COMPREHENSIVE POLYMER SCIENCE

Second Supplement

COMPREHENSIVE POLYMER SCIENCE

Second Supplement

CHAIRMAN OF THE EDTIORIAL BOARD

SIR GEOFFREY ALLEN, FRS

Kobe Steel Ltd., London, UK

VOLUME EDITORS

SUNDAR L. AGGARWAL

Global Polymer Technology Associates, Akron, OH, USA

&

SAVERIO RUSSO

Università di Genova, Italy

PERGAMON

To the memory of Mario Farina, who inspired and delighted so many alumni, colleagues, and friends by his keen intellect and great scientific enthusiasm, virtues able to permeate his manifold activities in polymer education and research.

UK	Elsevier Science Ltd., The Boulevard, Langford Lane, Kidlington, Oxford OX5 1GB, UK
USA	Elsevier Science Inc., 660 White Plains Road, Tarrytown, New York 10591-5153, USA
JAPAN	Elsevier Science Japan, 9-15 Higashi-Azabu 1-chome, Minato-ku, Tokyo 106, Japan

---

First edition 1996.

British Library Cataloguing in Publication Data
A catalogue record for this book is available from the British Library.

ISBN 0–08–042708–1

∞™ The paper used in this publication meets the minimum requirements of the American National Standard for Information Sciences—Permanence of Paper for Printed Library Materials, ANSI Z39.48–1984.

Typeset by The Universities Press (Belfast) Ltd., Belfast, UK.
Printed and bound in Great Britain by The Bath Press Ltd., Bath, UK.

Contents

Special Topics of Current Interest

Preface

The *Second Supplement* is the concluding volume of *Comprehensive Polymer Science*. The topics selected for this volume fall within two major categories:

(i) recent developments that justify authoritative critical reviews; and
(ii) polymer topics of current interest that could not be including the main seven-volume work and the *First Supplement*.

In the first of these categories we have included chapters on the following topics:

Advances in nontraditional catalysis and polymerization kinetics (Chapters 1 and 2);
Synthesis of polymers of novel structures (Chapters 3 and 4);
Advances in polymer characterization techniques (Chapters 5 and 6);
The present status of polymer crystallization and directions projected for future research (Chapter 7) and characterization of phase transitions by dynamic time-resolved scattering techniques (Chapter 8);
Structure–property relationships in perfluoropolyethers (Chapter 9) and wholly aromatic liquid crystalline copolyesters (Chapter 10); and
Recent developments in theories of rubber elasticity (Chapter 11) and phenomenological theories as a basis for determining material properties of rubbers needed for finite element analysis (FEA) and other engineering design applications (Chapter 12).

Under the second main category, we selected chapters on the following topics:

Statistical thermodynamics of phase separation in copolymers and blends of copolymers (Chapter 13) and self-assembly and morphology of block polymers and homopolymers (Chapter 14);
Single-chain polymers of high molecular weight that may separate under suitable conditions as particulates (Chapter 15);
Electrooptical properties of liquid crystalline polymers in a thermoplastic polymer matrix (Chapter 16);
Metal complexation with macromolecules (Chapter 17);
An overview of the status of current polymer recycling technologies and directions for future developments (Chapter 18); and
Microwave processing of polymers (Chapter 19), an area for several potential applications.

We wish to acknowledge the valuable assistance of Dr. Stefania Bisbano and Dr. Sabina Mazzanti, associates of Saverio Russo, for their meticulous checking of the accuracy of all the literature citations and references included in each chapter.

We would also like to thank several of our colleagues and professional associates for their reviews and comments on manuscripts of the chapters included in the *Second Supplement*. The guidance of Sir Geoffrey Allen, the Chairman of the Editorial Board, in the scope and organization of not only the main seven-volume work but especially of the supplementary volumes, has been invaluable to us and is gratefully acknowledged.

We also wish to thank Dr. Helen McPherson, Dr. Henri van Dorssen, Mr. Peter Frank and the editorial teams at Pergamon and Elsevier for their cooperation and work to ensure high publication standards and attention to a myriad of details necessary for the production of the main seven-volume work and the two supplementary volumes of *Comprehensive Polymer Science*.

SUNDAR L. AGGARWAL
Akron, OH, USA
April 1996

SAVERIO RUSSO
Genova, Italy
April 1996

Contributors

Dr. G. Ajroldi
Ausimont S.p.A., Research and Development Centre, Bollate (Milano), Italy

Professor J. Blackwell
Department of Macromolecular Science, Case Western Reserve University, Cleveland, OH 44106-7202, USA

Professor M. Brigodiot
Laboratoire de Synthèse Macromoleculaire, Boite 184, Université Pierre et Marie Curie, 4 place Jussieu, F-75252 Paris Cédex 05, France

Dr. A. Fradet
Laboratoire de Synthèse Macromoleculaire, Boite 184, Université Pierre et Marie Curie, 4 place Jussieu, F-75252 Paris Cédex 05, France

Professor J. M. J. Fréchet
Department of Chemistry, Baker Laboratory, Cornell University, Ithaca, NY 14853, USA

Dr. G. Goldbeck-Wood
Molecular Simulations, 245/250 The Quorum, Barnwell Road, Cambridge, CB5 8RE, UK

Professor H. Hasegawa
Department of Polymer Chemistry, Graduate School of Engineering, Kyoto University, Kyoto 606-01, Japan

Professor T. Hashimoto
Department of Polymer Chemistry, Graduate School of Engineering, Kyoto University, Kyoto 606-01, Japan

Dr. C. J. Hawker
IBM Almaden Research Center, Department K93-801, 650 Harry Road, San Jose, CA 95120-6099, USA

Professor A. D. Jenkins
School of Chemistry and Molecular Sciences, University of Sussex, Falmer, Brighton, BN1 9QJ, UK

Professor A. Keller
H. H. Wills Physics Laboratory, University of Bristol, Royal Fort, Tyndall Avenue, Bristol, BS8 1TL, UK

Professor S. I. Kuchanov
Keldysh Institute of Applied Mathematics of RAS, Miusskaya Square 4, 125047 Moscow, Russia

Professor T. Kyu
Institute of Polymer Engineering, The University of Akron, Akron, OH 44325-0301, USA

Professor T. Lalot
Laboratoire de Synthèse Macromoleculaire, Boite 184, Université Pierre et Marie Curie, 4 place Jussieu, F-75252 Paris Cédex 05, France

Dr. D. Mangaraj
Polymer Product Group, Battelle Columbus Division, 505 King Avenue, Columbus, OH 43201-2693, USA

Dr. G. Marchionni
Ausimont S.p.A., Research and Development Centre, Bollate (Milano), Italy

Professor E. Maréchal
Laboratoire de Synthèse Macromoleculaire, Boite 184, Université Pierre et Marie Curie, 4 place Jussieu, F-75252 Paris Cédex 05, France

Dr. C. M. McCullagh
Department of Macromolecular Science, Case Western Reserve University, Cleveland, OH 44106-7202, USA

Dr. F. Parodi
Saiag S.p.A., Corso Stati Uniti 61, I-10129 Torino, Italy

Dr. S. V. Panyukov
Keldysh Institute of Applied Mathematics of RAS, Miusskaya Square 4, 125047 Moscow, Russia

Professor G. Pezzin
Dipartimento di Chimica Fisica, Università di Padova, Via Loredan 2, I-35131 Padova, Italy

Professor R. Qian
Institute of Chemistry, The Chinese Academy of Sciences, PO Box 2709, Beijing 100080, People's Republic of China

Dr. G. Schnurpfeil
Institut für Organische und Makromolekulare Chemie, Universität Bremen, NW2, PO Box 330 440, D-28334 Bremen, Germany

Professor K. J. Smith, Jr.
Department of Chemistry, College of Environmental Science and Forestry, State University of New York, Syracuse, NY 13210-2786, USA

Dr. M. Srinivasarao
College of Textiles, North Carolina State University, Raleigh, NC 27695-8301, USA

Mr. D. D. Werstler
GenCorp Technology Center, 2990 Gilchrist Road, Akron, OH 44305, USA

Professor D. Wöhrle
Institut für Organische und Makromolekulare Chemie, Universität Bremen, NW2, PO Box 330 440, D-28334 Bremen, Germany

Dr. O.-H. Yeoh
Lord Corporation, Mechanical Products Division, 1635 West 12th Street, Erie, PA 16514, USA

1
Reactivity in Radical Copolymerization

AUBREY D. JENKINS
University of Sussex, Brighton, UK

1.1 INTRODUCTION

The quantitative study of radical copolymerization dates from 1944, when three papers on the relationship between the composition of a binary monomer mixture and that of the resulting copolymer were published almost simultaneously by Alfrey and Goldfinger,[1] Mayo and Lewis,[2] and Wall.[3] All were based on the premise of Norrish and Brookman[4] that four propagation reactions with different specific rate constants had to be taken into account, using the stationary-state hypothesis. Shortly afterwards, the treatment was extended to ternary systems[5] and to the general case of *n* monomers.[6] The subject then advanced so rapidly that by 1950 it was possible for Mayo and Walling[7] to publish a most scholarly and comprehensive review, which should still be compulsory reading for any student new to the topic; it was discussed in depth in books on radical polymerization by Bamford, Barb, Jenkins and Onyon in 1958,[8] by Bevington in 1961,[9] and in books devoted solely to copolymerization, by Alfrey, Bohrer and Mark in 1952[10] and by Ham in 1964.[11] The kinetic aspect of copolymerization was well summarized in a 1979 article by Wittmer[12] and, much more recently, by Fukuda, Kubo and Ma.[13] Good reviews are to be found in several textbooks of polymer chemistry, two of the best and most recent of which are those of Rudin[14] and Cowie,[15] while Volume 3 of *Comprehensive Polymer Science* contains a review of copolymerization in general,[16] covering the literature up to 1986. In 1987, an extremely successful conference on radical polymerization was held in Santa Margherita Ligure in Italy, the proceedings of which[17] contain several important articles on copolymerization. A review of the quantitative theory of many aspects of radical copolymerization appeared in 1992.[18]

The present article is restricted to copolymerization by radical mechanisms, and, in the main, seeks to summarize the more important developments since the mid-1980s. For one or two topics, it seemed best to present a complete review from the inception of the subject, but, for the most part, overlap with the article in *Comprehensive Polymer Science*[16] has been avoided; the symbolism and nomenclature used here are essentially the same as in that article, which should be regarded as preliminary reading for anyone who wishes to have a full review of the subject.

1.2 COPOLYMER COMPOSITION

1.2.1 Copolymer Microstructure

Earlier studies of copolymer composition were restricted to measurements of the overall composition of copolymers, that is, the proportions of the monomer units present in the product, usually related to the composition of the monomer feed, and some important consequences flowing from this work will be discussed below. Henceforth, the symbols A and B will be used to denote both the monomers and units derived from the monomers after incorporation into the copolymer. In fact, the manner in which the monomer units are distributed along a copolymer chain is far more informative than the mere mole-fraction composition with respect to A and B units, and this aspect of the subject will be summarized first because an appreciation of the statistical approach to copolymer structure is vital to an understanding of the current state-of-the-art.

The microstructure can be described by specifying the fractions of pairs of units (diads or dyads) of the possible kinds (AA, AB, BA, BB in the binary case) or triplets (triads) of consecutive units (AAA, AAB, ABA, BAA, ABB, BBA, BAB, BBB). Even higher groupings of four (tetrads) or five (pentads) or more units can be considered, but analysis beyond the pentad level is difficult and rarely attempted. The populations of diads (and the triads of which they form a part) are, of course, linked by mathematical formulae; for example, if $P(\mathrm{ABA})$ is the mole fraction of triads which have the structure ABA, one can write $P(\mathrm{AB}) = P(\mathrm{ABA}) + P(\mathrm{ABB})$, thus relating the mole fraction of a particular type of diad to the mole fractions of related triads, and so on. It is also possible to work in terms of sequences of all possible lengths composed exclusively of A units or of B units. All this is really an exercise in well-established statistics, and a very good account of the basic calculations is given in Koenig's comprehensive and clear discussion of the composition and microstructure of polymer chains[19] (at the state-of-the-art at the time), although this does not deal with the experimental side beyond 1980, and much relevant experimental work has been reported since then, mainly because NMR techniques have been refined tremendously. In this review, where Koenig's notation will be used, there is space to indicate only the most basic aspects of the study of microstructure but reference to more extensive reports will be included here.

The probability $P(\mathrm{A})$ of finding an A unit by random inspection at a point in a copolymer is equal to the mole fraction of A units present, and this is known as an *unconditional* probability because it is not dependent on any other circumstance. The same applies for B units. The probability $P(AB)$ of finding the pair AB at random depends not only on the mole fraction of A units present in the structure but also on the *conditional* probability $P(B/A)$ that a B unit will have added to an A unit. Thus the probability of finding the pair AB is given by $P(A)P(B/A)$. The probability $P(ABA)$ of finding an ABA triad is related to $P(AB)$ and also to the probability $P(A/AB)$ that an A unit will have added to a B unit that follows an A unit, so that $P(ABA) = P(AB)P(A/AB)$. By means of equations of this kind, it is possible to identify all the conditional probabilities as functions of mole fractions of component monads (isolated monomer units), diads and triads, etc. For the example chosen, then, we have $P(A/AB) = P(ABA)/P(AB)$. [$P(B/A)$ has already been identified as $P(AB)/P(A)$.] Further relationships are immediately available because the sum of all probabilities relating to a particular addition step must equal unity. Thus, $P(A/A) + P(B/A) = 1$ and $P(A/AA) + P(B/AA) = 1$ because either A or B must add at each stage. One especially interesting result of this analysis is that the ratio of mole fractions of A and B units in the polymer is given by a simple ratio of conditional probabilities

$$\frac{P(A)}{P(B)} = \frac{P(A/B)}{P(B/A)} \tag{1}$$

and this is a remarkably simple statement of the overall composition, one which can be derived in other ways, as will be seen later. The importance of these triad (and other) distributions is that they are related to monomer feed in different ways for the various kinetic models of copolymerization, and they provide a basis for discrimination among the models, as will be seen shortly.

1.2.2 The Terminal Model

The termination processes are not relevant to copolymer composition, but will be discussed separately below in Section 1.3.1. When copolymerization involving just two monomers (binary copolymerization) is under consideration, it is necessary to involve at least four propagation reactions, represented as follows:

$$P_{m,n,A} + M_A \xrightarrow{k_{AA}} P_{m+1,n,A} \tag{2}$$

$$P_{m,n,A} + M_B \xrightarrow{k_{AB}} P_{m,n+1,B} \tag{3}$$

$$P_{m,n,B} + M_A \xrightarrow{k_{BA}} P_{m+1,n,A} \tag{4}$$

$$P_{m,n,B} + M_B \xrightarrow{k_{BB}} P_{m,n+1,B} \tag{5}$$

$P_{m,n,A}$ denotes an active copolymer radical comprising a total of m units of monomer A (M_A) and n units of monomer B (M_B) in the chain, and with the terminal active centre located on a unit derived from monomer A; related symbols are defined similarly with the necessary changes. The velocity constant k_{BA} applies to a reaction in which a polymer radical with active terminal unit B adds to monomer A, and so on.

An assumption implicit in this scheme is that the reactivity of a polymer radical depends only on the nature of its terminal unit. This is the reason that the group of reactions represented by this scheme is known as the Terminal Model (TM). That being the case, it is a simple matter to deduce, on the basis of the stationary state assumption, that the relative rates of consumption of monomers A and B are given by the following differential equation

$$\frac{-d[A]}{-d[B]} = \frac{[A]\{r_A[A] + [B]\}}{[B]\{r_B[B] + [A]\}} \tag{6}$$

where $r_A = k_{AA}/k_{AB}$ and $r_B = k_{BB}/k_{BA}$. The parameters r_A and r_B were originally called monomer reactivity ratios by Mayo and Lewis,[2] but they are now (unfortunately) usually known simply as reactivity ratios. Equation (6) is the classical copolymer composition equation because the ratio $-d[A]/-d[B]$ corresponds to the composition of the copolymer being formed instantaneously from a monomer mixture containing molar concentrations [A] and [B] of the components. By dividing the numerator and denominator of the right-hand side of equation (6) throughout by [B], one obtains the ratio of the molar quantities of A and B in the copolymer, Q_p, as a function of the ratio $[A]/[B] = Q_m$, in the monomer mixture; thus

$$Q_p = \frac{Q_m\{r_A Q_m + 1\}}{\{r_B + Q_m\}} \tag{7}$$

This is actually identical to equation (1) because, in terms of the kinetics of TM, the probabilities involved are

$$P(A/B) = k_{BA}[A]/(k_{BA}[A] + k_{BB}[B])$$

and

$$P(B/A) = k_{AB}[B]/(k_{AB}[B] + k_{AA}[A])$$

and substitution in equation (1) produces equation (7). The kinetic derivation of equation (6) presumes that, in the stationary state, the rates of reactions (3) and (4) are equal, that is that B units are adding to A-terminated radicals at the same velocity that A units are adding to B-terminated radicals; the probabilistic equivalent statement is that $P(AB) = P(BA)$.

Although equation (7) has been used to a prolific extent, there are many examples of systems which it does not describe well. To account for these deviations from TM, two hypotheses have been advanced: (i) the reactivity of a polymer radical depends on the penultimate unit as well as the terminal unit [this is the Penultimate Model (PM)]; and (ii) the monomers may form a complex which behaves as a third monomer [this is the Complex Model (CM)]. Both these possible complications were envisaged at very early stages in the study of radical copolymerization; the PM can be traced back to Merz *et al.*[20] and the CM to Bartlett and Nozaki,[21] both suggestions being advanced in 1946.

1.2.3 The Penultimate Model

It has often been suggested that the reactivity of the terminal unit in a polymer radical is influenced by its predecessor in the chain through the agency of a so-called penultimate unit effect. If this idea is valid, it is necessary to double the number of propagation steps because the penultimate unit may be either A or B, whatever the nature of the terminal unit. The full set of propagation reactions then takes the following form, where the three subscripts to the rate constants denote the penultimate unit, the terminal unit and the monomer involved, in that order.

$$P_{m,n,AA} + M_A \xrightarrow{k_{AAA}} P_{m+1,n,AA} \tag{8}$$

$$P_{m,n,AA} + M_B \xrightarrow{k_{AAB}} P_{m,n+1,AB} \tag{9}$$

$$P_{m,n,BA} + M_A \xrightarrow{k_{BAA}} P_{m+1,n,AA} \tag{10}$$

$$P_{m,n,BA} + M_B \xrightarrow{k_{BAB}} P_{m,n+1,AB} \tag{11}$$

$$P_{m,n,AB} + M_A \xrightarrow{k_{ABA}} P_{m+1,n,BA} \tag{12}$$

$$P_{m,n,AB} + M_B \xrightarrow{k_{ABB}} P_{m,n+1,BB} \tag{13}$$

$$P_{m,n,BB} + M_A \xrightarrow{k_{BBA}} P_{m+1,n,BA} \tag{14}$$

$$P_{m,n,BB} + M_B \xrightarrow{k_{BBB}} P_{m,n+1,BB} \tag{15}$$

If an expression parallel to equation (7) is now calculated from equations (8)–(15), it is found that there are four (rather than two) ratios of rate constants that must be taken into consideration because there are four, rather than two, distinguishable types of radicals if the penultimate unit, in addition to the terminal unit, determines reactivity. Thus, it is necessary to define these four monomer reactivity ratios, as follows

$$r_A = k_{AAA}/k_{AAB} \qquad r'_A = k_{BAA}/k_{BAB}$$

$$r'_B = k_{ABB}/k_{ABA} \qquad r_B = k_{BBB}/k_{BBA}$$

Here, r_A and r_B are the monomer reactivity ratios for radicals in which the penultimate and terminal units are identical, while r'_A and r'_B are monomer reactivity ratios for radicals with terminal units indicated by the subscript but in which the penultimate unit is derived from the partner monomer. The instantaneous relationship between monomer feed and polymer composition is then given by equation (16), due to Merz *et al.*[20]

$$Q_p = \frac{\{1 + r'_A Q_m[(r_A Q_m + 1)/(r'_A Q_m + 1)]\}}{[1 + r'_B/Q_m\{[(r_B/Q_m) + 1]/[(r'_B/Q_m) + 1]\}]} \tag{16}$$

In recent times, this relationship has more commonly been employed in the following equivalent form

$$Q_p = Q_m(\bar{r}_A Q_m + 1)/(\bar{r}_B + Q_m) \tag{17}$$

where $\bar{r}_A = r'_A(r_A Q_m + 1)/(r'_A Q_m + 1)$ and $\bar{r}_B = r'_B(r_B + Q_m)/(r'_B + Q_m)$.

Problems arising out of penultimate unit effects can be overcome by working with a large excess of one of the monomers, so that it can be assumed that the monomer in excess always provides the penultimate unit.[22] Ordinarily, this would be impracticable because of the imprecision of the polymer analysis but the use of a radioactively labelled monomer solves this problem. Barson and Fenn went on to apply this technique to situations in which penultimate unit effects influence both of the monomers involved.[23]

1.2.4 The Complex Model

It has sometimes been suggested, particularly where charge transfer seems to be likely, that the monomers A and B might form a complex, C, which can enter into copolymerization along with its parents, and possibly in two different orientations (C′ and C″), each with its own reactivity parameters, according to whether it adds through its A end or its B end. In the simplest case, where radical reactivity is dependent only on the nature of the terminal unit, eight propagation steps are required, each with its own rate constant, as shown below.

$$P_{m,n,A} + M_A \xrightarrow{k_{AA}} P_{m+1,n,A} \tag{18}$$

$$P_{m,n,A} + M_B \xrightarrow{k_{AB}} P_{m,n+1,B} \tag{19}$$

$$P_{m,n,B} + M_A \xrightarrow{k_{BA}} P_{m+1,n,A} \tag{20}$$

$$P_{m,n,B} + M_B \xrightarrow{k_{BB}} P_{m,n+1,B} \tag{21}$$

$$P_{m,n,A} + M_{C'} \xrightarrow{k_{AC'}} P_{m+1,n+1,C'} \tag{22}$$

$$P_{m,n,A} + M_{C''} \xrightarrow{k_{AC''}} P_{m+1,n+1,C''} \tag{23}$$

$$P_{m,n,B} + M_{C'} \xrightarrow{k_{BC'}} P_{m+1,n+1,C'} \tag{24}$$

$$P_{m,n,B} + M_{C''} \xrightarrow{k_{BC''}} P_{m+1,n+1,C''} \tag{25}$$

An additional necessary parameter is the equilibrium constant K characterizing the formation of the complex C, according to equation (26), in the formulation used by Koenig.[19]

$$A + B \overset{K}{\rightleftharpoons} [\mathrm{AB}] \equiv \mathrm{C} \qquad (26)$$

Seiner and Litt[24] explored the system of eight propagation steps, and showed that the monomer reactivity ratios would be expected to change on dilution, as had already been observed in some cases. Pittman and Rounsefell[25] produced a computer program for the evaluation of the model, and pointed out that no data were available in the literature that were adequate to test it properly. An alternative treatment was advanced by Cais *et al.*,[26] including the derivation of sequence distributions and triad fractions. A complete analysis has more recently been published by Braun and Czerwinski.[27]

The expression for Q_p in terms of Q_m that follows from these postulates is considerably more complicated than equations (7) or (16), and is written as a function of six monomer reactivity ratios, which can be defined in a variety of ways. Of course, if there is sufficient attraction between A and B to form a complex, it would hardly be surprising if there were to be a penultimate unit effect as well. The combination of PM and CM would seem to imply a need for at least twelve monomer reactivity ratios, impossible to evaluate, but, in fact, Brown and Fujimori[28] have combined the penultimate and complex models in a scheme they call Comppen.

General reviews of the complex model have been published by Bartoň and Borsig[29] and by Rätzsch and Vogl.[30] Karad and Schneider[31] have considered the kinetics, distinguishing two cases where (i) the radical simply adds to the entire complex, *e.g.* A· + C → AAB·, and (ii) the radical displaces one of the component monomers from the complex, *e.g.* A· + C → AA· + B.

1.2.5 Determination of Monomer Reactivity Ratios

One of the principal topics of debate in the recent literature has been the question of the validity of TM, PM and CM, and of ways of distinguishing among them, a topic that will be discussed in Section 1.4. It has long been known that the manner in which monomer units are distributed along copolymer chains is much more informative than overall composition data,[32] but this concept was of theoretical importance only, prior to the development of high-frequency NMR. Any discussion of model discrimination can have validity only if reliable values of monomer reactivity ratios are available, and this is by no means something that can be taken for granted. It is therefore necessary to begin by examining the procedures required in arriving at dependable values of monomer reactivity ratios.

As early as 1941, Wall[33] demonstrated that a diagram in which F_A, the mole fraction of A units in the copolymer, is plotted against f_A, the mole fraction of A molecules in the monomer feed, reflects valuable information about the copolymerization process. It is clear from equation (7) that the shape of the curve depends on the values of r_A and r_B, and these values can be deduced by trial and error fitting to the curve. (Such figures may be referred to as Wall diagrams in the literature.) This was a very remarkable paper for its time; among other things, it clearly recognized the possibility of preparing tapered copolymers, although that term was not used in 1941.

Mayo and Lewis[2] suggested a method for the determination of monomer reactivity ratios based on the rearrangement of equation (7) to equation (27)

$$r_A = r_B/(Q_p/Q_m^2) + (Q_p - 1)/Q_m \qquad (27)$$

Linear plots of r_A versus r_B were then made using the sets of Q_p and Q_m data points to calculate the slopes and intercepts corresponding to this equation. In principle, this procedure should produce a family of straight lines, all intersecting at a single point corresponding to the true values of the monomer reactivity ratios.

In 1950, Fineman and Ross[34] offered a new version of equation (7) for determining monomer reactivity ratios from a single linear plot of the data set. They worked in terms of the mole fractions of monomers A and B in the polymer and in the monomer feed, that is, $Q_p = F_A/F_B$ and $Q_m = f_A/f_B$, respectively, leading to the equation

$$Q_m(1 - Q_p)/Q_p = r_B - r_A(Q_m^2/Q_p) \tag{28}$$

As the assignment of the symbols A and B to the monomers is arbitrary, it proved best to work out the results with both of the possible allocations to weight the data as evenly as possible.

Two milestone papers appeared in the mid-1960s, when Behnken[35] and Tidwell and Mortimer[36] pointed out that none of the then existing methods for determining monomer reactivity ratios was satisfactory, and they took the matter into the realm of the respectable statistical treatment of data, thus providing the essential basis for modern methods. The argument was that all the existing methods relied on subjective weighting of the data, and that the application of the linear least-squares routine to optimize the results is not appropriate. Both of these papers stressed the importance of estimating the confidence limits attached to the results (something that had been foreseen as far back as 1952[37]), and arrived at the conclusion that the best way to obtain accurate monomer reactivity ratios is to perform copolymerization at just two values of Q_m if TM is valid.

Computer experiments have demonstrated that the maximum information (*i.e.* the highest level of confidence) is obtained per experiment if half the experiments are conducted at each of the two chosen Q_m values.[35] To use the two-point technique it is necessary to have good prior estimates of the values of the monomer reactivity ratios, or to use an iterative procedure (or both). Thus, Tidwell and Mortimer suggested the following approximate values of f_B

$$\begin{aligned} f_B &= r_A/(2 + r_A) \\ f_B &= 2/(2 + r_B) \end{aligned} \tag{29}$$

and used the Fineman–Ross procedure to obtain the initial estimates of the monomer reactivity ratios.

A procedure was proposed by Chûjô *et al.*[38] for determining the monomer reactivity ratios from a single copolymer sample by analysis of the fractions of ABA, ABB and BBB triads in the NMR spectrum. This method was applied to the acrylonitrile/methyl methacrylate system, using the relative intensities of three OMe peaks. The ratios were found to be insensitive to monomer feed but slightly affected by the nature of the penultimate unit and strongly affected by inhomogeneity.

Several further linear least-squares methods were proposed (and later analysed[39]); of these, only the Kelen–Tüdős procedure will be discussed here because it has been used by many more authors than the others.

The novel feature introduced by Kelen and Tüdős[40] was an arbitrary constant α which was included in a particular rearrangement of the Fineman–Ross equation. They defined two new variables η and ξ as follows

$$\eta = Q_m(Q_p - 1)/Q_p[\alpha + (Q_m^2/Q_p)]$$

and

$$\xi = (Q_m^2/Q_p)/[\alpha + (Q_m^2/Q_p)] \tag{30}$$

and then rewrote the Fineman–Ross equation in the form

$$\eta = [r_A + (r_B/\alpha)]\xi - (r_B/\alpha) \tag{31}$$

It is clear that the parameter ξ has the extreme values 0 and 1, and then it follows that a plot of η versus ξ will extrapolate to $-(r_B/\alpha)$ at $\xi = 0$ and to r_A at $\xi = 1$. The value of α is arbitrary but the 'best' value is said to correspond to the square root of the product of the highest and lowest values of (Q_m^2/Q_p) used in the experiments; this is stated to afford optimum distribution of the data, that is, to secure symmetrical location of the data points in the interval $0 < \xi < 1$.

The Kelen–Tüdős method was applied to 8000 copolymerizations involving styrene as one of the monomers.[41] Only 649 sets of data were found suitable for analysis and, of these, 464 conformed to TM.

An interesting comparison of the linear and nonlinear least-squares methods was made by McFarlane *et al.*,[39] who simulated data, complete with error, for a hundred (computer) experiments for each of a number of pairs of monomer reactivity ratios, and used the available procedures to determine the monomer reactivity ratios in each of the experiments. They recommended that future determinations should use the two-point technique, with the values of f_B specified in equation (29). If this condition is observed, both the Kelen–Tüdős and the Yezrielev, Brokhina and Roskin[42] linear extrapolation methods give good results. If all the error can be attributed to inaccuracy in the polymer composition, and if the error distribution is 'normal', the Kelen–Tüdős and similar methods suffice; however, if there is significant error in the data for the composition of the monomer feed, as when there is a finite conversion with resulting drift in monomer concentrations, it is necessary to use another statistical technique, the Error-in-Variables Method (EVM).

The EVM method differs from the usual least-squares methods, which assume all the error to reside in data for the dependent variable (polymer composition), in allowing for error in all the data. It is therefore more demanding on the amount of initial data, requiring information about the error variances and covariances, but experience has shown that the results are not too sensitive to the precision of estimates of these quantities, although the evaluation of confidence limits does need good estimates of variances. For applications of the EVM method to copolymerization, the reader is referred to the work of van der Meer *et al.*,[43] Hautus *et al.*,[44] Patino-Leal *et al.*[45] and Chee and Ng.[46] Patino-Leal *et al.* pointed out that their procedure did not need the storage of large matrices, and neither did it necessitate the use of an optimization process like that of van der Meer *et al.* Petrak and Pitts[47] made it clear that the assumption that the relative error in determining the monomer feed composition by gas–liquid chromatography is independent of conversion is not always valid; such analysis is much less reliable in the presence of polymer. Chee and Ng showed that the EVM method is to be preferred to other nonlinear least-squares approaches if drift in composition is appreciable, even at low conversion.

Additional problems which can arise when momoner reactivity ratios are estimated using the integrated copolymer composition equation[14] have been discussed by Hautus *et al.*[48]

In reviewing methods for the determination of monomer reactivity ratios in 1987, O'Driscoll and Reilly[49] put forward a modified list of criteria for a satisfactory method, as follows.

(i) The method should utilize all (or nearly all) the information resident in the data with regard to the parameters to be estimated, thus providing precise estimates.

(ii) The parameter values calculated by the method should not depend upon arbitrary factors (such as which monomer is subscripted A or B, and the starting point of the calculation).

(iii) The method should supply a valid measure of the errors of the resulting estimates.

(iv) The method should be reasonably easy to use.

They concluded that these criteria were met by the nonlinear least-squares method in the EVM form, and regretted that the facility with which cheap and simple electronic calculators could perform the linear least-squares procedure continued to promote the use of the Fineman–Ross and similar methods. They also endorsed the two-point technique for the determination of values of monomer reactivity ratios but pointed out that more experiments at other compositions are necessary if the object of the exercise is to discriminate between models of the reaction, such as TM, PM and CM.

The use of the Monte Carlo procedure for the estimation of values of kinetic parameters has been discussed by Duever *et al.* and applied to both TM[50] and PM.[51] A program for evaluating monomer reactivity ratios, using EVM and based on a nonlinear minimization algorithm, has been described by Dubé *et al.*[52]

Another nonlinear procedure was suggested by Hagiopol *et al.*[53] in 1989, and, recently, a further iterative least-squares method, valid for data obtained for any percentage conversion, was put forward by Mao and Huglin.[54]

Equation (17), which applies when penultimate unit effects are important, cannot be converted into a linear relation between monomer reactivity ratios as it stands, but procedures for determining monomer reactivity ratios have been derived from it to cope with systems where one of the monomers does not homopolymerize, so that both (say) r_A and r'_A are zero. Proposals along these lines have been advanced by Kelen *et al.*,[55] Mao and Hughlin,[56] and by Bauduin and Boutevin.[57] Berger and Kuntz[32] showed in 1964 that copolymer composition information alone would have to be quite exceptionally accurate to permit one to decide whether a given copolymerization behaved according to the terminal or the penultimate model. Klumperman and Vonk[58] later showed that the Tidwell–Mortimer method could be used with PM as well as TM.

Two problems not addressed by Tidwell and Mortimer relating to the design of experiments for the determination of monomer reactivity ratios were examined by Burke *et al.*[59] One was the question of constraints on composition, and the other was the influence of the penultimate unit effect. In the first case, new equations were developed and, in the second case, an attempt was made to maximize precision.

Burke *et al.*[60] have used a computer simulation technique to reassess the application of statistical model-discrimination methods in copolymerization, and have concluded that composition data may be more useful than was previously thought; they highlight both the strengths and weaknesses of the available methods.

1.2.6 The Bootstrap Effect

It was often stated in the early literature that monomer reactivity ratios are insensitive to the solvent employed for a copolymerization, but it became apparent that there are important cases where this is by no means true or, perhaps more accurately, where it appears not to be true. Chapiro,[61] for example, described systems involving such monomers as methyl acrylate, acrylamide, acrylonitrile, acrylic acid and methacrylic acid, where the monomer reactivity ratios appeared to change dramatically with variation in the solvent medium, an effect which Chapiro ascribed to molecular association by hydrogen bonding. These results were reminiscent of some studies of solvent effects in homopolymerization, for example that of Gromov *et al.*[62] A comprehensive review was published by Plochocka.[63]

Harwood[64] produced a general explanation of solvent effects in copolymerization in terms of a phenomenon he called the 'Bootstrap Effect'. In considering competitive reactions in copolymerization, for example, the competition between monomers A and B to add to an A· radical, the ratio of rates of reaction is $k_{AA}[A\cdot][A]/k_{AB}[A\cdot][B]$ or $k_{AA}[A]/k_{AB}[B]$, which can be written as $r_A[A]/[B]$, where the ratio $[A]/[B]$ was always assumed to have the same value in the proximity of A· as it has in the bulk solution. Harwood's point was that this assumption may be false, and that the monomer concentration ratio at the site of reaction may vary from solvent to solvent as a result of a kind of partitioning process which may show some selectivity. In his words, 'a growing polymer radical can control its own environment; since monomer partitioning will be influenced by the composition of a growing radical, we can expect K (the partition coefficient) to vary with the overall conversion and even during the process of forming a single polymer chain'.

The essential experimental evidence which led Harwood to make this proposal was the finding that the Wall diagrams (see Section 1.2.5) for a number of binary copolymerization systems varied dramatically with change in solvent (at first implying a variation in monomer reactivity ratios) but that, nevertheless, copolymers obtained in different solvents but having the same overall composition had the same microstructure, expressed as triad fractions. This means that, regardless of the solvent used, the same conditional probabilities must have existed during all the copolymerizations. This fact indicates strongly that the monomer reactivity ratios were actually the same, and that the quantity responsible for the variation in behaviour was the ratio of monomer concentrations at the site of reaction in the various solvents. In fact, observations of a pertinent nature had already been made by San Roman, Madruga and Del Puerto[65,66] on the styrene/methyl methacrylate system. Solvent effects on the rate constant for propagation in homopolymerization were also well documented[67] and apparent dependence of monomer reactivity ratios on solvent dielectric constant had been reported by Bontà *et al.*[68] for the same pair of monomers. Kratochvíl *et al.*[69] had already conceived the idea of preferential solvation in systems containing monomer, polymer and solvent.

Confirmation of the solvent effect is found in the work of Semchikov *et al.*[70] who showed that, for 18 out of 24 binary systems examined, the Wall diagram displayed significantly different curves for experiments in different monomer mixtures. They reported that the copolymer composition is a function of molecular weight. They considered that the explanation for this correlation is a difference in the ratio of monomer concentrations in the neighbourhood of the polymer coil and in the bulk solution, *i.e.* the Bootstrap Effect. Further work led to the formulation of a quantitative thermodynamic criterion for the prediction of the deviation from ideality to be expected in any particular system.[71]

Harwood's finding that copolymers with identical overall compositions had identically the same microstructure, regardless of the solvent medium present during synthesis, was confirmed by Park *et al.*[72] by cyclizing pairs of acrylamide units in copolymers to imide units, an operation which depends critically on the adjacent placement of acrylamide residues. Further confirmation appeared to come from Davis,[73] who re-examined the microstructure of styrene/methyl methacrylate copolymers; however, doubts exist about the interpretation of NMR data accepted at that time.[74] Some anomalies in rates of copolymerization were being investigated by Fukuda and his colleagues, a topic to be discussed later (See Section 1.3.2), and Davis drew attention to the possibility that the two effects are interrelated. Possible, but less convincing, evidence for the Bootstrap Effect was reported by Hill *et al.*[75] for the styrene/acrylonitrile system, where the solvent effect is rather minimal.

The Bootstrap Effect took a large step forward in 1992 with a quantitative formulation presented by Klumperman and O'Driscoll.[76] First they treated the terminal model, starting with a consideration of the conditional probabilities $P(\mathrm{A/B})$ and $P(\mathrm{B/A})$, employing the relationship $F_A/F_B = P(\mathrm{A/B})/P(\mathrm{B/A})$, mentioned in Section 1.2.1, and the fractions of triads centred on an A unit, which are as follows

$$\begin{aligned} F_{AAA} &= [1 - P(\mathrm{B/A})]^2 \\ F_{AAB} + F_{BAA} &= 2P(\mathrm{B/A})[1 - P(\mathrm{B/A})] \\ F_{BAB} &= [P(\mathrm{B/A})]^2 \end{aligned} \tag{32}$$

Harwood[64] had already shown that $P(\mathrm{B/A}) = [1 + r_A K([\mathrm{A}]_b/[\mathrm{B}]_b)]^{-1}$ where the subscript b refers to concentrations in the initial monomer mixture, and K is a partition coefficient characterizing the difference in monomer concentrations at the radical site, $[\mathrm{A}]^*$ and $[\mathrm{B}]^*$, and in the bulk; thus $K = [\mathrm{A}]^*[\mathrm{B}]_b/[\mathrm{A}]_b[\mathrm{B}]^*$. The apparent monomer reactivity ratios in a given solvent will be designated r_A^s and r_B^s, and these must be related to the 'real' values by $r_A^s = r_A K$ and $r_B^s = r_B/K$; note that, if the result of the Bootstrap Effect is to enhance one monomer concentration at the radical site, it must, in bulk systems, simultaneously depress the concentration of the other.

Rather than enter into a philosophical discussion of the nature of 'real' monomer reactivity ratios, the usual thermodynamic practice of defining a reference state is adopted, and the chosen reference state is the bulk mixture of monomers, for which it may be assumed $K^b = 1$, even though it is not certain that the concentrations of monomers at the radical site will be the same as in the monomer mixture, even in this case. Denoting the monomer reactivity ratios determined in bulk monomers as r_A^b and r_B^b, it follows that $r_A^s/r_A^b = K$ and $r_B^b/r_B^s = K$, and, most importantly, that $r_A^s r_B^s = r_A^b r_B^b$. Algebraic manipulation of the relationships presented here leads to the following equation

$$P(\mathrm{B/A}) = \frac{1 - \{1 - [4F_A F_B(1 - r_A r_B)]\}^{1/2}}{2F_A(1 - r_A r_B)} \tag{33}$$

It is seen that the conditional probability $P(\mathrm{B/A})$, which alone governs the fractions of triads, is a function only of the polymer composition and the product $r_A r_B$, which is independent of the solvent. The microstructure will therefore not vary from solvent to solvent if the overall composition of the polymer is invariant. Experimentally, Bontà *et al.*[68] have shown that $r_A r_B$ does vary somewhat from solvent to solvent.

Subsequently, Klumperman and Vonk[58] looked at the styrene/maleic anhydride system, and found the Bootstrap Effect applicable but hard to quantify because of the lack of tendency of maleic anhydride to homopolymerize. Klumperman and Kraeger[77] extended the Klumperman–O'Driscoll theory by recognizing that K is not a constant for a given solvent but will vary with conversion if the composition of the (radical) copolymer chains changes as a result of the reaction; this would be expected from the definition of the partition coefficient *K*.

1.2.7 Copolymerization with Depropagation

In all the foregoing discussion, it has implicitly been assumed that reverse propagation steps can be neglected, an assumption that becomes less and less valid as the temperature is increased towards the ceiling temperature of the system. A variety of cases can be distinguished depending on how many of the terminal structures in the radical, AA·, AB·, BA·, BB· (invoking only the last two units), undergo appreciable depropagation. Some of these have been dealt with by Lowry,[78] Howell *et al.*[79] and Izu and O'Driscoll.[80]

1.3 THE RATE OF COPOLYMERIZATION

The discussion which follows is confined to issues specifically related to copolymerization but it should not be forgotten that termination in copolymerization is influenced by factors which are also operative in homopolymerization. It is not possible to assign a single k_t value to any polymerization system at a given temperature because k_t may depend, among other things, on the degree of polymerization of the radicals and the nature of the solvent (especially its viscosity). A quantitative treatment of many of these effects has been developed by Mahabadi and O'Driscoll, and a summary was provided by Mahabadi in 1987.[80] Further ideas on modelling termination have been advanced by Buback *et al.*[82] The problem of the influence of the size of the interacting radicals has been treated by Bamford *et al.*[83] and by Olaj *et al.*[84]

The polarity of the solvent can also affect the kinetics of polymerization to a significant extent, as demonstrated by Bontà *et al.*[85] and Madruga *et al.*[86] for example; the effect is not large but neither is it negligible.

In considering copolymer composition, only propagation reactions are of any significance so long as the long-chain assumption is valid, and this is almost always the case. The rate of copolymerization, by contrast, depends on both propagation and termination processes, and can be far more complicated to calculate. In the review below, it will be assumed that there is a constant rate of initiation, *I*, which is independent of the monomer and polymer concentrations. This assumption involves a modest approximation in some cases; thus Bontà *et al.*[87] have shown that *I* can vary significantly, and in nonlinear fashion, between $f_A = 0$ and $f_A = 1$, where f_A is the mole fraction of A in the monomer mixture.

The terminal model of propagation (Section 1.2.2) will be considered first, in combination with the various conceptions of termination, and then the implications, for the rate of reaction, of the penultimate model, described in Section 1.2.3, will be examined in some detail.

1.3.1 Using the Terminal Model for Propagation

1.3.1.1 The chemical model of termination

In addition to the propagation processes (2)–(5), it is now necessary to take into account the three possible termination steps (34)–(36) that follow from the basic premise of TM that the reactivity of a radical in termination depends solely on the nature of its terminal unit.

$$P_{m,n,A} + P_{m,n,A} \xrightarrow{k_{tAA}} \text{Polymer} \quad (34)$$

$$P_{m,n,A} + P_{m,n,B} \xrightarrow{k_{tAB}} \text{Polymer} \quad (35)$$

$$P_{m,n,B} + P_{m,n,B} \xrightarrow{k_{tBB}} \text{Polymer} \quad (36)$$

The stationary-state assumption stipulates that

$$I = k_{tAA}[A\cdot]^2 + 2k_{tAB}[A\cdot][B\cdot] + k_{tBB}[B\cdot]^2 \quad (37)$$

where [A·] represents the total concentration of all radicals with terminal A units, and likewise for [B·]. The stationary-state assumption must imply that there are constant concentrations of [A·] and [B·] radicals, so that the interchange processes (3) and (4) must maintain a balance, leading to equations (38) and (39).

$$k_{AB}[A\cdot][B] = k_{BA}[B\cdot][A] \tag{38}$$

$$[B\cdot] = [A\cdot]\{k_{AB}[B]/k_{BA}[A]\} \tag{39}$$

The rate, ω, is equal to the rate of consumption of the monomers in reactions (2)–(5), so the basic relation expressing the rate in terms of monomer and radical concentrations is equation (40).

$$\omega = \frac{-d\{[A] + [B]\}}{dt} = k_{AA}[A\cdot][A] + k_{AB}[A\cdot][B] + k_{BA}[B\cdot][A] + k_{BB}[B\cdot][B] \tag{40}$$

Substituting for [B·] from equation (39), this transforms into equation (41).

$$\omega = k_{AA}[A\cdot][A] + k_{AB}[A\cdot][B] + (k_{BA}[A] + k_{BB}[B])[A\cdot](k_{AB}[B]/k_{BA}[A])$$

or

$$\omega = [A\cdot]\{k_{AA}[A] + k_{AB}[B] + (k_{BA}[A] + k_{BB}[B])(k_{AB}[B]/k_{BA}[A])\} \tag{41}$$

In the same way, equation (37) turns into equation (42).

$$I = k_{tAA}[A\cdot]^2 + 2k_{tAB}[A\cdot]^2(k_{AB}[B]/k_{BA}[A]) + k_{tBB}\{[A\cdot](k_{AB}[B]/k_{BA}[A])\}^2$$

or

$$I = [A\cdot]^2\{k_{tAA} + 2k_{tAB}(k_{AB}[B]/k_{BA}[A]) + k_{tBB}(k_{AB}[B]/k_{BA}[A])\}^2 \tag{42}$$

If substitution for [A·] is now made from equation (42) into equation (41), the resulting expression for the rate of copolymerization in terms of the monomer concentrations and the rate constants of the component propagation and termination processes is equation (43).

$$\omega = \frac{I^{1/2}\{k_{AA}[A] + k_{AB}[B] + (k_{BA}[A] + k_{BB}[B])(k_{AB}[B]/k_{BA}[A])\}}{\{k_{tAA} + 2k_{tAB}(k_{AB}[B]/k_{BA}[A]) + k_{tBB}(k_{AB}[B]/k_{BA}[A])^2\}^{1/2}} \tag{43}$$

To proceed beyond this point, it is necessary to make an assumption about the value of the cross-termination rate constant, k_{AB}; equation (43) cannot be reduced to a more tractable form unless k_{tAB} can be written as a simple function of k_{tAA} and k_{tBB}. The usual solution to this difficulty[88,89] is to relate the cross-termination rate constant to the two homopolymerization rate constants through a geometric mean type of equation, as in equation (44)

$$k_{tAB} = \varphi(k_{tAA} \cdot k_{tBB})^{1/2} \tag{44}$$

where φ, the 'phi factor', is a parameter characteristic of the particular system of two monomers. If $\varphi = 1$, the simple geometric mean assumption is obeyed, but experience shows that this is rarely the case; in fact, for some systems, the value of φ required to fit the experimental data approaches 1000.

It is now helpful to define two new parameters, as in equation (45).

$$\delta_A = k_{tAA}^{1/2}/k_{AA}$$

and

$$\delta_B = k_{tBB}^{1/2}/k_{BB} \qquad (45)$$

With the substitutions of equations (44) and (45) into equation (43), the final expression for the rate of copolymerization becomes equation (46).

$$\omega = \frac{I^{1/2}(r_A[A]^2 + 2[A][B] + r_B[B]^2)}{(\delta_A^2 r_A^2[A]^2 + 2\varphi\delta_A\delta_B r_A r_B[A][B] + \delta_B^2 r_B^2[B]^2)^{1/2}} \qquad (46)$$

This equation gives a satisfactory account of copolymerization rate for a few systems, but all too often it fails, either because φ varies with the composition of the monomer feed or has to be allocated a value greatly in excess of unity, as mentioned above. Of course, the physical significance of φ becomes obscure if the rate of termination is not, in fact, governed by the chemical nature of the end units, as postulated in the section below.

1.3.1.2 The translational diffusion model

The chemical model treatment of copolymerization kinetics, described in Section 1.3.1.1, was widely used until about 1961 when it was pointed out by Allen and Patrick[90] and North and Reed[91] that the rate of termination in polymer systems must be controlled by diffusion of the radicals and not by the inherent chemical reactivity of their end units. It follows that only one rate constant for termination is necessary because the radicals can no longer be distinguished on chemical grounds. Of course, if diffusion is important, it also follows that the rate of termination may be a function of both the size and the chemical composition of a radical, a topic to be discussed below, but it may nevertheless be possible to use a single average rate constant to represent the behaviour of the system.

This concept was developed by Atherton and North[92] and applied to the copolymerization of methyl methacrylate and vinyl acetate. The stationary-state equation corresponding to equation (37) will now be equation (47), where the single rate constant for termination is k_t.

$$I = k_t([A\cdot] + [B\cdot])^2 \qquad (47)$$

Combination of this expression with equation (39) leads directly to equation (48) as the equation for the rate of reaction.

$$\omega = \frac{I^{1/2}(r_A[A]^2 + 2[A][B] + r_B[B]^2)}{(\varepsilon_A^2 r_A^2[A]^2 + 2\varepsilon_A\varepsilon_B r_A r_B[A][B] + \varepsilon_B^2 r_B^2[B]^2)^{1/2}} \qquad (48)$$

This equation is very close in form to equation (46) but the parameters ε_A and ε_B (equal to $k_t^{1/2}/k_{AA}$ and $k_t^{1/2}/k_{BB}$, respectively) replace δ_A and δ_B, and the phi factor has been eliminated.

It remains to express the single termination rate constant as a function of parameters relating to the two relevant homopolymerizations. Allen and Patrick[90] concluded that k_t must be the sum of k_{tAA} and k_{tBB} while Atherton and North[92] preferred to think that an 'ideal' copolymerization would correspond to the condition expressed in equation (49).

$$k_t = F_A k_{tAA} + F_B k_{tBB} \qquad (49)$$

(F_A and F_B are the mole fractions of A and B units in the copolymer.) A further proposal that

$$k_t = F_A^2 k_{tAA} + F_A F_B k_{tAB} + F_B^2 k_{tBB} \qquad (50)$$

was advanced by Chiang and Rudin,[93] who felt that this better represented the influence of the composition of the reacting radicals. They presented a thorough discussion of the Atherton and North model and of their own modification of it, which was shown to account well for the behaviour of several copolymerization systems. They realized that their model contains an additional (adjustable) parameter, similar to the phi factor, but recognized the advantage that its value can be predicted from a knowledge of the monomer reactivity ratios. Moreover, it was found that the value of their φ, required to bring about a fit to the experimental data, did not vary with monomer composition.

Experimental support for the Atherton–North concept came from the work of O'Driscoll *et al.*,[94] whose results were suggestive of the origin of diffusion control in the flexibility of the chain at least to the penultimate unit and probably beyond, an idea which is in harmony with the next model to be discussed.

1.3.1.3 The segmental diffusion model

As far back as 1959, Benson and North[95] had envisaged termination in radical copolymerization as a three-stage process, with (i) translational diffusion bringing the radicals into proximity, (ii) segmental diffusion allowing rearrangement of chains to permit (iii) chemical reaction to take place. It is generally agreed that the segmental diffusion element in this compound process would be the rate-controlling step. To devise a kinetic scheme that represented the role of segmental diffusion more realistically, Russo and Munari[96] attributed control of the relevant chain flexibility in a macroradical to the last four carbon–carbon bonds adjacent to the radical site. This is equivalent to saying that segmental diffusion is controlled by the last two monomer units in the radical, and it therefore constitutes a kind of penultimate unit effect.

According to this concept, there are essentially four types of radicals engaged in termination, which can be represented as AA·, AB·, BA· and BB·. Terminating in pairs, there are ten possible bimolecular reactions, of which six (cross-termination) take place between unlike radicals. To reduce the number of rate constants involved, it was assumed that the value of a rate constant for a cross-termination would be the geometric mean of those for the corresponding reactions between like radicals, *e.g.*

$$k_{tAABA} = 2(k_{tAAAA} \cdot k_{tABBA})^{1/2}$$

where k_{tAABA} characterizes the reaction between one AA· radical and one AB· radical. This leaves four termination rate constants to be handled, and these were incorporated into four δ parameters defined as follows

$$\delta_A = (k_{tAAAA})^{1/2}/k_{AA} \qquad \delta_B = (k_{tBBBB})^{1/2}/k_{BB}$$

$$\delta_{AB} = (k_{tABBA})^{1/2}/k_{BB} \qquad \delta_{BA} = (k_{tBAAB})^{1/2}/k_{AA}$$

where δ_{AB} and δ_{BA} are adjustable parameters, which are chosen to obtain the best possible fit to the data.

Russo and Munari used this ingenious basis to deduce an equation for the rate of copolymerization, which is best written in the form of equation (51), due to Procházka and Kratochvíl.[97]

$$\omega = \frac{I^{1/2}(r_A[A]^2 + 2[A][B] + r_B[B]^2)}{\left[\dfrac{(\delta_A r_A^2[A]^2 + \delta_{BA} r_A[A][B])}{(r_A[A] + [B])}\right] + \left[\dfrac{(\delta_B r_B^2[B]^2 + \delta_{AB} r_B[A][B])}{(r_B[B] + [A])}\right]} \tag{51}$$

(The rate of reaction is always represented by an equation in which the numerator depends only on propagation steps; the numerator is therefore always the same, as seen in equations (46), (48) and (51), whatever assumptions are made about termination.)

Further work on the styrene/methyl methacrylate system[85,98] showed that while δ_{BA}/δ_A shows no dependence on temperature, δ_{AB}/δ_B decreases with increasing temperature, a result which was explained on the basis of segmental flexibility. When the copolymerization was carried out in solvents (acetone, dimethylformamide, dioxane) rather than in bulk, k_p for styrene was reduced whereas the corresponding parameter for methyl methacrylate was enhanced. Similar effects had been reported with bromobenzene,[99] diethyl malonate[100] or benzonitrile[101] as solvent.

The Russo–Munari approach proved to be highly successful in accounting for the experimental rate-*versus*-composition data for many binary systems, but unfortunately it occasionally happens that the value to be attributed to one of the δ parameters is negative, and this is incompatible with the definition of δ. Although the Russo–Munari method is empirically very satisfactory, it has been suggested[102] that it cannot be accorded the full physical significance assumed in the derivation of equation (51), at least in those cases where negative values of δ arise; it may nevertheless be valid in general, but the reason for the occurrence of these exceptional cases is obscure.

Procházka and Kratochvíl[97] carried out a searching examination of all the termination models, taking into account the effect of the viscosity of the medium. They modified the Chiang and Rudin equation (equation 50) by introducing the geometric-mean assumption in the form of equation (44) with $\varphi = 2$, which leads to the formulation of the single termination constant as shown in equation (52).

$$k_t = (F_A k_{tAA}^{1/2} + F_B k_{tBB}^{1/2})^2 \tag{52}$$

Their thorough study showed that the chemical model is distinctly inferior to the diffusion models, and that the standard deviations of calculations based on the Russo–Munari treatment are always smaller, sometimes much smaller, than those for the other procedures.

The kinetic scheme of Russo and Munari has been extended by Jenkins and Mustafa[103] to include an additional termination reaction with a retarder or inhibitor for one of the monomers.

1.3.1.4 The 'overall' rate constants for propagation and termination

It is possible to view the rate of copolymerization as if it were parallel to homopolymerization by writing the rate of reaction as follows

$$\omega = \boldsymbol{k}_\mathbf{p}[\mathrm{X}\cdot][\mathrm{M}]$$

where $[\mathrm{X}\cdot] = ([\mathrm{A}\cdot] + [\mathrm{B}\cdot])$, $[\mathrm{M}] = ([\mathrm{A}] + [\mathrm{B}])$ and $\boldsymbol{k}_\mathbf{p}$ is regarded as the overall rate constant for propagation. Since this equation is equivalent to equation (41), the following relationship must be valid

$$\boldsymbol{k}_\mathbf{p}[\mathrm{X}\cdot][\mathrm{M}] = [\mathrm{A}\cdot]\{k_{AA}[\mathrm{A}] + k_{AB}[\mathrm{B}] + (k_{BA}[\mathrm{A}] + k_{BB}[\mathrm{B}])(k_{AB}[\mathrm{B}]/k_{BA}[\mathrm{A}])\}$$

Division throughout by $[\mathrm{X}\cdot]$ $\{= ([\mathrm{A}\cdot] + [\mathrm{B}\cdot])\}$, remembering equation (39), and additionally by $[\mathrm{M}]$ $\{= ([\mathrm{A}] + [\mathrm{B}])\}$, results in equation (53), where f_A and f_B are (as before) the mole fractions of A and B in the monomer feed.

$$\boldsymbol{k}_\mathbf{p} = [k_{AA} f_A + k_{AB} f_B + (k_{BA} f_A + k_{BB} f_B)(k_{AB} f_B / k_{BA} f_A)] \tag{53}$$

By rearrangement, and substitution of the monomer reactivity ratios, this is transformed into equation (54).

$$\boldsymbol{k}_\mathbf{p} = \frac{(r_A f_A^2 + 2 f_A f_B + r_B f_B^2)}{[(r_A f_A / k_{AA}) + (r_B f_B / k_{BB})]} \tag{54}$$

This equation defines an overall rate constant for propagation for a system obeying TM, regardless of the mechanism of the termination step.

If the overall rate equation for copolymerization is now written, by analogy with conventional homopolymerization, as

$$\omega = (\boldsymbol{k}_\mathbf{p} / \boldsymbol{k}_\mathbf{t}^{1/2}) I^{1/2} [\mathrm{M}] \tag{55}$$

substitution of equation (54) into the rate equation for a particular termination mechanism results in an expression for $\boldsymbol{k}_\mathbf{t}$ in terms of all the other parameters. To give an example, for the chemical termination mechanism the relevant rate equation is equation (46), from which $\boldsymbol{k}_t$ is found to be defined as follows

$$\boldsymbol{k}_t = \frac{(\delta_A^2 r_A^2[A]^2 + 2\varphi\delta_A\delta_B r_A r_B[A][B] + \delta_B^2 r_B^2[B]^2)}{[(r_A f_A/k_{AA}) + (r_B f_B/k_{BB})]^2} \tag{56}$$

Parallel expressions can be derived from equations (48) and (51).

1.3.2 Using the Penultimate Model for Propagation

It is almost universally accepted that the chemical model of termination has been superseded by some form of diffusion model, so the most realistic way to model the kinetics of copolymerization is to combine the penultimate model for propagation with the translational diffusion model or the segmental diffusion model for termination. Fukuda, Inagaki, Ma and their colleagues have worked in this way, using the translational diffusion option, and an account of their theory now follows.

When the eight propagation reactions (8)–(15) are taken into account, the overall rate constant for propagation $\boldsymbol{k}_p$ can be expressed as in equation (57).

$$\boldsymbol{k}_p = \frac{(\bar{r}_A f_A^2 + 2f_A f_B + \bar{r}_B f_B^2)}{[(\bar{r}_A f_A/\bar{k}_{AA}) + (\bar{r}_B f_B/\bar{k}_{BB})]} \tag{57}$$

In this equation, $\bar{r}_A$ and $\bar{r}_B$ have the meanings defined in relation to equation (17); if i,j = A,B and i$\neq$j throughout, these parameters can usually be formulated using the nomenclature $r_i' = k_{jii}/k_{jij}$ when, similarly, the terminal model monomer reactivity ratios become $r_i = k_{iii}/k_{iij}$. It is also necessary to define $\bar{k}_{AA}$ and $\bar{k}_{BB}$ by equation (58)

$$\bar{k}_{ii} = k_{iii}(r_i f_i + f_j)/[r_i f_i + f_j(1/s_i)] \tag{58}$$

and s_A and s_B by equation (59).

$$s_i = k_{jii}/k_{iii} \tag{59}$$

To ensure that the symbolism is clear, k_{ijm} is the rate constant for the reaction of a radical with penultimate unit i and ultimate unit j with monomer m.

It would be helpful to adopt O'Driscoll's suggestion[104] that these s parameters should be called 'chain-end reactivity ratios' in parallel with, but in contradistinction to, the monomer reactivity ratios.

In order to deal with termination controlled by diffusion, Fukuda *et al.*[105,106] chose to use the Procházka–Kratochvíl form of the Russo–Munari expression for termination, given in equation (52), and arrived at the following expression for the combined overall kinetic parameters

$$\boldsymbol{k}_p/\boldsymbol{k}_t^{1/2} = \frac{(\bar{r}_A f_A^2 + 2f_A f_B + \bar{r}_B f_B^2)}{[(\bar{r}_A \bar{\delta}_A f_A) + (\bar{r}_B \bar{\delta}_B f_B)]} \tag{60}$$

where $\bar{\delta}_i$ is defined by equation (61).

$$\bar{\delta}_i = \delta_i(r_i f_i + f_j)/[r_i f_i + f_j(1/s_i)] \tag{61}$$

Although lacking theoretical justification, the simplification $s_A = s_B$ $(=s)$ was introduced, and shown to be convenient in some circumstances; exceptions have been reported.[107] In this work, it was noted that the value of s^2 seemed to correlate with low values of $r_A r_B$; in fact, an argument based on the concept of stabilization energy was developed to show that equation (62) is to be expected.

$$r_A r_B = s_A s_B \ (=s^2) \tag{62}$$

Bamford,[108] by reference to transfer experiments, argued that the postulates of Fukuda *et al.* led to relations between ratios of rate constants that were unacceptable when considered in the context of the different types of transition states involved, and Ito[109] found the theory inappropriate to the acrylontrile/methyl methacrylate system. Responses to these criticisms[110] were made along the lines that, on the one hand, the assumptions are justified by their experimental success and that the origin of the effect may be different from that supposed, and, on the other hand, that Ito had reasoned on the basis of the incorrect chemical model of termination.

Fukuda *et al.*[110] went on to propose that the effects of penultimate units could be classified in two groups according to whether they showed: (i) an implicit penultimate unit effect (IPUE); or (ii) an explicit penultimate unit effect (EPUE). For an IPUE, composition studies reveal no evident penultimate unit effect on the monomer reactivity ratios, so $r_i' = r_i$; it is supposed that the origin of the phenomenon resides in stabilization of a radical by the penultimate unit. (This regime is sometimes referred to as the restricted penultimate unit effect.) By contrast, for an EPUE, both rate and monomer reactivity ratios are influenced, and it was proposed that specific interactions between the penultimate unit and the incoming monomer may be responsible. It must be noted that the s parameters do not figure in the expression for the overall composition of the copolymer nor in that for sequence distributions; they only affect the equation for the rate of reaction, a conclusion valid for the IPUE.

A phenomenological argument based on stabilization energies was invoked[111] to add a semiquantitative aspect to their theory. This leads to the conclusion that penultimate effects would disappear if the product of monomer reactivity ratios has a value close to unity; this idea was tested by Piton *et al.*[112] and found to be true for styrene and *p*-methoxystyrene. In the copolymerization of styrene and perdeutero-styrene and of methyl methacrylate and perdeutero-methyl methacrylate, Wittmer *et al.*[113] found no isotope effect in relation to copolymer composition but a definite effect on the rate of copolymerization; Davis and O'Driscoll[114] were later able to explain these results in terms of TM, in the first case, and the penultimate unit effect of Fukuda *et al.* (where all monomer reactivity ratios are equal to unity and $s_A = s_B = 0.73$), in the second case.

O'Driscoll and Davis[115] adapted a kinetic scheme, previously worked out to allow for depropagation, to show that the s parameters enter explicitly into the equation for rate of reaction but are always absent from the conditional probabilities which control composition. In fact, the relevant conditional probabilities are shown in equation (63).

$$P(\mathrm{B/A}) = [1 + r_A'(r_A + 1)/(r_A' + 1)]^{-1}$$

and

$$P(\mathrm{A/B}) = [1 + r_B'(r_B + 1)/(r_B' + 1)]^{-1} \tag{63}$$

If $r_A' = r_A$ and $r_B' = r_B$, these relations revert to the simple form seen in Section 1.2.2.

It can happen that the values of the s parameters derived by different methods show rather disappointing agreement; O'Driscoll and Huang[116] have shown that this is due to the inferior precision of the rotating-sector method compared to that of the pulsed-laser technique.

A possible explanation of the chemical mechanism of the penultimate unit effect has been put forward by Jenkins and O'Driscoll.[117]

1.3.3 A New General Model

Maxwell and Russell[118] have recently advanced a general treatment of diffusion-controlled copolymerization kinetics which attempts to treat the process even into the high-conversion region, taking into account the diffusion control of propagation as well as termination; the equations that result are no more complicated than those for homopolymerization, and the kinetics of propagation at high conversion are simpler than those at low conversion. The hope is expressed that this approach can be extended to systems of more than two components.

Their paper describes the concept of termination by 'reaction diffusion' or 'residual termination'[119,120] and also the idea of 'short–long' termination,[121] where most termination reactions are presumed to involve one long, relatively immobile, radical and one short, mobile, radical.

1.4 MODEL DISCRIMINATION

It has been seen above that, in studies of the composition of copolymers (that involve only propagation steps), it is necessary to decide, in each particular case, whether the behaviour of the system follows the Terminal Model (TM), the Penultimate Model (PM) or the Complex Model (CM). In corresponding studies of the rate of reaction, it is necessary to consider also the termination steps and, in each case, to distinguish between the Chemical Model, the Translational Diffusion Model and the Segmental Diffusion Model.

The first diagnosis is usually an examination of copolymer composition as a function of monomer feed composition to test for conformity with equation (7) or, failing that, with equation (17) or the corresponding equation for the complex model.

If the monomer reactivity ratios have constant values, *i.e.* they show no discernible trend with composition, it may seem to be reasonable to conclude that the system obeys TM. However, if a trend with composition is apparent, more complicated analysis is appropriate, and it may be that PM or CM will give a better account of the data. If the values of the monomer reactivity ratios show sensitivity to the solvent medium, CM is indicated as a possibility, particularly if the monomers have structures suggestive of charge-transfer interactions. However, the apparent constancy of reactivity ratios can be deceptive; Moad *et al.*[122] pointed out that this type of behaviour may very closely be approximated by systems in which a penultimate effect is operating, and stated that analysis of the sequence distribution would be required to decide between TM and PM for some cases.

To distinguish between PM and CM is far from easy, and it usually relies on high-resolution NMR analysis of sequence distributions; it seems that sequence distribution was first suggested as a diagnostic of kinetic models by Berger and Kuntz[32] in 1964, long before NMR had developed the necessary sensitivity. In 1966, Chûjô[123] discussed NMR as a model-discrimination technique, giving explicit functions for monomer reactivity ratios and persistence ratios in terms of NMR intensities. In recent years the NMR technique has been used extensively by groups in Brisbane and Eindhoven, and reference will be made to this work below.

Until very recently, all model-discrimination work relied on NMR triad analysis, but an additional tool is becoming available with the development of the mass spectrometry (MS) of polymers. The key innovation here was the use of soft ionization techniques, especially fast atom bombardment (FAB). This results in spectra in which the peak intensities of the molecualr ions derived from polymers are proportional to their relative amounts in the sample. Using FABMS, it has been established that decoding the information inherent in the MS peak intensities reveals the composition and the microstructure of copolymers. Low polymers can be studied directly but high polymers are examined after partial decomposition, *e.g.* by ozonolysis.[124]

Statistical analysis using Bernoullian, Markovian or sequential models can provide a means for characterizing the microstructure of a copolymer sample.[125] If partial decomposition is necessary, it is now possible to tell whether the cleavage process occurs by a selective or a nonselective process, and the method can be applied to polymers of both the step-growth and chain-growth types. The level of statistical modelling now practised enables FABMS to complement NMR in the elucidation of copolymer structure.[126]

1.4.1 Sequence Distributions

It was explained in Section 1.2.1 that the frequencies of occurrence of sequences of units in copolymers are governed by the conditional probabilities of the possible addition, *i.e.* propagation, steps; it is therefore not surprising to find that the conditional probabilities can be formulated in terms of monomer reactivity ratios combined with the ratio of monomer concentrations because the relative rates of competitive propagation reactions can be written as functions of these parameters. Thus, for example, according to TM, $P(\mathrm{B/A}) = (r_\mathrm{A}Q_\mathrm{m} + 1)^{-1}$ and $P(\mathrm{A/B}) = (r_\mathrm{B}/Q_\mathrm{m} + 1)^{-1}$, and it is easy to see why equations (1) and (7) are equivalent.

To explore models thoroughly on the basis of $Q_\mathrm{p} - Q_\mathrm{m}$ relationships, it is necessary to measure the frequency of occurrence of the various possible triads. Thus, the triads with a central A unit are AAA, AAB, BAA and BAB. Triads AAB and BAA are statistically equivalent and occur with the same frequency, so there are really only three types of A-centred triads to be considered, and similarly for B-centred triads. For TM, the expected fractions of A-centred triads are as follows

$$F_\mathrm{AAA} = r_\mathrm{A}^2Q_\mathrm{m}^2/(1 + 2r_\mathrm{A}Q_\mathrm{m} + r_\mathrm{A}^2Q_\mathrm{m}^2)$$

$$F_\mathrm{AAB} = F_\mathrm{BAA} = r_\mathrm{A}Q_\mathrm{m}/(1 + 2r_\mathrm{A}Q_\mathrm{m} + r_\mathrm{A}^2Q_\mathrm{m}^2)$$

$$F_\mathrm{BAB} = 1/(1 + 2r_\mathrm{A}Q_\mathrm{m} + r_\mathrm{A}^2Q_\mathrm{m}^2)$$

and the expected fractions of B-centred triads are

$$F_\mathrm{BBB} = r_\mathrm{B}^2/Q_\mathrm{m}^2(1 + 2r_\mathrm{B}/Q_\mathrm{m} + r_\mathrm{B}^2/Q_\mathrm{m}^2)$$

$$F_\mathrm{BBA} = F_\mathrm{ABB} = r_\mathrm{B}/Q_\mathrm{m}(1 + 2r_\mathrm{B}/Q_\mathrm{m} + r_\mathrm{B}^2/Q_\mathrm{m}^2)$$

$$F_\mathrm{ABA} = 1/(1 + 2r_\mathrm{B}/Q_\mathrm{m} + r_\mathrm{B}^2/Q_\mathrm{m}^2)$$

A different series of equations can be derived for PM, as quoted by Maxwell *et al.*,[127] in which the expected fractions of A-centred triads are

$$F_\mathrm{AAA} = r_\mathrm{A}r'_\mathrm{A}Q_\mathrm{m}^2/(1 + 2r'_\mathrm{A}Q_\mathrm{m} + r_\mathrm{A}r'_\mathrm{A}Q_\mathrm{m}^2)$$

$$F_\mathrm{AAB} = F_\mathrm{BAA} = r'_\mathrm{A}Q_\mathrm{m}/(1 + 2r'_\mathrm{A}Q_\mathrm{m} + r_\mathrm{A}r'_\mathrm{A}Q_\mathrm{m}^2)$$

$$F_\mathrm{BAB} = 1/(1 + 2r'_\mathrm{A}Q_\mathrm{m} + r_\mathrm{A}r'_\mathrm{A}Q_\mathrm{m}^2)$$

and the expected fractions of B-centred triads are

$$F_\mathrm{BBB} = r_\mathrm{B}r'_\mathrm{B}/Q_\mathrm{m}^2(1 + 2r'_\mathrm{B}/Q_\mathrm{m} + r_\mathrm{B}r'_\mathrm{B}/Q_\mathrm{m}^2)$$

$$F_\mathrm{BBA} = F_\mathrm{ABB} = r'_\mathrm{B}/Q_\mathrm{m}(1 + 2r'_\mathrm{B}/Q_\mathrm{m} + r_\mathrm{B}r'_\mathrm{B}/Q_\mathrm{m}^2)$$

$$F_\mathrm{ABA} = 1/(1 + 2r'_\mathrm{B}/Q_\mathrm{m} + r_\mathrm{B}r'_\mathrm{B}/Q_\mathrm{m}^2)$$

The CM gives rise to yet another group of expressions for the triad fractions. All three models are discussed by Hill and O'Donnell,[128] using a different formulation employing conditional probabilities rather than monomer reactivity ratios. They warn that discrimination between models becomes difficult if one of the triads is present in very low quantity because ratios of triad fractions are then subject to large uncertainty.

1.4.2 Individual Copolymerization Systems

1.4.2.1 Styrene and methyl methacrylate

This is the most extensively examined of all binary copolymer systems. The rate of reaction as a function of monomer feed composition at 60 °C had been studied over the greater part of the range by Walling,[89] and at the styrene-weak end of the range by Jenkins and Mustafa.[103]

Fukuda *et al.*[129] applied the rotating-sector technique in order to determine absolute values of $\boldsymbol{k}_p$ and $\boldsymbol{k}_t$ across the entire composition range; this study revealed that the termination step was completely inconsistent with the chemical model but followed the expectations of control by translation diffusion. At the same time, it was apparent that the kinetics of propagation did not obey the terminal model, even though that very model is adequate to explain the composition data. This state of affirs was explained by the same authors soon afterwards[105] as a consequence of a penultimate unit effect.

Fukuda *et al.*[130] made a comprehensive survey at 40 °C of the copolymer composition, the initiation rate, the volume contraction factor, the rate of polymerization, the molecular weight, and the radical lifetime; they found that the copolymer composition corresponded to TM for propagation but that the rate of reaction did not. Their data were compatible with the translational diffusion model for termination. For the monomer reactivity ratios, they found: $r_A = r'_A = 0.523$; $r_B = r'_B = 0.460$; $s_A = 0.30$ and $s_B = 0.53$. Clearly, these results place the system in the IPUE category.

A recent re-evaluation by van Herk[131] of the data of Fukuda *et al.* by a very convenient non-iterative method evaluates the s parameters as $s_A = 0.31$ and $s_B = 0.40$; it confirms that the model is very insensitive to the value of s_B.

In 1989, O'Driscoll and Huang[116] measured the rate of polymerization, the degree of polymerization and the composition of the copolymer formed as functions of conversion. Their results were in agreement with those of Fukuda *et al.* They also followed the rate of reaction up to 95% conversion;[132] they defined a 'gel-effect parameter', and analysed the progress of the reaction into three stages. The pulsed-laser technique of Olaj was also used to confirm the relation between $\boldsymbol{k}_p$ and composition.[133] An extensive re-evaluation of the kinetic data for this system has been presented by Schweer.[134]

Proper application of NMR techniques to the study of sequence distribution had to wait for the work of Maxwell *et al.*,[74] who made a complete re-assignment of the triad sequences in the ^{13}C NMR spectrum of this copolymer. They confirmed the conclusion of Fukuda *et al.* that a 'restricted' penultimate effect (IPUE) governs the behaviour. They also speculated that the basic reason for the apparent influence of non-terminal units may be found in the CM hypothesis, the Bootstrap Effect, or even in an effect of tacticity, which could mimic the penultimate unit effect but with different signficance attached to the monomer reactivity ratios. It should be noted that, when tacticity is to be taken into account, the number of possible types of triads in a binary polymer is 20, 10 centred on an A unit and 10 on a B unit. (Aerdts *et al.*[135] had previously shown that the existing ^1H NMR assignments for this monomer pair were incorrect; they made a new assignment, based on pentad sequence distribution — triads are inadequate.)

1.4.2.2 Styrene and ethyl, n*-butyl or lauryl methacrylate*

The pulsed-laser technique for the determination of $\boldsymbol{k}_p$ was used to examine the relationship of this parameter to the monomer feed composition.[136] None of these three systems obeys TM but all are fitted by the model of Fukuda *et al.* The monomer reactivity ratios and s parameters were determined.

1.4.2.3 Styrene and methyl or n*-butyl acrylate*

The pulsed-laser method[107] gave results broadly similar to the styrene/methacrylate systems. Monomer reactivity ratios were determined but the $s_A = s_B$ equality could not be made to fit the data. Clearly, though a penultimate acrylate group does not affect reactivity, a penultimate styrene unit does.

1.4.2.4 Styrene and acrylonitrile

After speculating about the use of NMR in model discrimination,[137] the technique was applied to this system at 60 °C.[127,138] From studies of the overall composition it was clear that TM was unsatisfactory but PM and CM were able to fit the data equally well. However, the triad fractions were in agreement with PM but incompatible with CM, so it was concluded that a penultimate effect exists with this monomer pair. This conclusion is in harmony with work on model compounds by Jones, Prementine and Tirrell,[139–141] who established that a penultimate acrylonitrile unit depresses the reactivity of a radical towards addition to acrylonitrile monomer by a factor of between 3 and 4.

The effect of using a solvent rather than bulk conditions is perceptible but small;[142] toluene mimics the behaviour of the bulk system at high styrene levels while acetonitrile similarly behaves as an analogue of acrylonitrile.

1.4.2.5 Styrene and maleic anhydride

A penultimate effect was first postulated in this system in 1953,[143] and several other workers used this model to fit their results. However, the known formation of complexes between these two monomers cast suspicion on PM, and suggested that CM should seriously be considered in this case. In fact, an NMR investigation[75] was unable to discriminate between the two models,[144] but recent work by Klumperman[145] and Amin Sanayei *et al.*[146] has settled the matter in favour of PM.

1.4.2.6 p-Chlorostyrene and methyl acrylate

This is one of the two systems which were described by Fukuda *et al.*[105] in introducing their penultimate unit theory.

1.5 COPOLYMERIZATION AND REACTIVITY

1.5.1 Experimental Studies of Radical and Monomer Reactivity

Copolymerization is capable of providing a wealth of information about the reactivities, particularly relative reactivities, of radicals and monomers. Monomer reactivity ratios provide the raw material to a large extent but, in recent years, both radioactive tracer and NMR techniques have been developed to augment simple composition studies in most useful ways, mostly as a result of excellent work by Bevington and Tirrell and their colleagues.

A feature of this work is the use of an initiator, labelled with ^{3}H, ^{13}C or ^{14}C, that decomposes to give labelled primary radicals which start the polymerization of a labelled monomer or of a selected pair of monomers. The relative rates of addition of the two monomers can then be deduced quantitatively from tracer assays or NMR assessment.

One phenomenon exposed by this kind of work is the influence of steric hindrance on radical addition. Thus, it was demonstrated that stilbene is reluctant to add to a radical that bears two substantial sustituents on the α-carbon atom,[147] and later work, comparing the reactivities of poly(methyl acrylate) and poly(methyl methacrylate) radicals, as well as poly(acrylonitrile) and poly(methacrylonitrile) radicals, confirmed the supposed hindrance of an α-methyl group.[148]

An interesting detail concerns the possible occurrence of 'head' addition, *i.e.* addition of a radical to the substituted end of the double bond in the monomer. This was shown to occur at the 20% level with vinyl acetate at 60 °C and to a greater extent at 100 °C, but to be absent in the polymerizations of styrene and methyl methacrylate.[149] Another intriguing detail is that addition of 1,4-butadiene to the 1-cyano-1-methylethyl radical takes place mostly in the 1,4 mode.[150]

It is important to know whether a primary radical with a structure resembling that of a monomer can be used as a model for the reactions of the polymer radicals terminating in a unit of that monomer. This has been tested for the 1-cyano-1-methylethyl radical and methacrylonitrile, for the 1-phenylethyl radical and styrene,[151] and for the 1-methoxycarbonyl-1-methylethyl radical and methyl methacrylate.[152] The correspondence is closer for the 1-phenylethyl radical and styrene than it is for the 1-cyano-1-methylethyl radical and methacrylonitrile; in the case of the 1-methoxycarbonyl-1-methylethyl radical and methyl methacrylate, it was found that the radicals concerned have very similar selectivities towards styrene and methyl methacrylate.

Further, the order or reactivity of several monomers towards the 1-phenylethyl and the 1-cyano-1-methylethyl radical has been established,[153] and the competitive addition of acrylonitrile and methyl methacrylate has been assessed.[154] Cywar and Tirrell[155] have also assessed the reactivities of the 1,3-diphenylethyl and the 3-cyano-1-phenylethyl radicals towards styrene and acrylonitrile; their results show that the rate of addition is indeed sensitive to the nature of the substituent on the γ-carbon atom in the radical, confirming the plausibility of the penultimate unit effect.

1.5.2 The Systematization of the Reactivity of Radicals, Monomers and Transfer Agents

It is generally accepted that reactivity in radical polymerization is governed by a combination of 'general' reactivity (*i.e.* stabilization energy) and polar effects. This concept was the basis of the well-known Q-e scheme of Alfrey and Price,[156] but there were several weaknesses in the form of this scheme. A more soundly based procedure, the 'Patterns of Reactivity' approach, based on the same principle but eliminating the arbitrary aspects of Q-e, was put forward in 1959, subsequently extended, and reviewed in 1967.[157] According to the Patterns scheme, the rate constant k_{iA} for reaction of a radical i with a monomer (or transfer agent) A is given by

$$\log k_{iA} = \log k_{3,T} + \alpha_A \sigma_i + \beta_A \tag{64}$$

where $k_{3,T}$ is the rate constant for the reaction of transfer to toluene for the same radical, and α_A and β_A are constants for the monomer which are evaluated experimentally; σ_i is the Hammett constant for the substituent(s) in the radical. Thus, all the parameters on the right-hand side of this equation are experimentally determined, and no arbitrary factors are necessary.

The main drawback to the use of the Patterns scheme has proved to be the need to measure the toluene transfer constants required for the evaluation of the $k_{3,T}$ vlaues. To overcome this obstacle, it has been suggested[158] that $k_{3,T}$ should be replaced in equation (64) by k_{iS}, the rate constant for addition of radical i to styrene. Although equation (64) is cast in terms of absolute rate constants, it should be noted that these quantities do not need to be determined because the equation can easily be rewritten so as to incorporate monomer reactivity ratios (or transfer constants), as follows

$$\log r_{ij} = \log r_{iS} - \alpha_j \sigma_i - \delta_j \tag{65}$$

In this equation, δ has replaced β because of the change of reference reaction described in the previous paragraph; δ is experimentally determined. When the reactions of a novel monomer are being investigated, its reactivity ratio in copolymerization with styrene is very likely to be determined, and hence the data necessary for use in the Patterns scheme are likely to be available. A further possible advance (and simplification) may be based on the use of the comparative reactivity of a radical with respect to a nonpolar monomer (styrene) and a polar monomer (acrylonitrile) to replace σ_i; if this is successful, all the necessary parameters will be obtained from polymerization studies.[159]

1.5.3 Moderated Copolymerization

Copolymerization can ingeniously be employed accurately to evaluate rate constants for very fast reactions of polymer radicals, *e.g.* transfer or retardation/inhibition processes. Consider a reaction with a very high transfer constant: normally working with a high concentration of transfer agent produces polymers too low in molecular weight to be isolated or characterized; on the other hand, the use of a low concentration of the transfer agent ensures that it will be depleted so rapidly that it is impossible to know what its effective concentration was during the course of the experiment.

Bamford solved this problem[160] by using the monomer of interest in low concentration, together with another monomer, with low reactivity towards the transfer agent, as a vehicle for the reaction. Although this technique successfully overcomes the inherent difficulty in evaluating the transfer constant, it introduces an uncertainty in the event that the process is subject to a penultimate unit effect because the penultimate unit in these experiments is derived from the vehicle monomer and not the monomer of prime interest, as would be the case in a simple transfer experiment. That such a penultimate effect does occur was established;[161] the transfer constant for the reaction of styrene radicals with tetrabromomethane was found to be 51, 169 and 368 $L\,mol^{-1}\,s^{-1}$ with, respectively, methyl methacrylate, methyl acrylate and styrene as the penultimate group. The origin of the penultimate unit effect was thought by Bamford and Basahel[161] to reside in local steric hindrance between bromine atoms in the transfer agent and the α-methyl group on the methyl methacrylate unit, the first time that such an effect had been invoked in transfer.

It was later shown that a similar approach works well in the evaluation of rate constants for reactions of polymer radicals with retarders or inhibitors, for the same reasons.[103]

1.6 MISCELLANEOUS

There are many aspects of copolymerization that cannot be covered in the space available here but, in some cases, recently published texts are available. A few notes are appended below to assist in obtaining access to the relevant research literature.

1.6.1 Emulsion Copolymerization

Much of the current research in this field is conducted by substantial groups in laboratories in Canada, Australia and The Netherlands. Reference is recommended to, among others, such papers as the following: Hamielec *et al.*,[162] Lacík *et al.*[163] and Maxwell *et al.*[164]

1.6.2 Alternating Copolymerization

A book on this topic was published by Cowie in 1985.[165]

1.6.3 Cyclocopolymerization

The discoverer of cyclopolymerization, Butler, published a book on this subject in 1992.[166]

1.6.4 Compositional Heterogeneity

Heterogeneity with respect to size is a familiar characteristic of polymers but, in the case of copolymers, it is necessary to consider also the statistical nature of the distribution of molecular composition. Considerable advances have been made in this area, treating block and graft copolymers[167,168] as well as statistical copolymers.[169,170]

1.6.5 Multi-component Copolymerization

Relatively infrequently, copolymerizations are performed with more than two monomers but analysis of terpolymerization is sometimes encountered, *e.g.* that of Fawcett and Smyth,[171] and some of the theoretical aspects of copolymerization have been extended to the general case of n monomers.[6] Ham and Lipman have developed a computer program for terpolymerization.[172] Many commercial polymerization systems involve more than two components.

Terpolymerization can usefully be employed to investigate some special situations, *e.g.* by incorporating a small molar percentage of an ionizable group. Thus, McCormick *et al.*[173] prepared three-component polymers from acrylamide, N-(4-butylphenyl)acrylamide and sodium acrylate (or an alternative ionizing monomer), and used them to study hydrophobic associations and structure–property relationships. McCormick and Salazar[174] went on to show that polyelectrolyte behaviour is detectable with as little as 0.5 mol % of charged groups present. Another use for terpolymerization is advantageously to modify the properties of a binary copolymer system; for example, Buback and Panten[175] added vinyl acetate to acrylonitrile and ethene in a mixture polymerizing under a pressure of 2000 bar at 170–240 °C. This enables a product with a high content of nitrile functional groups to be obtained from a reaction taking place in a single phase. The necessary six monomer reactivity ratios were obtained by application of first-order Markov chain statistics.

The EVM procedure (see Section 1.2.5) for the evaluation of monomer reactivity ratios has been extended to ternary systems by Duever *et al.*[176]

One of the recurrent questions in the terpolymerization field is that of the possibility of finding azeotropic conditions, where the copolymer formed has exactly the same composition as the monomer mixture from which it derives. Azeotropes are well known in binary systems[14,15] and the theory for ternary mixtures was worked out by Ring in 1967.[177] The conditions for ternary azeotropy were linked to the Q-e parameters (see Section 1.5.2) by Ham[178] following some proposals of Tarasov *et al.*,[179] but considerable dispute has revolved around this issue, involving Mayo,[180] O'Driscoll[181] and O'Driscoll and Ham[182] in the early days and, in the recent past, Quella[183] and Ham.[184] In view of the weaknesses inherent in the Q-e scheme and of the uncertain reliability of the majority of Q-e values in the literature, it may well be pointless to continue this argument at the present time.

Kinetic modelling, mostly applied to binary systems so far, is being developed.[185]

1.6.6 IUPAC Recommendations

The International Union of Pure and Applied Chemistry (IUPAC) has issued recommendations regarding the nomenclature of copolymers[186] and values of rate parameters for polymerization.[187]

ACKNOWLEDGEMENTS

The very helpful advice of Professors Kenneth O'Driscoll and Saverio Russo during the preparation of this review is most gratefully appreciated, as is the unfailingly helpful eye for detail of Ing. Jitka Jenkins.

1.7 REFERENCES

1. T. Alfrey and G. Goldfinger, *J. Chem. Phys.*, 1944, **12,** 205.
2. F. R. Mayo and F. M. Lewis, *J. Am. Chem. Soc.*, 1944, **66,** 1594.
3. F. T. Wall, *J. Am. Chem. Soc.*, 1944, **66,** 2050.
4. R. G. W. Norrish and E. F. Brookman, *Proc. Roy. Soc.*, 1939, **A171,** 147.
5. T. Alfrey and G. Goldfinger, *J. Chem. Phys.*, 1944, **12,** 322.
6. C. Walling and E. R. Briggs, *J. Am. Chem. Soc.*, 1945, **67,** 1774.
7. F. R. Mayo and C. Walling, *Chem. Rev.*, 1950, **46,** 191.
8. C. H. Bamford, W. G. Barb, A. D. Jenkins and P. F. Onyon, 'The Kinetics of Vinyl Polymerisation by Radical Mechanisms', Butterworths, London, 1958.
9. J. C. Bevington, 'Radical Polymerization', Academic Press, London, 1961.
10. T. Alfrey, J. J. Bohrer and H. Mark, 'Copolymerisation', Interscience, New York, 1952.
11. G. E. Ham, 'Copolymerisation', Interscience, New York, 1964.
12. P. Wittmer, *Makromol. Chem., Suppl.*, 1979, **3,** 129.
13. T. Fukuda, K. Kubo and Y-D. Ma, *Prog. Polym. Sci.*, 1992, **17,** 875.
14. A. Rudin, 'The Elements of Polymer Science and Engineering', Academic Press, New York, 1982.
15. J. M. G. Cowie, 'Polymers: Chemistry & Physics of Modern Materials', 2nd edn., Blackie, Glasgow, and Chapman and Hall, New York, 1991.
16. A. E. Hamielec, J. F. MacGregor and A. Penlidis, in 'Comprehensive Polymer Science', ed. G. Allen and J. C. Bevington, Pergamon Press, Oxford, 1989, vol. 3, p. 17.
17. K. F. O'Driscoll and S. Russo, eds., *Makromol. Chem., Makromol. Symp.*, 1987, **10/11.**
18. S. I. Kuchanov, *Adv. Polym. Sci.*, 1992, **103,** 1.
19. J. L. Koenig, 'Chemical Microstructure of Polymer Chains', Wiley, New York, 1980.
20. E. Merz, T. Alfrey and G. Goldfinger, *J. Polym. Sci.*, 1946, **1,** 75.
21. P. D. Bartlett and K. Nozaki, *J. Am. Chem. Soc.*, 1946, **68,** 1495.
22. C. A. Barson, *Eur. Polym. J.*, 1984, **20,** 125.
23. C. A. Barson and D. T. Fenn, *Eur. Polym. J.*, 1989, **25,** 719.
24. J. A. Seiner and M. Litt, *Macromolecules*, 1971, **4,** 308.
25. C. U. Pitman, Jr. and T. D. Rounsefell, *Macromolecules*, 1975, **8,** 46.
26. R. E. Cais, R. G. Farmer, D. J. T. Hill and J. H. O'Donnell, *Macromolecules*, 1979, **12,** 835.
27. D. Braun and W. Czerwinski, *Makromol. Chem.*, 1983, **184,** 1071.
28. P. F. Brown and K. Fujimori, *J. Polym. Sci., Polym. Chem. Ed.*, 1994, **32,** 2971.
29. J. Bartoň and E. Borsig, 'Complexes in Free-Radical Polymerisation', Elsevier, Amsterdam, 1988.
30. M. Rätzsch and O. Vogl, *Prog. Polym. Sci.*, 1991, **16,** 279.
31. P. Karad and C. Schneider, *J. Polym. Sci., Polym. Chem. Ed.*, 1978, **16,** 1137.
32. M. Berger and I. Kuntz, *J. Polym. Sci.*, 1964, **A2,** 1687.

33. F. T. Wall, *J. Am. Chem. Soc.*, 1941, **63,** 1862.
34. M. Fineman and S. D. Ross, *J. Polym. Sci.*, 1950, **5,** 259.
35. D. W. Behnken, *J. Polym. Sci.*, 1964, **A2,** 645.
36. P. W. Tidwell and G. A. Mortimer, *J. Polym. Sci.*, 1965, **A3,** 369.
37. T. Alfrey, J. J. Bohrer and H. Mark, 'Copolymerisation', Interscience, New York, 1952, p. 23.
38. R. Chûjô, H. Ubara and A. Nishioka, *Polym. J.*, 1972, **3,** 670.
39. R. McFarlane, P. M. Reilly and K. F. O'Driscoll, *J. Polym. Sci., Polym. Chem. Ed.*, 1980, **18,** 251.
40. T. Kelen and F. Tüdős, *J. Macromol. Sci.*, 1975, **A9,** 1.
41. W. Czerwinski, G. Disselhoff, F. Tüdős, D. Braun, T. Kelen and B. Turcsanyi, *Angew. Makromol. Chem.*, 1984, **125,** 161.
42. A. J. Yezrielev, E. L. Brokhina and Y. S. Roskin, *Vysokomol. Soedin.*, 1969, **A11,** 1670.
43. R. van der Meer, H. N. Linssen and A. L. German, *J. Polym. Sci., Polym. Chem. Ed.*, 1978, **16,** 2915.
44. F. L. M. Hautus, H. N. Linssen and A. L. German, *J. Polym. Sci., Polym. Chem. Ed.*, 1984, **22,** 3487, 3661.
45. H. Patino-Leal, P. M. Reilly and K. F. O'Driscoll, *J. Polym. Sci., Polym. Lett. Ed.*, 1980, **18,** 219.
46. K. K. Chee and S. C. Ng, *Macromolecules,* 1986, **19,** 2779.
47. K. L. Petrak and E. Pitts, *Polymer,* 1983, **24,** 729.
48. F. L. M. Hautus, A. L. German and H. N. Linssen, *J. Polym. Sci., Polym. Lett. Ed.*, 1985, **23,** 311.
49. K. F. O'Driscoll and P. M. Reilly, *Makromol. Chem., Makromol. Symp.*, 1987, **10/11,** 355.
50. T. A. Duever, K. F. O'Driscoll and P. M. Reilly, *J. Polym. Sci., Polym. Chem. Ed.*, 1988, **26,** 965.
51. T. A. Duever, K. F. O'Driscoll and P. M. Reilly, *Polym. Mat. Sci. Eng.*, 1988, **58,** 369.
52. M. Dubé, R. A. Sanayei, A. Penlidis, K. F. O'Driscoll and P. M. Reilly, *J. Polym. Sci., Polym. Chem. Ed.*, 1991, **29,** 703.
53. C. Hagiopol, O. Frangu and L. Dumitru, *J. Macromol. Sci., Chem.*, 1989, **A26,** 1363.
54. R. Mao and M. B. Huglin, *Polymer,* 1993, **34,** 1709.
55. T. Kelen, F. Tüdős, D. Braun and W. K. Czerwinski, *Makromol. Chem.*, 1990, **191,** 1853, 1863.
56. R. Mao and M. B. Huglin, *Polymer,* 1994, **35,** 3525.
57. G. Bauduin and B. Boutevin, *Makromol. Chem., Rapid Commun.*, 1992, **12,** 197.
58. B. Klumperman and G. Vonk, *Eur. Polym. J.*, 1994, **30,** 955.
59. A. L. Burke, T. A. Duever and A. Penlidis, *J. Polym. Sci., Polym. Chem. Ed.*, 1993, **31,** 3065.
60. A. L. Burke, T. A. Duever and A. Penlidis, *Macromolecules,* 1994, **27,** 386.
61. A. Chapiro, *Eur. Polym. J.*, 1989, **25,** 713.
62. V. F. Gromov, N. I. Galperina, T. O. Osmanov, P. M. Khomikovskii and A. D. Abkin, *Eur. Polym. J.*, 1980, **16,** 529.
63. K. Plochocka, *J. Macromol. Sci. Rev.*, 1981, **C20,** 67.
64. H. J. Harwood, *Makromol. Chem., Macromol. Symp.*, 1987, **10/11,** 331.
65. J. San Roman, E. L. Madruga and M. A. Del Puerto, *Angew. Makromol. Chem.*, 1979, **78,** 129; 1980, **86,** 1, 17.
66. E. L. Madruga, J. San Roman and M. A. Del Puerto, *J. Macromol. Sci., Chem.*, 1979, **A13,** 1105.
67. C. H. Bamford and S. Brumby, *Makromol. Chem.*, 1967, **105,** 122.
68. G. Bontà, B. M. Gallo and S. Russo, *Polymer,* 1975, **16,** 429.
69. P. Kratochvíl, D. Straková, J. Stejskal and Z. Tuzar, *Macromolecules,* 1983, **16,** 1136.
70. Yu. D. Semchikov, L. A. Smirnova, T. Ye. Knyazeva, S. A. Bulgakova and V. I. Sherstyanykh, *Eur. Polym. J.*, 1990, **26,** 883.
71. G. A. Egorochkin, Yu. D. Semchikov, L. A. Smirnova, N. V. Karyakin and A. M. Kut'in, *Eur. Polym. J.*, 1992, **28,** 681.
72. K. Y. Park, E. R. Santee and H. J. Harwood, *Eur. Polym. J.*, 1989, **25,** 651.
73. T. P. Davis, *Polym. Commun.*, 1990, **31,** 442.
74. I. A. Maxwell, A. M. Aerdts and A. L. German, *Macromolecules,* 1993, **26,** 1956.
75. D. J. T. Hill, A. P. Lang, P. D. Munro and J. H. O'Donnell, *Eur. Polym. J.*, 1992, **28,** 391.
76. B. Klumperman and K. F. O'Driscoll, *Polymer,* 1993, **34,** 1032.
77. B. Klumperman and I. R. Kraeger, *Macromolecules,* 1994, **27,** 1529.
78. G. G. Lowry, *J. Polym. Sci.*, 1960, **42,** 463.
79. J. A. Howell, M. Izu and K. F. O'Driscoll, *J. Polym. Sci. (A-1)*, 1970, **8,** 699.
80. M. Izu and K. F. O'Driscoll, *J. Polym. Sci. (A-1)*, 1970, **8,** 1687.
81. H. K. Mahabadi, *Makromol. Chem., Macromol. Symp.*, 1987, **10/11,** 127.
82. M. Buback, B. Huckestein and G. T. Russell, *Macromol. Chem. Phys.*, 1994, **195,** 539.
83. C. H. Bamford, E. Schofield and J. Eilers, *Polymer,* 1989, **30,** 540; *Eur. Polym. J.*, 1989, **25,** 683; *Eur. Polym. J.*, 1994, **30,** 1127.
84. O. F. Olaj, G. Zifferer and G. Gleixner, *Macromolecules,* 1987, **20,** 839.
85. G. Bontà, B. M. Gallo and S. Russo, *J. Chem. Soc., Faraday Trans. 1,* 1975, **71,** 1727.
86. E. L. Madruga, J. San Roman and M. A. Del Puerto, *Polymer,* 1981, **22,** 951.
87. G. Bontà, B. M. Gallo, S. Russo and C. Uliana, *Polymer,* 1976, **17,** 217.
88. H. W. Melville, B. Noble and W. F. Watson, *J. Polym. Sci.*, 1947, **2,** 229.
89. C. Walling, *J. Am. Chem. Soc.*, 1949, **71,** 1930.
90. P. E. M. Allen and C. R. Patrick, *Makromol. Chem.*, 1961, **47,** 154.
91. A. M. North and G. A. Reed, *Trans. Faraday Soc.*, 1961, **57,** 859.
92. J. N. Atherton and A. M. North, *Trans. Faraday Soc.*, 1962, **58,** 2049.
93. S. S. M. Chiang and A. Rudin, *J. Macromol. Sci., Chem.*, 1975, **A9,** 237.
94. K. F. O'Driscoll, W. Wertz and A. Husar, *J. Polym. Sci. (A-1)*, 1967, **5,** 2159.
95. S. W. Benson and A. M. North, *J. Am. Chem. Soc.*, 1959, **81,** 1339.
96. S. Russo and S. Munari, *J. Macromol. Sci., Chem.*, 1968, **A2,** 1321.
97. O. Procházka and P. Kratochvíl, *J. Polym. Sci., Polym. Chem. Ed.*, 1983, **21,** 3269.
98. G. Bontà, B. M. Gallo and S. Russo, *J. Chem. Soc., Faraday Trans. 1,* 1973, **69,** 328.
99. G. M. Burnett, G. G. Cameron and S. N. Joiner, *J. Chem. Soc., Faraday Trans. 1,* 1973, **69,** 322.

100. C. H. Bamford and S. Brumby, *Makromol. Chem.*, 1967, **105,** 122.
101. M. M. Zafar, *Makromol. Chem.*, 1972, **157,** 219.
102. K. Ito and K. F. O'Driscoll, *J. Polym. Sci., Polym. Chem. Ed.*, 1979, **A17,** 3913.
103. A. D. Jenkins and B. H. Mustafa, *Eur. Polym. J.*, 1983, **19,** 1009.
104. K. F. O'Driscoll, *Makromol. Chem., Macromol. Symp.*, 1992, **53,** 53.
105. T. Fukuda, Y.-D. Ma, K. Kubo and A. Takada, *Polym. J.*, 1989, **21,** 1003.
106. Y.-D. Ma, T. Fukuda and H. Inagaki, *Macromolecules,* 1985, **18,** 26.
107. T. P. Davis, K. F. O'Driscoll, M. C. Piton and M. A. Winnik, *Polym. Int.*, 1991, **24,** 65.
108. C. H. Bamford, *Polym. Commun.*, 1989, **30,** 36.
109. K. Ito, *Polym. Commun.*, 1988, **29,** 223.
110. T. Fukuda, Y.-D. Ma, K. Kubo and H. Inagaki, *Macromolecules,* 1991, **24,** 370.
111. T. Fukuda, Y.-D. Ma and H. Inagaki, *Makromol. Chem., Rapid. Commun.*, 1987, **8,** 495.
112. M. C. Piton, M. A. Winnik, T. P. Davis and K. F. O'Driscoll, *J. Polym. Sci., Polym. Chem. Ed.*, 1990, **28,** 2097.
113. P. Wittmer, H. Böck, H. Naarmann and B. J. Schmitt, *Makromol. Chem.*, 1981, **182,** 2505.
114. T. P. Davis and K. F. O'Driscoll, *Makromol. Chem., Rapid Commun.*, 1989, **10,** 509.
115. K. F. O'Driscoll and T. P. Davis, *J. Polym. Sci., Polym. Lett. Ed.*, 1989, **27,** 417.
116. K. F. O'Driscoll and J. Huang, *Eur. Polym. J.*, 1989, **25,** 629.
117. A. D. Jenkins and K. F. O'Driscoll, *J. Polym. Sci., Polym. Chem. Ed.*, 1995, **33,** 1189.
118. I. A. Maxwell and G. T. Russell, *Makromol. Chem., Theory Sim.*, 1993, **2,** 95.
119. W. J. Huang, T. S. Frick, M. R. Landry, J. A. Lee, T. P. Lodge and M. Tirrell, *AIChE J.*, 1987, **33,** 573.
120. G. T. Russell, D. H. Napper and R. G. Gilbert, *Macromolecules,* 1988, **21,** 2133.
121. G. T. Russell, R. G. Gilbert and D. H. Napper, *Macromolecules,* 1992, **25,** 2459.
122. G. Moad, D. H. Solomon, T. H. Spurling and R. A. Stone, *Macromolecules,* 1989, **22,** 1145.
123. R. Chûjô, *J. Phys. Soc. Jpn.*, 1966, **21,** 2669.
124. G. Montaudo, E. Scamporrino and D. Vitalini, *Macromolecules,* 1991, **24,** 376.
125. M. S. Montaudo and G. Montaudo, *Macromolecules,* 1992, **25,** 4264.
126. M. S. Montaudo and G. Montaudo, *Makromol. Chem., Macromol. Symp.*, 1993, **65,** 269.
127. I. A. Maxwell, A. M. Aerdts and A. L. German, *Macromolecules,* 1993, **26,** 1956.
128. D. J. T. Hill and J. H. O'Donnell, *Makromol. Chem., Macromol. Symp.*, 1987, **10/11,** 375.
129. T. Fukuda, Y.-D. Ma and H. Inagaki, *Makromol. Chem., Suppl.*, 1985, **12,** 125.
130. T. Fukuda, Y.-D. Ma and H. Inagaki, *Macromolecules,* 1985, **18,** 17.
131. A. M. van Herk, *J. Chem. Educ.*, 1995, in press.
132. K. F. O'Driscoll and J. Huang, *Eur. Polym. J.*, 1990, **26,** 643.
133. T. P. Davis, K. F. O'Driscoll, M. C. Piton and M. A. Winnik, *J. Polym. Sci., Polym. Lett. Ed.*, 1989, **27,** 181.
134. J. Schweer, *Makromol. Chem., Theory Sim.*, 1993, **2,** 485.
135. A. M. Aerdts, J. W. de Haan, A. L. German and G. P. M. van der Velden, *Macromolecules,* 1991, **24,** 1473.
136. T. P. Davis, K. F. O'Driscoll, M. C. Piton and M. A. Winnik, *Macromolecules,* 1990, **23,** 2113.
137. R. G. Farmer, D. J. T. Hill and J. H. O'Donnell, *J. Macromol. Sci., Chem.*, 1980, **A14,** 51.
138. D. J. T. Hill, J. H. O'Donnell and P. W. O'Sullivan, *Macromolecules,* 1982, **15,** 960.
139. S. A. Jones, G. S. Prementine and D. A. Tirrell, *J. Am. Chem. Soc.*, 1985, **107,** 5275.
140. G. S. Prementine and D. A. Tirrell, *Macromolecules,* 1987, **20,** 3034.
141. G. S. Prementine and D. A. Tirrell, *Macromolecules,* 1989, **22,** 52.
142. D. J. T. Hill, A. P. Lang, P. D. Munro and J. H. O'Donnell, *Eur. Polym. J.*, 1992, **28,** 391.
143. W. G. Barb, *J. Polym. Sci.*, 1953, **11,** 117.
144. J. R. Ebdon, C. R. Towns and K. Dodgson, *J. Macromol. Chem., Rev.*, 1986, **C26,** 523.
145. B. Klumperman, Ph.D. Thesis, Technical University of Eindhoven, 1994.
146. R. Amin Sanayei, K. F. O'Driscoll and B. Klumperman, *Macromolecules,* 1994, **27,** 5577.
147. C. A. Barson, J. C. Bevington and T. N. Huckerby, *Polym. Bull.*, 1986, **16,** 209.
148. K. Behari, J. C. Bevington and T. N. Huckerby, *Polymer,* 1988, **29,** 1867.
149. J. C. Bevington, S. W. Breuer, E. N. J. Heseltine, T. N. Huckerby and S. C. Varma, *J. Polym. Sci., Polym. Chem. Ed.*, 1987, **25,** 1085.
150. J. C. Bevington, D. A. Cywar, T. N. Huckerby, R. A. Lyons, E. Senogles and D. A. Tirrell, *Eur. Polym. J.*, 1991, **27,** 603.
151. J. C. Bevington, B. F. Bowden, D. A. Cywar, R. A. Lyons, E. Senogles and D. A. Tirrell, *Eur. Polym. J.*, 1991, **27,** 1239.
152. J. C. Bevington, R. A. Lyons and E. Senogles, *Eur. Polym. J.*, 1992, **28,** 283.
153. J. C. Bevington, D. A. Cywar, T. N. Huckerby, E. Senogles and D. A. Tirrell, *Eur. Polym. J.*, 1990, **26,** 41, 871.
154. D. A. Cywar and D. A. Tirrell, *Eur. Polym. J.*, 1989, **25,** 657.
155. D. A. Cywar and D. A. Tirrell, *J. Am. Chem. Soc.*, 1989, **111,** 7544.
156. T. Alfrey and C. C. Price, *J. Polym. Sci.*, 1947, **2,** 101.
157. A. D. Jenkins, *Adv. Free-Rad. Chem.*, 1967, **2,** 139.
158. A. D. Jenkins, *Eur. Polym. J.*, 1989, **25,** 721.
159. A. D. Jenkins, *Macromol. Chem. Phys., Rapid Commun.*, 1996, in press; *J. Polym. Sci.*, 1996, in press.
160. C. H. Bamford, *J. Chem. Soc., Faraday Trans. 1,* 1976, **72,** 2805.
161. C. H. Bamford and S. N. Basahel, *J. Chem. Soc., Faraday Trans. 1,* 1978, **74,** 1020.
162. A. E. Hamielec, J. F. MacGregor and A. Penlidis, *Makromol. Chem.*, 1987, **10/11,** 521.
163. I. Lacík, B. S. Casey, D. F. Sangster, R. G. Gilbert and D. H. Napper, *Macromolecules,* 1992, **25,** 4065.
164. I. A. Maxwell, L. F. J. Noel, H. A. S. Schoonbrood and A. L. German, *Makromol. Chem., Theory Sim.*, 1993, **2,** 269.
165. J. M. G. Cowie, 'Alternating Copolymers', Plenum Press, New York, 1985.
166. G. B. Butler, 'Cyclopolymerisation and Cyclopolymerisation', Dekker, New York, 1992.
167. J. Stejskal, J. Horská and P. Kratochvíl, *Macromolecules,* 1984, **17,** 920.
168. J. Stejskal, P. Kratochvíl and A. D. Jenkins, *Macromolecules,* 1987, **20,** 181.

169. J. Stejskal and P. Kratochvíl, *Macromolecules,* 1987, **20,** 2624.
170. J. Stejskal, P. Kratochvíl, D. Straková and O. Procházka, *Macromolecules,* 1986, **19,** 1575.
171. A. H. Fawcett and U. Smyth, *Eur. Polym. J.,* 1989, **25,** 791.
172. G. E. Ham and R. D. A. Lipman, *J. Macromol. Sci., Chem.,* 1967, **1,** 1005.
173. C. L. McCormick, J. C. Middleton and C. E. Grady, *Polymer,* 1993, **33,** 4184.
174. C. L. McCormick and L. C. Salazar, *Polymer,* 1993, **33,** 4384.
175. M. Buback and K. Panten, *Makromol. Chem.,* 1993, **194,** 2471.
176. T. A. Duever, K. F. O'Driscoll and P. M. Reilly, *J. Polym. Sci., Polym. Chem. Ed.,* 1983, **21,** 2003.
177. W. Ring, *Makromol. Chem.,* 1967, **101,** 145.
178. G. E. Ham, *J. Macromol. Chem.,* 1966, **1,** 93.
179. A. I. Tarasov, V. A. Tskhai and S. S. Spasskii, *Vysokomol. Soedin.,* 1960, **2,** 1601.
180. F. R. Mayo, *J. Polym. Sci.,* 1964, **A2,** 4207.
181. K. F. O'Driscoll, *J. Polym. Sci.,* 1965, **B3,** 305.
182. K. F. O'Driscoll and G. E. Ham, *J. Macromol. Sci., Chem.,* 1967, **A1,** 1365.
183. F. Quella, *Makromol. Chem.,* 1989, **190,** 1445.
184. G. E. Ham, *Makromol. Chem.,* 1992, **193,** 1123.
185. T. Xie and A. E. Hamielec, *Makromol. Chem., Theory Sim.,* 1993, **2,** 421, 455.
186. International Union of Pure and Applied Chemistry, Commission IV.1, *Pure Appl. Chem.,* 1985, **57,** 1427.
187. International Union of Pure and Applied Chemistry, Working Party on the Modelling of Polymerization Kinetics, *Pure Appl. Chem.,* 1992, **64,** 1563; *J. Polym. Sci., Polym. Chem. Ed.,* 1992, **30,** 851.

2

Enzymatic Catalysis in Organic Media for Polymer Synthesis

THIERRY LALOT, MARYVONNE BRIGODIOT and ERNEST MARÉCHAL
Université Pierre et Marie Curie, Paris, France

2.1 INTRODUCTION

The study and uses of enzymes are no longer limited to the domain of biologists.[1] Since the mid-1980s, chemists have used more and more of the enormous possibilities of these catalysts, particularly in organic chemistry. Enzymes exhibit a much higher selectivity than classical catalysts. An enzyme catalyzes only one type of reaction, for example hydrolysis, and it recognizes only one compound (the substrate) with a specific structure.

Many examples of enzymatic catalysis in aqueous media are reported in the literature,[2] but there are fewer references relating to the use of enzymes in organic media compared to those in water. However, fundamental and applied research on enzymatic catalysis in non-aqueous media have made important developments since the 1970s.

Examples show that the use of enzymes in organic media leads to well-controlled syntheses, with the following advantages.[3,4]

(i) Many non-polar substrates cannot be dissolved in water and this decreases drastically the efficiency of the enzymatic process. On the other hand, they are often soluble in organic solvents and in these conditions synthesis can be realized.

(ii) Enzymes are insoluble in organic media and can be recovered easily from the reaction mixture; consequently, immobilization is often unnecessary.

(iii) There is little possibility of microbial contamination.

(iv) Enzymes show their maximum activity at a temperature close to room temperature; this drastically decreases the contribution of side reactions which are observed in non-enzymatic processes, as the latter are most often carried out at high temperature. For example, the classical synthesis of unsaturated polyesters exhibits many side reactions leading to ill-defined structures;[5] on the other hand, Geresh[6] obtained unsaturated polyesters with a well-defined structure using enzyme-catalyzed synthesis in organic media (*Candida cylindracea* lipase in tetrahydrofuran or acetonitrile).

The use of enzymatic catalysis in organic media is rapidly expanding in polymer science and should lead to fantastic achievements.[7] Before analyzing the possible applications of enzymes in organic chemistry it is necessary to give a brief description of their structures, study their stability in media different from their natural environment, and study the kinetics of enzymatic processes.

Enzymes are proteins extracted from cells of living systems. They are linear copolymers whose monomer units are amino acid residues —(CO—CHR—NH)— linked by peptide linkages. Twenty different amino acids have been identified. The specificity and the efficiency of an enzyme depend on the microstructure of the chain, which is defined by the order and the nature of the amino acid residues and the conformation of the macromolecule.

There are several levels of structural order; each of them is responsible for a function in the formation and in the preservation of the catalytic activity. The primary structure corresponds to the sequence (order) of the monomer units in the chain. The secondary structure gives the respective position of each unit with respect to that of the adjacent ones. The chain can be *trans* zig-zag or helicoidal or a chain can contain both of these different configurations (Figure 1). The stability of these conformations results, for a large part, from the hydrogen bonds between CO and NH of the peptide links.

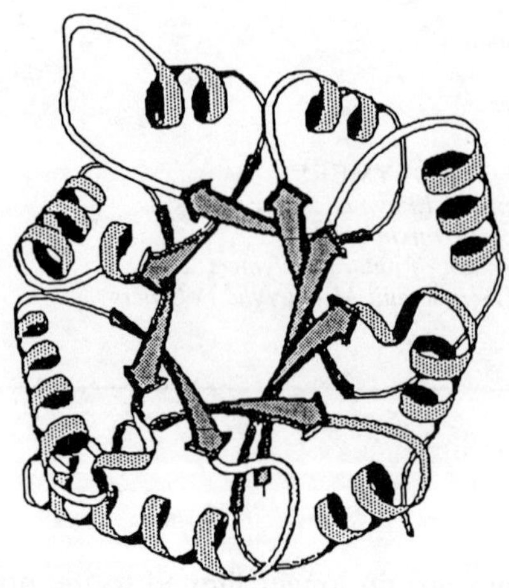

Figure 1 A part of the triose isomerase structure (an enzyme involved in glycolysis). The β sheets alternate with the α coils and form a 'tunnel' arrangement.[8]

The tertiary structure is responsible for the compactness of the globular proteins. About 30–50% of the substituents R are hydrophobic. When the enzyme is in a hydrophilic medium, the R groups are inside the coil and the polar groups are at the surface.

The stability of a protein results from the presence of bridges which link different units of the chain: hydrogen bonds, electrostatic and hydrophobic interactions, van der Waals interactions, disulfide bridges, various chelates. The three-dimensional structure results from folding of the main chain; this will make two units close even though they are separated by many units in the chain and this structural feature is responsible for the formation of several active sites.

According to Koshland,[9] the structure of the protein is not completely rigid and the active site conformation partially results from its association with the substrate. In some cases the formation of the catalytic site requires the contribution of several similar or different polypeptide chains. The stability of the quaternary structure is ensured by bonds analogous to those observed in the tertiary structure. The active sites, which have non-symmetrical structures, can be thought of as 'pockets'. Some areas of the active sites are formed of residues which are responsible for the attachment of the substrate to the enzyme, while other areas are directly responsible for the catalytic activity.

The formation of the enzyme–substrate complex decreases the energy of the complex of the reactants and this increases the interactions between the catalytic centre and the substrate. The conformation of the active site and that of the substrate are complementary in the transition state, as the enzyme is not a completely rigid macromolecule.

Some enzymes are active only in the presence of another component, which is termed a cofactor or coenzyme. Coenzymes are species derived from nucleotides which combine with the substrate and play the role of carrier with respect to different species such as redox systems, phosphate groups and protons. Coenzymes are regenerated by a second reaction which can be enzymatic, chemical or electrochemical. For example, the enzyme-catalyzed oxidation of a primary alcohol requires NAD^+ as cofactor and leads to an aldehyde and to the reduced form of the cofactor NADH. NAD^+ is regenerated using another aldehyde (Scheme 1).

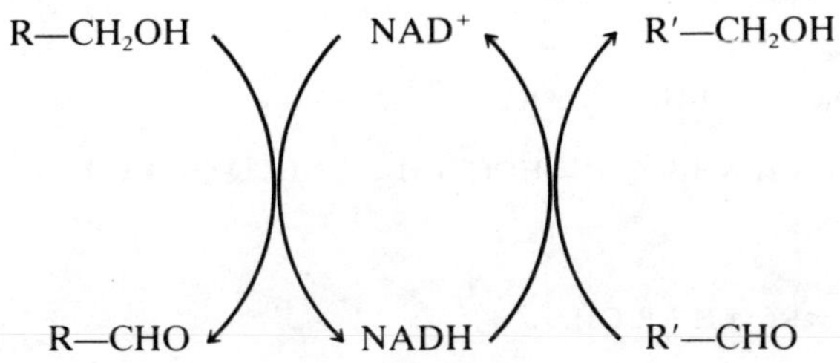

Scheme 1

2.2 NOMENCLATURE AND CLASSIFICATION

The high selectivity of enzymes is clearly shown by the rules of classification and of nomenclature which were established to identify each enzyme precisely. According to the recommendations of the Commission on Enzymes of the International Union of Biochemistry, each enzyme is characterized by a code number. Each enzyme number is formed of four elements, separated by points. The first figure refers to the group of enzymes which are responsible for a specific transformation (*e.g.* hydrolases), the second figure defines the sub-group related to a specific reaction (*e.g.* hydrolysis of ester function), the third figure refers to a well-defined family of compounds (*e.g.* carboxylic esters) and the last figure refers to the substrate which is used as the reference to define the activity units (*e.g.* glycerides).

The characteristics of an enzyme depend on its origin and it is necessary to specify the living organism from which it is extracted.

Example: EC 3.1.1.3	Lipase from *Rhizopus delemar*
EC 3	Hydrolase
EC 3.1	Hydrolysis of ester functions
EC 3.1.1	Carboxylic esters
EC 3.1.1.3	Glycerides

The enzymes of interest for polymer science are collected in Table 1; most of them are

Table 1 Enzymes of Interest for Polymer Science.

		Refs.
Group 1: Oxidoreductases		
1.1	Primary or secondary alcohol or hemiacetal $\rightleftharpoons$ aldehyde or ketone	
1.1.1	Donor–acceptor: $NAD^+ \rightleftharpoons NADH$ or $NADP^+ \rightleftharpoons NADPH$	
1.1.1.1	Large number of alcohols	
1.1.1.6	Glycerol or 1,2-propanediol $\rightleftharpoons$ dihydroxyacetone	
1.1.1.72	Glycerol $\rightleftharpoons$ glyceraldehyde	
1.1.3	Donor–acceptor: $O_2 \rightleftharpoons H_2O_2$	
1.1.3.13	Primary alcohols saturated or unsaturated	10
1.1.3.15	$RCH(OH)CO_2H \rightleftharpoons RCOCO_2H$	
1.3	Acting on CH—CH bond	
1.3.3	By O_2	
1.3.3.5	Bilirubine oxidase	11
1.4	Acting on amino acid or amine	
1.4.1	$NAD^+(NADP^+)$ + substrate + $H_2O \rightleftharpoons NADH(NADPH)$ + product + NH_3	
1.4.1.1	L-$MeCH(NH_2)CO_2H \rightleftharpoons MeCOCO_2H$	
1.4.1.9	Several amino acids: L-valine, L-norvaline, L-norleucine	
1.4.3	Substrate + $H_2O_2 + O_2 \rightleftharpoons$ product + $NH_3 + H_2O_2$	
1.4.3.2	L-$RCH(NH_2)CO_2H \rightleftharpoons$ L-$RCOCO_2H$	
1.4.3.2	D-$RCH(NH_2)CO_2H \rightleftharpoons$ D-$RCOCO_2H$	
1.4.3.6	$RCH_2NH_2 \rightleftharpoons RCO_2H$	
1.4.3.14	L-$H_2N(CH_2)_4CH(NH_2)CO_2H \rightleftharpoons H_2N(CH_2)_4COCO_2H$	
1.10	Acting on diphenols	
1.10.3	By O_2	
1.10.3.1	1,2-Benzenediol $\rightleftharpoons$ 1,2-benzoquinone	12
1.11	H_2O_2 as acceptor	
1.11.1.7	Peroxidase: by H_2O_2	10,13–22
	Aniline $\rightleftharpoons$ polyaniline	
	Phenol $\rightleftharpoons$ polyphenol	
Group 2: Transferases		
	Transfer of a group from one compound to another	
2.1	One-carbon groups	
2.1.1	—Me	
2.1.1.6	$1,2\text{-}(HO)_2C_6H_4 \rightleftharpoons 1\text{-}HO\text{-}2\text{-}MeOC_6H_4$	
2.1.3	—$CONH_2$	
2.1.3.2	$HOCOCH_2CH(NH_2)CO_2H \rightleftharpoons HOCOCH_2CH(NHCONH_2)CO_2H$	
Group 3: Hydrolases		
3.1	Ester functions	
3.1.1	Carboxylic esters	
3.1.1.1	$RCO_2R' \rightleftharpoons RCO_2H + R'OH$	23
3.1.1.3	Lipase	6,7,23–63
	$RCO_2R' \rightleftharpoons RCO_2H + R'OH$	
3.1.3	Phosphoric monoester	
3.1.3.1	In basic conditions	
	$ROPO_3^{2-} \rightleftharpoons ROH$	
3.1.3.2	In acid conditions	
	$ROPO_3^{2-} \rightleftharpoons ROH$	
3.1.6	Sulfuric esters	
	$ROSO_3^- \rightleftharpoons ROH$	
3.2	Glycosyl compounds	
3.2.1	*O*-Glycosyl bonds	
3.2.1.1	Polysaccharides (<3 D-glucose) $\rightleftharpoons$ maltose:α-amylase	59,64,65
3.2.1.4	Cellulose $\rightleftharpoons$ β-D-glucose:cellulase	59,65,66
3.2.1.21	β-Glucosidase	65
3.2.1.23	β-Galactosidase	67
3.2.2	*N*-Glycosyl bonds	
3.2.2.18	Glycopeptidase	
3.4	Peptide bonds	
3.4.11	Exopeptidases: *N*-terminus of peptide chains	
3.4.11.1 and 14	$RNHCH(Me)CONHR' \rightleftharpoons RNHCH(Me)CO_2H + R'OH$	
3.4.16	Exopeptidases: *C*-terminus of peptide chains	
3.4.16.1	Peptidyl-L-amino acid $\rightleftharpoons$ peptide + L-amino acid	
3.4.21	Endopeptidases: recognize peptides, amides and esters	
3.4.21.1	Chymotrypsin	36,68
3.4.21.4	Trypsin	
3.4.21.14	Subtilisin	69
3.4.22.2	Papain	70
3.4.24.4	Alkaline protease	71,72

Table 1 (continued)

	Refs.
3.5 C—N bonds other than peptide bonds	
3.5.1 Linear amides	
3.5.1.1 to 4 $RCONH_2 \rightleftharpoons RCO_2H$	
3.5.2 Cyclic amides	
3.5.2.6 β-Lactams	
3.8 Halide bonds	
3.8.1 C–halide bonds	
3.8.1.2 $RCHClCO_2H \rightleftharpoons RCH(OH)CO_2H$	
Group 4: Lyases	
Cleavage or formation of C—C, C—O, C—N bonds by process other than hydrolysis of oxidation (or reverse reaction). When the reverse reaction is used, the enzyme is considered as a synthase which is different from the synthetases registered in Group 6.	
4.1 C—C bonds	
4.1.1 CO_2 involved	
4.1.1.1 $RCOCO_2H \rightleftharpoons RCOH + CO_2$	
4.2 C—O bonds, unsaturated compounds involved	
4.2.1 $RCH_2CH(OH)CO_2H \rightleftharpoons RCH{=}CHCO_2H$	
4.2.2 Act on polysaccharides	
4.3 C—N bonds, unsaturated compounds involved	
4.3.1 $RCH_2CH(NH_2)CO_2H \rightleftharpoons RCH{=}CHCO_2H$	
Group 5: Isomerases	
5.2 *Cis–trans*	
5.2.2.1 Maleate $\rightleftharpoons$ fumarate	
5.3 Intramolecular oxidoreduction	
5.3.1 $RCH(OH)CHO \rightleftharpoons RCOCH_2OH$	
Group 6: Synthetases	
This group is most important in biosynthesis as the corresponding enzymes have a major contribution in the synthesis of proteins, including enzymes. As far as we know, there is no application of these enzymes in organic media at the present time.	

commercialized. When an enzyme has been used in macromolecular chemistry, a reference is given.

2.3 KINETICS

Enzyme catalysis in organic media has been the object of many investigations because of its exceptional performance in substrate selectivity. The determination of the rates and the structural analysis of the products have been widely studied; on the other hand, the literature gives very few examples of enzymatic kinetics in organic media. Some authors[4,15] made the assumption that many reactions have more or less similar kinetics in aqueous and non-aqueous media.

Enzyme-catalyzed reactions do not obey the classical kinetic laws that relate to chemical reactions carried out in homogeneous media. This is due to a change of the rate during the progress of the reaction because the activity of the enzyme decreases under the influence of various parameters (Figure 2); for this reason, the initial rate is the most significant value in the analysis of enzymatic kinetics.

The simplest kinetic scheme is that of Michaelis–Menten, which assumes the rapid formation of a complex between the substrate and the enzyme, followed by a slow step which corresponds to the formation of the products. The Michaelis model should be applied to the initial conditions where very few products are formed, which means that the second step can be considered as non-reversible.

$$E + S \rightleftharpoons ES \qquad (1)$$

Then

$$ES + (\text{a second reactant}) \xrightarrow{k_{cat}} E + \text{products} \qquad (2)$$

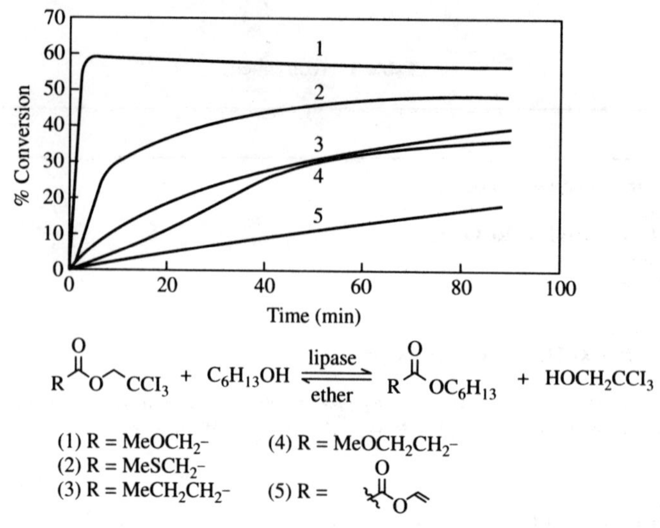

Figure 2 The relative rates and extents of conversion of lipase-mediated transacylations with various trichloromethyl esters and vinyl acetate.[73]

The thermodynamic equilibrium (equation 1) is rapidly reached and its dissociation constant is K_s

$$K_s = \frac{[E][S]}{[ES]} \tag{3}$$

Equation (2) corresponds to the reaction of complex ES with a second reactant; k_{cat} is the rate constant of the global process

$$v = k_{cat}[ES] \tag{4}$$

and

$$k_{cat} = k'[\text{second reactant}] \tag{5}$$

At the beginning of the reaction, the concentration of the second reactant can be considered constant. Let $[E]_T$, $[E]$ and $[ES]$ be the total concentration of the enzyme and of E and ES, respectively, at equilibrium.

$$[E]_T = [E] + [ES] \tag{6}$$

and

$$[ES] = \frac{[E]_T[S]}{K_s + [S]} \tag{7}$$

This description is based on the assumption that the first step is a thermodynamic equilibrium which is established rapidly.

A more accurate description permits the definition of the rate constant of each step.

$$E + S \underset{k_{-1}}{\overset{k_1}{\rightleftharpoons}} ES \xrightarrow{k_2} P \tag{8}$$

According to the above assumption, $k_{-1} \gg k_2$.

At the beginning of the reaction, $[E]_T \ll [S]$, and therefore [ES] is very low and the steady state can be applied to the chemical species ES.

$$\frac{d[ES]}{dt} = k_1[E][S] - k_2[ES] - k_{-1}[ES] = 0 \tag{9}$$

Taking equation (6) into account

$$[ES] = \frac{[E]_T[S]}{\frac{k_{-1}+k_2}{k_1} + [S]} \tag{10}$$

with

$$v = k_2[ES] = \frac{k_2[E]_T[S]}{\frac{k_{-1}+k_2}{k_1} + [S]} \tag{11}$$

Let us write

$$K_m = \frac{k_{-1}+k_2}{k_1} \tag{12}$$

$$v = \frac{k_2[E]_T[S]}{K_m + [S]} \tag{13}$$

K_m is an important kinetic parameter, which accounts for the specificity of an enzyme towards a given substrate in well-defined experimental conditions. The affinity of the substrate to the enzyme decreases with decreasing K_m and the concentration of the complex is low; to increase its value it is necessary to increase the concentration of the substrate.

The following cases are particularly interesting. When $k_2 \ll k_{-1}$ then

$$K_m = K_S \tag{14}$$

When $[S] \gg K_m$ (a high concentration of substrate is particularly true at the initial conditions) then

$$V_m = k_2[E]_T \tag{15}$$

(all the sites are occupied)

$$k_2 = k_{cat} \tag{16}$$

and

$$v = \frac{V_m[S]}{K_m + [S]} \tag{17}$$

Several plots of equation (17) were proposed (Figure 3); two of them are particularly important.

Several examples of synthesis carried out in organic solvents and following Michaelis–Menten kinetics have been reported; it was observed by Klibanov[4] for a pancreas lipase-catalyzed transesterification between tributyrine and heptanol in hexane, and by Drodick[15] who studied the horseradish (HPR)-catalyzed oxidation of *p*-cresol in dioxane (Figure 4).

2.4 THERMODYNAMIC ASPECTS

Most enzyme-catalyzed reactions are equilibria which must be shifted towards the formation of the desired products for good yields. This result can be achieved in several ways. Yamada *et al.*[41] studied the lipase-catalyzed macrolactonization of hydroxy acid esters in anhydrous isooctane (Scheme 2). To shift the equilibrium the authors trapped the methanol produced with molecular sieves.

Another example reports the use of an unstable low molar mass product (enol) which is eliminated in the form of the corresponding tautomer. This technique was used by Nicolosi *et al.*[49] for acetylating phenols in the presence of lipase; the acyl donor is vinyl acetate and the evolving enol is eliminated as ethanal (Scheme 3).

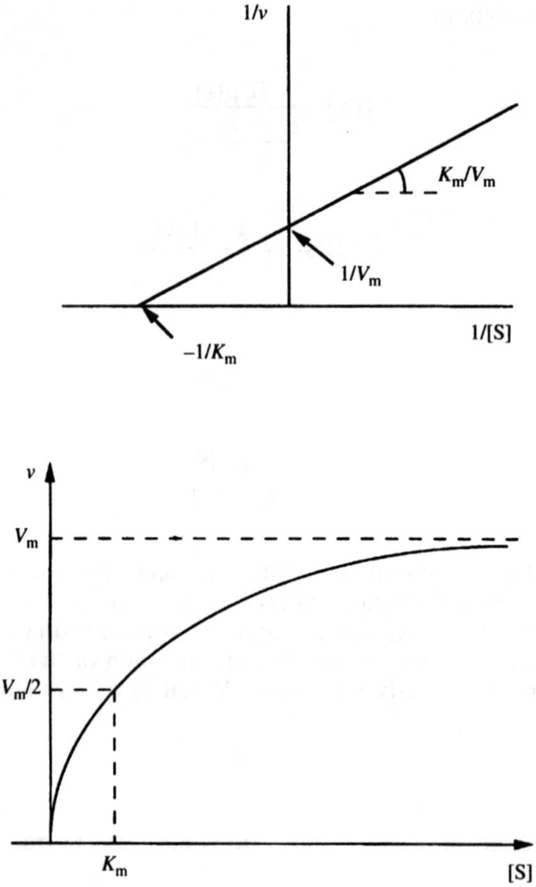

Figure 3 Several plots of equation (17).

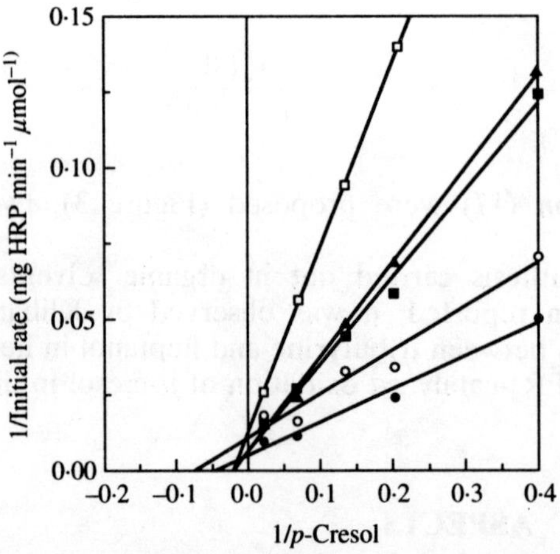

Figure 4 Saturation kinetics of peroxidase-catalyzed oxidation of *p*-cresol in different dioxane concentrations. (●) = 60% dioxane, (○) = 70%, (■) = 80%, (▲) = 85%, and (□) = 95%. [HPR] = 0.6 μg mL^{-1}; [H_2O_2] = 0.25 mM.[15]

The selective dissolution (or precipitation) of substrates, reactants and products is an efficient technique.

The effect of organic solvents on the equilibrium position is particularly shown in the enzymatic esterification of glycerol by decanoic acid, which leads to a mixture of three isomeric

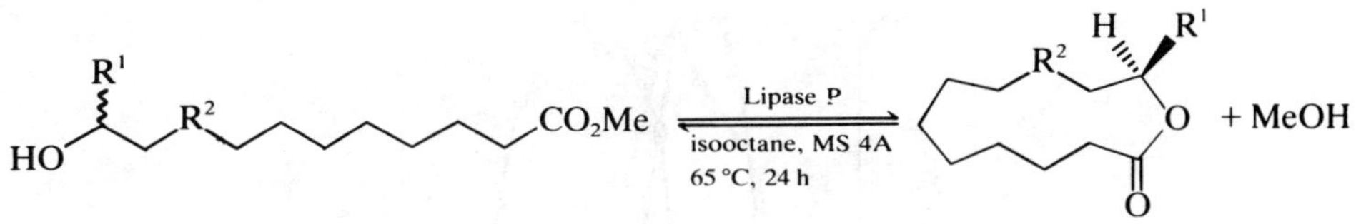

Scheme 2

$$ArOH + CH_2{=}CH{-}OAcyl \rightleftharpoons Ar{-}OAcyl + [CH_2{=}CH{-}OH] \rightleftharpoons MeCHO$$

Scheme 3

acylglycerols.[74] The reaction is carried out in a biphasic mixture of water with a solvent insoluble (or sparingly soluble) in water. The yields of the three isomers can be connected to the hydrophobicity or to the solubility of the water in the organic phase; both scales are equivalent, with a few exceptions. The position of the equilibrium can be determined by controlling the hydrophilicity of the medium.

More recent work accounts for the shift of the equilibrium by the elimination of a product of the reaction by nitrogen bubbling[63] or by using a high-boiling solvent under vacuum.[60]

2.5 FACTORS LIMITING ACTIVITY

The activity of the enzyme is connected to the tertiary structure or, possibly, to the quaternary structure. Any parameter able to modify these structures influences the enzyme's activity and generates a reversible or irreversible denaturation. The most important parameters are pH, temperature, inhibitors and hydrophilic solvents.

2.5.1 Effects of pH

An enzyme is active in a limited range of pH. The variation of activity with respect to pH is a bell-shaped curve, the top of which corresponds to the optimal activity. Enzymes are proteins which contain numerous ionizable groups and for this reason are sensitive to pH. A pH change can also modify the ionization state of the substrate and consequently the latter cannot be recognized by the enzyme (Figure 5).[75]

2.5.2 Effect of Temperature

Enzyme-catalyzed reactions follow the Arrhenius law when the temperature is low. The rate increases with increasing temperature and reaches a maximum. It then decreases rapidly as too high a temperature denatures the enzyme, modifies the stability of the substrate–enzyme complex and changes the ionization state of the active sites, as shown below.

$$\text{N (native)} \rightleftharpoons \text{D (denatured)} \rightarrow \text{I (inactive)} \qquad (18)$$

The first step is an equilibrium which depends on temperature through its thermodynamic and kinetic characteristics; on the other hand, the second step leads more or less rapidly to a completely inactive structure.

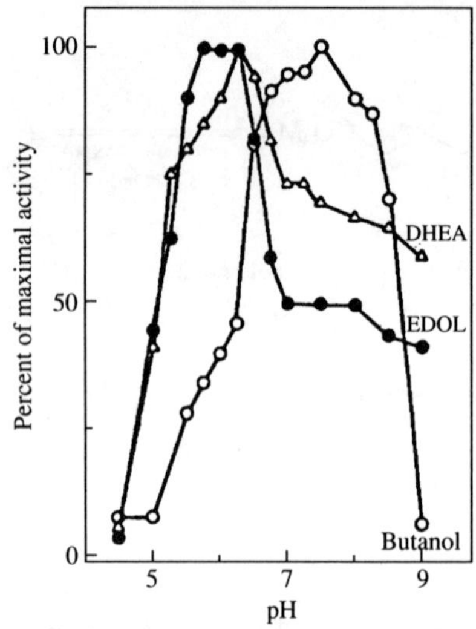

Figure 5 pH profile of purified hydroxysteroid sulfotransferase 1 measured with β-estradiol (●) [EDOL], butanol (○), and dehydroepiandrosterone (△) [DHEA]. Activity was measured with standard assay except that 0.5 M Tris-acetate buffer was used at the indicated pH. Data are presented as percentage of maximal activity for each substrate.[75]

2.5.3 Effect of Inhibitors

There are several kinds of inhibitors and they have different effects on V_m and K_m (Figures 6–8). Competitive inhibitors take the place of the substrate in the process. K_m is increased. Non-competitive inhibitors generate non-reactive species. As they are in equilibrium with the complex ES they affect V_m only. Uncompetitive inhibitors are bonded to the active site only when the substrate is linked. The whole system is locked. K_m is increased and V_m is decreased. Suicide inhibitors poison the active site by taking the place of substrate, and so prevent the enzyme–substrate interaction from taking place. They are largely used for structural analysis of the active site.

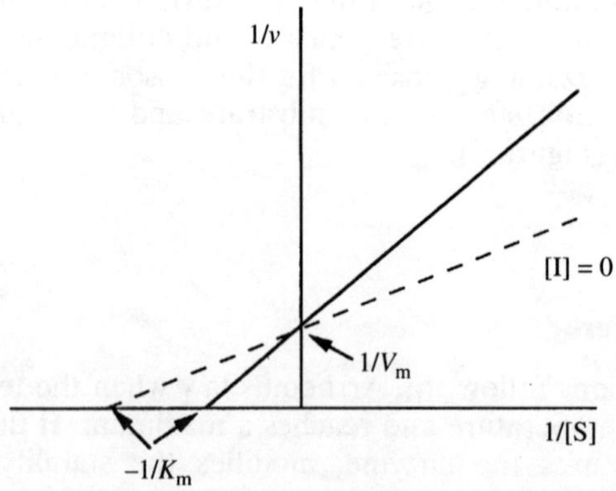

Figure 6 $1/v = f(1/[S])$ in competitive inhibition.

2.5.4 Effect of Solvent

The first attempts of enzymatic catalyzed reactions in non-aqueous media were not very promising as the denaturing power of some organic solvents is very high (Figure 9).

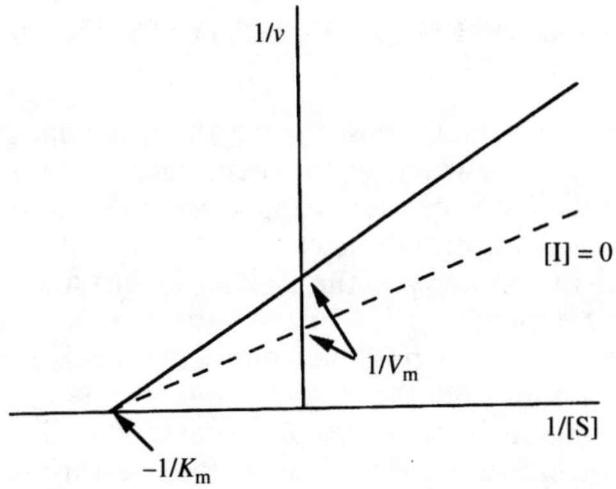

Figure 7 $1/v = f(1/[S])$ in non-competitive inhibition.

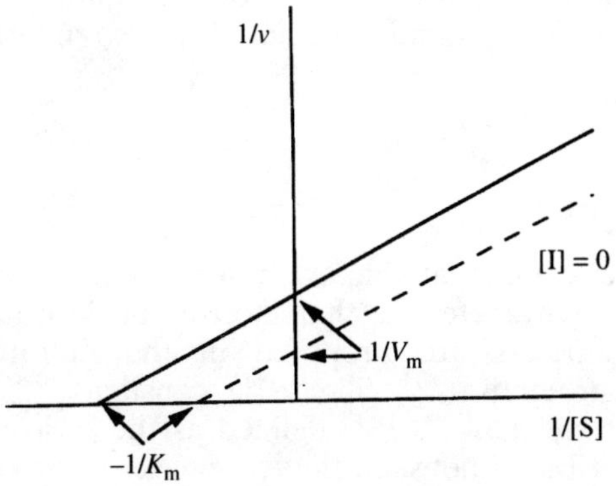

Figure 8 $1/v = f(1/[S])$ in uncompetitive inhibition.

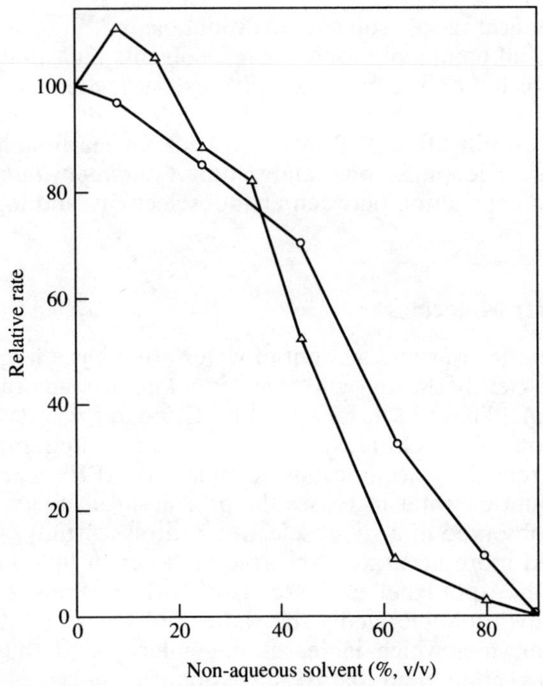

Figure 9 Effect of non-aqueous solvents on rate of hydrolysis of *p*-nitrophenyl phosphate by an acid phosphatase obtained from leaves of corn (*Zea mays*). (○) Ethylene glycol; (△) methanol.[76]

2.6 THE PRESERVATION OF ENZYME ACTIVITY IN ORGANIC MEDIA: THE MAIN FACTORS

The use of enzymes in organic media raises some thermodynamic and kinetic problems; the natural environment of enzymes is water, even when placed in a hydrophobic medium such as the cell wall. The constitutive water can be stripped away from an enzyme when it is used in organic media and its native conformation can be deeply modified, leading to a dramatic decrease in its original activity. Moreover, the rigidity of enzymes greatly affects their activity and their selectivity; rigidity depends on the internal electrostatic interactions and in particular on hydrogen bonding. The number of possible conformations is low when the solvent is polar and liable to hydrogen bonding and the enzyme internal bonds are decreased, leading to relaxation of the structure; this leads to a drastic decrease in activity and selectivity.[3,77,78]

An ideal solvent must not induce significant structural perturbation of an enzyme when it is used in an organic medium. Several assumptions were proposed to explain this change of activity in some experimental conditions. Solid state NMR showed that enzymes are not denatured when they are dispersed in non-aqueous solvent.[79] The loss of activity and selectivity can result from several side-effects specific to the non-aqueous solvents: competition with the substrate binding, dissociation of multimers, shift of the equilibrium between two conformations, and specific bonding. The organic solvent can also interfere or compete with the structural water molecules.[76]

2.6.1 Solvent Hydrophobicity

The solvent hydrophobicity plays an important role in enzymic activity.[80] However, the choice of a hydrophobicity parameter to characterize the biocatalytic activity is a difficult problem and several parameters are proposed in the literature: Hildebrand solubility parameters,[81] dye solvatochromatism,[80,81] dielectric constants,[80,82] dipole moments and the logarithm of the partition coefficient P. P is defined as the ratio of the concentrations of an organic solvent which is distributed between the two phases of an octanol–water mixture

$$P = \frac{[\text{solute}]_{\text{octanol}}}{[\text{solute}]_{\text{water}}} \tag{19}$$

Log P is a good indicator of solvent hydrophobicity.[80,81,83] It reveals differences in hydrophobicity between all commonly used organic solvents and, moreover, it can be used to estimate the hydrophobicity of mixtures of solvents, surfactant interphases and polymeric supports (Table 2).

The enantioselective esterification of 2-hydroxy acids in nearly anhydrous organic solvents with primary alcohols as nucleophiles was catalyzed by *Candida cylindracea* lipase (CCL):[84] the authors observed a good correlation between enantioselectivity and log P (Figure 10).

2.6.2 Constitutive Water Molecules

In many cases the presence of some amount of water often has a large effect on stereoactivity and selectivity. This was clearly shown by Monot[85] in a kinetic study of butyl butyrate synthesis catalyzed by *Mucor miehei* lipase in hexane, and by Gutman[86] who studied the effect of water on the transesterification of prochiral symmetrical disubstituted malonates (**1**) by benzoic alcohol using a lipase from liver acetone powder horse (LAPH) as catalyst (Scheme 4). The addition of a small amount of water increases the prochiral selectivity, as shown in Table 3.

The same effect was observed in a stereoselective hydrolysis study.[87]

Affleck *et al.*[88] defined more accurately the role of water in lipase-catalyzed transesterification; they used ESR with a spin-labelled lipase. This work confirms that the presence of water is necessary. In highly hydrophobic media the water molecules are strongly associated to the ionizable part of the enzyme, which increases the polarity and the flexibility of the active centre; it is necessary to keep in mind that freshly lyophilized enzymes are particularly rigid and their activity is limited. On the other hand, an excess of water (20 μL/mL) decreases enzymatic acvitivity drastically; this could result from the formation of water clusters in the immediate

Table 2 Log *P* Values of Commonly Used Organic Solvents.[80]

Solvents	*Log P*
Dimethyl sulfoxide	−1.3
Dioxane	−1.1
Methanol	−0.76
Acetone	−0.23
Acetic acid	−0.23
Tetrahydrofuran	0.49
Pyridine	0.71
Diethyl ether	0.85
Cyclohexanone	0.96
Dihydroxybenzene	1.0
Ethyl chloride	1.3
Phenol	1.5
Triethylamine	1.6
Acetophenone	1.8
Chloroform	2.0
Benzene	2.0
Toluene	2.5
Dibutyl ether	2.9
Tetrachloromethane	3.0
Pentane	3.0
Cyclohexane	3.2
Hexane	3.5
Diphenyl ether	4.3

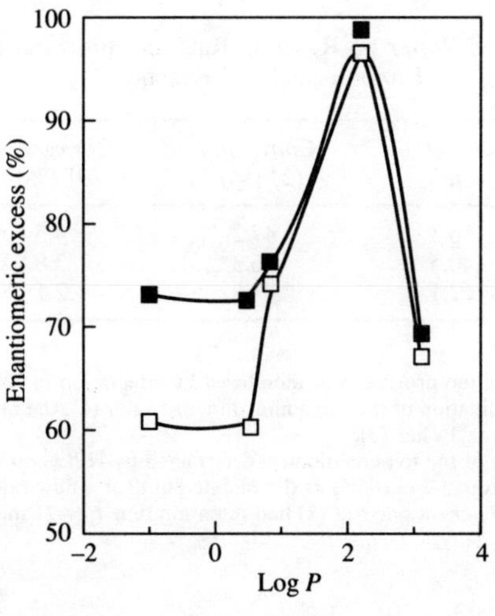

Figure 10 Influence of solvent hydrophobicity on the enantioselectivity of CCL catalysis for (*R,S*)-2-hydroxycaproic acid (□) and (*R,S*)-2-hydroxyisocaproic acid (■).[84]

neighbourhood of the active site and important modifications of the configuration are observed. These effects are clearly shown in Figure 11.[88]

In the case of less hydrophobic media, the distribution of the water molecules between the active centre and the bulk of the solvent is characterized by the thermodynamic activity (a_w). This is clearly shown by Valivety's study[89] relative to the esterification of dodecanol by decanoic acid in the presence of an immobilized lipase (Lipozyme®) and in different organic solvents.

When the initial rate is plotted against a_w, the curve shows a maximum at $a_w = 0.5$ whatever the solvent used (Figure 12). On the other hand, when the initial rate is plotted against the

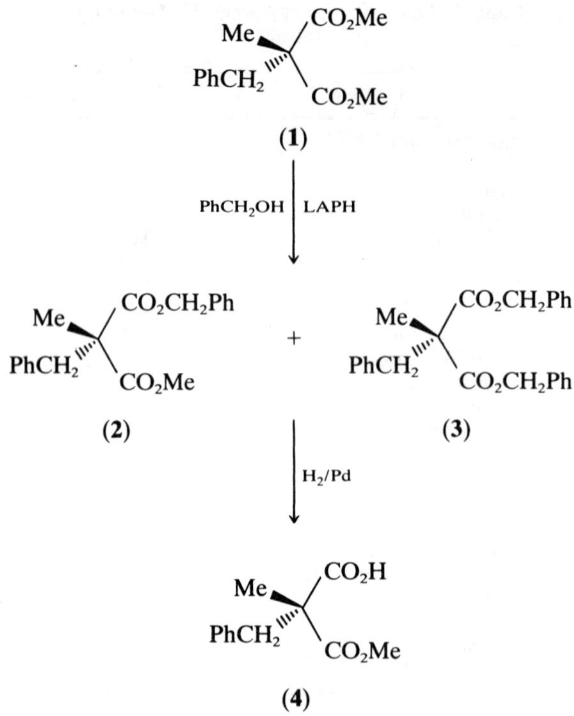

Scheme 4

Table 3 Effect of Added Water on Reaction Rate and Prochiral Stereoselectivity of the Enzyme-catalyzed Reaction.[86]

Added water (%)	*Initial rate*[a]	*Conversion* **(2)** *(%)*[b]	*Conversion* **(3)** *(%)*[b]	*e.e.* (%)[c]
0	56.3	47.2	4.5	39.1
1	140.5	76.3	3.8	51.0
5	162.1	81.6	2.1	61.8

[a] In $\mu mol\,h^{-1}$ (g enzyme)$^{-1}$.
[b] % formed after 45 h. Reaction progress was monitored by integration of ^1H NMR signals, which enabled quantitative determination of the remaining dimethyl ester (**1**), the mixed benzyl methyl diester (**2**) and the by-product dibenzyl ester (**3**).
[c] Enantiomeric excess (e.e.) of the (*S*)-enantiomer, determined by HPLC on a chiral column (Chiracel OJ, Daicel) with hexane–propan-2-ol (95:5) as the mobile phase at a flow rate of 0.9 mL min^{-1} using detection at 258 nm. The (*R*)-enantiomer of (**2**) had retention time $t_R = 21$ min, and the (*S*)-enantiomer had $t_R = 31$ min.

water concentration the curves have the same shape but the position of the maximum depends on the nature of the solvent (Figure 13).

2.6.3 Source and Conditioning

The source and the conditioning of the enzyme have the most relevant influence on its activity. Hogan *et al.*[73] showed that an enzyme can behave in different ways with respect to a given substrate, depending on its source.

The lyophilization history of an enzyme has an important influence on its activity, particularly the pH at which it was lyophilized. Gaertner[70] studied the papain-catalyzed peptide synthesis and oligomerization of amino acid amides in organic solvents and showed clearly the influence of the pH of the lyophilization medium (Figure 14).

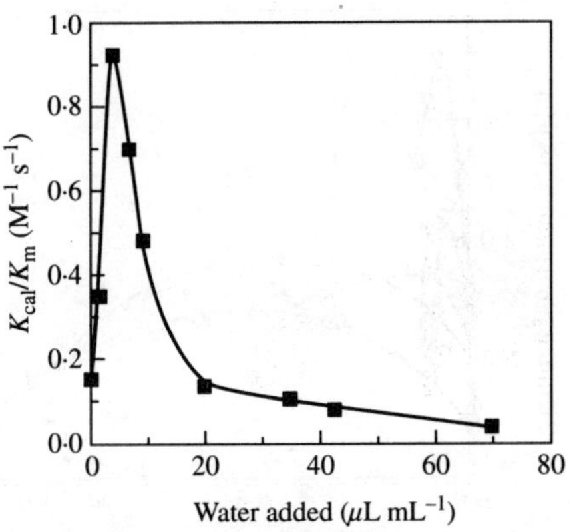

Figure 11 Catalytic efficiencies of subtilisin carlsberg in THF with different water content. The rate of *N*-acetyl-L-phenylalanine propyl ester formation from *N*-acetyl-L-phenylalanine 2-chloroethyl ester was determined by GC. Hydrolysis was not observed (as determined by the lack of appearance of *N*-acetyl-L-phenylalanine on reverse-phase HPLC) below 30 μL of added per mL of solvent. While the catalytic efficiencies of subtilisine are *ca.* 3 orders of magnitude below values in aqueous solutions, catalysis in an organic medium is a result of active-site chemistry as shown in two control experiments: (i) subtilisin preinactivated with phenylmethanesulfonyl fluoride was nearly completely inactive in THF (at least 4 orders of magnitude less active than the pH-adjusted enzyme – the sensitivity limit for detection of the ester product by GC); and (ii) the activity of an active site mutant S221A [whereby alanine replaces serine at the active site (position 221) of the subtilisin] for the related enzyme from *B. amyloliquefaciens* (subtilisin BPN′) is also at least 4 orders of magnitude less active than the pH-adjusted enzyme.[88]

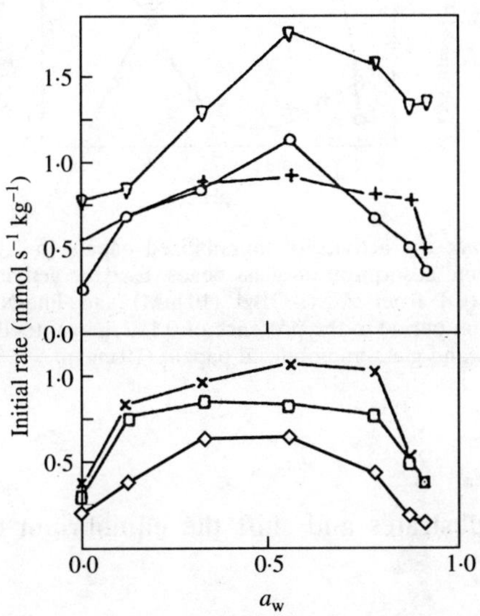

Figure 12 Activity of lipozyme catalyst as a function of water activity in a range of solvents. The solvents present were hexane (▽), toluene (X), trichloroethylene (○), diisopropyl ether (□), pentan-3-one (◇), and none, *i.e.* liquid reactant mixture (+).[89]

2.7 ENZYME PROTECTION

From the earliest work, authors underlined the necessity to protect the enzyme from the denaturing effect of the solvent. Several methods are used.

2.7.1 Biphasic Systems

To maintain the enzyme in a natural environment, one of the two phases is often a dilute aqueous solution (or water) emulsified in the organic solvent.[90] Moreover, the organic phase

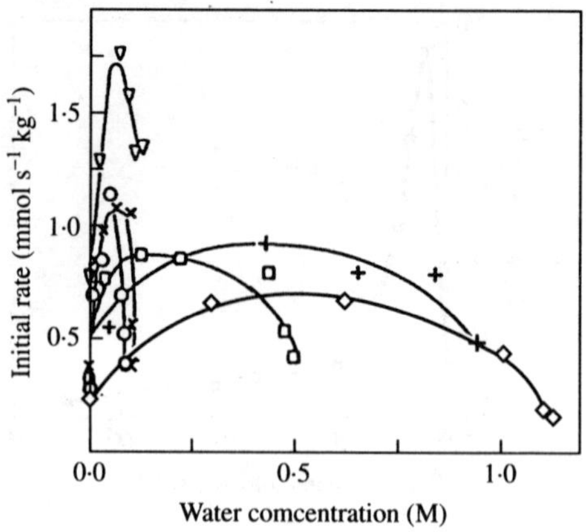

Figure 13 Activity of lipozyme catalyst as a function of water concentration dissolved in the organic phase. The solvents present are identified by the same symbols as for Figure 12. The curves are purely guide lines drawn by eye.[89]

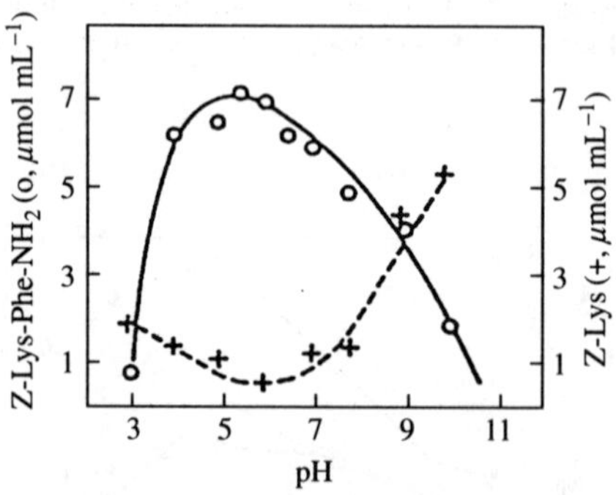

Figure 14 Dependence of the enzymatic activity of immobilized papain in 1,1,1-trichloroethane on the pH of the buffered solution used for the enzyme adsorption on glass beads. Enzyme activity was determined by measuring both the synthesis of Z-Lys-Phe-NH_2 (O) from Z-Lys-OBzl (10 mM) and Phe-NH_2 (20 mM) and the hydrolysis of Z-Lys-OBzl (+) over a 4 h incubation period in the presence of 0.1% mercaptoethanol (v/v), 0.09% added water (v/v) and glass-immobilized papain (10 mg mL^{-1}).[70]

can dissolve hydrophobic substrates and shift the equilibrium towards the formation of the product.

2.7.2 Immobilization

The immobilization of enzymes very often increases their resistance to denaturation factors[91] and they can be linked to different supports which are organic or inorganic and non-soluble in the reaction medium. Supported enzymes preserve favourable conformations and they are protected from any harmful effects of the solvent. Figure 15 shows clearly the behaviour of invertase in different experimental conditions.

2.7.3 Entrapment

The entrapment of the enzymes in reverse micelles[93] generates a microheterogeneous medium where the enzyme is protected from the denaturing solvent by the presence of the

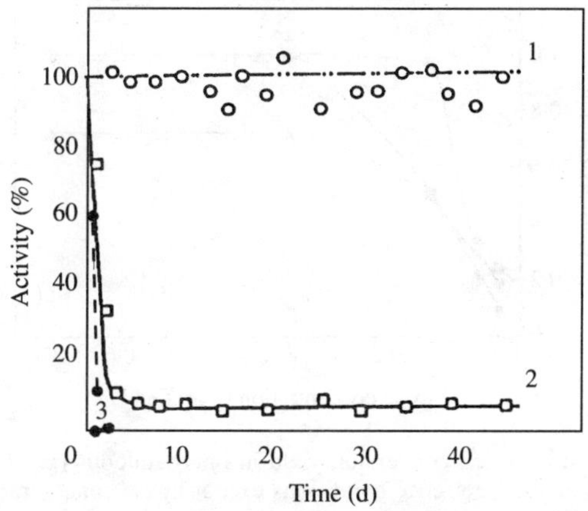

Figure 15 Stability of free (curve 2) and immobilized (curve 1) invertase under continuous operation at 50 °C. Curve 3 corresponds to free invertase without substrate. Invertase was covalently grafted to bentonite.[92]

surfactant monolayer. These micelles are microreactors which are hosts of the enzyme; the surfactants can be polymers instead of low molar mass species whose elimination is difficult (Figure 16).

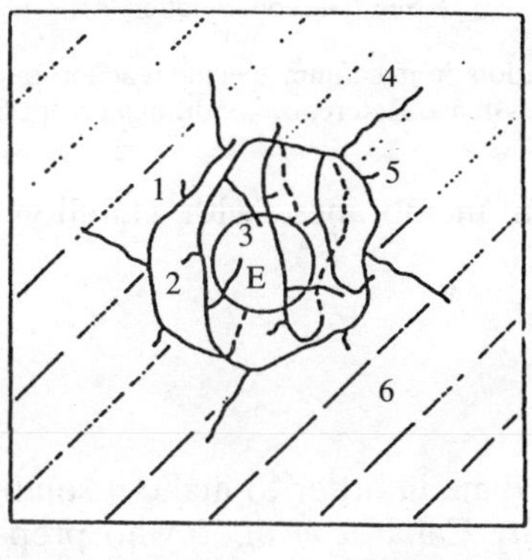

Figure 16 Assumed schematic structue of the enzyme-containing polymeric reversed micelle. (1) CEPEI [poly(ethyleneimine) modified by cetyl bromide and ethyl bromide] backbone bearing positively charged nitrogen atoms; (2) aqueous inner cavity of the reversed micelle; (3) enzyme molecule; (4) cetyl side chain; (5) ethyl side chain; (6) organic solvent.[94]

Khmelnitsky[94] studied the successive alkylation of polyethyleneimine with cetyl (or ethyl) bromide and used the resulting products as surfactants in the hydrolysis of a peptide in the presence of α-chymotrypsin. He observed a considerable increase in the catalytic activity and stability of α-chymotrypsin when it was entrapped in reverse micelles compared to its use in aqueous solutions.

Entrapment of enzymes was also carried out in liquid crystals;[95] such a system was used in the continuous production of (*R*)-cyanohydrins. Borzeix *et al.*[96] studied the esterification of butanol by butanoic acid, catalyzed by *Mucor miehei* lipase and carried out in three different media: a microaqueous medium containing the enzyme as a suspension in hexane, a water–hexane two-phase system, and reverse micelles. The best performances were obtained either in the microaqueous medium or in the two-phase system. The reverse micelles yielded lower performances from both kinetic (except at low enzyme concentration) and thermodynamic (conversion) points of view (Figures 17 and 18).

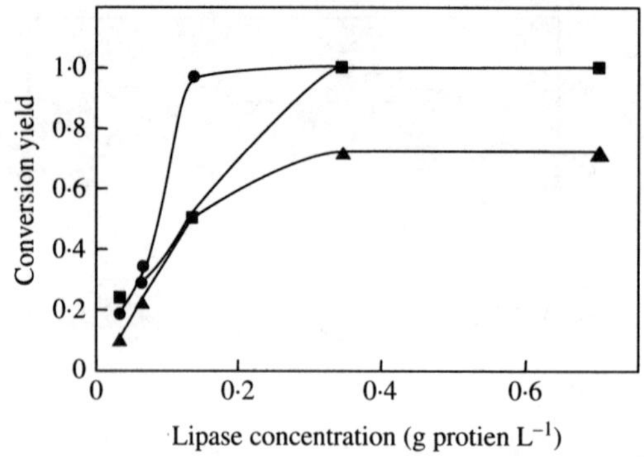

Figure 17 Effect of lipase concentration on conversion yield in microaqueous (■), biphasic (●) and reverse micellar (▲) systems. Reference conditions except lipase concentration.[96]

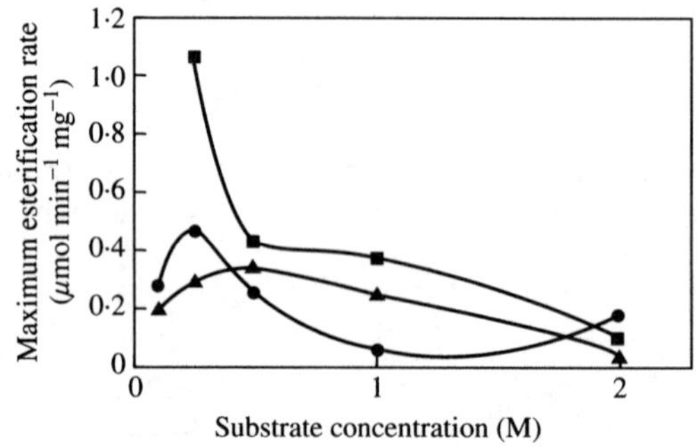

Figure 18 Effect of substrate concentration on maximum specific reaction rate in microaqueous (■), biphasic (●) and reverse micellar (▲) systems. Reference conditions except lipase concentration.[96]

Enzymes can be enclosed in membranes, with stability equal to or exceeding that of immobilized enzymes.[97]

2.7.4 Enzyme Modification

It is possible to modify the enzyme in order to make it soluble and active in organic solvents; some examples were reported by Cabaret *et al.*,[98] who prepared chymotrypsins modified by neutral or charged amphiphiles and used them as catalysts for ester synthesis in polar organic solvents.

Vazquez-Duhalt *et al.*[99] modified horseradish peroxidase in two ways: by deglycosylation and benzyl (or polyoxyethylene) modifications. In both cases the superficial hydrophobicity of the enzyme was increased, particularly around the active centre; hydrophobicity increases the activity and decreases its dependence with respect to the molecules present in the medium; however, Ohya[62] showed that a lipase and an esterase modified too much by PEG ($\bar{M}_n$ 5000) leads to deactivation of the enzyme; steric hindrance of PEG counteracts the positive amphiphilic effect of PEG.

2.8 ENZYMES IN CHAIN POLYMERIZATION

Derango[10] reported an example of radical polymerization through an enzymatic process by using the formation of free radicals from the interaction of peroxidases with xanthine

$$\text{xanthine oxidase} + \text{xanthine} + H_2O \rightarrow \text{urate} + O_2^{-}\cdot \qquad (20)$$

Some other peroxidases decompose hydrogen peroxide at room temperature, which leads to radical polymerizations of acrylates or of methacrylates under mild conditions.

2.9 ENZYMES IN POLYCONDENSATION AND CYCLIZATION

2.9.1 Lipases

2.9.1.1 Polycondensation

The great ease of use of lipases in polycondensation in organic solvents or in bulk permitted their use from the outset of enzymatic catalysis in non-aqueous media. Low molar mass oligomers were obtained from different AB + AB or AA + BB polycondensations. AB monomers include hydroxy acids,[24,62] hydroxy esters,[30,53,54] and lactones.[53,57,61] Stoichiometric AA + BB systems include several important reactions: diols + diacids,[25] diols + diesters[32,40,52] or diols + cyclic anhydrides.[55]

Enzyme-catalyzed reactions can be carried out under mild conditions, which is one of their main interests. This permits efficient control of both reactions and structures. Thus Geresh[6,52] did not observe any side reactions for polycondensations of various fumaric esters with 1,4-butanediol in the presence of lipases from different sources; no isomerization of the double bond took place. In the same way, Wallace and Morrow[39] described the polycondensation of bis(2,2,2-trichloroethyl) 3,4-epoxyadipate (**5**) with 1,4-butanediol (**6**) in anhydrous ether at room temperature and in the presence of porcine pancreas lipase; they observed that the epoxy group content is not modified during the polycondensation (Scheme 5).

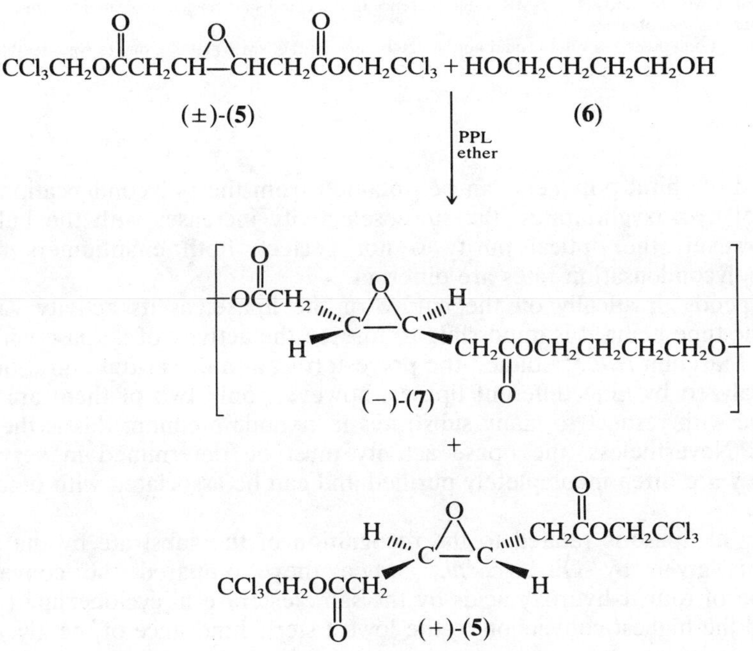

Scheme 5

Lipase-catalyzed reactions carried out in organic solvents are characterized by selective processes. Thus Margolin *et al.*[32] obtained chiral trimers and pentamers by transesterification of chiral diesters with achiral diols or of achiral diesters with chiral diols; the asymmetric carbon is α or β to the carboxylic group.

Gutman and Bravdo[36] polytransesterified a prochiral symmetrical dimethyl β-hydroxyglutarate in hexane and in the presence of different lipases: the liver acetone powder pig (LAPP), the liver acetone powder horse (LAPH) and the protease from *Streptomyces griseus* (SG). The resulting achiral polymer results from an asymmetric synthesis, as shown in equation (21) and Table 4.

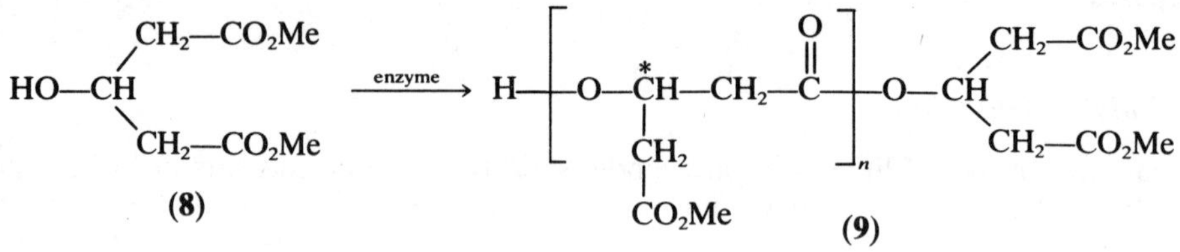

(21)

Table 4 Asymmetric Polymerization of Dimethyl β-Hydroxyglutarate with Enzymes in Organic Solvents.[36]

Enzymes	*Temperature* (°C)	*Chiral oligomer* (**9**)	$[\alpha]_D^{20}$ (deg) (*C in* CH_2Cl_2)	*e.e.* (%)[a]
LAPP	40	(+)-(**9**) ($n = 1$)	+1.4 (C 1.5)	37
LAPH	40	(+)-(**9**) ($n = 1$)	+2.0 (C 0.95)	35
SG	69	(+)-(**9**) ($n = 1$)[b]	—	30
LAPP	40	(+)-(**9**) ($n = 2$)[c]	+1.6 (C 0.5)	—

[a] Optical purity of dimers was determined by ^1H NMR in the presence of chiral shift reagent tris[3-((heptafluoropropyl)hydroxymethylene)-(+)-camphorato]europium(III) [(+)-Eu(tfc)$_3$].
[b] Because of the small quantity of isolated material, the authors failed to determine accurately the optical rotation of the dimer obtained with SG. However, NMR in the presence of the chiral shift reagent indicated that in this case the (+) enantiomer was also obtained.
[c] For trimers, enantiomeric excess values could not be determined by the NMR method, due to considerable line broadening.

The same kind of chiral polymers can be obtained from the polycondensation of a racemic mixture of chiral hydroxyglutarates; the stereoselectivity increases with the bulkiness of the substituent. However, the optical purity is not perfect; both enantiomers react but the corresponding polycondensation rates are different.

Selectivity depends drastically on the nature of the lipase, as its activity varies with the substrate. For the time being it is impossible to foresee the activity of a lipase with respect to a given substrate. Margolin *et al.*[32] studied the polyesterification of various chiral or achiral diols and diacids, catalyzed by nine different lipases; however, only two of them are active. Some lipases are active with respect to many substrates in organic medium; this is the case of pork pancreas lipase. Nevertheless, the lipase activity must be determined in very well-defined conditions as they are often incompletely purified and can be associated with other enzymes or with coenzymes.

The selectivity of lipase is related to the recognition of the substrate by the active site. A clear example is given by Ohya *et al.*[62] when they compared the conversion in the polycondensation of four α-hydroxy acids by the same esterase in cyclohexane (Table 5). The authors assigned the highest conversion to the lowest steric hindrance of the glycolic acid.

Table 5 Conversions for Polycondensation of Various α-Hydroxy Acids by Unmodified Esterase in Cyclohexane.[62]

Substrates	*Conversion* (%)
Glycolic acid	33
D,L-Lactic acid	16
L-Malic acid	9
2-Hydroxyisobutyric acid	12

In the following, we analyze the behaviour of lipases with respect to acyl groups and alcohols. The acyl group is very sensitive to its neighbourhood. As a result, there is no lipase able to catalyze reactions involving maleic acid or esters, at least for the time being; on the other hand, lipases are active with respect to fumarates.[52] The activity of many lipases is decreased when there is an aromatic function in the vicinity of the carboxylic group; *Mucor miehei* lipase is the only one which is active in many systems.[45] Several lipases are totally inactive with respect to some substrates; for example, lipase from *Pseudomonas fluorescens* is fairly active in the polycondensation of succinic anhydride with diols whereas it is totally inactive when glutaric anhydride is used.[55]

Alcohols are less sensitive to their neighbourhood than the carbonyl function; on the other hand, lipases do not recognize phenols. However, there is an exception: the lipase-catalyzed transesterification of diphenyl carbonate by phenols, particularly bisphenol A,[23] as shown in equation (22).

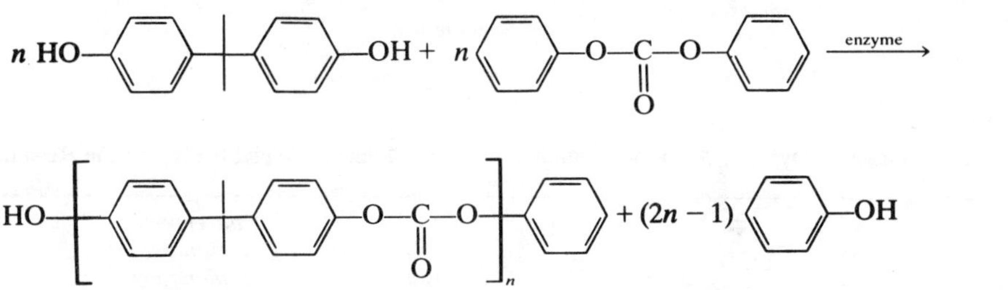

(22)

It is difficult to establish general rules governing the behaviour of lipases with respect to alcohols. Matsumura and Takahashi[27] state that secondary alcohols are more active than primary ones and that hydroxy acids with secondary alcohols lead to higher molar masses and broader distributions than those obtained with primary alcohols. On the other hand, several authors observed an inverse order for polytransesterifications of fumarates[6] and hydroxy esters.[54] The enantioselective polycondensation of hydroxy esters[54] is characterized by a selectivity increase and a rate decrease with the increasing volume of the substituent of the carbon α to the hydroxy function; these results are confirmed by those relative to hydroxy acids.[36]

The nature of the leaving group has an appreciable effect on the reaction kinetics; polytransesterifications are favoured by leaving groups with high electronegative charge, as shown by Geresh's results relative to the polyesterification of 1,4-butanediol with dichloroethyl fumarate or dicyanopropyl fumarate in the presence of different lipases.[6,45]

The characteristics of the polymers prepared by lipase-catalyzed polytransesterification (molar mass, distribution) depend greatly on experimental conditions. Molar masses depend directly on both the conversion and the thermodynamic equilibrium, as in the case of chemically catalyzed polycondensation; on the other hand, the accessibility of the active centre is an essential parameter in enzymatic polycondensations.

As already mentioned, the reactive medium has an important influence on the run of the process, particularly the distribution of water molecules which have an essential role in the structure of the active centre. When the reaction is carried out in an anhydrous and very hydrophilic solvent, the water molecules are abstracted from the active centre.

The analysis of the effects resulting from different parameters is complex and the early articles were limited to the description of the results with scanty interpretation. Matsumura[27] studied the polycondensation of several hydroxy acids in different media (Scheme 6 and Table 6).

The polymers (**11**) prepared in isooctane and those prepared in water with an added emulsifier have molar masses of the same order of magnitude (Table 6); the emulsifier probably plays the same role as isooctane: the solubilization of the products. However, Binns *et al.*[51] claimed that only oligomers are obtained when adipic acid is polycondensed by 1,4-butanediol in the presence of Lipozyme IM-20.

The partition coefficient P can help towards the choice of the solvent; $\log P$ expresses the competition between the active site and the solvent with respect to the water molecules and the competition between the substrate and the solvent with respect to the active site.

The relative solvating power of the solvent with respect to the different substrates is an important parameter of the reaction, particularly when these substrates are of a very different

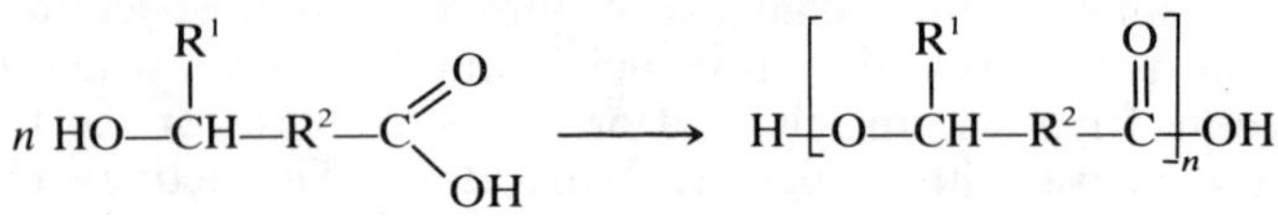

(10) (11)

	R^1	R^2
(10a), **(11a)**	$(CH_2)_5Me$	$—(CH_2)_{10}—$
(10b), **(11b)**	$(CH_2)_5Me$	$—CH_2—(CH\overset{cis}{=}CH—(CH_2)_7—$
(10c), **(11c)**	H	$—(CH_2)_{14}—$
(10d), **(11d)**	H	$—(CH_2)_{10}—$

Scheme 6

Table 6 Conversions of Hydroxy Acids and Number Average Molecular Weights ($\bar{M}_n$) of the Resulting Polymers.[27]

Run no.	*Sub-strate*	*Sol-vent*[a]	*Lipase*[b]	*Emulsifier*[c]	*Conversion*[d] *in % based on monomer*	$\bar{M}_n$ (g mol^{-1})	$\overline{DP_n}$
1	**(10a)**	IO	*C. rugosa*	–	99	1260	4.4
2	**(10c)**	W	*C. rugosa*	–	85	940	3.6
3	**(10c)**	W	*Chromobacterium*	–	93	896	3.5
4	**(10d)**	W	*C. rugosa*	–	99	860	4.3
5	**(10d)**	W	*Chromobacterium*	–	99	944	4.7
6	**(10b)**	W	*C. rugosa*	–	97	1149	4.0
7	**(10b)**	W	*Chromobacterium*	–	97	1124	4.0
8	**(10b)**	W	*C. rugosa*	POE-OE ($n = 10$)	97	1126	4.0
9	**(10b)**	W	*C. rugosa*	POE-OE ($n = 20$)	97	1307	4.6
10	**(10b)**	W	*C. rugosa*	POE-OE ($n = 50$)	97	1286	4.5
11	**(10b)**	IO	*C. rugosa*	–	97	1011	3.5
12	**(10b)**	O	*C. rugosa*	–	85	660	2.3
13	**(10b)**	CH	*C. rugosa*	–	95	772	2.7
14	**(10b)**	H	*C. rugosa*	–	65	710	2.5
15	**(10b)**	B	*C. rugosa*	–	15	600	2.1

[a] IO, isooctane; W, water; O, octane; CH, cyclohexane; H, hexane; B, benzene.
[b] Reaction mixture containing 3.0 g of substrate, 60 mL of solvent, 0.3 g of *Candida rugosa* lipase or *Chromobacterium viscosum* lipase and/or 0.3 g of emulsifier was incubated at 35 °C with stirring for 72 h.
[c] Poly(ethylene glycol) monooleyl ethers with average numbers of oxyethylene units of 10, 20 and 50.
[d] Conversion (%) was calculated on the amount of monomeric hydroxy acid consumed in the polymerization reaction.

nature, which is the case when esters and alcohols are involved in the same reaction. Geresh[6] studied the polyesterification of several fumarates with 1,4-butanediol in different solvents and compared their advantages and drawbacks; the choice of THF or acetonitrile was the best compromise. However, there is a striking difference between the results obtained with the two solvents: in THF the oligoester is soluble and its molar mass ranges from 600 to 800, whereas the one prepared in acetonitrile is not soluble and its molar mass is around 1200. It is important to mention that the lipase sample used in THF could be reused in acetonitrile and can give the same non-soluble oligomer; this shows that the activity of the lipase was not destroyed by THF. However, no clear explanation of the relative effect of THF and acetonitrile on the course of the reaction was given.

Knani *et al.*[53,54] compared different polycondensations of hydroxy esters carried out in several solvents and in the bulk (Table 7). When the reaction is carried out in good solvents, the oligomers are removed from the enzyme surface and do not permit the growing of the chain. The polymers with the highest molar masses were obtained in bulk and in poor solvents. According to these authors, the concentration of the soluble polymer around the active centre of the enzyme is low and does not permit the chain growing. The authors[53,54] compared polyesters prepared from hydroxy esters or from the corresponding lactones and the catalyst was porcine pancreatic lipase. In both cases the products contained linear oligomers and macrolactones.

Table 7 Enzyme-catalyzed Polymerization of Methyl 6-Hydroxyhexanoate.[53]

Solvent	*Temperature* (°C)	*Polymer* $\bar{M}_w$[a]	*Polymer* *Yield* (%)	*Macrolactone yield* (%)
– (in bulk)	70	1900 (16.4)	95	1.2
Isooctane	70	1130 (9.6)	62	7
Diisopropyl ether	70	1035 (8.8)	59	10.5
n-Hexane	70	850 (7.2)	61	8.5
CCl_4	70	395 (3.2)	69	8.8
Benzene	70	295 (2.3)	73	11.3
Acetonitrile	70	200 (1.5)	90	0
– (in bulk)	30	305 (2.4)	95	1.0
Isooctane	30	260 (2.0)	62	1.1
Diisopropyl ether	30	260 (2.0)	48	3
Diethyl ether	30	260 (2.0)	76	1.7
n-Hexane	30	260 (2.0)	69	1.3
CCl_4	30	260 (2.0)	46	1.9
Benzene	30	240 (1.8)	66	3
Acetonitrile	30	200 (1.5)	69	0

[a] Determined by NMR (number of mer units).

Recently, the enzyme-catalyzed polycondensation of 11-hydroxydecanoic acid in hexane with the lipase from *Candida cylindracea* was reported.[58] As in chemical polycondensation, runs were carried out under optimized conditions (pure monomer, added molecular sieve) and molecular masses up to 35 000 were obtained.

An advantage of the ring opening polymerization of lactones in the presence of a nucleophile (H_2O, alcohol) is that no leaving group is produced. In a general scheme, the serine group or the molecule of water strongly bound to the enzyme acts as the nucleophile in the first step of the initiation (Scheme 7).

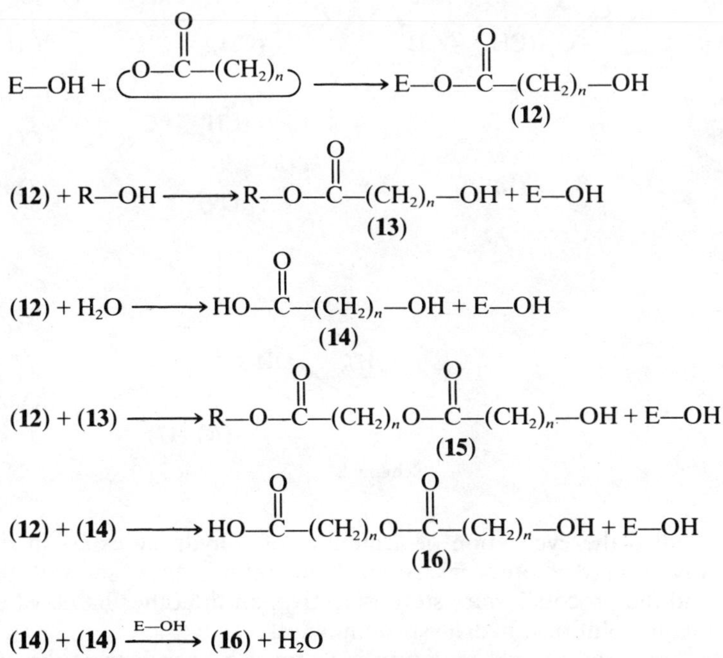

Scheme 7

By the structural analysis of the end group, MacDonald *et al.*[61] suggested that water is active in chain initiation even when an alcohol is added. It seems[59,61] that the ring opening of ε-caprolactone is more rapid than both the polymerization and the dilactone formation, because the molecular weight increases slowly with conversion.

2.9.1.2 *Lactonization*

Low molar mass polymers are obtained in many polymerizations even though most authors predetermined the optimal conditions. It is difficult to know whether $\overline{DP}_n$ depends on the extent of reaction (p) or on some other parameter; moreover, most authors give yields instead of extents of reaction, which are meaningless with respect to $\overline{DP}_n$ analysis. In most cases the conversion is low, which does not permit the formation of high polymers; it is important to recall that high $\overline{DP}_n$ is obtained only when p is very close to 1. Very often, low extents of reaction result from an inefficient control of the thermodynamic equilibrium.

It is interesting to compare p and $\overline{DP}_n$ values given in Table 6: experimental $\overline{DP}_n$ are around 2 to 5 compared to the theoretical value of 100 when $p = 0.99$. This discrepancy is due to side reactions, particularly intramolecular cyclizations. The formation of cycles was proposed for the first time by Ajima,[26] then confirmed by Gutman[30] and by Makita *et al.*[31] who prepared lactones by cyclization of various hydroxy esters in different organic solvents and in the presence of various lipases (equation 23).

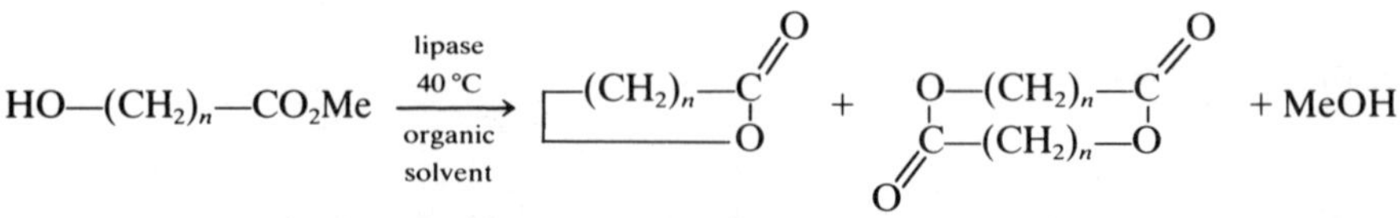

(23)

Among the lipases tested, only lipase P from *Pseudomonas* sp. and porcine pancreatic lipase catalyze lactonization efficiently, while other lipases show little or no activity, indicating that lactonization activity is not the common feature of lipases but more a specific character of some lipases. This procedure was also used to prepare asymmetric lactones from hydroxy esters such as racemic methyl 10-hydroxyundecanoate (**17**); methylation of (**18**) with diazomethane led to regenerated (*R*)-(**17**) with optical purity (Scheme 8).[31]

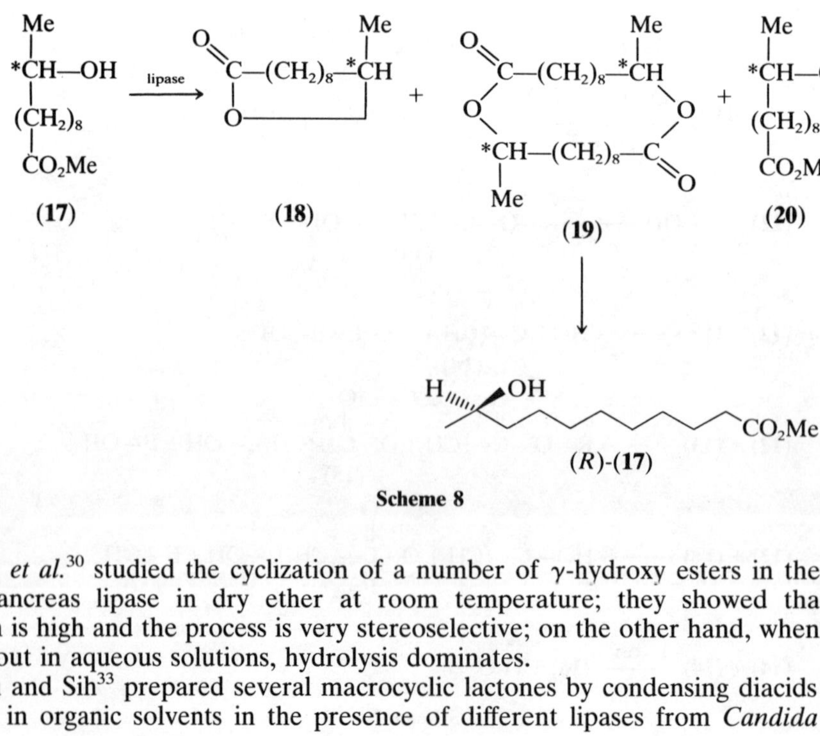

Scheme 8

Gutman *et al.*[30] studied the cyclization of a number of γ-hydroxy esters in the presence of porcine pancreas lipase in dry ether at room temperature; they showed that cyclization conversion is high and the process is very stereoselective; on the other hand, when the reaction is carried out in aqueous solutions, hydrolysis dominates.

Zhi-Wei and Sih[33] prepared several macrocyclic lactones by condensing diacids and diesters with diols in organic solvents in the presence of different lipases from *Candida cylindracea,*

Pseudomonas sp. and porcine pancreas. The results depend on several parameters. In any case, the cyclization is favoured by increasing temperature. The structure of the macrocycles depends on the size of the substrates. The system diester/diol leads to a mixture poorer in lactones. When α,ω-hydroxy acids are used (equation 24), the lactone distribution varies with the length of the substrate and the nature of the lipase (Table 8).

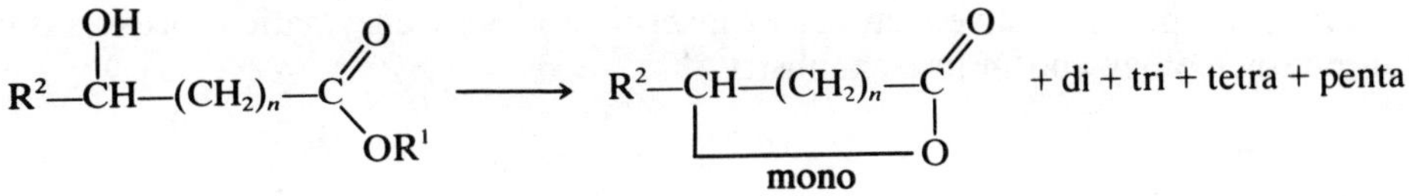

(24)

Table 8 Lipase-catalyzed Lactonization of ω-Hydroxy Acids.[34]

				Isolated yields of lactones (%)				
R^1	R^2	*n*	*Lipase*	*Mono*	*Di*	*Tri*	*Tetra*	*Penta*
H	H	8	*Pseudomonas* sp. *AK*	0	53	16	4	[a]
			Pseudomonas sp. *K-10*	0	33	14	15	7
			Porcine pancreas	0	57	20	6	2
			Candida cylindracea	0	4	26	11	[a]
			Mucor miehei	0	2	20	15	[a]
H	H	1	*Pseudomonas* sp. *AK*	66	26	[a]		
		4	*Pseudomonas* sp. *K-10*	62	30	[a]		
			Porcine pancreas	46	21	12		
			Candida cylindracea	19	15	9		
			Mucor miehei	3	43	15		

[a] Trace quantity (by thin layer chromatography).

The enantiospecificity of the inter- and intramolecular esterification was revealed by lactonization of the pair of diastereomeric dimers (**22**) and (**23**) in Table 9. No cyclization appears with the (*R*,*S*) (C-7′, C-7) diastereoisomer of (**22**). In lactones, the configuration of the asymmetrical carbon C-7′, prefers to be (*R*). The configuration of the second chiral centre influences the lactonization.

Table 9 Enantiospecificity of Lipase-catalyzed Intramolecular Esterification.[34]

		Isolated yields (%)		*Ratio of*
Substrate	*Reaction time* (h)	(*R*)-(*S*)	(*R*)-(*R*) & (*S*)-(*S*)	(*R*)-(*R*)/(*S*)-(*S*)
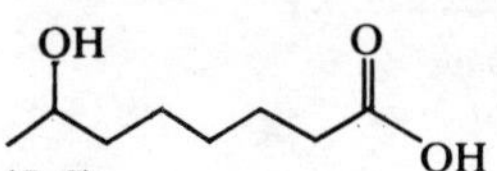 (*R*,*S*) (**21**)	144	5	12	96/4
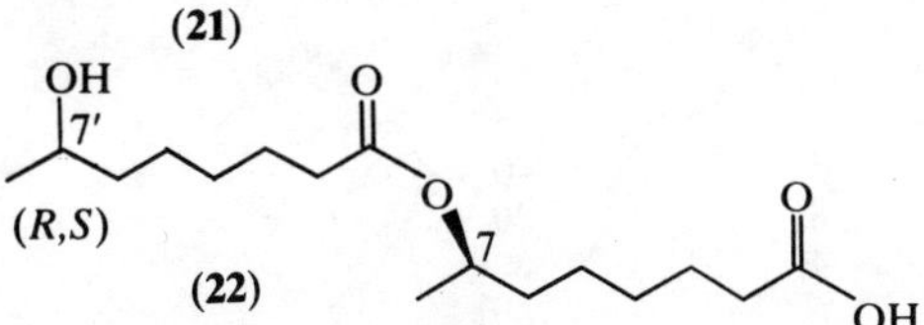 (*R*,*S*) (**22**) (*R*)	90	0	19	100/0
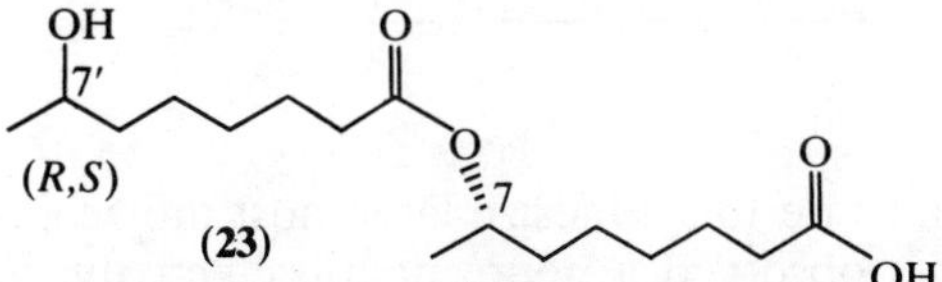 (*R*,*S*) (**23**) (*S*)	90	8	2	0/100

All these results were summarized in 1993 by Gutman *et al.*[53,54] The contribution of the rings during the polycondensation of hydroxy esters or during the polymerization of lactones increases with increasing temperature and reaction time; the cyclization is favoured when the reaction is carried out in the bulk or in a low polarity solvent. Finally, when a hydroxy ester and one of its oligomers are treated separately by the same lipase, the resulting lactone mixtures have exactly the same composition.

The synthesis of optically active lactones or macrolactones by enzymatic processes is of major interest in pharmacology and in fine chemistry.[29,35,37,43]

2.9.2 Proteases

2.9.2.1 Peptide bond formation

The formation of polypeptides in organic media in the presence of proteases is the subject of numerous articles. However, results depends strongly on experimental parameters: history of the enzyme, pH, nature of the substrates, *etc*. This was clearly shown by Gaertner[70] in his study of the peptide bond formation between *N*-acyl-L-amino acid esters and L-phenylalanine amide catalyzed by modified or supported papain. Conversion above 90% was obtained when either basic or acidic amino acids were used as acyl-donor residues, while neutral and aromatic residues are poor substrates and the conversion is low. In spite of this, the polymerization is initiated.

2.9.2.2 Lactamization

Lactams or dilactams can be prepared by aminolysis of various difunctional compounds: amino esters, diesters or diamines. Aminolysis can be catalyzed in organic media by lipases[4,100] or by proteases.[100,101] Gutman *et al.*[47] studied the formation of lactams in organic solvents in two ways: intramolecular aminolysis of amino esters (equation 25) and the condensation of 2-chloroethyl diesters with diamines. Conversion depends on the structure of the lactam (Table 10). The formation of bis-lactams is described by the reaction shown in Scheme 9.[47]

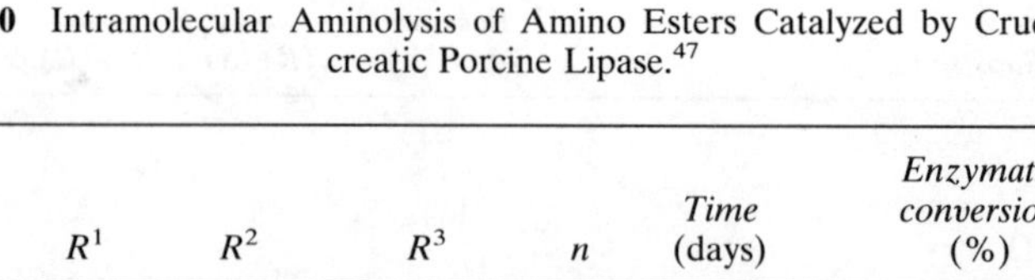

(25)

Table 10 Intramolecular Aminolysis of Amino Esters Catalyzed by Crude Pancreatic Porcine Lipase.[47]

No.	*R*^{1}	*R*^{2}	*R*^{3}	*n*	*Time* (days)	*Enzymatic conversion* (%)
1	H	H	Et	1	7	0
2	H	CO_2Et	Et	2	2	50
3	H	H	Pr^i	2	3	45
4	H	Me	Pr^i	2	6	40
5	Me	H	Pr^i	2	6	30
6	H	H	Pr^i	3	4	80
7	Me	H	Pr^i	3	5	40
8	H	H	Pr^i	4	7	10

The formation of macrocyclic bis-lactams provides a route to synthesize large host molecules in host–guest chemistry;[47] however, the above authors observed a poor enantioselectivity, in contrast to the results observed with lactones.

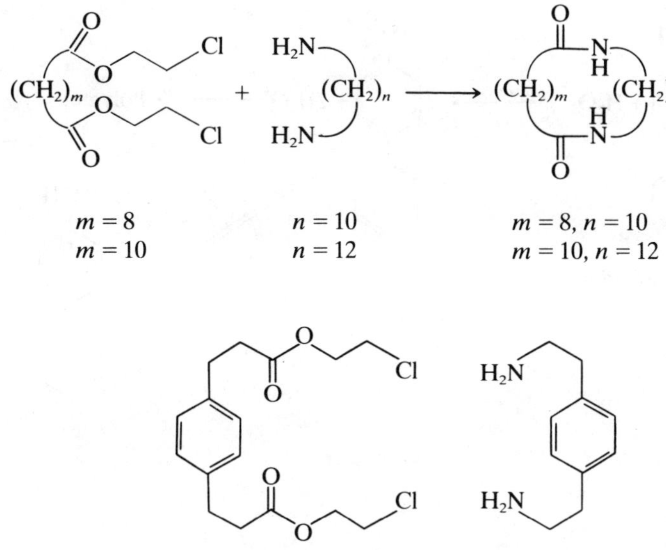

Scheme 9

2.9.3 Oxygenases and Peroxidases

Phenols and aniline derivatives can be polycondensed by the conjugated action of an oxidizing agent (O_2, H_2O_2) and of an enzyme (oxygenase, peroxidase). These enzymatic oxidations take place in aqueous media *in vivo*; however, the resulting polymers have low molar masses as they are insoluble.

The formation of oligomers and polymers depends on the nature of the reaction medium; it must permit an efficient contact between the substrate and the active site of the enzyme modified by the oxidizing agent.

Polyphenols exhibit some interest as substitutes for phenol–formaldehyde resins; examples are given in Schemes 10 and 11.[18]

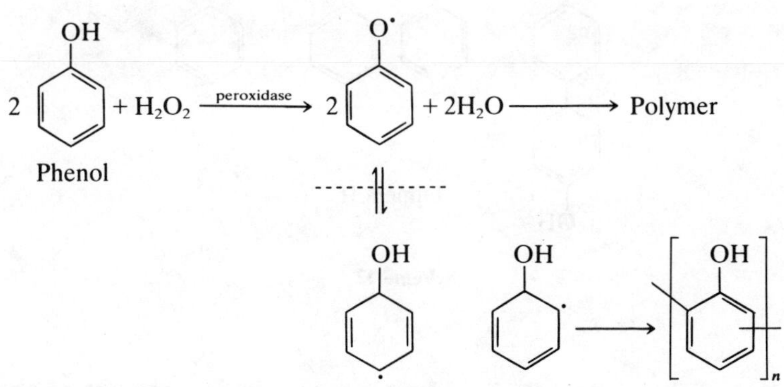

Scheme 10

In the same way, polyanilines are greatly studied as organic conductors.[19] Most of the chemical syntheses of these polymers give ill-defined materials; on the other hand, enzymatic processes are more regioselective and should permit a better control of the structure.

The reaction processes are very complex; they involve an oxidation step of the monomer giving free radicals, followed by dimerization of the radicals.[15] This is clearly shown in Scheme 12. Dimer II grows through oxidation steps which consume hydrogen peroxide, showing that it is a step-growth polymerization. Each step is activated by peroxidase cleavage of hydrogen peroxide; currently the relation between the stability of each mesomeric free radical and the catalytic process is not clear.

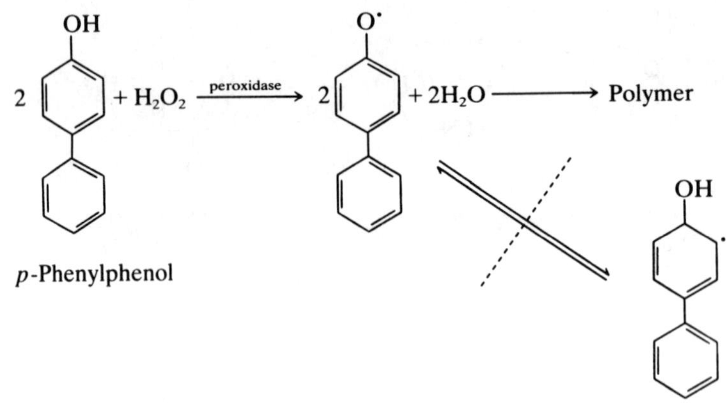

Scheme 11

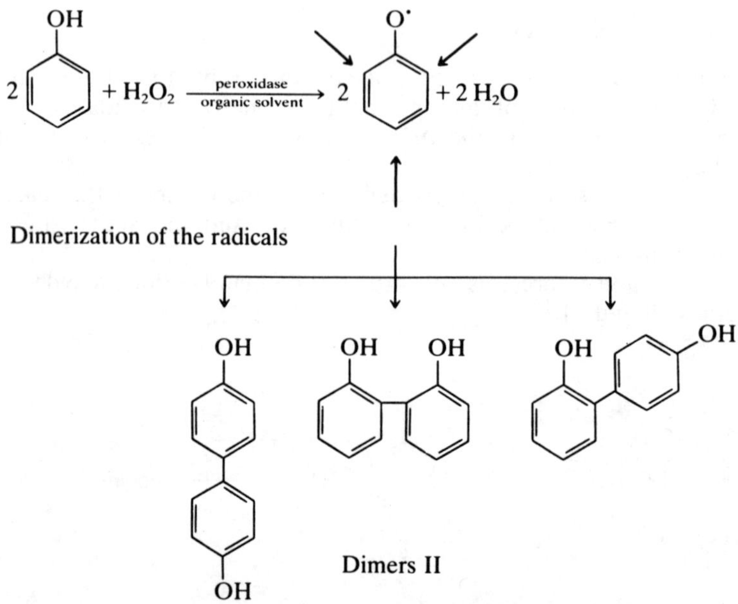

Scheme 12

Hammond[12] showed that polyphenol oxidase is a very active catalyst of the oxidation of phenols when it is carried in supercritical carbon dioxide or fluoroform containing oxygen. Oxidation of *p*-cresol and *p*-chlorophenol readily gives the corresponding *o*-benzoquinones which are polymerized to form poly(arylene oxides) whose molar mass ranges from 200 to 6500 (Figure 19).

When the enzyme was exposed to supercritical carbon dioxide (or fluoroform) containing oxygen but no phenolic substrate, there was little or no inactivation. However, the enzyme was inactivated during the oxidation process. With *p*-chlorophenol, 64% inactivation of the enzyme was observed at 23% conversion; with *p*-cresol, 30% conversion gave 36% inactivation and 90% conversion led to 90% inactivation.

Several studies relate almost exclusively to horseradish peroxidase (HRP).[14,15] When the reaction is carried out in an aqueous medium containing hydrogen peroxide, HRP oxidizes phenols but only very low oligomers are formed; this is due to their insolubility in water. On

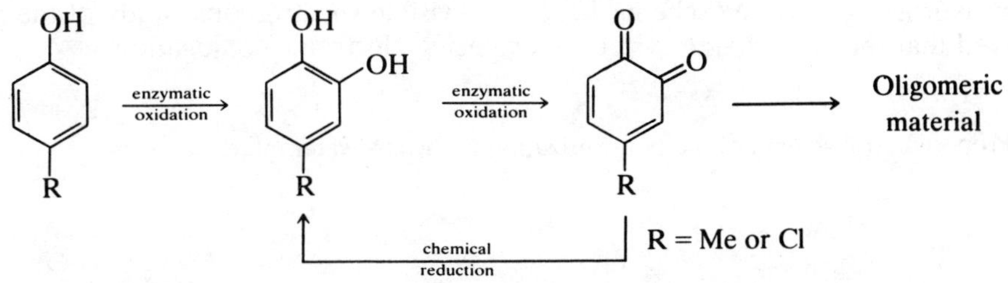

Figure 19 Diagram of the oxidation of phenols by polyphenol oxidase.[12]

the other hand, when the reaction is carried out in a water-miscible organic solvent (for instance dioxane), the polymer is soluble and its molar mass can be very high, as shown in Table 11.

Table 11 Average Molecular Weights of Phenolic Polymers Enzymatically Produced in 85% Dioxane.[14]

Compound	*Average molecular weight*
Phenol	1400
p-Methoxyphenol	2000
p-Cresol	1900
p-Chlorophenol	600
2,6-Dimethylphenol	500
4,4′-Biphenol	400
Aniline	1700
1-Naphthol	very high
2-Naphthol	2000
p-*t*-Butylphenol	1900
p-Phenylphenol	26000

The solvent/water ratio has an important influence on the results and it must be optimized in each case. In pure organic solvents the enzyme is inactive (Table 12).

Table 12 The Rates of Horseradish Peroxidase-catalyzed Oxidation of *p*-Phenylphenol in Aqueous Dioxane of Different Concentrations.[14]

Dioxane content (%)	*Reaction rate* (μmol min^{-1} (mg enzyme)$^{-1}$)
0	308
60	141
70	125
80	88
85	66
90	33
95	26
99	0
100	0

Reactions in inverse micelles are very promising; both molar mass and structure can be controlled by a good choice of the surfactant. Langmuir–Blodgett layers were prepared by HPR-catalyzed polycondensation of a mixture of 4-tetradecyloxyphenol and phenol in the presence of hydrogen peroxide (Scheme 13).[16] UV–visible spectroscopic study of the polymeric layers showed that they are characterized by extensive electronic conjugation.

Proposed arrangement of the polymerization at the air–water interface

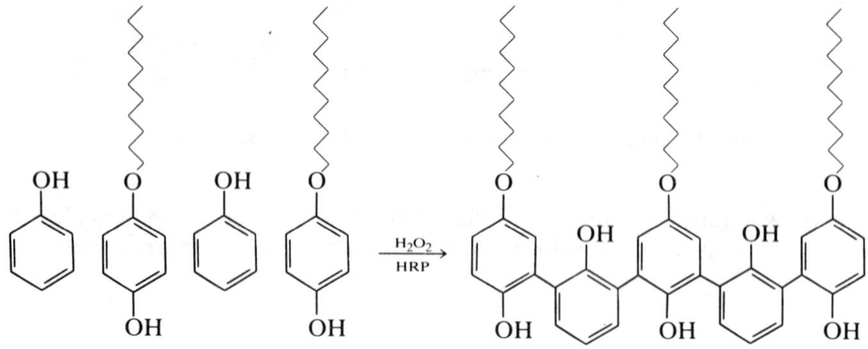

Scheme 13

The HPR-catalyzed polymerization of aniline was carried out in organic solvents, both water miscible and immiscible, and containing H_2O_2.[20] The results are nearly similar to those obtained with phenols; however, only monophasic systems lead to polymers. Polyaniline films were prepared from aqueous solutions of aniline, in the presence of an oxidase and in aerobic conditions.[11] The electrochemical properties of these polymers are different from those of the polymers which were chemically and electrochemically prepared; according to the authors, this results from structural differences. In fact, the structure of the polyanilines prepared by the enzymatic process is very complex: nitrogen is involved in secondary amino and imino groups as bonds in the chains and in primary amino groups as substituents of aromatic rings.

Kobayashi *et al.*[19] clearly pointed out the structural differences between the polymers obtained by oxidation of 1,2-diaminobenzene and by enzymatic catalysis or by classical oxidizing routes. When 1,2-diaminobenzene (**24**) is oxidized in 1,4-dioxane containing hydrogen peroxide, in the presence of HRP at pH 7 and at room temperature, the product is a polymer ($M = 20\,000$) whose chain is mainly formed of 2-aminophenylene units (**25**) (equation 26).

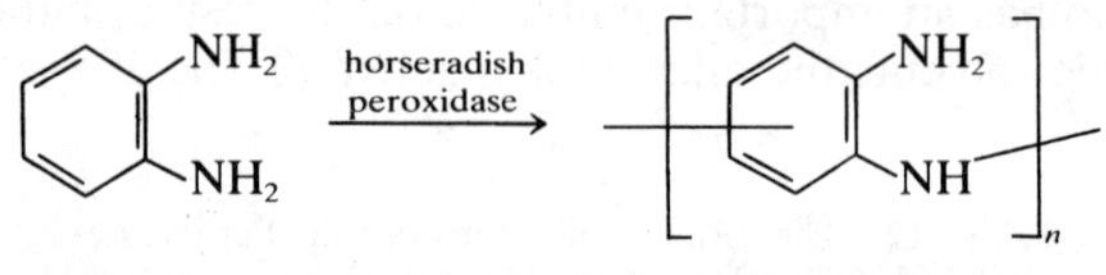

(26)

(**24**) (**25**)

These chains contain different kinds of monomer unit (**26**) and (**27**) (Scheme 14). On the other hand, a ladder polymer, mainly formed of phenazine rings, is obtained when 1,2-diaminobenzene is oxidized by an iron chelate complex[102] or by electrolysis.[103]

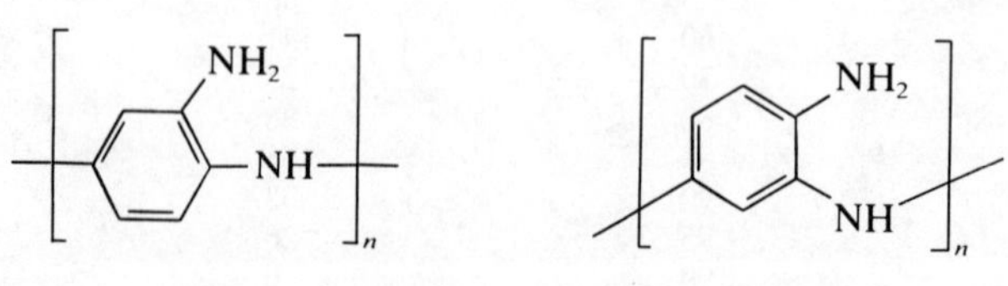

(**26**) (**27**)

Scheme 14

Various phenols and aromatic amines have been homopolymerized and copolymerized in very similar experimental conditions (HPR, dioxane/water). Most of the polymers are insoluble and thermostable; their structures are complex and difficult to analyze.[17,18,22]

The HPR-catalyzed copolymerization of a kraft lignine with phenols in the presence of H_2O_2 is always possible, but only in DMF;[21] a proposed reaction sequence is given in Figure 20.

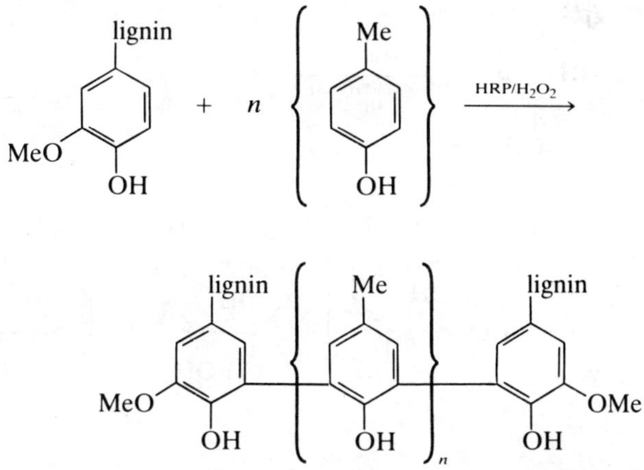

Figure 20 Proposed reaction sequence for lignin and multiple *p*-cresol molecules copolymerized by HRP in the presence of H_2O_2. The poly(*p*-cresol) forms a bridge between two lignin molecules, or two phenolic groups on the same lignin molecule. It is expected that the reactive moieties of lignin are the guaiacyl groups and these are shown in the figure.[21]

The characteristics of the copolymers (Table 13) permit their use as thermosets in place of classical phenol–formaldehyde resins. When the reaction is carried out in dioxane/water mixtures, some depolymerization of the lignin is observed.[13]

Table 13 Thermal Properties[21] of Lignin–Phenol Copolymers.[a]

Sample	T_g (°C)	ΔH (J g^{-1})	*30% Decomposition temperature* (°C)
Lignin	180.3	–	407
Lignin–cresol copolymer	119.7	117.2	425
Lignin–phenylphenol copolymer	98.0	357.5	438
Lignin–bisphenol A copolymer	178.5	89.7	433
Poly(*p*-cresol)	–	44.5 and 70.0[b]	441

[a] The copolymers were the DMF-insoluble fractions.
[b] The first exotherm occurred at 140 °C and the second exotherm at 405 °C.

2.9.4 Glycosidases

Amylases and cellulases belong to the class of hydrolysases and are known as catalysts of polysaccharide hydrolysis; however, they catalyze other reactions such as the glycosylation of C—F bond-containing substrates: the substrate behaves both as donor (C—F) and as acceptor (OH group).

As an example, Okada *et al.*[64] studied the behaviour of α-maltosyl fluoride in water and in the presence of α-amylase as catalyst and observed the conversion of α-maltosyl fluoride into maltooligosaccharides rather than its hydrolysis. According to these authors, α-amylases have a substantial capacity for binding two molecules of this small substrate in a distinctive way, with the C—F glycolysic bond of one and the free C-4 hydroxyl group of the other located in the region of the catalytic groups of the enzyme. This favours glycolysation of the suitably positioned acceptor over solvent water. It is noteworthy that enzymes recognize disaccharides more easily than monosaccharides.

The synthesis of cellulose (**29**) was achieved by Kobayashi *et al.* by cellulase-catalyzed polycondensation of β-D-cellobiosyl fluoride (**28**), carried out in a mixture of acetonitrile and water added to acetate buffer (pH 5) (equation 27).[65,66]

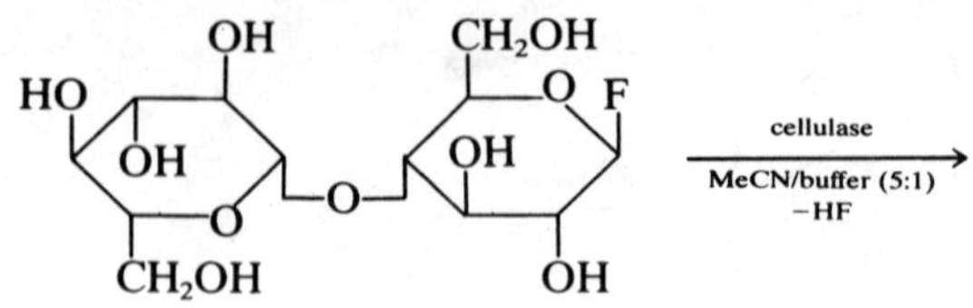

(27)

(**28**)

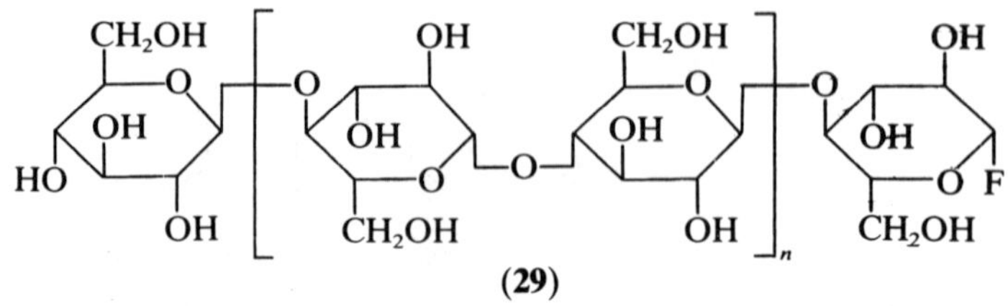

(**29**)

According to these authors, this synthesis would be the first *in vitro* synthesis of cellulose via a nonbiosynthetic path. Several solvents were studied; those which are miscible to the buffer gave the best results (Table 14).

Table 14 Reaction of D-Cellobiose Catalyzed by Cellulase[65] in Various Solvents.[a]

			w % of product[c]			
Entry	*Solvent*	*Time* (h)	*G1*	*G2*	*G3*	*G4*
1	Buffer[b]	24	76	22	2	0
2	MeCN/buffer (2:1)	24	53	38	18	1
3	MeCN/buffer (5:1)	48	19	60	18	3
4	Me_2CO/buffer (2:1)	24	26	62	10	2
5	Me_2CO/buffer (5:1)	48	10	84	5	1
6	DMF/buffer (2:1)	48	2	98	0	0
7	DMSO/buffer (2:1)	48	0	100	0	0

[a] The reaction was carried out at 30 °C: [cellobiose] = 2.5×10^{-2} mol L^{-1}, cellulase 7.5 wt % for cellobiose.
[b] Acetate buffer (0.05 M, pH 5).
[c] G1, G2, G3 and G4 denote glucose, cellobiose, cellotriose and cellotetraose, respectively.

An optimum organic solvent/water (buffer) ratio exists (Table 15): when the concentration of organic solvent is high, the polymers have low molar masses as the chains become insoluble when growing; on the other hand, when the water content is high the contribution of hydrolysis becomes predominant.

Table 15 Effect of Acetonitrile Concentration on Enzymatic Polymerization of (**28**)[65] Catalyzed by Cellulase.[a]

Entry	*Solvent*	*Yield* (%)[c]
1	Buffer	0 (0)
2	MeCN/buffer[b] (2:1)	16 (3)
3	MeCN/buffer (5:1)	64 (54)
4	MeCN/buffer (7.5:1)	77 (31)
5	MeCN/buffer (10:1)	67 (14)
6	MeCN/buffer (15:1)	50 (5)

[a] Polymerized at 30 °C for 12 h: [(**28**)] = 2.5×10^{-2} mol L^{-1}, cellulase 5 wt % for (**28**).
[b] Acetate buffer (0.05 M, pH 5).
[c] 5:1 Methanol/water-insoluble part. In parentheses, the yield of the water-insoluble part is given.

The cellulose obtained exhibits a very regular β(1–4) structure; a model of this reaction is given in Figure 21; however, it is more crystalline than natural cellulose and its crystalline structure is different. This study confirms the regio- and the stereoselectivity of α-amylase in transglycosylation: it recongizes β-D-fluorinated cellobiose but not the α-D anomer and no more than the β-D-fluorinated lactosyle.

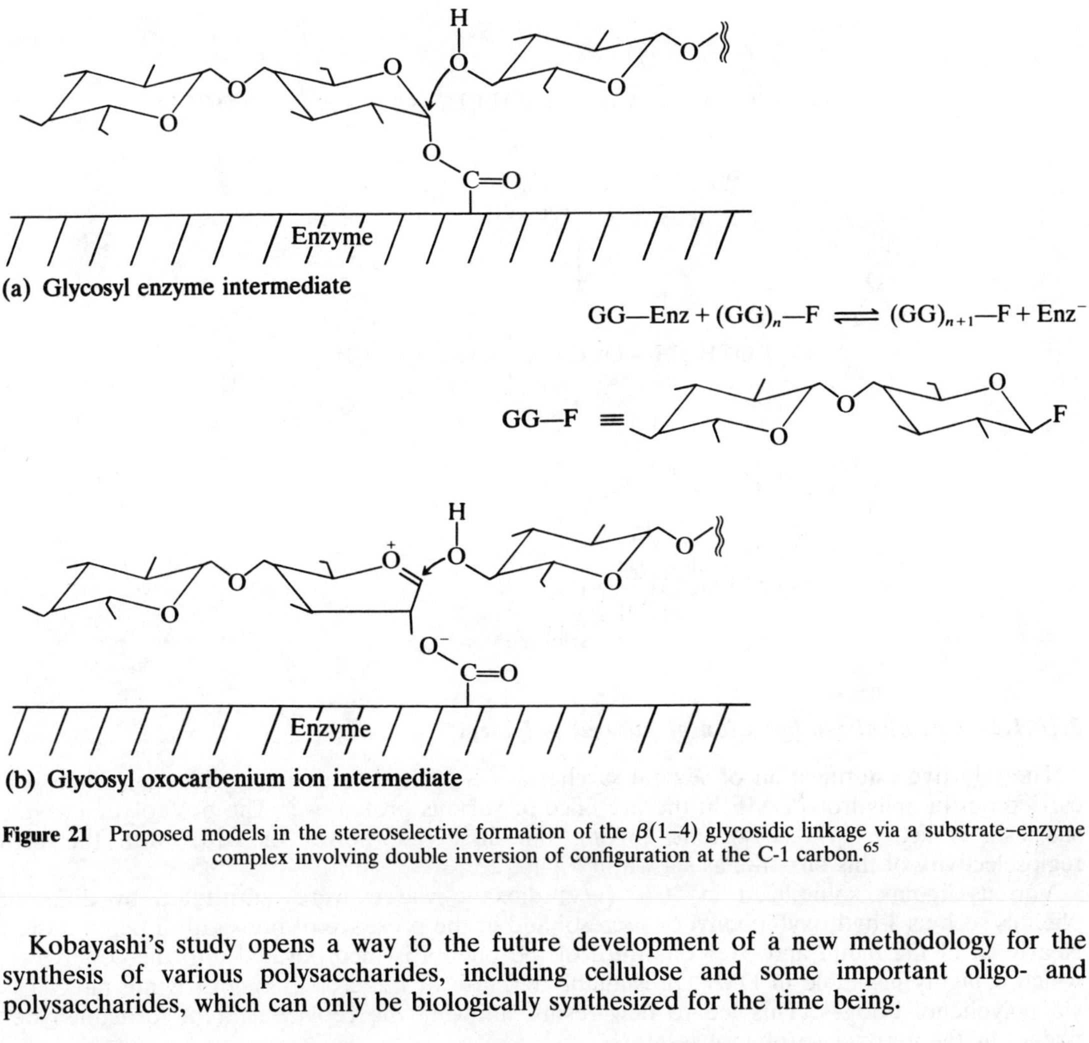

Figure 21 Proposed models in the stereoselective formation of the β(1–4) glycosidic linkage via a substrate–enzyme complex involving double inversion of configuration at the C-1 carbon.[65]

Kobayashi's study opens a way to the future development of a new methodology for the synthesis of various polysaccharides, including cellulose and some important oligo- and polysaccharides, which can only be biologically synthesized for the time being.

2.10 CHEMICAL MODIFICATION OF POLYMERS

Enzymatic catalysis is not only used in polycondensation but also in the chemical modification of polymers, where its regioselectivity and stereoselectivity are most useful.

In the following we will give two applications of the selectivity of enzymatic processes; then we will study particularly the chemical modification of acrylic and methacrylic polymers.

2.10.1 Application of the Selectivity of Enzymatic Processes

The polymer can be used as a product or as an intermediate agent in a synthetic process.

2.10.1.1 Resolution of racemic mixtures

One of the two enantiomers of a racemic mixture is selectively fixed on a polymer via an enzymatic process. This method was applied to the resolution of racemic 2,2,2-trichloroethyl

3,4-epoxybutanoate (**30**). When it is transesterified with α,ω-dihydroxy polyoxyethylene (**31**) in the presence of a porcine pancreas lipase (PPL), only the (*S*)-enantiomer reacts with the polymer (Scheme 15).[44] The modified polymer is easily purified by filtration at low temperature and the enantiomer can be regenerated by hydrolysis or by transesterification.

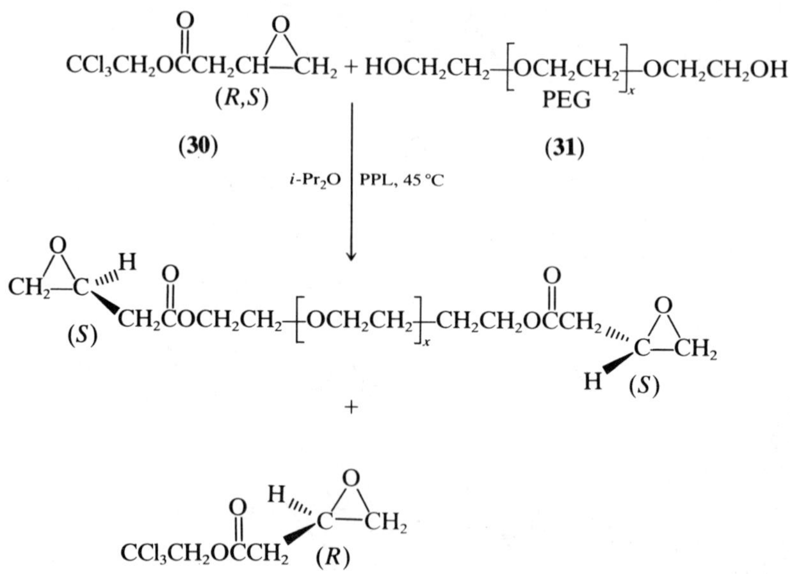

Scheme 15

2.10.1.2 *Chemical modification of natural polymers*

The selective esterification of several saccharides is limited to monoesterification when it is carried out in anhydrous DMF in the presence of various proteases.[69] The proteolytic enzyme subtilisin is active in this solvent, linking the dissolving power of DMF with the high regioselectivity of this enzyme, as shown in Figure 22.

Various lignins, solubilized in 50% (v/v) dioxane–water, were substituted by different phenols such as 4-hydroxybiphenyl or *p*-cresol and in the presence of horseradish peroxidase.[21] Nearly all of the lignin and over one-third of the phenol is incorporated into the copolymer, which is highly insoluble in DMF, presumably because of the crosslinking of lignin molecules via polyphenol bridges. This led to new resins, allowing the replacement of formaldehyde/phenol by the peroxidase/phenol system.

2.10.2 Chemical Modification of Acrylic and Methacrylic Polymers

Most of the enzymatic chemical modifications of macromolecular compounds in organic media concern acrylic and methacrylic polymers. Several examples relate to long side-chains bearing methacrylic polymers. For example, Rehse[68] studied the α-chymotrypsin-catalyzed reaction of alanine *t*-butyl ester (**32**) with a copolymer of methyl methacrylate and a long-chain-bearing methacrylate (**33a**) and (**33b**) (Scheme 16). This leads to polypeptide side-chain polymers and shows the importance of the spacer groups: only the acid groups of the long chains were transesterified by alanine *t*-butyl ester; the short methyl methacrylates were not modified.

In the same way, Pavel[38,50] studied the acylation of poly[*N*-(2-hydroxypropyl)-11-methacryloylaminoundecanamide-*co*-styrene] [(**36a**) and (**36b**)] by several esters in the presence of a lipase (Scheme 17). The monomer is more reactive than the copolymer and the reactivity of the latter depends on its composition; this is linked to steric and hydrophobic effects. The racemic

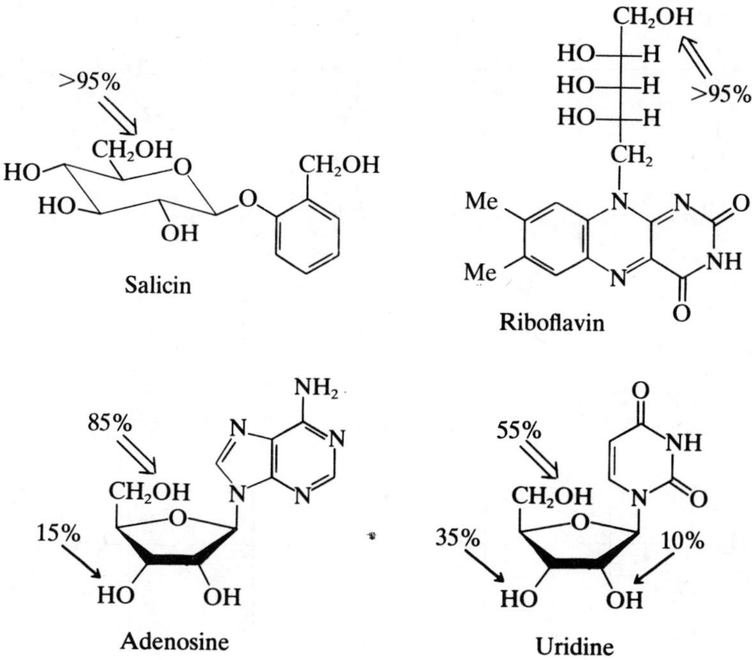

Figure 22 Structures of compounds containing both sugar and aglycon moieties used as targets for subtilisin-catalyzed acylations in DMF.[69]

secondary alcohol of the side-chains shows a slight resolution after acylation, which leads to optically active products. The highest optical rotation was observed when the acylating agent was phenyl stearate or benzoate.

From the early work of enzymatic chemistry in organic media it was observed that the efficiency of the enzyme depends to some extent on the structure of the substrate. The chemical modifications of poly(methyl acrylate) and of poly(methyl methacrylate) could be catalyzed by lipases. Pavel[50] observed the same phenomenon when studying the acetylation of copolymers of styrene with long side-chain methacrylamides which contain *CHOH groups. Yields and optical rotations are given in Table 16.

Lalot[46] showed that only the end groups of oligomethacrylates (PMA) were modified when these oligomers are transesterified by allylic alcohol; this observation results from the influence of two parameters: the steric hindrance close to the reactive site and the distance of the reactive site to the chain, that is to say the length of the spacer (Scheme 18). However, cleavages of the spacer are observed when it contains functions which can be recognized by the enzyme.[56]

Many difficulties are encountered in the course of chemical modifications of polymers; for this reason it is often easier to modify the monomer before the formation of the polymeric chain rather than to carry out the modification on the polymer. This involves two steps: (i) the enzymatic-catalyzed modification of the monomer; and (ii) the polymerization or polycondensation of the modified monomer. This method leads to higher conversions than the direct modification of the polymer.

Rehse[68] showed that the α-chymotrypsin-catalyzed formation of peptide bonds in methacrylic polymers is easier on the monomer than on the related polymer. This method was applied to chiral alcohols, which were then polymerized. Thus, Ghogare[42] transesterified *O*-acryloyloxime with several chiral alcohols in the presence of a lipase, and then the resulting monomers were polymerized by radical polymerization, leading to a chiral polymer (Scheme 19).

Methacroylaminoundecanoic acid was transesterified by various alcohols, including cyclohexanol, menthol, testerone and cholesterol; the reaction was carried out in a water/cyclohexane mixture and in the presence of a lipase. These monomers were polymerized, leading to biological active group-bearing polymers.[42]

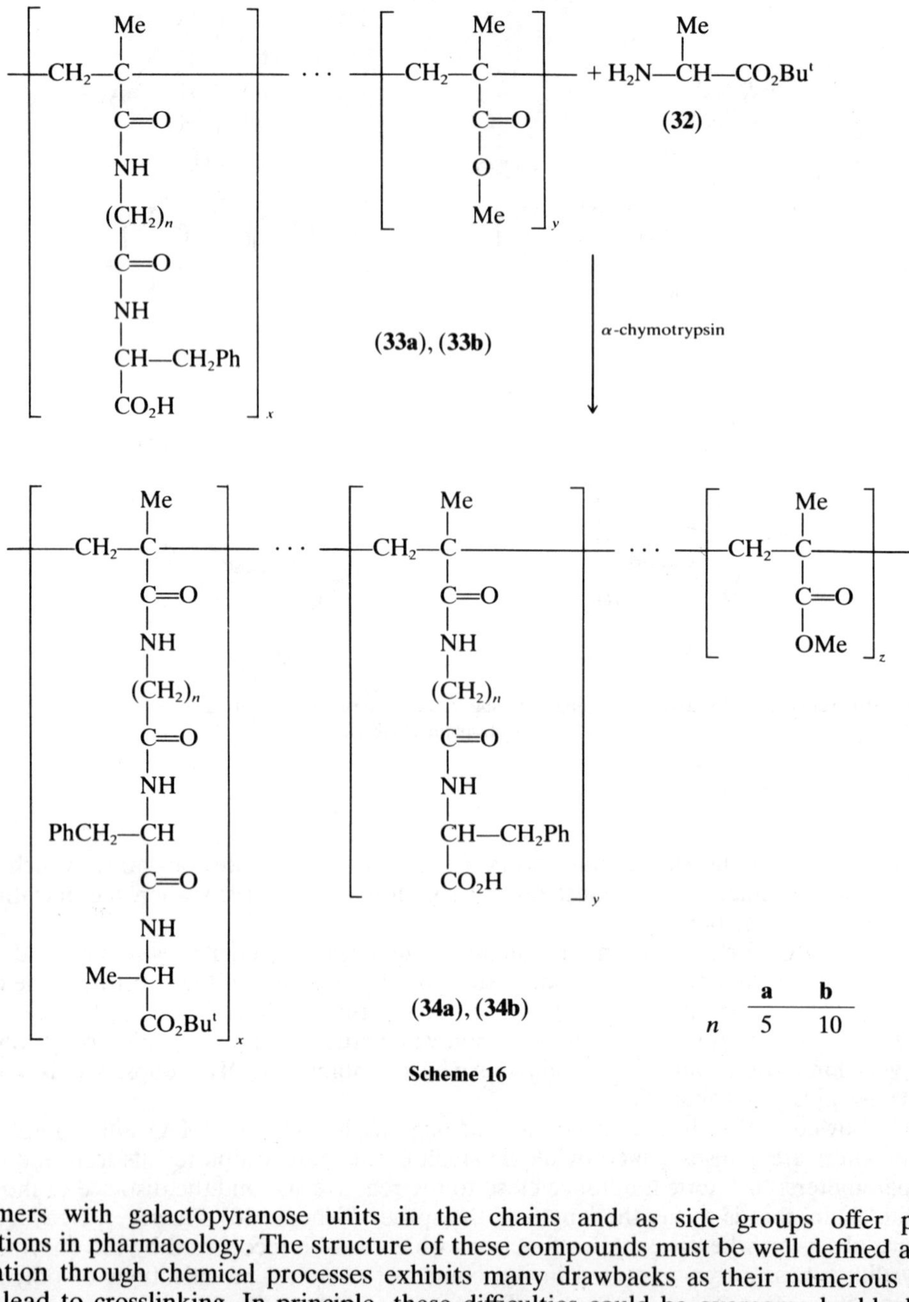

Scheme 16

Polymers with galactopyranose units in the chains and as side groups offer potential applications in pharmacology. The structure of these compounds must be well defined and their preparation through chemical processes exhibits many drawbacks as their numerous hydroxy groups lead to crosslinking. In principle, these difficulties could be overcome by blocking the functions which are sensitive to various reactions such as acylation; however, this requires cumbersome processes. Using a porcine pancreatic lipase in pyridine, Therisod[28] succeeded in preparing a monoacetylated α-glucose using 2,2,2-trichoroethyl butyrate; only the primary hydroxy reacted. The same process was applied to fructose and only the two primary alcohol groups were modified.

Polymers bearing sugar residues exhibit an increasing interest as pharmacological and biomedical materials. Matsumura[67] used β-galactosidase to catalyze the transesterification of lactose (**39**) or of *o*-nitrophenyl-β-D-galactopyranoside (ONPG) (**40**) with 2-hydroxyacrylate (**41**) or hydroxymethacrylate (**42**) in a phosphate buffer (Scheme 20).

It was found that both the yield and the rate of reaction of ONPG were higher than those for lactose. The transesterification preserved all the hydroxy functions.

Martin[48] prepared vinyl derivatives of several sugars, leading to saccharide-bearing hydrogels; the sugar was modified by vinyl acrylate in the presence of a lipase in pyridine.

Patil *et al.*[71,72] prepared polycondensable and polymerizable sugars by different enzymatic processes using a protease (Proleather) in pyridine. The process led to sugar-bearing

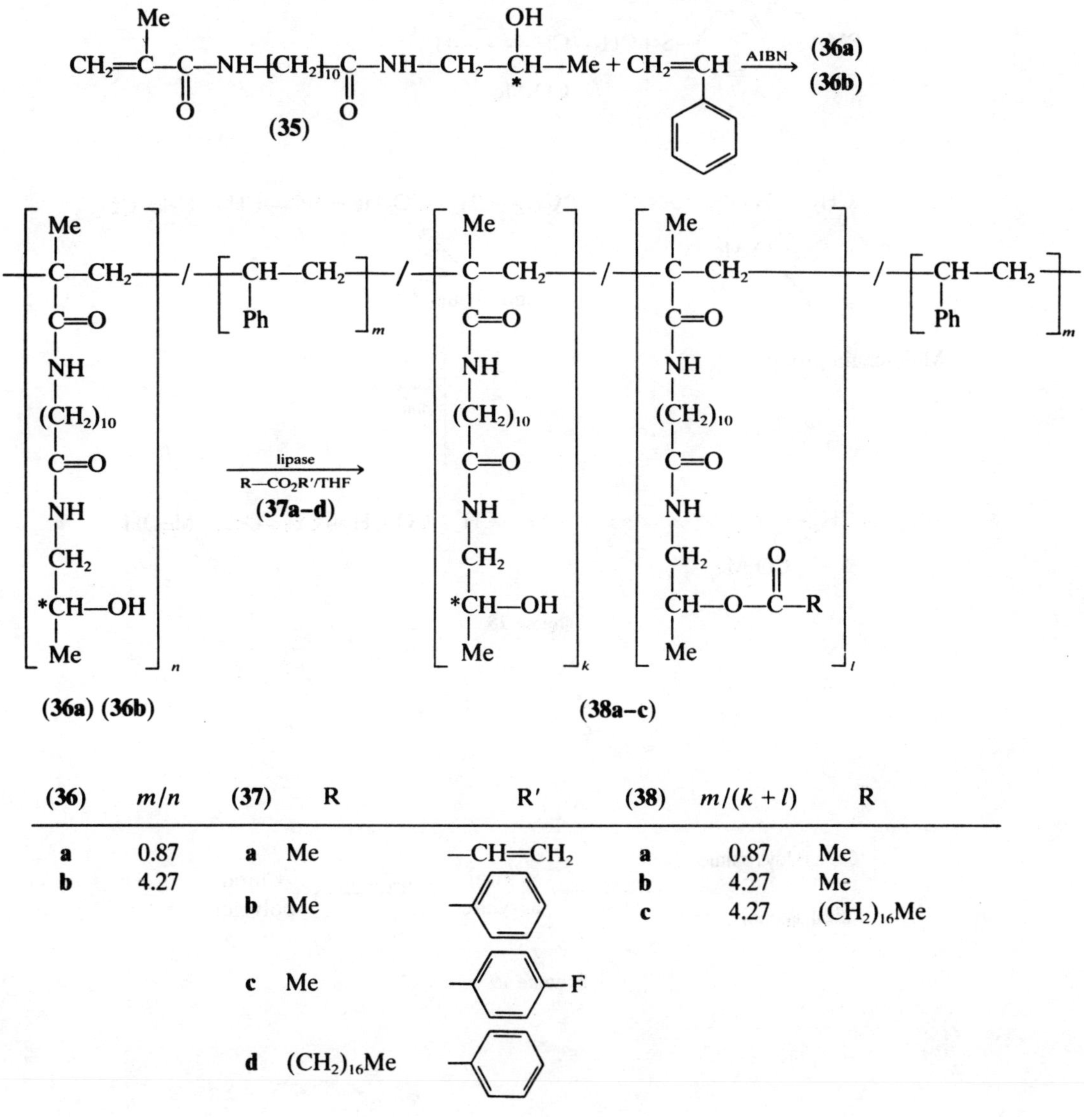

Scheme 17

Table 16 Yield of Acylation of Monomer **(35)** and the Copolymers **(36a)** and **(36b)** (Scheme 17) and Comparison of the Specific Optical Rotation $[\alpha]$.[50]

Substrate	*Acylating agent* **(37a–d)**	*Conversion after 6 days* (%)	k/l	$[\alpha]_{546}^{20}$ [a]
(35)	**a**	93	–	n.d.[b]
(35)	**b**	86	–	n.d.
(36a)	**a**	32	2.09	n.d.
(36a)	**b**	40	1.50	n.d.
(36b)	**a**	26	2.82	−0.63
(36b)	**b**	40	1.51	−1.20
(36b)	**c**	25	3.01	0.00
(36b)	**d**	20	4.03	−1.22

[a] Related to the content of methacrylic units in the copolymer.
[b] Not determined.

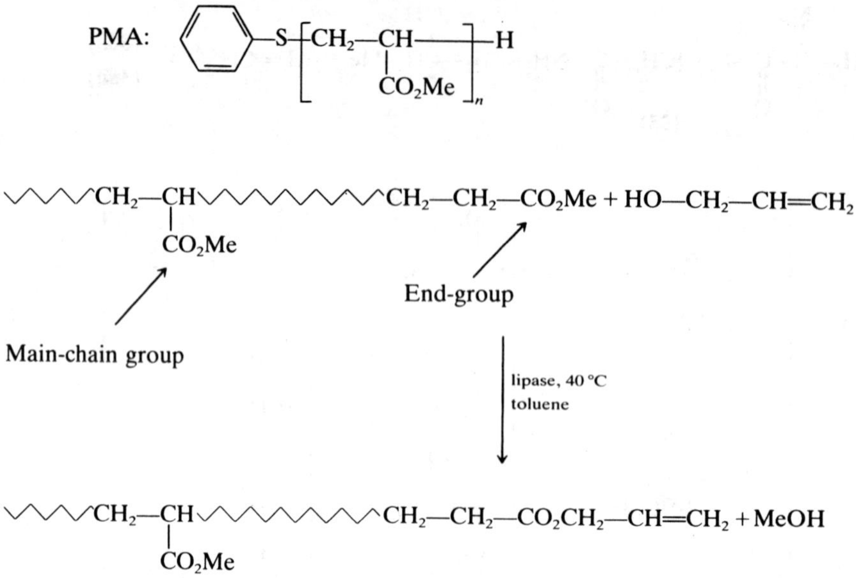

Scheme 18

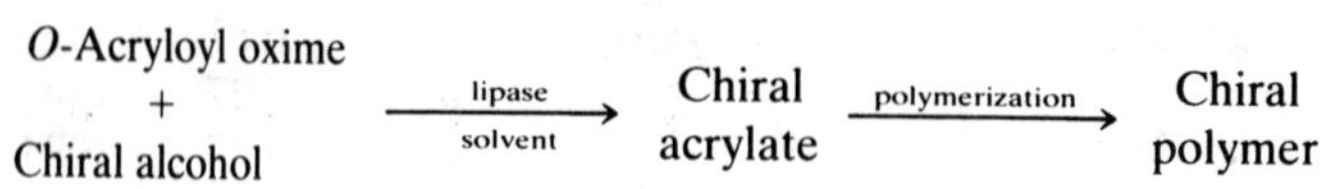

Scheme 19

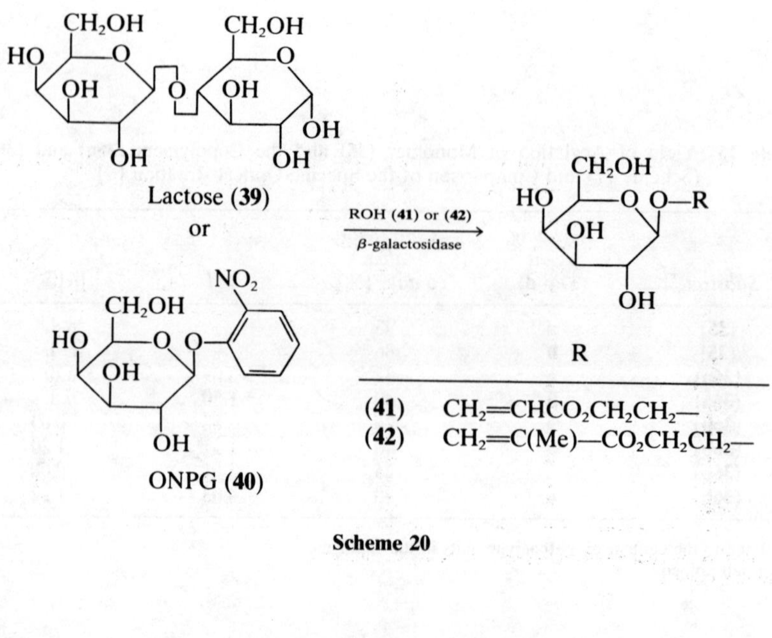

Scheme 20

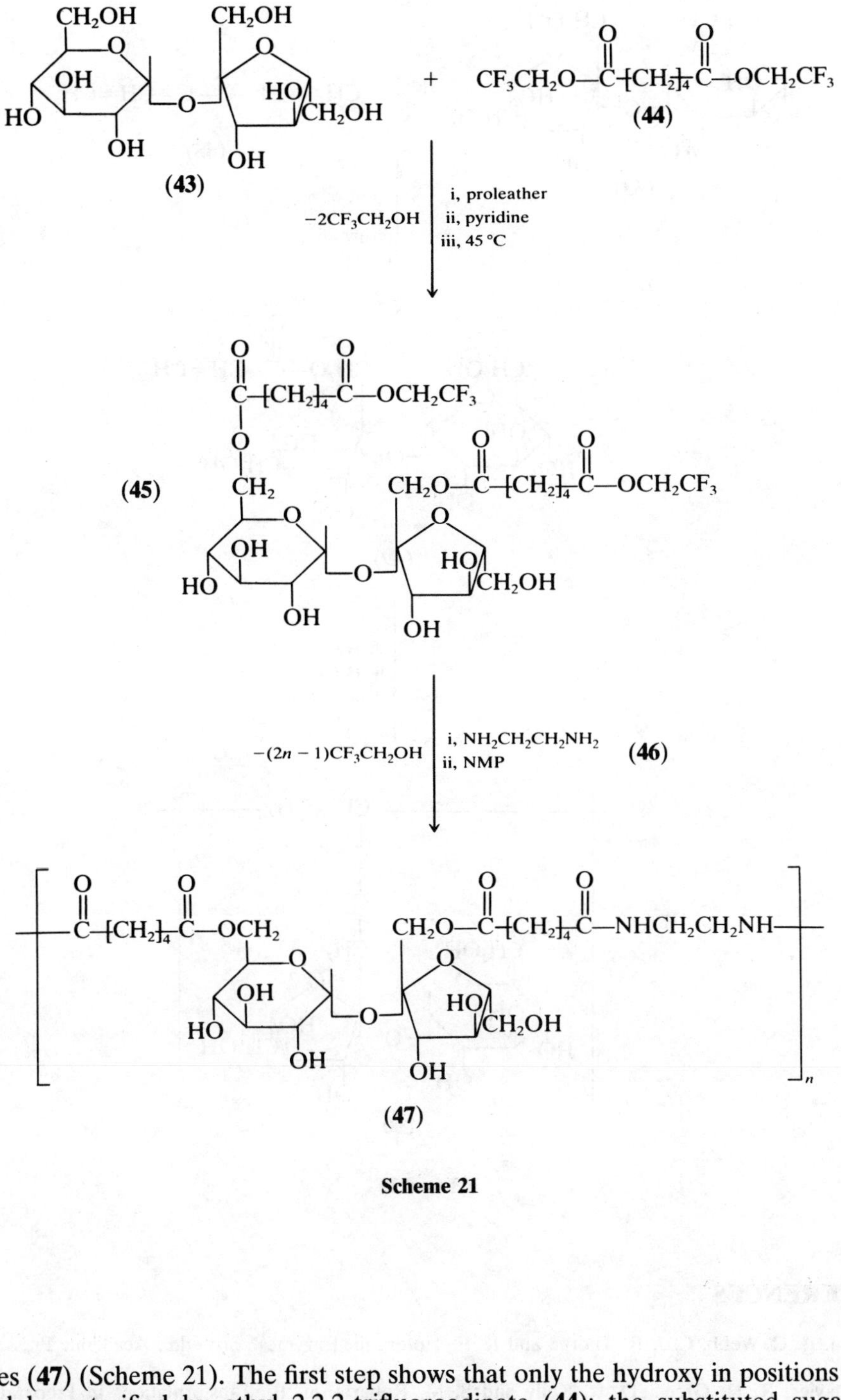

Scheme 21

polyamides (**47**) (Scheme 21). The first step shows that only the hydroxy in positions 1′ and 6 of (**43**) could be esterified by ethyl 2,2,2-trifluoroadipate (**44**); the substituted sugar (**45**) was polycondensed with a diamine (**46**).

The same sugar (**43**) was reacted with vinyl acrylate (**48**); only the hydroxy in position 1′ was modified (Scheme 22). This unsaturated monomer (**49**) was polymerized using a redox system as the radical initiator. It is interesting to underline that, depending on the process, the sugar residue is either in the chain (polycondensation) (**47**) or is a side group (chain polymerization) (**50**).

Enzymatic processes are particularly used for their very high selectivity; on the other hand, the low rate of most enzymatic reactions is their major drawback. The coupling of enzymatic and chemical processes permits the association of the selectivity of one with the kinetic efficiency of the other.

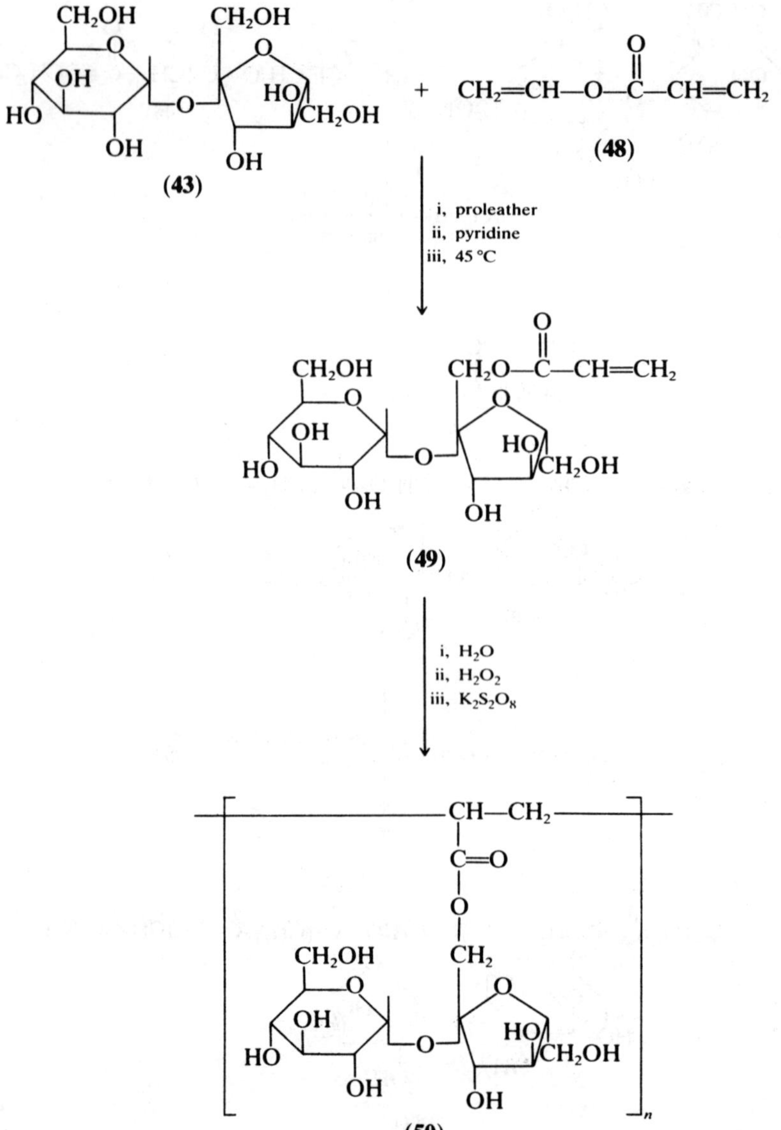

Scheme 22

2.11 REFERENCES

1. M. Dixon, E. C. Webb, C. J. R. Thorne and K. F. Tipton, in 'Enzymes', 3rd edn., Academic Press, New York, 1979.
2. H. G. Davies, R. H. Green, D. R. Kelly and S. M. Roberts, in 'Biotransformations in Preparative Organic Chemistry', Academic Press, London, 1989.
3. J. S. Dordick, *Biotechnol. Prog.*, 1992, **8,** 259.
4. A. Zaks and A. M. Klibanov, *Proc. Natl. Acad. Sci. USA*, 1985, **82,** 3192.
5. D. Judas, A. Fradet and E. Maréchal, *Makromol. Chem.*, 1984, **185,** 2583.
6. S. Geresh and Y. Gilboa, *Biotechnol. Bioeng.*, 1990, **36,** 270.
7. E. Maréchal, M. Brigodiot and T. Lalot, 1996, in press.
8. C. Chothia, *Nature,* 1988, **333,** 598.
9. D. E. Koshland, *Proc. Natl. Acad. Sci. USA*, 1958, **44,** 98.
10. R. A. Derango, L. C. Chiang, R. Dowbenko and J. G. Lash, *Biotechnol. Tech.*, 1992, **6,** 523.
11. M. Aizawa, L. Wang, H. Shinohara and Y. Ikariyama, *J. Biotechnol.*, 1990, **14,** 301.
12. D. A. Hammond, M. Karel, A. M. Klibanov and V. J. Krukonis, *Appl. Biochem. Biotechnol.*, 1985, **11,** 393.
13. J. S. Dordick, M. A. Marletta and A. M. Klibanov, *Proc. Natl. Acad. Sci. USA*, 1986, **83,** 6255.
14. J. S. Dordick, M. A. Marletta and A. M. Klibanov, *Biotechnol. Bioeng.*, 1987, **30,** 31.
15. K. Ryu, D. R. Stafford, J. S. Dordick, *ACS Symp. Ser.*, 1989, **389,** 141.

16. F. Bruno, J. A. Akkara, L. A. Samuelson, B. K. Mandal, D. L. Kaplan, K. A. Marx and S. Tripathy, *Polym. Preprints (Am. Chem. Soc., Div. Polym. Chem.)*, 1991, **32,** 232.
17. J. A. Akkara, K. J. Senecal and D. L. Kaplan, *J. Polym. Sci., Part A, Polym. Chem.*, 1991, **29,** 1561.
18. R. J. Patil and P. K. Kamath, *Paintindia,* 1991, **41,** 63.
19. S. Kobayashi, I. Kaneko and H. Uyama, *Chem. Lett.*, 1992, **3,** 393.
20. J. A. Akkara, P. Salapu and D. L. Kaplan, *Indian J. Chem.*, 1992, **31B,** 855.
21. A. M. Blinkovsky and J. S. Dordick, *J. Polym. Sci., Part A, Polym. Chem.*, 1993, **31,** 1839.
22. H. Uyama, H. Kurioka, I. Kaneko and S. Kobayashi, *Macromol. Rep.*, 1994, **A31,** 421.
23. D. A. Abramowicz and C. Keese, *Biotechnol. Bioeng.*, 1989, **33,** 149.
24. S. Okuruma, M. Iwai and Y. Tsujisaka, *Yukagaku,* 1983, **32,** 271.
25. S. Okuruma, M. Iwai and Y. Tominaga, *Agric. Biol. Chem.*, 1984, **48,** 2805.
26. A. Ajima, T. Yoshimoto, K. Takahashi, Y. Tamaura, Y. Saito and Y. Inada, *Biotechnol. Lett.,* 1985, **7,** 303.
27. S. Matsumura and J. Takahashi, *Makromol. Chem., Rapid Commun.*, 1986, **7,** 369.
28. M. Therisod and A. M. Klibanov, *J. Am. Chem. Soc.*, 1986, **108,** 5638.
29. A. L. Gutman, K. Zuobi and A. Boltansky, *Tetrahedron Lett.*, 1987, **28,** 3861.
30. A. L. Gutman, D. Oren, A. Boltanski and T. Bravdo, *Tetrahedron Lett.*, 1987, **28,** 5367.
31. A. Makita, T. Nihira and Y. Yamada, *Tetrahedron Lett.*, 1987, **28,** 805.
32. A. L. Margolin, J.-Y. Crenne and A. M. Klibanov, *Tetrahedron Lett.*, 1987, **28,** 1607.
33. G. Zhi-Wei and C. J. Sih, *J. Am. Chem. Soc.*, 1988, **110,** 1999.
34. G. Zhi-Wei, T. K. Ngooi, A. Scilimati, G. Fülling and C. J. Sih, *Tetrahedron Lett.,* 1988, **29,** 5583.
35. A. L. Gutman and T. Bravdo, *J. Org. Chem.*, 1989, **54,** 4263.
36. A. L. Gutman and T. Bravdo, *J. Org. Chem.*, 1989, **54,** 5645.
37. T. K. Ngooi, A. Scilimati, G. Zhi-Wei and C. J. Sih, *J. Org. Chem.*, 1989, **54,** 911.
38. K. Pavel, H. Rehse and H. Ritter, *Polym. Bull.*, 1989, **21,** 535.
39. J. S. Wallace and C. J. Morrow, *J. Polym. Sci., Part A, Polym. Chem.*, 1989, **27,** 2553.
40. J. S. Wallace and C. J. Morrow, *J. Polym. Sci., Part A, Polym. Chem.*, 1989, **27,** 3271.
41. H. Yamada, S. Ohsawa, T. Sugai, H. Ohta and S. Yoshikawa, *Chem. Lett.*, 1989, 1775.
42. A. Ghogare and G. S. Kumar, *J. Chem. Soc., Chem. Commun.*, 1990, 134.
43. A. L. Gutman, K. Zuobi and T. Bravdo, *J. Org. Chem.*, 1990, **55,** 3546.
44. J. S. Wallace, K. B. Reda, E. M. Williams and C. J. Morrow, *J. Org. Chem.,* 1990, **55,** 3544.
45. S. Geresh and Y. Gilboa, *Biotechnol. Bioeng.*, 1991, **37,** 883.
46. T. Lalot, M. Brigodiot and E. Maréchal, *Polym. Bull.*, 1991, **26,** 55.
47. A. L. Gutman, E. Meyer, X. Yue and C. Abell, *Tetrahedron Lett.*, 1992, **33,** 3943.
48. B. D. Martin, S. A. Ampofo, R. J. Linhardt and J. S. Dordick, *Macromolecules,* 1992, **25,** 7081.
49. G. Nicolosi, M. Piattelli and C. Sanfilippo, *Tetrahedron,* 1992, **48,** 2477.
50. K. Pavel and H. Ritter, *Makromol. Chem.*, 1992, **193,** 323.
51. F. Binns, S. M. Roberts, A. Taylor and C. F. Williams, *J. Chem. Soc., Perkin Trans. 1,* 1993, 899.
52. S. Geresh, Y. Gilboa, S. Abrahami and A. Bershadsky, *Polym. Eng. Sci.*, 1993, **33,** 311.
53. D. Knani, A. L. Gutman and D. H. Kohn, *J. Polym. Sci., Part A, Polym. Chem.*, 1993, **31,** 1221.
54. D. Knani and D. H. Kohn, *J. Polym. Sci., Part A, Polym. Chem.*, 1993, **31,** 2887.
55. S. Kobayashi and H. Uyama, *Makromol. Chem., Rapid Commun.*, 1993, **14,** 841.
56. K. Pavel and H. Ritter, *Makromol. Chem.*, 1993, **194,** 3369.
57. H. Uyama and S. Kobayashi, *Chem. Lett.*, 1993, 1149.
58. D. O'Hagan and N. A. Zaidi, *J. Chem. Soc., Perkin Trans. 1,* 1993, 2389.
59. S. Kobayashi, *Polym. Preprints (Am. Chem. Soc., Div. Polym. Chem.)*, 1994, **35** (1), 772.
60. E. M. Brazwell, D. Y. Filos and C. J. Morrow, *J. Polym. Sci., Part A, Polym. Chem.*, 1995, **33,** 89.
61. R. T. MacDonald, S. K. Pulapura, Y. Y. Svirkin, R. A. Gross, D. L. Kaplan, J. A. Akkara, G. Swift and S. Wolk, *Macromolecules,* 1995, **28,** 73.
62. Y. Ohya, T. Sugitou and T. Ouchi, *J.M.S. – Pure Appl. Chem.*, 1995, **A32,** 179.
63. G. Mezoul, T. Lalot, M. Brigodiot and E. Maréchal, *J. Polym. Sci., Part A, Polym. Chem.*, 1995, **33,** 2691.
64. G. Okada, D. S. Genghof and E. J. Herhe, *Carbohydr. Res.*, 1979, **71,** 287.
65. S. Kobayashi, K. Kashiwa, T. Kawasaki and S.-I. Shoda, *J. Am. Chem. Soc.*, 1991, **113**, 3079.
66. S. Kobayashi, K. Kashiwa, J. Shimada, T. Kawasaki and S. Shoda, *Makromol. Chem., Macromol. Symp.*, 1992, **54/55,** 509.
67. S. Matsumura, H. Kubokawa and K. Toshima, *Makromol. Chem., Rapid Commun.*, 1993, **14,** 55.
68. H. Rehse and H. Ritter, *Makromol. Chem.*, 1988, **189,** 529.
69. S. Riva, J. Chopineau, A. P. G. Kieboom and A. M. Klibanov, *J. Am. Chem. Soc.*, 1988, **110**, 584.
70. H. F. Gaertner, A. Ferjancic and A. J. Puigserver, *Biocatalysis,* 1990, **3,** 197.
71. D. L. Patil, J. S. Dordick and D. G. Rethwisch, *Macromolecules,* 1976, **24,** 3462.
72. D. R. Patil, D. G. Rethwisch and J. S. Dordick, *Biotechnol. Bioeng.*, 1991, **37,** 639.
73. V. F. Hogan, D. O'Hagan and J. Sanvoisin, *Indian J. Chem.*, 1992, **31B,** 883.
74. A. E. M. Janssen, A. Van der Padt, H. M. Van Sonsbeek and K. Van't Riet, *Biotechnol. Bioeng.,* 1993, **41,** 95.
75. E. S. Lyon and W. B. Jakoby, *Arch. Biochem. Biophys.*, 1980, **202,** 474.
76. L. G. Buttler, *Enzyme Microb. Technol.*, 1979, **1,** 253.
77. A. Zaks and A. M. Klibanov, *J. Biol. Chem.*, 1988, **263,** 3194.
78. A. Zaks and A. M. Klibanov, *J. Biol. Chem.*, 1988, **263,** 8017.
79. P. A. Burke, S. O. Smith, W. W. Bachovchin and A. M. Klibanov, *J. Am. Chem. Soc.,* 1989, **111,** 8290.
80. C. Laane, J. Tramper and M. D. Lilly, in 'Studies in Organic Chemistry 29', Elsevier, Amsterdam, 1987.
81. L. E. S. Brink and J. Tramper, *Biotechnol. Bioeng.*, 1985, **27,** 1258.
82. R. C. Weast (ed.), 'Handbook of Chemistry and Physics', 57th edn., CRC Press, Cleveland, OH, 1991.
83. R. F. Rekker and H. M. de Kort, *Eur. J. Med. Chim. Ther.*, 1979, **14,** 479.
84. S. Parida and J. S. Dordick, *J. Am. Chem. Soc.*, 1991, **113,** 2253.
85. F. Monot, F. Borzeix, M. Bardin and J.-P. Vandecasteele, *Appl. Microbiol. Biotechnol.,* 1991, **35,** 759.

86. A. L. Gutman and M. Shapira, *J. Chem. Soc., Chem. Commun.*, 1991, 1467.
87. K. Ohshima, Y. Tomiuchi and H. Kise, *J. Jpn. Oil Chem. Soc. (Yukagaku)*, 1992, **41,** 1141.
88. R. Affleck, Z.-F. Xu, V. Suzawa, K. Focht, D. S. Clark and J. S. Dordick, *Proc. Natl. Acad. Sci. USA*, 1992, **89,** 1100.
89. R. H. Valivety, P. J. Halling and A. R. Macrae, *Biochim. Biophys. Acta,* 1992, **1118,** 218.
90. P. J. Halling, *Biotech. Adv.*, 1987, **5,** 47.
91. P. Monsan and D. Combes, *Methods Enzymol.*, 1988, **137,** 584.
92. P. Monsan and G. Durand, *C. R. Acad. Sci., Ser. C,* 1971, **273,** 33.
93. M. P. Pileni, *J. Phys. Chem.*, 1993, **97,** 6961.
94. Y. L. Khmelnitsky, A. K. Gladilin, V. L. Roubailo, K. Martinek and A. V. Levashov, *Eur. J. Biochem.*, 1992, **206,** 737.
95. P. Miethe, R. Gruber and K.-H. Mohr, *BioEngineering,* 1992, **8** (3) 38.
96. F. Borzeix, F. Monot and J.-P. Vandecasteele, *Enzyme Microb. Technol.*, 1992, **14,** 791.
97. M. D. Bednarski, H. K. Chenault, E. S. Simon and G. M. Whitesides, *J. Am. Chem. Soc.*, 1987, **109,** 1283.
98. D. Cabaret, S. Bourcier, S. Maillot and M. Wakselman, *Biocatalysis,* 1992, **6,** 191.
99. R. Vazquez-Duhalt, P. M. Fedorak and D. W. S. Westlake, *Enzyme Microb. Technol.*, 1992, **14,** 837.
100. H. Kitaguchi, P. A. Fitzpatrick, J. E. Huber and A. M. Klibanov, *J. Am. Chem. Soc.*, 1989, **111,** 3094.
101. H. Kitaguchi and A. M. Klibanov, *J. Am. Chem. Soc.*, 1989, **111,** 9272.
102. E. Tscuchida, M. Kaneko and Y. Kurimura, *Kogyo Kagaku Zasshi,* 1969, **72,** 772.
103. K. Chida, T. Ohsaka, Y. Ohnuki and N. Oyama, *J. Electroanal. Chem.*, 1987, **219,** 117.

3
Synthesis and Properties of Dendrimers and Hyperbranched Polymers

JEAN M. J. FRÉCHET
Cornell University, Ithaca, NY, USA

and

CRAIG J. HAWKER
IBM Almaden Research Center, San Jose, CA, USA

3.1 INTRODUCTION

Because organic synthesis is a youthful science whose origins date back to the work of Michel Chevreul and Friedrich Wohler in the first part of the 19th century, it is not surprising that most molecules made by Nature have not been duplicated by Man. In contrast to the field of organic synthesis, where naturally occurring molecules have long remained important targets, the even younger field of polymer science has recently been dominated by wholly synthetic macromolecules that, although not found in living systems, have had a very significant impact on our daily lives. Despite the development of a wealth of synthetic reactions and processes, the fact remains that our best synthetic approaches cannot come, even close, to the ability of natural systems to accurately control variables such as molecular shape, size, and local microenvironment in large macromolecular assemblies. Furthermore, Nature controls all of these variables simultaneously and such ability has remained elusive for the synthetic chemist.

This review is concerned with the synthesis, properties, and applications of a very new family of polymers known as dendritic macromolecules. As their name indicates, these macromolecules have highly branched three-dimensional features that mimic the structure of trees. Once a certain size is reached, dendritic macromolecules adopt a globular shape that is quite different in its origin from the globular shape of enzymes, since it is the direct result of their branching pattern, as opposed to folding, intramolecular interactions, or bonding in enzymes.

Three broad types of dendritic macromolecules are recognized today: dendrimers, hyperbranched polymers, and hybrid dendritic–linear macromolecules.[1] While the first two families of dendritic molecules share common features such as extensive branching, they are quite different in that the former are highly regular structures that are frequently monodisperse, while the latter are irregular and just as polydisperse as other ordinary synthetic polymers. Hybrid dendritic–linear macromolecules are also less precisely defined than dendrimers, if only because their linear components are always polydisperse.

The term 'dendrimer' was originally coined by Tomalia[2] to describe a family of regularly branched poly(amidoamines) in which all bonds converge to a single point, each repeat unit contains a branch junction, and the final molecule features a very large number of identical chain ends. Although it is two- rather than three-dimensional, the 2D representation given in Scheme 1 conveys the essential notion that a dendrimer is a highly symmetrical layered molecule consisting of many identical building blocks emanating from a central core and radiating through a regular branching pattern towards numerous chain ends. Steric considerations, coupled with the branching geometry of the numerous internal building blocks, does not allow the dendrimer to assume a flat structure as shown in Scheme 1. Instead, the dendrimer fills space and adopts a globular shape that, in a fully extended, though perhaps energetically unfavorable conformation, may approximate a sphere. It is this postulated globular shape of dendrimers that has sparked considerable interest in recent years because it cannot be achieved by traditional one-dimensional linear polymers that adopt random coil or rigid rod structures. In particular, it is thought that the globular structure will have a major effect on the basic physical and chemical properties of these unusual macromolecules. For example, globular dendrimers are believed to be essentially free of chain entanglement, in direct contrast to most linear polymers where physical properties are dominated by chain entanglements. In addition, as a result of their layered structure, it is possible to imagine globular dendrimers in which specific functional groups are segregated as well defined locations within the nanoscopic macromolecule. Therefore, a unique nanoenvironment can be created around inner functionalities, leading to a structure that may mimic the active site of an enzyme. These unusual features, coupled with the ability to accurately control the structure, have stimulated much research as exciting new applications are being anticipated in areas ranging from drug delivery and diagnostic tools, to rheology control, adhesion, and nanofabrication.

While naturally occurring dendrimers are not known today, highly branched but irregular macromolecules such as amylopectin and glycogen have some structural analogies to a closely related class of materials, first named 'hyperbranched' polymers by Kim and Webster.[3] These are highly branched irregular polymers that are typically obtained, not by regular dendritic growth, but by the polycondensation of a multifunctional AB_x monomer containing one reactive group of type A and x of type B, where A and B can be made to combine with each other to form a chemical bond. Because of kinetic and steric considerations, the growth of such macromolecules is not regular and therefore reactive functionalities will be placed not only at chain ends, but also throughout the macromolecule.

Finally, it is possible to imagine synthetic macromolecules that are made up of dendritic and linear macromolecular fragments. Once again, these novel synthetic block copolymers have sparked the imagination of many scientists because the combination of properties associated with their separate blocks offers tantalizing development opportunities.

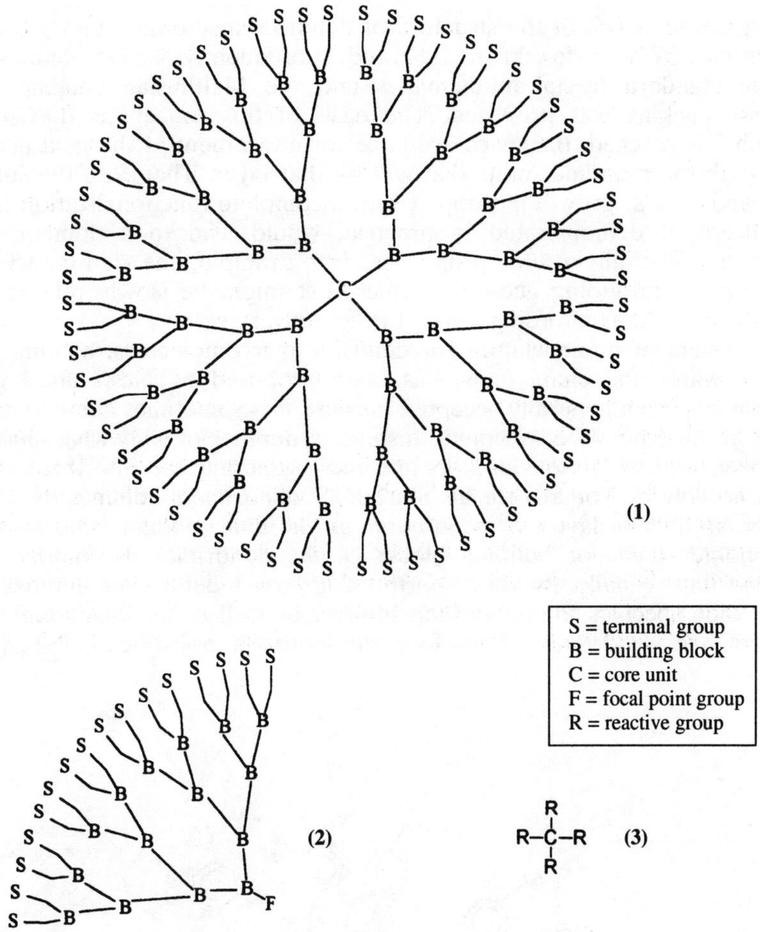

Scheme 1

3.2 DENDRIMERS

3.2.1 The Nomenclature of Dendrimers

Because dendrimers are so unlike linear polymers, it is useful to review the specialized terms and nomenclature that have been developed to describe their features and designate their components. This nomenclature is still evolving and many refinements are likely to be introduced in coming years.

A typical dendrimer such as structure (**1**), shown in Scheme 1, consists of three basic sections, the core moiety, C, the interior building blocks, also known as branch cells, B, and the terminal, or surface groups, S. The interior building blocks are arranged into layers, or generations, giving a fractal-like archiecture to the dendritic macromolecule. This fractal-like quality results in a highly symmetrical structure where the number of building blocks in each generation and the number of terminal groups are defined by geometrical progressions based on the functional multiplicity of both the core molecule and the building blocks. For example, dendrimer (**1**) has a core multiplicity of 4 and four layers of building blocks that have a branch multiplicity of 2. Therefore this dendritic macromolecule has 4×2^4 or 64 terminal groups and 60 building blocks arranged in four layers of 32, 16, 8, and 4 units, respectively.

The dendritic fragments that are attached to the core of (**1**) are termed dendrons. As will be seen below, the term dendron may be also used to describe dendrimers that will become part of a larger macromolecule once they are attached to a core moiety. For example, structure (**2**) represents a reactive dendron that may be attached to a core molecule such as (**3**) by reaction of its focal point[4] reactive group F with the reactive groups R of the core, affording the tetradendron dendrimer (**1**).

An interesting consequence of the structure of dendritic macromolecules was first proposed in theoretical studies by Maciejewski[5] in 1982 and, subsequently, by De Gennes and Hervet.[6] Borrowing from standard fractal mathematics and the Malthusian packing paradox,[7] the concept of dense packing was proposed. The basis of this concept is that once a specific generation number is reached, the space available for attachment of the next layer is less than that required by all the monomer units that will fill that layer. Therefore the surface becomes 'dense packed' and regular growth is stopped with incomplete functionalization of the terminal layer. This concept, if demonstrated in practice, would lead to a number of interesting consequences and potentially useful properties. For example, the dense-packed dendrimer would be capable of entrapping guest molecules that might be slowly released through the congested surface layer, thus affording a novel drug delivery system.

A systematic system of nomenclature for dendritic macromolecules, naming the molecules from the core towards the chain ends, has been proposed by Newkome.[8] However, this nomenclature has not been generally accepted because it is sometimes cumbersome and it may not apply easily to all types of dendrimers. Instead, a number of somewhat chaotic shorthand versions have been used by various authors and these generally fall into three categories. The first, used most notably by Tomalia for his Starburst® dendrimers,[2] defines the building blocks and specifies the number of layers or generations in the final product. Alternatively, the total number of monomer units, or building blocks of the dendrimer, is counted[9] and used to describe the dendrimer. Finally, the third shorthand approach defines the number and nature of the chain ends, then specifies the generation number as well as the functionality at the focal point or core of the dendrimer.[4] Therefore the notation polyether $(CN)_{16}$-[G-4]-OH has

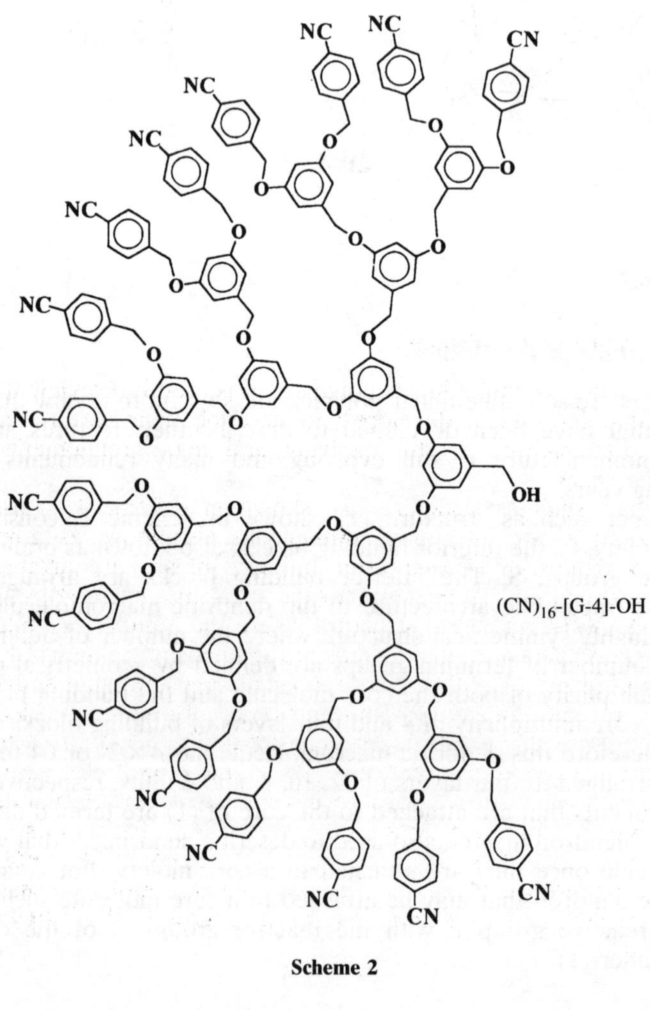

Scheme 2

been used[10] to describe a fourth generation dendrimer (or reactive dendron) consisting of four layers of ether-linked building blocks with 16 cyano terminal groups at the chain ends, and a single hydroxy group at the focal point, as shown in Scheme 2.

3.2.2 Synthesis of Regular Dendrimers

Two distinct synthetic strategies have been used for the preparation of dendrimers. The first, developed independently by Tomalia for Starburst dendrimers[2] and Newkome *et al.* for Cascade molecules,[11] is now recognized as the divergent growth approach. More recently, a novel convergent growth approach has been described by Hawker and Fréchet.[4] The basic difference between these two synthetic strategies is in the direction of growth. The divergent approach starts growth at a polyfunctional core, proceeding radially outwards with the stepwise addition of successive layers of building blocks. This process leads to larger and larger dendritic molecules with an ever increasing number of chain ends and reactive functional groups. In contrast, the convergent growth approach begins at the chain ends and proceeds inwards through successive additions of the growing dendritic molecules to a single building block or monomer unit. An optional final reaction is the attachment of several dendritic molecules (or dendrons) to a polyfunctional core. While it has been shown[12] that the two approaches can give exactly the same structure, the design of their growth process is essentially opposite and each has a number of advantages and disadvantages when compared to the other.

3.2.2.1 The concept of divergent growth

Historically, the initial attempts to prepare dendritic macromolecules have all employed the divergent approach. In 1978, Vögtle and co-workers[13] attempted the preparation of an oligometric nitrile-terminated polyamine by a series of Michael additions and subsequent reduction steps (Scheme 3). However, difficulties with the reduction step, the occurrence of cyclization side reactions, and incomplete functionalization of the terminal groups severely affected the growth scheme. In fact, it is this formation of unwanted side products that limited growth and restircted the synthesis to the preparation of a second generation derivative with a theoretical molecular weight (MW) of 873 daltons. As a result of their small size and uncontrolled growth the resultant molecules cannot be considered to be actual dendrimers. Subsequently, three patents by Denkewalter *et al.* disclosed[14] the synthesis of poly(lysine) via a stepwise repetitive procedure that used standard peptide coupling chemistry and the *p*-nitrophenyl ester of bis(*t*-Boc)-L-lysine as the monomer unit. Although the patents suggest that growth was continued to the 10th generation, affording truly polymeric materials with molecular weights up to 500 000 Da, little characterization data are available for these highly branched polymers. In a result that is still not fully explained, but may be related to the unsymmetrical nature of the lysine building blocks or to uncontrolled growth, the physical properties of the branched poly(lysine) were found to be rather unremarkable[15] and unlike those subsequently obtained for dendritic macromolecules produced using symmetrical building blocks. The behavior observed is actually analogous to that of some hyperbranched macromolecules,[16] suggesting that the poly(lysine) may assume an extended three-dimensional structure more closely resembling hyperbranched rather than dendritic macromolecules. Such a shape could easily be rationalized on the basis of either or both of the following factors: (i) the unsymmetrical nature of the monomer unit that leads to unequal branch segments; (ii) the occurrence of a large number of defects during the poly(lysine) synthesis. Because the standard peptide coupling steps used for the growth of these dendrimers are generally quite reliable, it is likely that the lack of symmetry[17] and conformational flexibility of the lysine building block are at the root of the seemingly disappointing physical properties of Denkewalter's poly(lysine).

Following these studies it was not until the mid-1980s that truly regular symmetrical dendritic macromolecules were first synthesized. Tomalia's Starburst dendrimers, prepared by a divergent approach, were first reported at several conferences in 1983[18] and 1984,[19] then fully disclosed in 1985.[2] Almost simultaneously, Newkome[11] reported in 1985 the preparation of a cascade molecule termed 'arborol' also using a divergent approach. Although these pioneering efforts employed different synthetic strategies, the fundamental characteristics required for a successful dendrimer synthesis are found in both cases. These characteristics demand that the chemistries used for both generation growth and activation give high yields with no significant side reactions. This allows the synthesis to be continued through a sufficient number of

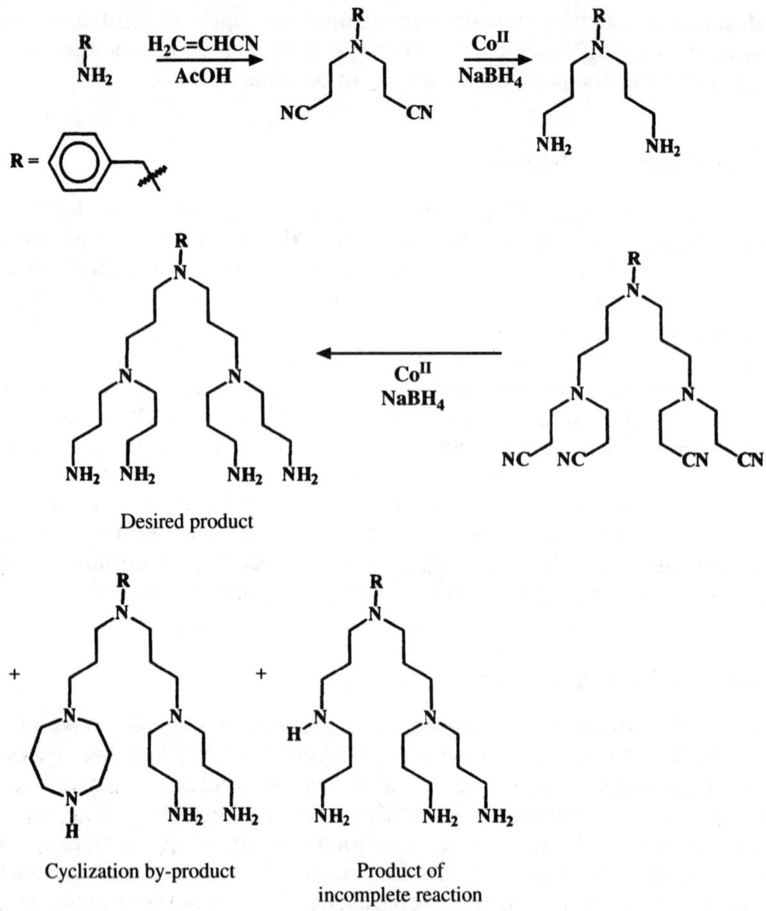

Scheme 3

iterations to obtain a macromolecule of respectable molecular weight. In addition, the monomer units were selected to afford symmetrical branch segments, which alleviates any possibility of asymmetric shape or unequal reactivity within the growing macromolecule. These basic principles have subsequently been used by a large number of researchers to prepare a variety of dendritic macromolecules with a symmetrical layering of building blocks by the divergent growth approach.[20,21]

In the synthesis of poly(amidoamine) (PAMAM or Starburst) dendrimers shown in Scheme 4, Tomalia[2,20] employed a repetitive sequence based on Michael additions similar in concept to that reported by Vögtle.[13] However, the troublesome reduction step was elegantly circumvented by using exhaustive amidation to regenerate the reactive amino groups. Reaction of the trifunctional core molecule, ammonia, with methyl acrylate results in formation of the triester (**4**). Regeneration of the reactive NH groups is then accomplished by exhaustive amidation using excess 1,2-diaminoethane. This process leads to a doubling in the number of reactive NH groups from three for ammonia to six for the first generation dendrimer (**5**). Significantly, all of the reactive NH groups of the growing dendrimer (**5**) are located at the chain ends. Repetition of this two-step reaction sequence, with purification at each stage to remove the large excess of reagents required to force the reactions to completion, then leads to larger and larger dendritic macromolecules. A characteristic of the divergent growth approach is that the number of terminal functional groups follows a strict geometrical progression. For example, Tomalia's PAMAM dendrimer synthesis utilizes a core moiety and a building block having multiplicities of three and two, respectively. Therefore, the number of terminal functional groups doubles at each growth step and while the fourth generation dendrimer (**6**) has only 48 active —NH end groups, the product obtained after 10 iterations, the 10th generation dendrimer, has 3072 (3×2^{10}) terminal groups. It must be emphasized here that for PAMAM dendrimers, because the two active hydrogens are located on the same nitrogen atom, this corresponds to 1536 primary amino, $—NH_2$, groups. Careful manipulation of the experimental conditions and purification procedures has allowed Tomalia and co-workers to produce PAMAM dendrimers on a kilogram scale and these products are now commercially available from a number of sources.[22]

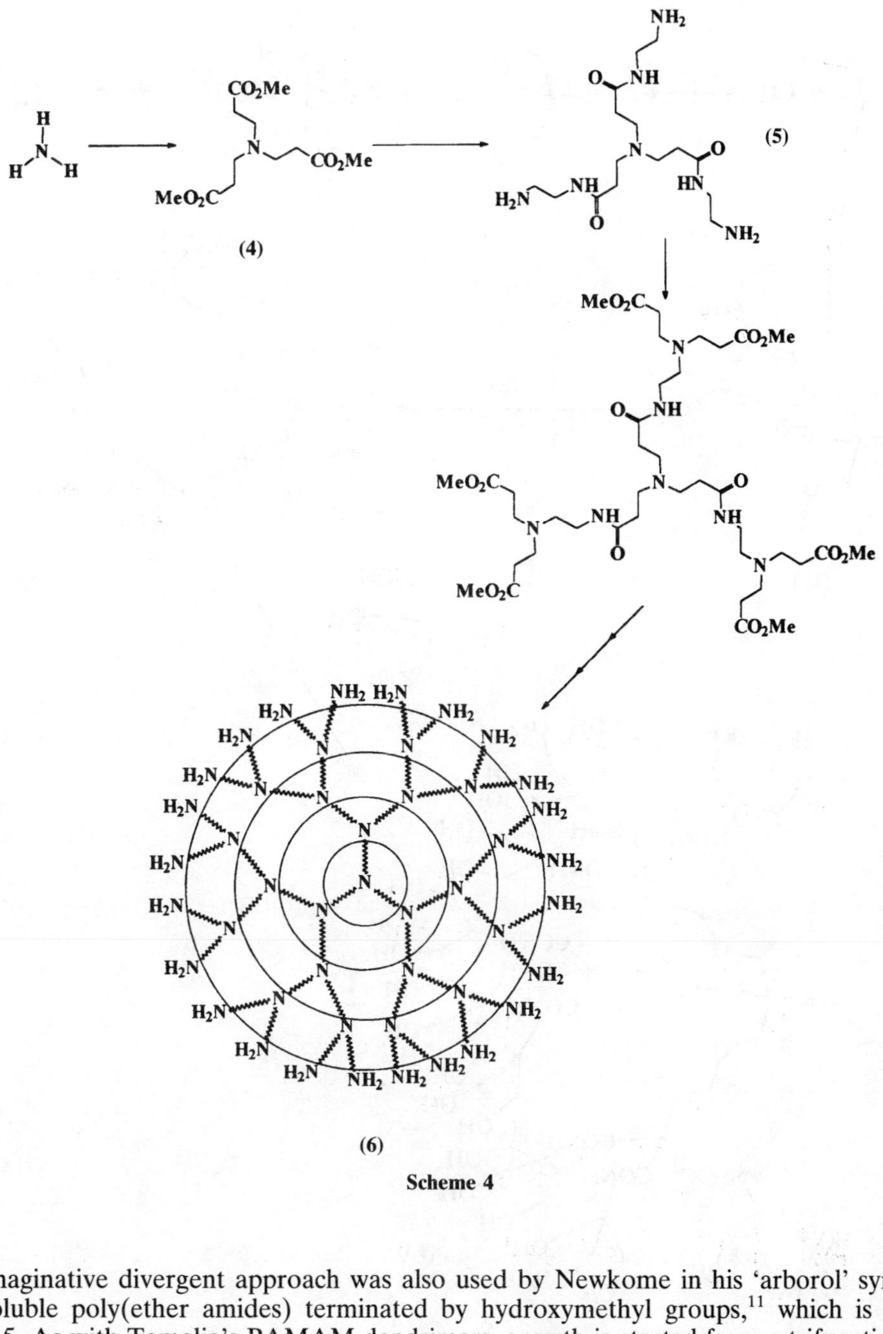

(6)

Scheme 4

An imaginative divergent approach was also used by Newkome in his 'arborol' synthesis of water-soluble poly(ether amides) terminated by hydroxymethyl groups,[11] which is shown in Scheme 5. As with Tomalia's PAMAM dendrimers, growth is started from a trifunctional core; however, to increase the branching density and hence the 'speed' of the synthesis, Newkome employed an AB_3 building block. This affords three branch points at each junction and triples the number of terminal functional groups at each step of growth. Initially, reaction of the tritosylate (**7**) with the sodium anion of triethyl methanetricarboxylate (**8**) could not be forced to completion, presumable due to steric congestion. To overcome this difficulty, a spacer group that alleviates the steric congestion around the quaternary carbon atom of (**9**) was introduced. Therefore, reaction of (**9**) with chloroacetic acid followed by reduction and tosylation afforded the tritosylate (**10**), which is similar to (**7**) except that a three-atom spacer group has been inserted between the tosylate group and the quaternary center. Reaction of (**10**) with (**8**) proceeded smoothly to give the nonaester (**11**) which, on exhaustive amidation with tris(hydroxymethyl)aminomethane (**12**), afforded the dendritic macromolecule (**13**) with 27 (3×3^2) terminal hydroxy groups.

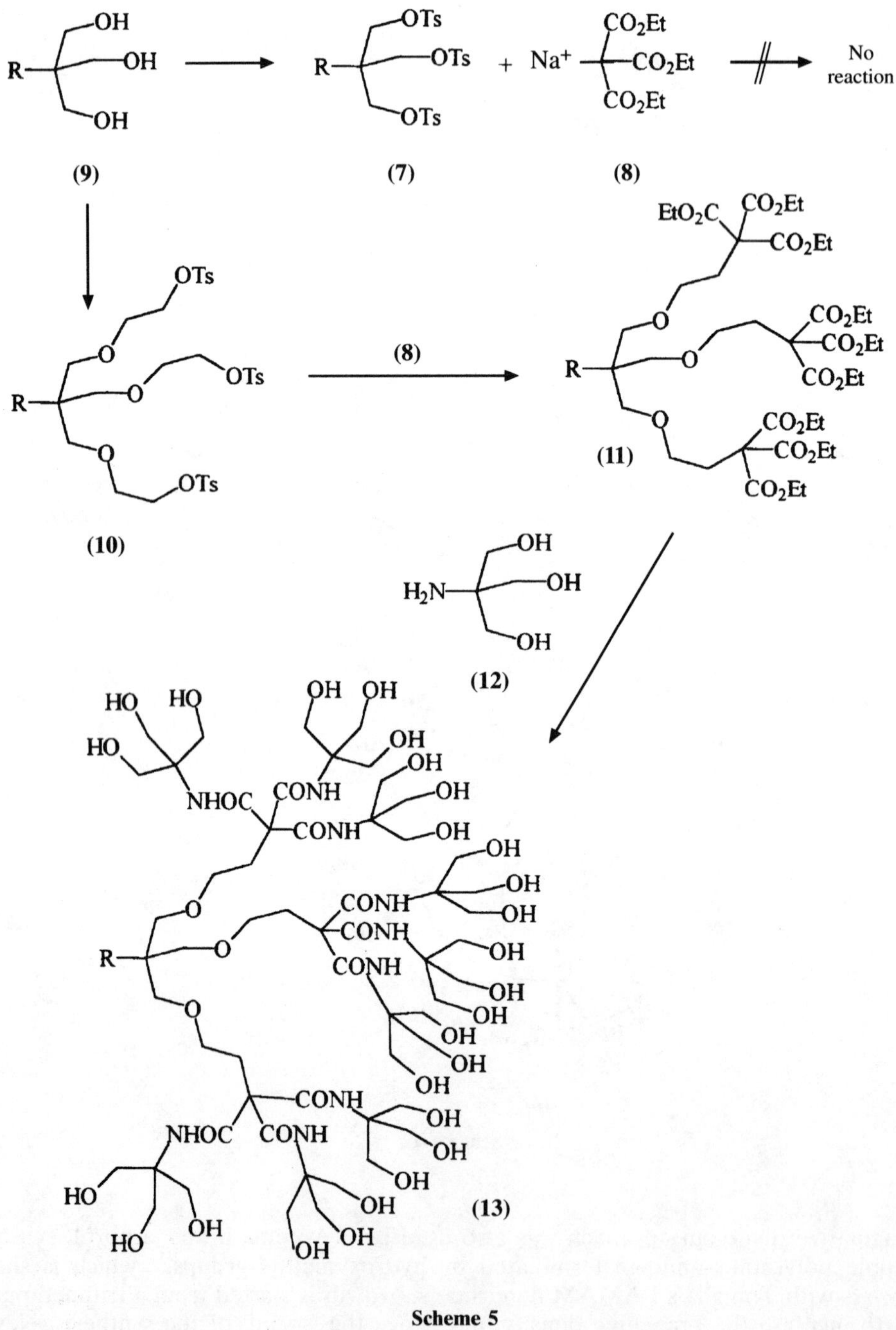

Scheme 5

As will be seen below, the fundamental principles that underlay these remarkable initial divergent syntheses have been used to prepare a large number of dendritic structures involving a great variety of building blocks.[22,23] With both Newkome's and Tomalia's synthetic approaches, success can be attributed to the careful selection of the high-yielding organic reactions that are used for growth, and also to the choice of symmetrical building blocks.

Because the divergent growth approach requires that reactions be carried out at the ever increasing number of chain ends of each growing dendrimer, it was initially feared that 'perfect' growth could not be sustained at high generation number, and therefore that the products obtained by the divergent approach might be highly contaminated.[24] However, careful engineering of the reactions with the judicious use of excess reagents and enhanced purification techniques has allowed the preparation of remarkably homogeneous products.[25,26] While some failure sequences do occur, it is generally acknowledged that the presence of a few imperfections in all but the most carefully purified divergent dendrimers has little effect on their ultimate properties.

3.2.2.2 The concept of convergent growth

Although the divergent growth approach has proven extremely successful for the preparation of dendritic macromolecules, it does not provide the level of versatility that would be desirable for the preparation of more complex globular macromolecules. This is especially true if accurate control over the placement of more than one type of end group at the periphery of the globular dendrimer is desired.

To overcome this and other difficulties, an alternative synthetic strategy for the production of dendritic macromolecules, the convergent growth approach, was demonstrated by Fréchet and Hawker and first disclosed in 1989.[27] A different, if conceptually similar, convergent synthesis of smaller dendrimers was outlined by Miller and Neenan in 1990.[31] Development of the convergent growth route shown in Scheme 6[4,28] relied on the application of the disconnection approach used in classical organic synthesis to the preparation of the highly branched dendritic macromolecules. Because of their fractal and highly symmetrical nature, disconnection of dendrimers eventually leads to the chains ends as the starting point of the synthesis. Therefore, growth is begun at the chain ends or 'surface' functional groups, S, and coupling with an AB_x monomer (*e.g.* **14**), where x is the multiplicity of the branch point, 2 or greater, leads to the next generation reactive dendron (*e.g.* **15**). While the number of chain ends has doubled, it should be noted that there is still only a single group P at the focal point of the reactive dendron. Activation of this latent functionality gives the reactive functional group R that can again be coupled with the monomer unit. Repetition of this two-step process results in an increase in generation number, or the number of layers of monomer units, and leads to larger and larger reactive dendrons (*e.g.* **16** or **17**). If desired, the final reaction step may be the attachment of several reactive dendrons to a polyfunctional core C to give a dendritic macromolecule (*e.g.* **18**) similar to that which might be obtained from a divergent strategy.

To demonstrate this new methodology, Hawker and Fréchet have reported[4,28] the synthesis of a series of dendritic polyether macromolecules based on 3,5-dihydroxybenzyl alcohol (**19**) as the monomer unit (Scheme 7). Preparation of the first generation reactive dendron (**21**) is started from the surface functional group, which, in this case, is a benzylic ether, by Williamson ether synthesis of a benzyl bromide with the monomer (**19**). This type of coupling was selected because it proceeds in high yields under mild conditions and does not suffer from significant side reactions. By design, alkylation of the monomer unit occurs regioselectively at the phenolic hydroxy groups, leaving the hydroxymethyl group available for further reactions. This hydroxymethyl group becomes the focal point of the growing dendrimer (**20**), and its activation with carbon tetrabromide and triphenylphosphine regenerates the bromomethyl group of reactive dendron (**21**). Repetition of this two-step alkylation/bromination procedures gives monodisperse reactive dendrons such as (**22**) or (**23**) that have a single functional group at the focal point and may be used in further reactions with monomer (**19**). Hawker *et al.*[4] carried this synthesis to the sixth generation reactive dendron which has 64 terminal groups, seven layers of aromatic rings, a molecular formula of $C_{889}H_{763}BrO_{126}$ and a molecular weight of 13 851. Key features of this entire process include the choice of monomer and end groups. In this case, the end groups may be varied by selecting different R substituents (Scheme 7) for the benzyl bromide used in the first step of the synthesis. More importantly perhaps, the choice of the monomer is such that purification of the intermediate dendrons at each step of the synthesis is easy. This ability to purify is an essential requirement of dendrimer syntheses if truly regular and monodisperse macromolecules are to be obtained.

The same coupling chemistry can also be used for attachment of the reactive dendrons to a polyfunctional core. For example, Scheme 8 shows the attachment of three molecules of a

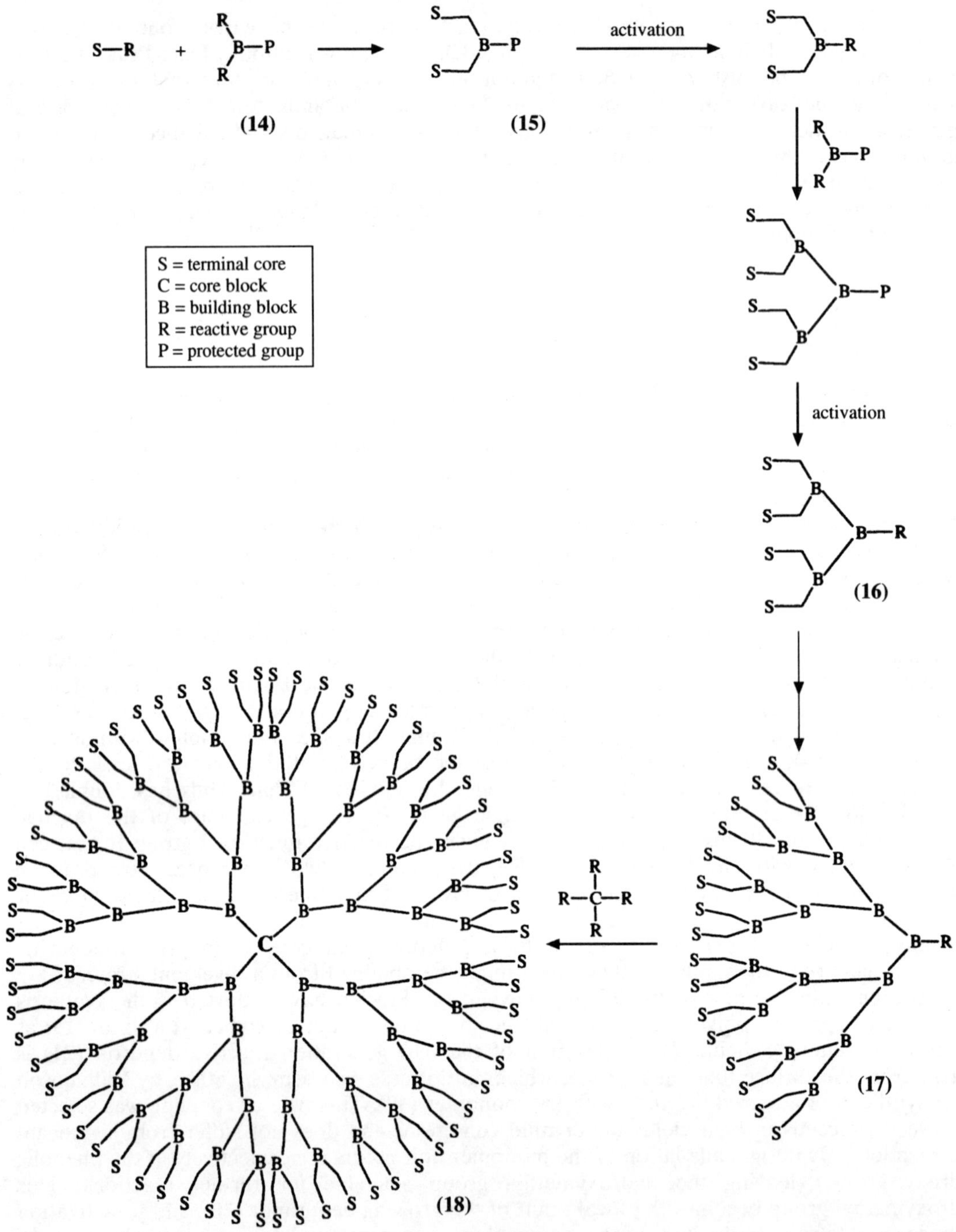

Scheme 6

fourth generation reactive dendron (**24**) to the trifunctional core 1,1,1-tris(4′-hydroxyphenyl)ethane to afford a large dendrimer (**25**) now devoid of focal point reactivity. This process, involving the various generations of reactive dendrons, (**21**), (**22**), (**24**), and larger (all with R = H), was used[4] to prepare a complementary series of monodisperse dendrimers (*e.g.* **25**) with molecular weights ranging from 576 to 40 689. Steric inhibition to reaction was not observed in any of these syntheses, and only a slight retardation and decrease in yield was observed in coupling steps involving dendrons of generation 5 and higher. Subsequently, the choice of core molecules has been extended to a host of other molecules and chemistries, and even buckminsterfullerene, C_{60},[29] or large phenol-terminated dendrimers[30] also known as 'hypercores' have proven to be useful core molecules.

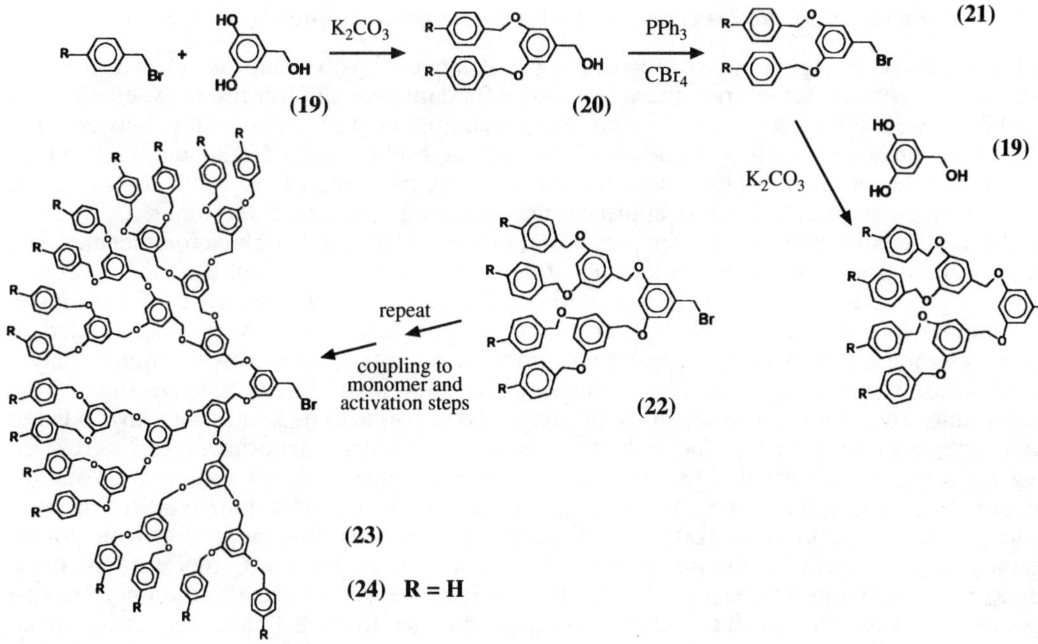

Scheme 7

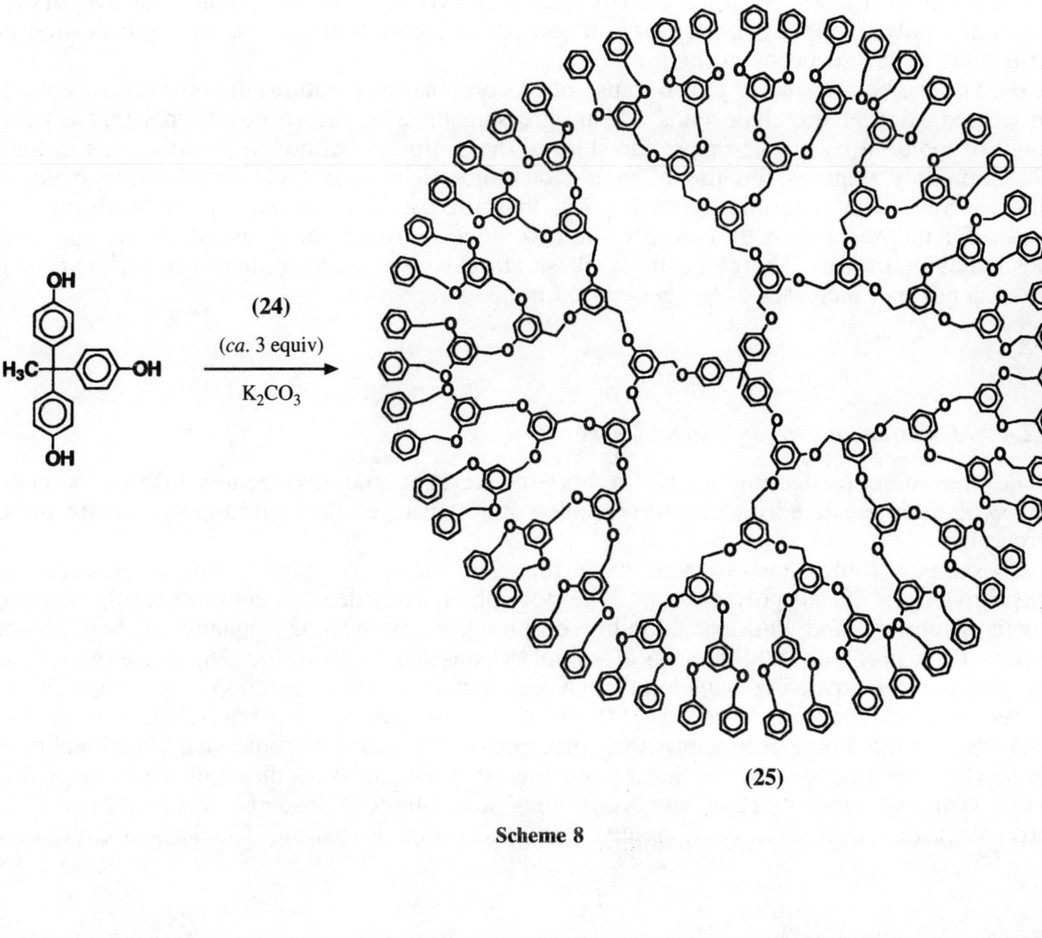

(25)

Scheme 8

3.2.2.3 A comparison of the divergent and convergent growth processes

While both the convergent and divergent growth strategies involve stepwise growth and can give the same dendritic structures, there are some fundamental differences between the two approaches. These differences result in an almost complementary relationship between the divergent and convergent methodologies, with neither method having a fundamental advantage over the other. However, the differences do influence the planning of the synthesis and for a specific dendritic macromolecule one approach may be more viable than the other.

For the synthesis of very regular or precisely functionalized dendritic macromolecules, the convergent growth approach allows a significantly greater degree of control than the divergent approach. This control is a result of the synthetic blueprint and the presence of the single, unique, functional group at the focal point of the intermediate dendrons. As was alluded to above, a consequence of this single focal point group is that generation growth requires only a single activation reaction and usually the coupling step only involves two reactive sites of the monomer unit. Therefore, the probability of unwanted side reactions is not only low, but the dramatic difference in the sizes and polarity between the desired dendrimers and unreacted starting material, or unwanted side products, frequently allows easy separation by standard techniques. This is in direct contrast to the divergent growth approach that requires an ever increasing number of both activation and coupling reactions for perfect growth to be maintained. The chances of failure sequences, or incomplete reactions, therefore increase rapidly as the generation numbers of the dendrimers increase and the small differences in size and polarity between the 'perfect' dendrimers and those containing failure sequences make their separation essentially impossible. Overall, the convergent growth approach provides easier access to 'perfect' dendritic macromolecules. In addition, it allows a greater degree of flexibility in controlling the number and nature of the terminal groups, 'interior' building blocks, and the focal point group.

In the production of a very large dendritic structure ($MW > 100\,000$) the divergent approach is the methodology of choice. Although the risk of irregular growth increases with the number of generations, the ability to use large excesses of reagents and techniques such as ultrafiltration and dialysis to remove these excesses allows the divergent approach to be continued to high generation numbers. Tomalia's PAMAM dendrimers are commercially available up to the tenth generation. In contrast, while the convergent growth approach maintains monodispersity, it eventually fails if coupling of very large fragments to a core is attempted, or if purification of intermediate dendrons becomes impractical.

It should also be recognized that because of its combinatorial nature, the convergent growth approach, starting at the chain ends and progressing towards the core, requires that a large amount of material be purified from the start of the synthesis. Although growth, even at high generation, only requires the use of near stoichiometric amounts of material, the mass of dendrimer after each generation growth is less than the combined masses of the dendrons that are used for its preparation. In contrast, the divergent approach starts growth at the core and builds radially outwards. Therefore the synthesis starts with a small amount of material and the mass of dendrimer increases as the generation number increases.

3.2.2.4 Miscellaneous organic dendrimers

Following these pioneering efforts in both convergent and divergent syntheses, a large number of alternate dendritic macromolecules and structures have been prepared by these approaches.

For example, Miller and Neenan have reported the convergent synthesis of dendritic poly(phenylenes)[31,32] and poly(esters).[33] The poly(phenylene) dendrimers are notable because of their symmetry and unusual solubility despite the absence of aliphatic or heteroatom linkages. Their preparation utilizes 3,5-dibromo-1-(trimethylsilyl)benzene (**26**) as the monomer, with growth steps involving palladium-catalyzed Suzuki coupling reactions with appropriate arylboronic acid dendrons such as (**27**), as shown in Scheme 9. Transformation of the trimethylsilyl focal point group into the corresponding reactive boronic acid functionality is then followed by repetition of the Suzuki coupling with another monomer unit. In the final step of the synthesis, the reactive dendrons that are obtained can be coupled to 1,3,5-tribromobenzene, used as a core moiety using the same procedure. The largest dendrimer obtained by this process, **(28)**, has 46 aromatic rings, a diameter of 31 Å, is stable to 500 °C, and is freely soluble in toluene.

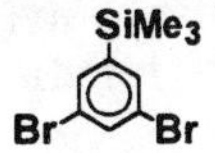

(26)

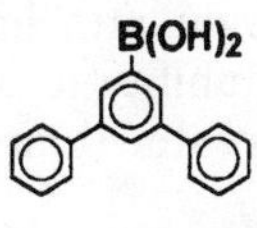

(27)

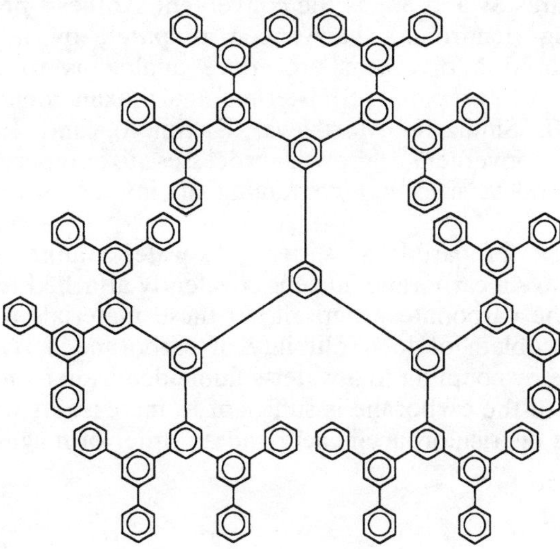

(28)

Scheme 9

Another interesting 'all hydrocarbon' family of dendrimers has been developed by Moore[34] using a convergent strategy with conformationally rigid building blocks. The synthesis utilizes 3,5-dibromo-1-(2-trimethylsilyl)ethnylbenzene **(29)** as the basic building block in a repetitive synthesis that involves a palladium-catalyzed cross-coupling with analogous moieties such as terminal dendron **(30)** containing an unprotected terminal alkyne (Scheme 10). Owing to the

(29)

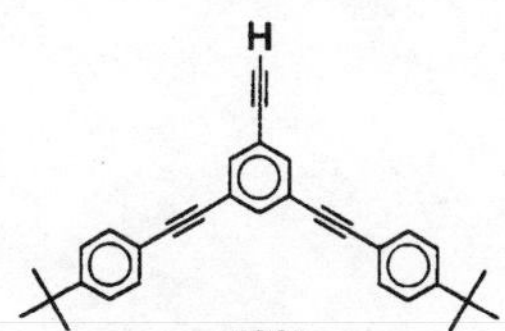

(30)

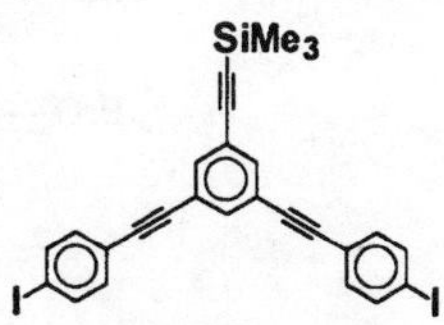

(31)

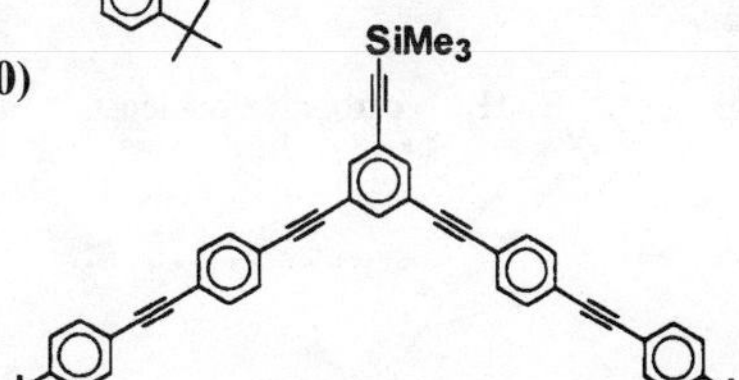

(32)

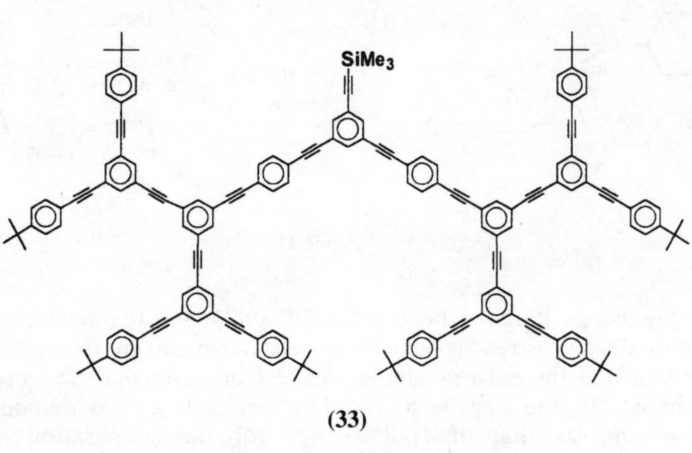

(33)

Scheme 10

rigidity of the building blocks and the resulting steric constraints, growth of dendrimers larger than the third generation (**33**) is not possible using this synthetic blueprint. A clever solution to this problem[35] involves the progressive enlargement of the monomer unit with rigid spacer groups (*e.g.* structures **31** and **32**) as the convergent synthesis proceeds to higher generations.

Silicon-containing dendrimers have been prepared by a number of research groups. Masamune[36] has used a divergent procedure analogous to that developed for Tomalia's Starburst dendrimers to prepare —SIH-terminated siloxane dendrimers with molecular weights as high as 15 000. Similarly, Morikawa, Kakimoto, and Imai have used well-designed divergent[37,38] and convergent[39] growth processes to prepare dendritic poly(siloxanes) of somewhat more modest sizes but containing an interesting variety of functionalized chain ends.[38]

Yamamoto[40] has prepared a series of water-soluble hydroxy-terminated dendritic poly(ethers) that have a carborane nucleus covalently attached to their focal point (Scheme 11). The presence of the carbonate group allows these materials to be used for neutron capture therapy and the problem of low solubility of carborane derivatives in aqueous solutions is elegantly overcome by coupling to a water-soluble dendrimer. Interestingly, attachment of even a small dendrimer to the carborane is sufficient to increase its water solubility to a level that is comparable to that of trichloroacetic acid and an order of magnitude greater than DL-valine.

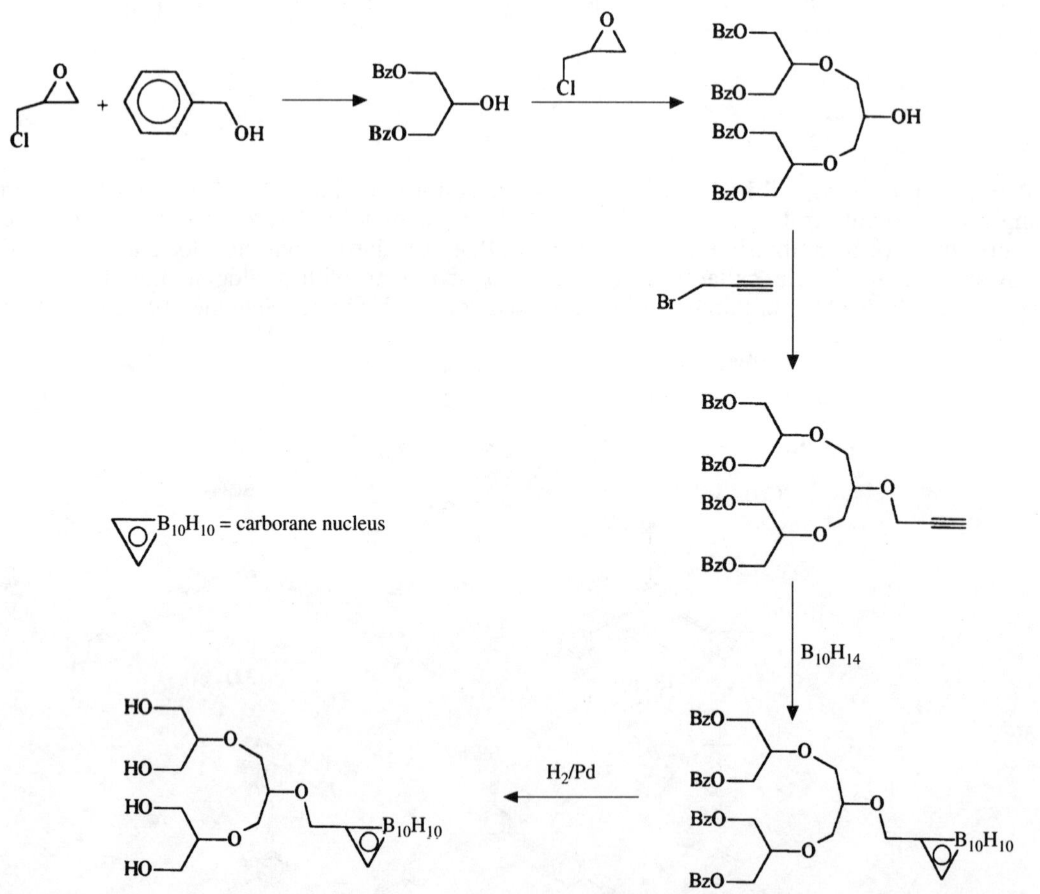

Scheme 11

The concept of dense packing has been discussed above and the generation number at which this elusive packing density is reached ought to be determined by the size and functionality of both the repeat unit and the core molecule. Aside from Tomalia's extensive studies with his Starburst dendrimers,[20,21] the approach that has come closest to demonstrating this dense packing phenomenon was that of Hall *et al.*[41] for the preparation of poly(ether) and poly(thioether) dendrimers. In this case a tetrafunctional core, pentaerythrityl tetrabromide

(**34**), and the compact AB_3 building block (**35**) were employed. Generation growth involved addition of the anion of the monomer unit to the polybromide (*i.e.* **34**) followed by regeneration of the bromide terminal groups by a deprotection, tosylation, and bromination procedure (Scheme 12). By using both a tetrafunctional core and a building block with a branch multiplicity of 3, the number of terminal groups increases at an extremely rapid rate. Therefore, as the generation number progresses from 1 to 2 to 3, the number of terminal groups increase from 12 to 36 to 108. Because the branch points are only separated by three covalent bonds, a very compact structure is obtained and growth beyond generation 3 appears to be inhibited, presumably as a result of extreme steric congestion in the vicnity of the chain ends. Tomalia has discussed the related concept of 'sterically induced stoichiometry'[21] that affects reactions for high generation Starburst dendrimers. Similarly, Hawker[4,28] has encountered difficulties in carrying out his dendrimer synthesis at the seventh generation, a finding that is not unexpected in view of the large steric requirements of convergent syntheses at high generations.

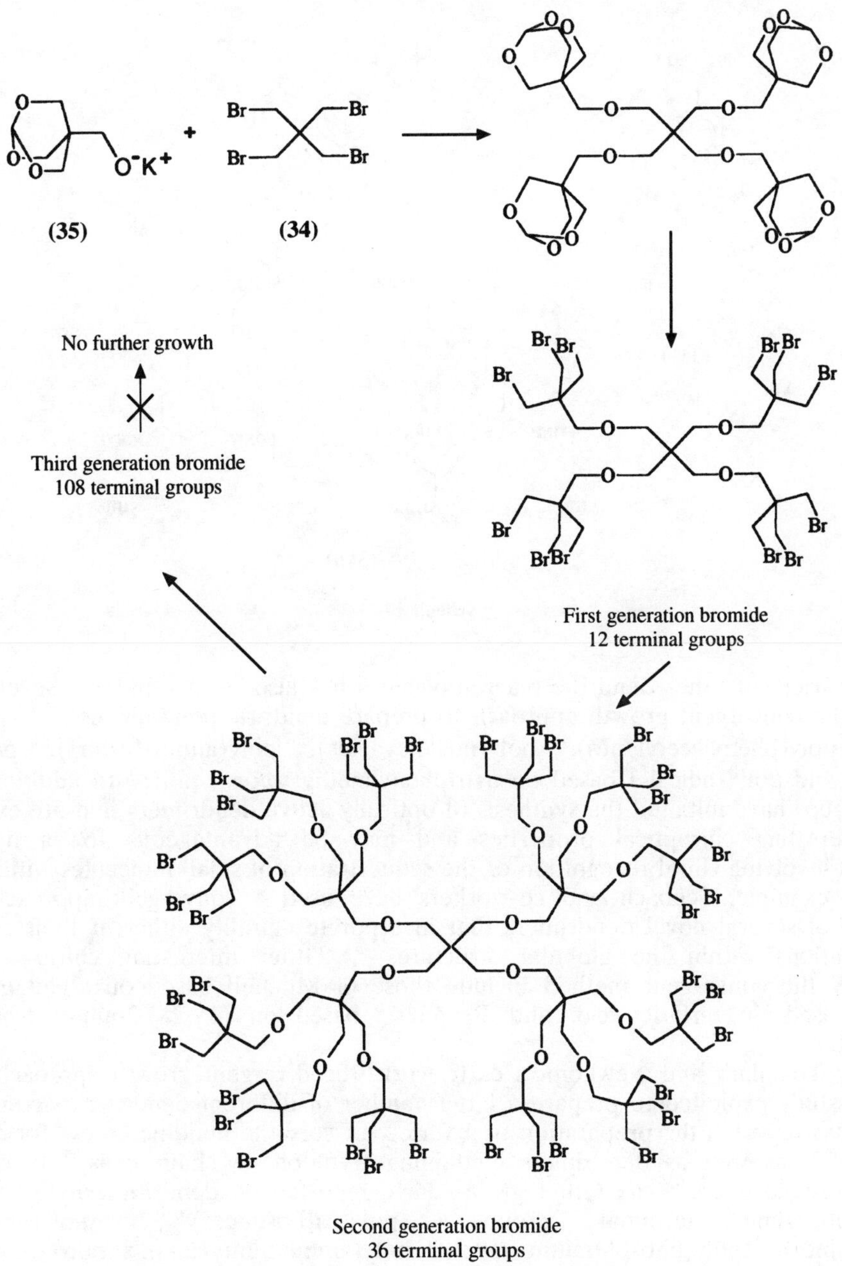

Scheme 12

In an attempt to simplify the purification of intermediate generations, the solid-phase synthesis of dendrimers has been attempted by Fréchet and co-workers.[42] The monomers tested in this approach to dendritic poly(amides) were based on 3,5-diaminobenzoic acid and a lightly crosslinked Merrifield-type chloromethylated poly(styrene) support, while classical peptide coupling chemistry was employed. The most efficient growth was observed with the chain-extended monomer (**36a**) shown in Scheme 13. Although the synthesis could be carried well past the second-generation polymer-bound dendrimer (**36b**) to the fourth generation dendrimer, steric inhibition by the matrix inhibited further growth. Nevertheless, this study demonstrates the feasibility of the solid-phase approach and it is likely that the use of enhanced supports, providing more space for growth of the compact globular molecule, could be used for the preparation of even larger dendrimers. Roy *et al.*[43] have also reported the preparation of second generation sialic acid dendrimers on a polymer support, while Moore employs a solid support to assist in the preparation of his dendritic phenylacetylenes.[44]

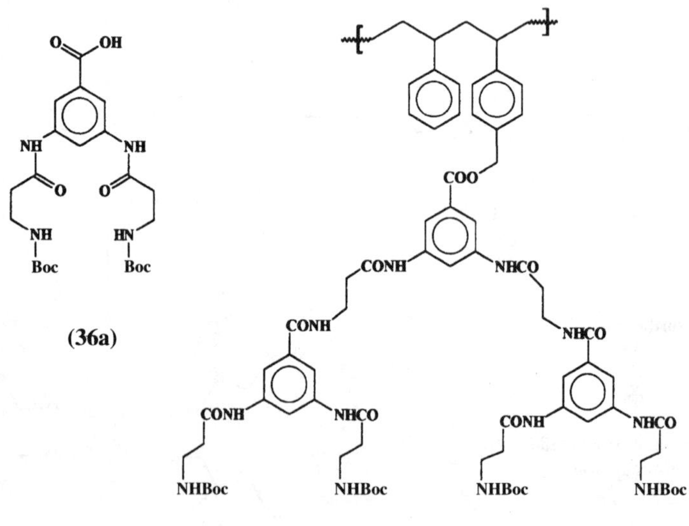

(36a)

(36b)

Scheme 13

A great variety of other dendritic macromolecules has also been reported. Several groups have used the convergent growth approach to prepare dendritic poly(amides),[45–47] poly(ether ketones),[48] poly(phenylacetylenes),[49] poly(nucleic acids),[50] poly(amino esters),[51] poly(crown ethers),[52,53] and poly(radicals) based on a triphenylmethyl repeat unit.[54] In addition, several research groups have initiated the synthesis of optically active dendrimers that are expected to possess interesting chiroptical properties and may be advantageous for a number of applications involving chiral recognition or the sequestration of small molecules within a chiral cavity. For example, Seebach and co-workers have used a convergent approach for the preparation of several novel dendrimers that incorporate chirality either at their core, or at several locations within the globular structure.[55–57] Other interesting chiral dendrimers prepared by the convergent method include those of Mitchell based on L-glutamic acid,[58] Chow[59,60] based on tartaric acid, and Brunner[61] based on (1*S*,2*S*)-2-amino-1-phenyl-1,3-propanediol.

Following Tomalia's and Newkome's early work, the divergent growth approach has also been successfully exploited to prepare a large number of different dendritic macromolecules. Newkome has reported the preparation of a variety of versatile building blocks for dendrimer synthesis,[23,62,63] as well as dendrimers containing tryptophane chain ends,[64] functionalized hydrocarbon dendrimers,[65] *etc*. Other groups have reported on dendritic poly(aryl sulfones) with azobenzene terminal groups,[66] poly(carbosilanes),[67–69] poly(amines),[70–72] poly(arylamines),[73] poly(phosphonium salts),[74] poly(ammonium salts),[75] poly(triptycenes),[76] and a novel dendrimer with a crown ether as the core.[77] Very recently, Majoral has reported the preparation of an interesting dendrimer with a pentavalent phosphorus atom at each branch point.[78] Finally, several reports on peptide dendrimers have appeared, including dendritic poly(amino acids) that are also termed 'multiple antigen peptides',[79] and a dense octabranched thiazolidinyl peptide dendrimer[80] with a molecular weight in excess of 25 000.

3.2.2.5 Organometallic dendrimers

The preparation of dendritic macromolecules has not been limited to those based on 'organic' units, or even covalent bonds. As was mentioned above, silicon-[36–39,81,82] and phosphorus-containing[74,78] dendrimers have been prepared by a number of groups, while transition metal centers have been introduced as the terminal groups in a variety of 'organic' dendrimers. The nature of the bonding in these transition metal-containing dendrimers range from 'ferrocene-like' sandwich complexes,[83,84] to metal–carbon σ-bonds,[85] and coordination complexes.[86,87] Examples of the latter include the dendritic structure shown in Scheme 14, which has an inner layer of 12 ruthenium atoms surrounded by 36 terminal groups.[86] As a result of the design of this macromolecule, the nature of these terminal groups can be modified to alter the physical properties of the organometallic dendrimer. It is interesting to note that this dendrimer is not built exclusively of classical covalent bonds. The very strong ruthenium terpyridyl coordination complex also acts as a connecting unit between the interior building blocks and the outermost branching layer of the dendrimer.

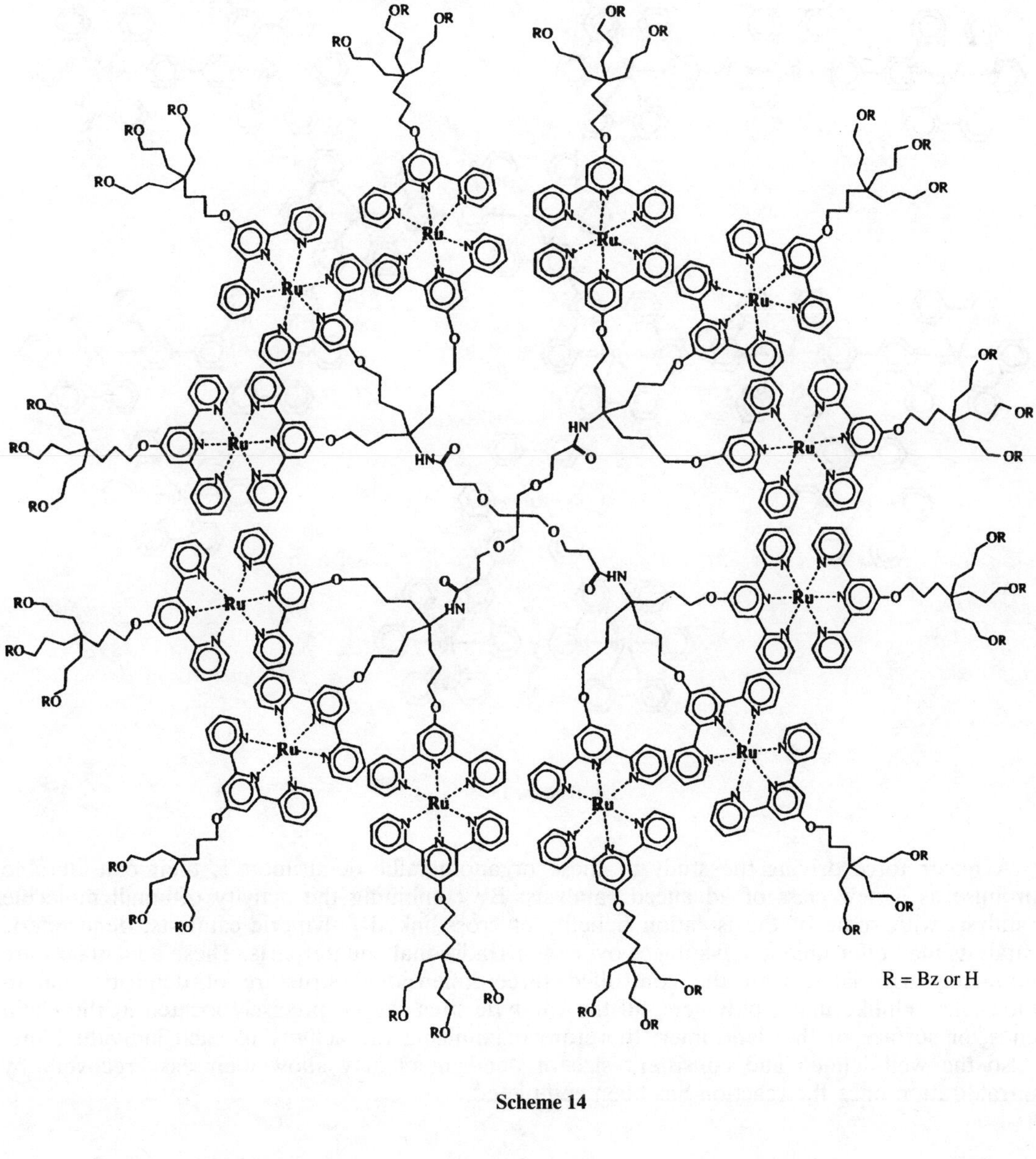

Scheme 14

An equally impressive use of coordination chemistry for the formation of dendrimers containing up to 22 metal atoms has been reported by a group of Italian researchers.[88,89] These organometallic macromolecules are the first examples of dendrimers where the multifunctional core and building blocks all rely on non-covalent bonding for their branching. By the use of hexacoordinate ruthenium or osmium metal centers and multifunctional ligands capable of chelating two different metal centers, dendritic structures can be built up using the divergent strategy with accurate control of the nature of metal atom in each generation level (Scheme 15).

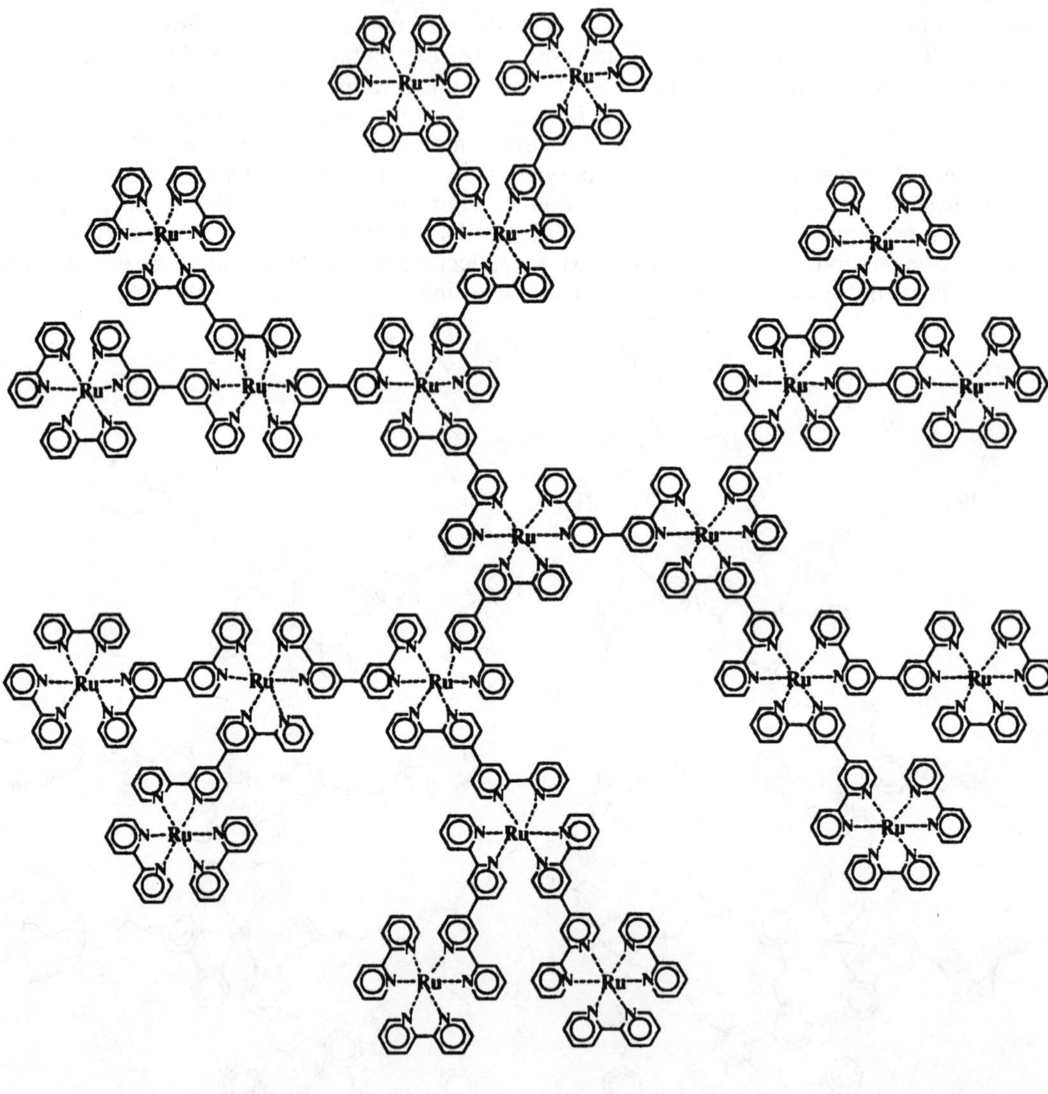

Scheme 15

A major force driving the study of these organometallic dendrimers is their considerable promise as a new class of advanced catalysts. By combining the activity of small molecule catalysts with some of the isolation benefits of cross-linked polymeric catalysts, dendrimeric catalysts may offer unique advantages over their traditional counterparts. These advantages are intimately associated with the controlled three-dimensional structure of dendritic macromolecules. Unlike linear polymers, all the catalytic sites can be precisely located at the chain ends, or surface of the dendrimer, therefore maximizing the activity of each individual site. Also the well-defined and consistent size of dendrimers may allow their easy recovery by ultrafiltration once the reaction has been completed.

The first example of such a dendrimeric catalysis with active metal sites present at the chain ends has recently been reported.[85] In this case, arylnickel(II) complexes are coupled to the chain ends of a silane dendrimer to give the dodecanickel(II) species shown in Scheme 16. This dendrimer demonstrated regiospecific catalytic ability for the Kharasch addition of polyhaloalkanes to carbon–carbon double bonds, and, significantly, it displayed a catalytic activity comparable to that of the analogous non-dendrimer-bound small molecule. While such organometallic systems hold great promise for catalyzing a variety of chemical processes, the potential is not limited to chemical reactions. Transition metal-based coordination dendrimers, similar to the structure in Scheme 15 but containing a precise arrangement of different transition metals, have been postulated as devices capable of harvesting solar energy.[90] In an important step towards obtaining this goal, Balzani has demonstrated that for the organometallic dendrimer of Scheme 17, excitation of the ruthenium centers by light causes an extremely efficient transfer of energy from the peripheral ruthenium atoms to the interior osmium center, which then dissipates the energy by luminescence. This ability to channel the energy in specific directions, coupled with the efficient nature of such transfer processes, suggest that a wide range of applications may exist for these organometallic dendrimers.

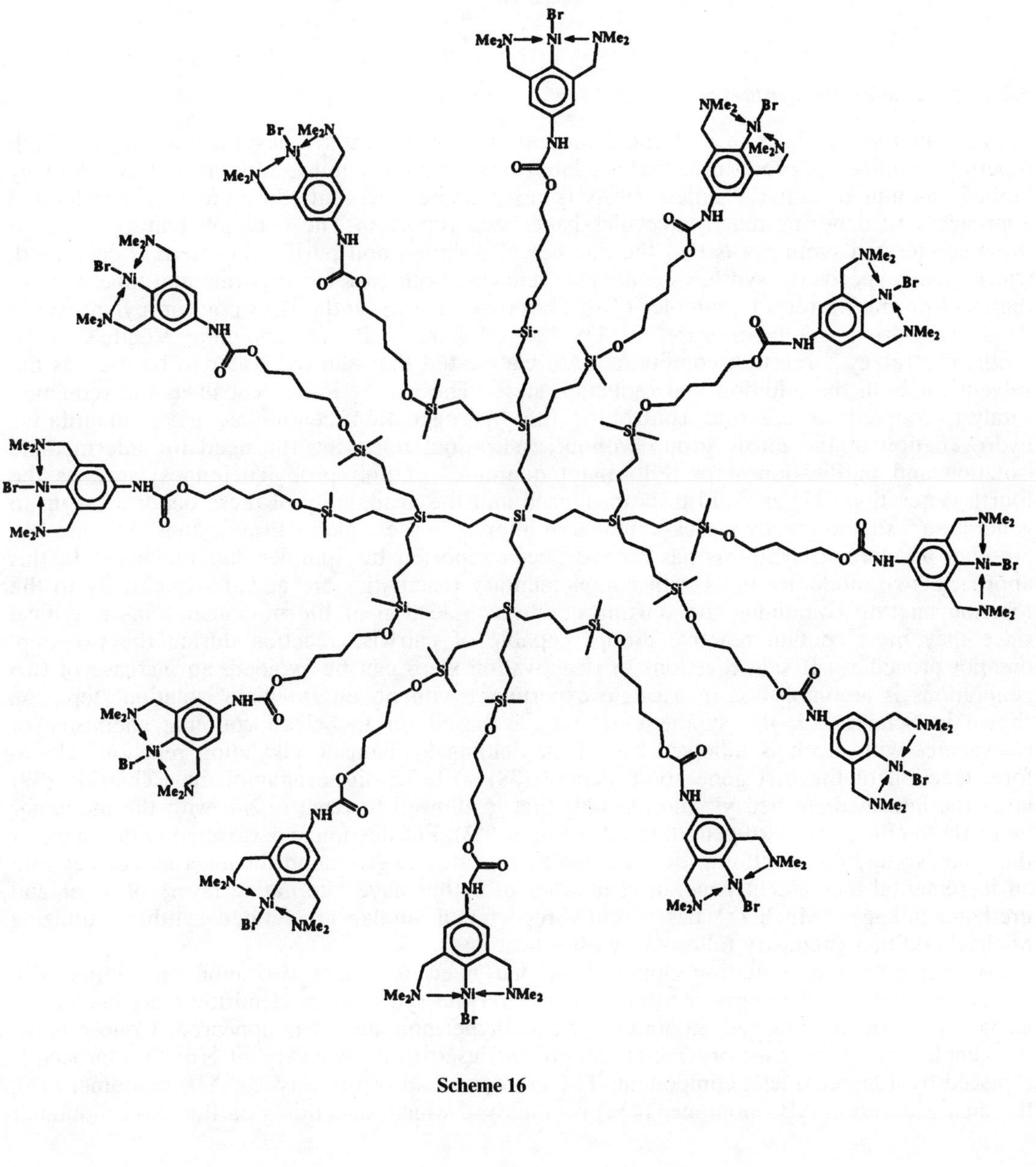

Scheme 16

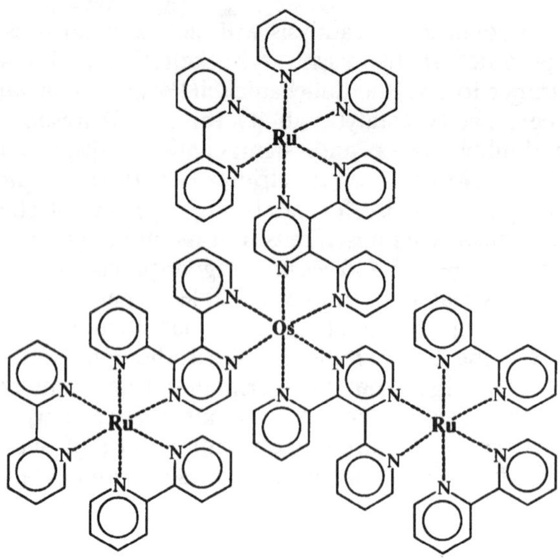

Scheme 17

3.2.2.6 Accelerated syntheses

One common drawback of both the divergent and convergent syntheses is that they are both repetitive, multi-step, procedures that are laborious, time consuming, and afford only relatively limited amounts of material unless purity is compromised. Recently, a number of accelerated approaches to dendritic macromolecules have been reported. The common feature of all of these accelerated syntheses is that the number of isolation and purification steps is decreased, which greatly speeds the synthesis while also reducing both cost and experimental time. One of the most promising recent examples of this approach is the synthesis of poly(propyleneimine) dendrimers from the laboratories of DSM in Holland.[72] By re-examining Vögtle's initial synthetic strategy,[13] reaction conditions were elaborated that allowed water to be used as the solvent for both the addition and reduction steps. The use of Raney cobalt as the reduction catalyst, coupled to accurate control of the hydrogenation conditions, gave quantitative hydrogenation of the nitrile group without cyclization, removing the need for intermediate isolation and purification steps. Pilot-plant quantities of poly(propyleneimines) such as the fourth generation (**37**) are said to be produced and the availability of these dendrimers up to generation 5 should greatly increase industrial interest in these globular molecules (Scheme 18). Another accelerated synthesis has recently been reported by Spindler and Fréchet.[91] In this approach, two monomer units with complementary reactivities are added sequentially to the reaction mixture containing the starting dendron. Selection of the monomer units is critical since they must contain reactive groups capable of pairwise reaction during the two-step, one-pot procedure. If side reactions or deactivation steps can be avoided, an increase of two generations is accomplished in a single experiment with no intermediate isolation steps. As shown in Scheme 19, the synthetic strategy is based on the clean coupling chemistry of isocyanates with alcohols followed by a high-yielding Williamson alkylation reaction. Therefore, reaction of the first generation alcohol (**38**) with 3,5-diisocyanantobenzyl chloride (**39**) gives the intermediate benzyl chloride (**40**) that is allowed to react *in situ* with the monomer unit (**41**) to afford the third generation dendrimer (**42**). Purification is performed at this stage of the synthesis and the reaction procedure can be repeated to give dendritic macromolecules with an incremental increase in generation number of 2 that have alternating layers of ether and urethane linkages. Mitchell[51] has recently reported a similar accelerated synthesis utilizing Michael addition chemistry followed by esterification.

To overcome the limitation imposed by the need to select two monomer units with compatible functional groups, a recent report[92] on the synthesis of dendritic macromolecules using a preformed branched monomer termed 'hypermonomer' has appeared. Conceptually, the simplest repeating unit or 'fractal component' used in the convergent growth approach is replaced by a larger fractal component. Therefore, instead of the classical AB_2 monomer (**19**), the next generation AB_4 monomer (**43a**) is employed which, depending on the exact monomer

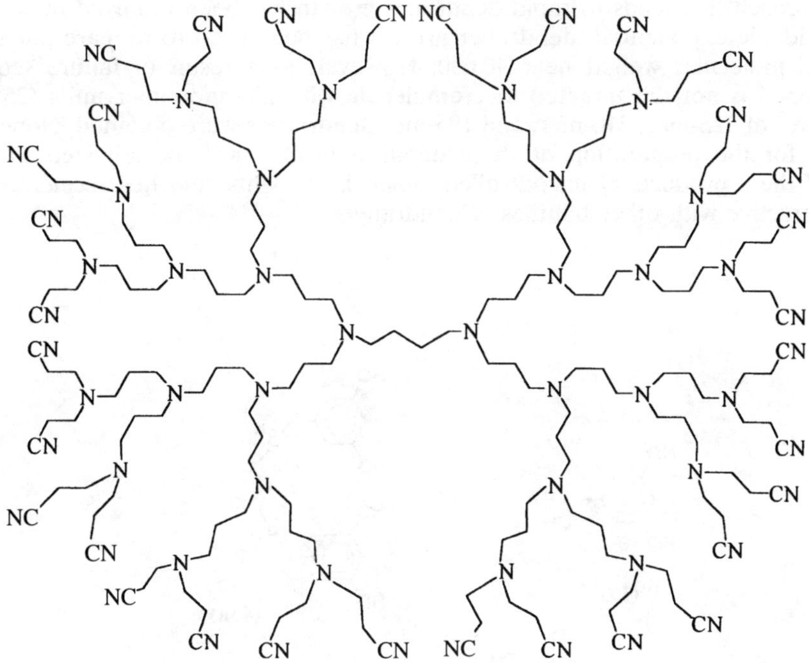

(37)

Scheme 18

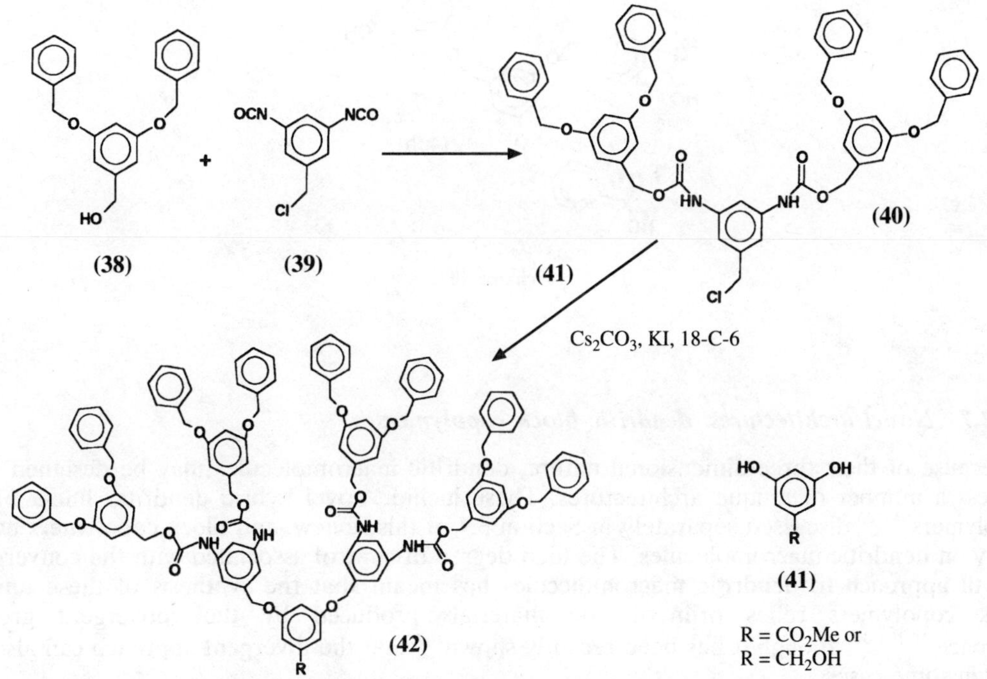

(38) (39) (41) (40)

Cs_2CO_3, KI, 18-C-6

(42) (41)

R = CO_2Me or
R = CH_2OH

Scheme 19

structure, allows the generation number to increase by 2 in a single coupling step (Scheme 20). In a nonoptimized procedure that involved hypermonomer (**43b**), an alternating dendritic polyether/ester macromolecule of molecular weight 9371 was obtained in only three steps from a starting material of molecular weight 381 with an overall yield of 31%. A somewhat analogous approach that leads to rapid dendrimer growth has been reported by Moore.[93] This imaginative 'double exponential' dendrimer growth has been used to prepare phenylacetylene dendrimers of molecular weight near 40 000. However, as a result of failure sequences the material obtained is not the targeted macromolecule with 255 monomer units (255-mer), but instead mixtures of 225-mer, 210-mer, and 195-mer dendrimers were obtained. Nonetheless, the method used for the preparation of these dendrimers may well be adjusted to reduce the occurrence of these products of uncontrolled molecular weights, and this accelerated synthesis may prove attractive with other families of dendrimers.

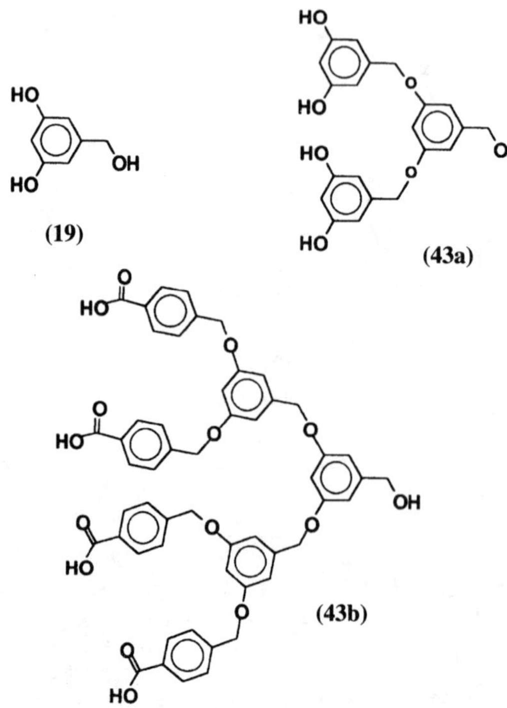

Scheme 20

3.2.2.7 *Novel architectures: dendritic block copolymers*

Because of their three-dimensional nature, dendritic macromolecules may be designed that possess a number of unique architectures. These include novel hybrid dendritic–linear block copolymers[94–103] discussed separately in Section 3.4 of this review, and block copolymers based solely on dendritic macromolecules. The high degree of control associated with the convergent growth approach to dendritic macromolecules has meant that the synthesis of these unique block copolymers relies primarily on materials produced by the convergent growth approach,[10,104–106] though it has been recently shown[107] that the divergent approach can also be used in some cases.

Traditional block copolymers are composed of linear blocks and therefore the one-dimensional nature of these units limits the number of possible arrangements. In contrast, three-dimensional systems, such as found in dendritic macromolecules, offer a substantially larger number of possibilities and a number of interesting and novel structures may be proposed when the concept of block copolymers is considered. At this time, three different arrangements of building blocks that lead to purely dendritic block copolymers have been studied. These arrangements, shown schematically in Scheme 21, have been termed dendritic *layer-block, segment-block,* and *surface-block* copolymers.[106] Significantly, initial studies of the properties of these block copolymers based on purely dendritic components have shown that a number of interesting and unique features are associated with these unusual materials.

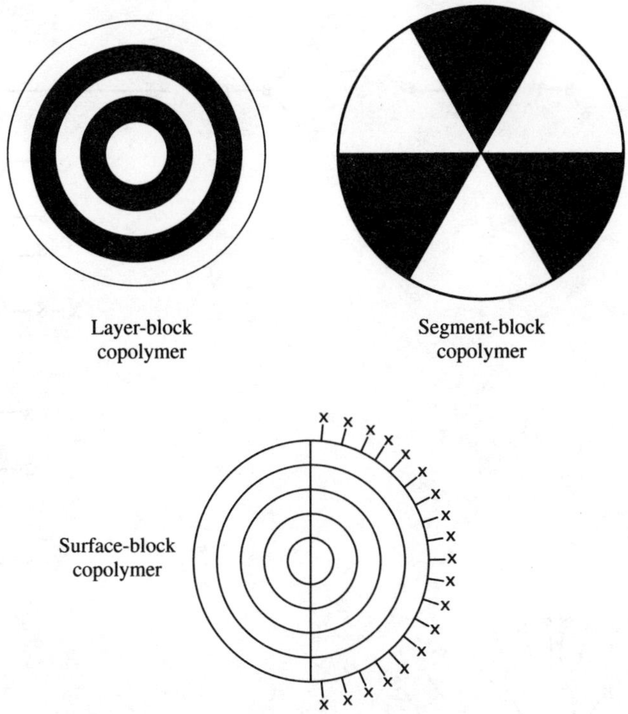

Scheme 21

Dendritic surface-block copolymers are characterized by the uniformity of their interior building blocks, with at least two different types of chain ends each localized in well-defined areas of the overall globular structure. To prepare such materials, a method for the accurate control over the number and nature of functional groups at the chain ends of dendritic macromolecules must be available. Owing to the limited number of reactions required for generation growth, the convergent growth approach offers such a methodology. To demonstrate this strategy for the control of functionalities at the chain ends, the preparation of two extreme cases has been reported.[104,105] In the first of these, a dendritic macromolecule containing 47 identical chain ends and one odd chain end that differs from all others has been prepared. In the second case, the dendritic macromolecules contain two different types of chain end functionalities, each type being localized on dendrons attached to a central core. The synthetic strategy used to prepare the 'monofunctionalized' moiety is illustrated in Scheme 22. The fundamental principle is that at each generation growth step the monomer unit is coupled to different dendritic fragments, one which carries a single surface functional group (*e.g.* **44**) and the other which is devoid of that functionality (*e.g.* **45**). In this way a dendritic macromolecule (**46**) can be prepared where only one of the exterior 64 terminal groups carries the desired functional group.

To prepare dendritic macromolecules having areas, or blocks, of different surface functional groups, a similar stepwise alkylation strategy is employed. This process has also been used to prepare dendritic surface-block copolymers[10] that have the surface of a well-defined 'hemisphere' substituted with electron-withdrawing cyano groups while the opposite half is substituted with electron-donating ether groups. Significantly, this partitioning of the surface functional groups into opposing areas results in a considerable increase in the dipole moment when compared to the analogous molecules with uniform chain ends. Conceptually, such structures may be useful in the design of switching or display devices based on dipolar globular dendritic macromolecules. The same approach has been used to prepare amphiphilic dendrimers made up of two hemispheres, one with terminal carboxylic acid or hydrophilic

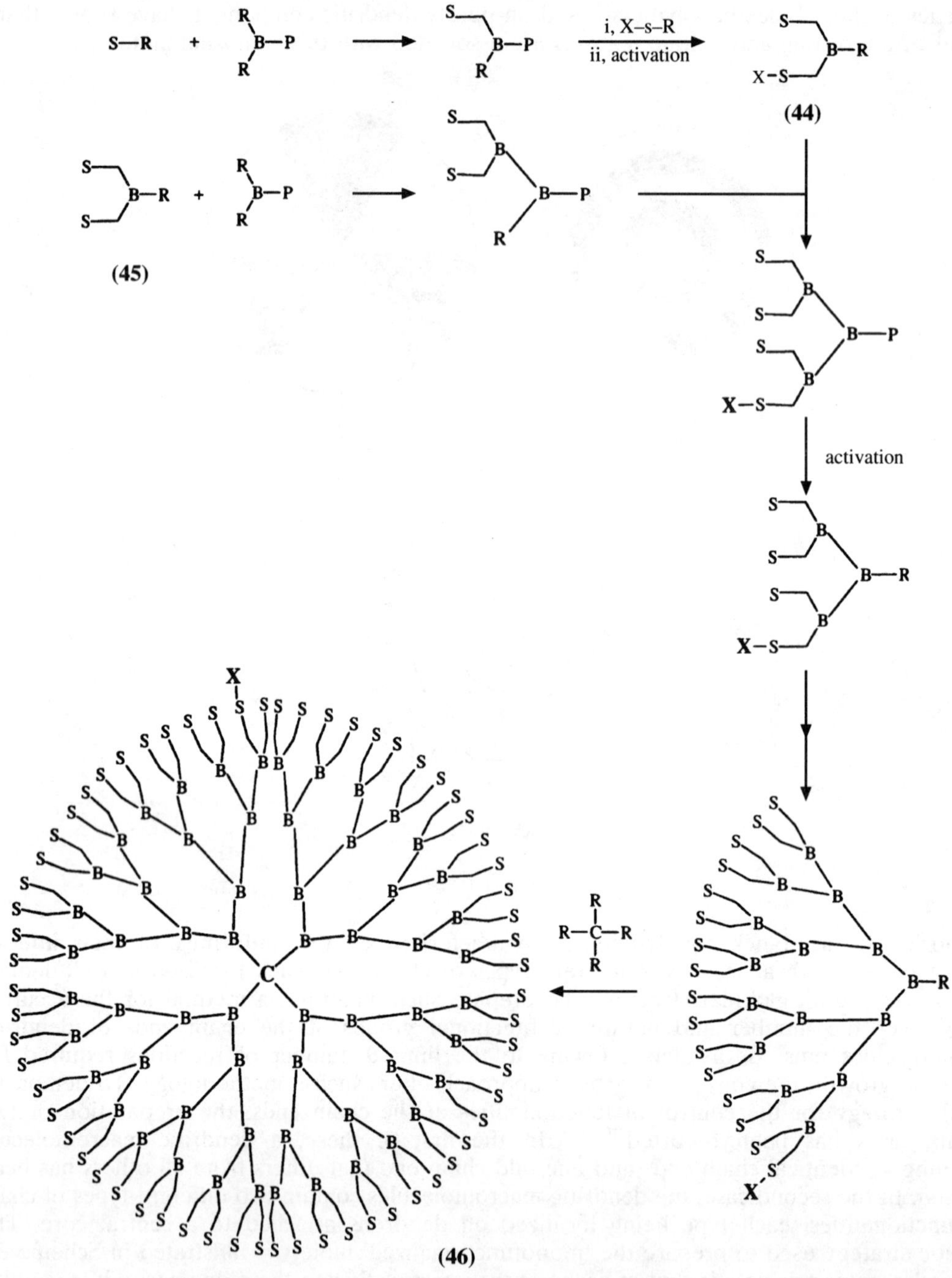

Scheme 22

carboxylate salt end groups, and the other with lipophilic benzyl ether chain ends.[108] By a combination of the methodologies used for the synthesis of monofunctionalized dendrimers and surface-block copolymers, dendritic macromolecules with any number of terminal functional groups can be prepared and a degree of control over their placement is also possible.

Unlike the surface-block copolymers, the preparation of dendritic segment-block and layer-block copolymers relies on the use of different building blocks. Therefore the chemistries used in the attachment of these blocks to each other and to a polyfunctional core must be compatible. In many divergent growth syntheses an alternating layer-block copolymer, where there are alternating layers of different building blocks or functional groups, is a natural consequence of the synthesis. Well-known examples are the PAMAM dendrimers of Tomalia that have alternating layers of amine and amide functional groups. Variations on this theme are dendritic macromolecules that have a terminal or chain end which is dramatically different in terms of functionality from the interior building blocks. Prime examples of these materials are dendritic micelles that have a hydrophobic core surrounded by a hydrophilic surface layer.[108–113]

A versatile synthesis of dendritic layer-block copolymers based on the convergent growth approach has recently been reported by Hawker and Fréchet.[106] In this approach, the number and size of each block can be varied by suitable changes in the synthetic blueprint. For the type of dendrimer shown in Scheme 23, growth is initially begun with linkages based on ether chemistry using 3,5-dihydroxybenzyl alcohol as the monomer unit. However, at the third generation, a different building block, 1,1,1-trichloroethyl 3,5-dihydroxybenzoate (**47**) (see Scheme 24) is introduced into the synthesis. This allows the coupling reaction to be changed from ether to ester linkages. The synthesis is continued with attachment to a trifunctional core using ester chemistry to give (**48**) that has two inner concentric layers of ester linkages surrounded by three outer concentric layers built with ether linkages. While the different blocks are both based on ether linkages, the synthesis of dendrimers using the hypercore concept[30] is a variant on this strategy; in this case a preformed dendritic core (the hypercore) is attached to dendritic fragments composed of different building blocks to give a dendritic layer-block copolymer.

The synthesis of dendritic segment-block copolymers employs a strategy similar to that for the control of surface functional groups. In this case a single monomer unit (*i.e.* **47**) is coupled in a stepwise fashion in two different dendritic fragments composed of different building blocks. As shown in Scheme 24, the monomer unit (**47**) is initially monoalkylated with the second generation ether fragment; the second phenolic group of the monomer unit is then esterified with a polyester fragment to give the third generation dendritic fragment (**49**). As defined by the synthetic blueprint, (**49**) is itself a block copolymer having an ether block and an ester block attached to the same unique monomer unit (Scheme 24). Deprotection of (**49**) followed by coupling with a trifunctional core then gives the dendritic segment-block macromolecule (**50**). Interestingly, while there are a number of possible conformations for (**50**), the constraints arising from the branching sequence do not allow a structural isomer where all three ester blocks are adjacent. Using a 'protected core' approach, Tomalia[107] has recently shown that the divergent growth approach can also be used for the preparation of analogous dendritic segment-block copolymers.

3.2.2.8 *Dendritic-graft architecture*

The synthesis of dendritic macromolecules also need not be confined to those derived from small molecules as building blocks. Both Tomalia[114] and Gauthier[115] have reported the synthesis of dendritic-graft or 'Combburst®' macromolecules where the building blocks are monodisperse linear polymers. Tomalia's original Combburst approach, which opened up the field, involves the grafting of a reactive polyoxazoline onto a polyethylene imine used as the core moiety. After removal of the amide groups from the side chains, the grafting process is repeated with more reactive polyoxazoline. This 'graft-on-graft' approach leads to a dramatic increase in the molecular weight of the resulting macromolecule as the growth of each generation is achieved.[114] Gauthier's approach to dendritic-graft polymers[115] is patterned after Tomalia's but uses a polystyrene of relatively low molecular weight (*ca.* 5000) and narrow molecular weight distribution as the core moiety. This core is activated by chloromethylation in a statistical process that involves approximately 30% of the pendant phenyl rings. Generation growth is achieved by reaction of the chloromethylated polymers (**51**) with living polystyrene to afford polymer (**52**) that contains the first level of branching. The two-step chloromethylation and subsequent coupling step with anionic polystyrene is then repeated to give the next generation macromolecule (**53**) that a has a high degree of layered branching and loosely resembles a dendritic macromolecule (Scheme 25). As was the case with Tomalia's pioneering approach to Combburst architecture, a rapid molecular weight increase results from this graft-on-graft process and, after only two iterations, the polymer that is isolated has a molecular weight of *ca.* 4 000 000 with a polydispersity of only 1.3. Comparison of the hydrodynamic volume of the highly branched polystyrene with linear polystyrene reveals that the former occupies 27 times less volume, clearly demonstrating the highly branched and compact nature of these dendritic-graft macromolecules.[115]

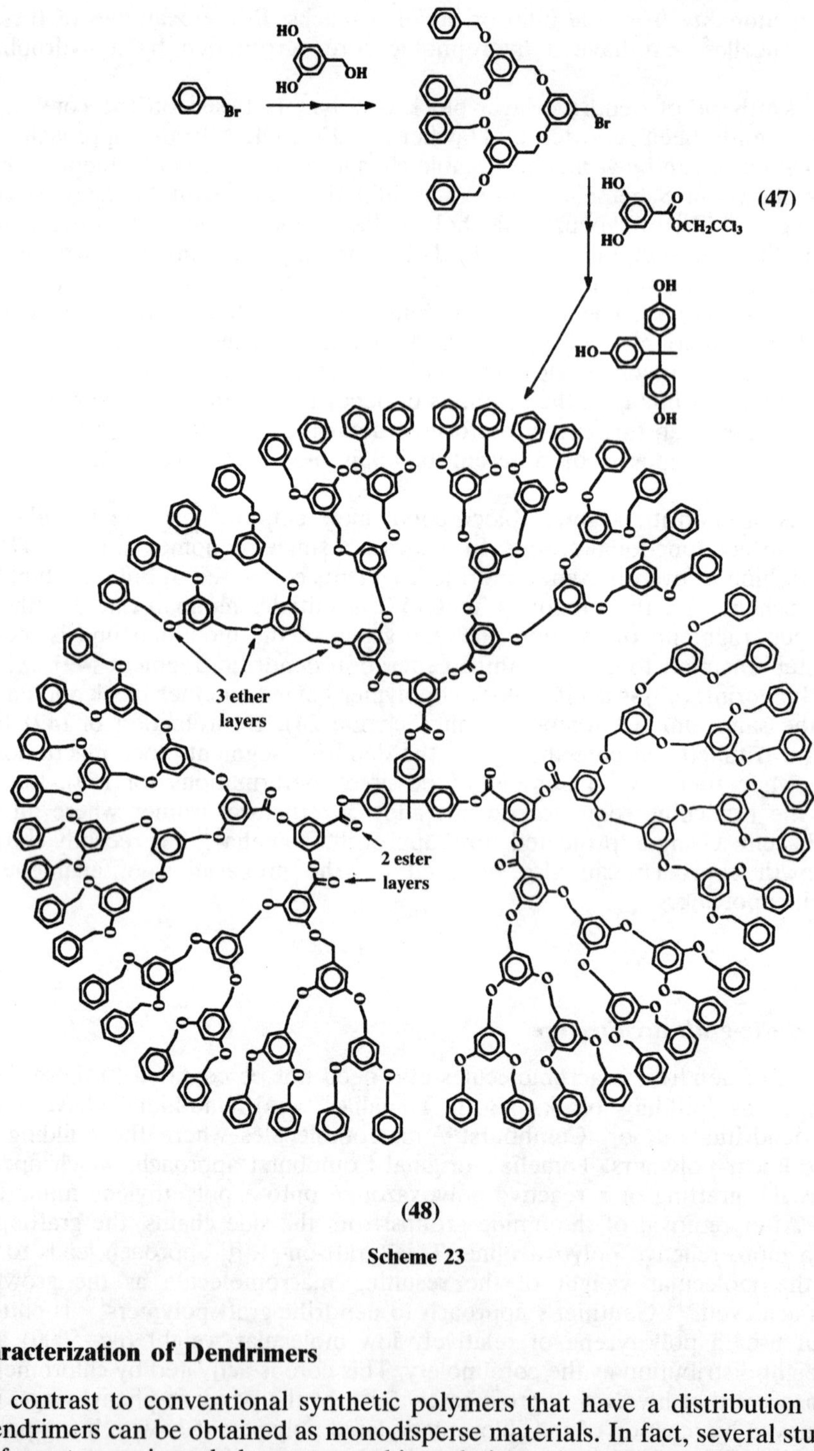

(47)

(48)

Scheme 23

3.2.3 Characterization of Dendrimers

In sharp contrast to conventional synthetic polymers that have a distribution of molecular weights, dendrimers can be obtained as monodisperse materials. In fact, several studies utilizing a variety of spectroscopic and chromatographic techniques have demonstrated that a carefully prepared dendrimer is essentially a single molecule. The unprecedented accuracy with which dendrimers can be characterized is the result of their unique features of high symmetry and monodispersity.

Several elegant demonstrations of the structural identification of dendritic macromolecules have been reported by various research groups.[4,21,32,37,116] Size exclusion chromatography (SEC) is a particularly useful routine tool for the monitoring of dendritic growth by the convergent approach since the molecular weights of the reactive dendrons essentially double at each generation growth step. This enables incomplete reactions to be observed as multimodal distributions, while the purified dendrons show single and narrow peaks. On comparison with

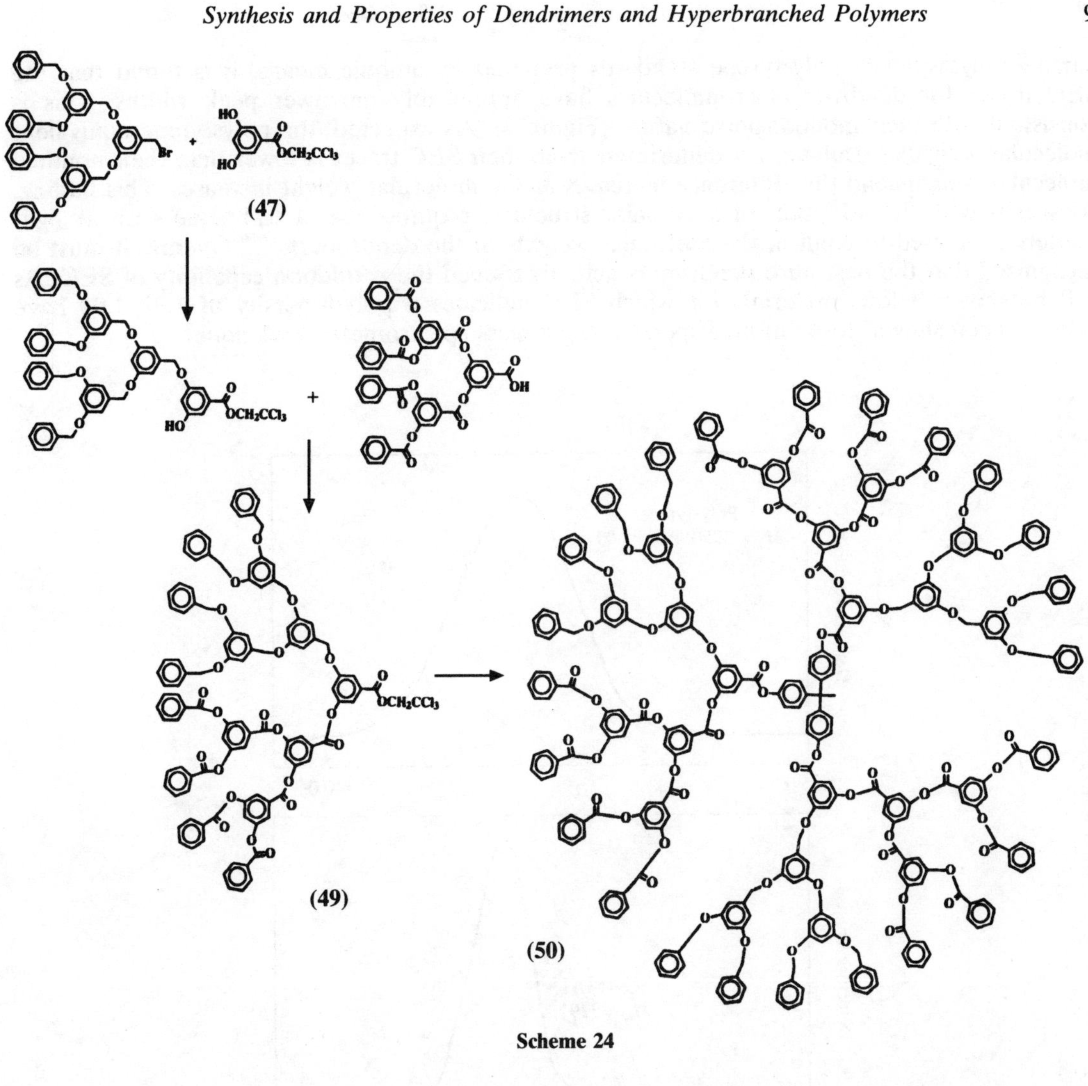

Scheme 24

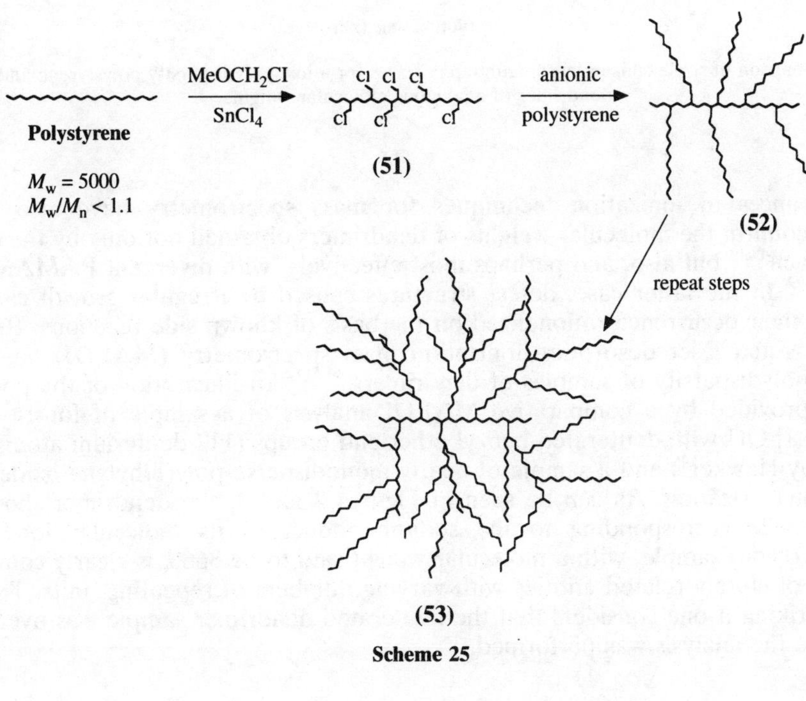

Scheme 25

narrow polydispersity polystyrene standards prepared by anionic means, it is found that the SEC traces for dendritic macromolecules have appreciably narrower peak widths. This is consistent with their monodisperse nature (Figure 1). As expected, the polystyrene equivalent molecular weights calculated for dendrimers from their SEC traces is lower than their nominal molecular weights and this difference increases as the molecular weight increases. This finding, consistent with the adoption of a globular structure, requires that a technique such as light scattering be used to confirm the molecular weights of the dendrimers.[4,116] Overall, it must be recognized that the best pure dendrimers actually exceed the resolution capability of SEC. As will be shown below, materials for which SEC indicates a polydispersity of 1.02–1.04 have actually been shown[4] to be monodisperse using a mass spectrometric technique.

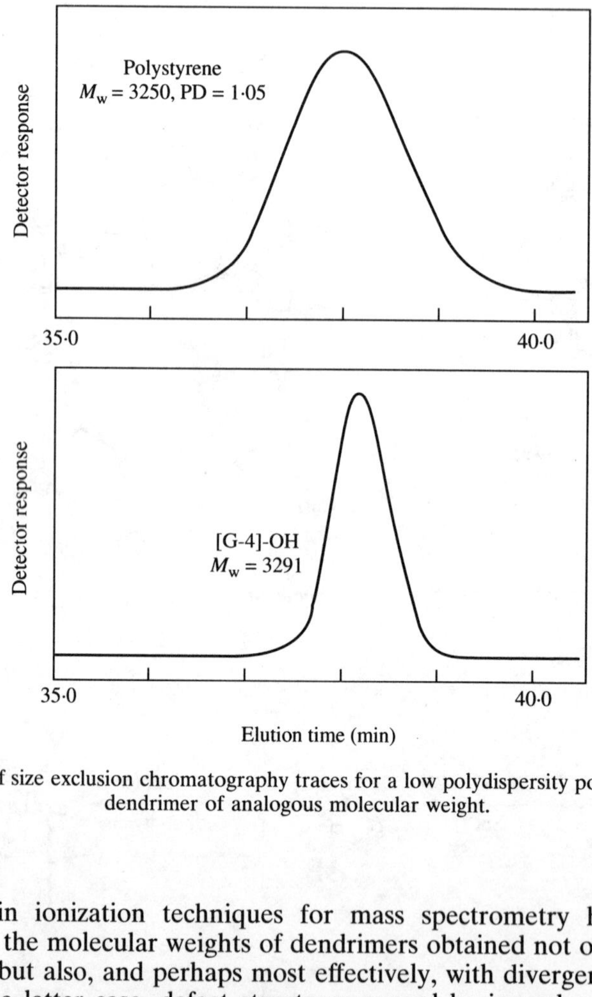

Figure 1 A comparison of size exclusion chromatography traces for a low polydispersity polystyrene and a convergent dendrimer of analogous molecular weight.

Recent advances in ionization techniques for mass spectrometry have also been used effectively to confirm the molecular weights of dendrimers obtained not only by the convergent growth approach[45,49] but also, and perhaps most effectively, with divergent PAMAM Starburst dendrimers.[21,22] In the latter case, defect structures caused by irregular growth can easily be observed and their occurrence rationalized on the basis of known side reactions. Recent work with matrix-assisted laser desorption/ionization mass spectrometry (MALDI) has confirmed the very low polydispersity of samples of dendrimers.[49,117] An illustration of the power of this technique is provided by a comparative MALDI analysis of a sample of fourth generation dendrimer [G-4]-OH with deuterated benzyl ether end groups (112 deuterium atoms, prepared as described by Hawker[4]) and a sample of nearly monodisperse poly(ethylene oxide) prepared by anionic polymerization. As can be seen in Figures 2 and 3, the dendrimer shows a sharp peak at m/z 3423 corresponding to the sodium adduct of its molecular ion, while the poly(ethylene oxide) sample, with a molecular weight said to be 3660, is clearly composed of a large number of closely related entities with varying numbers of repeating units. This result is particularly striking if one considers that the deuterated dendrimer sample was over two years old at the time the analysis was performed.[118]

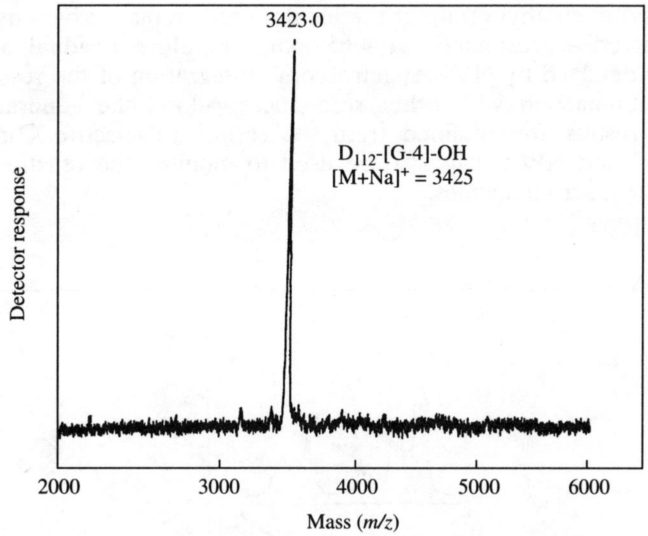

Figure 2 MALDI mass spectrum of a fourth generation convergent dendrimer.

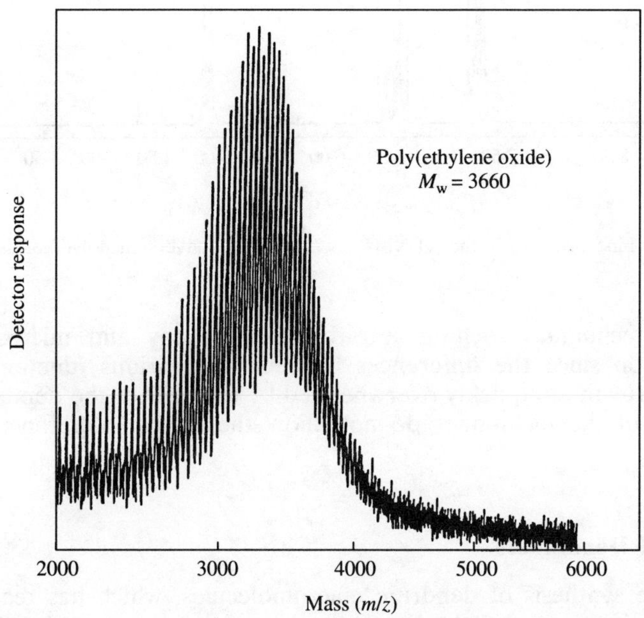

Figure 3 MALDI mass spectrum of a low polydispersity poly(ethylene oxide).

The structure and purity of dendritic macromolecules can also be confirmed by NMR spectroscopy since the highly ordered and symmetrical branching sequence allows small changes in the focal point group, terminal groups, and interior building blocks to be readily observed. For example, the third generation dendrimer derived from 3,5-dihydroxybenzyl alcohol gives a unique resonance for each layer of aromatic building blocks. As shown in Figure 4, the resonances for the aromatic protons of the terminal benzyl groups are observed as a multiplet between 7.50 and 7.25 ppm; the aromatic protons of the building blocks appear as a complex set of resonances between 6.45 and 6.70 ppm that, on expansion, reveal the individual resonances for each of the three layers of monomer units. A degree of fine structure is also observed for the benzylic methylene units at *ca.* 5.00 ppm and the unique resonance for the single bromomethyl group at the focal point appears at 4.30 ppm. On reaction with the monomer unit the bromomethyl group at the focal point is replaced by a hydroxymethyl group which has a characteristic resonance at 4.65 ppm. Therefore residual amounts of starting materials are easily detected by NMR spectroscopy. Integration of the resonance for the focal point group, and comparison with other resonances, allows the generation number to be confirmed. Similar results are obtained from the carbon-13 spectra. Chromatography techniques such as TLC and HPLC can also be used to monitor the reactions employed in the synthesis of dendritic macromolecules.

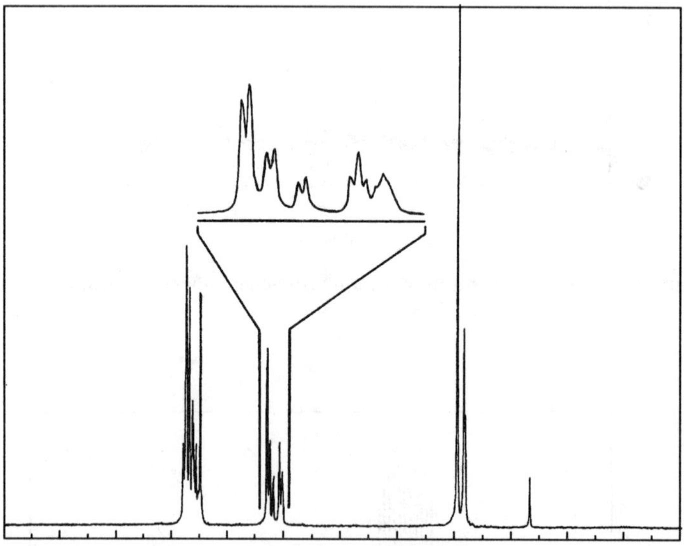

Figure 4 Fine structure in the ^{1}H NMR spectrum of a convergent polyether dendrimer.

Other chemical techniques such as infrared spectroscopy and microanalysis offer little structural information since the differences between the various functionalities involved at different stages of growth are quickly overwhelmed by the bulk of the dendritic macromolecule and the sensitivity of the techniques do not allow these small differences to be accurately observed.

3.2.4 Properties of Dendrimers

In contrast to the synthesis of dendritic macromolecules, which has reached a rather high level of maturity, the study of the physical properties of these unique globular macromolecules is still in its infancy, doubtless as a result of their lack of ready availability. However, the number of studies examining the physical properties of dendritic macromolecules is increasing steadily and these studies are gaining considerable attention. A particular aim has been the comparison of classical linear polymers and functionally analogous dendritic macromolecules, because it is expected that the shape and chain end functionality of dendrimers will result in new properties that will open new avenues for their applications.

To date, it is the solution properties of dendritic macromolecules that have been studied in the greatest detail, although many interesting findings will no doubt result from the study of their solid-state behavior. It is well known that as the molecular weight of linear polymers increases their intrinsic viscosity increases according to the Mark–Houwink–Sakurada equation: $[\eta] = KM^a$. This behavior is also observed for other nonlinear polymers such as stars or branched macromolecules. However, dendritic macromolecules do not obey this relationship once a threshold molecular weight is exceeded.[116] In the case of the regular polyether dendrimers shown in Scheme 8, a plot of log$[\eta]$ *vs.* log(molecular weight) gives a bell-shaped curve (Figure 5) with a maximum below 5000 amu instead of the expected fast rising curve expected for classical Mark–Houwink behavior shown as a dashed line in Figure 5. This unique behavior is readily explained if one considers the physical basis of intrinsic viscosity. Because inherent viscosity is concentration dependent, $[\eta]$ is obtained by extrapolation of the inherent viscosity to zero concentration. As a result, the value of $[\eta]$ is expressed in inverse concentration units, *i.e.* a volume divided by a mass. During generation growth the volume of a spherical dendrimer is expected to increase cubically ($V = 4/3\pi r^3$) while its mass doubles at each of the successive generations and increases exponentially [$MW \approx 2^{(G-1)}$]. Therefore, as one progresses from low to higher generation, the mass term dominates and a bell-shaped curve is indeed expected.

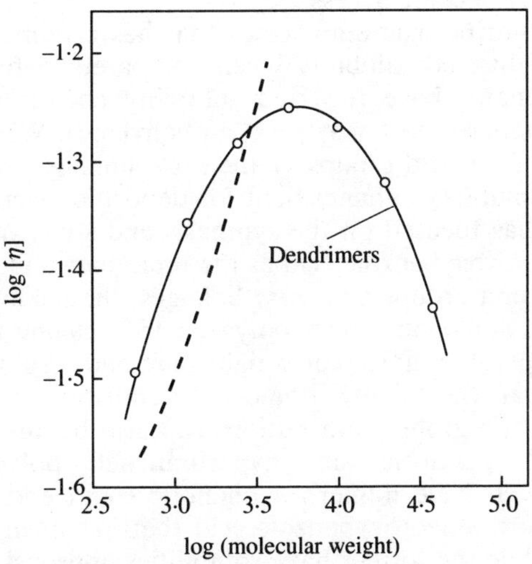

Figure 5 Intrinsic viscosity–molecular weight relationship for convergent polyether dendrimers (full line) and linear polystyrene (dashed line).

The maximum in intrinsic viscosity can also be correlated with a shape change from an extended to a more compact globular structure at higher molecular weights. Further evidence for this shape transition has been recently obtained from a series of solvatochromic measurements with uniquely functionalized convergent dendrimers.[119] A family of polyether dendrimers was prepared using the focal point to attach a solvatochromic probe, *N*-methyl-*p*-nitroaniline. Measurements made in a variety of low polarity solvents show that the absorption maximum of the solvatochromic probe increases with the molecular weight of the dendrimers as a consequence of the more effective shielding of the probe by the surrounding dendrimer at high generation. A plot of absorption maximum *vs.* generation displays a marked discontinuity between generations 3 and 4. This discontinuity can be directly correlated to the maximum in the intrinsic viscosity curve described above. This behavior suggests that the molecule undergoes a shape transition from an extended to a more globular shape as it grows larger, with the consequence that at a higher generation the solvatochromic probe is shielded to a greater degree.[119] These data confirm that dendrimers provide a unique microenvironment surrounding the solvatochromic probe, a concept that is similar to that which prevails for the active site of numerous enzymes. This interesting finding suggests that dendrimers may be of particular value as molecular catalysts of drug delivery systems. Molecular modeling experiments of Goddard[120] and photoinduced electron-transfer measurements on Tomalia's PAMAM Starburst dendrimers by Turro[121,122] also support the finding of a shape transition near generation 4.

This concept of a unique microenvironment for dendritic macromolecules has been elegantly exploited by Inoue and co-workers,[123] who attached four dendrons based on 3,5-dihydroxybenzyl alcohol to a central porphyrin core. The series of macromolecules that was obtained featured a porphyrin core surrounded by a dendrimer 'cage' of increasing size. Fluorescence quenching experiments showed that when fourth generation dendritic fragments were used, a size selective isolation of the porphyrin to external quenchers was achieved. Quenching by large molecules was not effective, while small molecules such as vitamin K_3 acted as efficient quenchers. Conversely, for small 'cages' made from the first generation fragments, it was found that quenching occurred at approximately the same rate for both small and large molecules. Significantly, the quenching rate for the [G-1] derivative was *ca.* three times smaller than for the quenching of vitamin K_3 with the G-4 cage. Therefore not only does the dendrimer cage act as an effective size discriminating barrier, at high molecular weight it acts a unique microenvironment in which vitamin K_3 quenching is enhanced, presumably by its selective attraction to the dendrimer's interior. Both of these properties are crucial to certain biological functions and illustrate clearly the intriguing possibilities opened by dendritic macromolecules.

Recently, Chujo[124] has reported the preparation of nanoporous silica by the pyrolysis of a dendrimer–inorganic polymer hybrid. Interestingly, the pore size could be controlled by the use of PAMAM dendrimers of specific generation and was essentially the same as the radius of the starting dendrimer. If this process can be accurately controlled, a host of zeolite-like materials with unusual properties may become accessible.

One characteristic of dendritic macromolecules that has become evident from a number of different studies is their enhanced solubility when compared to functionally analogous linear polymers. Miller and Neenan[32] have reported solubility enhancements of 10^5 for dendritic poly(phenylenes) when compared to linear poly(p-phenylene). While this study does not take into account contributions from end groups or the *meta* linkages in the dendritic molecules, it does suggest a significant solubility enhancement for dendritic macromolecules.

A more recent study[125] has focused on the synthesis and study of various isomeric aromatic polyesters based on 3,5-dihydroxybenzoic acid as the monomer unit. For the dendritic polyester (**54**) that has benzyl ether end groups and *meta* linkages, the solubility in THF was extremely high, 1.15 g mL^{-1}, while an analogous linear polyester (**55**) having the same number of benzyl ether end groups and comparable molecular weight only had a solubility of 0.025 g mL^{-1}. This study clearly confirms that the highly branched dendritic structure dramatically affects solubility. The question of end group contributions can also be answered through comparisons involving, on one hand, dendritic and hyperbranched polyesters derived from 3,5-dihydroxybenzoic acid, which have numerous phenolic chain ends, and, on the other hand, linear polyester derived from 3-hydroxybenzoic acid that has *meta* linkages but essentially no 'interacting' chain ends. While the former have solubilities approaching 1.0 g mL^{-1}, the latter is totally insoluble in THF. From this, it can be concluded that the increased solubility of dendritic macromolecules is due both to their branched architecture and to the solubilizing contributions made by their large number of end groups.

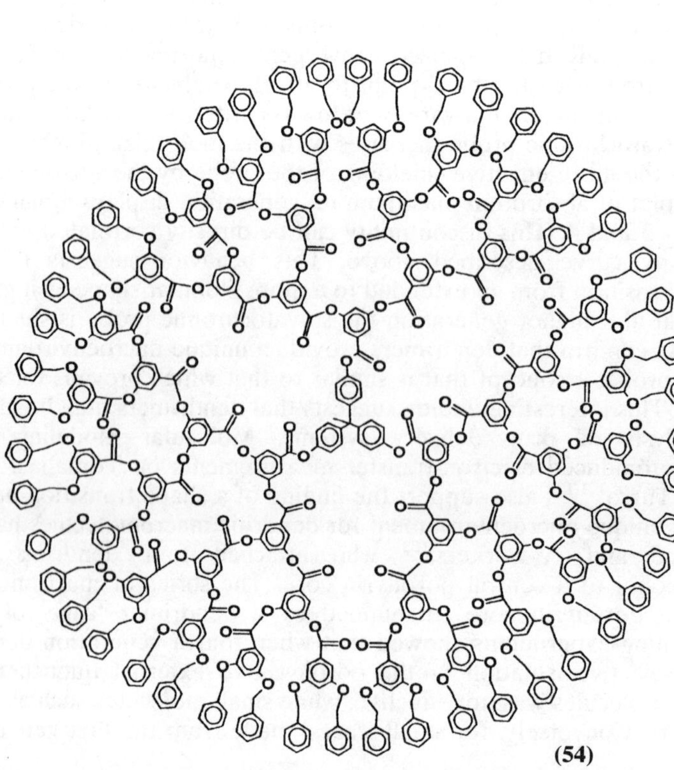

(54)

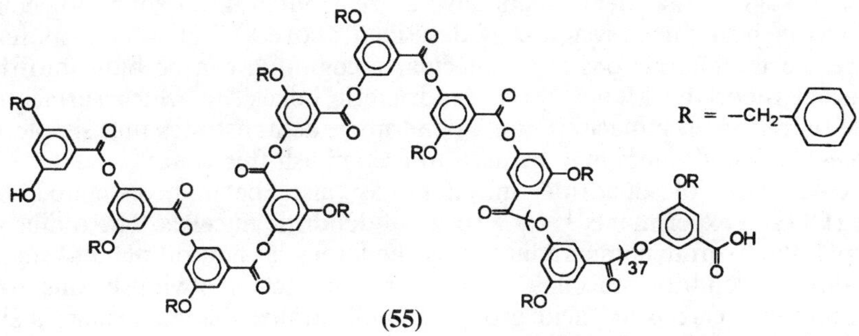

(55)

In contrast to the solution properties described above, the thermal properties of dendrimers and functionally analogous linear polymers are similar. For example, the variation of glass transition temperature with molecular weight and chain end composition has been examined[126] and the results follow a modified version of the chain-end free-volume theory if one takes into account the large number of chain ends. Similarly, and not unexpectedly, dendritic macromolecules have been shown to have essentially the same thermogravimetric behavior as linear and hyperbranched macromolecules composed of the same building blocks.[125]

The structure and properties of dendritic micelles[108–113] is a very promising area of research due to their similarity to natural systems. As mentioned previously, these unique materials have the basic features of traditional micelles except that they are static, covalently bound structures instead of dynamic associationas of molecules. Studies have shown that dendritic micelles such as **(56)** are not only capable of molecular inclusion but they can solubilize hydrophobic molecules in aqueous solutions to an extent equal to, if not greater than, traditional SDS micelles while not displaying any critical micelle concentration.[108] These characteristics have been employed in the design of a novel recyclable solubilization and extraction procedure that could be easily applied to the recovery of a great variety of organic substances from water.[108] Interestingly, the solubilization ability of dendritic micelles has been shown to be sensitive to the ionic strength of the aqueous solution and the internal chemistry of the dendrimer. For electron-rich aromatic polyether cores, noncovalent interactions lead to the preferential solubilization of electron-deficient hydrophobic molecules. The unique hydrophobic microenvironment in the interior of dendritic micelles based on alternate building blocks has also been confirmed by optical fluorescence and solvatochromic studies and provides further rationalization for the existence of molecular inclusions.[109,110]

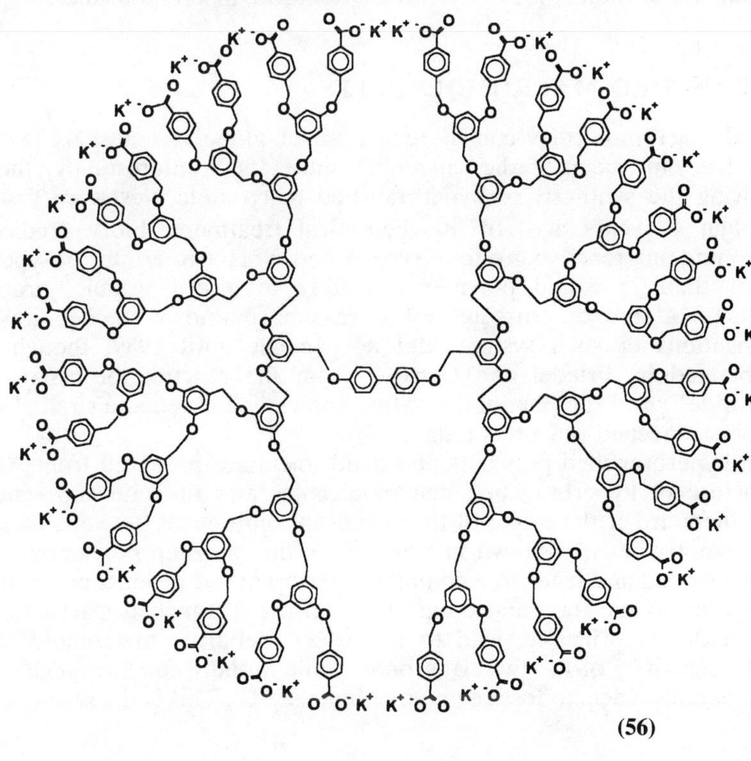

(56)

These findings suggest that dendrimers have a great potential in supramolecular chemistry, initially perhaps in host–guest systems, as described above,[108] but also in more finely tuned systems where specific interactions and molecular recognition can be built into the dendrimer moiety. A recent report by Meijer[130] on 'dendrimeric boxes', in which surface congestion is used as a tool to trap small molecules into a dendrimer, demonstrates that simple modifications of readily accessible dendrimers may be used to accomplish this goal.

The molecular size of dendritic micelles has also been investigated by 2D-NMR spectroscopy (DOSY) experiments[112] for a series of dendritic micelles. The radius was shown to change with pH, the hydrodynamic radius being the largest at neutral pH and smallest at acidic pH. This ability of dendritic micelles to swell or contract is obviously due to interactions between the terminal carboxylic acid groups and illustrates the molecular flexibility of this nonentangled, covalently bound, architecture.

Since one of the goals of supramolecular chemistry[175] is the accurate assembly of nanometer scale molecules, the structural and functionality control available for dendritic macromolecules makes these materials ideally suited for nanoscale assembly. A number of reports have recently appeared detailing initial experiments in this area. In terms of self-assembly, dendritic polyether macromolecules have been shown to form stable monolayers at the air–water interface with surfactant-like behavior being observed for [G-2]-OH to [G-4]-OH only (see analogous structures in Scheme 7, R = H). At higher generation, the surfactant-like behavior is lost and isotherms typical of hydrophobic molecules are observed. This finding is consistent with the solvatochromic results described above and provides further evidence that the focal point group is shielded by the surrounding dendrimer as the generation number increases. Similarly, all isotherms for tridendron dendrimers, $[G\text{-}X]_3\text{-}[C]$, such as shown in Scheme 8, reflect their hydrophobic character; it must be remembered that these dendrimers have no hydroxy group at their focal point for interaction with the aqueous subphase.[127,128]

Analysis of these monolayers by neutron reflectivity[128] revealed that for [G-4]-OH the initial monolayer collapses to a bilayer structure where the dendrimers in the top layer assume a spherical shape, while those in the bottom layer undergo further compression and assume a distorted ellipsoid shape with a ratio of height to width of approximately 2. This behavior, which has not been observed previously, indicates that the dendrimers are flexible enough to alter their shape in response to increased pressure and that entanglements between different molecules must be minimal. Recent atomic force microscopy studies[129] and the stabilization of aqueous–organic emulsions by amphiphilic dendrimers[108] support the ability of dendritic macromolecules to form stable self-assembled monolayers. This ability, coupled with the recent report of the orientation of a dendritic surface-block macromolecule in an electric field,[10] suggests promising applications for self-assembled dendritic macromolecules.

3.3 HYPERBRANCHED MACROMOLECULES

Hyperbranched macromolecules constitute a class of globular materials that is related to dendrimers but has only been studied in detail since 1989. Interestingly, the fundamental concepts underlying the synthesis of hyperbranched macromolecules were first discussed by Flory[131] more than 40 years ago. In his theoretical treatment, Flory predicted that AB_x monomers containing one reactive group of type A and x reactive groups of type B (where x is equal or greater than 2) would polymerize readily to afford soluble, processible three-dimensional structures free of cross-links. For reasons unknown, recognition and detailed synthetic investigations of such systems did not appear until 1989, though the insoluble hydrocarbon obtained by Friedel and Crafts[132] from the reaction of benzyl chloride with aluminum trichloride[133] and the phenol derivative known as Lautemann's red[134] would both be recognized as hyperbranched polymers today.

Whereas both hyperbranched polymers and dendrimers are prepared from AB_x monomers, the overall structure of hyperbranched macromolecules and the one-step synthetic strategy used to prepare them are at the origin of the significant differences between these two families of globular macromolecules. As shown in Scheme 26, the one-step polymerization of an AB_2 monomer is uncontrolled and leads to a complex hyperbranched product containing both linear and dendritic regions. While the one-step growth of a hyperbranched macromolecule from an AB_2 monomer could, in principle, lead to a 'perfect' dendritic macromolecule, it is never encountered. The reactivity of the two 'B' groups, while initially equal, changes as soon as one of the 'B' functionalities reacts to form a polymer linkage, 'C'. This is the result of an increase in steric constraints in the immediate vicinity of the remaining unreacted 'B', which, in turn, decreases its reactivity. Additional electronic factors may further reduce the reactivity of the second 'B' group. As a result, three different subunits are typically found in hyperbranched structures and they differ in the number of B functionalities that have undergone reactions to form the polymeric linkages, 'C'. Reaction of both B functionalities leads to 'dendritic' units that are analogous to the internal building blocks of dendritic macromolecules. Alternatively, if only one of the two B functionalities undergoes reaction, a 'linear' unit is obtained and these 'failure sequences' constitute the principal structural difference between hyperbranched macromolecules and dendrimers. Finally, if neither of the B functionalities react, a 'terminal' unit is obtained which can be considered the equivalent of the outer layer, or chain ends, of a dendritic macromolecule. Because of these incomplete reactions, the structures of hyperbranched macromolecules are intermediate between those of 'perfect' dendritic macromolecules and linear polymers. The term 'degree of branching', which is described more fully in the next section, has been developed to define this intermediate structure more accurately. For the sake of brevity the representation of numerous complex hyperbranched structures has been simplified using a shorthand version throughout the text. Therefore, the hyperbranched macromolecule shown as **(57)** in Scheme 26 can also be represented as a tri-bracketed building block **(58)** in which one of the bonds bears the functional group present at the chain ends. It should be noted that this notation does not imply that a linear structure is obtained.

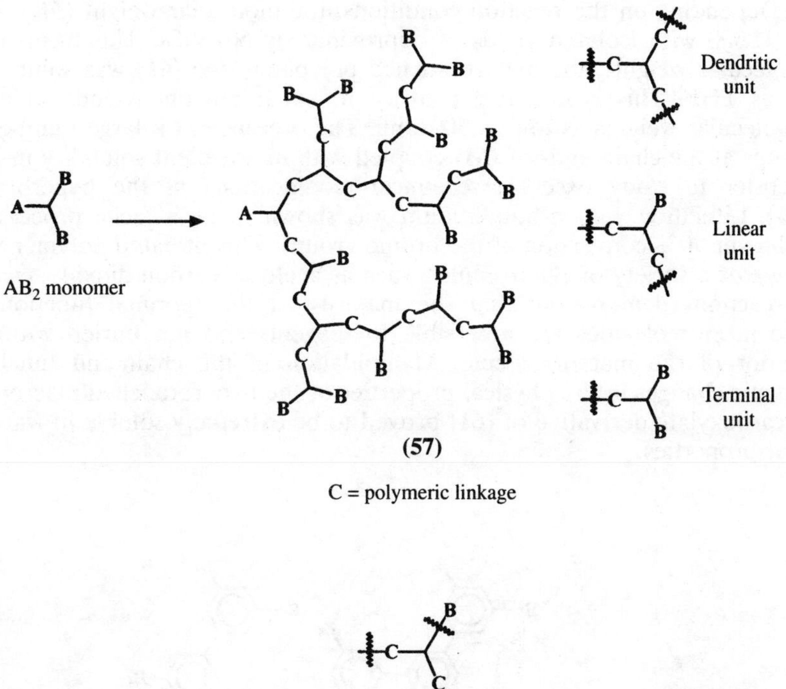

(57)

C = polymeric linkage

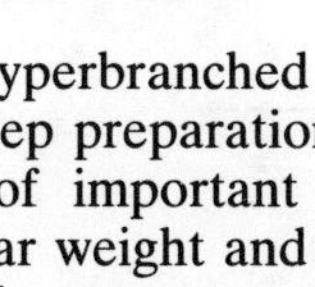

(58)

Scheme 26

The main distinguishing features of hyperbranched macromolecules, when compared to dendritic macromolecules, are their one-step preparation and the existence of 'linear' units or failure sequences. This has a number of important consequences because the materials obtained are polydisperse in both molecular weight and branching. Also, the molecular weight of hyperbranched macromolecules is to a large extent uncontrollable. This is in sharp contrast to dendrimers for which a well-defined molecular weight can be obtained. However, the one-step synthesis of hyperbranched macromolecules allows these materials to be produced on a much larger scale and at a significantly lower cost, which allows one to contemplate their use as specialty high-performance polymers, rheology modifiers, advanced coating systems, additives, lubricants, *etc.*

3.3.1 Synthesis of Hyperbranched Macromolecules

Several conditions must be met in order to obtain high molecular weight hyperbranched polymers from AB_x monomers. First, the reactive groups, A and B, should only react with each other in the presence of a catalyst, or after a suitable activation step has been carried out (*e.g.* the removal of a protecting group via a chemical, thermal, or photochemical process). Consequently the mutual reactivity of the functional groups, A and B, should be high to facilitate polymer formation. Side reactions should be kept to an absolute minimum to prevent deactivation or cross-linking of the growing polymer chain. The latter is an important consideration due to the large number of reactive B groups present at the chain ends of the hyperbranched macromolecule. Perhaps of greater importance is the potential deactivation of the single reactive A group at the focal point of the growing hyperbranched macromolecule; deactivation during polymerization would be expected to disturb the growth process severely and may lead to low molecular weight materials.

While early and nonconcerted attempts to prepare highly branched macromolecules date back to the 1880s,[132] the first rigorous treatment was a theoretical study on three-dimensional polymers from AB_x monomers by Flory in 1952.[131] A rigorous synthetic attempt at preparing these materials was not attempted for approximately 40 years until Kim and Webster reported their preparation of hyperbranched polyphenylenes from AB_2 monomers.[3,111,135] Starting from either 3,5-dibromophenylboronic acid (**59**) or 3,5-dibromophenylmagnesium bromide (**60**), polycondensation was found to give a bromo-terminated hyperbranched polyphenylene (**61**) (Scheme 27). Depending on the reaction conditions, the molecular weight (M_n) of (**61**) varied from 2000 to 32 000 with isolated yields of approximately 60–80%. The surprising feature is that, at all molecular weights, the hyperbranched polyphenylene (**61**) was soluble in common solvents such as THF. This is in direct contrast to linear polyphenylenes which are totally insoluble at molecular weights as low as 700 amu. The presence of a large number of bromine functional groups at the chain ends of (**61**) coupled with its excellent solubility in THF allowed Kim and Webster to study excessive chemical modifications of the hyperbranched polyphenylene (**61**). Lithiation with *n*-butyllithium was shown to be a facile process, affording a polyanion with over 90% conversion of the bromo groups. This lithiated polymer was shown to be reactive towards a variety of electrophiles, such as acetone, carbon dioxide, *etc.* All of these modification reactions demonstrate that the majority of the terminal functional groups in hyperbranched macromolecules are accessible to reagents and not buried within the highly branched interior of the macromolecule. Manipulation of the chain-end functional groups results in dramatic changes in the physical properties of the hyperbranched macromolecule. For example, the carboxylate derivative of (**61**) proved to be extremely soluble in water, displaying typical micellar properties.

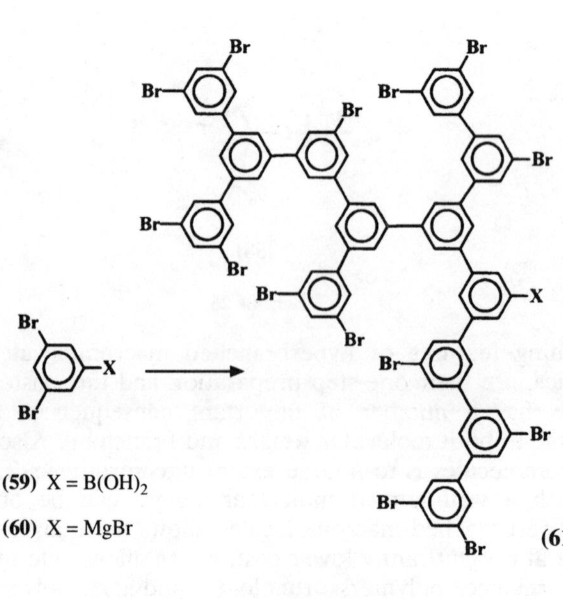

(**59**) X = $B(OH)_2$

(**60**) X = MgBr

(**61**)

Scheme 27

Following Kim and Webster's initial report, a large number of publications concerning the synthesis of hyperbranched macromolecules have appeared. As a consequence of their range of potential industrial applications, the family of hyperbranched polymers that has been studied in greatest detail is based on ester linkages. Hyperbranched polyesters with structures ranging from fully aromatic to fully aliphatic as well as others with partial aromatic/aliphatic character have been examined. Perhaps the most extensively studied of these hyperbranched polyesters are the fully aromatic systems based on a 3,5-dioxybenzoyl building block. Initial reports by Hawker and Fréchet[136] detailed the high yield (>80%), one-step synthesis of the hyperbranched aromatic polyester (**62**) from a thermally stable AB_2 monomer, 3,5-bis(trimethylsilyloxy)benzoyl chloride (**63**) (Scheme 28). The molecular weight of the polymer (**62**) was found to increase reproducibly with higher reaction temperatures, longer reaction times, and larger amounts of catalysts.[137] As with the hyperbranched polyphenylenes, thermal stability was similar to that observed for the corresponding commercial linear polymers, while the solubility of the hyperbranched polyester was dramatically increased. Previously, Kricheldorf[138] has used the AB_2 monomer (**63**) to prepare slightly branched polyesters by copolymerization with an AB monomer at low levels of incorporation of the AB_2 units. Turner[16] has also reported the preparation of aromatic hyperbranched polyesters from either the acid chloride (**63**) or 3,5-diacetoxybenzoic acid (**64**). The results obtained were similar to those outlined above, except that the polymerization of (**64**), which leads to the corresponding acetate-terminated hyperbranched polyesters, could be forced to significantly higher molecular weights (>1 000 000). The materials that were obtained not only retained high solubilities but had extremely low viscosities, such as $[\eta] = 0.34\ \mathrm{dL\,g^{-1}}$ at $MW = 10^6$. Subsequent work[137,139] has refined the synthetic procedures for both the preparation of the acid chloride (**63**) and its subsequent polymerization. These advances, combined with the recent commercial availability (Aldrich) of an advanced precursor to (**63**), significantly increase the opportunity to study these novel materials.

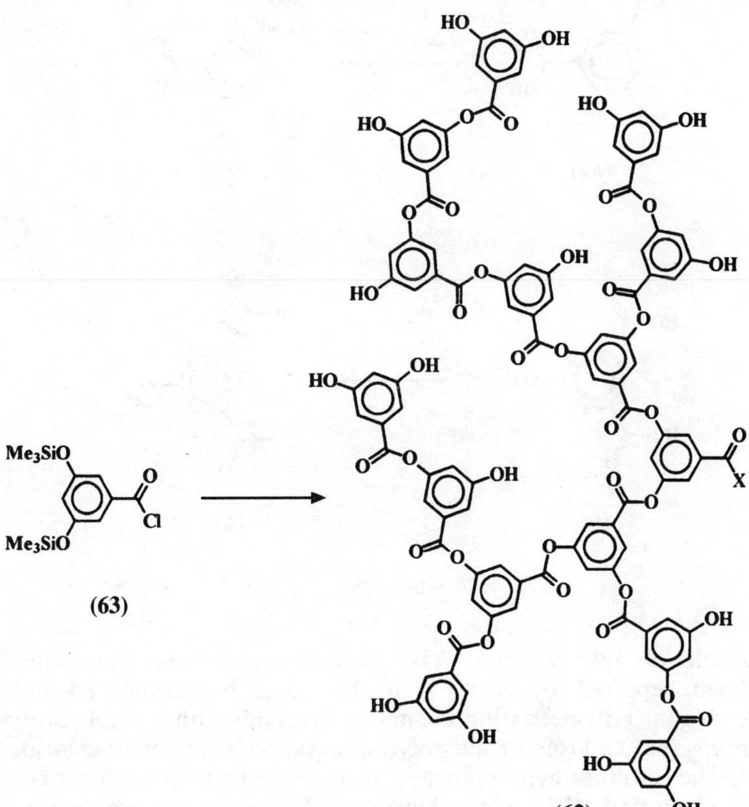

(63) **(62)**

Scheme 28

One distinctive feature of hyperbranched macromolecules, first demonstrated with aromatic polyesters such as (**62**), is that the choice of the A and B functionalities in the original AB_2

monomer is of crucial importance. Interchange of the A and B functional groups gives an A_2B monomer which, on polymerization, affords a material with significantly different properties. This behavior is totally unlike linear polymers prepared from either an AB or the corresponding BA monomers, which give essentially the same polymer containing a single A chain end and a single B chain end. This fundamental difference arises from the numerous B functionalities (their number equals DP + 1) that are present in the final hyperbranched macromolecule, and the large influence these functionalities have on the physical properties of hyperbranched macromolecules. It should be noted that theory predicts that hyperbranched macromolecules also contain a single A functionality.

A number of research groups have examined the significance of an interchange of the A and B functionalities in detail and analysis of the results demonstrates that this is a very important point because significant property changes arise from such an interchange. For example, polymerization of the AB_2 monomer 3,5-diacetoxybenzoic acid (**64**) ($A = CO_2H$; $B = OAc$) gives an acetoxy-terminated polyester (**65**) containing aromatic building blocks. If the functional groups, A and B, of the monomer (**64**) are exchanged, the corresponding A_2B monomer 5-acetoxybenzene-1,3-dicarboxylic acid (**66**) that is obtained may be used under similar polymerization conditions to afford another aromatic hyperbranched polyester (**67**) (Scheme 29).[140] However, while the building blocks are analogous for (**65**) and (**67**), the terminal functional groups are dramatically different, being carboxylic acid functionalities for (**67**) and acetate groups for (**65**). The physical consequences of this are that the glass transition temperature of (**65**) is 155 °C compared to 250 °C for (**67**). Significant differences are also observed in solubility, with (**67**) being soluble in aqueous base and alcoholic solvents, whereas (**65**) is completely insoluble. Recently, Kricheldorf[141] has also prepared the carboxylic acid-terminated polyester (**67**) using bis(trimethylsilyl) 5-acetoxylbenzene-1,3-dicarboxylate as an AB_2 monomer; the results obtained were similar.

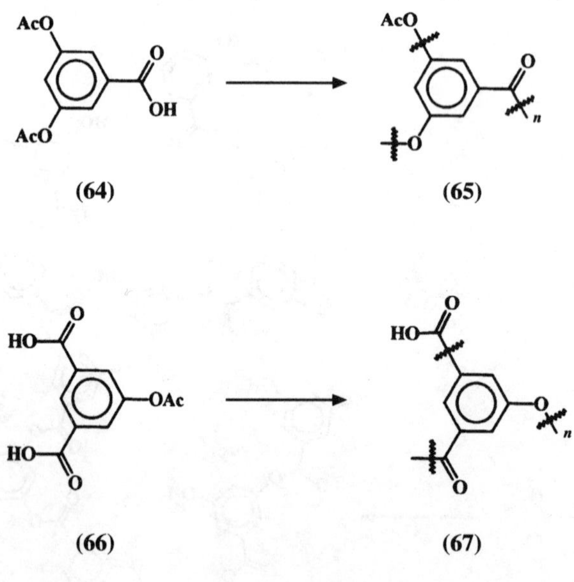

Scheme 29

A further example of 'interchanged' AB_2 monomer units and their effect on polymer properties has been reported by Kim.[142] In this case, hyperbranched polyamides were synthesized using elegant polymerization chemistry involving either the hydrolysis of sulfinyl-masked amino groups (**68**) and (**69**) or the polycondensation of the hydrochloride salts (**70**) and (**71**) (Scheme 30). The resultant hyperbranched aromatic polyamides contain either amino (**72**) or carboxylic acid chain ends (**73**). These hyperbranched polymers are similar to industrially important aromatic polyamides such as Kevlar, but they can be obtained as high molecular weight materials ($M_n = 24\,000–46\,000$) that have excellent solubility in a variety of solvents. Once again, differences are observed in the thermal and solubility behavior for these 'interchanged' hyperbranched polyamides. Surprisingly, size exclusion chromatography indicated that, in select solvents, supramolecular aggregates (MW = 700 000–1 000 000) were formed and amide solutions containing greater than 40 wt% of the carboxylic acid-terminated hyperbranched polyamide (**73**) displayed birefringence. This is the first reported example of hyperbranched or dendritic macromolecules that display lyotropic liquid crystalline properties.

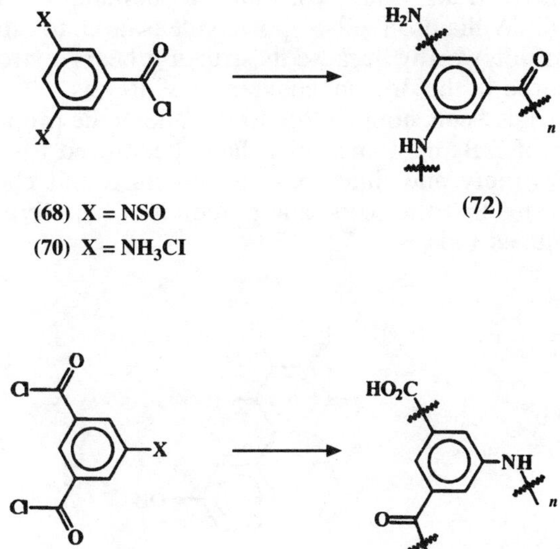

(68) X = NSO
(70) X = NH_3Cl
(72)
(69) X = NSO
(71) X = NH_3Cl
(73)

Scheme 30

A more rigorous examination of this aspect of hyperbranched macromolecules has recently been reported by Hawker.[143] In this case the interchangeable AB_2 monomers were designed so that the resulting hyperbranched macromolecules would contain exactly the same internal building blocks. Therefore, any differences in physical properties would only be due to the influence of the chain ends and not to subtle differences in the internal structure. 3,5-Difluoro-4′-hydroxybenzophenone (**74**) and 3,5-dihydroxy-4′-fluorobenzophenone (**75**) were chosen as the monomers and polymerization under standard conditions gave the hyperbranched poly(ether ketone)s (**76**) and (**77**) that contain the same 3,4′,5-trisubstituted benzophenone repeat units (Scheme 31). However, the chain ends of the hyperbranched macromolecules are very different, with fluoro groups in the case of (**76**) and hydroxy groups for (**77**). Again, this leads to significant differences in properties such as solubility, glass transition temperature, and melt viscosity for the hyperbranched macromolecules. Interestingly, the molecular weights and, perhaps more importantly, the degree of branching (see Section 3.3.2) of the two hyperbranched poly(ether ketone)s were also different. This finding demonstrates that the choice of the A and B functionalities can also influence the overall structure of hyperbranched macromolecules.

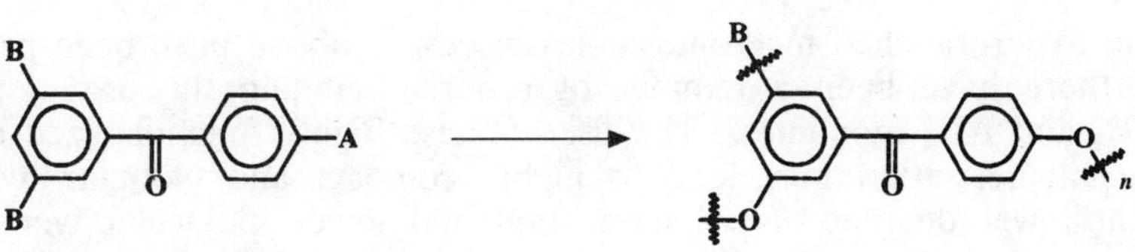

(74) B = F, A = OH
(75) B = OH, A = F
(76) B = F
(77) B = OH

Scheme 31

Classical condensation chemistry and AB_2 monomers have also been used by a number of other groups to prepare hyperbranched macromolecules. For example, the preparation of both phenolic-terminated[144] and isocyanate-terminated hyperbranched polyurethanes[145] using masked isocyanate chemistry has recently been independently described by two groups, while the synthesis of thermotropic hyperbranched liquid crystalline polyethers[146] and polyesters[147] has also been reported. Uhrich *et al.*[148] have used an AB_2 building block, 3,5-dihydroxybenzyl bromide, that can undergo Williamson ether polycondensation to afford a hyperbranched polyether. This polyether is noteworthy because its structure bears a strong resemblance to the dendrimers that were prepared in the original convergent synthesis.[4,27,28] However, the study of its reactivity and properties has been stopped due to the allergenic properties of the monomer. Figure 6 shows a selection of AB_2 monomers that have been used to prepare hyperbranched polymers.[148–153] The great variety and differences in structure and chemistry between these monomers amply demonstrates that the scope and potential of hyperbranched macromolecules is only starting to be recognized today.

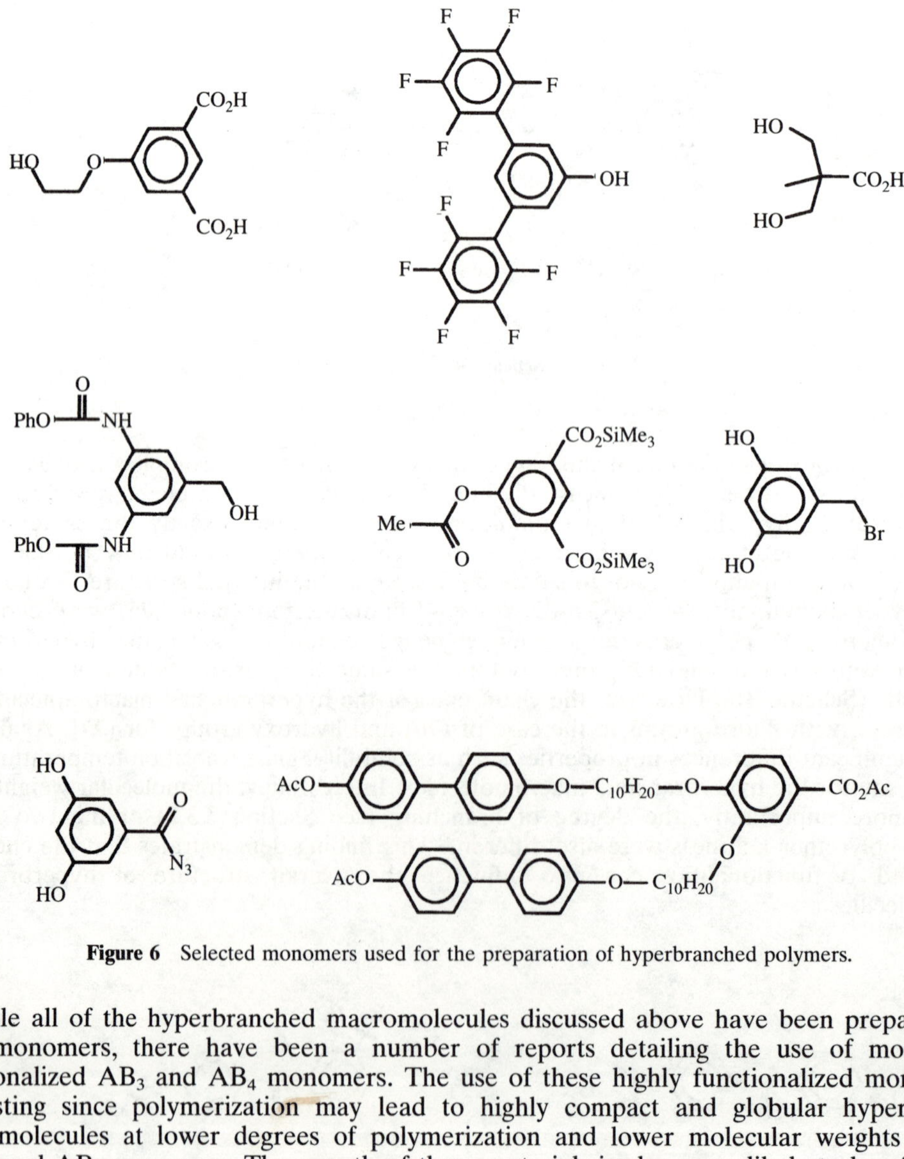

Figure 6 Selected monomers used for the preparation of hyperbranched polymers.

While all of the hyperbranched macromolecules discussed above have been prepared from AB_2 monomers, there have been a number of reports detailing the use of more highly functionalized AB_3 and AB_4 monomers. The use of these highly functionalized monomers is interesting since polymerization may lead to highly compact and globular hyperbranched macromolecules at lower degrees of polymerization and lower molecular weights than for traditional AB_2 monomers. The growth of these materials is also more likely to be affected by steric inhibition (or the dense packing phenomenon). To investigate this concept, the synthesis of the AB_3 monomer (**78**), containing an allyl group and three Si—H groups, and its subsequent platinum(IV)-catalyzed polymerization has been reported (Scheme 32).[154] Significantly, growth was observed to stop at a molecular weight of 19 000, a value which, for regular growth, would correspond to a fourth generation material. Further additions of more catalyst or monomer did not lead to any increase in molecular weight. This suggests that the chain ends are sterically inhibited to further dendritic growth, a finding that correlates well with the growth problems that are expected for such highly branched and congested structures. Subsequently, Hawker has shown that AB_3 and AB_4 monomers can also be used to control the degree of branching in hyperbranched poly(ether ketone)s.[155]

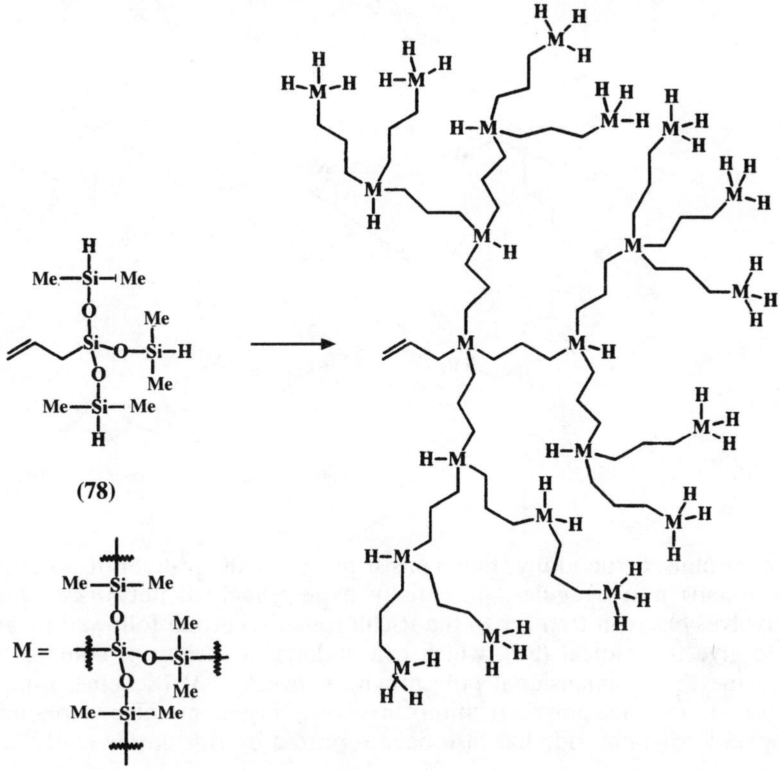

Scheme 32

The preparation of hyperbranched macromolecules has recently been extended to polymerizations other than those involving classical polycondensation chemistry. Suzuki[156] has reported an elegant application of ring-opening polymerization to the synthesis of hyperbranched polyamines with controlled molecular weights and low polydispersities (<1.3). Reaction of the cyclic carbamate (**79**), which in this case is a masked AB_2 monomer, with a palladium(0) catalyst activates the double bond and allows formation of a ring-opened species (**80**). Nucleophile attack by a primary amine, used as polymerization initiator, then leads to the loss of carbon dioxide and gives the diamino derivative (**81**). The unique feature of this synthesis is that diamine (**81**) is generated by ring opening of the monomer unit (**79**) and this diamine can react with additional monomer units at either of its primary or secondary amino sites. Formation of a dendritic unit results from reaction at the secondary amine (pathway A), while reaction at the primary amine leads to a linear unit (pathway B). Therefore, a combination of these two processes leads to a hyperbranched polyamine (Scheme 33). Despite the different reactivity of the primary and secondary amine sites, this elegant synthesis that affords products with low polydispersities and controlled molecular weights is very attractive and the preparation of hyperbranched macromolecules by either ring-opening or 'living' polymerizations[167] is a promising area of research.

An intriguing new class of three-dimensional polymeric materials, which may be considered hyperbranched, are the recently reported polycarbynes. These novel polymers are synthesized by the reductive condensation of the appropriate trichloromethyl monomers, $RCCl_3$ (R = Ph (**82**), Me, H) with highly reactive metals such as ultrasonically generated emulsions of sodium–potassium alloy. Bianconi[157] has elegantly shown that polycarbynes, such as poly(phenylcarbyne) (**83**) ($M_w = 4000$, PD = 1.3) are soluble in organic solvents and can be cast

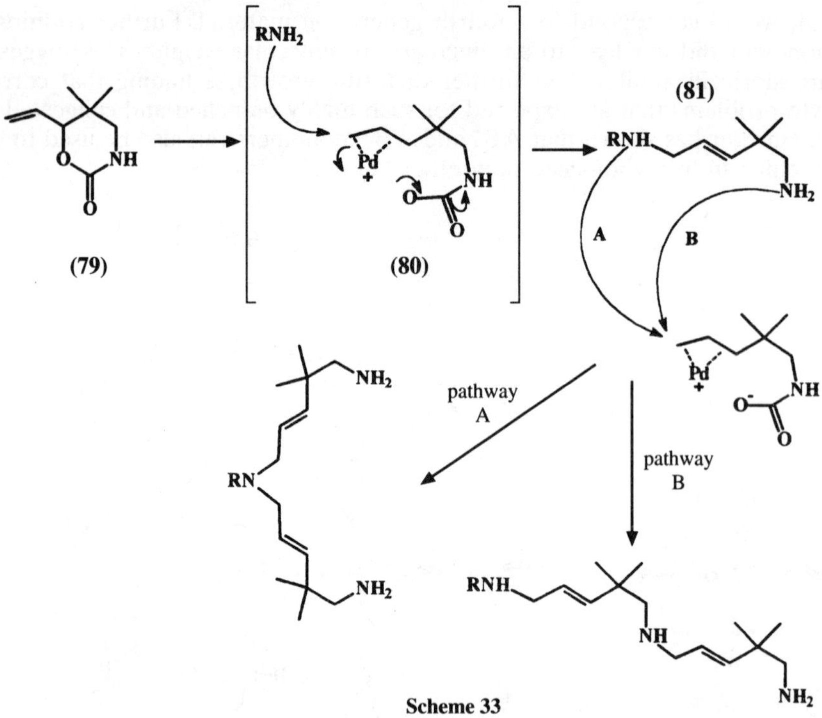

Scheme 33

to form transparent films. Structurally, they consist primarily of sp^3-hybridized units assembled via C—C single bonds into irregular, potentially hyperbranched, networks. One postulated mechanism[158] involves electron transfer to the trichloromethyl group followed by abstraction of a chloride ion to give the radical (**84**), which can undergo a variety of combination/addition reactions to give the three-dimensional polycarbyne network (**83**) (Scheme 34). A somewhat similar uncontrolled anionic polymerization involving organometallic monomers such as tris(pentafluorophenyl)silyl chloride has also been reported by Bochkarev *et al.*[159]

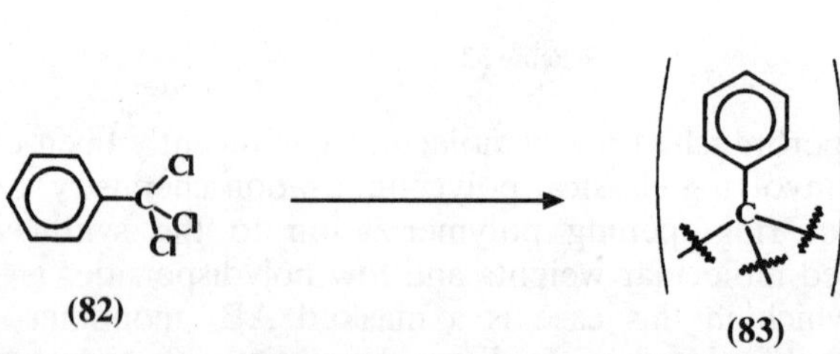

Scheme 34

3.3.2 Structure of Hyperbranched Macromolecules

As demonstrated above, hyperbranched macromolecules are fundamentally different from both linear polymers, because of their three-dimensional highly branched architecture, and dendritic macromolecules, because their branching is random and uncontrolled rather than highly regular. Therefore, traditional methods for structural analysis are frequently inadequate and initial examination of the proposed structures for hyperbranched macromolecules only suggests a complex arrangement of linear and branched segments of different sizes. More detailed examination shows that this complex arrangement can be simplified at the building block level since hyperbranched macromolecules, derived from AB_2 monomers, are composed of only three different types of subunits, or building blocks. These three subunits, termed linear, dendritic, or terminal, directly relate to the relative amount of branching in the hyperbranched macromolecule and it is this feature that allows structural analysis. The concept of degree of branching (DB) was therefore introduced to better define hyperbranched macromolecules and allow their comparison with linear or dendritic polymers.[136] As defined, $DB = 1 - f$, where f is the fraction of subunits linked through one of their B groups only (*i.e.* linear units).[136]

By default, a linear polymer would have a degree of branching of 0.0, while a perfectly branched dendritic macromolecule would have a DB of 1.0. When the random structure of hyperbranched macromolecules is considered, values for the degree of branching can range between 0.0 and 1.0, in effect between linear and perfectly dendritic structures. Since it has been conclusively demonstrated that the physical properties of linear polymers and dendritic macromolecules are frequently dramatically different, large variations in the physical properties of hyperbranched macromolecules may occur for different degrees of branching. For example, hyperbranched macromolecules with a low degree of branching (*e.g.* $DB = 0.1$) may be expected to have physical properties resembling those of linear polymers, while hyperbranched macromolecules with a high degree of branching (*e.g.* $DB = 0.9$) may be expected to have physical properties resembling those of dendrimers. Therefore, accurate determination of the degree of branching for hyperbranched macromolecules is required for understanding, predicting, and possibly manipulating the properties of these novel materials.

Two methods have been developed for the determination of the degree of branching for hyperbranched macromolecules. One method, based on NMR spectroscopy, relies on the ability to differentiate discrete resonances for the three different subunits present in the macromolecule. Integration of these resonances allows the relative percentage of each subunit to be determined and hence DB to be calculated. For example, the hyperbranched aromatic polyester (**62**) (Scheme 28) contains three different subunits: a terminal unit that has two phenolic groups (**85**), a linear unit that has one phenolic group and one acyloxy group (**86**), and a dendritic unit that has two acyloxy groups (**87**) (Figure 7).[136] This different substitution pattern leads to separate resonances for each subunit and, on comparison with model compounds, specific ^{1}H and ^{13}C resonances may be assigned to each of the subunits (Figure 8). Integration data from these values may then be used to calculate the relative percentage of each subunit and the degree of branching can be calculated (*e.g.* 0.6). The structural implication of a DB of 0.6 is that there is a branch point at approximately every second monomer unit, indicating that a highly branched structure is obtained even at such a relatively modest value of DB. Interestingly, the degree of branching for (**62**) was found to be independent of molecular weight, indicating that branching does not change as a function of conversion.[136]

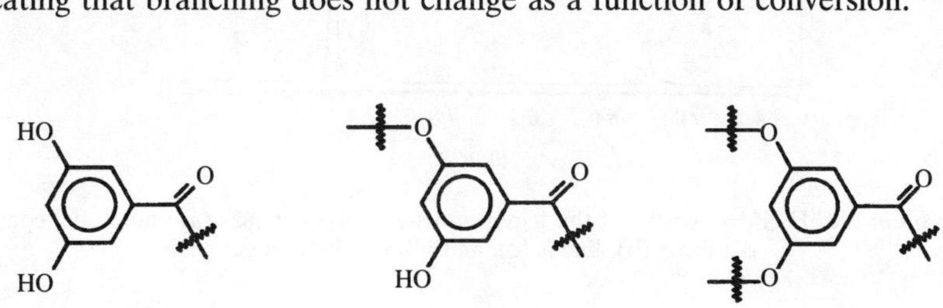

(85) Terminal unit　**(86)** Linear unit　**(87)** Dendritic unit

Figure 7 Three different subunits (**85**), (**86**), and (**87**) of hyperbranched polyester (**62**).

Other examples of the use of this NMR method for the determination of the degree of branching have been reported. For the hyperbranched polyphenylenes (**61**) prepared by Pd^0 catalysis, analysis of the ^{13}C spectrum revealed a DB of 0.7,[111] while the hyperbranched polyamines,[156] prepared by ring-opening polymerization of (**79**), were shown by ^{1}H NMR spectroscopy to have a DB of less than 0.5. A combination of ^{1}H, ^{13}C, and ^{19}F NMR

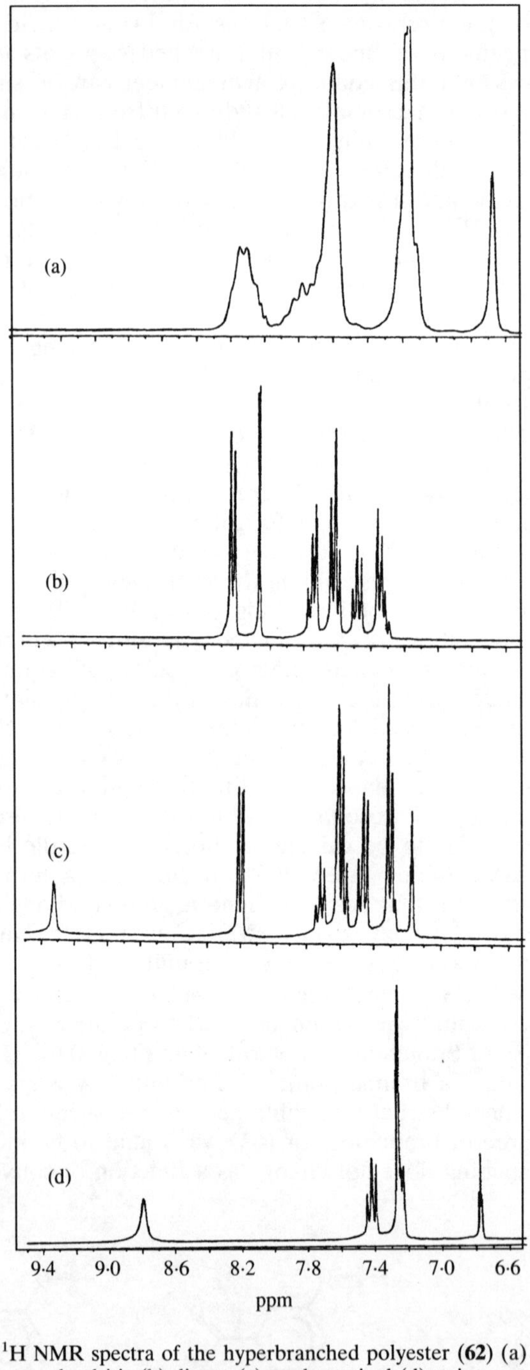

Figure 8 Comparison of the ^{1}H NMR spectra of the hyperbranched polyester (**62**) (a) and model compounds for the dendritic (b), linear (c), and terminal (d) units.

spectroscopy has also been used in a number of other cases and values for DB ranging from 0.15 to 0.85 have been determined.[155,162]

Because the degree of branching of hyperbranched polymers has a great effect on both the physical properties and the structure of these macromolecules, it is important to develop techniques that allow its determination. Unfortunately, there are a significant number of hyperbranched systems for which the NMR method of determination of DB fails. This is a result of the broadness and lack of resolution of the relevant peaks, or the absence of any significant differences between the resonances of the subunits. An example of this problem is the characterization of the hyperbranched polyester (**88**) that is obtained by polymerization of

4,4-bis(4′-hydroxyphenyl)pentanoate esters (**89**) (Scheme 35). As can be seen in Figure 9, the ^{1}H NMR spectrum of (**88**) shows no fine structure and comparisons with model compounds reveal essentially no differences between the aromatic protons of each of the subunits. This observation is not unreasonable given the fact that the phenolic groups are on different aromatic rings that are separated by a quaternary carbon atom.

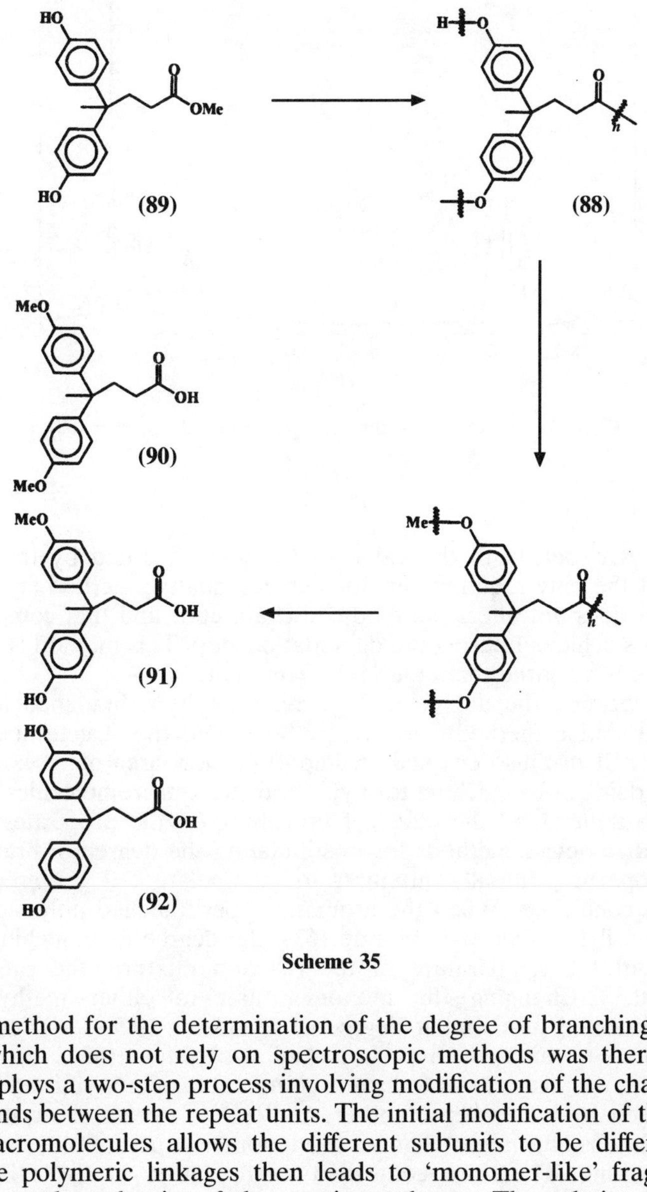

Scheme 35

An alternative method for the determination of the degree of branching for hyperbranched macromolecules which does not rely on spectroscopic methods was therefore introduced.[160] This technique employs a two-step process involving modification of the chain ends followed by cleavage of the bonds between the repeat units. The initial modification of the chain ends of the hyperbranched macromolecules allows the different subunits to be differentiated. Chemical degradation of the polymeric linkages then leads to 'monomer-like' fragments that can be directly correlated to the subunits of the starting polymer. The relative percentage of these 'monomer-like' fragments can be determined by either isolation or through a variety of chromatographic techniques, and then used to calculate DB. For example, the hyperbranched polyester (**88**), for which the degree of branching could not be determined by the conventional NMR method, is initially alkylated with silver oxide and methyl iodide to give the methyl ether terminated polyester. Each subunit is now substituted by a unique number of methyl ether groups and complete alkaline hydrolysis then leads to only three products, the dimethyl ether (**90**), the monomethyl ether (**91**), and the diphenol (**92**), and the relative percentage of each can be determined by chromatographic analysis (HPLC or GC) (Scheme 35). Since each of these fragments originates from the linear, dendritic, and terminal subunits found in the initial hyperbranched polyester (**88**), the degree of branching can be calculated for (**88**) (DB = 0.49). To confirm the validity of this method, it was applied to the aromatic hyperbranched polyester (**62**) discussed above. A degree of branching of 0.57 was determined by the degradative

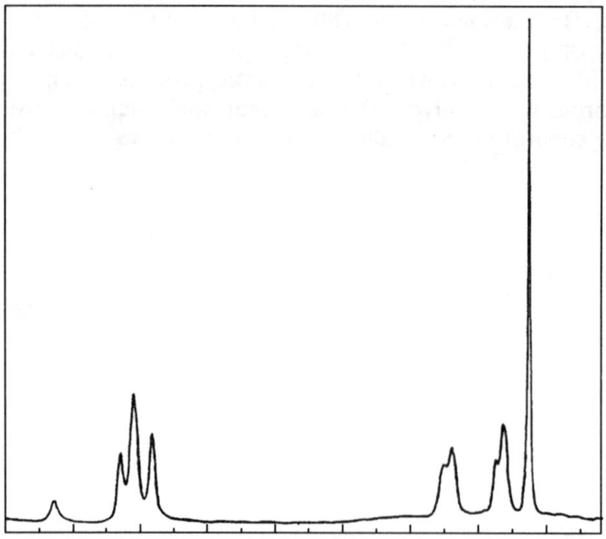

Figure 9 ^{1}H NMR spectrum of hyperbranched polyester (**88**).

method, which agrees closely with the value of 0.55–0.6 obtained by the NMR method. It should be noted that the only requirements for this degradative method are that the chemistry used for degradation does not affect the modified chain ends and that complete conversion to elementary subunits is achieved during the degradation step. This method is complementary to, and in some cases more versatile than, the NMR technique.

This ability to determine the degree of branching for hyperbranched macromolecules by either NMR or degradative methods has greatly facilitated the structural characterization of these novel materials. It has also enabled an important new area of research concerned with manipulation of the degree of branching for hyperbranched macromolecules to be probed. This is important both to understand the effect of branching on the properties of hyperbranched macromolecules, and to devise methods for manipulating the degree of branching in an effort to control these properties. Initially, attempts to manipulate DB centered on the effect of varying the reaction conditions. When the aromatic hyperbranched polyester (**62**) is prepared from 3,5-bis(trimethylsilyloxy)benzoyl chloride (**63**), the degree of branching was found to be 0.55–0.6 regardless of the temperature of the reaction mixture, the catalyst used, or the presence of solvent.[137] Changing the monomer unit to either methyl or phenyl 3,5-dihydroxybenzoate resulted in a similar degree of branching (0.5–0.55) for (**62**),[161] while the corresponding acetoxy-terminated hyperbranched polyester (**65**), prepared from 3,5-diacetoxybenzoic acid, was also shown to have a comparable degree of branching (0.5).[16] These findings are supported by a number of other studies[140,155] which show that changes in the polymerization conditions and minor changes in monomer structure have little or no effect on the degree of branching. Therefore controlled manipulation of the DB by changing the polymerization catalyst and/or polymerization conditions is, in most cases, unlikely to be a successful procedure.

Two alternate methods for successfully manipulating the degree of branching for hyperbranched macromolecules have been reported. The deliberate insertion of either a dendritic or linear fragment into a monomer gives an AB_4 or an AB_3 monomer, respectively. The presence of these pre-existing fragments would be expected to enhance growth and lead to hyperbranched macromolecules of different degrees of branching. Hawker[155] has examined this question in detail for hyperbranched poly(ether ketone)s based on a 3,4′,5-trioxy-substituted building block. For the standard AB_2 monomer, 3,5-difluoro-4′-hydroxybenzophenone (**74**), polymerization gives the hyperbranched poly(ether ketone) (**76**), which was shown by ^{19}F NMR to have a degree of branching of 0.5. Modification of (**74**) by insertion of a dendritic fragment leads to the AB_4 monomer (**93**), in which the presence of a pre-existing dendritic fragment in the monomer unit effectively enhances branching during polymerization (Scheme 36).

Therefore a hyperbranched poly(ether ketone) (**94**) with a degree of branching of 0.72 is obtained, a dramatic increase in the DB when compared to the polymerization of the corresponding AB_2 monomer. Conversely, introduction of a linear fragment gives the corresponding AB_3 monomer (**95**). In this case a bias against branching is present in the monomer structure and polymerization gives a hyperbranched poly(ether ketone) (**96**) that has a significantly lower degree of branching (0.39) (Scheme 37). It is worthwhile noting that the hyperbranched poly(ether ketone)s (**76**), (**94**), and (**96**) represent a special type of macromolecular isomer since they all have the same internal building blocks and chain ends but differ only in their degree of branching. These results suggest that changing the monomer structure, through the insertion of dendritic or linear units, can be used to control the degree of branching in a systematic manner.

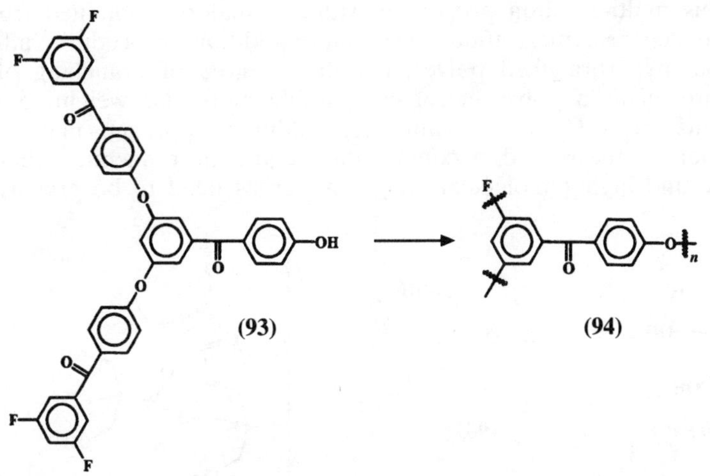

Scheme 36

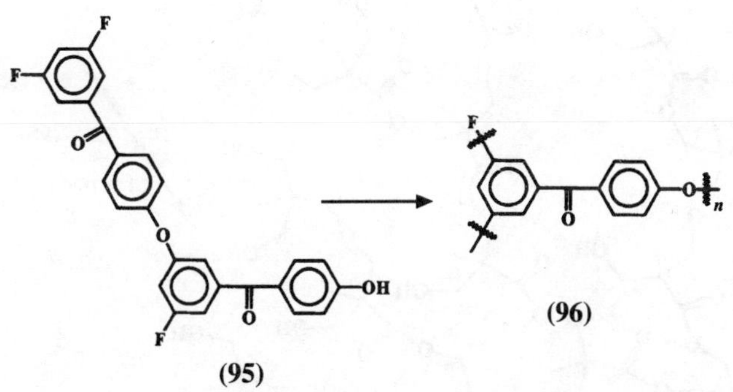

Scheme 37

An effective alternative to the above method is that of controlled monomer addition, first proposed by Malmstrom and Hult,[162] which essentially simulates the divergent growth approach used for dendritic macromolecules. In this procedure, a polyfunctional core, such as the trifunctional trihydroxymethylpropane (**97**), is reacted with 3 equiv of the AB_2 monomer 2,2-dihydroxymethylpropanoic acid (**98**), until complete disappearance of the acid functionality is observed. Assuming that the monomer unit preferentially reacts with the core molecule, and not with itself, dendritic growth would occur and the major product would be the next generation dendrimer (**99**) having six hydroxy groups at the chain ends. To ensure complete reaction of the chain ends of (**99**), 6 equiv of monomer are therefore added to the reaction mixture and the reaction is continued until completion. Repetition of the addition/polymerization process with incremental increases in the number of equivalents of monomer is then continued until a hyperbranched macromolecule of the desired molecular weight is obtained (Scheme 38). This clever procedure simulates the stepwise growth of dendrimers in a one-pot reaction mixture, and while the assumption that the monomer will react preferentially with the polyol and not with itself is unlikely to be correct, the results from this procedure warrant further investigations. Analysis of the hyperbranched oligoester shows that after the first addition step the degree of branching is very high, reaching 0.95, and decreases with each successive addition until a value of DB = 0.8 is obtained for the molecule (**100**) after four addition steps. As expected, the molecular weight of the polyester increases with each addition, though the ultimate molecular weight that has been achieved remains modest. This decrease in the degree of branching with increasing molecular weight, or number of addition steps, can be rationalized by the increasing number and cumulative effects of failure sequences and unwanted monomer–monomer couplings at each successive addition. Aside from its obvious usefulness, the most significant aspect of this important study is the comparison of the hyperbranched macromolecule obtained from this multi-addition procedure with the material obtained from the standard uncontrolled one-step polymerization. The multi-addition procedure affords a pseudo 'fourth generation' hyperbranched polyester with a degree of branching of 0.8, while the one-step procedure yields a polyester, of comparable molecular weight, but with a DB of 0.65. These results suggest that the multistep addition approach may well be used to simulate dendrimer synthesis and maximize the degree of branching, though more work needs to be done and higher molecular weight materials need to be prepared to prove this concept.

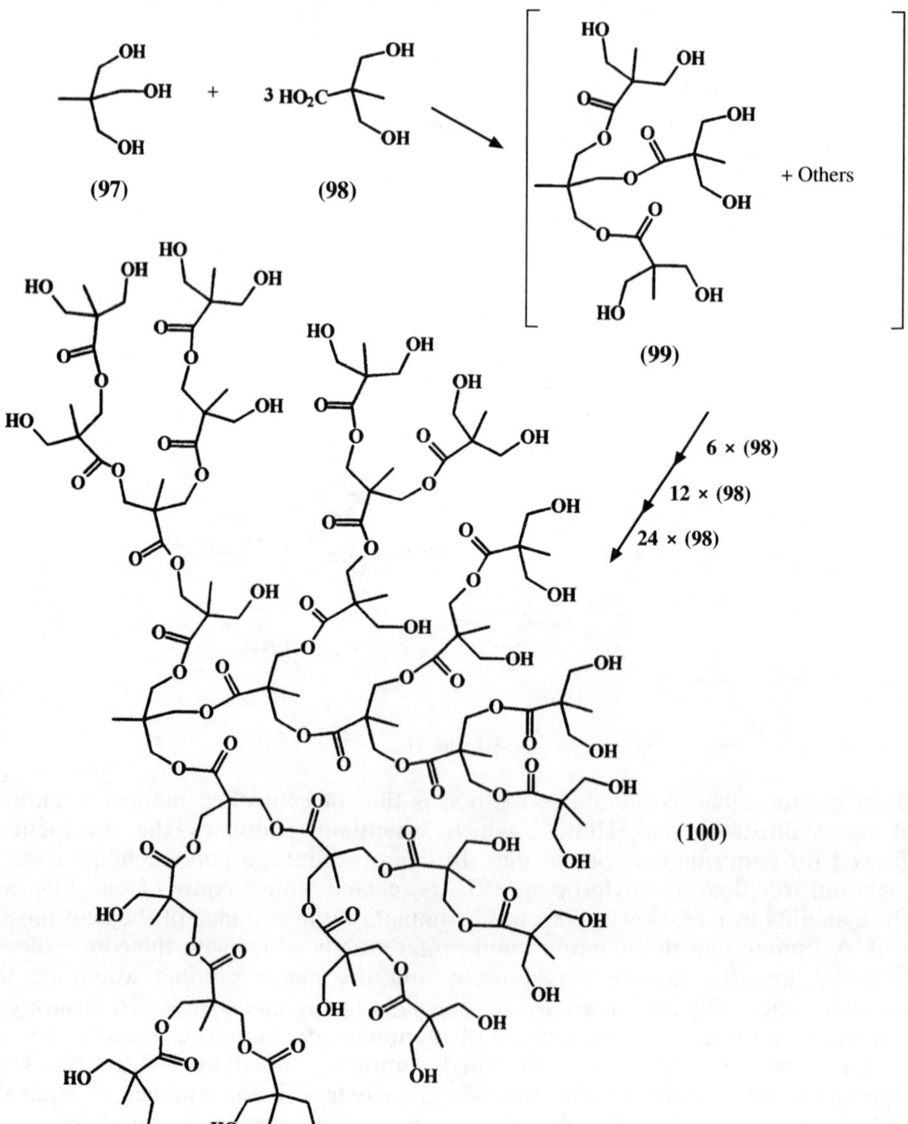

Scheme 38

One structural feature of hyperbranched macromolecules that has received little attention in the literature is the unique functionality which may be present at the focal point of these polymers. While the focal point group is a consequence of the convergent growth approach to dendritic macromolecules, its presence in hyperbranched macromolecules is questionable. This is due to the intramolecular cyclization reactions that may occur in hyperbranched systems but are not possible in the convergent synthesis of dendritic macromolecules. For hyperbranched macromolecules these cyclization reactions are facilitated by the presence of a large number of reactive B functionalities that can undergo an intramolecular coupling with the single A functionality of the same molecule. If this intramolecular cyclization does occur, the unique A functionality is lost. This premature termination may explain why some syntheses reach an upper molecular weight limit and cannot be pushed further.[163] Preliminary investigations into the fate of the focal point group in hyperbranched macromolecules has involved the preparation of labeled AB_2 monomers, such as methyl 4,4-bis(4′-hydroxyphenyl)pentanoate (**101**), in which the latent focal point group is 100% carbon-13. Polymerization leads to the hyperbranched polyester (**102**) and if ring formation is a dominant process the ^{13}C NMR resonance for the labeled methyl ester at the focal point is expected to disappear. Conversely, if ring formation does not occur, the relative size of the methyl ester resonance when compared to resonances for the repeat units should decrease with increasing conversion or molecular weight (Scheme 39). As shown in Figure 10, the ^{13}C NMR resonance for the methyl ester is observed for a range of molecular weights and a linear correlation between the decrease in the relative integral value of the labeled methyl ester resonance and the increase in molecular weight (M_n) of the hyperbranched polyester is observed. Therefore intramolecular cyclization leading to loss of the focal point group is either not occurring, or is a very minor process for this system. This is a surprising result since the monomer unit (**101**) has a relatively high degree of flexibility which would tend to favor ring formation.

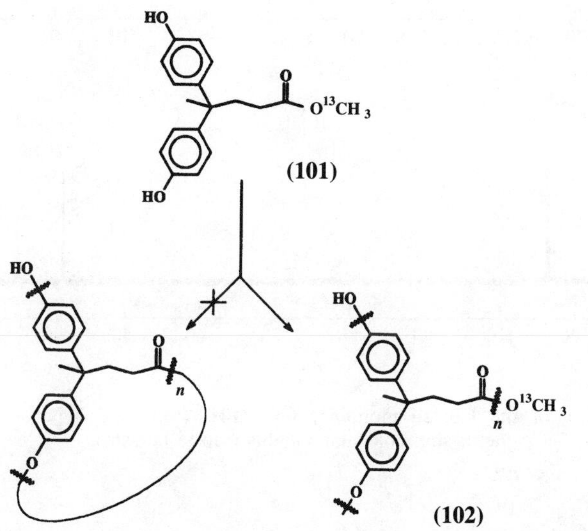

(**101**)

(**102**)

Scheme 39

Percec[164] has also examined this question in detail in an exquisite study that employed an even more flexible monomer unit (**103**) containing a long (C_{13}) alkyl chain. Polymerization of (**103**) under dilute phase-transfer conditions gave the hyperbranched polyether (**104**) (M_n = 11 800) in 56% yield, together with a small amount (5%) of the cyclized monomer (**105**) which contains an 18-membered ring (Scheme 40). ^{1}H NMR analysis of the purified hyperbranched polyether showed separate resonances for each of the three possible focal point groups, cyclic —CH_2O—Ar, —CH_2OH, and —CH_2Br in the molar ratio 0.56:0.15:0.29 (Figure 11). While this result clearly demonstrates that ring formation does occur, it does not shed any light on whether the rings arise from reaction of the cyclized monomer with (**103**) or are formed later in the polymerization process. The analysis also makes no distinction between the size of the rings formed by intramolecular cyclization of the hyperbranched polymer. From these results it can be concluded that ring formation is generally not a favored process for hyperbranched macromolecules and is only observed under extremely 'favorable' conditions such as found with a flexible monomer unit and with high dilution. Even under such conditions, cyclization only affects approximately 50% of the focal point groups.

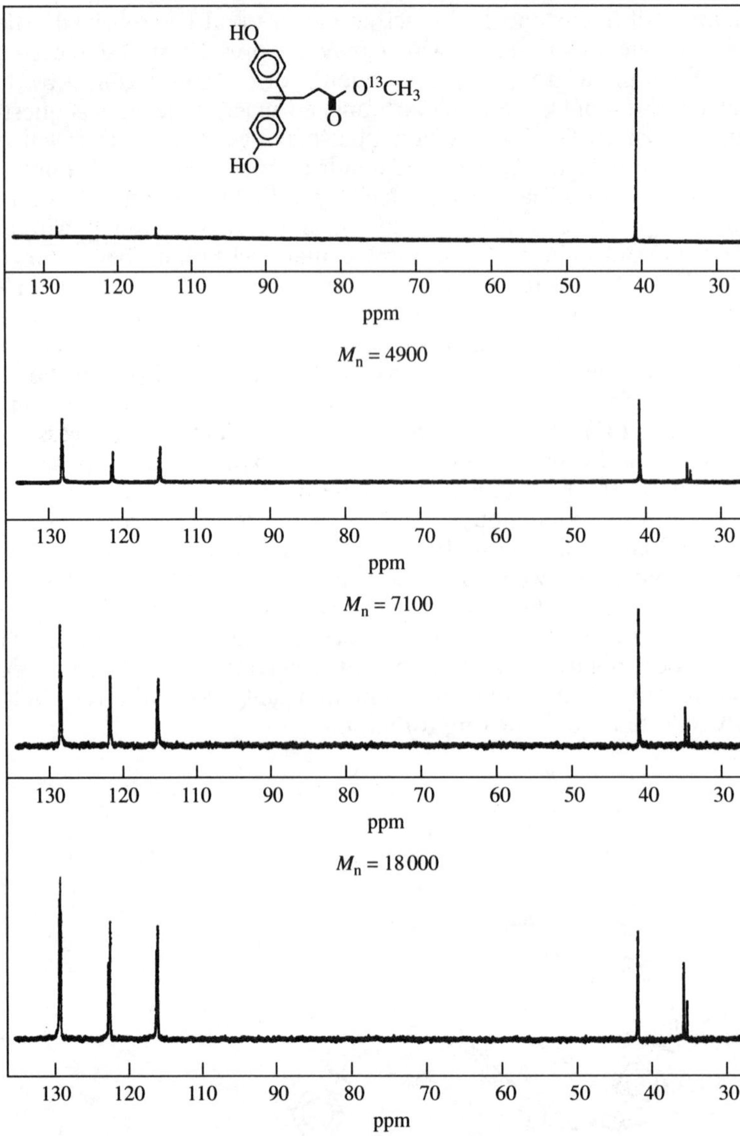

Figure 10 ^{13}C NMR spectra of the labeled monomer unit (**101**) (top) and hyperbranched polyesters (**102**) of increasing molecular weights (top to bottom).

3.3.3 Properties of Hyperbranched Macromolecules

As a result of their very recent development and intense industrial interest, very few thorough studies of the physical and other properties of hyperbranched polymers have been published to date. In general it can be said that many hyperbranched polymers possess the following characteristics.

- An irregular globular shape that is affected both by the flexibility of their components and their degree of branching.
- Low viscosity both in solution[16,139,140] and in the molten state.[135]
- High chemical reactivity;[125,137] this is particularly important with regards to the chemical modification of their chain ends, although this reactivity does not match that of regular dendrimers. This property may be of particular importance in processes such as reactive injection molding.
- Vastly enhanced solubility when compared to analogous linear polymers.[125]
- Enhanced compatibility with other polymers.[135]
- Mechanical properties that reflect their compact highly branched structures.

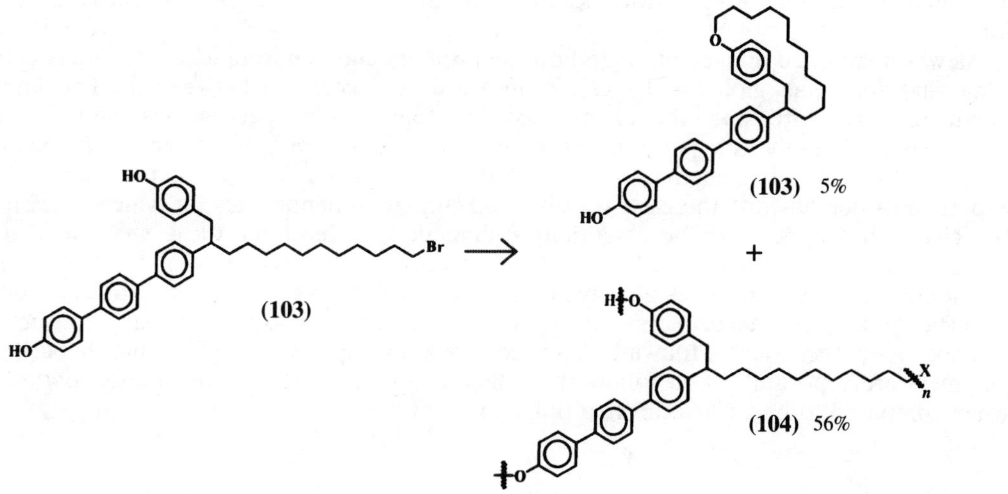

Scheme 40

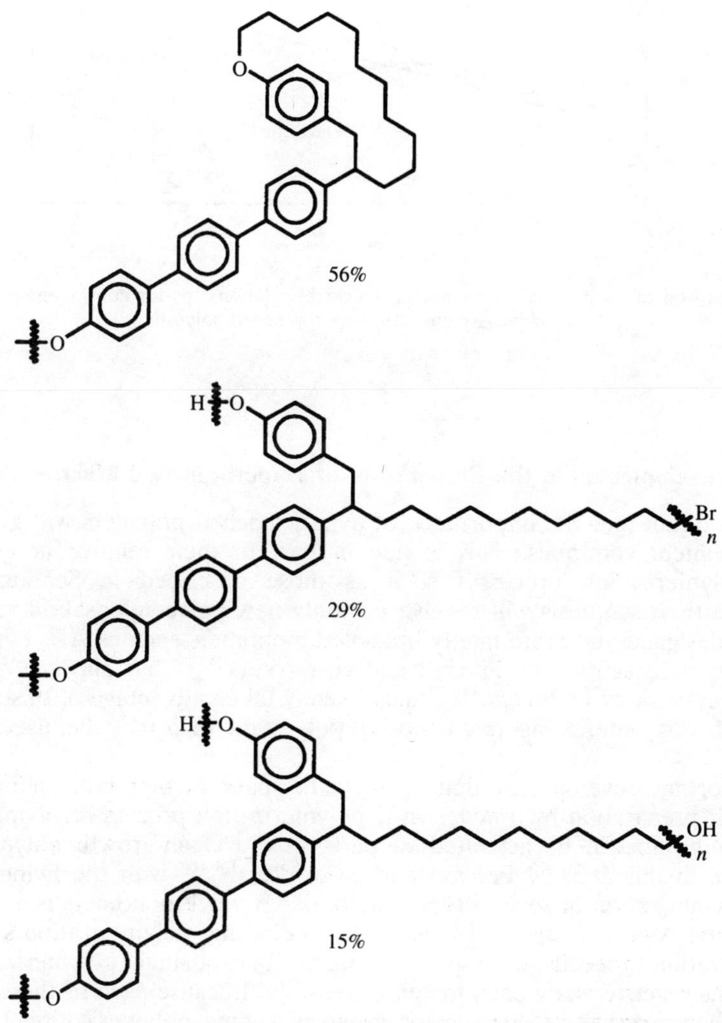

Figure 11 Nature of the focal end-groups for hyperbranched polymer (**104**).

This may lead to relative brittleness in some instances and high resistance to compression.

As was mentioned above, although both dendrimers and hyperbranched polymers can adopt somewhat analogous globular shapes, significant differences exist between the two families of polymers. Therefore one should not assume that the properties observed for regular dendrimers will apply to hyperbranched polymers. This is clearly demonstrated in experiments designed to probe the reactivity of the chain ends of both families of polymers.[1,125] These experiments demonstrate the clear reactivity advantage of dendrimers for which accessibility to the chain ends appears to be essentially unimpeded, at least for those structures in which 'dense-packing' is not a factor. Figure 12, illustrating the molecular weight–intrinsic viscosity relationship for different types of polymer architectures, points to another significant difference in behavior between dendrimers and hyperbranched polymers. Only regular dendrimers[116,139] do not obey the Mark–Houwink–Sakurada relationship $[\eta] = Km^a$, while hyperbranched polymers are reported[16,140] to follow the relationship, albeit with anomalously low 'a' values when compared to linear or branched polymers.

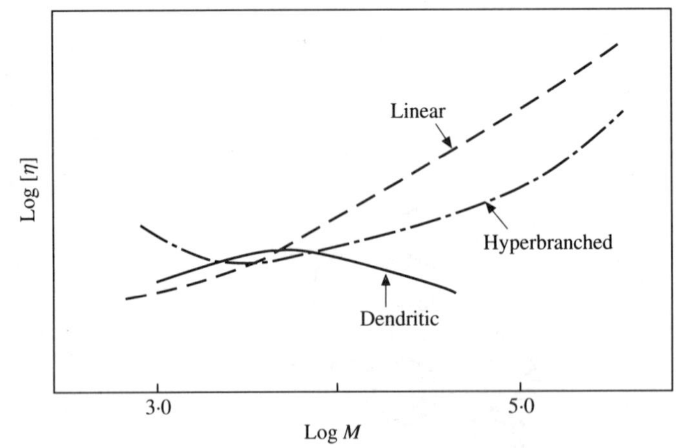

Figure 12 A comparison of intrinsic viscosity–molecular weight relationships for convergent polyether dendrimers, linear polystyrene, and hyperbranched polyester.

3.3.4 Future Developments in the Preparation of Hyperbranched Macromolecules

There is little doubt that the importance of hyperbranched polymers will grow considerably as their development continues. This is due in part to their relative accessibility through step-growth polymerization processes such as those described in Section 3.3.1. Further developments in their synthesis will involve not only new polycondensation reactions but also the use of well designed and more highly branched monomers such as AB_4 hypermonomers or higher AB_x systems described by Fréchet and co-workers[92,165] and Hawker.[155] As mentioned earlier, the recent work by Hult *et al.*[152,162] has already taken advantage of this approach for the development of very interesting low-viscosity polyesters that may be used in solvent-free coatings.

Another important development that is likely to spark further interest in hyperbranched polymers is their preparation by a novel vinyl polymerization process involving AB monomers in which the B group can be activated to perform the chain-growth polymerization of A. Pioneering work in this area by Fréchet and co-workers[166,167] with the living-like cationic or free-radical polymerization of several functionalized AB vinyl monomers is currently the focus of intense activity. A characteristic of these 'self-condensing' polymerization systems[167] is that, as the polymerization proceeds, each growing chain retains a single polymerizable group while the number of its reactive chain ends increases steadily. Because each of these reactive ends is capable of further propagation, the chains grow in a vinyl polymerization that involves the condensation of oligomeric species, affording highly branched structures, while network formation can be avoided.[167] A desirable feature of this novel type of polymerization is its versatility that allows a variety of copolymerization schemes leading to novel architectures, and the fact that high branching and high functionality are achieved in a living radical or ionic vinyl polymerization.

3.4 HYBRID DENDRITIC–LINEAR BLOCK COPOLYMERS

As was the case for the wholly dendritic block copolymers seen earlier in Section 3.2.2.7, the three-dimensional nature of dendrimers, coupled with the ability to perform site-selective functionalization, may be exploited for the design of block copolymers consisting of linear and dendritic components. For example, a linear polymer might conceivably be attached to some or all of the terminal groups of a dendrimer to give a multi-arm star with a globular core.[168] Alternately, attachment of a single linear polymer chain to the focal point of a dendrimer may lead to AB or ABA block copolymer arrangements in which the dendrimer and the linear polymer are the A and B blocks, respectively. Numerous other variations are possible, including some in which the linear polymer may have dendritic pendant groups, a combination that affords rather rigid and almost tubular structures,[95,169,170] or others in which macromolecular linear and dendritic components alternate in a 'string of pearls' arrangement.[171]

By far the largest number of hybrid dendritic–linear block copolymers prepared to date have been based on reactive dendrons obtained by the convergent growth approach. As was amply demonstrated above, the presence of a unique focal point reactive group in dendrons obtained by the convergent synthesis is a major advantage for accurately controlling the final macromolecular architecture. Therefore, most linear–dendritic block copolymers have been prepared by attachment of linear polymers to the focal point of dendritic moieties, although attachment of a linear polymer to a differentiated chain end of a dendrimer has also been reported.[171] A 1994 report[103] described a divergent synthesis in which a monofunctional linear polymer is used as the starting 'core' moiety from which the dendrimer is grown. This approach is interesting for specialized structures, but it lacks the versatility that is offered by the simple grafting processes which can be used with convergent dendrons.

3.4.1 Grafting of Linear Polymers onto Reactive Dendrons

The preparation of hybrid macromolecules that combine linear and globular blocks, disparate not only in terms of their structures but also in terms of their properties, is readily accomplished by simple coupling reactions. For example, Williamson ether syntheses[98,99,171] have been used to couple dendrons such as (**24**) having a single benzylic bromide group at their focal points with the alkoxide derivatives of poly(ethylene oxide) (PEO) or poly(ethylene glycol) (PEG) (Scheme 41). These reactions lead to AB diblock and AB triblock structures, respectively, in which the A block is dendritic and the B block is linear. Both types of hybrid macromolecules are amphiphilic and possess surfactant-like properties.

Two unusual features of these Williamson coupling reactions are their fast kinetics and their ability to reach completion with short reaction times, even in the absence of excess reagents.[94,99] These results, also indicating higher reaction rates for larger PEO or PEG blocks, are especially surprising since both reagents in the Williamson coupling are macromolecular. This is believed to be the result of a unique cooperative synergism between the two reactive fragments that provide a near ideal microenvironment for the mutual linking of their chain ends. Analogous diblock and triblock structures have also been prepared by a transesterification reaction involving methyl ester functionalities located at the focal point of reactive dendrons with the free alcohol end groups of either PEO or PEG.[99] This reaction, which is best done in the melt, provides the desired block copolymers in yields (50–80%) that are lower than those for the Williamson synthesis. It should be noted here that both methods afford essentially the same block copolymers since only the linkage between the blocks differ: an ester linkage resulting from the transesterification process instead of the ether linkage obtained from the Williamson synthesis.

Even more unusual macromolecules containing several linear and dendritic blocks in regular alternance (Scheme 42) can be obtained by polycondensation processes involving telechelic linear polymers and dendrons with two differentiated reactive chain ends.[171]

The combinations of a linear block that is flexible, hydrophilic, and crystalline with one or

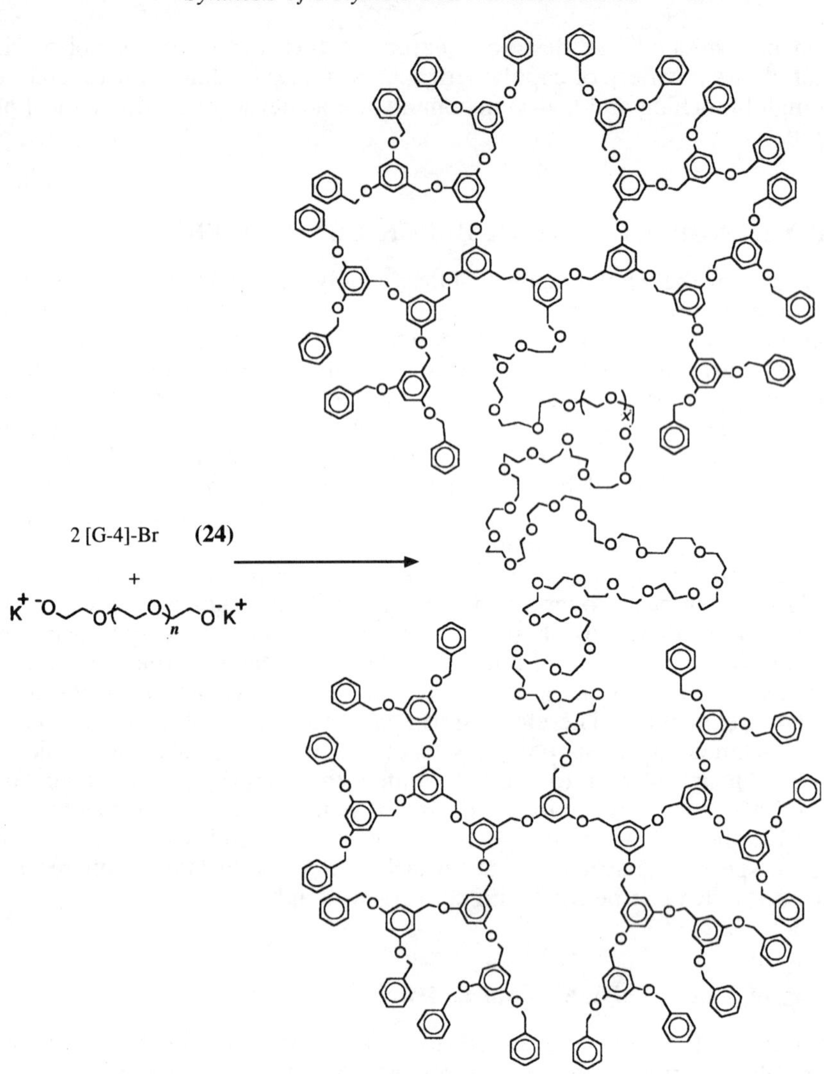

Scheme 41

more globular segments that are semi-rigid, hydrophobic, and amorphous leads to materials that have unusual solution as well as solid-state properties. For example, since chloroform is a good solvent for both types of blocks, the hybrid copolymers exist in their fully extended state in this solvent, allowing the observation of both the dendritic and linear blocks by NMR spectroscopy.[98] In contrast, NMR spectra taken in either THF, a good solvent for the dendritic block only, or in methanol, a good solvent for the linear block, show badly resolved spectra as one block or the other is in a collapsed state in each respective solvent. Further detailed studies by size exclusion chromatography (SEC) carried out in THF or in methanol demonstrate that each of the two dissimilar blocks has a great influence on the overall solution behavior of the hybrid molecules.[94,100] For example, with ABA triblock structures, a supramolecular arrangement of up to 12 hybrid molecules occurs in methanol–water, affording well-defined micelles that have the hydrophobic dendrons as their core and the hydrophilic PEG chains as their outer shells or corona.[100] These molecules, which have hydrophobic blocks connected to hydrophilic segments, behave both as surfactants and carriers for small molecules. As a result, they are the prototype of a new type of polymeric material that is likely to be extremely useful for biomedical applications.

Because the ABA hybrid copolymers contain a PEG central block, crystallization is possible when this block is of sufficient length. However, the melting points of the copolymers decrease as the percentage of their amorphous dendritic component increases. In the solid state, phase

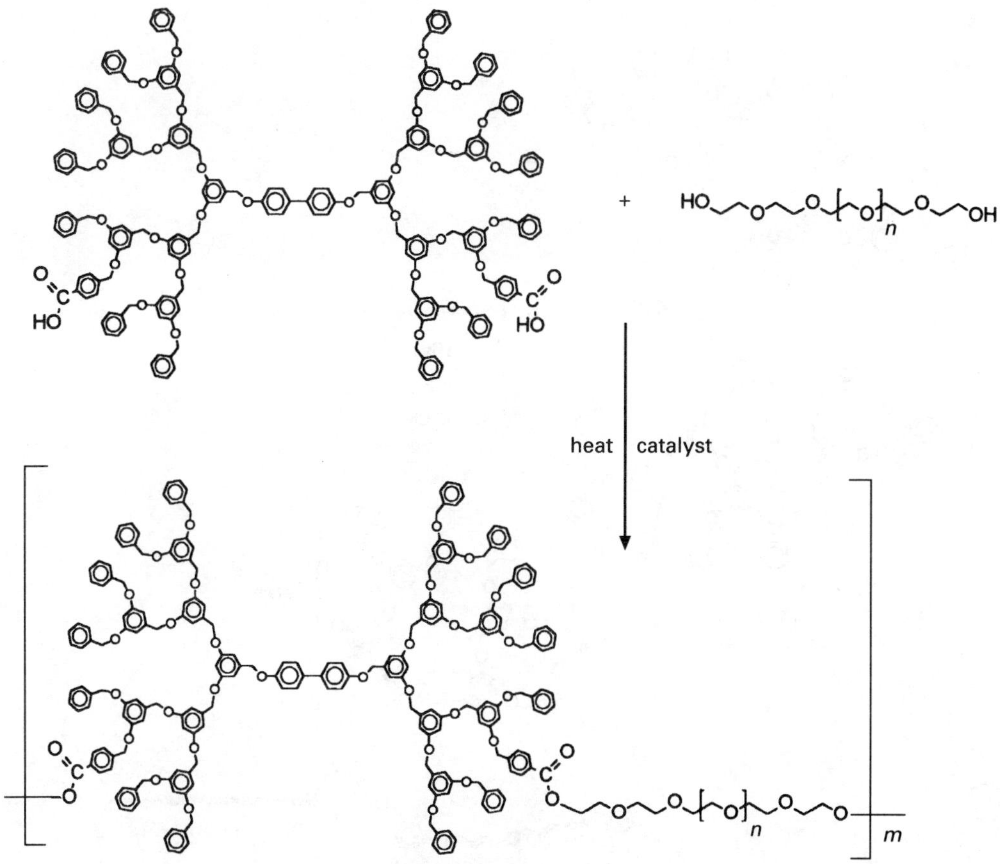

Scheme 42

separation leading to separate glass transition temperatures for each of the dissimilar blocks, as well as crystallization, has been shown to occur, but only after the linear polyether chains have reached a threshold molecular weight.[100]

Not unexpectedly, the unique functionality at the focal point of dendritic macromolecules can also be used to terminate living polymerizations in processes that also afford linear–dendritic block copolymers. For example, bifunctional living polystyrene may be terminated by reaction with dendrons having a benzylic bromide[97,172] or an aldehyde group at their focal point.[173] This gives rise to another type of 'barbell'-shaped ABA triblock structure in which the central B block is polystyrene. Similar AB diblock structures are obtained using monofunctional living polystyrene and an appropriate reactive dendron with benzylic bromide or an aldehyde group at its focal point. If a dendron having an ester group at its focal point is used, a BAB triblock structure in which the dendrimer A is the central block is obtained.[174] Alternately, Scheme 43 shows the preparation of a dendritic–linear diblock structure (**108**) by reaction of an amino-terminated polystyrene (**106**) with a convergent polyether dendron containing an aldehyde focal point group (**107**).[173]

Unfortunately, and quite unlike the Williamson syntheses of dendrimer–PEG–dendrimer triblock or dendrimer–PEO diblock structures, these reactions involving polystyrene and polyether dendrons are not easily pushed to completion. As a result, the product that is obtained is contaminated by the remaining individual blocks. In all cases, an excess of dendrimer is used to ensure that the kinetics of the process are adequate and that ABA triblock structures will indeed be obtained in the case of reactions involving bifunctional polystyrene. Purification by fractional precipitation is a laborious and challenging process, but the copolymers that are obtained again have unusual solution behavior.[97]

A very similar approach has been used by Kawaguchi and Moore[102] for the preparation of an AB block copolymer by quenching living poly(methyl methacrylate) with a phenylacetylene dendron that has an aldehyde group at its focal point.

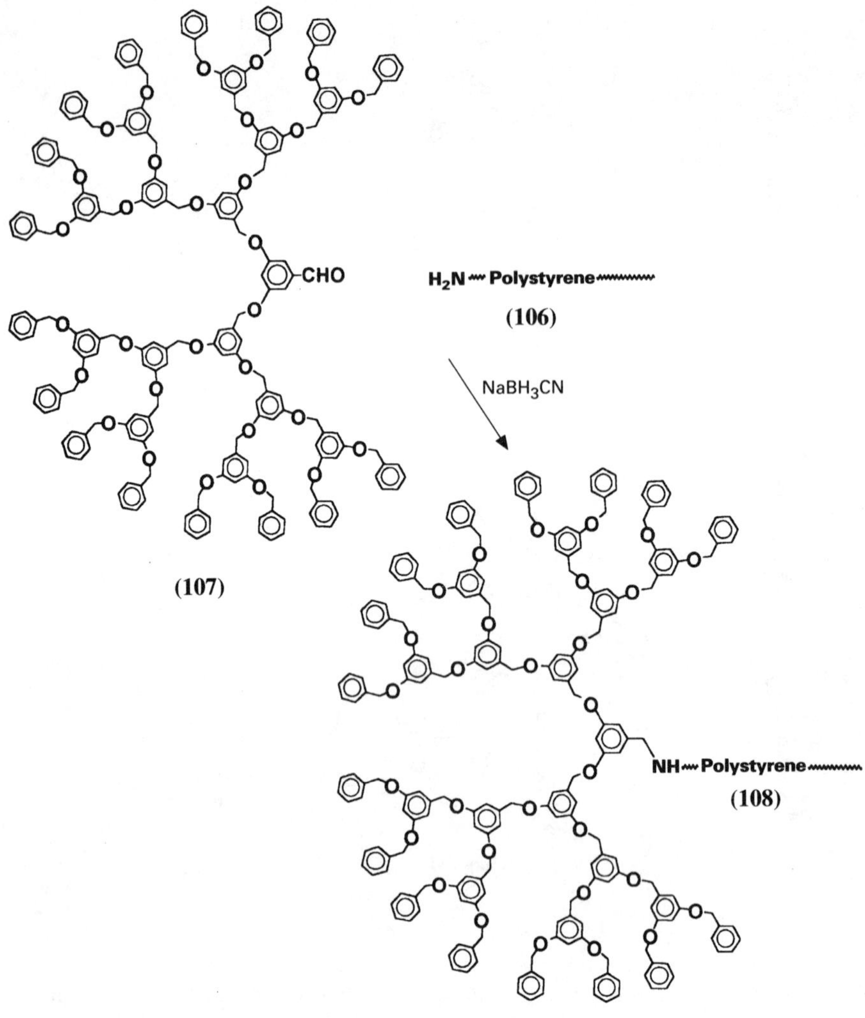

Scheme 43

3.4.2 Reactive Dendrons as Polymerization Initiators

Another interesting approach to hybrid globular–linear structures involves the use of the single reactive site located at the focal point of the dendrimer to initiate the chain-growth polymerization of a monomer. For example, the potassium alkoxide of a fourth generation dendritic alcohol [G-4]-OH (**109**) as been used to initiate the ring-opening polymerization of ε-caprolactone (Scheme 44). In contrast to similar polymerizations initiated with classical low molecular weight initiators such as potassium *t*-butoxide that only afford oligomeric products due to the occurrence of extensive back-biting, the use of the dendritic initiator allows controlled growth of the linear chain of poly(caprolactone) from its focal point to afford hybrid block copolymer (**110**). The molecular weight of the linear polyester block is readily controlled by adjusting the molar ratio of monomer to dendritic initiator, and therefore high molecular weight hybrid block copolymers (*e.g.* $M_n = 307\,000$) having narrow polydispersities (*e.g.* $M_w/M_n = 1.07$) can be obtained. In this instance, the dendrimer acts not only as the site of initiation for the polymerization of ε-caprolactone, but also as a nanoreactor that provides a unique microenvironment where the polymerization can proceed without the occurrence of deleterious back-biting. Previous solvatochromic studies[119] performed on these polyether dendrimers have shown that the polarity within the larger dendrons is quite high. As a result, the active propagating center remains localized within the favorable nanoenvironment of the dendritic globule through association of the alcoholate oxygen and its potassium counterion with the benzylic oxygens that are present in the inner cavity of the dendrimer (Figure 13). Back-biting is not a problem because the polyester that is formed does not remain within the nanoenvironment of the polyether dendrimer for which it has little thermodynamic affinity and is therefore separated from the alcoholate oxygen. Overall, the hybrid globular–linear macromolecule provides for its own assembly within a supramolecular arrangement of its components, and is another example of a unique synergistic relationship.

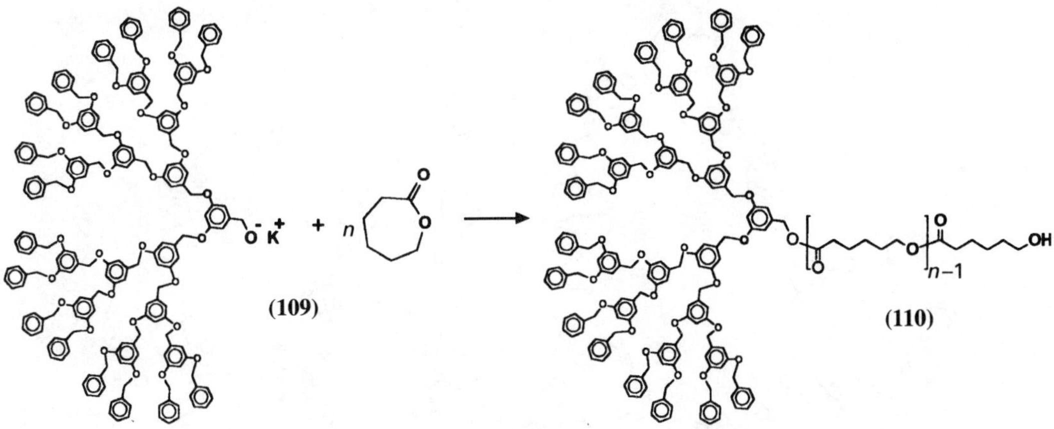

Scheme 44

These unusual findings suggest that dendrimers are likely to find an increasing role as catalysts, initiators, nanoreactors, *etc.* in applications where the interior of the dendrimer is used for the unique nanoenvironment it provides. A useful analogy is the increasingly important role played by a variety of complex cavities in supramolecular chemistry.[175–180]

3.4.3 Divergent Dendrimer Synthesis from a Linear Polymer

Chapman[103] has recently described another family of hybrid block copolymers that contains a dendritic poly(lysine) block connected to a poly(ethylene oxide) block. This hybrid copolymer is obtained by attaching a suitable reactive functionality onto preformed poly(ethylene oxide), then generating a poly(lysine) at the end of this functionalized PEO by a divergent dendritic growth procedure analogous to that used earlier by Denkewalter.[14] The new copolymers, termed 'hydraamphiphiles', also have surfactant properties and may be used to solubilize dyes.

3.4.4 Dendritic Monomers and Macromonomers in Copolymerizations

Because the focal point of convergent dendrimers can be fitted with almost any type of functional group, it is possible to transform reactive dendrons into polymerizable vinyl macromonomers[95,169] or co-monomers for condensation polymerizations.[170] Initial experiments[95,169] have detailed the preparation of dendrons containing a single polymerizable functional group at their focal point and their use as macromonomers. For example, reaction of the fourth generation alcohol [G-4]-OH (**111**) with *p*-chloromethylstyrene in the presence of base gives a polymerizable dendron (**112**) that has a single styrene unit attached to its focal point (Scheme 45). Copolymerization of (**112**) with styrene affords a hybrid block copolymer in which globular dendritic blocks are grafted to a linear polystyrene backbone. Similar vinyl-terminated dendrons with methacrylate or acrylate groups at their focal points have also been prepared[94] and used to terminate the propagation of a living polystyrene in much the same way as 1,1-diphenylethylene is used to cap living polymers.

Most recently, Schlüter[170] has prepared an interesting soluble, rigid rod, 'dendrimerized' poly(*p*-phenylene) by the Pd-catalyzed copolycondensation of a dibromobiphenyl derivative carrying small dendrons with an alkyl-substituted benzenediboronic acid. These approaches lead to novel cylindrically shaped dendritic structures.

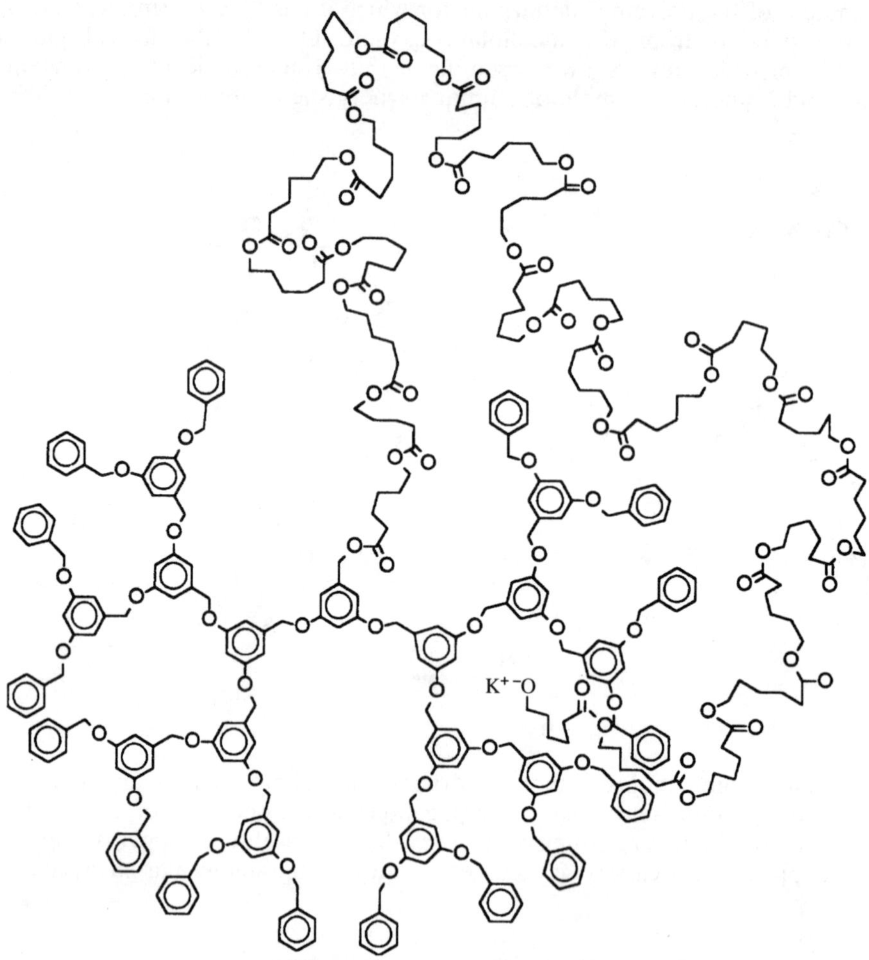

Figure 13 The nanoenvironment of the alcoholate chain-end in the polymerization of ε-caprolactone using a dendritic alcoholate initiator.

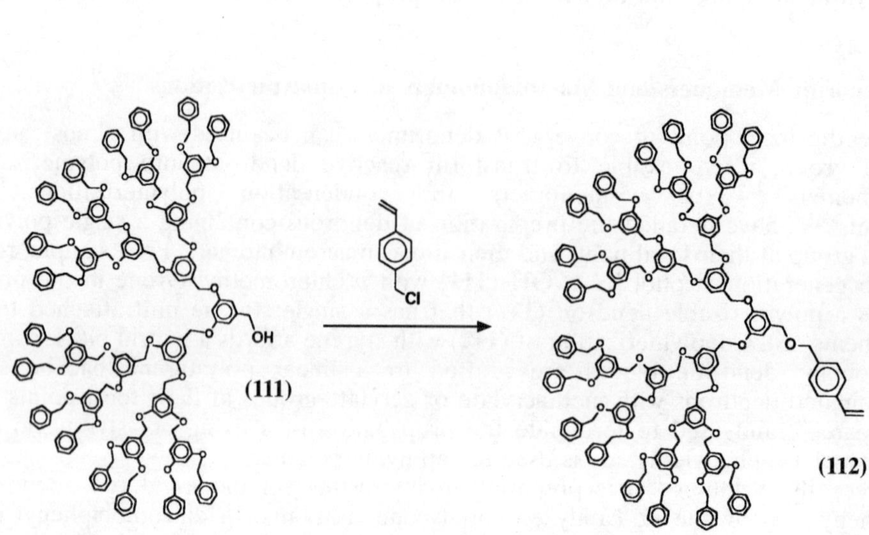

Scheme 45

3.5 CONCLUSIONS

The study of dendritic macromolecules has progressed extremely rapidly in recent years because these unusual macromolecules constitute a totally novel branch of polymer science. Some of their properties have been shown to be dominated, or at least strongly affected, by their three-dimensional shape that distinguishes them from all other polymers known today. In particular, their globular shape translates into viscoelastic and solution behaviors that appear to be free of the entanglement dominated features that are so characteristic of linear or other classical branched polymers.

In order to continue the fundamental study of this entirely novel class of polymers, it is necessary to aim for synthetic approaches that afford essentially 'perfect' molecules that have no defects or molecular weight dispersity. Much can be learned from the behavior of these molecules and by comparing them to other less perfect globular structures such as hyperbranched polymers or dendrimers prepared using less controlled accelerated syntheses.

While there is no doubt that dendrimers as well as hyperbranched polymers will find numerous applications, both improved synthetic techniques and a full understanding of these unique materials and their associated structure–property relationships are required before their impact can be fully measured. Unfortunately, the rather tedious multistep syntheses and repetitive purifications that are required for the preparation of 'perfect' dendrimers have limited their availability. As a result, many researchers have been tempted to use more accessible materials obtained by processes that do not ensure regular growth. This trend may be beneficial because it provides for faster progress of the field, yet it is also worrisome because the seemingly small differences that result from the presence of structural imperfections or impurities may in fact translate into totally different ultimate behavior. As discussed earlier, this has already been demonstrated by experiments such as those comparing the relationship between intrinsic viscosity and molecular weight for dendrimers and hyperbranched polymers (Figure 12). Only the former show a distinct bell-shaped curve, while the latter exhibit a significant deviation from the behavior that is characteristic of linear polymers but still obey the Mark–Houwink–Sakurada equation.

Although the 'perfect' dendrimers known today may be mostly valuable for their critical contribution to the fundamental understanding of synthetic globular polymers, their use in high added-value applications will continue to be explored. The purity of the materials coupled to their controlled sizes, chemical reactivity, and other unique features and properties make them ideal for use in the medical and high technology fields. Indeed, Tomalia's pioneering and now famous Starburst dendrimers are already finding applications in the medical and biotechnology fields, where the purity of the materials, their controlled size, and the biocompatible nature of their building blocks is of particular significance.[21,22,181,182] Very recent (1995) measurements of the rheological properties of convergent polyether dendrimers in the melt[183] have confirmed that they are non-entangled macromolecules that may well deserve their informal designation as 'molecular ball bearings'. While the low melt viscosity of hyperbranched polymers has also been noticed, no experiments have been reported that clarify their state of entanglement as a function of their degree of branching. It is certain that hyperbranched polymers that have a low degree of branching are subject to significant entanglement. Other applications in adhesion, catalysis, or uses as carriers, nanoreactors, membranes, surfactants, coatings, additives, *etc.* may well proceed quite satisfactorily with less perfect dendrimers or hyperbranched polymers.[184]

ACKNOWLEDGEMENTS

The authors thank the National Science Foundation (Division of Materials Research) and IBM Corporation (Materials and Processing Sciences Program) for their support of the early stages of research on convergent dendrimers at Cornell University. Additional support by The Eastman Kodak Company and The Australian Research Council is also acknowledged.

3.6 REFERENCES

1. J. M. J. Fréchet, *Science*, 1994, **263**, 1710.
2. D. A. Tomalia, H. Baker, J. Dewald, M. Hall, G. Kallos, S. Martin, J. Roeck, J. Ryder and P. Smith, *Polym. J.*, 1985, **17**, 117.
3. Y. H. Kim and O. W. Webster, *Polym. Prepr.*, 1988, **29**, 310.

4. C. J. Hawker and J. M. J. Fréchet, *J. Am. Chem. Soc.*, 1990, **112**, 7638.
5. M. Maciejewski, *J. Macromol. Sci., Chem.*, 1982, **A17**, 689.
6. P. G. De Gennes and H. Hervet, *J. Phys. Lett.*, 1983, **44**, L-351.
7. M. Gordon, *Pure Appl. Chem.*, 1973, **43**, 1.
8. G. R. Newkome, G. R. Baker, J. K. Young and J. G. Traynham, *J. Polym. Sci., Polym. Chem.*, 1993, **31**, 641.
9. E. W. Kwock, T. X. Neenan and T. M. Miller, *Chem. Mater.*, 1991, **3**, 775.
10. K. L. Wooley, C. J. Hawker and J. M. J. Fréchet, *J. Am. Chem. Soc.*, 1993, **115**, 11 496.
11. G. R. Newkome, Z. Yao, G. R. Baker and V. K. Gupta, *J. Org. Chem.*, 1985, **50**, 2003.
12. T. Nagasaki, M. Ukon, S. Arimori and S. Shinkai, *J. Chem. Soc., Chem. Commun.*, 1992, 608.
13. E. Buhleier, W. Wehner and F. Vögtle, *Synthesis*, 1978, 155.
14. R. G. Denkewalter, J. F. Kolc and W. J. Lukasavage, *US Pat.*, 1981, 4 289 872; *US Pat.*, 1982, 4 360 646; *US Pat.*, 1983, 4 410 688.
15. S. M. Aharoni, C. R. Crosby, III and E. K. Walsh, *Macromolecules*, 1982, **15**, 1093.
16. S. R. Turner, B. I. Voit and T. H. Mourey, *Macromolecules*, 1993, **26**, 4617.
17. D. A. Tomalia, M. Hall and D. M. Hedstrand, *J. Am. Chem. Soc.*, 1987, **109**, 1601.
18. See reference 1 cited in P. G. De Gennes and H. Hervet, *J. Phys. Lett.*, 1983, **44**, L-351.
19. D. A. Tomalia presented at 1st Int. Polym. Conf., Society Polymer Science Japan, August 1984, Kyoto, Japan.
20. D. A. Tomalia, A. M. Naylor and W. A. Goddard, III, *Angew. Chem., Int. Ed. Engl.*, 1990, **29**, 138.
21. D. A. Tomalia and H. D. Durst, *Top. Curr. Chem.*, 1993, **165**, 193.
22. D. A. Tomalia, *Aldrichim. Acta*, 1993, **26**, 91.
23. G. R. Newkome, C. N. Moorefield and G. R. Baker, *Aldrichim. Acta*, 1992, **25**, 31.
24. P. B. Smith, S. J. Martin, M. J. Hall and D. A. Tomalia, in 'Applied Polymer Analysis and Characterization', ed. J. Mitchell, Jr, Chapman & Hall, London 1987, Chapter III-D, p. 357.
25. D. A. Tomalia, personnal communication.
26. D. A. Tomalia, *Adv. Mater.*, 1994, **6**, 529.
27. J. M. J. Fréchet, Y. Jiang, C. J. Hawker and A. E. Philippides, in 'Proc. IUPAC Int. Symp. "Molecular Design of Functional Polymers", Seoul', 1989, p. 19; J. M. J. Fréchet, C. J. Hawker and A. E. Philippides, *US Pat.*, 1991, 5 041 516 (filed June 21, 1989).
28. C. J. Hawker and J. M. J. Fréchet, *J. Chem. Soc., Chem. Commun.*, 1990, 1010.
29. C. J. Hawker, K. L. Wooley and J. M. J. Fréchet, *J. Chem. Soc., Chem. Commun.*, 1994, 925.
30. K. L. Wooley, C. J. Hawker and J. M. J. Fréchet, *J. Am. Chem. Soc.*, 1991, **113**, 4252.
31. T. M. Miller and T. X. Neenan, *Chem. Mater.*, 1990, **2**, 346.
32. T. M. Miller, T. X. Neenan, R. Zayas and H. E. Bair, *J. Am. Chem. Soc.*, 1992, **114**, 1018.
33. T. M. Miller, E. W. Kwock and T. X. Neenan, *Macromolecules*, 1992, **25**, 3143.
34. J. S. Moore and Z. Xu, *Macromolecules*, 1991, **24**, 5893.
35. Z. Xu and J. S. Moore, *Angew. Chem., Int. Ed. Engl.*, 1993, **32**, 1354.
36. H. Uchida, Y. Kabe, K. Yoshino, A. Kawamata, T. Tsumuraya and S. Masamune, *J. Am. Chem. Soc.*, 1990, **112**, 7077.
37. A. Morikawa, M. Kakimoto and Y. Imai, *Macromolecules*, 1991, **24**, 3469.
38. A. Morikawa, M. Kakimoto and Y. Imai, *Polym. J.*, 1992, **24**, 573.
39. A. Morikawa, M. Kakimoto and Y. Imai, *Macromolecules*, 1992, **25**, 3247.
40. H. Nemoto, J. G. Wilson, H. Nakamura and Y. Yamamoto, *J. Org. Chem.*, 1992, **57**, 435.
41. A. B. Padias, H. K. Hall, D. A. Tomalia and J. R. McConnell, *J. Org. Chem.*, 1987, **52**, 5305.
42. K. E. Uhrich, S. Boegeman, J. M. J. Fréchet and S. R. Turner, *Polym. Bull.*, 1991, **25**, 551.
43. R. Roy, D. Zanini, S. J. Meunier and A. Romanowska, *J. Chem. Soc., Chem. Commun.*, 1993, 1869; R. Roy, D. Zanini, S. J. Meunier and A. Romanowska, *ACS Symp. Ser.*, 1994, **560**, 104.
44. J. K. Young, J. C. Nelson and J. S. Moore, *J. Am. Chem. Soc.*, 1994, **116**, 10 841.
45. K. E. Uhrich and J. M. J. Fréchet, *J. Chem. Soc., Perkin Trans. 1*, 1992, 1623.
46. P. M. Bayliff, W. J. Feast and D. Parker, *Polym. Bull.*, 1992, **29**, 265.
47. S. C. E. Backson, P. M. Bayliff, W. J. Feast, A. M. Kenwright, D. Parker and R. W. Richards, *Makromol. Chem., Macromol. Symp.*, 1994, **77**, 1.
48. A. Morikawa, M. Kakimoto and Y. Imai, *Macromolecules*, 1993, **26**, 6324.
49. Z. Xu, M. Kahr, K. L. Walker, C. L. Wilkins and J. S. Moore, *J. Am. Chem. Soc.*, 1994, **116**, 4537.
50. R. H. E. Hudson and M. J. Damha, *J. Am. Chem. Soc.*, 1993, **115**, 2119.
51. L. J. Twyman, A. E. Beezer and J. C. Mitchell, *J. Chem. Soc., Perkin Trans. 1*, 1994, 407.
52. T. Nagasaki, M. Ukon, S. Arimori and S. Shinkai, *J. Chem. Soc., Chem. Commun.*, 1992, 608.
53. T. Nagasaki, O. Kimura, M. Ukon, S. Arimori, I. Hamachi and S. Shinkai, *J. Chem. Soc., Perkin Trans. 1*, 1994, 75.
54. A. Rajca and S. Utamapanya, *J. Am. Chem. Soc.*, 1993, **115**, 10 688.
55. J. M. Lapierre, K. Skobridis and D. Seebach, *Helv. Chim. Acta*, 1993, **76**, 2419.
56. D. Seebach, J. M. Lapierre, K. Skobridis and G. Greiveldinger, *Angew. Chem., Int. Ed. Engl.*, 1994, **33**, 440.
57. D. Seebach, J. M. Lapierre, G. Greiveldinger and K. Skobridis, *Helv. Chim. Acta*, 1994, **77**, 1673.
58. L. J. Twyman, A. E. Beezer and J. C. Mitchell, *Tetrahedron Lett.*, 1994, **35**, 4423.
59. H. F. Chow, L. F. Fok and C. C. Mak, *Tetrahedron Lett.*, 1994, **35**, 3547.
60. H. F. Chow and C. C. Mak, *J. Chem. Soc., Perkin Trans. 1*, 1994, 2223.
61. H. Brunner and S. Altmann, *Chem. Ber.*, 1994, **127**, 2285.
62. G. R. Newkome, X. Lin and J. K. Young, *Synlett*, 1992, 53.
63. G. R. Newkome, S. Arai, F. R. Fronczek, C. N. Moorefield, X. Lin and C. D. Weis, *J. Org. Chem.*, 1993, **58**, 898.
64. G. R. Newkome, X. Lin and C. D. Weis, *Tetrahedron*: *Asymmetry*, 1991, **2**, 957.
65. G. R. Newkome, C. N. Moorefield, G. R. Baker, A. L. Johnson and R. K. Behera, *Angew. Chem., Int. Ed. Engl.*, 1991, **30**, 1176.
66. H. B. Mekelburger, K. Rissanen and F. Vögtle, *Chem. Ber.*, 1993, **126**, 1161.
67. A. W. van der Made and P. W. N. M. van Leewen, *J. Chem. Soc., Chem. Commun.*, 1992, 1400.

68. A. W. van der Made, P. W. N. M. van Leewen, J. C. de Wilde and R. A. C. Brandes, *Adv. Mater.*, 1993, **5,** 466.
69. J. Roovers, P. M. Toporowski and L. L. Zhou, *Polym. Prepr.*, 1992, **33,** 182.
70. C. Wörner and R. Mülhaupt, *Angew. Chem., Int. Ed. Engl.*, 1993, **32,** 1306.
71. R. Moors and F. Vögtle, *Chem. Ber.*, 1993, **126,** 2133.
72. E. M. M. de Brabander van den Berg and E. W. Meijer, *Angew. Chem., Int. Ed. Engl.*, 1993, **32,** 1308.
73. H. K. Hall and D. W. Polis, *Polym. Bull.*, 1987, **17,** 409.
74. K. Rengan and R. Engel, *J. Chem. Soc., Perkin Trans. 1,* 1991, 987; K. Rengan and R. Engel, *J. Chem. Soc., Chem. Commun.*, 1990, 1084.
75. K. Rengan and R. Engel, *J. Chem. Soc., Chem. Commun.*, 1992, 757.
76 .K. Shahlai and H. Hart, *J. Org. Chem.*, 1991, **56,** 6905.
77. K. Kadei, R. Moors and F. Vögtle, *Chem. Ber.*, 1994, **127,** 897.
78. N. Launay, A. M. Caminade, R. Lahana and J. P. Majoral, *Angew. Chem., Int. Ed. Engl.*, 1994, **33,** 1589.
79. A. Pessi, E. Bianchi, F. Bonelli and L. Chiappinelli, *J. Chem. Soc., Chem. Commun.*, 1990, 8.
80. C. Rao and J. P. Tam, *J. Am. Chem. Soc.*, 1994, **116,** 6975.
81. B. Alonso, I. Cuadrado, M. Moran and J. Losada, *J. Chem. Soc., Chem. Commun.*, 1994, 2575.
82. D. Seyferth, D. Y. Son, A. L. Rheingold and R. L. Ostrander, *Organometallics,* 1994, **13,** 2686.
83. Y. H. Laio and J. R. Moss, *J. Chem. Soc., Chem. Commun.*, 1993, 1774.
84. J. L. Fillaut and D. Astruc, *J. Chem. Soc., Chem. Commun.*, 1993, 1320.
85. J. W. J. Knapen, A. W. van der Made, J. C. de Wilde, P. W. N. M. van Leeuwen, P. Wijkens, D. M. Grove and G. van Koten, *Nature,* 1994, **372,** 659.
86. G. R. Newkome, F. Cardullo, E. C. Constable, C. N. Moorefield and A. M. W. C. Thompson, *J. Chem. Soc., Chem. Commun.*, 1993, 925.
87. P. Lange, H. Beruda, W. Hiller and H. Schmidbaur, *Z. Naturforsch., B: Chem. Sci.*, 1994, **49,** 781.
88. G. Denti, S. Campagna, S. Serroni, M. Ciano and V. Balzani, *J. Am. Chem. Soc.*, 1992, **114,** 2944.
89. S. Campagna, G. Denti, S. Serroni, M. Ciano, A. Juris and V. Balzani, *Inorg. Chem.*, 1992, **31,** 2982.
90. V. Balzani, *New Sci.*, 1994, 12 November, 32.
91. R. Spindler and J. M. J. Fréchet, *J. Chem. Soc., Perkin Trans. 1,* 1993, 913.
92. K. L. Wooley, C. J. Hawker and J. M. J. Fréchet, *Angew. Chem., Int. Ed. Engl.*, 1994, **33,** 82.
93. T. Kawaguchi and J. S. Moore, *Polym. Prepr.*, 1994, **35,** 669.
94. J. M. J. Fréchet and I. Gitsov, *Makromol. Chem., Macromol. Symp.*, 1996, in press.
95. C. J. Hawker and J. M. J. Fréchet, *Polymer,* 1992, **33,** 1507.
96. I. Gitsov, K. L. Wooley, C. J. Hawker and J. M. J. Fréchet, *Polym. Prepr.*, 1991, **32,** 631.
97. I. Gitsov and J. M. J. Fréchet, *Macromolecules,* 1994, **27,** 7309.
98. I. Gitsov, K. L. Wooley and J. M. J. Fréchet, *Angew. Chem., Int. Ed. Engl.*, 1992, **31,** 1200.
99. I. Gitsov, K. L. Wooley, C. J. Hawker, P. T. Ivanova and J. M. J. Fréchet, *Macromolecules,* 1993, **26,** 5621.
100. I. Gitsov and J. M. J. Fréchet, *Macromolecules,* 1993, **26,** 6536.
101. I. Gitsov, P. T. Ivanova and J. M. J. Fréchet, *Makromol. Chem., Rapid Commun.*, 1994, **15,** 387.
102. T. Kawaguchi and J. S. Moore, *Polym. Prepr.*, 1994, **35,** 872.
103. T. M. Chapman, G. L. Hillyer, E. J. Mahan and K. A. Schaffer, *J. Am. Chem. Soc.*, 1994, **116,** 11 195.
104. C. J. Hawker and J. M. J. Fréchet, *Macromolecules,* 1990, **23,** 4726.
105. K. L. Wooley, C. J. Hawker and J. M. J. Fréchet, *J. Chem. Soc., Perkin Trans. 1,* 1991, 1059.
106. C. J. Hawker and J. M. J. Fréchet, *J. Am. Chem. Soc.*, 1992, **114,** 8405.
107. D. A. Tomalia, D. R. Swanson, J. W. Klimash and H. M. Brothers, *Polym. Prepr.*, 1993, **34,** 52.
108. C. J. Hawker, K. L. Wooley and J. M. J. Fréchet, *J. Chem. Soc., Perkin Trans. 1,* 1993, 1287.
109. D. A. Tomalia, V. Berry, M. Hall and D. M. Hedstrand, *Macromolecules,* 1987, **20,** 1164.
110. G. R. Newkome, C. N. Moorefield, G. R. Baker, M. J. Saunders and S. H. Grossman, *Angew. Chem., Int. Ed. Engl.*, 1991, **30,** 1178.
111. Y. H. Kim and O. W. Webster, *J. Am. Chem. Soc.*, 1990, **112,** 4592.
112. G. R. Newkome, J. K. Young, G. R. Baker, R. L. Potter, L. Audoly, D. Cooper, C. D. Weis, K. Morris and C. S. Johnson, Jr., *Macromolecules,* 1993, **26,** 2394.
113. D. A. Tomalia, H. M. Brothers, II, L. T. Piehler and Y. Hsu, *Polym. Mat. Sci. Eng.*, 1995, **73,** 75.
114. D. A. Tomalia, D. M. Hedstrand and M. S. Ferritto, *Macromolecules,* 1991, **24,** 1435.
115. M. Gauthier and M. Moller, *Macromolecules,* 1991, **24,** 4548.
116. T. H. Mourey, S. R. Turner, M. Rubinstein, J. M. J. Fréchet, C. J. Hawker and K. L. Wooley, *Macromolecules,* 1992, **25,** 2401.
117. K. L. Walker, M. S. Kahr, C. L. Wilkins, Z. F. Xu and J. S. Moore, *J. Am. Soc. Mass Spectrom.*, 1994, **5,** 731.
118. J. W. Leon and J. M. J. Fréchet, *Polym. Bull.*, 1995, in press; J. W. Leon, C. J. Hawker and J. M. J. Fréchet, unpublished data.
119. C. J. Hawker, K. L. Wooley and J. M. J. Fréchet, *J. Am. Chem. Soc.*, 1993, **115,** 4375.
120. A. M. Naylor, W. A. Goddard, III, G. E. Kiefer and D. A. Tomalia, *J. Am. Chem. Soc.*, 1989, **111,** 2339.
121. M. C. Moreno-Bondi, G. Orellana, N. J. Turro and D. A. Tomalia, *Macromolecules,* 1990, **23,** 910.
122. K. R. Godipas, A. R. Leheny, G. Caminati, N. J. Turro and D. A. Tomalia, *J. Am. Chem. Soc.*, 1991, **113,** 7335.
123. R. Jin, T. Aida and S. Inoue, *J. Chem. Soc., Chem. Commun.*, 1993, 1260.
124. Y. Chujo, H. Matsuki, S. Kure, T. Saegusa and T. Yazawa, *J. Chem. Soc., Chem. Commun.*, 1994, 635.
125. K. L. Wooley, J. M. J. Fréchet and C. J. Hawker, *Polymer,* 1994, **35,** 4489.
126. K. L. Wooley, C. J. Hawker, J. M. Pochan and J. M. J. Fréchet, *Macromolecules,* 1993, **26,** 1514.
127. P. M. Saville, J. W. White, C. J. Hawker, K. L. Wooley and J. M. J. Fréchet, *J. Phys. Chem.*, 1993, **97,** 293.
128. P. M. Saville, P. A. Reynolds, J. W. White, C. J. Hawker, J. M. J. Fréchet, K. L. Wooley, J. Penfold and J. R. P. Webster, *J. Phys. Chem.*, 1995, **99,** 8283.
129. T. Kowalewski, K. L. Wooley, C. J. Hawker and J. M. J. Fréchet, in 'Proc. 35th IUPAC Int. Symp. Macromolecules, MacroAkron '94', 1994, p. 327.
130. J. F. G. A. Jansen, E. M. M. de Brabander van den Berg and E. W. Meijer, *Science,* 1994, **266,** 1226.
131. P. J. Flory, *J. Am. Chem. Soc.*, 1952, **74,** 2718.

132. C. Friedel and J. M. Crafts, *Bull. Soc. Chim.*, 1885, **43,** 53.
133. R. A. Jacobsen, *J. Am. Chem. Soc.*, 1932, **54,** 1513.
134. W. H. Hunter and G. H. Woollett, *J. Am. Chem. Soc.*, 1921, **43,** 131.
135. Y. H. Kim, *US Pat.*, 1989, 4 857 630.
136. C. J. Hawker, R. Lee and J. M. J. Fréchet, *J. Am. Chem. Soc.*, 1991, **113,** 4583.
137. K. L. Wooley, C. J. Hawker, R. Lee and J. M. J. Fréchet, *Polym. J.*, 1994, **26,** 187.
138. H. R. Kricheldorf, Q. Zang and G. Schwarz, *Polymer,* 1982, **23,** 1821.
139. J. M. J. Fréchet and C. J. Hawker, *React. Funct. Polym.*, 1995, **26,** 127.
140. S. R. Turner, F. Walter, B. I. Voit and T. H. Mourey, *Macromolecules,* 1994, **27,** 1611.
141. H. R. Kricheldorf and O. Stober, *Makromol. Chem., Rapid Commun.*, 1994, **15,** 87.
142. Y. H. Kim, *J. Am. Chem. Soc.*, 1992, **114,** 4947.
143. F. Chu and C. J. Hawker, *Polym. Bull.*, 1993, **30,** 265.
144. A. Kumar and S. Ramakrishnan, *J. Chem. Soc., Chem. Commun.*, 1993, 1453.
145. R. Spindler and J. M. J. Fréchet, *Macromolecules,* 1993, **26,** 4809.
146. V. Percec and M. Kawasumi, *Macromolecules,* 1992, **25,** 3843.
147. S. Bauer, H. Fischer and H. Ringsdorf, *Angew. Chem., Int. Ed. Engl.*, 1993, **32,** 1589.
148. K. E. Uhrich, C. J. Hawker, J. M. J. Fréchet and S. R. Turner, *Macromolecules,* 1992, **25,** 4583.
149. T. M. Miller, T. X. Neenan, E. W. Kwock and S. M. Stein, *J. Am. Chem. Soc.*, 1993, **115,** 356.
150. G. D. Figuly, *US Pat.* 1992, 5 136 014.
151. Y. H. Kim and O. W. Webster, *Macromolecules,* 1992, **25,** 5561.
152. M. Johansson, E. Malmstrom and A. Hult, *J. Polym. Sci., Polym. Chem.*, 1993, **31,** 619.
153. N. Ventosa, D. Ruiz, C. Rovira and J. Veciana, *Mol. Cryst. Liq. Cryst.*, 1993, **232,** 333.
154. L. J. Mathias and T. W. Carothers, *J. Am. Chem. Soc.*, 1991, **113,** 4043.
155. C. J. Hawker and F. Chu, *Macromolecules,* 1996, in press.
156. M. Suzuki, A. Ii and T. Saegusa, *Macrmolecules,* 1992, **25,** 7071.
157. G. T. Visscher and P. A. Bianconi, *J. Am. Chem. Soc.*, 1994, **116,** 1805.
158. R. D. Rieke and T. Chen, *Chem. Mater.*, 1994, **6,** 576.
159. M. N. Bockarev, V. B. Cilkin, L. B. Mayorova, G. A. Razuvaev, U. D. Cemchkov and V. E. Sherstyanux, *J. Organomet. Chem. USSR,* 1988, **1,** 115.
160. P. Kambouris and C. J. Hawker, *J. Chem. Soc., Perkin Trans. 1,* 1993, 2717.
161. C. J. Hawker, P. Kambouris and F. Chu, *Macromolecules,* 1996, in press.
162. E. Malmstrom and A. Hult, in 'Proc. 35th IUPAC Int. Symp. Macromolecules, MacroAkron '94', 1994, p. 338; M. Johansson, A. Hult and J. A. E. Manson, in 'Proc. 35th IUPAC Int. Symp. Macromolecules, MacroAkron '94', 1994, p. 333; H. Ihre and A. Hult, in 'Proc. 35th IUPAC Int. Symp. Macromolecules, MacroAkron '94', p. 337.
163. Y. H. Kim, personal communication.
164. V. Percec, P. Chu and M. Kawasumi, *Macromolecules,* 1994, **257,** 4441.
165. E. M. Sanford, J. M. J. Fréchet, K. L. Wooley and C. J. Hawker, *Polym. Prepr.*, 1993, **34,** 654.
166. S. Aoshima, J. M. J. Fréchet, R. B. Grubbs, M. Henmi and M. A. Leduc, *Polym. Prepr.*, 1995, **36,** 531.
167. J. M. J. Fréchet, M. Henmi, I. Gitsov, S. Aoshima, M. Leduc and R. B. Grubbs, *Science,* 1995, **269,** 1080; J. M. J. Fréchet and S. Aoshima. *US* and *PCT Pats.* pending.
168. L. Lochmann and J. M. J. Fréchet, unpublished data.
169. C. J. Hawker, K. L. Wooley, R. Lee and J. M. J. Fréchet, *Polym. Mat. Sci. Eng.*, 1991, **64,** 73.
170. W. Claussen, N. Schulte and A. D. Schlüter, *Makromol. Chem., Rapid Commun.*, 1995, **16,** 89.
171. K. L. Wooley and J. M. J. Fréchet, *Polym. Mat. Sci. Eng.*, 1992, **67,** 90; C. J. Hawker, K. L. Wooley and J. M. J. Fréchet, unpublished data.
172. I. Gitsov, K. L. Wooley, C. J. Hawker and J. M. J. Fréchet, *Polym. Prepr.*, 1991, **32,** 631.
173. I. Schipor, I. Gitsov and J. M. J. Fréchet, publication in preparation.
174. J. Schipor, I. Gitsov and J. M. J. Fréchet, in 'Proc. 35th IUPAC Int. Symp. Macromolecules, MacroAkron '94', 1994, p. 339.
175. J. M. Lehn, *Angew. Chem., Int. Ed. Engl.*, 1990, **29,** 1304; J. M. Lehn, *Angew. Chem., Int. Ed. Engl.*, 1988, **27,** 90.
176. D. B. Amabilino, J. F. Stoddart and D. J. Williams, *Chem. Mater.*, 1994, **6,** 1159.
177. B. J. Whitlock and H. W. Whitlock, *J. Am. Chem. Soc.*, 1994, **116,** 2301.
178. F. Diederich and D. R. Carcanague, *Helv. Chim. Acta,* 1994, **77,** 800.
179. G. Wenz, *Angew. Chem., Int. Ed. Engl.*, 1994, **33,** 803.
180. F. C. J. M. Vanveggel, W. Verboom and D. N. Reinhoudt, *Chem. Rev.*, 1994, **94,** 279.
181. E. C. Wiener, M. W. Brechbiel, H. Brothers, R. L. Magin, O. A. Gansow, D. A. Tomalia and P. C. Lauterbur. *Magn. Reson. Med.*, 1984, **31,** 1.
182. D. A. Tomalia, *Sci. Am.*, May 1995, 62.
183. C. J. Hawker, P. J. Farrington, M. E. Mackay, K. L. Wooley and J. M. J. Fréchet, *J. Am. Chem. Soc.*, 1995, **117,** 4409.
184. B. I. Voit, *Acta Polym.*, 1995, **46,** 87.

4

Coupling Reactions in Polymer Synthesis

ALAIN FRADET

Université P. et M. Curie, Paris, France

4.1 INTRODUCTION

The number of references related to 'coupling' reactions published each year in the area of synthetic polymer chemistry has shown a steep increase between 1984 and 1994 (250%), much more than the increase of the total number of references in the same field during the same period (25%).[1] This increase is mainly due to a growing interest in two fields of polymer science, which are the synthesis of conjugated polymers and reactions between polymers. Although both are related to step polymerization, the chemistry involved in these two fields is completely different from each other. The term 'coupling' is actually used in polymer chemistry for a great number of reactions which have very little similarities.

(i) In the first set of reactions, coupling reactions are involved in the building of the polymer backbone. These reactions can be referred to as coupling step polymerizations. In this case, the definition of coupling is very similar to that in organic chemistry. It involves reactions between chemical functions of a similar nature promoted by a compound called a coupling agent, which normally does not participate in the newly formed bond. Wurtz (equation 1) and Ullmann (equation 2) reactions are examples of coupling reactions in which the coupling agents are sodium and copper, respectively. The coupling reactions of Grignard reagents are also frequently used and find application in the synthesis of poly(p-phenylenes) (equation 3).

$$2\ R{-}Cl \xrightarrow{Na} R{-}R \quad (1)$$

$$2\ Ar{-}Cl \xrightarrow[\Delta]{Cu} Ar{-}Ar \quad (2)$$

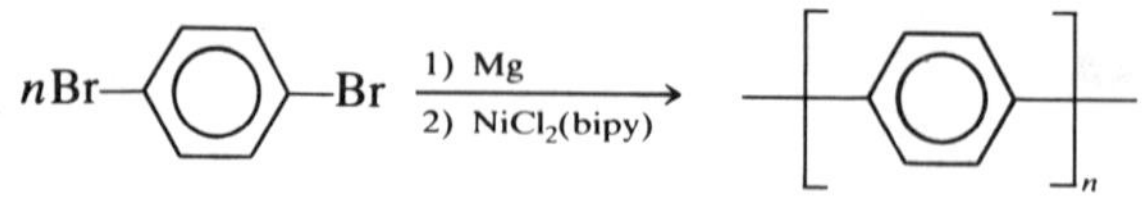

(3)

Coupling reactions are not restricted to the formation of symmetrical compounds, and the terms 'cross-coupling' or 'unsymmetrical coupling' apply when mixtures of starting compounds are involved, or when two-step reactions are carried out, *e.g.* the reaction between a Grignard reagent and an alkyl halide (equation 4). Moreover, some reactions are often called coupling reactions although they actually do not involve reactions between functional groups of similar nature. This is the case in the Heck (equation 5) coupling reaction, which finds a number of applications in the synthesis of polyarylenevinylenes.

$$R^1MgCl + R^2Br \longrightarrow R^1{-}R^2 \quad (4)$$

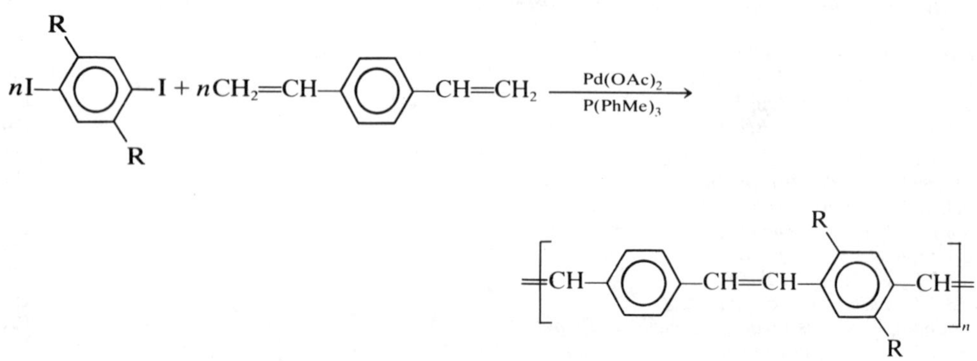

(5)

Coupling reactions mainly relate to the formation of C—C bonds;[2–4] however, a great number of references can be found where the term 'coupling' applies to the formation of other bonds, *e.g.* the C—O bond in the Williamson synthesis (equation 6), the oxidative coupling polymerization of phenols (equation 7), where oxygen is the coupling agent, and the C—N bond in diazonium coupling reactions, which are widely used in the synthesis of dyes, but find only few applications in polymer synthesis.

$$2\ R{-}OH \xrightarrow{Na} R{-}O{-}R \quad (6)$$

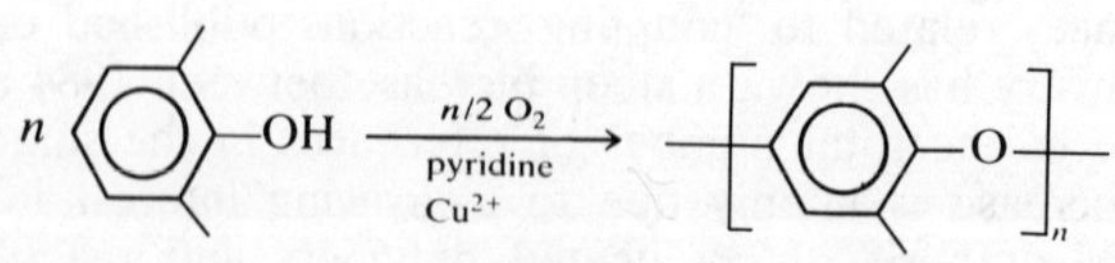

(7)

(ii) The second set of reactions involves reactions between polymers with functional groups and monomers. These monomers are termed chain-coupling agents or reagents, chain extenders or linking agents. The chemical reaction is not a 'coupling' reaction in the meaning of organic chemistry, and can be of any type. After the reaction the coupling agent forms a linkage between two polymer chains and is therefore included in the final polymer backbone.

From a chemical point of view, this is a step polymerization between a monomer and one or more prepolymers. Such reactions are termed as coupling reactions because: (a) coupling agents promote reactions between oligomers with similar end groups, which would not normally react together, and (b) the molecular weight of the coupling agent is generally 'small' in comparison to that of the reactive polymer, so that the influence of its chemical nature on the final polymer properties can be neglected. Typical examples of polymer coupling reactions are the chain extension of polymers (molecular weight boosting) (equation 8), and the synthesis of block copolymers with a random distribution of blocks (equation 9).

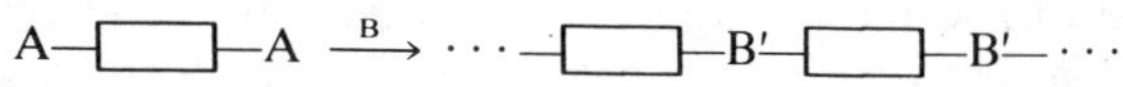

(8)

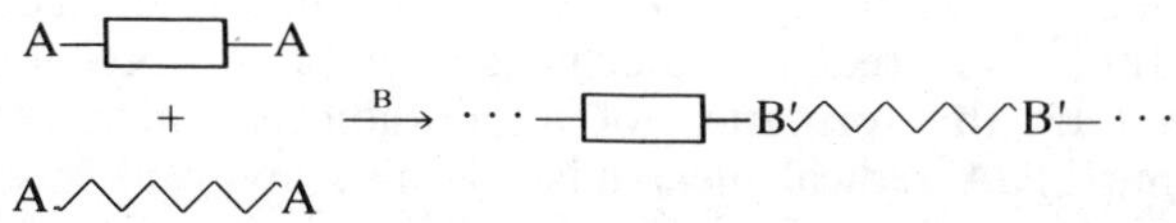

(9)

Reactions between prepolymers bearing mutually reactive end groups, which are sometimes called coupling reactions in the literature, will not be reviewed here. They cannot be considered as true polymer coupling reactions as they lead to alternating instead of random block copolymers. The same holds for the synthesis of polyurethane block copolymers, where the short-chain diol comonomer is often called a 'chain extender'.

(iii) 'Coupling agents' are often used to modify surfaces of fillers or fibers. However, in most references related to this point, the role of the coupling agent is only to enhance the adhesion between filler or fiber and polymer matrix.[5] Since there is no chemical reaction related to polymer synthesis, this topic will not be reviewed here.

(iv) Peptide coupling reactions are widely used in biochemistry, and sometimes in the synthesis of polyamides. Such reactions should be better regarded as activated amidification reactions, and will not be reviewed here.

(v) The last group of reactions related to coupling are side reactions or termination reactions of chain polymerization such as radical polymerization. They are reviewed in chapters of Volumes 3 and 4 of *Comprehensive Polymer Science.*

4.2 COUPLING STEP POLYMERIZATION

4.2.1 Carbon–Carbon Coupling Reactions

4.2.1.1 Coupling reactions between sp³ carbon centers

The classical C—C coupling reaction between sp^3 carbon centers is the treatment of haloalkanes with metals such as Na and K (Wurtz reaction) (equation 10).[2,6] Since a number of side reactions take place, *i.e.* eliminations or rearrangements, this reaction has a number of limitations and, with the exception of polysilylene synthesis (see Section 4.2.5), has not been widely used in polymer synthesis.

$$2\ \mathrm{R{-}X} \xrightarrow{\mathrm{M}} \mathrm{R{-}R} \quad \mathrm{R} = \mathrm{sp^3\ center},\ \mathrm{M} = \mathrm{metal},\ \mathrm{X} = \mathrm{halogen} \qquad (10)$$

Grignard reagents react with alkyl halides in the presence of transition metal compounds, usually copper, silver or iron salts,[7,8] with a much higher selectivity. The reaction of an alkyl dihalide with 1 equiv of magnesium gives a monohalide/mono-Grignard compound, which reacts in a second step in the presence of a catalyst. The intermediate Grignard reagent, which is not isolated, is actually a mixture of unreacted dihalide and of mono- and di-Grignard derivatives. Grignard coupling has been applied by Yamamoto *et al.*[9] to the synthesis of polymethylene in the 2000–3700 molecular weight range (equation 11). The mechanism is believed to proceed via the formation of an alkylcopper intermediate. The solubility of the final polymer in THF is a controlling factor of the chain growth, and the molecular weight of the polymer is low (2000–3700). As expected for step polymerization, the molecular weight distribution agrees with the most probable distribution. Somewhat higher molecular weights ($\overline{DP}_n = 3000–10\,000$) but lower yields (7–31%) were obtained using the same scheme in the synthesis of head-to-head polyisobutylene[10] and poly(1-vinylnaphthalene)[11] from substituted bromobutanes.

$$n\mathrm{Br{-}(CH_2)_6{-}Br} + n\mathrm{Mg} \xrightarrow{\mathrm{THF}} [n\mathrm{Br{-}(CH_2)_6{-}MgBr}] \xrightarrow[\mathrm{rt,\ 1{-}2\ h}]{\mathrm{CuBr(PPh_3)}} \mathrm{{-}(CH_2)_{6n}{-}}$$

$M_{visco} = 3400$
Yield = 80% (11)

The coupling reaction of alkylboron compounds induced by silver nitrate is another method of C—C bond formation between sp^3 centers.[12] Thus, the coupling reactions of benzylic boranes obtained from substituted divinylbenzene leads to high molecular weight polymers in good yields (equation 12).[12] The reaction mechanism involves alkyl group migration from boron to silver, followed by the formation of alkyl radicals which recombine, giving the coupling product (Scheme 1). A radical mechanism is also involved in the high-temperature oxidative coupling copolymerization of diphenylmethane and *p*-xylene in the presence of di-*t*-butyl peroxide (DBPO). Diphenylmethane and *p*-xylene units are probably linked in an alternate manner, and the composition of the final polymer does not depend on the initial monomer ratio (equation 13).[13]

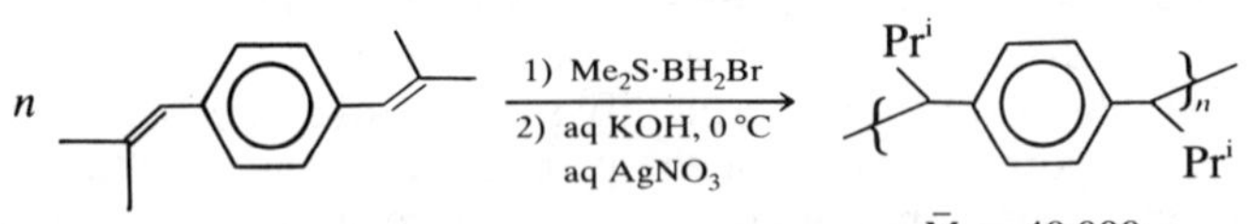

$\bar{M}_n = 40\,000$
Yield = 80% (12)

$$\mathrm{R{-}B{<}} \xrightarrow{\mathrm{OH^-}} \mathrm{R{-}\overset{|}{\underset{|}{B^-}}{-}OH} \xrightarrow{\mathrm{Ag^+}} \mathrm{R{-}Ag + HO{-}B{<}}$$

$$2\ \mathrm{R{-}Ag} \xrightarrow[-2\mathrm{Ag}]{} \mathrm{R{-}R}$$

Scheme 1

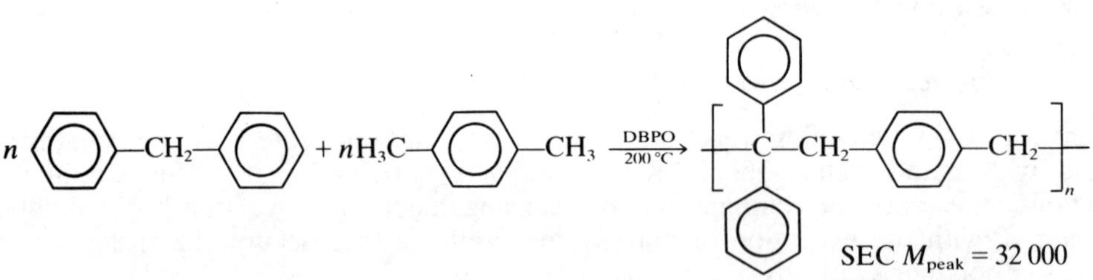

SEC $M_{peak} = 32\,000$
Yield = 27% (13)

4.2.1.2 *Coupling reactions between sp^3 and sp^2 carbon centers*

These coupling reactions are of little importance in polymer synthesis, although most of the methods described below for sp^2–sp^2 coupling by reaction between organometallic reagents and halogenated compounds can be applied. Two examples can be reported. One is the synthesis of polybenzyls by step polymerization of benzyl halides in the presence of Friedel–Crafts catalysts. This reaction leads to the formation of a C—C bond between methylene and phenyl groups (equation 14). The structure of the resulting polymers, which present a variable amount of branching and cross-linking, and the reaction mechanism are described in Volume 5, Chapter 26 of *Comprehensive Polymer Science.* The other example is the palladium(0)-catalyzed cross-coupling reaction between alkylboron compounds and aromatic halides[14]

(Suzuki coupling reaction[15]) (Scheme 2). The alkylboron compounds are obtained by hydroboration of the corresponding alkene with 9-borabicyclo[3.3.1]nonane (9-BBN). The reaction mechanism is depicted in Scheme 3.

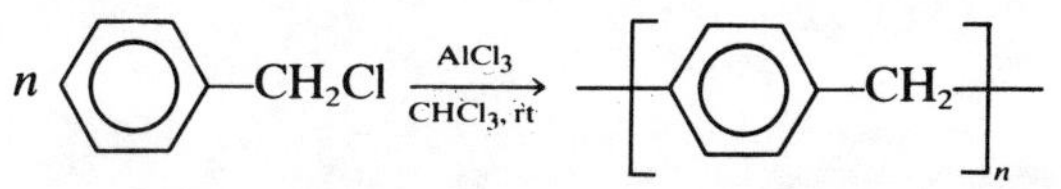

(14)

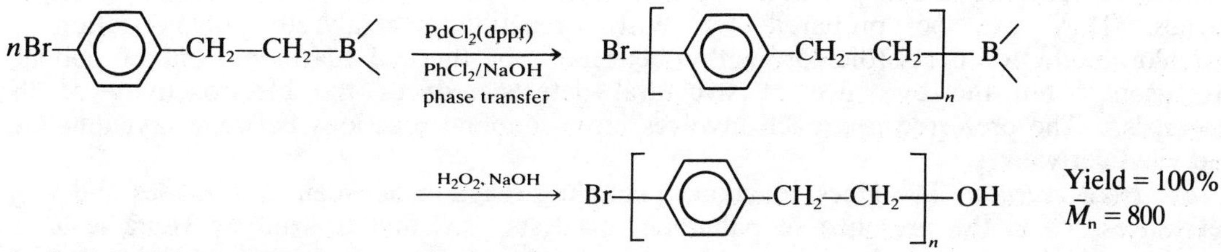

dppf = 1,1′-bis(diphenylphosphino)ferrocene

Scheme 2

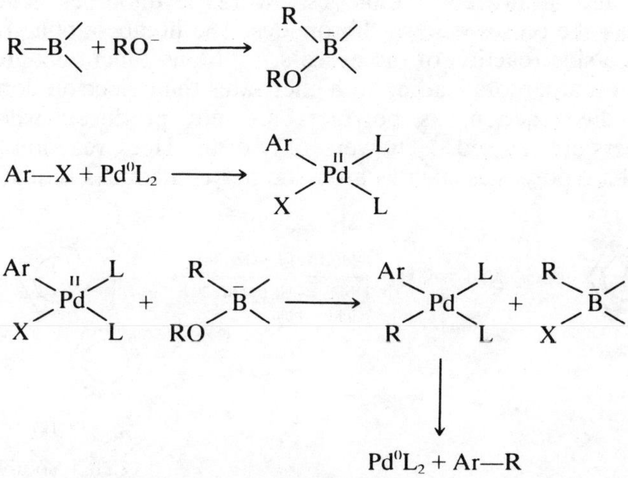

L_2 = 1,1′-bis(diphenylphosphino)ferrocene

Scheme 3

4.2.1.3 Coupling reactions between sp² carbon centers

Since the introduction of nickel(0) and palladium(0) catalysts in the early 1970s, the coupling reactions between sp^2 carbons have made enormous progress. Formerly restricted to the classical copper-mediated Ullmann reaction between aryl halides, it has now become an efficient and widely used route in organic chemistry (equation 15).

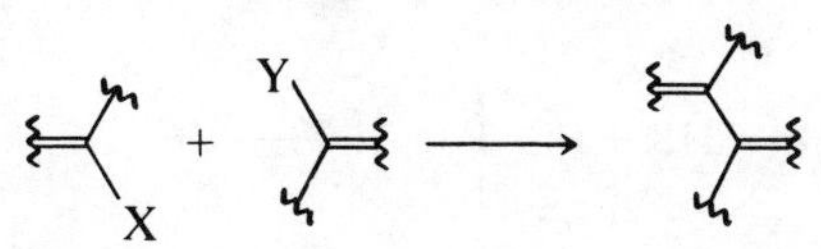

(15)

It can immediately be seen from equation (15) that the application of such chemistry to macromolecular synthesis will produce polymers with extended π systems. Conjugated polymers have considerable potential applications as conducting, magnetic or photoactive organic materials. As a consequence, during the last few years a considerable effort has been made by polymer chemists to apply these reactions to the synthesis of polymers.

(i) Aryl–alkenyl coupling reactions

Polyarylenevinylenes and polyheteroarylenevinylenes exhibit interesting electrooptical properties. They can be prepared by Wittig reaction,[16] catalyzed polymerization of bis(chloromethyl)-[17] or bis(dichloromethyl)benzene,[18] or thermal rearrangement of polymer precursors,[19] but the existence of structural defects reduces the electroactivity of the materials.[20] The preferred approach involves cross-coupling reactions between aryl dihalides and vinyl derivatives.

(a) Heck reaction. The Heck reaction, a coupling reaction between aryl halides and vinyl derivatives,[21,22] in the presence of palladium catalysts, was first utilized by Heitz *et al.* to prepare polyphenylenevinylenes by reaction between ethylene and aryl dihalides, mainly dibromides (equation 16).[23–25] The organic base acts as a hydrogen halide acceptor, and a radical inhibitor is sometimes added.[26] Instead of ethylene, divinylbenzene may be used. The reaction is assumed to proceed via the formation of an arylpalladium halide followed by insertion of the alkene (Scheme 4).[23,27] The predominant structure is that of the *trans* isomer.[23,28] This reaction has been applied to the synthesis of various substituted polyphenylenevinylenes,[28–30] and to the synthesis of polyarylenevinylenes with cinnamoyl,[31] distyrylthiophene,[26] and anthracene[32] moieties. Aromatic diiodides generally give a higher molecular weight than the corresponding dibromides. The homocoupling reaction between aryl halide end groups is a side reaction of the process.[26,31] Many functional groups are tolerated at the haloarene, electron acceptors leading to higher rates than electron donors. However, steric hindrance disturbs the reaction, as polymers are not produced when α,β-disubstituted unsaturated monomers are reacted.[31] The versatility of the Heck reaction permits its use in the synthesis of heterochain polymers such as aromatic polycinnamides (equation 17).[33]

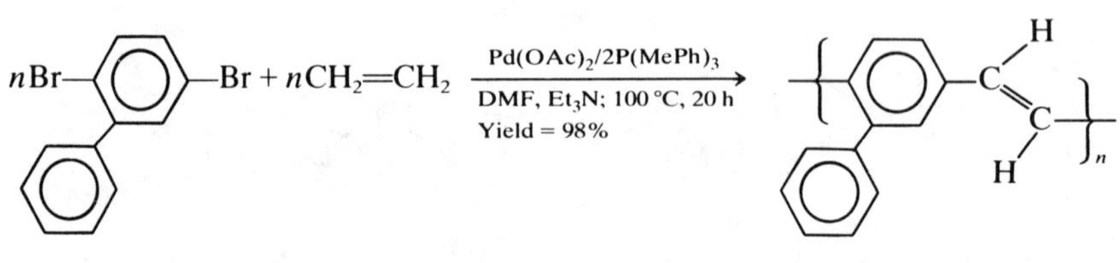

$\bar{M}_w = 7100$

($CHCl_3$ soluble fraction) (16)

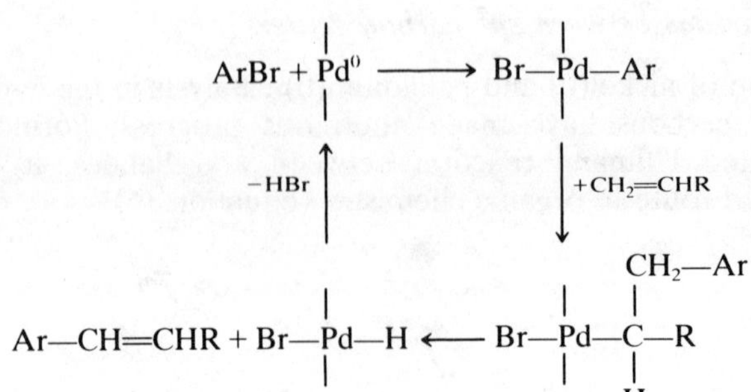

Scheme 4

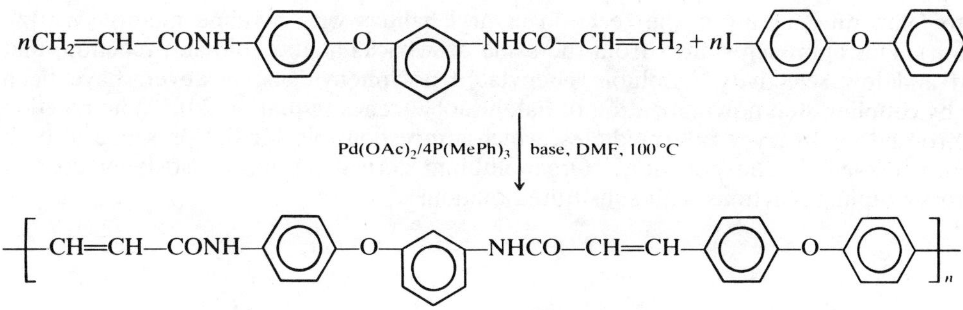

(17)

Yield = 99%
$\bar{M}_n = 93\,200$

(b) Reactions involving organometallic compounds. The reaction of arylmagnesium halides with alkenyl halides using nickel derivatives as catalysts is another way of achieving aryl–alkenyl coupling. Thus, polyheteroarylenevinylenes have been prepared by nickel-catalyzed cross-coupling between the Grignard derivatives of substituted thiophenes and 1,2-dichloroethylene.[34] Organotin derivatives can also couple with 1,4-bis(trifluoromethanesulfonyl)benzene[35] or dihalothiophenes in the presence of palladium phosphine complexes (equation 18)[36,37] (Stille coupling reaction;[38,39] see also Section 4.2.1.3(b)).

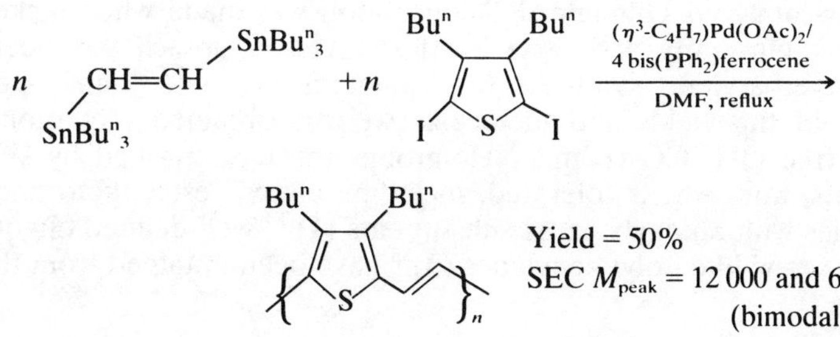

(18)

Yield = 50%
SEC M_{peak} = 12 000 and 6000 (bimodal)

(ii) Aryl–aryl coupling reactions

Polyarylenes (Volume 5, Chapter 27 of *Comprehensive Polymer Science*) or polyheteroarylenes are materials known for their high thermal and thermooxidative stability, and for their conducting properties. They are mainly prepared by coupling reactions, either oxidative coupling of aromatic compounds or coupling reactions between aromatic dihalides and organometallic derivatives.

(a) Ullmann-type reactions. One of the most employed approaches for the synthesis of biaryls is probably the classical Ullmann reaction (equation 2).[40,41] The reaction is usually carried out in solvents such as pyridine or DMF, but can also take place without solvent at temperatures above 200 °C. The catalyst used is copper powder[40] or activated copper powder obtained by reduction of copper salts.[42,43] The reaction fails in the presence of OH, CO_2H or NH_2 substituents, which must be protected before reaction by trimethylsilylation. The Ullmann reaction probably proceeds via the formation of arylcopper species.[40,41] It is not a highly selective reaction, as a number of by-products are formed, such as branched structures,[44,45] and consequently it has not been extensively used in polymer synthesis. A recent example is the synthesis of polypyrrole oligomers from N-protected dibromopyrrole (equation 19).[46] A reductive debromination side reaction, giving unreactive H end groups, limits chain growth.

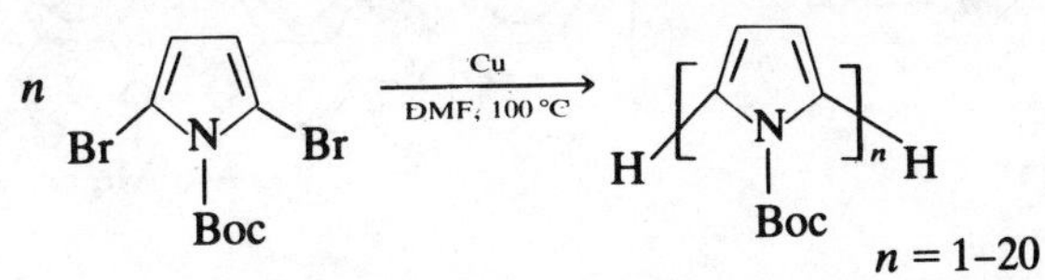

(19)

Related to Ullmann coupling is the reaction of aryl halides with alkaline metals (Wurtz–Fittig reaction). This approach suffers from the same drawbacks as the Ullmann reaction, that is low yield and low selectivity.[45] Soluble phenylated polyphenylenes, however, have been synthesized by coupling step polymerization of halolithiobenzenes (equation 20).[47] The reaction appears to proceed via benzyne intermediates, which are responsible for the presence of both *meta* and *para* linkages in the polymer.[47] Organolithium derivatives have also been used in reductive cross-coupling reactions with substituted quinones.[48]

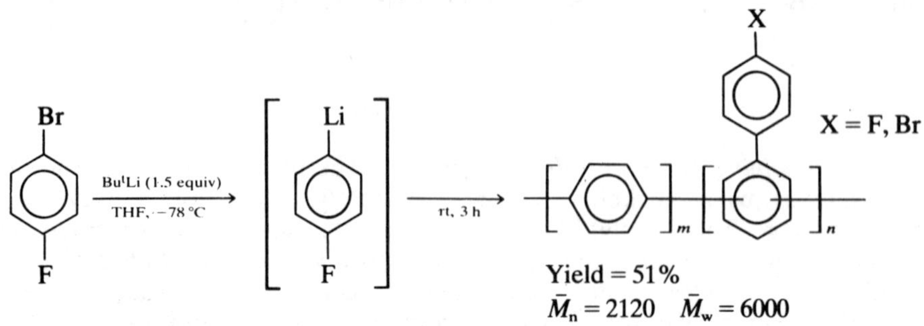

(20)

(b) Nickel(0)-catalyzed coupling reactions of aryl dihalides. A major improvement compared to the copper-catalyzed Ullmann dehalogenation was made when nickel(0) compounds were found to react efficiently with aryl halides.[49] This approach has been transposed to polyarylene and heteroarylene synthesis by Yamamoto *et al.*[50–52] and seems to be very promising, in view of the yields and molecular weights obtained (equation 21). As in the Ullmann reaction, free OH, CO_2H and NH_2 groups must be masked by silylation. A large range of substituents, however, is tolerated, including alkoxy, ester, keto and formyl groups. Thus, polyphenylenes with alkoxybenzoyl substituents (**1**),[53] well-defined oligothiophenes (**2**),[54] and optically active screw-like polyphenylenes (**3**)[55] have been obtained from the corresponding dibromo monomers.

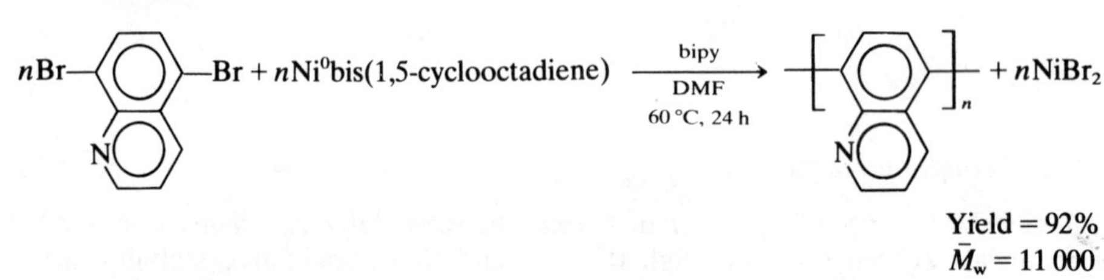

(21)

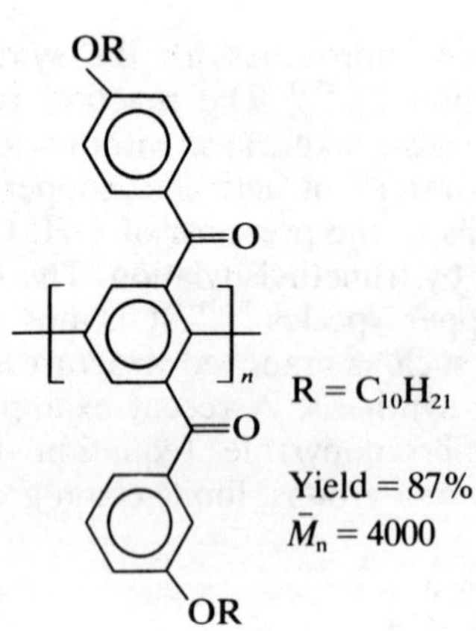

(**1**)

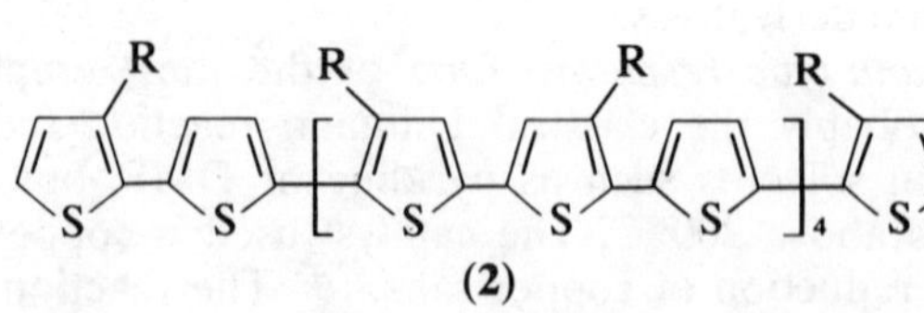

(**2**)

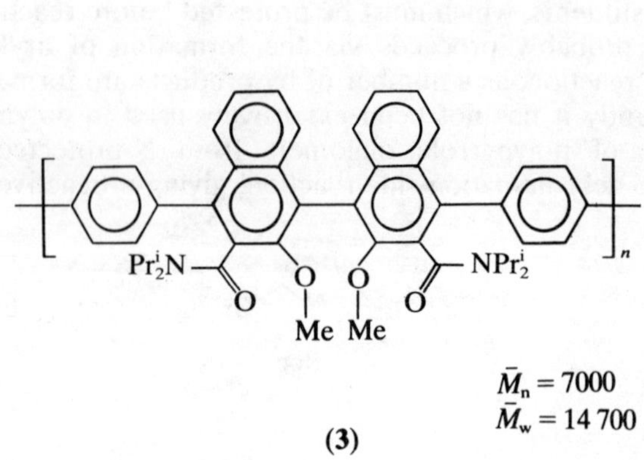

(**3**)

The stoichiometric amount of nickel complex can be replaced by catalytic quantities of nickel(0) species generated in situ by chemical[56] or electrochemical[57–59] reduction of nickel salts. The mechanism of the coupling reaction of dibromo-*N*-ethylcarbazole in the presence of a nickel(0) complex, generated in situ by electroreduction of a $NiBr_2$/bipyridine complex (equation 22), has been examined by Siove *et al.*[59] In the presence of monomer, nickel(0) is instantaneously inserted into a C—Br bond,[60] giving rise to C—Ni—Br* active species which are involved in the step polymerization (Scheme 5).

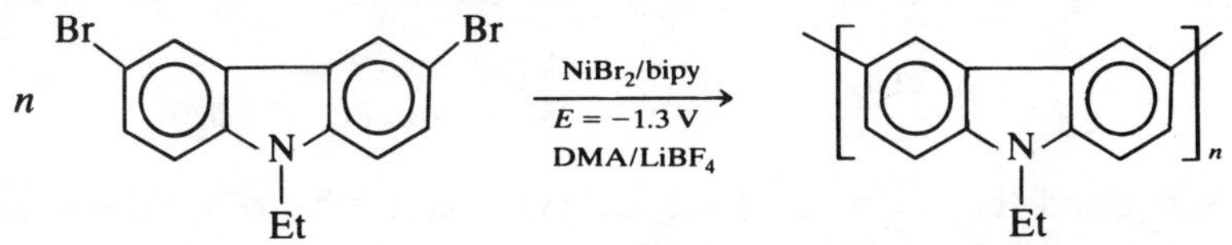

Yield = 95%
$\bar{M}_n = 1800$ (22)

$$NiBr_2/bipy + 2e^- \xrightarrow[E = -1.3\ V \text{ bipy}]{\text{electrolysis}} Ni^0/(bipy)_2 + 2Br^-$$

$$Ni^0 + \text{—Ar—Br} \xrightarrow{\text{insertion}} \text{—Ar—Ni—Br*}$$

$$\text{—Ar—Ni—Br*} + \text{—Ar—Br} \xrightarrow{\text{coupling}} \text{—Ar—Ar—} + NiBr_2$$

Scheme 5

The scope of nickel(0)-catalyzed dehalogenative coupling is not limited to the synthesis of conducting polymers. It is an alternative method towards aromatic poly(ether sulfones) and poly(ether ketones), which are usually prepared by electrophilic (Friedel–Crafts) or nucleophilic substitution reactions (C—O bond formation). Thus, high molecular weight aromatic poly(ether sulfones)[56,61] and poly(ether ketones)[62] have been prepared by nickel-catalyzed coupling reactions of activated aromatic dichloride monomers in the presence of a reducing metal. Aromatic polyketones have been obtained using the same scheme (equation 23).[63]

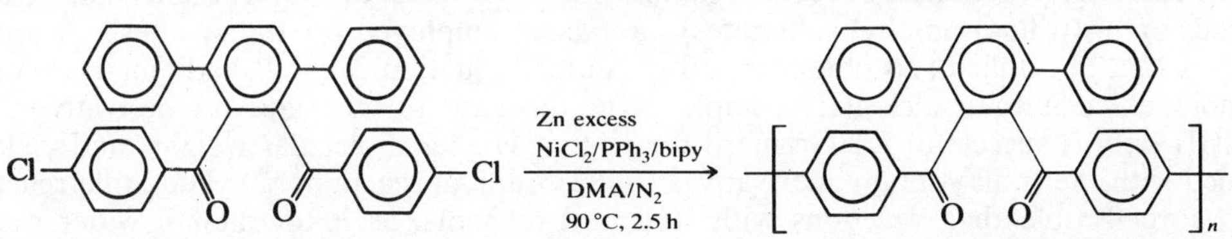

$\bar{M}_n = 21\,000$, $\bar{M}_w = 46\,500$ (23)

(c) Coupling reactions of preformed organometallic compounds. Polyphenylene is produced when the Grignard derivative of a dihalobenzene is heated in the presence of catalytic amounts of transition metal salts, such as iron(III), cobalt(II) or nickel(II) salts.[45] Nickel(II) derivatives are the most utilized (equation 24).[64] The reaction is believed to proceed via a transmetallation reaction and the formation of C—Ni species.[65] Not only polyphenylenes, but also poly(2,5-thienylene),[65,67] hyperbranched polyphenylenes,[68] and thermotropic polyalkylenephenylenes[69] (equation 25) have been synthesized. A stoichiometric amount of magnesium must be employed, and the intermediate Grignard reagent is of course a mixture of unreacted starting material and di- and mono-Grignard compounds. The study of the coupling reaction of alkyl-substituted 2,5-dibromothiophenes reveals a very low regioselectivity, resulting in lower conducting properties with respect to that of the polymers obtained by the oxidative coupling of aromatic compounds.[67]

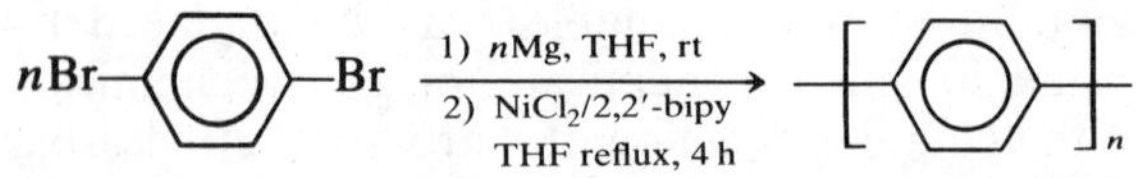

Yield = 95% (24)

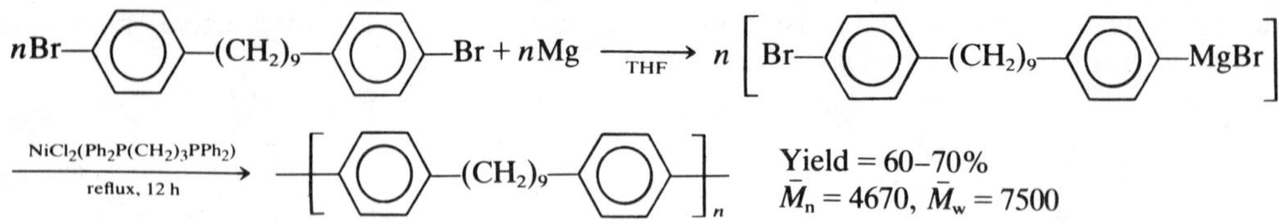

Yield = 60–70%
$\bar{M}_n = 4670$, $\bar{M}_w = 7500$ (25)

Taylor *et al.*[70] have effected the coupling of the Grignard derivative of dibromobenzene in very high yields (92–100%) using 1,4-dichloro-2-butene as the coupling agent. A mechanism involving electron transfer from the Grignard reagent to the allylic dichloride has been suggested. This reaction has been applied to the synthesis of soluble poly(1-butylperylene), for which the nickel(0)-induced coupling of dibromo monomers and the palladium(0)-catalyzed cross-coupling reaction between bis(trimethylstannyl)- and dioodo-1-butylperylene were unsuccessful (equation 26).[71]

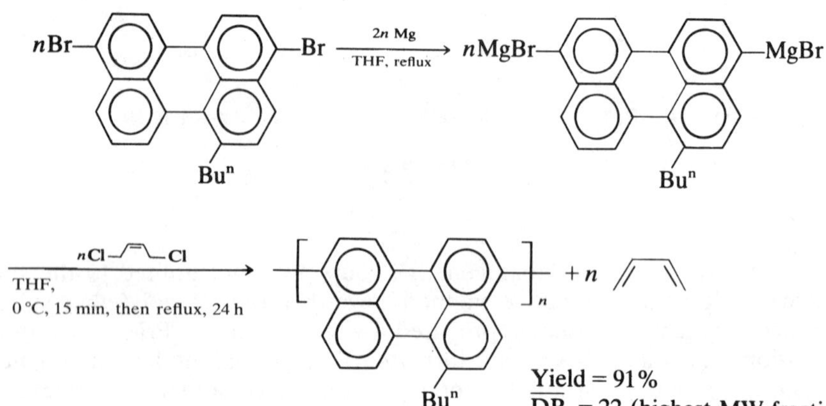

Yield = 91%
$\overline{DP}_n = 22$ (highest MW fraction) (26)

Cross-coupling reactions between organomagnesium, -zinc or -tin reagents and aromatic dihalides or bis(trifluoromethyl sulfonates) have been employed for the synthesis of aromatic polymers that are difficult to obtain by other routes (equation 27).[72] Palladium(II) complexes are more efficient than nickel(II) complexes in the case of zinc and tin derivatives, while nickel(II) only is successful for Grignard reagents.[72] Higher molecular weights and yields are obtained with the trialkylstannyl derivatives (Stille coupling reaction[38,39]). The Stille reaction is much more flexible than reactions with Grignard reagents, as it tolerates a wider range of functional groups such as ester, keto amine and hydroxy groups. Besides, tin reagents offer the advantage of greater solubility in a range of solvents, and a consequently easier chain growth reaction in homogeneous media.[73] Thus, thermotropic polyphenylenethienylenes can be synthesized in refluxing dioxane by reacting 1,5-bis(tributylstannyl)thiophene with alkoxy-substituted 1,4-dihalo- or 1,4-bis(trifluoromethylsulfonyl)benzene in the presence of a palladium catalyst.[74,75] Thermally and chemically stable poly(biphenylene ether ketones) have been synthesized by a similar reaction.[76,77] Molecular weight is controlled, however, by the limited solubility of the growing polymer, which precipitates from the reaction medium. This can be overcome by converting the ketone function into a diacetal, which gives the polymer better solubility (Scheme 6).

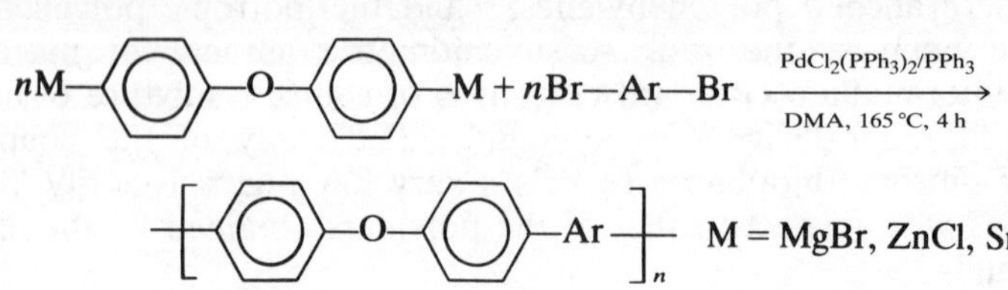

M = MgBr, ZnCl, $SnBu^n_3$ (27)

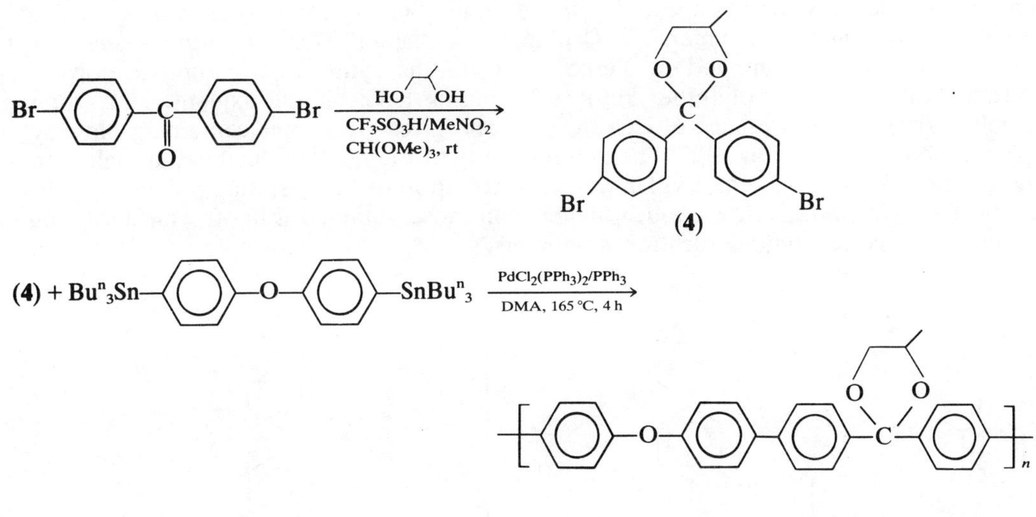

Scheme 6

(d) Coupling reactions between halo and boron derivatives. The palladium-catalyzed coupling reaction between arylboronic acid and bromobenzene described by Suzuki *et al.*[78] (see Section 4.2.1.2) proceeds in good yields and has been applied to the synthesis of polyphenylenes. Thus, a substituted benzene bearing both boronic acid and bromo substituents reacts regiospecifically, affording a relatively high molecular weight substituted polyphenylene (equation 28).[79] In the same way, polypyrrole oligomers,[80] polyphenylene derivatives with functional end groups,[81] and copolymers with *m*-phenylene and 2,5-thiophene units[82] have been synthesized. This approach has also been employed for the synthesis of rigid rod polyimides (equation 29).[83] The properties of the polymer are identical to those of a polymer of similar structure synthesized by the conventional pyromellitic dianhydride/*p*-terphenyldiamine reaction.

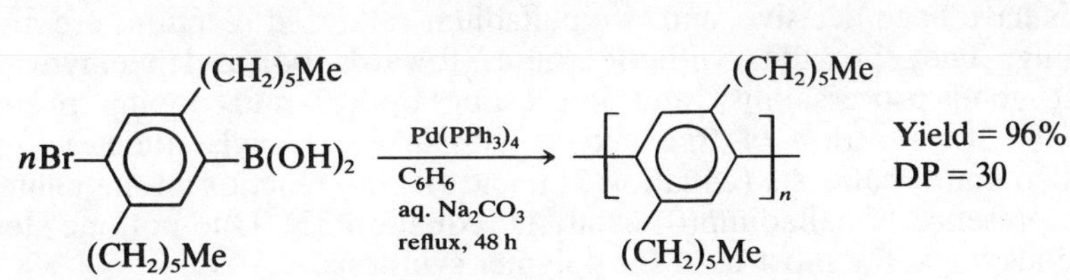

(28)

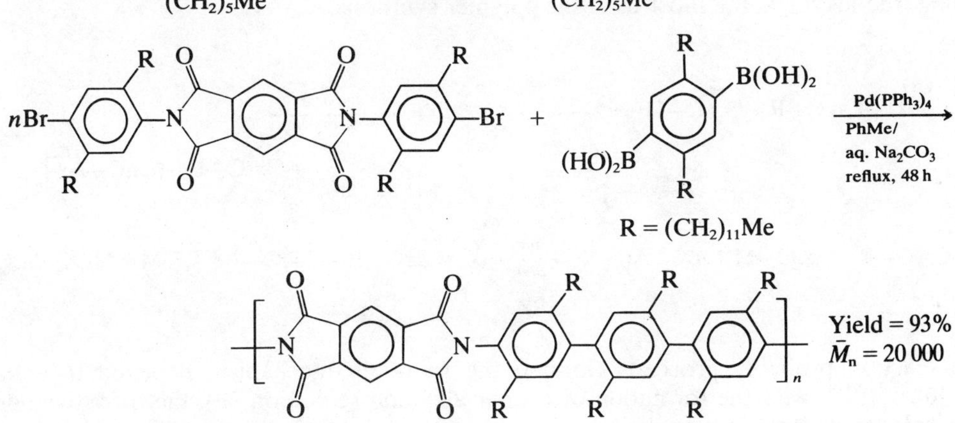

(29)

(e) Oxidative coupling of aromatic compounds. This approach involves the treatment of aromatic compounds with a system consisting of a Lewis acid (*e.g.* $AlCl_3$) and an oxidant (*e.g.* $CuCl_2$) under mild conditions.[84,85] A mechanism involving the formation of a radical cation

seems to be the most acceptable.[85,86] This reaction (Scholl reaction) is widely used in the synthesis of conjugated polymers[87,88] (Volume 5, Chapter 27 of *Comprehensive Polymer Science*), and has been applied by Percec *et al.* to the synthesis of aromatic polyethers,[86] poly(ether sulfones) and poly(ether ketones)[89–92] using $FeCl_3$ as the oxidant. Thus, monomer (**5**) polymerizes in the presence of excess $FeCl_3$ and gives a polymer containing biphenyl and naphthyl moieties (equation 30).[86] The heterogeneity of the reaction medium at high monomer concentration broadens the molecular weight distribution of the resulting polymer. Yields and molecular weights, however, are much higher than those obtained using the nickel(0) coupling reaction of the corresponding dichloro monomers.

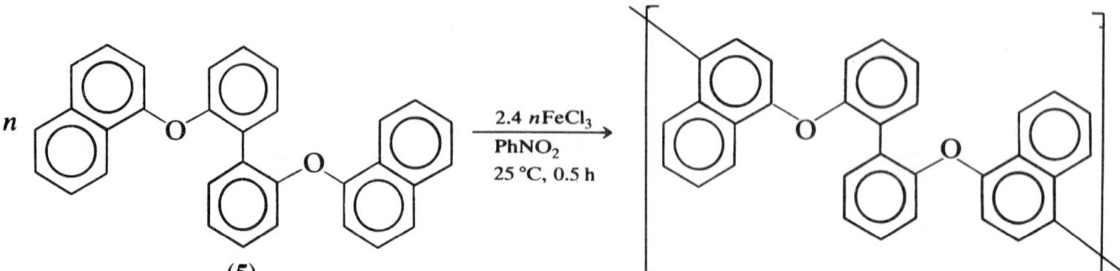

(**5**)

Yield = 99%
$\bar{M}_n = 90\,500$ (30)

The oxidative coupling of aromatic or heteroaromatic compounds can also be carried out by electrochemical means. The anodic oxidation of benzene or biphenyl leads to conducting films of poly(*p*-phenylene) and is reviewed in Volume 5, Chapter 27 of *Comprehensive Polymer Science.*

4.2.1.4 *Coupling reactions between sp and sp² carbon centers*

Copper(I) alkynides undergo cross-coupling reactions with halogen-bearing sp^2 carbon compounds.[93,94] However, the need for a stoichiometric amount of copper limits the scope of this reaction. As in the case of sp^2–sp^2 coupling, the advances made in the area of transition metal catalysis have been decisive, and two palladium-catalyzed reactions are now available for sp–sp^2 coupling. They provide synthetic routes towards polyaryleneethynylenes, polymers which exhibit good processability and interesting optical and doping properties.[95] These reactions are (i) the reaction of free alkyne with aryl or vinyl halides in the presence of copper(I)/palladium(0) catalysts (equation 31), and (ii) the reaction of metallated alkynes with halides in the presence of palladium(0) catalysts (equation 32). One-pot one-step reaction (31) involving free alkyne is the most used for polymer synthesis.

$$n\mathrm{H{-}C{\equiv}C{-}R{-}C{\equiv}C{-}H} + n\mathrm{X{-}Ar{-}X} + 2n\mathrm{NR_3} \xrightarrow[-2n\mathrm{NR_3HX}]{\mathrm{Pd^0/CuI}} \mathrm{-\!\!\!\![C{\equiv}C{-}R{-}C{\equiv}C{-}Ar]\!\!\!\!-}_n \qquad (31)$$

$$n\mathrm{M{-}C{\equiv}C{-}R{-}C{\equiv}C{-}M} + n\mathrm{X{-}Ar{-}X} \xrightarrow[-2n\mathrm{MX}]{\mathrm{Pd^0}} \mathrm{-\!\!\!\![C{\equiv}C{-}R{-}C{\equiv}C{-}Ar]\!\!\!\!-}_n \quad \mathrm{M = MgX, ZnX, SnR_3} \qquad (32)$$

The reaction probably proceeds through the four-step mechanism depicted in equations (33)–(36),[27,73,95–97] with the formation of copper alkynide (equation 33), the oxidative addition of the halogenated compound to the palladium(0) complex (equation 34), transmetallation (equation 35) and the reductive elimination of the palladium(0) complex, which liberates the coupling product (equation 36).

$$R{-}C{\equiv}C{-}H + CuX \rightarrow Cu{-}C{\equiv}C{-}R \quad (33)$$

$$(PPh_3)_2Pd + Ar{-}X \rightarrow PPh_3{-}\underset{\displaystyle PPh_3}{\overset{\displaystyle Ar}{Pd}}{-}X \quad (34)$$

$$Cu{-}C{\equiv}C{-}R + PPh_3{-}\underset{\displaystyle PPh_3}{\overset{\displaystyle Ar}{Pd}}{-}X \rightarrow PPh_3{-}\underset{\displaystyle PPh_3}{\overset{\displaystyle Ar}{Pd}}{-}C{\equiv}C{-}R \quad (35)$$

$$PPh_3{-}\underset{\displaystyle PPh_3}{\overset{\displaystyle Ar}{Pd}}{-}C{\equiv}C{-}R \rightarrow (PPh_3)_2Pd + Ar{-}C{\equiv}C{-}R \quad (36)$$

The use of a palladium(0) complex circumvents the need for the preliminary reductive steps (equations 37 and 38) that are required to enter the catalytic cycle when palladium(II) derivatives are used.

$$(PPh_3)_2PdY_2 + R{-}C{\equiv}C{-}H \xrightarrow[-2Y^-]{} (PPh_3)_2Pd(C{\equiv}C{-}R)_2 \longrightarrow R{-}C{\equiv}C{-}C{\equiv}C{-}R + (PPh_3)_2Pd \quad (37)$$

$$(PPh_3)_2PdY_2 + 2PPh_3 \xrightarrow[-2HY]{} (PPh_3)_4)Pd \underset{2PPh_3}{\overset{-2PPh_3}{\rightleftharpoons}} (PPh_3)_2Pd \quad (38)$$

This approach has been applied to the synthesis of low to moderate molecular weight polyethynylene(arylene ethers) and polyethynylene(arylene ketones) in dimethylacetamide.[73] Nickel catalysts were found to be inactive in these conditions. Homopolymerization of the diacetylenic units is a side reaction of the process, especially at high temperature, leading to the formation of highly branched structures above 140 °C, and to the beginning of thermal cross-linking above 170–180 °C.[73] Thermotropic hydrocarbon oligomers ($\bar{M}_n = 3000$–4000) containing a tolan mesogen connected by a flexible polymethylene spacer have been synthesized by reaction between bis(4-iodophenyl)alkanes and bis(4-ethynylphenyl)alkanes in benzene, in the presence of $PdCl_2(PPh_3)_2$/CuI. The insolubility of the polymers, which precipitate during reaction, limits the molecular weight.[98] Higher molecular weights have been obtained with heteroaromatic compounds, probably due to both better reactivity and better solubility. Thus, the reaction of substituted diiodothiophenes and diethynylpyridine yielded polymers with molecular weights that are uncommon for step polymerization polymers ($\bar{M}_w = 100\,000$–$700\,000$) (equation 39).[95] Besides their interesting optical and n-doping properties, such polymers are usable as starting materials for the synthesis of poly(arylenevinylenes) by addition reaction to the triple bond.[95]

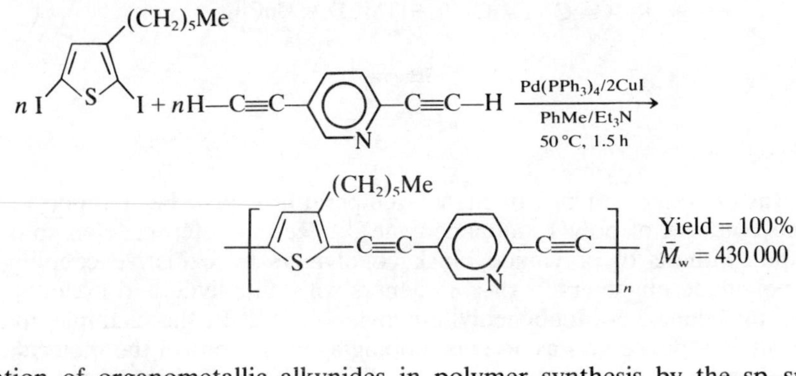

(39)

The application of organometallic alkynides in polymer synthesis by the sp–sp^2 coupling reaction is mostly restricted to tin alkynides.[35,73,76] Reaction takes place in the presence of palladium catalysts, and the mechanism involves the same kind of transmetallation reaction as proposed for palladium/copper catalysis. In contrast to the latter method, where the organometallic (organocopper) derivative is synthesized in situ, tin alkynides must be synthesized in a preliminary step. These reagents impart the growing chains higher solubility and enhanced thermal stability with respect to copper derivatives. They have been found to yield higher molecular weight polymers than the palladium/copper reaction when reacted with bis(bromophenyl) compounds.[73]

4.2.1.5 Coupling reactions between sp carbon centers

1-Bromoalkynes undergo cross-coupling reactions with terminal alkynes or with alkyne Grignard derivatives in the presence of copper catalysts.[99,100] However, the applications of sp–sp coupling to polymer synthesis mainly relate to the Glaser oxidative homocoupling reactions of diethynyl monomers introduced in polymer chemistry by Hay in the 1960s.[101–103] Dioxygen is bubbled into a solution of the diethynyl compound in a complex-forming solvent such as pyridine or tetramethylethylenediamine (TMEDA), in presence of a catalytic amount of copper(I) salts. The yield is almost quantitative (equation 40).[102] The oxidative coupling step may take place from a complexed dimeric alkynylcopper derivative (Scheme 7).[104]

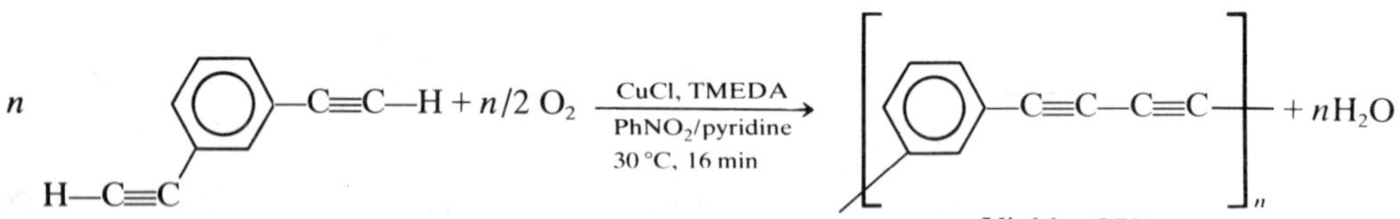

(40)

Yield = 95%
$\bar{M}_n = 30\,000$

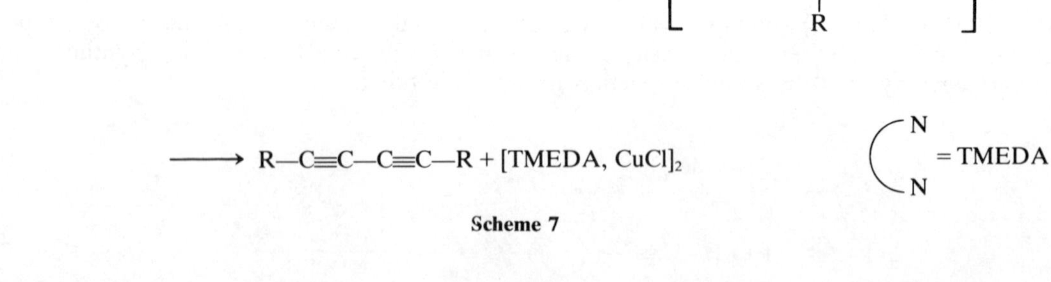

Scheme 7

The removal of water and use of higher temperatures have been reported to give better results in the synthesis of poly(1,8-nonanediyne).[105] Recent references on sp–sp coupling are related to the synthesis of polyimide block copolymers by oxidative coupling of acetylene-terminated polyimide oligomers,[106] of copolymers with butadiyne and cyclobutadiene units,[107] and of poly(ethynylene-2,5-thiophenediyl-ethynylene).[108,109] In the example given in equation (41),[110] 2-ethynylthiophene acts as an end-capping agent to control the molecular weight. Such polymers undergo cross-linking on heating, giving thermally stable materials.[108] Another application is the reaction of the butadiyne linkage with aniline[108,109] or sodium sulfide,[109] leading respectively to alternating polymers (**6**) and (**7**), which have potential applications as conducting and nonlinear optics materials.

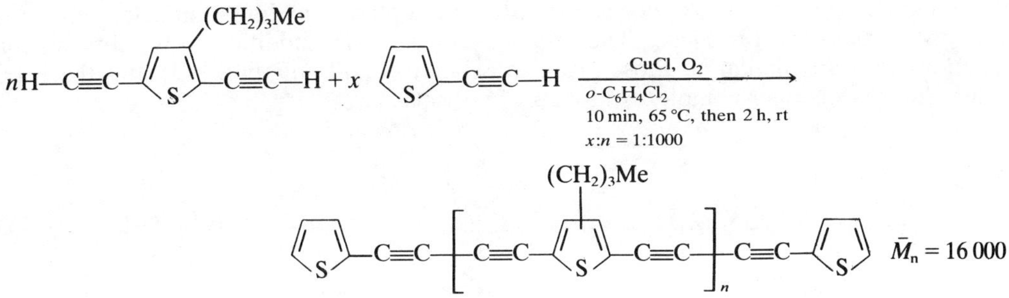

(41)

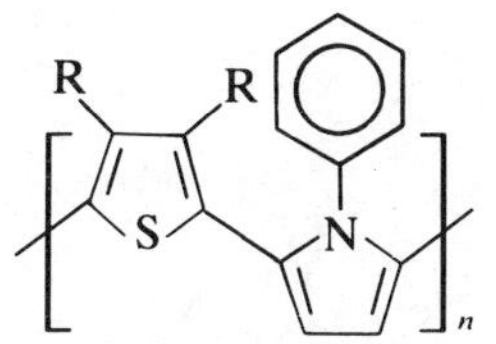

(6)

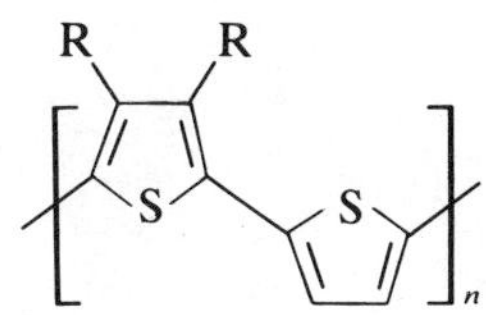

(7)

4.2.2 Carbon–Oyxgen Coupling Reactions

Carbon–oxygen coupling reactions have been mainly used for the synthesis of poly(phenylene ethers). In addition to the nucleophilic halogen displacement reaction of *ortho*-substituted 4-halophenols in the presence of bases,[111] these polymers can be synthesized by two coupling-related reactions: Ullmann ether synthesis and oxidative coupling of phenols. Ullmann reaction of 4-halophenols or phenolates in the presence of copper or copper salts can be carried out on phenols without *ortho* substituents, but is accompanied by a number of side reactions.[112] The preferred synthetic method is the oxidative coupling of phenols. The reaction has been known for a long time, as Hay first reported in 1959 the synthesis of poly(oxy-2,6-dimethyl-1,4-phenylene) (PPO) by oxidative coupling of 2,6-dimethylphenol (DMP) with dioxygen.[113] The reaction mechanism of PPO synthesis is detailed in Volume 3, Chapter 28 of *Comprehensive Polymer Science,* and will only be briefly discussed here, covering recent results. The reaction is usually carried out in solvents such as toluene, in the presence of a copper(II) catalyst and aliphatic or heterocyclic amines. The formation of diphenoquinone (**8**) by C—C coupling is a side reaction of the process, especially at the beginning of the reaction (equation 42).[114]

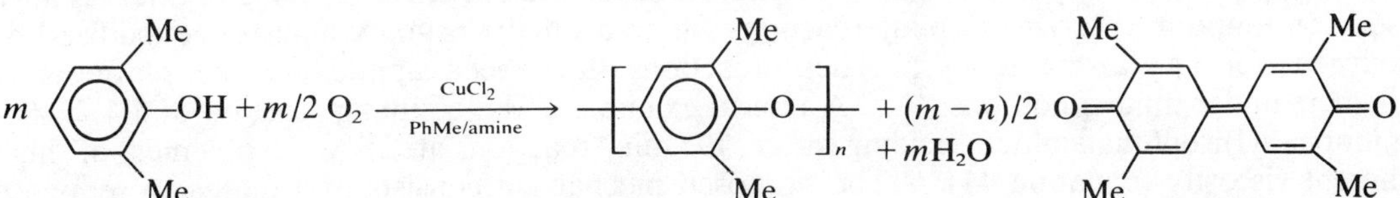

(8)

(42)

Polymerization appears to proceed following two distinct steps: (i) successive additions of DMP phenoxy radicals to oligomeric radicals with formation of quinol ether (**9**) (equation 43) and (ii) reaction between oligomeric phenoxy radicals with formation of ketals, followed by the dissociation/redistribution steps shown in Scheme 8. Not only copper salts, but also compounds such as bis(acetylacetonato)copper(II) and especially bis(ethylacetoacetato)copper(II) are efficient catalysts in the presence of amines.[115] Polymer-bound imidazole–copper(II) complexes also show good specificity for C—O coupling and very good stability during prolonged use in continuous processes.[116] In addition to its role of ligand in the copper/amine complex, the amine increases the pH of the solution, thus lowering the oxidation potential of the phenol.[117] Higher reaction rates and higher C—O/C—C coupling ratios are obtained when the basicity of the amine increases.[115] The reaction rate is significantly enhanced by the addition of strong bases such as KOH, resulting in a maximum rate at an OH/Cu ratio = 1.[118] The kinetics of the reaction catalyzed by the copper/TMEDA complex in the presence of KOH has been examined by Challa *et al.*[114,116,118–121] The results have been explained by an equilibrium leading to the formation of a bridged copper/phenolate/phenol complex, followed by a slow rate-determining step where a two-electron transfer takes place, yielding a dinuclear copper(I) complex and oxidation products. The chain growth with addition and dissociation/redistribution steps are similar to those given in Scheme 8 and equation (43), but the active species are dinuclear copper complexes instead of radicals.[119]

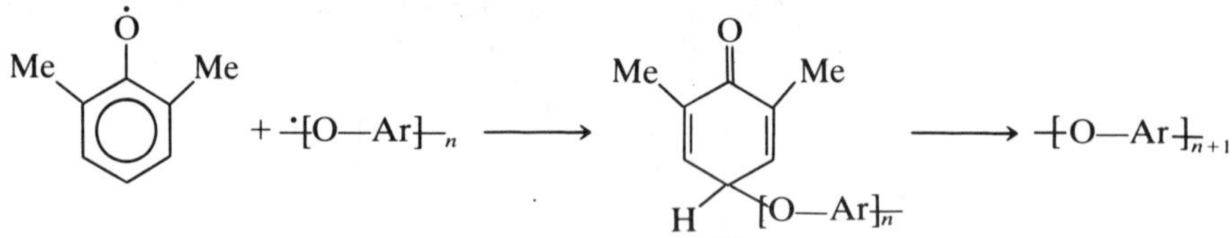

(9) (43)

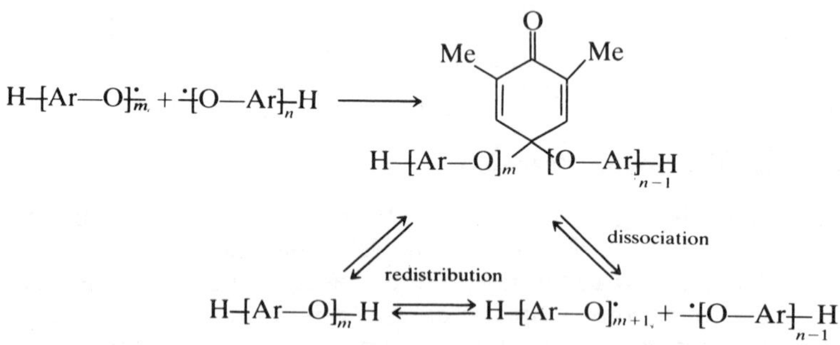

Scheme 8

SEC[122] and kinetic[114] studies have evidenced the much higher reaction rates of dimer and trimer with respect to that of starting monomer, resulting in a narrow Poisson molecular weight distribution at low conversion. At high conversion, the redistribution reaction becomes predominant and Flory-type distributions are obtained.[114]

4.2.3 Carbon–Nitrogen Coupling Reactions

The copper(II)/ternary amine complexes that catalyze the oxidative coupling of phenols have also been found to catalyze coupling reactions where aromatic primary amines are oxidized by dioxygen, forming azobenzenes.[123] These reactions have been applied to the synthesis of polymers under mild conditions.[124,125] A typical example is the coupling reaction of 4,4′-bis(*p*-aminophenyl)isophthalamide, yielding after 195 min reaction at 25 °C a polymer of high inherent viscosity (equation 44).[124] The proposed mechanism consists of a multistep sequence of copper complex formation, electron transfer, coupling and catalyst reoxidation.[124] The scope of this reaction is very wide, as diamines $H_2NC_6H_4$—X—$C_6H_4NH_2$ with X = —O—, —S—, —CH=CH—, —SO_2—, or —CONH—C_6H_4—NHCO—, or diamines with thiazole or oxazoline moieties, can be successfully polymerized.[124] Aminoaryl-terminated aromatic poly(ether sulfone), aromatic poly(ether amide), aliphatic polyethers or aliphatic polyesters also easily couple in DMF/pyridine at 50 °C following the same route.[126]

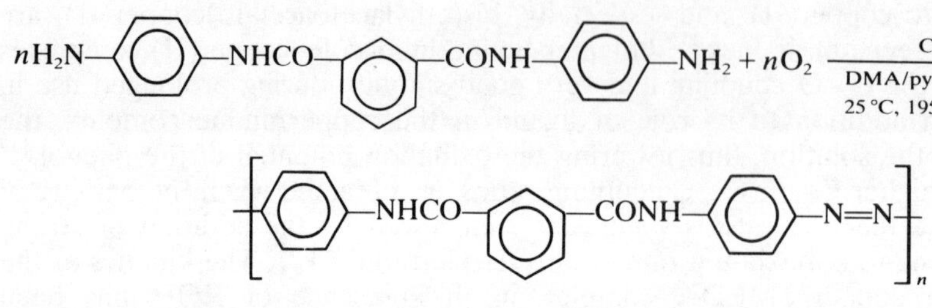

Yield = 100%
inh. visc. = 2.1 dL g^{-1} (44)

When oxidants such as ammonium persulfate[127] or ammonium perchlorate[128] are used, aniline polymerizes to polyaniline, a polymer exhibiting remarkable electrooptical, electrical, and tensile properties. The polymerization of aniline, of 1-aminonaphthalene and of 1-aminoanthracene by H_2O_2 in the presence of iron(II) catalysts has recently been reported (equation 45).[129,130] Two mechanisms are suggested. The first one proceeds through the formation of an ArNH· radical and the ensuing attack of this species on another amine. The second one, which is proposed as the more plausible, involves electron transfer from the monomer to an iron(III) species to give an $ArNH_2^{+}$· radical cation that is expected to undergo coupling with $ArNH_2$.[130]

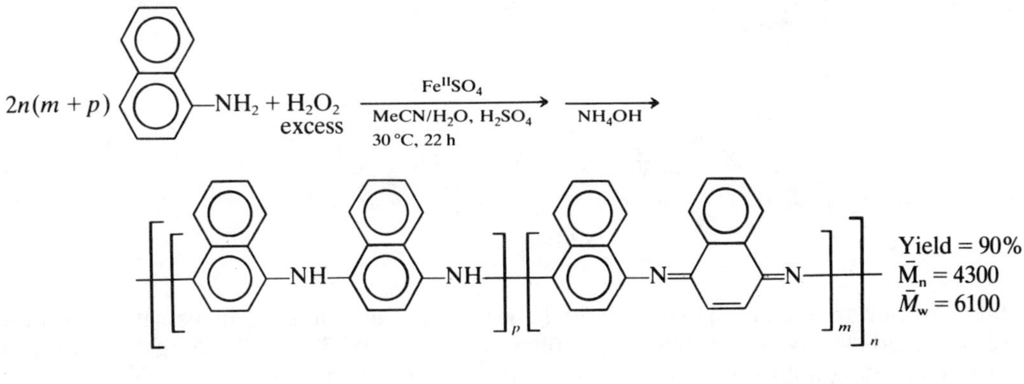

(45)

4.2.4 Carbonylation Coupling Reactions

The reaction of carbon monoxide with σ-bonded organopalladium complexes produces acylpalladium complexes which can be reacted with a wide range of functional groups, *e.g.* alcohols and amines.[27] Since organopalladium complexes are easily obtained from organic halides, this provides a general method to effect a 'coupling' polymerization of dihalides and diols or diamines with CO. Imai *et al.* have reported the synthesis of polyaramides by reaction between aromatic diamines, aromatic dibromides and CO in the presence of palladium(II), palladium(0)[131,132] or nickel[133] catalysis and a strong base (equation 46).

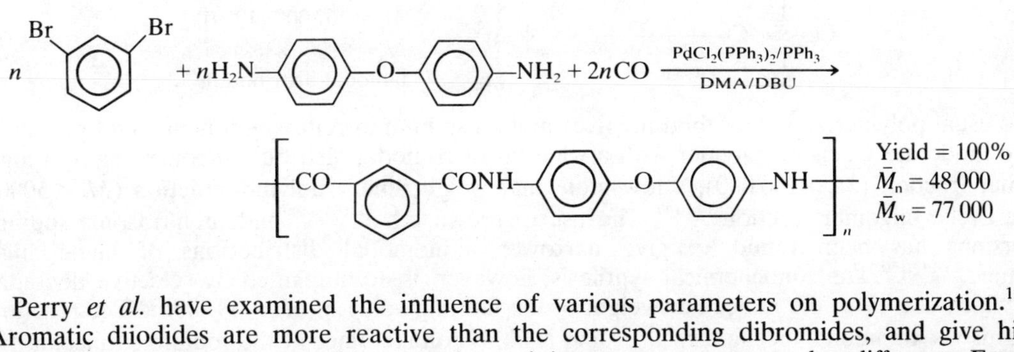

(46)

Perry *et al.* have examined the influence of various parameters on polymerization.[134–136] Aromatic diiodides are more reactive than the corresponding dibromides, and give higher molecular weight polymers. The rate-determining step appears to be different. For aryl bromides the rate-determining step is the oxidative addition of palladium(0), which forms arylpalladium bromide. For aryl iodides, oxidative addition is much more rapid, and the rate-determining step is CO insertion (Scheme 9). This is reflected in the influence of CO pressure on the carbonylation reaction. At high CO pressure the palladium(0) species is coordinated by a greater number of CO molecules, rendering palladium less nucleophilic. Oxidative addition is therefore disfavoured. As a result, aryl bromide carbonylation is inhibited and low molecular weight polymer is obtained. On the other hand, higher CO pressure favours the insertion rate-determining step in the case of aryl iodides, leading to a higher reaction rate and high molecular weight aromatic polyamide (up to $\bar{M}_n = 100\,000$).[136] A slight amine excess is necessary to minimize the reduction of aryl iodide end groups to the corresponding nonreactive arene by palladium compounds. A wide range of substituents, from electron donating to moderately electron withdrawing, are tolerated. With strongly electron-withdrawing substituents, however, a noticeable decrease of reaction rate is observed.[135,136] This reaction has been applied to the synthesis of four-armed star and branched aramids,[137] of poly(amide-imines),[138] and of poly(benzoxazole) precursors.[139]

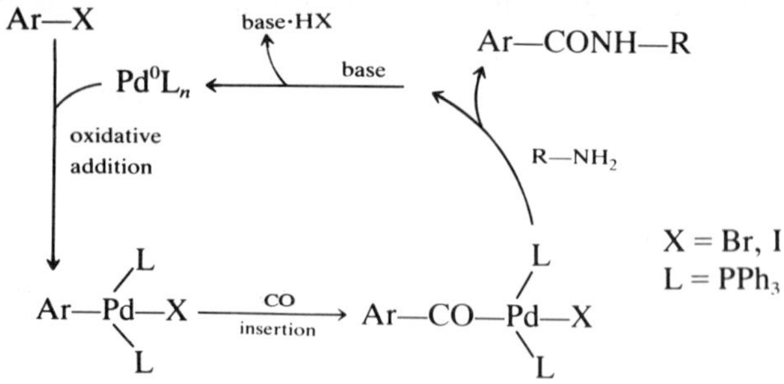

Scheme 9

Another carbonylation coupling reaction is the reaction of aliphatic diamines with CO in the presence of palladium acetate, triphenylphosphine and iodine. The synthesis of polyureas or copolymers containing both urea and oxamide units has thus been reported.[140]

4.2.5 Silicon–Silicon Coupling Reactions

The synthesis of polysilylenes has been the subject of recent and increasing interest, as a consequence of a number of important potential technological applications, *i.e.* photoactive and photoconductive materials, precursors to ceramic materials, and radiation-sensitive materials in microlithography.[141] Polysilylenes have been prepared by the dehydrogenative coupling of monosubstituted silanes using transition metal complexes,[142,143] and by the ring-opening polymerization of strained cyclosilanes.[144] The Wurtz coupling reaction, however, remains the most effective general route for polysilylene preparation (equation 47).[145]

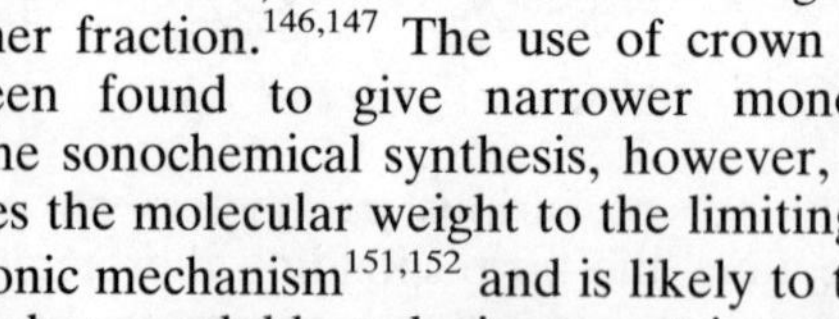

Yield = 25%
$\bar{M}_n = 267\,000$ (12%)
$\bar{M}_n = 8100$ (88%)
bimodal distribution (47)

The usual polymerization method involves molten sodium in refluxing toluene, and produces low to moderate yields of polymer with a wide bi- or trimodal distribution consisting of a high polymer fraction ($\bar{M}_n > 100\,000$), a low molecular weight linear polymer fraction ($\bar{M}_n < 5000$) and a cyclic oligomer fraction.[146,147] The use of crown ethers[145,148] and/or ultrasonic sodium dispersions has been found to give narrower monomodal distributions of linear high polymer.[145,149,150] The sonochemical synthesis, however, is accompanied by selective degradation, which decreases the molecular weight to the limiting value of $\bar{M}_n = 50\,000$.[153] The reaction proceeds via an anionic mechanism[151,152] and is likely to take place at the sodium surface, as no polymer is formed when a soluble reducing agent is used.[154] Although any polymer chain will still have reactive end groups,[150] *i.e.* a characteristic of a step-growth mechanism, polymerization resembles the classical chain-growth process[153] with formation of high polymer from the very beginning of reaction. The higher reactivity of monomer towards the growing chain than towards sodium,[153] and the heterophasic nature of reaction medium, may be involved in this phenomenon.

4.2.6 Dehydrative Coupling Reactions

Alcohols or carboxylic acids can undergo coupling reactions in the presence of dehydrative coupling agents, leading to ethers or anhydrides. As polyethers can be synthesized by much more convenient routes, this chemistry has been mainly used for the synthesis of polyanhydrides. Polyanhydrides are biodegradable materials which find potential applications in controlled drug delivery.[155–157] Compounds such as phenyl *N*-phenylphosphonamidoyl chloride, *N,N*-bis(2-oxo-3-oxazolidinyl)phosphoramidoyl chloride, phosgene or diphosgene are efficient dehydrative coupling agents in solution, but the best results have been obtained using a melt procedure and acetic anhydride (equation 48).[158]

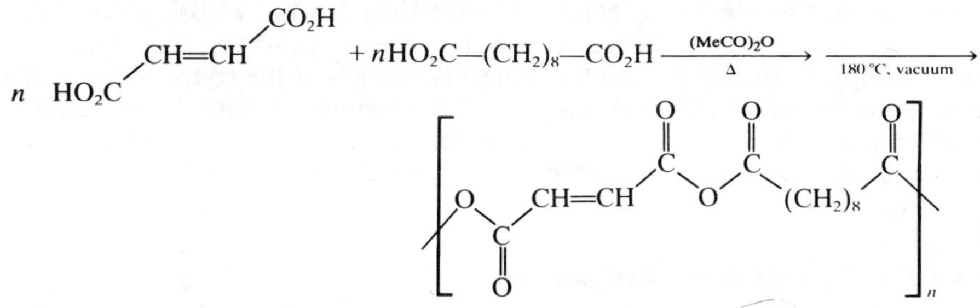

Yield = 100%
$\bar{M}_n = 13\,250$ (48)

4.3 POLYMER COUPLING REACTIONS

As mentioned in the introduction (Section 4.1), the reactions reviewed below involve the linking of polymer chains by a coupling agent. Any chemical reaction can theoretically be applied to polymer-chain coupling provided the coupling agent can react with at least two chain ends. When applied to di- or more functional polymers, the process is typical of step-growth polymerization. However, it is often applied to monofunctional polymers, for the synthesis of ABA-type block copolymers or of star-shaped polymers. This is more related to simple polymer reactions than to step polymerization. However, in both cases the process is governed by similar statistical considerations.

4.3.1 Molecular Weight Distribution

4.3.1.1 Reactions involving difunctional polymers

(i) Polydispersity

The classical equations of step polymerization obviously apply to the reaction of a monodispersed polymer A with a coupling agent, provided the usual conditions concerning equireactivity and the absence of intramolecular reactions are fulfilled. These equations lead to a final polydispersity index (PDI) of 2 for stoichiometric reactions (see Volume 5, Chapter 4 of *Comprehensive Polymer Science*). If coupling agent moieties are not considered as monomer units, but only as links between A units, number- and weight-average degrees of polymerization ($\overline{DP_n}$ and $\overline{DP_w}$) become the corresponding average numbers of A units per chain. They are given by equations (49) and (50),[159] where r is the stoichiometric ratio ($r \leq 1$) and p is the conversion with respect to functional groups not in excess.

$$\overline{DP_n} = 1/(1 - rp^2) \quad (49)$$

$$\overline{DP_w} = (1 + rp^2)/(1 - rp^2) \quad (50)$$

If polydispersed polymers are used, they can be regarded as mixtures of 'monomers' with various molecular weights. In this case, the general Stockmayer's equation[160] still applies, leading to the limiting value of 2 for PDI at total conversion. Therefore, whatever the polydispersity of the initial mixture, it gradually approaches that of the Flory distribution in the course of the reaction. If, in some cases, very broad distributions are obtained, this should not be assigned to the polydispersity of initial prepolymers, but to parameters such as non-equireactivity of functional groups, or specific reaction conditions, *e.g.* the heterophasic reaction medium.

(ii) Block length distribution

When two difunctional polymers A and B react simultaneously with a coupling agent, a multiblock copolymer is obtained. If A and B are equireactive towards coupling agent moieties, copolymer chains present a random distribution of blocks A and B, separated by coupling agent moieties. Constitutional regularity is therefore lower than for the $(AB)_n$ poly(polyA-*block*-polyB) alternating copolymers obtained using a reaction between oligomers with mutually reactive end groups. In the ideal case, the molecular weight of the coupling agent can be neglected in comparison to that of blocks A and B, and the copolymer can be considered as a poly(polyA-*stat*-polyB), with a random succession of blocks A and B. The number- and weight-average lengths of A- and B-type sequences can then be calculated from step-growth copolymerization statistics.[161–164]

4.3.1.2 Reactions involving monofunctional polymers

Reactions between monofunctional polymers and multifunctional coupling agents (functionality $= f$) lead to linear (when $f = 2$) or star-shaped (when $f > 2$) copolymers. Statistical[165–167] or kinetic[168] considerations permit the calculation of molecular weight distributions. In the case of the reaction of a monofunctional polymer with a f-functional coupling agent, $\overline{DP_n}$ and $\overline{DP_w}$ before any fractionation are given by equations (51) and (52)[168]

$$\overline{DP_n} = \frac{f}{f + r\left[1 - \frac{fp}{r} - \left(1 - \frac{p}{r}\right)^f\right]}\overline{DP_{n0}} \qquad (51)$$

$$\overline{DP_w} = \overline{DP_{w0}} + (f-1)\frac{p^2}{r}\overline{DP_{n0}} \qquad (52)$$

where $\overline{DP_{n0}}$ and $\overline{DP_{w0}}$ are the number- and weight-average degrees of polymerization of the starting polymer, p is the conversion in polymer end groups and r is the stoichiometric ratio of coupling agent to polymer reactive groups. For a stoichiometric reaction at total conversion, the polydispersity becomes (equation 53)

$$\overline{DP_w}/\overline{DP_n} = 1 + (\overline{DP_{w0}}//\overline{DP_{n0}} - 1)/f \qquad (53)$$

Therefore the situation is quite different from the case of difunctional polymers, since the reaction always results in a decrease of polydispersity. The more arms in the star-shaped polymer, the more homogeneous is the molecular weight distribution. Moreover, it can be demonstrated[168] that the type of final distribution is the same as that of the starting polymer (*e.g.* Flory or Poisson).

4.3.2 Coupling Reactions Involving Living Chain Ends

4.3.2.1 Carbanionic chain ends

Anionic polymerization is ideally suited for the synthesis of AB block copolymers, as they can be prepared by the simple sequential addition of monomers. Therefore the main purpose of anionic living-end coupling reactions will be: (i) the synthesis of $(AB)_n$ and, in some cases, of ABA block copolymers; (ii) the introduction of a given functionality in the chains; and (iii) the synthesis of polymers with controlled architecture such as macrocyclic or star-shaped polymers and copolymers. Some examples are reported below.

Carbanionic chain ends can react with a number of electrophilic coupling agents such as chlorosilanes or halogenomethyl benzenes (see Volume 7, Chapter 1 of *Comprehensive Polymer Science*). Terephthaldehyde has recently been reported to react with a poly(*t*-butyl acrylate) (block A)/polystyrene (block B) ABA living triblock copolymer, leading to a well-defined $(AB)_n$ block copolymer (equation 54).[169]

$$\mathrm{Li^+,\ ^-PBu^tA{-}PS{-}PBu^tA^-,\ Li^+} \xrightarrow[\text{2) } H^+]{\text{1) } H{-}C(=O){-}C_6H_4{-}C(=O){-}H} \left[\mathrm{PBu^tA{-}PS{-}PBu^tA{-}CH(OH){-}C_6H_4{-}CH(OH)}\right]_n \quad \bar{M}_n = 123\,000 \qquad (54)$$

Azo-containing polystyrene has been prepared by coupling living carbanionic polystyrene with azobisisobutyronitrile.[170] Coupling proceeds with the elimination of CN^- and the formation of azo links in the coupled polymer, which can afterwards be used as a polymeric radical initiator.

Cyclic polymers can be easily prepared from the reaction of living bicarbanionic polymers with difunctional coupling agents in dilute solution. Thus, cyclic polystyrene,[171] cyclic poly(2-vinylpyridine),[172,173] cyclic polystyrene-*block*-poly(2-vinylpyridine) and polystyrene-*block*-poly(dimethylsiloxane)[174] have been synthesized. The synthesis of multicyclic and of 'figure-of-eight-shaped' polyisoprenes has recently been reported: a bicarbanionic polyisoprene reacts with a diethylenic coupling agent, giving a bicarbanionic macrocycle able to reinitiate the polymerization of isoprene. Subsequent coupling with dichlorodimethylsilane yields the bicyclic polyisoprene (Scheme 10).[175,176]

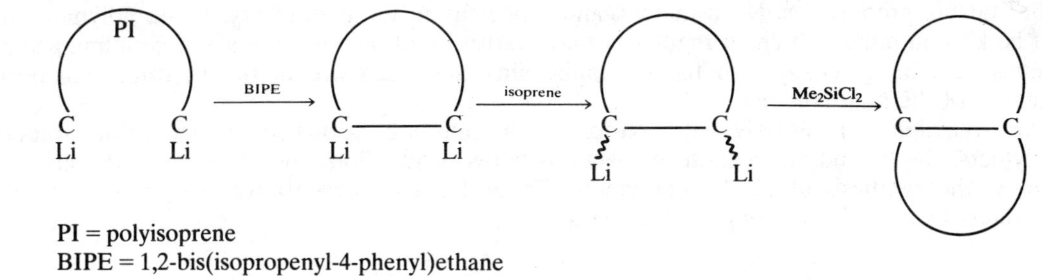

Scheme 10

The synthesis of star-shaped polymers and the most commonly used coupling agents are reviewed in Volume 7, Chapter 1, Section 1.3.1.3 of *Comprehensive Polymer Science.* Among recent developments, the synthesis of star polybutadienes with 32,[177] 64 or 128[178] arms can be noted. The coupling agents are carbosilane dendrimers with 32, 64 or 128 surface Si—Cl groups, respectively.

4.3.2.2 Carbocationic chain ends

The scope of living cationic polymerization is much more limited, since it practically concerns only the vinyl ether polymerization by the HI/I_2 or HI/ZnI_2 initiating systems[179,180] (see Volume 3, Chapter 42 of *Comprehensive Polymer Science*). Some multifunctional coupling agents have been described for these polymerizations, such as hexamethylenediamine[179] and multifunctional malonate salts.[181] Multifunctional silyl enol ethers, which are nonionic nucleophiles, are much more soluble and have proven to be very efficient.[182,183] The coupling reaction involves an electrophilic addition of the living end onto the enol ether double bond of the coupling agent to form a ketone (Scheme 11).

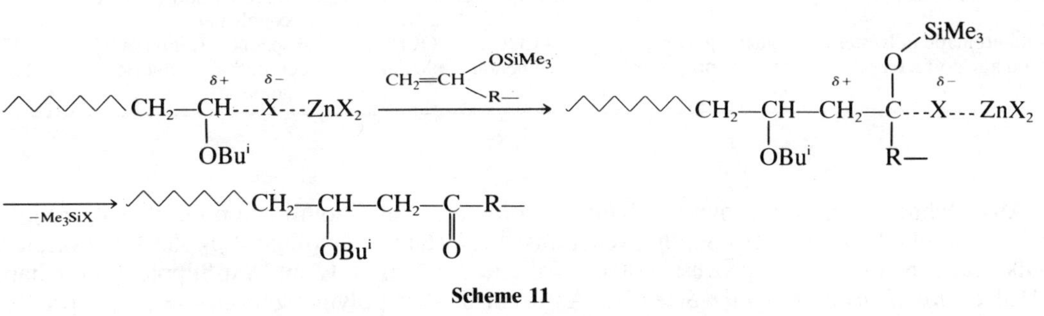

Scheme 11

4.3.3 Coupling Reactions Involving Functional Prepolymers

The synthesis of high molecular weight polymers by step-growth polymerization requires high degrees of conversion. In many cases, this means high temperatures, long reaction times and low pressures to distil off the last traces of reaction by-product, *e.g.* water. Moreover, in such conditions the existence of some undesired side reactions is unavoidable. To circumvent these drawbacks, a coupling agent may be added in the last steps of the reaction. This coupling agent should be able to give a rapid polyaddition with the end groups of the growing chains, so that high molecular weight copolymers can be obtained using lower temperatures, shorter reaction times and no vacuum. With coupling agents of functionality >2 the process leads to highly branched or cross-linked products. With difunctional coupling agents, high molecular weight linear copolymers are produced, and the process is often termed 'chain extension'. In order to find similar properties in the final linear copolymer and in the corresponding high molecular weight homopolymer, the MW of the coupling agent is usually much lower than that of the starting prepolymer. Nevertheless, and especially in the case of crystalline polymers, it must be kept in mind that chain regularity may be disturbed by the presence of coupling agent moieties, leading to changes in the crystallizability and therefore in the thermomechanical behaviour of the final polymer.

Since reactions of prepolymer mixtures with coupling agents obviously afford block copolymers, the second application of reactions between functional prepolymers and coupling agents is the synthesis of block copolymers. Table 1 lists representative syntheses of block copolymers by prepolymer coupling reactions.

Table 1 Representative Syntheses of Block Copolymers by Oligomer Coupling Reactions.

Oligomer A	*Oligomer B*	*A reactive groups, B reactive groups*	*Coupling agent and reaction conditions*	*Ref.*
Aliphatic polyether	Aliphatic polyamide	$—NH_2$, $—NH_2$	Bisoxazolinone, bulk	216
Aliphatic polyether	Aliphatic polyamide	$—CO_2H$, $—CO_2H$	Bisoxazoline, bulk	231
Aliphatic polyether	Aliphatic polyanhydride	$—CO_2COMe$, $—CO_2COMe$	Self coupling, bulk	232
Polyoxyethylene + polyoxypropylene	Polystyrene	(—OH + OH), —phenol	Diisocyanate, solution, three-constituent block copolymer	190
Triblock polyether/ polystyrene/ polyether	—	—ONa	Phosgene, solution	194
Polyformal	Aromatic polyether	—OH, —phenol	Isophthaloyl chloride, interfacial	196
Aliphatic polyester	Aliphatic polyester	—OH, —OH	Diisocyanate, solution	187
Aliphatic polyester	Aliphatic esteramide	—OH, —phenol	Phosgene, solution	195
Poly(dimethylsiloxane)	Aliphatic polyamide	(—OH, $—NH_2$, or $—CO_2H$), ($—NH_2$ or CO_2H)	Diisocyanate, solution	188
Polyisobutylene	Polycarbonate	—OH, —phenol	Diisocyanate, solution	233
Polyether + polybutadiene	Poly(ether sulfone)	(—OH, —OH), —phenol	Diisocyanate, solution, three-constituent block copolymer	234
Semi-aromatic polyester	Aromatic polyester	—COCl, —COCl	Bisphenol A, interfacial	235
Aromatic polyester	Aromatic polyester	—phenol, —phenol	Terephthaloyl chloride, solution	198

Any technique of step-growth polymerizations, *e.g.* bulk, solution or interfacial reactions, may be applied to polymer-coupling reactions. Important technological applications often use bulk reactions or reactive processing (see Volume 7, Chapter 12 and 1st Supplement, Chapter 21 of *Comprehensive Polymer Science*). As usual in step polymerization, the major problems arise from: (i) the control of the stoichiometry and therefore control of the functionality and purity of the starting reactive oligomers; and (ii) control of side reactions, which lead to undesired structures and to a decrease of the molecular weight, especially at high temperatures.

4.3.3.1 *Self-coupling reactions*

There are only few references relating to oligomer self-coupling reactions, *i.e.* reactions where the coupling agent is not included in the chains of the final polymer. The oxidative coupling reaction of amino-terminated prepolymers has already been mentioned in Section 4.2.3.[126] Heating dihydroxypoly(oxytetramethylene) in the presence of acid catalysts leads to a molecular weight increase due to dehydrative chain coupling.[184] The carbenoid polycondensation of dichloroacetate-terminated polyethers affords segmented copolymers (equation 55).[185]

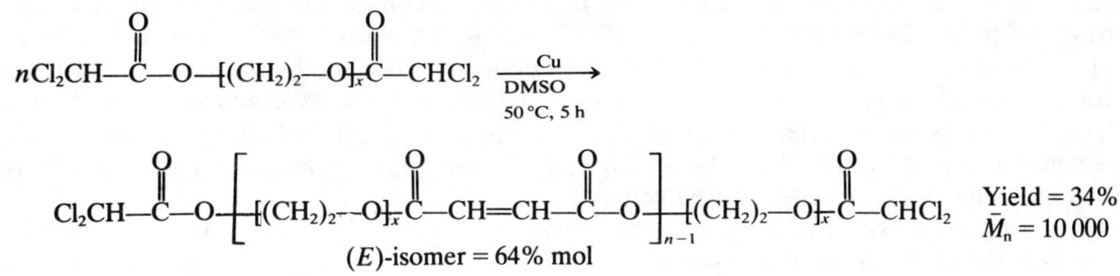

(55)

4.3.3.2 *Polymers with OH end groups*

(i) Coupling reactions with isocyanates

Isocyanates are probably the most used coupling agent, as they give rapid reactions with a number of protic nucleophiles, particularly hydroxy, amino or carboxy end groups, which are very common polymer end groups. The reaction between dihydroxy aliphatic polyethers or polyesters, diols and diiisocyanates is the basis of the preparation of polyurethane elastomers (see Volume 5, Chapter 24 of *Comprehensive Polymer Science*). In this case, diisocyanate does not act as a chain coupling agent only, but confers specific properties on the final block copolymer through the formation of polyurethane hard blocks. Consequently, this will not be regarded as a coupling reaction. A number of coupling reactions involving the alcohol/isocyanate solution reaction were described during the 1960s and early 1970s[186,187] for the synthesis of polyester–polyester block copolymers. Such copolymers would undergo randomization by ester interchange if high-temperature bulk reactions were used. More recent references are related to the coupling reaction of OH-terminated polysiloxanes with NH_2-terminated polyamides,[188] to the syntheses of polyisobutylene/polyoxyethylene di- and triblock copolymers,[189] and of polystryene/polyoxypropylene/polyoxyethylene multiblock copolymers.[190,191] by solution coupling reactions involving OH-terminated prepolymers. Other examples are given in Table 1.

Diisocyanates have also been used in a number of chain extension reactions, either in solution, *e.g.* for dihydroxypoly(methyl methacrylate),[192] or in the bulk, *e.g.* for the preparation of a high molecular weight aliphatic polyester.[193] In this last case, chain coupling probably involves both OH/NCO and CO_2H/NCO reactions.

(ii) Coupling reactions with acid chlorides

Phosgene[194,195] has been used for the synthesis of various block copolymers by reaction with OH end groups, but aromatic diacid chlorides, such as isophthaloyl chloride or terephthaloyl chloride, are now the most used coupling agents of this type.[196,197] In some cases, coupling agent moieties are also constituent units of one block and are therefore not anomalous units in the final block copolymer. This occurs in the coupling reactions of two OH-terminated homopolymers of *cis*- or *trans*-cyclohexanedimethanol and terephthalic acid with terephthaloyl chloride.[198]

(iii) High-temperature coupling reactions

At high temperatures, side reaction problems arise with diisocyanates and acid chlorides. Diisocyanates undergo a trimerization reaction, which produces branched polymers,[199] and the hydrogen chloride released during the OH/acid chloride bulk reactions can promote a number of side reactions.

Diesters, such as diphenyl carbonate or diphenyl terephthalate, react with the OH end groups of poly(ethylene terephthalate) (PET) at high temperatures,[199] but with the formation of undesirable by-products such as phenol or ethylene carbonate. Dianhydrides have been used for the coupling reactions of dihydroxypolybutadiene, -polyoxyethylene or -polyoxypropylene,[200] and for chain extension of PET.[201] However, their use in high-temperature bulk reactions may result in branched products.[199] The preferred coupling agents for such reactions are bis(cyclic iminoesters) and bis(*N*-acyllactams).

Bis(cyclic iminoesters) exhibit good reactivity with the OH end groups of molten poly(alkylene terephthalates) at temperatures above 240 °C. The most efficient are bis(4,4′-dimethyl-5(4*H*)-oxazolinones) (**10**), bis(4*H*-3,1-benzoxazin-4-ones) (**11**) and (**12**) and especially those presenting electron-withdrawing effects in the 2 position, such as 2,2′-bis(4*H*-3,1-benzoxazin-4-one) (**13**) (equation 56).[202]

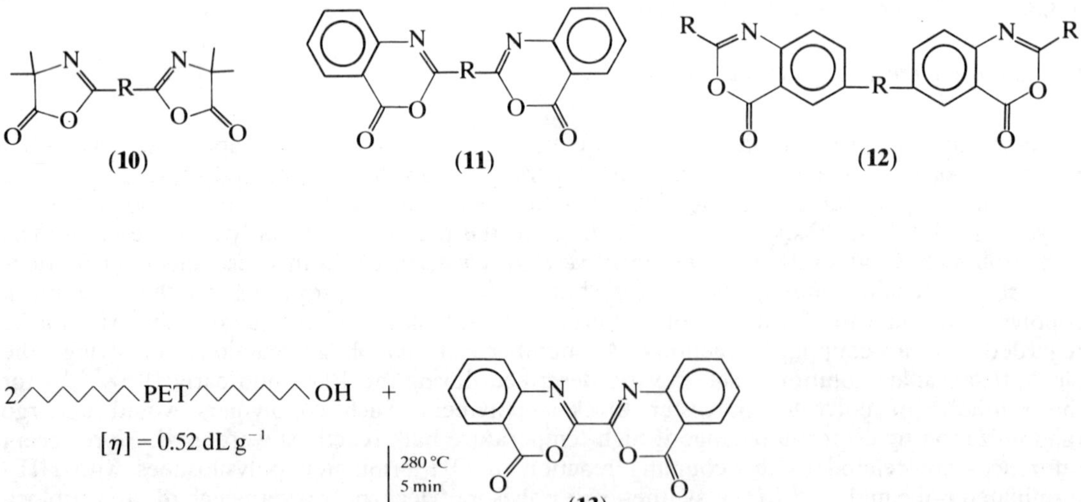

(**10**) (**11**) (**12**)

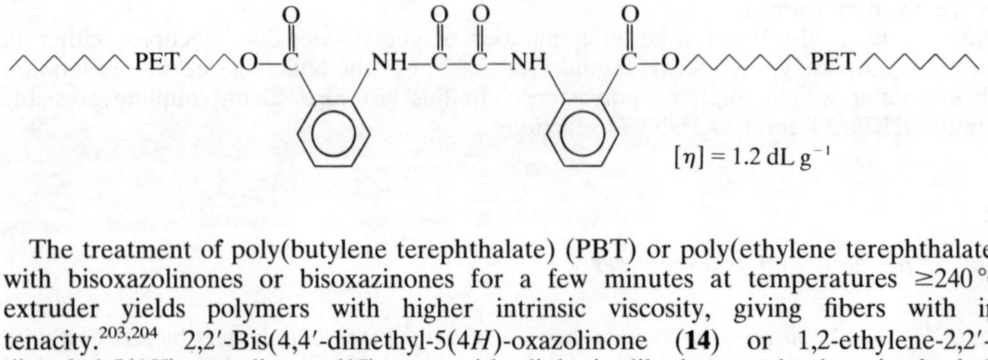

(56)

The treatment of poly(butylene terephthalate) (PBT) or poly(ethylene terephthalate) (PET) with bisoxazolinones or bisoxazinones for a few minutes at temperatures ≥240 °C in an extruder yields polymers with higher intrinsic viscosity, giving fibers with improved tenacity.[203,204] 2,2′-Bis(4,4′-dimethyl-5(4*H*)-oxazolinone (**14**) or 1,2-ethylene-2,2′-bis(4,4′-dimethyl-5(4*H*)-oxazolinone (**15**) react with aliphatic dihydroxy polyethers in the bulk in the presence of catalysts which increase the reaction rate and limit the importance of some side reactions (equation 57).[205]

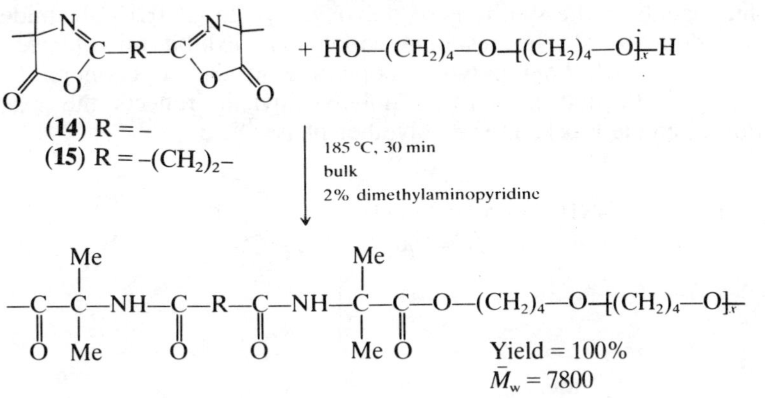

(57)

Terephthaloyl bis(laurolactam) has been found to be quite effective in increasing the molecular weight of PET by reactive extrusion.[206] The efficiency of the chain extension has been explained by the predominating ring opening process over the lactam elimination/ester–lactam interchange route (Scheme 12).

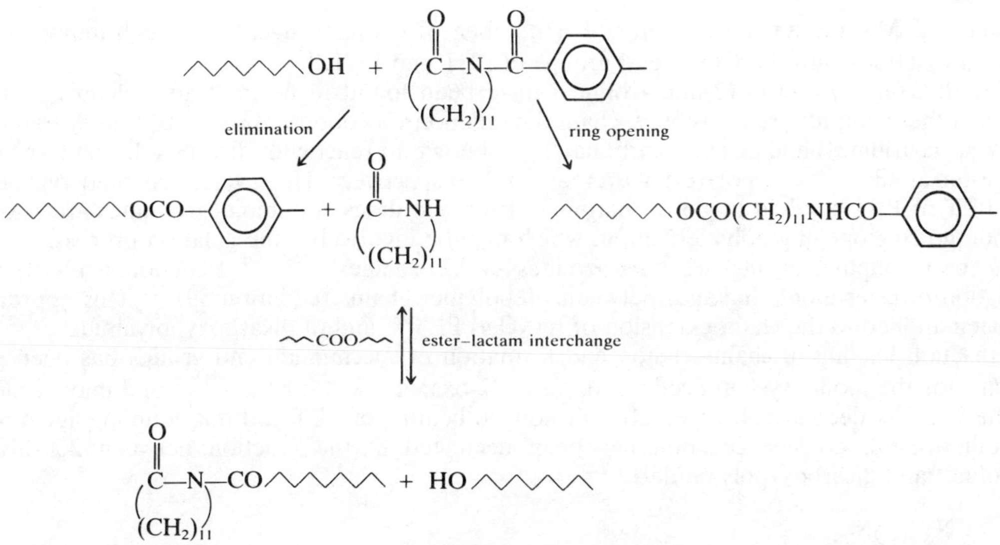

Scheme 12

Other coupling agents have been reported for reaction with OH end groups, such as phenoxy-*s*-triazine[207] and diisopropylidene methylenedimalonate[208] for PET high-temperature bulk reactions, 4,4′-difluorobenzophenone for poly(arylene ether sulfone) solution reactions[209] and 1,3-dioxolane for polystyrene solution reactions.[210]

4.3.3.3 *Polymers with NH_2 end groups*

Most coupling agents used for OH groups give efficient reactions with amino end groups. Isocyanates are more reactive with amino than with hydroxy groups, leading to the rapid formation of urea linkages (see Volume 5, Chapter 25 of *Comprehensive Polymer Science*). They have been used in the coupling reactions of OH-terminated prepolymers with NH_2-terminated prepolymers,[188] for the synthesis of networks with controlled structure[211] and for the synthesis of block copolymers from NH_2-terminated prepolymers.[212,213] Bis(*N*-acyllactams) readily react with the amino groups of aliphatic polyamides, leading to an effective chain coupling reaction.[206] The reaction of bisoxazolinones with amino groups[214,215] has been applied to the synthesis of block copolymers in the bulk at high temperatures. Thus, bisoxazolinones acted as coupling agents in the synthesis of polyoxyethylene (POE)/polyamide-12 (PA) block copolymers (equation 58).[216] The resulting copolymers exhibit microphase separation, the thermal transitions of each homopolymer appearing in the copolymers. However, the T_g increase with respect to that of starting polyoxyethylene reflects the existence of some compatibilization of amide blocks in the polyether phase.[216]

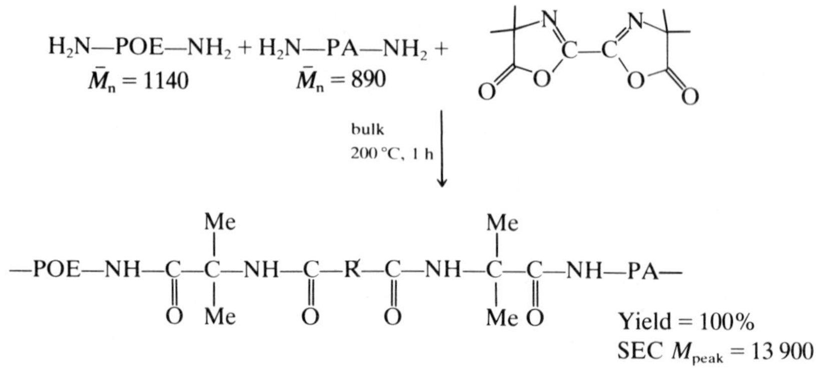

(58)

4.3.3.4 Polymers with CO_2H end groups

Inata and Matsumura have examined a number of coupling agents for high-temperature addition reactions with the CO_2H end groups of PET and PBT.[217]

Bis(2-thiazolines) and bis(2-imidazolines) have been found to be inefficient chain coupling agents in these conditions. Moreover, chain degradation was observed with bis(2-imidazolines). Bis(*N*-acylaziridines) and bisiminocarbonates are known to react with dicarboxylic acids, giving poly(ester amides)[218] and poly(ester urethanes),[219] respectively. However, their melt reactions with PET or PBT resulted in gel formation.[217] Both aziridines and iminocarbonates are known to undergo ring opening polymerization, which may be involved in the gelation process.

The best coupling agents are bisoxazolines or bisoxazines.[217,220,221] Reaction leads to the formation of ester amide linkages between prepolymer chains (equation 59).[221] This approach has been applied to the chain extension of recycled PET[222] and of dicarboxypolyamide-12.[223] A side reaction leading to chain scission and formation of succinimide end groups has been put forward for the model system acetic acid/1,2-bis(2-oxazolin-2-yl)ethane,[224,225] and may account for the viscosity decrease observed after prolonged heating of PET and this coupling agent. On the other hand, no side reaction has been dectected in the reaction between 2,2′-bis(2-oxazoline) and dicarboxypolyamide-12.[223]

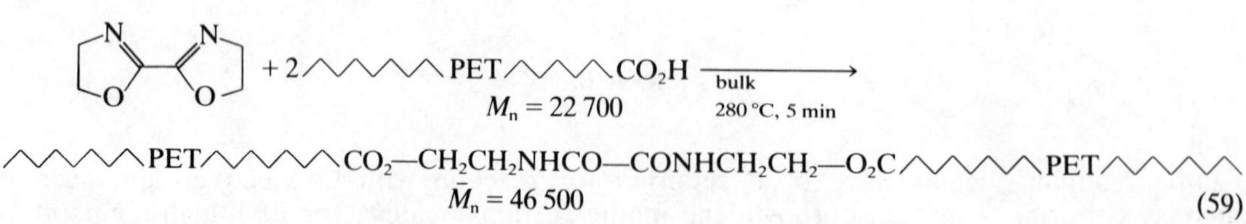

(59)

4.3.3.5 Miscellaneous coupling reactions

Halo-terminated prepolymers may be coupled by reaction with dicarboxylic acid salts.[226,227] For example, potassium succinate has been used for the solution coupling reaction between dibromo-terminated aliphatic polyesters.[227]

Bis(silylcyclopentadiene)-terminated polyisobutylenes may be chain extended by reaction with bismaleimides,[228] and vinyl-terminated polymers by reaction with *p*-bis(dimethylsilyl)benzene in the presence of chloroplatinic acid catalyst.[229]

A final example is the synthesis of star polyoxytetramethylene by a coupling reaction of a monofunctional azetidinium-terminated polyoxytetramethylene with tetrafunctional carboxylates.[230]

4.4 REFERENCES

1. Source: *Chemical Abstracts,* Chemistry of Synthetic High Polymers, reference period 1984–1994.
2. J. March, 'Advanced Organic Chemistry', 4th edn., Wiley, New York, 1992.
3. B. M. Trost and I. Fleming (eds.), 'Comprehensive Organic Synthesis', Pergamon Press, Oxford, 1991, vol. 3.
4. R. C. Larock, 'Comprehensive Organic Transformations', VCH, Weinheim, 1989, p. 45.
5. E. P. Plueddemann, in 'Encyclopedia of Polymer Science and Engineering', ed. H. F. Mark, N. M. Bikales, C. G. Overberger and G. Menges, Wiley, New York, 1988, vol. 4, p. 284.
6. A. Wurtz, *Ann. Chim. Phys.,* 1855, **44,** 275.
7. M. Tamura and J. K. Kochi, *J. Organomet. Chem.,* 1972, **42,** 205.
8. J. K. Kochi, 'Organometallic Mechanisms and Catalysis', Academic Press, New York, 1978, p. 372.
9. T. Yamamoto, T. Taguchi, K. Sanechika, Y. Hayashi and A. Yamamoto, *Macromolecules,* 1983, **16,** 1555.
10. M. Malanga and O. Vogl, *J. Polym. Sci., Polym. Chem. Ed.,* 1983, **21,** 2629.
11. M. Nanasawa, L. Hu and O. Vogl, *Polymer,* 1987, **28,** 514.
12. Y. Chujo, I. Tomita, Y. Kozawa and T. Saegusa, *Macromolecules,* 1993, **26,** 2643.
13. M. Inoue, S. Cantua, F. Medrano, K. Asano and M. Nakamura, *Polym. Bull.,* 1992, **28,** 29.
14. E. Cramer and V. Percec, *J. Polym. Sci., Part A: Polym. Chem.,* 1990, **28,** 3029.
15. A. Suzuki, *Acc. Chem. Res.,* 1982, **15,** 178.
16. R. N. McDonald and T. W. Campbell, *J. Am. Chem. Soc.,* 1960, **82,** 4669.
17. H.-H. Hörhold and J. Opfermann, *Makromol. Chem.,* 1970, **131,** 105.
18. H.-H. Hörhold and D. Raabe, *Acta Polym.,* 1979, **30,** 86.
19. D. R. Gagnon, J. D. Capistran, F. E. Karasz and R. W. Lenz, *Polym. Bull.,* 1984, **12,** 293.
20. I. Murase, T. Ohnishi, T. Noguchi and M. Hirooka, *Polym. Commun.,* 1984, **25,** 327.
21. R. F. Heck, *Acc. Chem. Res.,* 1979, **12,** 146.
22. R. F. Heck, *Org. React.,* 1982, **27,** 345.
23. W. Heitz, W. Brügging, L. Freund, M. Gailberger, A. Greiner, H. Jung, U. Kampschulte, N. Nießner, F. Osan, H.-W. Schmidt and M. Wicker, *Makromol. Chem.,* 1988, **189,** 119.
24. A. Greiner and W. Heitz, *Makromol. Chem., Rapid. Commun.,* 1988, **9,** 581.
25. H. Martelock, A. Greiner and W. Heitz, *Makromol. Chem.,* 1991, **192,** 967.
26. J.-C. Lim, M. Suzuki and T. Saegusa, *Polym. Bull.,* 1993, **31,** 651.
27. R. F. Heck, 'Palladium Reagents in Organic Synthesis', Academic Press, London, 1985.
28. M. Suzuki, J.-C. Lim and T. Saegusa, *Macromolecules,* 1990, **23,** 1574.
29. Z. Bao, Y. Chen, R. Cai and L. Yu, *Macromolecules,* 1993, **26,** 5281.
30. Z. Bao, Y. Chen and L. Yu, *Macromolecules,* 1994, **27,** 4629.
31. M. Suzuki, K. Sho, J.-C. Lim and T. Saegusa, *Polym. Bull.,* 1989, **21,** 415.
32. H.-P. Weitzel and K. Müllen, *Makromol. Chem.,* 1990, **191,** 2837.
33. M. Yoneyama, M. Kakimoto and Y. Imai, *Macromolecules,* 1989, **22,** 4148.
34. K. Y. Chen, H. Eckardt, T. R. Jow, L. W. Shacklette and R. L. Elsenbaumer, *J. Chem. Soc., Chem. Commun.,* 1988, 215.
35. M. E. Wright, *Polym. Prepr., Am. Chem. Soc., Div. Polym. Chem.,* 1992, **33** (1), 1127.
36. R. Galarini, A. Musco, R. Pontellini, A. Bolognesi, S. Destri, M. Castellani, M. Mascherpa and G. Zhuo, *J. Chem. Soc., Chem. Commun.,* 1991, 364.
37. A. Bolognesi, M. Catellani, A. Musco and R. Pontellini, *Synth. Met.,* 1993, **55–57,** 1255.
38. D. Milstein and J. K. Stille, *J. Am. Chem. Soc.,* 1979, **101,** 4992.
39. W. J. Scott and J. K. Stille, *J. Am. Chem. Soc.,* 1986, **108,** 3033.
40. P. E. Fanta, *Synthesis,* 1974, 9.
41. M. Sainsbury, *Tetrahedron,* 1980, **36,** 3327.
42. P. H. Gore and G. K. Hugues, *J. Chem. Soc.,* 1959, 1615.
43. R. D. Rieke and L. D. Rhyne, *J. Org. Chem.,* 1979, **44,** 3445.
44. S. Ozasa, N. Hatada, Y. Fujioka and E. Ibuki, *Bull. Chem. Soc. Jpn.,* 1980, **53,** 2610.
45. J. G. Speight, P. Kovacic and F. W. Koch, *J. Macromol. Sci., Rev. Macromol. Chem.,* 1971, **C5,** 295.
46. L. Groenendaal, H. W. I. Peerlings, J. L.-J. van Dongen, E. E. Havinga, J. A. J. M. Vekemans and E. W. Meijer, *Polym. Prepr., Am. Chem. Soc., Div. Polym. Chem.,* 1994, **35**(1), 194.
47. E. B. Stephens, K. E. Kinsey, J. F. Davis and J. E. Tour, *Macromolecules,* 1993, **26,** 3519.
48. U. Müller, T. Mangel and K. Müllen, *Makromol. Chem., Rapid Commun.,* 1994, **15,** 45.
49. M. F. Semmelhack, P. M. Helquist and L. D. Jones, *J. Am. Chem. Soc.,* 1971, **93,** 5908.
50. T. Kanbara, N. Saito, T. Yamamoto and K. Kubota, *Macromolecules,* 1991, **24,** 5883.
51. T. Yamamoto, A. Morita, Y. Miyazaki, T. Maruyama, H. Wakayama, Z. Zhou, Y. Nakumura, T. Kanbara, S. Sasaki and K. Kubota, *Macromolecules,* 1992, **25,** 1214.
52. T. Yamamoto, A. Morita, T. Maruyama, Z. H. Zhou, T. Kanbara and K. Sanechika, *Polym. J. (Tokyo),* 1990, **22,** 187.
53. K. Chmil and U. Scherf, *Makromol. Chem., Rapid Commun.,* 1993, **14,** 217.
54. M. Sato and M. Hiroi, *Chem. Lett.,* 1994, 985.
55. P. V. Bedworth and J. M. Tour, *Macromolecules,* 1994, **27,** 622.
56. M. Ueda and T. Ito, *Polym. J.,* 1991, **23,** 297.
57. A. Siove, D. Adès, C. Chevrot and G. Froyer, *Makromol. Chem.,* 1989, **190,** 1361.
58. J.-F. Fauvarque, M.-A. Petit, G. Pfluger, A. Jutand, C. Chevrot and M. Troupel, *Makromol. Chem., Rapid Commun.,* 1983, **4,** 455.
59. A. Aboulkassim, K. Faïd and A. Siove, *Makromol. Chem.,* 1993, **194,** 29.
60. M. Troupel, Y. Rollin, O. Sock, G. Myer and J. Perichon, *Nouv. J. Chim.,* 1986, **10,** 593.
61. I. Colon and G. T. Kwiatkowski, *J. Polym. Sci., Part A: Polym. Chem.,* 1990, **28,** 367.
62. M. Ueda and F. Ichikawa, *Macromolecules,* 1990, **23,** 926.
63. R. Singh and A. S. Hay, *Makromol. Chem., Macromol. Symp.,* 1992, **54/55,** 357.

64. T. Yamamoto, Y. Hayashi and A. Yamamoto, *Bull. Chem. Soc. Jpn.*, 1978, **51,** 2091.
65. M. B. Jones and P. Kovacic in 'Comprehensive Polymer Science', ed. G. Allen and J. C. Bevington, Pergamon Press, Oxford, 1989, vol. 5, chap. 27, p. 465.
66. J. W.-P. Lin and L. P. Dudek, *J. Polym. Sci., Polym. Chem. Ed.*, 1980, **18,** 2869.
67. H. Mao, B. Xu and S. Holdcroft, *Macromolecules,* 1993, **26,** 1163.
68. Y. H. Kim and O. W. Webster, *Macromolecules,* 1992, **25,** 5561.
69. S. W. Kantor, T.-C. Sung and E. D. T. Atkins, *Macromolecules,* 1992, **25,** 2789.
70. S. K. Taylor, S. G. Bennett, I. Khoury and P. Kovacic, *J. Polym. Sci., Polym. Lett. Ed.*, 1981, **19,** 85.
71. U. Anton and K. Müllen, *Makromol. Chem., Rapid Commun.*, 1993, **14,** 223.
72. M. Bochmann, K. Kelly and J. Lu, *J. Polym. Sci., Part A: Polym. Chem. Ed.*, 1992, **30,** 2511.
73. M. Bochmann and K. Kelly, *J. Polym. Sci., Part A: Polym. Chem. Ed.*, 1992, **30,** 2503.
74. L. Yu, W.-K. Chan, Y. Chen, Z. Peng, Z. Bao and D. Yu, *Polym. Prepr., Am. Chem. Soc., Div. Polym. Chem.*, 1993, **34** (2), 536.
75. Z. Bao, R. Cai and L. Yu, *Polym. Prepr., Am. Chem. Soc., Div. Polym. Chem.*, 1993, **34** (2), 749.
76. M. Bochmann and K. Kelly, *J. Chem. Soc., Chem. Commun.*, 1989, 532.
77. M. Bochmann and J. Lu, *J. Polym. Sci., Part A: Polym. Chem.*, 1994, **32,** 2493.
78. N. Miyaura, T. Yanagi and A. Suzuki, *Synth. Commun.*, 1981, **11,** 513.
79. M. Rehahn, A.-D. Schlüter, G. Wegner and W. J. Feast, *Polymer,* 1989, **30,** 1060.
80. S. Martina, V. Enkelmann, A.-D. Schlueter and G. Wegner, *Synth. Met.*, 1991, **41–43,** 403.
81. J. K. Kallitsis, M. Rehahn and G. Wegner, *Makromol. Chem.*, 1992, **193,** 1021.
82. M. M. Murray, P. Kaszynski, D. A. Kaisaki, W. Chang and D. A. Dougherty, *J. Am. Chem. Soc.*, 1994, **116,** 8152.
83. F. Helmer-Metzmann, M. Rehahn, L. Schmitz, M. Ballauf and G. Wegner, *Makromol. Chem.*, 1992, **193,** 1847.
84. M. B. Jones and P. Kovacic, in 'Encyclopedia of Polymer Science and Engineering', ed. H. F. Mark, N. M. Bikales, C. G. Overberger and G. Menges, Wiley, New York, 1987, vol. 10, p. 670.
85. P. Kovacic and M. B. Jones, *Chem. Rev.*, 1987, **87,** 357.
86. V. Percec, S. Okita and J. H. Wang, *Macromolecules,* 1992, **25,** 64.
87. M. Bragadin, P. Cescon, A. Berlin and F. Sannicolò, *Makromol. Chem.*, 1987, **188,** 1425.
88. M. L. Blohm, J. E. Pickett and P. C. Van Dort, *Macromolecules,* 1993, **26,** 2704.
89. V. Percec, J. H. Wang, Y. Oishi and A. E. Feiring, *J. Polym. Sci., Polym. Chem. Ed.*, 1991, **29,** 965.
90. V. Percec, J. H. Wang and Y. Oishi, *J. Polym. Sci., Polym. Chem. Ed.*, 1991, **29,** 949.
91. V. Percec and J. H. Wang, *Polym. Bull.*, 1991, **25,** 9.
92. V. Percec, J. H. Wang and S. Okita, *J. Polym. Sci., Polym. Chem. Ed.*, 1991, **29,** 1789.
93. R. D. Stephens and C. E. Castro, *J. Org. Chem.*, 1963, **28,** 3313.
94. G. H. Posner, *Org. React.*, 1975, **22,** 253.
95. T. Yamamoto, W. Yamada, M. Takagi, K. Kiza, T. Maruyama, N. Ooba, S. Tomaru, T. Kurihara, T. Kaino and K. Kubota, *Macromolecules,* 1994, **27,** 6620.
96. C. Pugh and V. Percec, *J. Polym. Sci., Part A: Polym. Chem.*, 1990, **28,** 1101.
97. V. Percec, C. Pugh, E. Cramer and R. Weiss, *Polym., Prepr., Am. Chem. Soc., Div. Polym. Chem.*, 1991, **32** (1), 329.
98. S. W. Kantor and T.-C. Sung, *Macromolecules,* 1993, **26,** 3758.
99. H. K. Black, D. H. S. Horn and B. C. L. Weedon, *J. Chem. Soc.*, 1954, 1704.
100. H. H. Schlubach and V. Franzen, *Justus Liebig Ann. Chem.*, 1951, **572,** 116.
101. A. S. Hay, *J. Org. Chem.*, 1961, **27,** 3320.
102. A. S. Hay, *J. Polym. Sci., Part A1,* 1969, **7,** 1625.
103. A. S. Hay, D. A. Bolon, K. R. Leimer and R. F. Clark, *J. Polym. Sci., Part B,* 1970, **8,** 97.
104. H. J. Kevelam, K. P. de Jong, H. C. Meinders and G. Challa, *Makromol. Chem.*, 1975, **176,** 1369.
105. K. E. Knol, L. W. Van Horssen, G. Challa and E. E. Havinga, *Polym. Commun.*, 1985, **26,** 71.
106. H. S.-I. Chao and M. A. Vallance, *J. Polym. Sci., Part A: Polym. Chem.*, 1990, **28,** 1209.
107. M. Altmann and U. H. F. Bunz, *Makromol. Chem., Rapid Commun.*, 1994, **15,** 785.
108. D. R. Rutherford and J. K. Stille, *Macromolecules,* 1988, **21,** 3530.
109. B. R. McKellar and W. A. Feld, *Polym. Prepr., Am. Chem. Soc., Div. Polym. Chem.*, 1993, **34** (2), 380.
110. M. R. Callstrom, T. X. Neenan and G. M. Whitesides, *Macromolecules,* 1988, **21,** 3528.
111. M. A. Semsarzadeh and C. C. Price, *Macromolecules,* 1977, **10,** 482.
112. H. M. Van Dort, C. A. M. Hoefs, E. P. Magre, A. J. Schopf and K. Yntema, *Eur. Polym. J.*, 1968, **4,** 275.
113. A. S. Hay, H. S. Blanchard, G. F. Endres and J. W. Eustance, *J. Am. Chem. Soc.*, 1959, **81,** 6335.
114. F. J. Viersen, J. Renkema, G. Challa and J. Reedijk, *J. Polym. Sci., Part A: Polym. Chem.*, 1992, **30,** 901.
115. S. Tsuruya, K. Nakagawa and M. Masai, *J. Polym. Sci., Part A: Polym. Chem.*, 1987, **25,** 995.
116. G. Challa, W. Chen and J. Reedijk, *Makromol. Chem., Macromol. Symp.*, 1992, **59,** 59.
117. D. Aycock, V. Abolins and D. M. White, in 'Encyclopedia of Polymer Science and Engineering', ed. H. F. Mark, N. M. Bikales, C. G. Overberger and G. Menges, Wiley, New York, 1988, vol. 13, p. 1.
118. F. J. Viersen, G. Challa and J. Reedijk, *Polymer,* 1990, **31,** 1361.
119. F. J. Viersen, G. Challa and J. Reedijk, *Polymer,* 1990, **31,** 1368.
120. F. J. Viersen, G. Challa and J. Reedijk, *Recl. Trav. Chim. Pays-Bas,* 1990, **109,** 97.
121. F. J. Viersen, G. Challa and J. Reedijk, *Recl. Trav. Chim. Pays-Bas,* 1989, **108,** 163.
122. W. Risse, W. Heitz, D. Freitag and L. Bottenbruch, *Makromol. Chem.*, 1985, **186,** 1835.
123. A. P. Terent'ev and Ya. D. Mogilyanskii, *Dokl. Akad. Nauk SSSR,* 1955, **103,** 91.
124. H. C. Bach and W. B. Black, *J. Polym. Sci., Part C,* 1969, **22,** 799.
125. I. L. Kotlyarevskii, M. P. Terpugova and E. K. Andrievskaya, *Izv. Akad. Nauk SSSR, Ser. Khim.*, 1964, 1854.
126. E. Radlmann and G. Nischk, *Makromol. Chem.*, 1971, **145,** 39.
127. E. M. Geniès, A. Boyle, M. Lapkowski and C. Tsintavis, *Synth. Met.*, 1990, **36,** 139, and references therein.
128. M. Inoue, F. Brown, I. C. Muñoz and F. Ofelia Muñoz, *Polym. Bull.*, 1991, **26,** 403.
129. D. K. Moon, K. Osakada, T. Maruyama and T. Yamamoto, *Makromol. Chem.*, 1992, **193,** 1723.
130. D. K. Moon, K. Osakada, T. Maruyama, K. Kubota and T. Yamamoto, *Macromolecules,* 1993, **26,** 6992.

131. M. Yoneyama, M. Kakimoto and Y. Imai, *Macromolecules,* 1988, **21,** 1908.
132. M. Yoneyama, M. Kakimoto and Y. Imai, *J. Polym. Sci., Part A: Polym. Chem.,* 1989, **27,** 1985.
133. M. Yoneyama, T. Konishi, M. Kakimoto and Y. Imai, *Makromol. Chem., Rapid Commun.,* 1990, **11,** 381.
134. S. R. Turner, R. J. Perry and R. W. Blevins, *Macromolecules,* 1992, **25,** 4819.
135. R. J. Perry and B. D. Wilson, *Macromolecules,* 1993, **26,** 1503.
136. R. J. Perry, S. R. Turner and R. W. Blevins, *Macromolecules,* 1993, **26,** 1509.
137. V. R. Reichert and L. J. Mathias, *Macromolecules,* 1994, **27,** 7024.
138. R. J. Perry, S. R. Turner, R. W. Blevins and B. D. Wilson, *Polym. Prepr., Am. Chem. Soc., Div. Polym. Sci.,* 1994, **35** (1), 706.
139. R. J. Perry and B. D. Wilson, *Macromolecules,* 1994, **27,** 40.
140. G. M. Dahlen and A. Sen, *Polym. Prepr., Am. Chem. Soc., Div. Polym. Sci.,* 1992, **33** (2), 180.
141. R. D. Miller and J. Michl, *Chem. Rev.,* 1989, **89,** 1359.
142. C. Aitken, J. F. Harrod and E. Samuel, *J. Organomet. Chem.,* 1985, **279,** C11.
143. C. T. Aitken, J. F. Harrod and E. Samuel, *J. Am. Chem. Soc.,* 1986, **108,** 4059.
144. M. Cypryk, Y. Gupta and K. Matyjaszewski, *J. Am. Chem. Soc.,* 1991, **113,** 1046.
145. R. D. Miller, D. Thompson, R. Sooriyakumaran and G. N. Fickes, *J. Polym. Sci., Part A: Polym. Chem.,* 1991, **29,** 813.
146. R. West, *J. Organomet. Chem.,* 1986, **300,** 327.
147. J. Devaux, J. Sledz, F. Schue, L. Giral and H. Naarmann, *Eur. Polym J.,* 1989, **25,** 263.
148. M. Fujino and H. Isaka, *J. Chem. Soc., Chem. Commun.,* 1989, 466.
149. H. K. Kim and K. Matyjaszewski, *J. Am. Chem. Soc.,* 1988, **110,** 3321.
150. G. J. Price, *J. Chem. Soc., Chem. Commun.,* 1992, 1209.
151. H. K. Kim and K. Matyjaszewski, *Polym. Prepr., Am. Chem. Soc., Div. Polym. Chem.,* 1990, **31** (2), 278.
152. K. Matyjaszewski, M. Cypryk, H. Frey, J. Hrkach, H. K. Kim, M. Moeller, K. Ruehl and M. White, *J. Macromol. Sci., Chem.,* 1991, **A28,** 1151.
153. K. Matyjaszewski, D. Greszta, J. Hrkach and H. K. Kim, *Macromolecules,* 1995, **28,** 59.
154. J. M. Zeigler, *Polym. Prepr., Am. Chem. Soc., Div. Polym. Chem.,* 1986, **27** (1), 109.
155. K. W. Leong, J. Kost, E. Mathiowitz and R. Langer, *Biomaterials,* 1986, **7,** 364.
156. K. W. Leong, V. Simonte and R. Langer, *Macromolecules,* 1987, **20,** 705.
157. K. W. Leong, V. Simonte and R. Langer, *Macromolecules,* 1988, **21,** 1925.
158. A. J. Domb, E. Mathiowitz, E. Ron, S. Giannos and R. Langer, *J. Polym. Sci., Part A: Polym. Chem.,* 1991, **29,** 571.
159. Y. Gnanou, P. Chaumont, P. Lutz, G. Hild and P. Rempp, *Makromol. Chem.,* 1984, **185,** 2647.
160. W. H. Stockmayer, *J. Polym. Sci.,* 1952, **9,** 69.
161. F. Lopez-Serrano, J. M. Castro, C. W. Macosko and M. Tirrell, *Polymer,* 1980, **21,** 263.
162. E. Sorta and A. Melis, *Polymer,* 1978, **19,** 1153.
163. L. H. Peebles, Jr., *Macromolecules,* 1974, **7,** 872.
164. L. H. Peebles, Jr., *Macromolecules,* 1976, **9,** 58.
165. M. Morton, T. E. Helminiak, S. D. Gadkary and F. Bueche, *J. Polym. Sci.,* 1962, **57,** 471.
166. G. Kraus and J. T. Gruver, *J. Polym. Sci., Part A,* 1965, **3,** 105.
167. X. Chen, A. Tang and D. Yan, *Sci. Sin., Engl. Ed.,* 1981, **24,** 1379.
168. D. Y. Yan, *J. Macromol. Sci., Chem.,* 1986, **A23,** 59.
169. Y. Gnanou, *Polymer,* 1994, **35,** 4011.
170. G. Riess and R. Reeb, *Polym. Prepr., Am. Chem. Soc., Div. Polym. Chem.,* 1980, **21** (1), 55.
171. P. Lutz, C. Strazielle and P. Rempp, *Polym. Prepr., Am. Chem. Soc., Div. Polym. Chem.,* 1986, **27** (1), 190.
172. T. E. Hogen-Esch, J. Sundararajan and W. Toreki, *Makromol. Chem., Macromol. Symp.,* 1991, **47,** 23.
173. J. Sundararajan and T. E. Hogen-Esch, *Polym. Prepr., Am. Chem. Soc., Div. Polym. Chem.,* 1992, **33** (1), 162.
174. Y. D. Gan, J. Zoller, R. Yin and T. E. Hogen-Esch, *Makromol. Chem., Macromol. Symp.,* 1994, **77,** 93.
175. A. El Madani, J.-C. Favier, P. Hémery and P. Sigwalt, *Polym. Int.,* 1992, **27,** 353.
176. P. Hémery, J.-M. Boutillier, A. El Madani, J.-C. Favier and P. Sigwalt, *Polym. Prepr., Am. Chem. Soc., Div. Polym. Chem.,* 1994, **35** (2), 478.
177. L. L. Zhou, N. Hadjichristidis, P. M. Toporowski and J. Roovers, *Rubber Chem. Technol.,* 1992, **65,** 303.
178. J. Roovers, L. L. Zhou, P. M. Toporowski, M. van der Zwan, H. Iatrou and N. Hadjichristidis, *Macromolecules,* 1993, **26,** 4324.
179. M. Miyamoto, M. Sawamoto and T. Higashimura, *Macromolecules,* 1984, **17,** 265.
180. M. Sawamoto and T. Higashimura, *Makromol. Chem., Macromol. Symp.,* 1986, **3,** 83, and references therein.
181. H. Fukui, M. Sawamoto and T. Higashimura, *J. Polym. Sci., Part A: Polym. Chem.,* 1993, **31,** 1531.
182. H. Fukui, M. Sawamoto and T. Higashimura, *Macromolecules,* 1993, **26,** 7315.
183. H. Fukui, M. Sawamoto and T. Higashimura, *Macromolecules,* 1994, **27,** 1297.
184. G. Pruckmayr and R. B. Osborne, *U.S. Pat.* 5 284 980 (1992) (*Chem. Abstr.,* 1994, **121,** 36 483).
185. Y. Tezuka and C. Hoshino, *J. Polym. Sci., Part A: Polym. Chem.,* 1994, **32,** 897.
186. A. Noshay and J. E. McGrath, 'Block Copolymers, Overview and Critical Survey', Academic Press, New York, 1977, chap. 7, and references therein.
187. J. J. O'Malley, *J. Polym. Sci., Polym. Phys. Ed.,* 1975, **13,** 1353.
188. T. Kanno, Y. Toga and N. Ikeda (Daicel Huels Ltd), *Ger. Offen.,* 3 432 509 (1984) (*Chem. Abstr.,* 1985, **103,** 6892).
189. J. P. Kennedy, *Polym. J.,* 1985, **17,** 29.
190. H. Xie, X. Chen and J. Guo, *Angew. Makromol. Chem.,* 1992, **200,** 49.
191. H. Xie and X. Chen, *Makromol. Chem., Macromol. Symp.,* 1988, **20/21,** 19.
192. G. M. Gordon, *Polym. Prepr., Am. Chem. Soc., Div. Polym. Chem.,* 1988, **29** (2), 46.
193. E. Takiyama, T. Fujimaki, S. Seki, T. Hokari and Y. Hatano (Showa High Polymer Co. Ltd.), *Eur. Pat. Appl.,* 572 256 (1993) (*Chem. Abstr.,* 1994, **120,** 299 584).
194. D. H. Richards and M. Szwarc, *Trans. Faraday. Soc.,* 1959, **55,** 16.

195. H. Kobayashi, K. Sasaguri and M. Makita (Asahi Chemical Co), *Jpn. Pat.* 69 15 431 (1969) (*Chem. Abstr.*, 1969, **71,** 102 905).
196. D. M. White (General Electric Co.), *U.S. Pat.*, 4 356 290 (1982) (*Chem. Abstr.*, 1983, **98,** 161 341).
197. Y. Qing and H. Jiasong, *Polym. Bull.*, 1992, **29,** 633.
198. R. W. Lenz and S. Go, *J. Polym. Sci., Polym. Chem. Ed.*, 1973, **11,** 2927.
199. H. Inata and S. Matsumura, *J. Appl. Polym. Sci.*, 1986, **32,** 4581.
200. R. A. Rhein and J. D. Ingham, *Polymer,* 1973, **14,** 466.
201. H. Al Ghatta (M. and G. Ricerche S.p.A.), *PCT Int. Appl.*, 92-17 521 (1992) (*Chem. Abstr.*, 1993, **118,** 125 780).
202. H. Inata and S. Matsumura, *J. Appl. Polym. Sci.*, 1987, **34,** 2609.
203. Teijin Ltd, *Jpn. Kokai Tokkyo Koho,* 57 082 519 (1982) (*Chem. Abstr.*, 1983, **98,** 35 976).
204. Teijin Ltd, *Jpn. Kokai Tokkyo Koho,* 57 002 330 (1982) (*Chem. Abstr.*, 1982, **96,** 163 390).
205. M. Acevedo and A. Fradet, *Polym. Prepr., Am. Chem. Soc., Div. Polym. Sci.*, 1993, **34** (1), 457.
206. M. K. Akkapeddi and J. Gervasi, *Polym. Prepr., Am. Chem. Soc., Div. Polym. Sci.*, 1988, **29** (1), 567.
207. R. L. Wear (Minnesota Mining and Mfg. Co), *U.S. Pat.*, 3 966 680 (1976) (*Chem. Abstr.*, 1976, **85,** 109 487).
208. C. Serres, Jr. (Standard Oil Co), *U.S. Pat.*, 4 065 440 (1977) (*Chem. Abstr.*, 1978, **89,** 110 858).
209. S. D. Wu, J. L. Hedrick, B. K. Carter, D. K. Mohanty, E. Yilgor, G. L. Wilkes and J. E. McGrath, *Polym. Prepr., Am. Chem. Soc., Div. Polym. Chem.*, 1985, **26** (2), 277.
210. E. Franta, P. Kubisa, J. Refai, S. O. Kada and L. Reibel, *Makromol. Chem., Macromol. Symp.*, 1988, **13–14,** 127.
211. A. Shefer, A. J. Grodzinsky, K. L. Prime and J. P. Busnel, *Macromolecules,* 1993, **26,** 5009.
212. H. H. Wang and W. L. Chen, *J. Polym. Sci., Part A: Polym. Chem.*, 1989, **27,** 1359.
213. I. Vulic, A. J. B. Loman, J. Feijen, T. Okano and S. W. Kim, *J. Polym. Sci., Part A: Polym. Chem.*, 1990, **28,** 1693.
214. M. Ueda, K. Kino, K. Yamaki and Y. Imai, *J. Polym. Sci., Polym. Chem. Ed.*, 1978, **16,** 155.
215. J. K. Rasmussen, S. M. Heilmann, L. R. Krepski, H. K. Smith, A. R. Katritzky and K. Sakizadeh, *J. Polym. Sci., Polym. Chem. Ed.*, 1986, **24,** 2739.
216. M. Acevedo and A. Fradet, *J. Polym. Sci., Part A: Polym. Chem.*, 1993, **31,** 1579.
217. H. Inata and S. Matsumura, *J. Appl. Polym. Sci.*, 1985, **30,** 3325.
218. G. J. Del Franco, M. Guagliardo, P. Loewrigkeit and N. Georgalas, *J. Polym. Sci., Part B,* 1967, **5,** 487.
219. T. Mukaiyama, T. Fujiasawa and T. Hyugaji, *Bull. Chem. Soc. Jpn.*, 1962, **35,** 687.
220. H. Inata and S. Matsumura, *J. Appl. Polym. Sci.*, 1986, **32,** 5193.
221. H. Inata and S. Matsumura, *J. Appl. Polym. Sci.*, 1987, **33,** 3069.
222. N. Cardi, R. Po, G. Giannotta, E. Occhiello, F. Garbassi and G. Messina, *J. Appl. Polym. Sci.*, 1993, **50,** 1501.
223. A. Douhi and A. Fradet, *J. Polym. Sci., Part A: Polym. Chem.*, 1995, **33,** 691.
224. T. Loontjens, W. Belt, D. Stanssens and P. Weerts, *Makromol. Chem., Macromol. Symp.*, 1993, **75,** 211.
225. T. Loontjens, W. Belt, D. Stanssens and P. Weerts, *Polym. Bull.*, 1993, **30,** 13.
226. Q. Han and G. Wu, *Gaofenzi Cailiao Kexue Yu Gongcheng,* 1989, **5,** 21 (*Chem. Abstr.*, 1990, **113,** 232 135).
227. J. Djonlagic, M.-O. Sepulchre, M. Sepulchre, N. Spassky, B. Dunjic and M. S. Jacovic, *Makromol. Chem.*, 1990, **191,** 1529.
228. J. P. Kennedy and G. M. Carlson, *J. Polym. Sci., Polym. Chem. Ed.*, 1983, **21,** 2973.
229. M. A. Sharaf and J. E. Mark, *Polym. Mater. Sci. Eng.*, 1988, **58,** 518.
230. Y. Tezuka and E. J. Goethals, *Makromol. Chem.*, 1987, **188,** 791.
231. A. Douhi, A. Fradet and D. Judas (Elf Atochem S. A.), *Eur. Pat. Appl.*, 581 642 (1994) (*Chem. Abstr.*, 1994, **120,** 246 095).
232. A.-C. Albertsson and S. Lundmark, *Br. Polym. J.* 1990, **23,** 205.
233. T. P. Liao and J. P. Kennedy, *Polym. Bull.*, 1982, **7,** 233.
234. I. A. Volegova, Yu. K. Godovskii, A. N. Aksenov, I. P. Storozhuk and V. V. Korshak, *Vysokomol. Soedin., Ser. B,* 1983, **25,** 792.
235. T. Y. Tseng, N. J. Chu and Y. D. Lee, *J. Appl. Polym. Sci.*, 1990, **41,** 1651.

5

Recent Advances in Optical Microscopy and its Application to the Characterization of Polymers

MOHAN SRINIVASARAO

North Carolina State University, Raleigh, NC, USA

5.1 INTRODUCTION

An optical microscope is perhaps one of the most used imaging instruments in a variety of disciplines to 'see' things at high magnification. One of the limitations of traditional optical microscopy techniques (far-field techniques) is its diffraction-limited resolution.[1] Despite the limited resolution, its widespread use as an imaging instrument is due to its simplicity and a unique variety of available contrast mechanisms. Examples of such contrast mechanisms are absorption, polarization, phase contrast, dark-field, and fluorescence, to name just a few. Fluorescence microscopy in particular has revolutionized cell biology, allowing one to obtain images of cells and organelles in exquisite detail. Such advances have been made by combining lasers, electronic cameras, and digital image analysis. In view of the advantages provided by various contrast mechanisms and the ease of use, efforts to push the resolution (both spatial and axial) of the optical microscope well beyond that of the diffraction limit have produced a number of very useful techniques, such as confocal microscopy,[2] standing-wave fluorescence microscopy (SWFM)[3] (where one is able to visualize structure in 3D by optical sectioning of the object being viewed), solid immersion microscopy (SIM),[4] photon tunneling microscopy (PTM),[5,6] and, at a higher spatial resolution, near-field optical microscopy (NSOM).[7–10] All of these techniques, by virtue of using light at visible wavelengths, provide information that is not accessible by other high-resolution techniques such as electron microscopy and the recent family of probe microscopes (scanning tunneling microscopy and force microscopy).[11–14] We will not dwell on such forms of microscopy and will refer the reader to excellent reviews and books that have been written.[14]

The purpose of this article is to discuss the progress that has been made in optical microscopy since the mid-1980s. It has been recognized[15,16] that the limitations of optical microscopy are not fundamental but arise from the fact that the detector is placed many wavelengths away (far-field) from the object of interest. Much progress has been made by recognizing this fact and this has lead to the development of methods which probe the near-field. Probing the near-field is accomplished by laterally scanning a light source in close proximity to the sample, where the light source is made smaller than the wavelength. This article will focus on methods that use the information in the near-field to obtain higher spatial resolution and will also discuss methods that are diffraction-limited spatially but nevertheless have achieved an increased axial resolution.

5.2 HISTORICAL ASPECTS OF MICROSCOPY

Historically, the need to 'see' things better has led to the development of tools that have accomplished that purpose. Apparently the earliest of imaging devices was a glass lens found in the ruins of Nineveh, dating back to 610 BC. A Roman philosopher later (about 100 AD) noted that a globe of water held properly would magnify objects.[17] The optical microscope itself dates back to the early 1600s and was primarily considered a *philosophical* instrument.[18] It is believed that the optical microscope made its transformation from a philosophical instrument to a research/diagnostic tool around the 1840s. This was primarily due to the increased effort to improve public health as a result of the outbreak of cholera epidemics in Britain around the 1830s.[18] The optical microscope still continues to play a very important role as a research/diagnostic tool and the picture of a 'researcher' peering into a microscope has become a symbolic representation of the sciences in general.

The 1830s and 1840s also saw an increase in the quality of the microscope in terms of small features that could be seen. The fact that smaller and smaller features were resolved with progression of time is shown graphically in Figure 1. This graph clearly shows the increase in resolution, reaching its limit (see below) around 1880. This coincided with the large increase in the manufacture of microscopes during the same period.[18]

The limit of resolution (the ability to distinguish closely spaced objects) of an optical microscope was recognized by Abbe with his fundamental work on diffraction and image formation.[1,19] He recognized that light of wavelength λ cannot be focused to a size below $\lambda/2$, which in turn implies that one cannot distinguish two particles that are spaced less than $\lambda/2$ apart. This implies that if one uses green light, then the resolution is about 0.25 μm. This resolution was obtained as far back as 1880 with the use of oil immersion objectives, as is clearly seen from Figure 1. We will come back to the issue of resolution in later sections, since this forms the subject matter of this review. The term *super-resolution* will be used to describe any (optical) imaging technique that is able to produce resolution beyond this diffraction limit of $\lambda/2$.

Not much progress was made in increasing the resolution up until 1928 or so, when a method for improving the resolution of the optical microscope was put forward in an elegant paper by Synge.[20,21] The aim of the proposal was to overcome the diffraction limit on resolution, and has led to the practice of near-field optical microscopy. The essence of the idea was to produce a very small light source by getting the light to an aperture smaller than the wavelength of light, which was to be placed very close to the sample and raster-scanned to obtain an image. This was to be accomplished by having subwavelength apertures in an otherwise opaque (to visible

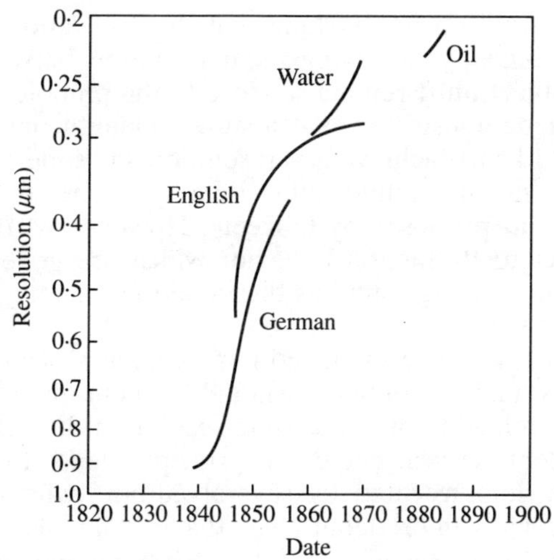

Figure 1 Improvements in resolution during the nineteenth century.[18]

light) screen, as is schematically shown in Figure 2. Light incident on one side of the opaque screen is transmitted through the aperture, and if the object of interest is placed in the near-field, it illuminates only a small region of the object. Synge also went on to list the technical difficulties that would have to be solved in order for the realization of the idea and to make it a reality. Among the difficulties were: (a) the amount of light available for imaging, (b) problems associated with moving very small distances, (c) preparing samples that are extremely flat, and (d) the construction of subwavelength apertures for the light source. The fact that several of these problems have been solved in making the near-field optical microscope a reality, rather than an abstract idea that was presented quite elegantly some 60 years ago, will be seen in later sections of this article. This also serves to illustrate the point that it is quite worthwhile to look at the historical aspects of developments and that awareness of the early literature is quite beneficial to the practicing scientist.

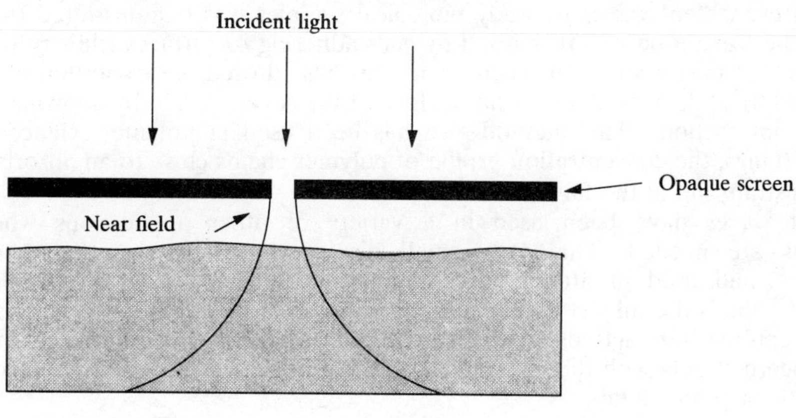

Figure 2 Schematic diagram showing the concept of obtaining high resolution by having the sample in the 'near-field' of a scanned probe.

Later, in 1956, O'Keefe,[22] almost certainly without knowledge of Synge's ideas, proposed a scanning near-field microscope and deemed it possible to 'go around the resolving power of visible light'. However, he went on to conclude that the realization of such a microscope was rather remote due to the difficulty of having relative motion between the submicron pinhole and the object, while the object must remain so close to the pinhole. Following up on this idea, Baez[23] later the same year demonstrated that a subwavelength object (his finger) was clearly resolved using sound ($\lambda = 14$ cm), achieving a resolution far superior to half the wavelength. Because the imaging was done with sound rather than light, one could argue that this does not constitute a test of the idea as proposed by O'Keefe. However, as Baez points out, since both light and sound are subject to diffraction behavior which are governed by the same laws of physical optics, the experiment using sound as the radiation for imaging is in fact quite valid to test the proposed idea of O'Keefe.

For most part in the above we have discussed how 'scanning' devices can increase resolution beyond that of conventional far-field optics. It should be mentioned that it was recognized that nonscanning devices can also lead to an increase in resolution. It was recognized fairly early on that the use of evanescent waves could in principle lead to enhanced resolution or super-resolution. This was demonstrated by several authors who used evanescent waves to record grating-like objects with finer detail than that obtainable by the classical resolution limit.[24–26] The existence of evanescent waves has been known at least since the time of Newton.[27]

Evanescent waves are formed in phenomena commonly encountered in optics. One of the most common ways of creating evanescent waves is by creating conditions for total internal reflection (TIR). The use of evanescent waves to observe topographical features was reported as early as 1930.[28] It appears as though this method did not enjoy widespread use until evanescent waves were demonstrated as being useful in the measurement of film thickness and surface relief by Harrick in the early 1960s. Harrick[29] demonstrated the use of frustration of total internal reflection (FTIR) for recording fingerprints and the contrast of the image obtained is rather remarkable (see Figure 3 in Harrick's[29] paper). At about the same time, use of FTIR as a tool for metrology in proximity problems was demonstrated by Young and Rothrock.[30] The use of evanescent waves as a spectroscopic tool to probe chemisorbed molecules at a surface was also demonstrated by Harrick.[31,32]

The optical system designed by Harrick to observe the surface relief or 'the topography' is not without its problems. One of the problems is having an image that is both undistorted and in focus everywhere. In an effort to circumvent this, an axially symmetric system of illumination was developed by McCutchen.[33,34] Such an illumination system provides an image of surface relief or the topography of the surface. In this review we will describe recent progress in using a simple configuration to obtain images of surface topography in three dimensions.

The use of evanescent waves to study biological samples was demonstrated by Ambrose,[35] who termed the frustration of TIR caused by cells adhering to surfaces (thus ruining the TIR) as 'surface contact microscopy'. In recent years this has allowed the excitation of fluorescence of species (mainly biological samples) quite close to the surface, thereby allowing one to study cell/substrate interactions. This method also has been used in polymer science to measure, among other things, the concentration profile of polymer chains close to an absorbing surface[36] and to demonstrate slip in the flow of polymers.[37]

Evanescent waves have been used in a variety of other applications where distance measurements are needed. That very small displacements can be measured has been demonstrated[38] and used in atomic force microscopes.[39] Its use in holography has been demonstrated[24] and is the subject of an excellent review by Bryngdahl.[40,41] It has also been used to measure repulsive interactions between a colloidal particle and a surface.[42–45] As we are primarily concerned only with the imaging aspects of this subject, we shall not discuss many of these applications in any detail.

It is the purpose of this article to present recent advances in optical microscopy and discuss some applications where these techniques find use in the characterization of polymers. Techniques are described here that have been developed in the past few years which have achieved the goal of enhancing both the lateral (spatial) and axial resolution. Our discussion here will be limited only to optical microscopy techniques and we shall categorize these methods in terms of spatial resolution and axial resolution.

This article is structured in the following manner. The limits of resolution of an imaging system are discussed, primarily based on the classical Rayleigh criterion for resolution, for both spatial and axial resolution. This will be followed by a section on various techniques that have appeared since the mid-1980s to enhance both types of resolution. This will be followed by a discussion of various applications of these methods to the characterization or the study of polymers. The methods that have shown an increase in spatial resolution which will be discussed are (a) photon tunneling microscopy (PTM), (b) photon scanning tunneling microscopy (PSTM), and (c) near-field scanning optical microscopy. Methods that have shown a considerable increase in axial resolution which will be discussed are (a) standing-wave fluorescence microscopy (SWFM) and (b) 4pi confocal fluorescence microscopy. The general use of confocal microscopy has been in biology and there are many very well written books and review papers to which the reader is referred.[2,46,47] The use of confocal microscopy in polymer characterization is increasing,[48] but one has to admit that its use has been under-utilized in polymer science. The article ends by describing the use of an old but useful technique, optical crystallography or conoscopy, in the study of liquid crystals. Its use in the study of static properties of liquid crystals will be discussed. The static properties include the measurement of birefringence of uniaxial media and individual refractive indices. This method can also be used to study dynamic properties.

5.3 IMAGE FORMATION IN A MICROSCOPE AND RESOLUTION

There are many well-written books on image formation in a microscope to which the reader is referred for a detailed description.[1,16,49,50] The purpose of this section is to provide a brief account of how an image is formed and the limits on resolution — the so-called 'diffraction limit'. It should be recognized that there is a clear distinction between resolving an image and visualizing an object. Visualization of objects that are far below the resolution limit of an optical microscope is not a new event but resolution below the diffraction limit is novel. These two issues must not be confused. For any image-forming device the resolution is generally expressed as a measure of the ability of the device to separate images of two neighboring points, given by the Rayleigh criterion.

The Rayleigh criterion is often used to describe the resolution of optical instruments and image-forming devices. There have been several, rather elegant, discussions on the resolution limit in the literature, claiming that this Rayleigh limit is only a convenient measure of the resolution that is obtainable by optical instruments, and that it should be possible, in principle, to obtain resolution without limit.[16] In a very elegant paper, Toraldo di Francia[51] has shown that even though it should be possible to obtain resolution without any limits based on mathematical grounds, the practical realization of the mathematical ideal is quite impossible. Hence, it is worthwhile to remind ourselves of the arguments that have been presented.

We begin by considering an imaging system as shown schematically in Figure 3. Image formation in a microscope is well understood in terms of diffraction of the incident beam on passing through the sample and recombination thereafter to give a real (and magnified) image of the object.[52] Image formation in a microscope can then be considered as a double diffraction process. In order to illustrate this in a qualitative fashion, consider an object which is a diffraction grating G. A parallel beam of light, on passing through this object, produces several orders of diffraction in the back focal plane (BFP) of the objective (first lens). The diffracted beams emerge at angles θ given by $nd \sin \theta = m\lambda$, where n is the refractive index in which the grating of period d is placed and m is an integer. Each diffracted order can be considered as a plane wave and the set of plane waves produced by the diffraction grating can be collected by a lens or an objective having a sufficiently large numerical aperture (NA), so that they converge individually to a set of points in the focal plane F (Fourier transform plane) of the objective. The diffracted light passes on to interfere, resulting in an accurate image of the diffraction grating in a plane conjugate to that of the grating (the object), and only in that plane, which is the image plane I.

A diffraction grating was chosen as the object for several reasons: historically, a grating was used to demonstrate Abbe's theory[50,53] and it also serves the purpose of illustrating that image formation in a microscope can be considered as two distinct stages: (a) production of the Fraunhoffer diffraction pattern in the rear focal plane and (b) a Fourier transform of the Fraunhoffer diffraction pattern to form a magnified image. In the focal plane of the lens F, having produced the Fraunhofer diffraction pattern of the object, the various diffraction orders S behave like a set of equally spaced point sources and the image is their diffraction pattern. The finest structure observable is determined by the highest order of diffraction that is transmitted by the lens used for the purpose of imaging, and is a function of the numerical aperture of the objective used in the microscope. The numerical aperture (NA) of an objective

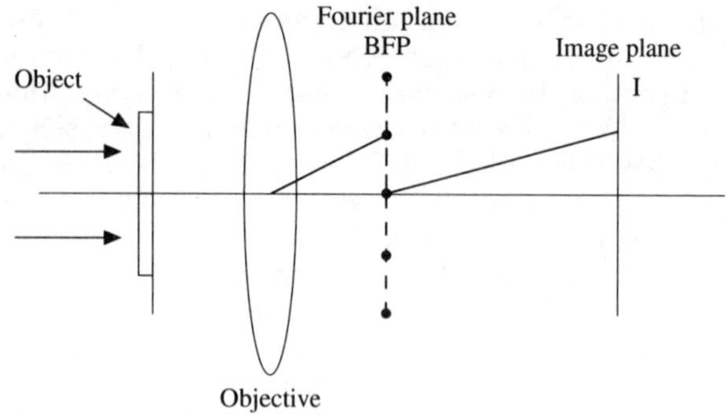

Figure 3 Schematic diagram showing image formation in a microscope.

is the light gathering power and is defined as $n \sin \theta$, where n is the refractive index of the medium in which the light is collected and θ is the half-angle of the cone of light entering the objective.

In a light microscope, the part of the Fraunhofer diffraction most likely to be lost is the outer part, which corresponds to larger angles of diffraction, and it will be lost if the numerical aperture (the light gathering power) of the objective is too small. This results in a loss of the sharpness in the image, leading to a deterioration of the resolution and a loss of the finer detail since it is the sharper boundaries and finer detail which diffract light at larger angles. It is seen that it is useful to collect as many orders of diffraction as possible, since this contributes to the detailed structure of the image.

In order to understand the image formation better, one can resort to Fourier optics.[1,16] Consider the image of the grating as described above. It is seen that in the back focal plane (which is the Fourier plane) we have several orders of diffraction behaving as a set of equally spaced point sources. The zeroth order beam in the diffraction plane produces a uniform amplitude (and by itself carries no information about the period of the grating), which provides for a uniform illumination across the image plane, while the higher order beams produce a sinusoidal amplitude distribution. These various amplitudes are superposed in the image plane to produce the image. The image or the amplitude distribution across the image plane can then be represented in terms of various Fourier components or spatial frequencies. To observe the periodicity of the grating it is enough to reproduce the fundamental or the lowest spatial frequency component in the amplitude distribution of the image. This is accomplished if the objective of the microscope can admit the zeroth order beam and one of the first order beams, giving two effective sources in the Fourier plane, thus giving maxima in the intensity distribution across the image plane which correspond to the grating spaces, leading to a resolved grating. This leads us to the Abbe principle: the need to transmit the zeroth order beam and one of the first order beams. The limit to resolution is then given by the finest grating that can be resolved in this way. It becomes clear that finer gratings give rise to larger angles between the zeroth and first orders, thus requiring lenses of larger numerical aperture in order to be resolved. Hence the resolution of a microscope is a function of the numerical aperture used and the theoretical limit is given by the expression $0.61\lambda/\mathrm{NA}$, where λ is the wavelength of the light.[52] This expression is often approximated by $0.5\lambda/\mathrm{NA}$ and comes about when one takes account of the numerical aperture of the condenser. We shall not go into the details, as they can be found elsewhere.[54]

One has to realize that the above discussion represents an approximate theory of image formation. It is assumed that light associated with a given diffraction order is in a single direction and all light associated with this order is either accepted or rejected by the collecting objective. The theory does not take into account the possibility of part of the light in a given diffracted order being admitted nor the finite amount of light which is diffracted between the principal maxima.

The above discussion has been for nonluminous objects, as Abbe described in his theory for the microscope. Rayleigh[55] took a different approach to the problem. Light from each point in an object that is illuminated by a condenser produces a set of circular rings (called the Airy pattern) in the conjugate image plane and the final image is a superposition of these Airy patterns. The effects of such superposition can be calcualted if the phase relations and the degree of coherence are known, which are determined by the structure of the object and the form of illumination. The Airy pattern then results from diffraction by the objective and the structure of the object is taken into account by the phase relations between the Airy patterns. This leads to another definition of resolution, the Rayleigh criterion: the separation of two just resolvable points. This is given when the maxima of one Airy pattern is located at the minimum of the other Airy pattern. The resultant two-point intensity will then have a minimum, between two just resolved objects, that has about 20% less intensity than the maximum. In all of this, nothing has been said about the signal-to-noise ratio (SNR) in the measurements. As with any measurement, the SNR plays an important role. Hence, if one can have a good SNR in the measurement of the convoluted intensity distribution from two just resolved Airy patterns, a smaller separation can be estimated with adequate certainty.

We now return to the issue of resolution and its limits. As pointed out earlier, it has been claimed in a series of elegant papers that the Rayleigh criterion is not a fundamental limit.[16,51,56–60] It is known that an object of finite extent has a Fourier transform that is an analytic function. Given enough information, even if a part of the Fourier transform is lost (due to the presence of effective apertures) in the Fourier transform plane (back focal plane), it must, in principle, be possible to recover the entire function. This has been the topic of many elegant papers, presenting arguments that have been given to establish that the Rayleigh criterion is just a practical limit and not a fundamental one. This argument is mathematically quite correct, as was pointed out by Marcuse,[61] among others. It should be recognized that these arguments are of little use in terms of going around the diffraction limit due to the issue of noise entering into the detection scheme. It is inevitable that one will be limited by the noise in any measuring or detection scheme, and hence there is really very little hope of being able to reproduce precisely the object field from a knowledge of the image field. Resolution issues for the scanning forms of optical microscopy techniques (NSOM) will be discussed later. We will just mention that the Rayleigh criterion is useful to think about when considering any scheme for enhancement of lateral resolution.

Thus far the discussion has been on lateral resolution. Since optical sectioning microscopy (OSM) has in a sense revolutionized biology,[2,3,46] there has been considerable effort to obtain thinner and thinner sections, and hence a number of methods have been developed in order to achieve high axial resolution. This form of microscopy has been primarily used in biology to study the structure of cells in three dimensions. The image formed in an optical microscope contains information from the area that is in clear focus in a thin focal plane, as well as out-of-focus information. This is especially true if the object being imaged is much thicker than the depth of field. The depth of field can be as low as 0.15–0.20 μm when a high magnification, high NA objective is used. A vast majority of experiments use fluorescence microscopy to visualize the 3D structure of cells, and the imaging is usually done in low light conditions. This decreases the SNR and limits the axial resolution attainable.

Axial resolution is defined as the point at which two objects separated only axially (along the optical axis of the microscope) can be identified not to be in the same plane (plane normal to the optical axis of the microscope). As there is no single agreed formula that defines the limit to axial resolution, we will use the expression given by Francon (1961) for the axial setting.[62] This limit is given by the expression

$$\Delta z = \lambda\{8n \sin^2(\theta/2)\}^{-1}$$

where n is the refractive index of the sample and θ is the half-angle of the cone of light accepted by the objective. Here, Δz is defined as the minimum axial separation when two objects can be distinguished as such by changing the focal plane. This expression would predict an axial setting accuracy or resolution of about 0.18–0.14 μm for high NA lenses, which has been demonstrated by Inoue[63] in his experiments. We will return later to the issue of axial resolution and the forms of microscopy that have demonstrated an enhancement in the axial resolution.

5.4 PHOTON TUNNELING MICROSCOPY

In many situations it is desirable to 'see' selectively at surfaces, ignoring the structure in the bulk. Examples of such situations abound in a variety of disciplines: examination of cell adhesion to surfaces;[35,64] observing or obtaining the concentration of a polymer that is in contact with a surface;[36] obtaining the orientation of liquid crystals at a surface;[32] and observing the motion of diatoms on a surface. A useful way to study these processes is by the use of selective illumination provided by the use of evanescent waves created by total internal reflection (TIR) at a surface. Photon tunneling microscopy (PTM)[5,65] is a form of microscopy that uses the properties of evanescent waves to study the topographical features on a surface or index variations at the surface of interest which is otherwise flat. In this section are described the uses of PTM in the study of polymeric surfaces. Its utility in the characterization of static as well as dynamic structures will be illustrated.

PTM, described fully elsewhere,[5,6,65] has all the benefits of reflected light microscopy. These include whole field imaging with direct visual access, and in real time, with higher (than diffraction-limited) lateral resolution (while less than its scanning probe cousins), benefiting from the super-numerical aperture afforded by the tunneling or near-field information that is normally lost in a light microscope. PTM also shares the subnanometer vertical resolution and profiling ability of its scanning probe microscopy cousins, and has proved to be a valuable tool for imaging a variety of polymeric surfaces. Related to and descended from frustrated TIR microscopy (FTIRM)[66] (also referred to as evanescent wave or surface contact microscopy[35] when used in the transmission mode), PTM is differentiated from its relatives at the tunneling interface by a contiguous transducer that acts like a close-packed array of probes, allowing more kinds of samples to be imaged in real time without probe errors or damage, and by its unique 3D analog display of the photometric video image that facilitates real-time imaging with continuous and immediate image rotation about all axes, as well as profiling.

Most importantly for polymer applications, PTM images dielectric surfaces directly without metallization, carbon shadowing, vacuum, or damaging electrons. Also, surface alteration is negligible. While the transducer is in optical proximity to the sample surface, it is usually separated by a tunneling gap (or else it would not function), moves with the sample, is flexible enough to accommodate particulates, and is of a hardness comparable to most polymers.

5.4.1 Brief Description of PTM

The photon tunneling microscope is based on the behavior of light at a TIR surface, or the boundary between a medium 1 with refractive index n_1 and medium 2 with a lesser refractive index n_2 where TIR occurs in medium 1 if the angle of incidence is equal to or greater than the critical angle θ_c

$$\theta_c = \sin^{-1}(n_2/n_1) \qquad (1)$$

Probing the less dense medium (medium 2) near the TIR surface (hence 'near-field') with a dielectric surface of n_3 greater than n_2 reveals energy coupling from the first medium into the third medium, indicating the presence in the less dense medium, when unperturbed, of an electromagnetic wave with the same frequency as its parent wave in the denser medium, and 'evanescent' because it does not carry any energy from the TIR surface. The field strength of the evanescent wave decays exponentially as

$$E_{\text{evanescent}} = E_0 \exp(-z/d_p) \qquad (2)$$

where z is the distance normal to the TIR boundary and d_p is the penetration depth at which the field strength falls to $1/e$ of its initial value E_0 and is given by

$$d_p = \frac{\lambda_1}{2\pi(\sin^2\theta - n_{21}^2)^{1/2}} \qquad (3)$$

where $n_{21} = n_2/n_1$ is the ratio of refractive indices of media 1 and 2 and λ_1 is the illumination wavelength in medium 1. Figure 4 shows schematically the penetration depth and the exponential decay discussed above.

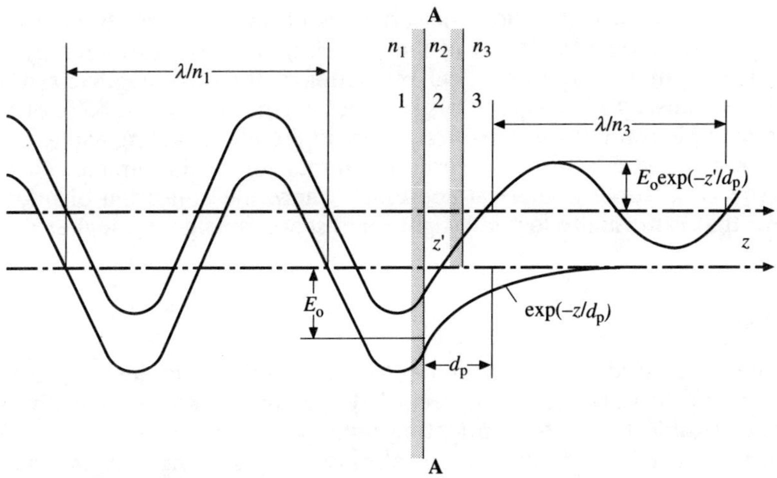

Figure 4 Schematic diagram showing the energetics and geometry of photon tunneling.

Evanescent waves are fully explained with classical optics[41,50] and one need not invoke the quantum theory, but doing so adds to the intuitive understanding of the phenomenon. The analogy to the electron barrier tunneling experiment has been elegantly made[10,67] and the effect termed as 'optical' or 'photon tunneling'.[68] Equation (2) then describes the probability of finding a photon from medium 1 in the tunneling gap at some distance z normal to the TIR boundary. Perhaps one of the first 'devices' to make use of 'optical tunneling' is a beam splitter based on the phenomenon of frustration of TIR, shown schematically in Figure 5.

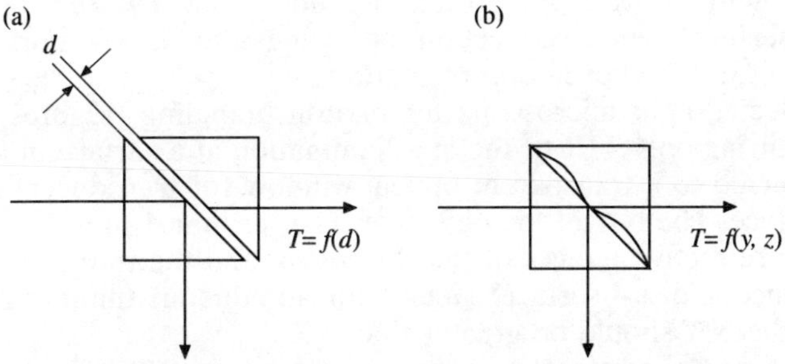

Figure 5 Schematic diagram of a beam splitter based on the principle of frustration of total internal reflection.

Depending on the separation between the two prisms, a certain fraction of light will be transmitted and the rest reflected (Figure 5a). However, if the air gap is not uniform, as is the case in Figure 5(b), then the transmission and reflectance are a function of the thickness and position, giving rise to an intensity distribution.[29,34]

The use of tunneling photons or evanescent waves brings with it several remarkable properties that are of use to microscopy. First, each tunneling photon serves as a height probe of a third medium ($n_3 > n_2$) as TIR is locally frustrated to a degree, exponentially and inversely dependent on z. Thus when the third medium is a dielectric sample with homogeneous optical properties, the reflected light will be modulated in intensity corresponding spatially to the modulation of the tunneling gap by the sample surface topography, and an image of the surface topography in grayscale is formed wherein height information is mapped into an intensity distribution.

Second, photometry of this tunneling image yields quantified height measurement directly and without ambiguity, as bright is down and dark is up in the intensity mapped topography. More importantly, the exponential decay of the tunneling probabilty (potential) with distance z normal to the TIR boundary frees vertical resolution from wavelength constraints. Vertical resolution is limited only by photometric resolution, determined by the SNR of the detector.

Third, a tunneling photon has the same frequency in the air tunneling gap as in medium 1, so that high lateral resolution is achieved without oil immersion of the sample. This has also been described in terms of a 'super-numerical aperture', where the collection of information in the evanescent wave that is normally lost in a light microscope results in a higher resolution.

5.4.2 Resolution

It has been demonstrated that increased resolution can be obtained by using evanescent waves, as was done by Nassenstein,[24] who recorded information holographically with resolution greater than was possible within the diffraction limit. The explanation for this lies in the fact that the wavelength to be considered for resolution λ_1 (wavelength in the medium) is λ/n, where n is the refractive index of the medium in which TIR is occurring.

PTM has been shown to have resolution better than $\lambda/4$,[65] demonstrating resolution beyond the diffraction limit. Examination of Figure 6(b) shows that the image in PTM is formed using totally internally reflected light, which leads to an annular illumination, although there is no central stop to block the central beam. However, central obstruction is unnecessary since TIR is so intense that it overwhelms the central illumination. Such an illumination is known to enhance spatial resolution by a factor of 1.3–1.6,[52] and is used to advantage in applications ranging from microlithography to astronomy. By having a centrally obstructed aperture, the diameter of the central peak of the diffraction image can be made smaller than the standard Airy disk, which leads to enhanced spatial resolution.[52,63] This annular illumination can be made arbitrarily small with the hope of attaining higher resolution. However, as the diameter of the central obstruction is increased, the amount of light available for imaging is decreased, and hence the resolving power is limited by the amount of light available (see Section 5.3). The possibility that evanescent waves generated by the sample topography or microfeatures can be converted to image-forming rays by the proximity of the transducer, leading to an additional factor for enhancement of lateral resolution, cannot be ignored. This will lead to a 'super-numerical aperture' where collection of evanescent waves normally lost in a light microscope may result in a higher lateral resolution.

Configuring a reflected-light microscope for photon tunneling (Figures 6a and 6b) is rather straightforward, requiring only TIR of the epi-illumination at a surface in the object plane. The objective is oil immersed to a transparent optical window (or transducer) and is focused at the transducer–air interface. The NA of the objective is larger than 1 in order to contain the critical angle defined by the refractive indices of the transducer and the tunneling air gap, so that TIR occurs at the transducer's distal surface. Hence, for an aqueous tunneling gap, as with *in vivo* biological samples, the NA should be greater than 1.33.

For photon tunneling to occur, the sample is placed in the proximity of the transducer (so-called due to the fact that the window facilitates conversion of height modulation present in the sample topography into intensity modulation) before oil immersion to the objective (the sample remains dry). In addition to providing the required optical continuity for TIR (now frustrated by local photon tunneling into the sample, Figures 5a, 5b, and 6b), the immersion oil allows the sample and transducer to travel together. This eliminates damage to the sample by sliding contact and allows complete freedom of movement about the surface. A recent innovation is the use of a proprietary (Polaroid) flexible optical membrane as the transducer: it is stiff and flat over the field of view of the microscope, but flexible enough on a larger scale to accommodate curved surfaces and even dirt on the sample.

The vertical tunneling range is about 0.75λ, or about 0.3 μm for green light. Where the gap grows larger than this, the transducer serves as a soft contact interference reference surface, with only refocusing and suppression of TIR illumination required for interferometry. Unlike the scanning tunneling microscopes, whether electron or photon, which operate at constant tunneling current, the photon tunneling microscope forms an image with the variations in tunneling referenced to a constant surface (the transducer). The grayscale tunneling image of the surface thus formed is viewed directly through the microscope.

The PTM gleans height information from the tunneling modulation caused by the (optically homogeneous) sample topography referenced to the transducer. This simple arrangement

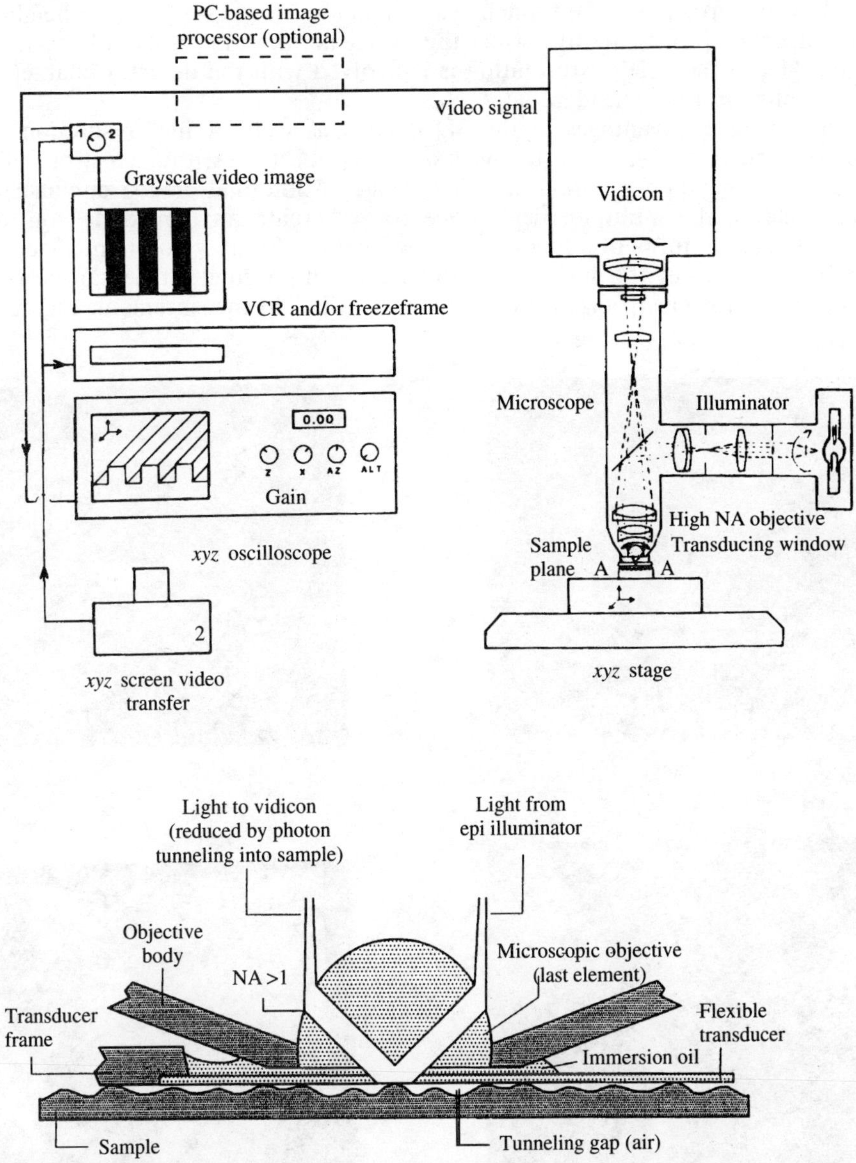

Figure 6 Schematic diagram of (a) the photon tunneling microscope and (b) the transducer, objective and sample.

eliminates the need for separate vertical control instrumentation, as is necessary in PSTM (see below). However, in order to measure height (and not optical variations as in latent images or biological samples), the tunneling response must be calibrated to the particular optical property of the sample. There are a number of possibilities for doing this, but a straightforward calibration method is to use a material with a known geometry, such as a spherical surface, allowing the relation of the sample height to the grayscale tunneling image to be measured directly, and also providing a check on the optical homogeneity. For polymers which cannot be fashioned into a sphere, spin coating a spherical substrate works well. Alternately, an occasional dirt particle under the flexible transducer offers the opportunity to calibrate directly on the sample by using the interference mode.

For height quantification, a charge-coupled device (CCD) or vidicon detector views the tunneling image at the microscope's phototube in the normal manner, and this output is readily analyzed and displayed in any number of ways, including 3D, by any number of image processing programs now available for PCs. Detector trade-offs must be considered, in that while the CCD has the better SNR and therefore higher vertical resolution and vertical range, the vidicon has higher lateral resolution at lower cost, with real-time response; with a pasicon tube, the vertical resolution is about a nanometer. Fortuitously, the nonlinear vidicon response to intensity largely corrects for the tunneling's exponential response to linear height, so that in the absence of image processing to correct the grayscale, the 3D display of the topography is not distorted. Hence the vertical resolution is convolved with the detector characteristics, and detector trade-offs must be considered.

Analog signals have advantages in the 3D display as well. A high-resolution x/y oscilloscope displays each of the video raster lines as amplitude traces, thus electronically mapping intensity back into height for a real-time 3D image. Additional analog circuits, to shear or expand and collapse the multiple oscilloscope traces, create an immediate control over 3D perspective with simple turns of a few knobs. The quality of reality and immediacy this display provides helps to interpret the surface and understand it. A single trace can be isolated anywhere in the image for electronic cross-sectioning and height measurement. Figure 7 provides

Figure 7 Comparison images obtained of an optical storage disc. Left column, top to bottom: SEM, WYKO and AFM images; right column: PTM images.

comparative images of a molded optical storage structure using PTM, scanning electron microscopy (SEM), micro interferometer (Wyko Corporation), and atomic force microscope (AFM).

5.4.3 Imaging Polymers with PTM

5.4.3.1 Film formation of polymer latex

Film formation from polymer latexes and the mechanism of film formation is of considerable industrial importance and has received a lot of attention in recent years.[69–71] Polymer latexes are water-based coating formulations which contain the polymer in the form of spheres, forming a colloidal suspension. When allowed to dry, some latexes produce transparent continuous, crack-free, mechanically rigid films, while some others give rise to opaque powders. This process has been found to be dependent, among other things, on the latex particle size and the temperature at which the film was formed, the so-called minimum film-forming temperature (MFT). Films formed above the MFT are usually found to give rise to a transparent, continuous film.

Several multistep mechanisms have been proposed in order to understand the film formation process• and to take account of several variables that affect the final film. Film formation is generally thought to occur in a number of successive steps and the most widely accepted mechanism is as follows. In stage 1 the water evaporates, leading to close contact of particles. In stage 2 it is believed that the particles pack and deform. In the final stage, stage 3, annealing the film at temperatures greater than the T_g of the polymer latex leads to a mechanically rigid film. All of these stages and the respective temperature ranges depend on a number of parameters, which include the latex structure, size, and composition as well as dispersion formulation.

We used 50 nm spheres of polystyrene suspended in water. A drop of the suspension was air dried on a glass plate, leading to very interesting surface topography of the film. The film is rather turbid and is not mechanically rigid. Normarski imaging at 100× magnification reveals a beautiful system of radial and tangential cracks (Figure 8a). The grayscale tunneling image from the PTM (Figure 8b) shows these surface height variations in high contrast, which when viewed in 3D (Figures 8c and 8d) are revealed to be ridges running diagonally, with occasional orthogonal flips across tiles defined by cracks. These features are also seen in Nomarski imaging, but with much less resolution than with the PTM. Further, the cusping of the tile edges is made obvious in PTM. The spheres themselves are just beyond PTM's lateral resolution, but structure is indicated.

It is quite well known that as water evaporates the latex particles come close together and pack to give structures (as examined by SEM and TEM) consistent with *fcc* packing. Although an extensive analysis of the structures observed has not been made, it appears that the observed topography of the latex film is consistent with an incomplete close-packed structure for the latex particles. This is evidenced by the fact that the observed structures change as a function of time, suggesting an incomplete film formation, yet to reach its final stage.

5.4.3.2 Imaging of polymer single crystals

Many synthetic polymers can be crystallized from dilute solutions to form what are usually referred to as 'polymer single crystals'. The discovery of such 'single crystals', made at almost the same time by Keller,[72] Till,[73] and Fischer[74] in 1957, has had a profound impact on the understanding of crystallization and self-assembly of macromolecules (recently, proteins). Their electron diffraction data showed that the polymer chains (polyethylene) were normal to crystal platelets formed from dilute solution, and most remarkable was the fact that the platelets were only about 10 nm thick. This led Keller and O'Connor[75] to conclude that the molecules must

Figure 8 (a) Grayscale image of radial cracks formed during film formation of 50 mm latex spheres. (b) Nomarski image of the same cracks. (c) and (d) Two 3D images of the same film with PTM.

fold sharply on themselves to give rise to lamellar periods of only 10 nm. Since then several studies have been carried out on solution-grown crystals.[76]

The principal tool that has been used to study the morphology of single crystals is electron microscopy and diffraction either from electrons or X-rays. Despite its obvious importance, electron microscopy is not without its shortcomings. These include the relatively high cost, the need for high vacuum, loss of the ability to manipulate samples, and possible damage to samples upon irradiation with the electron beam. These limitations are overridden in many cases by the resolution achievable and by the fact that one can perform selected area diffraction to determine the unit cell structure and orientation of crystalline polymers. Nevertheless, in an effort to circumvent many of the limitations of electron microscopy, Wunderlich and co-workers[77] and others[78,79] have used several forms of interference microscopy to visualize the single crystals formed from solution. All forms of interference microscopy use the phase difference between a ray of light split into two or more rays which on passing through a polymer crystal are made to recombine at a later stage, in the microscope, into a single ray. This phase difference is then converted into an intensity difference, thus making the polymer crystal visible and in some cases even leading to quantitative measurements of the thickness of the lamellae.[77]

Polyethylene (Marlex-50) was crystallized from a dilute solution in *p*-xylene (0.5% solution) at 80 °C to obtain large single crystals. A small amount of the suspension of *p*-xylene containing the single crystals was deposited on a microscope slide and the solvent was removed by evaporation under atmospheric conditions.

Figure 9 shows the grayscale image of a polyethylene single crystal as viewed directly under the PTM, where the individual single crystals are visible as intensity differences in the grayscale image. The high resolution and contrast of this image is rather remarkable, as viewed through the microscope (although the quality of this image is considerably reduced by the video chain). The analog 3D display of the grayscale image (Figure 10) clearly shows the individual lamellae of the polyethylene crystals, with step heights of ~7.0 nm. Manipulation of the perspective is relatively simple and helps reveal the structure of the surface. It should be pointed out that the apparent height in the 3D image is exaggerated in comparison with the lateral dimensions.

While the long side of the image is 25 μm across, the step heights are only on the order of 7–8 nm.

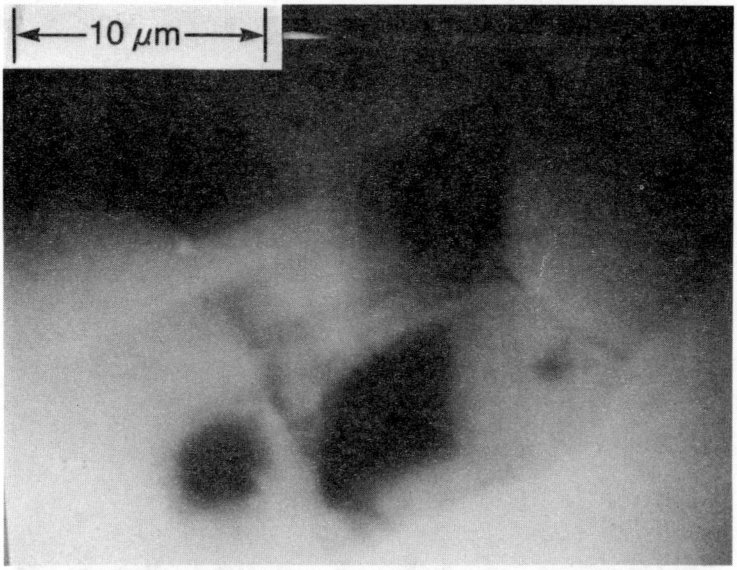

Figure 9 Tunneling image of polyethylene single crystals as seen through the PTM.

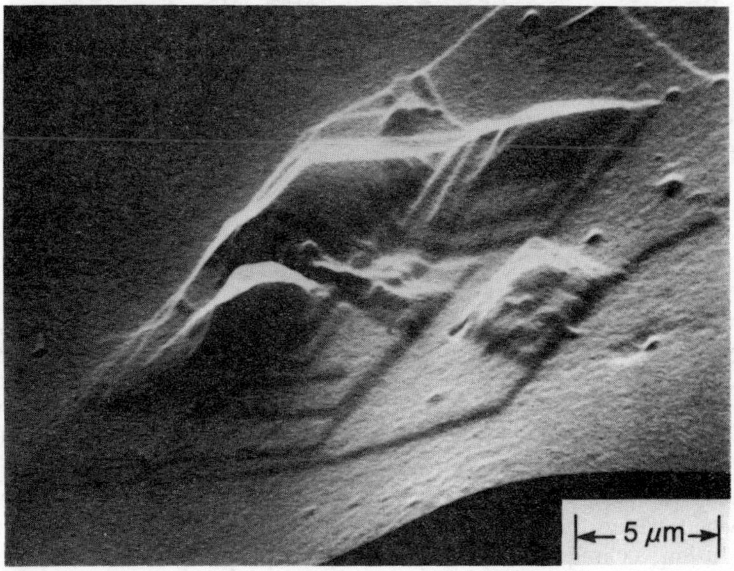

Figure 10 Topography (3D image) of a family of polyethylene single crystals.

Frequently the topography of the single crystals exceeded the tunneling range. This was due primarily to the fact that many of the crystals had a center of the growth spiral that was much thicker than the rest of the crystal. This lifted the flexible transducer much beyond the tunneling range. A limit to the tunneling range exists since the probability with which one frustrates the TIR decays exponentially. This is typically 0.75λ, which for green light is about 0.38 μm. In such a situation, one uses the other attributes of the microscope to image the thicker crystal, for example conversion of PTM to an interference microscope. This is accomplished by suppression of the TIR illumination, required for interferometry, and refocusing the microscope to image the crystal in the interference mode. Now, the flexible transducer serves as a contact interference reference surface. An interference image of a single crystal consisting of several lamellae obtained using the PTM in the interference mode is shown in Figure 11. Use of monochromatic light allows for quantification of the height of each of the lamellae.

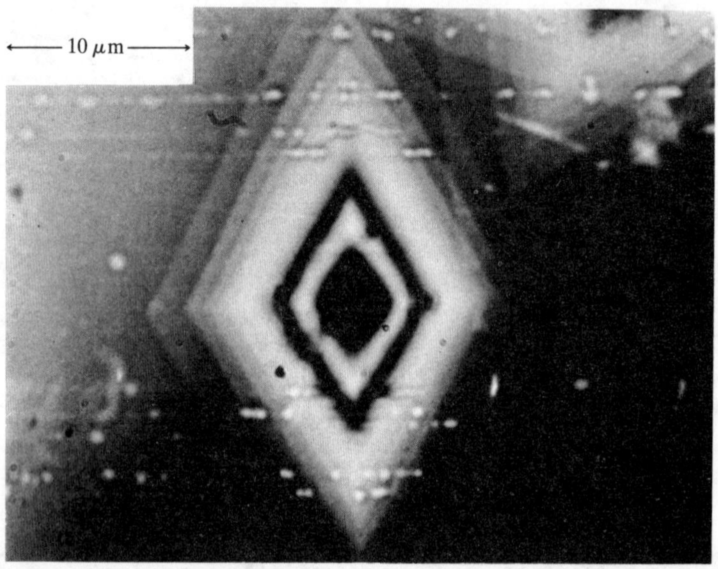

Figure 11 A contact interference image of the polyethylene single crystals obtained by suppression of the TIR illumination of the PTM.

5.4.3.3 Dewetting studies of polystyrene

The process of a liquid wetting a surface or the drying of a liquid film from a solid surface are important in many practical applications, for example in paints and coatings, detergents, and flow through porous media. Thin polymer films are used in a variety of applications where their thermal stability is important; in many instances a smooth and stable surface is desirable. The process of a liquid film wetting or dewetting a solid surface is dependent on the nature of long-range interactions between the liquid and the solid. The case of a liquid wetting and spreading on a surface is relatively well understood.[80–82] Here we describe experiments designed to study the inverse process, retraction/rupture or dewetting of a thin polymer film on a glass substrate. A common example of such a process can be seen in a film of water spread on a waxed automobile, where the film spontaneously ruptures and is drawn up into small beads.

It is possible using techniques like spin coating to prepare smooth surfaces of controlled thickness of polymer films even on nonwettable (defined by a nonzero contact angle between the substrate and the polymer film) surfaces. Films prepared on nonwettable surfaces, however, are not stable above their glass transition temperature and will rupture spontaneously, giving rise to beads of the polymers.[83] Experimental studies of thick films of a nonvolatile fluid dewetting[84] and of thin films ($<$100 nm) of polystyrene have been reported.[85] Thick films undergo the dewetting process by a nucleation and growth mechanism,[84] while thin films ($<$100 nm) are expected to undergo dewetting via a spinodal type decomposition. Such a study was reported in 1992[85] and evidence for spinodal decomposition was given based on the size of the droplets observed on completion of dewetting.

The experiments described below were motivated by our desire to directly observe the undulations of the film which grows and eventually ruptures the film, leading to droplets of size h^2/a where h is the initial thickness of the film and a is a molecular size.[83] For the experiments we used polystyrene ($M_w = 19\,000$, $M_w/M_n < 1.05$) obtained from Polymer Laboratories, UK. Thin films of polystyrene were coated onto a glass substrate from dilute solutions of polystyrene in toluene. The concentration and spinning speed were chosen so as to yield a film about 70 nm thick. The thickness of the films were measured by a profilometer to be 71 nm (within 1 nm). The glass surface was a plano-convex lens with a focal length of 1000 mm. The uniformity of the spin-coated films was checked by differential interference contrast microscopy and was found to be uniform over the entire disc (2 inches–51 mm–in diameter). A glass transducer in contact with the polystyrene film was used. The observations were made 2 mm away from the point of contact, a distance large compared to the thickness of the film.

When the thin polystyrene film is heated to above its glass transition temperature, a variety of patterns of ripples and periodic lines arise, and the merging of these structures, giving rise to droplets, is dramatically revealed in Figure 12. The structures by now are about 0.2 μm high and the periodicity displayed is remarkable.

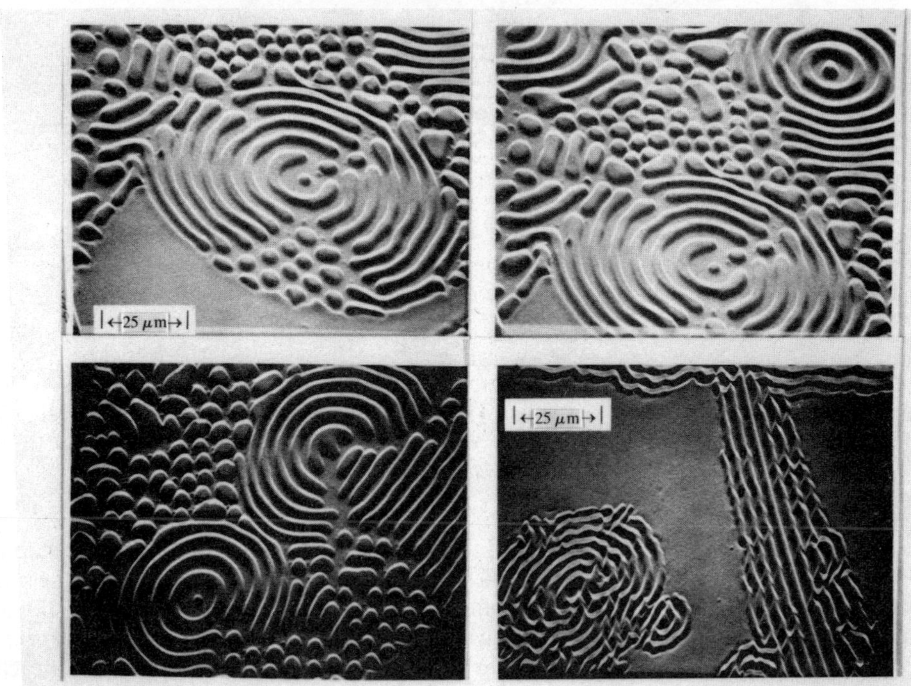

Figure 12 A variety of structures formed during the dewetting of polystyrene films on a glass surface, with a typical peak-to-valley distance of 0.2 μm.

Figure 13 shows a time sequence of the dewetting process. The polystyrene films in this case were heated to 165 °C, well above the glass transition temperature. As can be seen from the figure, the process of dewetting starts out producing a ripple-like structure which grows in amplitude as a function of time. The amplitude at the start of the breakup is on the order of a few nanometers, eventually reaching a peak to valley height of 0.2 μm. The spacing between the ripples is about 5 μm. Several such ripples start at the same instant, leading to the observed structure. Should the film break up by a spinodal decomposition mechanism, one would expect to find ripples or waves, as is observed, but one does not expect the ordered structure that is also observed. The ripples eventually grow in amplitude and spatial extent, colliding with similar structures formed a distance away. Each filament of the polymer making up the ripple undergoes a Rayleigh-type instability, giving rise to discrete droplets. In time the entire surface of the glass substrate is filled with little beads of the polymer. We believe that this is the first report of direct observation of the path that is taken by a thin polymer film as it starts to dewet by a spinodal decomposition mechanism. The ripples are attributed to the amplifications of thermal fluctuations prior to dewetting of the film.

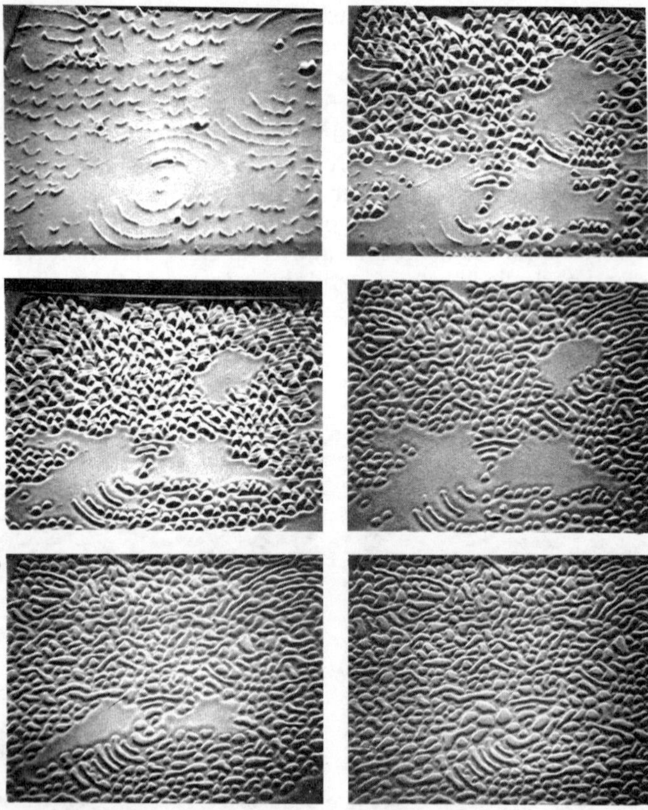

Figure 13 A time sequence of the various stages in the dewetting process of polystyrene films: (a) 60 s, (b) 456 s, (c) 540 s, (d) 564 s, (e) 864 s, and (f) 1200 s. Images (c) and (d) were made almost simultaneously to show the demagnification and for calibration of the entire sequence. Peak-to-valley height is typically 0.2 μm.

It can be argued that the structure observed is due to a Rayleigh–Benard-type instability as there is a temperature gradient in a direction normal to the film plane (along the viewing direction of the microscope). The situation then becomes identical to the case of a liquid film heated from below, giving rise to buoyancy-driven convection. In this case, above a certain critical temperature gradient, an instability due to convection will be initiated. In the case of polymeric fluids such instabilities do not occur due to the high viscosities involved.[86] Hence we attribute the observed structures to the dewetting through a spinodal mechanism, rather than to a Rayleigh–Benard-type instability.

Experiments were also done with a higher molecular weight polystyrene ($M_w = 80\,000$, $M_w/M_n < 1.04$) which reveal structures similar to those shown in Figure 12. The circular, ripple-like structure does eventually break up into little droplets, as is expected[83] for a film breaking up via a spinodal decomposition process. It appears from the results presented that the film breaks up with a complicated mechanism that has characteristics of spinodal decomposition.

5.4.3.4 Imaging monolayers

Monolayers of a 'hairy rod-like polymer' formed by Langmuir–Blodgett deposition have been studied both by PTM and polarization modulation laser scanning microscopy (PM-LSM).[87,88] Multilayers of this polymer are formed on a glass plate by repeatedly dipping the glass slide through a monolayer in a Langmuir trough, as described elsewhere.[88] PTM was used to study the topographic variations that are seen in the film as the number of layers are varied spatially along the length of the slide. The boundary between 21 (~46.2 nm) and 11 (~26.0 nm) layers was chosen to study the topographic variation. Figure 14 shows the variation in topography at this boundary, which ideally would be sharp if the dipping process had been carried out precisely to the same depth. It is apparent that the topography at the boundary is far from sharp, and perhaps its origins can be traced to how the material becomes deposited during the dipping process. Figure 15 shows the uniformity of orientation of the two layers where the image has been formed by implementing polarization modulation into a laser scanning microscope.[87,88] This allows one to map the spatial variation of orientation of very thin films.

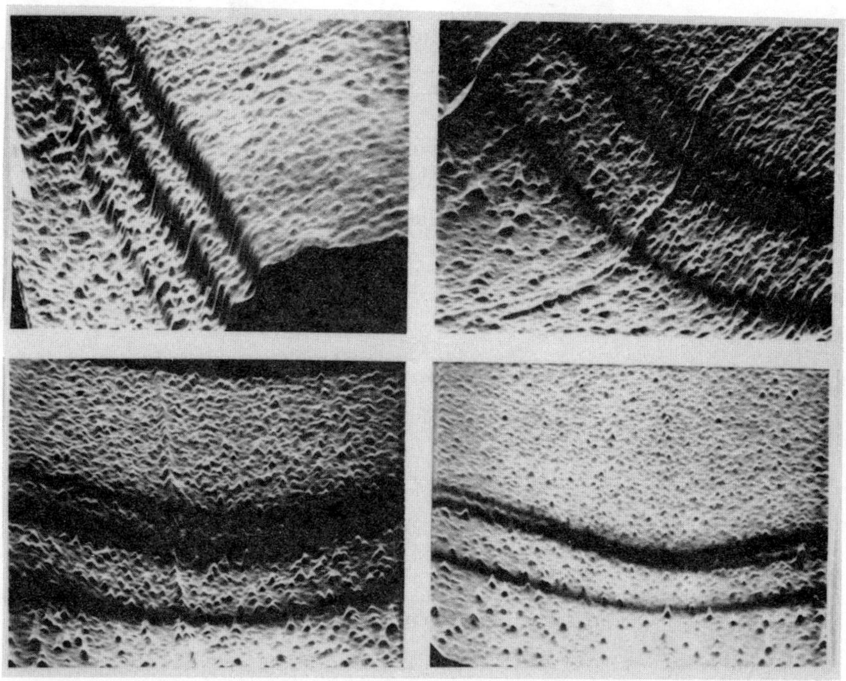

Figure 14 Various views of the topography of LB films of a rod-like polymer on a glass surface. The transition is from 21 layers to 11 layers.

5.4.3.5 Latent images of photoresists

So far, the use of PTM to visualize the topography of polymeric surfaces has been discussed. PTM is also sensitive to variations in the local refractive index. Often this feature of the microscope is not used to advantage. A situation where it can be used is in visualizing the latent images that are formed on photoresists, before they are developed. With the devices moving towards smaller and smaller dimensions (as was envisioned by Feynman over 30 years ago[89]), use can be made of the enhanced lateral resolution of the PTM in characterizing the photoresist. Figure 16 demonstrates the use of PTM to visualize latent images. The grating has a spacing of about 2 μm. The topography that is represented is not the true topography, but rather due to a refractive index variation of the exposed and unexposed areas of the photoresist. This false topography can be used to measure the change in the refractive index or used to measure the size of features before they are developed to have topographic variations (metrology of photoresists, as was suggested by Nyyssonen and Larrabee[90]).

5.5 SCANNING OPTICAL MICROSCOPY

The quest to resolve objects that are ever smaller has produced two optical microscopy techniques, near-field scanning optical microscopy (NSOM, also referred to as SNOM) and photon scanning tunneling microscopy (PSTM),[91–98] that use a probe that is raster-scanned over the sample to obtain an image at higher spatial resolution. The principle of NSOM was

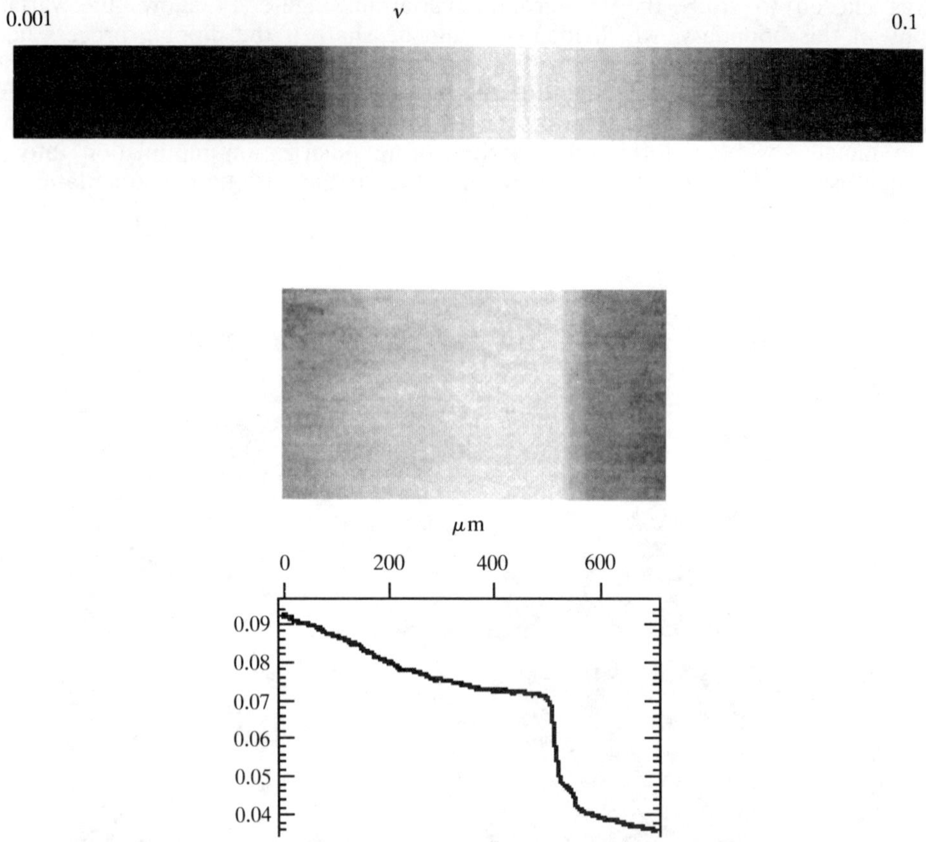

Figure 15 The linear dichroism of the same transition zone obtained using a PM-LSM. The plot shows the magnitude of the dichroism as a function of position along the image.

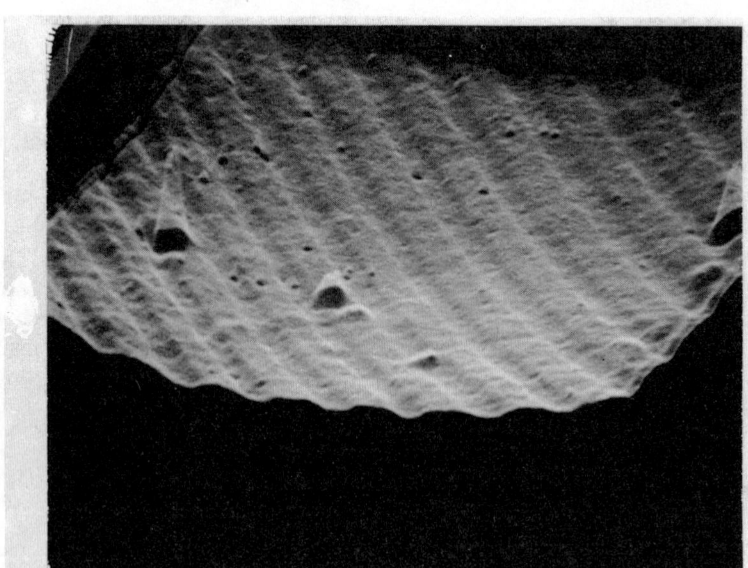

Figure 16 PTM image of 'false topography' of a latent image. The apparent topography is due to refractive index variation and not to height variation.

described earlier (Section 5.2, see Figure 2) and we will return to this form of microscopy. In the following are described the principle involved and the attributes of PSTM to imaging objects at higher spatial resolution.

5.5.1 Photon Scanning Tunneling Microscopy

The principle of operation of PSTM is essentially the same as that described for PTM. One creates an evanescent wave by TIR at the surface of a transparent object and the perturbation or the frustration of the TIR is detected by having a dielectric stylus, an optical fiber drawn down to subwavelength dimensions. The optical fiber is raster-scanned in a constant height mode or constant intensity mode, provided by a feedback circuit, with the light coupled into the fiber being fed to a photomultiplier tube. Hence the image is built by probing the evanescent field locally. A schematic diagram of the PSTM is shown in Figure 17. The light source is usually a low-power He–Ne laser, but white light sources may be used as well.[91] The dielectric probe is usually an optical fiber sharpened to about 100 nm radius with an apex of about ~20° by chemical etching with a hydrofluoric acid solution.

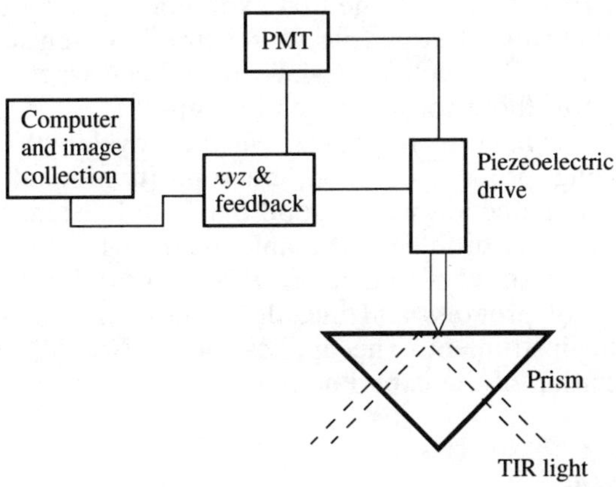

Figure 17 Schematic diagram of photon scanning tunneling microscopy.

A sample placed on the prism (TIR) surface will cause a spatial variation of the evanescent field that is characteristic of its dielectric and topographic properties, as was explained in some detail for the operation of PTM. In the case of PSTM, the method of detecting the frustration relies on raster-scanning a dielectric probe over the sample. Interpretation of the images are not quite unambiguous, as opposed to the case in PTM. This is so for at least two reasons: polarization effects need to taken into account due to the illumination and artifacts arise due to the tip–sample interaction.

In all forms of SOM, issues of resolution need to be looked at rather carefully and the experimenter needs to be quite critical in establishing the 'true' lateral resolution. Taking PSTM as an example, studies in the past few years have claimed resolution ranging from 100 nm to 10 nm. Often this resolution comes from the edge steepness of an object that is imaged.[92] Edge steepness is not adequate in itself for testing the resolution of an instrument. In order to put limits on the resolution it is quite obvious that one needs to image a close-packed structure of some sort, especially without topography. A comparison then needs to be made of the same structure imaged by some other well-established technique, like electron microscopy for example, to determine the limits to the resolution of the microscope. It appears that this has not been done for many claims of the resolution for PSTM.

The difficulties come from a number of factors that are perhaps unique to problems associated with scanning a probe in close proximity to the sample. It has been pointed out[99] that it is quite easy to generate high spatial frequencies which, on further analysis, turn out not to be due to the lateral imaging of the scanned probe. The complication arises from the evanescent field which, as was pointed out earlier, has a high degree of sensitivity in a direction normal to the sample. In order to minimize this effect, the relative distance of the sample to the probe must be kept constant over the course of a scan to obtain the image. In order to ascertain that a given form of SOM provides true lateral resolution, it is necessary to use test samples that have no topography (or small compared to the resolution limits claimed) and that no distance regulation be used.[99] In almost all cases of PSTM, to the author's knowledge, images presented do not provide for an unequivocal demonstration of the true enhanced lateral resolution that is claimed. A theoretical investigation[100] into issues of lateral resolution for PSTM sets the limit at $\lambda/3$. This is primarily due to diffraction effects, which become significant for increasingly fine tips that are not coated with a metal to contain the light (bound modes).

Although there have been several papers published in this area, this form of microscopy is still in its infancy. Not much work has been done to characterize material properties. PSTM has been recently adapted for spectroscopic measurements, the details of which can be found elsewhere.[91]

5.5.2 Near-field Scanning Optical Microscopy

In this section we review a method near-field scanning optical microscopy (NSOM), which has been demonstrated to have 'true' enhanced lateral resolution down to about $\lambda/40$.[99] The basic principle involves scanning a light source of subwavelength dimensions over the sample that is to be imaged (see Figures 2 and 18). This method of circumventing the diffraction limit was proposed by Synge in 1928.[20] In 1972, the first experimental demonstration of resolution of $\lambda/60$ was achieved using microwave frequencies ($\lambda = 3$ cm).[101] Extensions to optical frequencies have followed more recently.[7–9,99] Excellent reviews have been written on this subject and the reader is well advised to go to those for a detailed account.[10,99]

In order to extend the near-field microscope to visible wavelengths, several technical issues had to be solved, including probe formation, micropositioning of the probe, mechanical stability of the system, signal detection with a good SNR, and distance regulation of the probe from the sample. Among several problems, the amount of light delivered to the sample is of great importance and a variety of ways to increase this intensity have been developed. We will outline the various aspects of progress and include some images that have been obtained to show the versatility of the instrument. The application of NSOM to problems seems to be limited only by the imagination of the experimenter.

5.5.2.1 Probe development

One of the most important aspects of NSOM is the fabrication of a probe of subwavelength size and this probe has to be able to deliver sufficient amount of light. The initial approaches were to fabricate an aperture array by e-beam lithography[7] in a thin, planar silicon nitride membrane. Once the arrays were formed, the membranes were made opaque by evaporating aluminum. These arrays are useful for the characterization of the optical properties of submicron apertures. Since the extent of the near-field is rather limited, such arrays do not form a convenient subwavelength source in practical situations. Another early approach was to use glass pipettes.[7] In this instance, the pipettes were drawn down to diameters of about 100 nm, after which they were metallized to form the probe. Betzig *et al.*,[7] using such probes (well characterized by electron microscopy), demonstrated that light can be transmitted through the aperture. Using such apertures a resolution of $\lambda/3.6$ was demonstrated in 1986.

Pohl *et al.*,[8] at about the same time, used a quartz rod, cut and polished to the shape of a pointed tip, which was then covered with an opaque metallic film. The tip covered with a metallic film was then cold-deformed until a weak spot of light became visible at the tip of the would-be probe. It was estimated that under these conditions a hole with a diameter of about 10–20 nm was created, which was then used for imaging. Details of a variety of other methods of obtaining illumination for NSOM can be found elsewhere.[10,102]

An alternative approach to obtain a subwavelength light source was proposed in which one relied on crystals that would emit fluorescence at the end of a pipette in which the crystals were grown.[103] Crystals of anthracene were grown at the end of a pipette tip which can then be used as a light source. It was demonstrated that such a probe can be incorporated into an inverted light microscope to obtain images with enhanced lateral resolution.[104] A scheme that would allow the excitation of the crystal by electroluminescence was also suggested as an alternative.[103]

The most significant development has been the use of an adiabatically tapered (*i.e.* the taper is sufficiently gradual that energy is not lost to radiation modes) single mode optical fiber coated with aluminum. This allows a significantly high throughput of light to be used for near-field imaging[99] due to the ability of the fiber to transfer light to the aperture via waveguiding. A nice illustration can be found in Figure 1 of Betzig.[105] The fiber also preserves the incident polarization very well, with attainable extinction in excess of 2000:1.[99] Another more recent innovation to obtain more light through the aperture is the use of a near-field/fiber laser probe.[106] This probe has been demonstrated to lead to a higher throughput (by about a factor of 10^3) combined with high spatial resolution ($\lambda/10$).

Thus far, nothing has been said about aluminum and what limits the ultimate resolution obtainable in NSOM. Aluminum is used to contain the light to subwavelength dimensions, which would otherwise not be contained in the fiber probe, and energy would be lost to other evanescent modes. A coating of about 80 nm thick aluminium is found sufficient to contain the light to the probe dimensions. The limit to the attainable spatial resolution is given by the optical penetration depth (skin depth) of the opaque screen. The prevalent use of aluminum as the 'opaque screen' is due to the fact that it has the shortest skin depth of any material for visible radiation.

One obvious solution to increased flux is to increase the power level of the input laser source to the probe. One has to remember that doing so is not helpful due to the fact that a finite amount of light is absorbed by aluminum and dissipated as heat. Hence, increasing the input power level to obtain higher throughput will work only when one has a perfect reflector. One can use multilayer dielectric stacks (as for laser cavity applications),[107] but their reflectivities do not reach 100% for thin films (<80 nm). The construction of better probes has come a long way since 1990, and it is certain that better probes will be made with further refinements.

5.5.2.2 Near-field imaging

One of the things that is most important is that the sample to be imaged should be in the near-field. A useful definition of near-field would be the region in which the radiation transmitted from the aperture remains confined to an area of roughly the size of the aperture. This region will extend to approximately z size of the aperture.[10,99] The optical field in this region (the near-field) does not propagate due to the fact that it has a large evanescent component, in much same way as the optical field near a TIR surface does not propagate. Similar to FTIR, if a sample should be placed in this region, the normal optical processes can be observed. Since this field changes rather rapidly in both the propagation direction and in a plane normal to it, the resolution attainable in an NSOM is dependent on the distance of the sample from the aperture, as was demonstrated quite clearly by Betzig and Trautman.[99]

In order to keep the distance of the aperture to the sample constant and be able to manipulate the probe over samples with varied topography, a method was needed to accomplish this function. A shear force mechanism was incorporated in NSOM to enhance the reliability and versatility of the microscope.[108] In this method the probe functions as a cantilever (at its resonant frequency) to measure small shear forces between the tip of the probe and the sample (Figure 18). This provides for topographic mapping as well as near-field imaging of the sample, although the two images are coupled. One can also detect the lateral dither motion of the probe by external focusing of light onto the probe and detection of the scattering from the tip.[105] It has been pointed out that the scattering method provides a topographic signal that is not coupled to the near-field image (see Figure 3 in ref. 105). While the optical effects on the topographical signal can be decoupled, it is frequently not possible to accomplish the reverse. This is primarily due to how light couples to the sample while the probe is scanned over a rough sample. Hence, any true test for resolution of NSOM should present images that are virtually uncorrelated with the topographical features of the sample. In very few instances are such criteria met. Often such correlations are taken to be a test for resolution![91]

The microscope can be adapted to perform imaging in the so-called transmission geometry or reflection geometry, as is schematically shown in Figure 19. To demonstrate that NSOM is capable of imaging at high resolution, a state-of-the-art diffraction-limited image of a metal test pattern on a silica plate and a transmission geometry NSOM image are shown in Figure 20. The advantages of using NSOM are quite clear, where resolution is improved to at least $\lambda/43$.[99] That a variety of contrast mechanisms unique to optical microscopy can still be retained with high spatial resolution is demonstrated in Figure 21 (a color version of this can be found elsewhere[99]). It is clear from the images presented that this form of microscopy is quite useful for a variety of applications, making use of all the contrast mechanisms available. The reader is referred to the original literature for a detailed account of the images that are shown in Figure 21.[99]

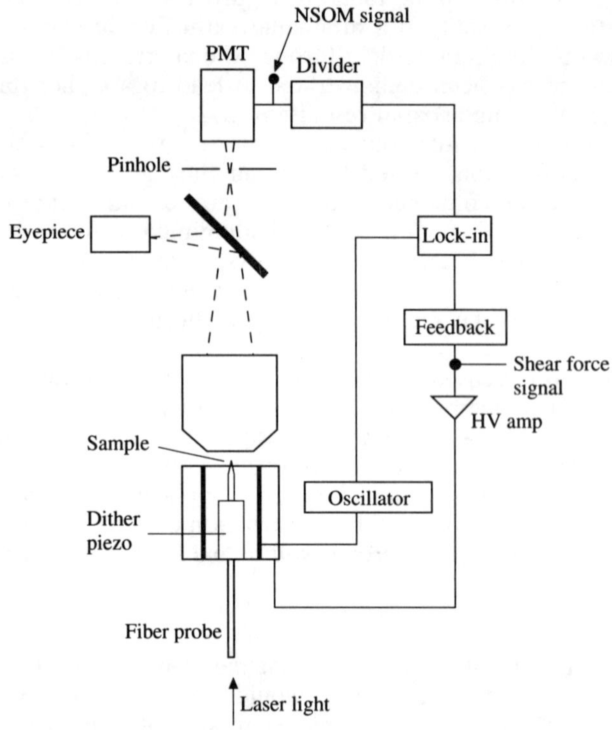

Figure 18 Schematic diagram of a combined shear force and near-field scanning optical microscope (redrawn after Betzig *et al.*[108]).

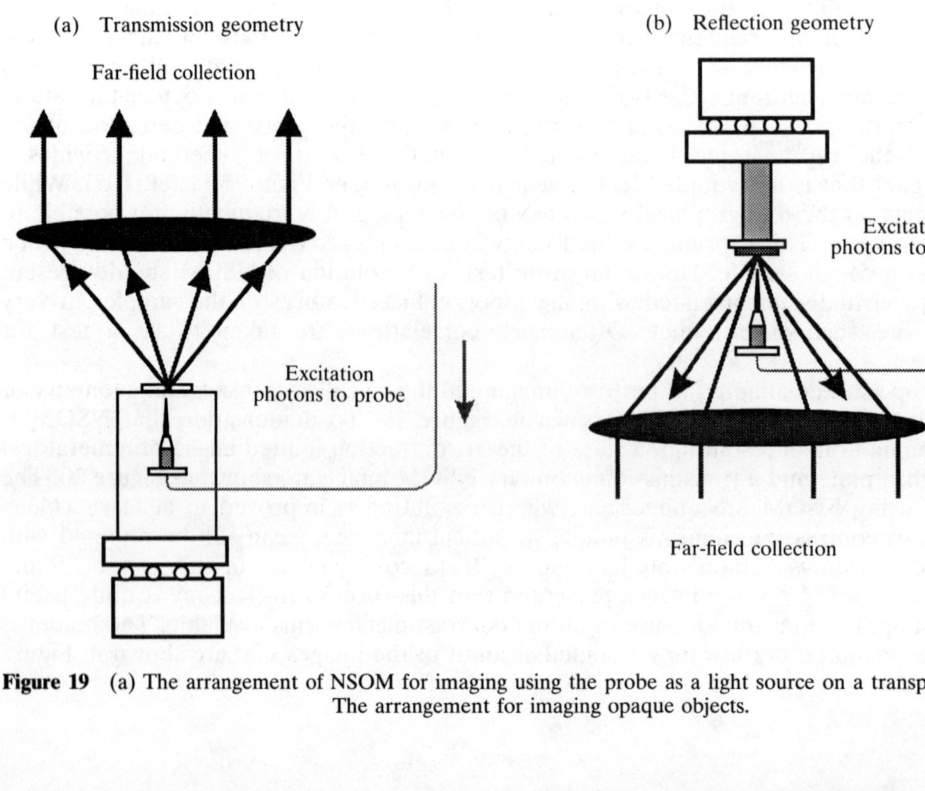

Figure 19 (a) The arrangement of NSOM for imaging using the probe as a light source on a transparent sample. (b) The arrangement for imaging opaque objects.

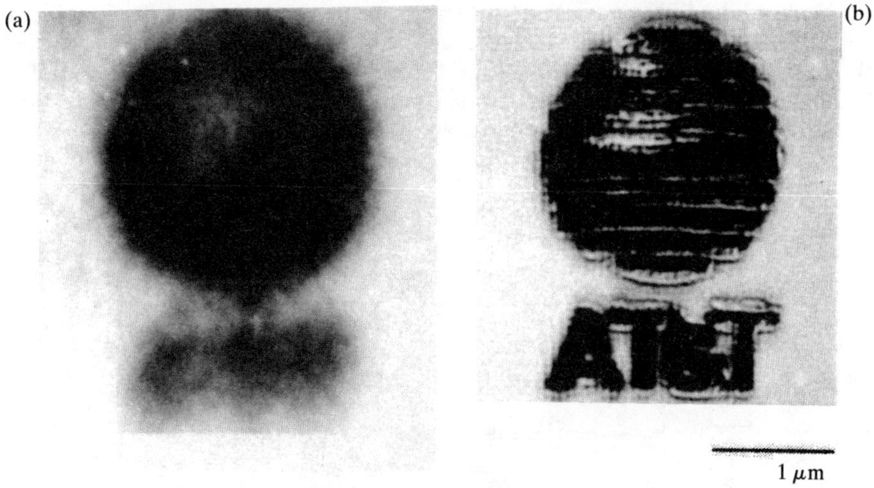

Figure 20 (a) State-of-the-art diffraction-limited image of a metal pattern. (b) NSOM image of the same pattern in the transmission geometry (Figure 19a) showing resolution of at least $\lambda/43$.

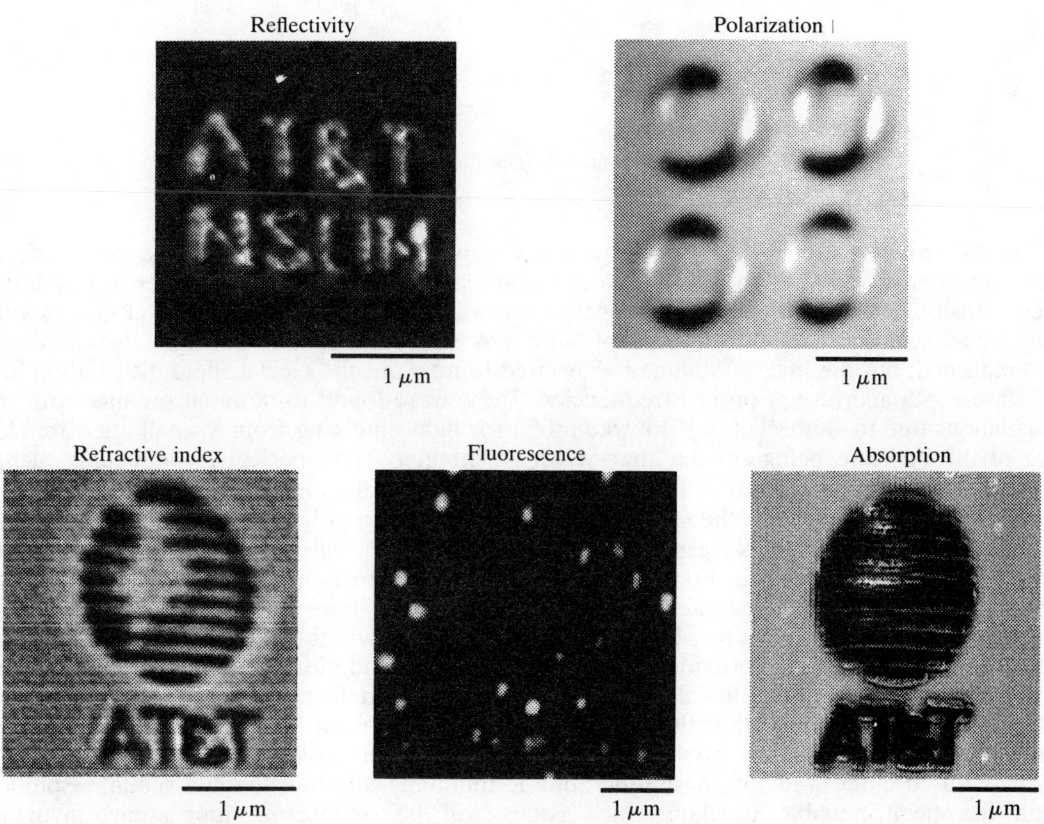

Figure 21 Montage of micrographs obtained using the various contrast mechanisms available with optical imaging with the use of an NSOM (courtesy of Jay Trautman, Bell Labs.).

NSOM has been applied to a variety of material problems in various disciplines, the details of which can be found elsewhere. One of the applications that has not been touched upon is spectroscopic applications (Figure 22). NSOM has been very successfully applied to the detection of single molecules. Much can be learned from single molecule detection (SMD). SMD represents the ultimate goal in chemical analysis[109] and has been suggested as a tool for base sequencing of DNA.[110]

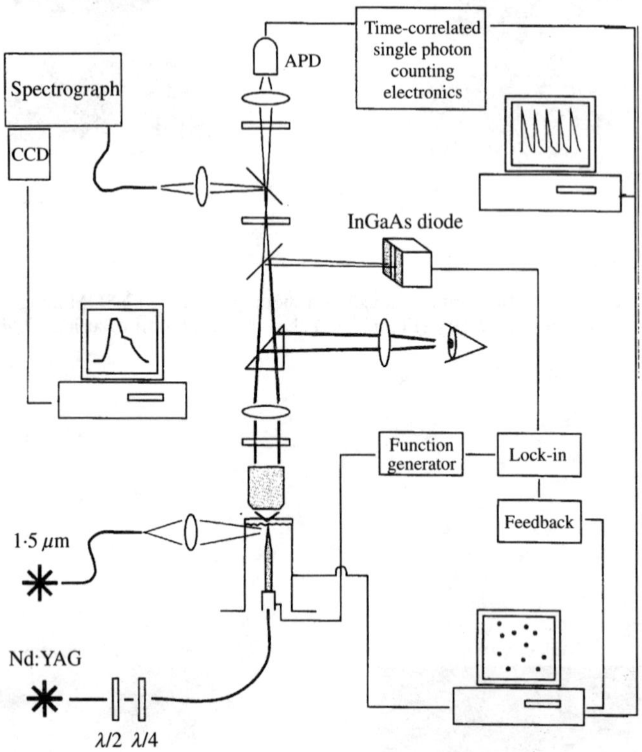

Figure 22 Schematic diagram of NSOM for spectroscopic measurements.

The experiments using NSOM consisted of placing a dye on the surface of a polymer that was spun coated to a glass coverslip. The dye solution was spread on the polymer to give a low areal density. The sample was then imaged using NSOM, which provided several images with polarized excitation and detection. Not only was the detection of a single dye molecule accomplished, but the images obtained were used to map out the electric field distribution in a subwavelength aperture at optical frequencies. These were found to be in accordance with the calculations due to Bethe[111] (and Bouwkamp[112]) for light emerging from a small aperture (the size of the aperture being smaller than λ) in an infinitely thin, perfectly conducting, planar screen. The reader is referred to the original publication for the elegant experiments.[113]

Applications of NSOM to the study of problems relating to polymer science are nonexistent to the author's knowledge. Nevertheless, there are many areas where NSOM can be applied to problems in polymer science. For instance, the study of defects in liquid crystals will benefit from NSOM. There are several models that describe the core of orientational defects in nematic liquid crystals. Thus far, EM has been applied to study the core structure. As pointed out earlier, EM does not provide information on the orientational state of molecules, and hence will not be of use in elucidating the core structure of defects. For example, EM will not provide information on whether the core of a defect is isotropic or biaxial. NSOM, with its high spatial resolution, can in fact provide valuable information on the structure of the defect core. It can be used in the study of phase separation in thin films with much higher spatial resolution than has been possible to date. Such issues will be of use in applications involving photolithography, where the true reproduction of lithographically encoded patterns seems to depend on the state of miscibility of two polymers.[114] As the making of devices, and hence feature sizes,[115] moves to ever smaller sizes, NSOM with its high spatial resolution will prove to be a valuable tool in the characterization and measurement of these structures (submicron metrology).[90]

5.6 STANDING WAVE FLUORESCENCE MICROSCOPY

In an optical microscope the image that is formed contains the in-focus and out-of-focus light, which does not allow for a 3D visualization of the structure of the object under study. Three-dimensional microscopy has revolutionized biology and is becoming a widely used form of microscopy in many areas of biology and medicine. It has been used to study, among other things, the 3D organization of single cells, cells in tissues, *in vivo* microscopy of the human eye,[116] and colloidal crystals.[117,118] An important attribute of this form of microscopy is that it can probe the 3D structure by optical sectioning of the object. The difference between a conventional microscope and a confocal microscope can be understood from the schematic diagram of a confocal microscope shown in Figure 23. While both the in-focus and out-of-focus light is used in a conventional microscope, only the in-focus light is used in confocal microscopy.

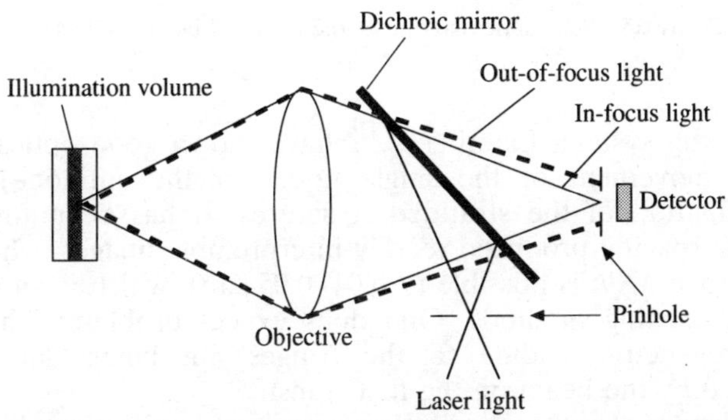

Figure 23 Schematic diagram of a confocal fluorescence microscope.

The axial resolution of a microscope, using the Rayleigh quarter-wave criterion[3] based on the phase shift suffered by the marginal rays compared to that of the central ray passing through the lens, is given by

$$\Delta z = \lambda\{8n \sin 2(\theta/2)\}^{-1}$$

where n is the refractive index. As pointed out earlier, this expression would give an axial resolution of 0.17–0.13 μm for a high-NA objective. While this has been demonstrated for transmitted light imaging,[63] the limit reached in the fluorescence mode of a confocal microscope is usually on the order of 0.7–1.0 μm. This is perhaps due to the lower SNR.

A way to overcome this limitation was demonstrated by Lanni *et al.*[119] and Bailey *et al.*[3] The approach was to use the three-dimensional standing waves (SW) created in the sample volume by crossing two coherent light beams of equal amplitude at complementary angles relative to the optical axis of the microscope. This excitation field varies sinusoidally along only the axial direction given by

$$I_{ex} = 1 + \cos(Kz + \Phi) \qquad (4)$$

Where $K = (4\pi n/\lambda)\cos\theta$, and the electric field nodal plane spacing is given by $\lambda/2n\cos\theta$. The minimum node spacing occurs when the two beams are counter propagating ($\theta = 0$). By varying the phase of one of the beams, the nodal planes can be moved axially relative to the specimen, thus selectively illuminating the stratified structures in a sample. A schematic diagram of the reflector system is shown in Figure 24. A confocal scanner has also been developed where the principle of standing wave excitation has been used,[120] and is referred to as the 4Pi confocal fluorescence microscope.

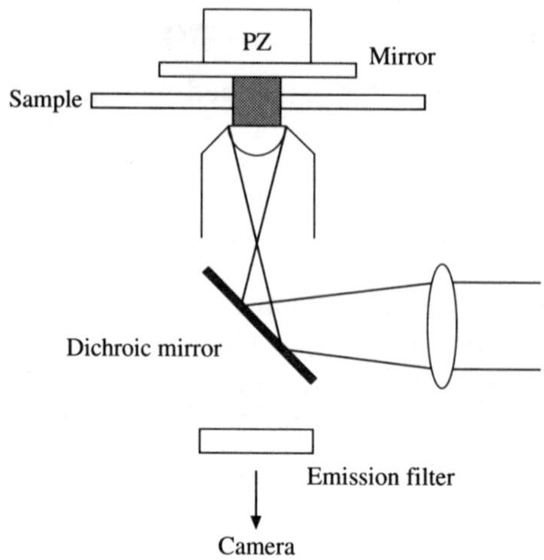

Figure 24 Schematic diagram of the reflected-beam SWFM.

Using the reflector beam system, Lanni *et al.*[119] have shown good optical sectioning of thin specimens. Controlled movement of the single node (or the antinode) within the object provides for the illumination of the stratified structures. It has been found that a shift of one-quarter of the node spacing produced readily interpretable images, which suggests that an axial resolution better than $\lambda/8n$ is possible ($\sim$0.04–0.05 μm), which is an order of magnitude improvement over the existing methods. One does expect problems when the sample has refractive index heterogeneity, leading to the fringes not being flat due to the phase information accumulated by the beam on the first transit.

This development is rather recent and will have important implications for imaging biological structures. Its application to polymers will await the time when confocal microscopy becomes a routne instrument for studying the morphology of polymers. Confocal microscopy should prove to be a valuable tool to learn about the 3D structure development during phase separation. It has been applied to problems in colloidal crystals.[117,118] Although this form of microscopy has not found widespread use in polymer science, its use is growing.

5.7 CONOSCOPY OF NEMATIC LIQUID CRYSTALS

A nematic liquid crystalline phase has uniaxial symmetry and possesses two refractive indices, the ordinary refractive index n_o and the extraordinary refractive index n_e.[121] Hence, optical characterization of these materials requires the measurement of two refractive indices. In the nematic phase, the measured birefringence Δn ($=n_e - n_o$) is directly related to the order parameter S defined as $(1/2)[\langle 3\cos^2\theta\rangle - 1]$. Therefore measurement of Δn affords a simple way to characterize the order of the nematic phase.

We discuss here an optical crystallographic technique, conoscopy, and demonstrate its usefulness in the measurement of Δn of nematic solutions of a rod-like polymer (poly(1,4-phenylene-2,6-benzobisthiazole); PBT) in methanesulfonic acid (MSA). This method is, of course, useful for any material that has uniaxial (or biaxial) symmetry.

5.7.1 Conoscopy

In using this method for optical characterization, the nematic single crystal is illuminated by a highly convergent laser beam. In doing so, one is able to probe the sample at many different propagating angles of light simultaneously. This method is especially useful for samples with large birefringence, since orthoscopic polarized light microscopy will not reveal small distortions of the optical axis, due to the so-called adiabatic limit.[121] By contrast, small distortions of the director field are revealed by rotation of the conoscopic interference figures about its center of symmetry as the director field is rotated in the sample plane, or by translation of the pattern if the director field tilts out of the sample plane. Figure 25 is a schematic diagram of an optical microscope for conoscopic measurements. The conoscopic figures are observed in the rear focal plane of the objective (see Figures 3 and 25).

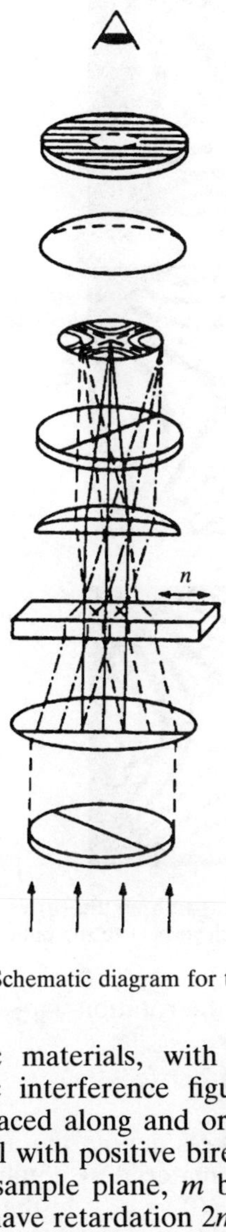

Figure 25 Schematic diagram for the observation of conoscopic figures.

For uniaxially symmetric materials, with the optical axis normal to the axis of the microscope, the conoscopic interference figures are hyperbola-like principal isochromates (curves of zero intensity) placed along and orthogonal to the optical axis (or director) of the material.[122,123] For a material with positive birefringence the retardation $\delta = 2m\pi + \delta'$ for light propagating normal to the sample plane, m being an integer and $0 < \delta' < 2\pi$; the successive fringes along the optic axis have retardation $2m\pi$, $2(m-1)\pi, \ldots, 2\pi$ on moving away from the center, and those along the orthogonal axis have retardation $2(m+1)\pi$, $2(m+2)\pi, \ldots$ on moving outward. For the samples studied here, $m = 20$ or more, depending on the solution concentration and the cell thickness. The birefringence of the samples can be determined by measuring the distance between successive fringes that are placed normal to the director or the optical axis. Experimentally, the distance between the fringes can be measured using a calibrated eyepiece on the microscope or by using a photograph of the interference figures. A set of interference figures are shown in Figure 26 for a quartz crystal having a thickness of 4096 μm. Figure 26(a) shows the interference figure for the crystal with its optical axis in the plane of the microscope stage where the center of symmetry of the interference figure is centered along the optical axis of the microscope. Figure 26(c) shows the same crystal tilted with respect to the optical axis of the microscope. It is easy to see that the center of symmetry is no longer at the center of the field of view, thus providing the valuable information that the optical axis of the system is tilted, which is of importance for the shear flow of nematic liquid crystals.[126] Figure 26(b) shows the quartz crystal rotated in the plane of the microscope stage and this is apparent from the rotation of the conoscopic figure. The rotation of the figure is related to the rotation of the optical axis of the crystal, as is discussed below. This is the principle that is used to measure the twist viscosity of the nematic liquid crystal.[124,125]

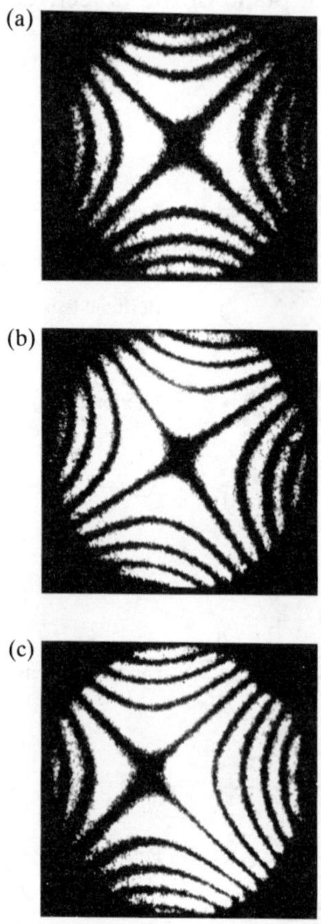

Figure 26 Conoscopic figures obtained for a quartz crystal (a) with no tilt, (b) with a small rotation about the optical axis of the microscope, and (c) with a small tilt with respect to the optical axis of the microscope.

For a continuous rotation of the director field, the rotation Φ_{ROT} of the interference figures is given by[124]

$$\tan 2\Phi_{ROT} = \langle \sin 2\varphi \rangle / \langle \cos 2\varphi \rangle \tag{5}$$

where φ is the twist angle, and the angular brackets denote the average of twist $\varphi(z)$ through the sample thickness z $(-d/2 \le z \le d/2)$. For a material at equilibrium in a field imposing a continuous twist distortion, with maximum twist $\varphi(0) = \varphi_{max}$ in the midplane of the sample and zero twist $\varphi(d/2) = \varphi(-d/2) = 0$ at the surfaces, Φ_{ROT} may be expressed as[121,124,125]

$$\tan 2\Phi_{ROT} = \frac{2\sin(\varphi_{max})}{2E(\pi/2, \sin\varphi_{max}) - F(\pi/2, \sin\varphi_{max})} \tag{6}$$

where F and E are the complete elliptic integrals of the first and the second kinds, respectively. In this analysis, $\varphi(z)$ is essentially given by $\varphi_{max}\cos(\pi z/d)$ for small φ_{max}, though the rigorous expression is more complicated.[127] For the range of φ_{max} of interest here, calculations using the above equation show that φ_{max} may be approximated from the measured Φ_{ROT} by the simple expression $\varphi_{max} = 1.57\Phi_{ROT}/(1 + 0.14\Phi_{ROT}^2)$ within 3% error (with Φ_{ROT} in radians). By comparison, a bulk rotation of the director field, such that $\varphi(z) = \varphi_{UNF}$ at all z, gives $\Phi_{ROT} = \varphi_{UNF}$ for the range of retardations of interest here, as is the case in Figure 26(c). For the sake of completeness, the translation Δ of the center of symmetry of the fringe pattern from the optical axis of the microscope is related to an average of the tilt $\theta(z)$ of the director out of the sample plane. For a uniform tilt ($\theta(z) = \theta_{UNF}$), $2\Delta/D\Omega_{NA} = n_l(\sin\theta_{UNF})$, where D is the diameter of the field of view, Ω_{NA} is the numerical aperture of the objective (here 0.60), and n_l is the average refractive index of the shape.[128]

5.7.2 Measurement of Birefringence

A typical conoscopic figure for a nematic monodomain with planar alignment is shown in Figure 27 (solution with $w = 0.0483$ (wt) or 0.0715 (g cm^{-3})). Several nematic monodomains with different concentrations were studied. The values of the birefringence as a function of concentration are shown in Figure 28. It is seen that the birefringence increases linearly with the concentration of the polymer in solution. The details of the calculations are given elsewhere.[129] The order parameter of the solution mainly depends on the concentration of the solute polymer, PBT in this case. The birefringence of the system $|n_e - n_o|$ can be related to the order parameter through the relation

$$|n_e - n_o| = |n_e - n_o|_{MOL} S\phi \tag{7}$$

where ϕ is the solute volume fraction, $|n_e - n_o|_{MOL}$ is the molecular birefringence, and S is the order parameter. Using certain approximations,[128] the birefringence can be written as

$$|n_e - n_o| = 1.46S(\phi - 0.002) \tag{8}$$

for $\phi > 1.1\phi^*$.

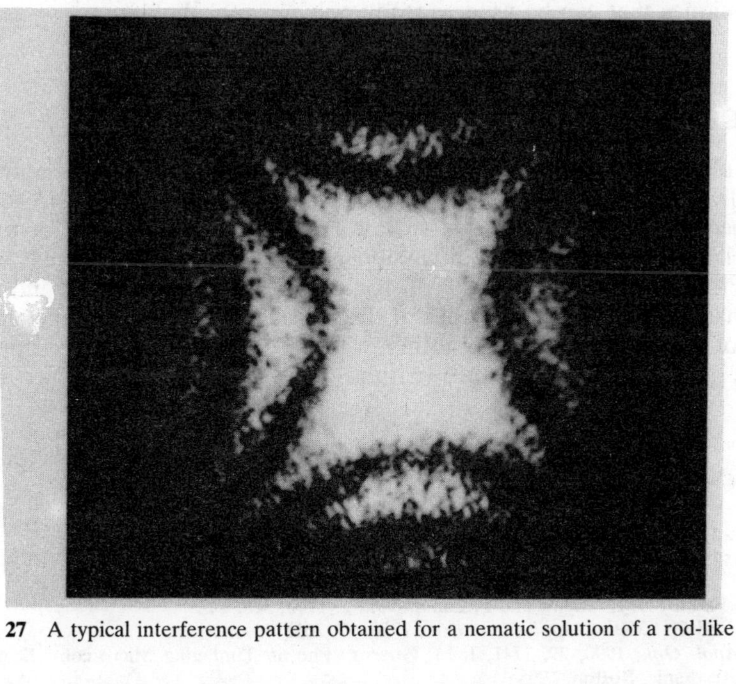

Figure 27 A typical interference pattern obtained for a nematic solution of a rod-like polymer.

Based on this it is clear that the birefringence increases as the concentration increases, and that S at ϕ^* is equal to about 0.94, which is rather high. Also the data suggests that S is not affected by an increase in concentration. The reason for this is under further investigation, since this is a rather surprising result.[129]

It is apparent that application of an optical crystallographic technique, conoscopy, to liquid crystals can lend itself to the measurements of static properties of the medium, *i.e.* one can measure the birefringence and the two individual refractive indices of the medium (in some cases as a function of wavelength[129]). Although not described here, measurement of the dynamic properties is also possible, as has been demonstrated elsewhere.[124,125]

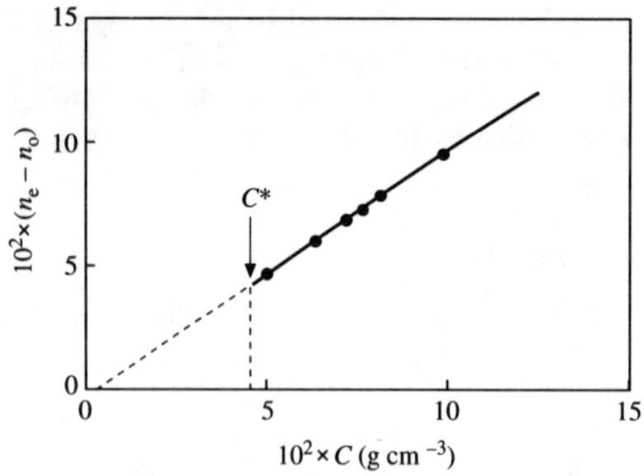

Figure 28 Concentration dependence of birefringence for PBT solutions.

5.8 CONCLUDING REMARKS

In this review we have tried to give an overview of many of the newer methods of imaging using light. Owing to the choices that had to be made due to constraints of space, we have not been able to discuss a few other techniques which are quite important and which have shed new insights into the behavior of polymers, both on their static properties and their behavior in a flow field.[130–132] Single DNA molecules labeled by a fluorescent dye manipulated by means of optical tweezers have been observed by fluorescence microscopy, thus providing direct evidence for many of the assumptions used in the reptation model.[133,134] Use of many of these new forms of microscopy will certainly lead to a better understanding of the behavior of polymers. It is the hope of this author that this article serves as a useful source of reference for many of the techniques that can be used to study polymer morphology.

ACKNOWLEDGEMENTS

It is a great pleasure to acknowledge many people who helped make this manuscript possible, especially John Guerra at Polaroid Corporation with whom I have collaborated for several years. Special thanks are also due to Jay Trautman at Bell Labs, for providing some figures and especially for many (useful) discussions, some lasting hours! Thanks are also due to Dr Fred Lanni, Prof. Douglas Keith and Prof. Richard Stein for discussions and suggestions. It is delightful to thank Dr J. O. Park for useful discussions on many occasions and for being a constant source of encouragement. The author wishes to acknowledge the support of AT&T Bell Labs, where parts of this review were written.

5.9 REFERENCES

1. R. S. Longhurst, 'Geometrical and Physical Optics', Wiley, New York, 1968.
2. C. J. R. Sheppard and T. Wilson, 'Theory and Practise of Optical Scanning Microscopy', Academic Press, New York, 1985.
3. B. Bailey, D. L. Farkas, D. L. Taylor and F. Lanni, *Nature,* 1993, **366,** 44.
4. S. M. Mansfield and G. S. Kino, *Appl. Phys. Lett.,* 1990, **57,** 2615.
5. J. M. Guerra, *Appl. Opt.,* 1990, **29,** 3741; J. M. Guerra, 'Photon Tunneling Microscopy: Seeing in Forbidden Light', Kluwer Academic, Boston, 1995.
6. J. M. Guerra, *SPIE Proc.,* 1988, **1009,** 254.
7. E. Betzig, A. Lewis, A. Harootunian and E. Kratschmer, *Biophys. J.,* 1986, **49,** 269.
8. D. W. Pohl, W. Denk and M. Lanz, *Appl. Phys. Lett.,* 1984, **44,** 652.
9. E. Betzig, J. K. Trautman, T. D. Harris, J. S. Weiner and R. L. Kostelak, *Science,* 1991, **251,** 1468.
10. D. W. Pohl, in 'Advances in Optical and Electron Microscopy', ed. C. J. R. Sheppard and T. Mulvey, Academic Press, London, 1990, p. 243.
11. G. Binnig, H. Rohrer, Ch. Gerber and E. Weibel, *Phys. Rev. Lett.,* 1982, **49,** 57.
12. G. Binnig and H. Rohrer, *Physica,* 1984, **127B,** 37.
13. D. M. Eigler and E. K. Schweizer, *Nature* 1990, **344,** 524.
14. H. K. Wickrasmasinghe, ed., 'Scanned Probe Microscopy', American Institute of Physics, New York, 1992.
15. G. A. Massey, *Appl. Opt.,* 1984, **23,** 658.

16. J. W. Goodman, 'Introduction to Fourier Optics', McGraw-Hill, New York, 1968, chap. 6.
17. G. E. Schuster, *Am. Lab.*, April 1970, 53.
18. G. L. E. Turner, 'The Great Age of the Microscope', Adam Hilger, New York, 1989, p. 7.
19. E. Abbe, *Arch. Mikros. Anat.*, 1873, **9,** 413.
20. E. H. Synge, *Philos. Mag.*, 1928, **6,** 356.
21. D. McMullen, *Proc. R. Microsc. Soc.*, 1990, **25,** 127.
22. J. A. O'Keefe, *J. Opt. Soc. Am.*, 1956, **46,** 359.
23. A. V. Baez, *J. Opt. Soc. Am.*, 1956, **46,** 901.
24. H. Nassenstein, *Opt. Commun.*, 1970, **2,** 231.
25. H. Nassenstein, *Optik,* 1969, **29,** 597.
26. H. Nassenstein, *Optik,* 1969, **30,** 44.
27. I. Newton, 'Opticks', Dover, New York, 1730 and 1979, Part I, p. 193.
28. O. Bryngdhal, *Prog. Opt.*, 1973, **11,** 189.
29. N. J. Harrick, *J. Appl. Phys.*, 1962, **33,** 2774.
30. T. R. Young and B. Rothrock, *J. Opt. Soc. Am.*, 1961, **51,** 1038.
31. N. J. Harrick, *Phys. Rev. Lett.*, 1960, **4,** 224.
32. N. J. Harrick, 'Internal Reflection Spectroscopy', Harrick Scientific, New York, 1979.
33. C. W. McCutchen, *Appl. Opt.*, 1962, **1,** 253.
34. C. W. McCutchen, *Rev. Sci. Instr.*, 1964, **35,** 1340.
35. E. J. Ambrose, *Nature,* 1956, **24,** 1194.
36. D. Ausserre, H. Hervet and F. Rondelez, *Phys. Rev. Lett.*, 1985, **54,** 1948.
37. K. B. Migler, H. Hervet and L. Leger, *Phys. Rev. Lett.*, 1993, **70,** 287.
38. M. Allegrini, C. Ascoli and A. Gozzini, *Opt. Commun.*, 1971, **2,** 435.
39. A. Diaspro and M. Aguilar, *Ultramicroscopy,* 1992, **42–44,** 1668.
40. O. Bryngdhal, *J. Opt. Soc. Am.*, 1969, **59,** 1645.
41. O. Bryngdhal, *Prog. Opt.*, 1973, **11,** 169.
42. M. A. Brown and E. J. Staples, *Faraday Discuss. Chem. Soc.*, 1990, **90,** 193.
43. D. C. Previe, F. Luo and F. Lanni, *Faraday Discuss Chem. Soc.*, 1987, **83,** 297.
44. D. C. Previe and N. A. Frej, *Langmuir,* 1990, **6,** 396.
45. H. Chew, D-S. Wang and M. Kerker, *Appl. Opt.*, 1979, **18,** 2679.
46. G. J. Brakenhoff, E. A. van Spronsen, H. T. M. van der Voort and N. Nanninga, 'Methods in Cell Biology', ed. D. L. Taylor and Y.-L. Wang, Academic Press, New York, 1989, vol. 30, part B, chap. 14.
47. A. Lewis, in 'New Techniques of Optical Microscopy and Microspectroscopy', ed. R. J. Cherry, CRC Press, Boca Raton, FL, 1991, chap. 2.
48. L. Li *et al.*, *Langmuir,* 1994, **10,** 2495.
49. R. W. Ditchburn, 'Light', Dover, New York, 1991, chap. 8.
50. S. G. Lipson and H. Lipson, 'Optical Physics', Cambridge University Press, London, 1981, chaps. 3 & 9.
51. G. Toraldo di Francia, *J. Opt. Soc. Am.*, 1969, **59,** 799.
52. M. Born and E. Wolf, 'Principles of Optics', Pergamon, New York, 1986, p. 418.
53. A. B. Porter, *Philos. Mag.*, 1906, **11,** 154.
54. Ref. 1, p. 310.
55. Lord Rayleigh, *Philos. Mag.*, 1896, **42,** 167.
56. W. Lukosz, *J. Opt. Soc. Am.*, 1966, **56,** 1463.
57. G. Toraldo di Francia, *J. Opt. Soc. Am.*, 1955, **45,** 497.
58. C. Pask, *J. Opt. Soc. Am.*, 1977, **66,** 68.
59. C. W. McCutchen, *J. Opt. Soc. Am.*, 1967, **57,** 1190.
60. J. L. Harris, *J. Opt. Soc. Am.*, 1964, **54,** 931.
61. D. Marcuse, 'Light Transmission Optics', Van Nostrand Reinhold, New York, 1972, p. 164.
62. M. Francon, 'Progress in Microscopy', Row, Peterson, Evanston, IL, 1961.
63. S. Inoue, in 'Methods in Cell Biology', ed. D. L. Taylor and Y.-L. Wang, Academic Press, New York, 1989, vol. 30, part B, chap. 3.
64. D. Axelrod, in 'Methods in Cell Biology', ed. D. L. Taylor and Y.-L. Wang, Academic Press, New York, 1989, vol. 30, part B, chaps. 15 & 9.
65. J. M. Guerra, M. Srinivasarao and R. S. Stein, *Science,* 1993, **262,** 1395.
66. P. A. Temple, *Appl. Opt.*, 1981, **20,** 2656.
67. S. Zhu, A. W. Yu, D. Hawley and R. Roy, *Am. J. Phys.*, 1986, **54,** 601.
68. P. W. Baumeister, *Appl. Opt.*, 1967, **6,** 897.
69. M. A. Winnik, Y. Wang and F. Haley, *J. Coat. Tech.*, 1992, **64,** 51.
70. Y. Wang *et al.*, *Langmuir,* 1992, **8,** 1435.
71. S. T. Eckersley and A. Rudin, *J. Coat. Tech.*, 1990, **62,** 89.
72. A. Keller, *Philos. Mag.*, 1957, **2,** 1171.
73. P. H. Till, Jr., *J. Polym. Sci.*, 1957, **24,** 301.
74. E. W. Fischer, *Z. Naturforsch.*, 1957, **120,** 753.
75. A. Keller and A. O'Connor, *Discuss. Faraday Soc.*, 1958, **25,** 114.
76. P. H. Geil, 'Polymer Single Crystals', Wiley, New York, 1963.
77. P. Sullivan and B. Wunderlich, *SPE Trans.*, 1964, **4,** 2.
78. P. Sullivan and B. Wunderlich, *Polym. Lett.*, 1964, **2,** 537.
79. D. H. Reneker and P. H. Geil, *J. Appl. Phys.*, 1960, **31,** 1916.
80. G. F. Teletzke, H. T. Davis and L. E. Scriven, *Chem. Eng. Commun.*, 1987, **55,** 41.
81. P. G. de Gennes, *Rev. Mod. Phys.*, 1985, **57,** 827.
82. J. Daillant, J. J. Bennatar and L. Leger, *Phys. Rev.*, 1990, **A41,** 1963.
83. F. Brochard-Wyart and J. Daillant, *Can. J. Phys.*, 1990, **68,** 1084.
84. C. Redon, F. Brochard-Wyart and F. Rondelez, *Phys. Rev. Lett.*, 1991, **66,** 715.

85. G. Reiter, *Phys. Rev. Lett.*, 1992, **68,** 75.
86. R. G. Larson, *Rheol. Acta,* 1992, **31,** 213.
87. V. K. Gupta and J. A. Kornfield, *Rev. Sci. Instr.*, 1994, **65,** 2823.
88. V. K. Gupta, J. A. Kornfield, A. Ferencz and G. Wegner, *Science,* 1994, **265,** 940.
89. R. P. Feynman, in 'Miniaturization', ed. H. D. Gilbert, Reinhold, New York, 1962, chap. 16.
90. D. Nyyssonen and R. B. Larrabee, *J. Res. Nat. Bur. Stand.*, 1987, **92,** 187.
91. S. L. Sharp, R. J. Warmack, J. P. Goudonnet, I. Lee and T. L. Ferrell, *Acc. Chem. Res.*, 1993, **26,** 377.
92. N. F. van Hulst, N. de Boer and B. Bolger, *J. Microsc.*, 1991, **163,** 117.
93. D. Courjon, K. Sarayeddine and M. Spajer, *Opt. Commun.*, 1989, **71,** 23.
94. J. Cites *et al.*, *J. Appl. Phys.*, 1992, **71,** 7.
95. N. F. van Hulst, F. B. Segerink and B. Bolger, *Opt. Commun.*, 1992, **87,** 212.
96. F. de Fornel *et al.*, *Ultramicroscopy,* 1992, **42–44,** 422.
97. N. F. van Hulst, M. H. P. Moers and B. Bolger, *J. Microsc.*, 1993, **171,** 95.
98. T. L. Ferrell, S. L. Sharp and R. J. Warmack, *Ultramicroscopy,* 1992, **42–44,** 408.
99. E. Betzig and J. K. Trautman, *Science,* 1992, **257,** 189.
100. E. L. Buckland, P. J. Moyer and M. A. Paesler, *J. Appl. Phys.*, 1993, **73,** 1018.
101. E. A. Ash and G. Nicholls, *Nature,* 1972, **237**, 510.
102. L. Malmqvist and H. M. Hertz, *Opt. Commun.*, 1992, **94,** 19.
103. K. Lieberman, S. Harush, A. Lewis and R. Kopelman, *Science,* 1990, **247,** 59.
104. A. Lewis and K. Lieberman, *Nature* 1991, **354**, 214.
105. E. Betzig, in 'Near Field Optics', ed. D. W. Pohl and D. Courjon, Kluwer Academic, London, 1992, p. 7.
106. E. Betzig, S. G. Grubb, R. J. Chichester, D. J. DiGiovanni and J. S. Weiner, *Appl. Phys. Lett.*, 1993, **63,** 3550.
107. I. N. Court and F. K. von Willisen, *Appl. Opt.*, 1964, **3,** 719.
108. E. Betzig, P. L. Finn and J. S. Weiner, *Appl. Phys. Lett.*, 1992, **60,** 2484.
109. W. B. Whitten, J. M. Ramsey, S. Arnold and B. V. Bronk, *Anal. Chem.*, 1991, **63,** 1027.
110. E. B. Shera, N. K. Seitzinger, L. M. Davis, R. A. Keller and S. A. Soper, *Chem. Phys. Lett.*, 1990, **174,** 553.
111. H. A. Bethe, *Phys. Rev.*, 1944, **66,** 163.
112. C. J. Bouwkamp, *Philips Res. Rep.*, 1950, **5,** 321.
113. E. Betzig and R. J. Chichester, *Science,* 1993, **262,** 1422.
114. E. Reichmanis and A. E. Novembre, *Annu. Rev. Mater. Sci.*, 1993, **23,** 11.
115. M. J. Bowden, in 'Introduction to Microlithography', ed. L. F. Thompson, C. Grant Wilson and M. J. Bowden, ACS, Washington, 1994.
116. B. R. Masters, in 'Confocal Microscopy', ed. T. Wilson, Academic Press, London, 1990.
117. S. Dosho *et al.*, *Langmuir,* 1993, **9,** 394.
118. K. Ito, H. Yoshida and N. Ise, *Science,* 1994, **263,** 66.
119. F. Lanni, B. Bailey, D. Farkas and D. L. Taylor, *Bioimaging,* 1993, **1,** 187.
120. S. Hell and E. H. K. Stelzer, *J. Opt. Soc. Am.*, 1992, **A9,** 2159.
121. P. G. deGennes, 'The Physics of Liquid Crystals', Clarendon Press, Oxford, 1974.
122. M. Born and E. Wolf, 'Principles of Optics', Pergamon, New York, 1986, p. 691.
123. A. Sommerfeld, 'Optics', Academic Press, San Diego, 1992, p. 172.
124. M. Srinivasarao and G. C. Berry, *Mol. Cryst. Liq. Cryst.*, 1992, **223,** 99.
125. P. E. Cladis, *Phys. Rev. Lett.*, 1972, **28,** 1629.
126. M. Srinivasarao and G. C. Berry, *J. Rheol.*, 1991, **35,** 379.
127. H. Zocher, *Trans. Faraday Soc.*, 1933, **29,** 945.
128. F. D. Bloss, 'An Introduction to the Methods of Optical Crystallography', Holt, Rinehart and Winston, New York, 1961, p. 120.
129. H. Mattoussi, M. Srinivasarao, P. G. Kaatz and G. C. Berry, *Macromolecules,* 1992, **25,** 2860.
130. T. T. Perkins, D. E. Smith and S. Chu, *Science,* 1994, **264,** 819.
131. T. T. Perkins, S. R. Quake, D. E. Smith and S. Chu, *Science,* 1994, **264,** 822.
132. S. B. Smith, L. Finzi and C. Bustamante, *Science,* 1992, **258,** 1122.
133. P. G. de Gennes, 'Scaling Concepts in Polymer Physics', Cornell University Press, New York, 1979.
134. M. Doi and S. F. Edwards, 'The Theory of Polymer Dynamics', Clarendon Press, Oxford, 1986.

6

Two-dimensional NMR Spectroscopy for the Structural Characterization of Polymers

DENVER D. WERSTLER
GenCorp, Akron, OH, USA

6.1 INTRODUCTION

The recent use of two-dimensional (2D) nuclear magnetic resonance (NMR) in the study of polymers has contributed significantly to a better understanding of polymer structure–physical property relationships. 2D NMR experiments have provided detailed structural and dynamic information in synthetic polymers and polymer blends that often could not be obtained in any other way. This has been made possible due to recent significant advances in 2D NMR experimentation.

Prior to this, 1D NMR had long provided valuable information on the microstructure of polymers, particularly in terms of stereochemical configuration and sequencing of monomer units.[1] Carbon-13 (^{13}C) NMR was often the nucleus of choice because of its large spectral range and relative spectral simplification in terms of single spectral lines resulting from proton (^{1}H) decoupling. However, spectral complexity was still often a problem due to sensitivity to numerous unique nuclear environments in complex polymers. ^{1}H NMR was potentially even more applicable to such studies because it could provide information on through-bond scalar J-coupling of protons. However, it suffered from a small chemical shift range and broad, overlapped, and often uninterpretable spectra, particularly for polymers. This often resulted in the loss of significant data. Both ^{13}C and ^{1}H 1D microstructure studies of polymers often required the synthesis of specific oligomers and/or polymers, estimation of chemical shifts via empirical additivity rules,[2,3] and/or estimation of relative intensities assuming Bernoullian or Markovian statistics to identify structural species.

2D NMR was initially developed and used for the study of small molecules, mostly biological. The biological molecules were excellent candidates for this type of study because they had small and relatively well-defined structures, resulting in sharp and often well-separated peaks. 2D NMR has been used for making spectral resonance (peak) assignments in many proteins, peptides, and nucleic acids.[4–15] It was initially felt that the size and complexity of synthetic polymers would prohibit significant use of 2D experiments for their structural characterization. Nuclei in synthetic polymers with random coil configurations are generally located in a variety of environments and orientations, resulting in extensive peak broadening. This lack of chemical shift specificity, the high viscosity of polymer solutions, and short polymer nuclear spin–spin relaxation times (T_2) were all factors working against this type of application.[16] Improved NMR instrumentation and 2D techniques and the superior resolution of 2D data have overcome many of these problems. It has repeatedly been demonstrated that resonances for individual species that are 'lost' in badly overlapped spectral regions in 1D spectra of polymers can be isolated and identified in a 2D spectrum.

Bovey *et al.* were among the first to realize the value of 2D NMR in the characterization of polymers.[17–19] Much of the early work in this area involved the study of detailed stereochemical configurations and regiochemical structures in homopolymers. Sequences from diads to heptads have been assigned in this way and this type of study is still providing useful information today.

2D NMR experiments can also be used to provide information on intra- and intermolecular spatial relationships of polymer chains. This is particularly useful in the study of miscibility in polymer blends at the molecular level. In this application, polymer chains which are within 0.5 nm of one another can be identified. These experiments have been carried out both in solution and in bulk.

More recent 2D studies have also probed the dynamics of homopolymers and polymer blends. Motional rearrangements near T_g have been detected in homopolymers or in one or both of the components in polymer blends. The result has been a better understanding of the motional character of homopolymers and polymer blends during this transition period. The use of 2D NMR in the study of molecular motion in crystalline regions of semicrystalline polymers and of molecular order in orientated polymers has also resulted in important information.

While most of the 2D literature deals with formal studies of highly specific, detailed structure, the value of 2D NMR data goes far beyond this. 2D NMR experiments are now used as routine probes in many analytical laboratories in the characterization of homopolymers, copolymers, blends of polymers, and related additives. The 2D data are often even more fundamental in these applications in that they can be used to first identify individual polymeric components whose stereostructure, regiostructure, *etc.* can then be characterized by the 2D experiments.

The 2D NMR experiments take advantage of nuclear spin interactions to characterize polymeric structure and dynamics. As there are many different types of nuclear spin interactions resulting in different types of spin-related NMR data (chemical shifts, coupling constants, nuclear spacial distances, *etc.*), there are varieties of 2D NMR techniques that can be applied to extract each type of information. The different types of structural and dynamic characterization data resulting from these techniques will be discussed after a brief description of the basic 2D experiment. Each 2D technique is reviewed where its application is most useful. Structure–property relationships developed as a result of the 2D NMR data are discussed where appropriate. Detailed mathematical relationships are not presented and concepts are emphasized at the expense of tedious detail.

6.2 THE BASIC 2D EXPERIMENT

A 1D NMR spectrum of a polymer is obtained by applying a series of short radiofrequency (rf) pulses at right angles to the spectrometer's magnetic field. All of the polymeric nuclei in different environments are simultaneously excited and absorb individual frequency components, reflecting their different environments. The signals resulting from these excitations are instrumentally detected and amplified. The result is a free induction decay (FID) pattern which plots intensity as a function of time. An FID of a single nucleus is an exponentially decaying sine wave pattern. Since polymers typically contain numerous nuclei in different environments, the FID of a polymer is typically the superimposition of many different exponentially decaying sine wave patterns. The FID is Fourier-transformed (FTed) from the time domain to the frequency domain to put the data in a format that can be better interpreted. The net result is intensity plotted as a function of frequency.[20]

A 2D spectrum is characterized by two frequency axes located perpendicular to one another and one intensity axis. Application of a single pulse for the 1D experiment is replaced by application of a pulse sequence interrupted by two time intervals (t_1 and t_2). The two frequency axes result from the introduction of the two time variables into the pulse sequence as explained below. The 2D spectrum resulting from the pulse sequence is a map of internuclear interactions occurring in the polymer under study.

Almost all 2D experiments are divided into at least three periods separated by pulses, *e.g.* preparation period – 1st pulse – evolution period – 2nd pulse – detection period. The preparation period is a time delay between pulse sequences to allow the nuclei to equilibrate. The first pulse perturbs the nuclei from their equilibrium positions and the nuclei become established at their Larmor (resonance) frequencies during the evolution period (t_1). The second pulse again perturbs the nuclei and those nuclei which are interacting in some manner (*e.g.* magnetization exchange between nuclei which are *J*-coupled to one another) become established at frequencies different from their Larmor frequencies. The detection period (t_2) allows the observation of the changes in frequencies resulting from the second pulse.

An FID is, therefore, detected during the detection period (t_2) after the second pulse for each t_1 value. The experiment is repeated many times, increasing the t_1 value incrementally. An FID is acquired in t_2 for each t_1 value. The FID for each increment is different because the interacting spins modulate each other's response. The result of this is a data matrix $D(t_1, t_2)$. To convert the data to a frequency format, two FTs are carried out. First, each row in the data matrix is FTed in the same way as a typical 1D FID with respect to t_2. The result is $D(t_1, F_2)$. The matrix is transposed (interchange of rows and columns) and each row of the transposed matrix is then FTed with respect to t_1. The result is a 2D spectrum as a function of two frequency axes $S(F_1, F_2)$. The 2D spectrum is a map of the interactions that have taken place as manifested in the changes in frequencies resulting from the 2D pulse sequence. The specific pulse sequence employed depends on the type of interaction being studied.

The 2D experiments themselves can be divided into three types: 2D correlated, 2D *J*-resolved, and 2D exchange. 2D correlation experiments correlate homonuclear or heteronuclear spins interacting with one another via through-bond scalar *J*-coupling. 2D *J*-resolved experiments separate different types of interactions, usually coupling constants and chemical shifts, along the two frequency axes. 2D exchange experiments provide information resulting from internuclear cross relaxation between spatially close nuclei via through-space dipole–dipole interactions or from chemical exchange.

6.3 STEREOSTRUCTURE, REGIOSTRUCTURE, AND POLYMER IDENTIFICATION

6.3.1 2D Correlated Spectroscopy

6.3.1.1 Homonuclear 2D COrrelated SpectroscopyY (COSY)

The first 2D experiment was the COSY experiment proposed by Jeener in 1971.[22] The first successful experiment using 2D COSY was conducted by Ernst *et al.* in 1976.[23] The COSY experiment provides information on the protons in the polymer system that are interacting through two-bond, three-bond, or four-bond (*e.g.* for $^{19}F–^{19}F$ bonds) scalar *J*-coupling between nuclear spins via the pulse sequence shown in Figure 1.[24,25]

The preparation period in the 2D COSY experiment is actually a time delay between the application of consecutive pulse sequences to allow the nuclei to equilibrate. Nuclei which have

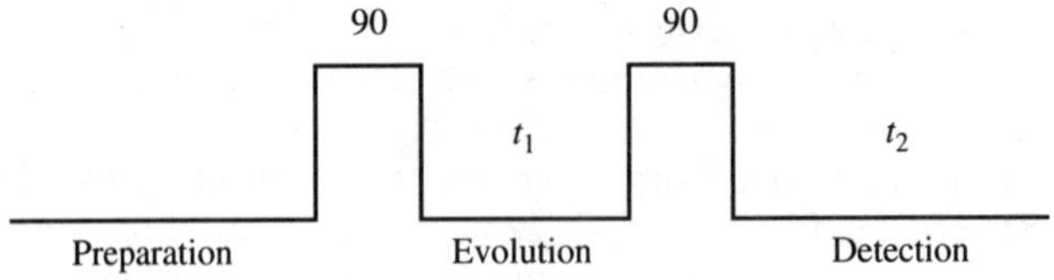

Figure 1 2D COSY pulse sequence.

been flipped from the magnetization component along the z-axis into the $x-y$ plane return back to equilibrium during this period via well-known relaxation processes. Nuclei which would not come to equilibrium between pulse sequences would have their peak intensities truncated, thus distorting their contributions. The first 90° pulse causes each nuclear spin to precess at its characteristic resonance frequency during the evolution period. This frequency is referred to as F_1. The evolution time is referred to as the t_1 time. The second 90° pulse is a mixing pulse. Magnetization exchange occurs between nuclear spins that are J-coupled to one another. As a result of the second 90° pulse, the precessional frequencies of nuclei whose spins are J-coupled are altered. The final precessional frequencies (F_2) are measured during the detection period (t_2). For those nuclear spins which do not exchange magnetization, the final frequency is the same as the initial frequency, *i.e.* $F_1 = F_2$, and their resonances lie along the diagonal of the 1D spectrum. For those nuclear spins which exchange magnetization, off-diagonal or cross peaks occur in the 2D spectrum, reflecting the change in frequency that has taken place. These cross peaks are symmetrically located around the diagonal in (F_i, F_j) and (F_{-i}, F_{-j}) pairs.

The information obtained from a 2D COSY experiment is routinely used for spectral identification, detailed stereosequence assignments, and for the determination of J-coupling constants in many polymers and copolymers. Spectral identification may involve identifying the components in a polymer blend, copolymer, or terpolymer. It may also involve determination of polymer sequencing of monomer units and end group analysis (these are usually small for higher molecular weight polymers), or even distinguishing polymers and polymeric additives from one another. Often, for many vinyl polymers, spectral resonances resulting from regiostructural effects complicated by stereochemical effects can be sorted out using this technique.

A very simple example of an application of the 2D COSY pulse sequence to a synthetic polymer is the partial ^1H 2D COSY spectrum shown in Figure 2 of a hydroxy-terminated, highly branched polyalkene which has been capped with MDI which in turn has been capped with ε-caprolactam. Excess unreacted caprolactam is also present. This is a particularly simple example because most of the peaks of interest in the ^1H spectrum are actually the more mobile polymeric end groups. The 1D spectrum lying along the diagonal results from those nuclei which did not exchange magnetization and $F_1 = F_2$. The other peaks are the cross peaks symmetrically located around the diagonal, which result from nuclei which have exchanged magnetization via J-coupling and where the final frequencies and the initial frequencies are different. While aesthetically pleasing, it is usually difficult to make all of the appropriate correlations with the data in this form. For example, a cross peak is hidden behind a diagonal peak at 2.77 ppm. A contour plot of the data (Figure 3) is much more practical for making cross peak correlations.

The 2D contour plot allows one to observe several of the superimposed resonances in the 1.5–1.9 ppm region. It is often important to analyze spectral regions such as this to fully understand the polymer structure and to be comfortable with quantitative data obtained from the ^1H 1D data. Seven types of methylene protons are superimposed in this area. Five of these resonances (U4, R4, U6, R6, and β-PO) are isolated and easily distinguished as cross peaks in this 2D COSY spectrum. The U3 and U7 resonances are easily identified from an unreacted caprolactam spectrum. U4 and U6 protons are respectively coupled with U3 and U7 and are identified via the connectivity lines. Protons on reacted caprolactam, R3 and R7, are shifted downfield as indicated. Protons R4 and R6 are then identified as they are J-coupled with peaks R3 and R7. Unreacted and reacted protons R5 and U5 are too far removed from the reaction sites to be influenced by the presence or absence of reaction there and superimpose at 1.75 ppm on the R6 peak. The polyalkene methylene group α to the reaction site (4.23 ppm) is J-coupled with the methylene group β to this site and the protons from the latter superimpose in the 1.5–1.9 ppm region as well. The MDI methylene protons are equivalent and are not directly

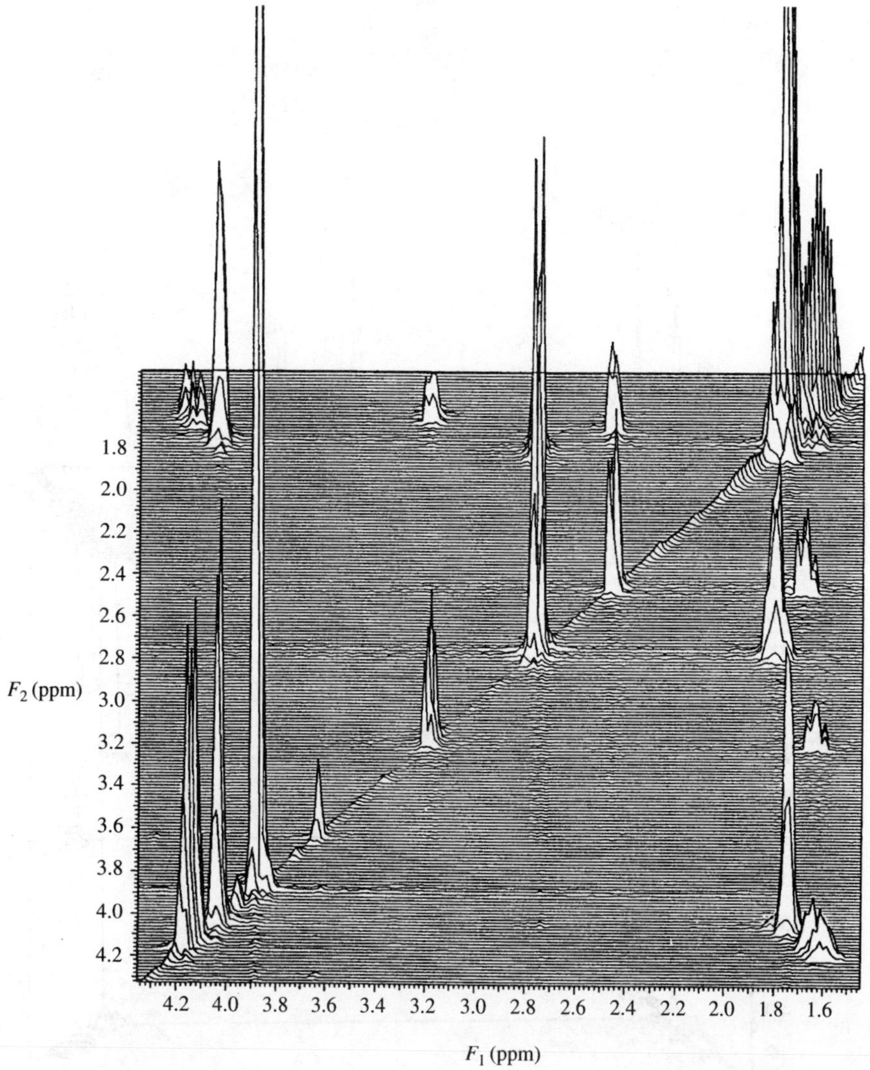

Figure 2 Partial ^{1}H 2D COSY spectrum of a hydroxy-terminated polyalkene capped with MDI and ε-caprolactam.

coupled with any protons. They resonate at 3.84 ppm (relative to TMS) and have no cross peaks.

From the aromatic region (Figure 4), the protons in MDI units adjacent to the polyalkene reaction site can be distinguished from those in MDI units adjacent to the caprolactam reaction site. Proton resonances from each type of linkage superimpose at 7.12 ppm in the ^{1}H 1D spectrum but are isolated in the 2D spectrum. This allows distinction of the urethane and urea linkages from the aromatic proton data. All of the above assignments are relevant in themselves but, more importantly, eventually lead to the assignment of all of the corresponding carbon resonances in the ^{13}C spectrum, as explained in Section 6.3.1.2.

A more involved example is the ^{1}H 2D COSY spectrum of a complex resin system shown in Figure 5. This system contains a propylene glycol–fumaric acid (PG–FA) unsaturated polyester, a propylene glycol–ethylene glycol–adipic acid–MDI (PG–EG–AA–MDI) polyester polyurethane, and low molecular weight poly(propylene glycol) (PPG) polyether. The presence of propylene glycol in all three components, as well as the presence of terminal and internal PG units, severely complicates the ^{1}H 1D spectrum due to peak overlap. The 2D COSY spectrum allows one to observe all of the monomer units in these components, *e.g.* the nine different PG methyl resonances which resonate in the 1.0–1.4 ppm region. By studying the ^{1}H 1D spectrum

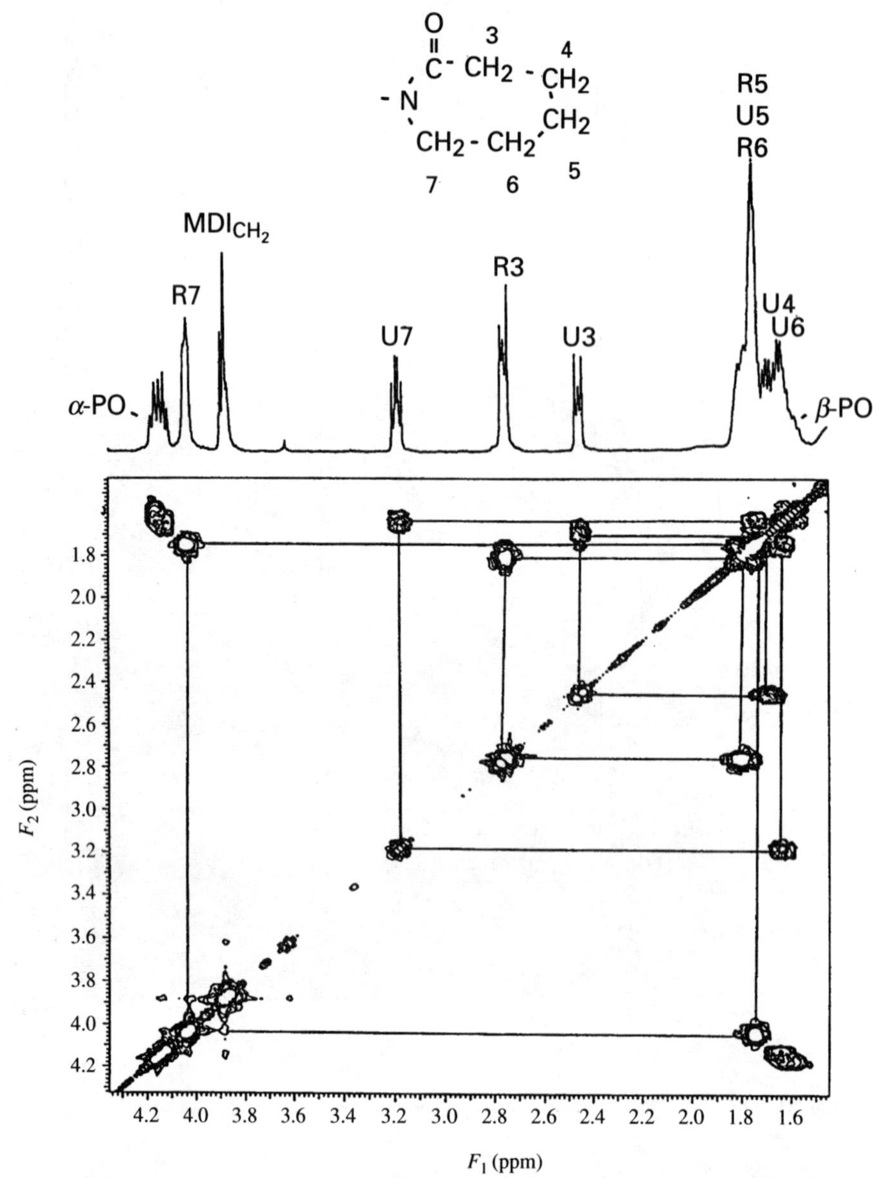

Figure 3 Contour plot of the partial ^{1}H 2D COSY spectrum of the hydroxy-terminated polyalkene capped with MDI and ε-caprolactam shown in Figure 2. U = unreacted caprolactam, R = reacted caprolactam, α-PO = PO protons in PO methylene group α to reaction site, and β-PO = protons in PO methylene group β to reaction site.

alone, there is no way of knowing that all of the species responsible for these resonances exist in the polymer. Each of the methine protons to which the methyl protons are *J*-coupled can be identified via methyl–methine cross-peak correlations (for simplicity, only the PG_{CH_3}–PG_{CH} coupling from internal PG–FA units is shown). Then, with appropriate spectral expansions, the methylene protons to which the respective methine protons are coupled can be observed in the 3.3–4.4 ppm region by mapping methine–methylene cross-peak correlations (the PG_{CH}–PG_{CH_2} coupling from the internal PG units is indicated by the correlation lines in the lower left corner of the spectrum – the methylene protons are not equivalent, resulting in two sets of correlation lines).

In the process of making the cross correlations, the proton resonances from each of the internal PG units in different environments in the three different components are isolated in the 2D spectrum in Figure 5. The proton resonances from each PG unit in different types of end groups, *i.e.* oriented in either a secondary or primary configuration, are also isolated in the 2D

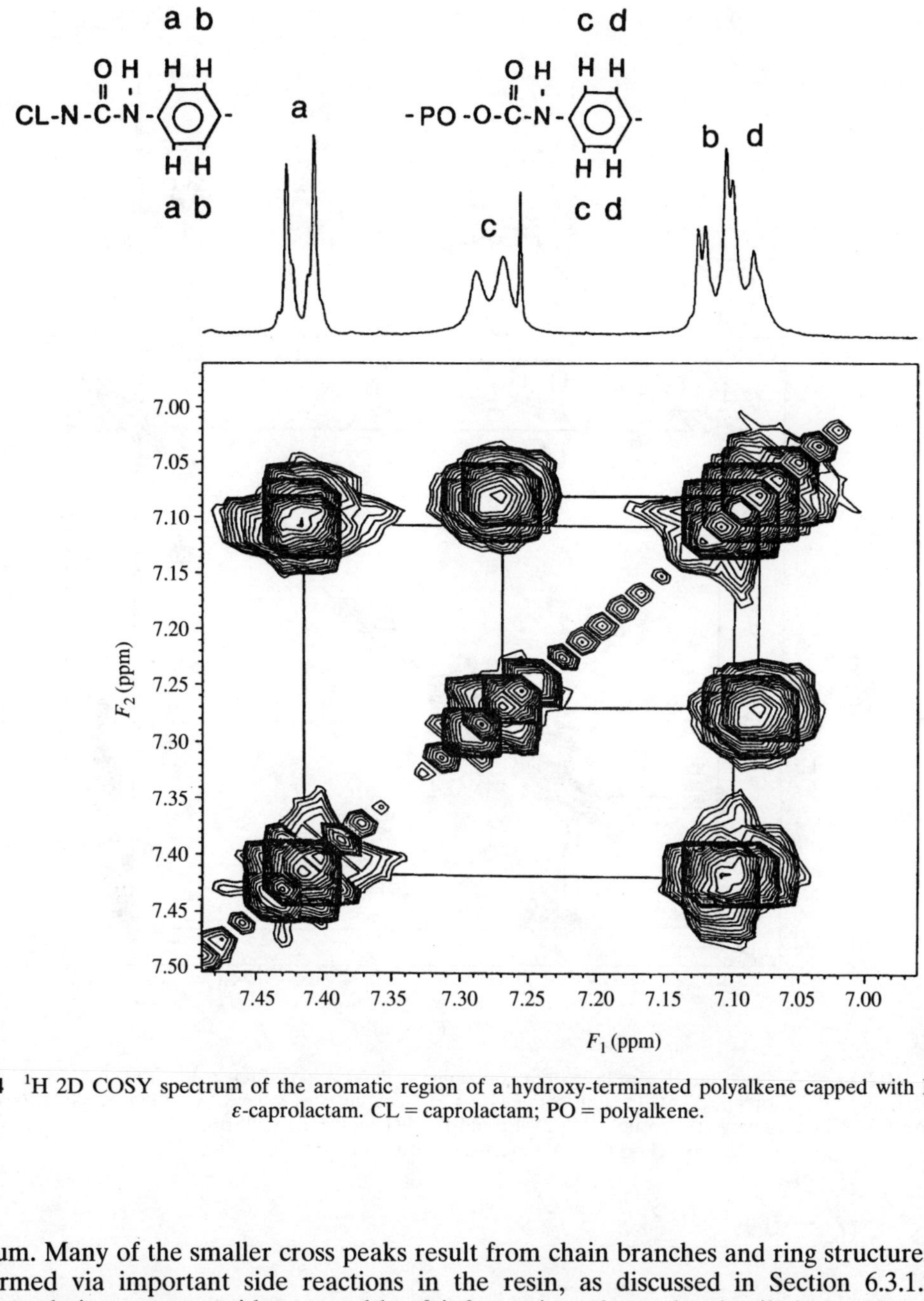

Figure 4 ^{1}H 2D COSY spectrum of the aromatic region of a hydroxy-terminated polyalkene capped with MDI and ε-caprolactam. CL = caprolactam; PO = polyalkene.

spectrum. Many of the smaller cross peaks result from chain branches and ring structures which are formed via important side reactions in the resin, as discussed in Section 6.3.1.2. This cross-correlation map provides a wealth of information about the detailed structures in this complex polymer blend which is certainly not available from the 1D data.

One of the most rewarding applications of 2D COSY has been in the assignment of stereochemical sequences in homopolymers. The first application of 2D COSY to a synthetic polymer was by Ferguson and Ovenall to make a stereochemical assignment in poly(vinyl fluoride) (PVF).[26] This initial work was extended by Bruch *et al.* who made comprehensive assignments of stereosequences to the pentad level in PVF using ^{19}F 2D COSY (Figure 6).[18] Assignments were made by identifying and correlating cross peaks resulting from the central pair of four-bond J-coupled fluorines in pentad sequences which share a common hexad. For example, the central fluorines in the hexad shown in Scheme 1 (one in an r$\underline{\text{mm}}$r pentad sequence and one in an m$\underline{\text{mr}}$r pentad sequence) are J-coupled to one another and have connecting cross peaks in the ^{19}F 2D COSY spectrum.

Correlation of the assignments in the mr region provides an example of how stereosequence assignments are made. The rmrr and m$\underline{\text{mr}}$r peaks are coupled with both the rrrm and mrrm peaks from the rr region. The peaks for mmrm and m$\underline{\text{mr}}$r are coupled with both the mmmr

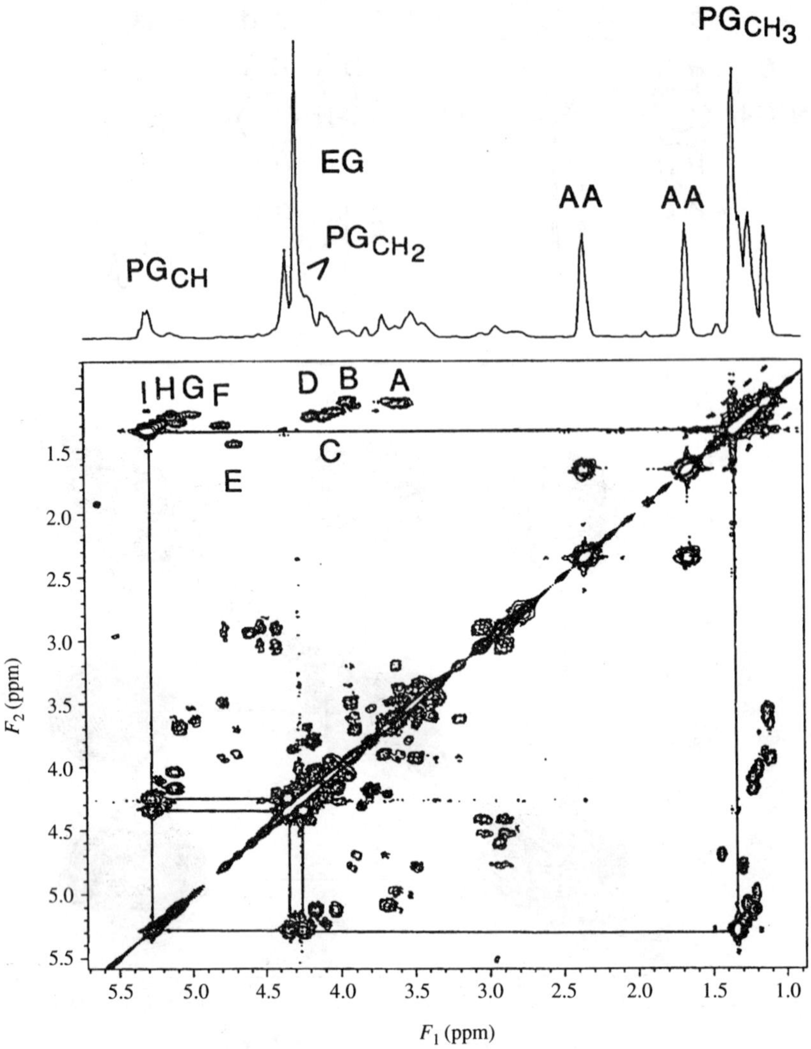

Figure 5 ^{1}H 2D COSY spectrum of a resin system containing a PG–FA polyester, a PG–EG–AA–MDI polyester polyurethane and PPG. Cross peaks A–I result from CH_3—CH *J*-couplings as follows: A = PPG; B = PPG (secondary end group); C = PG (secondary end group)–FA; D, E, and F = PG branching or lactone; G = PG–FA (primary end group); H = PG–AA (internal); and I = PG–FA (internal).

and rmmr peaks in the mm region. The mmrr peak is identified as the only peak that couples with peaks in both the rr and mm regions. The rmrr peak is then assigned as it is the only other peak coupled with the rr region. The mmrm peak is then assigned as it is the only other peak coupled with a peak from the mm region. The remaining peak is rmrm which can only couple within the mr region.[18]

500 MHz ^{1}H 2D COSY data have also confirmed the chemical shift assignments of stereochemical sequences in poly(methyl methacrylate) (PMMA) via the study of isotactic and atactic PMMA.[19,27] Advantage was taken of the fact that the methylene protons in a syndiotactic or racemic diad (r) are equivalent magnetically, while methylene protons in an isotactic or meso diad (m) are not, as shown in Scheme 2 (R = CH_3).

The r diads result in singlets in the ^{1}H spectrum of PMMA while the m diads result in a pair of doublets with geminal coupling (coupling of protons attached to the same carbon) of approximately 15 Hz. Assignments were made using cross-peak matching of the methylene–methyl 2D resonances.[19,27]

Spectral data from the isotactic PMMA also indicated that the preferred conformation of the isotactic chain is *trans–trans*. Cross-peak correlation between the α-methyl proton resonance

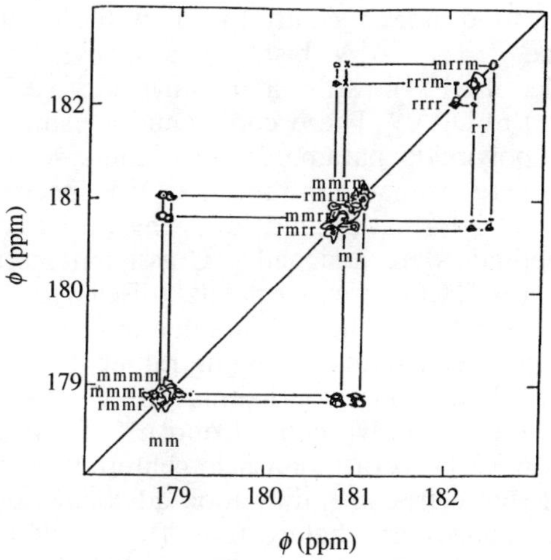

Figure 6 ^{19}F 2D COSY spectrum of an 11% solution of isoregic poly(vinyl fluoride) in DMF at 130 °C. X denotes peaks that are visible at lower contour levels (reproduced by permission of the American Chemical Society from Bruch *et al.*[18]).

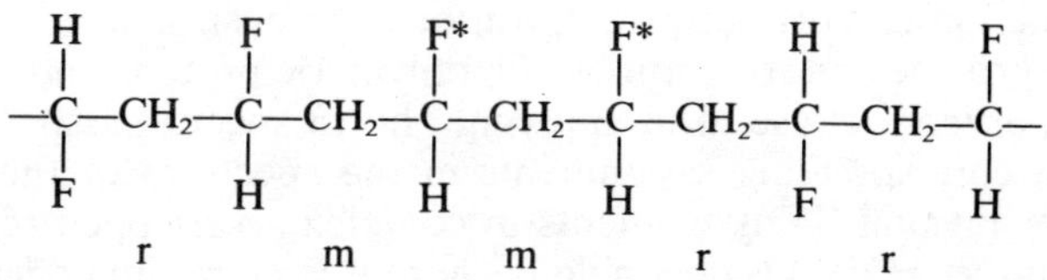

Scheme 1

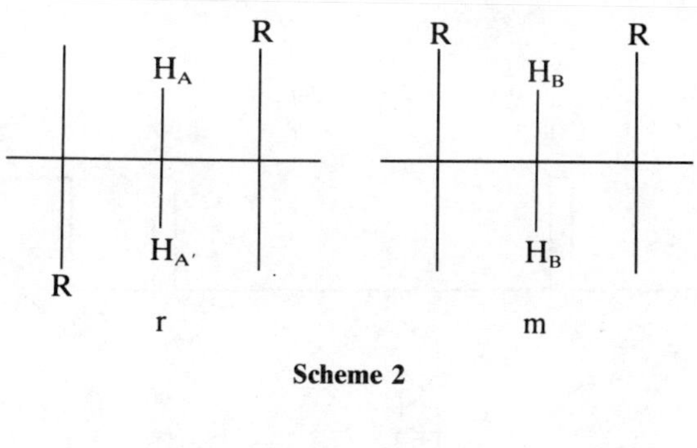

Scheme 2

and the *erythro* methylene proton resonance but not the *threo* proton resonance was observed. This results from a very weak four-bond coupling of approximately 1–2 Hz. Only the *erythro* proton can form a 'W'-shaped four-bond path favorable for such long-range coupling with the α-methyl proton in this conformation.[19,27]

^{19}F 2D COSY experiments were also successfully applied to the study of configuration in aregic poly(vinylidene fluoride) (PVF_2).[28] Polymers with high defect levels (18%) were used to observe connectivities between isoregic and aregic signals. Data from the aregic polymers were coupled with data from isoregic and syndioregic polymers to assign all of the sequences involving head-to-head (h–h) and tail-to-tail (t–t) units.

^1H 2D COSY data were used more recently to follow the reaction of a series of precursors for an imido diamine (*e.g.* *N,N'*-bis(4-aminophenyl)-2,5-bis[(isopropyloxy)carbonyl]-benzene-1,4-dicarboxamide; PDPM) with a dianhydride (2,2-bis(3,4-carboxyphenyl)hexafluoropropane dianhydride; 6FDA).[29] These copolyimides exhibited enhanced solubility relative to known random copolyimide parameters in organic solvents without loss of thermal stability. ^1H–^1H couplings were evaluated in the 2D COSY spectrum of unknown copolyimides based on background ^1H 1D information and following the connectivity pathways in the 2D spectra. Regiosequence heptads were assigned. ^{13}C assignments were then made via ^{13}C–^1H heteronuclear correlation (HETCOR) experiments (discussed in Section 6.3.1.2) based on knowledge of the proton assignments. The 2D data demonstrated that the copolyimide reaction products with superior properties had an alternating rather than a random structure.

Gippert and Brown used 2D COSY experiments with F_1 scaling to assign triad and tetrad stereosequence configurations in poly(vinyl alcohol).[30] They also used homonuclear 2D *J*-resolved NMR (Section 6.3.2.1) to distinguish m-centered and r-centered sequences. As in many 2D NMR studies of stereostructure, they took advantage of the fact that the methylene protons in m-centered sequences are heterosteric (have different chemical environments) while the methylene protons in r-centered sequences are homosteric (have identical environments).

6.3.1.2 2D HETeronuclear shift CORrelated spectroscopy (HETCOR)

The 2D HETCOR experiment correlates heteronuclear coupling, *e.g.* the coupling that takes place between a proton or protons attached to a carbon. It is an invaluable analytical tool used daily in the determination of polymeric compositions, stereostructure, and regiostructure. It is important because it provides information on the polymeric protons and carbons from the same experiment. This experiment is useful in applying known spectral assignments for a particular nucleus to make the corresponding assignments in the spectrum of the 'other' nucleus. It is particularly useful for making ^1H assignments in complex, overlapped regions where quantitative information from the ^1H data is desirable because it is more time efficient and one does not have to account for different NOE and relaxation effects. It is often imperative to account for the contribution from each species in an overlapped region before quantitative data from that or other regions of the spectrum are valid. In many cases, this experiment is essential in the study of complex polymers, copolymers, and polymer blends for isolating and identifying ^{13}C resonances which cannot be identified by any other means. The HETCOR pulse sequence is shown in Figure 7.[31–33]

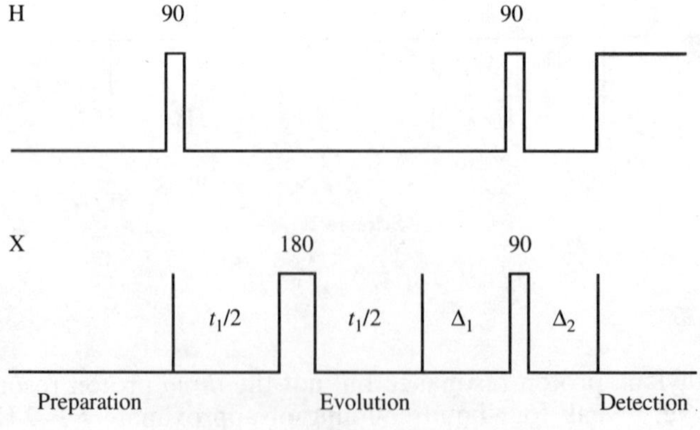

Figure 7 2D HETCOR pulse sequence.

As for the COSY experiment, the preparation period allows the nuclei to come to equilibrium. The first proton pulse causes the protons to precess at their characteristic frequencies. The heteroatom (^{13}C will be demonstrated) 180° pulse refocuses the affects arising from ^{13}C–^{1}H coupling. The net result is that the precessional frequency during t_1 depends only on the ^{1}H chemical shift. The simultaneous 90° pulses cause exchange of magnetization between the coupled carbons and protons. The cross-correlated signals are allowed to build up during the delay times Δ_1 and Δ_2. Δ_1 is set to $1/(2J_{CH})$ (J_{CH} is the carbon–proton coupling constant) for optimum magnetization and Δ_2 is set to approximately $1/(4J_{CH})$ to observe CH, CH_2, and CH_3 signals equally well. The result of this is a series of heteronuclear signals which are detected during the detection period (t_2). ^{1}H decoupling is applied during the detection period. As a result, the final frequencies depend only on the heteronuclear chemical shift. All peaks in a 2D HETCOR spectrum are cross peaks. Each cross peak provides information about the carbons and directly attached protons for each hydrocarbon group.

A simple example of the application of 2D HETCOR to the structural characterization of polymers is the ^{13}C–^{1}H 2D HETCOR spectrum in Figure 8 of the double-capped hydroxy-terminated polyalkene discussed in Section 6.3.1.1. The 1D resonance position for each type of nucleus can be obtained by moving at right angles to the ^{13}C and ^{1}H 1D spectra plotted along the F_1 and F_2 axes. This allows one to observe all of the different species that superimpose in both spectral regions. Each of the six caprolactam methylene resonances (U4–U6, R4–R6)

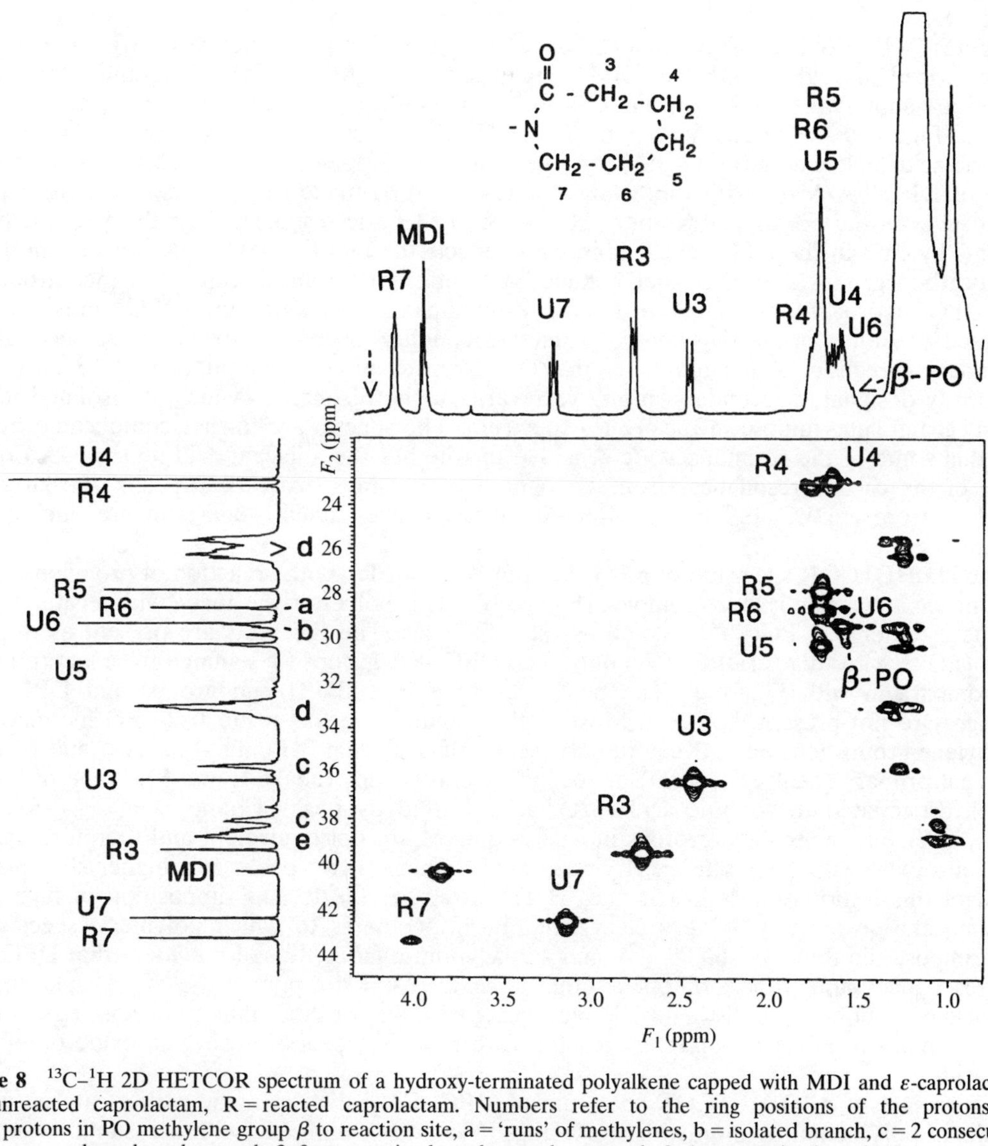

Figure 8 ^{13}C–^{1}H 2D HETCOR spectrum of a hydroxy-terminated polyalkene capped with MDI and ε-caprolactam. U = unreacted caprolactam, R = reacted caprolactam. Numbers refer to the ring positions of the protons. β-PO = protons in PO methylene group β to reaction site, a = 'runs' of methylenes, b = isolated branch, c = 2 consecutive branches, d = mostly 2–3 consecutive branches, and e = mostly 3–4 consecutive branches.

which are superimposed in the 1.5–1.9 ppm region of the ^{1}H 1D spectrum are isolated in the 2D HETCOR spectrum and in the ^{13}C 1D spectrum. The polyalkene methylene protons β to the reaction site (β-PO) also are superimposed in this region but have isolated 2D and 1D ^{13}C resonances. One has a clear understanding from the 2D data that the R5, R6, and U5 resonances superimpose in the ^{1}H 1D spectrum, but that the corresponding ^{13}C resonances are isolated in the ^{13}C 1D spectrum. In like fashion, assignments for the ^{13}C 1D resonances for U6, U4, and R4 are easily made. It is also possible to observe that alkenic resonances superimpose on the U6 and U5 resonances in the ^{13}C 1D spectrum. These peaks can be distinguished in the ^{13}C 1D spectrum as the caprolactam resonances are much sharper than the alkenic resonances.

The assignment of the remaining reacted and unreacted caprolactam resonances as well as confirmation of the MDI methylene resonance is very easy as these spectral peaks are well isolated. There is no cross peak for the polyalkene methylene group α to the reaction site (arrow) as its ^{13}C resonance is split into two peaks resonating at 63.1 ppm (α to a primary end group) and 64.6 ppm (α to a secondary end group), which is beyond the range of the F_2 (^{13}C) axis shown in Figure 7. The different polyalkene ^{13}C resonances reflecting different sequences of monomer units overlap severely in the ^{1}H 1D spectrum, but most are isolated in the ^{13}C 1D spectrum and the 2D HETCOR spectrum. The distribution of alkenic peaks reflects a high level of chain branching. Peaks c, d, and e which result from $\geq$2 consecutive branches are far more dominant than isolated branches (peak b) and 'runs' of methylene units (peak a).

Figures 9–12 provide a dramatic example of how complex polymeric resin structures can be characterized through ^{13}C–^{1}H 2D HETCOR experiments. Many of the assignments for these figures are summarized in Melby and Castro.[34] Others have resulted from more recent internal studies. Figure 9 is the relatively simple 2D HETCOR spectrum of a dipropylene glycol–fumaric acid/maleic acid (DPG–FA/MA) polyester. This polymer has a 1.05/0.63/0.37 weight ratio of DPG/FA/MA. With the catalyst system used to make this polymer, the DPG units alternate with the FA and MA units. There are eight basic resonances for the four methines and methylenes in the DPG repeat unit. This reflects the fact that (i) the carbons in the DPG hydrocarbon groups α to the ether linkage have different chemical shifts than the carbons in the DPG hydrocarbon groups α to the ester linkage, and (ii) the DPG units can be oriented in either primary–primary, primary–secondary, secondary–primary, or secondary–secondary direction. An example is the DPG repeat unit shown in Scheme 3, which is arbitrarily designated secondary–primary. Several of the resonances which are isolated in the 2D spectrum superimpose in the ^{1}H 1D spectrum. The spectra are further complicated by the fact that some of the resonances are sensitive to whether the adjacent acid unit is MA or FA. Each of the carbon resonances from 71–74 ppm is split into two carbon peaks, the larger of which is from a DPG–FA linkage. The small resonances at 66.1–66.3 ppm are due to end groups.

The 2D HETCOR spectrum of a PG–FA polyester made via the reaction of propylene oxide and maleic anhydride is more complex (Figure 10). This polyester was made with a catalyst that results in ether (*i.e.* PG–PG) as well as ester linkages. The PG units are present as isolated units (PG–FA) or diads (DPG–FA) only. The DPG–FA groups (designated by ★) are obvious by comparison with Figure 9. The peaks resulting from DPG end groups and DPG–MA linkages are not present, however. Most of the remaining peaks result from PG methine and methylene groups located between fumaric acid units (internal PG units – peaks A and B) or in terminal primary (peaks C and D) or secondary end groups (peaks E and F). Some of he FA was hydrogenated to succinic acid (SA), as indicated by peaks G and H (PG_{CH_2}–SA and PG_{CH}–SA). The methylene protons in each sequence are not equivalent and therefore each is split into two peaks (peaks B, F, H, and peak D, which is two peaks at a higher slice plane). Most of this information is lost in the ^{1}H 1D spectrum due to superimposition of numerous resonances. However, it is very clear from the 2D plot as to which polymeric species are superimposed in the 3.4–3.8, 3.9–4.4, and 4.9–5.3 ppm regions. It is also evident that DPG and PG resonances superimpose on one another at *ca.* 68.0 and 70.5 ppm in the ^{13}C 1D spectrum. It has also been possible to distinguish longer runs ($\geq$3 units) of PPG units from isolated PG units and PG diads in PG–FA polyesters reacted with a large excess of propylene oxide relative to the acid.

Figure 11 is the 2D HETCOR spectrum of a PG–FA polyester made with a catalyst which results in polyester formation with alternating PG and FA units but with occasional branching

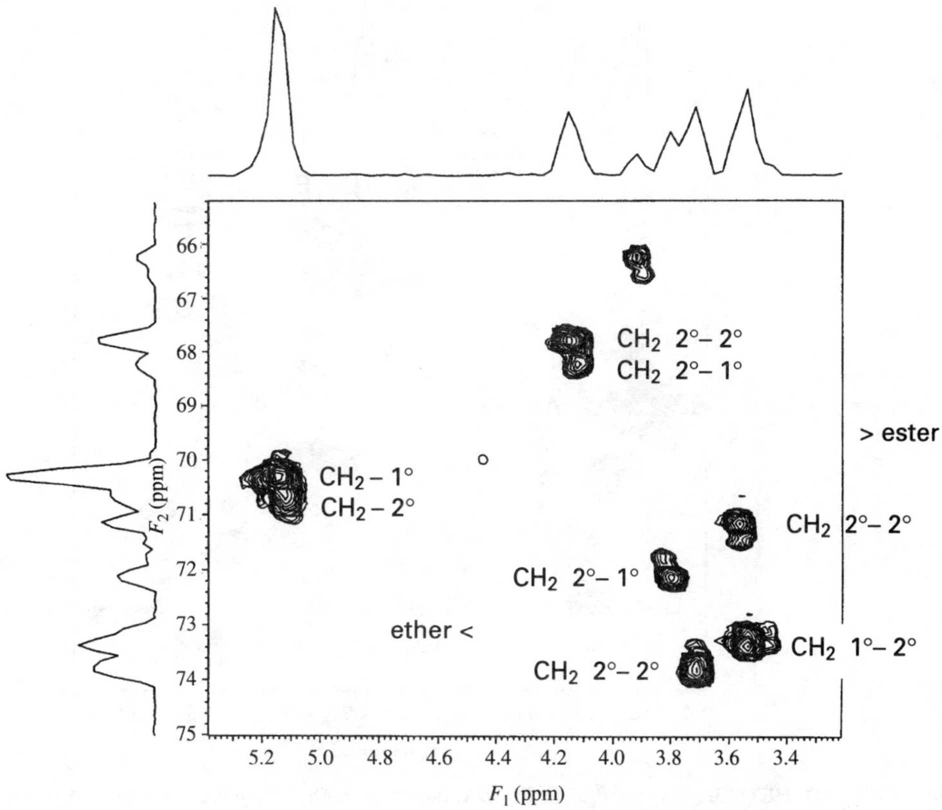

Figure 9 ^{13}C–^{1}H 2D HETCOR spectrum of a 1.05/0.63/0.37 mole ratio DPG–(FA/MA) polyester resin. Numbers refer to the orientation of PG units within each DPG unit.

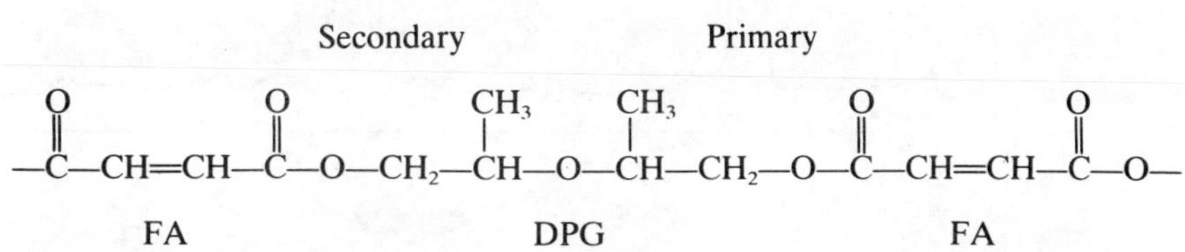

Scheme 3

(reaction of a PG unit with an internal FA double bond) and lactone formation (reaction of a terminal FA double bond with a PG unit to form a terminal six-membered ring). The branched and lactone structures are important in that the former increases chain functionality while the latter decreases chain functionality. Both reactions also decrease fumarate double bonds available for styrene cross-linking reactions where applicable. The 2D data isolate PG_{CH}–FA and PG_{CH_2}–FA internal groups and the PG methylene and methine end groups in both secondary and primary orientations for easy identification in the same manner as in Figure 10. It also establishes the fact that polyether (*i.e.* 'runs' of PG > 1) is not present in this polymer. Branching and lactone content can also be determined: 7.7% of the FA units are involved in branching; 2.2% are involved in lactone rings. This information was not available from the 1D spectra due to severe peak overlap until these data were observed in the 2D spectrum.

The 2D HETCOR spectrum in Figure 12 is from the very complex resin system discussed in

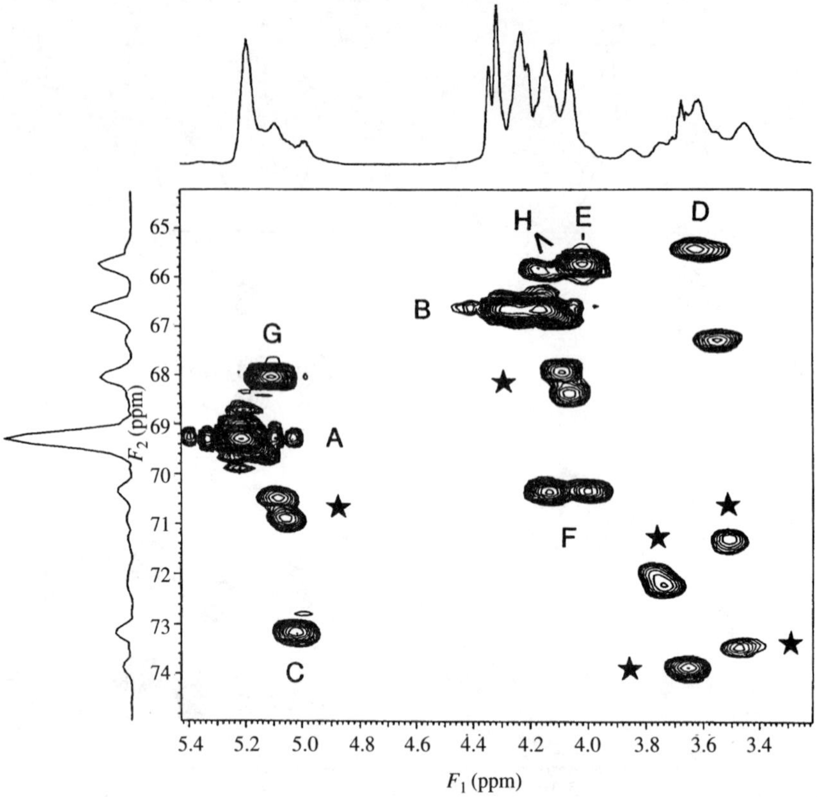

Figure 10 ^{13}C–^{1}H 2D HETCOR spectrum of a PG–FA polyester resin containing polyether. A = PG_{CH}–FA, B = PG_{CH_2}–FA, C = PG_{CH} in terminal primary unit, D = PG_{CH_2} in terminal primary unit, E = PG_{CH} in terminal secondary unit, F = PG_{CH_2} in terminal secondary unit, G = PG_{CH}–SA, H = PG_{CH_2}–SA, ★ = DPG–FA.

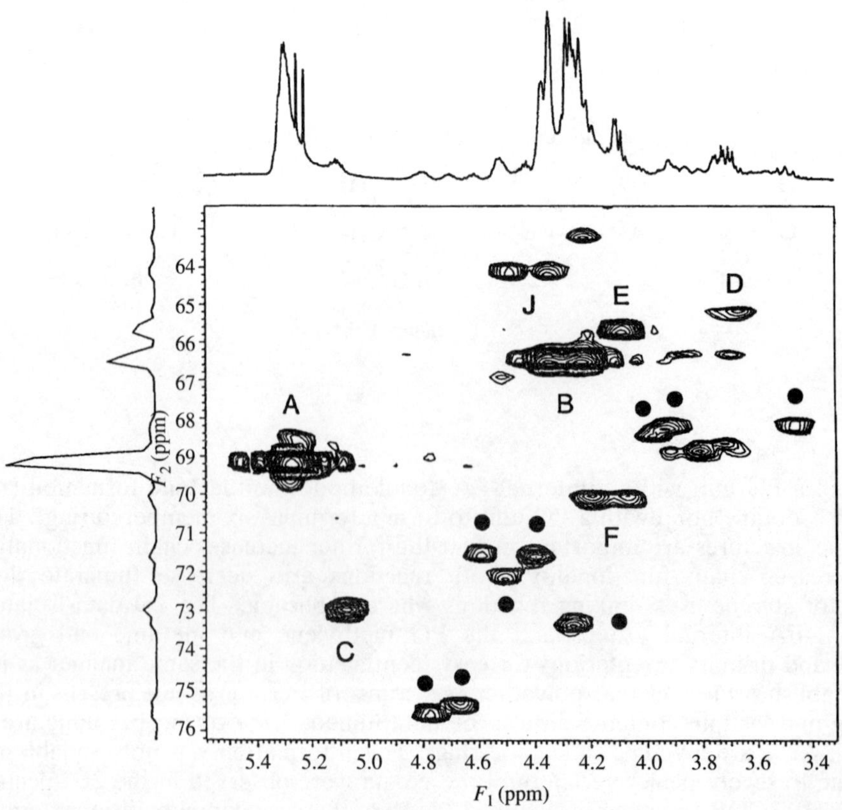

Figure 11 ^{13}C–^{1}H 2D HETCOR spectrum of a PG–FA polyester resin containing branching and lactone. A–F are PG–FA cross peaks as in Figure 10. ● = branched or lactone species. J = 1,5-pentanediol initiator methylene.

Section 6.3.1.1. It is even more complicated than the previous 2D plots in that it contains a PG–FA polyester, a PG–EG–AA–MDI polyester polyurethane, and PPG. The catalyst used in both polyesters results in polyester formation only (no ether formation) with chain branches and lactone present. This system is complicated by the presence of PG in all three components. These PG sequences can be distinguished because ^{13}C and, to a lesser extent, ^{1}H NMR are sensitive to the neighboring units around the PG units, *i.e.* both the methine and methylene PG units in PG–FA, PG–AA, and PG–PG sequences have unique peaks which are easily observed in the 2D spectrum (peaks A–F, K, L, and those marked with △ or ▲). All of the branched, lactone, and end group carbons and protons can be identified in much the same way as in Figure 11. The EG–AA resonance (peak I) from the polyester polyurethane is well isolated and easily identified. The absence of PG end groups from the PG–AA polyester component is consistent with this being a polyester polyurethane with no residual unreacted hydroxyls. There would be no hope of unraveling all of these overlapped spectral data, particularly in the ^{1}H spectrum, without the 2D data. Once all of the individual species have been isolated, a combination of ^{13}C and ^{1}H data allows the quantitative determination of the monomer contents in each of the components, secondary/primary end group ratio, and total branching and lactone contents.

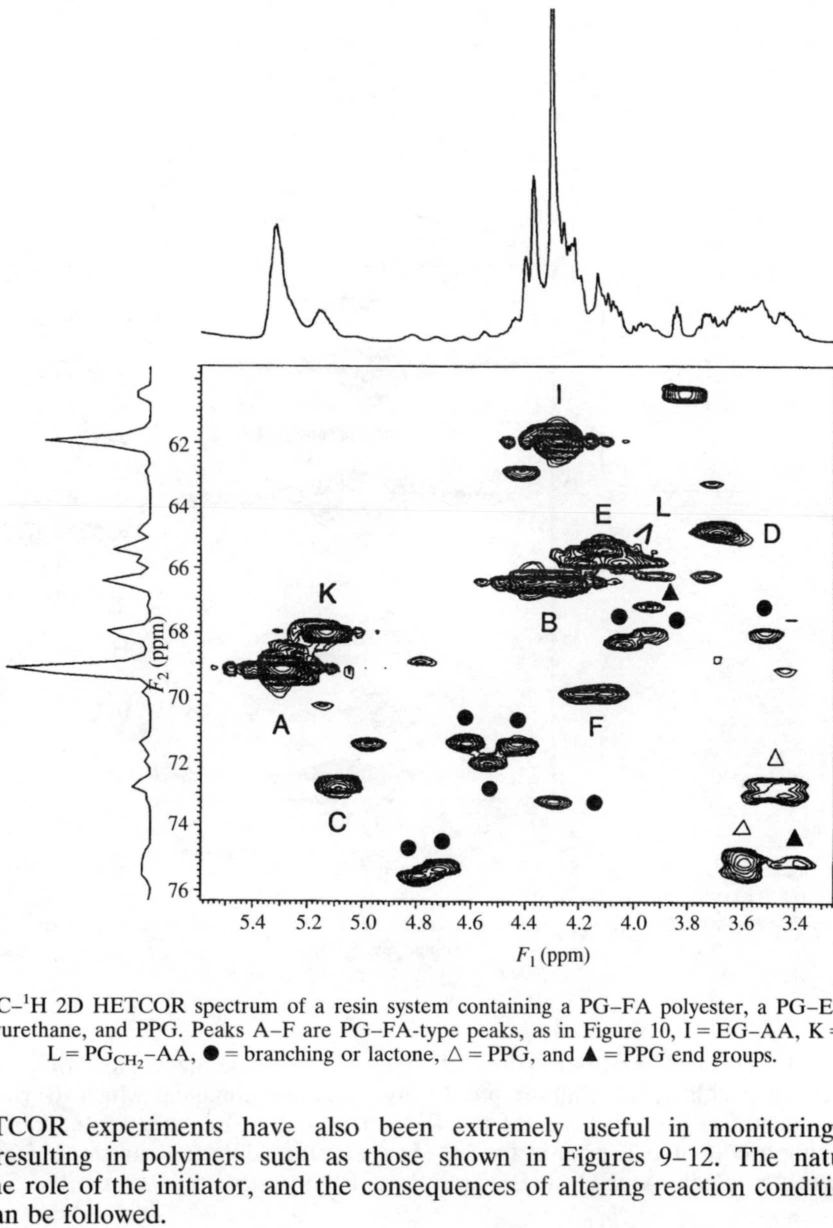

Figure 12 ^{13}C–^{1}H 2D HETCOR spectrum of a resin system containing a PG–FA polyester, a PG–EG–AA–MDI polyester polyurethane, and PPG. Peaks A–F are PG–FA-type peaks, as in Figure 10, I = EG–AA, K = PG_{CH}–AA, L = PG_{CH_2}–AA, ● = branching or lactone, △ = PPG, and ▲ = PPG end groups.

2D HETCOR experiments have also been extremely useful in monitoring polyester reactions, resulting in polymers such as those shown in Figures 9–12. The nature of the reaction, the role of the initiator, and the consequences of altering reaction conditions on the structure can be followed.

2D HECTOR techniques have also been applied to the analysis of the sequence distribution of monomer units in polybutadienes (PBDs), styrene–butadiene (SBD) copolymers, and SBD–X terpolymers. The aliphatic region of the ^{13}C–^{1}H 2D HETCOR spectrum of a 53/21/26 *trans*/*cis*/vinyl (TCV) PBD is shown in Figure 13. While the vinyl 1,4 methylene protons in a V_{CH_2}–1,4 linkage are well isolated from the *trans* and *cis* 1,4 resonances of this PBD (1.8–2.2 ppm region), the resonances for a vinyl methine proton in a V_{CH}–1,4 linkage and vinyl diad and triad protons superimpose on the 1,4 resonances in this region of the ^{1}H spectrum. This spectrum also serves to demonsrate just one of the reasons why peak broadening in ^{1}H 1D spectra of homopolymers occurs and is a problem in many ^{1}H spectra. While both ^{13}C and ^{1}H are sensitive to monomer sequencing differences, the carbon peaks are decoupled (resulting in single lines), and are spread out over approximately 200 ppm, while the proton peaks are not decoupled (resulting in multiple lines) and are restricted to usually about 10 ppm spectral width. This is why the resonances for the C–V_{CH_2}, T–V_{CH_2}, T–V_{CH}, V_{CH_2}–1,4, V_{CH}–1,4, V–V, V–V–V, C–C,C–T, and T–T,T–C linkages are well isolated in the ^{13}C spectrum,[35–37] but have only slightly different chemical shifts in the ^{1}H spectrum. They superimpose in the 2.0–2.1 ppm region and broaden the ^{1}H resonance.

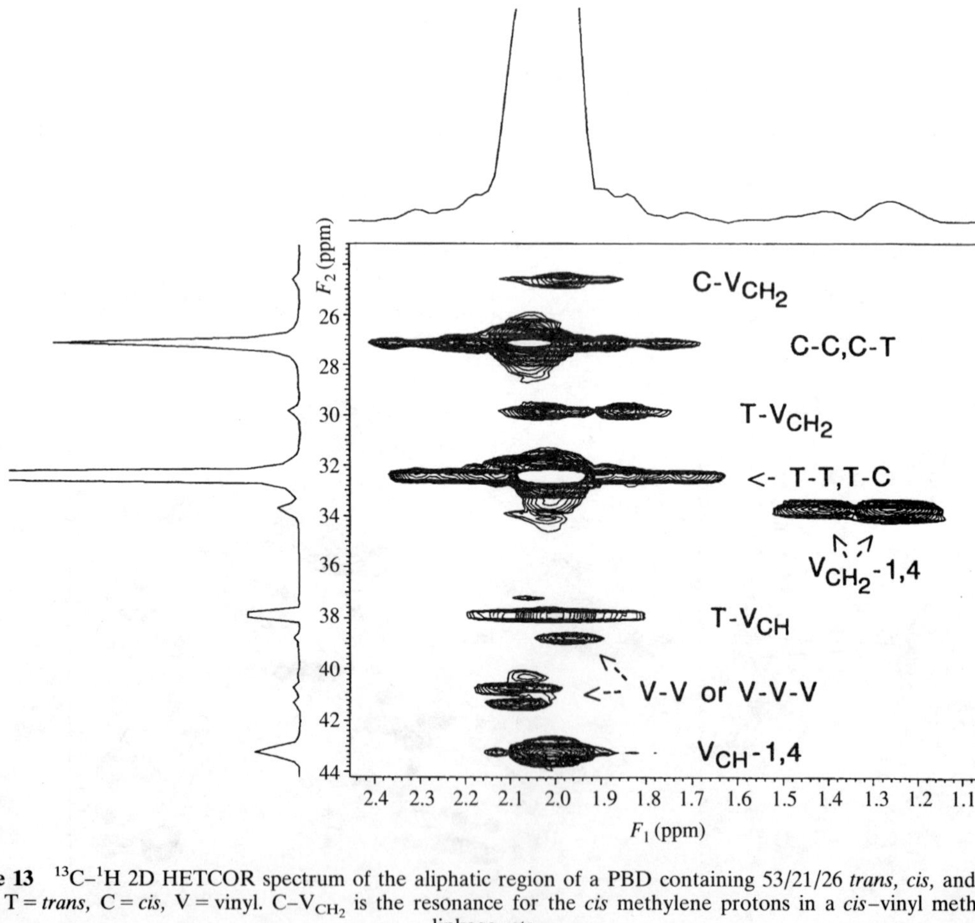

Figure 13 ^{13}C–^{1}H 2D HETCOR spectrum of the aliphatic region of a PBD containing 53/21/26 *trans, cis,* and vinyl units. T = *trans,* C = *cis,* V = vinyl. C–V_{CH_2} is the resonance for the *cis* methylene protons in a *cis*–vinyl methylene linkage, *etc.*

The ^{13}C–^{1}H 2D HETCOR spectrum of the aromatic/alkenic region of a commercial SBD–VP (2-vinylpyridine) latex allows one to observe the resonances which overlap in the ^{1}H and ^{13}C 1D spectra (Figure 14). All of the $BD_{1,4}$ resonances for unsaturated carbons (peaks A–D) superimpose at *ca.* 5.25 ppm in the 1D ^{1}H spectrum, while they are more isolated in the ^{13}C 1D spectrum. Peak A [$BD_{1,4}$–(BD_V, S, or VP)$_{CH_2}$] superimposes on the $S_{2,6}$ and $S_{3,5}$

resonances in the ^{13}C 1D spectrum at 127.8–128.1 ppm. The VP_3 and VP_5 1D carbon resonances are well isolated in the 1D ^{13}C spectrum but their protons superimpose on the $S_{2,6}$, $S_{3,5}$, and S_4 resonances in the 6.8–7.3 ppm region. The VP_4 and VP_6 carbon and proton resonances are well isolated in both 1D spectra. There are no heteronuclear resonances for S_1 or VP_1, of course, because these are not heteronuclear groups. Accurate quantitative data can be obtained from both regions of the 1D spectra, but care must be taken to account for all of the superimposed resonances.

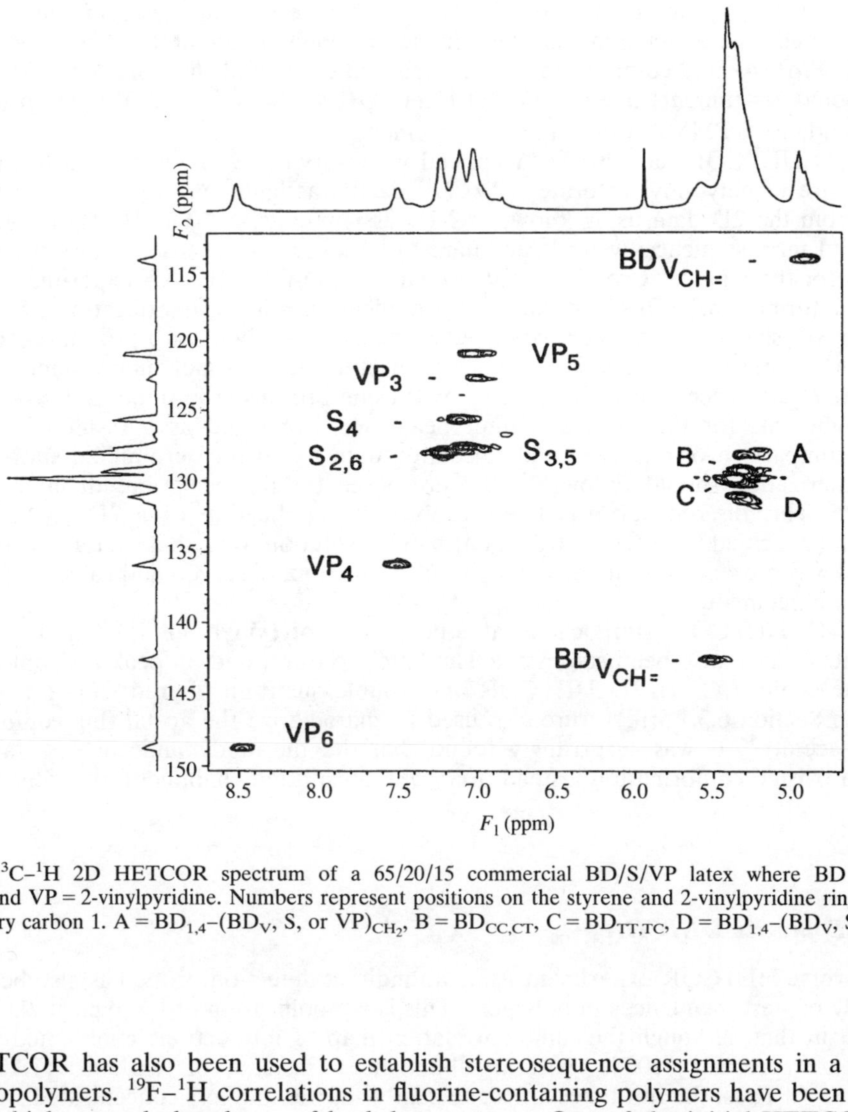

Figure 14 ^{13}C–^1H 2D HETCOR spectrum of a 65/20/15 commercial BD/S/VP latex where BD = butadiene, S = styrene and VP = 2-vinylpyridine. Numbers represent positions on the styrene and 2-vinylpyridine rings relative to the quaternary carbon 1. A = $BD_{1,4}$–(BD_V, S, or VP)$_{CH_2}$, B = $BD_{CC,CT}$, C = $BD_{TT,TC}$, D = $BD_{1,4}$–(BD_V, S, or VP)$_{CH}$.

2D HETCOR has also been used to establish stereosequence assignments in a variety of vinyl homopolymers. ^{19}F–^1H correlations in fluorine-containing polymers have been successful due to the high natural abundance of both heteroatoms. One of the initial HETCOR studies involved ^{19}F–^1H correlations in poly(*p*-fluorostyrene) made by Gerig, who observed that ^{19}F–^1H 2D correlation spectra separated resonances which were superimposed in the ^1H 1D spectra.[38]

^{13}C–^1H 2D HETCOR has been used to clarify tactic placement assignments in ^1H and ^{13}C spectra of isotactic, atactic, and stereoirregular polypropylene.[39] Ten pentad ^{13}C methyl resonances were correlated with the corresponding ^1H resonances largely from previous ^{13}C assignments. Several of the methylene proton assignments were made by taking advantage of equivalence/nonequivalence differences in the central methylene protons of tetrad sequences. By establishing assignments for obvious sequences, *e.g.* those in the isotactic polymer, and then using peak connectivity correlations, relative intensity information, and deductive reasoning, all of the ^{13}C hexad and ^{1}H tetrad methylene assignments from the 2D HETCOR data were established or confirmed.

^{13}C–^{1}H 2D HETCOR studies of stereoregular ethylene–propylene copolymers have been used to interpret ^{1}H assignments for sequencing in the ^{1}H spectrum using well-established ^{13}C assignments.[40] This study provided evidence that seven previous ^{1}H assignments were incorrect. It was also determined that a direct measure of copolymer composition from ^{1}H data was not possible. The authors, however, used the corrected assignments resulting from the 2D data coupled with a first-order Markovian probability model[41–43] to determine copolymer composition and comonomer reaction probabilities.

Assignment of the α-methyl, quaternary, and carbonyl carbon sequences in random PMMA up to the heptad level has also been made with substantial input from ^{13}C–^{1}H 2D HETCOR correlation data.[44,45] Assignment of the PMMA methylene carbons has also been achieved to the tetrad level. These assignments were made by comparison of data from isotactic and syndiotactic PMMAs and comparison of observed and calculated (first-order Markov statistics) data. Carbonyl assignments made from 2D HETCOR studies of ^{13}C carbonyl enriched[46] and natural abundance[45] PMMA were in good agreement.

^{13}C–^{1}H 2D HETCOR has also been applied with varying levels of success to the study of stereostructure in poly(vinyl chloride) (PVC).[47] Partial assignments were made in the ^{1}H 1D spectrum from the 2D data using known ^{13}C 1D assignments.[48–51] The ^{1}H 1D assignments for the mmr and mrm sequences were later found to be incorrect because of an incorrect ^{13}C 1D assignment for these peaks. Crowther *et al.* used a 2D spin-lock RELAY experiment to provide additional information.[52] This experimental modification allows identification of carbons by recognition of shared pairs. The observed nucleus is carbon but proton chemical shift information is preserved. This is particularly useful in vinyl polymers, where distinction between the equivalence/nonequivalence of methylene protons is so important.

The assignments for the mrm and mmr peaks were reversed as a result of this work.[52] However, comparison of observed and calculated data based on Bernoullian statistics (which PVC polymerization should follow[53,54]) did not agree for the peaks based on the corrected assignments. This discrepancy was later resolved by application of a 2D INADEQUATE experiment (discussed in Section 6.3.1.5(i)).[55] The solution was that stereostructure at the hexad level was present in both peaks. With this correction, observed and calculated data were in complete agreement.

^{13}C–^{1}H 2D HETCOR stereochemical studies of poly(vinyl amine)[58] and poly(*n*-butyl methacrylate)[57] have also been successful. The latter concentrated on peak assignments for the *n*-alkyl side chain. ^{13}C–^{1}H 2D HETCOR and double-quantum filtered COSY (DQFCOSY, discussed in Section 6.3.1.5(iii)) were also used to characterize the acetal ring conformation in poly(vinyl acetal).[58] It was surprisingly found that the thermodynamically less favored ring exhibited a twisted conformation instead of a *trans*-substituted chain conformation.

6.3.1.3 2D inverse HETCOR

A 2D inverse HETCOR experiment using an indirect detection probe has also been applied to the study of stereosequences in polymers. This is a modification of the typical 2D HETCOR experiment in that, although the same correlation map of interactions is obtained, the nuclei observed are protons (100% natural abundance) instead of carbons (1.1% natural abundance).[59,60] Thus, the run times are significantly shorter and less concentrated solutions are required. Sample quantities required for analysis are in the mg range. This technique overcomes one of the major disadvantages of the HETCOR experiment – lengthy data acquisition times.

The 2D inverse HETCOR experiment has been successfully applied to the 500 MHz NMR study of tactic placement distributions in poly(methacrylonitrile).[61] Tonelli had shown that the γ effect on chemical shifts used to assign tactic placements in monosubstituted vinyl polymers is not applicable to disubstituted vinyl polymers such as poly(methacrylonitrile).[62] The authors took advantage of the nonequivalence of the methylene protons in r-centered sequences and the equivalence of the methylene protons in m-centered sequences coupled with Bernoullian statistics to assign the methylene tactic placements to the hexad level. The DQFCOSY experiment (discussed in Section 6.1.3.5(iii)) was necessary to assign cross-peak correlations near the diagonal in the 2D COSY spectrum. The ^{1}H assignments were then used to make ^{13}C assignments. The results demonstrated that there was considerable overlap of r and m tactic peaks (which were separated in the 2D data) and that previous ^{1}H 1D diad assignments, made assuming that each tactic group had a unique chemical shift, were incorrect.

A modified 2D inverse HETCOR experiment was also used to make long-range carbon–proton correlations in polyamide-4,6 (PA-4,6).[63] The ^{1}H-detected heteronuclear multiple-quantum coherence sequence of Bax and Subramanian was used for one-bond correlations.[64] A 2D inverse HETCOR experiment developed by Bax and Summers,[65] designed for making long-range correlations, was successful in establishing C_4–H_3 and C_2–H_1 correlations.

6.3.1.4 2D HETCOR–TOCSY

The 2D inverse HETCOR–TOCSY (TOtal Correlation SpectroscopY) experiment is an example of a pseudo three-dimensional experiment which is complementary to other 2D correlation data. This experiment results in heteronuclear and homonuclear correlations in a single 2D experiment. This technique exploits the high sensitivity of protons and the high resolution of the ^{13}C dimension. A direct correlation of neighboring species is possible. The net result is that complete characterization of the sequencing of monomer units in complex polymers and copolymers can be made. In this experiment a second polarization transfer step (TOCSY, Figure 15) is introduced among the proton spins.

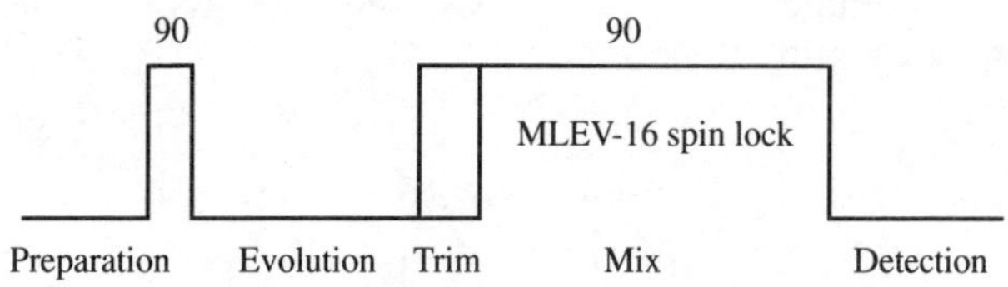

Figure 15 TOCSY pulse sequence.

Introduction of the second polarization transfer step results in signals from three-bond ^{1}H–^{1}H correlations as well as from single-bond ^{13}C–^{1}H correlations in the 2D spectrum. The spectrum of a typical vinyl homopolymer, therefore, results in two pairs of cross peaks for each of the methylene and methine carbons. For the methylene carbon, one set correlates it with the directly attached methylene protons, while the other set correlates it with the neighboring methine protons. This effectively establishes the sequence of units along a polymer chain. For example, the ^{13}C resonance for the carbon in the central methylene of a tetrad sequence is assigned via correlation with the directly attached protons and correlations with nearest-neighbor methine protons. The latter are located at the center of the two triad sequences which make up the tetrad. It is necessary to obtain a 2D inverse HETCOR spectrum to compare with the 2D inverse HETCOR–TOCSY spectrum. The cross peaks present in the latter that are not present in the former are due to the ^{1}H–^{1}H three-bond interactions. This technique is independent of chemical shift and peak intensity predictions based on an empirical approach. Thus, complications resulting from peak shifts due to temperature, pH, concentration, and impurity variations do not need to be considered.

This method has been applied to the assignment of sequencing and stereochemistry in poly(acrylic acid).[66] Disagreements of stereochemical configuration assignments in the literature[67–70] attributed to the sensitivity of spectral chemical shifts to pH effects were resolved from this study, due in large part to the independence of this experiment to variables such as pH. Unambiguous stereochemical assignments were made from the triad to the pentad level. Other minor, but relevant, resonances of previously unknown origin were attributed to end groups.

The HETCOR–TOCSY method has also been applied to the determination of sequencing in ethylene–vinyl acetate copolymers.[71] Assignments were to the pentad level for the methylenes and to the triad and tetrad level for the methines. Cross peaks resulting from carbons in obvious unique sequences were assigned first from the 2D spectrum. For example, a secondary carbon in the 2D inverse HETCOR–TOCSY spectrum did not show correlation to any of the other secondary units. Therefore this had to be the resonance for the $S_{\alpha\alpha}$ carbon (nomenclature of Carmen[72] and of Cheng[73]), which is the only carbon located between two tertiary carbons. A methine carbon from the same spectrum had only a strong correlation to $S_{\alpha\alpha}$ protons. Since $S_{\alpha\alpha}$ is the only secondary carbon next to a $T_{\beta\beta}$ carbon, the methine carbon had to be $T_{\beta\beta}$. In a similar fashion, the other sequence assignments were established.

6.3.1.5 Double quantum spectroscopy

Double quantum coherence transfer can occur through application of 2D pulse sequences because two coupled spins possess states with a double quantum energy gap.[21] Single quantum coherence transfer, such as is present in a typical 2D COSY experiment, is suppressed by a phase-cycling routine in this experiment. Both homonuclear and heteronuclear double quantum experiments are possible. Both suffer from some inherent limitations, but important results have been obtained from these experiments.

(i) 2D INADEQUATE

The 2D INADEQUATE (Incredible Natural Abundance DoublE QUAntum Transfer Experiment) technique uses double quantum coherence to detect signals from homonuclear coupled spins selectively, usually ^{13}C–^{13}C, while suppressing signals from isolated spins. The net result is a map of ^{13}C–^{13}C connectivities.[74] The pulse sequence typically used was developed by Bax *et al.*[75] and is shown in Figure 16.

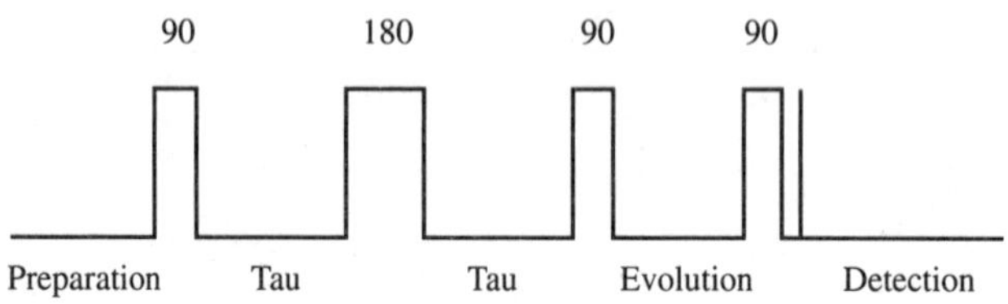

Figure 16 2D INADEQUATE pulse sequence.

An obvious limitation of this technique is its insensitivity due to the low natural abundance of ^{13}C nuclei (1.1%). This means that only 1 of every 10 000 carbon bonds is ^{13}C–^{13}C. The result is extremely long acquisition times, which are very often prohibitive in practical terms. Selective ^{13}C enrichment has been effectively used to partially eliminate this problem. This technique has typically been applied where all other NMR techniques have been explored and characterization data are still incomplete. However, when used, it has produced some very important and unique information.

2D INADEQUATE has been used recently in the study of the termination mechanism of styrene radical polymerization.[76] 99% ^{13}C enrichment of the α-methine carbon (C_α) was used to partially overcome the problem of insensitivity.

^{13}C 1D and ^{13}C–^{13}C 2D INADEQUATE techniques were applied to a low molecular weight PS polymer and ^{13}C 1D techniques to a high molecular weight PS polymer to distinguish cross peaks from C_α pairs resulting from h–h propagation reactions from those resulting from termination via coupling of propagating radicals (Scheme 4). The (C_α–C_α) junctions from random h–h propagation had resonances of equal relative intensity in the ^{13}C 1D spectra of the two polymers, while the resonances for C_α–C_α junctions resulting from termination via combination were much larger in the ^{13}C 1D spectrum of the low molecular weight polymer. To confirm this, C_α–C_α, C_α–C_β (h–h segment), and C_β–C_α (h–t segment) connectivities were distinguished in a ^{13}C–^{13}C 2D INADEQUATE spectrum of the low molecular weight polymer (Figure 17). The C_α–C_α resonances from h–h chain termination by combination coincided with the significantly reduced 1D spectral peaks in the low molecular weight spectrum relative to the high molecular weight spectrum. Consequently, this study produced the first direct evidence that polystyrene radicals are terminated largely by combination rather than by disproportionation.

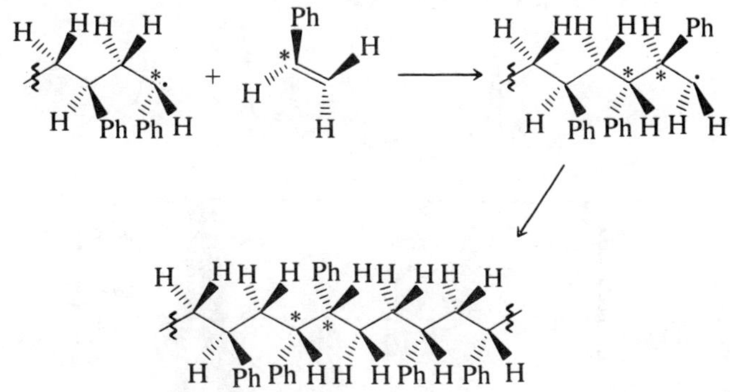

Head-to-head propagation reaction

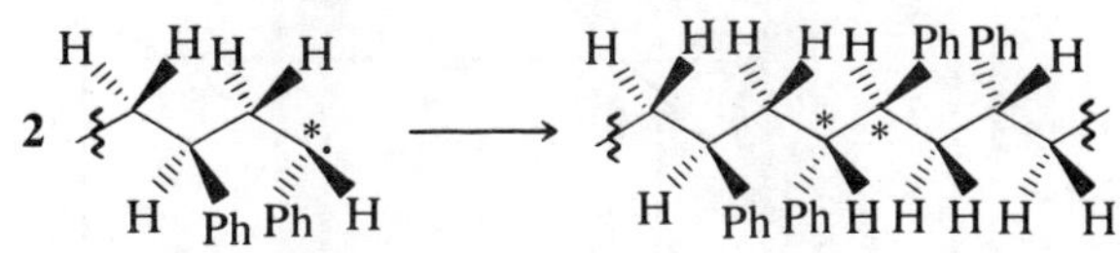

Termination via combination (head-to-head termination reaction)

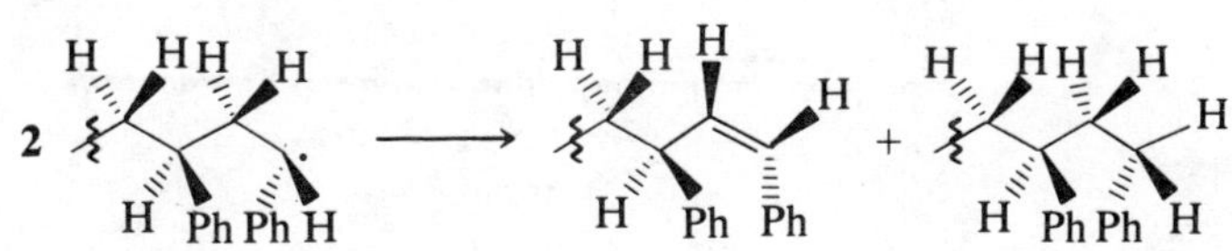

Termination via disproportionation (note: terminal methylene and alkenic carbons)

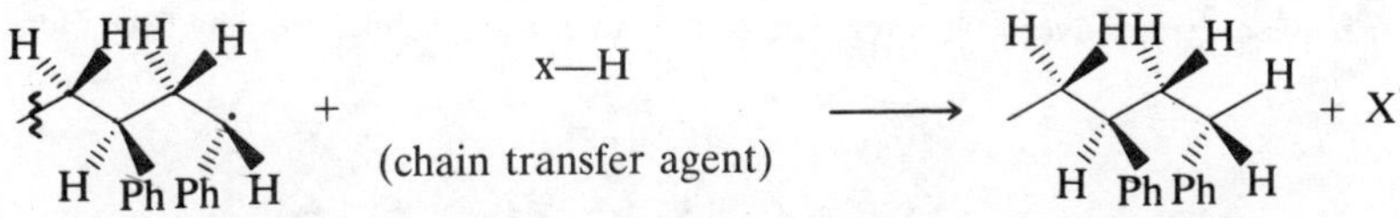

Termination via chain transfer reactions (terminal methylene carbon)

Scheme 4

^{13}C–^{13}C 2D INADEQUATE data have also been used to correct ^{13}C chemical shift assignments in poly(vinyl chloride) (PVC).[77] Previous 2D HETCOR,[47] COSY,[47] DQCOSY,[47] and spin–lock relay[52] NMR experiments had been used to make ^{13}C chemical shift assignments of stereosequences in PVC (discussed in Section 6.3.1.2). A contradiction arose when the relative areas of methylene tetrad peaks based on these assignments did not agree with relative areas based on Bernoullian propagation statistics[52,77] which should apply to PVC propagation.[53,54] The spectral peaks in question were those assigned to mmr and mrm peaks at 46.0–46.3 ppm. The 2D INADEQUATE connectivity map revealed that the previous assignments for this region were in error in that the peak for the central carbon in one of the mrm-centered hexad sequences superimposed on the mmr-centered hexad peaks at 46.3 ppm. The central carbons from the other mrm-centered hexads resonate at 46.1 ppm. This accounted for the disparity of observed and calculated data.[77]

^{13}C–^{13}C 2D INADEQUATE techniques have also been successfully applied to the ^{13}C chemical shift assignments of carbons in a stereoirregular ethylene–propylene copolymer,[78,79] stereoirregular polyalkenes,[80,81] regioirregular polypropylene[82] and poly(1-butene),[83] stereoregular ethylene–propylene,[78,79] and 1-butene–propylene copolymers.[84] In several of these studies the 2D INADEQUATE data were combined with the following types of complementary data to make detailed assignments: INEPT (Insensitive Nuclei Enhancement by Polarization Transfer) technique for distinguishing methine, methylene, methyl, and quaternary carbons;

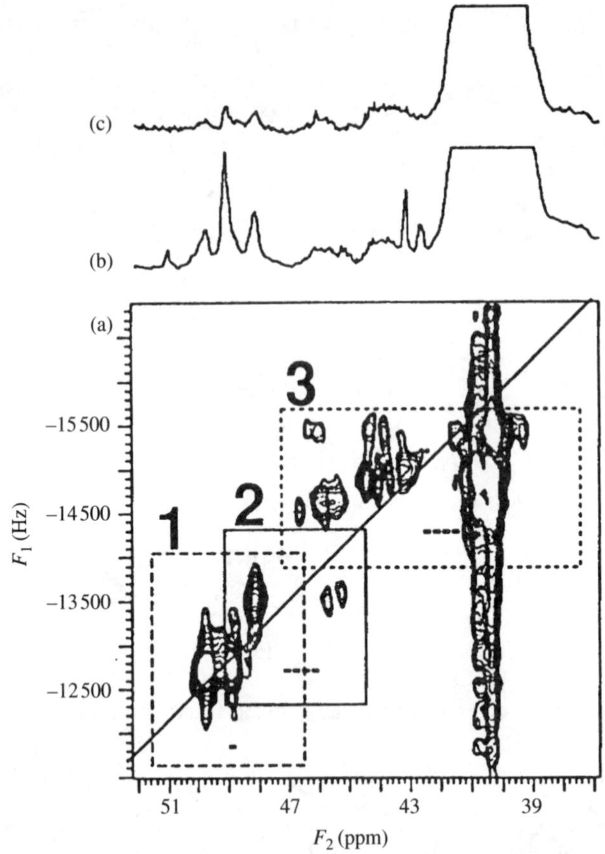

Figure 17 (a) 2D INADEQUATE spectrum of low molecular weight PS with corresponding ^{13}C 1D NMR spectra of the low (b) and high (c) molecular weight polymers. Cross peaks in region 1 arise from C_α–C_α bonds, those in region 2 correspond to C_β–C_α bonds (C_α from a h–h segment), and those in region 3 correspond to C_β–C_α bonds (C_α from a h–t segment) (reproduced by permission of the American Chemical Society from Hensley *et al.*[76]).

empirical data for making ^{13}C chemical shifts via the γ-effect on the ^{13}C chemical shift using the rotational isomeric state (RIS) model;[85] and empirical data using either the additivity rules of Cheng and Bennett[86] for predicting the ^{13}C chemical shifts of methyl-substituted alkanes or the additivity rules of Linderman and Adams.[87] All of these techniques were necessary because the presence of the numerous different types of regular and irregular sequences complicated by different tactic distributions resulted in very complex structures and extremely complex NMR spectra to interpret.

(ii) Homonuclear double quantum COSY

The homonuclear double quantum COSY (DQCOSY) experiment results in a map of the coupling network but without the diagonal peaks found in a typical 1D COSY spectrum as a result of suppression of single quantum coherence transfer.[21,47] The spectral layout is unique in that the typical ^1H 1D spectrum appears on the F_2 axis while the F_1 axis is the double quantum frequency axis. The cross peaks appear as the sum of the frequencies of the coupled spins on the F_1 axis, increasing the spectral resolution by spreading the peaks further apart. The cross peaks are correlated horizontally and the intensities vary with the coupling constants and the delay time τ_m. A disadvantage is that the spectrum can be very complex and difficult to interpret. It has, therefore, had only limited applications.

The 2D DQCOSY technique has been applied to the study of PVC stereochemistry.[47] Two ^1H–^1H DQCOSY spectra were obtained at delay times of 8 ms and 18 ms. The 8 ms delay emphasized longer couplings, such as the geminal coupling (15 Hz) of the nonequivalent methylene protons and $J_{\alpha-\beta}$ coupling (11 Hz) of r-centered stereosequences. With an 18 ms delay, the smaller $J_{\alpha-\beta}$ couplings (6.5 and 7.5 Hz for m-centered stereosequences) were emphasized. These data complemented 2D COSY,[47] 2D spin-lock RELAY,[52] and 2D INADEQUATE[77] data in the assignment of tetrad PVC stereosequences.

(iii) Homonuclear double quantum filtered COSY

Another modification of the 2D COSY experiment is the 2D double quantum filtered COSY (DQFCOSY) experiment.[21,58,61] The advantage of this experiment is that the diagonal peaks appear in the absorption mode instead of the dispersion mode that is found in a typical 2D COSY spectrum. The result of this is the elimination of 'tailing' of the diagonal peaks, making it possible to observe correlation between protons which resonate close to one another in the ^{1}H 1D spectrum. Another advantage is the suppression of single lines since double quantum coherence cannot be created for spins that are not coupled.

This technique has been successfully applied to the assignment of stereochemical structure in poly(methacrylontrile).[61] Many cross correlations near the diagonal were made in the 2D DQFCOSY contour plot that would have been 'swamped out' by dispersion-mode peak-tailing in a 2D COSY spectrum. The strategy for making these assignments was the same as that used in a 2D COSY experiment, only there was more information available. This technique has also been used in the study of the geometry of the ring systems in poly(vinylbutyral) and poly(vinylformal) at 400 MHz.[58]

6.3.2 2D *J*-Resolved Spectroscopy

6.3.2.1 Homonuclear 2D J-resolved

The strength of the homonuclear 2D *J*-resolved experiment is that multiplets resulting from complex *J*-couplings (through-bond scalar coupling) which overlap in a 1D spectrum (*e.g.* ^{1}H) are resolved in the 2D *J*-resolved spectrum. This results in information such as homonuclear chemical shifts and coupling constants which appear on the separate frequency axes in the 2D spectrum. It also allows identification of polymeric species in complex superimposed spectral areas. On many occasions this has prevented misinterpretation of the data and has resulted in valid quantitative data. This experiment, first proposed by Ernst in 1976, employs a spin-echo pulse sequence shown in Figure 18.[88,89]

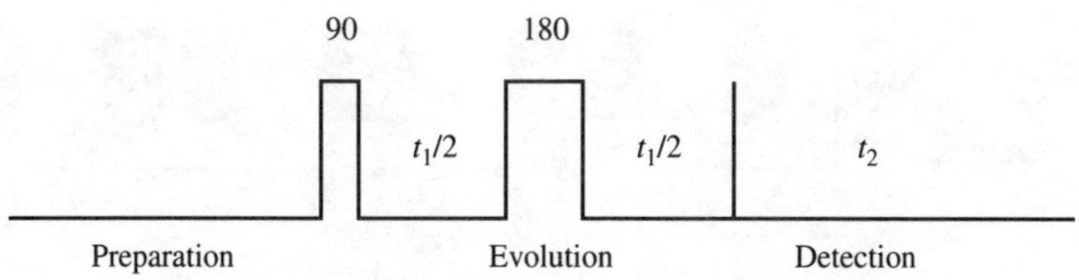

Figure 18 Homonuclear 2D *J*-resolved pulse sequence.

As in all 2D experiments, the preparation period allows the spin system to equilibrate. The 90° pulse perturbs the nuclei from their equilibrium positions. The individual components dephase during the first $t_1/2$ period. The chemical shift effects are refocused as a result of the 180° pulse. However, the effect of the *J*-coupling is not refocused because of the counteracting exchange of components in the same multiplet. The components continue to defocus during the second $t_1/2$ time period. The net result is that the frequency during t_1 is determined by the scalar coupling constant associated with each multiplet. The chemical shift components are completely refocused at the end of the t_1 interval and a 'spin echo' is detected during the t_2 detection period. F_2 is determined by both the chemical shift and the coupling effects, while F_1 is determined only by the coupling effects. The individual multiplets lie along parallel lines at a 45° angle with respect to the F_2 axis. For simplicity of interpretation, the 45° lines are typically rotated by −45° so that the multiplets are horizontal with respect to F_2. In this form the chemical shift is plotted as a function of the coupling constant. Each multiplet can be easily isolated by taking a trace (slice) through the multiplet. The distance between the spectral peaks of the coupled protons is the *J*-coupling constant.[88–90]

An application of this technique is the 2D *J*-resolved spectrum of poly(1,2-butylene glycol) (BG) shown in Figure 19. This spectrum demonstrates *J*-couplings involving protons in the main-chain methylene group (A) and in the side-chain methylene group (B) in the BG repeat unit. The information is limited to each repeat unit as the ether linkages prevent longer range couplings. The main-chain methylene protons are not equivalent, resulting in two resonances (F_2 axis) which are further split due to stereosequence effects. The 2D *J*-resolved spectrum of this region is a doublet of doublets, reflecting geminal coupling of −9.9 Hz and vicinal coupling (coupling of protons on adjacent carbons) of 5.7 Hz. All of the peak splittings have similar coupling values, demonstrating that their coupling is independent of the polymer stereochemistry. A projection of the main-chain methylene proton resonance at 3.62 ppm is shown at the top of Figure 19.

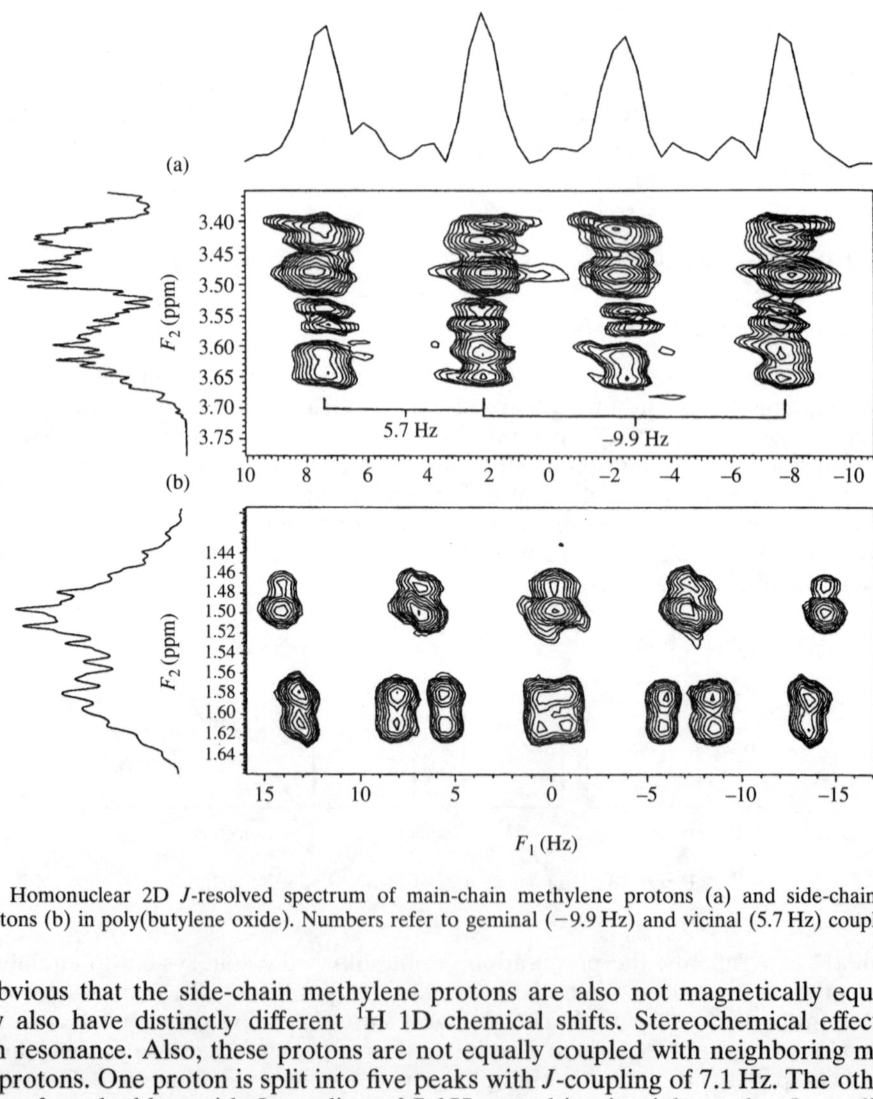

Figure 19 Homonuclear 2D *J*-resolved spectrum of main-chain methylene protons (a) and side-chain methylene protons (b) in poly(butylene oxide). Numbers refer to geminal (−9.9 Hz) and vicinal (5.7 Hz) coupling.

It is obvious that the side-chain methylene protons are also not magnetically equivalent in that they also have distinctly different ^{1}H 1D chemical shifts. Stereochemical effects further split each resonance. Also, these protons are not equally coupled with neighboring methyl and methine protons. One proton is split into five peaks with *J*-coupling of 7.1 Hz. The other proton is split into four doublets with *J*-coupling of 7.6 Hz, resulting in eight peaks. *J*-coupling of the methyl resonance (not shown) was observed to be 7.4 Hz.

An earlier homonuclear 2D *J*-resolved analysis of poly(propylene glycol) (PPG) by Bruch *et al.* yielded similar results, *i.e.* *J*-couplings of −9.9 (geminal) and 5.0 Hz (vicinal) for the methylene protons and 6.5 Hz for the methyl protons.[90] *J*-coupling for the methine protons was too complex to interpret. The coupling values were also independent of the stereochemistry of the polymer. Another early study of PVC by 2D *J*-resolved techniques was only partially successful because of complications from longer range coupling.[91] The repeat units in PVC are not separated by ether linkages as in the BG and PPG polymers.

Homonuclear 2D *J*-resolved experiments have also been used to determine tactic placement assignments in urea canal polymerized poly(acrylonitrile) (PAN).[92] Nonequivalence of the methylene protons was observed, indicating that the predominant stereosequence was isotactic. An ionically polymerized PAN resulted in three multiplets in the 2D *J*-resolved spectrum for the methine proton. These were assigned to rr, mr, and mm sequences, *i.e.* an atactic distribuiton of tactic placements.

Homonuclear 2D *J*-resolved experiments have been used to distinguish several regioregular and regioirregular species in cationic poly(epichlorohydrin) (PECH).[93] Previous tacticity studies required degradation of this polymer to PPG, which could then be partially analyzed by NMR.[94,95] Curve deconvolution analysis was also applied.[96] 2D *J*-resolved slices through five areas of the complex spectral region isolated the peaks for several regioregular and regioirregular species. These species were identified and their coupling constants were determined. These data were combined with other advanced 1D and 2D data (HETCOR) (discussed in Section 6.3.1.2) to establish that this polymer contained *ca.* 63% h–t and *ca.* 37% h–h and t–t sequences.

Homonuclear 2D *J*-resolved, ^{1}H 2D COSY, and ^{13}C–^{1}H 2D HETCOR data were complementarily used in the analysis of the chain ends of low molecular weight, highly isotactic PMMA prepared in Bu^{t}MgBr in toluene at −78 °C.[97] The data allowed the complete assignment of stereochemical sequences of the first (α_1, α_2) and last (ω_1, ω_2) MMA units in the polymer chains. Meso/racemic (m/r) ratios of 10/1 and 6/4 for the α and ω ends were observed, demonstrating high stereospecificity at the initiation but not the termination site. The two ω_1 r methylene protons were each observed as doublets of doublets in the 2D *J*-resolved spectrum, with $J_{gem} = -14.0$ and J_{vic} of 7.9 Hz. The corresponding ω_1 m methylene protons both had $J_{gem} = 14.3$ and J_{vic} of 3.5. A complementary study showed the tacticity to be independent of the molecular weight.

6.3.2.2 Heteronuclear 2D J-resolved

The heteronuclear 2D *J*-resolved experiment provides information on heteronuclear coupling constants. As a ‘bonus’, its splitting patterns, *e.g.* for a hydrocarbon group, provide information on the number of directly attached protons for each carbon in the spectrum. Other NMR experiments such as DEPT (Distortionless Enhancement of Polarization Transfer) provide the same information. The pulse sequence is shown in Figure 20.[98,99]

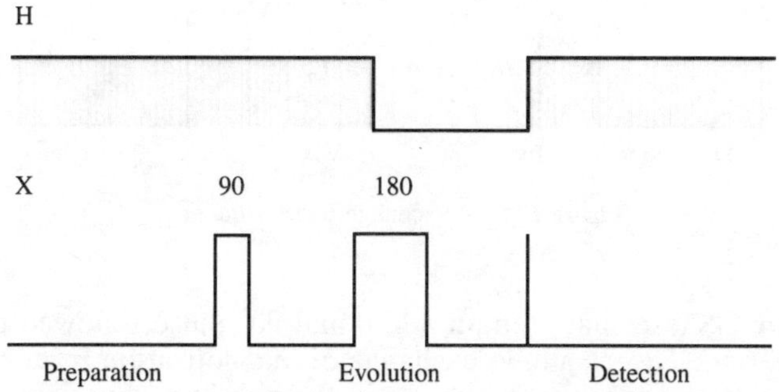

Figure 20 Heteronuclear 2D *J*-resolved pulse sequence.

An application of this technique is to the complex three-component resin system discussed previously in Sections 6.3.1.1 and 6.3.1.2. Each ^{13}C resonance in the heteronuclear 2D *J*-resolved spectrum (Figure 21) is split into $N+1$ peaks, where N is the number of directly attached protons. This spectrum was beneficial in distinguishing PG methines and methylenes in this complex spectral region, *e.g.* a methine carbon (the doublet peak E from PG_{CH} in a secondary end group) and a methylene carbon (the triplet peak L from internal PG_{CH_2}–AA) resonate very close to one another at *ca.* 65.6 ppm in the ^{13}C spectrum. The coupling constant is the distance between the centers of each of the peaks in the splitting patterns multiplied by 2.

The coupling constant of PG hydrocarbons in PG–FA sequences is 148.5 Hz, slightly larger than the coupling constant (145.7 Hz) of PG hydrocarbons in PG–AA sequences.

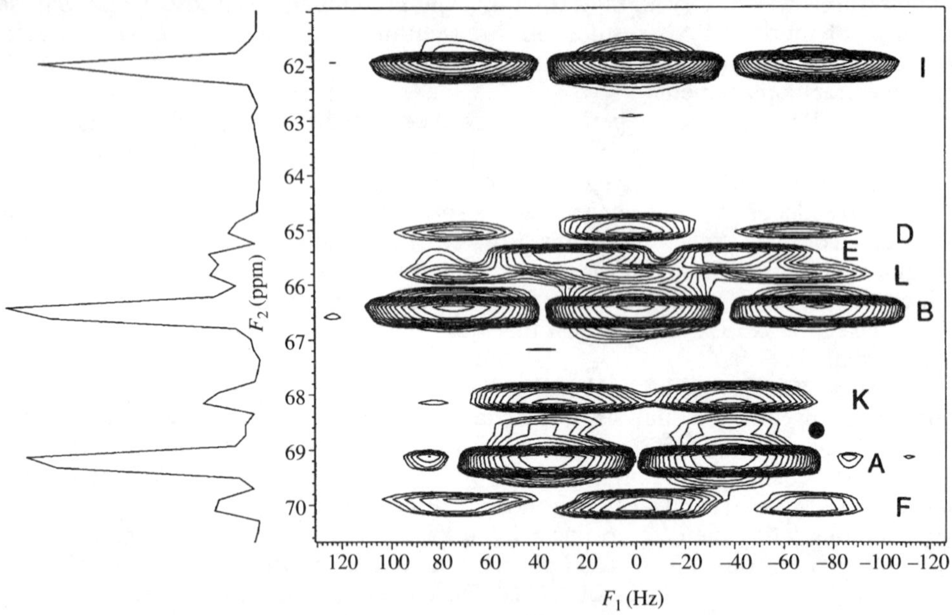

Figure 21 Portion of a heteronuclear 2D *J*-resolved spectrum of a resin containing a PG–FA polyester, a PG–EG–AA–MDI polyester polyurethane, and PPG. Peak assignments are the same as those designated in Figure 12.

6.3.3 2D Exchange Spectroscopy

The interaction studied in a 2D exchange spectroscopy experiment is chemical exchange or magnetization exchange from the NOE (Nuclear Overhauser Effect). The 2D exchange pulse sequence is shown in Figure 22.[100–103]

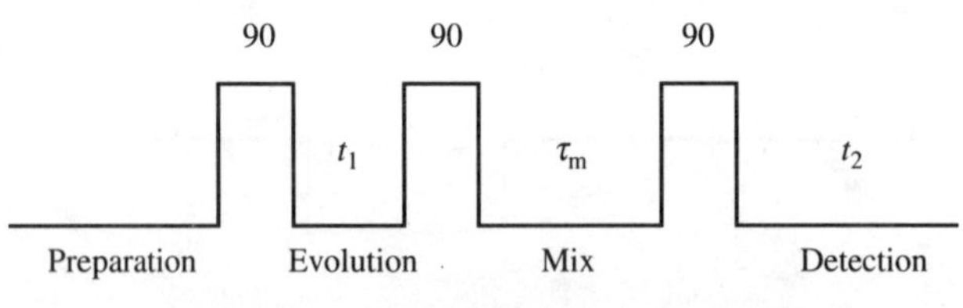

Figure 22 2D exchange pulse sequence.

This is basically a COSY format with an additional 90° pulse followed by a mixing time τ_m after the evolution period. The τ_m allows exchange of magnetization from NOE (via incoherent interactions) or chemical exchange to occur. As in the previous experiments, the results of the chemical or magnetization exchange are frequency changes which are detected during the detection period. Chemical shifts (usually ^1H) are plotted along both axes with the exchange information present as cross peaks in the 2D spectrum. Cross peak intensities reflect the amount of exchange that has taken place. By varying τ_m, under certain conditions one is able to obtain the build-up rate of cross peak intensities, which is a measure of the exchange rate.[100–102]

6.3.3.1 2D Nuclear Overhauser Effect SpectroscopY (NOESY)

In a 2D NOESY experiment, through-space, dipole–dipole coupled protons exchange magnetization via the NOE effect. The mixing time approximates the spin–lattice relaxation time (T_1). The exchange that takes place is proportional to the inverse of the difference in distance ($1/r^6$) between the dipole-coupled protons. In practical terms, magnetization exchange will occur only when the protons are within 0.5 nm of one another. This is an excellent experiment for determining intimate spatial relationships of closely packed polymer chains, particularly in polymer blends, and the specific nuclei which are involved in these interactions. It therefore has particularly important implications for the study of miscibility in polymer blends (discussed in detail in Section 6.4). The 2D NOESY experiment also provides the means to observe nuclei which are located very close to one another along polymer chains. This allows the determination of specific types of sequencing patterns in many polymers.

The first application of 2D NOESY to synthetic polymers was the assignment of ^{1}H tetrad sequences in a 65/35 mole ratio vinylidene chloride/isobutylene copolymer by Bovey and Bruch.[104] The polymer structure in this copolymer is relatively simple in that there are no stereoconfigurations or scalar coupling to complicate the spectrum. Only adjacent methylenes are close enough together for magnetization exchange, resulting in spectral cross peaks, to take place. It was also observed in this study that the isobutylene methyl protons were correlated with not only the adjacent methylene protons but also the isobutylene methylene protons further away, suggesting tight packing of the polymer chains.

2D NOESY has also been successfully applied to the study of hydrogen bonds in polyurethanes. This is important in that it is well established that different types of hydrogen bonds in polyurethanes result in different physical properties, such as tensile strength and strength/elongation relationships.

The reaction product of poly(EG–PG)–AA (PEPA), 4,4′-diphenylmethyl diisocyanate (MDI), and *N*-methyldiethanolamine (MDEA) was characterized using this method.[105,106] The MDI–MDEA–MDI sequences are the hard segments and the PEPA are the soft segments in this polymer. A strong cross correlation was found between the MDI *ortho* aromatic protons and the N—H proton from the urethane linkage in another chain, a result of hydrogen bonding between the urethane carbonyl oxygen from one chain and the N—H proton from another chain. The data suggested that the carbonyl, N atom, and MDI phenyl group were in the same plane. This meant that the hard segments were highly aligned. Only a small amount of hydrogen bonding of soft segment carbonyl oxygens with N—H protons from another chain was observed. The authors also came to the important conclusion that the hydrogen bonding interactions of the concentrated solutions were similar to those in the bulk polyurethane.

6.4 MISCIBILITY IN POLYMER BLENDS

The physical properties of polymer blends are directly related to their miscibilities at the molecular level. Several 2D experiments have been developed to determine the degree of intimacy of the polymer chains at this level. It is possible using 2D NMR techniques to determine whether intermolecular interactions are taking place between spatially close protons in polymer blends and, if so, the specific nuclei that are involved. The latter capability provides a distinct advantage over other analytical techniques. In certain cases it is possible to determine the strength of these interactions.

Most polymers are prepared for this type of study as films cast from a variety of solvents. Some have been redissolved in an appropriate solvent and run by solution NMR techniques (Section 6.4.1); others have been run in bulk by solid-state NMR techniques (Section 6.4.2).

6.4.1 2D Solution Studies of Polymer Blend Miscibility

The most commonly used solution experiment is 2D NOESY, which provides a map of all nuclei (usually protons) in chains which are spatially located within 0.5 nm of one another. Intra- and intermolecular interactions are observed as cross peaks in the 2D NOESY spectrum. The intramolecular interactions are between nuclei on the same polymer chains, either within a repeat unit or on another portion of the chain that has folded back onto itself. Intermolecular interactions in miscible polymer blends can be between similar or dissimilar polymer chains. The latter is an indication of polymer miscibility at the molecular level. Cross peak intensities provide a means for determining the closeness of the polymer chains in that they vary with both the inverse power of the distance ($1/r^6$) between the nuclei and the mixing time. By varying the mixing time over a range of values (all greater than the correlation times but less than the spin–lattice relaxation time) the maximum cross peak intensities can be determined. This is a measure of the strengths of the interactions and therefore an approximation of the distances between the interacting nuclei. The lower limit of detection is 0.18 nm at this time.[107] Comparison of the internuclear distances between several nuclei from dissimilar polymer chains can also allow the determination of the geometry of the monomer units involved in the interaction. Finally, there are several means by which chains can become closely packed in a polymer blend: hydrogen bonding, ion pair formation, van der Waal forces, *etc.* Data from the 2D NMR experiment can sometimes provide information on which type of interaction is responsible for the polymer blend miscibility.

Polymer chain entanglements in synthetic polymers are considered to be necessary to restrict local chain motions which would attenuate the NOEs by moving the interacting nuclei away from one another.[108] For polymers with low to moderate molecular weight, the 2D Rotating frame Overhauser Effect SpectroscopY (ROESY) experiment has advantages over the NOESY experiment.[109] An unfortunate characteristic of NOESY is that, due to its relationship to τ_c, its sign may change for an intermediate τ_c (which can occur in polymers at *ca.* 5–20 K), resulting in a small cross-correlation constant. NOE cross peaks may be reduced significantly or eliminated as a result. The cross-correlation rate of the ROESY experiment is less sensitive to τ_c and is positive for polymers of all sizes. The ROESY pulse sequence involves the application of a field gradient pulse or a series of 180° pulses during the mixing time. For a detailed description of the ROESY experiment, see Ernst.[109]

A 2D NOESY study of complex formation at the molecular level in blends of poly(acrylic acid) (PAA) and poly(ethylene oxide) (PEO) has confirmed that an intermolecular complex was formed via hydrogen bonding between the PEO ether oxygen and the PAA hydroxyl group (*i.e.* O—H—O hydrogen bonds).[110]

Intermolecular PAA–PEO interactions between the PEO methylene protons and all of the PAA protons (particularly the α-protons) were observed. Cross peaks between the PEO methylene protons and the nonequivalent PAA β-protons had approximately the same intensity, demonstrating a lack of stereosequence preference in the complex formation. Intermolecular cross peak intensities decreased with increasing pH. Intramolecular PAA cross peaks for a PAA homopolymer and for the PAA/PEO complex were very similar, indicating little change in the average chain conformation upon complex formation. Hydrogen bonding in these blends was very strong and the polymers were extremely miscible at the molecular level, as indicated by the observation of cross peaks in PAA/PEO blends at concentrations as low as 0.01 wt.%.[110]

500 MHz 2D NOESY studies of PVC/s-PMMA blends have found much weaker polymer miscibility than that observed in the PEO/PAA blends.[111] The data supported complex formation via weak hydrogen bonding of the s-PMMA carbonyl and the PVC methine. This is consistent with the fact that O—H—C hydrogen bonds are typically weaker than N—H—N, O—H—N, or O—H—O hydrogen bonds.

Short intrachain distances were observed between the PMMA methoxyl protons and the methylene and methyl protons, indicating that the methoxyl group was not extended away from the main chain. Intermolecular chain distances varied from 0.31 nm for PMMA methoxyl–PVA methylene interactions to 0.35–0.4 nm for PMMA methylene–PVC methine interactions, and 0.4–0.5 nm for PMMA methyl–PVC interactions. The authors concluded that the PVC surface defined by the methylene protons is oriented toward the PMMA surface covered by the methoxyl protons (0.31 nm intermolecular distance) and the PMMA methyl protons are oriented away from the surface of molecular contact (0.4–0.5 nm intermolecular distance). Rate matrix analysis of the 2D NOESY spectra also showed that the internuclear interactions were not strongly influenced by polymer tacticity or temperature.[111]

Intermolecular interactions have been observed in 500 MHz 2D NOESY studies of poly(*N,N*-dimethylacrylamide) (PDMA) with polystyrene-*co*-styrenesulfonic acid [P(S–SA)] and with poly(4-vinylphenol) (PVPh).[108] The nature of the solvent was an important consideration in that certain polymer–solvent (pyridine-d_5) interactions effectively competed with polymer–polymer interactions via hydrogen abstraction.

Relatively weak internuclear interactions of the methyl protons of PDMA with PS aromatic protons (Figure 23) occurred as a result of ion pair formation between the sulfonic acid group of the P(SA) and the amide group from the PDMA, as shown in Scheme 5. The latter was largely concluded because of the lack of cross peaks connecting the aromatic protons *ortho* to the sulfonic group to the acidic proton. This interaction would be present if the acidic proton were still attached to the sulfonic acid group.[108]

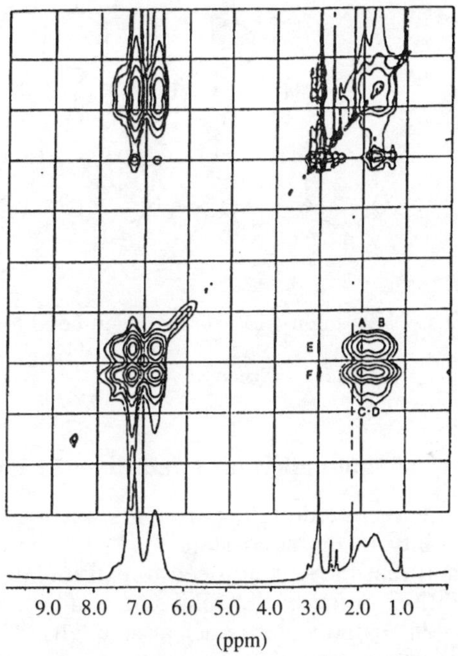

Figure 23 2D NOESY spectrum of 25 wt.% PMDA/P(PS–SA) in DMSO-d_6. Peaks A and B are from intramolecular PS–PS interactions, peaks C and D represent intermolecular PS–PS interactions, and peaks E and F represent intermolecular PS (*ortho, meta,* and *para,* respectively) ring proton–PMDA methyl interactions (reproduced by permission of the American Chemical Society from Kelts *et al.*[108]).

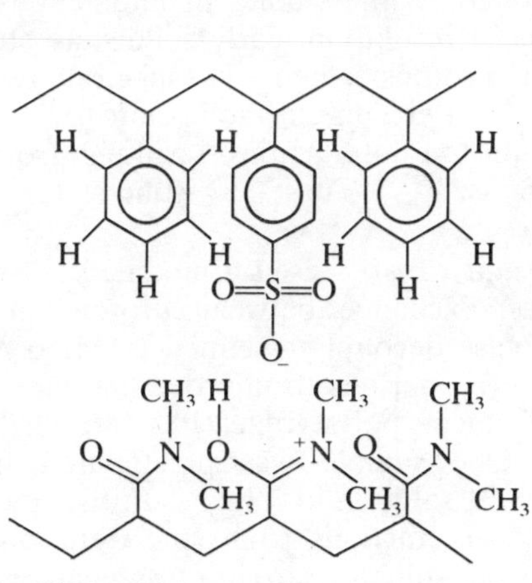

PMDA–P(PS–SA) interaction via ion pair formation

Scheme 5

Intermolecular interactions in PDMA/PVPh blends in DMSO-d_6 solvent were also observed (PMDA methyl–PVPh hydroxyl and PDMA methyl–PVPh aromatic proton). Intramolecular PVPh aromatic proton–PVPh hydroxyl proton interactions were also observed, demonstrating that the hydroxyl group was intact and that the interaction bringing the polymer chains close to one another was hydrogen bonding (Scheme 6).[108]

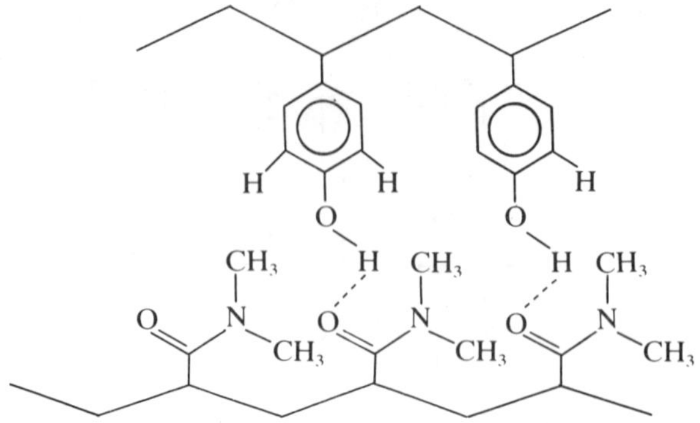

PMDA–PVPh interaction via hydrogen bond formation

Scheme 6

6.4.2 2D Solid-state Studies of Intermolecular Interactions in Polymer Blends

Several 2D experiments involving unique experimental techniques have been developed for the study of polymer miscibility in the solid state.[112–119] This is usually preferred to solution studies for the obvious reason that there is no need to extrapolate the solution study results to the real life solid-state condition. Some of the solid-state studies use 2D exchange experiments to study spin diffusion between spatially close nuclear spins (usually protons) via magnetization exchange. It has been well established that spin diffusion in solids is a primary relaxation mechanism.[120–125] Spin diffusion in heterogeneous solids has been studied by the Goldman–Shen experiment, where differences in domain mobilities are exploited.[126] An advantage of the 2D exchange experiments is that it can provide information on polymer miscibility, or the lack thereof, in polymers regardless of the mobilities of the nuclei being studied. The NMR experiments also have the advantage that information on the specific nuclei interacting via spin diffusion can be identified. While studies of interactions between carbons would be preferred due to superior spectral resolution, early solid-state studies involved spin diffusion between the more abundant protons when ^{1}H spectral resolution was good enough to isolate these interactions.[112] Later, unique solid-state experiments were developed involving the study of carbon–proton interactions resulting from ^{1}H–^{1}H spin diffusion via solid-state HETCOR experiments.[113–119] These 2D solid-state experiments are summarized below.

Caravatti, Neuenschwander, and Ernst[112] used a modified 2D exchange experiment for the determination of magnetization exchange between protons in polymer blends cast from different solvents. A multiple-pulse decoupling sequence (MREV-8) was applied during both the evolution (t_1) and detection (t_2) periods to improve line narrowing. It was applied to two 0.61/1.00 mole ratio blends of atactic polystyrene (PS) ($M_n = 279\,000$) and atactic poly(vinyl methyl ether) ($M_n = 53\,500$). One sample was precipitated from a 5 wt.% solution in chloroform, another from a 5 wt.% solution in toluene. In the spectrum of the blend cast from chloroform, only intramolecular interactions from spin diffusion were observed. In the 2D spectrum of the polymer cast from toluene, strong cross peaks were present between the PS aromatic protons and the PVME methoxy and/or methine protons (which superimpose). Thus, polymer misciblity was observed in the blend cast from toluene but not in the blend cast from chloroform. These data confirmed thermal and NMR relaxation data that the physical properties of the blend depend on the solvent from which it was precipitated or the films from which they were cast.

Modified ^{13}C–^{1}H 2D HETCOR experiments have recently been developed to allow the study of polymer miscibility by observing interchain carbon–proton correlation interactions which depend on the rate of proton–proton spin diffusion.[113–119] This involves proton–proton dipolar mixing followed by proton–carbon dipolar mixing, where transfer of magnetization from protons to carbons occurs. This type of experiment, therefore, has many advantages over the earlier study of spin diffusion via homonuclear interactions in terms of spectral resolution and the instrumentation needed to do the experiment. It also more frequently allows the identification of the specific sites where the interactions (often due to hydrogen bonding) occur between polymer chains. The pulse sequence is shown in Figure 24.

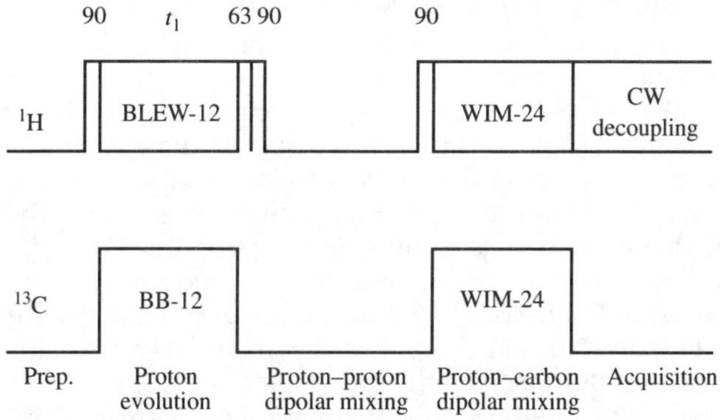

Figure 24 2D solid-state HETCOR pulse sequence.

BLEW-12 and BB-12 pulse sequences (or variations thereof) are applied during the evolution period to suppress homonuclear and heteronuclear dipolar couplings, respectively. A mixing time is introduced after the evolution period to allow proton–proton dipolar mixing. A WIM-24 (Windowless Isotropic Mixing)[115] pulse sequence is then applied (as opposed to a typical cross polarization sequence) to allow proton–carbon dipolar mixing. Proton–proton spin diffusion is eliminated during this time as a result of the WIM mixing, in contrast to only being scaled down by a factor of 0.5 in the typical cross polarization technique. This is important, because the proton chemical shift information resulting from the evolution period should not be complicated by proton spin diffusion during the mixing time if carbon–proton correlations are to be made. This modified experiment provides much better resolution than the 2D ^{1}H–^{1}H method and the correlation intensities depend on the rate of abundant ^{1}H–^{1}H spin diffusion mixing. The 2D HETCOR method does not require simultaneous multiphase ^{1}H detection, and therefore the expensive 'CRAMPS' (Combined Rotation and Multiple Pulse NMR) capability is not required.[113–119]

The general approach is to obtain first a 'normal' 2D solid-state HETCOR spectrum to establish proton–carbon couplings between directly attached nuclei. Then a modified 2D solid-state HETCOR spectrum is obtained to identify heteronuclear interchain interactions. Two different types of cross peaks are present in the 'modified' 2D HETCOR spectrum. One type results from the same proton–carbon couplings of directly attached nuclei observed in a 'normal' 2D HETCOR spectrum. The other type correlates a proton resonance from one chain with a carbon resonance from another chain. This type of cross peak can be an indication of internuclear interaction of dissimilar polymer chains. Introduction of the proton–proton dipolar mixing time τ_m in the modified experiment allows one to distinguish short and longer range interactions by varying this parameter and observing changes in the 2D spectrum. Short values result in largely nearest neighbor interactions; longer τ_m values result in longer range interactions as it takes more time for these interactions to occur.

The modified 2D HETCOR method has been applied to the study of deuterated 1/1 weight ratio PS-d_5/poly(2,6-dimethyl-*p*-phenylene oxide) (PXE).[114] The PS was deuterated to simplify the ^{1}H dimension of the 2D spectrum. At $\tau_m = 0.3$ ms, only intramolecular correlations in each component were observed. Because of the deuteration, intrasegmental aliphatic and aromatic spin diffusion cannot occur within the PS-d_5. The only way an aromatic proton–aliphatic PS correlation could take place is through heterosegmental contact of PS-d_5 and PXE nuclei. At $\tau_m = 10.0$ ms, interchain PXE aromatic carbon–aliphatic PS proton interactions resulting from spin diffusion were observed.

Application of this method showed that the PXE and PS-d_5 were miscible at the molecular level but with weakly interacting chains. This study demonstrated that the modified 2D HETCOR method is particularly useful for polymer blends which result in only one resolvable

^{1}H chemical shift in one component. Then, protons on the other chain can be distinguished by magnetization transfer to ^{13}C, where the resolution is much better. Previous homonuclear methods required resolvable ^{1}H resonances from both components.[114]

Kaplan has applied a modified 2D solid-state HETCOR experiment to the study of the miscibility of a blend of *N,N'*-diphenyl-*N,N'*-bis(3-methylphenyl)-1,1'-biphenyl-4,4'-diamine (TPD) in bisphenol A polycarbonate (BPA).[127] TPD is highly crystalline with $\tau_m = 167\,°C$ and $T_g = 63\,°C$; BPA has a T_g of 137 °C. A 50% (w/w) blend of TPD/BPA cast from methylene chloride is amorphous with a single $T_g = ca.$ 89 °C.

At longer τ_m values, magnetization was transferred from the aromatic protons to the methyl protons via spin diffusion. The methyl protons then isotropically cross polarized with the methyl carbons during the WIM-24 sequence. Although distinguishable in the ^{13}C dimension, there was insufficient chemical shift resolution in the proton dimension to resolve effectively the TPD and BPA aliphatic resonances in this blend. However, nearly all of the polymeric protons achieved a common spin temperature via the spin diffusion process. By plotting the volume integral fraction of aliphatic protons as a function of spin diffusion time and by plotting their difference, a curve was observed which was interpreted as an initial rise from both inter- and intramolecular spin diffusion followed by a slower, long-term delay from intermolecular proton spin diffusion. The data indicated that the polymer compounds were miscible at the molecular level in spite of the fact that the specific interaction sites could not be observed.[127]

White and Mirau have also studied polymer miscibility with a slightly different modified solid-state HETCOR experiment for a PVPh/poly(methyl acrylate) (PMA) blend.[128] A sideband suppression technique, TOSS (TOtal Sideband Suppression),[129] was applied to eliminate superimposition of sidebands on polymeric peaks.

The 2D HETCOR/TOSS spectrum of a 1:1 mole PVPh/PMA blend showed a correlation between the PMA carbonyl carbon (176 ppm) and the PVPh hydroxyl proton (5 ppm) (Figure 25). This demonstrated the presence of carbonyl–hydroxyl hydrogen bonding in this system. This cross peak was not present in a 2D HETCOR/TOSS spectrum where the hydroxyl proton had been exchanged with a deuterium. A similar study of a PVPh/PMMA blend demonstrated the hydroxyl bonding of the PMMA carbonyl and the PVPh hydroxyl proton.[128]

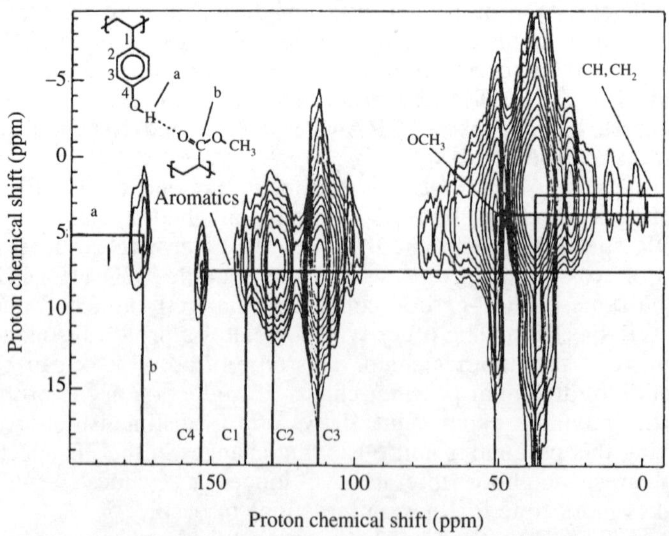

Figure 25 2D solid-state HETCOR/TOSS spectrum for a 1:1 PVPh/PMA blend. The inset plot shows the chemical structures of the polymers, and the assignments are indicated. The intermolecular correlation between the PVPh hydroxyl protons (a) and the PMA carbonyl carbon (b) are also shown (reproduced by permission of the American Chemical Society from White and Mirau[128]).

Heffner and Mirau have recently reported weak intermolecular interactions between the *cis*-polyisoprene (PIR) methyl group and the 1,2-polybutadiene (PBD) vinyl side chain (particularly the methylene) in 50/50 weight ratio PIR/PBD blends via ^{1}H 2D solid-state magic angle spinning (MAS) NOESY experiments.[130] Internuclear interactions involving the main-chain PBD protons were not observed. Under these experimental conditions the chains are sufficiently mobile and the resolution is good enough to observe these interactions. It was noted that the PIR methyl group is electron deficient while the PBD vinyl side chain is electron rich, a possible explanation for the preferential interactions.

6.5 POLYMER DYNAMICS

6.5.1 2D Solid-state Exchange

This topic is worthy of more mention than the current space allows. Several impressive ^{2}H and ^{13}C solid-state exchange NMR studies of dynamic behavior in homopolymers and polymer blends have been carried out. Schmidt-Rohr and Spiess have documented many of these studies in a book 'Multidimensional Solid-State NMR and Polymers'.[131] Blümich and Spiess have also covered this topic, as well as an overview of solid-state NMR techniques, very effectively in an earlier article.[132] This review has a comprehensive set of references for 2D studies prior to 1989. For a detailed presentation of the work done in this field of study, these publications and the literature cited below are recommended. The following is a cursory overview of important and interesting 2D applications in this area.

Molecular motions of polymer chains have a significant influence on the macroscopic properties of polymers.[133] Dynamic properties, in turn, are intimately related to the arrangement of the chains in the amorphous and crystalline regions. Pulsed ^{2}H and ^{13}C 1D solid-state NMR have been used to study molecular dynamics in solids for many years, providing very useful information concerning segmental motions.[132,134–147] However, these techniques have not been able to distinguish between different types of molecular motions and the timescales were not in the range comparable to the much slower macroscopic properties. 2D exchange experiments are now able to provide the means to distinguish among different types of motions. For example, the 2D spectral pattern resulting from Brownian motion can be distinguished from the pattern resulting from discrete reorientational jumps. 2D exchange experiments are essential for the study of slow and ultraslow rotational chain dynamics such as those that occur near the T_g in amorphous polymers and in crystalline regions of semicrystalline polymers.[131,132,148–165] Spiess *et al.* have developed 2D techniques which effectively monitor the dynamic process via study of motions in this range while ultimately attempting to account for faster motions which may occur in the evolution and detection periods.[148–151] These experiments correlate changes in molecular motion and orientation with changes in frequency. Unique geometric information is also provided since the angle by which the molecular motion is measured can be determined directly and a model is not needed. The timescale is also comparable to that employed for mechanical and dielectric measurements so direct comparisons are valid. With this experiment, it is possible to study the effect that such variables as chain configuration, conformation, and packing have on the chain dynamics. For detailed descriptions of this procedure, the assumptions made, and mathematics involved, see Spiess *et al.*[132,148–151]

2D exchange experiments have been used to study segmental reorientation dynamics in homopolymers[148,149,153–159] and in each component in miscible polymer blends.[160,162–164] Backbone reorientation can be quantitatively determined as a function of temperature for each component in a blend near the T_g. Many of these studies involve the observation of ^{2}H nuclei in powdered, deuterated homopolymers and polymer blends exploiting ^{2}H quadruple coupling of the ^{2}H nucleus with the electric field gradient of the C—^{2}H bond.[148,149,153–155,157–160,164] In this experiment, the resonance frequency depends on the reorientation angle of the C—^{2}H bond or the threefold axis of a rotating methyl group in isotopically labeled polymers relative to the magnetic field of the spectrometer. ^{13}C and ^{31}P 2D solid-state exchange experiments exploiting the chemical shift tensor (discussed in Section 6.5.2) have also been successful in this type of study.[156,162,163,165]

6.5.1.1 ^{2}H 2D solid-state exchange

The ^{2}H 2D solid-state experiment for the study of powdered, deuterated polymers initially used a three-pulse sequence. It recently has been modified to the four-pulse sequence shown in

Figure 26. The fourth pulse was added during the detection period to overcome receiver dead-time.[149] This effectively results in the addition of two more time periods (Δs) during the detection period. A five-pulse sequence has also been used.[160]

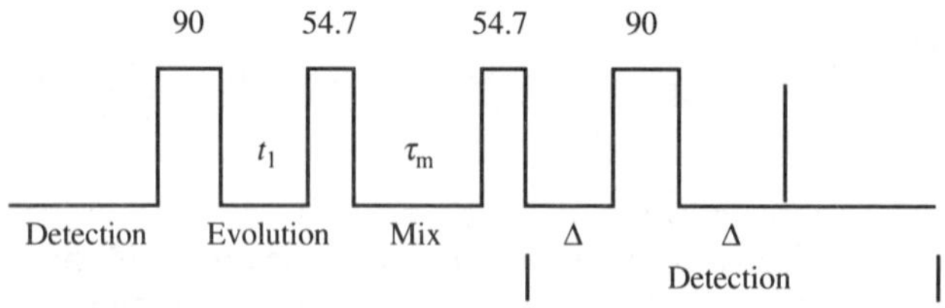

Figure 26 ^{2}H 2D solid-state exchange pulse sequence.

The first pulse in the sequence establishes an initial frequency during the evolution period (reflecting the initial orientation). The mixing time τ_m allows time for molecular reorientation to take place. If $\tau_m > \tau_c$, many C—^{2}H bonds undergo reorientation. The frequency at this point is different from the initial frequency because of molecular reorientation. The new frequency is detected during the detection period. Cross peaks in this spectrum reflect the changes in frequency that have taken place as a result of changes in molecular orientation. As in previous 2D experiments, t_1 is incrementally increased, generating a 2D exchange spectrum.

The angular dependence of the NMR frequency is given by equation (1)[120,166]

$$F = F_0 + 1/2\,\delta(3\cos^2\theta - 1) \tag{1}$$

where F_0 is the Larmor frequency, θ is the orientational angle that the field gradient tensor (FGT) axis (C—^{2}H bond direction) makes with the static magnetic field, and δ is a measure of the anisotropic quadrupolar interaction (for nuclei such as ^{2}H with a spin number of 1). A condition which is met is that the electric field gradient of the ^{2}H must be essentially uniaxial about the C—^{2}H axis.[148,149,154,160]

Since the spin number of ^{2}H is 1, reorientation of a C—^{2}H bond results in two transitions and four spectral peaks. Peaks occur at $(F_1, F_2) = [F^{\pm}(\theta_e)F^{\pm}(\theta_d)]$, where θ_e and θ_d are the orientational angles during the evolution and detection times, respectively. These peaks are symmetrically located around the diagonal. These signals can be very broad and inhomogeneous and cover the entire frequency range because of the numerous reorientations that take place in many polymers.[148,149,154,160]

In amorphous polymers, those polymer chains not involved in reorientation appear as a Pake pattern along the diagonal. The overall cross peak map of reorientation frequencies shows the type and extent of reorientation that has taken place in the polymer chains. By systematically varying the τ_m, all of the reorientation angle distributions (RADs) can be observed. The dynamic process can be monitored in real time and different motional processes can be identified and characterized. The upper limit of τ_m is the T_1 spin–lattice relaxation time.[148,149,154,160]

In polymers in which the all of the C—^{2}H bond directions change by the same reorientation angle θ, *e.g.* as the result of a discreet jump, an ellipse forms in the 2D spectrum. The reorientation angle θ in this case is calculated by equation (2)[132,148]

$$|\tan\theta| = b/a \tag{2}$$

where a and b are the principle axes of the ellipse, parallel and perpendicular to the diagonal of the spectrum. Evaluation of 2D exchange spectra of diffusive motions by small angles near the diagonal is more complex. These motions result in the superimposition of the spectra of an infinite number of reorientation angles. A broad distribution of reorientation angles results. These data can be best fit using a log Gaussian distribution of correlation times.[132,148,149,154,160]

Distinguishing orientational exchange due to molecular motion from that due to spin diffusion is generally not a problem in ^{2}H 2D exchange experiments because spin diffusion between deuterons is slow and ineffiicient.

The ^{2}H 2D exchange experiment has been applied to the study of ultraslow rotational dynamics in deuterated atactic PP (aPP) and deuterated isotactic PP (iPP).[148] Evolution of the dynamic process for the aPP at 270 K (well above the T_g of 253 K) is shown in Figure 27. Both diffusive motion by small angles of parts of the chains (broadening of the pattern near the diagonal) and large-angle reorientation of the other chain segments (ridges parallel to the frequency axes) have occurred. Figure 28 shows the reduction in RAD resulting from a decrease in temperature from 264 to 258 K. Changes in the spectra and in the RADs dramatically demonstrate the slowing down of the motional process. At 258 K, slightly above T_g, only rotational diffusive motions at small reorientation angles with a distribution of correlation times were observed. These data fit a log Gaussian distribution of correlation times and followed WLF behavior over 11 orders of magnitude (10^{-10}–10^1 s).

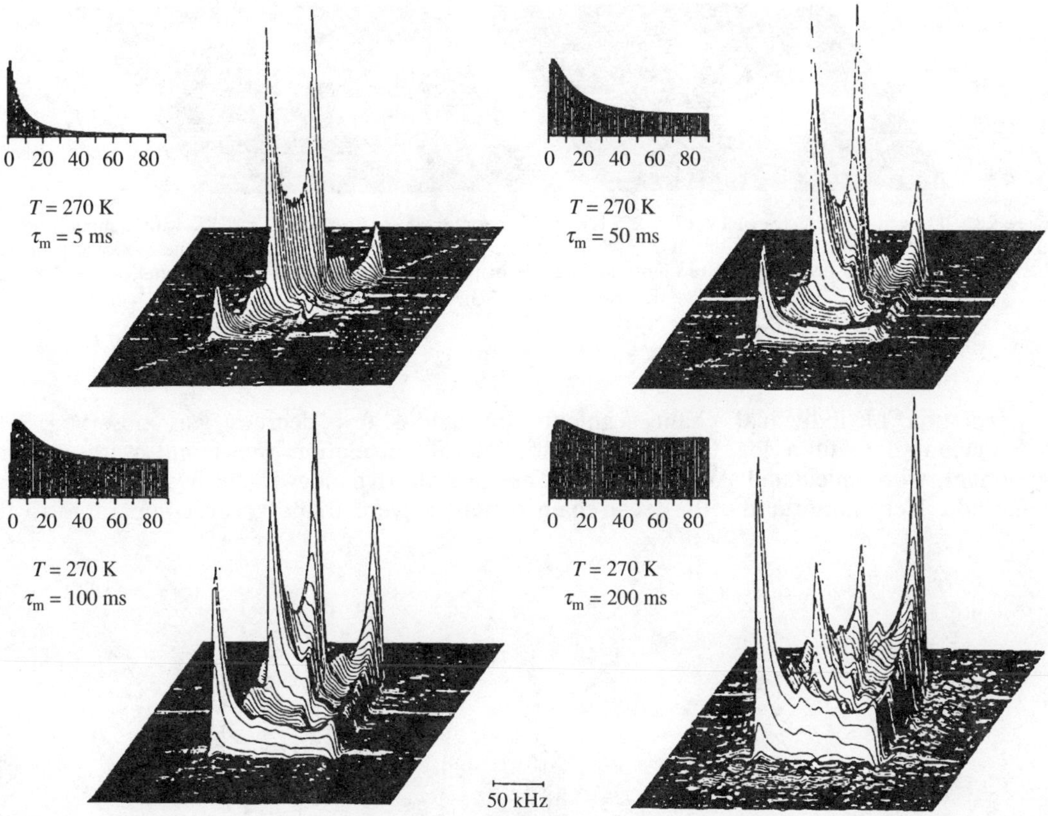

Figure 27 2D exchange spectra of aPP (T_g = 253 K) at T = 270 K with varying mixing time τ_m. The evolution of the dynamic process can be followed by inspection of the spectra or the reorientational angle distributions (reproduced by permission of the American Chemical Society from Schaefer *et al.*[148]).

Molecular motion in the deuterated iPP was totally different. Chain motions in the crystalline region (387 K) involved a limited number of helical jumps caused by 120° rotations around the 3_1 helix axis (Figure 29). A reorientational angle of 113° was calculated, which was consistent with this observation. The different behavior between aPP and iPP was attributed to different conformation and packing in the polymers.[148]

Helical jump motions have also been observed in the crystalline regions of poly(oxymethylene)[153] and PVF_2.[154] The chains rotated primarily via translation of one repeat unit in each case. A study of polyethylene has also shown that, at temperatures above the T_g of the amorphous region but below the T_m of the crystalline region, this motional process can lead to chain diffusion between crystalline and amorphous regions.[155]

The 2D exchange experiment has also been applied to the study of PS-d_3 (T_g = 373 K).[149,157] The results were very similar to the behavior observed in aPP. At 391 K (well above the T_g), both diffusive motion by small angles and large-angle reorientation were observed. At 375 K,

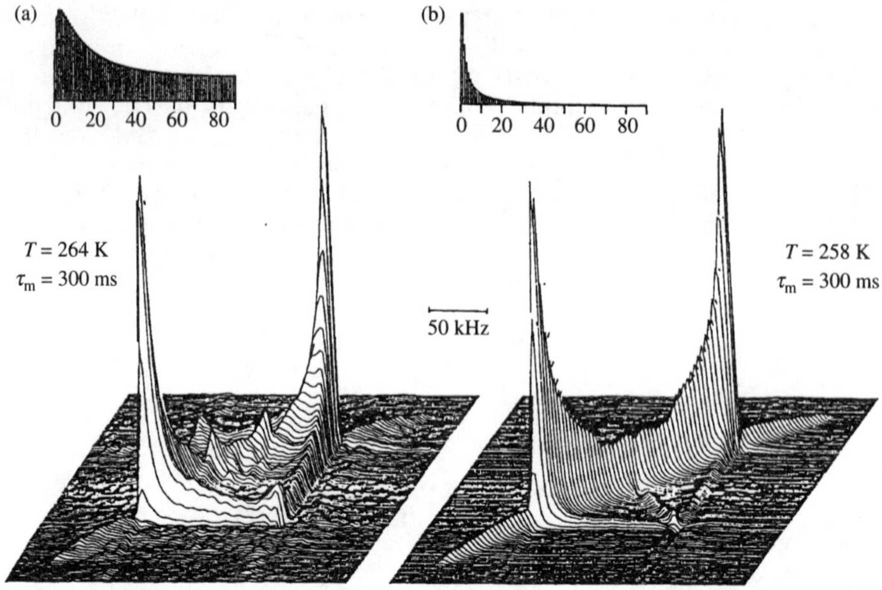

Figure 28 2D exchange spectra of PP ($T_g = 253$ K). (a) Experimental spectrum at $T = 264$ K with a mixing time of $\tau_m = 300$ ms. (b) Experimental spectrum at $T = 258$ K with $\tau_m = 300$ ms. The slowing down of the motional process is clearly visible from the spectra or the reorientational angle distributions (reproduced by permission of the American Chemical Society from Schaefer *et al.*[148]).

only rotation of individual chain segments through a few degrees was observed. The experimental data fit a log Gaussian distribution of correlation times and were in good agreement with calculated WLF values. The overall dynamic range was 11 orders of magnitude. Very importantly, the local chain dynamics were found to be comparable to the

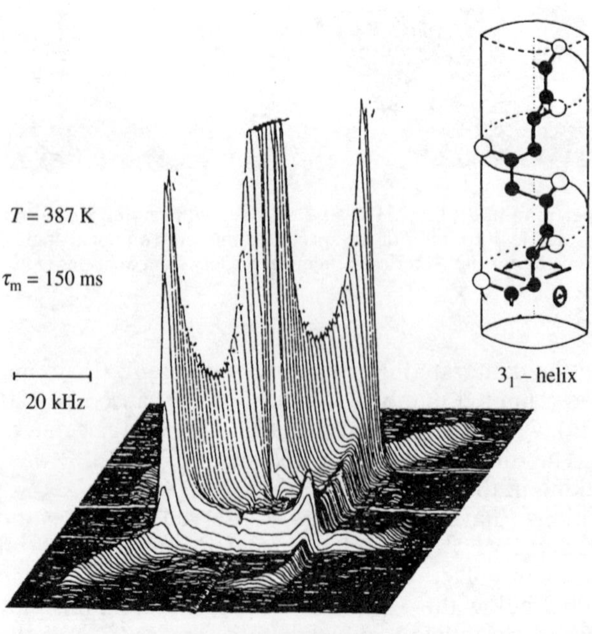

Figure 29 2D exchange spectrum of iPP. Experimental spectrum at $T = 387$ K with a mixing time $\tau_m = 150$ ms. The inset shows the 3_1 helix of iPP. The elliptical ridges in the 2D plane correspond to a reorientational angle of 113°. This value results from discrete jumps in 120° steps about the helix axis (reproduced by permission of the American Chemical Society from Schaefer *et al.*[148]).

macroscopic viscoelastic properties of this homopolymer, allowing a direct comparison of these data.

While the initial static studies of aPP and PS failed to show the expected *trans–gauche* transitions as would be manifested by sharp rotational angles near the melt, more recent 2D solid-state ^{13}C MAS exchange studies (discussed in Section 6.5.2) resulted in observations of ultraslow *trans–gauche* conformational transitions (10^{-3}–1 s) involving the aPP methylene group.[156] Their correlation times also had WLF behavior. The data showed that conformational transitions occur on the same timescale as rotational motions and must be accompanied by significant angular displacement. This supported the hypothesis that conformational changes can arise as a result of relaxation processes involving 'longer' chain units as opposed to strictly localized motions about a fixed C—C axis.

^2H 2D exchange NMR study of ultraslow reorientation of the phenylene groups in phenylene-deuterated bisphenol A polycarbonate (PC-d_4 and PC-d_8) under tensile stress has provided information about the geometry of these flip motions.[158] At temperatures below 273 K, magnetization exchange resulting from slow motions was observed in the unstretched polymer with a τ_m of 100 ms. The exchange increased with decreased temperature, with a maximum at 233 K and $\tau_m = 500$ ms. From the 2D pattern a mean reorientational angle of 120°, corresponding to a 180° flip of the phenylene ring about the 1,4 axis, was calculated. A distribution of flip angles was observed. Application of stress up to 50 MPa (3% strain) from 273 to 253 K resulted in an enlarged distribution of flip angles around 180° (*ca.* twice that of the unstretched state). It was concluded that tensile stress reduces the angular restriction of the slow phenylene flip motion (due to increased free volume of 1% at 3% strain).

2D and 3D techniques have been combined in the study of β-relaxation dynamics in PMMA.[159] ^{13}C 2D exchange NMR (Section 6.5.2) analysis of the PMMA carbonyl and ^2H 2D exchange NMR analysis of the PMMA methoxy group of ^{13}C- and ^2H-enriched polymers showed that the O—C—O plane of *ca.* 50% of the side groups undergo 180° ($\pm 20°$) flips at 330 K. The flips are accompanied by a random rotation around the local chain axis, with a 20° root-mean-square amplitude. The authors hypothesized that this was due to the fact that the side group no longer fits into its original environment after the flip, resulting in an average displacement of *ca.* 12°. The remaining side groups undergo a rocking motion with *ca.* 7° amplitude. It was also concluded that rotation of the methoxy methyl group around the O—CH_3 bond occurs, but rotation of the methoxy group around the C—OCH_3 bond does not.

^2H 2D exchange experiments have also been successfully applied to reorientation dynamic studies of each component in 1,4-polyisoprene (PIR)/1,2-polybutadiene (PBD) miscible blends.[160] In separate experiments, one component in each blend was deuterated. A significant difference in the reorientation of the two components was observed in 50/50 mole ratio blends. Mean correlation times (τ_c) of 0.025 and 0.55 were obtained for the dPIR/PBD and PIR/dPBD blends, respectively, while both had distribution widths (σ) of 3 decades. Compared to the homopolymers, the distributions of correlation times of the copolymers were much broader in the copolymer spectra (Figure 30). The experimental data fit a simple log Gaussian distribution of correlation times with no systematic deviation. The authors concluded that the observed broad T_g arose from both a wide distribution of segmental motional rates for each species and from intrinsic differences in the motional rate between the two species. The data suggested that the two components underwent distinct T_g values, *i.e.* dynamic heterogeneity in a miscible blend. These heterogeneities were explained via local compositional variations in addition to intrinsic differences in chain mobilities.[160]

6.5.1.2 ^{13}C and ^{31}P 2D solid-state exchange

Frequency–molecular motion correlations exploiting ^{13}C and ^{31}P chemical shift anisotropy interactions can be made using ^{13}C and ^{31}P solid-state exchange experiments. The pulse sequence is shown in Figure 31.[148–151,161,163]

Cross polarization in this sequence (Figure 31) allows magnetization exchange between the abundant ^1H nuclei and the much less abundant ^{13}C nuclei. The mixing pulse allows time for molecular reorientation to occur. Changes in orientation result in cross peaks in the 2D spectrum, just as in the ^2H experiment. Since each of these nuclei has a spin number of $\frac{1}{2}$, only one transition takes place and one set of cross peaks per exchange occur. Portions of chains which do not undergo reorientation resonate in a Pake diagram along the diagonal.

Orientational exchange from this experiment can result from either molecular motion or

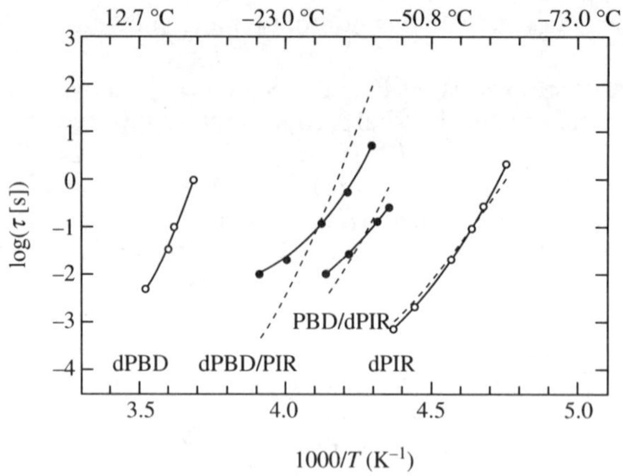

Figure 30 Log mean correlation times as a function of temperature τ_c (T). The dashed lines are obtained by using WLF parameters estimated by mechanical measurement well above T_g. The solid curves are obtained by restricted WLF fits, constraining the T_g alone for pure PI and constraining $\tau_g \equiv \tau_c(T_g) = 0.32$ s for both species in the blend (reproduced by permission of the American Chemical Society from Chung *et al.*[160]).

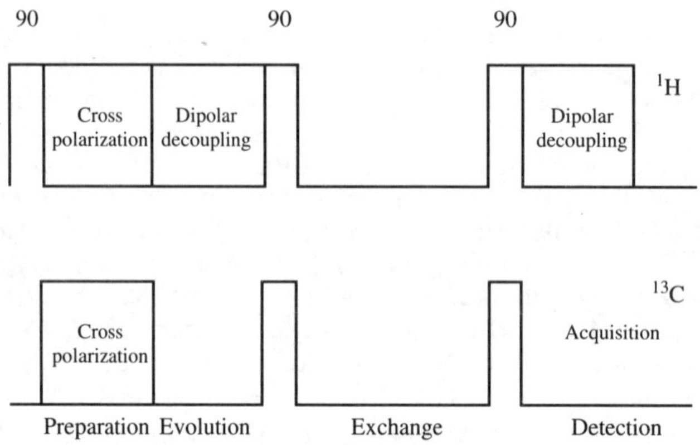

Figure 31 ^{13}C 2D solid-state exchange pulse sequence.

spin diffusion. These can be distinguished via temperature dependence characteristics, as molecular motion is strongly temperature dependent while spin diffusion is weakly temperature dependent. Other means to make this distinction are through concentration variance of ^{13}C-labeled nuclei and through variation of τ_m (shorter τ_m values emphasize exchange due to molecular motion).[162]

The Williams–Watts stretched exponential correlation function shown in equation (3) has been used to interpret quantitatively the 2D patterns characteristic of Brownian rotational diffusion resulting from this experiment. In this application, this function is actually written as a weighted sum of exponential correlation functions[137]

$$\phi(t) = \exp[-(t/\tau)^{\alpha}] \qquad 0 < \alpha < 1 \tag{3}$$

where τ is the center correlation time and α is a measure of the breadth of the distribution of correlation times.[137]

This method has recently complemented ^{2}H 2D exchange experiments in the study of reorientational motions near the T_g in the individual components of 75/25 weight ratio miscible blends of PS-d_3–^{13}C-labeled PXE[162,163] and PS-d_3–PXE.[164] The T_g of the PXE homopolymer is 210 °C, the T_g of PS is 100 °C, and the T_g of the polymer blend is approximately 128 °C. In one study,[162,163] the motion of methyl ^{13}C-labeled PXE repeat units was monitored near the T_g as a function of temperature and τ_m using ^{13}C 2D solid-state exchange techniques. Deuteration of the PS served to emphasize the PXE spectral resonances. In the other study,[164] the reorientational motion of the PS-d_3 polymer repeat units was monitored, also as a function of temperature and mixing time, near the T_g using ^2H 2D solid-state exchange NMR techniques.

Rotational Brownian diffusion was observed in both the PS-d_3 and PXE components, with a broadening of correlation times (considerably broader than for typical single-component polymeric glasses). Onset of motion was observed well below (10 °C) the T_g. Both the ^{13}C and ^2H spectral patterns were best simulated by a bimodal distribution, with each mode having a distribution of correlation times similar to one another. The relative weighting of the modes corresponded to local concentration fluctuations of the nearest neighbors as calculated from a lattice model with an assumption of random mixing.[163,167–170] This model counted nearest-neighbor contacts where the lattice sites are occupied by either PS or PXE units. It was concluded that the observed dynamic heterogeneity (as indicated by the distribution of correlation times) for each component was determined largely by the surroundings and not as much by the intrinsic dynamics of the repeat unit. In essence, the mobility of each unit influences the mobilities of the surrounding units. Therefore, a PXE unit surrounded by PS units (lower T_g, more mobility) is more mobile than a PXE unit surrounded by other PXE units and only slightly less mobile than a PS unit surrounded by other PS units. The data demonstrated that, although the polymer chains are intimately mixed and have a much narrowed T_g relative to the individual components, heterogeneities produced by the surroundings persist in the blend.[162–164]

^{31}P 2D solid-state exchange NMR has been used to study the dynamics of cross-links in tris(4-isocyanatophenyl) thiophosphate interconnected with telechelic PPG. The motion of the cross-links at temperatures near the T_g (*ca.* 30 °C) was observed to be isotropic Brownian reorientational diffusion. τ_c was 1.0 s at 30°C and 0.05 s at 35 °C.[165]

As Spiess has pointed out, a limitation of the 2D methods for these types of studies is that the orientation of molecules is measured only twice. Therefore, this does not provide information on the path that a unit follows in rotating from one orientation to another. 3D and even 4D techniques have been developed to provide additional information.

6.6 MOLECULAR ORDER

6.6.1 ROtor-Synchronized 2D ^{13}C/CP MAS NMR (ROS-MAS)

A relatively new 2D NMR experiment has been shown to be an excellent tool for the study of molecular order in partially oriented polymers. This is the 2D (ROS-MAS) NMR experiment developed by Harbison *et al.* where magnetization exchange is driven by rotor spinning.[171,172]

In the ROS-MAS experiment, an ordered polymer is spun with the order axis not parallel to the rotor axis. A series of ROS-MAS NMR spectra of the oriented polymer is obtained with the rotor located at different positions at the beginning of each experiment. Because of the dependance of resonance frequency on orientation, the peaks in each spectrum depend on the position of the rotor when the spectrum is obtained. By synchronizing data acquisition with the rotor position, a sideband pattern in two dimensions is obtained for each spectral resonance. This leads to phase and intensity modulations of spectral resonances and their sidebands. Sidebands in the F_1 dimension of a 2D ROS-MAS spectrum are modulated by the orientation order of the polymer in the F_2 dimension ($F_2 = 0$ is a typical CP MAS spectrum). Crystalline samples have side bands in the second dimension reflecting their order. The intensities of these side bands vary with the degree of orientational order for each resonance. Amorphous polymers will not have side bands in the second dimension because it does not matter where the experiment starts. It is possible to use the 2D ROS-MAS technique to study polymers containing different morphological components (amorphous and crystalline) and to provide information on the orientational order (or lack of it) in each component.[132,134,173–175,178]

2D ROS-MAS techniques have been successfully applied to the study of molecular orientation in undrawn and hot-drawn Nylon G films and fibers with draw ratios (λ values) of 2.5, 3.5, 4.5, and 5.5.[175] Sidebands for the carbonyl and the superimposed aliphatic C-4 and C-5 carbons (Figure 32) were used to determine the second and fourth moments of the orientational distribution function (ODF). The undrawn sample had no order. An aggregate model was used and modified to correlate order parameter $\langle p_2 \rangle$ and λ data in the drawn samples. By way of the modification, it was shown that the order parameters varied with draw temperature and thermal history. $\langle p_2 \rangle_{max}$ values, the maximum order parameters obtained by complete alignment of the disordered chains, of *ca.* 0.83 for the films and *ca.* 0.69 for the fibers were obtained. The reduction of $\langle p_2 \rangle_{max}$ from 1.00 (complete order) to these values represents the disorder in the chains.

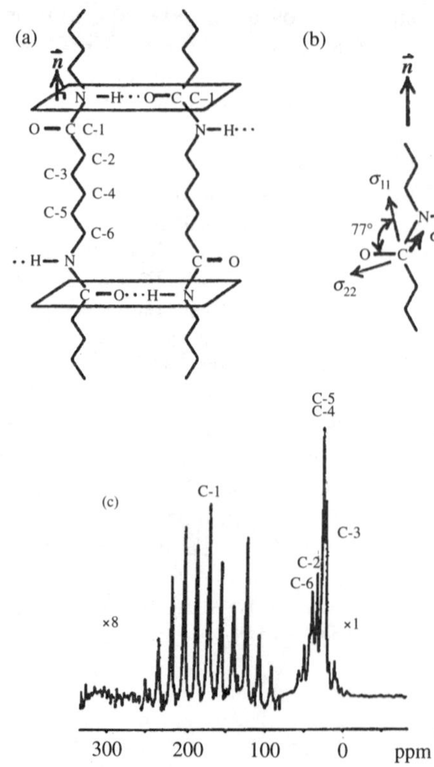

Figure 32 (a) Sketch of nylon 6 layer structure.[176] (*b*) Carbonyl carbon tensor orientation; σ_{33} is perpendicular to the O═C—N plane.[177] (c) ^{13}C 1D CP/MAS spectrum of nylon 6 in ppm scale referenced to TMS (reproduced by permission of Wiley from Tzov *et al.*[175]).

The data demonstrated that, as λ increased, $\langle p_2 \rangle$ increased to a maximum at $\lambda = 5.5$. Crystalline orientation increased from $\lambda = 1$ to $\lambda = 3$ (crystalline content increased from 15% to 45%) and then plateaued. The amorphous orientation did not plateau at $\lambda = 3$, but increased monotonically from that point. Crystalline form changed from primarily γ to primarliy α when λ increased from 1 to 3. Tensile strength and modulus increased primarily after the crystalline content was maximized ($\lambda > 3$) and were, therefore, correlated with changes in amorphous and not crystalline orientation.[175]

Orientation in two ethylene fibers gel-spun with ultrahigh molecular weight (UHMPE, $>10^6\ g\ mol^{-1}$, draw ratio > 40), one melt spun ($10^5\ g\ mol^{-1}$), and one hot-drawn (draw ratio = 3) has also been studied by 2D ROS-MAS.[178,179] Analysis of the spectral sideband patterns was done mathematically (procedure of Hentschel *et al.*[180]) to determine ODFs assuming transverse isotropy for uniaxially spun fibers.[181]

Two crystalline (C_1, C_2) regions and one amorphous (A) region were observed in each spectrum. Order parameters were determined mathematically. $\beta_{1/2}$, the width at half-height (the angle where the distribution function is half its maximum) values for the largest crystalline component, C_2, were determined: 18° and 21° for the UHMPE polymers, 31° for the MW = 10^5 polymer, and 50° for the hot-drawn polymer.[178,179]

The authors concluded that since the C_1 and C_2 components in the UHMPE polymers were relatively narrow, observed broadness of the ODFs was due to long-range disorder in the macrofibers and not disorder of the bundle of polymer chains within the microfibrils. Also, since the spectral peaks for the moderate molecular weight and hot-drawn polymers were much broader, disorder must be present in these polymers in both the microfibers and macrofibrils.

Order differences in the C_2 major component in all four polymers and in the C_1 versus the C_2 components of one UHMPE fiber were observed.[178,179]

The 2D ROS-MAS technique has also been applied to the determination of the orientational distribution in drawn amorphous PMMA up to a draw ratio of 3.[182] The carbonyl carbons were used to determine the orientational distribution as this was the only nonoverlapped spectral region. The ODF was obtained from lineshape analysis of ^{13}C static spectra. 2D ROS-MAS data were used to obtain an order parameter $\langle p_2 \rangle$ of 0.22 for this polymer at $\lambda = 3$ and 100 °C. This agreed perfectly with X-ray data from a comparable sample.

6.7 REFERENCES

1. F. A. Bovey, 'High Resolution NMR of Macromolecules', Academic Press, New York, 1972.
2. D. M. Grant and E. G. Paul, *J. Am. Chem. Soc.*, 1964, **86,** 2984.
3. A. E. Tonelli, *Macromolecules,* 1978, **11,** 565.
4. K. Nagayama, K. Wüthrich and R. R. Ernst, *Biochem. Biophys. Res. Commun.*, 1979, **90,** 305.
5. A. Kumar, G. Wagner, R. R. Ernst and K. Wüthrich, *Biochem. Biophys. Res. Commun.*, 1980, **96,** 1156.
6. K. Nagayama and K. Wüthrich, *Eur. J. Biochem.*, 1981, **114,** 365.
7. K. Wüthrich, G. Wider, G. Wagner and W. Braun, *J. Mol. Biol.*, 1982, **155,** 311.
8. G. Wagner and K. Wüthrich, *J. Mol. Biol.*, 1982, **155,** 347.
9. A. S. Arseniev, G. Wider, F. J. Joubert and K. Wüthrich, *J. Mol. Biol.*, 1982, **159,** 323.
10. G. Wagner and K. Wüthrich, *J. Mol. Biol.*, 1982, **160,** 343.
11. Y. Kobayashi, Y. Kyogoku, J. Emura and S. Sakakibara, *Biopolymers,* 1981, **20,** 2021.
12. H. Kessler, R. Schuck and R. Siegmeier, *J. Am. Chem. Soc.*, 1982, **104,** 4486.
13. D. Davoust, B. Bodo, S. Rebuffat and N. Platzer, *Biochem. Biophys. Res. Commun.*, 1983, **116,** 1.
14. L. S. Kan, D. M. Cheng and J. Cadet, *J. Magn. Reson.*, 1982, **48,** 86.
15. J. Feigon, J. M. Wright, W. Leupin, W. A. Denny and D. R. Kearns, *J. Am. Chem. Soc.*, 1982, **104,** 5540.
16. D. Y. Gagnaire, F. R. Taravel and M. R. Vignon, *Macromolecules,* 1982, **15,** 126.
17. P. A. Mirau and F. A. Bovey, *Macromolecules,* 1986, **19,** 210.
18. M. D. Bruch, F. A. Bovey and R. E. Cais, *Macromolecules,* 1984, **17,** 2547.
19. F. C. Schilling, F. A. Bovey, M. D. Bruch and S. A. Kozlowski, *Macromolecules,* 1985, **18,** 1418.
20. G. C. Levy and G. L. Nelson, 'Carbon-13 Nuclear Magnetic Resonance for Organic Chemists', Wiley, New York, 1972.
21. M. D. Bruch, 'NMR Spectroscopy Techniques', ed. C. Dybowski and R. L. Lichter, Marcel Dekker, New York, 1987.
22. J. Jeener, Ampère International Summer School, Basko Polje, Yugoslavia, 1971.
23. W. P. Aue, E. Bartholdi and R. R. Ernst, *J. Chem. Phys.*, 1976, **64,** 2229.
24. G. Bodenhausen, R. Freeman, R. Niedermeyer and D. L. Turner, *J. Magn. Reson.*, 1977, **26,** 133.
25. P. Bachmann, W. P. Aue, L. Muller and R. R. Ernst, *J. Magn. Reson.*, 1977, **28,** 29.
26. R. C. Ferguson and D. W. Ovenall, *Polym. Prepr., Am. Chem. Soc., Div. Polym. Chem.*, 1984, **25** (1), 340.
27. F. A. Bovey and P. A. Mirau, *Makromol. Chem., Macromol. Symp.*, 1990, **34,** 1.
28. R. E. Cais and J. M. Kometani, *Macromolecules,* 1985, **18,** 1354.
29. J. W. Park, M. Lee, M.-H. Lee, J. W. Liu, S. D. Kim, J. Y. Chang and S. B. Rhee, *Macromolecules,* 1994, **27,** 3459.
30. G. R. Gippert and L. R. Brown, *Polym. Bull.*, 1984, **11,** 585.
31. A. Bax and G. A. Morris, *J. Magn. Reson.*, 1981, **42,** 501.
32. A. Bax, *J. Magn. Reson.*, 1983, **53,** 517.
33. V. Rutar, *J. Magn. Reson.*, 1984, **58,** 306.
34. E. G. Melby and J. M. Castro, in 'Comprehensive Polymer Science', ed. G. Allen and J. C. Bevington, Pergamon Press, Oxford, 1989, vol. 7, chap. 3.
35. J. C. Randall, *J. Polym. Sci., Polym. Phys. Ed.*, 1973, **11,** 275.
36. A. D. H. Clague, J. A. M. van Broekhoven, L. P. Blaauw, *Macromolecules,* 1974, **7,** 348.
37. D. D. Werstler, *Rubber Chem. Technol.*, 1981, **8,** 114.
38. J. T. Gerig, *Macromolecules,* 1983, **16,** 1797.
39. H. N. Cheng and G. H. Lee, *Polym. Bull.*, 1985, **13,** 549.
40. H. N. Cheng and G. H. Lee, *Polym. Bull.*, 1984, **12,** 463.
41. H. N. Cheng, *Anal. Chem.*, 1982, **54,** 1828.
42. H. N. Cheng, in 'Transition Metal Catalyzed Polymerizations. Alkenes and Dienes', ed. R. P. Quirk, Harwood, New York, 1983, p. 617.
43. H. N. Cheng, *J. Polym. Sci., Polym. Phys. Ed.*, 1983, **21,** 573.
44. T. Kawamura, N. Toshima and K. Matsuzaki, *Macromol. Chem. Phys.*, 1994, **195,** 2677.
45. T. Kawamura, N. Toshima and K. Matsuzaki, *Makromol. Chem., Rapid Commun.*, 1993, **14,** 719.
46. G. Moad, E. Rizzardo, D. H. Solomon, S. R. Johns and R. I. Willing, *Macromolecules,* 1986, **19,** 2494.
47. P. A. Mirau and F. A. Bovey, *Macromolecules,* 1986, **19,** 210.
48. F. A. Bovey, 'Chain Structure and Conformation of Macromolecules', Academic Press, New York, 1982.
49. F. A. Bovey, 'High Resolution NMR of Macromolecules', Academic Press, New York, 1972.
50. C. J. Carman, A. R. Tarpley and J. H. Goldstein, *J. Am. Chem. Soc.*, 1971, **93,** 2864.
51. F. Heatley and F. A. Bovey, *Macromolecules,* 1969, **2,** 241.
52. M. W. Crowther, N. M. Szeverenyi and G. C. Levy, *Macromolecules,* 1986, **19,** 1333.

53. N. I. Liu, S. N. Tong and J. L. Koenig, *J. Appl. Polym. Sci.*, 1980, **25,** 2205.
54. Y. Inoue, K. Koyama, R. Chujo and A. Nishioka, *J. Polym. Sci., Polym. Lett. Ed.*, 1973, **11,** 55.
55. N. Nakayama, A. Aoki and T. Hayashi, *Macromolecules,* 1994, **27,** 63.
56. C. Chang, D. D. Muccio and T. St. Pierre, *Macromolecules,* 1985, **18,** 2334.
57. G. R. Quinting and R. Cai, *Macromolecules,* 1994, **27,** 6301.
58. P. A. Berger, E. E. Remsen, G. C. Leo and D. J. David, *Macromolecules,* 1991, **24,** 2189.
59. A. Bax and S. Subramanian, *J. Magn. Reson.*, 1986, **67,** 565.
60. A. A. Maudsley, L. Müller and R. R. Ernst, *J. Magn. Reson.*, 1977, **28,** 463.
61. L. Dong, D. J. T. Hill, J. H. O'Donnell and A. K. Whittaker, *Macromolecules,* 1994, **27,** 1830.
62. A. E. Tonelli 'NMR Spectroscopy and Polymer Microstructure', VCH, New York, 1989.
63. N. K. de Vries, *Polym. Bull.*, 1991, **26,** 451.
64. A. Bax and S. Subramanian, *J. Magn. Reson.*, 1986, **67,** 565.
65. A. Bax and M. F. J. Summers, *J. Am. Chem. Soc.*, 1986, **108,** 2093.
66. K. Beshah, *Makromol. Chem.*, 1993, **194,** 3311.
67. J. Schaefer, *Macromolecules,* 1971, **4,** 98.
68. D. C. Watts, *J. Biomed. Mater. Res.*, 1979, **13,** 423.
69. H. J. Prosser, C. P. Richards and A. D. Wilson, *J. Biomed. Mater. Res.*, 1982, **16,** 431.
70. C. Chang, D. D. Muccio and T. St. Pierre, *Macromolecules,* 1985, **18,** 2154.
71. K. Beshah, *Macromolecules,* 1992, **25,** 5597.
72. C. J. Carman, R. A. Harrington and C. E. Wilkes, *Macromolecules,* 1977, **10,** 536.
73. H. N. Cheng and G. H. Lee, *Macromolecules,* 1988, **21,** 3164.
74. J. K. M. Sanders and B. K. Hunter, 'Modern NMR Spectroscopy. A Guide for Chemists', Oxford University Press, New York, 1987.
75. A. Bax, R. Freeman, T. A. Frenkiel and M. H. Levitt, *J. Magn. Reson.*, 1981, **43,** 478.
76. D. R. Hensley, S. D. Goodrich, H. J. Harwood and P. L. Rinaldi, *Macromolecules,* 1994, **27,** 2351.
77. N. Nakayama, A. Aoki and T. Hayashi, *Macromolecules,* 1994, **27,** 63.
78. K. Hikichi, T. Hirai, M. Ikura, K. Higuchi, K. Eguchi and M. Ohuchi, *Polym. J.*, 1987, **19,** 1317.
79. T. Hayashi, Y. Inoue, R. Chûjô and T. Asakura, *Polym. J.*, 1988, **20,** 895.
80. I. Ando, Y. Yamanobe and T. Asakura, *Prog. NMR Spectrosc.*, 1990, **22,** 349.
81. T. Asakura, K. Hirano, M. Demura and K. Kato, *Makromol. Chem., Rapid Commun.*, 1991, **12,** 215.
82. T. Asakura, N. Nakayama, M. Demura and A. Asano, *Macromolecules,* 1992, **25,** 4876.
83. A. Aoki, T. Hayashi and T. Asakura, *Macromolecules,* 1992, **25,** 155.
84. T. Asakura and N. Nakayama, *Polymer,* 1992, **33,** 650.
85. F. C. Schilling and A. E. Tonelli, *Macromolecules,* 1980, **13,** 270.
86. H. N. Cheng and M. A. Bennett, *Makromol. Chem.*, 1987, **188,** 135.
87. L. P. Linderman and J. Q. Adams, *Anal. Chem.*, 1971, **43,** 1245.
88. W. P. Aue, J. Karhan and R. R. Ernst, *J. Chem. Phys.*, 1976, **64,** 4226.
89. K. Nagayama, P. Bachmann, K. Wüthrich and R. R. Ernst, *J. Magn. Reson.*, 1978, **31,** 133.
90. M. D. Bruch, F. A. Bovey, R. E. Cais and J. H. Noggle, *Macromolecules,* 1985, **18,** 1253.
91. S. Macura and L. R. Brown, *J. Magn. Reson.*, 1983, **53,** 529.
92. M. Minagawa, K. Ute, T. Kitayama and K. Hatada, *Macromolecules,* 1994, **27,** 369.
93. K. R. Lindfors, S. Pan and P. Dreyfuss, *Macromolecules,* 1993, **26,** 2919.
94. K. E. Steller, *ACS Symp. Ser.*, 1974, **6,** 136.
95. A. Dworak, *Makromol. Chem., Rapid Commun.*, 1985, **6,** 665.
96. H. N. Cheng and D. A. Smith, *J. Appl. Polym. Sci.*, 1987, **34,** 909.
97. K. Hatada, K. Ute, K. Tanaka, M. Imanari and N. Fujii, *Polym. J.*, 1987, **19,** 425.
98. G. Bodenhausen, R. Freeman and D. Turner, *J. Magn. Reson.*, 1977, **27,** 511.
99. R. Freeman and J. Keeler, *J. Magn. Reson.*, 1981, **43,** 484.
100. J. Jeener, B. H. Meier, P. Bachmann and R. R. Ernst, *J. Chem. Phys.*, 1979, **71,** 4546.
101. S. Macura and R. R. Ernst, *Mol. Phys.*, 1980, **41,** 95.
102. S. Macura, Y. Huang, D. Suter and R. R. Ernst, *J. Magn. Reson.*, 1981, **43,** 259.
103. D. J. States, R. A. Haberkorn and D. J. Ruben, *J. Magn. Reson.*, 1982, **48,** 286.
104. M. D. Bruch and F. A. Bovey, *Macromolecules,* 1984, **17,** 978.
105. X. Lu, Y. Wang and X. Wu, *Polymer,* 1994, **35,** 2315.
106. X. Lu, Y. Wang and X. Wu, *Polymer,* 1993, **34,** 56.
107. A. Bax, *Annu. Rev. Biochem.*, 1989, **58,** 223.
108. L. W. Kelts, C. J. T. Landry and D. M. Teegarden, *Macromolecules,* 1993, **26,** 2941.
109. R. R. Ernst, 'Nuclear Magnetic Resonance Fourier Transform Spectroscopy', Reprint from *Les Prix Nobel,* 1991.
110. S. Heffner and P. Mirau, *Polym. Prepr., Am. Chem. Soc., Div. Polym. Chem.*, 1992, **33** (2), 549.
111. P. A. Mirau and G. Kogler, *Polym. Prepr., Am. Chem. Soc., Div. Polym. Chem.*, 1992, **33** (2), 578.
112. P. Caravatti, P. Neuenschwander and R. R. Ernst, *Macromolecules,* 1985, **18,** 119.
113. A. Bielecki, D. P. Burum, D. M. Rice and F. E. Karasz, *Macromolecules,* 1991, **24,** 4820.
114. S. Li, D. M. Rice and F. E. Karasz, *Macromolecules,* 1994, **27,** 2211.
115. P. Caravatti, L. Braunschweiler and R. R. Ernst, *Chem. Phys. Lett.*, 1983, **100,** 305.
116. D. P. Burum and A. Bielecki, *J. Magn. Reson.*, 1991, **94,** 645.
117. D. P. Burum, M. Linder and R. R. Ernst, *J. Magn. Reson.*, 1981, **44,** 173.
118. P. Caravatti, G. Bodenhausen and R. R. Ernst, *Chem. Phys. Lett.*, 1982, **89,** 363.
119. J. E. Roberts, S. Vega and R. G. Griffin, *J. Am. Chem. Soc.*, 1984, **106,** 2506.
120. A. Abragam, 'The Principles of Nuclear Magnetism', Clarendon Press, Oxford, 1961.
121. N. Bloembergen, *Physica,* 1949, **15,** 386.
122. D. W. McCall and D. C. Douglass, *Polymer,* 1963, **4,** 433.
123. D. C. Douglass and G. P. Jones, *J. Chem. Phys.*, 1966, **45,** 956.
124. V. J. McBrierty, D. C. Douglass and T. K. Kwei, *Macromolecules,* 1978, **11,** 1265.

125. V. J. McBrierty, *Faraday Discuss. Chem. Soc.,* 1979, **68,** 78.
126. M. Goldman and L. Shen, *Phys. Rev.,* 1966, **144,** 321.
127. S. Kaplan, *Macromolecules,* 1993, **26,** 1060.
128. J. L. White and P. A. Mirau, *Macromolecules,* 1994, **27,** 1648.
129. W. T. Dixon, J. Schaefer, M. D. Sefcik, E. O. Stejskal and R. A. McKay, *J. Magn. Reson.,* 1982, **49,** 341.
130. S. A. Heffner and P. A. Mirau, *Macromolecules,* 1994, **27,** 7283.
131. K. Schmidt-Rohr and H. W. Spiess, 'Multidimensional Solid-State NMR and Polymers', Academic Press, New York, 1994.
132. B. Blümich and H. W. Spiess, *Angew. Chem., Int. Ed. Engl.,* 1988, **27,** 1655.
133. J. D. Ferry, 'Viscoelastic Properties of Polymers', Wiley, New York, 1980.
134. M. M. Maricq and J. S. Waugh, *J. Chem. Phys.,* 1979, **70,** 3300.
135. A. K. Roy, A. A. Jones and P. T. Inglefield, *Macromolecules,* 1986, **19,** 1356.
136. A. A. Jones, in 'High Resolution NMR Spectroscopy of Synthetic Polymers in Bulk', ed. R. A. Komoroski, VCH, Deerfield Beach, FL, 1986, p. 247.
137. A. A. Jones, J. F. O'Gara, P. T. Inglefield, J. T. Bendler, A. F. Yee and K. L. Ngai, *Macromolecules,* 1983, **16,** 658.
138. L. C. Dickinson, P. Morganelli, C. W. Chu, Z. Petrovic, W. J. MacKnight and J. C. W. Chien, *Macromolecules,* 1988, **21,** 338.
139. L. C. Dickinson, J. F. Shi and J. C. W. Chien, *Macromolecules,* 1992, **25,** 1224.
140. R. Voelkel, *Angew. Chem., Int. Ed. Engl.,* 1988, **27,** 1468.
141. H. W. Spiess, *Adv. Polym. Sci.,* 1985, **66,** 23.
142. J. Schaefer, E. O. Stejskal, R. A. McKay and W. T. Dixon, *Macromolecules,* 1984, **17,** 1479.
143. H. W. Spiess, *Colloid Polym. Sci.,* 1983, **261,** 193.
144. L. W. Jelinski, 'High Resolution NMR Spectroscopy of Synthetic Polymers in Bulk', ed. R. A. Komoroski, VCH, Deerfield Beach, FL, 1986, p. 335.
145. H. W. Spiess, *Adv. Polym. Sci.,* 1985, **66,** 23.
146. A. Schmidt, S. O. Smith, D. P. Raleigh, J. E. Roberts, R. G. Griffin and S. Vega, *J. Chem. Phys.,* 1988, **85,** 4248.
147. M. S. Greenfield, A. D. Ronemus, R. L. Vold, R. R. Vold, P. D. Ellis and T. E. Raidy, *J. Magn. Reson.,* 1987, **72,** 89.
148. D. Schaefer, H. W. Spiess, U. W. Suter and W. W. Fleming, *Macromolecules,* 1990, **23,** 3431.
149. S. Kaufmann, S. Wefing, D. Schaefer and H. W. Spiess, *J. Chem. Phys.,* 1990, **93,** 197.
150. S. Wefing and H. W. Spiess, *J. Chem. Phys.,* 1988, **89,** 1219.
151. S. Wefing, S. Kaufmann and H. W. Spiess, *J. Chem. Phys.,* 1988, **89,** 1234.
152. C. Schmidt, S. Wefing, B. Blümich and H. W. Spiess, *Chem. Phys. Lett.,* 1986, **130,** 84.
153. A. Hagemeyer, K. Schmidt-Rohr and H. W. Spiess, *Adv. Magn. Reson.,* 1989, **13,** 85.
154. J. Hirschinger, D. Schaefer, H. W. Spiess and A. J. Lovinger, *Macromolecules,* 1991, **24,** 2428.
155. K. Schmidt-Rohr and H. W. Spiess, *Macromolecules,* 1991, **24,** 5288.
156. K. Zemke, B. F. Chmelka, K. Schmidt-Rohr and H. W. Spiess, *Macromolecules,* 1991, **24,** 6874.
157. U. Pschorn, E. Rössler, H. Sillescu, S. Kaufmann, D. Schaefer and H. W. Spiess, *Macromolecules,* 1991, **24,** 398.
158. M. T. Hansen, B. Blümich, C. Boeffel, H. W. Spiess, L. Morbitzer and A. Zembrod, *Macromolecules,* 1991, **25,** 5542.
159. K. Schmidt-Rohr, A. S. Kulik, H. W. Beckham, A. Ohlemacher, U. Pawelzik, C. Boeffel and H. W. Spiess, *Macromolecules,* 1994, **27,** 4733.
160. G.-C. Chung, J. A. Kornfield and S. D. Smith, *Macromolecules,* 1994, **27,** 964.
161. H. T. Edzes and J. P. C. Bernards, *J. Am. Chem. Soc.,* 1984, **106,** 1515.
162. P. T. Inglefield, A. A. Jones, Y. H. Chin and C. Zhang, *Polym. Prepr., Am. Chem. Soc., Div. Polym. Chem.,* 1992, **33** (1), 76.
163. Y. H. Chin, C. Zhang, P. Wang, P. T. Inglefield, A. A. Jones, R. P. Kambour, J. T. Bendleer and D. M. White, *Macromolecules,* 1992, **25,** 3031.
164. Y. H. Chin, P. T. Inglefield and A. A. Jones, *Macromolecules,* 1993, **26,** 5372.
165. J.-F. Shi, L. C. Dickinson, W. J. MacKnight, J. C. W. Chien, C. Zhang, Y. Liu, Y. H. Chin, A. A. Jones and P. T. Inglefield, *Macromolecules,* 1993, **26,** 1008.
166. H. W. Spiess, in 'NMR Basic Principles and Progress', Springer, Heidelberg, 1978, vol. 15, p. 55.
167. C. Zheng, P. Wang, A. A. Jones and P. T. Inglefield, *Macromolecules,* 1991, **24,** 338.
168. A. A. Jones, P. T. Inglefield, Y. Liu, A. K. Roy and B. J. Cauley, *J. Non-Cryst. Solids,* 1991, **131,** 556.
169. B. J. Cauley, C. Cipriani, K. Ellis, A. K. Roy, A. A. Jones, P. T. Inglefield, B. J. McKinley and R. P. Kambour, *Macromolecules*, 1991, **24,** 403.
170. Y. Liu, A. K. Roy, A. A. Jones, P. T. Inglefield and P. Ogden, *Macromolecules,* 1990, **23,** 968.
171. G. S. Harbison and H. W. Spiess, *Chem. Phys. Lett.,* 1986, **124,** 128.
172. G. S. Harbinson, V.-D. Vogt and H. W. Spiess, *J. Chem. Phys.,* 1987, **86,** 1206.
173. M. Maricq and J. S. Waugh, *Chem. Phys. Lett.,* 1977, **47,** 327.
174. J. S. Waugh, M. M. Maricq and R. Cantor, *J. Magn. Reson.,* 1978, **29,** 183.
175. D. L. Tzou, H. W. Spiess and S. Curran, *J. Polym. Sci., Part B: Polym. Phys.,* 1994, **32,** 1521.
176. D. R. Holmes, C. W. Bunn and D. J. Smith, *J. Polym. Sci.,* 1955, **17,** 159.
177. R. E. Stark, L. W. Jelinski, D. J. Ruben, D. A. Torchia and R. G. Griffin, *J. Magn. Reson.,* 1983, **55,** 266.
178. D. L. Tzou, T.-H. Huang, P. Desai and A. S. Abhiraman, *J. Polym. Sci., Part B: Polym. Phys.,* 1993, **31,** 1005.
179. D. L. Tzou, P. Desai, A. S. Abhiraman and T.-H. Huang, *J. Polym. Sci., Polym. Phys. Ed.,* 1991, **29,** 49.
180. R. Hentschel, J. Schlitter, H. Sillescu and H. W. Spiess, *J. Chem. Phys.,* 1978, **68,** 56.
181. V. J. McBrierty, in 'Comprehensive Polymer Science', ed. G. Allen and J. C. Bevington, Pergamon Press, Oxford, 1989, vol. 1, chap. 19.
182. A. S. Kulik and H. W. Spiess, *Macromol. Chem. Phys.,* 1994, **195,** 1755.

7

Polymer Crystallization: Fundamentals of Structure and Crystal Growth of Flexible Chains

ANDREW KELLER and GERHARD GOLDBECK-WOOD
University of Bristol, UK

7.1 INTRODUCTION

Polymer crystallization is a subject of many facets where, in addition, each of them lends itself to a variety of modes of presentation. For an article of the present restricted length, clearly choices have to be made which will first be defined. The basis of the present review will be the experimental observation itself. As there exists a large multitude of observational facts, again choices have to be made. Here we contend that the multitudinous manifestations of crystallization can be reduced to a comparatively few which are both characteristic and basic, from which most of the rest derive. Such we shall select to serve as core material for the present review. Having placed what we consider basic observations on the map, together with such inferences as are arising from them directly, we shall proceed to present theories set up to account for them, including attempted assessments of their comparative merits. In all the above, special emphasis will be given to the open endedness of the subject and we shall conclude the review with pointers towards new departures as seen from our present perspectives.

Of the large variety of crystalline polymers, we shall confine ourselves to such as are (1) perfectly periodic, (2) highly flexible and (3) chemically aspecific. By restriction (1) we exclude, amongst others, copolymers; by that of (2), rigid rod-like molecules; and by that of (3), polar and hydrogen bonding group containing molecules. The adoption of such severe restrictions is in the service of bringing down the subject to its basic essentials, namely to approximate, as closely as possible, the idealized physicist's model of a flexible string of identical beads without specific interactions between them. Amongst available polymers, the closest approximation to such an idealization is polyethylene (PE) which, accordingly, will feature most prominently throughout this article.

A further restriction will be the mode of crystallization. Only crystallization from stationary unstrained systems will be considered and this as induced thermally: in specific terms, the crystallization of chain molecules from the random coil state on cooling. This means that the whole specifically polymeric phenomenon of strain and chain extension induced or promoted crystallization will fall outside the orbit of the present review, as will also crystallization concurrent with polymerization (nascent crystallization).

In proceeding along the above declared path we shall obviously touch upon material featuring in the four previous articles on crystallization in the present series of 'Comprehensive Polymer Science';[1–4] reference to these will be made where appropriate. Yet the present paper is not to be regarded either as an extension, or an updating or a replacement of these preceding papers, but as a different, self-contained treatment standing on its own. Even so, to avoid repetition we shall frequently quote from the illustrative material already presented specially by Vaughan and Bassett.[2]

7.2 THERMODYNAMIC AND KINETIC FRAMEWORK

Crystallization, like any phase transition, is subject to the laws of thermodynamics, in this case expressed in the form of phase diagrams. In the first place these tell us whether, under a specific circumstance, crystals can exist or not. Then, within a given existence regime, whether crystallization will take place and how fast is determined by the kinetics of the process.

The kinetics not only tells us how fast crystallization progresses, but also how far it proceeds ultimately. In polymers, specifically, we never reach the state of ultimate stability; there is nearly always amorphous like material left; hence the much referred to 'amorphous–crystalline' ratio. Further, even amorphous content apart, we never attain the stablest crystal state, that of infinite (in practice macroscopic) size, as the crystals usually remain in the submicroscopic size regime, which in the above sense is a metastable state. The texture, hence morphology, is therefore an expression of this size-determined metastability.

The same applies to liquid–liquid (L–L) phase separation, which in the case of polymers never goes to completeness, *e.g.* to the formation of two separate macroscopic liquid layers, and the phase separated liquids remain in some kind of dispersion. Thus the otherwise amorphous system acquires a morphology of its own, which may be termed 'phase morphology' as distinct from the 'crystal morphology' (our main concern in what follows). If then one or more of the phase-segregated liquid components crystallizes, a crystal morphology will arise within the previously established phase morphology leading to a whole morphological hierarchy wihin such a sample. Such situations can arise in solution crystallization and become particularly important for blends, a line which we shall not pursue here further. (For this subject area including illustrations, see reviews by Keller.[5,6])

The systems of concern here will be single component, *i.e.* melts, and two-component (solutions). For a fully miscible solution system the phase diagram (L–C, *i.e.* liquid–crystal) will be of the kind as Figure 1, with a phase line defining the temperature (T) below which the crystals are stable at a given concentration (c). The abscissa at $c = 1$ is the phase diagram of the melt, here a temperature scale with a single crystal–liquid coexistence point, the melting point T_m, beneath which the crystals are stable.

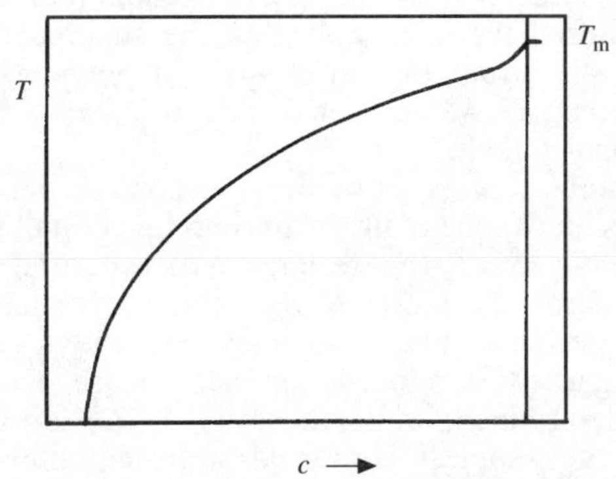

Figure 1 Schematic liquid–solid (crystal) phase diagram; temperature (T) versus concentration (c), for a fully miscible polymer solution.

If a crystallizable polymer in the melt state is not fully miscible with the solvent, L–L separation will take place in a certain c range. In such a case the phase diagram will be compounded from those of L–L and L–C in a way represented schematically by Figure 2. The significant feature here is an invariant line A*–B*–C*, above which the system will behave as L–L or L–C according to the c range. At and above this invariant line, and at c between A* and B*, L–L and L–C phase segregation will occur consecutively, leading to a phase and crystal morphology, respectively; for illustrations, we refer to Keller.[5,6]

Figures 1 and 2 are only two examples of phase diagram types relevant to the present review; there are many others. One important point to make here is that in the sense of equilibrium thermodynamics they refer to infinite phase size, which is the state of ultimate stability. As polymer crystals cannot be obtained in macroscopic sizes, the phase lines in such phase diagrams need to be obtained by extrapolation from small crystal sizes. Such extrapolated values for melting and dissolution temperatures are not undisputed, and in any case their determination has only been attempted in a very few cases. Accordingly, most existing phase

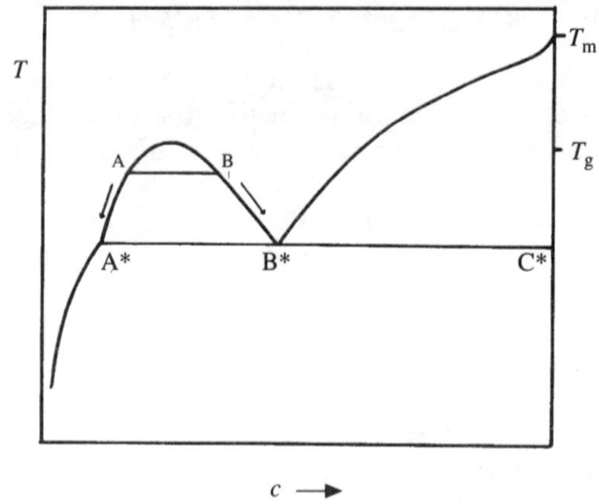

Figure 2 Schematic phase diagram; temperature (T) versus concentration (c), showing liquid–liquid phase separation plus crystallization from solution.

diagrams involving the crystal phase are approximate, semi-quantitative at their best. They are usually adequate for practical purposes but seldom for more in-depth considerations. The salient exception is PE as a single component melt with a theoretical melting point, *i.e.* for infinite size, T_m^0, of 145 °C, a value largely agreed upon.

For the above stated reasons, phase diagrams can only serve as guides as to what is theoretically possible. In practice, polymer crystals, which are always small and imperfect, hence less than ideally stable, can only exist well below the phase line (in terms of T), a point to which we give special attention in Section 7.11.2.2 as the source of new developments.

In addition to thermodynamic conditions, there are kinetic reasons for the L–C conversion not taking place at the phase line itself. Namely, crystallization is unrealizably slow there and for practicable rates we need to go to much lower temperatures; hence we need substantial supercooling ($\Delta T = T_m^0 - T_c$ where T_c is the crystallization temperature). While a common feature in many other material types, in polymers the supercoolings required are specially large. It is essential to realize from the onset that, in polymers, all crystallization studies conducted are at substantial ΔT values for which the phase diagram provides only an unrealizable upper theoretical T limit.

The ΔT required for crystallization is set by the rates, or vice versa (for a given polymer, ΔT determines the rate R). R is an exponentially increasing function of ΔT (see later) but is limited by vitrification at a sufficiently low T_c (hence large ΔT), where at the glass transition (T_g) all crystallization stops. In general, therefore, R versus T_c curves have a maximum (Figure 3), where $T_m^0 - T_g$ is the temperature interval of crystallizability. Here the temperature scale of the ordinate can be regarded as a one-component phase diagram with two points, one (T_m^0) thermodynamically determined, while the second (T_g) is not a thermodynamic but a kinetic quantity. This is the simplest combined representation of stability and rates. It would be desirable to carry this over to multicomponent phase diagrams, say Figures 1 and 2 for a full description of crystallizability (needed, for example, to predict where a particular cooling pathway would take the system), but we know of no example of this having been done. Also, in the bicomponent system, T_g would have a decisive influence, for the assessment of which curves of T_g vs. c would need inserting into the phase diagram. Here T_g arrests not only crystallization but also L–L phase separation, with particularly interesting consequences for the combined phase–crystal morphology, a point we cannot pursue here further (see Keller[5,6]).

Reverting to single component (melt) systems, it is important to appreciate the difference between fast and slowly crystallizing polymers. Fast crystallizing polymers, such as PE, cannot be cooled to T values even approaching the rate maximum before crystallization is complete even for the fastest cooling rates. In contrast, slowly crystallizing polymers (such as poly(ethylene terephthalate) (PET), poly(phenyl ether ether ketone) (PEEK) and isotactic polystyrene (i-PS)) can be cooled down to T_g without noticeable crystallization, and hence can be quenched amorphous. This is of consequence for the whole subject area surveyed here; namely since PE is the main model substance it confines investigations to the low ΔT side of the temperature interval of crystallizability, *i.e.* comparatively close below T_m^0. This also means that, rate studies apart, crystallization over much of the realizable ΔT range remains largely

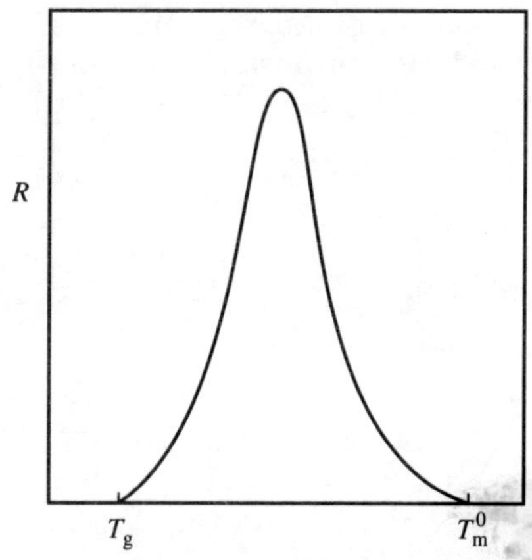

Figure 3 Schematic representation of the rate of crystallization as a function of temperature as commonly pertaining for polymer melts.

unexplored territory. This needs to be kept in mind as at such very high supercoolings, which can amount to 100–200 K, mechanisms which are even qualitatively different may come into play. In this respect, for example, in a quite recent study on PET a complete change in the nucleation mechanism was indicated.[7,8]

As a further important point, the requirement (or possibility) of large ΔT for crystallization can result in reaching down to the stability region of altogether new metastable phases. Such can be crystal polymorphs or even liquid crystal phases, if such exist. It is in the nature of metastability (*i.e.* phases of stability which are intermediate between the liquid and the crystal of ultimate stability) that they develop faster for the identical supercoolings than the phase of ultimate stability, or what amounts to the same, require lower ΔT for the attainment of practicable rates. Consequently, once their existence regime is entered they will dominate the phase transition, altering altogether the process and the resulting product. More of this will be discussed in Section 7.11.2.2. A particularly interesting possibility arises in bicomponents, *i.e.* solvent–polymer systems in the case of slowly crystallizing polymers. Here, such high ΔT values can be reached that, by latest indications, the chain conformation of the dissolved isolated chains seems to be affected, possibly through a coil–helix transformation. In this case the latter is metastable with respect to the crystal, which is the state of maximum stability, while stable with respect to the dissolved coil molecule. Crystallization, when it finally occurs, takes place not from the random coil but from the transformed chain molecule in its new conformation (see also Section 7.13.10). This possibility, indicated by ongoing works on syndiotactic PS[9] and poly(methyl methacrylate) (PMMA),[10] would be the ultimate manifestation of metastability, here on a molecular level, with potentially far reaching consequences (see Keller[6]).

Having brought all these issues arising at very high ΔT to attention, the material to be reviewed will be largely confined to crystallizations at low ΔT.

7.3 HISTORICAL

The existence of crystallinity in appropriate crystallizable polymers has long been established by X-ray diffraction.[11] From X-ray diffraction patterns the crystal structure could then be determined by traditional crystallographic procedures,[12] the example for polyethylene (PE) being shown by Figure 4. This should all be too familiar to require further elaboration. From the point of view of the present subject, the essential point is that here the repeating motif, the unit cell, comprises only a small portion of a full chain, either a single chemical repeat or a small multiple of it, leaving the question open as to what is the trajectory of the full chain within a crystalline polymer. There is no single answer to the latter question, neither is there a single uniquely decisive generally applicable method for attacking this problem which, on the molecular level, is central to the subject of the present review. Both the complexity and also the richness of the topic are the consequence of the open endedness of the above issue, and so are the innumerable controversies, and also surprises, which punctuate the development of the subject.

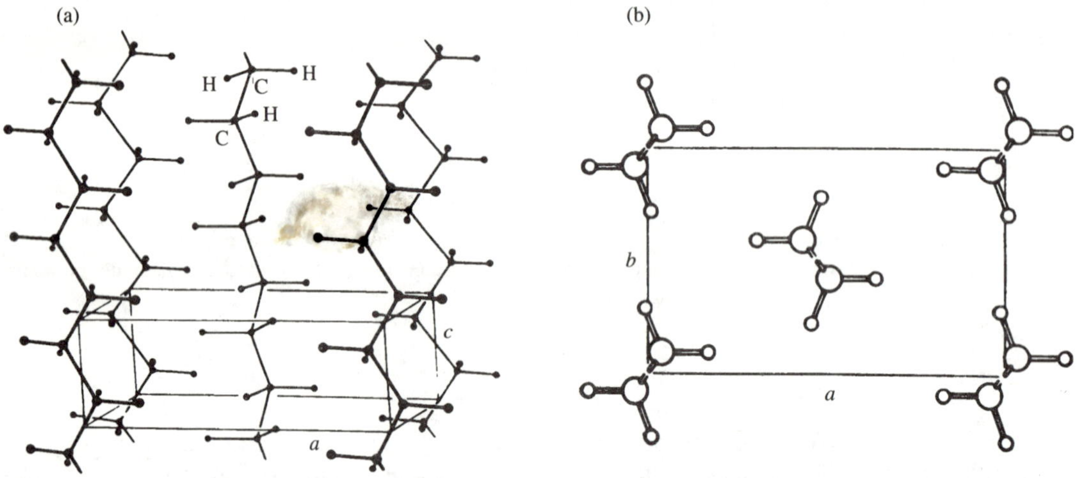

Figure 4 The crystal structure of polyethylene: (a) a general view; (b) projection along *c* (chain direction) (after Bunn[13]).

Historically, on the basis of the assumed organizational capabilities of random coils, chain molecules were first believed to crystallize by statistical parallelization of chain portions to form small interconnected micelles (the 'fringed micelle' model: see Figure 1 in Vaughan and Bassett[2]) without any identifiable faceting or any other identifiable morphology. Events, however, proved otherwise by revealing a previously unsuspected wealth of morphological features which form the present basis for the understanding, such as we have, of polymer crystallization. It is worth reflecting upon the failure of *a priori* predictions in this subject; in the light of it, subsequent theoretical efforts had to content themselves with constructing models to account for what had been revealed by direct observation in the first place.

Even still within the era of the fringed micelle there was an increasing number of pointers to the fact that this conception in itself is inadequate, and that the subject is much richer than implied by it. Morphologically such were the spherulites identifiable by polarizing light microscopy.[2] Further, by X-ray diffraction, discrete reflections were observed at small angles corresponding to structure elements in the range of 10–30 nm. None of these fitted the picture at the time but they fell readily in place in the light of subsequent developments. Spherulites were much of the subject matter of Vaughan and Bassett[2] and will here only be briefly touched upon. The small-angle X-ray scattering effect (SAXS), however, will be central to what is to follow. They were found to arise from lamellae,[14] which have emerged as the universal features of flexible crystalline polymers, the subject to be surveyed in the present review.

7.4 THE LAMELLAR MORPHOLOGY: BASIC FEATURES

7.4.1 The Isolated Regular Lamella; Recognition of Chain Folding

As stated at the end of Section 7.3, the lamellar morphology is the basic and universal feature of a crystalline polymer and is also the source of the ubiquitously observed SAXS reflections (see later). Lamellae arise in their clearest form (as first discovered) when crystallized from solution, in which case they may appear as isolated suspended entities that are readily accessible for microscopic examinations (Figure 5).[15–17] Lamellae also arise on crystallization from the melt, in which case their identification requires more elaborate sample-preparation methods (etching, staining, etc.; see Bassett[3]). These methods took about two decades to develop with, in retrospect, much futile argument about the existence, or otherwise, of such lamellar structures taking place in the intervening period.

(a) (b)

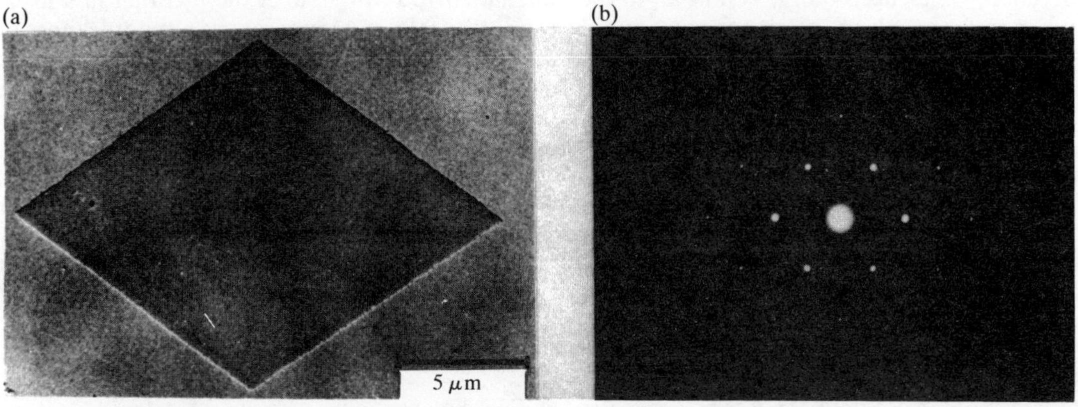

Figure 5 (a) Electron micrograph of a monolayer single crystal of polyethylene crystallized from solution, and (b) corresponding electron diffraction pattern (in identical orientation) (after Keller[18]).

The methodology for revealing lamellae within the melt crystallized interior, together with numerous examples, is described in the article by Basset.[3] Here we only illustrate it with a recent example (Figure 6) of an isolated layer of a specially well formed crystal as extracted from the interior of a bulk sample.[19]

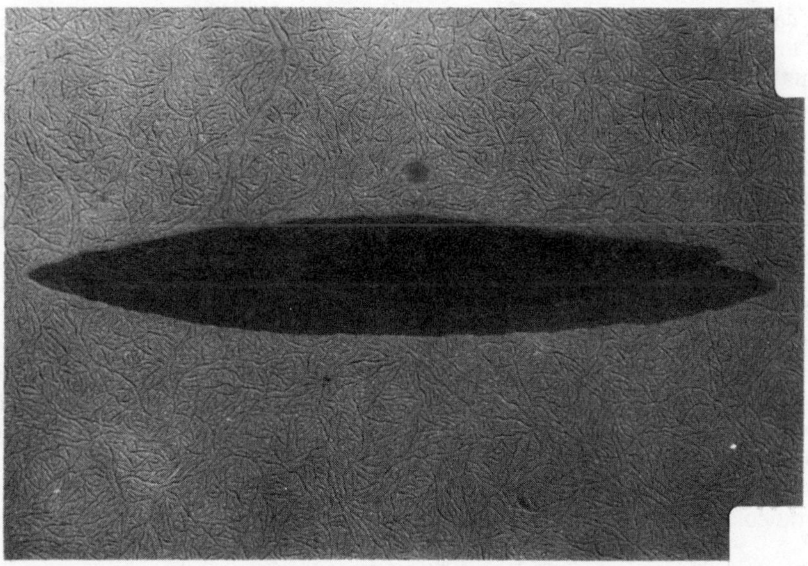

Figure 6 Electron micrograph of a crystal of polyethylene formed from the melt, extracted from the solidified sample (after Toda and Keller[19]).

Taking the isolated layer such as in Figure 5(a) as a starting point, we see that such layers are usually bounded by lateral facets indexable crystallographically, a feature on which all interpretations have relied up to the mid-1980s. In some instances the lateral faces can also be curved (see Figure 6), a feature which, after it had been duly taken note of, had a major influence on theoretical developments over the last few years (see Section 7.13.5).

However, of prime importance has always been the thinness and uniformity of the lamellae themselves, at least as formed under 'normal' conditions, which in the case of polyethylene means at atmospheric pressure. By 'thin' we mean a few tens of nanometres, as in cases like

Figure 5. Electron diffraction patterns (Figure 5(b)) of individual layers such as Figure 5(a) revealed that the molecules are either perpendicular or, in appropriate other cases, at a specifiable large angle to the lamellar surface. The one in Figure 5(b) corresponds to the projection of the PE structure as represented by Figure 4(b), hence as viewed along the chain direction, which accordingly is normal to the lamellar surface. In view of the fact that in a usual polymer the length of the chains is many times (*i.e.* 10–100 times) greater than the thickness of the lamella, it follows, by the now familiar argument, that the chains must repeatedly fold back upon themselves so as to be accommodated by the lamella (Figure 7). In the light of this, the high degree of uniformity then must mean that crystal growth proceeds by lamellae growing laterally with unaltered thickness, which in turn then means constant fold length. Figure 7 is a schematic model of a chain folded crystal established on the above basis. For the history of this important recognition we refer to an article by Keller;[20] for summing up of all relevant earlier facts, still valid today and contained in abbreviated form by the present article, see the review by Keller.[21] (For other treatises of the broader subject area we refer to a review,[18] recent short summary,[22] recent review,[23] monographs,[24,25] textbooks,[26–27] one most recent,[28] and an incomparable collection of annotated micrographs.[29]

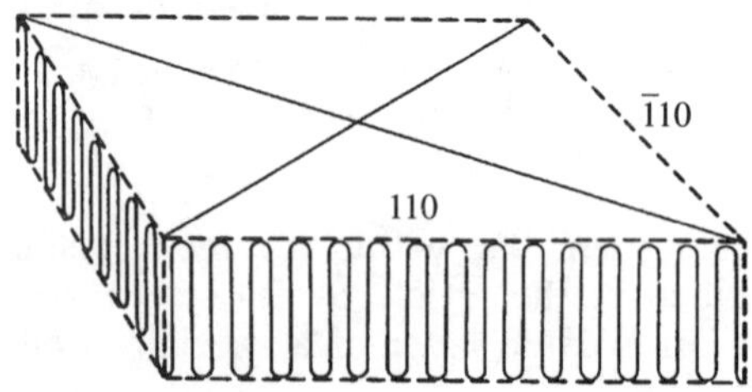

Figure 7 Schematic illustration of a chain folded crystal structure to account for the evidence of Figure 5.

It needs recalling that the above described findings went in the face of all existing ideas and expectations at the time. Long chain molecules, being intrinsically fibre forming, were expected to give fibrous entities with the chains in a more or less extended conformation, as opposed to the lamellae actually observed, with the chains folded up within them. The simplicity of the argument leading to chain folding should be particularly noted: it is totally straightforward, in fact common sense, and at the same time totally conclusive and irrefutable.

The last point needs to be kept in mind when coming to the numerous controversies surrounding the subject of chain folded crystals. Even if sometimes obscured by all the arguments, the basic concept of chain folded crystallization itself cannot be in any doubt; the arguable points, such as there are, relate to particular features of the chain folded structure (and not to chain folding itself), which from the global viewpoint, important as they may be in themselves, are rather matters of detail.

The above conclusion on chain folding is not confined to PE but has been extended to virtually all flexible chain polymers. The lateral habit types will vary from polymer to polymer (*e.g.* see hexagonal crystals of polyoxymethylene and square crystals of poly(4-methyl-1-pentene), Figures 8(a) and 8(b), respectively, of Vaughan and Bassett[2]), according to the intrinsic symmetry of the crystal lattice, but as long as there are distinct and thin lamellae with chains perpendicular (or closely so) to the lamellar surface the same argument for chain folding holds.

So far the argument has been based on isolated, well-formed lamellae. In most cases of a 'real life' polymer the lamellae are not in isolation and, in addition, may not be well formed. It is to this issue we turn next.

7.5 MULTILAYERS AND IRREGULAR LAMELLAE

7.5.1 Multilayer Crystals

In most usual situations the lamellar crystals are in the form of multilayer assemblies. In fact, in the course of growth, single lamellae 'multiply' via the familiar screw dislocation growth mechanism and branch out in various, including dendritic-type, growth patterns. These give

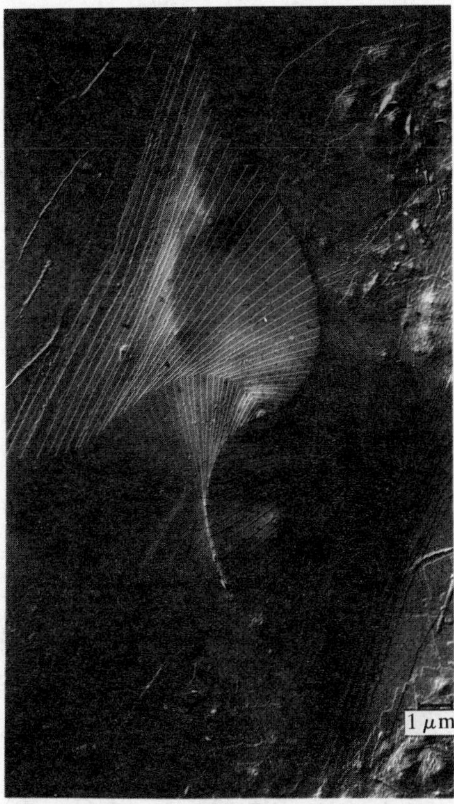

Figure 8 Multilayer solution-grown polyethylene crystal with regularly rotated terraces. Electron micrograph replica (after Keller[32]).

rise to further, larger scale assemblies, such as sheaves and ultimately spherulites, the light-optically identifiable structure elements of a usual crystalline thermoplastic. Such larger scale elements of the polymeric crystal structure hierarchy are important for the final properties of polymeric matter, and they raise also many still unclarified issues in crystal growth. The latter pertains not only to polymers but also to simpler substances which can often display identical or similar features. Such features, *e.g.* spherulitic structure and growth in particular, remain a more general and still unsolved issue in the wider field of crystal growth.[30,31] Having placed the issue of multilayers and its ultimate end product on the map, it will not concern us further at this place, particularly as some aspects of it have already received attention in the present series,[2] where also numerous illustrations, including that of spherulites, are displayed which will not be repeated here. Two examples nevertheless will be shown and commented on.

The first is Figure 8 which, in addition to its obvious aesthetic appeal, serves to convey the impression of the conspicuous self-organizational ability of macromolecules (in face of *a priori* predictions to the contrary), given the right conditions. This is just one of the many examples that can be quoted; for others we refer to the extensive pictorial material available elsewhere.[18,21–29]

The second illustration of multilayering, Figure 9, has a rather more explicit purpose in the service of the issue of the main theme of chain folding. Namely, in its original and most indisputable form the postulate of chain folding arose in connection with the isolated layer where the chain has no other possible option but to re-enter the same layer. The argument, it may seem, could lose some of its strength when the layers are stacked, as in that case the chain could then also traverse from one lamella to the next. Indeed, such interlamellar ties are being inferred through a variety of means (see later), but it is easy to see that if they become the dominant feature then not only the overall chain folded character of the crystals but also their lamellar nature on the thickness level of the fold length would become suppressed.

Observations on multilayers such as provided by Figure 9, and on numerous other analogous examples,[2,3] including solution grown multilayers,[21] where no special etching and/or extraction treatment is needed for seeing the relevant structural features, largely eliminate such a possibility. Namely, it shows that the layers in the multilayer crystals are not spatially contiguous but are splaying apart. Without entering the question of the possible origin of such a splay,[2] its mere existence asserts the individuality of the layers within the assembly, thus safeguarding the concept of chain folding. Even in the case when such a splay effect is not as conspicuous or not as readily assessable (*e.g.* in Figure 8 where the layers seem more contiguous), the distinctness of the layers on the 10 nm scale is brought out by the distinctness of the SAXS maxima on the corresponding length scale, reaffirming that chain folding is still at least a dominant feature throughout a representative portion of the preparation.

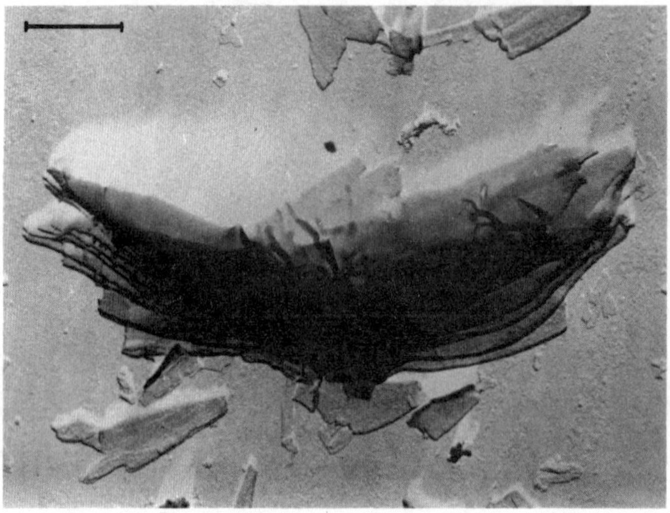

Figure 9 Splaying multilayer crystal of polyethylene crystallized from the melt and dispersed by fuming nitric acid digestion (after Keller and Sawada[33]).

7.5.2 Ill-formed Lamellae or No Lamellae

In many instances the lamellae are not as well and regularly formed as those under discussion so far. This feature can either be intrinsic to a particular polymer (as far as we know) or can be due to particular crystallization conditions in the case of polymers which under different circumstances may yield well-formed lamellae.

In the 'intrinsic' category belong polymers which display low degrees of crystallinity by conventional crystallinity determination. These are of two kinds: polymers which contain chemical irregularities, *e.g.* branched polyethylene or copolymers of polyethylene.[34] Important as such substances are, by our own self-declared choice they fall outside the scope of this review. The second kind, within the orbit of this article, are polymers with chemically regular chains which nevertheless, for various reasons (or for still unknown reasons), are incapable of attaining high degrees of crystallinity, the attainable levels of crystallinity being below 50% compared to 70–90% in polymers like PE. Examples are PET, PEEK, poly(phenylene ether ketone) (PEK) and poly(phenyl sulfide) (PPS).[35,36] Figure 10 is an example in the case of PEK. Its edges are jagged, its internal structure knobbly and its overall attainable size small compared to crystals of the regular lamellar category, such as PE. Yet the lamellar nature is apparent and by electron diffraction evidence, analogous so that in Figure 5(b), the chains must be in a folded configuration. Even so, because of the poor definition of the crystals, many of the considerations to follow in this review will not be applicable. A reason for this habit degeneration has been suggested:[37] it could be due to the fact that the chains in the corresponding polymers are not quite as flexible as in PE (due to aromatic groups in PEEK,

PEK, *etc.*), so that folding cannot be as sharp, from which the habit degeneration and also low crystallinity could follow. For further particulars at this place we need to refer to the original article;[37] see also Section 7.8 where we shall briefly return to this issue.

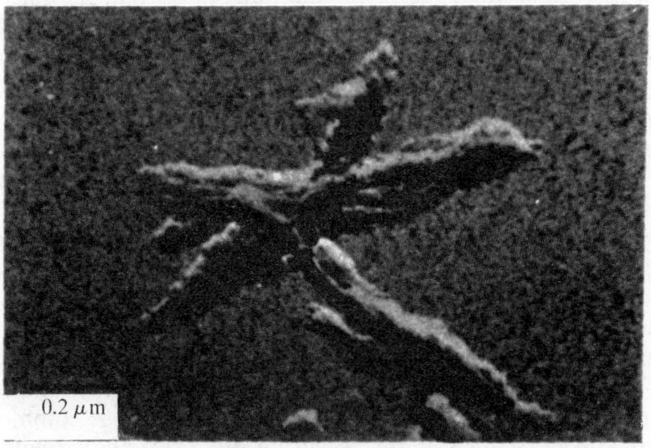

Figure 10 Solution grown crystal of PEK (poly(phenyl ether ketone)) in illustration of irregular degenerate lamellar habit (after Waddon *et al.*[35])

The 'non-intrinsic' category of irregular habit may arise when polymers, which are otherwise obtainable in regular lamellae, are crystallized very fast, say by sudden precipitation (from solution) or by quenching (from the melt). Altogether, this is not surprising and in keeping with general experience in the crystallization of any substance, not only polymers, namely that too rapid crystallization leads to irregular and malformed habits. In polymers the issue may, nevertheless, acquire additional significance. Namely, it leads to the question as to whether, in cases where lamellae are poorly defined, or even become indiscernible, other modes of crystallization may set in, altogether different from that of chain folding, *e.g.* a local coming together of chain segments into small bundles, thus harking back to the historical precedent of the fringed micelle. As in practice many thermoplastics are being used in the rapidly crystallized form (say as obtained by quenching), this issue continues to remain on the cards.

Having broadly outlined the wider landscape, we shall restrict ourselves to well-defined regular lamellae and to those mostly present in isolation, *i.e.* as monolayers, in the rest of this review. This is in keeping with the general tradition of crystal growth studies, where the best developed special cases are selected for closer examination in the conviction that they display the basic principles in their clearest form, which otherwise become overlaid and partly obscured in the more general 'real life' situation. We believe that many disputes in the field originate from the failure to realize the above declared distinction, which in our opinion pertains not only to crystal growth but to all experimental science.

Despite the above truism, it needs stating that, in polymers, imperfect crystallinity can acquire importance of its own, often regarded as intrinsic to the 'polymeric matter of state'. Namely, it finds expression in the amorphous material trapped between crystals and in the intimate and multiple connectedness of crystals through noncrystalline chain portions. In fact, in certain treatises such factors can even form the starting point.[38] In the present review, by focusing on the well-developed individual crystal lamella, we are taking the opposite route in the belief that this entity is the truest expression of the intrinsic organizational capabilities of long-chain molecules, from which other aspects follow. As already noted, at this stage it leads to crystal entities which have most of the features of traditional crystals, thus linking polymers to the general body of crystal studies, with the all-important additional feature of the underlying chain folding. We believe that this approach will eventually provide a basis also for the treatment and understanding of the more conventionally regarded 'polymeric matter of state' of a semicrystalline polymer in the sense defined above, a point to be taken up again in the Concluding Remarks (Section 7.14) of the present review.

7.6 THE CHARACTERISTICS OF LAMELLAE

7.6.1 The Principal Features

The lamellae are characterized by two main features, thickness and lateral habit, which will be taken in turn.

As implied by the foregoing, thickness is far the most important because it corresponds, or is simply related, to the fold length. The fold length is a molecular characteristic intimately connected with the process of crystallization and as such will receive much further attention in what follows. At this point we may take note of the quite unique situation with which polymers are presenting us, namely, here we have an intimate connection between external morphology (and crystal dimensions) and structure on a molecular level. These are two aspects which are separate facets in traditional crystal studies but which in polymers are closely related. The molecular feature, *i.e.* the fold length, and the underlying process of chain folding determine a dimension, *i.e.* the thickness of the crystal, and conversely, by measuring the lamellar thickness we obtain an important structural parameter on the molecular level. It is to this item to which we shall turn in much of what follows.

Although conspicuous by inspection, the lateral habit is, with some exceptions, only of secondary consequence for the basic issues in polymer crystallization. The lateral habit of a PE crystal, such as in Figure 5(a), is identical to that of the familiar *n*-alkane (or fatty acid) crystals. In this case it reflects the symmetry of the lattice which is common to both classes of system. For different polymers this symmetry will be different, and so will the resulting crystal habit in its most ideal form (*e.g.* the POM lattice along the chain axis projection is hexagonal and so is the habit of the most regular crystals; see Figure 8a of Vaughan and Bassett[2]). Under different crystallization conditions the lateral habit may also contain other faces, high and low index ones, it may become more complex, acircular, dendritic, *etc.* This is in line with crystal growth in substances also other than polymers and can be the consequence of special conditions of solute diffusion (in solutions), the influence of impurities, local differences in supersaturation and/or undercooling, *etc.*, all aspects given attention in the practical control and interpretation of crystal growth in general. Important as such matters are for general crystal growth, we shall not be concerned with them in the present review on polymers except in the special cases where features of the lateral habit suggest a connection with the core issue of chain folded crystal growth. One such connection is that between the external lateral habit and the fold plane direction within the crystal, leading to sectorization, to which a separate section (Section 7.7) will be devoted. Another one is the rounding of lateral facets, such as in Figure 6 which usually (but not exclusively) develops in crystallization at increasingly higher temperatures.[39] As will be shown in Section 7.13.5, such curvilinear faces, after having been taken note of in the mid-1980s, presented a challenge to polymer crystallization theories as they stood at the time.

7.6.2 Lamellar Thickness; Fold Length

7.6.2.1 Methods of measurement

With the lamellar thickness (l) acquiring such basic importance, the methods of measurement require special attention.

The primary information comes from the electron microscope (EM) image itself, where the layer thickness is directly assessable from the shadow cast by the lamellar edge in samples coated with a suitable metal at a specified oblique angle, and also in sections of bulk samples by staining methods.[3]

A totally different method is small-angle X-ray scattering (SAXS), where a lamellar stack gives rise to discrete reflections which by Bragg's law then give a numerical value for the lamellar periodicity. Historically, this identification of discrete SAXS reflections, widely observed in semicrystalline polymers, with the lamellar periodicity has been one of the most important items in the whole subject.[14] Of course, it relies on the existence of a periodicity. Multilayer crystals with layers in contact and lamellae closely impinging on each other, as they do in a sample fully crystallized from the melt, automatically provide this. When lamellae are separate, as frequently in the suspended state in solution crystallization, they need to be sedimented first, thus producing an appropriately layered mat.

Note the complementary nature of the two methods: electron microscopy provides the real space information of what the entity actually is (together with light optical microscopy when applicable[40] and since recently AFM – see Figure 17), the only method capable of doing so. The thickness measurement it yields is direct but only relates to the specific selected entities in the field of view. In contrast, the SAXS method requires no special sample preparation and provides information on a representative amount of sample.

Another method, applicable only to certain polymers and specifically to PE, is low-frequency Raman spectroscopy (LAM), where the frequency of the first mode of longitudinal acoustical vibration relates to the extended chain portion traversing the crystal.[41] By appropriate conversion of frequency to length, good correspondence with the lamellar thickness, as determined along other routes, was found, enabling this vibrational spectroscopic technique to be applied for assessing the lamellar thickness. The precise numerical value depends on the conversion formula used and on the model adopted for the decoupling of the coherently vibrating straight-chain portion. Rather satisfactorily, in our hands at any rate, good agreement was obtained with the SAXS (and EM) method when using the simplest possible model and corresponding conversion formulae,[40] that of a vibrating isolated rod. Accordingly, this LAM method has been utilized productively for our work on PE, including results to be referred to in this review (see a critical survey[42]). This simple approach is not being shared by all concerned,[43] a point to which we shall return later. Note that, irrespective of interpretative details, the LAM method provides information on individual lamellae and does not rely on a perodicity as does SAXS.

Finally, calorimetry, specifically melting point (T_m) determination, can also be used. It relies on the fact that when the crystal is small, T_m is a function of crystal size, smaller crystals melting at a lower temperature (T). Quantitatively the relation between the lamellar thickness and T_m is given by the Hoffman–Weeks formula[44] as a special case of the general Gibbs–Thomson equation relating T_m to phase size, to feature prominently later on in this review (Sections 7.13.3 and 7.13.8). This method of melting point depression is applicable quantitatively as long as only the relevant crystal dimension, *i.e.* the lamellar thickness l, is the dominant variable, with other parameters (heat of fusion — hence internal perfection — and surface free energy) remaining approximately constant. In fact, as it turns out this calorimetric approach to determine l works remarkably well in most (even if not all) cases, in which respect polymers are providing a lead to thermal studies of materials on a nanometre scale, providing support for the Gibbs–Thomson relation not readily verifiable experimentally otherwise, a fact not widely recognized or appreciated.

In summary, the lamellar thickness, and hence fold length, can be reliably measured by a variety of methods based on entirely different physical principles, generally in good agreement with each other, thus placing the whole subject on a firm, quantitative experimental basis.

7.6.2.2 The lamellar thickness (*l*) and its functional dependence

It was amongst the earliest observations that l is not invariant but a function of crystallization conditions: it varies with crystallization temperature T_c, higher T_c yielding larger l values.[45] When comparing different solvents (in the case of solution crystallization) it emerged that l actually is a function of the supercooling ΔT and not of the absolute temperature of crystallization.[46] This means that allowing for variations of T_m^0, or rather dissolution temperature (T_d^0), all l versus T_c curves can be superposed onto a single master curve, of which Figure 11 is the latest and most definitive example, displaying the functional relationship

$$l \propto \frac{K_l}{\Delta T} \tag{1}$$

with the constant K_l containing the parameters characteristic to a particular polymer.[47] Equation (1) is widely obeyed and is basic to the whole subject. In fact, if the temperature is being changed in the course of crystallization, either upwards or downwards, l will change discontinuously according to the altered ΔT, leading to corresponding steps in the lamellar thickness.[48,49] It follows that changes in the growth temperature become permanently recorded in the crystal, Figure 12 being a particularly instructive example.

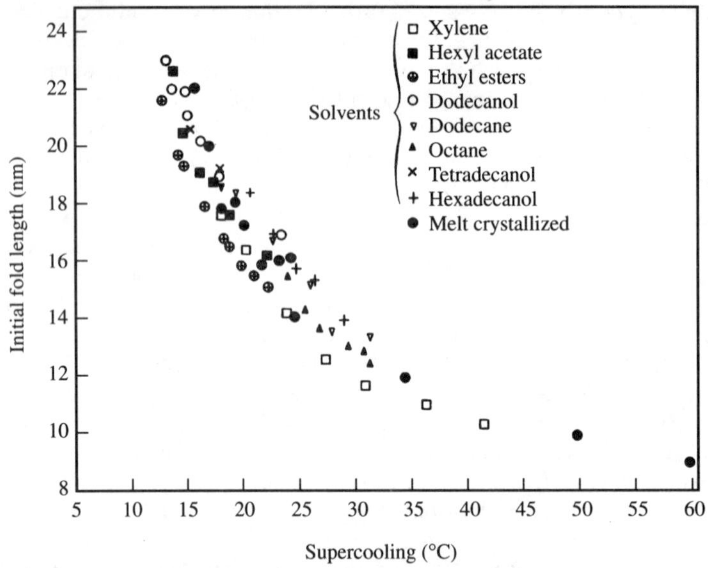

Figure 11 Initial fold length l_g^* against supercooling for both solution crystallization (involving a variety of solvents) and melt crystallization, with isothermal thickening in the latter case having been accounted for (after Barham *et al.*[47]).

Figure 12 Isochronically decorated polyethylene crystal. The up and down steps correspond to the response of the fold length to the respective raising and lowering of the temperature during crystallization. Electron micrograph (after Dosière *et al.*[49]).

In case of PE, and of other similarly fast crystallizing polymers, equation (1) acounts for the full ΔT range that can be covered. For slowly crystallizing polymers, crystallization can be conducted in a controlled manner up to higher ΔT values, in which case there is a leveling off of the l versus ΔT curve down to a constant lower plateau value, as first observed for isotactic polystyrene.[50] In the case of a homologous series, as in polyamides with different numbers of CH_2 groups in the monomer, such a lower plateau value for l was found to relate to the chemical constitution in a systematic manner: specifically, it corresponded to fixed numbers of H-bonds along the straight segments.[51,52]

In the case of solution crystallization, the concentration is a variable in addition to the temperature. Although there is no comprehensive investigation embracing the full concentration range from $c = 0$ to $c = 1$ for a given solvent, partial studies (in the range of $c = 0.01–1.00\%$[48,53] and isolated data points at higher concentrations (unpublished)) have failed to register any change of l with c. Thus, in contrast to supercooling, l seems to be invariant to supersaturation (Δc). This is a most puzzling feature of the subject which has received little attention and is seemingly at variance with expectations from presently held theories. It has been suggested[54] that the invariance of l on Δc could be due to the fact that the growing crystal only responds to the local concentration along the growth face, which is determined by the segment density of the attaching chain that is shielding the face from the solution interior, a point to which we shall return later in connection with crystal growth rates (Section 7.10).

7.6.2.3 *Increase in l subsequent to initial crystal formation*

It was an early experimental observation that the crystal lamellae could thicken after they have formed.[48] Such an increase in l corresponds to refolding to greater thickness, the process having been found irreversible.[55] In the course of it, the initial chain orientation within the crystal is retained, and in cases where the process can be followed in individual crystal layers, even the lateral register corresponding to that of the original crystal is seen to be retained, in spite of the severe disruption in the external morphology (see Figure 13). From the present view point, we identify this process as secondary crystallization leading to crystal perfectioning, a kind of maturation common to all crystallizing materials, which in the present case of polymers is accomplished by the rather unique molecular process of refolding.

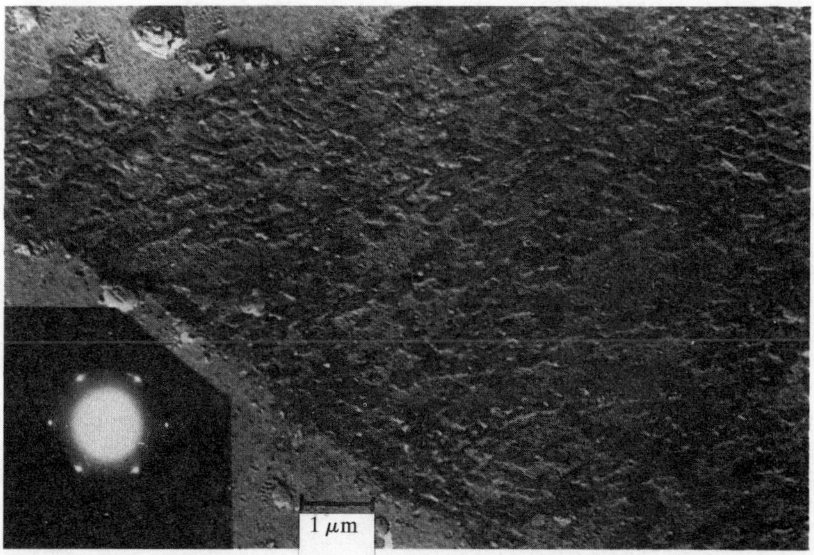

Figure 13 Solution-grown single crystal of polyethylene which has thickened due to heat annealing in the dried state after crystallization. Electron micrograph. Insert is the electron diffraction pattern of the same annealed crystal in corresponding orientation, showing that lattice coherence is preserved in spite of the crystal having broken up morphologically (the formation of the holes and lacework textures is the consequence of the layer being in isolation; in a stack the layers would intermesh with essentially unaltered overall volume) (after Keller[21]).

There are two circumstances under which such subsequent increase in l can arise. One is through heating the crystal above its original crystallization temperature T_c, when it is referred to as 'annealing'. The second situation arises when l starts to increase at T_c itself, either on prolonged storage at T_c or even while the crystal is growing, *i.e* behind the growth front itself, in which cases the effect is usually referred to as 'isothermal thickening'. The latter is more likely to occur when T_c (absolute) is high, which means that in the case of PE it is absent when crystallizing from solution but becomes pronounced in crystallization from the melt.

The latter is of consequence for the determination of curves such as in Figure 11, and thus for the underlying relation of equation (1). Namely, we need to distinguish between the initial l value which corresponds to primary crystallization, commonly denoted as l_g^* (see later) and the

$l > l_g^*$ which results from subsequent thickening. In the light of what was said above, the problem does not arise for crystallization from solution (in the case of PE) where for unannealed crystals the l values, as measured, correspond to l_g^* directly. However, this is not so for melt crystallization, where the conventionally determined l values exceed l_g^*. Here, obtaining l_g^* required a major research programme on its own. It involved 'catching' l at an early stage of crystallization with the newly forming crystals still 'young' before any thickening could set in (details[47]). Only then could the points for melt crystallization in Figure 11 be obtained and the universality of equation (1) (with the same constants for a given polymer) be asserted.

The time (t) dependence of the increase in l on annealing (secondary crystallization) is expressed by the Fischer–Schmidt relation[56]

$$l = l_0 + B(T)\log\left(\frac{t}{t_0} + 1\right) \tag{2}$$

where l_0 is the value of l at time t_0 and B is a constant for a given T.

However, in general this logarithmic relation does not extend down to, for the present purposes, the all important shortest times. In cases where this could be assessed, the thickening process starts off with a sudden discontinuous jump, which can correspond to an increase in l by an integer factor, *i.e.* 2×, 3×, *etc.*, of l with significant implications for the underlying molecular processes.[57,58] Such a jump with a subsequent logarithmic increase with time (see equation (2) above) could, more recently, also be registered in situ in the course of isothermal thickening.[59]

7.7 STRUCTURAL CONSEQUENCES OF CHAIN FOLDING

7.7.1 Sectorization and its Origin

Chain folding can have some explicit consequences for the crystal morphology, and conversely, certain morphological features can be unmistakable indications of the existence of chain folding, even, as we shall see, of some specific attributes of chain folding. The most conspicuous one is sectorization. Although abundantly covered in early reviews, we are giving it renewed emphasis here, partly to draw attention to the salient and clear-cut nature of the effect, seemingly not always fully appreciated (together with its implications), and partly because recent results, as we shall show, are bringing it to the forefront again.

Sectorization has already been implied in Figure 7. Here folding is represented as occurring along the growing lateral faces, a very plausible assumption which is in fact borne out by all subsequent experiments. This fold plane direction parallel to the growth face will be retained behind the growth front, hence within the crystal interior, leading to domains of different fold plane direction corresponding to each growth face within the crystal layer. Such a fourfold sectorization corresponding to the faces of the simplest PE crystal is indicated in Figure 7. By the same argument, crystals with other and further lateral faces should display a correspondingly increased number of sectors, there being as many sectors as there are lateral faces with the fold plane direction parallel to the lateral face which bounds each particular sector, *e.g.* as shown by Figure 14.

The above *a priori* considerations have been amply verified by experiments of many different kinds, involving different properties and techniques, to which we shall turn in what follows. The works will not all be referenced individually; further summaries together with the original references are contained in general[21] and specific reviews.[60]

7.7.2 Experimental Evidence for Sectorization

7.7.2.1 Manifestation of sectors by geometry

This most conspicuous effect arises from the fact that often a given crystal layer is not planar but has a three-dimensional configuration displaying the existence of sectors. The simplest is the hollow pyramid where the crystals are to be visualized as 'tents', with the different panels corresponding to the different sectors, which accordingly are manifest through their distinct

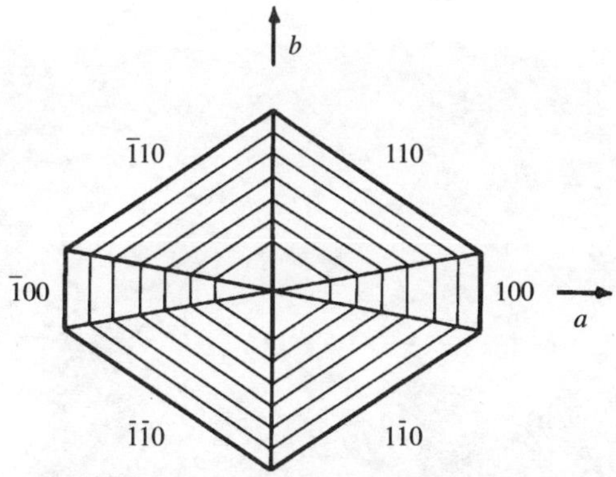

Figure 14 Diagram of a truncated lozenge polyethylene crystal with {110} and {100} lateral faces. The thin concentric lines represent the crystal outlines at distinct stages of lateral growth, revealing the origin of, in this case, six-fold sectorization (after Keller[60]).

and specific inclination to the common central pyramidal axis. Under special conditions, such pyramids can be observed in their undistorted three-dimensional form, specifically as suspended in their mother liquor, when formed in solution, as shown by Figure 15. More commonly, such pyramidal and other three-dimensional forms are seen in their flattened version, arising in the course of sample preparation for microscopy (*e.g.* sedimentation of the suspension) when the underlying sectorization becomes recognizable through specific collapse features. Such are pleats (see the central pleat in Figures 16 and 17) and the diffraction contrast in Figure 16 which displays the sectors directly, the latter arising from the different chain inclinations within the different portions of the flattened lamellae (for collapse modes, see Figure 23 later). (Sectorizations as revealed by diffraction contrast can also arise in *n*-alkanes without any chain folding.[61] Such effects do not detract from the present evidence associated with chain folding as these authors[61] seem to imply. While true that in the light of these results[61] the mere detectability of sectors *per se* need not signify chain folding, the presently discussed sectorization associated with chain folding is also backed by much further evidence of qualitatively different kinds: see below. Even within the scope of the diffraction-based evidence alone, the inclination differences involved can be huge, 30–40°, in contrast to that in the *n*-alkanes where it is very small and unspecified. In the latter case, sectorization is likely to be due to a fourfold twin arrangement and associated stresses in the central crystal nucleus.[61]) The origin of the visibility of sectors in Figure 17 will be commented on further below.

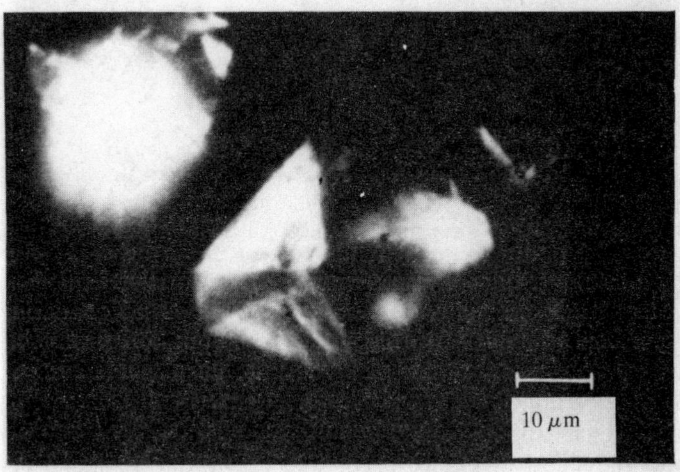

Figure 15 Truncated lozenge solution-crystallized polyethylene crystal while floating in the liquid, revealing its three-dimensional hollow pyramidal configuration. Dark field optical micrograph (after Keller *et al.*[60,70]).

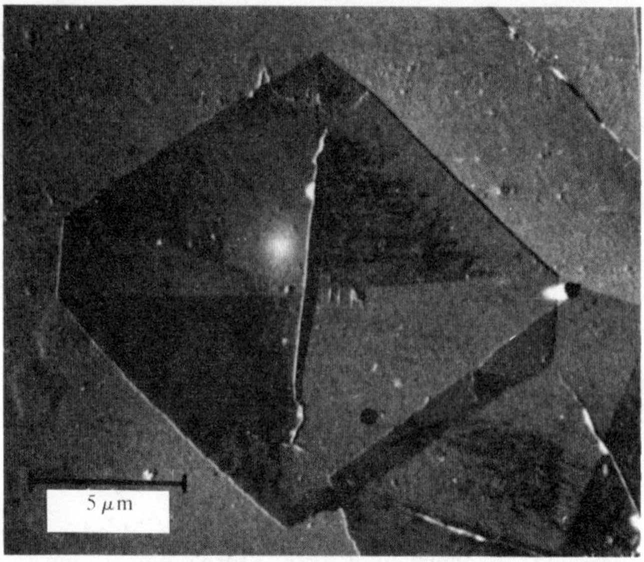

Figure 16 Hollow pyramidal polyethylene single crystal, such as in Figure 15, collapsed on a substrate, showing central pleat and sectorization by diffraction contrast (after Keller[60]).

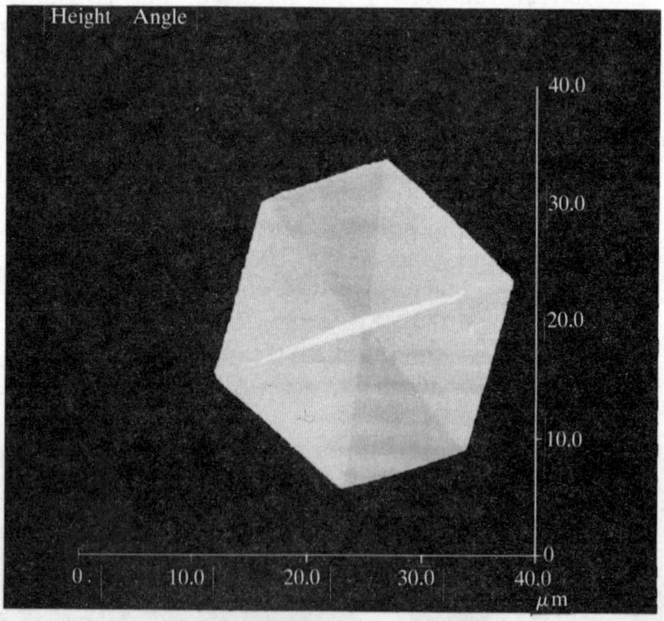

Figure 17 Similar flattened hollow pyramidal crystal as in Figure 16 but imaged by atomic force microscopy, revealing sectorization due to thickness differences (some foreshortening is indicated by the differences in scales along the horizontal and vertical directions) (after Nakagawa *et al.*[67]).

There can be more complex three-dimensional arrangements, similarly displaying the existence of sectors which are derivatives of the simple hollow pyramid. Such are the corrugated structures modelled in Figure 19(b) and illustrated as actual crystals by Figure 9 of Vaughan and Bassett.[2] A further version observed simultaneously with the pyramids, having come into prominence again recently in connection with studies on melt crystallized material,[19] are the 'chair' crystals (Figure 18 and model Figure 19(d)). Their relation to the pyramids are best visualized by considering a pyramid such as in Figure 19(a) first cut in half along its short diagonal, then one half rotated through 180° with respect to the other, the two halves being joined together again, which then necessarily will involve the central distortion associated with the 'chair' shape (see Figure 19(d)). The underlying physical significance of these three-dimensional shapes will be commented on in Section 7.7.3.

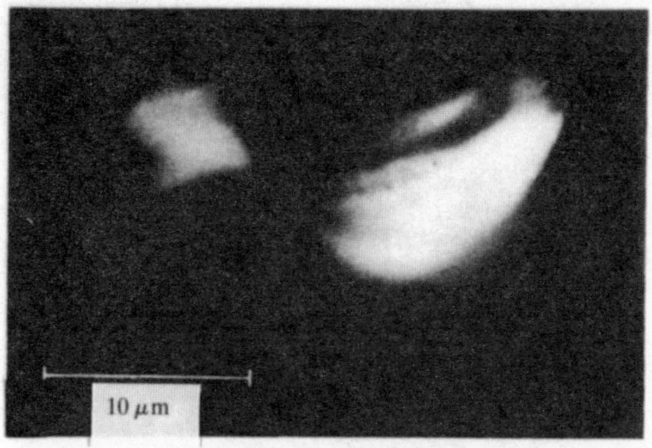

Figure 18 Truncated lozenge solution-crystallized polyethylene crystal while floating in the liquid revealing, in this case, a 'three-dimensional chair' conformation. Dark field optical micrograph (after Keller *et al.*[60,70]).

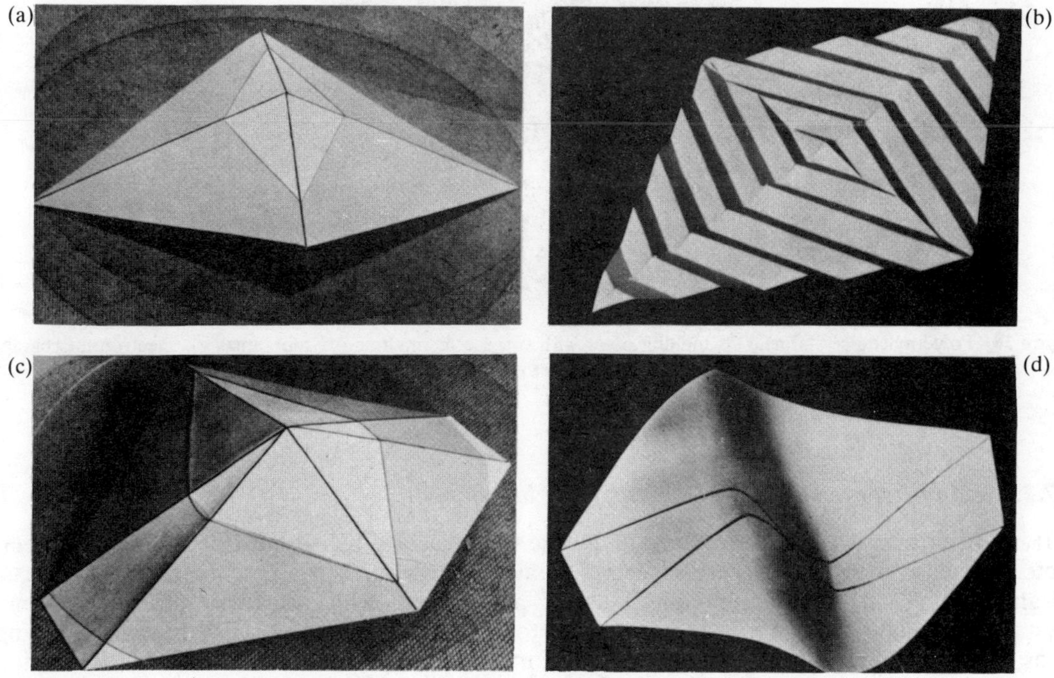

Figure 19 Models of the main monolayer crystal configuration in three dimensions for polyethylene. (a) Hollow pyramid (of unequal base) with {110} lateral faces. (b) Corrugated layers with {110} lateral faces (for corresponding crystals in this series, see Vaughan and Bassett[2]). (c) Hollow pyramid with {110} and {100} lateral faces. (d) 'Chair' crystal with {110} and {100} faces (after Keller *et al.*[60,70]).

7.7.2.2 *Manifestation of sectors by thermal properties*

In sectors which are distinct not only by symmetry (*e.g.* (110) and (1$\bar{1}$0) as in Figure 7) but also by actual index *type* (such as in the six-sided crystal of Figure 14 where the indices are {110} and {100}), the crystallographically distinct sectors may melt (or undergo thermal transformation) at different temperatures as shown by Figure 20. Accordingly, the different portions of the crystals can be distinct also as regards thermal stability. (The extent to which this effect is relying on a preceding collapse on a substrate, as in Figure 20, or whether it is also demonstrable while in suspension, has quite recently become an issue of topicality.[63] The issue raised concerns the interpretation of multiple melting peaks in DSC and does not affect the essence of the present argument, namely that whenever such effects as in Figure 20 are seen, manifestation of sectorization by thermal properties pertains, whatever the circumstance necessary for its appearance may be. In this context we recall that morphologically observable sectorization due to differential thermal stability is on record even on crystals which are in suspension.[64])

Figure 20 Polyethylene crystal near its melting point with two sectors melting (or reorganizing). Electron microgram (after Keller *et al.*[60,62]).

7.7.2.3 *Manifestation of sectors by cleavage behaviour*

It has been a basic observation for a long time that cleavage behaviour distinguishes between sectors, or conversely that different sectors cleave differently.[65] In a given sector, cleavage is 'clean' parallel to the corresponding lateral prism face, while cleavage along the same crystallographic direction but in the 'wrong sector' produces cracks with fine threads spanning it, as clearly demonstrated by Figure 21. The former case is consistent with the crack running along the prism face, hence between fold planes, while in the latter case the crack would cut across the fold plane thus pulling out threads, in full support of the underlying sectorization concept. This effect is particularly significant because it demonstrates the intrinsically different directionality of the molecular trajectory in the different sectors, and further that this embraces the full crystal interior.

(a)

(b)

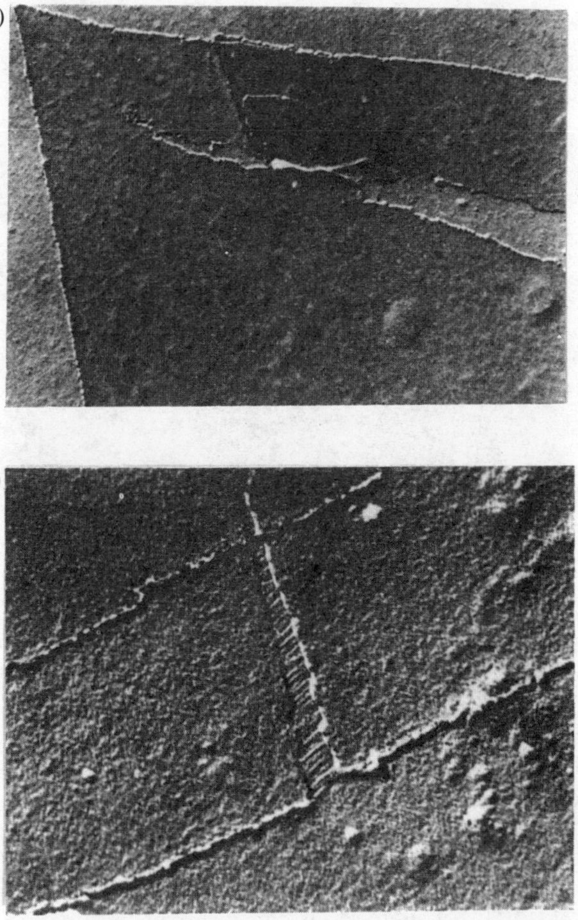

Figure 21 Fracturing in polyethylene crystals: (a) fracture parallel and (b) perpendicular to the (110) plane bounding the corresponding sector. The threads pulled in (b) should be noted. Electron micrographs (after Lindenmeyer[65]).

7.7.2.4 Manifestation of sectorization by surface decoration

A particular decoration technique involving vapour deposition of paraffin crystals clearly reveals sectorization (Figure 22).[66] Here the tiny rod-like paraffin crystals are normal to the lateral prism plane, and hence must lie across the envisaged fold plane associated with each sector, revealing sectorization through the particular surface texture pattern which thus arises. Hence, even without any further specific assumption, this method implies that the crystal surfaces themselves are distinct according to sector type, a point which acquires special significance in discussions on the nature of the crystal surface and on the structure of the fold itself. This surface decoration method is particularly powerful and of general applicability. It is applicable not only to PE but to chain folded crystals of all the polymers examined so for this particular behaviour. Clearly, some generalized epitaxy must be involved where the fold plane direction and not the particular crystal lattice as such is the determining factor.

7.7.2.5 Sectorization by atomic force microscopy (AFM)

Figure 17, already referred to, is a particularly revealing example of the application of this most powerful latest technique to the issue in question. In this case we have a low-resolution application of AFM utilizing its great sensitivity to thickness differences.[67] The contrast by which the sectors here become visible is the smaller thickness of the {100} (the truncated) sectors compared to the {110} sectors, for reasons which we shall discuss below.

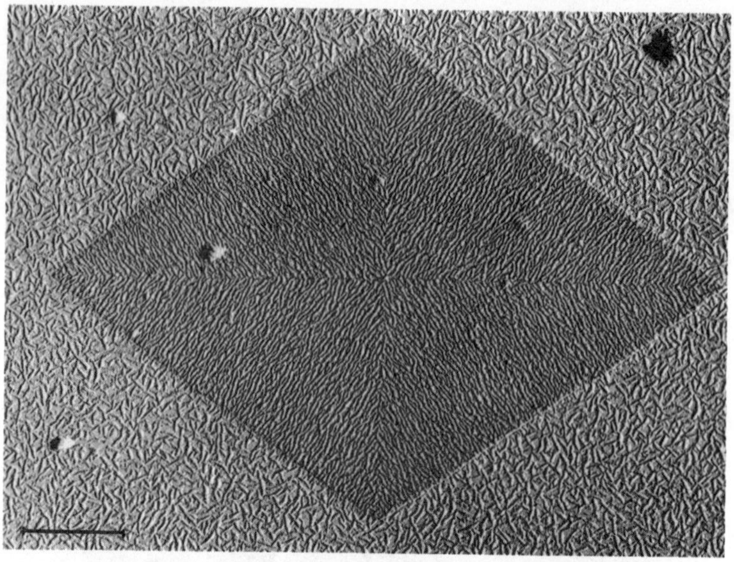

Figure 22 Solution-grown monolayer polyethylene crystal with four-fold sectorization revealed by surface decoration. The decorating particles are paraffin crystals resulting from vapour deposition of evaporated (and degraded) polyethylene, in epitaxial relation with respect to the fold plane direction (after Wittmann and Lotz[66]).

A quite recent and different application of AFM consists of measuring the 'surface viscosity' by sliding the probe tip along different directions with respect to the sectors. It was observed that in a given sector the frictional resistance is significantly lower parallel to the corresponding prism face compared to what it is normal to it.[68,69] This 'anisotropy of surface viscosity' is fully consistent with the concept that the frictional resistance is expected to be lower when sliding the probe parallel to the fold plane, *i.e.* along the ridges or grooves, compared to normal to it when it should traverse the hills and valleys. Most significantly, the degree of anisotropy is affected by molecular weight:[68] for high M_w it can disappear altogether, a point to which we shall return (see Section 7.8.3.4).

7.7.3 Physical Implication of Sectorization

As repeatedly stressed, the sector effect in itself implies folding along the lateral prism face. The specific nature of the three-dimensional configurations of the crystals (pyramids, chairs), together with the associated sectorization, has further additional information and pointers to convey.

7.7.3.1 Pyramids

By electron diffraction evidence the straight chain segment, *i.e.* *c*-axis orientation, within the pyramid is as in Figure 23(a). This may change on flattening onto a substrate in either of two ways (Figure 23(b) or Figure 23(c)), mode (c) being the one appropriate to Figures 16 and 17. In Figure 16 the diffraction contrast by which the sectors are manifest is a consequence of the different chain inclination (see the sketch in Figure 23(c)). In Figure 17 the visibility of the sectors is due to the {100} sectors being thinner than the rest, which in turn is largely (even if not entirely) due to the larger chain inclination in these sectors of closely identical stem lengths.[67]

The principal point, implicit in the uncollapsed pyramidal conformation, is that the straight segment ends, and hence the folds, must be staggered, and this in a different sense in diametrically opposed sectors. This structure is sufficiently regular to be definable in

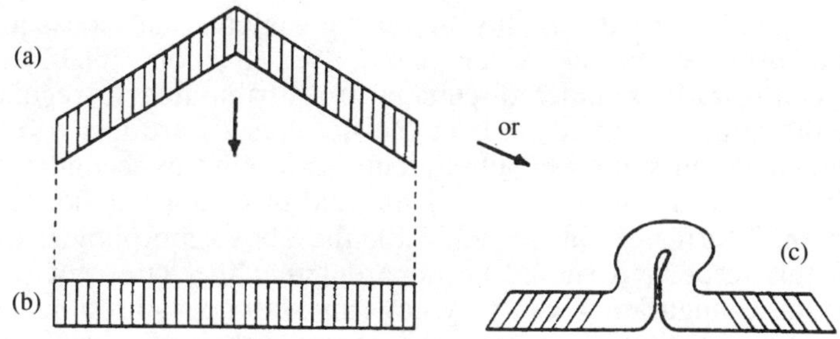

Figure 23 Schematic representation of two modes of crystal collapse: (a) section through unflattened crystal or portion thereof; (b) flattening by plastic deformation only, with no change in straight segment orientation; (c) flattening without any plastic deformation leading to chain (c-axis) tilts, which differ according to sectors (after Keller[60]).

crystallographic terms, with the important implication of a corresponding regularity of the fold itself for causing such a stagger (see reviews[21,60,71]). For this stagger to be of opposed sign in opposite sectors, an asymmetry of the fold is required. In fact, the pyramid is not flat based, in the sense apparent from Figure 19(a), which means that the {100} faces are sloping with respect to the horizontal. This, in turn, implies stagger of folds *within* these planes, in addition to the stagger *between* fold planes. Figure 24 is an attempt to convey the two-fold asymmetry implied, (a) in plane, and (b) out of plane. In fact, attempts to construct fold structures on the basis of conformation considerations have indeed arrived at such a double asymmetry (see reviews[21,60]). To sum up, the regular three-dimensional layer configurations as associated with sectorization are indicative of the nature of the fold structure itself, and hence their physical significance.

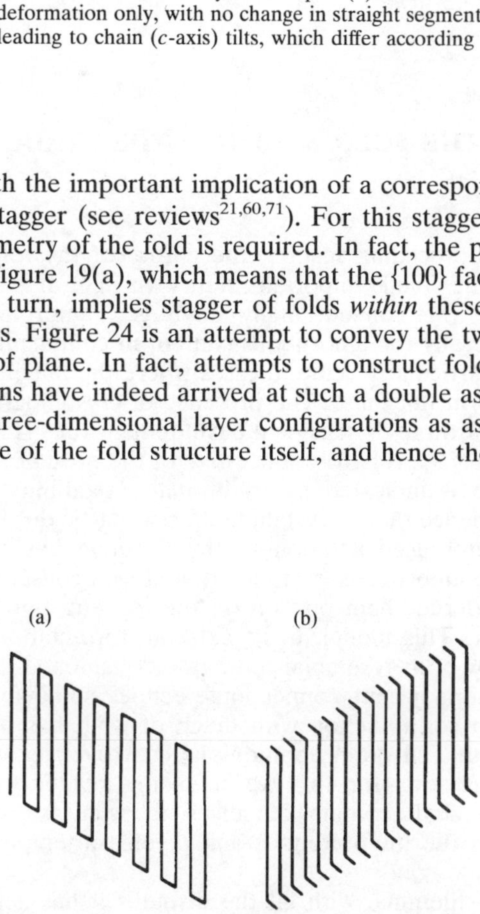

Figure 24 Sketch illustrating the principles of fold staggering: (a) ribbon of folded molecule seen normal to ribbon plane; (b) stacking of consecutive ribbons seen along the ribbon plane (after Keller *et al.*[60,70]).

7.7.3.2 Chair crystals

The chair shape is related to the pyramid as described earlier, from which it follows that, the central curving region apart, the stem, and hence the fold stagger in the diametrically opposed sectors, is in the *same* sense (and not opposite as in the pyramids). The possibility (and co-existence) of these two variants were attributed early on to the consequence of rearrangement subsequent to primary growth. Accordingly, the crystals would first grow as flat lamellae, with the stagger leading to the appropriate three-dimensional configuration setting in subsequently. By statistical chance, and/or whether there is any coordination between the sense of the stagger, either pyramids or chair crystals, or both together, would result. This early hypothesis, having received little attention at the time, became reinforced by recent developments, both in a specific study of melt crystallized morphology[19] and also by several other developments pointing to the reality of rearrangements and consequent perfectioning subsequent to the primary crystal development (see Section 7.9.3.1). As the regular three-dimensional configurations under discussion are attributable to regular fold stagger, it would follow that the initially formed folds in the flat crystals are less regular, possibly even quite irregular, regularization setting in subsequently as a kind of secondary perfectioning. In view of the fact that regularity or otherwise of the fold has been a major issue in the field (a topic to which we shall turn next in Section 7.8), the above morphological findings acquire signficance also in this respect. It should be apparent that the nature of the fold surface, as reflected by the crystal configuration, can vary not only with crystallization conditions but even within the same crystal with time. This has implications both for the structural issue as to the nature of the fold and, by pointing to the role of rearrangements, for the issue of how the crystals actually form.

7.8 THE NATURE OF THE FOLD SURFACE: THE STRUCTURE OF THE FOLD

7.8.1 The General Issue

As already raised in the preceding section, the nature of the fold in molecular terms is a salient but still unsettled issue. While important in its own right, it is linked to the more general issue of how to visualize the conventionally envisaged amorphous–crystalline two-phase structure, or still more generally the amorphous content in conjunction with crystal lamellae.

The issue here became particularly acute in cases where the lamellae are crystallographically defined and of highly uniform thickness. The presence of crystallographically indexable facets, laterally and also in terms of three-dimensional development for PE in terms of $\{hk0\}$ and $\{hkl\}$ planes, respectively (Section 7.7), is hardly suggestive of the presence of a significant amount of amorphous material, which is indicated by traditional crystallinity determinations and reinforced by the general experience that a crystalline thermoplastic displays plastic and viscoelastic characteristics. A widely embraced solution to the problem has been to visualize the fold surface as the source of the amorphous component, and as a consequence to envisage the fold as a largely irregular disordered chain portion on the top and bottom of the crystalline core within the lamellar interior. This model, in its extreme formulation often referred to as the 'switchboard' model, in view of its resemblance to an old-fashioned telephone switchboard with loosely dangling, largely nonadjacent connections, echoes some features of the old fringed micelle and seems to be in contradiction with much of what has been said in the preceding section. The long drawn out and often acrimonious controversies which resulted will not be followed through in this review (for this we refer the reader to the Discussion Meeting on record[72]). Instead, we shall confine ourselves to what we consider forward looking physical issues arising from the multitudinous and diverse attempts to find a solution to the problem.

In the widest sense, the dilemma, with all the disputes it has generated, has already been implied by the foregoing. It could be formulated as follows. In a high crystallinity polymer like PE the morphology of the crystal elements, whenever such could be identified, displayed crystallographic features similar to those in conventional polycrystalline materials with the added richness of specifically 'polymeric' effects arising from chain folding. There is nothing obvious in the direct structural examination of such entities which would suggest that they are partially amorphous (see interim reviews[71,73]). On the other hand, measurements of bulk properties carried out on macroscopic assemblies of such entities (*e.g.* a piece of melt crystallized material or even a sedimented mat of solution grown crystals) display substantial amounts of crystallinity deficiency (say 20–25% in PE), traditionally attributed to amorphous material (see interim review[74]). This in essence is the dilemma, unresolved to the present day.

7.8.2 Assessment of the Fold Surface from Macroscopic Properties: Scattering Effects

The most basic properties used for assessing the degree of crystallinity are the intensive thermodynamic quantities of density and heat of fusion. More intricate measurements on macroscopic samples, leading broadly to the same result, relate to the detailed analysis of SAXS intensities aimed at determining the electron density profiles of stacked layers, usually as formed in melt crystallized samples but also on mats of sedimented single crystals. These usually lead to a distinction between a high (electron) density crystal core and a low (electron) density interlamellar layer, interpreted as disordered amorphous material in keeping with the traditional amorphous crystalline two-phase model dividing the l value, the total periodicity arising from Bragg's Law, into crystal (l_c) and amorphous (l_a) components where $l = l_c + l_a$. Also, more detailed analyses are increasingly leading to the postulate of a transition layer with order intermediate between the crystal and the amorphous portions (for a comprehensive source, see a textbook[75]). Similar conclusions have also been arrived at by Raman LAM analysis in the case of PE. This, in addition to any amorphous crystalline subdivision, can also lead to a substantial transition layer in between[76] through the use of models and corresponding conversion formulae considering end effects, added to the simple picture of an isolated vibrating rod. (Here we refer to Section 7.6 and Section 7.9 to follow.)

A methodology representing a class of its own which has been extensively applied to the above problem is neutron scattering. As a full article is devoted to this subject within the present series,[4] it will only be included here in its briefest outline. It relies on the difference in the ability to scatter neurons elastically between hydrogen and deuterium. This raised the expectation that by mixing a small amount of deuterated material (guests) with a largely hydrogen containing (host) polymer — or vice versa — the scattering of neutrons will give information about the trajectory of the guest molecule, and hence, in a first approximation, of a polymer molecule in its own environment. This should also tell us what happens to the molecule on crystallization, and hence should provide an answer to the most central issues in the whole field.

While powerful in many ways, in actual application the method did not prove to be as straightforward as initially expected. One of the major problems arose from the tendency of the isotopically distinct chains to segregate on crystallization, negating the main purpose of the investigation. This handicap eventually restricted the technique to samples which had been crystallized fast enough to prevent such a segregation. However, such fast crystallization leads to less than ideally developed, in fact mostly highly imperfect, crystals, which is counter to the general practice in the study of fundamental features of crystals and crystallization, namely to use crystal samples in their maximum obtainable perfection. Even so, what did emerge was that the overall chain trajectory was much less affected by the crystallization process on crystallization from the melt, where it is in the state of the random coil, compared to that from solution. This implies that at least for (by necessity) fast crystallization there must be considerable randomness, hence disorder, in the resulting crystal, seemingly in line with the 'switchboard' type model, while crystallization from solution (apparently much less prone to isotopic segregation) leads to structures closer to the idealized model such as Figure 7.

The global features of the chain trajectory, as characterizable through the radius of gyration (R_g), can be assessed from the scattering pattern at the smallest angles. While no or little change in R_g was found in (fast) crystallization from the melt, there was a large contraction in R_g when crystallized from solution. Further features of the scattering curves, those at larger scattering angles, allowed conclusions to be drawn about the overall 'shape' of the chain envelope within solution-grown single crystals which, thus derived, was consistent with ribbons, in turn consistent with the picture of chain folded deposition along the growing lateral faces as visualized by *a priori* considerations. An increase in molecular weight M_w caused the ribbons to become thicker but hardly longer (assessable from the scattering curves), indicating stacking of ribbon planes ('stacked ribbon' arrangement[77]) which in molecular terms we interpreted as the ribbon constituted by a given molecule folding up on itself ('superfolding', see Figure 25). Certain deductions could also be made as regards the arrangement of stems belonging to a given chain within such a ribbon, most informatively in combination with IR spectroscopy, which through analyzing fine splitting of certain crystallinity sensitive vibrational bands can give information on the isotopic environment of stems, *i.e.* how far a given stem has neighbours of the same or different isotopic species, and hence belonging to the same or different molecules.[79,80] Figure 26 shows the outcome of such an analysis. It is the computation of a stem arrangement that can account for the fine splitting (up to 11 components) of the IR spectra and is also consistent with the neutron scattering results. It shows that while stems from a given molecule are not all adjacent, and hence the folds are not all adjacently reentrant, they cluster in groups in a nonrandom fashion, implying more or less extended runs of adjacent folds interspersed with nonadjacent ones. For the specifics we refer to a recent review on this particular subject.[81]

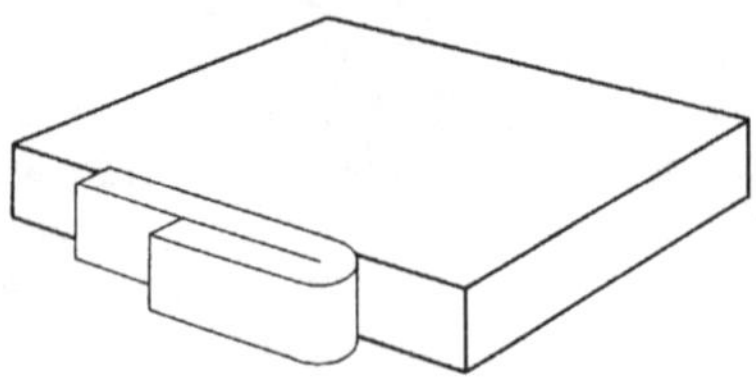

Figure 25 The overall envelope of a chain depositing along the growth face of a chain-folded single crystal. Neutron scattering (at medium small angles) registers it as a ribbon, which with increasing molecular weight does not increase proportionally in length only in thickness; hence the concept of 'superfolding' represented schematically in this figure (after Sadler and Keller[78]).

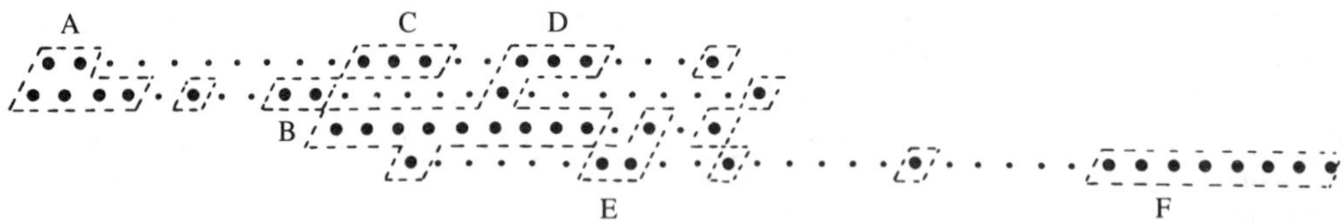

Figure 26 Stem projections of a single molecule — heavy dots — as deposited along the growth face of a solution-grown single crystal. Computed so as to be compatible with both the complex multiple peak structure of the IR spectrum and with information from neutron scattering measurements, both obtained on isotopically doped material (*e.g.* the heavy dots could correspond to the deuterated guest molecule). It incorporates the conception of superfolding (Figure 25) but with clusters of adjacent stems interspersed with stems from other molecules (after Spells *et al.*[80]).

Finally, even if outside the fold structure topic, the technique of neutron scattering has revealed a further and most remarkable effect needs placing on the map. It relates not to the formation but to the melting of crystals. As stated previously, in the case of solution-grown single crystals, R_g contracts drastically, in keeping with the more compact ribbon shape adopted by the initially random coil on crystallization. On melting, the inverse process takes place, as assessed by neutron scattering. The chain expands and reverts to its initial random coil state,[82] the scale of the effect being apparent from Figure 27. The most remarkable feature is that such a huge effect is taking place practically instantaneously (on the time scale set by the heat transfer in the sample) and is, at least within wide limits, independent of molecular weight. We seem to have a kind of an 'explosion' of the molecule. The driving force clearly must be the entropy (we have a kind of an 'entropy spring'), but the molecular process and the dynamics by which this is achieved, however, seems quite inexplicable at present. (For latest results, see 'Notes Added in Proof', p. 302, no. 1.) As a practical sequitur it follows that the memory of the crystal should not be maintainable in the melt on any time scale which could be of consequence. This in fact is borne out by technological experience (*e.g.* the advantages of the 'disentangled' state of solution processed material for ultra-drawing of fibres is immediately lost on melting).[83]

At this stage of the discussion of the fold surface issue, two general points are appropriate. The first is the limit to the amorphous–crystalline two-phase model in combination with lamellar crystals as set by geometry. *All* the chains emerging from a crystal just cannot pass into a zone of complete randomness: there is simply no room for it. Randomization requires excess volume for which, given the constraint of the crystal, there is no space laterally (see papers[72]). Such space can only be created by terminating some of the 'chain flux', which is achieved by folding the chain back into the crystal. In fact, for this reason, folding can be regarded as a necessity for forming lamellar crystals. The question which remains is numerical: what is the minimum amount of backfolding required by such geometric constraints? By calculation, at least half of the chains would need to reenter the same crystal, hence fold, thus setting a geometric limit to a switchboard-like model, there being some differences between existing estimates.[84–87]

Figure 27 Typical molecular trajectories before and after melting a solution-grown polyethylene single crystal of molecular weight 390 000. Final radius of gyration: 64 nm (after Barham and Sadler[82]).

The second point will be apparent from some of the foregoing. There is no single answer to the fold surface problem. The fold surface and fold structure can vary widely with crystallization conditions and even, in the same crystal, with time. Any quest for a single decisive experiment to decide the issue once and for all, as pursued in many past instances, is therefore basically flawed. Therefore instead of pursuing the controversies generated by such misplaced attitudes, we shall proceed to survey such variations and identify some of their causes.

7.8.3 On the Variability of the Fold Surface

7.8.3.1 Decoration of crystal surfaces

One morphological method for the study of surface structure was by decorating the crystals with vapour deposited gold.[88–90] The gold particles seen help to reveal steps and other surface features, for which the method is usually employed in the field of simple crystals (*e.g.* alkali halides). In addition, however, in the case of polymer single crystals it can show differences in gold particle size and density of coverage, in crystals prepared in different ways and even in different portions of the same crystal (Figure 28). While such effects cannot be interpreted quantitatively, they serve to show that surface structures can be, and in fact usually are, different according to the way in which the crystal arises even in the course of growth of the same crystal. (It is salutary to reflect that all such variations would be averaged out by any macroscopic measurement on an aggregate sample.) One specific feature is the response of the decoration pattern to exposure of a swelling agent. The latter can disrupt the initial decoration pattern. We see from Figure 28 that such a disruption (mobilization) is confined to the crystal interior but does not occur nearer the edge. In that case the interior was crystallized at a higher temperature than the edge (beyond the step) and also, due to fractionation,[91] certainly contained longer molecules. In fact the higher 'mobilizability', hence swelling ability, of crystal surfaces formed from longer molecules could be substantiated in a subsequent study (see below).

At the same time, a class of separate investigations was performed along different lines to identify the location of the chain ends. These relied on the accessibility of terminal double

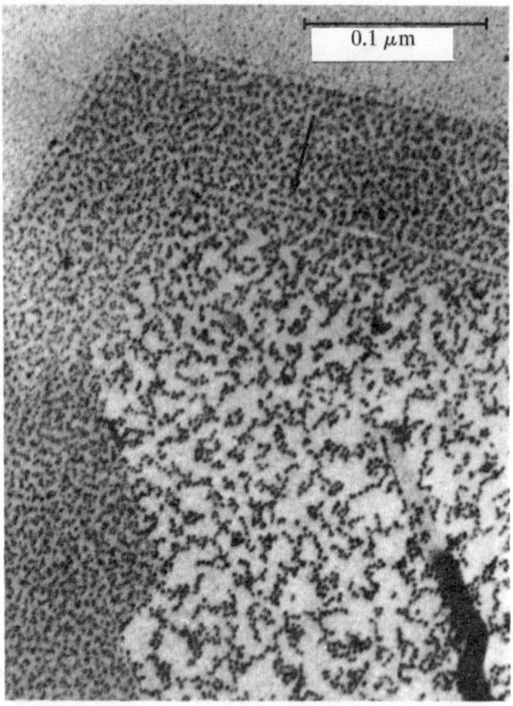

Figure 28 Polyethylene crystal decorated with gold and subsequently flooded with xylene, revealing mobilization of the fold surface in the layer interior (coarser decoration pattern). The finer decoration pattern at the periphery and in the overgrowth layer characterizes the unmobilizable surface; this was the pattern over the entire crystal before flooding. A step due to deliberate changes in growth temperature is also brought out by the decoration (arrowed) (after Blundell and Keller[89]).

bonds (present in the particular type of PE) to ozone. It could be established that the majority of ends must lie at the crystal surface, and hence are ejected in the course of crystallization,[92] a finding of note in itself. When combined with the above-mentioned decoration effects, plus the swelling-induced mobilization, it could be inferred that the mobilizable material contains largely chain ends, and hence must be, at least to a large extent, loose cilia overlaying the chain-folded crystal layer itself,[90] where the higher M_w material (and also higher T_c) is leading to more pronounced ciliation at the crystal surface.

7.8.3.2 Selective chain cutting

A particularly informative method for the study of chain folded structures was the exposure of crystalline polymers to strong oxidizing agents such as, in the case of PE, fuming nitric acid and ozone. In the case of the melt crystallized bulk, nitric acid, in particular, penetrated the sample and digested all interlamellar material including tie molecules, leading to a dispersion of lamellae.[93] In fact, this led to the first conclusive demonstration of the existence of lamellae within the bulk interior,[33,93] and is one of the methods for exploring bulk morphology (see Figure 9). The folds are also cut in the process, converting PE into closely uniform dicarboxylic acids of lengths which correspond to, yet are somewhat smaller than, the initial lamellar thickness.[94–98]

When applying this method under milder oxidative conditions, including using ozone as the oxidizing agent, and combining with the measurement of M_w distribution by GPC, it proved to be possible to follow the progressive cutting of the folds. It was found that the M_w distribution at the low end became quantized, corresponding to simple multiples of the layer thickness identifiable as 1, 2, 3 and higher numbers of chain traverses, and hence correspondingly folded chain fragments (Figure 29).[94–98] In fact, on its own this represents a chemical proof of chain folding. By refining the quantitative analysis of the peak values in the GPC traces such as Figure 29, and including also the variation of peak area ratios as arising with progressive degradation, further information was obtained on the fold surface structure.[94–98] It could be inferred that there was a distribution in fold stem lengths within the outermost zones of the lamellae and that folds terminating deeper down in the crystal are increasingly sharper (the M_w peaks corresponding increasingly to integer chain length ratios). This led to the picture that the outer lamellar zones are less ordered than the more deeply lying but still fold-containing zones where the order becomes closely crystallographic. This is fully consistent with a 'buried' fold conception arising from the combined space requirement considerations and the neutron scattering plus isotopic 'tagging' based IR results, and can at the same time account for the existence of a transition layer between the ordered crystal core and the outermost highly disordered (amorphous?) layer (see Figure 5 of Sadler[4]). Even if such selective degradation studies were not pursued further, so as to include variations with the initial molecular weight, it has become apparent from subsequent studies by other techniques that the existence and degree of surface disorder depends on M_w, higher M_w resulting in more disorder, a point which ties up with the results of the decoration studies quoted above.

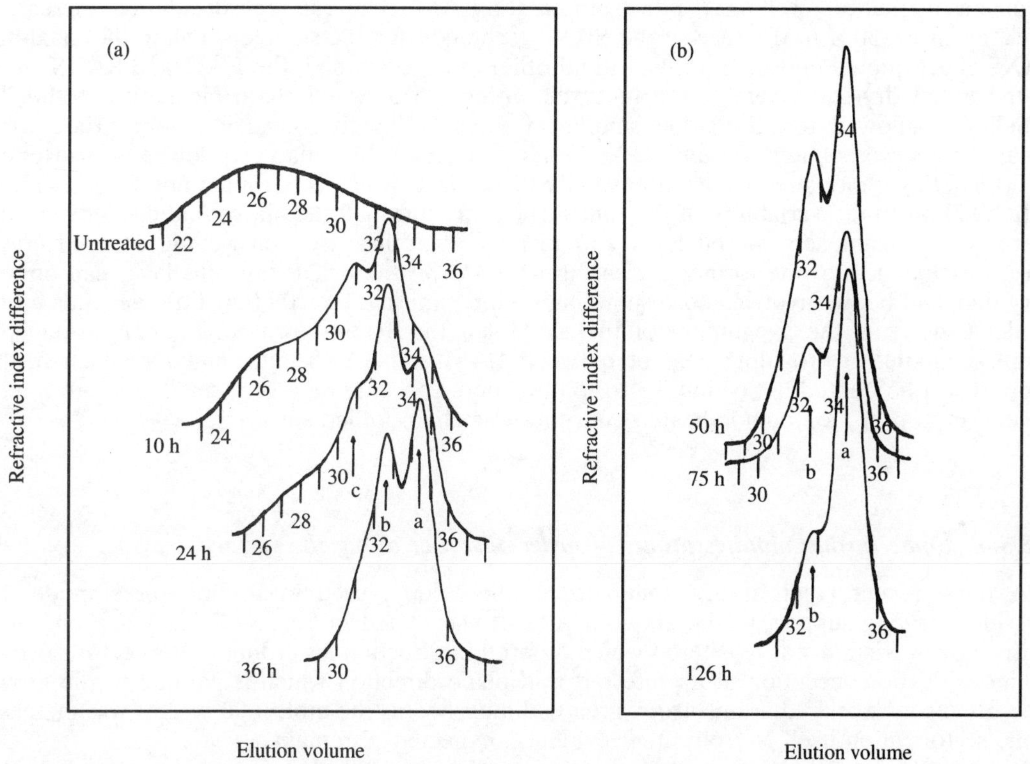

Figure 29 A series of gel permeation chromatograms obtained from polyethylene crystals at increasing stages of degradation with nitric acid. The detected refractive index difference is plotted against elution volume (larger elution volume corresponds to lower molecular weight (after Blundell *et al.*[94]).

7.8.3.3 *Swelling of single crystals*

The influence of molecular weight on the surface structure has become directly apparent in several ways. One is through swelling of the crystal surface, this time quantitatively assessed by the SAXS technique.[99,100] Exposing solution-grown mats of single crystals to swelling agents (such as are solvents for the crystals at appropriately higher temperatures), the SAXS spacings were found to increase to specific terminal values which were a function of molecular weight in a sense that higher M_w gave larger spacings. This terminal value is to a lesser extent also affected by crystallization conditions. In isothermal crystallization, a higher T_c leads to a larger terminal spacing than a lower T_c,[99] but, on the other hand, quenching again gives larger terminal values.[100] The effect itself is reversible, the spacing reducing to its original unswollen value on drying even if, on repeated cycling, the swelling effect, *i.e.* the spacing increase on solvent addition, becomes reduced,[101] and increasingly so for swellings at higher temperatures. The above hysteresis with cycling means that the swellable surface layer can become progressively immobilized, which would indicate subsequent perfectioning of the surface structure under the influence of the swelling agent. Such immobilization, as indicated by the reduced swellability, possibly associated with belated ordering, can occur also without any swelling agent merely on storage of the mats under ambient conditions on the time scale of weeks and months, being another indication of the variability of the fold surface in this instance with time. It follows that in all these studies the time factor also needs to be considered (otherwise the notion of apparent irreproducibility can arise, which could possibly account for some differences between the results from different sources[100]). Specifically, in order to have the swelling effect in its most pronounced form, freshly prepared mats are advantageous. For high M_w material ($M_w \approx 10^6$) with fresh mats, the spacing increase on swelling can be substantial: at room temperature, ~35%; at 80 °C, nearly 50%.

All the above is in line with the message of the gold decoration technique referred to above, where the swelling effect, while unquantifiable, was revealed by the real space image of the crystal. It is significant that, in contrast to SAXS, the Raman LAM mode registers hardly any change as a result of swelling.[101] For example, the LAM technique records a length increase of ~5% against ~50% in the case of the SAXS technique for the sample showing the maximum SAXS effect quoted above. Here, and in all other cases examined, the LAM and SAXS values in the initial dry state were found to agree within ~5%, when chain inclination within the lamellae is allowed for, using the simplest frequency–length conversion appropriate to an isolated rod (see Sections 7.6 and 7.9). (Other conversion formulae give lower values for the LAM length without affecting its insensitivity to swelling, which is our main point at this stage.)

In addition to the variability of the fold surface structure, all the above findings indicate that there is a swellable disordered layer spread over the otherwise nonswelling ordered crystal core, the thickness of the former increasing with M_w. In the dried state this layer can only be very thin, and hence involving correspondingly little material. Exactly how little depends on the model underlying the assignment of the LAM length. By our own assignment through the simplest model[40,42] providing near fit between LAM and SAXS data, and for which further support will be quoted in Section 7.9 to follow, there can only be a very small amount, a mere 'smear' on top of the nonmobilizable but otherwise chain-folded surface.

7.8.3.4 Some further manifestations of order–disorder along the surface

A quite recent result using atomic force microscopy, used in friction force mode, has provided striking support to the above model. It was stated in Section 7.7.2 that pronounced anisotropy in friction was registered which differed in direction according to the sector selected, in line with the conception of a preferred fold plane direction which is parallel to the growth face. By recent work[68] this anisotropy greatly diminished as the molecular weight was increased while sectorization itself is, from other evidence, expected to remain.

It follows from all the foregoing that we would expect little, if any, disordered matter, specifiable in the form of the above indicated overlay, for material which is of very low molecular weight but with chains still long enough to crystallize by folding. Section 7.9 will in fact be devoted explicitly to the crystallization of such short chains. At this point, merely a very striking finding of long standing will be illustrated in the form of Figure 30. The fringe and network structure in Figure 30 arises through diffraction contrast and is attributable to a dislocation network[102–104] along the interface where two overlaying layers are in contact along their basal planes. Dislocations are crystal imperfections which in this case arise through mismatch of the crystal lattices of the two rotationally slightly misoriented layers. Even if they are defects in the crystallographic sense, they rely on interaction between lattices in contact, in other words on crystallographic regularity along the surfaces. The underlying misfits and associated energies can in fact be analyzed,[103,104] but this is beyond the scope of this review. What is more pertinent from the present point of view is the fact that the occurrence of such dislocation networks could be associated with short but still folding chains, which is consistent with a crystallographically definable fold surface and the absence of substantial looseness such as, by the foregoing, we have attributed to the influence of longer chains. (Assertions[105] that dislocation networks are not relevant to chain folding as they are confined to crystals with fully extended chains are not tenable in view of the evidence[104] to the contrary: the nitric acid degradation experiments there yielded, amongst others, double and treble traverse chain lengths. This is not to deny that crystals with shorter chains crystallizing in the extended mode — Section 7.9 to follow — can also give rise to dislocation networks, without detracting from the argument relating to chain folded crystals.)

Figure 30 Polyethylene crystal showing dislocation network between two overlaying layers. Electron micrograph (after Sadler and Keller[103]).

Finally, there is the ongoing quest to see the fold directly. In principle, the new technique of atomic force microscopy should be able to achieve this. As matters stand at the time of writing, periodicities appropriate to certain lattice spacings of PE were made visible along the basal surface of single crystals.[106] In one crystal area (so far) there were additional directional features visible which were qualitatively and quantitatively consistent with their representing actual chain folds. This is an important pointer which invites further following up. For this, in addition to the technical difficulties involved, the variability of the fold surface with preparation conditions, and even between different regions in the same crystal, as laid out in this section, will need to be taken note of. Even, apart from the issue of seeing the folds themselves, the mere fact that the lattice, a feature of the crystal interior, can be seen by a technique specific to the surface only is of significance in the present context. Namely, it reveals that the lattice periodicity is at least imprinted onto the surface, which must have bearing on any structural disorder that can be visualized there. That directionality is present along the surface has already been shown by the epitaxial surface decoration method (Section 7.7.2).[66] The new findings now extend this theme to the much higher resolution of the crystal lattice.

Another route for diagnosing the fold directly is by giving special attention to crystals of very short chains. This is a subject in its own right and will be covered by the section to follow. As will be seen, in addition to much else, directly or indirectly, such works also provide further information on the fold.

7.9 CRYSTALLIZATION OF SHORT CHAINS; THE ONSET OF CHAIN FOLDING

7.9.1 The Issues, Background and Scope

It has long been familiar that short, chain-type nonpolymeric organic molecules crystallize by forming lamellae with the chains perpendicular, or at a specific large angle to the lamellar surface. Here the chains are extended and the smallest lamellar thickness is equal, or simply related to, the length of the molecule, corresponding to the basic crystallographic repeat, *i.e.* to the unit cell dimension. Thus, here the basic repeat, and hence the crystal structure, and molecular identity are closely related. The question clearly arises as to the relation between such extended chain 'oligomeric' crystals and their chain folded polymeric counterparts. Or more specifically, at what chain length does chain folding set in and how? Or looking at it from the other direction, what are the structural characteristics and mode of crystallization of short polymeric chains of length at, or closely above which, chain folding sets in?

In order to address the above questions, chain molecules are needed which are not only in the required length range but are also of uniform length. Ideally, to investigate the transition from nonpolymeric to polymeric behaviour, chain molecules of increasing length would need to be synthesized through controlled step by step addition of monomers (or small well defined multiples of it) by traditional methods of organic chemistry, a very demanding procedure in the length range in question. Instead, the subject of polymers developed along the familiar route of polymer chemistry, yielding products with a statistical distribution of chain lengths, in principle unsuited for addressing the problems posed in this section. Yet it deserves a record that even the earliest attempts, using low M_w polyethylenes with appreciable M_w distribution, placed the chain length where folding should set in, in the range of 100–200 carbon atoms.[107] Oligomers of well-defined uniform chain lengths did become available in the polyamide family, where the onset of chain folding was inferred to set in beyond a specific fixed length at which the SAXS spacing became independent and unrelated to the length of the molecule.[108–110] The inference at that time was that, beyond a certain chain length, the chain ends start folding back into the crystal. This inference is at variance with the later findings on other oligomers to be reviewed below, calling for a reexamination of the polyamide system.

The first stage of the developments to be outlined was using poly(ethylene oxide) (PEO) grades that, thanks to new polymerization procedures at the time, became available for this purpose. The chain length was not strictly uniform yet of sufficiently narrow M_w distribution to provide the distinctive effects which emerged. The next stage was work on the *n*-alkane–PE class materials, in this instance of strictly uniform lengths having become available through traditional organic synthesis along two different routes.[111,112] The results of at least some of these latter investigations then fed back into the work with PEO, leading to a renewal of activity with this polymer combining some of the distinctive features of each of the two systems.

7.9.2 Investigations on Short-chain Narrow Molecular Weight Distribution PEO

The first work on short-chain, very narrow distribution PEO was concerned with the SAXS spacings given by materials of different chain lengths as crystallized at different temperatures. In the course of extensive works[113,114] it emerged that l varied in discrete steps with T_c, with the spacing values corresponding to the mean chain length and integer fractions thereof. For the highest T_c it corresponded to the extended chain and for lower T_c values progressively to 1/2, 1/3, *etc.*, of this length. The latter corresponds to once, twice, *etc.*, folded chains. Clearly, this means that the chain ends are outside the crystal, a conclusion already reached with PE of conventional chain length[92] but in this case with the additional feature of it needing to lie flush along the surface.

The major groups of work[115,116] which followed were concerned with the morphology (see review[117]). Rather remarkably, crystals were grown within thin layers of melt, where the crystal lamellae could be observed to develop in their flat-on basal plane view under the light microscope. They could be made visible by a 'self-decoration' method, through which the crystal edges became bright when viewed between crossed polaroids, together with steps, asperities and other irregularities within the confines of the basal surface (Figure 31). In this way, many features of the crystal could be registered at any selected stage of growth. Amongst much else it was observed that within a growing crystal, at some stage of growth, thickening sets in, usually in the centre of the first grown lamellae (*e.g.* the top pair in Figure 31). This thickening itself was by a multiple fraction of the chain length (say if the crystal first grows as a once-folded chain, then the thickened part will be an extended chain, and so on). The growth of the thickened portion can then be followed simultaneously with that of the original layer further out. The two growth rates are in general different, and the innermost thicker layer can catch up with the outer one (see a recent simulation[118] to be quoted again later). It can also take up different lateral habits and, as seen from the example in Figure 31, it can give rise to a variety of morphological effects. As implied already, the lateral growth can be followed from micrographs such as Figure 31 pertaining to a particular temperature, and thus the lateral growth rate (G) can be determined as a function of temperature for each thickness. The particular point to be mentioned here is that while the rate increases with decreasing temperature, there are singularities in the form of cusps along the log G vs. T curves (Figure 32). It was found that the T values associated with these discontinuities correspond to the stability limit of the next shortest fold length. Thus, for example, when following the rate of extended chain crystallization downwards in T, the discontinuity in the G vs. T curves occurs at the T value where the once-folded crystallization sets in, there being a new branch of the G vs. T curve for each fold length. This foreshadows the much more pronounced effect resulting in an actual reversal in growth rates observed in the even more homogeneous system of the next section.

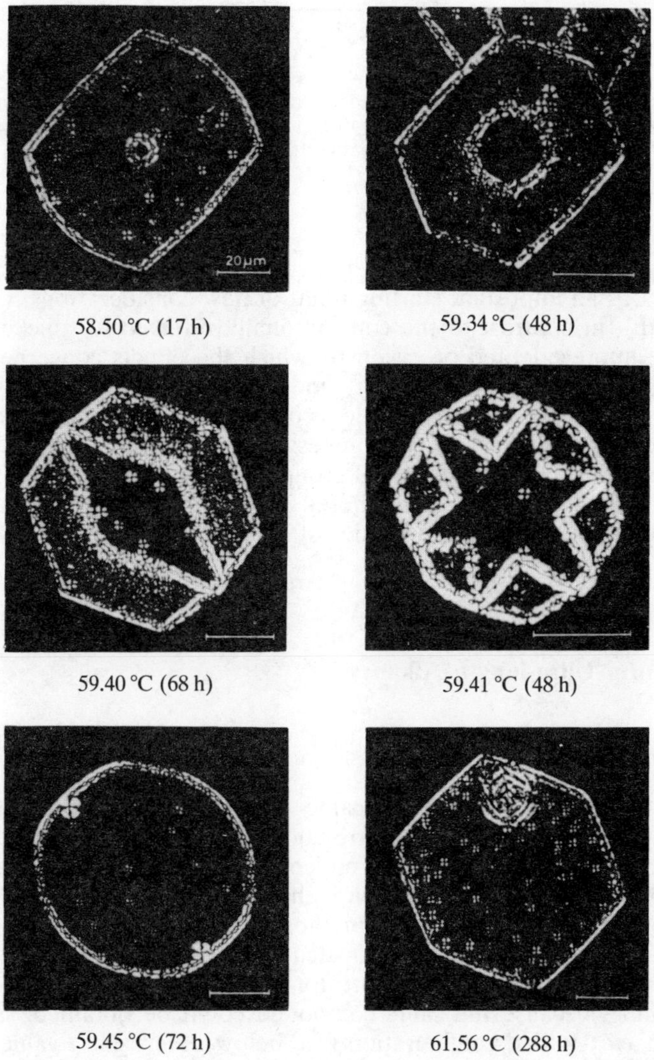

Figure 31 Single crystals of polyethylene oxide of $M_w = 6000$ grown from the melt at a range of closely spaced temperatures in the temperature range where the transition from once-folded to extended-chain growth occurs. Bars represent 20 μm (after Kovacs *et al.*[115]).

Out of the numerous features, conclusions and inferences which arise from the above experiments, we mention two here. First, that the formation of the thicker interior layer is not equivalent to the thickening processes referred to under Section 7.6.2.3. The latter corresponded to rearrangement within crystals already formed, hence to secondary perfectioning, without more crystal arising. In contrast, the effect presently discussed involves the

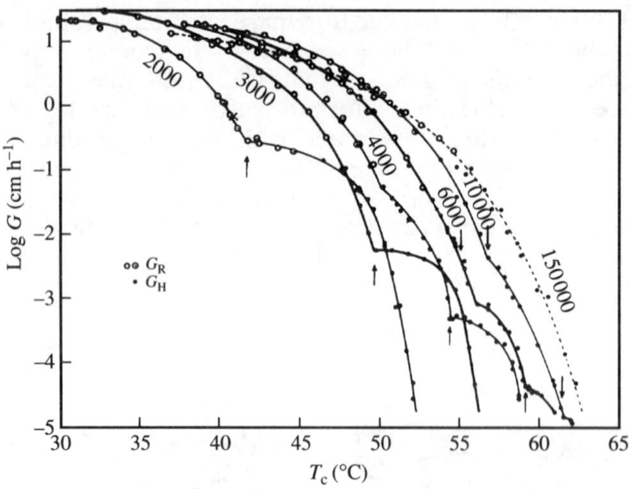

Figure 32 Growth rate (G) versus crystallization temperature (T_c) plots of six PEO fractions (M_w as indicated by numbers) of melt grown crystals. Transition temperatures are indicated by vertical arrows (after Kovacs and Gonthier[116]).

creation of new crystals, as part of primary growth. This distinction is basic and will emerge again in Section 7.11 as an important starting point of new considerations, where we shall term it 'thickening growth' (in contrast to the conventionally referred to 'thickening'). The second point of note is the unprecedented precision by which the effects concerned are manifest. As apparent from Figure 31, the effects seen, and hence also the underlying chain folding behaviour, respond to changes in T of as little as 0.01 °C, and this in a reproducibly controlled fashion. With reference to the issue of randomness versus regularity, in this special instance at least there seems to be no room for statistical grouping: the chains seem to 'know' exactly what to do and when, a point to be remembered in any more general quest towards the understanding of the organizational capabilities of macromolecules.

7.9.3 Strictly Uniform Ultra-long *n*-Alkanes

7.9.3.1 Structural aspects

Once the strictly uniform *n*-alkanes became available along synthetic routes,[111,112] the research to follow set out to probe the correlation between lamellar thickness (in this case related to the crystallographic unit cell) and molecular length, or alternatively, to establish the point of cessation of such a correlation, which is the signal of the onset of chain folding. SAXS was the principal tool, but the Raman LAM method was also used. The agreement between the two techniques was practically complete provided the simplest model of a vibrating rod was used (see Section 7.6.2.1) in strong support for this model. In this case the argument is particularly compelling, because the same compound could be obtained, say in the extended and the once, twice, *etc.*, folded configuration (see below). Here the l values as obtained from the Raman LAM mode all fell on the same line in a length vs. reciprocal frequency plot, conforming to the simplest rod model.[119]

The results showed that the chains did indeed fold, starting at around 100–150 carbon atom lengths, the precise value depending on crystallization temperature.[120] The fold lengths in the final stages of crystallization were found to be integer fractions of the chain length, changing by such fractional lengths with changes in crystallization temperatures and/or annealing conditions. This falls in line with the preceding results on PEO, but in this case, thanks to the strict uniformity of the chains and in the knowledge of their exact lengths, the portion of chain within the fold region could be assessed. This, in the most regular cases at least, amounted to no more than 2–4 carbon atoms.[121] Such cases also revealed specific vibrational bands in the IR spectrum, providing access to the study of the structure of the fold itself.[122] Further, by analysis of the SAXS intensities the interlayer could be assessed; this, for once folded crystals, could be as narrow as 0.4 nm per fold surface, in the most regular case. This is actually narrower than the interlayer of 0.7 nm found for melt-crystallized extended chains.[123]

It was recognized by synchrotron-based X-ray work[124] that the folds were not always sharp by the above-described experimental criterion. In fact, it was observed that crystallization starts with rather imperfect crystals, with significant disorder along the fold surface, where the fold length itself is not closely of integer fractional value. However, during isothermal crystal growth the crystal can still perfect itself, in the course of which the fold length either increases or, rather surprisingly, decreases to its nearest lying integer fractional value, with a concomitant regularization of the fold surface. The process, as envisaged presently for a once-folded chain, is illustrated schematically in Figure 33, the meaning of which, with the help of the caption, should be self-explanatory.

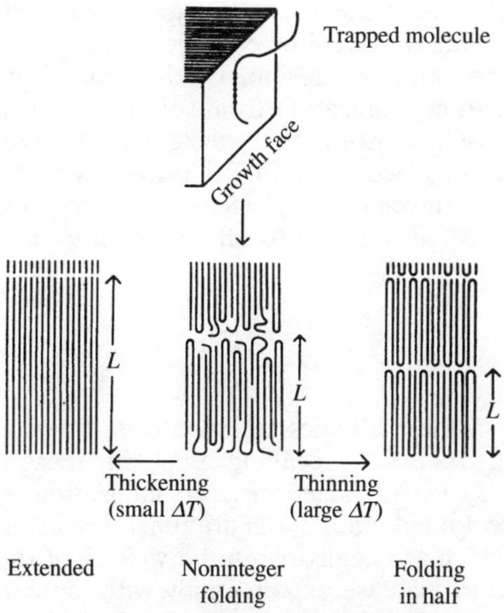

Figure 33 Schematic representation of crystallization of a long *n*-alkane of uniform length. Top: initial deposition, in general unfavourable by probability, for formation of either fully extended or exactly once-folded conformation. The optimum kinetic pathway leads to a once-folded crystal with disordered fold and internal defects (bottom row middle). The latter can then rearrange subsequently in two ways: either fully chain extend by thickening (bottom left) or achieve a precisely once-folded state with ends at the surface and folds regular through thinning (bottom right) (adapted from Ungar[119] based on Ungar and Keller[124]).

At this point the following features deserve highlighting, particularly with reference to the previous sections. First, consider the fold surface problem. It seems that the folds can be sharp with identifiable structure features and, at least for up to four folds per chain (the maximum number achievable with these compounds), must be adjacently re-entrant. (Note that this compares with the average degree of stem adjacency in Figure 26 derived for a 'real life' polymer.) However, as just seen, there can be disorder even for adjacent re-entry and even for such a short chain. Here the nature of the final product will depend on the stage at which crystallization and rearrangement is interrupted by cooling. Thus the practically achieved fold surface is not unique and can span a spectrum of order from near amorphousness to crystallographic. This is also reflected by the interlayer thickness as assessed by analysis of the SAXS intensities[123] quoted above, which showed a spread of 0.4–0.75 nm according to crystallization conditions, the largest value being from a fast-cooled melt-crystallized specimen. The most important point, however, is the pronounced drive for the chains to fold and the trend towards maximum regularity in the folded state which, depending on the degree of rearrangement allowed, they may achieve to varying degrees. Note that the variability of the fold surface structure and the role of rearrangements have been points of emphasis throughout this review, here emerging forcefully with these model compounds.

The second and third points relate to the mode of deposition and subsequent rearrangements, the two being in a way interrelated. The scheme in Figure 33 will reveal the emphasis on probability-controlled deposition, and hence on the entropic nature of the initial act of growth, a concept to feature again in the theoretical section in a different, yet not quite unrelated, context. The chain will attach itself according to the laws of probability and will 'stick' provided that the minimum thermodynamic stability criterion is satisfied. Optimization, and hence the resulting maximum rate, will not correspond to the most perfect structure even within the limits of a particular fold number (one fold, in Figure 33). Perfectioning will then be a further stage which may be subsequent to growth, or may still be part of the rate determining step.

At this place we can only draw attention to the numerous and noteworthy morphological features emerging from the study of *n*-alkane crystals.[125,132] Amongst these is the issue of sectorization. As assessed by the surface decoration method (epitaxial deposition of paraffin crystals with directionality corresponding to sectorization;[66] see Section 7.7.2), there is no sectorization in crystals formed by extended chains, but the sectors appear conspicuously in all chain-folded crystals[125] even where the chains are only folded once. The crystals are all capable of displaying interfacial dislocation networks[125] which, as stated in Section 7.8.3.4, are indicative, amongst much else, of some crystallographic regularity along the local surfaces, irrespective of whether consisting of folded or extended chains, the emphasis here being on its significance in the case when folding is present. Further notable features arise in the case of chain refolding in the course of crystallization from solutions. It is noteworthy that, in the case of regular *n*-alkanes, such refolding (isothermal thickening) does take place even in the course of crystallization from solution, in contrast to PE proper where the solution-grown crystals always appear in their initial unthickened form (Section 7.6.2.3). In the case of *n*-alkanes, such isothermal thickening effects are also linked to rate considerations, the point to which we turn next.

7.9.3.2 Rate effects

A further unusual feature of the *n*-alkanes emerged from the study of crystallization rates as a function of temperature. It needs recalling that, in these alkanes, on lowering the crystallization temperature T_c (*i.e.* raising ΔT), crystallization will occur consecutively in extended, once folded, twice folded, *etc.*, configurations. The overall rates, G, would (in the absence of transport terms which are negligible in the vicinity of the melting point, the T range here concerned) be expected to increase exponentially with increasing ΔT. In fact, G, after an initial increase, was found to reach a maximum, then to dip into a sharp minimum and to rise again thereafter (Figure 34).[126,127] This was observed to apply for growth and primary nucleation both from melts and solutions, as assessed by X-ray diffraction, calorimetry and visually by microscopy.[126–130] At growth temperatures above the minimum the chains were extended, whereas below they were in a once-folded conformation, the changeover occurring at the start of the second rise. This seems to be the same effect as the discontinuity in Figure 32 observed with PEO, which in the strictly homodisperse *n*-alkane system proceeds further to a complete inversion of rates. This effect is unprecedented in any crystal growth; it is extremely sharp and strictly reproducible.

A proposed explanation for the rate inversion and consequent minimum in growth rate is as follows.[119,126] Downwards (in terms of T) until the minimum, only the extended chain deposition is stable but not that of the once-folded chains. However, even if transiently, once-folded depositions will still occur and their probability will increase along the growth face with decreasing T. Being subcritical, these once-folded nuclei cannot grow, yet they block the deposition of the chains in their stable extended conformation, and hence the growth of the crystal (and also of the primary nucleation) itself. This novel phenomenon, seemingly characteristic of polymers alone, was termed 'self-poisoning'. Indirectly it is also apparent in annealing/recrystallization phenomena in solutions, with some striking morphological consequences.[131,132] Clearly it is a manifestation of the general fact that the chain has to deposit in the 'right' conformtion to perpetuate the crystal, and hence of the probabilistic control of chain-folded crystallization (see Section 7.13.6 on theories below). Concurrent computer simulation[133] in fact did contain the above growth minimum, which became fully appreciated only after the actual experimental observation itself. In a more recent simulation, modelling the actual *n*-alkane system in question, in fact a close match with the observed curves for the rates was attained.[118] Reference to these points will be made again in the theoretical section below.

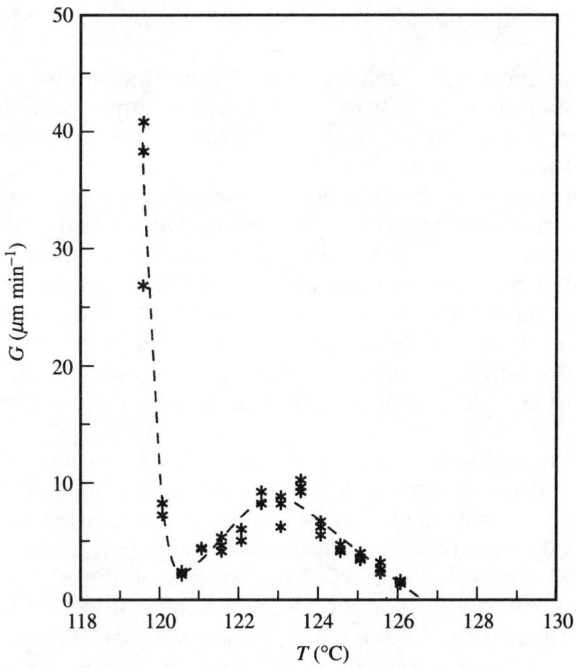

Figure 34 Linear growth rate (G) of crystals of $C_{246}H_{494}$ as a function of crystallization temperature (T_c) displaying a maximum and a minimum (after Organ *et al.*[129,130]).

There are current indications that rate inversion and associated growth minimum are not confined to the present n-alkanes. The likely connection with PEO in the form of Figure 32 has already been indicated. In fact, more recently an actual minimum has been observed also in this class of material (see Section 7.9.4).[134] In wider generality, a minimum in rate with T, when observed, is indicative of two competing processes being operative, where the one setting in at a lower temperature can block the other. In this wider sense the concept of 'self poisoning' has proved to be even of predictive power, productively being applied to account for an observed rate minimum in a different system (a copolymer of PEEK) to which here only a reference can be made.[135]

Having placed all the above emphasis on the rate inversion effect, it needs stating that in the work on uniform n-alkanes from the only other available source,[112] the actual existence of the effect has been negated.[136] There, in a calorimetric study of crystallization from solution, no rate inversion, only a kink in the rate curve, was obtained, on the basis of which the rate inversion effect itself, with all else it entails, is dismissed. Specifically, all previous observations to the contrary, it was attributed to erroneous interpretation of the actual experimental record throughout the entire wide ranging material. The present review is not the place for arguing out the rights and wrongs of the respective cases; for this the interested reader is referred to the original papers for forming judgement.

There are several further works based on uniform alkanes obtained by the alternative synthetic route mentioned above.[112] While broadly in line with the overall picture here presented (with the exception of the disputed minimum in growth rate), they concern the detailed analysis of the fusion process,[137] isothermal thickening,[138] and some aspects of morphology and crystallization kinetics[139] to which within the confines of this review only reference can be made.

7.9.4 Other Uniform Oligomers and Return to Narrow Molecular Weight Distribution PEO

The informative results on the uniform n-alkanes gave impetus to the synthesis and subsequent crystallization studies of similar oligomers of other chemical species. All round, the outcome was broadly similar, with some special features appropriate to the specific materials.

Here we merely place the appropriate references on record, which should be consulted for particulars.

One such material was the strictly uniform oxyethylene/methylene block oligomer.[140] Here chain folding set in at the chain length of ca. 150 chain atoms, with integral fractional fold length. Special attention was given to the refolding/annealing behaviour and the underlying intricacies of the crystal structure.

Quite recently the oligomers of PHB (poly-β-hydroxybutyrate) were synthesized.[141] While the original interest was biological (PHB is a bacterial polyester also usable as a thermoplastic), the oligomers proved useful for the study of polymer crystallization.[142] Namely, PHB gives chain-folded single crystals with considerably shorter fold lengths than PE, down to a minimum of 4 nm.[142] The latter means that the stage of folding is attained at a correspondingly lower degree of polymerization of the monomer. This stage, which in fact has been attained leading to integral folding of the corresponding oligomer, is now opening the way to fundamental crystallization studies using strictly uniform PHB as a model substance, with certain specific advantages over PE.

As a follow up of the *n*-alkane work, investigations were resumed again on narrow M_w distribution PEO. As already mentioned, a rate inversion such as in Figure 34 with *n*-alkanes could be registered, even if in a much less pronounced form, using the methoxy terminated variant of PEO.[134] This demonstrates, amongst other things, the connection between the rate curves such as Figures 32 and 34, in support of the contention that the latter rate inversion is a more pronounced variant of the former, no doubt brought about by the strict uniformity in chain length, the underlying retardation effect, which we attribute to self poisoning, being common to both.[126]

Other renewed studies on narrow M_w distribution PEO include an investigation of the primary stages of crystallization during which the initial formation of non-integer folding (NIF) followed by rearrangement to integer folding (IF) was also observed in the case of this polymer, in the form of either thickening or thinning as in the case of *n*-alkanes. This recognition was followed up by in-depth studies of this class of effects as a function of crystallization temperature, molecular weight and end groups.[143,144] The earlier morphological studies on this system (such as yielding crystals as in Figure 31) were extended in these renewed investigations, in the course of which certain habit changes could be associated with discontinuities in the growth rates.[145] Amongst much else, a rounding of facets at the discontinuities, rate reversal in particular, was seen, a point to be commented on further in the theoretical section below.

7.10 CRYSTAL GROWTH RATES

The kinetics of crystallization is a major subject on its own, which has been treated more explicitly elsewhere.[25,146] Here we are confining ourselves to some selected aspects that have a bearing on the core issues of this article, specifically on chain folded crystallization. Such aspects are the rate of growth of crystals as a function of temperature (T) and concentration (c), the latter in the case of solutions.

7.10.1 On Growing Uniform Crystals

Observing the growth of crystals individually *in situ* is only possible under some special circumstances. Such are the short-chain PEOs in Section 7.9.2, where the growth of crystals, as forming in the melt, can be followed (Figure 31) and the rates recorded (Figure 32). Another such case (growth of extended chain-type crystals from the melt) will be referred to in Section 7.11 below. In the more general situation, such as in standard crystallization from solution, a method of sampling at given stages of crystal growth is needed where it can be ensured that the crystals selected are representative of the crystal population as a whole at the particular stage of growth. This can be achieved by the self-seeding crystal growth method that yields crystals which, in a given preparation, are all of uniform lateral size, and if grown isothermally, of uniform thickness (*i.e.* having a fixed l value; Section 7.6.2). Also they are, to a good approximation, all monolayers, which ensures that the single parameter of a lateral dimension is an adequate representation of a given stage of crystal development.

The essence of the method[147] is to initiate crystal growth through seed crystals, which ensures that all crystals start growing at closely the same time, yielding crystals all of the same age at any given instance. If the seeds are plentiful the crystals will be restricted to small sizes over large portions of the total crystallization period, during which, in the best explored and utilized case of solution crystallization, they will have little chance to develop the usual sheaving and other overgrowth features, thus remaining essentially monolayers. While in principle this should be achievable through a suitable external nucleating agent, in practice it is most simply and reproducibly accomplished by relying on the last traces of the same material which have stayed behind on a preceding melting or dissolution process now acting as seeds, hence the term 'self seeding'. The particulars are contained in the original publications.[147,148] Here it will only be stated that the seed-forming residues contain the highest M_w end of the M_w distribution which, coupled with perfectioning on heating, have become stabilized against complete melting or dissolution. By choosing the dissolution temperature within certain specificable limits, the number of 'self seeds', and hence the size of the crystals which result on subsequent crystallization, can be predesigned, examples of which are shown in Figure 35.

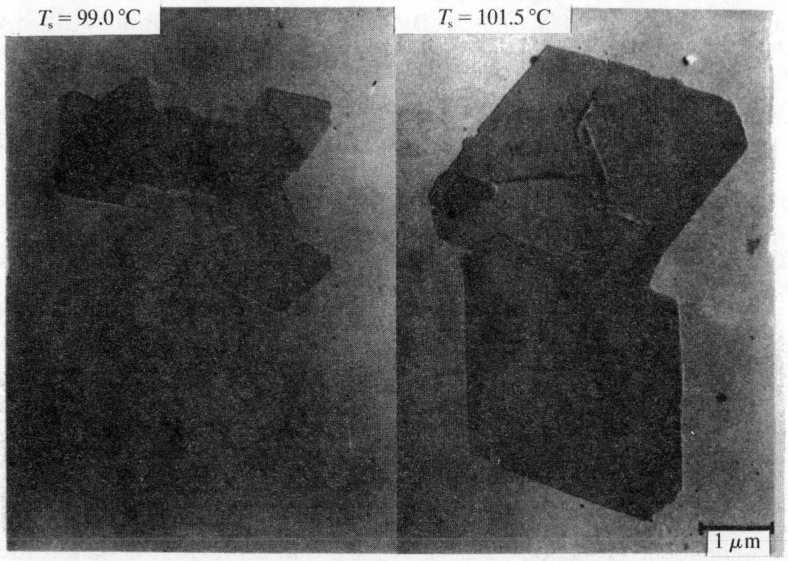

Figure 35 Uniform monolayer crystals of polyethylene grown from solution by the self-seeding method. The smaller crystals (left) originate from more seeds, the number of which can be controlled by the self-seeding temperature (T_s). Electron micrograph (after Blundell *et al.*[147]).

The last point may well give cause for reflection. Complicated as polymer crystallization may appear, and disputed many of the issues, in several respects they display a simplicity unrealizable in even the simplest crystals. Thus, they can be grown so that all crystals in a given preparation are closely identical in shape, thickness and size, where both the thickness (l) (by choosing the appropriate ΔT) and the lateral size (by the control of the number of nuclei through appropriate control in the 'self seeding' process) can be predesigned.

In possession of a uniform crystal population, the growth rate (G) can be determined. In most work this is done by sampling the suspension at different stages of growth, where the interruption of crystal growth at the intended T is imprinted on the crystal through a step, signalling a change in layer thickness. Further, with crystals all uniform and of known shape, the change in lateral size can also be followed through *in situ* turbidimetry.[149] It was found unanimously in the course of numerous studies that, with the exception of the late stages of the crystallization, where material depletion and pronounced fractionation can set in, isothermal lateral growth is linear with time. This allows the growth process to be characterized by a single, in the first approximation, size-independent rate parameter (G), the basis of the further studies to be surveyed. Also, the finding of a size-independent G parameter has an important impact on the theories (see Section 7.13.4.1).

7.10.2 The Temperature (T) and Concentration (c) Dependence of Growth Rates in Single Crystals

Over 30 years ago, using preparations which must have been accidentally of the self-seeded type, a growth law of the form

$$G \propto \exp\left(\frac{-K_G}{T_c \Delta T}\right) \tag{3}$$

was established (the constant K_G containing the material parameters).[150] Equation (3) was found to be universally obeyed in all subsequent work. Also it connects with the identical functional relation found earlier with spherulites.[151] Amongst other things in a study on PEO, complete identity between solution-grown single crystals and melt-grown spherulites could be established not only for the functional relation (equation (3)) but even for the numerical values of the constants.[152] Equation (3) is basic and together with equation (1) for l forms the two fundamental quantitative relations underlying the whole subject under review here.

The concentration dependence of G is also of note. It was found to obey a power law such as $G \propto c^{\alpha}$ where $\alpha \ll 1$. In fact, by the first work,[153] $\alpha = 0.3$ was systematically registered over a 100-fold concentrations range (0.008–0.8%). By subsequent investigations[154,155] it was found to be molecular weight dependent, higher M_w yielding lower values. Thus for very high M_w ($\geq 2.5 \times 10^5$), $\alpha = 0.2$ was found, while for low M_w fractions ($M_w \approx 10\,000$), α could be 1 or even higher.

The above expressed insensitivity of growth rates to concentration is unprecedented in crystal growth. Effects of diffusion could certainly be excluded.[153,156] The explanation must rest in the fact that the still unattached molecules, in the form of random coils, accumulate along the lateral growth face in the course of their much slower incorporation into the crystal. Accordingly, the face will become shielded from the solution interior and thus would not 'respond' to changes therein, 'sensing' only the concentration within the local environment which would be largely that corresponding to the segment density of the random coil. At extremely low concentrations, this shielding effect is still incomplete and its development can be followed.[157] In fact it was found recently[158] that it is describable by a Langmuir adsorption isotherm. For a discussion of the implication of this evidence for crystallization theories, see a review.[159]

The issue of the local concentration as a controlling factor in chain folded crystallization has come up previously in a different context, namely to account for the invariance of l with concentration, and hence supersaturation (Δc), seemingly in contradiction with its observed pronounced ΔT dependence (Section 7.6.2.3). We see that a previous apparent thermodynamic anomaly could find explanation by the concept (*i.e.* local concentration) which, quite independently, has also arisen in search for an explanation of an unusual kinetic effect. Thus the thermodynamic and the possibly better substantiable kinetic reasonings are mutually supportive.

7.11 EXTENDED CHAIN-TYPE CRYSTALLIZATION

7.11.1 Basic Features

A separate line of approach in polymer crystallization, still centred on PE, has arisen from investigation of the crystallization of melts under high hydrostatic pressure ($P \geq 4$ kbar) (see textbooks[24,27] and a review[160]). Here in a material of distinctly different consistency, lamellae of unusually large thickness (up to several μm) can be observed, compatible with full chain extension (crystals can even be thicker than this, see below) (Figure 36). This 'extended chain-type' crystallization arises through the chains first folding but extending continuously immediately behind the lateral growth front. In this case the crystals, as they grow, are wedge shaped (at least before they impinge on others) (Figure 37), in contrast to growth with uniform (and much smaller) thickness in the main stream of polymer crystallization studies. In the case of PE, this type of growth was found to occur in the hexagonal (as opposed to the usual orthorhombic) structure, where the chains are highly mobile, and correspond to a phase with liquid-crystal-like characteristics.

Figure 36 Extended chain-type morphology in polyethylene crystallized under pressure. Electron micrograph of fracture surface replica (after Wunderlich *et al.*[161]).

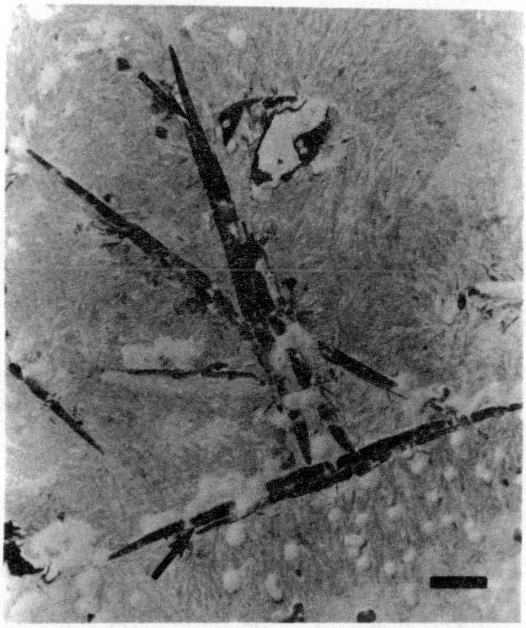

Figure 37 Cross sectional view of extended-chain-type crystals of PE while growing in the hexagonal phase at 3.2 kbar pressure. Where in isolation, thickening growth proceeds unimpeded concurrently with lateral growth, giving rise to straight wedge-shaped profiles. Where layers impinge face-to-face, primary thickening growth stops and (the much slower) secondary thickening takes over. Electron micrograph of a fracture surface from a pressure quenched sample. Scale bar = 3.5 μm (after Hikosaka *et al.*[162]).

This hexagonal phase, intermediate between the usual orthorhombic crystal and the melt, is metastable at atmospheric pressure but becomes the stable one beyond a triple point at elevated pressure, allowing the type of crystallization just described (Figure 38). High pressure is not the prime condition in itself, merely as far as it opens up a mobile crystal-phase regime. In some other, less well studied and less tractable polymers (*e.g.* polytetrafluoroethylene), such a mobile phase may exist even at atmospheric pressure with the consequent distinct extended chain-type morphology. This whole extensive field cannot be surveyed in further detail within the space restrictions of the present review. In what follows, a few items will be selected from recent and still-ongoing research that promise to have forward looking consequences, to be taken up again in the last section of this review.

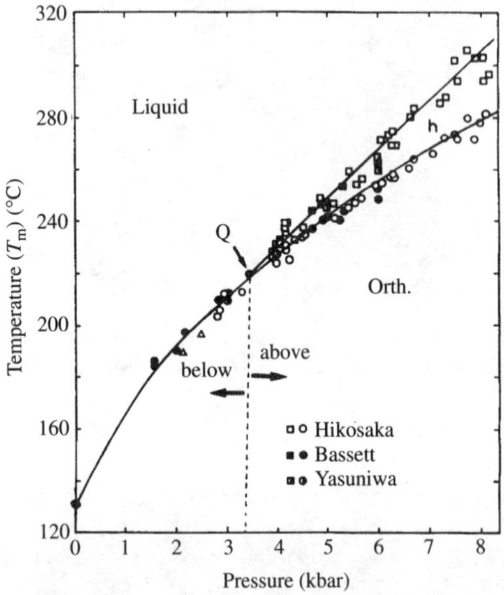

Figure 38 p–T phase diagram of PE displaying the appearance of a hexagonal (h) phase region at elevated pressure beyond triple point Q. Experimental points are from different sources (after Hikosaka *et al.*[163]).

7.11.2 Some Facts of Recent Emphasis

Against the above general background, the following two further factual recognitions need to be added, particularly as they provided the impetus for the new scheme to be presented in Section 7.13.8 of this review.

7.11.2.1 Thickening growth

First, in the case of the hexagonal phase, the crystals, when in isolation, grow continuously in the thickness direction from the folded to the extended-chain state and, most significantly, beyond, until stopped or slowed down by impingement on other layers (Figure 37).[162] It is important to recognize that this is primary crystal growth (first mentioned in Section 7.9.2 with PEO). We call this 'thickening growth' to distinguish it from the long familiar 'lamellar thickening' which, as referred to in the past, usually corresponded to rearrangement within lamellar stacks, *i.e.* after impingement, consequently to secondary crystallization (see caption to Figure 37). The above distinction is fundamental, as the two processes obey fundamentally different laws.[162] In molecular terms it involves a sliding diffusion mechanism by which chains, deposited first in the folded form along the leading edge, can refold towards chain extension. This process forms the basis of a new theoretical approach (see Section 7.13.7).

Originally it was thought that chains must also be 'sucked' in along the basal planes (Hikosaka, as quoted by Keller[20]). However, latest results suggest differently. Here[163] lateral (V) and thickening (U) growth rates were measured separately *in situ* and both were found to obey equation (3), V being much larger (by a factor of 10). It follows, amongst others, that unobstructed extended chain-type crystals grow with constant shape. Whatever the ultimate explanation, the consequences of the new finding could well be far reaching for our understanding of crystal growth and molecular dynamics within the crystal. (See also Notes Added in Proof, p.302, no. 2.)

7.11.2.2 The role of metastable phases

The second group of new facts centred around the prominence of metastable states. Namely, it is observed that crystallization will proceed in the hexagonal phase even when the system is in the orthorhombic phase stability regime, where the hexagonal crystal is metastable.[162,164] While such metastability has been reported previously,[165] it has now re-emerged particularly forcefully. At least in the part of the phase diagram considered, the crystal in the metastable hexagonal phase (h) was the only one capable of formation and growth. Growth in fact stopped on transformation into the stable orthorhombic (o) phase (at least as seen on the timescale of those experiments). This h$\rightarrow$o transformation and consequent stopping (or drastic slowing down) of growth (both thickening and lateral growth) was found to set in after a certain lamellar thickness had been attained, more precisely depending on the undercooling,[164] an illuminating but no doubt extreme case of competition between the two phases first mentioned in 1974.[165] The significance attached to this preference for metastable phases should become apparent in Sections 7.13.7–7.13.9.

7.12 AN INTERIM ASSESSMENT

The totality of the material presented so far deserves some reflecting. The overwhelming impression is the remarkable organizational capability of a chain molecule like that of PE, given the appropriate circumstances. This is borne out by the variety, distinctness and regularity of the resulting morphological features, of which only a small sample has been reproduced here. This is seemingly at variance with the traditional conception of an interlocked assembly of chains with organizational capability confined to within the entanglement points, at least in the case of melts (or concentrated solutions) of conventional molecular weights, say above 10 000 in PE, which is above the entanglement threshold. We see that not only can the chains fold, with some uncertainties and arguments about the regularity of the folding, but the chains when in the hexagonal mesophase can even completely extend within the crystal in the course of thickening growth. This means that either they have never been truly entangled or if they were, they can slip out of their mutual entanglement.

The question arises therefore whether there is any circumstance where entanglements are in evidence in the crystallization behaviour? One indication is that very high M_w material gives poorly defined crystals. While this is a general trend, it is not clear whether it is truly intrinsic and that slow enough crystallization would not re-establish the morphological regularity. Unpublished studies in Bristol (R. Law and P. J. Barham, 1990) show that, at least in the case of solutions, the latter can take place. More conclusive evidence for the existence of entanglements in crystallization behaviour comes from kinetic studies. Here, for the model polymer poly(tetramethyl-*p*-phenylenesiloxane) investigated for this purpose, it was found that above a certain M_w the crystallization rate became M_w independent.[25] (Magill[25] only asserts this for identical T_c values, but the same conclusion is also supposed to hold when referred to identical ΔT, thus taking account of the M_w dependence of T_m (discussion contribution by Magill at Faraday Discusison 68, which remained unrecorded).) This means that here the chain does not 'know' its length on crystallization any more, which is at least consistent with it being confined by entanglement points. Clearly at this stage the effect is merely a pointer for further work along such lines. In the light of it all the theoretical considerations to be described in what follows will regard the chains as in isolation, unimpeded by entanglements, neighbouring chains such as in high concentrations or in the melt merely providing viscous hindrance expressable by an appropriate transport term with which we shall not be otherwise concerned in specific detail.

7.13 CRYSTALLIZATION THEORY

7.13.1 Introductory Remarks

In the previous section the experimental background has been laid for the discussion of the theoretical understanding of polymer crystallization, as it has developed over the past 40 years or so and presents itself to us today.

The description and analysis of polymer crystallization is generally based on concepts developed in the much older sciences of thermodynamics and crystal growth of low molecular weight materials. Their application to the case of polymeric materials has often been fruitful, but sometimes at the cost of drastic assumptions and simplifications.

The difficulties arise usually because of the long-chain nature of the molecules, which is no longer identical with the basic crystallizing unit, the latter comprising only one or a few monomers. Furthermore, the connectivity of the crystallizing units along a chain acts as a hindrance inherent to the system.

Such fundamental but also much more specific questions have led to often controversial and intricate discussions within the community of active researchers. We would like to use this opportunity to provide an overview of the different theories and models, as well as pointing out some possible future directions. In particular, after introduction of the basic thermodynamics pertaining to chain-folded crystals, the main theories of the kinetics of chain-folded crystallization will be outlined, followed by a discussion of the issue of crystal shapes (or habits) (see Section 7.13.5) which, though originally regarded as of rather peripheral interest, has in recent years become a focus for the development and testing of crystallization models. In a somewhat similar fashion, the special case of crystallization of short chains (see Section 7.13.6) has evoked a strong debate between proponents and opponents of kinetic theories based on surface nucleation.[166] As well as outlining the facts at the centre of the controversies, we shall aim to point out some common ground and possible synthesis. We shall then go on to discuss some issues which have come into the fore in recent years. First, the description of thickening growth (see Section 7.13.7) and then the case where two or more crystal phases compete kinetically as a result of their relative thermodynamic stability (Sections 7.13.8 and 7.13.9). Finally, we shall return to the chain molecule itself, considering its conformation in the liquid phase, in particular at high undercooling, already mentioned in Section 7.2. First of all, however, the following section will summarize a number of salient experimental facts, all of which have been discussed in more detail above.

7.13.2 Summary of Experimental Evidence

The following summary of experimental facts should both serve to act as a guide through the huge amount of experimental evidence on polymer crystals and crystallization as well as laying out what any theory must aim to explain.

(1) Polymer crystals typically grow only laterally to form very thin (~10 nm) lamellae which, under given growth conditions, have a uniform thickness (Section 7.6).

(2) In folded-chain crystal growth there is only a limited amount of thickening as a secondary perfectioning process. However, under certain conditions of temperature and pressure a dramatically different mode of growth is observed which we term as 'thickening growth' (see Figure 37[22]). This happens in a so-called 'mobile' phase (*e.g.* the hexagonal phase in polyethylene), which allows substantial longitudinal chain movement. Wedge-shaped crystals are formed which can have a thickness larger than the length of an extended chain (Section 7.11.2.1).

(3) Crystallization has been found to proceed in the mobile phase even below the phase boundary to the proper crystal phase, thus raising questions about relative stability and kinetic competition (Section 7.11.2.2).

(4) In folded-chain growth the 'fold-length', or lamellar thickness l, has been found by a number of methods to be inversely related to the degree of undercooling ΔT, as given by equation (1) in Section 7.6.2.2, except for the highest undercoolings where l tends towards a constant value.

(5) The above relation applies at any time, independent of substrate or crystallization history (see Figure 5).

(6) The lateral growth of folded-chain crystals proceeds at a rate which depends exponentially on undercooling according to equation (3) in Section 7.10.2. The front factor in that

equation contains mainly the effects of the rate of transport of polymer to the crystal surface.[167] Since we are considering the actual crystal growth process, this is not our immediate concern. We must note, however, that as the glass transition temperature is approached, the rate of transport becomes the dominant factor, slowing down crystallization, and any possible error in correcting for this effect will greatly affect the crystal growth data. Hence caution must be applied in the interpretation of the commonly used plots of polymer crystal growth data as $\log G/G_0$ versus $(T_c\Delta T)^{-1}$ (following from equation (3)). Finally, the constant K_G derived from such plots may have different values in different ΔT ranges (*e.g.* Figure 39). Hence there appear to be certain growth 'regimes'. Their theoretical interpretation has caused substantial controversy (see Section 7.13.4.1).

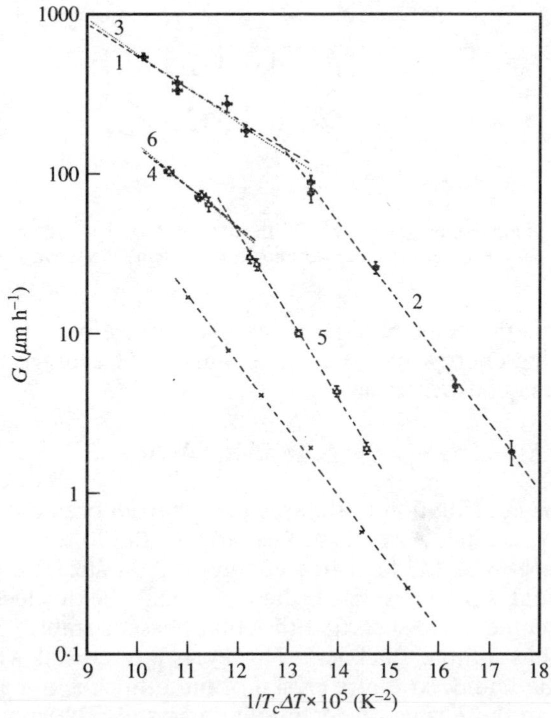

Figure 39 Growth rate plot of solution-crystallized polyethylene single crystals according to the expected functional dependency given by equation (3). The data fall on straight lines with a break in the slope, indicating the existence of regimes (after Organ and Keller[149]).

(7) Where crystals are morphologically well defined, the plane of folding is parallel to the growth face (sectorization, see Section 7.7).

(8) The lateral shape (or habit) of folded-chain polymer single crystals tends to be facetted at low temperatures and curved-edged at high crystallization temperatures (see Section 7.6.1).

The following subsections will be concerned with the thermodynamics and kinetics of chain-folded crystallization, which takes the chain from its random state in the liquid phase directly into a folded-chain state in a stable crystal phase. This will be followed by the more general case involving the possibility of an intermediate (mestastable) phase, and finally by consideration of the chain conformation prior to crystallization.

7.13.3 Basic Thermodynamics of Chain-folded Crystals

A basic requirement of any theory involving a change of state is the evaluation of the difference in Gibbs free energy. In the present case this is related to the driving force for crystallization and establishes relationships between the physical parameters, such as temperature and thickness, which must be obeyed if a lamella is to be stable compared with the liquid state.

The following considerations are based on the idealized model of a lamellar crystal of homogeneous thickness *l*, as shown in Figure 40, surrounded by the liquid phase. Further assumptions arise from the fact that most real polymers are multicomponent systems, consisting of chains of a range of lengths and hence a range of lowest free energy states, melting points, *etc.* For that reason, phase boundaries in real polymer systems are never sharp, but rather there are regions of phase transition. Here we will consider only the simplest case of a lamellar crystal grown from a homogeneous melt or solution of infinitely long chains (*i.e.* having no ends). For a discussion of the effects of finite chain length, see a review.[159]

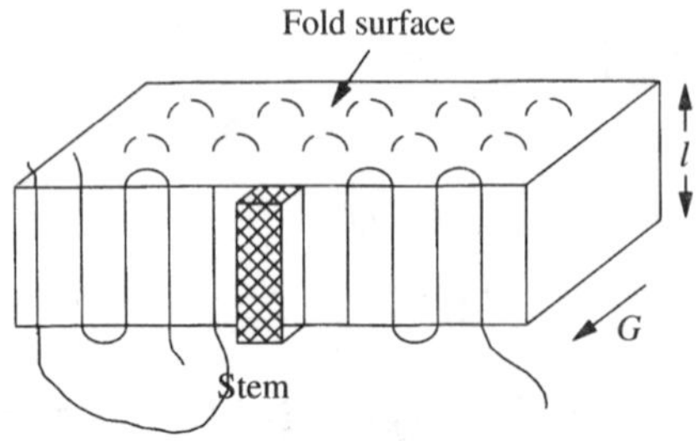

Figure 40 Schematic model of a polymer single crystal with chain folding. The attachment of a section of a molecule on the growth face, forming a new 'stem', is shown.

Let the surface free energy per unit area of the fold surface of a lamellar crystal be σ_e, then the increase in Gibbs free energy due to a single lamella, of average thickness *l*, immersed in the surrounding liquid may be written as

$$\Delta\mathcal{G} = \sum\nolimits_{lat} + 2A\sigma_e - \Delta F A l \tag{4}$$

where Σ_{lat} represents the contribution to the free energy from both the thin lateral surfaces and from edges bounding the crystal, A is the surface area of the fold surface ($2A$ is the total area of both fold surfaces) and ΔF is the bulk free energy of crystallization per unit volume.

If the area of the fold surface is large then Σ_{lat} may be neglected compared with the remaining terms. At a temperature such that the solid phase is stable, ΔF is necessarily positive and the stable crystal has infinite thickness. However, any crystal which has $\Delta\mathcal{G} < 0$ will be stable compared with the liquid, so that a crystal of finite thickness may be metastable if there is a free energy barrier to the formation of an infinite crystal. The condition for $\Delta\mathcal{G} < 0$ is

$$l > \frac{2\sigma_e}{\Delta F(T)} \tag{5}$$

The lower limit of this inequality is the minimum stable thickness l_{min} of a lamella for a given ΔF. The latter can be expressed, in a first approximation, as

$$\Delta F(T) = \Delta H(T_m^0)\Delta T / T_m^0 \tag{6}$$

where $\Delta H(T_m^0)$ is the bulk enthalpy of melting per unit volume. Hence

$$l_{min} = \frac{2\sigma_e T_m^0}{\Delta H(T_m^0)\Delta T} \tag{7}$$

which shows that the minimum thickness of a lamella varies inversely with the undercooling. Equivalently, the melting point of such a lamella is depressed to

$$T_m = T_m^0\left\{1 - \frac{2\sigma_e}{l\Delta H(T_m^0)}\right\} \tag{8}$$

This equation, known as the Hoffman–Weeks equation,[44] is a particular form of the widely known Gibbs–Thomson relation which describes the melting point depression due to small size or curved surface.

7.13.4 Theories of Chain-folded Crystallization

Having discussed some equilibrium aspects, we now proceed to the crystallization theories themselves. It is now widely accepted that the experimental observations listed above can best be accommodated by theories based on the consideration of kinetic processes on the lateral growth faces. Nevertheless, for completeness we would like to mention other approaches. Firstly, there are the equilibrium theories[159] which are based on free energy calculations of the crystal, including linear vibrations and torsional oscillations. They predict that below a certain critical temperature there are two minima in the free energy, one as a finite thickness and one at infinite thickness. However, the growth kinetic observations, as well as the strong dependence of thickness on supercooling, cannot be explained by this approach. Secondly, a rarely quoted but nevertheless interesting approach[168] assumes that the chains form metastable folded clusters already in the liquid phase. These so-called bundles subsequently aggregate to the side-surfaces of the crystal. Although the lamellar thickness, determined by the average contour length of the loops within the bundles, is in reasonable agreement with experimental data, the model again does not address the growth kinetics. A third approach, in recognition of the strong negative temperature coefficients observed in the crystallization, primary nucleation proposes as the relevant process.[1] However, this approach does not account for the lamellar growth, and in particular the observation that the lamellar thickness does not depend on the substrate (point (5) above). Finally, the so-called 'molecular nucleation theory'[169] assumes that each molecule must independently form a stable nucleus before crystallization. This model is based on the observation of fractionation during crystallization,[170] a mechanism by which lower molecular weight components tend to remain in the melt (or solution). In particular, a temperature dependent 'critical molecular length' was found, above which molecules are accepted into the crystalline phase. This length diverges significantly above that predicted by equilibrium thermodynamics, particularly at low undercoolings. Likewise, the growth rate decreases with molecular weight at a given undercooling,[171] a trend which was reproduced by computer simulations which incorporated the molecular nucleation idea.[172] We shall come back to this point in Section 7.13.6, but now turn to the main stream theories of polymer crystallization, which are built on considering the kinetic factors operative on the lateral growth face.

Kinetic theories assume that the observed growth faces would have a range of possible thicknesses, each having a growth rate which depends on thickness. The preferred thickness is then that which maximizes the growth rate. Going one step further, the growth rate itself is regarded as the product of two opposing forces. As depicted in Figure 41, the thermodynamic driving force (related to $\Delta\mathcal{G}$) increases with the thickness of the crystal, whereas some 'barrier' term in the rates decreases exponentially with the thickness, leading to a maximum growth rate not much above the minimum stable one. Accordingly, one should expect to observe crystals corresponding to the conditions of this maximum, *i.e.* with a thickness $l = l_{min} + \delta l$, where δl is a small term. It is hence the underlying dependence of l_{min} on the supercooling (equation (7)) which is at the origin of the observed thickness dependence.

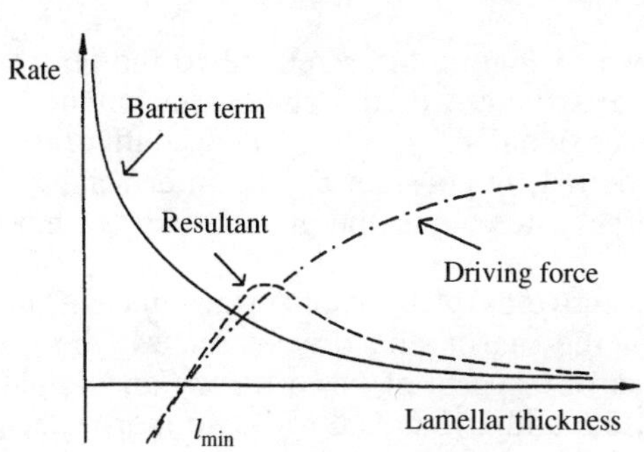

Figure 41 Crystal growth rate as a result of the product of a barrier term and a driving force term.

Logically, the next step must then be to find the origin of the barrier term. This is an issue which is still controversial today. The so-called 'surface nucleation' theories[166] invoke a barrier due to the increase in surface free energy as a molecule attaches itself to a flat portion of the growth face. Naturally this surface free energy and hence the barrier increases with thickness. The 'entropic barrier' theories,[173–175] on the other hand, argue that in addition to any surface free energy terms the attachment of a long-chain molecule in an 'orderly fashion' is slowed down because the molecule will explore many partially attached configurations, only a few of which are viable for further growth. As the number of unfavourable configurations increases exponentially with thickness, a barrier with the same functional dependency as that of surface nucleation is found.

Before discussing the surface nucleation and entropic barrier theories and their consequences in some more detail, we would like to introduce three abstract 'images' of the growth face and attachment kinetics. The first two, shown in Figures 42 and 43, are related to the surface nucleation model. Figure 42 represents the surface nucleation model proper: the kinetics on the side surface involves the addition or removal of so-called stems, *i.e.* molecular traverses of the whole face. The heat-of-fusion and surface free energies of the stems are considered as basic parameters of the model. Figure 43 is a further abstraction, namely a projection view looking down along the stems. The growth process involves simply the addition or removal of squares, by some abstract rates not *a priori* related to free energies or any other material parameters. We shall call models based on this image 'projection models'. Both of the above have been used in deriving the equations which constitute surface nucleation theory as we know it today; the former relates growth rate and lamellar thickness to bulk and surface free energies, the latter leads to growth rate regimes.

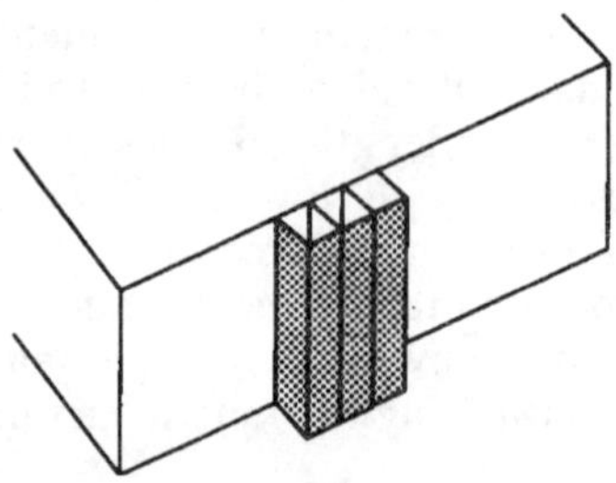

Figure 42 Polymer surface nucleation model. Straight stems form a nucleus on a smooth section of the lateral growth face.

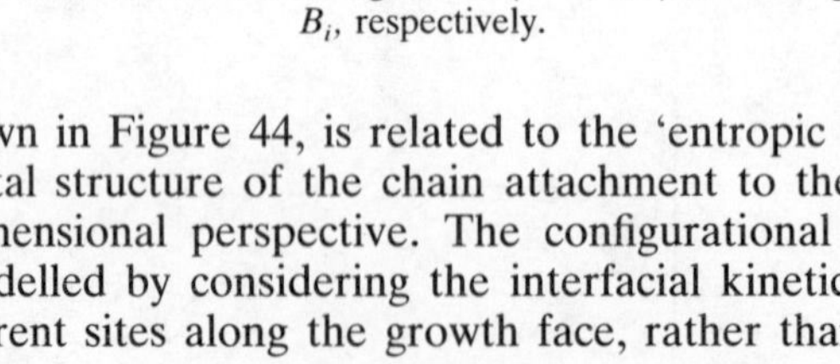

Figure 43 'Projection model' of crystallization. It is basically a top view of Figure 42, the stems now being shown as squares. Attachment and detachments with i lateral neighbours ($i = 0, 1, 2$) are distinguished by rate constants A_i and B_i, respectively.

The third 'image', shown in Figure 44, is related to the 'entropic barrier' theory. The main point is that the segmental structure of the chain attachment to the growth face is resolved, giving a truely three-dimensional perspective. The configurational degeneracy of the chain attachment is thence modelled by considering the interfacial kinetics of small chain sections, which can attach at different sites along the growth face, rather than building up 'stems' in a sequential way.

Some details of the respective derivations of surface nucleation and entropic barrier theories, as well as a discussion of the merits and drawbacks, will be given in the following sections: 7.13.4.1 and 7.13.4.2 focus on aspects of basic theory and applicability to basic experimental facts, whereas Sections 7.13.5 and 7.13.6 deal with two more specific applications of the theories to experimental evidence. These sections aim to reveal the basic facts as well as do justice to at least some of the often intricate arguments and counterarguments between proponents and opponents of surface nucleation theory. Unfortunately, the latter often draw the attention away from the achievements of polymer crystallization theories, namely on the basis of barrier-type (nucleation or entropy, or both) theories to account for the experimental facts listed in Section 7.13.2.

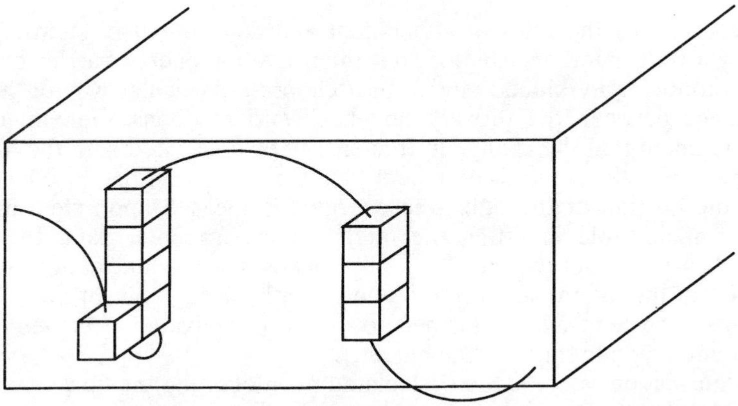

Figure 44 Attachment of the chain molecule as envisaged by the 'entropic barrier' model. Small sections of the chain come into crystallographic register at different locations on the growth face, forming conformations unfavourable to further growth.

The reader content with this brief overview, and not interested in further detail, may proceed to Section 7.13.7, where we pick up the thread again with a summary of recent theoretical developments related to the thickening growth mode.

7.13.4.1 Polymer surface nucleation theory

(i) Basic theory

The classical concept of surface nucleation envisages the formation of a two-dimensional patch on the crystal surface which corresponds to a maximum free energy state. Since, in chain-folded polymer crystals, growth is mainly lateral, and the fold length is constant, an adapted surface nucleation model was proposed,[166] shown schematically in Figure 42. A single traverse of the molecule across the growth face, a so-called stem, takes the place of the critical surface nucleus. Successive fold stem depositions next to this nucleus must lead to a reduction in free energy, and hence to a stable crystal. This, employing the Hoffman–Weeks equation (8), sets a minimum thickness level, as shown in Figure 45. The basic deficiency of this adaptation of the surface nucleation concept to the polymer case arises from the fact that the 'surface nucleus' here can only grow laterally, *i.e.* in one dimension, and hence the critical nucleus is not well defined (see Figure 45).[22,176]

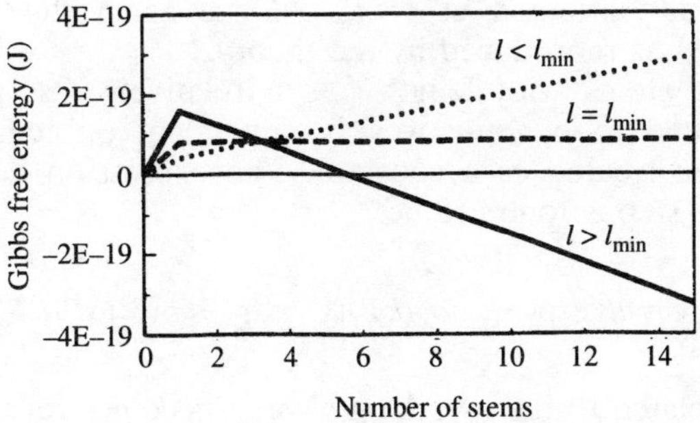

Figure 45 Gibbs free energy of a surface nucleus such as shown in Figure 42. It is the result of bulk and surface free energy of stems of fixed length l (given by equation (12)), and hence only a linear function of the number of adjacent stems. For $l > l_{min}$ (equation (6)), the deposition of the first stem represents the maximum, and hence the critical nucleus is always a single stem. In the case of the δl catastrophe, the Gibbs free energy becomes negative for the first stem already. (Numbers in this diagram apply to PE.)

With the above thermodynamic considerations in mind, we now proceed to the kinetic argument. The rate determining step in nucleation theory is the laying down of the first stem of a surface nucleus. For any given thickness the flux of this process, $S(l)$, can be calculated within the model by considering the rates of attachment and detachment of stems. The ratio of the latter is given by a Boltzmann distribution, but the activation energy barrier between the states remains a free variable. Any kinetic model must choose a particular way of 'apportioning' the free energy change between the forward and backward reactions. General considerations[177] lead to the requirement that the change in free energy be associated with the 'upward' step, *i.e.* in the present case with the removal of a stem. Surface nucleation theory introduces a factor ψ which denotes the portion of the bulk free energy ΔF released upon stem attachment. (The factor ψ plays a major role in fitting the model to experimental data. In the most recent versions of the theory, ψ is set to zero.[187–189] This means that the 'nucleation' step involves first of all an increase in the lateral surface free energy without any gain of bulk free energy. The relevant section of the molecule is deemed to attain an activated stretched-out state on the surface without any crystallographic attachment.)

The average 'nucleation rate' i is then calculated by integrating the fluxes over all thicknesses larger than $l_{\min}$. For $\psi = 0$ the result is:

$$i = i_0 P \exp(-4b\sigma\sigma_e/\Delta F kT) \tag{9}$$

where

$$P = \frac{kT}{2b\sigma} - \frac{kT}{2b\sigma + ab\Delta F} \tag{10}$$

and σ is the lateral surface free energy per unit area. Hence, if the approximate expression (6) for ΔF is inserted, the theory indeed reproduces the observed growth rate dependence on temperature, as given by equation (3).

The predicted average thickness prior to any secondary thickening (*cf.* Section 7.6.2) corresponds to the first moment of the flux distribution:

$$l_g^* = \frac{\int_{l_{\min}}^{\infty} l S(l)\,\mathrm{d}l}{\int_{l_{\min}}^{\infty} S(l)\,\mathrm{d}l} \tag{11}$$

which yields

$$l_g^* = \frac{2\sigma_e}{\Delta F} + \delta l \tag{12}$$

where for $\psi = 0$

$$\delta l = \frac{kT}{2b\sigma} \frac{\Delta F + 4\sigma/a}{\Delta F + 2\sigma/a} \tag{13}$$

the latter 'overlength' being a small term in general, which represents the net gain in stability over the marginal stability attained at $l = l_{\min}$. Once again the empirical relationship, as expressed by equation (1), is reproduced by the theory.

The spreading of the stems with length l is a relatively fast process, which in surface nucleation theory is found to be only weakly dependent on temperature in the range of interest, and hence often treated as a constant. The expression derived for the net rate of attachment of stems to a step is found to be:

$$g = a\beta \exp(-2ab\sigma_e/kT)[1 - \exp(-ab(\delta l)\Delta F/kT)] \tag{14}$$

hence net spreading is related to the 'overlength' δl; g becomes zero for $\delta l = 0$.

The above relationships, in particular (12) and (14), constitute the heart of the surface nucleation model. Despite some basic criticisms, it has proven remarkably successful in accounting for the observed trends in growth rate and lamellar thickness.

(ii) *Projection view and regimes*

The actual growth rate expressions and the theory which is put forward to explain crystallization in different regimes is a further development which is best understood by considering the kinetics of the stem attachment in the 'projection' view. As much of the literature on polymer crystallization theory is concerned with such 'regimes', we shall go into the matter in some detail.

Consider, to the above stated end, the interfacial processes in the projection models (see Figure 43). In the general case one must distinguish between additions and removals involving either no, one or two nearest neighbours. The kinetics in such a case can be simulated on a computer,[178,179] but much can be learnt from a 'mean-field' version based on a net nucleation rate i and a spreading rate g (see Figure 46), concrete expressions for which can then be inserted from the surface nucleation treatment above.

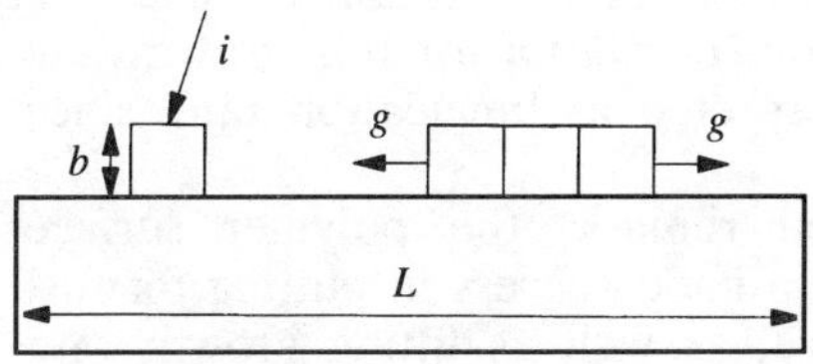

Figure 46 Representation of the Seto/Frank model, a mean-field version of the projection model in Figure 43, based on a nucleation rate i and a spreading rate g.

A mathematical formulation of this model was given independently by Seto[180] and by Frank,[181] who considered a pair of kinetics equations for the densities of 'right' and 'left' moving steps $r(x, t)$ and $l(x, t)$, respectively:

$$\frac{\partial r}{\partial t} + g\frac{\partial r}{\partial x} = -2grl + i \tag{15}$$

$$\frac{\partial l}{\partial t} - g\frac{\partial l}{\partial x} = -2grl + i \tag{16}$$

on a substrate of width L with 'absorbing' boundary conditions:

$$r(-L/2, t) = 0 \tag{17}$$

$$l(L/2, t) = 0 \tag{18}$$

i.e. no steps come in from outside the boundary or are reflected.

Once again by analogy to previous work on small molecules, two limiting cases have been identified.[159] The first is that of large g so that a nucleus, having formed, spreads rapidly to cover the substrate before another nucleation event occcurs. If the substrate has a width L, then the resulting growth rate is given by

$$G_M = biL \tag{19}$$

where b is the thickness of the stem and the subscript M stands for 'mononucleation'. Notice that the growth rate is predicted to be proportional to the size L of the face but independent of g. As i increases, the separation between nuclei, also known as the kinetic length $L_k = (g/2i)^{1/2}$, decreases and eventually becomes smaller than L, at which point the 'polynucleation' regime is entered. There are several derivations of the growth rate in this case.[159] We shall only quote the result, which is

$$G_P = b(2ig)^{1/2} \tag{20}$$

Equations (19) and (20) are at the heart of nucleation theory's explanation of the change in slope by a factor two observed in some plots of log G versus $T^0_m/(T\Delta T)$ (*e.g.* Figure 39). The underlying exponential undercooling dependence is due to that of the nucleation rate; the spreading rate varies only slowly with temperature (see above). Hence a transition from mono- to polynucleation is expected to lead to a change in the exponential by a factor of two. However, the also ensuing size dependence of the growth rate within the mononucleation regime (equation (19)) was never observed.[182] Hence it was argued that a real crystal face contains imperfections which limit the spreading to an effective length L_p, often called 'persistence length' (not to be confused with that describing the chain stiffness in the rotational isomeric state model of polymers). With L_p replacing L in equation (19), this regime becomes known as Regime I, and that for polynucleation as Regime II.

The justification for, or indeed existence of, L_p is a matter of intense dispute.[183,184] The issue, albeit often intricate, is central to the actual existence or not of a Regime I in the sense described above, and hence to the interpretation of breaks in the growth rate curve by regime theory, and last not least to the calculation of material parameters such as σ on that basis.

Finally, if the kinetic length L_k becomes as small as a few stem widths, no space is left for step spreading. At this point a third regime (Regime III) sets in, in which growth once again becomes dominated by the primary stem attachment, and hence the growth rate is usually expressed as $G_{III} \propto i$ (for a strict derivation of this growth law, see Gates and Westcott[185]), although i should not be interpreted as 'nucleation rate' since the surface is serrated on a molecular scale.

Notwithstanding our critical remark, the polymer surface nucleation theory offers a comprehensive treatment with a good success in putting forward explanations for most of the experimental facts listed above, as well as fitting growth rate data for a range of specific polymers.[184]

(iii) Critique and recent developments

Apart from the basic criticism mentioned above, concerning the oversimplified application of the classical surface nucleation concept to the polymer case, a number of specific deficiencies and inconsistencies of the model have been raised.[159,186] The best known is probably the so-called δl catastrophe, *i.e.* a divergence of the predicted thickness if a certain critical undercooling is exceeded. This arises basically as a consequence of the 'one-dimensional', *i.e.* single-stem, nucleus. Beyond the critical undercooling the Gibbs free energy becomes a decreasing function of the length of this first stem; hence there is no barrier to stems of any thickness.

The second major criticism follows from the fact that at least in Regimes I and II the crystals are expected to exhibit straight faces, since the completion of new growth layers is a fast process. However, in polyethylene, curved faces have been observed at high temperatures.

On the side of surface nucleation theory, the above criticisms have been addressed by (1) putting forward more forcefully the '$\psi = 0$' version of the model, in which case the δl catastrophe is blanked out,[187] and (2) the introduction of surface strain,[188] supposedly due to the folds, as an additional parameter which leads to a reduction and undercooling dependence of the substrate completion rate, and according to further work on the projection models (see Section 7.13.5 below) to rounded faces. Although promising at a first look, the dependence of g on ΔT in this approach is in conflict with the regime interpretation discussed above, which relies on constant g.

A further recent development of surface nucleation theory[187,189] is also based on the '$\psi = 0$' picture of chain alignment prior to crystallographic attachment. This can in general terms be associated with a loss of entropy of the chain relative to the melt. In the context of the surface nucleation model, on the other hand, it is associated with the lateral surface free energy σ. This recognition leads to the following relationship between σ and the effective segment length, represented by the characteristic ratio C_∞ of the rotational isomeric state model

$$\sigma = \frac{T \Delta H_f a}{T_m} \frac{l_b}{2 l_u} \frac{1}{C_\infty} \tag{21}$$

where σ, ΔH_f, and T_m are as defined above, a is the width of the chain, l_b is the bond length, and l_u is the projected length per atom along the chain axis. Good agreement is claimed between values of C_∞ derived from chain dimensions and via equation (21), employing values for σ derived from fitting the nucleation theory to experimental growth rates.

At the same time, further specific deficiencies of the model[186] have been pointed out more and more forcefully. Apart from the δl catastrophe these are: (1) the 'artificial' substrate length L_p, (2) the need to rely on adjusting the apportioning factor ψ, (3) the observation that the regime transitions are usually sharp (Figure 39), whereas they are predicted by the Seto/Frank model to be gradual, (4) the observation that, in some cases at least, the regime transition is accompanied by morphological or even crystallographic changes, and hence different values of the lattice and fold energy parameters would have to be applied, and (5) the fact that the length scales involved (kinetic length L_k, 'persistence length' L_p, width of growth face L) are at times inconsistent with the growth rate dependence and regime transitions.

Arguments and counter-arguments have been following each other for many years. We do not want to enter further into this debate, and leave the interested reader to consult the references provided. Rather, the following section will introduce an alternative picture of interfacial kinetics, and we ask the readers to judge the basic facts for themselves.

7.13.4.2 Entropic barrier models

The above discrepancies led to a search for alternative or more complete treatments of polymer crystallization. Here we can only mention a few. It was first suggested that the multitude of ways in which the chain can attach to the growth face should not be neglected.[173] This naturally leads to an 'entropic' barrier to the formation of thick crystals,[174] and to a well-defined and limited l, thereby remedying the δl catastrophe. Models based on this principle have become known as 'entropic barrier models' in order to distinguish them from the nucleation models which are based on an enthalpic barrier.

Entropic barrier models envisage the chain to be made up of segments comprising a certain number of monomers (depending on the chain stiffness, *etc.*) which act as 'growth units' (Figure 44). Their addition to and removal from the growing crystal is limited to solid-on-solid sites,[190] *i.e.* without leaving gaps in the c direction along the lateral growth faces in order to mimic the chain connectivity as well as the presumed chain folds on the basal planes. Rate constants, based on the principle of detailed balance, govern the transition processes. As in any kinetic model, the actual rate constants require an apportioning of the free energy change. However, in contrast to surface nucleation theory, the value of ψ has no effect on the predicted steady state growth rates and lamellar thicknesses. This result hence leaves some doubts about the 'physical' interpretation of ψ within polymer surface nucleation theory.

We also emphasize the difference in the views as regards the entropy. Whereas the change in entropy associated with the alignment of the chain as proposed by surface nucleation theory is a part of the heat of fusion, the present model proposes an additional configurational entropy associated with the multiplicity of attachments of the chain segments after and apart from losing the freedom of movement in the liquid phase.

On the basis of the model of addition and removal of small chain segments described above, the crystal growth can be simulated. In the full three-dimensional case this is accomplished by a kinetic Monte Carlo method (Figures 47 and 48(a)).[175,191] In simpler versions, which consider either a slice cut out of the crystal perpendicularly to the growth face as a two-dimensional representation of the whole lamella (Figure 48(b)), or only the spreading of a step along a smooth growth face (Figure 48(c)), the evolution of the system can alternatively be formulated by rate (or Master) equations which can be solved numerically.

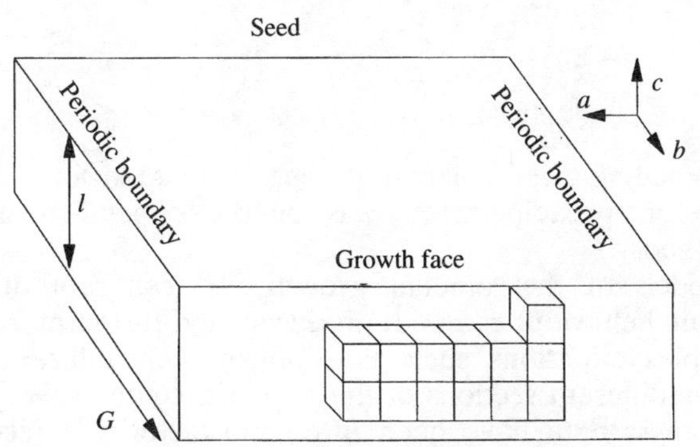

Figure 47 Schematic representation of the three-dimensional 'entropic barrier' model according to Sadler and Gilmer. Growth is simulated by the addition and removal of small 'growth units' (shown as cubes) on one lateral growth face. Periodic boundary conditions apply in the other directions as indicated.

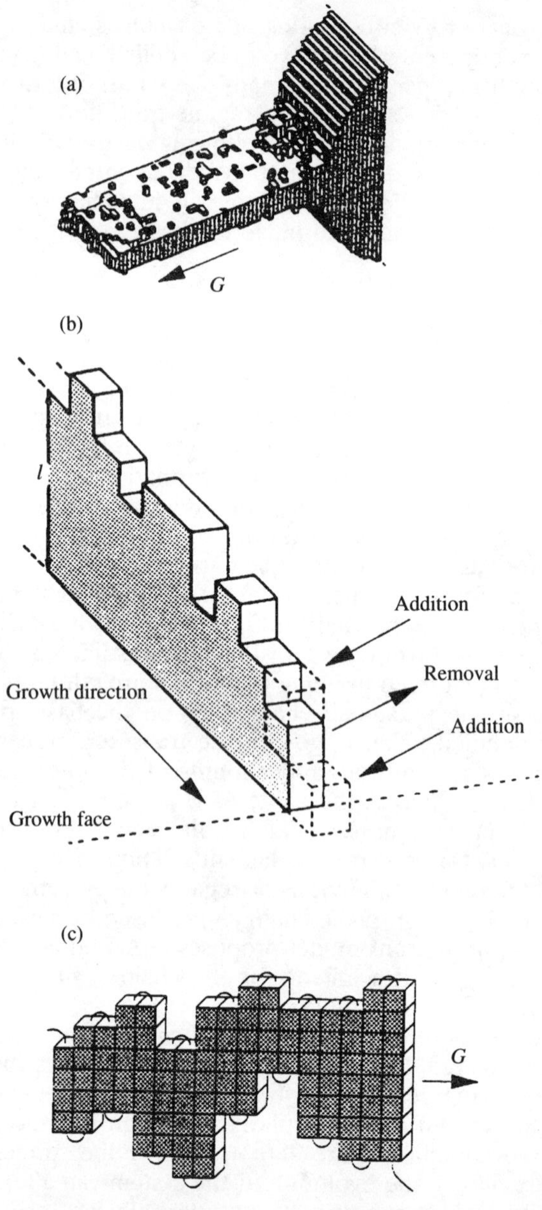

Figure 48 Three different versions of the entropic barrier model: (a) lamella simulated with the three-dimensional version (see also Figure 47); (b) a two-dimensional slice cut out of the lamella perpendicular to the growth face; and (c) the spreading of a layer on an otherwise smooth growth face.

The model of step spreading (Figure 48(c)) has been studied in some detail,[192] with the result that the rate of advance or 'spreading rate' g was found to depend strongly on supercooling, in contrast to nucleation theory.

Entropic barrier models of the lamellar growth[175,191] can reproduce the typical crystal thickness and growth rate behaviour can be reproduced, and furthermore that the model can be applied to more complex situations such as copolymer crystallization,[193] where the units represent the chemically different sections of the chain. Recently, specific experimental growth rate and lamellar thickness data have been fitted satisfactorily[192] (see Figure 49) by using literature data for the model parameters, which are the heat of fusion, the surface free energies, the melting point, and the lattice parameters. Moreover, it was recognized that the only 'free' parameter of the model, that is the length of one growth unit in terms of the number of monomers contained within it, can be related to the chain persistence length in the melt. In fact, the length of chain in one growth unit amounts to five persistence lengths for vinyl polymers.[192]

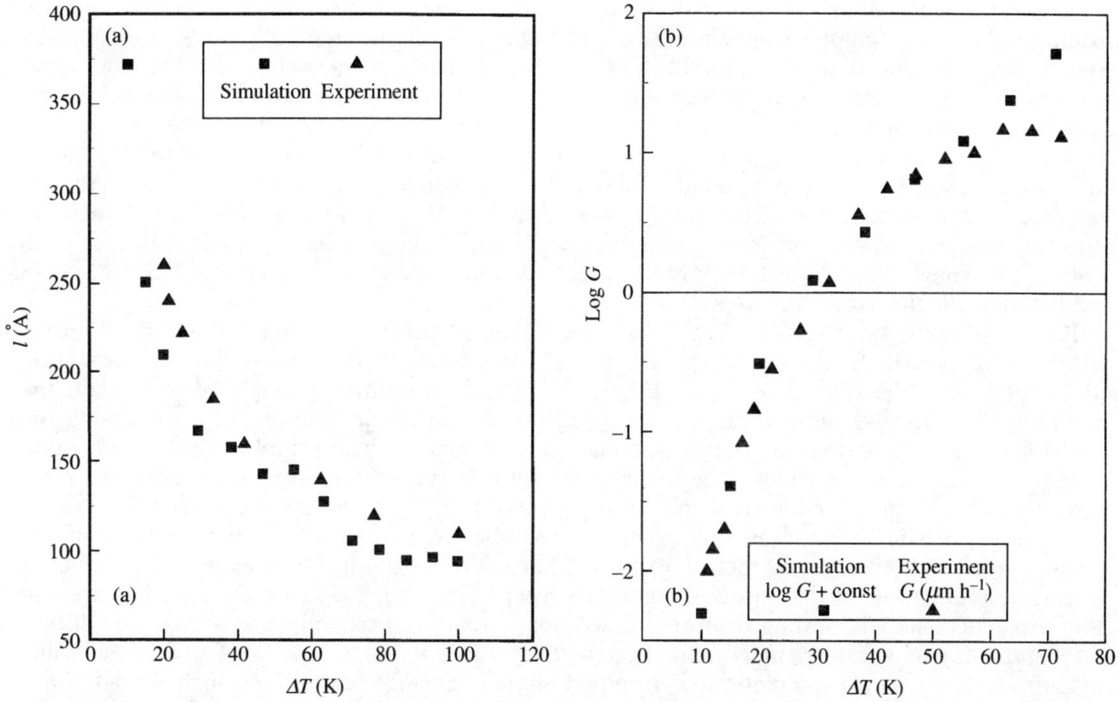

Figure 49 Application of the two-dimensional 'entropic barrier' model (Figure 48(b)), to polystyrene: comparison of experiment and simulation results for (a) lamellar thickness (experimental data from Edwards and Phillips[194]), and (b) growth rates (experimental data from Vaughan[195]).

Hence we conclude that both nucleation and entropic barrier models have led to the recognition that the experimental knowledge about polymer crystallization can be accounted for on the basis of kinetic models which feature a strongly thickness-dependent barrier, leading to a sharp maximum in the predicted flux or growth rate not far above the minimum thickness. Both models have their own strengths and weaknesses, which have been discussed in detail in a previous review.[159] In principle, the entropic barrier approach, simulating by a computer the multitude of processes which lead to the actual formation of patches on the surface, is only a generalization of the nucleation approach.[173] As argued by Binsbergen,[173] the latter was devised originally by Becker and Döring out of necessity, since the complete range of possible processes could not be computed at the time. We believe it is important to keep this common background in mind, in order to put more effort into further development and improvement of theories, rather than in pointing out deficiencies. It should be understood as a pointer that, lately, both models have also come to realize relations between the chain conformation in the liquid phase and the crystallization kinetics. Although the actual relationships are different, further work into this direction could be fruitful for our understanding of polymer crystallization as well as for the modelling of materials in a wider sense.

7.13.5 Discussion of Factors Affecting the Shapes of Crystals

As discussed in Section 7.6.1, the lateral shapes of polymer crystals, which have formed under controlled conditions, have become recognized as indicators of the underlying interfacial growth processes. Given the fact that the predictions of both surface nucleation and entropic barrier theories concerning G and l are very similar (due to similar functional dependencies of the respective activation barriers), further predictions of the respective approaches, such as to shapes, have become central issues of discussion in recent years.

Historically, the discussion relating shapes to the kinetics begins with the observation of increasingly rounded growth faces in polyethylene single crystals with increasing crystallization temperatures. At such low supercoolings, if surface nucleation was operative, it would be expected that the substrate completion rate g was far larger than the nucleation rate i and hence faceted crystals should be expected.

This issue was taken up with reference to a 'roughening theory' of polymer crystallization.[175,190] If the length of the chain segments which can be treated as 'rigid' growth units is significantly smaller than the complete 'stems' of the surface nucleation model, the 'bond energy' per unit would equally be smaller, possibly reduced to values for which an equilibrium roughening transition can be expected.[190] Above the transition temperature, surface nucleation is no longer necessary for growth, hence as known from the crystal growth of simple substances[196] the faces become rounded above the roughening transition, and the growth rate becomes proportional to ΔT. However, as was shown in the previous section, in polymers an entropic barrier is always active, retaining the exponential dependence of the growth rate even above any roughening transition. Hence the 'entropic barrier' picture offered a consistent explanation for the observed shapes.

It remained to be checked that it is indeed the structure (or roughness) of the two-dimensional growth face, *i.e.* incorporating the c direction, in contrast to just the serration obtained in a 'projection model' (see Figure 43), which is ultimately required to explain the rounding. To this end, simulations were carried out of the 'projection model',[178] avoiding the mean-field approximations of the Frank and Seto model. It was found that the projection model was unable to reproduce the experimentally observed rounding; hence the need for a model which accounts for the structure during growth in the c direction of the growth face.

However, in simulation[178] as well as in the analytical treatments,[180,181] the width L of the growth face was kept constant, hence assuming that any variation in L during growth would not affect the predictions of the model. This has in recent years been refuted. Renewed interest in the 'projection models' sprung from the realization[197] that an increasing size of the growth face L (Figure 46) at a rate h in combination with a step spreading rate g of about the same magnitude leads to the prediction of rounded shapes. Several variants of such models have been proposed since. One model considered the case $g > h$,[197] consistent with a truncated lozenge shape with curved {100} faces, another the case $g < h$,[198] leading to a lenticular shape. Both models still employ absorbing boundary conditions (see above), and depend on the existence of at least a small neighbouring growth facet or sector to provide the widening of the growth face in question. In contrast, yet another model[199] envisages reflection of the steps at the boundaries and does not depend on a leading facet. The most general case so far has been studied in a recent computer simulation,[179] leading to the conclusion that nearly circular shapes must be expected if the spreading rate is sufficiently small.

The success of these models in reproducing the whole range of observed rounded and lenticular shapes has certainly been impressive. However, it must not be forgotten that the origins of the required dependencies of the parameters, in particular g on the undercooling, remains beyond the scope of these mathematical models.

As for the reason for the reduction in spreading rate, effects such as impurity adsorption on the growth face,[200] and surface strain due to the folding[201] have been suggested. However, there is still no conclusive evidence for such phenomena. Furthermore, as to the implications of the above finding (reduction of g) for the validity, or otherwise, of surface nucleation theory, the challenge remains. Either the classical version outlined in the last section, with a constant spreading rate g and a constant effective substrate width L_p is retained with the ensuing Regime explanation, or a small and ΔT-dependent g and an increasing substrate width L are built into the theory, in which case the traditional Regime explanation is lost.

In contrast, entropic barrier models seem to provide a more natural explanation of the required functional dependency of g (see above). The issue of crystal shapes hence shows a way in which both nucleation-type (as projection models for the shape calculation) and entropic barrier ideas (as origin of the slow net step spreading rate) could be combined, thereby enhancing each other to find a better description of polymer crystallization.

7.13.6 Some Aspects Pertaining to the Crystallization of Short Chains

The experimental evidence relating to the crystallization of short chains was outlined in Section 7.9. One of the salient facts is the observation of distinct branches of the growth rate related to extended and n times ($n = 1, 2, \ldots$) folded-chain crystal formations (Figure 32). As in the main stream of crystallization theory, entropic barrier and surface nucleation models compete in their efforts to account for the experimental facts.

As for the entropic barrier models, such crystallization at a fixed thickness over a range of temperatures provides an ideal test case, since the activation barrier in these models is in the first instance dependent on the thickness, and only indirectly on the supercooling. Hence at constant thickness the activation barrier remains constant and the growth rate is expected to be dominated by the driving force term, and hence in a first approximation,[159] proportional to ΔT. This was indeed found[202] to be the case, except for temperatures close to the transition points.

Furthermore, a particular version of the 'entropic barrier' model, which included a free energy preference for adjacent stems of particular length (*e.g.* corresponding to extended or once-folded chain lengths), reproduced branches with constant thickness and a linear growth rate dependence.[133] Moreover, close to the transition points the model predicted minima in the growth rate. This is one of the few occasions in polymer crystallization theory where a model actually predated the experimental observation, outlined in Section 7.9.3.2. Recently, the 'entropic barrier' modelling of the once-folded to extended growth transition has been taken up again.[118] The modelling simply considers two types of growth unit, one representing a fully extended chain, the other one half of a folded chain. The latter are allowed to attach onto the extended segments but not vice versa; hence growth can be blocked, an effect also known as 'self-poisoning'. The result is a sharp minimum in the growth rate at the temperature where folded chains are marginally stable. Best agreement with the experimental evidence was obtained when, in addition to the simple attachment mechanism described above, a small degree of thickening of the folded into extended chains was allowed, a finding with repercussions in the recent evidence pertaining to thickening growth (Section 7.11.2.1). Finally, since this self-poisoning should also affect any step spreading across the growth face, a much reduced g is expected and hence, according to the models outlined in the previous section, rounded or possibly circular crystals. This indeed agrees with the experimental evidence available so far (Figure 31 and Section 7.9.4).

The surface nucleation treatment, on the other hand, was adapted to the case of extended chain growth[203] by replacing the usual flux integral over all stem lengths greater than $l_{\min}$ (see equation (9)) by the flux of stems of extended-chain length only. Then, if ψ is once again chosen to be close to zero, a maximum in the growth rate is obtained. The folded chain case at lower temperatures is treated with the full flux integral over all lengths between $l_{\min}$ and infinity. However, this seems in conflict with the limited chain length. Furthermore, it is unsatisfactory that the model proposes very different growth mechanisms on either side of the minimum.

Finally, at this point a new approach deserves mentioning,[204] which is based on ideas akin to those of the 'molecular nucleation' concept introduced in Section 7.13.4. It is assumed, in contrast to both surface nucleation and entropic barrier models, that the attachment of a molecule to a crystal implies 'size, shape and force recognition', leading to molecular weight segregation during crystallization. In the calculation of growth rates the latter fact finds its expression in a dependency on the length of the crystallizing molecules. Whether a molecule can crystallize or not in a particular configuration (*e.g.* extended or once folded) is decided by a criterion similar to relation (5). The result of this segregation–crystallization approach are growth rate curves which fit the experimental data, such as shown in Figure 32, remarkably well.

7.13.7 Thickening Growth Models

The observations of polymer crystals with thicknesses larger than the extended chain length, as summarized in points (2), (3) and (7) in Section 7.13.2, clearly demonstrates that the kinetic theories outlined above are not able to deal with all the variations of polymer crystal growth. Hence the proposal of a so-called 'unified model' of polymer crystallization[205] based on the premise that in principle the polymer crystal can grow not only laterally but also longitudinally, *i.e.* in the chain direction. In the main-stream case discussed above the rate for this latter process is very small, hence folded-chain crystals result. In the case that a 'mobile' phase (*cf.* Section 7.11.2.2) is present, however, the longitudinal rate becomes significant due to the process of chain sliding diffusion, the latter with an activation free energy ΔE_e. The approximate analytical treatment,[205] as well as subsequent computer simulation of the model,[193] agree well with the experimental observations. In its most recent development the theory explains the experimentally observed supercooling dependence of the thickness of pressure-crystallized polyethylene[163] as being the result of the halting of thickening growth by a phase transformation from mobile hexagonal (h) to the stable orthorhombic (o) phase (for a discussion of the such phase behaviour, see Section 7.13.8 below). This opens up the discussion about the factors governing the lamellar thickness in polymer crystals in general, with consequences still to be fully appreciated.

7.13.8 Thermodynamics of Polymorphous Systems

In the light of the evidence that some intermediate phase (which is often a mesophase, and we shall refer to it as such) displaying thickening growth can be involved in the crystallization process, both the thermodynamic and kinetic treatments outlined above need to be considered and extended. As above, we shall first deal with the thermodynamic aspects.

The emphasis in the thermodynamic treatment of the liquid–mesophase–crystal system lies once again on the effects of the limited crystal thickness. The size dependence of the triple point is a well-known effect in solid–liquid–gas systems. However, it has until recently not been properly appreciated in the context of polymer crystallization which involves two solid phases and the melt, with direct consequences for their growth.

In the polymer case, such a system was first studied in detail in the context of the hexagonal phase of PE occurring at elevated pressure (see Section 7.11). Hence a full exploration of the pressure–temperature (p–T) phase space, common in simple materials, was called for. This led not only to the experimental work outlined in Section 7.11 but also to theoretical efforts, *inter alia* concerning the triple point depression with size, which in fact is a generalization of the Gibbs–Thomson equation (see equation (8)) with pressure as an additional variable. The expected behaviour of the system can be represented in the form of a three-dimensional p–T–$1/l$ diagram (Figure 50). This is an extension of a p–T phase diagram such as in Figure 38 so as to incorporate size (l) dependence, where the high p phase (hexagonal h) is metastable (as referred to infinite size) at $p = 0$; see below and Figure 51. As the schematic representation in Figure 50 shows, the stability regime of the hexagonal phase must be expected to extend down to ambient pressure for sufficiently small thicknesses l. A full analytical treatment of this triple point depression has been given.[206]

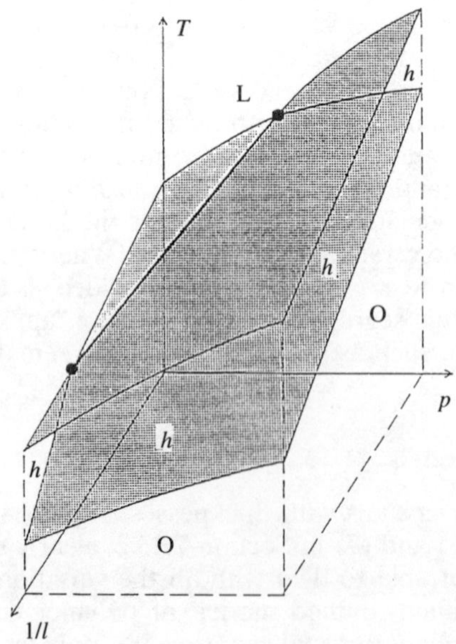

Figure 50 Schematic three-dimensional p–T–$(1/l)$ phase stability diagram, showing how the stability regime of the hexagonal extends for small phase size.

Here we consider a simplified approach,[206] in which the three-phase system is simply split up into three competing but independent two-phase systems, namely (a) liquid–mesophase, (b)

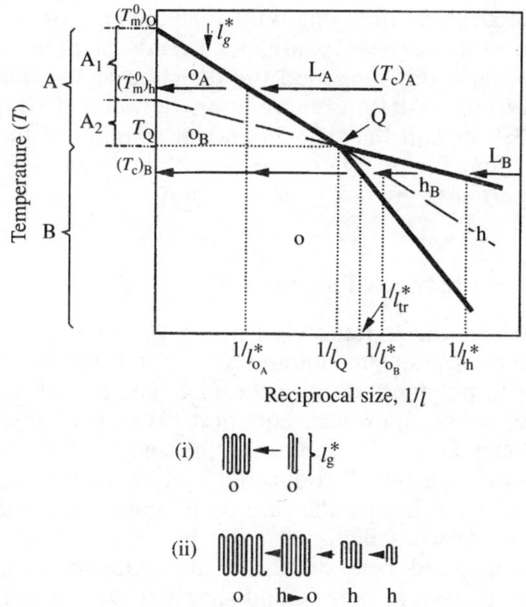

Figure 51 Phase growth in terms of a 'phase stability' diagram with notation referring to crystal form (h and o) in PE with L standing for melt (liquid). The two sets of horizontal arrows, pointing towards $1/l = 0$, denote isothermal growth pathways at the two selected (crystallization) temperatures, here chosen to be in the two temperature regimes A and B. Shown below are representations of chain-folded polymer crystal growth: (i) in region A leading to lamellae of a specific restricted thickness l_g^*, at which they continue to grow laterally through direct growth in phase o, and (ii) in region B where crystals arise in the h phase and develop by simultaneous lateral and thickening growth with the latter stopping (or slowing down) on h→o transformation and/or impingement.

liquid–crystal and (c) mesophase–crystal. The size-dependent stability of aggregates of one phase with respect to the other phase is governed by the respective Hoffman–Weeks equations (8).

Now consider the three plots of equation (8) compounded into one graph (Figure 51), which we call a 'phase stability diagram'.[22,206] Again, we take the mesophase (here h) as metastable (as referred to infinite size), in which case its melting point will be lower than that of the crystal phase, *i.e.* $(T_m^0)_{meso} < (T_m^0)_{crystal}$. The melting transition lines will intersect if

$$\left(\frac{\sigma_e}{\Delta H}\right)_{meso} < \left(\frac{\sigma_e}{\Delta H}\right)_{crystal} \tag{22}$$

Such an intersection represents a triple point at which the three phases can coexist. Moreover, below this triple point the phase stability is inverted, the mesophase becoming the most stable one. Hence point (3) of Section 7.13.2 may be a result of the small size of the crystals.

In order to elucidate this point we invoke Figure 51, which can be regarded as a T, $1/l$ section of the p–T–$1/l$ phase stability diagram (Figure 50) for $p < p_Q$ ($p_Q \equiv$ triple point pressure), say at $p = 0$. Figure 51 shows two possible pathways of isothermal growth (horizontal arrows). Pathway A is above the triple point temperature, and hence never enters the stability regime of the mesophase, whereas pathway B actually passes through a region of mesophase stability before entering the region where the crystal phase is the stable one, with the mesophase becoming metastable. Such differences in the thermodynamics can, kinetics permitting of course (see below), have strong effects on the growth and the ensuing morphology. Pathway A is likely to lead to the typical chain-folded polymer crystal since we know that the crystal phase, *e.g.* the orthorhombic phase in polyethylene, features closely packed chains with little longitudinal mobility, and hence the size l will remain close to the minimum stable one. Pathway B, on the other hand, has the nucleus first appearing in the mesophase, characterized by much greater mobility. Hence thickening growth can set in, which is likely to stop, however, as the stability regime of the crystal phase is entered as a result of increasing size (here thickness). This scheme is in line with the observations described in

Section 7.11, which indeed show that not only thickening growth but growth altogether is halted on transformation to the 'crystal' phase. As already pointed out in the previous section, the realization of this scenario has reopened the question as to what determines the lamellar thickness of polymer crystals: is it the competition between driving force and barrier term described in Section 7.13.4, or is it the size-dependent stability of the mesophase? There need be no conflict. As seen from Figure 51, it can be either depending on the undercooling, *i.e.* whether region A or B pertains.

7.13.9 Kinetic Competition between Polymorphs

In this section we consider the kinetics concerning the case of two solid phases competing crystallization. This situation, not uncommon to crystal growth in general, has received relatively little attention in polymers despite the fact that crystal polymorphs have been well known for some time. In a first approach, equation (3) can be applied to each crystal phase using the respective pre-exponential and exponential constants. In this way, general graphs of the range of possible cases have been produced,[206] providing a basis for the interpretation of experimental data. In Figure 52 the specific case pertaining to the orthorhombic and hexagonal phases of polyethylene is shown, calculated[207] by inserting literature data for the transport factor, surface free energies and heat of fusion into equation (3), irrespective of its strict validity over such large temperature ranges, and simply to demonstrate the likely magnitude of the effect. It is obvious that the kinetics strongly favour the growth of the mesophase. Hence for small phase sizes, both thermodynamics and kinetics work in the same direction, favouring the phase which ultimately is only metastable.

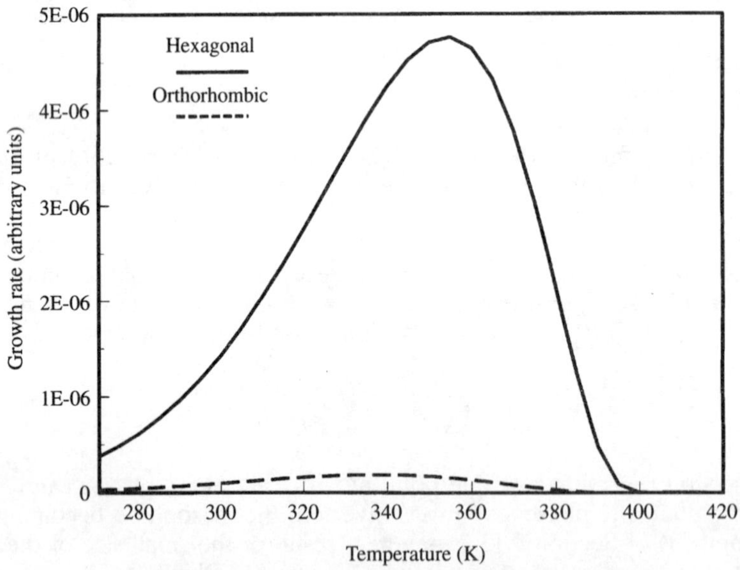

Figure 52 Growth rate of orthorhombic and hexagonal phase of polyethylene, calculated[207] on the basis of equation (3). The hexagonal 'mesophase' is clearly favoured kinetically.

7.13.10 Conformations of the Single Chain

Finally we return to the chain molecule itself, motivated by recent evidence, partly outlined in Section 7.2, which questions the hitherto common, although not always clearly spelt out, belief that the flexible, chemically nonspecific polymer chain, which has been our main concern here, approaches its minimum free energy state only on crystallization.

The typical minimum free energy configuration is a helical structure, which in the simplest case of polyethylene degenerates to a simple zig-zag. These structures are of course well known, particularly in biological macromolecules where specific interchain interactions lead to the formation of stable helical structures independent of crystallization. Crystallization is only a possible further stage in a structural hierarchy beginning at the molecule itself. In the opposite extreme of the flexible, chemically nonspecific chain, the polymer is usually found in a so-called random coil configuration, and attains a state close to its minimum of free energy only in the crystal. Hence, in the latter case the ordering of the individual chain conformation and the cooperative process of crystallization are seemingly inseparable events.

However, the recent evidence pertaining to the crystallization of syndiotactic and also isotactic polystyrene from dilute solutions at very high undercoolings[9] displays behaviour which, as stated in Section 7.2, is associated with a conformational change in solution, with strong inference of a coil–helix transition. Furthermore, independent evidence from computer simulation of polyethylene[208] indicates that at high undercooling the chains in a vacuum assume a predominantly *trans* conformation, with mostly sharp chain-folds. Although large external driving forces are required to bring about such conformational changes, these are not uncommon in polymers, where often high degrees of supercooling are reached in the laboratory as well as in industrial processing. If, indeed, such changes occur in the liquid phase prior to crystallization, this will of course change the thermodynamics and have consequences for the kinetics of crystal growth. This raises new issues for future research, and possibly some older approaches to polymer crystallization should be revisited, *e.g.* the 'bundle' model[168] (see Section 7.13.4), which was based on an equilibrium distribution of intramolecular clusters reminiscent of the structures found in the recent computer simulations.

7.13.11 Conclusion

In this section on theories we have tried to summarize the fundamental basis as well as topical issues in the theory of polymer crystallization. The description of folded-chain growth is, despite many years of discussion, still broadly split into nucleation and entropic-barrier type models. The focus on the particular issues of the relationship between kinetics and melt structure, as well as the crystal shapes, seems to point towards some possible synthesis between the different approaches. We hope that the present review can make a contribution towards this. Beyond these 'classical' questions the more recent developments concerning the mode of thickening growth, and the effects of kinetic as well as phase stability competition between polymorphs, have been discussed on the basis of the size-dependent stability for small phase sizes, opening up wider scenarios, in particular via the use of phase stability diagrams. The final section on the conformations of the single chain can be considered to be the most forward looking and speculative part in the potential overall phase scheme.

7.14 CONCLUDING REMARKS

As stated at the onset, within the given restrictions in space, this review had to be highly selective. Even so, we hope to have given a representative cross section of the subject, recognizing that this was only one of many ways to traverse the field. Throughout the particular selection adopted, we hope that we succeeded to convey the message that the subject is of virtually infinite variety, and further, that it is by no means closed. On the contrary, even some of its most basic aspects are still open to queries and even to radically new approaches.

It will have been noticed that some of the material chosen is comparatively old on the standards of the subject and has been reviewed repeatedly before. However, for those in the know, such items should not serve as mere repetition but as reminders of what the salient effects are which have stood the test of time and, simultaneously, keep showing the way ahead. In the latter respect, gratifyingly, some most recent, in fact concurrent, material seems to connect with past knowledge in a highly constructive way. Hopefully, it will have been apparent that this review put special emphasis on such linking up of the very old with the very new, in an attempt to create a unity in this ramifying subject while spanning its full life history of about four decades.

If there is an overall message, it is that the organizational capability of long-chain molecules is much greater than usually envisaged, and possibly with still new surprises. Interpretation of the effects must clearly lie within the basic laws of physical science except that the application of these laws to long-chain molecules seems to extend much further than envisaged by traditional projections. In a nutshell, the last statement expresses the past history of our subject, a lesson to be heeded when looking ahead to the future.

Finally, we end on the note raised at the end of Section 7.5 and in Section 7.12. All through this review we considered the crystallization of chains acting as on their own, producing crystal entities essentially in isolation, where interconnections, when they arise, are rather accidental. In doing so, we have been probing the basic organizational ability of long chains *per se*. For this purpose we left out of consideration interchain interactions, entanglements in particular, and in fact, as stated in Section 7.12, the experimental evidence considered does not require invoking such interactions. Nevertheless, it is the common experience that a 'real life' semicrystalline polymer *is* a molecularly interconnected network of crystals where the nature of the connections, in the form of some kind of constrained amorphous intercrystalline material, acquires intrinsic significance, possibly linked to preexisting entanglements in the preceding isotropic state (melt). The existence of such an amorphous state should not detract from the importance of what we know about the unconstrained crystal, the subject of the present review; it is to be regarded as a subsequent factor arising from it. Under appropriate circumstances it is created by the crystal itself and can eventually arrest the crystallization which has brought it into being. The exploration of these circumstances, the resulting structures with the crystal–amorphous interactions generated, is clearly the next task ahead.

Of course, the subject just raised does feature in many preceeding works ever since the crystallization of polymers has received attention. Thus, it has gradually emerged that such a connecting, constrained amorphous phase is promoted by very long chains,[25] by the lack of mobility in specific polymer species (poorly and slowly crystallizing polymers, such as PET, i-PS, PEEK, *etc.*) and by crystallization close to T_g (*i.e.* on the low T side of rate curves such as in Figure 3), areas intentionally left uncovered by the present review. However, a more systematic attack on the problem, and this in the light of what we know of the formation and structure of the isolated crystal, is only now beginning to take shape. An expression of the latter is a preview[209] which has appeared as the present article is being finallized, to which we are now drawing attention.

7.15 NOTES ADDED TO PROOFS

1. Since the submission of the manuscript, an explanation for the 'coil explosion' has been proposed.[210] It is based on a model of crystal melting in which each chain successively detaches itself from the crystal starting at one of its ends while remaining attached along the rest of its length, 'inventing its own tube' in the process. This model accounts for the experimental observation of a radius of gyration which, at early stages at least, is independant of chain length;[82] also in our view it is consistent with the majoirty of chain ends being located outside the crystal[92] and thus available to start off detachment.
2. For documentation of th effect of thickening growth, see Hikosaki *et al.*[211]

7.16 REFERENCES

1. L. Mandelkern, in 'Comprehensive Polymer Science', ed. G. Allen and J. C. Bevington, Pergamon, Oxford, 1989, vol. 2, p. 363.
2. A. S. Vaughan and D. C. Bassett, in 'Comprehensive Polymer Science', ed. G. Allen and J. C. Bevington, Pergamon, Oxford, 1989, vol. 2, 415.
3. D. C. Bassett, in 'Comprehensive Polymer Science', ed. G. Allen and J. C. Bevington, Pergamon, Oxford, 1989, vol. 1, p. 841.
4. D. M. Sadler, in 'Comprehensive Polymer Science', ed. G. Allen and J. C. Bevington, Pergamon, Oxford, 1989, vol. 1, p. 731.
5. A. Keller, *Pure Appl. Chem.*, 1992, **64,** 193.
6. A. Keller, *Makromol. Chem., Macromol. Symp.*, 1995, **98,** 1.
7. I. Gutzov and A. Dobreva, *Ceram. Trans.*, 1993, **30,** 151.
8. A. Dobreva, T. Stoyani, S. Tsuparska and I. Gutzov, *Thermochim. Acta,* 1996, in the press.
9. H. Berghmans and F. Deberdt, *Philos. Trans. R. Soc. London, Ser. A,* 1994, **348,** 117.
10. M. Berghmans, S. Thijs, M. Cornette, H. Berghmans, F. C. de Schryver, P. Moldenaers and J. Mewis, *Macromolecules,* 1994, **27,** 7669.
11. M. Katz, *Naturwissenschaften,* 1925, **13,** 410.
12. H. Tadokoro, 'Structure of Crystalline Polymers', Wiley, New York, 1979.
13. C. W. Bunn, *Trans. Farday Soc.*, 1939, **35,** 482.
14. A. Keller and A. O'Connor, *Nature,* 1957, **180,** 1289.
15. P. H. Till, *J. Polym. Sci.*, 1957, **24,** 301.
16. A. Keller, *Philos. Mag.*, 1957, **2,** 1171.
17. E. W. Fischer, *Z. Naturforsch.*, 1957, **12a,** 753.
18. A. Keller, in 'Polymers, Liquid Crystals and Low-Dimensional Solids', ed. N. March and M. Tosi, Plenum, New York, 1984, p. 3.
19. A. Toda and A. Keller, *Colloid Polym. Sci.*, 1993, **271,** 328.

20. A. Keller, in 'Sir Charles Frank, OBE, FRS: An Eightieth Birthday Tribute', ed C. G. Chambers, J. E. Enderby, A. Keller, A. R. Lang and J. W. Steeds, Adam Hilger, Bristol, 1991, p. 265.
21. A. Keller, *Rep. Prog. Phys.*, 1968, **31,** 623.
22. A. Keller, G. Goldbeck-Wood and M. Hikosaka, *Faraday Discuss.*, 1993, **95,** 109.
23. P. J. Barham, in 'Materials Science and Technology', ed. E. L. Thomas, VCH, Weinheim, 1993, vol. 12, p. 153.
24. D. C. Bassett, 'Principles of Polymer Morphology', Cambridge University Press, Cambridge, 1981.
25. J. H. Magill, in 'Treatise on Materials Science and Technology', ed. J. M. Schultz, Academic Press, 1977, vol. 10, Part A, p. 1.
26. P. H. Geil, 'Polymer Single Crystals', Interscience, New York, 1963.
27. B. Wunderlich, 'Macromolecular Physics', Academic Press, New York, 1973, vol. 1.
28. U. W. Gedde, 'Polymer Physics', Chapman and Hall, London, 1995.
29. A. E. Woodward, 'Atlas of Polymer Morphology', Hanser, Munich, 1988.
30. A. Keller, in 'Growth and Perfection of Crystals' (Proc. Int. Conf. Crystal Growth, Cooperstown), ed. R. H. Doremus, B. W. Roberts and D. Turnbull, Wiley, New York, 1968, p. 499.
31. J. Bisault, G. Ryschenkow and G. Faivre, *J. Crystal Growth,* 1991, **110,** 889.
32. A. Keller, *Kolloid Z. Z. Polym.*, 1967, **219,** 118.
33. A. Keller and S. Sawada, *Makromol. Chem.*, 1964, **74,** 190.
34. P. J. Holdsworth and A. Keller, *Polym. Lett.*, 1967, **5,** 605.
35. A. J. Waddon, M. J. Hill, A. Keller and D. J. Blundell, *J. Mater. Sci.*, 1987, **22,** 1773.
36. A. J. Lovinger and D. D. Davies, *Polymer Commun.*, 1985, **26,** 322.
37. A. J. Waddon, A. Keller and D. J. Blundell, *Polymer,* 1992, **33,** 27.
38. R. K. Bayer, *Colloid Polym. Sci.*, 1994, **272,** 910.
39. S. J. Organ and A. Keller, *J. Mater. Sci.*, 1985, **20,** 1571.
40. A. Keller, in 'Structural Order in Polymers' (International Symposium on Macromolecules, Florence, 1980), ed. F. Ciardelli and P. Giusti, Pergamon, Oxford, 1980, p. 135.
41. W. L. Peticolas, G. W. Hibler, J. L. Lippert, A. Peterlin and H. Olf, *Appl. Phys. Lett.*, 1971, **18,** 87.
42. G. V. Fraser, *Indian J. Pure Appl. Phys.*, 1978, **16,** 344.
43. G. R. Strobl and R. Eckel, *J. Polym. Sci., Polym. Phys. Ed.*, 1976, **14,** 913.
44. J. D. Hoffman and J. J. Weeks, *J. Chem. Phys.*, 1965, **42,** 4301.
45. A. Keller and A. O'Connor, *Discuss. Faraday Soc.*, 1958, **25,** 114.
46. T. Kawai and A. Keller, *Philos. Mag.*, 1965, **11,** 1165.
47. P. J. Barham, R. A. Chivers, A. Keller, J. Martinez-Salazar and S. J. Organ, *J. Mater. Sci.*, 1985, **20,** 1625.
48. D. C. Bassett and A. Keller, *Philos. Mag.*, 1961, **6,** 1053.
49. M. Dosière, M.-Ch. Colet and J. J. Point, in 'Morphology of Polymers', ed. B. Sedlacek, de Gruyter, Berlin, 1986, p. 171.
50. D. J. Jones, A. J. Latham, A. Keller and M. Girolamo, *J. Polym. Sci., Polym. Phys. Ed.*, 1973, **11,** 1759.
51. P. Dreyfuss and A. Keller, *J. Macromol. Sci. Phys.*, 1970, **B4,** 811.
52. P. Dreyfuss and A. Keller, *J. Polym. Sci., Polym. Phys. Ed.*, 1973, **11,** 201.
53. T. Kawai, T. Hama and K. Ehara, *Makromol. Chem.*, 1968, **113,** 282.
54. T. Kawai, *Polym. Lett.*, 1964, **2,** 965.
55. W. O. Statton and P. H. Geil, *J. Appl. Polym. Sci.*, 1960, **3,** 357.
56. E. W. Fischer and G. F. Schmidt, *Angew. Chem.*, 1962, **74,** 551.
57. P. Dreyfuss and A. Keller, *Polym. Lett.*, 1970, **8,** 253.
58. A. Keller and D. J. Priest, *Polym. Lett.*, 1970, **8,** 13.
59. P. J. Barham and A. Keller, *J. Polym. Sci., Polym. Phys. Ed.*, 1989, **27,** 1029.
60. A. Keller, *Kolloid Z. Z. Polym.*, 1964, **197,** 98.
61. D. L. Dorset, R. G. Alamo and L. Mandelkern, *Macromolecules,* 1993, **26,** 3143.
62. A. Keller and D. C. Bassett, *J. R. Microsc. Soc.*, 1960, **79,** 243.
63. G. C. Alfonso, M. Ceschina, V. Chiappa and E. Pedemonte in 'Crystallization of Polymers', NATO ASI Series, ed. M. Dosière, Kluwer, Dordrecht, 1992, vol. C405, p. 455.
64. R. J. Roe, C. Gieniewski and R. G. Vadimsky, *J. Polym. Sci., Polym. Phys. Ed.*, 1973, **11,** 1653.
65. P. H. Lindenmeyer, *J. Polym. Sci. C,* 1963, **1,** 5.
66. J. C. Wittmann and B. Lotz, *J. Polym. Sci., Polym. Phys. Ed.*, 1985, **23,** 205.
67. Y. Nakagawa, H. Hayashi, T. Takahagi, F. Soeda, A. Ishitani, A. Toda and H. Miyaji, *Jpn. J. Appl. Phys.*, 1994, **33,** 3771.
68. T. Kajiyama, I. Ohki and A. Takahara, *Macromolecules,* 1995, **28,** 4768.
69. *R. Nisman, P. Smith and G. J. Vancso, Langmuir,* 1994, **10,** 1667.
70. D. C. Bassett, F. C. Frank and A. Keller, *Philos. Mag.*, 1963, **8,** 1739 and 1753.
71. A. Keller, in 'MTP International Review of Science', Physical Chemistry, Series One, ed. C. E. H. Bawn, Butterworth, London, 1972, vol. 8, p. 105.
72. Faraday Discussion No. 68, 'Organization of Macromolecules in the Condensed Phase', 1979, entire issue.
73. A. Keller, *Kolloid Z. Z. Polym.*, 1969, **231,** 386.
74. E. W. Fischer, *Kolloid Z. Z. Polym.*, 1969, **231,** 458.
75. F. J. Baltá-Calleja and C. G. Vonk, 'X-ray Scattering of Synthetic Polymers', Elsevier, Amsterdam, 1989.
76. G. R. Strobl, *Colloid Polym. Sci.*, 1979, **257,** 584.
77. M. Stamm, E. W. Fischer, M. Dettenmaier and P. Convert, *Faraday Discuss.*, 1979, **68,** 263.
78. D. M. Sadler and A. Keller, *Science,* 1979, **203,** 263.
79. M. Tasumi and S. Krimm, *J. Polym. Sci. A-2,* 1968, **6,** 995.
80. S. J. Spells, A. Keller and D. M. Sadler, *Polymer,* 1984, **25,** 749.
81. S. J. Spells, in 'Characterization of Solid Polymers', ed. S. J. Spells, Chapman and Hall, London, 1994, p. 166.
82. P. J. Barham and D. M. Sadler, *Polymer,* 1991, **32,** 395.
83. P. J. Lemstra, N. A. M. J. van Aerle and C. W. M. Bastiaansen, *Polym. J.*, 1987, **19,** 85.
84. F. C. Frank, *Faraday Discuss.*, 1979, **68,** 7.

85. J. D. Hoffman, C. M. Guttman and E. A. DiMarzio, *Faraday Discuss.*, 1979, **68,** 179.
86. P. J. Flory, D. Y. Yoon and K. A. Dill, *Macromolecules,* 1984, **17,** 862.
87. C. G. Vonk, *J. Polym. Sci., Polym. Lett.*, 1986, **24,** 305.
88. G. A. Bassett, D. J. Blundell and A. Keller, *J. Macromol. Sci.*, 1967, **B1,** 161.
89. D. J. Blundell and A. Keller, *J. Macromol. Sci., Phys.*, 1973, **B7,** 253.
90. A. Keller and D. M. Sadler, *J. Macromol. Sci., Phys.*, 1973, **B7,** 263.
91. D. M. Sadler, *J. Polym. Sci. A-2,* 1971, **9,** 779.
92. A. Keller and D. J. Priest, *J. Macromol. Sci., Phys.*, 1968, **B2,** 479.
93. R. P. Palmer and A. J. Cobbold, *Makromol. Chem.*, 1964, **74,** 174.
94. D. J. Blundell, A. Keller, I. M. Ward and I. J. Grant, *J. Polym. Sci. B,* 1966, **4,** 781.
95. D. J. Blundell and A. Keller, *J. Polym. Sci. A-2,* 1967, **5,** 991.
96. Y. Udagawa and A. Keller, *J. Polym. Sci. A-2,* 1971, **9,** 437.
97. A. Keller, E. Martuscelli, D. J. Priest and Y. Udagawa, *J. Polym. Sci. A-2,* 1971, **9,** 1807.
98. G. N. Patel and A. Keller, *J. Polym. Sci., Polym. Phys. Ed.*, 1975, **13,** 2259.
99. A. Keller and Y. Udagawa, *J. Polym. Sci. A-2,* 1971, **9,** 437.
100. E. Ergos and L. Mandelkern, *J. Polym. Sci., Polym. Lett. Ed.*, 1972, **10,** 631.
101. G. V. Fraser, A. Keller, E. J. George and D. Dreyfuss, *J. Macromol. Sci., Phys.*, 1979, **B16,** 295.
102. V. F. Holland and P. H. Lindenmeyer, *J. Appl. Phys.*, 1965, **36,** 3049.
103. D. M. Sadler and A. Keller, *Kolloid Z. Z. Polym.*, 1970, **239,** 641.
104. D. M. Sadler and A. Keller, *Kolloid Z. Z. Polym.*, 1970, **242,** 1081.
105. F. Kloos, S. Go and L. Mandelkern, *J. Polym. Sci., Polym. Phys. Ed.*, 1974, **12,** 1145.
106. R. Patil and D. H. Reneker, *Polymer,* 1994, **35,** 1909.
107. A. Keller and A. O'Connor, *Polymer,* 1960, **1,** 163.
108. H. Zahn and W. Pieper, *Kolloid Z. Z. Polym.*, 1962, **180,** 97.
109. W. Kern, J. Davidovits, K. J. Rauterkus and G. F. Schmidt, *Makromol. Chem.*, 1961, **43,** 106.
110. F. J. Baltá-Calleja and A. Keller, *J. Polym. Sci. A,* 1964, **2,** 2151.
111. I. Bidd, D. W. Holdup and M. C. Whiting, *J. Chem. Soc., Perkin Trans. 1,* 1987, 2455.
112. K. S. Lee and G. Wegner, *Makromol. Chem., Rapid Commun.*, 1985, **6,** 203.
113. J. P. Arlie, P. Spegt and A. Skoulios, *Makromol. Chem.*, 1967, **104,** 212.
114. B. Gilg, P. Spegt, J. Terrisse and A. Skoulios, *Makromol. Chem.*, 1967, **107,** 39.
115. A. J. Kovacs and A. Gonthier and C. Straupe, *J. Polym. Sci., Symp.*, 1975, **50,** 283.
116. A. J. Kovacs and A. Gonthier, *Kolloid Z. Z. Polym.*, 1972, **250,** 530.
117. C. P. Buckley and A. J. Kovacs, in 'Structure of Crystalline Polymers', ed. I. H. Hall, Elsevier Applied Science, London, 1984, p. 261.
118. P. G. Higgs and G. Ungar, *J. Chem. Phys.*, 1994, **100,** 640.
119. G. Ungar, in 'Integration of Fundamental Polymer Science and Technology', ed. P. J. Lemstra and L. A. Kleintjens', Elsevier Applied Science, London, 1988, vol. 2, p. 346.
120. G. Ungar, J. Stejny, A. Keller, I. Bidd and M. C. Whiting, *Science,* 1985, **229,** 386.
121. G. Ungar, S. J. Organ and A. Keller, *J. Polym. Sci., Polym. Lett.*, 1988, **26,** 259.
122. G. Ungar and S. J. Organ, *Polym. Commun.*, 1987, **28,** 232.
123. G. Ungar, in 'Characterization of Solid Polymers', ed. S. J. Spells, Chapman and Hall, London, 1994, p. 56.
124. G. Ungar and A. Keller, *Polymer,* 1987, **27,** 1835.
125. S. J. Organ and A. Keller, *J. Polym. Sci., Polym. Phys. Ed.*, 1987, **25,** 2409.
126. G. Ungar and A. Keller, *Polymer,* 1987, **28,** 1899.
127. S. J. Organ, G. Ungar and A. Keller, *Macromolecules,* 1989, **22,** 1995.
128. A. Keller, S. J. Organ and G. Ungar, *Makromol. Chem., Macromol. Symp.*, 1991, **48/49,** 93.
129. S. J. Organ, A. Keller and G. Ungar, in: 'Crystallization of Polymers', NATO ASI Series, ed. M. Dosière, Kluwer, Dordrecht, 1992, vol. C405, p. 81.
130. S. J. Organ, G. Ungar, A. Keller and M. Hikosaka, *Polymer,* 1996, in press.
131. G. Ungar and S. J. Organ, *J. Polym. Sci., Polym. Phys. Ed.*, 1990, **28,** 2353.
132. S. J. Organ, G. Ungar and A. Keller, *J. Polym. Sci., Polym. Phys. Ed.*, 1990, **28,** 2365.
133. D. M. Sadler and G. H. Gilmer, *Polym. Commun.*, 1987, **28,** 242.
134. S. Z. D. Cheng and J. Chen, *J. Polym. Sci., Polym. Phys. Ed.*, 1991, **29,** 311.
135. D. J. Blundell, J. J. Liggat and A. Flory, *Polymer,* 1992, **33,** 2475.
136. R. G. Alamo, L. Mandelkern, G. M. Stack, C. Kröhnke and G. Wegner, *Macromolecules,* 1994, **27,** 147.
137. L. Mandelkern, A. Prasad, R. G. Alamo and G. M. Stack, *Macromolecules,* 1990, **23,** 3696.
138. R. G. Alamo, L. Mandelkern, G. M. Stack, C. Kröhnke and G. Wegner, *Macromolecules,* 1993, **26,** 2743.
139. G. M. Stack, L. Mandelkern, C. Kröhnke and G. Wegner, *Macromolecules,* 1989, **22,** 4351.
140. S. G. Yeates and C. Booth, *Makromol. Chem.*, 1985, **186,** 2663.
141. D. Seebach, H. Michael Burger, H. M. Müller, U. D. Lengweiler, A. Beck, K. E. Sykes, P. A. Barker and P. J. Barham, *Helv. Chim. Acta,* 1994, **77,** 1099.
142. P. J. Barham, A. Keller, E. L. Otun and P. A. Holmes, *J. Mater. Sci.*, 1984, **19,** 2781.
143. S. Z. D. Cheng, A. Q. Zhang, J. S. Barley, J. H. Chen and A. Habenschuss, *Macromolecules* 1991, **24,** 3937.
144. S. Z. D. Cheng, J. H. Chen, J. S. Barley, A. Q. Zhang, A. Habenschuss and P. R. Zschack, *Macromolecules,* 1992, **25,** 1453.
145. S. Z. D. Cheng, S. S. Wu, J. H. Chen, Q. H. Zhou, R. P. Quirk, E. D. Meerwall, B. S. Hsiao, A. Habenschuss and P. R. Zschack, *Macromolecules,* 1993, **26,** 5105.
146. P. J. Phillips, *Rep. Prog. Phys.*, 1990, **53,** 549.
147. D. J. Blundell, A. Keller and A. J. Kovacs, *J. Polym. Sci. B,* 1966, **4,** 481.
148. D. J. Blundell and A. Keller, *J. Macromol. Sci., Phys.*, 1968, **B2,** 301.
149. S. J. Organ and A. Keller, *J. Polym. Sci., Polym. Phys. Ed.*, 1986, **24,** 2319.
150. V. F. Holland and P. H. Lindenmeyer, *J. Polym. Sci.*, 1962, **57,** 589.
151. L. Mandelkern, 'Crystallization of Polymers', McGraw-Hill, New York, 1964.

152. M. J. Nardini and F. P. Price, in 'Crystal Growth', (Proc. Int. Conf. Crystal Growth, Boston), Pergamon, Oxford, 1966, p. 395.
153. D. J. Blundell and A. Keller, *J. Polym. Sci. B,* 1968, **6,** 433.
154. A. Keller and E. Pedemonte, *J. Cryst. Growth,* 1973, **18,** 111.
155. M. Cooper and R. St. J. Manley, *Macromolecules,* 1975, **8,** 219.
156. S. J. Organ and A. Keller, *J. Polym. Sci., Polym. Lett.,* 1987, **25,** 67.
157. A. Toda, H. Miyaji and H. Kiho, *Polymer,* 1986, **27,** 1505.
158. A. Toda, *J. Chem. Soc., Faraday Trans.,* 1995, **91,** 2581.
159. K. Armitstead and G. Goldbeck-Wood, *Adv. Polym. Sci.,* 1992, **100,** 219.
160. D. C. Bassett, in 'Developments in Crystalline Polymers — 1', ed. D. C. Bassett, Elsevier Applied Science, London, 1982, p. 115.
161. B. Wunderlich, L. Melillo, C. M. Cormier, T. Davidson and G. Snyder, *J. Macromol. Sci., Phys.,* 1967, **B1,** 485.
162. M. Hikosaka, S. Rastogi, A. Keller and H. Kawabata, *J. Macromol. Sci. Phys.,* 1992, **B31,** 87.
163. M. Hikosaka, H. Okada, S. Rastogi and A. Keller, *J. Chem. Soc., Faraday Trans.,* 1995, **91,** 2573.
164. S. Rastogi, M. Hikosaka, H. Kawabata and A. Keller, *Macrmolecules,* 1991, **24,** 6384.
165. D. C. Bassett and B. Turner, *Philos. Mag.,* 1974, **29,** 925.
166. J. D. Hoffman, G. T. Davies and J. I. Lauritzen, in 'Treatise on Solid-State Chemistry', ed. N. B. Hannay, Plenum, New York, 1976, vol. 3, p. 497.
167. N. Okui, in 'Crystallization of Polymers', NATO ASI Series, ed. M. Dosière, Kluwer, Dordrecht, 1992, vol. C405, p. 593.
168. G. Allegra, *Ferroelectrics,* 1980, **30,** 195.
169. B. Wunderlich and A. Mehta, *J. Polym. Sci., Polym. Phys. Ed.,* 1974, **12,** 255.
170. D. M. Sadler, *J. Polym. Sci., A-2,* 1971, **9,** 779.
171. S. Z. D. Cheng and B. Wunderlich, *J. Polym. Sci., Polym. Phys. Ed.,* 1986, **24,** 595.
172. S. Z. D. Cheng, D. W. Noid and B. Wunderlich, *J. Polym. Sci., Polym. Phys. Ed.,* 1989, **27,** 1149.
173. F. L. Binsbergen, *J. Cryst. Growth,* 1972, **13/14,** 44.
174. J. J. Point, *Faraday Discuss.,* 1979, **68,** 167.
175. D. M. Sadler and G. H. Gilmer, *Polymer,* 1984, **25,** 1446.
176. G. Goldbeck-Wood, in 'Science and Technology of Crystal Growth', ed. J. P. van der Eerden and O. S. L. Bruinsma, Kluwer, Dordrecht, 1995, chap. 6.3.
177. A. Ziabicki and L. Jarecki, *J. Chem. Phys.,* 1984, **80,** 5751.
178. D. M. Sadler, *J. Chem. Phys.,* 1987, **87,** 1771.
179. M. L. Mansfield and L. I. Klushin, *Polymer,* 1994, **35,** 2937.
180. T. Seto, *Rep. Prog. Polym. Phys. Jpn.,* 1964, **7,** 67.
181. F. C. Frank, *J. Cryst. Growth,* 1974, **22,** 233.
182. J. J. Point, M. C. Colet and M. Dosière, *J. Polym. Sci., Polym. Phys. Ed.,* 1986, **24,** 357.
183. J. J. Point and M. Dosière, *Macromolecules,* 1989, **22,** 3501.
184. J. D. Hoffman and R. L. Miller, *Macromolecules,* 1989, **22,** 3502.
185. D. J. Gates and M. Westcott, *Proc. R. Soc. London, Ser. A,* 1988, **416,** 443 and 463.
186. J. J. Point and J. J. Janimak, *J. Cryst. Growth.,* 1993, **131,** 501.
187. J. D. Hoffman, R. L. Miller, H. Marand and D. B. Roitman, *Macromolecules,* 1992, **25,** 2221.
188. J. D. Hoffman, *Polymer,* 1991, **32,** 963.
189. J. D. Hoffman, *Polymer,* 1992, **33,** 2643.
190. J. P. van der Eerden, in 'Science and Technology of Crystal Growth', ed. J. P. van der Eerden and O. S. L. Bruinsma, Kluwer, Dordrecht, 1995, chap. 1.2.
191. D. M. Sadler and G. H. Gilmer, *Phys. Rev. B,* 1988, **38,** 5684.
192. G. Goldbeck-Wood, *Makromol. Chem., Macromol. Symp.,* 1994, **81,** 221.
193. G. Goldbeck-Wood, in 'Crystallization of Polymers', ed. M. Dosière, Kluwer, Dordrecht, 1993, p. 249.
194. B. C. Edwards and P. J. Phillips, *Polymer,* 1974, **15,** 351.
195. A. Vaughan, University of Reading, personal communication, 1992.
196. L. A. M. J. Jetten, H. J. Human, P. Bennema and J. P. van der Eerden, *J. Cryst. Growth,* 1984, **68,** 503.
197. M. L. Mansfield, *Polymer,* 1988, **29,** 1755.
198. A. Toda, *Faraday Discuss.,* 1993, **95,** 129.
199. J. J. Point and D. Villers, *J. Cryst. Growth,* 1991, **114,** 228.
200. A. Toda, *J. Phys. Soc. Jpn.,* 1986, **55,** 3419.
201. J. D. Hoffman and R. L. Miller, *Macromolecules,* 1989, **22,** 3038.
202. D. M. Sadler, *J. Polym. Sci., Polym. Phys. Ed.,* 1985, **23,** 1533.
203. J. D. Hoffman, *Polymer,* 1991, **32,** 2828.
204. J. J. Point, *J. Chem. Soc., Faraday Trans.,* 1995, **91,** 2565.
205. M. Hikosaka, *Polymer,* 1987, **28,** 1257; 1990, **31,** 458.
206. A. Keller, M. Hikosaka, S. Rastogi, A. Toda, P. J. Barham and G. Goldbeck-Wood, *J. Mater. Sci.,* 1994, **29,** 2579.
207. G. Goldbeck-Wood, 1994, unpublished results.
208. T. A. Kavassalis and P. R. Sundararajan, *Macromolecules,* 1993, **26,** 4144.
209. H. Marand, V. Velikov, R. K. Verma, P. M. Cham, V. Prablu and D. Dillard, *Polym. Prepr., Am. Chem. Soc., Div. Polym. Chem.,* 1995, **36** (1), 263.
210. P. G. de Gennes, *C. R. Acad. Sci., Ser. IIb,* 1995, **321,** 363
211. M. Hikosaka, K. Amano, S. Rastagi and A. Keller, *Macromolecules,* 1996, in press.

8

Time-resolved Characterization of Polymer Phase Transitions

THEIN KYU

University of Akron, OH, USA

8.1 INTRODUCTION

Recent advances in time-resolved scattering and spectroscopy have been reviewed with emphasis on their application to the characterization of the dynamical aspects of polymer phase transitions involving polymer relaxation, crystallization, phase separation, and ordering transitions in multicomponent polymers, and interplay among various processes. The time-dependent scattering method, often known as a real time approach, basically refers to the determination of the evolution of a static (elastic) structure factor as a function of elapsed time during phase transitions under quiescent conditions and/or subjected to flow or deformation.

The application of on-line nondestructive optical techniques to the investigation of rheological and viscoelastic properties[1] such as flow and deformation mechanisms of polymers, known as rheo-optics, can be dated back to the early 1960s.[2–5] The primary focus of rheo-optical studies was to identify the structural origin of the viscoelastic responses of polymers subjected to deformation. The first generation of the rheo-optical technique was based on the principle of a dynamic mechanical method by simultaneous employment of optical and X-ray techniques.[2–8] The sinusoidal stress and optical responses were monitored simultaneously while the sample was subjected to the oscillatory strain. On-line birefringence, small-angle light scattering (SALS), infrared (IR) dichroism, wide-angle X-ray diffraction

(WAXD), or small-angle X-ray scattering (SAXS) were employed by synchronizing with the external applied field to elucidate the relaxation mechanisms of structural units involving polymer segments, amorphous chains, and crystals.[2–8] Later, an alternative technique based on the principle of stress relaxation was introduced,[9] where a polymer sample was rapidly stretched (typically less than a second) within the linear elastic limit of the materials, then subsequently monitoring the time decay of stress while keeping the specimen at a constant strain. The responses of internal structural units may be analyzed through time- or frequency-dependent behavior of stress–optical or strain–optical coefficients. By virtue of the specificity of the individual optical techniques utilized, the rheo-optical techniques are, in principle, capable of probing the structural responses at various length scales from amorphous chains all the way to lamellae and spherulites.[2–10]

Newer rheo-optical techniques such as shear optical birefringence[11–13] shear small-angle light scattering,[14] and shear small-angle X-ray scattering,[15] have emerged to characterize the rheological behavior of liquid crystalline polymer (LCP) solutions. Similar progress has been made in studies of the structural transformation in solutions of block copolymers and polymer blends subjected to flow.[16–19] This progress may be attributed to the recent emergence of advanced scattering techniques such as small-angle neutron scattering (SANS)[18,19] and synchrotron radiation for X-ray scattering[17] that make static and dynamic studies on polymer phase transitions more attractive and exert significant impact on polymer research in unprecedented ways. At present, the application of SANS to time-resolved studies of polymer phase transformations may be still in its infancy, but the number of such studies have increased considerably. In this review we focus on the recent advances in the applications of time-resolved SANS, SAXS, and SALS techniques to polymer phase transitions. Some pertinent spectroscopic studies are also discussed.

8.2 PRINCIPLES AND TECHNIQUES

8.2.1 Scattering Techniques

The static (elastic) scattering by neutrons, X-rays, and light occurs in accordance with a common principle based on the interaction between radiation and matter.[20–22] The intensity of the incident radiation (I_0) and the scattered intensity (I) by a scattering element may be expressed in terms of a correlation function (or scattering power) and the phase difference between the position vector ($\boldsymbol{r}$) and the scattering wavevector ($\boldsymbol{q}$):

$$I(q) = I_0(q)v\langle\eta^2\rangle \int \gamma(\boldsymbol{r})\exp[-i(\boldsymbol{q}\cdot\boldsymbol{r})]\,\mathrm{d}r^3 \tag{1}$$

where $\gamma(\boldsymbol{r})$ is the correlation function given by

$$\gamma(\boldsymbol{r}) = \langle\eta_1(\boldsymbol{r})\eta_2(\boldsymbol{r})\rangle/\langle\eta^2(\boldsymbol{r})\rangle \tag{2}$$

where the scattering wavenumber is defined as $q = (4\pi/\lambda)\sin\theta/2$, λ and θ being the wavelength of the radiation and the scattering angle measured in the medium, respectively. $\eta_i = p_i - p_0$ where p_i represents the scattering quantities such as electron density or dielectric constant of the component i (=1 or 2), and p_0 is the mean value. $\langle\eta^2(\boldsymbol{r})\rangle$ is the mean square fluctuations. The difference among the three techniques is originated in their scattering power, *e.g.* the contrast $(p_1 - p_2)$ or the fluctuation (η_i) in the neutron scattering length of the components 1 and 2 for neutron scattering or in the electron densities for X-rays. The intensity of light scattering arises due to the contrast of the refractive index or the polarizabilities of the components. The static scattering technique can be utilized to study the time-dependent processes underlying the phase transitions of polymers, provided that the dynamical process is sufficiently slow. Such a methodology is called time-resolved scattering, which is basically a measure of the temporal evolution of a static structure factor (or scattered intensity) during the course of the phase transitions. The time-dependent scattering of light, X-rays, or neutrons may be described as

$$I(q, t) = I_0(q, 0) \int \langle \eta_1(r, t)\eta_2(r', t')\rangle \exp[-i(\mathbf{q} \cdot \mathbf{r})]\, dr^3 \quad (3)$$

where $\eta(r, t)$ is the time-dependent fluctuations of electron density or neutron scattering length or refractive index of the correlated scattering entities.

8.2.1.1 Light scattering

Time-resolved scattering of light is essentially the same as the conventional static (elastic) light scattering technique which permits the simultaneous acquisition of angular dependence of scattering data in one-dimension or in two-dimensions in relatively short time intervals. The development of a one-dimensional position sensitive detector (PSD) and a two-dimensional charge couple detector (CCD) camera coupled with dedicated personal computer (PC) has made scattering measurements in real time possible for studying time-dependent phenomena. These PSD detectors are powerful, especially for monitoring the rapid evolution of scattering profiles at relatively small angular ranges.

Time-resolved light scattering apparatus reported thus far were mostly custom-built, which vary from one laboratory to another. The SALS apparatus, built at the University of Akron, is depicted schematically in Figure 1. It consists of a He–Ne laser light source (wavelength of 632.8 nm), a heating stage sample holder, a set of a polarizer and analyzer, a neutral density filter, a screen (or a lens), and a PSD. A Reticon one-dimensional detector (EG & G Princeton Applied Research) is controlled by an Optical Multichannel Analyzer (OMA III) interlinked with a dedicated computer (IBM PS-2/30). The detector controller is equipped with a dual 16-bit ADC (analog to digital converter), but the dynamic range is 14-bit. The data sampling rate is 1024 pixels per 60 μs. The data can be stored in RAM (random access memory) directly. They may be unloaded to a floppy disc for a permanent record or further data reduction. A 2 mm diameter hole was drilled in the heating stage sampler holder to collimate the incident laser light. A couple of heating stages are placed side by side for temperature jump/quench experiments; one is used for preheating a sample at a single phase temperature, while the other is controlled at an experimental temperature. The sample is rapidly transferred from one heating stage to the other.

The two-dimensional CCD camera (EG & G Co) has a detector area of $25 \times 25\ mm^2$ with a resolution of 576 pixels $\times$ 384 lines and 16-bit ADC. While the data information available by the 2D CCD detector far exceeds that of the 1D Reticon, the data handling for 2D scattering patterns is time-consuming and tedious. A 2D data analysis program, dedicated to scattering and image analysis (SASIMA©), has been written in visual C^{++} on Windows. It offers a quick and easy way for angle corrections, background subtraction, data smoothing, circular and quadrant averaging, and area integration with quality graphic outputs. Figure 2 shows a typical time-evolution of scattering patterns in 2D and 3D perspective plots of spinodal decomposition for a mixture of side-chain liquid crystal polymer (ethylene glycol diglycidyl ether-*graft*-4-(ω-aminohexyloxy)-4′-cyanobiphenyl) and low molecular weight liquid crystal (cyanobiphenyls: E7)[23] as observed by the CCD detector. The time required to acquire individual scattering images is 1–2 s, depending on the pixel size.

In recent years, time-resolved scattering of light and neutrons has been applied to polymer melts and/or solutions under shear flow or deformation. Again, the detailed instrumentation may vary from one laboratory to another, but the essential feature remains the same. Figure 3 shows a schematic diagram of shear flow–SALS apparatus developed at Kyoto University.[24,25] The steady-state Couette-type shear device is designed with a set of an optically transparent cone (E) and plate (G) made of fused quartz. The radius of the cone and plate is 40 mm and the cone angle is 0.91°. A laser light scattering apparatus consisting of a He–Ne laser ($\lambda = 632.8$ nm) and associated optics is mounted on the shear device such that light can pass through the shear-sample cell in the x direction. The polarization direction of the incident beam (D) and the analyzer (K) is set parallel to the flow direction (along the z axis). The velocity gradient occurs in the plane parallel to the x–y plane during shearing. The scattered image

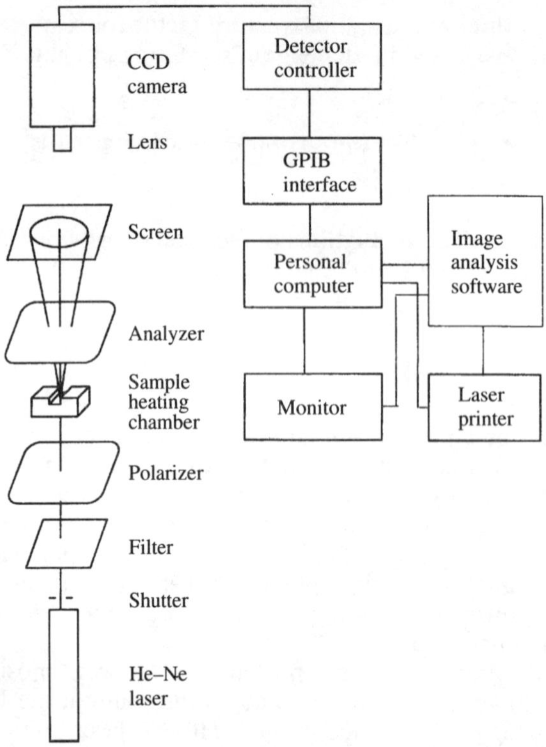

Figure 1 A schematic diagram of a small-angle light scattering set-up.

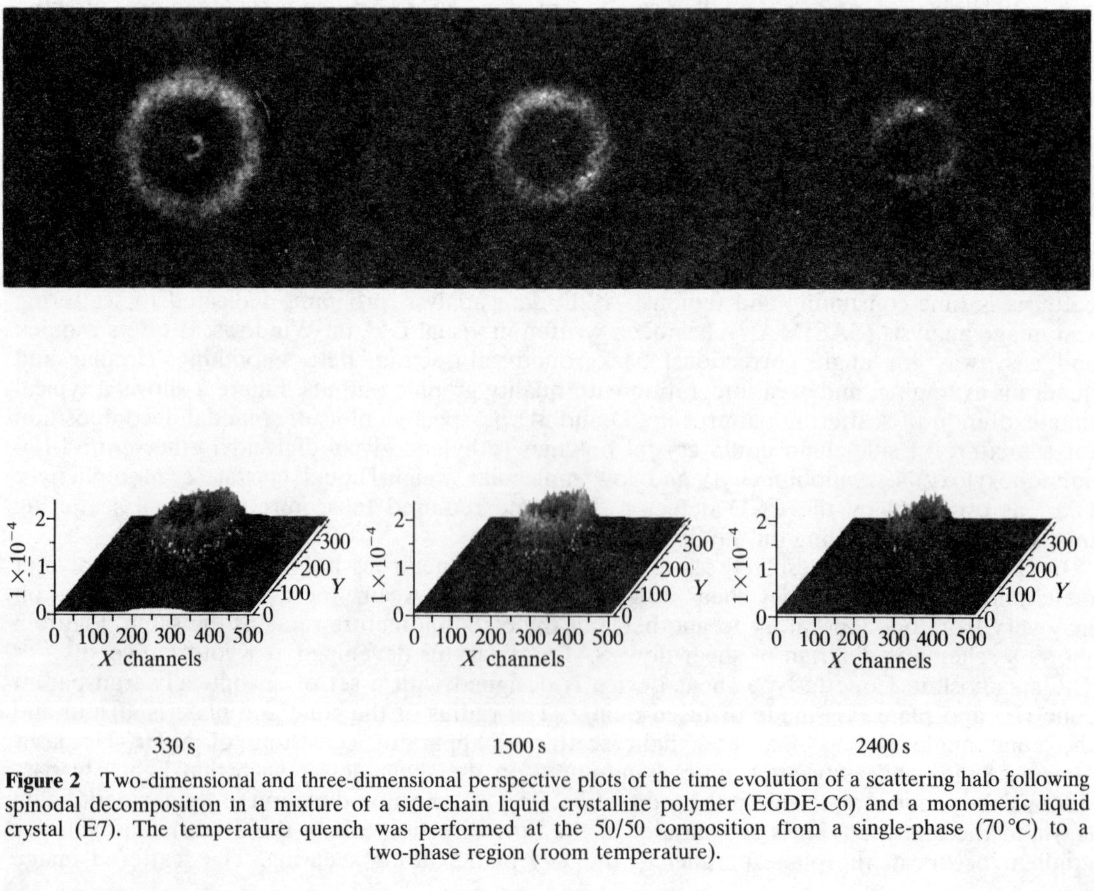

Figure 2 Two-dimensional and three-dimensional perspective plots of the time evolution of a scattering halo following spinodal decomposition in a mixture of a side-chain liquid crystalline polymer (EGDE-C6) and a monomeric liquid crystal (E7). The temperature quench was performed at the 50/50 composition from a single-phase (70 °C) to a two-phase region (room temperature).

may be recorded by using one of the following detectors: (i) a photographic film, (ii) a TV camera with a VTR monitor system or a dynamic image analyzer, or (iii) a position sensitive photodiode array.

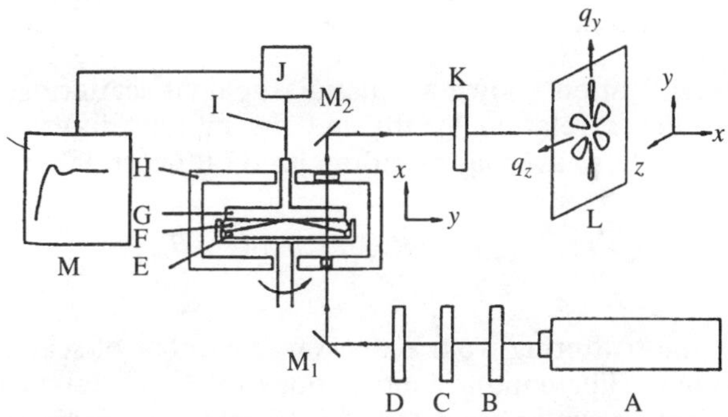

Figure 3 A schematic drawing of a cone and plate type viscometry–SALS apparatus consisting of (A) He–Ne laser, (B) polarizer, (C) chopper, (D) half-wave plate, (E) cone, (F) sample, (G) plate, (H) thermal chamber, (I) shaft, (J) load cell, (M) recorder, (K) analyzer, (M_i) mirrors, and (L) detector.

8.2.1.2 X-ray scattering

The small-angle X-ray technique, developed at the Oak Ridge National Laboratory, renders the measurement of the angular (azimuthal and Bragg angles) dependence of scattering in two dimensions.[26] The 10 m SAXS apparatus comprises a two-dimensional position sensitive detector (2D-PSD), a rotating anode X-ray generator, and associated X-ray optics. The fast data acquisition and the increased X-ray flux permit time-resolved SAXS measurements possible on polymer crystallization, structural relaxation, and deformation. However, it has limited use for dynamical studies of fast transitions such as polymer–polymer phase separation because of the poor contrast of electron densities of the constituent polymer pairs, requiring prolonged accumulation of the scattered intensity to improve the signal-to-noise ratio.

Recently, synchrotron small-angle X-ray scattering has been extensively employed for characterizing the kinetics of polymer crystallization and polymer–polymer phase separation. The high-flux synchrotron X-rays make time-resolved SAXS studies on polymer phase separation feasible. The detailed principles of synchrotron radiation along with instrumentation, such as detection systems and associated optics, may be found in a recent review paper.[27] The shear or tensile deformation device used in time-resolved X-ray scattering is similar in principle to that of light scattering and thus will not be repeated here. We focus on the applications of synchrotron SAXS to time-resolved studies of polymer phase transitions.

8.2.1.3 Neutron scattering

The use of various kinds of radiation and wavelengths has its own merit. Neutron scattering originates from atomic nuclei, as opposed to electron clouds for X-ray scattering and atomic/bond polarizability for light scattering. While the wavelengths of neutrons and X-rays are in a comparable range of a few ångstroms, their scattering powers show different sensitivities to a given atom. For instance, while hydrogen is known to be a poor scatterer for X-rays, it is a strong scattering center for neutrons. Consequently, the substitution of hydrogen by deuterium exerts a profound effect on neutron scattering, but it has insignificant influence on X-ray scattering. Neutron scattering has a clear advantage over X-rays by virtue of the possible labeling of polymeric chains with radioisotopes. The wavelength of neutrons is about 3–4-fold larger than that of X-rays. Hence SANS permits measurements at intermediate scattering angles which are not accessible by the conventional small-angle X-ray scattering approach.

As in the case of SALS and SAXS, the SANS equipment for time-resolved scattering experiments is essentially the same as the static SANS set-up. The detailed descriptions of the static SANS apparatus may be found in the original literature.[22,28] The neutron flux in the existing SANS facilities has improved over the years, thereby permitting time-resolved SANS studies possible for relatively slow phase-separating systems.

8.2.1.4 Dynamic light scattering

Dynamic light scattering,[29] often known as quasielastic light scattering, has been customarily employed to determine the correlation length and the mutual diffusion coefficient of dilute binary solutions in their single phase via the normalized autocorrelation function, $g(\tau)$, of the scattered electric field, $E(\tau)$;

$$g^{(1)}(\tau) = \langle E(0) \cdot E(\tau) \rangle = \exp(-q^2 D_m \tau) \tag{4}$$

Here, it is assumed that the scattering from a very large number of scatterers is simultaneously detected so that the 'number fluctuation' contributions to the correlation function vanish and multiple scattering does not occur. In practice, $g^{(1)}(\tau)$ is obtained indirectly from the measured homodyne intensity autocorrelation function, $G^{(2)}(\tau)$;

$$G^{(2)}(\tau) = \langle I(0) \cdot I(\tau) \rangle = B(1 + f\,|g^{(1)}(\tau)|^2) \tag{5}$$

where B is a baseline and f an instrumental parameter. It is assumed that the intensity is a Gaussian random variable so that the normalized intensity autocorrelation function, $g^{(2)}(\tau)$, is given by $g^{(2)}(\tau) = 1 + |g^{(1)}(\tau)|$. The data of $|g^{(1)}(\tau)|$ may be analyzed according to a cumulate method that gives the average line width, $\bar{\Gamma}$. This dynamic light scattering technique is particularly useful for determining the correlation length and diffusivity in the single phase of a binary system.[36]

8.2.2 Spectroscopic Techniques

Spectroscopic techniques such as nuclear magnetic resonance (NMR),[30,31] infrared (IR),[32,33] fluorescence,[34] and electron spin resonance (ESR)[35] have been widely used for characterizing organic and biomaterials more than metals or ceramics. One advantage of the spectroscopic studies of polymer phase transitions over other classes of materials is the possibility of tailoring macromolecules with specific architecture via established synthetic routes. The possible control of molecular weight and intricate molecular structure provides a new avenue for fine-tuning of polymer phase diagrams, as well as testing with various scaling and dynamical theories of polymer phase transitions. The spectroscopic technique could be informative as it is specific in selectively probing macromolecules during phase transitions by labeling the chains fully or partially, *e.g.* by substituting with deuterium or a fluorescence chromophore. Additionally, spectroscopic techniques could permit insights into investigations at a microstructural level (typically <5 nm) which is too small to be seen by other techniques, such as the cloud point method.

Solid state NMR was the first spectroscopic technique[30] utilized for characterizing polymer miscibility through the measurement of chain mobility subject to its environment. The NMR quantity $T_{1\rho}$ is basically a measure of the chain or segmental mobility of constituent polymers. This technique is capable of probing at a few nanometers length in scale, and therefore it has a greater sensitivity for mimicking microheterogeneity at an early stage of polymer–polymer phase separation. ESR spectroscopy[35] has a similar sensitivity for probing small length scales of local heterogeneity via the motion of labeled nitroxide radicals. The ESR technique has excellent sensitivity to molecular mobility with correlation times of 10^{-6}–10^{-11} s, thus covering the temporal correlation range not accessible by the NMR technique. Most spectroscopic techniques thus far have been applied to the characterization of structures, molecular interactions, and equilibrium properties of multicomponent polymers, but to a lesser extent to dynamic aspects of phase transitions. There are only a few instances where spectroscopic techniques have been employed to study time-resolved experiments.[34] Another major advantage of the spectroscopic techniques compared to the scattering techniques is the possible determination of the concentration of one component in mixed phases during the course of polymer–polymer phase separation.[34,35]

8.3 APPLICATIONS TO POLYMER PHASE SEPARATION AND PHASE TRANSITION

8.3.1 Polymer–Polymer Phase Separation

The progress of time-resolved scattering studies on dynamical properties of phase transformations is particularly significant in polymers relative to small molecule systems such as metals, ceramics, or low molar mass solutions.[36] Such remarkable progress is partly due to the inherent slow diffusivity or high effective viscosity of macromolecules.[37,38] That is to say, the diffusion coefficient (D) of a macromolecule is inversely proportional to the square of the degree of polymerization (N), *viz.* $D \approx N^{-2}$. The degree of polymerization of a typical polymer is in the range of several thousands, whereas it is unity for small molecules; thus polymer diffusivity could be many orders of magnitude slower. The slow diffusion of polymers renders the determination of the dynamic processes of phase transitions possible relatively easily by time-resolved scattering or spectroscopic equipment accessible in many laboratories.

The dynamics of liquid–liquid phase separation in metal alloys had been customarily studied in the past by the so-called aging technique, by annealing the alloys for varying times and taking electron micrographs or scattering measurements of the aged samples at ambient temperature.[36] Recent development of high-energy radiation sources such as synchrotron radiation, coupled with enhanced detection systems, has facilitated the on-line measurement of the dynamical processes of alloys in real time.[39]

8.3.1.1 Phase equilibria

Polymer–polymer phase separation reveals a unique phase behavior originating from the interconnectivity of long-chain molecules and stereoregular arrangements that do not exist in atomic or small molecule systems.[36] Most small molecule mixtures such as metal alloys, inorganic glasses, or solution mixtures display an upper critical solution temperature (UCST). On the other hand, most high-polymer blends exhibit an inverted phase diagram with a feature known as a lower critical solution temperature (LCST). Some oligomers and rubbery blends, however, show a UCST behavior.[37]

A classical theory for describing polymer–polymer miscibility is the Flory–Huggins lattice theory.[37,38] The FH theory is based on the mean field approach, where the change in the free energy on mixing per lattice site may be expressed as

$$f/k_BT = [\phi \ln \phi/N_1 + (1-\phi)\ln(1-\phi)/N_2] + \chi\phi(1-\phi) \tag{6}$$

where k_B is the Boltzmann constant, T the absolute temperature, ϕ the volume fraction of the component 1, and N the degree of polymerization; χ is the interaction parameter per unit segment of constituent polymers. The bracket term on the left hand side of equation (6) represents the combinatory free energy associated with entropy, whereas the second term arises from the segmental interaction of constituent polymers which is commonly inferred as the enthalpic contribution. For a symmetric polymer pair with an equivalent degree of polymerization, $N_1 = N_2 = N$, and the effective interaction parameter between the two polymers at the critical point is reduced to $(\chi N)_c = 2$. This relationship implies that a small change of the segmental interaction parameter χ due to a subtle variation of polymer stereospecificity or conformation would be amplified by a factor of N; thus the overall effect of χ on the polymer blend phase diagram could be significant. In high molecular weight polymer blends, a combinatory free energy will be reduced with increasing molecular weight (or N), which implies that most polymer pairs would be immiscible unless the systems have favorable attractive interactions.

The original FH theory captures some general trends of UCST systems, but it is not capable of predicting the LCST unless one assumes the empirical relationship of χ to be inversely proportional to absolute temperature, *i.e.*

$$\chi = a + \frac{b}{T} \tag{7}$$

where a and b are constants. Another drawback of the FH theory is the assumption that the volume does not change upon mixing, which is obviously an oversimplification of real polymer systems. It has been known experimentally that χ is dependent on the temperature, concentration, and microstructure of polymers.[37] Newer theories,[40,41] such as equations of state and lattice fluid theories, were subsequently introduced to account for compressibility and free volume effects. These theories certainly not only predict both UCST and LCST behavior, but also capture the trends of complex phase diagrams such as those with combined LCST and UCST, e.g. immiscibility loop or hourglass phase diagrams.

Experimentally, LCST and UCST could be observed to coexist in some polymer blends. Figure 4 shows a virtual immiscibility loop phase diagram of polycarbonate (PC) and syndiotactic poly(methyl methacrylate) (s-PMMA).[42] There are some concerns about thermal reversibility of the phases for the LCST, due to the proximity to the glass transition temperature of the constituent polymers.[40,41] Upon prolonged annealing at elevated temperatures, an irreversible interchange reaction occurs in this blend; therefore the UCST is not reversible and thus it should be regarded as a virtual one. When the blend was heated above the UCST, phase dissolution occurred first and was completed within 15 min at 230 °C, which has been attributed to copolymer formation. The PC–PMMA copolymer probably takes place preferentially at the interface and acts like a compatibilizing agent to favor the homogenization of PC and PMMA. Upon annealing at 250 °C for 60 min, a crosslinking reaction occurs as the reaction becomes severe, and then the blend becomes intractable. Recently, Debier and co-workers[43] identified the reaction mechanism to be due to the hydrolysis of the ester bonds of PMMA leading to acid pendent groups, which in turn reacted with carbonate bonds of PC to form copolymers. Subsequently, benzoic acid produced by PC degradation could further react with PMMA, leading to the crosslinking reaction. They admitted that sufficient contact between the reacting units was necessary for this reaction to occur. This is exactly what was pointed out by Kyu *et al.*:[42] that phase dissolution must occur so that PC and PMMA can reside in close proximity in the melt so that an interchange reaction can take place. This phase mixing is believed to be triggered by the chemical reaction leading to the copolymer formation; thus the UCST is strictly not a real one in the thermodynamic sense.

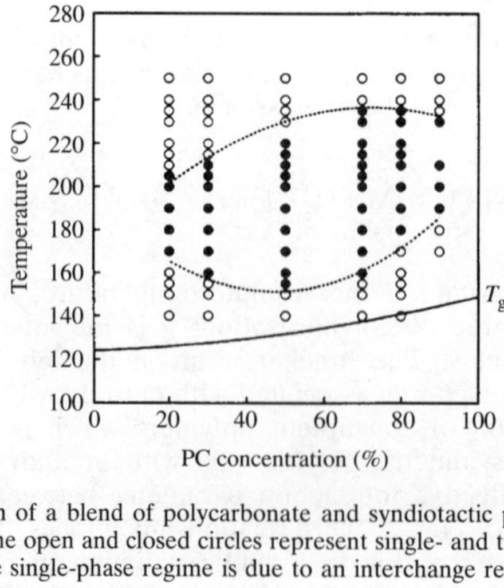

Figure 4 A virtual phase diagram of a blend of polycarbonate and syndiotactic poly(methyl methacrylate) obtained via annealing the blends for 2 h. The open and closed circles represent single- and two-phase regimes, respectively. The high-temperature single-phase regime is due to an interchange reaction of the pair.

Another type of the combined LCST/UCST phase diagram was observed by Tomura *et al.*[44] for poly(vinylidene fluoride) (PVDF)/poly(methyl methacrylate) (PMMA) blends in which the LCST is located above the UCST (Figure 5). The LCST temperature was located in the vicinity of the polymer degradation temperatures of the constituent polymers, while the UCST appeared near the onset of the crystallization temperatures of the blends. Shimada and co-workers,[32] who studied the same PVDF/PMMA system using the ESR technique, found a UCST (Figure 6). The actual concentrations of PMMA in mixed phases, such as the PMMA-rich and PMMA-poor regions, can be determined by the ESR technique.[35] In the 11/89 composition, the annealed temperature vs. PMMA composition curve appears comparable to that of the cloud point phase diagram, but the annealing of the 50/50 composition shows phase separation at appreciable higher temperatures in that the UCST point seemingly appears around 170 °C. It is unclear what kind of effect the spin probe might exert on the UCST phase diagram. In view of the greater sensitivity of ESR to a smaller length scale, it is possible to detect the local heterogeneity in the blends at temperatures appreciably higher than the UCST temperatures obtained by light scattering. To our astonishment, the authors chose to identify the UCST at 100 °C, which is puzzling.

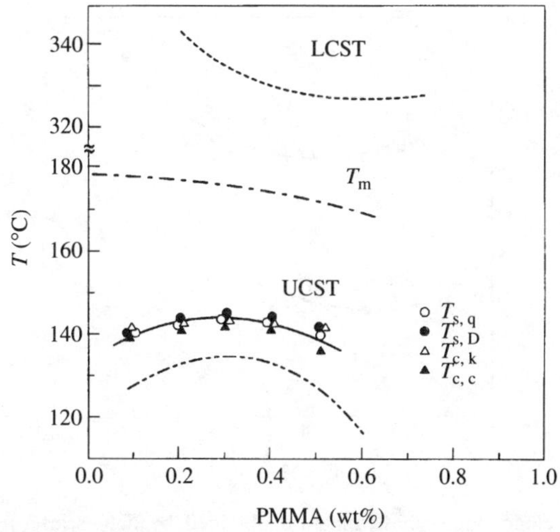

Figure 5 A combined LCST + UCST phase diagram of a blend of poly(vinylidene fluoride) and poly(methyl methacrylate) obtained by cloud point determination and crystallization experiments.

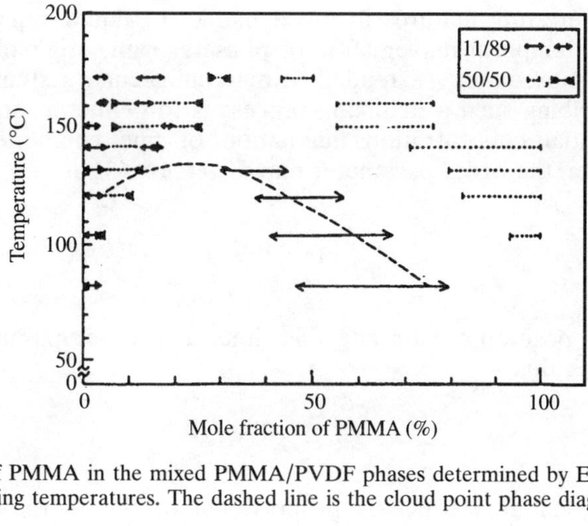

Figure 6 Concentrations of PMMA in the mixed PMMA/PVDF phases determined by ESR during phase separation at various annealing temperatures. The dashed line is the cloud point phase diagram of Figure 5.

Muller *et al.*[45] studied the blends poly(styrene-*co*-maleic anhydride)/poly(vinyl methyl ether). Their ESR studies showed that the LCST, as characterized by the appearance of the fast component, is approximately 70 °C lower than that obtained by the cloud point measurement (Figure 7). The detection of the lower coexistence temperatures in the LCST systems by ESR is consistent with the general perception. The possible existence of a combined LCST and UCST phase diagram in hydrogen bonding polymer blends has been predicted by Painter and co-workers,[41,46] who modified the FH theory by adding a free energy contribution term arising from strong hydrogen bonding between electron donor and acceptor groups of the dissimilar polymer pair. However, the predicted temperature gap between the LCST and UCST is unrealistically large, and would exceed the degradation temperatures of most polymers. Recently, Hammouda *et al.*[47] reported the existence of the combined LCST and UCST in the low molecular weight blends of d-PS ($M_w \approx 10\,400$; $M_n \approx 5020$) and poly(butyl methacrylate) ($M_w \approx 18\,800$; $M_n = 9600$). According to their SANS studies, both LCST and UCST are thermally reversible.

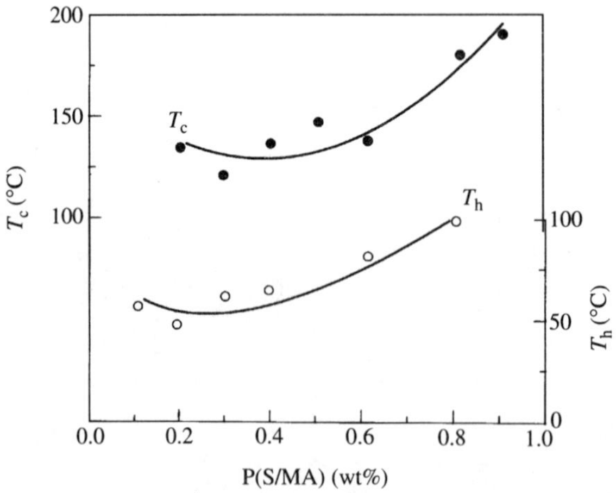

Figure 7 Comparison of the cloud point phase diagram obtained by light scattering and the temperature of the appearance of the fast component versus composition curve obtained by ESR for the blends of poly(styrene-*co*-maleic anhydride) and poly(vinyl methyl ether).

8.3.1.2 Dynamics of polymer–polymer phase separation

One of the most interesting features in the dynamics of liquid–liquid phase separation in binary mixtures is the 'temporal universality' of phase growth, originally developed for small molecule systems and subsequently extended to macromolecular systems.[33,48–55] A convenient starting point for describing such a demixing process is to define an order parameter, $\varphi(r, t)$, associated with a spatial concentration fluctuation of one component with time. For a conserved binary system, the order parameter may be expressed as:

$$\frac{\partial}{\partial t}\int \varphi(\boldsymbol{r}, t)\,\mathrm{d}r = 0 \tag{8}$$

where $\boldsymbol{r}$ represent the position vector and t is time. The order parameter must satisfy the continuity equation such that

$$\frac{\partial\varphi(r, t)}{\partial t} = -\mathrm{div}\,\boldsymbol{J} \tag{9}$$

where $\boldsymbol{J}$ is the diffusional flux which is proportional to the thermodynamic driving force $(X(r, t))$ as follows:

$$\boldsymbol{J} = M\boldsymbol{X}(\boldsymbol{r}, t) \tag{10}$$

where M is the Onsager-type mobility given by $M = \Lambda/k_B T$, in which the diffusion coefficient

(Λ) satisfies the Onsager reciprocity. By definition, the thermodynamic driving force is the gradient of a potential, *i.e.*

$$\boldsymbol{X}(\boldsymbol{r}, t) = -\boldsymbol{\nabla}\mu \tag{11}$$

where $\boldsymbol{\nabla}$ is a vector operator given by $\boldsymbol{\nabla} = \left(\frac{\partial}{\partial x}\boldsymbol{i} + \frac{\partial}{\partial y}\boldsymbol{j} + \frac{\partial}{\partial z}\boldsymbol{k}\right)$ and μ is the local chemical potential given by

$$\mu = \frac{\delta F}{\delta \varphi} \tag{12}$$

where F is the total free energy of the system. From equations (9)–(11) one obtains[48]

$$\frac{\partial \varphi(\boldsymbol{r}, t)}{\partial t} = M\boldsymbol{\nabla}^2 \mu \tag{13}$$

For a system subject to thermal fluctuations,[49] equation (13) may be modified as follows:

$$\frac{\partial \varphi(\boldsymbol{r}, t)}{\partial t} = M\boldsymbol{\nabla}^2 \frac{\delta F}{\delta \varphi} + \eta(\boldsymbol{r}, t) \tag{14}$$

where $\eta(\boldsymbol{r}, t)$ is thermal noise. According to the fluctuation–dissipation theorem, thermal fluctuation is related to the mobility by the following equation:

$$\langle \eta(\boldsymbol{r}, t)\eta(\boldsymbol{r}', t')\rangle = -2k_{\mathrm{B}}TM\nabla^2\delta(\boldsymbol{r} + \boldsymbol{r}')\delta(t - t') \tag{15}$$

where the angle brackets denote the ensemble average and $\delta(\boldsymbol{r} - \boldsymbol{r}')$ and $\delta(t - t')$ are the Dirac delta functions.

For a binary polymer blend of species 1 and 2 having chain lengths $N_1 = N_2 = N$ and the characteristic length $a_1 = a_2 = 1$, de Gennes[50] modified the Cahn-Hilliard theory[48] by incorporating the FH free energy density and the gradient term κ given by the random phase approximation (RPA). The gradient term for a symmetric blend may be expressed as $\kappa = a^2/36\phi(1 - \phi)$ where $a = 1$.[50,51] This leads to the following equation:

$$\frac{F(\varphi)}{k_{\mathrm{B}}T} = \int \left\{\frac{1}{N}[\phi \ln \phi + (1 - \phi)\ln(1 - \phi)] + \chi\phi(1 - \phi) + \frac{1}{36\phi(1 - \phi)}(\nabla\phi)^2\right\} \mathrm{d}r \tag{16}$$

For an asymmetric blend, the gradient term may be modified as $\kappa \approx [(a_1^2/\phi) + \{a_2^2/(1 - \phi)\}]$.[52] Substituting equation (16) into equation (14), one obtains

$$\frac{\partial \varphi(r, t)}{\partial t} = M\nabla^2\left[\frac{1}{N}\ln\left(\frac{\phi}{1 - \phi}\right) - 2\chi\phi + \frac{1 - 2\phi}{36\phi^2(1 - \phi)^2}(\nabla\phi)^2 - \frac{\nabla^2\phi}{18\phi(1 - \phi)}\right] + \eta(r, t) \tag{17}$$

It is usually a demanding task to simulate the time evolution of the order parameter in three dimensions. To illustrate the time evolution of the order parameter, the simulation was performed in two dimensions using a finite difference scheme for the spatial derivative and a semi-implicit method for the temporal derivative with a periodic boundary condition.[53–55] The spatial discretization was made by replacing the continuous space of the position vector by a simple square lattice having a grid size of 64×64. The simulated SD patterns[56] of a symmetric polymer blend as a function of the phase separation time (in arbitrary units) are shown in Figure 8. The coalescence of the domains can be discerned during growth. The Fourier transform of these SD domains reveals the characteristic scattering halo corresponding to the average periodic distance of the SD domains (Figure 9). Reflecting the growth of the domains, the ring collapses to a smaller diameter with elapsed time while the intensity increases.

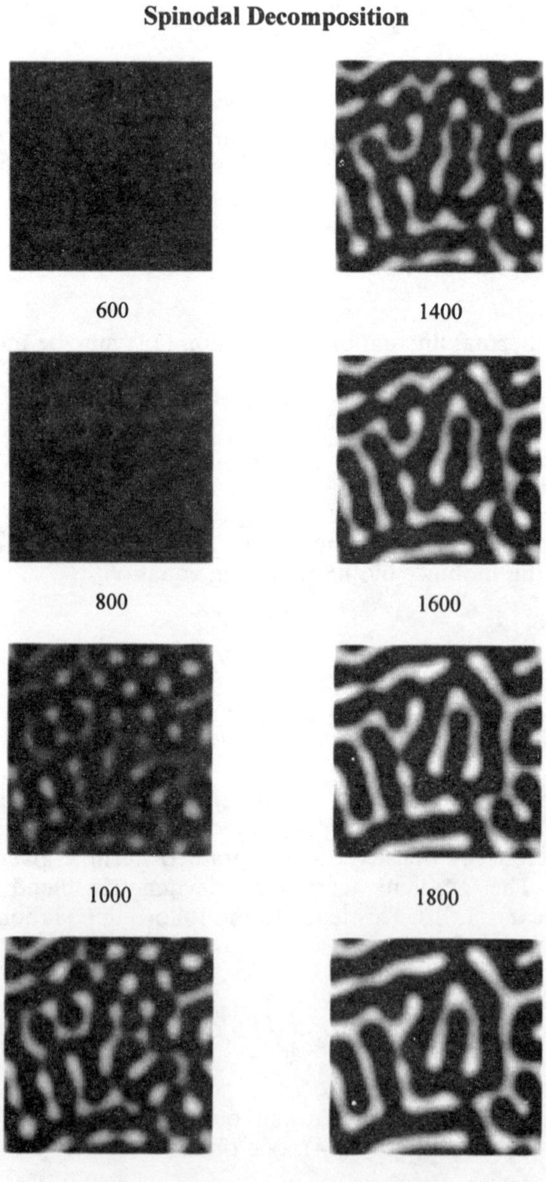

Figure 8 Simulated two-dimensional structures of spinodal decomposition in a symmetric blend coarsening with elapsed time (in arbitrary units) via coalescence.

(i) *Early Stage of SD*

Cahn and Hilliard[48] were the first to derive the time-evolution equation for the ordering process by neglecting the higher-order terms. In the standard linearization approximation, the order parameter $\varphi(r, t) = \phi(r, t) - \phi_0$ must be very small, *i.e.* the concentration is very close to the average value. In practice, the linearization of the CH theory may be valid only in the very

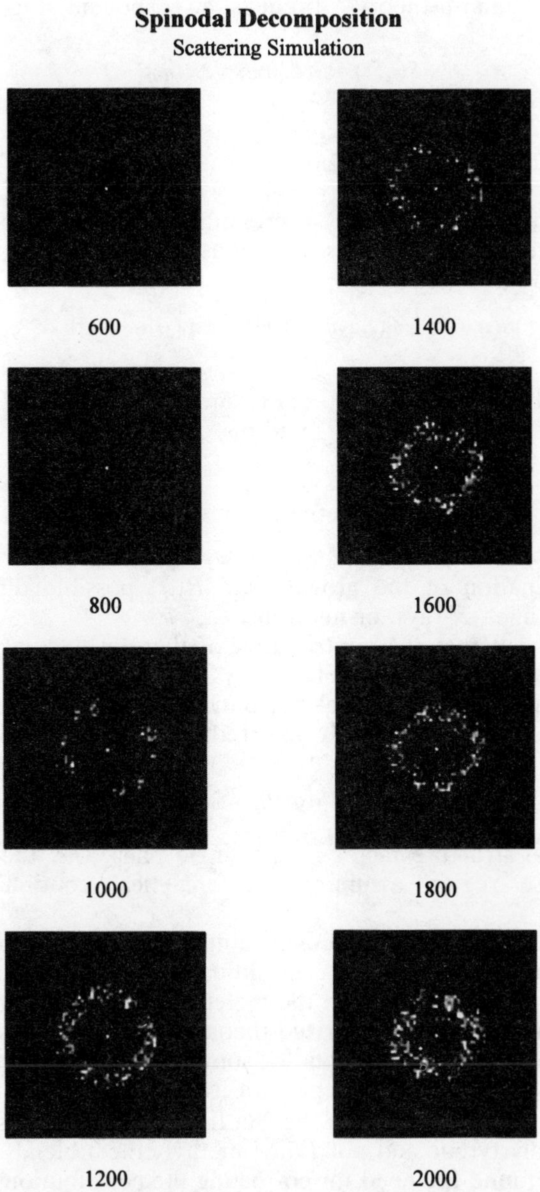

Figure 9 The corresponding Fourier transformed SD patterns of the same symmetric blend of Figure 6, displaying the collapse of the scattering halo to a smaller diameter due to coarsening.

early stages of SD. If one keeps only terms linear in the small-order parameter limit, the linearization gives

$$\frac{\partial \varphi(r, t)}{\partial t} = M\nabla^2[2(\chi_s - \chi) - \tfrac{2}{9}\nabla^2]\varphi(r, t) + \eta(r, t) \tag{18}$$

Equation (18) may be solved in the q Fourier space to give

$$\frac{\partial \varphi_q(t)}{\partial t} = -Mq^2[2(\chi_s - \chi) + \tfrac{2}{9}q^2]\varphi_q(t) + \eta_q(t) \tag{19}$$

At small q ranges, $M(q \rightarrow 0) = N\phi(1-\phi)D_c$,[50] where D_c is the translational coefficient while the thermal noise is given as

$$\langle \eta_q(t)\eta_q(t')\rangle = 2Mq^2\delta(t - t') \tag{20}$$

The linearized Cahn–Hilliard theory[48] predicts an exponential growth of concentration fluctuations:

$$S(q, t) = S(q, 0)\exp[2R(q)t] \tag{21}$$

with

$$R(q) = -2Mq^2[(\chi_s - \chi) + \tfrac{1}{9}q^2] \tag{22}$$

where $R(q)$ is the amplification factor representing the growth rate of concentration fluctuations. The structure factor, $S(q, t)$, is defined as

$$S(\boldsymbol{q}, t) \propto \int \phi(\boldsymbol{r}, t)\phi(\boldsymbol{r} + \boldsymbol{r}', t')\exp(-i\boldsymbol{q} \cdot \boldsymbol{r}')\, dr'^3 \tag{23}$$

where $\phi(r, t)$ is the local concentration. $S(q, t)$ is directly proportional to the time-dependent scattering function $I(q, t)$, which is given in terms of the Fourier inversion of the electron density $\rho(r, t)$:

$$I(\boldsymbol{q}, t) \propto \int \rho(\boldsymbol{r}, t)\rho(r + \boldsymbol{r}', t')\exp(-i\boldsymbol{q} \cdot \boldsymbol{r}')\, dr'^3 \tag{24}$$

This makes the determination of the growth rate $R(q)$ possible directly by time-resolved small-angle scattering of light, X-rays, or neutrons.

Later, Cook[49] pointed out that the contribution of thermal fluctuation is not negligible as originally assumed in the linearized CH theory for the early stages of SD. Binder[52] incorporated Cook's correction for thermal fluctuations into a polymer-based derivation as a modification of the CH theory, commonly inferred as Cahn–Hilliard–Cook (CHC) theory, which reads:

$$S(q, t) = S_0(q) + [S(q, 0) - S_0(q)]\exp\{2R(q)t\} \tag{25}$$

where $S_0(q)$ is the virtual structure factor in the single phase, *i.e.* the extrapolated structure factor into a two-phase region assuming the analytical continuity in the mean-field approximation.

In the small molecule system it is extremely difficult to confirm the validity of the CHC theory because of the fast diffusion rates. The diffusion process is naturally slowed down in polymer systems in a manner depending on the molecular weights of the constituent polymers. The experimental task for testing the linearized theory in the early stage of SD becomes easier for polymer blends. However, early attempts for some polymer systems by time-resolved light and small-angle X-ray scattering show a departure from the linear theory.[57–59] Hashimoto and co-workers[60,61] demonstrated the validity of the CH theory in the early stages of SD in the near critical composition of polystyrene and poly(vinyl methyl ether) blends. For the same polymer blend, Okada and Han[62] found the need for correcting the contribution of thermal noise to the structure factor for shallow quenches. Kumaki and Hashimoto[61] contended that the effect of thermal noise is negligible so long as the temperature quenched depth is sufficiently large. Later, several polymer blends[63–66] were found to follow the linearized CH theory for the early stages of SD. As an illustration, a typical time evolution of scattering profiles for a polycarbonate (PC)/isotactic poly(methyl methacrylate) (i-PMMA) blend following a temperature jump from ambient to 220 °C is depicted in Figure 10.[66] The scattering peak remains invariant for some initial periods up to 10 min. The plot of $R(q)/q^2$ vs. q^2 shows a straight line, suggesting the validity of the linearized CH theory (Figure 11).

Excimer fluorescence techniques have been applied to gain insight into the mechanism associated with phase separation in polymer systems. Gelles and Frank[67] employed excimer fluorescence to study spinodal decomposition in blends of PS and PVME. Subsequently, they developed a two-phase model to describe the time-dependent behavior through the ratio of excimer and monomer fluorescence intensity, I_E/I_M, and to determine the concentrations of mixed phases such as PS-rich and PS-poor. Their model requires simultaneous measurement of excimer and monomer fluorescence. Figures 12(a) and 12(b) show the variation of I_E/I_M and the PS concentration in the mixture as a function of elapsed time, respectively. An exponential increase of the concentration of the phases generated during phase separation is consistent with the mechanism of spinodal decomposition. Their analysis showed that the early stage of SD of their blend obeyed the linearized Cahn–Hilliard theory.

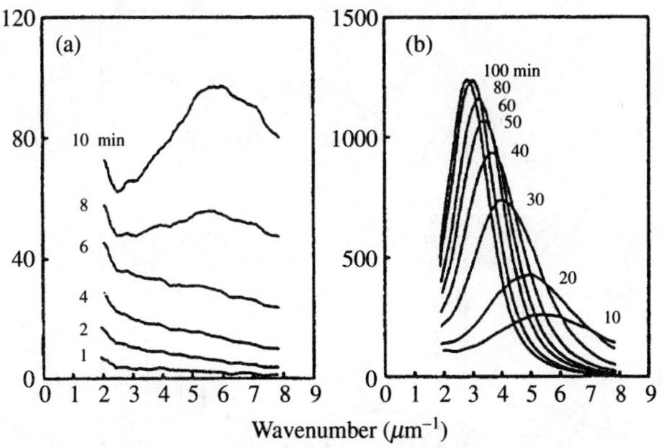

Figure 10 Time evolution of light scattering profiles of a 40/60 PC/i-PMMA blend following a temperature jump from ambient to 220 °C.

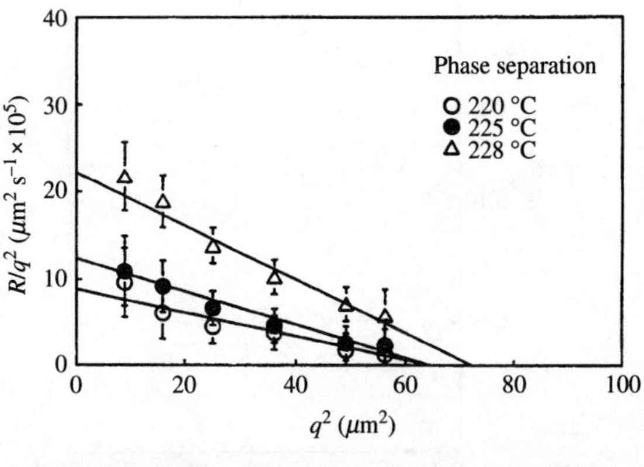

Figure 11 Plots of $R(q)/q^2$ vs. q^2 for the 60/40 PC/i-PMMA blend at several T jumps from ambient to 220, 225 and 228 °C, showing the validity of the linearized CH theory.

Torkelson *et al.*[68] investigated phase separation in PS–methylcyclohexane solutions using excimer fluorescence and found that the Gelles–Frank model[34,67] correctly predicted the magnitude of the change in I_E/I_M during phase separation. However, the complications due to convective flow during macroscopic phase separation precluded a quantitative test with the SD theories. In a subsequent study on blends of oligomeric PS ($M_w \approx 2500$; $M_w/M_n = 1.13$)/polybutadiene (PB), it was found that the Gelles–Frank model predicted qualitatively the behavior of I_E/I_M at equilibrium for blends that were initially homogeneous. The kinetics of SD were found to follow the Cahn–Hilliard linearized theory with an apparent mutual diffusion coefficient of the order of $10^{-12}\,s^{-1}\,cm^2$, consistent with those obtained by time-resolved light scattering studies.

There is no assurance that the early stage of SD can be discerned by time-resolved light scattering, in which case the correlation length and the mutual diffusivity must be determined by alternative techniques such as dynamic light scattering,[69] or time-resolved static SANS[70,71] or SAXS[72] in the single phase. The correlation length, ξ, of the binary fluid in the single phase may be obtained by assuming the scattered intensity to have the Ornstein–Zernike form:

$$\frac{I(\theta)}{I(0)} \propto \frac{1}{1+\xi^2 q^2} \tag{26}$$

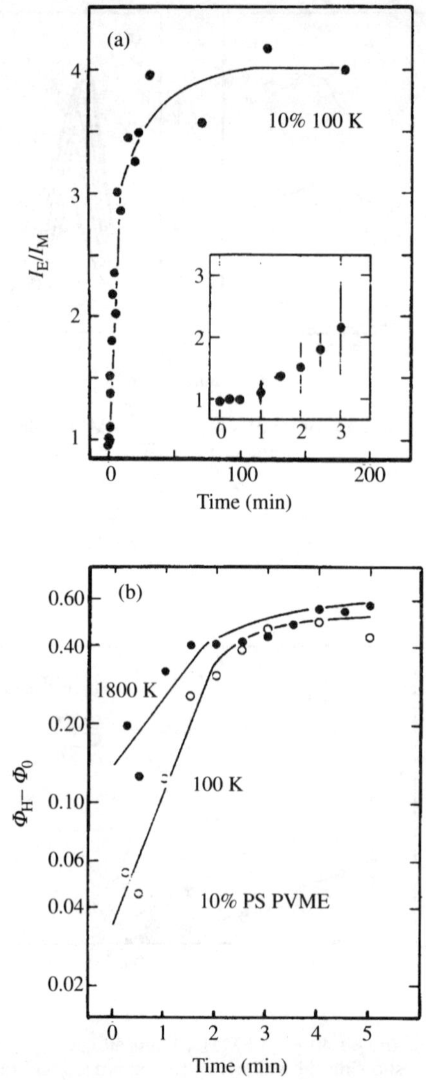

Figure 12 Variation of (a) I_E/I_M and (b) the concentration of the PS phase in the PS/PVME mixture as a function of elapsed time.

The spinodal temperature T_s can be estimated from the following relation

$$I(0) \propto (|T - T_s|/T_s)^{-\gamma} \tag{27}$$

by extrapolating $I(0)^{-1/\gamma}$ vs. T to $I(0)^{-1/\gamma} = 0$. The correlation length (ξ) near T_s can be expressed by

$$\xi = \xi_0(|T - T_s|/T_s)^{-\nu} \tag{28}$$

where ν and ξ represent the critical exponent and the amplitude, respectively. The diffusion coefficient D can be measured from equation (29) by $D = \bar{\Gamma}/q^2$, which may be assumed to have the following relationship near T_s

$$D = D_0(|T - T_s|/T_s)^{\nu} \tag{29}$$

Here D_0 is usually constant provided that the experimental temperature is far removed from the glass transition temperature of the polymer.

Schwahn *et al.*[70] studied the relaxation behavior of blends of deuterated PS (d-PS) and PVME in the early stages of spinodal decomposition using time-resolved small-angle neutron scattering. Several temperature jumps were undertaken from a single phase to both single phase and two-phase temperatures. Figure 13 shows a typical time evolution of the structure factor following a T jump to 134.1 °C. In all T jumps, a pronounced interference peak appears and remains stationary for some time, suggestive of the early stages of SD. The analysis of the results obtained within the homogeneous phase appears to obey the Cahn–Hilliard–Cook theory. At very early times, <10 s, a small deviation from the predicted exponential growth of fluctuation was observed, which has been attributed to the proximity to the glass transition temperature and/or to the internal 'Rouse' modes of the polymer segment. This effect is more accentuated in the PS/d-PS blends.[71] When the T jump was undertaken inside the unstable region, the 'quasi-equilibrium' structure factor may be analyzed according to an extrapolation method proposed by Cook[49] at $q > q_c$, where the modes are still stable. The analysis revealed that the early stages of decomposition within the miscibility gap follow the CHC theory.

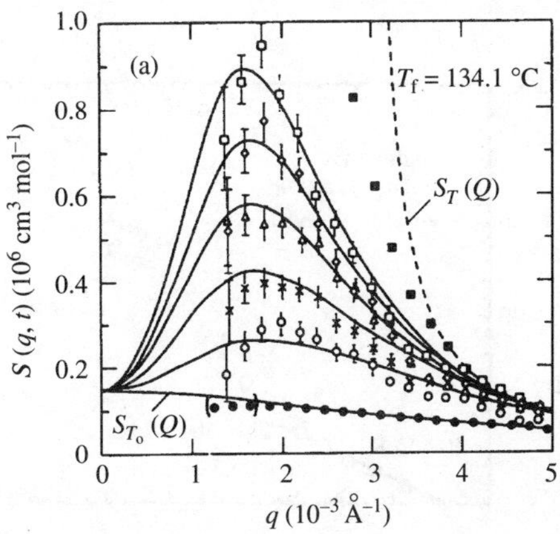

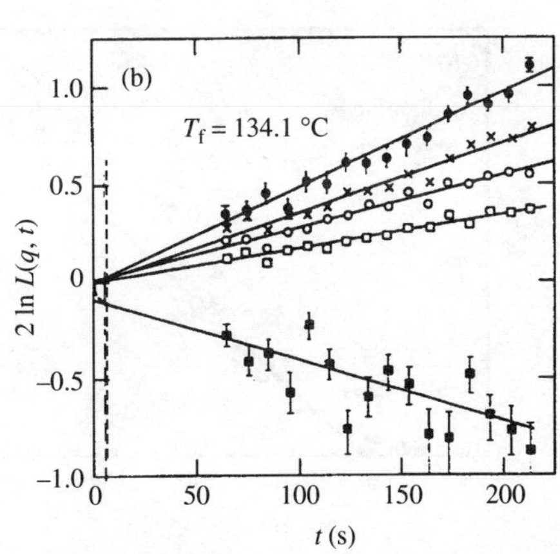

Figure 13 (a) Time-dependent structure factor of the 13/87 d-PS/PVME blend following the 134.1 °C jump at various time intervals of 60, 110, 150, 180, 210 s (shown by open symbols in the order from bottom to top), and (b) the corresponding relaxation function where $L = \exp(-R(q) \cdot t)$, displaying a small deviation from exponential behavior for $t < \tau_R$ at various q: 1.79×10^{-3}, 2.2×10^{-3}, 2.4×10^{-3}, 2.6×10^{-3} and 3.6×10^{-3} (from top to bottom).

The Onsager coefficient as determined from the time-dependent scattering data showed strong dependence on q (or q^{-2} at large q) and the corresponding length scale was found to be much larger than the radius of gyration of the polymer. In the comparison with light scattering experiments, the q range covered by light and neutron scattering has little or no overlap. They found that the SANS data are consistent with the light scattering data if extrapolated to $q = 0$.

It is indeed encouraging to learn that the SANS and SALS experiments are complementary and permit investigations covering a wide range of length scales. When the structure factor was extrapolated into the miscibility gap near the SD and the critical temperature, the observed susceptibility $S(q=0)$ reveals a critical behavior above and below the spinodal that follows the 3D Ising exponents.[70] The crossover to meanfield behavior was observed within a few degrees span above and below the extrapolated spinodal, suggesting that the spinodal, as defined by the extrapolation based on the mean-field approach, is bridged by a 3D Ising regime of the structure factor $S(q=0)$.

The crossover from meanfield to Ising critical behavior was explored by Chu *et al.*[72] on blends of poly(2-chlorostyrene)/PS using synchrotron SAXS. Both polymers have a narrow molecular weight distribution (P2ClS: $M_w \approx 233\,000$; $M_w/M_n \approx 1.19$; PS: $M_w \approx 370\,000$; $M_w/M_n \approx 1.06$). This P2ClS/PS system exhibits a lower critical solution temperature with $T_c = 134.3\,°C$. SAXS studies were conducted at various temperatures in the single phase. The scattered intensity showed a monotonic decay behavior with increasing scattering wavenumber, which can be analyzed in accordance with the Ornstein–Zernike relationship (equation 26). The crossover was found to occur 2.6 °C below the critical temperature in the static susceptibility and the correlation length, with exponent γ changing from 1.00 (meanfield) to 1.23 (Ising). As can be seen in Figure 14, the correlation length (ξ) showed a crossover with a slope (ν) changing from 0.50 (meanfield) to 0.63 (Ising).[72]

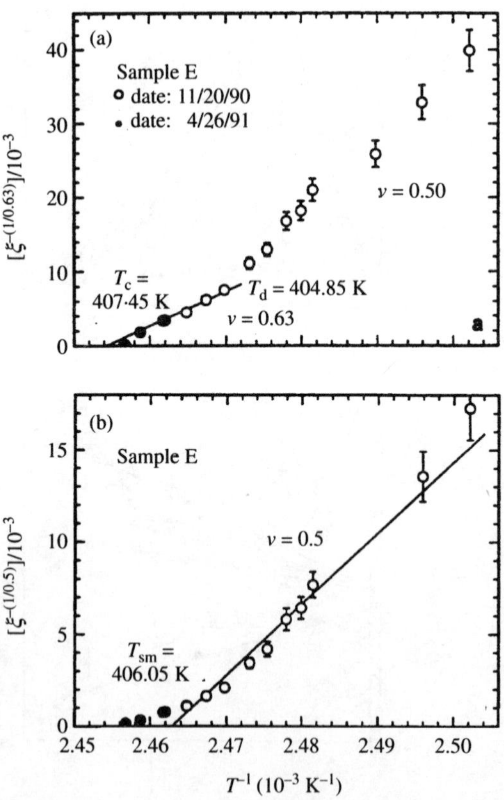

Figure 14 Inverse correlation length $\xi^{-1/\nu}$ versus $1/T$ for the 48/52 P2ClS/PS blend, exhibiting (a) $\nu = 0.63$ and (b) $\nu = 0.50$.

(ii) Intermediate and late stages of SD

When phase separation proceeds beyond the linear limit, the amplitude of the fluctuations have increased due to the coupling of additional modes arising from the intermediate and late stages of SD.[36,73] The linearized equation becomes inadequate to account for the late stages. There is no exact analytical solution for the growth of spatial concentration fluctuations that occurs through a sequence of nonlinear and nonequilibrium processes. The SD process may be best analyzed by dividing it into three regimes such as early, intermediate, and late stages

(Figure 15).[73] Wiltzius and Cumming[74] suggested a four-stage model by refining the late stage into transition and final regions of SD.

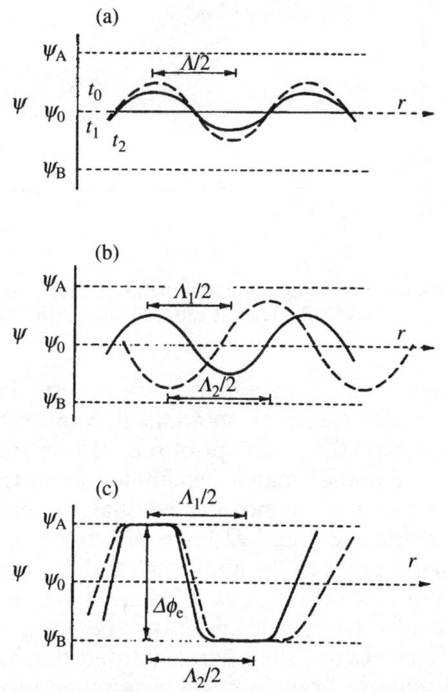

Figure 15 Schematic representation of the spatial concentration fluctuation in the (a) early, (b) intermediate and (c) late stages of spinodal decomposition.

A question arises as to what is the common rule that governs the nonequilibrium dynamics of the SD in approaching thermodynamic equilibrium. A common approach to describe this dynamical process is a scaling law which starts by noting the similarity of the structure factor $S(q, t)$ at later times to that at earlier times. The structural growth involves only the change in size or length scale while preserving the self-similarity of the system. The length scales at various times can be scaled as:

$$q_m(t) \approx \Lambda(t)^{-1} \propto t^{-\alpha} \tag{30}$$

and

$$I_m(t) \approx \langle \eta(t)^2 \rangle \Lambda(t)^3 \propto t^{-\beta} \tag{31}$$

where $\Lambda(t)$ is the length scale of periodic fluctuations and $\langle \eta(t) \rangle^2$ is the mean square fluctuations of dielectric constant or electron density. These exponents have been predicted to have various values with the $\beta = 3\alpha$ relationship in the late stages of SD, but for intermediate stages, $\beta > 3\alpha$ as $\langle \eta(t)^2 \rangle$ has not reached its limiting values.[74] The typical values of α as predicted by simulations based on the Monte Carlo method, Ising model, and Landau–Ginzburg model vary from 1/5 to 1.[36,73,74] Experimentally, the α value was found to be 1/3 in most high-polymer blends at shallow quenches, but it could cross over to 1 at late stages if the quench depths were large.[60–66] Typical quench depth dependences ($\log q_m$ versus $\log t$) for the polycarbonate (PC)/isotactic PMMA blend are depicted in Figure 16.[66]

It appears that the observed power law relationship depends strongly on the polymer pairs, molecular weight, blend compositions (critical versus noncritical) and quench depths. The α and β values themselves are often time dependent and cross over among various coarsening processes. Hence, it is difficult to recognize the temporal universality without converting to dimensionless quantities such as reduced time (τ) and reduced scattering maximum $Q_m(\tau)$

$$\tau = t/t_c = D\xi^{-2}t \tag{32}$$

while

$$Q_m(\tau) = q_m(\tau)/q_m(0) \tag{33}$$

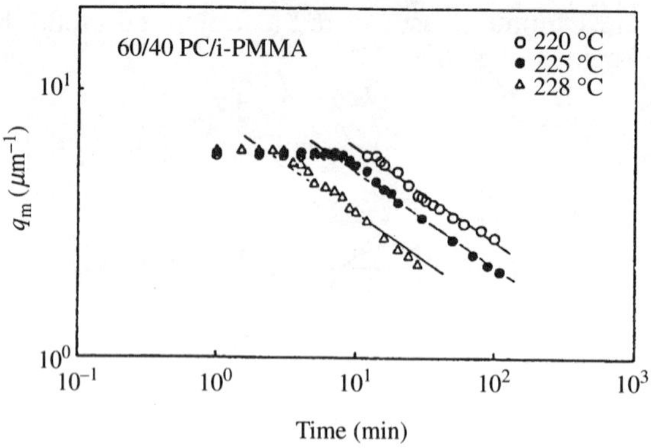

Figure 16 Variation of the scattering maximum versus phase separation time in double logarithmic plots for the 60/40 PC/i-PMMA blend at 220, 225 and 228 °C.

where ξ is the correlation length and D the apparent diffusivity. These dimensionless quantities can be determined from the early stages of spinodal decomposition using the linearized CH theory, *viz.* $\xi = 1/q_m(0)$ and D from the intercept of the $R(q)/q^2$ versus q^2 plot. When the early stage of SD is not measurable, ξ and D may be estimated from the dynamic light scattering in the single phase regime. Nose *et al.*[69] demonstrated that the temporal universal scaling can indeed be established by determining ξ and D from the dynamic light scattering experiments. Snyder and Meakin[57] demonstrated that the universal scaling is valid in polymer blends as in the case of metal alloys and solution mixtures. Hashimoto *et al.*[60] contended that the chain entanglement effect could lead to a deviation from the temporal universality, which they attributed to the 'polymer effect'. Later, they demonstrated that the reduced master curve, the $Q_m(\tau)$ versus τ plot, showed massive branching when some additives such as block copolymers are added to homopolymer blends. Bates and Wiltzius[65] reported that the master curve cannot be established in their deuterated and protonated polybutadiene blends. Ko *et al.*[42] found the failure of time universality in the blends of PC and s-PMMA due to the formation of PC–PMMA copolymers driven by the interchange reaction at the interphase.

Wiltzius and Cumming[74] observed the influence of surface wetting exerted by an interface on the kinetics of the coarsening of two coexisting liquid phases. Their experiments showed dual scattering maxima (Figure 17). One characteristic length coarsens with time with an α value of 3/2 and the other varies with a smaller value ($\alpha = 1/3 \sim 1$), typical for the bulk (Figure 18). Siggia[75] predicted that the dominant length scale of bicontinuous phases grows linearly with time in the late stages of SD, where hydrodynamic flow is dominant. This has been observed under the condition that the effect of gravity could be avoided.[76] Goldburg *et al.*[77,78] showed that gravity indeed affects the growth law $q_m \approx t^{-\alpha}$ in which the value of α changes from 0.3 to 2.

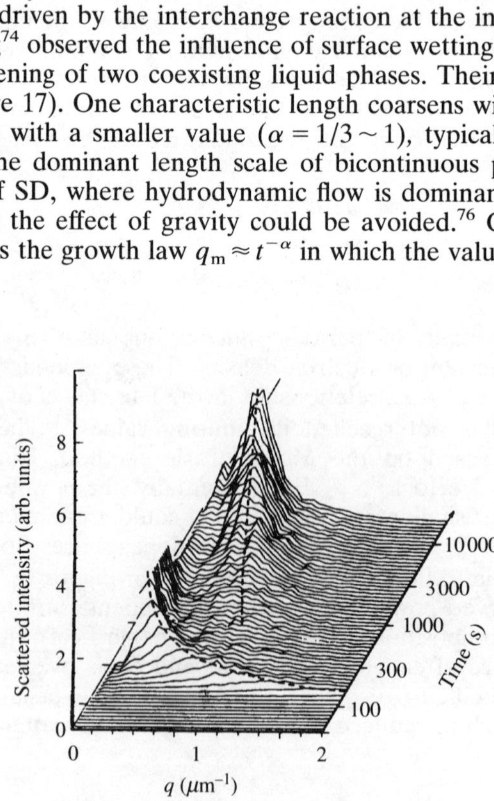

Figure 17 The time evolution of scattering profiles after a quench below the critical temperature. The locus of the maxima of the fast mode and slow mode are indicated by dashed lines.

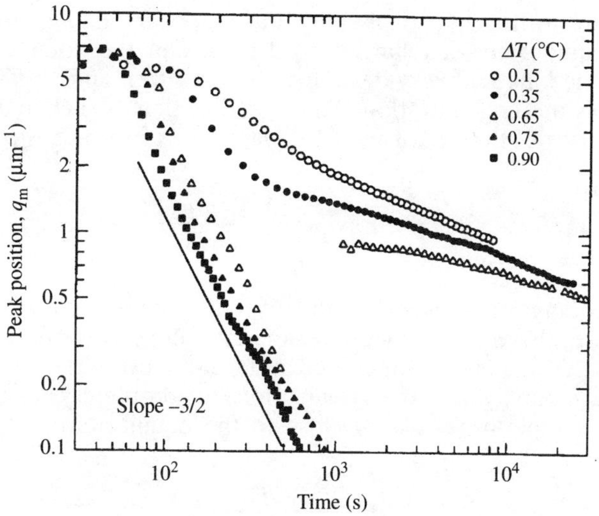

Figure 18 Temporal variation of the peak position of the fast and slow mode for the blend of polyisoprene and poly(ethylene–propylene) at various T quenches.

8.3.2 Polymer Ordering Transitions

The first-order phase transitions such as liquid–vapor or liquid–solid phase transitions occur primarily through nucleation and growth (NG).[79] The characteristic time, ξ^2/D, of liquid–vapor equilibrium, where ξ is the correlation length and D the diffusivity, is much faster than the liquid–solid or liquid–liquid phase transitions. It is technically difficult to follow the dynamical processes of liquid–vapor phase transitions due to the fast diffusion of the liquid–vapor systems. We shall focus on the NG process of the solid–liquid phase transition, which is unstable at infinitesimal fluctuations but stable against large fluctuations. The initial stable fragments act like a nucleus, upon which growth occurs through coalescence or ripening until a thermodynamic equilibrium has been reached.[79]

Polymer crystallization from the melt generally occurs via a solid–liquid phase transition process.[80] One major feature that distinguishes polymer crystallization from small molecule systems is that the lamellar crystals thus formed exhibit a plate-like morphology in which polymer chains fold back and forth while some chains traverse from one lamellar crystal to another. Such plate-like lamellae have unusually high surface free energy due to the chain folding at the surface, and thus are far from equilibrium. When polymer crystals are annealed near their crystallization temperature, crystal lamellae tend to grow thicker when approaching equilibrium. This lamellar thickening process may be characterized as a solid–liquid transition, if there is an increase in the overall crystallinity, or otherwise a solid–solid transition. Another crystallization process that can be characterized as a solid–liquid phase transition is melting–recrystallization, in which the preformed (unstable) crystals melt first and then recrystallize to a more stable form. Such melting–recrystallization phenomena are prevalent in aromatic polycondensates due to the inherently slow crystallization rate relative to flexible chains and thus equilibrium can hardly be achieved in those systems. The incomplete crystallization occasionally results in recrystallization during the course of reheating, and it is called 'cold crystallization'.

It has been well documented that crystalline polymers often exhibit polymorphic crystalline forms in which one crystal form may be more stable than the other. Annealing of an unstable crystal structure could lead to a transformation to a more stable crystalline form while approaching equilibrium. This kind of crystal-to-crystal transformation can be undoubtedly categorized as a solid–solid phase transition.

8.3.2.1 Polymer crystallization

There are two notable theories in describing polymer crystallization; the first one is the equilibrium theory introduced by Peterlin and Fischer.[81,82] The equilibrium theory captures some qualitative features of lamellar thickening, but it appears inadequate to describe polymer crystallization since equilibrium can hardly be achieved in the complex crystalline polymer systems. The other is the kinetic theory developed by Lauritzen and Hoffman,[83] which has been used extensively in polymer crystallization. The crystallization of such anisotropic lamellae may be best described in terms of the Lauritzen–Hoffman theory in which the rate of crystal growth (v) is expressed by

$$v = v^0 \exp\frac{-\Delta f}{k_B T}\exp\frac{-4b_0\sigma_s\sigma_e T_m^0}{\Delta h_f \Delta T k_B T} \tag{34}$$

where Δf is the free energy for transport of the crystallizing unit across the amorphous–crystal interface, v^0 is related to the volume fraction of the crystalline portion, σ_s and σ_e represent the surface free energy at the side and at the folded surface of a cylindrical embryo, respectively, b_0 is the thickness of the monolayer of the growing nucleus on the crystal surface, and Δh_f is the heat of fusion. If the temperature is very close to the equilibrium melting temperature, one finds[83]

$$l^* = \frac{4\sigma_e T_m^0}{\Delta h_f(T_m^0 - T)} \tag{35}$$

where l^* is the characteristic length scale, representing the thickness of the straight crystal chain length. It is obvious from the above equation that this lamellar thickness or size is primarily governed by the undercooling ($\Delta T = T_m^0 - T$). With continued crystallization, the lamellar structures further organize to form a spherulite. The spherulitic growth is by no means straightforward. A bundle of anisotropic crystals forms initially, and crystal growth presumably occurs through diffusion-controlled interfacial growth to a dendritic shape by gradually branching out, ultimately forming a spherulitic structure.[84,85] The growing spherulites impinge into neighboring spherulites, leading to truncation. The average length scale (or size) of spherulites can no longer grow, but the crystallization can continue internally within spherulites. This process is known to be a secondary crystallization, where crystal perfection or reorganization takes place.

(i) *Crystallization of homopolymers*

As pointed out above, the initial thickness of the crystal lamellae is inversely proportional to the undercooling, *viz.* the difference of the crystal melting temperature and the crystallization temperature. Time-resolved SAXS and WAXD studies have been customarily undertaken to determine the crystallization kinetics by rapidly cooling a polymer from its melt state to an experimental temperature significantly lower than the equilibrium crystallization temperature. An alternative approach is a self-seeding technique in which a polymer melt (typically well above its equilibrium melting temperature) is cooled rapidly below the crystallization temperature and is allowed to complete crystallization. The specimen is then reheated very slowly at a controlled rate to one or two degrees below the equilibrium melting temperature and kept for a certain time (typically a few tens of minutes). The major portion of the polymer other than the seeds melts away. The specimen is then exposed to the X-ray beam for SAXS measurements at preset crystallization temperatures. This self-seeding approach permits possible control of the number of nuclei generated, so the crystal growth can be followed in a reproducible manner.

The kinetics of crystallization of linear polyethylene were studied by Schultz[86] using the 10 m SAXS facility at the Oak Ridge National Laboratory. At low crystallization temperatures, the lamellar spacing of linear polyethylene was invariant with time. At high crystallization temperatures, the lamellar spacing decreased initially with time. Since then, time-resolved SAXS and WAXD by synchrotron radiation have been extensively employed to probe the time evolution of the lamellar long period of single crystal mats. Grubb *et al.*[87,88] investigated the polyethylene single crystals that were crystallized at 80 °C by rapidly changing the temperature to various annealing temperatures of 121–129 °C. As shown in Figure 19(a), the WAXD integrated intensity of the (110) reflection decreases drastically at first for a few seconds, then it increases until it reaches the limiting magnitude in a manner depending on the annealing temperature. The SAXS long period shows little or no change at the crystallization temperature of 121 °C (Figure 19(b)). At 123 and 125 °C the long period increases initially for some periods, then levels off. The crystallization at 127 °C shows a decrease in the long period initially, then remains constant with continued crystallization. An initial decrease in the long period during crystallization at high temperatures (a few degrees below their equilibrium melting transition) was observed by Kawaguchi *et al.*[89] in the time-resolved SAXS and WAXD studies of conventional PE single-crystal mats. These observations point to the fact that the crystal melting occurs first, and is then succeeded by the recrystallization. This initial decrease in the lamellar spacing has been interpreted in terms of a lamellar insertion model by Strobl *et al.*[90] and the car parking model by Hashimoto and Saijo,[91] *i.e.* the new lamellae form in the preformed interlamellar region so that the distance between the lamellae becomes smaller initially. Later, the continued increase of the WAXD intensity may be attributed to crystal perfection, such as lamellar thickening.

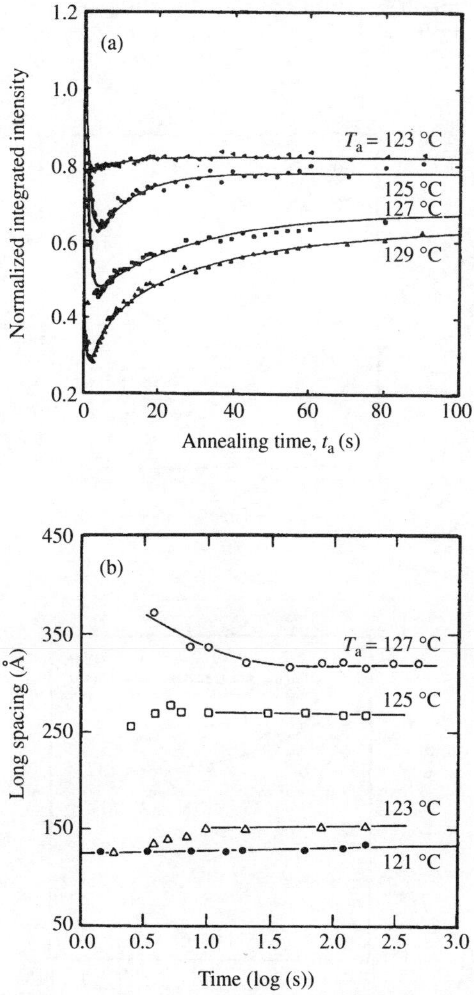

Figure 19 Variation of (a) integrated intensity of the (110) WAXD reflection and (b) the SAXS long period representing the average center-to-center distance of adjacent polyethylene crystals as a function of annealing time.

Barham and Keller[92] employed an enhanced self-nucleation technique by first lowering the temperature of the PE melt from 160 °C to 125 °C instantaneously and allowing the polymer to crystallize, and then gradually reheating the specimen to various experimental temperatures in the synchrotron SAXS beamline. Their results showed that the initial crystal thickness is nearly a factor of two smaller than the quasi-equilibrium crystal thickness obtained by the static SAXS measurement. The authors[92] claimed that the lamellar thickness could double or triple or quadruple during the course of crystallization. The doubling of the initial lamellar thickness was also suggested by Martinez-Salazar *et al.*,[93] based on Raman acoustic mode spectroscopy.

The lamellar thickening process during isothermal annealing of a gelation-crystallized ultrahigh molecular weight polyethylene ($M_w \approx 6 \times 10^6$) has been investigated by time-resolved SAXS and WAXD.[94] The gel-crystallized UHMWPE films exhibit a two-point SAXS pattern characteristic of a folded chain lamellar crystal. At low isothermal annealing temperatures (below 130 °C), the long period increases slightly and then levels off (Figure 20(a)), while the degree of crystallinity (as estimated from the time-resolved WAXD studies) shows little or no change (Figure 20(b)). This observation has been attributed to the surface melting of crystallites without appreciably perturbing the core crystal stems. At 134 °C there is a slight decrease of the apparent crystallinity initially, but it remained unchanged thereafter. This crystallinity decrease is more pronounced during annealing at 137 °C in the initial period, but the crystallinity remains unchanged with continued annealing time. The SAXS long period at a comparable temperature (135 °C) continues to increase gradually. The initial decrease of WAXD crystallinity may be due to the surface melting, which may be counter-balanced by the recrystallization (or transformation to a more stable crystal form) so that there is little or no appreciable change in the net crystallinity. On annealing at a higher temperature near the melting transition, terminal melting occurs without crystallization.

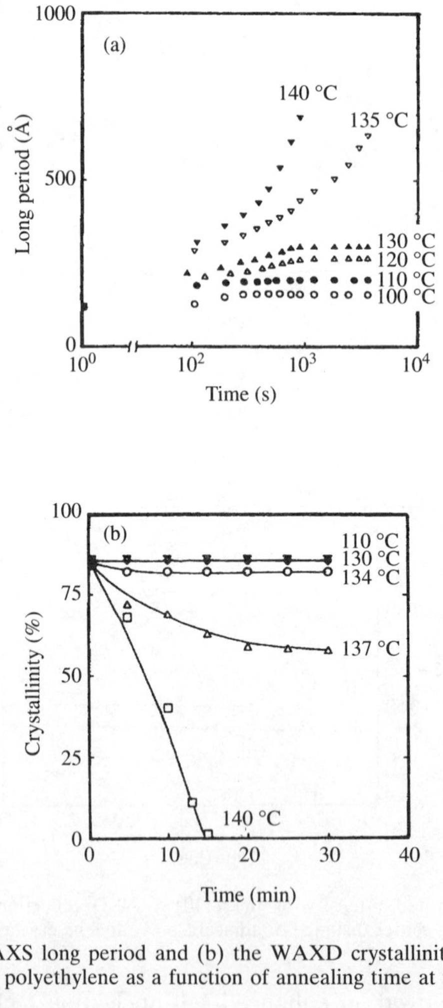

Figure 20 Variation of (a) the SAXS long period and (b) the WAXD crystallinity of gelation-crystallized ultrahigh molecular weight polyethylene as a function of annealing time at various T jumps.

The increase in the SAXS long period has been interpreted with the melting of neighboring (or alternate) lamellae, which results in an increase of interlamellar distances. This phenomenon is exactly the reverse sequence of crystallization, where the long period is large at the initial stage of crystallization, but it reduces as crystallization proceeds. The decrease in the long period during crystallization may be explained in the context of a lamellar insertion model in which new lamellae form in the preformed interlamellar spaces.

Recently, Rastogi and Ungar[95] studied the monoclinic–hexagonal transition of 1,4-*trans*-butadiene(1,4-t-PB) by synchrotron SAXS. Their experiment revealed that 1,4-t-PB can form several hundred nanometer-thick lamellae of the extended or nearly extended chain type, which may be formed either by direct growth from the melt or through annealing of thinner chain-folded lamellae so long as they are in the columnar phase. One important aspect of the columnar phase is that it provides the main avenue to extended chain crystals. Figure 21 shows time-resolved synchrotron WAXD traces, exhibiting a phase change from monoclinic to hexagonal at the beginning and a subsequent phase reversal from hexagonal to monoclinic as the low-temperature crystal phase reappears spontaneously on annealing above the temperature of the initial transition to the high-temperature columnar phase.

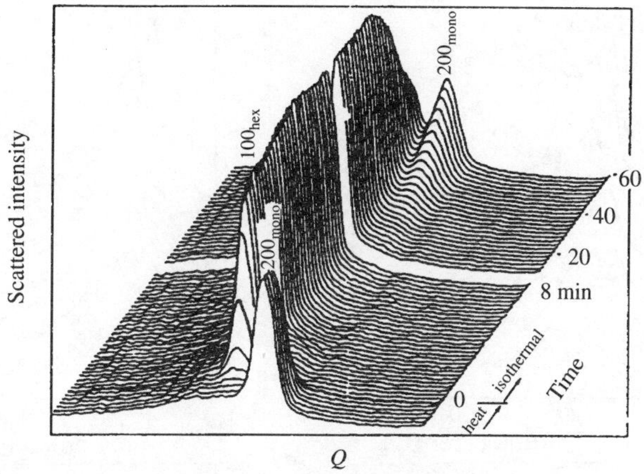

Figure 21 Time-resolved WAXD traces in the angular range of the main monoclinic (200) and hexagonal (100) reflections obtained at 15 s intervals for the first 37 curves and 2 min intervals for the later 26 scans, exhibiting isothermal phase reversal. The monoclinic → hexagonal transition is seen at the beginning and the reverse hexagonal → monoclinic transition toward the end of the run.

(ii) *Crystallization in liquid crystalline polymers and blends*

The occurrence of solid–solid crystal transformations and solid–liquid transitions is not unique to flexible chain polymers; it occurs in rigid chain crystals. Thermal transitions of liquid crystalline polymers (LCP) are generally characterized by a glass transition temperature and dual melting transitions.[96] These dual endotherms were attributed to fast and slow mode transitions associated with the melting of two types of crystals to a mesophase such as nematic. At low-temperature annealing above the glass transition, Cheng *et al.*[97] found solid–solid crystal transitions from hexagonal to a more stable orthogonal structure in a polyester-based LCP as well as in Vectra (a copolymer of hydroxybenzoic acid and 6-hydroxy-2-naphthoic acid). In the case of hydroxyquinone-based LCP (Granlar), the transition was found to occur from monoclinic I to monoclinic II when the LCPs were annealed at temperatures appreciably lower than the melting temperatures. On the other hand, Lin and Winter[98] observed a high-temperature melting peak during annealing above the original melting temperature, which is obviously a manifestation of melting–recrystallization phenomenon.

Nam *et al.*[99] applied the time-resolved WAXD technique to characterize the dynamics of phase transitions in a LCP prepared from bisphenol E diacetate, isophthalic acid, and 2,6-naphthalenedicarboxylic acid. Figure 22 depicts the time-dependent WAXD scans of the LCP below and above the melting temperatures. Three small peaks corresponding to *d* spacings of 5.7, 4.0, and 3.2 Å appear in the equatorial WAXD scan of the original LCP. Annealing at 300 and 310 °C the three WAXD peaks first disappear completely, suggestive of crystal melting to a nematic mesophase. With continued annealing isothermally for some 20 min, WAXD peaks reappear but at different Bragg angles. This process is nothing but recrystallization. Upon cooling after extended annealing for several hundred minutes, the LCP shows four distinct peaks, suggesting that a new type of crystal co-exists with the original crystal. Upon annealing at well above the melting temperature (320 °C), the crystalline WAXD peaks vanished completely and reappeared only when the LCP was cooled down. There was no memory of melting–recrystallization of the LCP at 320 °C. Annealing at 270 °C the three peaks were weaker, but did not disappear although they moved to low angles due to thermal expansion. With continued annealing the peaks became more distinct and a new peak also appeared. These observations imply that the partial melting of the original crystals occur first, followed by the subsequent formation of new crystals. The sequence of melting (or partial melting) and recrystallization cannot be known in the static WAXD characterization (at ambient temperature) of the LCPs annealed under various conditions. It is plausible that the solid–solid crystal transformation hitherto reported undergoes partial melting (disordering) before crystal ordering takes place again.

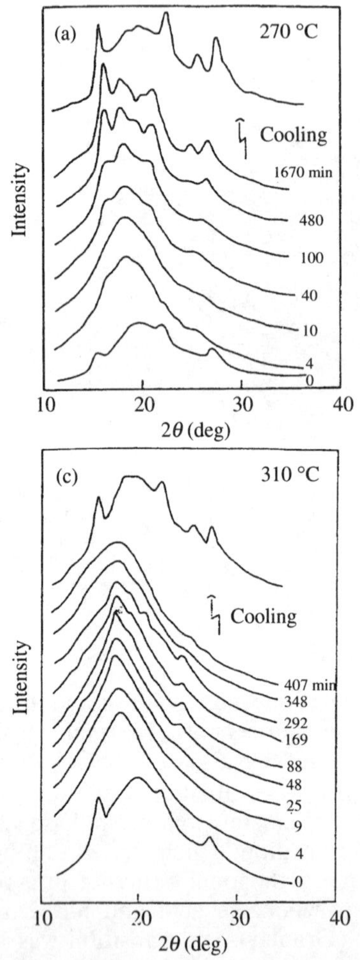

Figure 22 Time evolution of WAXD scans of the bulk LCP at 270, 300, 310 and 320 °C, exhibiting the phase transition of solid crystal–nematic–new crystal as a result of melting–recrystallization.

Nojima and co-workers[100] studied the crystallization kinetics in blends of poly(ε-caprolactone) (PCL) and poly(vinyl chloride) (PVC) using the time-resolved SAXS technique. For both pure PCL and its blends with PVC the crystallization at half-time, as measured by the SAXS peak characteristics of the center-to-center distance of neighboring lamellae, was found to increase with increasing temperature. The half-time for crystallization at a given isothermal crystallization temperature appears to slow down, as the concentration of the PVC increases. It is interesting to point out that this crystallization half-time by time-resolved SAXS correlates with the reciprocal linear growth rate of spherulites as seen by optical microscopic investigations. It seems to suggest that the kinetics of crystallization at the lamellar length scale and at the macroscopic spherulitic level are closely linked.

Synchrotron SAXS has been applied to mimic the effect of undercooling on crystallization of a 50/50 blend of high-density (linear) polyethylene and a low-density (branched) polyethylene.[101] As shown in Figure 23, the normalized integrated SAXS intensity varies in a sigmoidal manner with the crystallization temperature for the pure HDPE and LDPE. However, for the 50/50 blend the normalized integrated SAXS intensity shows a two-step increase. This behavior is due to the separate crystallization of HDPE and LDPE. The Q/Q_{max} versus temperature curve for the 50/50 blend may be separated into three regions. (1) In the first region of 120–110 °C, Q/Q_{max} rapidly rises to the first plateau due to the crystallization of HDPE; (2) in the second region (110–100 °C) the crystallization of HDPE continues while LDPE starts to crystallize; and (3) the third region of 100–90 °C is dominated by the LDPE crystallization. The behavior of the stepwise variation of Q/Q_{max} versus temperature suggests that no co-crystallization has occurred in the HDPE/LDPE blend at the slow cooling at a rate of 0.3 °C min^{-1}. Phase separation presumably occurs at an interfibrillar scale within the spherulites that contain both HDPE and LDPE. On the other hand, the rapid cooling of the same blend at a fast cooling rate of 110 °C min^{-1} showed that HDPE and LDPE crystallize separately on an interlamellar scale, suggesting that phase separation has occurred before crystallization has taken place. When the blend was rapidly cooled to two successive temperatures, *i.e.* the first temperature is below the crystallization temperature of the HDPE but higher than the crystallization temperature of the LDPE, and the second temperature is below the crystallization temperature of the LDPE, the scattering profiles are similar to those obtained from the directly quenched specimen.

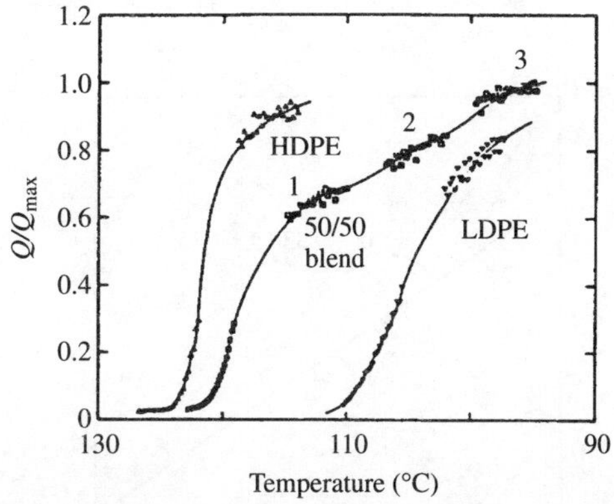

Figure 23 Variation of the normalized scattered SAXS intensity as a function of crystallization temperature for pure HDPE, pure LDPE, and their 50/50 blend.

Time-resolved Fourier transform infrared spectroscopy (FTIR) has been employed by Tashiro *et al.*[102] to follow the detailed structural evolution during isothermal crystallization of deuterated HDPE and hydrogenous linear low-density polyethylene (LLDPE) having various levels of branching. In the blend of d-HDPE and h-LLDPE (with 17 ethyl branches per 1000 carbons), co-crystallization occurs in which the d-HDPE and h-LLDPE chains are thoroughly mixed with a probability determined by the relative content of the d- and h-species. When the ethyl branching increases to 41 branches per 1000 carbons, the blend results in phase separation, but some portion of the h-LLDPE co-crystallizes with the d-HDPE chains.

8.3.2.2 Polymer gelation

Chowdhury and Russo[103] applied time-resolved video optical microscopy to investigate the late stages of phase separation-induced gelation in concentrated solutions of poly(γ-benzyl α-L-glutamate) (PBLG) in *N,N*-dimethylformamide (DMF) containing traces of water (2 wt%). Upon instantaneous cooling of the 10 wt% PBLG/DMF/H_2O from its homogeneous state to a two-phase region, phase separation occurs with a periodic length scale. Fourier transforms of the real space images, equivalent to scattering patterns, show a radially symmetric diffuse ring, which collapses to a smaller diameter as gelation proceeds. As shown in Figure 24, the wavenumber maximum (q_m) of the structure factor obeys a scaling relationship, $q_m \sim t^{-1/3}$, and is consistent with the prediction of the Lifshift–Slyozov evaporation/condensation model as well as the Binder–Stauffer cluster dynamics model.[36] Although the spinodal decomposition is likely the dominant mechanism, the optical microscopic technique is unable to establish the early stage of SD.

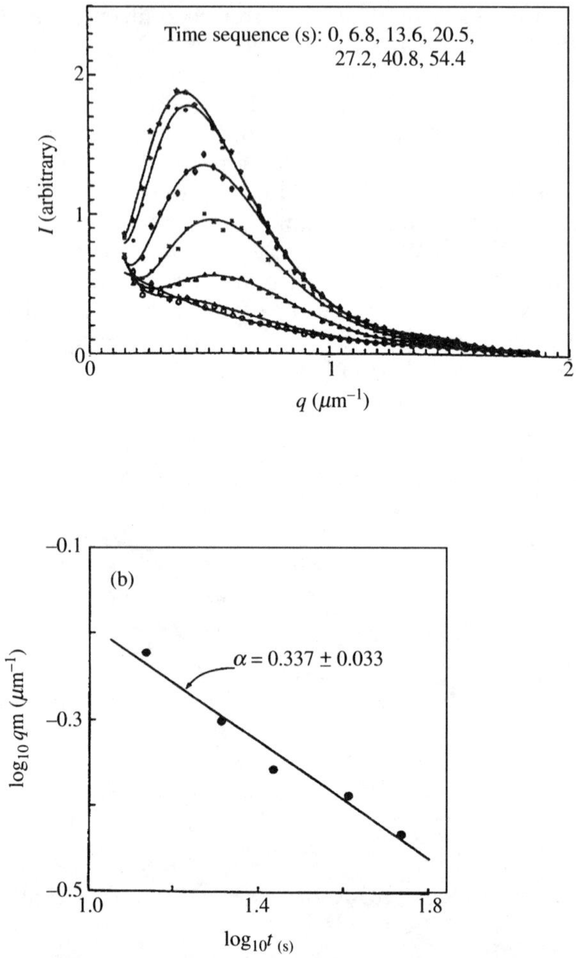

Figure 24 (a) Time evolution of scattering profiles of 10% PBLG/DMF with traces of water, obtained from the Fourier transform of the corresponding optical micrographs following a T quench from 54 to 50 °C and (b) the log–log plot of q_m versus elapsed time.

Otake *et al.*[104] studied the early stage of spinodal decomposition in *N*-isopropylacrylamide hydrogel during phase transition using time-resolved light scattering. The phase transition temperature of the hydrogel was determined to be 34 °C. Time evolution of the scattered intensity profiles was measured at various time intervals following temperature jumps to 35.0, 35.5, 36.0 and 36.5 °C. Their results were tested with the CH theory by plotting $R(q)/q^2$ versus q^2 (Figure 25). The linear slopes appear to verify the applicability of the linearized theory to the early stage of SD in this hydrogel system. They concluded that the volume change transition in the *N*-isopropylacrylamide takes place through the spinodal decomposition mechanism.

Tanaka *et al.*[105,106] utilized time-resolved optical microscopy to study the kinetics and morphological evolution of surface patterns during the swelling of gels prepared from a copolymer of acrylamide and sodium acrylate (AA-SA) lightly cross-linked by *N,N'*-methylenebisacrylamide and water. The morphological change was discerned on the gel surface in the late stage of pattern formation. The real space images were Fourier transformed so as to correspond to the scattering patterns, from which the wavenumber maxima were determined to evaluate the kinetic exponents of gel swelling. As shown in Figure 26, the temporal change of

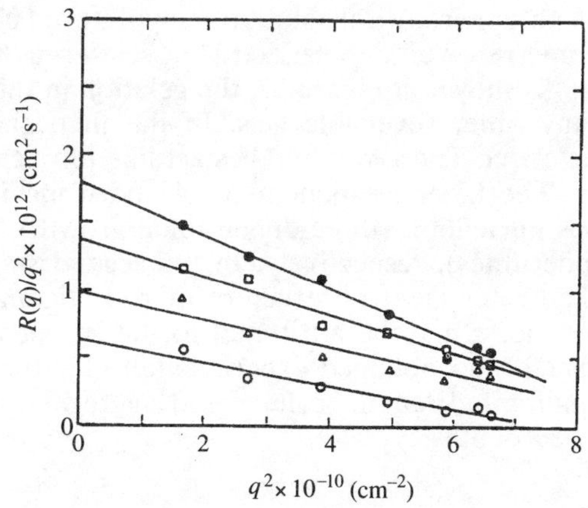

Figure 25 Cahn–Hilliard plots of $R(q)/q^2$ versus q^2 in a N-isopropylacrylamide hydrogel following T jumps to 35.0, 35.5, 36.0 and 36.5 °C.

the wavenumber maxima for three different specimens of varying thickness can be approximated by a straight line with a slope of $-1/2$, but deviations were seen at longer times. The late stage of gel swelling can be categorized into two regimes: (1) the formation of short-range order (tree-like structure) between cusps, and (2) the build-up of long-range order (honeycomb-like structure). In the regime I, the characteristic length scale $L(t)$ scaled according to 1/2 law. The regime II is prevalent only under strong geometrical constraint, when the cusps start to interact strongly with each other through the locally stored deformation energy. The morphological change occurs when the kinetics change from the diffusive growth regime ($t^{1/2}$ growth) to the relaxation region. This kinetic change can be well described by a simple one-dimensional diffusion in a finite system, indicating that the finite thickness of the system is responsible for this kinetic change.

Figure 26 Temporal change of the scattering maximum of a copolymer of acrylamide with sodium acrylate cross-linked by N,N'-methylenebisacrylamide with different initial thicknesses, showing a temporal scaling of $-\frac{1}{2}$.

Kyu *et al.*[107] investigated thermoreversible gelation in biphenyltetracarboxylic dianhydride–(perfluoromethyl)benzidine (BPDA–PFMB) polyimide/*m*-cresol solutions induced by phase separation and subsequent nematic formation. A temperature–composition phase diagram of

BPDA–PFMB/*m*-cresol is an upper critical solution temperature (UCST) type. The kinetics of gelation were studied by time-resolved depolarized light scattering by quenching the system at an off-critical composition. As shown in Figure 27, the gelation in this system was found to be a two-step process, like many other reversible gels. In the first stage, the rods coalesce into bundles due to an entropic effect. These rod bundles act like physical junctions upon which the aggregate structure grows. The kinetic exponents of 4 and 6 incidentally correspond to that predicted for homogeneous nucleation (three-dimensional growth) of the solid crystal and the solid sheaf (disordered spherulites), respectively. In the second stage, the aggregate domains impinge on each other. The refinement of structure or rod alignment takes place within the preformed domains. This process may be attributed to the secondary NG, in which a kinetic exponent of 0.5 was predicted and obtained experimentally for the 8 wt% PBDA–PFMB/*m*-cresol. The kinetically equilibrated length scale of rod aggregates was found to be inversely proportional to the quenched depths.

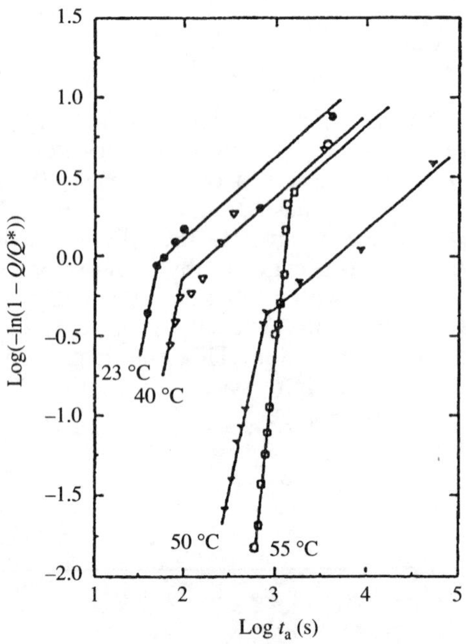

Figure 27 Log–log plots of scattering invariant versus annealing time for the 8 wt% PBDA–PFMB/*m*-cresol solution at the various temperatures indicated.

8.3.3 Interplay of Phase Separation and Ordering Transition

When one or both components of polymer blends are crystalline or liquid crystalline, there can be mutual interference of phase separation on ordering processes such as crystallization or nematic ordering. Flory and co-workers[38] were the first to show the existence of a UCST-type liquid–liquid phase separation above the melting transitions of polymer/solvent systems. This UCST curve depends strongly on the kinds of solvent used and their solubility parameters. A similar observation was reported by Tanaka and Nishi[108] for the blend of polycaprolactone (PCL)/poly(methyl methacrylate) in which the UCST occurred in the vicinity of the crystallization temperature of PCL (Figure 28). Because of the influence of phase separation on the crystallization of high PEO compositions, a new pattern formation was claimed to occur.

Kyu and Yang[109] studied the miscibility in crystalline polymer blends consisting of poly(vinylidene fluoride) (PVDF) and Nafion (perfluorinated ion exchange resins). A LCST appeared about 20 °C above the crystal melting temperature of PVDF (Figure 29). Nafions are semicrystalline polymers in nature, but the degree of crystallinity is extremely low. Recently, a LCST was observed in a crystalline–crystalline blend of PVDF/(poly(butyl adipate) (PBA)

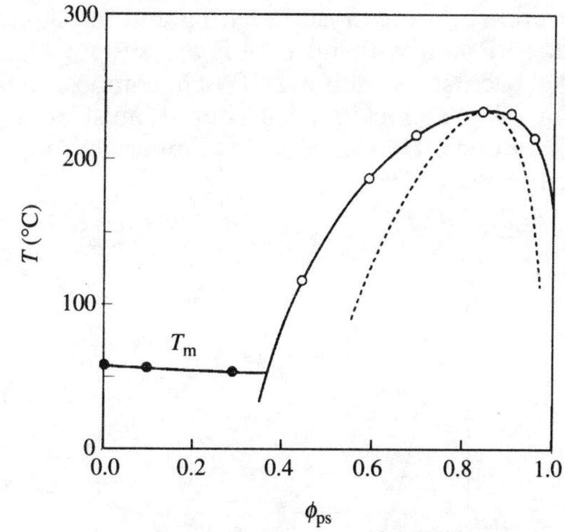

Figure 28 A UCST-type coexistence curve of polycaprolactone (PCL) and polystyrene influenced by the crystal melting of PCL. The solid and dashed lines were fitted to the experimental data using the Flory–Huggins theory.

which is located over 60 °C higher than the melting temperature of PVDF.[110] The phase diagrams in both systems are thermally reversible.[109,110] Kinetic studies by time-resolved light scattering on those systems show the regular growth behavior of liquid–liquid phase separation by spinodal decomposition normally observed in the amorphous–amorphous polymer blends.

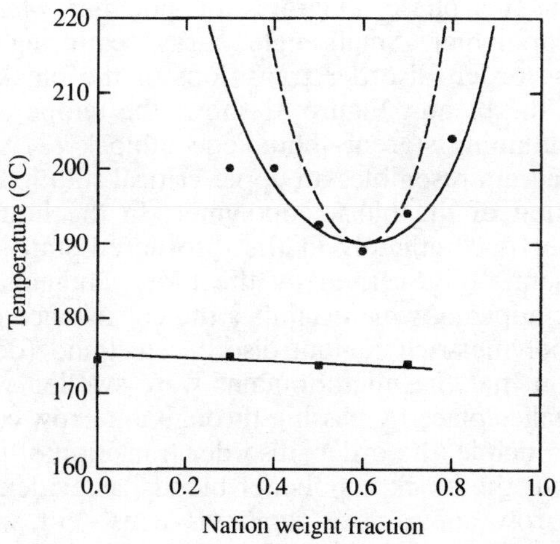

Figure 29 A LCST-type cloud point phase diagram and melting transitions of Nafion (perfluorinated ion exchange resin) and poly(vinylidene fluoride) blends, showing no mutual interference.

The blends of isotactic polypropylene with ethylene–propylene–diene terpolymer (EPDM), known for its thermoplastic elastomeric properties, were perceived to be completely immiscible. Recently, a liquid–liquid phase separation with LCST behavior was found to occur in the i-PP/EPDM (ENP-type with ethylene content of 70 wt%) blend.[111] Figure 30 shows the LCST appearing in the vicinity of the onset of crystal melting of i-PP, which is thermally reversible. At low temperatures below LCST, the amorphous phase of i-PP and EPDM appears miscible. When i-PP crystals start to melt, liquid–liquid phase separation occurs in the intermediate compositions, revealing a LCST. The uniqueness of this system from the above crystallizable blends is the mutual interference of the crystal melting and the liquid–liquid phase separation. At high concentrations of i-PP and with pure i-PP the system transforms to a single phase, when i-PP crystal melting occurs. At some i-PP-rich compositions (>80 wt% i-PP), phase separation occurs as soon as the crystals melt, but it must revert to a single phase with continued increase of temperature. Hence, a narrow immiscibility gap could exist at high i-PP compositions, as illustrated in Figure 30.

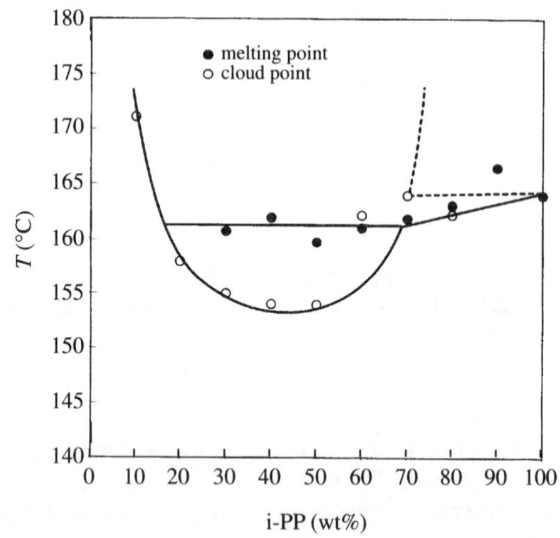

Figure 30 A LCST-type cloud point phase diagram of i-PP/EPDM blends overlapping with a melting transition of i-PP.

Roe *et al.*[112,113] studied the phase diagrams of polymer blends in which one of the components is a block copolymer. Small-angle X-ray scattering and light scattering were employed to determine the order–disorder transitions of the block copolymer and the phase separation temperature of the blends. Figure 31 shows the temperature vs. composition phase diagram of the blends containing styrene-butadiene diblock (27% styrene) and polystyrene ($M_w \approx 2400$). The phase diagram resembles an upper critical solution temperature influenced by the order–disorder transition of the block copolymer. In the liquid–liquid phase separation region the block copolymer is completely in the disordered state and molecularly dispersed. This $L_1 + L_2$ region is essentially governed by the Flory–Huggins free energy of mixing. A peritectic temperature line appears, which signifies the coexistence of the three phases L_1, M_1, and M_2. At the block copolymer-rich region, disordered liquid (denoted L_1) transforms to a mesophase (denoted M_1) in that the microdomains were swollen with homopolystyrene. This order–disorder transition takes place by passing through a narrow coexistence gap consisting of the L_1 and M_1 phases. In principle, the order–disorder transition of the pure block copolymer is the first-order transition, but the block copolymer blends, as evident by SAXS studies, show a broad transition. This narrow coexistence region is consistent with that of the crystalline polymer blends shown in Figure 30. At high concentrations below the first peritectic temperature, M_1 and L_2 coexist. The authors[112] suggested the existence of a second peritectic temperature characterized by the coexistence of M_1, M_2, and L_2 phases. The origin of the M_2 phase is not clearly understood, but is speculated to be due to the block copolymer micelles suspended in the continuum of PS.

The coupling of liquid–liquid phase separation and another type of first-order phase transition involving nematic ordering was observed by Shen and Kyu[114] for a mixture of low molar mass main-chain liquid crystal and a polymer. They employed the combination of Flory–Huggins free energy of mixing and Maier–Saupe free energy of nematic ordering to describe the complex phase diagram. Figure 32 shows a cloud point temperature versus composition phase diagram for a mixture of cyanobiphenyl eutectic liquid crystals (commercially known as E7) and poly(methyl methacrylate) ($M_w \approx 22\,000$). The coexistence curve is basically a UCST type overlapped by a narrow coexistence N + L region at high E7

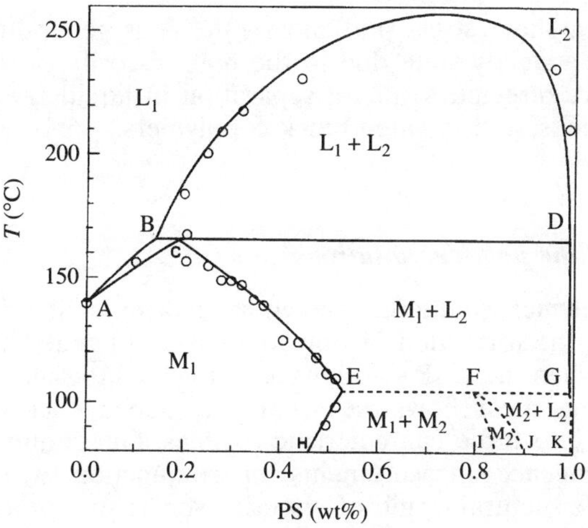

Figure 31 A coexistence curve of a blend of PS–PB block copolymer and homo-PS, exhibiting liquid–liquid phase separation and mesophase ordering transitions of block copolymers. The BCD and EFG lines represent the peritectic temperatures.

concentrations. This narrow coexistence region is consistent with those reported for liquid–liquid phase separation and phase transitions of binary systems such as crystalline polymer blends and block copolymers containing blends. The experimental cloud points corroborate favorably well with the theoretical curve.

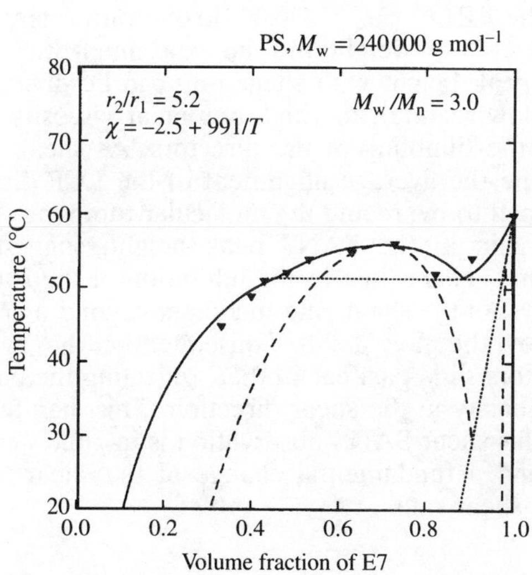

Figure 32 A coexistence curve of a mixture of a polydispersed polystyrene and E7 liquid crystal, showing liquid–liquid phase separation overlapped by a nematic ordering. The data points were obtained by cloud point experiments while the solid and dashed lines are the theoretical binodal and spinodal curves calculated from the combined free energies of Flory–Huggins for phase mixing and Maier–Saupe for nematic ordering.

8.3.4 Polymer Phase Separation under Shear Flow

Shear flow-induced phase transitions in multicomponent polymer systems have been the subject of recent interest. The melt of polymer blends is customarily sheared between two parallel quartz plates (or cone-and-plate) while the scattered intensity of light or neutrons is monitored *in situ* with a shear stress and shear rate. It is generally difficult for the sheared polymer melts to reach a steady state due to the high viscosity of the polymer molecules. At present, most studies are directed to phase separation in liquid crystalline polymer solutions, polymer blends in solutions, and solvated block copolymers.

8.3.4.1 Liquid crystalline polymer solutions

Liquid crystalline polymer solutions, known as lyotropic liquid crystals, display unique characteristics including the formation of polydomains and banded structure, director orientation and tumbling, *etc.* when the LCP solutions are subjected to shear flow. In response to these structural changes during shear flow, the shear rate dependence of the viscosity has been analyzed by dividing it into three characteristic regimes. Onogi and Asada[11] were the first to employ *in situ* birefringence measurements in conjunction with a cone-and-plate type rheometer to study the structural origin of viscosity versus shear rate behavior of poly(benzyl L-glutamate) (PBLG) solutions. They proposed a domain-based model comprising three shear rate regimes to account for the shear rate dependence of the viscosity, by correlating the response of the well-known polydomain nature of a LCP solution to shear flow. In the low shear rate (regime I), the domains, while preserving their integrity, flow past each other. The viscosity versus shear rate exhibits non-Newtonian behavior characteristic of two-phase flow. In the intermediate shear rate (regime II), the viscosity is independent of shear rate, suggesting that LCPs are flowing within the incipient domains, but the domains are starting to break up due to increasing shear. The high shear rate (shear thinning) region (regime III) occurs when the domains begin to break down to a continuous phase as the shear rate exceeds the critical value.

This structural transition behavior of LCP solutions during shear flow has been investigated at a microscopic molecular level by small-angle neutron scattering in order to gain further insight into the molecular origin.[114] Figure 33 shows the variation of shear viscosity and SANS peak height of the azimuthal intensity (a qualitative measure of molecular orientation) as a function of shear rate of the PBLG ($M_w \approx 236\,000$) solution in deuterated benzyl alcohol. The polymer concentration was 21% by weight and the experiment was carried out at 65 and 75 °C. The change of the SANS peak height with shear rate can be divided into three regions. At a low shear rate regime (a slow regime), the high nonlinear viscosity (not covered in Figure 33) may be a consequence of the tumbling of the directors. As shear rate increases, the director tumbling is suppressed while the average alignment of the LCP directors increases. However, the shear rate is still too small to overcome the molecular tumbling. In the intermediate regime, there is little or no net gain in the SANS peak height, suggesting that the net director orientation remains the same. This is due to the inhibition of the rotation of the LCP directors by the increased shear flow. As the shear rate increases beyond a critical value, the local chain dynamics are hampered by the flow field. Consequently, the average director orientation increases as the LCP directors slide past each other, reflecting the narrowing of the distribution of the SANS azimuthal intensity in the shear direction. This manifests itself in the viscosity as shear thinning behavior. This shear SANS observation is in good agreement with that obtained by Ernst *et al.*,[14] who found a fundamental change in the shear SALS pattern of PBLG in *m*-cresol above the critical shear rate.

8.3.4.2 Solution of polymer blends

Hashimoto and co-workers[115–118] studied the critical phenomena of a semidilute solution of a polystryene/polybutadiene blend under a quiescent condition and subject to the Couette-type shear flow using time-resolved light scattering and a shear device. The shear rate dependent scattering results were analyzed with emphasis on temporal universality and uniqueness of the polymer blend solution in comparison with the behavior of the simple liquid mixtures. They proposed to categorize the solution behavior into five regimes in ascending order of shear rate (Figure 34). Regime I corresponds to a region where the system undergoes macroscopic phase separation into two coexisting solution phases. In regime II the macroscopic interface is broken due to increased shearing and transforms to a percolated domain structure with a characteristic length scale R being essentially independent of the flow direction. With further increase of

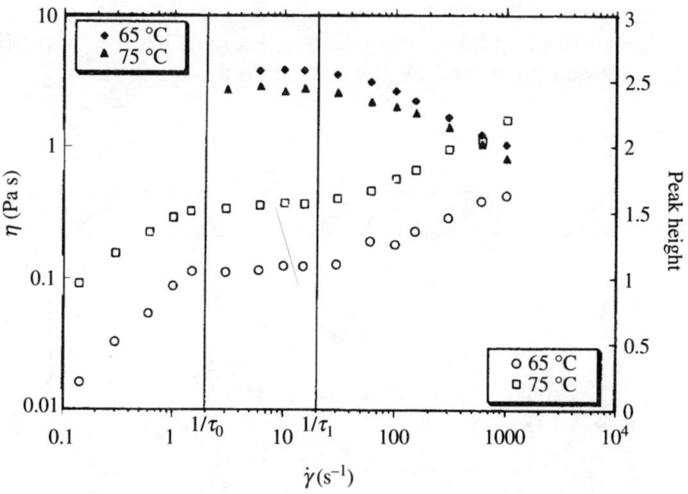

Figure 33 Variation of viscosity and the SANS peak height of the azimuthal intensity (a qualitative measure of molecular orientation) with shear, correlating the microscopic structural response (molecular orientation) of a LCP in solution to the macroscopic response (viscosity).

shear rate the domain structure is highly elongated toward the shear direction in regime III. In regime IV the amplitude of the concentration fluctuations becomes smaller with a further increase of shear, resulting in suppression of the scattering contrast and hence the intensity. Finally, in regime V the system becomes homogenized. The behavior of the crossover among the various regimes appears to depend on polymer concentration and quench depth ($\Delta T = T_c - T$), where T_c and T are the coexistence temperature under the quiescent state and the temperature of the measurement, respectively. In the shear-induced homogenized state the scattered intensity in the transverse direction of the flow (shear direction) was shown to follow the Ornstein–Zernike equation. The correlation length ξ and the intensity at zero angle were scaled with the shear rate, *viz.* $\xi^{-2} \sim \dot{\gamma}^{1/2} - \dot{\gamma}_c^{1/2}$ and $I(0)^{-1} \sim \dot{\gamma}^{1/2} - \dot{\gamma}_c^{1/2}$, and diverge at the critical shear rate $\dot{\gamma}_c$. At $\dot{\gamma}_c > \dot{\gamma}$ the hydrodynamic interaction is greatly suppressed, but is not completely screened out, thereby giving rise to the small mode–mode coupling.

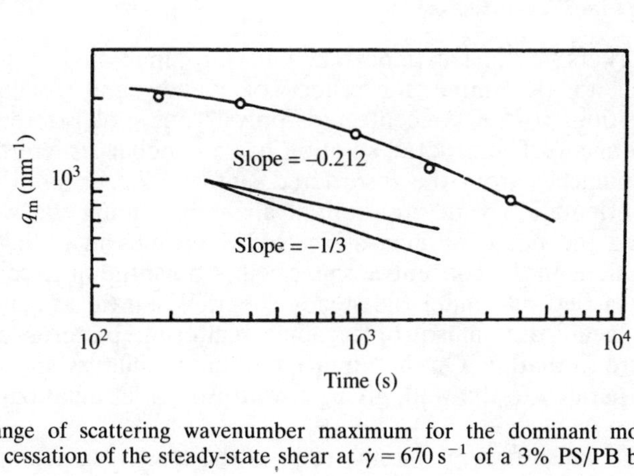

Figure 34 Temporal change of scattering wavenumber maximum for the dominant mode of the concentration fluctuations following a cessation of the steady-state shear at $\dot{\gamma} = 670\,s^{-1}$ of a 3% PS/PB blend in DOP solution.

The authors extended their studies to demixing following the cessation of shear at the near critical composition of PB/PS at the steady-state shear rate of $\dot{\gamma} = 670\,s^{-1}$. A diffuse scattering halo that developed was circularly symmetric, suggesting that the polymer chains were relaxed completely. The ring then collapsed to a smaller diameter with elapsed time, due to the phase growth. As shown in Figure 35, the shift of the scattering wavenumber maximum was found to scale in accordance with the theoretical prediction of Langer, Bar-on, and Miller,[36] *i.e.* $q_m \approx t^{-\alpha}$ and $\alpha = 0.212$ in the timescale covered by the experiment.

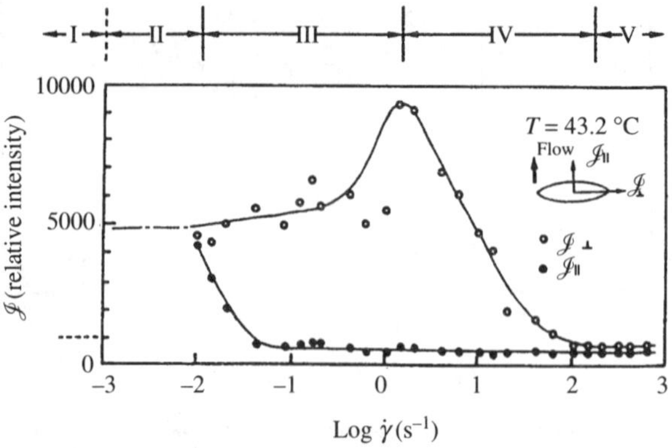

Figure 35 Variation of scattered intensities parallel to the flow direction and in the transverse direction versus shear rate of PS/PB/DOP (3%). The flow behavior was divided into five regimes.

Hammouda *et al.*[119] applied small-angle neutron scattering to determine the behavior of shear-induced phase separation in a solution of deuterated polystyrene in dioctyl phthalate (DOP) solution. The shear device was a Couette-type concentric quartz cylinder in which the 3 wt% PS in DOP solution was sheared in the neutron beam line. The neutron cloud point appeared to be about 14 °C lower than that by light scattering, which may be attributed to SANS's sensitivity to a smaller length scale. The scattered intensity at zero scattering angle increased when the shear rate increased beyond its critical value. The fact that the scattering pattern is isotropic regardless of the applied shear rates is puzzling. The polymer–solvent effective parameter χ_{ps} was analyzed in terms of the random phase approximation (RPA) theory, assuming that shear effects were contained exclusively in χ_{ps}. The validity of its application to polymer solutions is questionable. The critical shear rate values determined by SANS appear to be consistent with those obtained by light scattering.

8.3.4.3 Solution of block copolymers

Balsara and co-workers[120,121] have undertaken *in situ* small-angle neutron scattering in a concentric shear cell to determine the effect of steady and oscillatory shear on the order–disorder transitions of a concentrated polystryene–polyisoprene block copolymer solution in dioctyl phthalate. The ordered solution has a lamellar structure under the quiescent condition. Thermal quenches from the disordered state to the ordered state always give an anisotropic lamellar orientation with preferential alignment along the wall of the shear cell. Their results indicated the need for measuring SANS profiles in both tangential and radial (equatorial) configurations in the concentric shear cell. Above the quiescent ODT, the majority of experiments were carried out under the steady shear. When the steady shear rate increases beyond the critical shear rate, anisotropic radial scattering patterns appear, indicative of shear-induced structure formation. On the other hand, the oscillatory shear above the quiescent ODT yields lamellae parallel to the wall, giving rise to isotropic radial patterns and anisotropic tangential patterns.

The plots of the peak anisotropy, ζ (*i.e.* the ratio of the scattering peak position transverse to and along the shear direction), in the radial configuration versus the shear rate are depicted in Figure 36, at various temperatures. The scattering peak anisotropy shows a scaling $\zeta \sim (\dot{\gamma}/\dot{\gamma}_c)^{0.19\pm0.03}$ for $\zeta < 0.95$. The characteristic shear rate, $\dot{\gamma}_c$, which represents the onset of shear-induced order, may be estimated by extrapolating the log ζ versus log $\dot{\gamma}$ linear relation to $\zeta = 1$ (disordered state). The scaling exponents appear to be the same regardless of the shear temperatures investigated, suggesting that the scattering peak anisotropy at various temperatures follows a universal scaling law. If the shear rate is rescaled to $\dot{\gamma}/\dot{\gamma}_c$, the anisotropy versus shear rate curves corresponding to various temperatures can be superimposed on a master curve (Figure 36). This phenomenon is reminiscent of the shear rate–temperature superposition principle of polymer rheology.[1]

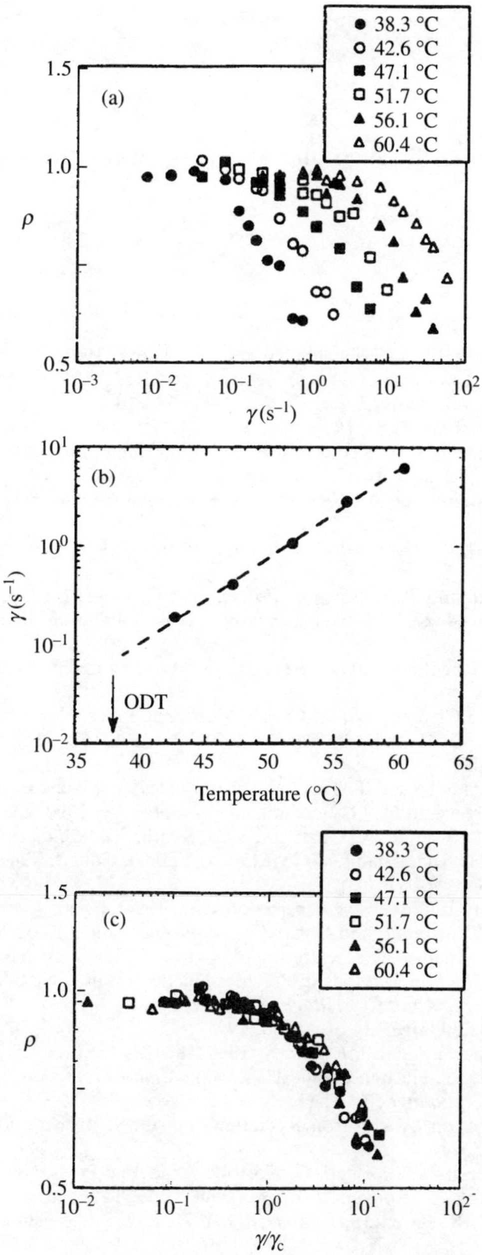

Figure 36 (a) Variation of the peak anisotropy ratio, ρ (*i.e.* the ratio of the peak height along the vertical and horizontal directions) as a function of shear rate at various temperatures above the order–disorder transition; (b) variation of the characteristic shear rate (onset of shear-induced order) with sample temperature; (c) reduced master curve for the variation of the peak anisotropy ratio with rescaled shear rate γ/γ_c for all temperatures.

In conclusion, the use of on-line time-resolved scattering of X-rays, neutrons, and light to flow or shear deformation of multicomponent polymer melts is anticipated to blossom in the coming decade. At this writing, the applications of spectroscopic techniques to time-resolved characterization of dynamical properties of polymer phase transitions are not abundant, but the number continues to increase. By virtue of their specificity in probing labeled molecules and segments, and their capability for determining the actual compositions of individual components in the mixed state, the field of static and time-dependent spectroscopic characterizations of polymer phase separations is expected to grow in the near future. Of particular interest is the time-dependent self-assembly of block copolymers because of possible control of the nanostructure of the blocks and the cross-fertilization with other areas such as the statistical mechanics of complex fluids.

ACKNOWLEDGMENT

The author is grateful to C. S. Shen and H. W. Chiu for their assistance in preparing this manuscript.

8.4 REFERENCES

1. J. D. Ferry, 'Viscoelastic Properties of Polymers', Wiley, New York, 1961.
2. S. Onogi, D. A. Keedy and R. S. Stein, *J. Polym. Sci.*, 1961, **50**, 153.
3. R. S. Stein, S. Onogi and D. A. Keedy, *J. Polym. Sci.*, 1962, **57**, 801.
4. B. E. Read, *J. Polym. Sci. C*, 1963, **5**, 87; *Polymer*, 1964, **5**, 1.
5. T. Kawaguchi, T. Ito, H. Kawai, D. A. Keedy and R. S. Stein, *Macromolecules*, 1968, **1**, 126.
6. R. S. Stein, in 'Rheology', ed. F. R. Eirich, Academic Press, New York, 1969, vol. 5, p. 297.
7. T. Hashimoto, R. E. Prud'homme, D. A. Keedy and R. S. Stein, *J. Polym. Sci., Polym. Phys. Ed.*, 1973, **11**, 693; 1973, **11**, 709.
8. P. Young, T. Kyu, S. Suehiro, J. S. Lin and R. S. Stein, *J. Polym. Sci., Polym. Phys. Ed.*, 1983, **21**, 881; 1988, **26**, 745.
9. S. Onogi, T. Asada, Y. Fukui and T. Fujisawa, *J. Polym. Sci., Part A-2*, 1967, **5**, 1067; 1970, **8**, 1195.
10. H. Kawai, *Rheol. Acta*, 1975, **14**, 27; H. Kawai, S. Suehiro, T. Kyu and A. Shimomura, *Polym. Eng. Rev.*, 1983, **3**, 109.
11. S. Onogi and T. Asada, in 'Rheology', ed. G. Astarita, G. Marrucci and L. Nicholias, Plenum, New York, 1980, vol. 1, p. 127.
12. T. Asada, in 'Polymer Liquid Crystals', ed. A. Ciferri, Academic Press, New York, 1982, p. 247.
13. W. R. Burghardt and G. G. Fuller, *J. Rheol.*, 1991, **35**, 379; K. Hogladarom and W. R. Burghardt, *Macromolecules*, 1993, **26**, 785.
14. B. Ernst, P. Navard, T. Hashimoto and T. Takebe, *Macromolecules*, 1990, **23**, 1370.
15. S. J. Picken, J. Aerts, R. Visser and M. G. Northolt, *Macromolecules*, 1990, **23**, 3849; 1991, **24**, 1366.
16. K. Amundson, E. Helfand, S. S. Patel, X. Quan and S. D. Smith, *Macromolecules*, 1992, **25**, 1935.
17. T. Jian, A. N. Semenov, S. H. Anastasiadis, G. Fytas, F.-J. Yeh, B. Chu, S. Vogt, F. Wang and J. E. L. Rovers, *J. Chem. Phys.*, 1994, **100**, 3286.
18. K. Mortensen, W. Brown and E. Jorgensen, *Macromolecules*, 1994, **27**, 5654.
19. H. Jinnai, H. Hasegawa, T. Hashimoto and C. C. Han, *Macromolecules*, 1991, **24**, 282.
20. A. Guinier and G. Fournet, 'Small-Angle Scattering of X-rays', Wiley, New York, 1955.
21. L. E. Alexander, 'X-ray Diffraction Methods in Polymer Science', Wiley, New York, 1969.
22. J. S. Higgins and R. S. Stein, *J. Appl. Crystallogr.*, 1978, **11**, 346.
23. H. W. Chiu, M. S. Thesis, The University of Akron, 1995.
24. T. Hashimoto, T. Takebe and S. Suehiro, *Polym. J.*, 1986, **18**, 123; *J. Chem. Phys.*, 1988, **88**, 5874.
25. S. Suehiro, K. Saijo, Y. Ohta, T. Hashimoto and H. Kawai, *Anal. Chim. Acta*, 1986, **189**, 41.
26. R. W. Hendricks, *J. Appl. Crystallogr.*, 1978, **11**, 15.
27. T. P. Russell, in 'Hand Book on Synchrotron Radiation', ed. G. S. Brown and D. E. Moncton, North-Holland, Amsterdam, 1991, vol. 3, p. 379.
28. 'Neutron Scattering in Materials Science', ed. G. Kostorz, Academic Press, New York, 1978.
29. A. K. Rizos, G. Fytas and J. E. L. Roovers, *J. Chem. Phys.*, 1992, **97**, 6925.
30. T. Nishi, T. T. Wang and T. K. Kwei, *Macromolecules*, 1975, **8**, 227; T. Nishi and T. T. Wang, *Macromolecules*, 1975, **8**, 909.
31. H. Tanaka and T. Nishi, *Phys. Rev. B*, 1986, **33**, 32.
32. E. J. Moskala, S. E. Howe, P. C. Painter and M. M. Coleman, *Macromolecules*, 1984, **17**, 1671.
33. P. C. Painter, Y. Park and M. M. Coleman, *Macromolecules*, 1989, **22**, 570; 1989, **22**, 580.
34. R. Gelles and C. W. Frank, *Macromolecules*, 1982, **15**, 1486.
35. S. Shimada, Y. Hori and H. Kashiwabara, *Macromolecules*, 1988, **21**, 3454.
36. G. Gunton, M. San Miguel and P. S. Sahni, in 'Phase Transitions and Critical Phenomena', ed. C. Domb and J. L. Lebowitz, Academic Press, New York, 1979, vol. 8, p. 269.
37. O. Olabisi, L. M. Robeson and M. T. Shaw, 'Polymer–Polymer Miscibility', Academic Press, New York, 1979.
38. P. J. Flory, 'Principles of Polymer Chemistry', Cornell University Press, New York, 1953.
39. J. J. Hoyt, B. Clark, D. de Fontaine, J. P. Simon and O. Lyon, *Acta Metall.*, 1989, **37**, 1597; 1989, **37**, 1611.
40. I. C. Sanchez and R. H. Lacomb, *J. Chem. Phys.*, 1976, **80**, 2352; 1976, **80**, 2568; I. C. Sanchez and A. C. Balasz, *Macromolecules*, 1989, **22**, 2325.
41. M. M. Coleman, J. F. Graf and P. C. Painter, 'Specific Interactions and the Miscibility of Polymer Blends', Technomic Publishing, Lancaster, PA, 1991.

42. T. Kyu, C.-C. Ko, D. S. Lim, S. D. Smith and I. Noda, *J. Polym. Sci., Part B: Polym. Phys.*, 1993, **31**, 1641; C.-C. Ko, T. Kyu and S. D. Smith, *ibid.*, 1995, **33**, 517.
43. D. Debier, J. Devaux and R. Legras, *J. Polym. Sci., Part A: Polym. Chem.*, 1995, **33**, 407.
44. H. Tomura, H. Saito and T. Inoue, *Macromolecules*, 1992, **25**, 1611.
45. G. Muller, R. Stadler and S. Schlick, *Makromol. Chem., Rapid Commun.*, 1992, **13**, 117.
46. P. C. Painter, J. Graf and M. M. Coleman, *J. Chem. Phys.*, 1990, **92**, 6166.
47. B. Hammouda, B. J. Bauer and T. P. Russell, *Macromolecules*, 1994, **27**, 2357.
48. J. W. Cahn, *Trans. Metall. Soc., AIME*, 1968, **242**, 166; J. W. Cahn and J. E. Hilliard, *J. Chem. Phys.*, 1958, **28**, 258.
49. H. E. Cook, *Acta Metall.*, 1970, **18**, 297.
50. P. G. de Gennes, *J. Chem. Phys.*, 1980, **72**, 4756.
51. P. Pincus, *J. Chem. Phys.*, 1981, **75**, 1996.
52. K. Binder, *J. Chem. Phys.*, 1983, **79**, 6387.
53. R. Petschek and H. Metiu, *J. Chem. Phys.*, 1983, **79**, 3443.
54. A. Chakrabarti, R. Toral, J. D. Gunton and M. Muthukumar, *J. Chem. Phys.*, 1990, **92**, 6899.
55. M. A. Kotnis and M. Muthukumar, *Macromolecules*, 1992, **25**, 1716.
56. C. S. Shen, Ph. D. Dissertation, The University of Akron, 1995.
57. H. L. Synder, P. Meakin and S. Reich, *Macromolecules*, 1983, **16**, 757; H. L. Snyder and P. Meakin, *J. Chem. Phys.*, 1983, **79**, 5588.
58. T. P. Russell, G. Hadziioannou and W. K. Warburton, *Macromolecules*, 1985, **18**, 78.
59. R. G. Hill, P. E. Tomlins and J. S. Higgins, *Macromolecules*, 1986, **18**, 2555.
60. T. Hashimoto, J. Kumaki and H. Kawai, *Macromolecules*, 1988, **16**, 641.
61. J. Kumaki and T. Hashimoto, *Macromolecules*, 1986, **19**, 763.
62. M. Okada and C. C. Han, *J. Chem. Phys.*, 1986, **85**, 5317.
63. T. Kyu and J. M. Saldanha, *Macromolecules*, 1988, **21**, 1021.
64. T. Kyu and P. Mukherjee, *Liq. Cryst.*, 1988, **3**, 631.
65. F. S. Bates and P. Wiltzius, *J. Chem. Phys.*, 1989, **91**, 3258.
66. T. Kyu and D.-S. Lim, *Macromolecules*, 1991, **24**, 3645.
67. R. Gelles and G. W. Frank, *Macromolecules*, 1983, **16**, 1448.
68. J. M. Torkelson, M. Tirrel and C. W. Frank, *Macromolecules*, 1984, **17**, 1505.
69. S. Nojima, K. Tsutsumi and T. Nose, *Polym. J.*, 1982, **14**, 225; S. Nojiima and T. Nose, *ibid.*, 1982, **14**, 269; S. Nojima, Y. Ohyama, M. Yamaguchi and T. Nose, *ibid.*, 1982, **14**, 907.
70. D. Schwahn, S. Jansen and T. Springer, *J. Chem. Phys.*, 1992, **97**, 8775.
71. D. Schwahn, H. Eckerlebe, E. Hädicke and T. Springer, *J. Phys. IV*, 1993, **3**, 13.
72. B. Chu, Q. Ying, K. Linliu, P. Xie, T. Gao, Y. Li, T. Nose and M. Okada, *Macromolecules*, 1992, **25**, 7382.
73. T. Hashimoto, in 'Current Topics in Polymer Science', ed. R. M. Otenbrite, L. A. Utracki and S. Inoue, Hanser, Munich, 1987, vol. 2, p. 199.
74. P. Wiltzius and A. Cumming, *Phys. Rev. Lett.*, 1991, **66**, 3000.
75. E. Siggia, *Phys. Rev. A*, 1979, **20**, 595.
76. D. Beysens, P. Guenoun and F. Perrot, in 'Dynamics of Ordering Process', ed. S. Komura and H. Furukawa, Plenum, New York, 1988, p. 373.
77. W. I. Goldburg and J. S. Huang, in 'Fluctuations, Instability and Phase Transitions', ed. T. Riste, Plenum, New York, 1975, p. 87.
78. W. I. Goldburg, in 'Light Scattering Near Phase Transitions', ed. H. Z. Cummins and A. P. Levanyuk, North-Holland, Amsterdam, 1983, p. 531.
79. F. F. Abraham, 'Homogeneous Nucleation Theories: Pretransition Theories of Vapor Condensation', Academic Press, New York, 1974.
80. K. Armistead and G. Goldbeck-Wood, *Adv. Polym. Sci.*, 1992, **100**, 219.
81. A. Peterlin and E. W. Fischer, *Z. Phys.*, 1960, **159**, 272.
82. A. Peterlin, E. W. Fischer and C. Reinhold, *J. Chem. Phys.*, 1962, **37**, 1403.
83. J. I. Lauritzen, Jr. and J. D. Hoffman, *J. Chem. Phys.*, 1959, **31**, 1680; *J. Res. Nat. Bur. Stand., Sect. A*, 1961, **65**, 297.
84. P. H. Geil, 'Polymer Single Crystals', Wiley, New York, 1963.
85. J. M. Schultz, 'Polymer Materials Science', Prentice-Hall, Englewood Cliffs, NJ, 1974.
86. J. M. Schultz, *J. Polym. Sci., Polym. Phys. Ed.*, 1976, **14**, 2291.
87. D. T. Grubb, J.-H. Liu, M. Caffrey and D. H. Bilderback, *J. Polym. Sci., Polym. Phys. Ed.*, 1984, **22**, 367.
88. D. T. Grubb and J. J. Liu, *J. Appl. Phys.*, 1985, **58**, 2822.
89. A. Kawaguchi, T. Ishida, S. Murakami and K. Katayama, *Colloid Polym. Sci.*, 1984, **262**, 597.
90. G. R. Strobl, M. J. Schneider and I. G. Voight-Martin, *J. Polym. Sci., Polym. Phys. Ed.*, 1980, **18**, 1361.
91. T. Hashimoto and K. Saijo, *Mater. Res. Soc., Symp. Proc. Ser.*, 1987, **79**, 267.
92. P. J. Barham and A. Keller, *J. Polym. Sci., Part B: Polym. Phys.*, 1989, **27**, 1029.
93. J. Martinez-Salazar, P. J. Barham and A. Keller, *J. Mater. Sci.*, 1985, **20**, 1616.
94. M. H. Cho, T. Kyu, J. S. Lin, K. Saijo and T. Hashimoto, *Polymer*, 1992, **33**, 4152.
95. S. Rastogi and G. Ungar, *Macromolecules*, 1992, **25**, 1445.
96. G. D. Butzbach, J. H. Wendorff and H. J. Zimmermann, *Polymer*, 1986, **27**, 1337; T. H. Sauer, H. J. Zimmermann and J. H. Wendorff, *Colloid Polym. Sci.*, 1987, **265**, 210.
97. S. Z. D. Cheng, J. J. Janimak, A. Zhang and Z. Zhou, *Macromolecules*, 1989, **22**, 4240.
98. Y. G. Lin and H. H. Winter, *Macromolecules*, 1991, **24**, 2877.
99. J. Nam, T. Fukai and T. Kyu, *Macromolecules*, 1991, **24**, 6250.
100. S. Nojima, W. Kosaka, M. Urushihara, W. Kosaka, N. Kato and T. Ashida, *Polym. J.*, 1986, **18**, 451; S. Nojima, K. Kosaka, T. Ashida, Y. Muroga, T. Ueki, Y. Hiragi, M. Kataoka, Y. Izumi, H. Tagawa and Y. Amemiya, *ibid.*, 1985, **17**, 1229.

101. H. H. Song, R. S. Stein, D. Q. Wu, M. Ree, J. C. Phillips, A. Legrand and B. Chu, *Macromolecules*, 1990, **23,** 1280.
102. K. Tashiro, M. Izuchi, F. Kaneuchi, C. Jin, M. Kobayashi and R. S. Stein, *Macromolecules*, 1994, **27,** 1240.
103. A. H. Chowdhury and P. S. Russo, *J. Chem. Phys.*, 1990, **92,** 5744.
104. K. Otake, H. Inomata, Y. Yagi, M. Konno and S. Saito, *Polym. Commun.*, 1989, **30,** 203.
105. T. Hayashi, H. Tanaka, T. Nishi, T. Hirose, T. Amiya and T. Tanaka, *J. Appl. Polym. Sci.: Appl. Polym. Symp.*, 1989, **44,** 195.
106. H. Tanaka, H. Tomita, A. Takasu, T. Hayashi and T. Nishio, *Phys. Rev. Lett.*, 1992, **68,** 2794.
107. T. Kyu, J. C. Yang, S. Z. D. Cheng, S. L. C. Hsu and F. W. Harris, *Macromolecules*, 1994, **27,** 1861.
108. H. Tanaka and T. Nishi, *Phys. Rev. A: Gen. Phys.*, 1989, **39,** 783.
109. T. Kyu and J.-C. Yang, *Macromolecules*, 1990, **23,** 176; J.-C. Yang and T. Kyu, *ibid.*, 1990, **23,** 182.
110. K. Fujita, T. Kyu, J. P. Penning and R. St. J. Manley, *Polym. Prepr., Am. Chem. Soc., Div. Polym. Sci.*, 1995, **36,** 332.
111. T. Kyu, C.-Y. Chen and W. M. Z. M. Yunus, unpublished results.
112. R.-J. Roe and W.-C. Zin, *Macromolecules*, 1984, **17,** 189.
113. R.-J. Roe, *Polym. Eng. Sci.*, 1985, **25,** 1103.
114. C. Shen and T. Kyu, *J. Chem. Phys.*, 1995, **102,** 556.
115. M. D. Dadmun and C. C. Han, *Macromolecules*, 1994, **27,** 7522.
116. T. Hashimoto, T. Takebe and S. Suehiro, *J. Chem. Phys.*, 1988, **88,** 5874; T. Takebe and T. Hashimoto, *Polym. Commun.*, 1988, **29,** 227.
117. T. Takebe, R. Sawaoka and T. Hashimoto, *J. Chem. Phys.*, 1989, **91,** 4369.
118. S. Okamoto, K. Saijo and T. Hashimoto, *Macromolecules*, 1994, **27,** 5547.
119. B. Hammouda, A. I. Nakatani, D. A. Waldow and C. C. Han, *Macromolecules*, 1992, **25,** 2903.
120. N. P. Balsara and B. Hammouda, *Phys. Rev. Lett.*, 1994, **72,** 360.
121. N. P. Balsara, B. Hammouda, P. K. Kesani, S. V. Jonnalagadda and G. C. Straty, *Macromolecules*, 1994, **27,** 2566.

9

Structure–Property Relationships in Perfluoropolyethers: A Family of Polymeric Oils

GIUSEPPE MARCHIONNI and GIUSEPPE AJROLDI

Ausimont S.p.A., Research and Development Centre, Bollate (Milano), Italy

and

GIOVANNI PEZZIN

University of Padova, Padova and National Research Council, Bologna, Italy

9.1 INTRODUCTION

Among the different classes of fluorinated organic molecules the perfluoropolyethers (PFPEs), composed entirely of carbon, fluorine and oxygen, represent a well known family that is of interest to the researcher and the user for the novelty and uniqueness of their chemical and physical properties. PFPEs have therefore become, in the past two decades, very important technically and commercially for a variety of applications, the most diffuse of which is related to their use as high quality lubricants.

As known, PFPE oils under normal conditions (but also at relatively high or low temperatures) are transparent fluids, both odorless and tasteless, that from the toxicological point of view are considered not harmful. They possess high thermal, oxidative and hydrolytic stability, are chemically inert and nonflammable, show very low volatility, which, together with their exceptional rheological and tribological properties, makes them useful in applications areas such as high temperature, high speed and aerospace lubrication, as well as in lubrication of pumps exposed to an aggressive environment and/or to radiation. Several other applications, including their use as barrier creams and cosmetics, are also described.

Details of the molecular structure of PFPE oils are known for the four major distinct types of PFPE polymers that are produced commercially. Ausimont in Italy, Du Pont in the US and Daikin in Japan produce several distinct grades of different oils, whose schematic structures and trade names will be given in the following pages. They are basically linear polymers in which the chain atoms are carbon and oxygen only. The main structural differences among the various oils are related to the different chemical structure of the repeating units, whereas the differences in the length of the chain are seen as changes of 'grade' of the given product. In the repeating structures, the number of C atoms present between two consecutive O atoms varies between 1 and 3. Among the four types, two are homopolymers and two are random copolymers. In addition, two types, which contain pendant trifluoromethyl —CF_3 groups, are often indicated as 'branched' or nonlinear PFPE oils. The chemical and physical properties of the four types of commercial oils, as well as those of other experimental samples prepared in the laboratory, have much in common when compared with those of other classes of lubricants, but their variations within the PFPE class are extremely important, being tied to the details of the relative structures (including chain length), and it is the scope of the present review chapter to describe the relationships between structure and properties of PFPE oils. Only limited attention will be given to the applications.

Historically, it appears that one of the first mentions of liquid perfluorinated peroxides of polymeric structure is due to Caglioti *et al.*[1] who obtained them in 1964 by the radiation-induced oxidation of tetrafluoroethylene. At the same time, in a series of patents in 1964–67, Du Pont researchers[2–5] described PFPE oils, useful as high temperature lubricants and hydraulic fluids, that were obtained by the catalytic polymerization of perfluoroepoxides. In the same years, a different route to the preparation of PFPE lubricants via the photochemical oxidation of perfluoroalkenes was described by Sianesi *et al.*[6–10] of Montedison in several patents and papers. Gozzo *et al.*[11,12] of the same company also obtained them by other reaction conditions.

9.2 SYNTHESIS OF PERFLUOROPOLYETHERS

The first synthesis of unspecified perfluorinated polyether mixtures was reported in 1953 by Haszeldine,[13] who described an oily product obtained in the course of the photooligomerization of hexafluoropropene when the monomer was inadequately purified. Since then, a number of different perfluorinated polyethers have been synthesized and described in the literature, but up to now there are only four different classes of commercial perfluoropolyethers, all of which have essentially linear chains but differ in their chemical structures. The latter are shown schematically as (**1**)–(**4**), together with their respective tradenames.

$$CF_3CF_2CF_2O[CF(CF_3)CF_2O]_wCF_2CF_3 \qquad \text{Krytox}^{®}$$

(1)

$$CF_3O[(CF_2O)_n(CF(CF_3)CF_2O)_m]CF_3 \qquad \text{Fomblin Y}^{®}\text{, Galden}^{®}$$

(2)

$CF_3O[(CF_2CF_2O)_p(CF_2O)_q]CF_3$ Fomblin Z®

(3)

$CF_3CF_2CF_2O(CF_2CF_2CF_2O)_rCF_2CF_3$ Demnum®

(4)

where *w* for Krytox and *r* for Demnun indicate the degree of polymerization of the homopolymers, whereas the letters *n, m* and *p, q* for Fomblin and Galden indicate the content of the various moieties in the chain, whose distribution is random.

The first perfluoropolyether oils to appear on the market under the tradename Krytox were the homopolymers of hexafluoropropene oxide, produced by Du Pont via the base-catalyzed polymerization of hexafluoropene oxide (equation 1).[14,15] The terminal acyl fluoride group is removed by fluorination, and chemically inert polyethers are produced whose structure is shown in (**1**). These fluids, actually available in the polymerization degree range from $w = 2$ to $w \approx 50$, are obtained by controlling the oligomerization conditions (temperature and solvent type).

$$CF_3CF\overset{O}{—}CF_2 \xrightarrow[\text{solv.}]{F^-} CF_3CF_2CF_2O[CF(CF_3)CF_2O]_{w'}CF(CF_3)COF \quad (1)$$

Recently, the PFPE fluids of structure (**5**), where $j = 0$, 1 and 2 only, produced by a similar reaction but modified by successive electrochemical Kolbe decarboxylation and dimerization (equation 2), have become available commercially from Hoechst[16] under the tradename Hostinert®.

$$[CF_3CF_2CF_2O—(CF(CF_3)CF_2O)_j—CF(CF_3)]_2$$

(5)

$$2R_fCOOH \longrightarrow R_f—R_f + 2CO_2 + H_2 \quad (2)$$

The commercial products appearing under the tradename Demnum are homopolymers of perfluorotrimethylene oxide. They have structure (**4**) and are obtained via the base-catalyzed ring-opening polymerization of 2,2,3,3-tetrafluoroxetane, the successive fluorination and final stabilization of the chain ends (equation 3).[17,18]

$$\begin{array}{cc} CF_2—CF_2 \\ | \quad\quad | \\ CH_2—O \end{array} \xrightarrow[\text{solv.}]{F^-} F(CH_2CF_2CF_2O)_rCH_2CF_2COF \quad (3)$$

The second commercially available fluids to appear were the copolymers Fomblin Z and Y,[19] prepared by the photooxidation of tetrafluoroethylene and hexafluoropropene, respectively, whose synthesis will be described in detail.

The copolymer of structure $—(CF_2O)_n(CF(CF_3)CF_2O)_m—$ is produced under the tradename Fomblin Y by photooxidation,[19,20] introducing oxygen into liquid hexafluoropropene (HFP) at −40 °C under UV irradiation (<300 nm). From the reaction one obtains an oligomeric mixture of polyether chains containing peroxide units, as shown in equation (4).

$$CF_3CF{=}CF_2 + O_2 \xrightarrow[< -40\,°C]{h\nu} TO(CF_2CF(CF_3)O)_m(CF_2O)_n(O)_vT' \quad (4)$$

$$T, T' = CF_3,\ COF\ (\text{major}),\ CF_2COF\ (\text{minor})$$

Under favourable conditions the reaction yield exceeds 95% (based on HFP) and the by-products are COF_2 and CF_3COF. When the reaction temperature is above −40 °C, hexafluoropropene oxide becomes the major by-product. Reaction rate and product composition are independent of oxygen concentration, but are sensitive to temperature, HFP concentration and light intensity. Quantum yields based on HFP or oxygen change according to the reaction conditions but rarely fall under 20, indicating free radical chain processes.

The products obtained consist, structurally, of randomly distributed fluoroether repeating units randomly interspaced with peroxide units. The terminal groups are neutral $—CF_3$ and two other functional groups, the fluoroformate —OCOF (major) and the acyl fluoride —CF_2COF (minor). The average molecular weight varies between 500 and 10 000, depending on the reaction temperature: low temperatures favour high molecular weight.

The in-chain peroxide units are relatively unstable and can be eliminated by thermal and photochemical reduction. The weight loss due to peroxide decomposition is always less than 5%, and is a function of peroxide content (which is normally 0.5% by weight). The by-products of the reaction are predominantly COF_2 and CF_3COF. Fluoroformates are converted to acyl fluorides and ketones; the ratio between the two species depends on the structure, as shown by the examples in equations (5) and (6). Ketones are generally prevalent under normal reaction conditions.

$$\text{—OCF(CF}_3\text{)CF}_2\text{OCOF} \longrightarrow \text{—OCF(CF}_3\text{)COF} + \text{COF}_2 \tag{5}$$

$$\text{—OCF}_2\text{CF(CF}_3\text{)OCOF} \longrightarrow \text{—OCF}_2\text{COCF}_3 + \text{COF}_2 \tag{6}$$

Both the acyl fluorides and the ketones are then converted, by direct fluorination, completely to fluoro end-caps (equations 7 and 8). The fluorination proceeds through hypofluorite intermediates, which are detectable and can be isolated. The reaction can be performed either thermally, at temperatures of 200–250 °C, or photochemically below 100 °C.[21] The perfluoro copolymeric fluids have the final prevalent structure (**2**), where $m/n > 40$.

$$\text{—OCF(CF}_3\text{)COF} \xrightarrow{F_2} \text{—OCF}_2\text{CF}_3 + \text{COF}_2 \tag{7}$$

$$\text{—OCF}_2\text{COCF}_3 \xrightarrow{F_2} \text{—OCF}_3 + \text{CF}_3\text{COF} \tag{8}$$

Samples of Fomblin Y with viscosities at 20 °C ranging from 40 cSt to 1800 cSt became commercially available in the late 1960s. Products of lower molecular weight (viscosity at 20 °C from 0.5 to 40 cSt) are commercialized under the tradename Galden. They are obtained by catalyzed cracking reactions.[22–25]

The linear PFPE copolymer (**3**), commercialized by Ausimont under the tradename Fomblin Z, is prepared by photo-oxidation of tetrafluoroethylene (TFE) (equation 9).[19,26,27] The reaction is carried out at low temperatures (<−40 °C) in the solvent CF_2Cl_2. The by-products are COF_2, small amounts of perfluoroethylene oxide and CF_3COF. The yields are dependent on temperature and on TFE concentration; under favourable conditions the yield is greater than 70%. The p/q ratio ranges from 0.5 to 2, and the average molecular weight ranges from 1000 to 200 000. Peroxide removal and fluorination convert polyperoxides to neutral PFPEs. The products have the linear structure (**3**).

$$\text{CF}_2\text{=CF}_2 + \text{O}_2 \xrightarrow[<-40\,^\circ\text{C, solv.}]{h\nu} \text{TO(CF}_2\text{CF}_2\text{O)}_p\text{(CF}_2\text{O)}_q\text{(O)}_v\text{T}' \tag{9}$$

Two PFPE products which combine the characteristics of the Y and Z fluids described above, and which are named Fomblin K and Fomblin J, have been developed recently by Ausimont. The first, Fomblin K, has the structure (**6**), where $m/p = 1–2$ and $n/(m+p) = 0.03–0.05$, and is obtained by the photo-oxidative polymerization of TFE/HFP mixtures.[28,29] The second product, Fomblin J, has the structure (**7**), where $l/u = 1.1$ and $q/(l+u) = 0.7$, and is obtained by using a perfluoro(methyl vinyl ether) as monomer.[30] As expected, the characteristics of these fluids lie between those of the Y and Z structures.

$$\text{CF}_3\text{O(CF}_2\text{O)}_n\text{(CF}_2\text{CF}_2\text{O)}_p\text{(CF}_2\text{CF(CF}_3\text{)O)}_m\text{CF}_3$$

(6)

$$\text{CF}_3\text{O(CF}_2\text{O)}_q\text{(CF}_2\text{CF(OCF}_3\text{))}_l\text{(CF(OCF}_3\text{)O)}_u\text{CF}_3$$

(7)

An advantage of the photo-oxidation technology described above is the relative ease with which the chemistry responds to external variables to produce desired structural variations of the PFPE–polyperoxide intermediates. In such a context, an innovative technological development appears to be the production of functionalized PFPEs. The availability of PFPE with functional end groups, coupled with the use of conventional techniques for further chemical transformations, has opened up a vast new field to PFPE chemistry and applications.

Thus, the polyperoxide series (equation 9) can be converted directly to diacid products by chemically induced reductive cleavage of the peroxide bonds,[31] as exemplified in Scheme 1, where structure (**8**) is a specific version of the generalized formulae of equation (9). The acid group can then be converted into salts, amides, nitriles, alcohols and alcohol methacrylates, opening up new possibilities for introducing fluorine into conventional products to improve their performance, and to open up new applications.

$$\underset{(\mathbf{8})}{\text{—}OOCF_2O(CF_2CF_2O)_pCF_2CF_2OO\text{—}} \xrightarrow{[H]_{aq}} [HOCF_2O(CF_2CF_2O)_pCF_2CF_2OH]$$

$$\downarrow -2HF$$

$$[FOCO(CF_2CF_2O)_pCF_2COF]$$

$$\downarrow 2H_2O \mid -2HF$$

$$[HO_2CO(CF_2CF_2O)_pCF_2CO_2H]$$

$$\downarrow -CO_2$$

$$\underset{(\mathbf{9})}{[HO_2CCF_2O(CF_2CF_2O)_{p-1}CF_2CO_2H]} \xleftarrow[\text{(ii) } H_2O\ (-HF)]{\text{(i) } -HF} [HOCF_2CF_2O(CF_2CF_2O)_{p-1}CF_2CO_2H]$$

Scheme 1

An extension of the photo-oxidation technology in which tetrafluoroethylene or hexafluoropropene are irradiated in the presence of chlorotrifluoroethylene[32] gives monofunctional derivatives having structures (**10**) and (**11**):

$$AO(CF_2CF_2O)_p(CF_2O)_qCF_2CO_2H \qquad \text{(from TFE)}$$

(**10**)

$$A'O[CF_2CF(CF_3)O]_m(CF_2O)_nCF_2CO_2H \qquad \text{(from HFP)}$$

(**11**)

$$A = \text{—}CClF_2, \text{—}CF_2CF_2Cl$$

$$A' = \text{—}CF(CF_3)CF_2Cl, \text{—}CF_2CFClCF_3, \text{—}CF_2Cl$$

When the starting products are tetrafluoroethylene and perfluorobutadiene, a unique polyfunctional PFPE has been obtained.[33–35] As usual, the co-photo-oxidation is followed by peroxide group elimination. The product is a polyether containing terminal acid fluorides and epoxy pendant groups, as shown in Scheme 2. The average molecular weight, the structure and the end group functionality of the product can be controlled by the reaction conditions. As found for bifunctional derivatives, conventional chemistry can be used to transform the COF group into any other group. The polyfunctional products have been used to obtain cross-linked polymers by UV treatment.[35]

Two alternative technologies for the synthesis of PFPEs are the oxidation of alkenes induced by ionizing radiation,[36–38] and the direct fluorination (using elemental fluorine) of hydrocarbon polyethers.[39] In addition, recent patents by Sianesi *et al.*[40,41] describe a new synthetic method in which the polymerization of fluoroalkenes and oxygen is obtained by initiation with specific radical initiators.

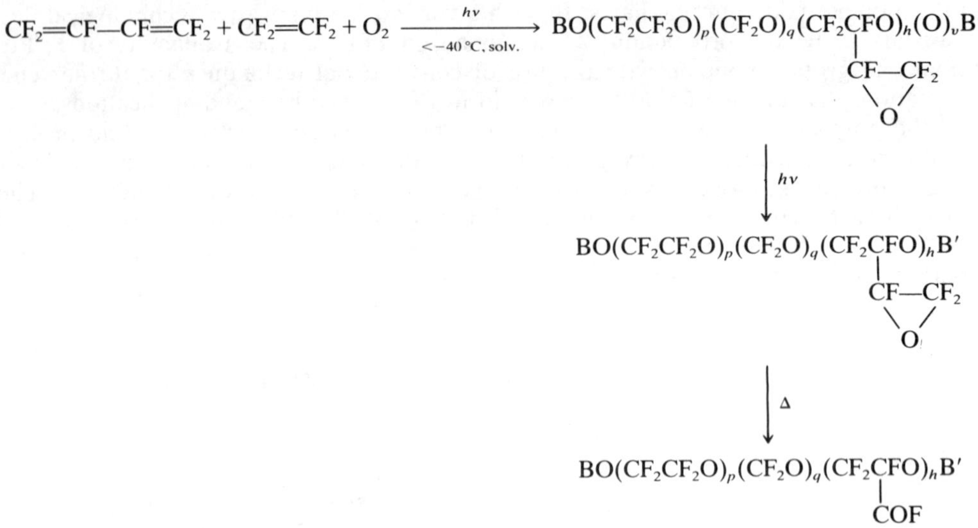

where B, B′ = —CF_2COF, —CF_3 (minor)

Scheme 2

9.3 STRUCTURE AND CONFORMATION

Historically, the determination of the chemical structure of the various classes of perfluoropolyethers has accompanied and has been strictly related to the understanding of the reaction mechanism of the respective syntheses. An example is the synthesis of Fomblin Z and Y polyethers, which were obtained by UV initiated reaction of fluoroalkenes with oxygen, and is described schematically in Section 9.2. The basic polymerization reaction, as well as the subsequent removal of the peroxide bonds and modification of the reactive terminal groups, have been understood and described structurally and kinetically by determing the chemical structure and concentration of the many different chemical species present in the reaction vessels under the different reaction conditions.

Sianesi *et al.*[19] have reported the use of the NMR technique, and particularly of the already known ^{19}F NMR spectral data,[42] to determine the concentration and the succession of the CF_2 and CF_2CF_2 groups present in the peroxidic perfluoropolyether obtained by peroxidation of tetrafluoroethylene, as well as the nature and concentration of several by-products. There are no problems in the interpretation of the spectra, since the characteristic NMR band position varies in a range of about 40 ppm when the structure of the C_2 and C_1 (CF_2CF_2O and CF_2O) proximal groups changes. NMR data are collected in Table 1. The ratio of the so-called C_2 and C_1 chain groups, easily and precisely determined, was found to depend on monomer concentration and polymerization temperature. The most important structural conclusion of the above studies, once the nature of the chain groups has been established, was of course the random nature of the distribution of the C_1 and C_2 units in Fomblin Z, and correspondingly, of the C_3 and C_1 units in Fomblin Y.[19,42,43]

An example of the NMR spectrum of a typical perfluoropolyether sample is given in Figure 1. More recently, ^{19}F NMR data have been used to control the chemistry of the synthesis of a very large number of novel perfluoroethers, and of other interesting perfluoroether classes (such as spherical perfluoroethers, perfluoro crown ethers, spiro compounds and cryptands) obtained by direct fluorination of hydrocarbon polymers (and in general of hydrocarbon materials) to fluorocarbon analogues, without alteration of the original structures, as well as of compounds obtained via a second synthetic route, which starts from hydrocarbon polyesters that are firstly fluorinated to perfluoropolyesters, and then treated with sulfur tetrafluoride to produce the final perfluoropolyethers.[39] Needless to say, the ^{19}F NMR spectroscopic studies are invaluable in any structural investigation of perfluoropolyether chemical modification or degradation reaction.[44]

Table 1 Assignment of the NMR ^{19}F Peaks of PFPE.

Fluid	*Structure*	*Group*	*Chemical shift, δ (ppm)*[a]
	1 2 3 4 5 6 7	1	−82
Krytox	$CF_3CF_2CF_2O[CF(CF_3)CF_2O]_nCF_2CF_3$	2	−130
		3	−145
		4	−80
		5	−80
		6	−89
		7	−87
	1 2 3 4 5 6 7 8	1	−82
Demnum	$CF_3CF_2CF_2O[CF_2CF_2CF_2O]_mCF_2CF_3$	2	−130
		3	−84
		4	−83
		5	−129
		6	−83
		7	−89
		8	−87
Fomblin Y	$—O\underline{CF}_2O—$		−49.5
	$—OCF_2CF(CF_3)O\underline{CF}_3$		−54.0
	$—OCF(CF_3)CF_2O\underline{CF}_3$		−56.2
	$—OCF(\underline{CF}_3)\underline{CF}_2O—$		−80.0
	$—OCF(\underline{CF}_3)\underline{CF}_3$		−81.2
	$—OCF_2CF_2\underline{CF}_3$		−82.0
	$—OCF_2\underline{CF}_3$		−87.3
	$—O\underline{CF}(CF_3)O—$		−89.0
	$—OCF_2\underline{CF}_2CF_3$		−130.0
	$—O\underline{CF}(CF_3)CF_2O—$		−145.0
Fomblin Z	$—OCF_2CF_2O\underline{CF}_2OCF_2CF_2O—$		−52.0
	$—OCF_2O\underline{CF}_2OCF_2CF_2O—$		−53.6
	$—OCF_2O\underline{CF}_2OCF_2O—$		−55.3
	$—OCF_2CF_2O\underline{CF}_3$		−56.3
	$—OCF_2O\underline{CF}_3$		−58.0
	$—OCF_2CF_2O\underline{CF}_2CF_2CF_2O—$		−83.2
	$—OCF_2O\underline{CF}_2CF_2CF_2O—$		−84.9
	$—OCF_2CF_2O\underline{CF}_2\underline{CF}_2OCF_2CF_2O—$		−89.0
	$—OCF_2O\underline{CF}_2\underline{CF}_2OCF_2O—$		−90.5
	$—OCF_2\underline{CF}_2\underline{CF}_2CF_2O—$		−126.0
	$—OCF_2\underline{CF}_2CF_2O—$		−130.0

[a] With reference to $CFCl_3$.

As far as the conformation of perfluoropolyethers is concerned, there is paucity of direct experimental results, a situation which is easily understood if one keeps in mind the fact that even the conformational preferences and dynamic behaviour of simple fluorocarbons are little known, and that, still lacking experimental data, 'the different possible isomers and even the existence of rotational isomerism in some perfluoroalkanes are still matter of controversy'.[45]

The most important information has been obtained by Pacansky *et al.*[46] who investigated simple perfluoroethers (CF_3—O—CF_3, CF_3—O—CF_2CF_3 and CF_3CF_2—O—CF_2CF_3) and diethers (CF_3—O—CF_2CF_2—O—CF_3) by *ab initio* SCF calculations, and studied the structure of perfluoropolyether chains by the molecular mechanics method. They obtained optimized geometries, net charges, molecular orbital energies, bond stretching force constants, vibrational frequencies and infrared intensities which they used to understand the polymer structure and assign the IR band centers. According to their results, cumulative rotations about the successive C—O bonds imparted by interaction between lone pairs of electrons on fluorine and oxygen lead to a helical nature of the geometry of the (CF_2O) and (CF_2CF_2O) polymer chains. The structures calculated for the model compounds CF_3—O—CF_2CF_2—O—CF_3 and CF_3—O—$CF(CF_3)CF_2$—O—CF_3 show that the O—C—C—O unit retains a roughly planar *trans* form.

The characteristic IR spectra of the four different commercial PFPE oils, as obtained in our laboratory, are given as examples in Figure 2. Numerical data are to be found in the Kasai paper,[44] in which the molecular mechanics method has been again applied to elucidate the

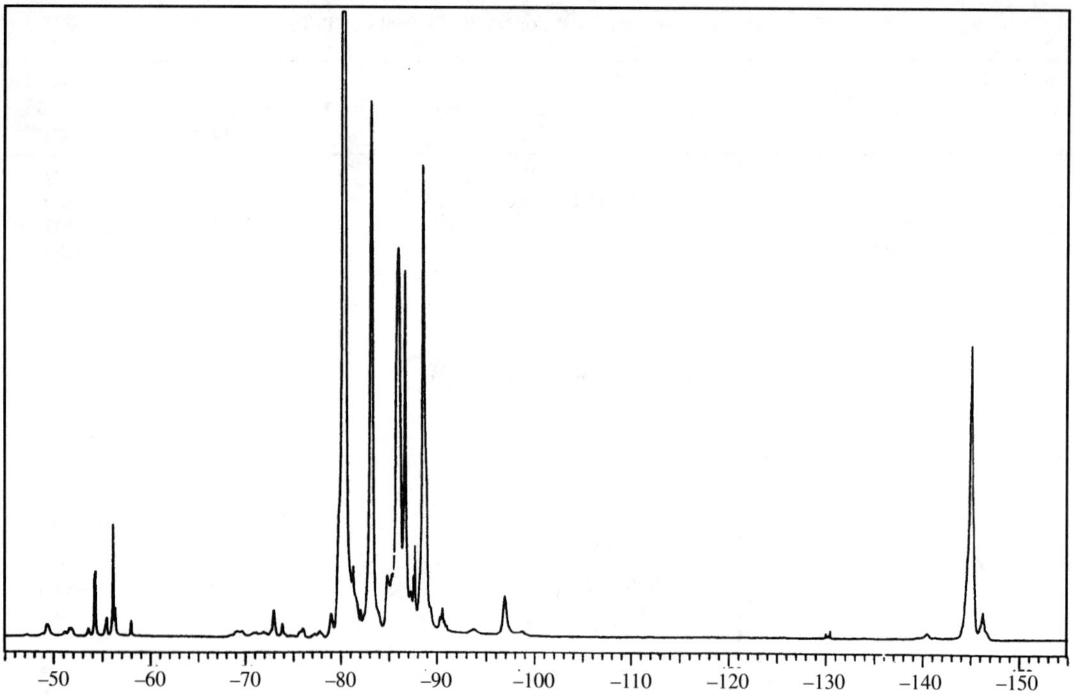

Figure 1 ^{19}F NMR spectrum of a PFPE oil (Fomblin K type).

conformation of short models of Fomblin Z, Krytox and Demnun perfluoropolyethers, and of the (noncommercial) homopolymer perfluoro(ethylene oxide), of structure (CF_2CF_2O). Figure 3 shows the space-filled entering of the computed forms. In the short-chain model of Fomblin Z ether, in which a central acetal group O—CF_2—O is flanked by two C_2 groups, the two acetal oxygen atoms are found to be close and rather exposed. In the Demnum model, the two oxygen atoms are also positioned on the same side of the chain whereas in ethylene oxide and propylene oxide (Krytox) models they are positioned on opposite sides of the chain. This '*trans*' disposition of the successive ethereal oxygen atoms has been invoked to explain the different characteristics of the perfluoropolyether thermal degradation.[44]

9.4 THERMODYNAMIC PROPERTIES OF PERFLUOROPOLYETHERS

9.4.1 *P–V–T* (Pressure, Volume, Temperature) Behaviour

Specific volume data for selected PFPE fluids of the Fomblin Y series, of various molecular weights, were first published by Sianesi *et al.*[47,48] Similar data, reported as density *vs.* temperature, for Krytox samples of different molecular weight are found in a paper by Ouano *et al.*[40] and in a recent review paper of Del Pesco.[50] Ohsaka has reported data for Krytox, as well as for Fomblin Y and Demnum samples.[17] Extensive numerical data on Fomblin Y and Z fluids are found in two papers by Marchionni *et al.*[51,52] and are shown in Figure 4. It is clear that, for a given sample, and in a temperature range of practical interest, the density ρ is a linearly decreasing function of temperature T.

The molecular weight dependence of the specific volume, at constant temperature T, is shown in Figure 5 to be given, for a variety of materials, by the equation

$$v = a + b/M_n \tag{10}$$

where M_n is the number average molecular weight of the sample. The equation can be derived, for monodisperse samples having molecular weight M and degree of polymerization $n = M/M_0$ (M_0 = molecular weight of the repeating unit of molar volume V_0), in terms of molar volume V.

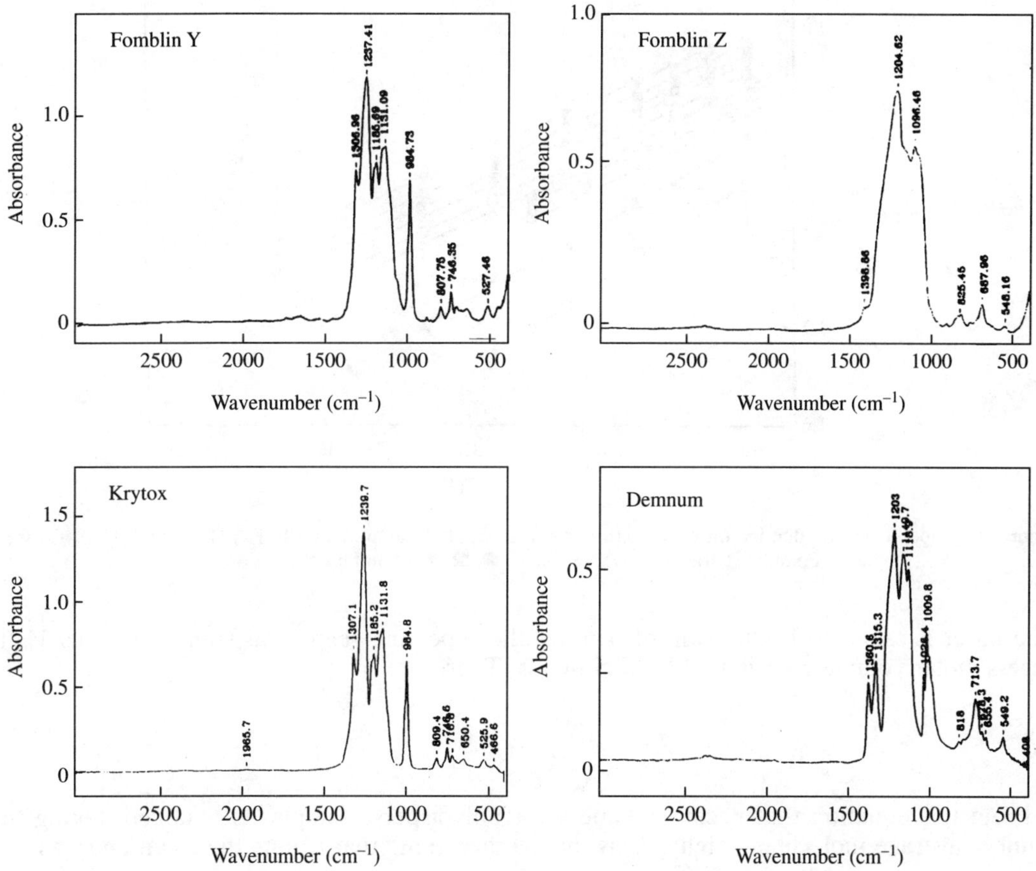

Figure 2 Typical IR spectra for some commercial PFPE copolymers and homopolymers.

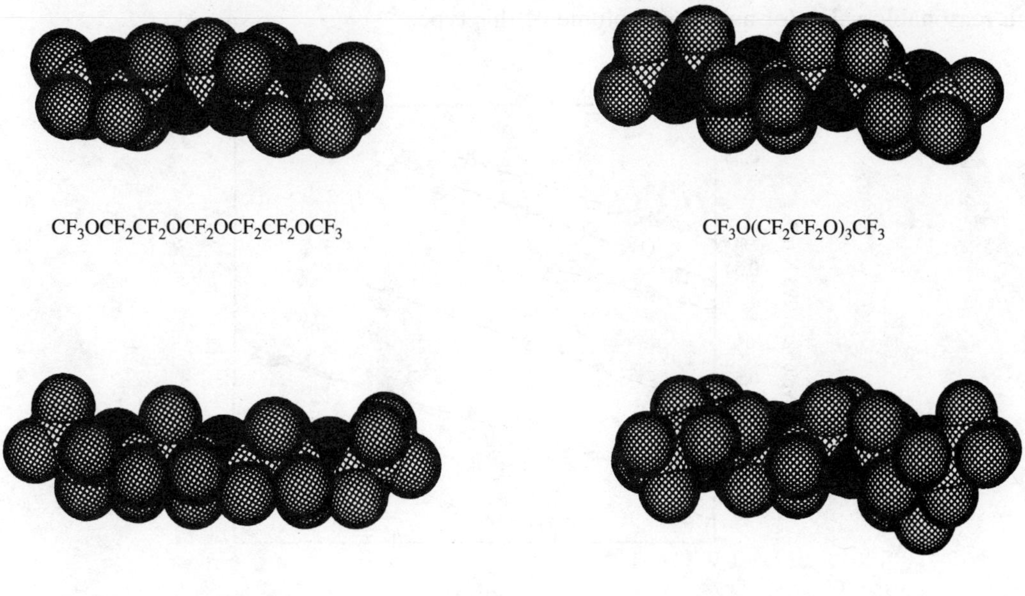

$CF_3OCF_2CF_2OCF_2OCF_2CF_2OCF_3$

$CF_3O(CF_2CF_2O)_3CF_3$

$CF_3O(CF_2CF_2CF_2O)_2CF_2CF_3CF_3$

$CF_2O[CF(CF_3)CF_2O]_3CF_2CF_3$

Figure 3 Conformations of PFPE homopolymers and copolymers obtained by the molecular mechanics method (reproduced by permission of the American Chemical Society from *Macromolecules,* 1992, **25,** 6791).

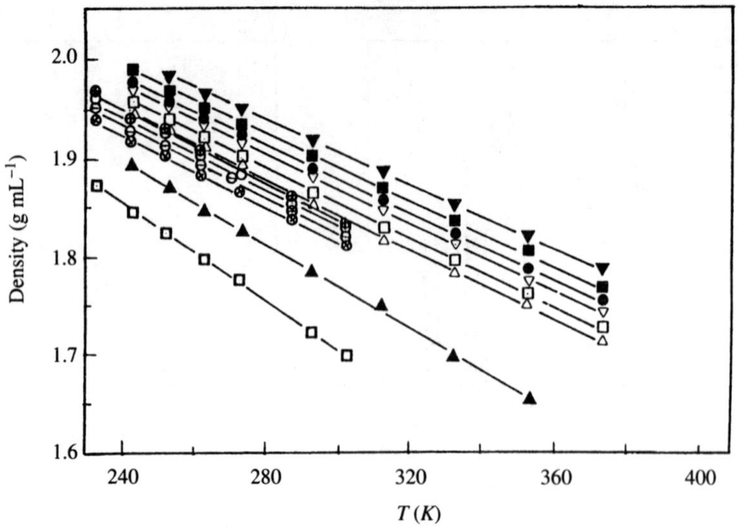

Figure 4 Dependence of density on temperature for fractionated samples of PFPE polymers: (□, ⊕, ○, ⊗, ⊖) Fomblin Z fractions; (△, ▲, □, ▽, ●, ■, ▼). Fomblin Y fractions.

The latter is taken to be the sum of n times the repeating segmental volume V_0 plus V_e, the excess molar volume contributed by chain ends. Therefore

$$V = nV_0 + 2V_e \tag{11}$$

From this equation the specific volume v for polydisperse samples is obtained, taking their number average molecular weight M_n as the average value that counts the chain ends, as

$$v = V_0/M_0 + 2V_e/M_n \tag{12}$$

which gives a physical interpretation of the constants of the empirical equation (10). From Figure 5 the excess molar volume V_e is found to be, at 25 °C, about 20% of the molar volume V_0, a reasonable value for an excess volume of this type.[52]

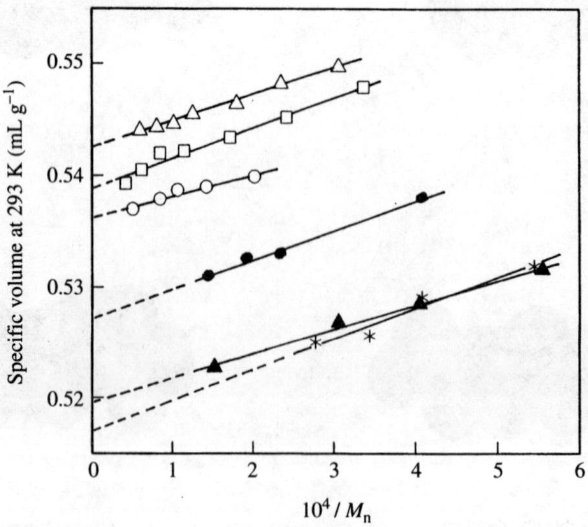

Figure 5 Specific volume at 293 K as a function of reciprocal number average molecular weight M_n for fractions of Z-type copolymers with different p/q ratios (△, □, ○) and for Denmum (●), Fomblin Y (▲) and Krytox (∗) (reproduced by permission of the Society of Plastic Engineers from *Polym Eng. Sci.*, 1990, **10,** 829).

In a 1994 study, Danusso *et al.*[53] considered the v–M relationship for bifunctional Fomblin Z structures, whose chain ends were either CH_2OH or CH_2OCOCF_3 groups. While the ratio V_0/M_0 was approximately the same found for Fomblin Z, the excess volume V_e was larger, *i.e.* 20 and 40 mL mol^{-1} in comparison with the average value of 11.3 mL mol^{-1} found for perfluorinated chain ends.

The thermal expansion coefficients $\alpha = (1/V)(dV/dT)_P$ at normal pressure (0.1 MPa) are also reported in some of the mentioned papers. The present authors found that, at 293 K, α is a function of both structure and chain length.[54] Z polymers have a thermal expansion coefficient higher than Y polymers (1.04×10^{-3} and 0.86×10^{-3} K^{-1}, respectively, for M_n of about 7000) and such a difference was attributed to the bulky CF_3 side group, which also affects strongly the glass transition temperature T_g. As far as the molecular weight dependence is concerned, it has been found that α is a function of the reciprocal molecular weight M_n;[54] similar results can be also found for Krytox fluids on the basis of the data reported by Del Pesco.[50]

The influence of pressure on the volume of PFPE fluids has been recently studied, for selected samples of Demnum fluids, by Tanaka *et al.*,[55] for an unspecified Krytox fluid by Del Pesco,[50] and for two Fomblin polymers (Y and Z) by Wolf *et al.*[56]

Unpublished data from the present authors, who carried out an extensive study on the two Fomblin families,[57] are shown as example in Figure 6 for temperatures ranging from 0 to 141 °C.

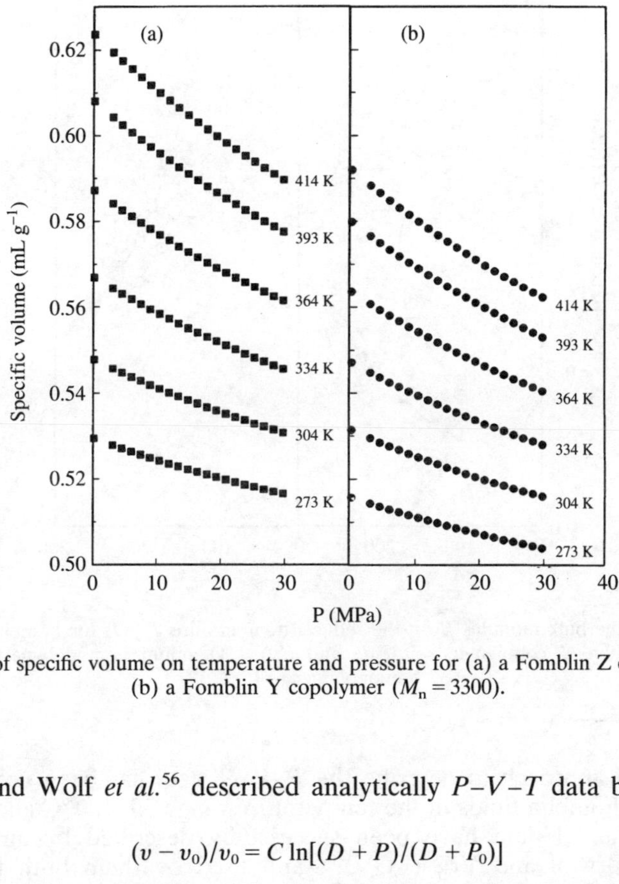

Figure 6 Dependence of specific volume on temperature and pressure for (a) a Fomblin Z copolymer ($M = 9700$) and (b) a Fomblin Y copolymer ($M_n = 3300$).

Tanaka et al.[55] and Wolf *et al.*[56] described analytically P–V–T data by means of the Tait equation

$$(v - v_0)/v_0 = C \ln[(D + P)/(D + P_0)] \tag{13}$$

where v_0 and v are the specific volume at P_0 and P (P_0 being the atmospheric pressure). The constants C and D are characteristic of the selected fluid and temperature. The isothermal compressibility β is related to them by the equation

$$\beta = (d \ln \rho/dP)_T = C/(D + P_0 + P) \tag{14}$$

It is interesting to observe that the Tait equation, which was proposed at the very beginning of this century[58] when synthetic polymers were unknown, has proven to hold also for polymeric liquids; as a matter of fact the basis of this equation lies in the assumption that the bulk equilibrium modulus B, *i.e.* the reciprocal of compressibility, varies linearly with pressure. Such an assumption was found to be valid for both a theoretical point of view[59] and for a number of experimental observations.[60]

The bulk modulus data, deduced from the known literature, are shown in Figure 7 both at normal pressure (0.1 MPa) and at a pressure of 30 MPa as a function of the reduced temperature $T - T_g$; it is in fact known that the temperature dependence of the zero pressure bulk modulus is a function of the difference $T - T_0$, where T_0 is a reference temperature.[60,61] It is seen that data pertaining to the same family lie on the same curve independently of the molecular weight; in other words, the molecular weight dependence seems to be mainly related to the molecular weight dependence of T_g. Surprisingly, the scatter is higher at 0.1 MPa than a 30 MPa. Differences among families are larger at 30 than at 0.1 MPa; this is due to the fact that the pressure dependence of T_g, which is at present unknown, is presumably different for each family.

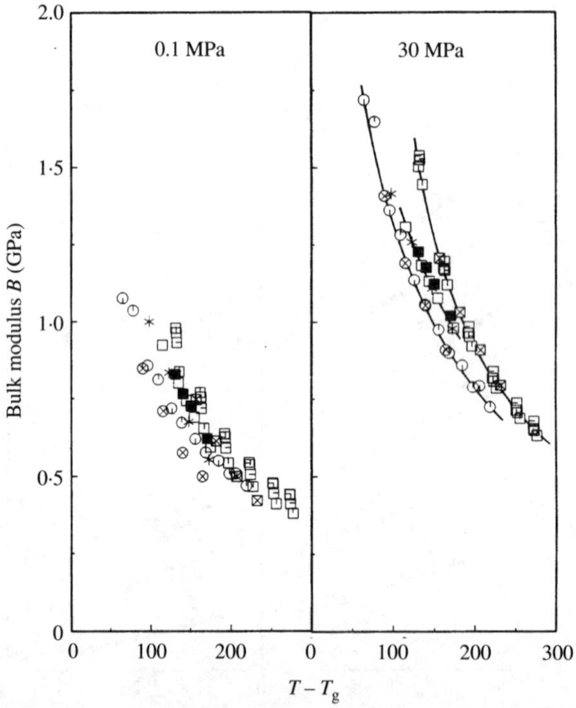

Figure 7 Dependence of the bulk modulus B on the reduced temperature $T - T_g$ for Fomblin Z copolymers (◫, ⊟, ⊟, ◫)[57] and (⊠),[56] Fomblin Y copolymers (⦶, ⦶)[57] and (⊗),[56] Demnum homopolymers (□, ■)[55] and Krytox homopolymer (∗).[50]

Another theoretical approach to describe the P–V–T data has been used by Wolf *et al.*,[56] who investigated two Fomblin fluids in the temperature range 25–100 °C and at pressures up to 40 MPa. The experimental data have been successfully described by an equation of state proposed by Flory, Orwoll and Vrij (FOV),[62] and used by them both for different linear polymers and low molecular weight compounds. The FOV equation is

$$\tilde{p}\tilde{v}/\tilde{T} = \tilde{v}^{1/3}/(\tilde{v}^{1/3} - 1 - 1/\tilde{v}\tilde{T}) \tag{15}$$

where the tilde over letters indicates the reduced variable of state (*e.g.* $\tilde{v} = v/v^*$, with v^* the reduction volume, or specific hard core volume). Figure 8 shows, by comparing calculated densities with experimental ones, how the Flory, Orwoll and Vrij equation describes the

P–V–T data of the PFPEs investigated over the P and T ranges explored. In Table 2 the reduction parameters of the two PFPE fluids obtained by Wolf *et al.* are compared with those of other fluids, perfluorinated and hydrogenated, of low and high molecular weight.

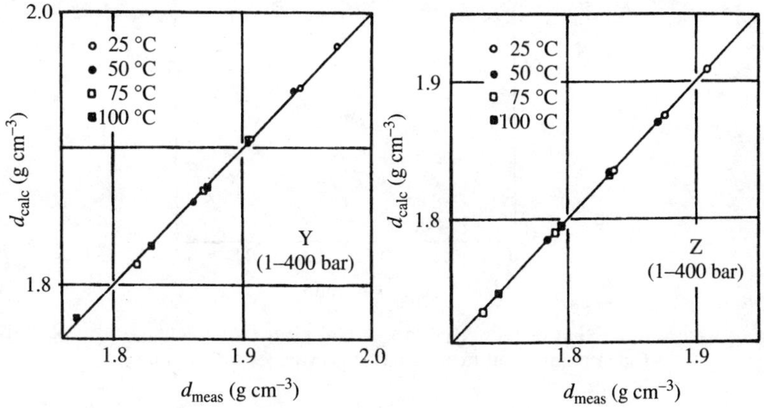

Figure 8 Comparison of measured densities d with those calculated on the basis of equation (15) by means of the reduction parameters collected in Table 2; the values given for the indicated temperatures refer to 1, 200 and 400 bar (reproduced by permission of the American Chemical Society from *J. Phys. Chem.*, 1989, **93,** 2682).

Table 2 Reduction Parameters of the Lubricants and of Some Polymers and Solvents.[56]

	v^* (mL g^{-1})	T^* (K)	p^* (bar)	Z^{*a}	*Ref.*
Fomblin Y	0.4266	5505	3536	5.4	56
Fomblin Z	0.4318	5035	3675	7.2	56
Polyisobutene	0.9493	7580	4483	7.8	63
Polystyrene	0.8131	7592	5320	4.5	64
Poly(dimethylsiloxane)	0.8468	5700	3300	4.5	65
Perfluoroheptane	0.4330	4292	3578	2.5	66
Perfluoro(methylcyclohexane)	0.4171	4299	3892	4.1	66
Benzene	0.8860	4709	6285	5.3	63
Cyclohexane	1.0012	4720	5320	5.2	67
Carbon tetrachloride	0.4881	4675	5607	0.9	68

[a] Number of backbone atoms contributing three external degrees of freedom.

Further information about the role of fluorine and oxygen comes from the packing density $\rho^* = V_w/V$, where V_w is the van der Waals volume; molar volume data for PFPEs were those extrapolated to infinite molecular weight by equation (12), data for amorphous PTFE, PE and polyoxides were taken from Van Krevelen's book,[69] while the van der Waals volumes were calculated by the group contributions reported by Bondi.[59] The data reported in Figure 9, although somewhat scattered, show that the introduction of an ethereal oxygen reduces the packing density more for the perfluorinated chains than for the hydrogenated series.

At 298 K and 0.1 MPa the specific volume v has been shown by the present authors[52] to be described successfully by the method of group contribution. In a 1992 paper[54] the same authors have used density data for a variety of perfluorinated ethers to obtain the group contributions to molar volume V at 298 K (0.1 MPa) that are collected, together with other data, in Table 3. The excess terminal group volume V_e has been found to be about 10.1 mL mol^{-1}, close to the oxygen contribution. The above volume values differ substantially from those reported earlier by Lawson *et al.*,[70] obtained on unspecified low molecular weight fluorocarbon–hydrocarbon hybrid compounds.

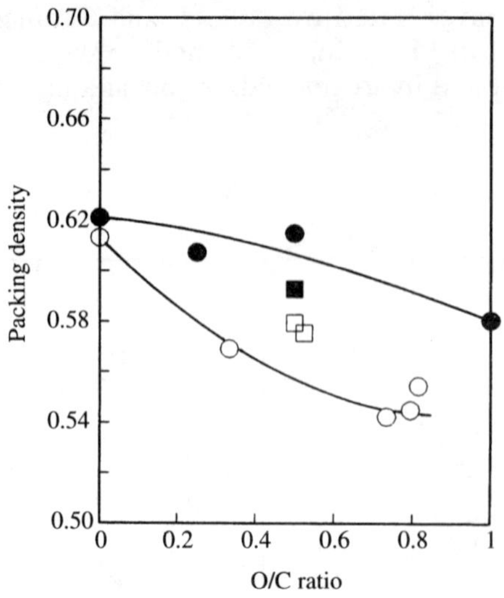

Figure 9 Packing density as a function of the oxygen to carbon ratio O/C for linear (○) and (□) branched PFPEs (open points) and hydrogenated polyoxides (filled points).

Table 3 Group Contributions at 298 K to Volume, Vaporization Energy and Glass Transition Temperature.

	V_i (mL mol^{-1})		ΔE_i° (kJ mol^{-1})		Y_{gi} (K g mol^{-1})	
Group	*X* = *F*	*H*	*F*	*H*	*F*	*H*
$—CX_2—$	25.51	16.45	3.65	4.81	10 250	2700
$—CX(CX_3)—$	49.9	32.70	6.40	10.60	27 600	8000
—O—	11.0	8.50	−0.47	4.80	−1950	4000

9.4.2 Vaporization Energy and Cohesive Energy Density

The important thermodynamic parameter ΔH_v (vaporization enthalpy) of PFPE fluids has been investigated,[71] and extensive literature data on low molecular weight perfluoroalkanes and ethers have been reported. Comparison with the corresponding hydrogenated compounds has also been given. The composition and chain length dependence of ΔH_v^{298} is plotted in Figure 10, where it is seen that the linear dependence of ΔH_v on chain length *N*, typical of perfluoroalkanes, is severely changed when the O/C (oxygen to carbon ratio) of the chain is progressively increased. It is worth noting that the majority of the data in Figure 10 have experimental origin, only a third of them being obtained from the boiling point T_b by using a Hildebrand-type $\Delta H_v^{298} - T_b$ equation.[71]

Strictly related to ΔH_v^{298} is the vaporization energy ΔE^0, which is given, as reported by the same authors,[71] by an 'additivity' equation, *i.e.* through a simple group contribution method. As an example, for Fomblin Z fluids the equation is

$$\Delta E^0 = 2\Delta E_e^0(—CF_3) + X_1 \Delta E^0(—CF_2—) + X_2 \Delta E^0(—O—) \qquad (16)$$

where X_1 and X_2 are the number of CF_2 and O groups and ΔE_e^0 is the CF_3 chain end contribution to the molecular ΔE^0. The data obtained for linear PFPE fluids at 298 K are shown in Table 3.

Of interest appears to be the negative (albeit small) ΔE^0 contribution of the oxygen atoms, which in hydrogenated polyethers is positive and has the value $\Delta E^0(—O—) = 4.80 \pm 0.28$ kJ mol^{-1} (a value not too different from that of the $—CH_2—$ and $—CF_2—$ groups).[69]

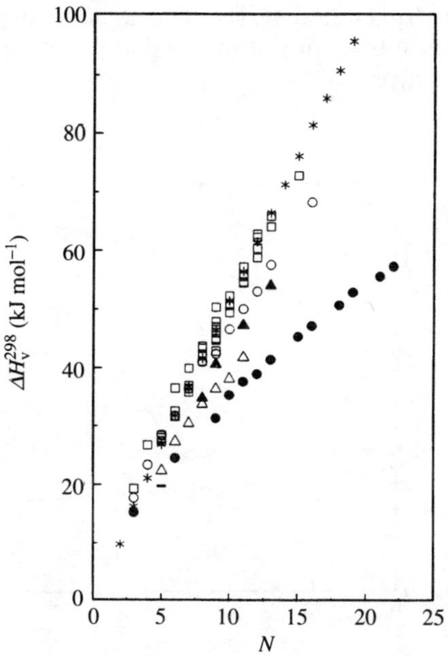

Figure 10 Standard vaporization enthalpy (ΔH_v^{298}) as a function of the chain atoms number N for 96 linear molecules: (∗) paraffins, (□) ethers, (○) perfluoroparaffins and perfluoroethers having an O/C ratio of 0.11 (▲), 0.25 (△), 0.44 (●), 0.66 (–); 52 points are experimental and 34 are calculated from the $\Delta H_v^{298} - T_b$ relationship (reproduced by permission of the American Chemical Society from *Macromolecules,* 1993, **26,** 1751).

In perfluorinated molecules the contribution of ethereal oxygen to intermolecular interactions is small (and negative), a situation that can be interpreted on the basis of the structure work of Pacansky *et al.*[46] and that can, in part, explain the peculiar physical properties of PFPE fluids, when they are compared with the corresponding fluoroalkanes and hydrogenated ethers.[71]

From both experimental data and group contribution analysis, the Cohesive Energy Density (CED) defined as

$$\mathrm{CED} = \Delta E^0/V \tag{17}$$

was calculated for linear PFPEs;[71] numerical data for CED and its square root, the so-called solubility parameter δ, are collected in Table 4. In the same table the value of δ found by Sianesi[47] for Fomblin Y is also reported.

Table 4 Solubility Parameter of Different PFPE Structures.

Structure	*O/C*	*CED* (J mL^{-1})	δ (J mL^{-1})$^{1/2}$	*Ref.*
Demnum	0.33	135	11.6	71
Fomblin Z ($p/q = 0.54$)	0.83	121–142	11–11.9	71
Polyperfluoro(oxymethylene)	1.0	108–146	10.4–12.1	71
Fomblin Y	0.5[a]	—	11.2	47

[a] Calculated on the basis of backbone atom number.

It is worth noting that the solubility parameter is very low (one of the lowest known), that it is in practice independent of the structure of the polymeric chain, and that it is always lower than the value of 12.1 found for polytetrafluoroethylene.[69,71] The effect of the structure on CED is shown in Figure 11, where CED is plotted against the O/C ratio; comparison is also made with the homologous hydrogenated series. The introduction of oxygen in the backbone has an opposite effect in the sense that for perfluorinated polyethers CED is slightly decreasing, while it strongly increases for polyoxides.

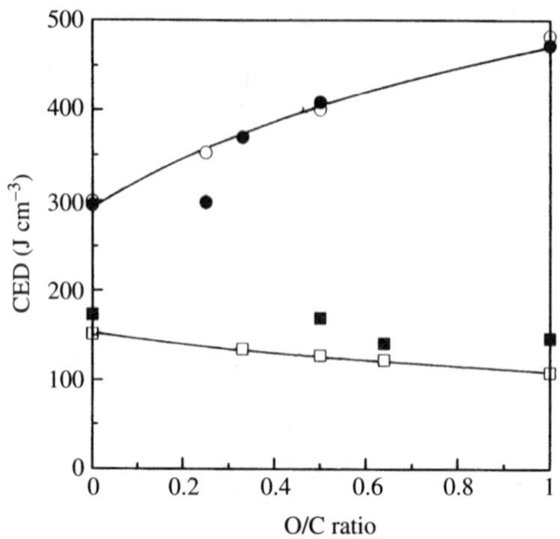

Figure 11 CED dependence on the oxygen to carbon ratio (O/C) for hydrogenated (○, ●) and perfluorinated (□, ■) polymers of high molecular weight. Open symbols: calculated by the group contribution. Full symbols: experimental data (reproduced by permission of the American Chemical Society from *Macromolecules,* 1993, **26,** 1751).

9.4.3 Specific Heat

Specific heat data at constant pressure and at 298 K are collected in Table 5 for some PFPE structures; the molar heat capacities C_p are first of all compared with values calculated by group contribution data reported by van Krevelen[69] and by Wunderlich and co-workers;[72,73] it appears that data reported by the last authors is, on average, more apt to predict the molar heat capacities of PFPEs. In the last column of the same table the ratio between the experimental molar heat capacities and the number of atoms n in the monomeric unit (or in the average monomeric unit for copolymers) is also shown; the ratio is close to $2R$ for both PFPEs and PTFE, while van Krevelen reports that for liquid hydrocarbon polymers this ratio is close to R; another difference lies in the fact that while the C_p/n ratio seems to be only slightly dependent on the O/C ratio, data of Wunderlich and co-workers on liquid hydrogenated polyoxides[73] shows a sensible increase of C_p/n with increasing the ethereal group concentration.

Table 5 Specific, Molar, and Reduced Heat Capacities for Some PFPEs and Amorphous Polytetrafluoroethylene (PTFE): Comparison with Values Calculated[69,72,73] from Group Contributions.

			C_p (J mol^{-1} K^{-1})				
Polymer	*M*	c_p (J g^{-1}K^{-1})	*Exp.*	*Calc*[72,73]	*Calc.*[69]	C_p/nR (atom^{-1})	*Ref.*
Fomblin Y25	3 260	1.00	160.4	174.5	178.9	1.99	57
Fomblin Z25	9 520	1.01	88.4	97.8	104.8	2.05	57
Demnum S200	7 050	1.04	172.4	179.8	183.9	2.07	57
Krytox	—	0.96	159.4	179.8	183.9	1.92	50
PTFE	>10^6	0.96	96.0	102.5	99.0	1.92	69

The temperature dependence of the specific heat c_p is reported in a recent review paper of Del Pesco[50] for Krytox; c_p increases linearly with temperature, as found for several polymeric liquids by Wunderlich and co-workers[74,75] and by van Krevelen.[69]

9.4.4 Solution Properties

The molecular properties in solutions of PFPEs have been investigated practically by means of all available techniques: light scattering, vapour pressure osmometry, intrinsic viscosity and gel permeation chromatography (GPC). However, the low solubility parameter and the perfluorinated nature do not allow the use of common laboratory solvents (Table 6). The solubilities of various gases in PFPEs are given in Table 7. Known solvents are trichlorotrifluoroethane (CFC 113) and its mixtures with acetone and perfluoroalkanes.

Table 6 Solubility of Common Solvents and PFPE Oils.[a]

Solvent	δ^b $(kJ\,g^{-1})^{1/2}$	*Solubility of PFPE in solvent*	*Solubility of solvent in PFPE*	*Solvent (g in 100 g PFPE)*
n-Hexane	14.9	SS	PS	2.0
n-Heptane	15.1	PS	SS	1.6
n-Octane	15.6	I	I	—
Cyclohexane	16.8	I	I	—
Benzene	18.8	I	SS	1.3
Toluene	18.2	I	I	—
Xylene	18.0	I	I	—
Ethyl ether	15.1	SS	PS	2.6
Tetrahydrofuran	18.6	I	SS	1.4
Ethyl acetate	18.6	I	I	1.5
Methyl formate	20.9	I	I	—
Dimethyl ketone	20.3	I	I	—
Methyl ethyl ketone	19.0	I	I	—
Trichloroethylene	18.8	I	PS	3.0
Monochlorobenzene	19.4	I	I	—
Chloroform	19.4	SS	PS	4.1
Carbon tetrachloride	17.6	SS	PS	2.9
1,1,2-Trichloro-1,2,2-trifluoroethane	14.8	Miscible	Miscible	—
Perfluorohexane	12.0	Miscible	Miscible	—
Perfluoroheptane	12.3	Miscible	Miscible	—
Perfluorooctane	12.5	Miscible	Miscible	—

[a] PS, partially soluble (3–5% by weight); SS, slightly soluble (1–2% by weight); I, practically insoluble (<1% by weight).
[b] Ref. 76.

Table 7 Solubilities of Various Gases[57] in PFPE at 296 K.

Gas	*Bunsen coefficient*[a] *Fomblin Y25*	*Fomblin Z25*
He	0.059	0.061
H_2	0.186	0.100
N_2	0.189	0.232
O_2	0.296	0.359
CO_2	1.310	1.500
F_2	0.197	—
CF_4	0.658	—
C_2F_6	2.120	—
SF_6	3.510	4.290
$C_2H_2ClF_3$	9.997	12.555
$CHCl_2F$	9.602	12.696
$CHClF_2$	3.590	4.325

[a] Bunsen coefficient is defined as the volume of gas, reduced to 0 °C and 760 Torr, which is absorbed by the unit volume of solvent (at the temperature of the measurement) under a gas pressure of 760 Torr.

The weight average molecular weight M_w of PFPEs was determined in CFC 113 by means of low-angle laser light scattering (LALLS) on practically all the known structures.[49,77–79] The measurements are rather difficult, the difference in refractive index between solvent and solute being small. Literature data are summarized in Table 8. It is worthy to observe that according to Sanguineti *et al.*[79] there is no influence of structure on (dn/dc) for Fomblin Z with different O/C or p/q ratios. From dynamic light scattering measurements the hydrodynamic radius R_h was determined as a function of the molecular weight and the following relationship was found[79]

$$R_h = 1.74 \times 10^{-2} M^{0.5} \tag{18}$$

The above plot is shown in Figure 12.

Table 8 Refractive Index Increments for PFPEs.

Polymer	*Wavelength* (nm)	dn/dc (mL g^{-1})	*M range* (g mol^{-1})	*Ref.*
Krytox	—	−0.04	1600–5200	49
Fomblin Y	—	$-0.037 - 12.9/M_n$	1000–9000	77
Fomblin Z	546	-0.046 ± 0.003	9000–380 000	79

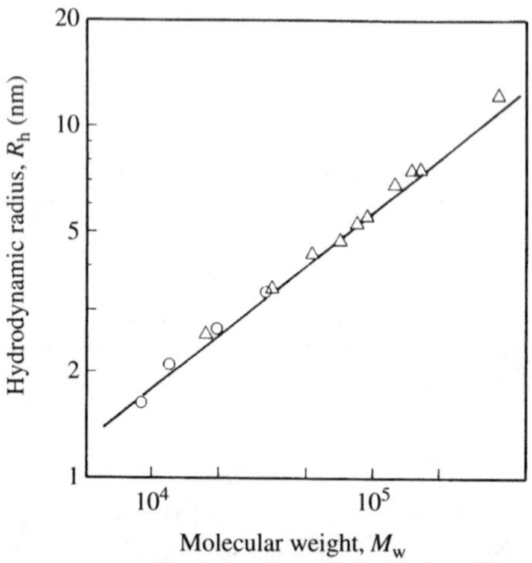

Figure 12 Hydrodynamic radius as a function of the weight average molecular weight for Fomblin Z fractionated samples with $p/q = 0.54$ (○) and $p/q = 1.27$ (△).

Number average molecular weights have been determined by vapour pressure osmometry in CFC 113 for molecular weights lower than 20 000, *i.e.* over all the molecular weight range of practical interest for the majority of applications.[49,51,52,79] The LALLS and osmometry data so obtained have been used to calibrate the GPC apparatus. Ouano *et al.*[49] and Cantow *et al.*[77] used CFC 113 as solvent and porous silica as support, while Marchionni *et al.*,[51,52] Sanguineti *et al.*[79] and Cotts[78] used successfully a PL-GEL styrene–divinylbenzene column set. The universal GPC calibration method was also used by Cantow *et al.*[77] and Cotts[78] to determine the root-mean-square end-to-end distance $\langle r^2 \rangle^{1/2}$.

Intrinsic viscosity measurements carried out at room temperature were also used by Cantow *et al.*[77] and Sanguineti *et al.*[79] to gain more fundamental information about the chain conformation. Firstly, the coefficients of the Mark–Houwink–Sakurada relationship $[\eta] = KM^a$ have been determined (Table 9). Data[77,79] show that, for PFPEs, CFC 113 is a theta solvent, while the perfluorinated paraffin is a good solvent;[79] it seems also, on the basis of the same references, that the K constant is practically independent of the chain structure, at least for the above linear polymers and copolymers. The fact that CFC 113 is a theta solvent is also supported by the second virial coefficient measured from light scattering data; Cantow *et al.*[77] found values (derived from A_2) of about 0.5 for a 10^4 molecular weight Fomblin Y; Sanguineti *et al.*,[79] who investigated much higher molecular weight samples of Z copolymers, obtained a value of 0.49. On the other hand, the value for perfluoroheptane ranges between 0.17 and 0.28, depending on the value of a constant used in the calculation of the Stockmayer–Fixman–Burchardt equation.[80,81] The different behaviour of the linear PFPEs in the two solvents is better illustrated in Figure 13, where the ratio $[\eta]/M_w^{1/2}$ is plotted against $M_w^{1/2}$. The fact that data in both solvents extrapolate to a common K value of $(3.8 \pm 0.3) \times 10^{-2}$ indicates that specific solvent effects are negligible, so that K_Θ can be safely used to obtain structural information on the chain configuration of PFPEs in solution. Therefore the characteristic ratio C_∞ defined as

$$C_\infty = (\langle r_0^2 \rangle / M)/(\langle l^2 \rangle / M_0) \tag{19}$$

can be calculated. In equation (19), $\langle r_0^2 \rangle$ is the unperturbed root-mean-square end-to-end distance calculated from the relationship:

$$\langle r_0^2 \rangle / M = (K_\Theta / \Psi_0)^{2/3} \tag{20}$$

M being the molecular weight and Ψ_0 a universal constant $= (2.6 \pm 0.1) \times 10^{21} \cdot M_0$ and $\langle l^2 \rangle$ are the average molecular weight and the average quadratic bond length of the structural unit, respectively. C_∞ values calculated for both Fomblin Z and Demnum are reported in Table 10, where they are compared with literature data for other polymers. As shown in this table, PFPEs appear to be very flexible polymers, and their flexibility compares well with those of some polyoxides and polydimethylsiloxanes. The introduction in the perfluorinated backbone of polytetrafluoroethylene ($C_\infty = 7$–9) of an ethereal oxygen greatly increases the flexibility.

Table 9 Intrinsic Viscosity and Molecular Weight: Mark–Houwink–Sakurada Coefficients.

Polymer	*M range* (g mol^{-1})	*T* (°C)	*Solvent*	$K \times 10^2$ (mL g^{-1})	*a*	*Ref.*
Fomblin Y	1100–9000	25	CFC 113	4.8	0.5	77
Fomblin Z	7500–25 000	25	CFC 113	4.0	0.5	77
Fomblin Z	9000–380 000	20	CFC 113	3.8 ± 0.3	0.51 ± 0.03	79
Fomblin Z	9000–380 000	20	f-heptane	0.95 ± 0.02	0.68 ± 0.02	79
Fomblin Z	medium–low[a]	27	CFC 113	0.75	0.69	78
Demnum	5000–9000	20	CFC 113	3.8	0.51	79
Demnum	5000–9000	20	f-heptane	0.95	0.68	79
Demnum	medium–low[a]	27	CFC 113	1.01	0.63	78
PF-PEO[b]	medium–low[a]	27	CFC 113	2.0	0.56	78

[a] Not reported. [b] PF-PEO = polyperfluoro(ethylene oxide).

9.5 TRANSITIONS AND RELAXATION PROCESSES

Commercial PFPE polymers do not crystallize at ambient pressure when cooled down to low temperatures, and the glass transition temperature T_g is the major transition observed. It has been reported by Alper *et al.*,[83] however, that a Krytox sample crystallizes at low temperature at a pressure of 200 MPa. It is not surprising that disordered structures such as those of Z copolymers and K terpolymers would not crystallize, and that a side-substituted chain like that of Krytox, which is presumably atactic from a steric point of view, crystallizes only under particular conditions. On the other hand, it is known since the early 1960s that a regular homopolymeric structure such as that of perfluoro(ethylene oxide) appears to be able to crystallize;[2] in some way it is therefore surprising that the regular structure of perfluoro(trimethylene oxide) does not, when it is well known that the hydrogenated homologue poly(trimethylene oxide) is a crystalline material which melts at 308 K.[84,85] Up to now a reasonable interpretation of the different behaviour is not found in the literature, to the best of the present authors' knowledge.

The following section is devoted to the transitions occurring in the amorphous state.

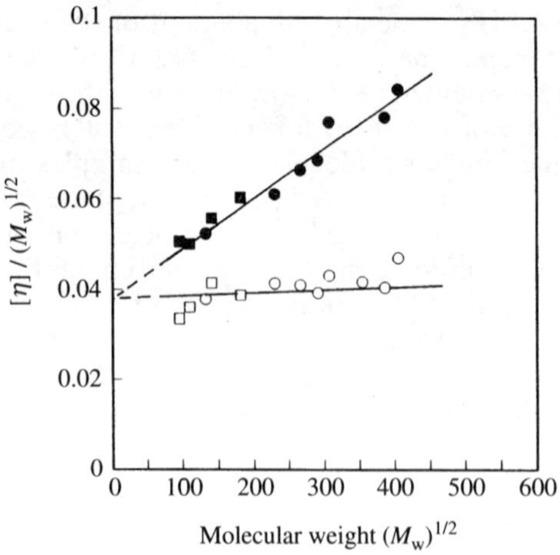

Figure 13 The Stockmayer–Fixman–Burchardt plot for Fomblin Z fractionated samples with $p/q = 0.54$ (○) and $p/q = 1.27$ (□) in 1,1,2-trichloro-1,2,2-trifluoroethane (open points) and in perfluoroheptane (filled points).

Table 10 Characteristic Ratio C_∞ Data from Intrinsic Viscosity.

Polymer	C_∞	*Ref.*
Poly(tetrafluoroethylene)	7–9	82
Demnum	5.3 ± 0.5	79
Demnum	5.4	78
Polyperfluoro(ethylene oxide)	4.9	78
Fomblin Z	4.8 ± 0.5	79
Fomblin Z	7.5	78
Polyethylene	5.2–10.3	76
Poly(trimethylene oxide)	3.9–4.2	76
Poly(ethylene oxide)	3.8–4.8	76
Poly(methylene oxide)	10.5	76
Poly(dimethylsiloxane)	5.2–7.6	76

9.5.1 Glass Transition

The glass transition T_g of PFPE fluids of the Y type, having different molecular weights, was first reported by Sianesi *et al.*,[48] who demonstrated that T_g depends on the number-average molecular weight M_n according to the well known equation

$$T_g = T_g^\infty - K_g/M_n \tag{21}$$

where T_g^∞ is the T_g of a chain of infinite length. Similar data on both Y and Z oils were reported by Marchionni *et al.*,[51] who also investigated Krytox and Demnum samples[52] and Fomblin K oils.[54]

The molecular weight influence on T_g is shown in Figure 14, and it is given for all types of PFPE oils by the simple equation (21). In Table 11 are collected the relevant T_g^∞ and K_g data for all series investigated. The general empirical relationship between T_g and K_g for different polymers given by Boyer[86]

$$K_g = 3.9 \times 10^{-11}(T_g^\infty)^{4.4} \tag{22}$$

is applicable to the above data.

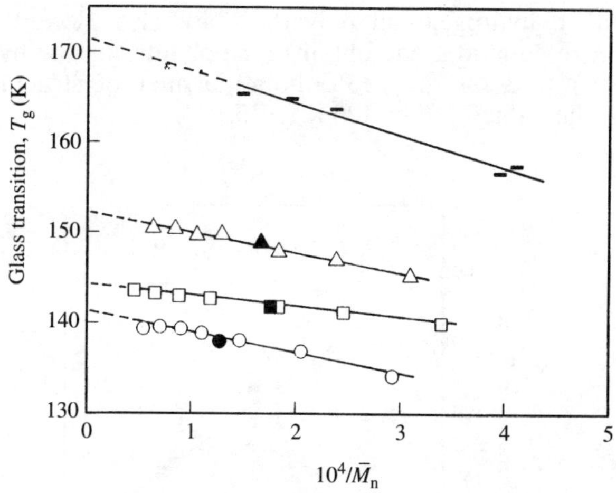

Figure 14 Glass transition temperature for fractions of Fomblin Z copolymers with different p/q ratios (○) 0.54, (□) 0.69, (△) 1.15 and for unfractionated Demnum samples (–) as a function of recpirocal M_n; filled symbols: unfractionated copolymers (reproduced by permission of the Society of Plastic Engineers from *Polym. Eng. Sci.*, 1990, **10,** 829).

Table 11 T_g^∞ and K_g Data for Different Series of PFPE Oils.

Series	T_g^∞ (K)	$10^4\ K_g$ (K)	*Ref.*
Fomblin Y	216	5.0	48
Fomblin Y	216	5.4	52
Fomblin Z	142	1.5	52
Fomblin K	199	3.6	54
Demnum	172	3.3	52
Krytox	219	5.5	52
Fomblin Z DOL[a]	150	−1.8	53
Fomblin Z ACF[b]	150	−0.7	53

[a] $—CH_2OH$ end groups. [b] $—CH_2OCOCF_3$ end groups.

The T_g^∞ data have been used by Marchionni *et al.*[52] to obtain, by using a group contribution method, the relevant relationship between the chemical structure of the repeating unit and the glass transition temperature (at infinite chain length). The group contribution equation to be used has been proposed by Van Krevelen[69] and is

$$T_g^\infty = (\sum Y_{gi})/M \tag{23}$$

where M is the molar weight of the repeating units, and Y_{gi} the relative characteristic transitional parameter of the groups. The numerical results obtained for PFPE polymers are reported, together with other data, in Table 3. It has to be noted that the oxygen group contribution to T_g^∞ is again negative (albeit small). For comparison, the oxygen Y_g group contribution for a hydrogenated polymer is reported to be $Y_g(—O—) = 4000\ \mathrm{K\ g\ mol^{-1}}$ by Van Krevelen.[69] From these data one can calculate, on the basis of equation (23), the T_g^∞ of any type of PFPE oil composed of the above groups.

The T_g^∞ dependence on chemical composition of the homopolymers and copolymers in which the side chain group CF_3 is absent has been directly related to the overall oxygen content by Marchionni *et al.*,[54] who obtained the equation

$$T_g^\infty = 200 - 80(\mathrm{O/C})\ \mathrm{K} \tag{24}$$

where (O/C) is the ratio between oxygen and carbon atoms content. The data are plotted in

Figure 15, where data on hydrogenated polyethers are also plotted for a comparison. By extrapolation of the above equation, one obtains for polytetrafluoroethylene (O/C = 0) the T_g^∞ value of 200 K (−73 °C), and for the PFPE homopolymer of structure (—CF_2O—), never synthesized up to now, the value of $T_g^\infty = 120$ K (−153 °C).

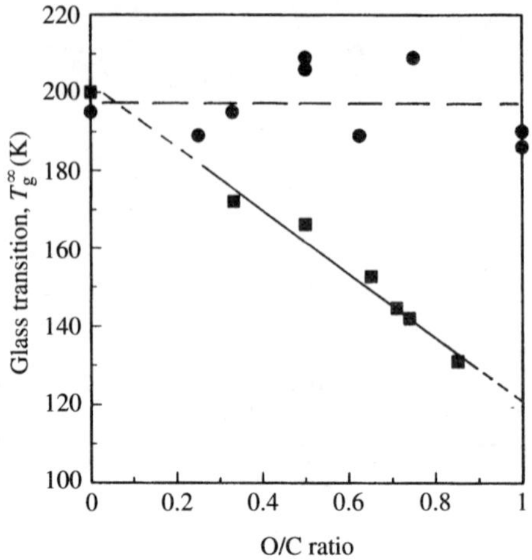

Figure 15 Composition dependence of the glass transition temperature T_g^∞ for two series of polyethers: (●) hydrogenated polyethers, copolyethers and polyethylene; (■) perfluorinated polyethers, copolyethers and polytetrafluoroethylene (reproduced by permission of the American Chemical Society from *Macromolecules,* 1993, **26,** 1751).

When the terminal groups are different from the normal perfluorinated ones, and in particular when they are ionic in character, the glass transition temperature is strongly changed, as shown by Danusso *et al.*[53,87] These authors have studies samples having the Z structure of various chain length, the chain ends being —CH_2OCH_3, —$CH_2OSi(CH_3)_2$, —CH_2OCOCF_3, —CH_2OH and —CH_2OK. The T_g obtained for the above samples are plotted as a function of the number of chain atoms X_M (number average degree of polymerization) in Figure 16. The T_g–X_M dependence is seen to be strongly affected by the nature of the chain ends, as expected. Moreover, while the T_g–M_n equation (21) holds also for functional PFPEs, positive instead of negative slopes are observed[53] for bifunctional Fomblin Z having —CH_2OH and —CH_2-$OCOCF_3$ end groups. According to the above authors, the polarity of chain ends and their mutual interaction is such that the intrinsic mobility of the backbone is gradually restricted by lowering the molecular weight. The negative K_g values found for the mentioned series are reported in Table 11.

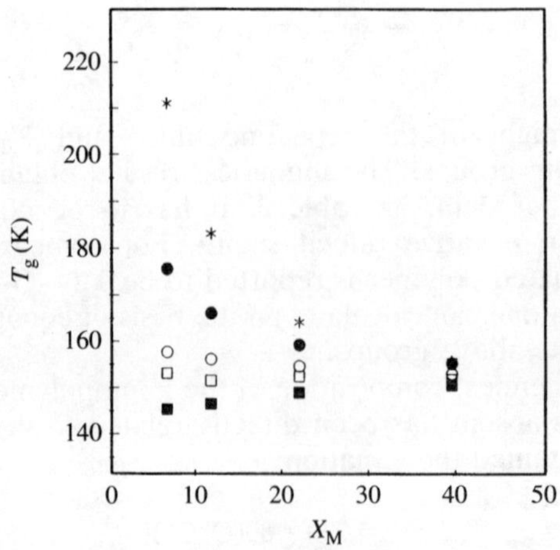

Figure 16 Relation between T_g and the degree of polymerization X_M for bifunctional Z copolymers: (●) —CH_2OH, (■) —CH_2OMe, (□) —CH_2OSiMe_3, (○) —CH_2OCOCF_3, and (∗) —CH_2OK end groups (reproduced by permission of Butterworth-Heinemann from *Polymer,* 1993, **34,** 3687).

The dependence of T_g^∞ on the O/C ratio has been recently analyzed by Marchionni *et al.*[71] on the basis of cohesive energy density (CED) (equation 17). The relationship between T_g^∞ and CED for selected linear Z-type homopolymers and copolymers is shown in Figure 17. In the figure, data on hydrogenated polyethers are also shown, to stress the difference between perfluorinated and hydrogenated polymers. It is seen that whereas oxygen atoms increase the CED of hydrogenated ethers but does not change T_g, the reverse is true for perfluorinated ones, in the sense that CED decreases together with T_g as O/C increases. The oxygen contribution to both T_g and CED is therefore found to be negative in fluorine surroundings, but positive in hydrogenated ones.[71]

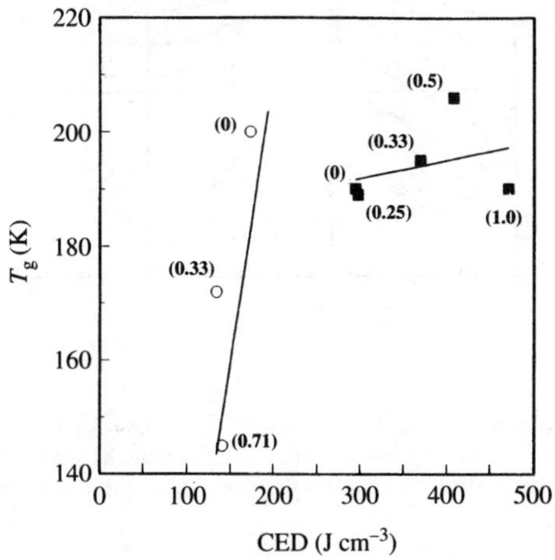

Figure 17 T_g^∞ as a function of CED for perfluoropolyethers (○) and hydrogenated polyoxides (■), having various O/C ratios.

9.5.2 Relaxation Processes

The dynamic-mechanical properties of several perfluoropolyether fluids have been studied as a function of molecular weight and structure by Ajroldi *et al.*[88] and an example is shown in Figure 18. Measurements were carried out at a frequency of about 1 Hz by means of a free oscillation torsion pendulum, between −180 °C and room temperature, with the inert support technique. Three relaxation processes are present.

The first, which occurs at very low temperatures, is related to local motions of short backbone chain segments; the transition temperature is dependent upon structure, being higher for perfluoromethyl-branched polymers, but practically independent of chain length (the minimum chain length investigated being about 55 backbone chain atoms).

The second relaxation is related to the glass transition, and depends strongly upon both molecular weight and structure; the CF_3 side group increases the transition temperature with respect to that of the linear chains, at a constant oxygen-to-carbon ratio, while an increase of the O/C ratio, at constant molecular weight and composition, lowers the transition. These results agree well with those relative to the T_g obtained by calorimetric methods and discussed before.

Finally, the third relaxation process, which occurs at temperatures higher than T_g, seems to be typical of an isoviscous state.[89]

A low-temperature noncooperative motion was found by Starkweather *et al.*[90] for the Krytox branched structure, by measuring the dielectric behaviour; the relaxation maximum lies at about 100 K at a frequency of 1 kHz; its activation energy was found to be 36.6 kJ mol^{-1}. This relaxation process was tentatively related to the out-of-chain motion of the side group; as a matter of fact, similar relaxation maxima have been found by the same authors in the same temperature range and at the same frequency for copolymers of tetrafluoroethylene with

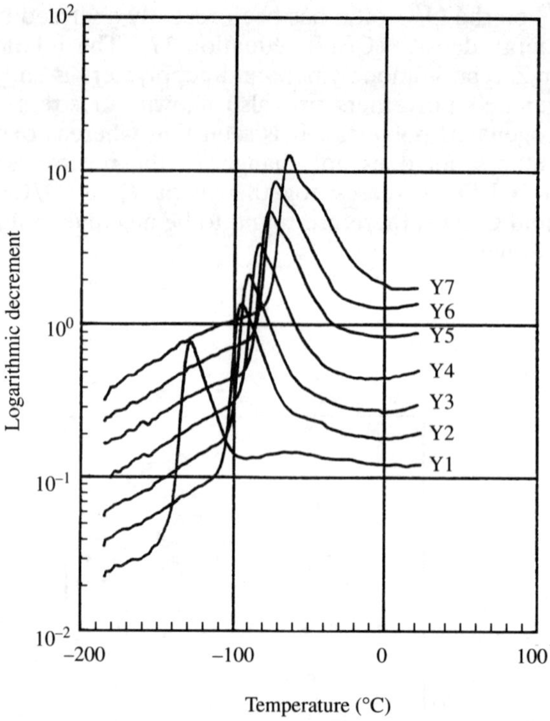

Figure 18 The influence of molecular weight on the mechanical damping for Fomblin Y copolymers (M_n = 750, 1110, 1200, 1800, 2470, 3260 and 6530 from Y1 to Y7). Curves are shifted vertically and progressively by 0.2 decade for clarity (reproduced by permission of The Institute of Materials from *Plast. Rubber Comp. Proc. Appl.*, 1992, **17**, 307).

hexafluoropropene (side group —CF_3) and with perfluoro(propyl vinyl ether) (side group —$OCF_2CF_2CF_3$).[91]

The dynamic mechanical analysis was also used to determine the glass transition temperature of mixtures of Y and Z fractions having practically the same bulk viscosity of about 0.45 Pa s at room temperature, and T_g values of 195 and 142 K, respectively.[92] The PFPE fluids of different structure were miscible and the composition dependence of T_g is shown in Figure 19; the experimental points can be fitted by the well known Fox equation.

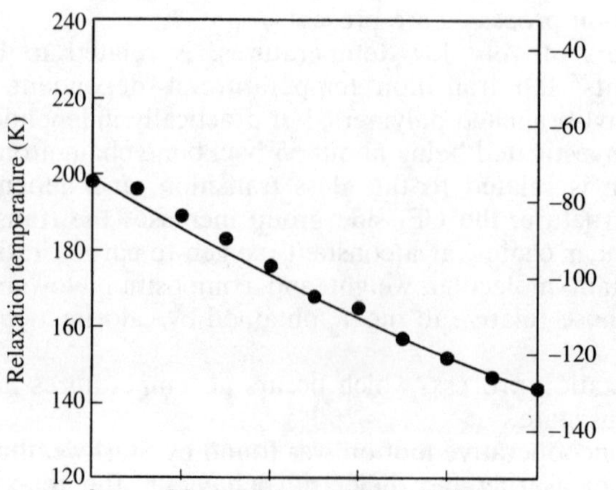

Figure 19 Composition dependence of the glass transition temperature T_g of mixtures of Fomblin Z and Y copolymers. The continuous line is the Fox equation calculated on the basis of the T_g values of pure components.

9.6 FLOW PROPERTIES

For macromolecular chains in the liquid state, *i.e.* for polymeric melts, the most important flow characteristics are both the elasticity and the fact that the viscosity η is a function not only of the state variables temperatures T and pressure P (as typical of any fluid), but also of the so-called shear rate (or velocity gradient) $\dot{\gamma}$ and of the molecular weight M. For linear monodisperse polymeric samples, therefore, η is usually expressed as a function of four main variables T, P, $\dot{\gamma}$ and M

$$\eta = \mathrm{f}(T, P, \dot{\gamma}, M) \tag{25}$$

As far as the perfluoropolyether oils are concerned, the elastic behaviour does not play, under many conditions, a major role in their flow, and the shear rate dependence of η, at constant M, T and P, has been found to be very small, in the temperature and shear rate ranges of practical interest. Only at very low T and at high values of η does the viscosity decrease appreciably with increasing $\dot{\gamma}$, a manifestation of non-Newtonian behaviour.[93] For many practical purposes the PFPE oils can be therefore considered inelastic Newtonian fluids, *i.e.* fluids whose viscosity only depends on chemical structure, on (average) molecular weight, and on T and P.

Up to now it has not been experimentally assessed what average value of M has to be used for the polydisperse samples, but the practice derived from other polymeric fluids and the related theoretical models suggests the weight-average molecular weight M_w. Often the number-average value M_n, easier to be measured, is used. Since the molecular weight distribution (MWD) of commercial samples is substantially constant for all of them, the above practice does not introduce serious errors in the treatment of the experimental data.

9.6.1 Viscosity and Molecular Weight

The molecular weight dependence of η has been investigated by several groups, and extensive results have been published by the present authors.[51,52] Typical of linear macromolecular chains are isothermal and isobaric $\eta(M)$ curves which in a bilogarithmic plot, at low values of M, have a slope of about 1, and then bend to reach a slope of about 3.5 when M passes through a critical molecular weight, M_c. The physical meaning of the above transition is that below M_c the 'flow units' are represented by entire macromolecules, and above M_c by portions (segments) of the macromolecules, being the chains connected by 'entanglements'. M_c is consequently interpreted as the critical molecular weight for influencing steady flow viscosity by entanglement coupling, and its measure represents an essential step of the rheological investigation.

For most polymer melts the critical chain atoms length $N_c = n_0 M_c / M_0$ (where n_0 is the number of chain atoms in the polymer repeating unit and M_0 the repeating unit molecular weight) has been found to be in the range of values from 300 to 600, and in view of some peculiarities of the PFPE properties[51,52] it seems interesting to determine the N_c of these macromolecular oils. The task has been found to be rather complex, due to the fact that the samples that are obtained from the polymerization reaction have molecular weight M close to M_c, so that from the log η–log M plots it is difficult to locate precisely the slope change. This is shown for example in Figure 20, where different slopes are found for different oils.

Various workers have explored the subject. An estimated value of M_c, based on bulk viscosity data, has been given by Monfort *et al.*[94] as $M_c = 19\,000$ for the Fomblin Z oils. On the other hand, Ouano *et al.*[49] have reported the value $M_c = 3000$ for Krytox fluids, and Cantow *et al.*[77] have estimated for the same oils, from dilute solution data, the exceptionally high value of $M_c = 75\,000$. The first value estimated by Marchionni *et al.*[51] for Fomblin Y oils, which are chemically almost identical to the Krytox ones, was much lower ($M_c \geq 5000$). The dispersion of the above results indicates the complexity of the problem, which was solved[51,52,95] by using two different approaches. In the first,[95] M_c was evaluated following the method suggested by Lee and Williams,[96] according to which the dependence of T_g^{-1} on ln M shows a sudden bend at M_c.

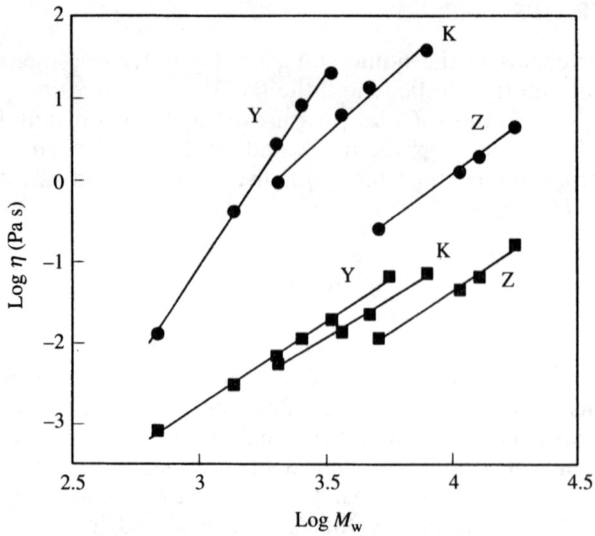

Figure 20 Relationship between zero shear viscosity and molecular weight M_w for Fomblin Y, K and Z ($p/q = 0.67$) oils measured at various temperatures: (●) 253 K; (■) 373 K (reproduced by permission of the Society of Automotive Engineers from SAE Paper #922289).

The method, applied to three series of fractionated Fomblin Z oils of slightly variable composition (see Figure 21), gives $M_c \approx 7000$, *i.e.* $N_c \approx 200$. The second method used by the present author,[51,52] based on the usual bilogarithmic η–M plot[97] (see Figure 22), gives a very close result, $M_c \approx 8000$–9000, *i.e.* $N_c \approx 230$.

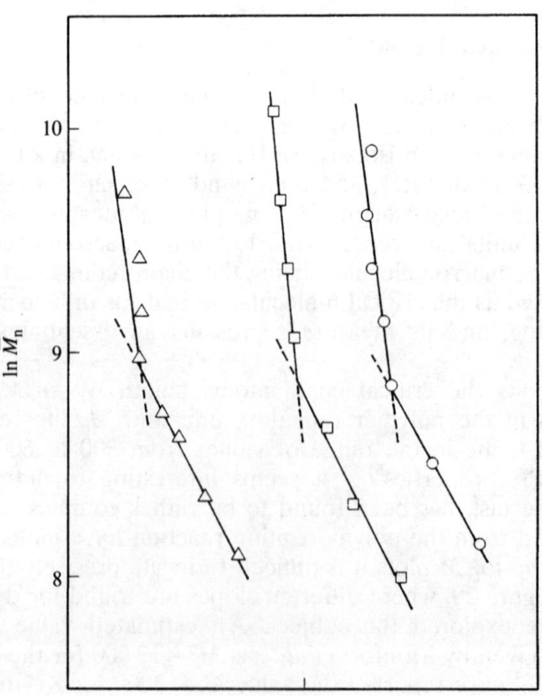

Figure 21 Lee–Williams plot for fractionated Fomblin Z copolymers: (○) $p/q = 0.54$, (□), $p/q = 0.69$ and (△) $p/q = 1.15$ (reproduced by permission of Butterworth-Heinemann from *Polym. Commun.*, 1991, **32,** 71).

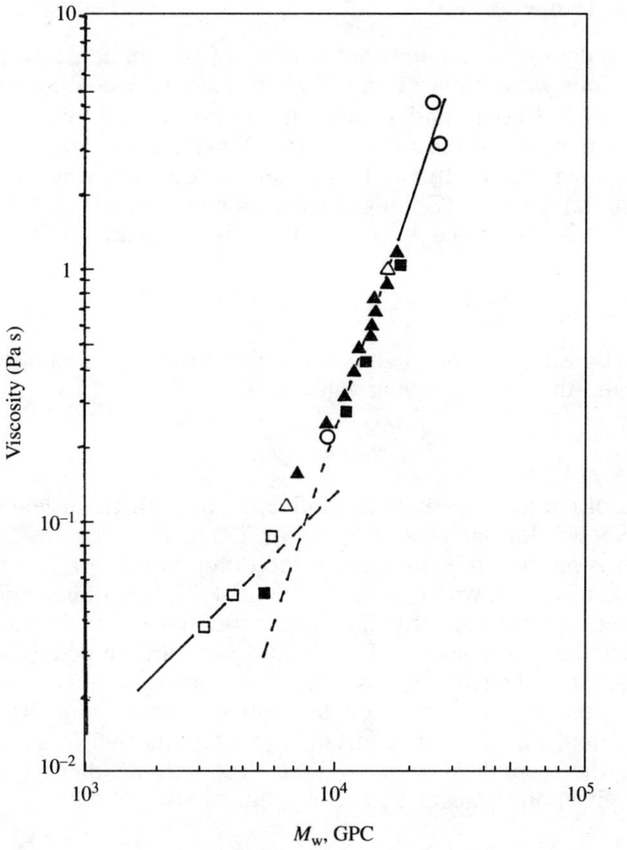

Figure 22 Relationship between zero shear viscosity and M_w for polymers and fractionated Z polymers at 293 K.

The M_c and N_c values for a variety of different polymeric melts are collected for comparison in Table 12. It is seen that the critical chain length values reported by Marchionni *et al.*[51,52,95] for the Fomblin Z fluids are close to those of polymer chains of similar structures, as the polyoxides and polydimethylsiloxane. If it is assumed that the value $N_c \approx 200$ of the Fomblin Z fluids can be applicable even to the similar chain structures of the Fomblin Y and Krytox oils, one finds that the value for them is $M_c \approx 11\,000$.

Table 12 Critical Molecular Weights M_c and M_e and Number of Chain Atoms N_c and N_e for Perfluoropolyethers and Related Polymers.

Polymer	M_c	M_e	N_c	N_e	*Method*	*Ref.*
Polyethylene	3800	—	—	270	viscosity	97
Polyethylene	—	1250	—	90	compliance	97
Poly(tetrafluoroethylene)	—	6600	—	132	modulus	98
Poly(tetrafluoroethylene)	—	7500	—	150	compliance	99
Poly(tetramethylene oxide)	2500	—	175	—	viscosity	100
Poly(ethylene oxide)	3500	—	240	—	viscosity	101
Poly(ethylene glycol)	4400	—	300	—	viscosity	97
Poly(dimethylsiloxane)	—	10 000	—	225	compliance	97
Fomblin Z	8000	—	230	100	viscosity	51, 52
Fomblin Z	7000	—	200	—	T_g	95

9.6.2 Viscosity and Temperature

The first extensive data on the important subject of the influence of temperature T on the viscosity η of PFPE oils were published in 1971 by Sianesi *et al.*[48] Successively, several other workers have reported $\eta(T)$ data under the form of curves or of tables.[48,49,51,52,55,56,102]

All normal polymers have, in the molten state, Newtonian viscosities whose T dependence obeys, at sufficiently high temperatures, the Arrhenius equation, whereas in the T range closer to the glass transition temperature T_g (in particular between T_g and $T_g + 100\,°C$), the Newtonian viscosity data have to be described by the well-known Williams, Landel and Ferry (WLF)[97] equation

$$\ln \eta_r = C_1(T - T_g)/(C_2 + T - T_g) \tag{26}$$

where the relative viscosity η_r is defined as the ratio between the value of $\eta\rho T$ of the fluid at the temperature T and the corresponding value at T_g

$$\eta_r = \eta\rho T / \eta_g \rho_g T_g \tag{27}$$

Typical examples of curves that show the influence of both chain length and temperature T on viscosity η are shown, for samples of Fomblin Y and Z, in Figure 23 where, according to equation (26), log η is plotted as a function of the difference $T - T_g$. It is clear that the curves belong to a unique family, and, whereas an Arrhenius-type equation cannot be used to describe them, they are fitted satisfactorily by the WLF equation, the constants C_1 and C_2 being considered as adjustable parameters. Their values are determined for each oil by suitable regression methods, after which, by assuming the simple free volume treatment[97] to be applicable, the two constants can be used to evaluate other parameters, the so-called 'free volume constants' f_g and α_f. The first constant f_g represents the 'free volume fraction' of the polymer at T_g, and the second constant, α_f, the 'free volume thermal expansion coefficient' above T_g. They are given, in terms of C_1 and C_2, by

$$f_g = 0.43/C_1 \qquad \alpha_f = f_g/C_2 \tag{28}$$

The average values of C_1 and of C_2 have been reported as about 11 and about 60, respectively, for both the Z and Y series of samples,[48,51] although relatively large deviations from the averages are found for selected oils, the reasons for this being unclear. From the above data, one obtains the average value $f_g = 0.040$, which is substantially larger than the 'universal' value $f_g = 0.025$, applicable to almost all polymeric species[97] (with the exception of polydimethylsiloxane and of *cis*-polybutadiene, which also show high values of f_g). As far as the free volume expansion coefficients α_f are concerned, the average value for the Y and Z oils is about $6.5 \times 10^{-4}\,°C^{-1}$, again larger than the universal value of $4.8 \times 10^{-4}\,°C^{-1}$ found to be applicable to most polymeric systems.[97]

Above approximately $T_g + 100\,°C$, as mentioned before, the temperature dependence of the Newtonian viscosity of polymeric melts can be described by the simple Arrhenius equation

$$\ln \eta = A + E_a/RT \tag{29}$$

where E_a, the so-called 'flow activation energy', has the physical meaning of a temperature coefficient of viscosity. In principle, E_a could be calculated by suitable differentiation of the WLF equation, which however cannot be applied confidently to high temperatures; since PFPE oils are often used as high T lubricants, the actual values of the above mentioned C_1 and C_2 constants, derived from η data at relatively low T, are in practice of limited utility as a source of reliable temperature coefficients, and there is therefore the need to obtain E_a directly from η measurements carried out in the temperature ranges of practical interest. E_a data on PFPE Fomblin samples have been reported in the literature by Cantow *et al.*[102] and on Demnum oils by Tanaka *et al.*[55]

By carrying out measurements of η at conditions close and above room temperature, it is found that E_a increases slowly with chain length, levelling off to a constant value when the number N of backbone atoms is larger than about 40–50. This is shown in Figure 24 for the Z oils and in Figure 25 for the Y oils.[54] It is interesting to compare the above data with the curves, plotted in the same figures, of the corresponding enthalpy of vaporization ΔH_v which, as expected, increases almost linearly with N, as already found for the two other systems

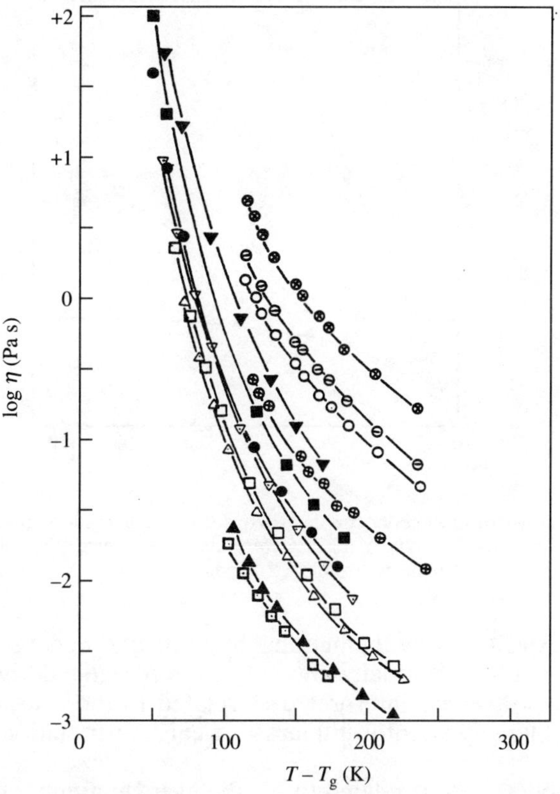

Figure 23 Dependence of the zero shear viscosity on the reduced temperature $T - T_g$ for (□, ⊕, ○, ⊗, ⊖) Z and (△, ▲, □, ▽, ●, ■, ▼) Y copolymers.

investigated, normal paraffins[103] and polydimethylsiloxanes.[104] These data, together with the results shown in the figures, appear to be perhaps the evidence of the transition from a substantially molecular flow mechanism to a segmental one with increasing the length of linear polymeric chains.

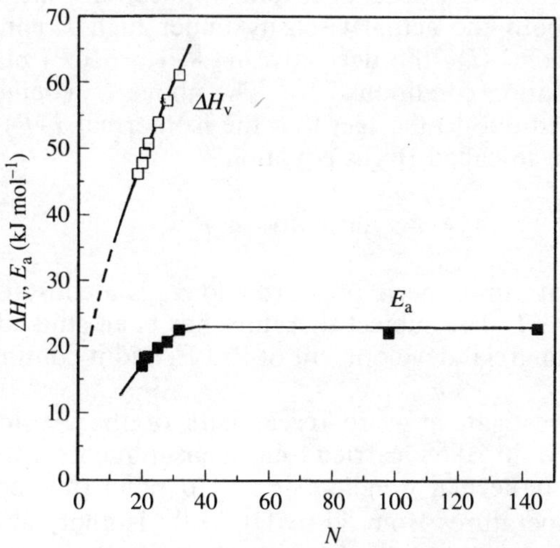

Figure 24 Dependence of the activation energy for viscous flow E_a and the vaporization enthalpy ΔH_v on the number of chain atoms for Fomblin Z PFPEs (reproduced by permission of the Society of Automotive Engineers from SAE Paper #922289).

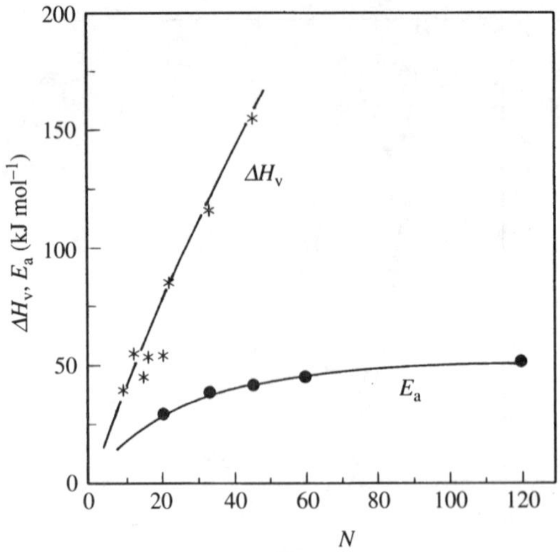

Figure 25 Dependence of the activation energy for the viscous flow E_a and the vaporization enthalpy ΔH_v on the number of backbone chain atoms for Fomblin Y PFPEs (reproduced by permission of the Society of Automotive Engineers from SAE Paper #922289).

There remains to be explained why the limiting flow activation energy E_a value is reached, in the above figures, at N values substantially lower than the N_c value derived from the isothermal $\eta(M)$ curves, since both values are interpreted as related to the transition from molecular to segmental flow. The problem apparently still lacks a clear interpretation in the scarce literature on the subject.

According to some models,[97] N_c is related to N_e, the average number of chain atoms between 'coupling loci' (*i.e.* between entanglements), by the approximate relation $N_c \approx 2N_e$. The values of N_e are found to be, for almost all polymers,[97] between 100 and 350 (Table 12). For the PFPE oils, $N_c \approx 200$ and if the critical N indicated by the curves of Figures 24 and 25, above which E_a remains constant, could be identified with N_e rather than with N_c (which appears not too meaningful), it would result that this experimental N_e is approximately 4 times, rather than 2 times, smaller than the N_c obtained from the $\eta(M)$ curves. This problem remains, up to now, unresolved.

9.6.3 Viscosity and Pressure

The pressure dependence of viscosity of perfluoroether oils, as well as that of other classes of lubricants, has obvious important aspects in several applications, including those of tribological relevance. In particular, both the actual viscosity under high P conditions and the so-called 'pressure–viscosity coefficient' (*i.e.* the derivative $\alpha_\eta = \mathrm{d}\ln\eta/\mathrm{d}P$) play important roles in the elastohydrodynamic lubrication conditions.[105,106] The above α_η coefficient (or a more complex form of it[106]) is widely used due to the fact that the isothermal $\eta(P)$ are often described, to a first approximation, by the so-called Barus equation[107]

$$\ln(\eta/\eta_0) = \alpha_\eta P \tag{30}$$

where η_0 is the viscosity at atmospheric pressure and α_η is a constant that depends on the oil nature and on temperature T. The subject therefore has been studied experimentally since the very beginning of the commercial development of PFPE, and it continues to be of interest up to the present.

Among the first to investigate it were researchers of the Ausimont Company (formerly Montecatini Edison), who in 1968 carried out measurements of viscosity and density of Fomblin Y oils of various molecular weight, using a Hoeppler rheometer, in the pressure range up to 30 MPa and at temperatures from 50 to 110 °C.[108] Isothermal data were described with sufficient accuracy by the well-known Doolittle free volume treatment, as modified by Matheson,[109] *i.e.* by the equation

$$\ln\eta(P) = A + B(v_0/(v(P) - v_0)) \tag{31}$$

where $v(P)$ is the sample specific volume at temperature T and pressure P, and v_0, a limiting value of v, is a more or less empirical parameter that can be determined, under suitable conditions, from the experimental $\eta(T)$ curves at normal pressure. The reciprocal of the volumetric ratio of equation (31) has the physical meaning of the 'free volume fraction' of the sample at pressure P and temperature T. An important result of the investigation was the absence, in the T and P ranges investigated, of molecular weight effects on the $\eta(P)$ function, for the restricted range of molecular weight investigated (M varying from 1400 to 7200).

Viscosity–pressure data, obtained under a variety of T and P conditions, are found in several papers.[55,56,83,102,110–113] They are, however, described from the various authors by means of different analytical equations (some of which are very complex[106,110,111]), a fact that complicates the overall analysis. Only some important results will therefore be reported in the present review.

Alper *et al.*,[83] for example, have investigated a Krytox sample of molecular weight 4220 in the temperature range from −30 to 61 °C and up to pressures of 300 MPa. The isothermal $\eta(P)$ curves (not described analytically by the authors) are seen to be, at all temperatures, quite linear up to about 30 MPa on a semilogarithmic plot, and moreover, below the temperature of 15 °C, essentially linear up to 300 MPa. Their slope, *i.e.* the pressure–viscosity coefficient α_η, increases by a factor of 2 when T decreases from 61 to −30°C. The increase of the pressure coefficient of viscosity with decreasing T seems to be a general feature.[55,102,108,109] The Krytox oil was found by Alper *et al.* to crystallize at pressures above 200 MPa whenever its viscosity approached 10^6 Pa s.

Tanaka *et al.*[55] investigated, between 0 and 60 °C and up to 180 MPa, two Demnum samples of molecular weight 2500 and 4500 and compared the results with those obtained previously by Cantow *et al.*[102] on Fomblin Y and Fomblin Z samples, both of slightly higher molecular weight. They found that the best analytical description of the isothermal $\eta(P)$ Demnum data could be obtained not by equation (30) but by a polynomial cubic equation. Plotting their data as log η against P gives essentially linear plots, however, from which approximate values of the pressure–viscosity coefficient α_η can be obtained. The α_η value is of course a function of T, as reported above, but when the temperature is kept constant (in the range from 25 to 50 °C), the coefficient is the lowest for the Fomblin Z perfluoropolyether, has an intermediate value for the Demnum samples, and is the highest for the Fomblin Y (their approximate ratios being 1.0:1.6:2.3). It appears therefore that the structure of the PFPE molecules plays a role in determining the value of the pressure–viscosity coefficient at a given value of T, a fact that has important practical implications. In particular, the high O/C ratio of Fomblin Z and its parallel low glass transitions temperature T_g (in comparison with those of the Demnum and Fomblin Y samples) appear to be clearly related to the low value of the α_η coefficient.

Similar results are obtained when $\eta(P, T)$ data are described by Arrnenius-type equations. The apparent activation energy

$$E_a = R(\partial \ln \eta / \partial(1/T))_P \tag{32}$$

and the activation volume for the viscous flow

$$V^{\#} = RT(\partial \ln \eta / \partial P)_T \tag{33}$$

obtained by these equations are collected for the same PFPEs in Table 13.

Table 13 Activation Energy and Activation Volume for the Viscous Flow of PFPEs.

Polymer	*Activation energy* E_a (kJ mol^{-1}) *@ 0.1 Pa and 298 K*	*Activation volume* V (mL mol^{-1}) *@ 298 K and P* (Pa) = 0.1	50	100	*Ref.*
Fomblin Z ($M \to \infty$)	21	—	—	—	54
Fomblin Z25	19	62	46	39	102
Fomblin Y ($M \to \infty$)	46	—	—	—	54
Fomblin YR	40	117	95	—	102
Fomblin K ($M \to \infty$)	33	—	—	—	54
Demnum S20	30	81	65	48	55

A novel approach to the description of the $\eta(P)$ data for the above two Fomblin oils investigated by Cantow *et al.*[102] has been presented by Wolf *et al.*[56] They used the Flory, Orwoll and Vrij reduced equation of state (equation 15) to describe the oil densities, measured from 25 to 100 °C and from 0.1 to 40 MPa, and thereafter demonstrated that the Doolittle equation is not obeyed, *i.e.* that 'the logarithm of the viscosity does not depend linearly on the inverse of the fraction of free volume'. Their analysis suggests that, when the P and T conditions are sufficiently distant from the glassy solidification, the viscosity is only a function of reduced pressure and temperature, and that the pressure influences can be calculated from density measurements and $\eta(T)$ data alone. The above conditions are applicable to the Fomblin Z sample, but not to the Fomblin Y sample, whose T_g is about 60 °C higher than that of the Z sample.

Summing up the literature results on the subjects of the present section, it can be said that the rheological behaviour of PFPE fluids does not differ substantially from that of typical polymeric melts, except for the low elasticity and the substantially Newtonian flow behaviour. This can be related to the flexibility of their chains (revealed by the low T_g values) at ambient and at higher temperatures, as well as to the fact that the average molecular weights M of both the commercial oils and of the other samples investigated in the various laboratories are relatively low and close to M_c.

9.7 MISCELLANEOUS PROPERTIES

9.7.1 Optical Properties

The refractive index n of many PFPE polymers is known and varies as expected with molecular weight M, increasing with M up to the asymptotic value n_∞. The latter can be easily obtained by extrapolation by plotting, for example, n_D^{20} as a function of reciprocal molecular weight M (Figure 26). The values of $n_{D\infty}^{20}$ (where the index ∞ has the meaning of infinite molecular weight) are shown in Table 14 together with the angular coefficient of the linear regression analysis. It is evident that $n_{D\infty}^{20}$ is quite insensitive to the linearity of the backbone since it is practically the same for C_3 branched and unbranched structures. However, as observed in previous sections for T_g and CED, an increase in the O/C ratio seems to bring about a small decrease in the refractive index. As a matter of fact, the refractive index of rubbery amorphous PTFE, calculated by means of group contribution analysis from low molecular weight perfluoroalkanes data, is 1.36, which compares well with the value of 1.35 reported in the literature.[69,76] It is worth pointing out that the trend experimentally observed cannot be predicted by means of the group contributions to the molar refractivity method published by Groh and Zimmermann.[115] According to these authors, the minimum value of n_D^{20} that can be calculated for polymers containing only C, F and O atoms is close to 1.29 ± 0.03.

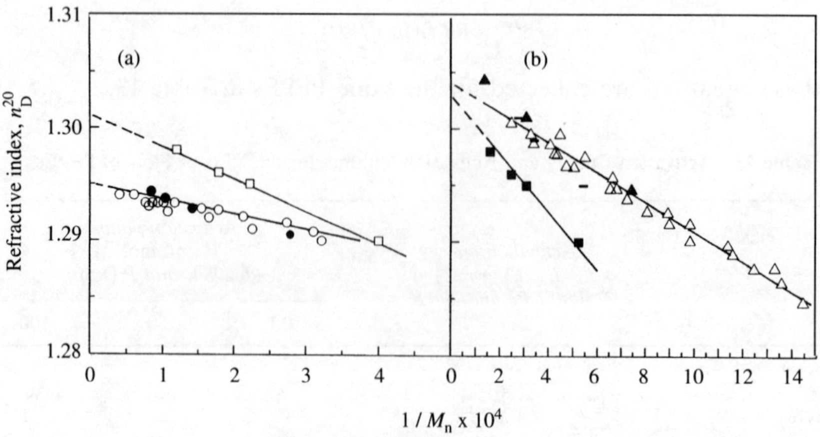

Figure 26 Refractive index n_D as a function of the reciprocal M_n for several PFPE polymers and copolymers: (a) (○) fractionated and (●) unfractionated Fomblin Z copolymers, and (□) Demnum; (b) (△) fractionated and (▲) unfractionated Fomblin Y copolymers, (–) Krytox and (■) Fomblin K.

Table 14 Refractive Index at 20 °C and its Dependence on Molecular Weight.[114]

Structure	$n^{20}_{D\infty}$	$dn_D/d(1/M_n)$ (mol g^{-1})
Fomblin Y[a] and Krytox[b]	1.304	−13.16
Fomblin Z[a]	1.295	−12.92
Fomblin K[b]	1.303	−25.68
Demnum[b]	1.301	−28.11

[a] Fractions. [b] Unfractionated materials.

9.7.2 Electrical Properties

The determination of the electrical properties of perfluoropolyethers requires some care because the materials easily adsorb moisture and gases; the volume resistivity and the dielectric loss factor are affected by impurity concentration.[116] The dielectric constant of PFPEs at standard temperature is very similar to that of rubbery amorphous PTFE, as shown in Table 15. The loss factor tan δ at the same temperature is about 10^{-3}. These data point out again that the introduction of an ethereal oxygen in the perfluorinated chain has a very small influence (if any). Such behaviour is completely different from that observed for hydrogenated polyoxides, where the introduction of the oxide group increases strongly the dielectric constant (Table 15).

Table 15 Dielectric Constant at 298 K of PFEPs, Hydrogenated Polyoxides and Related Materials.

Polymer	*Status*[a]	*O/C*	ε	Hz	*Ref.*
Poly(tetrafluoroethylene)	l	0.00	2.11	static	69
Krytox	l	0.50[b]	2.1–2.2	static	117
Krytox	l	0.50[b]	2.18	static	83
Fomblin Y	l	0.50[b]	2.15	static	116
Demnum	l	0.33	1.99	static	116
Fomblin Z	l	0.70	2.04	static	116
Polyethylene	l	0.00	2.30	static	69
Poly(propylene oxide)	l	0.50	4.80	static	118
Poly(tetramethylene oxide)	s	0.25	4.00	10^3	119
Poly(ethylene oxide)	s	0.50	4.60	10^3	120
Poly(methylene oxide)	s	1.00	3.5–3.7	0–10^3	76, 121

[a] l = liquid; s = solid. [b] Calculated on the basis of backbone atom number.

Below room temperature the dielectric behaviour is influenced by the main relaxation related to T_g, as shown in Figure 27 for two Fomblin Y of different molecular weights.

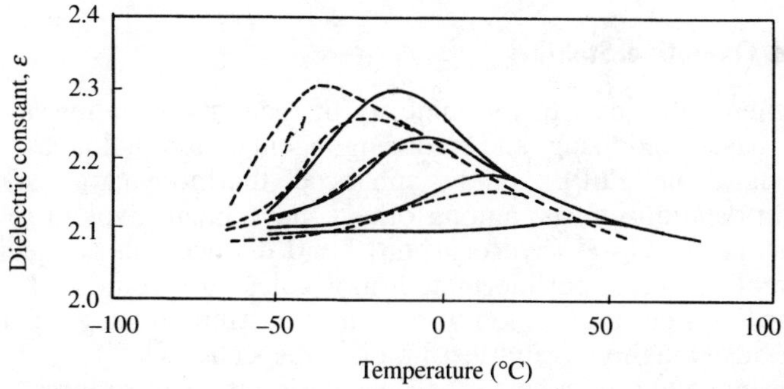

Figure 27 Dielectric constant as a function of temperature and frequency for two Fomblin Y copolymers: $M_n = 3300$ (dotted line), $M_n = 6500$ (continuous line); frequencies are in the order from left to right: 10^2, 10^3, 10^4 and 10^5 Hz.

A careful study of the dielectric relaxation behaviour of a Krytox sample ($M_n = 4220$) over a wide frequency and temperature range has been carried out by Alper *et al.*,[83] who investigated also the parallel viscoelastic relaxation. The dielectric behaviour is related to the relaxation of dipoles lying perpendicular to the polymer main chain. Alper *et al.* found that both the dielectric and viscoelastic characteristic relaxation times were dependent on temperature, following the WLF equation.

Another electrical property which has a relevant interest, especially for applications in the electric and electronic field, is volume resistivity (VR). Data reported in Figure 28 for a Fomblin Z oil and two samples of Y oils[116] show that PFPEs are good insulators and that the insulation capability decreases with decreasing molecular weight, due to the larger mobility of the macromolecule, so that for electrical purposes high molecular weight materials have to be preferred.

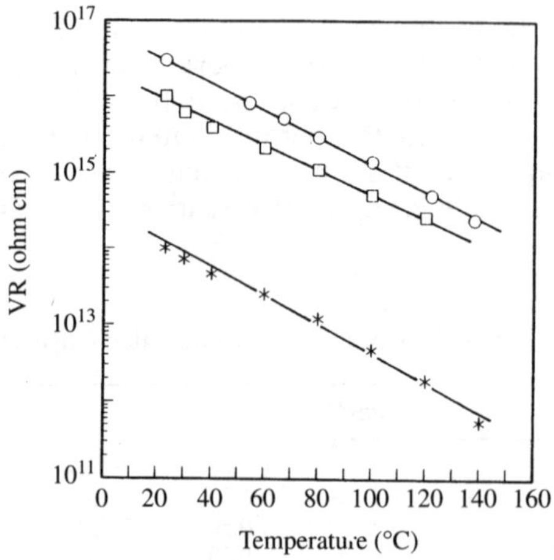

Figure 28 Dependence of the volume resistivity on temperature for a Fomblin Z copolymer (□) ($M_n = 9700$) and two Fomblin Y copolymers: (∗) $M_n = 3300$, (○) $M_n = 6550$. Oils were tested after degassing under vacuum for 2 h at 155 °C.

9.7.3 Surface Properties

As reported in the preceding sections, the CED of perfluoropolyethers is one of the lowest found for any class of compounds so that, because of the known empirical relation between CED and the surface tension,[69] low values for this property are to be expected. This is indeed the case, as shown in Figure 29, where surface tension data are plotted for several PFPEs as functions of the molecular weight. The low surface tension of PFPEs is important in applications which require wettability and low adhesion properties.

9.7.4 Thermal and Oxidative Stability

The PFPEs are chemically inert to the majority of reagents including mineral acids, alkalis, organic acids and bases, oxidizing and reducing agents, and halogens. In comparison to hydrocarbon materials, the PFPEs enjoy enhanced thermodynamic stability and kinetic invulnerability. Hybridization effects, among others, have been invoked to explain the larger C—F (fluorocarbon) *versus* C—H (hydrocarbon) bond dissociation energies (the difference is about 100–120 kJ mol^{-1}, when perfluorinated molecules are compared with hydrogenated ones). Along similar lines, perfluorination brings about a strengthening influence on both C—C and C—O single bonds (roughly 35 and 90 kJ mol^{-1}, respectively).[45]

As far as the thermal stability is concerned, perfluorination introduces a network of highly electronegative fluorine substituents, each having dimensionally protruding p-electron clouds that form a veritable barrier toward reagent aggression upon the carbon backbone. All PFPEs

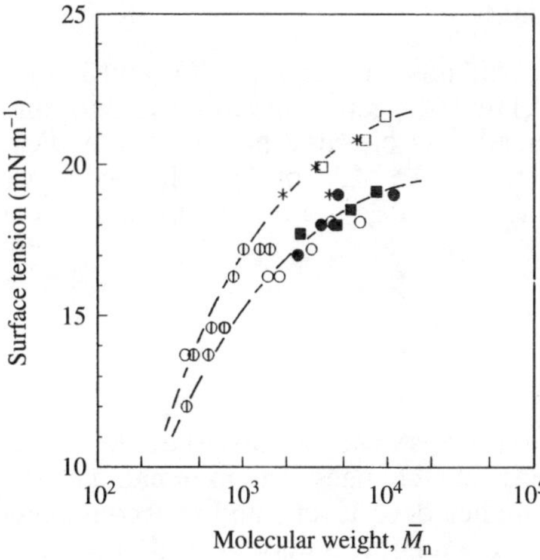

Figure 29 Surface tension at 298 K for different PFPE families as a function of the molecular weight: (⦶) Galden, (○) Fomblin Y, (□) Fomblin Z, (∗) Fomblin K, (●) Krytox, and (■) Demnum.

containing a fluorine in an α position to ethereal oxygen undergo halogen exchange and ether cleavage on contact with strong Lewis acids at elevated temperatures.[122] Reaction rates depend on substituent (steric and electronic) effects and reaction conditions (homogeneity/stirring rates), but generally proceed at measurable rates according to Lewis acid strength (in the series SbF_5 (*ca.* 100 °C), $AlCl_3$ (*ca.* 120 °C), $SbCl_3$ (*ca.* 200 °C), $FeCl_3$ (*ca.* 220 °C)). Schemes involving carbocation intermediates appear reasonable (Scheme 3).[123]

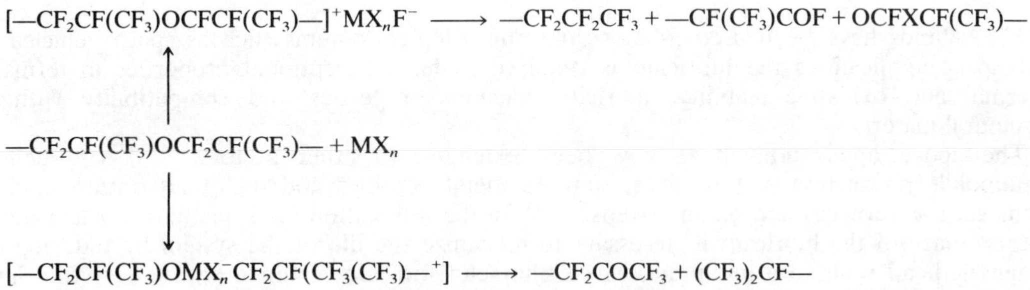

Scheme 3

The PFPEs are responsive to reactions with strong electron transfer agents as well, *e.g.* Na, K, Mg and Al over 100 °C, when shear (in lubrication applications) produces freshly exposed metal surfaces. Reactions proceed exothermically and can be rather capricious. The thermal oxidative stability of PFPEs was determined by thermogravimetric analysis (TGA) of Fomblin Y[48] and Fomblin Z[112] fluids; the decomposition onset occurs at temperatures exceeding 350 °C. The presence of metals and oxygen negatively influences decomposition temperatures, as functions of metal and fluid structure.[124] In the most reactive combinations, Fomblin Y–titanium/aluminum alloy and Fomblin Z–aluminum, the TGA-determined threshold temperatures in oxygen reduce to 325 and 270 °C, respectively. The greater susceptibility of Z fluids was verified by the microoxidation test,[125] whereby aggressive metal/oxygen combinations reduce operating temperatures to 240 and 315 °C for Fomblin Z and Y, respectively.

In the course of thermal or metal/oxygen induced decomposition, volatile gases such as fluoroalkene monomers as well as toxic COF_2 and CF_3COF are evolved, so that in an open system the undecomposed fluid fundamentally retains its physical properties.

9.7.5 Stability against Oxidants

The resistance to extreme oxidants (for example, O_2 and F_2) is another characteristic of the PFPEs. Test results have related oxygen compatibility, ignition temperature and impact of various fluorinated ether fluids.[112] At elevated pressure the indicative ignition temperature for Y fluid ($M_w = 3000$) is 395 °C at 23.5 MPa of O_2. This value varies slightly with molecular weight and oxygen concentration. PFPEs are not flammable and show no self-ignition in air at atmospheric pressure.

9.7.6 Radiation Resistance

PFPEs show good radiation resistance; at absorbed levels of 50 MRad (from a ^{60}Co γ source), weight loss and viscosity changes are minimal, *i.e.* 0.1% and 0.5%, respectively. Decomposition increases at higher dose levels, and is accompanied by the formation of lower molecular weight fragments containing acid fluoride end groups.[126]

9.8 APPLICATIONS

Although a section on applications of PFPEs is not (strictly speaking) pertinent to the title and content of the present review, the authors believe that it can be useful to some readers, and therefore described here, in a very concise way, are some of the established applications and some of the novelties recently published on new PFPE structures. A short list of application references includes not only lubrication, but cosmetics,[127,128] monument protection,[129,130] electronics,[131,132] the nuclear industry[133,134] and the aerospace industry.[124,135] Some of the many applications are discussed below.

9.8.1 Lubrication

PFPE fluids have been used as lubricants for complex systems such as space vehicles. In aerospace applications the lubricant is required to have exceptional properties in terms of thermal and oxidative stability, lubricity, vacuum properties and compatibility with all structural materials.

The above applications have now been extended to other sectors[112,113,136,137] such as automobile production, automotives, sintered metal bearings and high-temperature applications such as furnaces and baking ovens.[138–140] In the lubrication of magnetic recording media the presence of the lubricant is necessary to maximize the life of the system by reducing the magnetic head wear to a minimum. These characteristics are obtained with a very thin layer (20–50 Å) of fluids based, for example, on Fomblin Z.[141–143]

9.8.2 Electronics

The Galden fluids, which are available with molecular weights ranging from about 350 to 1500 (which corresponds to boiling point ranges from *ca.* 60 °C to 270 °C and to viscosities from 0.5 to 14 cSt), show excellent dielectric properties and compatibility with all electronic materials, and have found applications in electronics quality-control testing such as gross leak, thermal shock and liquid burn-in.[131]

Single-fluid thermal shock technology fully exploits the high flexibility of the molecular chain of PFPEs. For example, selected samples of Galden, in spite of the high boiling point (165 °C), have viscosity low enough at −65 °C (normal test temperature) to make possible this application. Single-fluid technology has indeed greatly reduced the fluid consumption, and hence the cost per device tested,[131] since the use of other perfluoro compounds, not containing ether groups, requires two separate baths with two different products.

Direct contact cooling of high-watt-density electronic devices is an important sector where the good dielectric properties and inertness of the perfluoro compounds are utilized.[144]

Vapour phase heating is also a field of application.[145,146] PFPEs are used in vapour phase soldering and to some extent in vapour phase curing. Only PFPEs have the right selection of properties for the above application, *i.e.* high vapour density, inertness to materials of construction, thermal stability, low toxicity, nonflammability and no-residue evaporation. The characteristics are very rapid heat transfer, minimum thermal stress and minimal oxidation of metal contacts, due to the absence of oxygen.

9.8.3 Biomedical

Some application of PFPEs have been found in the biomedicals sector, where the high gas solubility (see Table 7), low surface tension and low toxicity of the PFPEs are important. Blood analysis is an established sector.[147,148]

9.8.4 Vacuum Fluids

The chemical inertness and the lubrication properties of PFPEs make them some of the best vacuum pump fluids to be used in semiconductor manufacturing,[149–152] where corrosive chemicals enter the process. The reaction products, the gases and the solid particles produced, are in fact taken into the pump. Under these conditions, conventional oils experience degradation and polymerization, so that frequent oil changes and careful maintenance are needed. Personel safety can be often at risk. The use of PFPEs as fluids in vacuum pumps reduces frequent oil changes and frequency of maintenance. Their inertness and nonflammability increase the safety and the production cost is reduced.

Some types of low-cost PFPE fluids are designed for use in solvent recovery and in chemical industries where harsh environments are encountered. The oxygen production industry is another industrial field where PFPEs have found application. When liquid or gaseous oxygen comes in contact with fluids or greases that are not absolutely inert in the operating conditions, fire and explosion danger can occur. PFPEs are used in transfer pumps for liquid oxygen, in compressors, in valve lubrication and in measuring instruments. Oxygen producers and users have been using PFPEs, with excellent technical results and safety records, for years.

9.8.5 Functional PFPEs

An additional and recent innovative development is the production of functionalized PFPEs. The presence of functional end groups has opened up large new fields to PFPE chemistry.[153,154] Some properties of commercial products already known are listed in Table 16.

Table 16 Typical Properties of Fomblin Derivatives $XCF_2O(CF_2CF_2O)_p(CF_2O)_qCF_2X$.

	X					
Product	$—CO_2H$	$—CO_2Me$	$—CH_2OH$	$—CH_2(OCH_2CH_2)_kOH^c$	$—CH_2OCH_2CH(OH)CH_2OH$	$—CONHC_6H_3$-3-NCO-4-Me
Molecular weight	2000	2000	2000	2200	2000	3000
Difunctional (%)	>94	>94	>94	>94	>95[a]	>94
Viscosity @ 20 °C (cSt)	60	20	80	140	2000	160[b]
Specific gravity ($g\,mL^{-1}$)	1.82	1.77	1.81	1.73	1.74	1.72
Refractive index	1.300	1.298	1.300	1.316	1.324	—
Vapour pressure						
@ 20 °C (mm Hg)	5×10^{-5}	3×10^{-4}	2×10^{-5}	6×10^{-6}	1×10^{-7}	—
@ 100 °C (mmHg)	2×10^{-3}	2×10^{-3}	2×10^{-3}	1×10^{-3}	5×10^{-3}	—

[a] Tetrafunctional. [b] 75% by wt. soluble in CF 113. [c] $k = 1–2$.

Bifunctional derivatives were the first products to appear, and they found unchallenged application in thin layer lubrication of magnetic recording media. The hydrogenated end groups anchor themselves to the coating, providing long-life lubrication.[155–158]

Some versatile condensation prepolymer intermediates can be converted into thermoplastics, thermosets and elastomers.[159–165] The Z-type backbone present in the polymer allows the attainment of low T_g's, of hydro- and oleophobicity, of chemical and thermal resistance, of low surface energy and refractive index (Table 16). The potential applications of these polymers are too vast to be described in detail here.

9.8.6 Functional Derivatives in Lubrication

When used as lubricants,[124,134] PFPEs do not attack or corrode materials, but neither give complete protection against certain types of corrosion. As a matter of fact they have excellent thermal and thermo-oxidative stabilities, but future applications will demand improvements. The antiwear properties, too, have to be improved to the level of conventional mineral and synthetic oils containing suitable additives. These shortcomings of the PFPEs can be corrected by using additives; however, the conventional nonfluorinated additives are insoluble in PFPEs and do not have satisfactory stability characteristics (as required under the temperature and harsh environmental conditions where PFPEs are or will be used). Development of soluble additives for PFPEs represents a great challenge.[166–169] The solution appears to lie in the functional PFPEs described above.

The use of conventional chemistry with nonconventional preparative methods has been used to produce fluorinated soluble additives which improve anticorrosion, antiwear and thermal stability characteristics of PFPEs.

9.8.7 Microemulsions

The PFPE monoacid and its derivatives can be converted into surfactants, which can be used in a novel and unique emulsion and microemulsion technology.[170–174] The introduction of fluorine in desired quantities into nonfluorinated systems, to impart special characteristics in a controlled manner, is made possible by the use of microemulsions. The technology has opened up new applications, such as protection of marble against air-borne corrosive chemicals. In countries where historical monuments are abundant and are a part of the national heritage, this has been a significant development.[129,130]

The emulsion technology is still in its initial stage and the potential uses cover surface treatment to impart oleo/hydrophobicity and lubricity to various materials, such as metals, polymers and fibres. Car wax formulations containing PFPE have been commercialized.

The use of neutral PFPE fluids in cosmetic creams,[127,128] foaming and defoaming agents are possible applications of the emulsion technology.

9.9 CONCLUSIONS

The perfluoropolyethers investigated until now, reviewed in the present work, are polymers of relatively low molecular weight. They constitute a variegated family comprising structurally different classes, all characterized by the fact that the main chain is linear and is composed of carbon and oxygen only. Additional structural variations, such as chain branching and closed chain structures, have not been reported up to now in the vast literature on PFPEs explored in the present work, and are not therefore treated in it.

Within a given class the chain length is obviously variable, as required for investigations and specific applications purposes. Among the classes the differences are mostly related to structural aspects such as:

- the number of main chain oxygen atoms per carbon atom;
- the presence or absence along the chain of CF_3 side groups;
- the homopolymeric or copolymeric structure;
- the presence of specific functional groups at one or two of the chain ends.

Are there in the PFPE family important or unexpected pecularities in the relationship between structure and properties? Are their properties related or not to structure in a way similar to that deducible from the general knowledge of polymeric systems? Those are some of the questions that should be answered at the conclusive stage of the present review, in which the subject has been explored in several specific sections with due relevance given to the thermodynamic, transitional and rheological properties of perfluoropolyethers.

Broadly speaking, the answer to the main question is negative. The structure–property relationships do not appear in PFPEs to be diverse from those existing in other polymeric systems, and when one introduces modifications of the main chain length, of the very short side chains, and of the monomer composition, the property changes introduced are basically those expected. Moreover, several properties can be forecast when the chemical composition is known, since they largely obey to the rule of group contribution.

The exceptional characteristics of the PFPE molecules are therefore largely due to the fact that in these compounds the hydrogen atoms are all substituted by fluorine. In other words, what really makes the difference is the fluorine substitution, and as far as the PFPE macromolecules are concerned, their coil conformation in dilute solution is normal, the thermal characteristics are those usually found for amorphous solids and nonassociated fluids, and the flow behaviour does not present odd aspects (when a reduced temperature such as $T - T_g$ is used, as required — but often omitted — in the data analysis of the literature).

On the other hand, when the PFPEs are compared closely with several common macromolecular systems, some differences in the structure–property relationship do indeed appear. The prominent one, and the unique to which emphasis is given here, is the effect that changes in the oxygen to carbon ratio of hydrogenated and perfluorinated polyethers have on some thermodynamic and thermal properties. Packing density, glass transition temperature, vaporization energy and cohesive energy density of the PFPEs are strongly influenced by the above ratio, and the present review has stressed the fact that in perfluoropolyethers the influence of an increase of oxygen is markedly different, and often opposite, to that found in analogous hydrogenated polymers.

9.10 REFERENCES

1. V. Caglioti, A. Delle Site, M. Lenzi and A. Mele, *J. Chem. Soc.*, 1964, 5430.
2. J. L. Warnell (Du Pont), *US Pat.* 3 125 599 (1964) (*Chem. Abstr.*, 1964, **60,** 13 407h).
3. W. T. Miller (Du Pont), *Fr. Pat* 1 366 119 (1964) (*Chem. Abstr.*, 1965, **62,** 9313b).
4. E. P. Moore, A. S. Milian, Jr. and A. S. Eleuterio (Du Pont), *US Pat.* 3 250 808 (1966) (*Chem. Abstr.*, 1966, **65,** 13 554b).
5. E. P. Moore (Du Pont), *US Pat.* 3 322 826 (1967) (*Chem. Abstr.*, 1967, **67,** 44 292j).
6. D. Sianesi, A. Pasetti and C. Corti, *Makromol. Chem.*, 1965, **86,** 308.
7. D. Sianesi, A. Pasetti and C. Corti (Montecatini), *Neth. Appl.* 6 504 428 (1965) (*Chem. Abstr.* 1966, **64,** 14 360g).
8. D. Sianesi, G. C. Bernardi and A. Pasetti (Montecatini Edison), *Fr. Pat* 1 499 094 (1967) (*Chem. Abstr.*, 1968, **69,** 51 649a).
9. D. Sianesi and R. Fontanelli (Montecatini Edison), *Ital. Pat.* 793 154 (1967) (*Chem. Abstr.*, 1969, **70,** 20 513t).
10. D. Sianesi, A. Pasetti and G. Belardinelli (Montecatini Edison), *Ital. Pat.* 790 651 (1973) (*Chem. Abstr.*, 1969, **70,** 20 511r).
11. F. Gozzo and G. Camaggi, *Tetrahedron,* 1966, **22,** 2181.
12. F. Gozzo and V. Oprandi (Montecatini Edison), *Br. Pat.* 1 204 748 (1970) (*Chem. Abstr.*, 1970, **73,** 110 319a).
13. R. N. Haszeldine, *J. Chem. Soc.*, 1953, 3559.
14. W. H. Gumprecht, *ASLE Trans.*, 1966, **9,** 24.
15. J. T. Hill, *J. Macromol. Sci., Chem.*, 1974, **A8,** 499.
16. P. Bickle and W. Schwertfeger (Hoechst), *Ger. Offen.* 3 828 848 (1990) (*Chem. Abstr.*, 1990, **113,** 67 438u).
17. Y. Ohsaka, *Petrotech.*, 1985, **8,** 840.
18. Y. Ohsaka, T. Tozuka and S. Takaki (Daikin), *Eur. Pat. Appl.* 148 482 (1985) (*Chem. Abstr.*, 1986, **104,** 69 315q).
19. D. Sianesi, A. Pasetti, R. Fontanelli, G. C. Bernardi and G. Caporiccio, *Chim. Ind.* (*Milan*), 1973, **55,** 208.
20. A. Faucitano, A. Buttafava, V. Comincioli, G. Marchionni and R. J. De Pasquale, *J. Phys. Org. Chem.*, 1991, **4,** 293.
21. G. Marchionni, A. Staccione and G. Gregorio, *J. Fluorine Chem.*, 1990, **47,** 515.
22. G. T. Viola, G. Marchionni and G. Caporiccio (Ausimont), *Eur. Pat. Appl.* 167 258 (1986) (*Chem. Abstr.*, 1986, **104,** 187 128d).
23. G. Caporiccio, G. T. Viola and G. Marchionni (Ausimont), *Eur. Pat. Appl.* 224 201 (1987) (*Chem. Abstr.*, 1987, **107,** 199 200w).
24. G. Caporiccio, G. T. Viola, G. Marchionni, P. De Iorio and A. Tasca (Montedison), *Eur. Pat. Appl.* 223 238 (1987) (*Chem. Abstr.*, 1987, **107,** 134 897j).
25. G. Marchionni, G. Guglielmo, U. De Patto and M. Padovan (Ausimont), *Eur. Pat. Appl.* 440 199 (1991) (*Chem. Abstr.*, 1991, **115,** 256 982u).
26. A. Faucitano, B. Buttafava, F. Martinotti, G. Marchionni, A. Staccione and R. J. De Pasquale, ACS 9th Winter Fluorine Conference, St. Petersburg, FL, USA, 1989, p. 52.
27. A. Faucitano, B. Buttafava, P. A. Guarda and G. Marchionni, 3rd International Conference on Polymer Photochemistry, Sestri Levante, Italy, 1993, p. 178.
28. D. Sianesi, A. Pasetti and C. Corti (Montedison), *Ital. Pat.* 792 673 (1967) (*Chem. Abstr.*, 1969, **70,** 29 557t).
29. G. Marchionni and U. De Patto (Ausimont), *Eur. Pat. Appl.* 344 547 (1989) (*Chem. Abstr.*, 1990, **112,** 180 167x).

30. A. Marracini, A. Pasquale, A. M. Staccione and G. Marchionni (Ausimont), *Eur. Pat. Appl.* 445 738 (1991) (*Chem. Abstr.*, 1991, **115,** 256 985x).
31. D. Sianesi, A. Pasetti and G. Belardinelli (Montedison), *US Pat.* 4 451 646 (1984) (*Chem. Abstr.*, 1984, **101,** 91 718h).
32. G. Marchionni and A. M. Staccione (Ausimont), *Eur. Pat. Appl.* 340 740 (1989) (*Chem. Abstr.*, 1990, **112,** 159 247p).
33. G. Marchionni, U. De Patto and E. Strepparola (Ausimont), *Eur. Pat. Appl.* 244 839 (1987) (*Chem. Abstr.*, 1988, **108,** 168 200n).
34. G. Marchionni, E. Strepparola and G. Spataro (Ausimont), *Eur. Pat. Appl.* 337 346 (1989) (*Chem. Abstr.*, 1990, **112,** 8373a).
35. G. Martini, M. F. Ottaviani, S. Ristori, G. Marchionni and M. Visca, *Macromolecules,* 1991, **24,** 1050.
36. V. Caglioti, M. Lenzi and A. Mele, *Nature,* 1964, **201,** 610.
37. V. V. Shapovalov and V. A. Poluektov, *Khim. Vys. Energ.*, 1991, **25,** 325.
38. M. Loeffelholz, A. Pasetti and D. Sianesi (Montecatini Edison), *Fr. Addn.* 90 845 (1968) (*Chem. Abstr.*, 1969, **70,** 58 632u).
39. R. J. Lagow, T. R. Bierschenk, T. J. Juhlke and H. Kawa, in 'Synthetic Fluorine Chemistry', ed. A. Olah, R. D. Chambers and J. K. S. Prakash, Wiley, New York, 1992, chap. 5.
40. D. Sianesi, A. Marracini and G. Marchionni (Ausimont), *Eur. Pat. Appl.* 393 700 (1990) (*Chem. Abstr.*, 1991, **114,** 123 289f).
41. D. Sianesi, A. Marracini and G. Marchionni (Ausimont), *Eur. Pat. Appl.* 393 705 (1989) (*Chem. Abstr.*, 1991, **114,** 144 242a).
42. F. Ciampelli, M. Tacchi Venturi and D. Sianesi, *Org. Magn. Reson.*, 1969, **1,** 281.
43. M. Tacchi Venturi and C. Tosi, *J. Mol. Struct.*, 1972, **14,** 293.
44. P. H. Kasai, *Macromolecules,* 1992, **25,** 6791.
45. B. E. Smart, in 'Molecular Structure and Energetics', ed. J. F. Liebmann and A. Greenberg, VCH, New York, 1986, vol. 3, chap. 4.
46. J. Pacansky, M. Miller, W. Hatton, B. Liu and A. Scheiner, *J. Am. Chem. Soc.*, 1991, **113,** 329.
47. D. Sianesi, *Chim. Ind.* (*Milan*), 1968, **50,** 206.
48. D. Sianesi, V. Zamboni, R. Fontanelli and M. Binaghi, *Wear,* 1971, **18,** 85.
49. A. C. Ouano, B. Appelt and T. D. Watson, *Prepr.—Am. Chem. Soc., Org. Coating Appl. Polym. Sci. Div.*, 1981, **46,** 230.
50. T. W. Del Pesco, in 'Synthetic Lubricants and High-Performance Functional Fluids', ed. R. L. Shubkin, Dekker, New York, 1993, chap 6.
51. G. Marchionni, G. Ajroldi and G. Pezzin, *Eur. Polym. J.*, 1988, **24,** 1211.
52. G. Marchionni, G. Ajroldi, P. Cinquina, E. Tampellini and G. Pezzin, *Polym. Eng. Sci.*, 1990, **30,** 829.
53. F. Danusso, M. Levi, G. Gianotti and S. Turri, *Eur. Polym. J.*, 1994, **30,** 647.
54. G. Marchionni, G. Ajroldi and G. Pezzin, *Soc. Automot. Eng.*, 1992, **SP-936,** 87.
55. Y. Tanaka, N. Noijri, K. Ohta, H. Kubota and T. Makita, *Int. J. Thermophys.*, 1989, **10,** 857.
56. B. A. Wolf, M. Klimiuk and M. J. R. Cantow, *J. Phys. Chem.*, 1989, **93,** 2672.
57. Ausimont, unpublished data.
58. P. Tait, 'Physics and Chemistry of the Voyage of H.M. Ship Challenger', Cambridge University Press, Cambridge, 1900, vol. II.
59. A. Bondi, 'Physical Properties of Molecular Crystals, Liquids and Glasses', Wiley, New York, 1968.
60. R. W. Fillers and N. W. Tschoegl, *Trans. Soc. Rheol.*, 1977, **21,** 51.
61. P. J. Blatz, in 'Polymer Networks: Structural and Mechanical Properties', ed. A. J. Chompff, Plenum, New York, 1971.
62. P. J. Flory, R. A. Orwoll and A. Vrij, *J. Am. Chem. Soc.*, 1964, **86,** 3507.
63. B. E. Eichinger and P. J. Flory, *Trans. Faraday Soc.*, 1968, **64,** 2053.
64. H. Hoecker, G. J. Blake and P. J. Flory, *Trans. Faraday Soc.*, 1971, **67,** 2251.
65. H. Shih and P. J. Flory, *Macromolecules,* 1972, **5,** 758.
66. A. Abe and P. J. Flory, *J. Am. Chem. Soc.*, 1966, **88,** 2887.
67. B. E. Eichinger and P. J. Flory, *Trans. Faraday Soc.*, 1968, **64,** 2061.
68. A. Abe and P. J. Flory, *J. Am. Chem. Soc.*, 1965, **87,** 1838.
69. D. W. Van Krevelen, 'Properties of Polymers', Elsevier, Amsterdam, 1976.
70. D. D. Lawson, J. Moacanin, K. V. Scherer, Jr., J. F. Terranova and J. D. Ingham, *J. Fluorine Chem.*, 1971, **12,** 221.
71. G. Marchionni, G. Ajroldi, M. C. Righetti and G. Pezzin, *Macromolecules,* 1993, **26,** 1751.
72. S. F. Lau, H. Suzuki and B. Wunderlich, *J. Polym. Sci., Polym. Phys. Ed.*, 1984, **22,** 379.
73. H. Suzuki and B. Wunderlich, *J. Polym. Sci., Polym. Phys. Ed.*, 1985, **23,** 1671.
74. S. Z. D. Cheng and B. Wunderlich, *Thermochim. Acta,* 1988, **134,** 161.
75. K. Loufakis and B. Wunderlich, *J. Phys. Chem.*, 1988, **92,** 4205.
76. J. Brandrup and E. H. Immergut, 'Polymer Handbook', 3rd edn., Wiley, New York, 1989, chap. 7.
77. M. J. R. Cantow, R. B. Larrabee, E. M. Barral II, R. S. Butner, P. Cotts, F. Levy and T. Y. Ting, *Makromol. Chem.*, 1986, **187,** 2475.
78. P. M. Cotts, *Polym. Prepr., Am. Chem. Soc., Div. Polym. Chem.*, 1994, **35,** 108.
79. A. Sanguinetti, P. A. Guarda, G. Marchionni and G. Ajroldi, *Polymer,* 1995, **36,** 3697.
80. W. H. Burchardt, *Makromol. Chem.*, 1961, **50,** 20.
81. W. H. Stockmayer and M. Fixman, *J. Polym. Sci., Polym. Symp.*, 1963, **1,** 137.
82. B. Chu, C. Wu and J. Zuo, *Macromolecules,* 1987, **20,** 700.
83. T. Alper, A. J. Barlow, R. W. Gray, M. G. Kim, R. J. MacLachlan and J. Lamb, *J. Chem. Soc., Faraday Trans. 2,* 1980, **76,** 205.
84. U. Gaur and B. Wunderlich, *J. Phys. Chem. Ref. Data,* 1981, **10,** 1001.
85. S. Yoshida, H. Suga and S. Seki, *Polym. J.*, 1973, **5,** 11.

86. R. F. Boyer, *Macromolecules,* 1974, **7,** 142.
87. F. Danusso, M. Levi, G. Giannotti and S. Turri, *Polymer,* 1993, **34,** 3687.
88. G. Ajroldi, G. Marchionni, M. Fumagalli and G. Pezzin, *Plast. Rubber Compos. Proc. Appl.,* 1992, **17,** 307.
89. R. F. Boyer, in 'Order in the Amorphous State of Polymers', ed. S. Keinath, R. E. Miller and J. K. Rieke, Plenum, New York, 1987.
90. H. W. Starkweather, Jr., P. Avakian, J. J. Fontanella and M. C. Wintersgill, *Macromolecules,* 1992, **25,** 3815.
91. H. W. Starkweather, Jr., P. Avakian, R. R. Matheson, Jr., J. J. Fontanella and M. C. Wintersgill, *Macromolecules,* 1991, **24,** 3853.
92. G. Ajroldi, unpublished data.
93. M. J. R. Cantow, T. Y. Ting, E. M. Barral II, R. S. Porter and E. R. George, *Rheol. Acta,* 1986, **25,** 69.
94. J. P. Montfort and G. Hadziioannou, *J. Chem. Phys.,* 1988, **88,** 7187.
95. G. Marchionni, G. Ajroldi, M. C. Righetti and G. Pezzin, *Polym. Commun.,* 1991, **32,** 71.
96. M. C. H. Lee and M. C. Williams, *J. Macromol. Sci., Phys.,* 1987, **B26,** 145.
97. J. D. Ferry, 'Viscoelastic Properties of Polymers', 3rd edn, Wiley, New York, 1980.
98. A. V. Tobolsky, D. Katz and A. Eisenberg, *J. Appl. Polym. Sci.,* 1963, **7,** 469.
99. G. Ajroldi, C. Garbuglio and M. Ragazzini, *J. Appl. Polym. Sci.,* 1970, **14,** 79.
100. V. P. Privalko, Y. S. Lipatov and A. P. Lobodina, *J. Macromol. Sci., Phys.,* 1975, **B11,** 441.
101. G. C. Berry and T. G. Fox, *Adv. Polym. Sci.,* 1968, **5,** 261.
102. M. J. R. Cantow, E. M. Barral II, B. A. Wolf and H. Geerissen, *J. Polym. Sci., Polym. Phys. Ed.,* 1987, **25,** 603.
103. R. E. Powell, W. E. Roseveare and H. Eyring, *Ind. Eng. Chem.,* 1941, **33,** 430.
104. M. J. Hunter, E. L. Warrick, J. F. Hoyle and C. C. Currie, *J. Am. Chem. Soc.,* 1946, **68,** 2284.
105. M. Aderin, G. J. Johnston, H. A. Spikes and G. Caporiccio, *Lubr. Eng.,* 1992, **48,** 633.
106. H. S. Hsiao, S. K. Sharma and B. J. Hamrock, *SAE Techn. Paper Series,* 1992, 922 344.
107. C. Barus, *Am. J. Sci.,* 1893, **45,** 87.
108. A. Arcozzi and F. Branzoli, 'Caratteristiche viscosimetriche dei fluidi fluorurati Fomblin Y', Montecatini Edison, Internal Report, Milan, 1968.
109. A. J. Matheson, *J. Chem. Phys.,* 1966, **44,** 695.
110. S. Yasutomi, S. Bair and W. O. Winer, *J. Tribol.,* 1984, **106,** 291.
111. W. R. Jones, R. L. Johnson, W. O. Winer and D. M. Sanborn, *ASLE Trans.,* 1975, **18,** 249.
112. C. Corti and P. Savelli, *J. Synth. Lubr.,* 1993, **9,** 311.
113. H. A. Spikes, P. Cann and G. Caporiccio, *J. Synth. Lubr.,* 1984, **1,** 73.
114. G. Marchionni, unpublished data.
115. W. Groh, A. Zimmermann, *Macromolecules,* 1991, **24,** 6660.
116. Ausimont, unpublished data.
117. Du Pont literature.
118. T. Alper, A. J. Barlow and R. W. Gray, *Polymer,* 1976, **17,** 665.
119. R. E. Wetton and G. Williams, *Trans. Faraday Soc.,* 1965, **61,** 2132.
120. T. M. Connor, B. E. Read and G. Williams, *J. Appl. Chem.,* 1964, **14,** 74.
121. B. E. Read and G. Williams, *Polymer,* 1961, **2,** 239.
122. D. Sianesi and R. Fontanelli, *Makromol. Chem.,* 1967, **102,** 115.
123. B. E. Smart and D. A. Dixon, *J. Fluorine Chem.,* 1992, **57,** 251.
124. G. Caporiccio, L. Flabbi, G. Marchionni and G. C. Viola, *J. Synth. Lubr.,* 1989, **6,** 133.
125. C. E. Snyder, Jr. and R. E. Dolle, Jr., *ASLE Trans.,* 1976, **19,** 171.
126. A. Faucitano, A. Buttafava, P. A. Guarda and G. Marchionni, *J. Fluorine Chem.,* 1993, **64,** 189.
127. G. Pantini, C. Corti, G. Marchionni and M. Pianca, *Tecnol. Chim.,* 1991, **6,** 154.
128. F. Brunetta and G. Pantini, *Seifen, Oele, Fette, Wachse,* 1992, **11,** 3.
129. F. Piacenti, L. Manganelli Del Fa', V. Matteoli, P. Tiano and A. Scala, Proc. VI Int. Cong. Deterioration and Conservation of Stone, Torun, Hungary, 1988, pp. 492 and 881.
130. G. Moggi, *Adv. Org. Coat. Sci. Technol. Ser.,* 1991, **13,** 472.
131. D. S. L. Slinn and S. W. Green, 'Preparation, Properties and Industrial Applications of Organofluorine Compounds', ed. R. E. Banks, Horwood, Chichester, 1982.
132. G. Marchionni and P. Srinivasan, Fluoropolymer Conference '92, UMIST, Manchester, 1992, paper 14.
133. G. Caporiccio, *J. Fluorine Chem.,* 1986, **33,** 227.
134. A. Ciancia, A. Ascensioni, C. Corti and G. Caporiccio, *Nucl. Sci. Eng.,* 1984, **86,** 232.
135. E. Falcone, *Chim. Oggi,* 1983, **8,** 27.
136. C. E. Snyder, Jr., L. J. Gschwender and C. Tamborski, *Lubr. Eng.,* 1981, **37,** 344.
137. S. Mori and W. Morales, *Wear,* 1989, **132,** 111.
138. J. C. Anderson, L. Flabbi and G. Caporiccio, *J. Synth. Lubr.,* 1988, **5,** 199.
139. V. D'Agostino, V. Niola and G. Caporiccio, *Tribol. Int.,* 1988, **21,** 105.
140. L. J. Gschwender and C. E. Snyder, *J. Synth. Lubr.,* 1992, **9,** 115.
141. P. Buttafava, V. Bretti, G. Ciardiello, G. Caporiccio and A. M. Scarati, *IEEE Trans. Magn.,* 1985, **21,** 1533.
142. K. Tanimoto and E. Rabinowicz, *Tribol. Trans.,* 1992, **35,** 537.
143. B. D. Strom, D. B. Bogy, R. G. Walmsley, J. Brandt and C. S. Bhatia, *IEEE Trans. Magn.,* 1991, **29,** 253.
144. A. Luches and L. Provenzano, *J. Phys. D,* 1977, **10,** 339.
145. C. Lea and K. W. E. Johns, *Brazing Soldering,* 1987, **12,** 34.
146. R. S. Briggs, P. Newton and P. Srinivasan, *Hybrid Circuits,* 1989, **18,** 42.
147. G. Pantini and P. L. Bencini, *Tecnol. Chim.,* 1991, **5,** 130.
148. G. Pantini, R. Forestieri, F. Brunetta and P. L. Bencini, *Int. J. Cosmet. Sci.,* 1990, (December), 273.
149. G. Caporiccio, C. Corti, S. Soldini and G. Carniselli, *Ind. Eng. Chem., Prod. Res. Dev.,* 1982, **21,** 515.
150. L. Laurenson, N. T. M. Dennis and J. Newton, *Vacuum,* 1979, **29,** 433.
151. M. Masuko, I. Fujinami and H. Okabe, *Wear,* 1992, **159,** 249.
152. M. Vertes, *Vacuum,* 1993, **44,** 769.
153. J. G. Newman and K. V. Viswanathan, *J. Vac. Sci. Technol.,* 1990, **A8,** 2388.

154. C. Tonelli, G. Marchionni, P. Gavezzotti and R. J. De Pasquale, Soc. Plast Ind., Fluoropolymer Div., Proceed. Fall Meeting, Asheville, North Carolina, September 1992.
155. G. Caporiccio, E. Strepparola and M. A. Scarati (Montedison), *Eur. Pat. Appl.* 165 649 (1988) (*Chem. Abstr.*, 1986, **104,** 130 510t).
156. A. M. Scarati and G. Caporiccio, *IEEE Trans. Magn.*, 1987, **23,** 106.
157. D. D. Saperstein and L. J. Lin, *Langmuir,* 1990, **6,** 1522.
158. T. H. D. Lee, *J. Vac. Sci. Technol.*, 1991, **A9,** 1287.
159. D. Sianesi, Int. Symp. Cent. Disc. Fluorine, Paris, 1985, p. 223.
160. F. Pilati, M. Toselli, A. Re, F. A. Bottino, A. Pollicino and A. Recca, *Macromolecules,* 1990, **23,** 348.
161. F. Pilati, V. Bonora, P. Manaresi, A. Munari, M. Toselli, A. Re and M. De Giorgi, *J. Polym. Sci., Polym. Chem. Ed.*, 1989, **27,** 951.
162. G. Caporiccio, E. Strepparola, G. Bargigia, G. Novaira and E. Peveri, *Makromol. Chem.*, 1983, **184,** 935.
163. D. Sianesi, III Eur. Polym. Fed. Symp. Polymeric Materials, Sorrento, Italy, October 1990, vol. 1, p. 85.
164. G. Bargigia, P. Gavezzotti, C. Tonelli and G. Gavazzi, *Poliplasti,* 1992, **420,** 106.
165. F. Pilati, M. Toselli, A. Vallieri and C. Tonelli, *Polym. Bull.*, 1992, **28,** 151.
166. S. K. Sharma, L. J. Gschwender and C. E. Snyder, Jr., *J. Synth Lubr.*, 1990, **7,** 15.
167. P. Srinivasan, C. Corti, L. Montagna and P. Savelli, *J. Synth. Lubr.*, 1993, **10,** 143.
168. V. G. Melnikov, V. I. Nazarov, M. S. Khots, L. B. Kapustina and T. V. Popkova, *Corrosion,* 1992, **48,** 846.
169. K. Johns, C. Corti, L. Montagna and P. Srinivasan, *J. Phys. D,* 1992, **25,** A141.
170. G. Caporiccio, F. Burzio, G. Carniselli and V. Biancardi, *J. Colloid Interface Sci.*, 1984, **98,** 202.
171. A. Chittofrati, D. Lenti, A. Sanguineti, M. Visca, C. Gambi, D. Senatra and Z. Zhen, *Colloids Surf.*, 1989, **41,** 45.
172. A. Chittofrati, D. Lenti, A. Sanguineti, M. Visca, C. Gambi, D. Senatra and Z. Zhen, *Prog. Colloid Polym. Sci.*, 1989, **79,** 218.
173. M. Monduzzi, A. Chittofrati and M. Visca, *Langmuir,* 1992, **8,** 1278.
174. A. Chittofrati, A. Sanguineti, M. Visca and N. Kallay, *Colloids Surf.*, 1992, **63,** 219.

10

Structure and Properties of Aromatic Liquid Crystalline Copolyesters

C. M. McCULLAGH and JOHN BLACKWELL
Case Western Reserve University, Cleveland, OH, USA

10.1 INTRODUCTION

Liquid crystalline materials have the ability to form anisotropic mesophases with order intermediate between the crystalline solid state and the isotropic liquid. In 1888, Reinitzer reported two melting points for cholesteryl ethers, between which a birefringent phase was observed.[1] Since that time, the existence of such mesophases has been reported for an enormous range of compounds, and their structures and properties have received extensive study. Vorlander[2] studied the properties of the oligomers of LC compounds, setting the stage for work on LC polymers (LCPs). These are classified as *lyotropic* or *thermotropic,* depending on whether the LC behavior is seen in solution or in the melt. Early work on LCPs focused on the concentration-dependent lyotropic phases formed by rod-like polymers, notably poly(γ-benzyl-L-glutamate)[3] and tobacco mosaic virus.[4] The first commercial LCP products were Kevlar® fibers, based on poly(phenylene terephthalamide), which were developed by DuPont in the 1960s.[5] The fibers were produced by dry-jet wet spinning from an LC solution in hot concentrated sulfuric acid. The first well characterized thermotropic LCP was the ethylene terephthalate/oxybenzoate copolyester developed in the early 1970s by Eastman Kodak.[6] This copolymer, which is often referred to in the literature as X7G, could be melt-spun as fibers, but the tensile properties were greatly inferior to those of Kevlar. Nevertheless, this discovery prompted the development of a range of wholly aromatic LC copolyesters, which will be the topic of this review. Two such LCPs that are now commercially available are Xydar®, produced by Amoco,[7] and Vectra®, produced by Hoechst Celanese.[8] Xydar is based on the copolymer of *p*-hydroxybenzoic acid (HBA), biphenol (BP), and terephthalic acid (TPA); Vectra is a copolymer of HBA and 2-hydroxy-6-naphthoic acid (HNA).

The large increase in understanding of the structure–property relationships of LC polyesters (LCPEs) over the last 20 years can be directly attributed to the commercial importance of these materials as high-strength fibers and precision molded parts.[9–11] The liquid crystalline character of the melt leads to orientation of the fully extended chains along the fiber axis, resulting in high strength and high modulus. The melts have low shear viscosity resulting from shear alignment, such that it is much easier to mold complex shapes than is the case for equivalent isotropic polymer melts. In addition, the densities of the LC melt and the solid state are very similar, with the effect that there is very little mold shrinkage. These properties, plus excellent surface characteristics and solvent resistance, make the thermotropic polyesters ideal for precision moldings, especially for connectors used in the electronics industry. However, the inherent orientation in these systems also leads to disadvantages, including highly anisotropic mechanical properties and weaknesses at weld lines.

Liquid crystallinity in low molar mass compounds often derives from the presence of rod-, disc-, plate-, or board-like groups; typically these groups are aromatic, and are known as *mesogens.* In LCPs the mesogens may be attached as side chains to a flexible backbone (side chain LCPs). Generally, LC behavior requires the existence of a flexible spacer group, such as $(CH_2)_n$ or $(CH_2CH_2O)_n$, between the backbone and mesogen.[12] Alternatively, main-chain LCPs are formed when the mesogens are linked either directly or via flexible spacers.

Main-chain LCPs are relatively rigid and tend to have extended conformations in the LC state. The chains preferentially align parallel to their neighbors on a local scale, and are often easily oriented on a larger scale by flow. In the solid state the chain segments in crystalline regions tend to be rod-like, whereas those in noncrystalline regions are less ordered but still maintain uniaxial orientation. Consequently, the densities of the crystalline and noncrystalline regions can be very similar. This is especially the case for the thermotropic random copolyesters, where the ordered regions are relatively distorted compared to the crystallites formed by homopolymers. Semi-rigid main-chain LCPs may undergo thermal transitions to one or more liquid crystalline mesophases before becoming an isotropic melt. (The high temperatures typically required to observe isotropization may result in degradation of the polymer before this phase is achieved.)

The entropy of a mesophase is lowest when all mesogens are aligned along the same axis, known as the *director.* In the *nematic* phase the molecules possess one-dimensional order due to the alignment of the mesogenic groups in the backbone. However, there is random stagger of the molecules (and mesogens) along the director axis, and also random orientation about that axis. Two-dimensional order is present in the *smectic* phase, such that the mesogens are aligned in layers, but otherwise have random orientation about their axes. There are numerous subdivisions of the smectic classification, including smectic A, in which the director is perpendicular to the layers, and smectic C, in which the director is tilted. There are also higher ordered smectic structures with more limited rotational freedom for the mesogens, such as smectic E, which approach the order found in the crystalline state. The *cholesteric* phase is similar to the nematic, but the presence of a chiral center in the mesogen results in a helicoidal twist of successive layers of molecules. In all these structures the alignment of the mesogens is generally imperfect. The degree of alignment is measured by the order parameter, S, which is defined as:

$$S = \tfrac{1}{2}(3\langle\cos^2\theta\rangle - 1) \tag{1}$$

where θ is the angle between the molecular axis and the director. For an isotropic system $S = 0$, whereas $S = 1$ for a perfectly aligned system.

10.2 WHOLLY AROMATIC LCPEs

The basic principle in the design of semi-rigid wholly aromatic LCPEs is to reduce the processing temperature of wholly aromatic homopolyesters by the introduction of comonomers while retaining the excellent properties of the homopolymers. Poly(phenylene terephthalamide) cannot be melt-processed because it decomposes before the melting temperature is reached. The polymer is highly crystalline, and the crystal structure is further stabilized by hydrogen bonding between the amide groups. These strong intermolecular interactions can be disrupted in only a very few solvents, allowing for wet-spinning as fibers. The situation for similar homopolyesters, such as poly(phenylene terephthalate) or poly(oxybenzoate) (hereafter referred to as poly(HBA)), is if anything worse, in that the highly crystalline polymers are also insoluble. The melting point, T_m, which will generally be a solid state-to-nematic transition in LCPEs, is a thermodynamic transition defined by:

$$T_m = \Delta H_f / \Delta S_f \tag{2}$$

T_m can be reduced by decreasing ΔH_f, either by incorporating defects into the polymer structure or reducing interchain interactions. Alternatively, we can increase ΔS_f by reducing the chain stiffness. The various approaches used to lower T_m by introducing comonomers are summarized below.

10.3 CHAIN DISCONTINUITIES

The introduction of comonomers into a homopolymer can break the regular interactions of the crystalline structure, thereby lowering the heat of fusion and reducing T_m. In the Vectra series of resins, HBA and HNA monomer units are interspersed. The HNA ester linkages are 'off-set', such that the chain conformation is less linear than that for poly(HBA). Examples of this type of 'crankshaft' discontinuity are shown in Figure 1. Figure 2 shows the effect on the melt transition temperature for copoly(HBA/TPA/hydoquinone (HQ)) when the TPA units are progressively replaced by 2,6-naphthalenedicarboxylic acid (NDA), or when HQ is replaced by 2,6-dihydroxynaphthalene (DHN). An even greater reduction in T_m occurs when TPA/HQ is replaced by HNA. Figures 3(b) and 3(c) show the structures of copoly(HBA/HNA) and copoly(HBA/TPA/DHN), respectively.

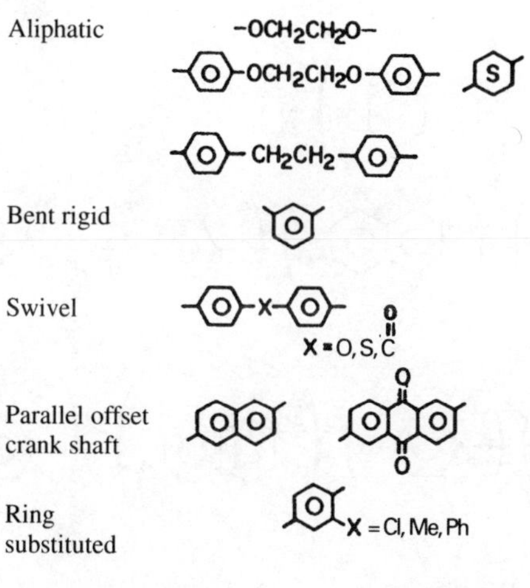

Figure 1 Types of comonomers which reduce T_m in LC copolyesters (after Calundann[11]).

T_m for the wholly *p*-phenylene esters can also be reduced by copolymerizing 'bent' monomers, such as resorcinol (RC), *m*-HBA, or isophthalic acid (IPA). The 1,3-phenylenes lead to a 60° change in the direction of propagation of the chain. This can be compensated almost immediately by modification of the adjacent phenylene ester torsion angles, but the result is still a more substantial defect than is caused by a 2,6-naphthylene unit. Thus incorporation of 1,3-phenylenes reduces T_m more rapidly, but higher mole fractions may lead to a loss of LC properties. For example, 33/33/33 copoly(HBA/TPA/RC) forms isotropic melts, even though the terpolymers are nematic at higher HBA contents. Carbonate units (CO) are similar to IPA and RC in their ability to reduce T_m, but 60/40/40 copolymers of HBA/RC/CO exhibit no LC characteristics.[13] Examples of patented copolymers incorporating both crankshaft and bent monomers are shown in Figures 3(d) and 3(e).

Incorporation of flexible comonomers, such as $(CH_2)_n$ or $(CH_2CH_2O)_n$, reduce T_m by

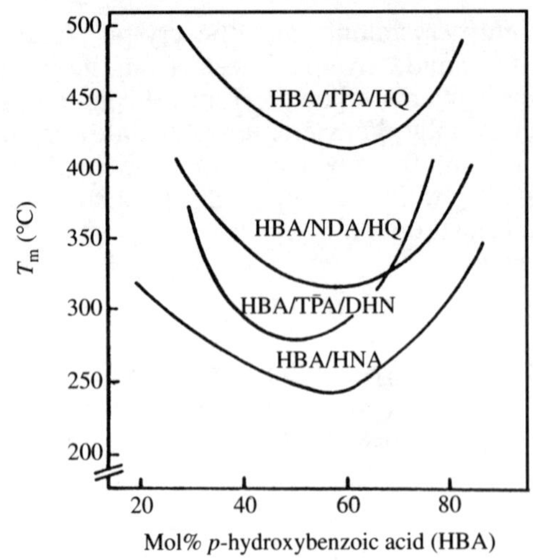

Figure 2 Effect of incorporating symmetrical and nonsymmetrical crankshaft units into a LC polyester (after Calundann[11]).

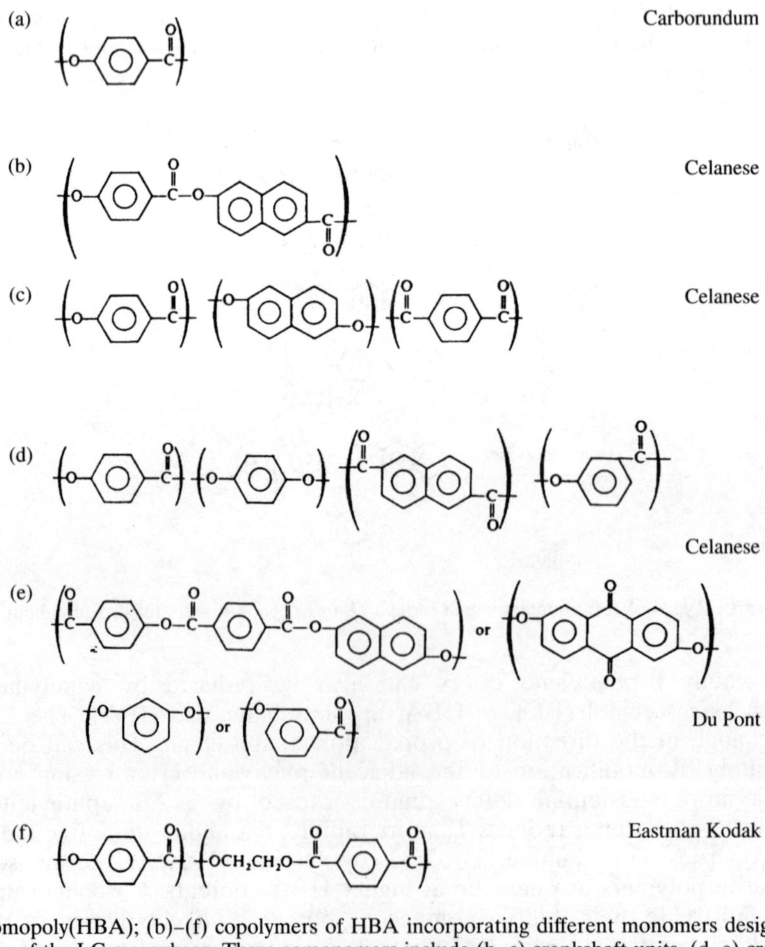

Figure 3 (a) Homopoly(HBA); (b)–(f) copolymers of HBA incorporating different monomers designed to widen the temperature range of the LC mesophase. These comonomers include (b, c) crankshaft units, (d, e) crankshaft and bent units, and (f) aliphatic units (after Blackwell and Biswas[9]).

increasing the entropy of the transition. The best example of this class of copolymer is X7G, *i.e.* copoly(HBA/ethylene terephthalate (ET)) shown in Figure 3(f), which is prepared by the transesterification reaction of acetoxy-HBA and PET. Approximately 50% HBA is necessary to convert isotropic PET to a nematic melt. The presence of the flexible ethylene spacers leads to tensile moduli for melt-spun fibers that are much lower than those for wholly aromatic copolyesters. Overall, T_m decreases with increasing spacer length, and there are pronounced odd–even effects for n in the homologous series. The spacers exist predominantly in the fully extended *trans* conformation. Hence, when n is even, such as in copoly(HBA/ET), the axes of the mesogens are parallel. However, odd values of n result in nonlinear minimum energy conformations, and the distortions necessary in a structure with a nematic arrangement of the mesogenic groups leads to lowering of T_m.

Probably the most effective way of disrupting the interactions between rigid wholly aromatic polyester backbones is to introduce side chains, which reduce T_m by interfering with chain packing. Substituents such as a halide, methyl, or phenyl group can sufficiently hinder intermolecular packing but still allow the formation of a stable mesophase. Long alkyl groups reduce the stability of the LC phase,[14] but systematic attachment of these groups can result in a second ordered phase perpendicular to the polymer backbone if the number of carbon atoms is greater than ~12.[15–17]

10.4 HOMOPOLYMER STRUCTURES AND PROPERTIES

The chemical structure of poly(HBA) is shown in Figure 3(a). The polymer exhibits two thermal transitions, at ~340 and 450 °C. The first of these corresponds to a crystal-to-plastic crystal transition; it has been argued that the latter state corresponds to a smectic E structure.[18] The higher temperature transition may be conversion to a nematic structure, but the data in this region are complicated by thermal degradation. Several polymorphic structures of poly(HBA) have been characterized by X-ray and electron diffraction.[19–21] At room temperature, high molecular weight poly(HBA) has an orthorhombic unit cell with dimensions $a = 7.52$ Å, $b = 5.70$ Å, and $c = 12.49$ Å, which has been designated form I.[19] However, a second structure is seen at room temperature for oligomers (form II), which have an orthorhombic unit cell with dimensions $a = 3.77$ Å, $b = 11.06$ Å, and $c = 12.89$ Å. Heating the oligomers results in partial conversion to form I, which coexists with form II.[19] Above the transition at ~350 °C, poly(HBA) can be processed by sintering the powder to form shaped objects. Both forms I and II are converted to the pseudohexagonal form III at this temperature: the orthorhombic unit cell dimensions are $a = 9.2$ Å, $b = 5.3$ Å, and $c = 12.4$ Å.[20] The calculated density of form III (1.32 g cm^{-3}) is ~13% lower than the equivalent values for forms I and II (1.48 g cm^{-3}) and 1.49 g cm^{-3}, respectively). The c dimension corresponds to a dimer repeat in which successive monomers along the chain are related by a glide plane, and mutually inclined by ~60°.[22,23] Similar unit cells for forms I and III have been determined by Yoon.[21] In the proposed smectic E structure for form III the chains are in three-dimensional register, and the HBA units are packed in a herring bone array, but there is disorder that is probably due to random selection of two conformations of the HBA unit separated by a 180° rotation about the chain axis.

Information about the crystalline structure of poly(HBA) has also been obtained indirectly from the X-ray fiber diagrams of copoly(HBA/ET). Unlike wholly aromatic copolyesters, such as copoly(HBA/HNA), the X-ray data for copoly(HBA/ET) are not strongly composition dependent.[20] Although NMR indicates that the sequence distribution is approximately random, it is possible that thermal treatment may result in solid state transesterification, leading to blocky sequences of HBA. This is suggested by the observation of biphasic domains by optical and electron microscopy, multiple thermal transitions, and the existence of fractions with different solubilities.[24–26] Blackwell *et al.*[20] have examined the X-ray diffraction patterns of 60/40 and 80/20 copoly(HBA/ET) and have compared the results with those for poly(HBA). The 80/20 copolymer showed equatorial reflections at $d \approx 4.4$ Å (very strong) and $d \approx 2.5$ Å (weak), an off-equatorial reflection at $d \approx 3.2$ Å (medium), and meridonial reflections at $d \approx 6.2$ Å (weak), 3.15 Å (strong) and 2.1 Å (strong). These data indicated an extended chain conformation with two aromatic units within a fiber repeat of 12.4 ± 0.2 Å. Thus the structure of the 80/20 copolymer was determined to be very similar to that of the high-temperature form (form III) of poly(HBA) determined by electron diffraction. The features of the X-ray diffraction pattern of the 60/40 copolymer were similar, but the reflections were considerably less sharp and the first meridional peak was absent. It was concluded that these random copolymers are capable of crystallizing as the less dense form III, in which there is sufficient space available for the ethylene terephthalate units to exist as defects. It is possible that the crystalline regions contain predominantly HBA-rich sequences, although this is by no means essential in the light of the packing studies for the ordered regions of wholly aromatic copolyesters, as will be discussed below.

Studies of the structure of poly(HNA) indicate that this polymer has physical characteristics similar to those of poly(HBA).[22,27] Single crystals showed a slab morphology similar to that of poly(HBA), and a nematic-like structure was evident in quenched specimens.[27] X-ray diffraction indicated a two-monomer repeat along the chain axis direction of the unit cell. The X-ray powder pattern obtained at room temperature was indexed by an orthorhombic unit cell with dimensions $a = 7.66$ Å, $b = 5.98$ Å, and $c = 17.12$ Å ($\rho_{calc} = 1.44\ g\ cm^{-3}$).[22] Two thermal transitions were detected at approximately 338 and 453 °C, corresponding to the solid-to-smectic E and smectic E-to-nematic transitions at 340 and 455 °C in poly(HBA). The enthalpy of the lower transition is similar for both polymers, but that for the higher transition is ~15 times greater than that for the corresponding transition in poly(HBA), suggesting that the crankshaft linkages in poly(HNA) result in a lower order parameter in the nematic melt.[27] The unit cell dimensions of the smectic E phase observed at 370 °C are $a = 9.28$ Å, $b = 5.64$ Å, and $c = 17.04$ Å ($\rho_{calc} = 1.27\ g\ cm^{-3}$),[22] which correlate with those for the high-temperature phase of poly(HBA).

10.5 STRUCTURE AND PROPERTIES OF COPOLY(HBA/HNA)

The remainder of this review will focus on the structure and properties of copoly(HBA/HNA). This system of wholly aromatic LCPEs has received the most extensive study, and is in many ways typical of the entire class.

10.5.1 Molecular Weight and Persistence Length

Molecular weight, polydispersity, and molecular dimensions of 70/30 copoly(HBA/HNA) have been determined in solution in 3,5-bis(trifluoromethyl)phenol (BTFMP) at 60 °C by light scattering, size-exclusion chromatography (SEC), and viscometry.[28] The preparation studied had $M_w = 36\,000\ g\ mol^{-1}$ as determined by light scattering and a polydispersity index of ~3.2 calculated from SEC. Earlier, Calundann had reported an average molecular weight of $\sim 25\,000\ g\ mol^{-1}$ for copoly(HBA/HNA) dissolved in pentafluorophenol.[29] Fractionation of the 70/30 copolymer[28] reduced the polydispersity to $M_w/M_n = 1.2$–2.3, and intrinsic viscosity measurements on these fractions yielded a Mark–Houwink exponent of 0.84, indicative of a highly extended chain conformation. (The Mark–Houwink exponent ranges from 0.5 to 0.8 depending on solvent conditions for a flexible polymer coil, and approaches ~0.98 for more rigid wholly aromatic polyesters.[30]) In addition, the ratio of the radius of gyration to the hydrodynamic radius ($\rho = R_g/R_h$) varied between 2.2 and 2.9, compared to $\rho < 1.7$ for flexible polymer coils. A persistence length of $l = 120$ Å was calculated from R_g, R_h, and M_w/M_n, using the Kratky–Porod worm-like coil model. The chain conformations for copoly(HBA/HNA) were also investigated using rotational isomeric state (RIS) theory to calculate the persistence lengths and characteristic ratios, C_∞.[31] The copoly(HBA/HNA) chains all had characteristic ratios similar to that for poly(HBA) ($C_\infty = 236$). The persistence length, however, depended on the average monomer length and increased with increasing HNA content. For a 70/30 copolymer, $l = 810$ Å at $T = 0$ K. However, when the temperature dependence of the chain conformation was taken into account, $l = 160$ Å at 60 °C, in reasonable agreement with the experimental results of Kromer *et al.*[28]

10.5.2 Sequence Distribution, Conformation, and Chain Packing

The X-ray diffraction patterns of oriented fibers of copoly(HBA/HNA) have been studied in considerable detail.[32–37] The meridional maxima are aperiodic, *i.e.* the d spacings are not orders of a simple repeat, and are explained by a structure consisting of parallel, extended chains of completely random comonomer sequence, which will be explained further below. The number and positions and the maxima are dependent on comonomer ratio. Intense equatorial Bragg reflections are observed for the as-spun fibers at $d \approx 4.6$, 2.6, and 2.3 Å, and there is also an off-equatorial reflection on the first layer line at $d \approx 3.3$ Å. These data indicate that the structure has three-dimensional order despite the random monomer sequence. The reflections are indexed with a two-dimensional orthorhombic unit cell with dimensions $a = 9.2$ Å and $b = 5.2$ Å (pseudohexagonal packing), similar to those for form III of poly(HBA).

Figure 4(a) shows the observed diffractometer scans along the chain axis direction for six monomer ratios of copoly(HBA/HNA). Within the scattering range of $2\theta = 5$–$50°$, we see either three or four maxima, with d spacings listed in Table 1. The maximum at $d \approx 2.1$ Å is seen for all compositions, but the other maxima shift with composition: the first maximum moves to progressively higher 2θ with increasing HBA content, while the third maximum similarly moves to lower 2θ; the second maximum seen at high HNA content weakens as the proportion of HBA is increased, and disappears after ~50/50. These data effectively rule out a block copolymer microstructure: if this was the case we would see the maxima characteristic of both blocks, but the d spacings would not change with the monomer ratio.

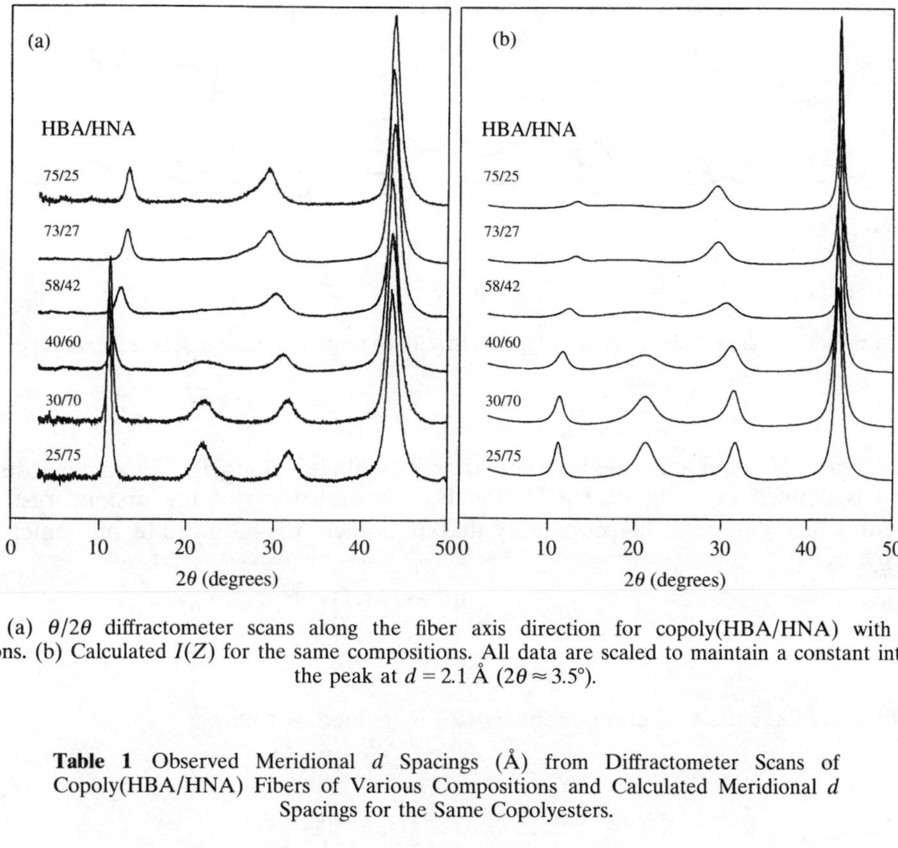

Figure 4 (a) $\theta/2\theta$ diffractometer scans along the fiber axis direction for copoly(HBA/HNA) with indicated compositions. (b) Calculated $I(Z)$ for the same compositions. All data are scaled to maintain a constant intensity for the peak at $d = 2.1$ Å ($2\theta \approx 3.5°$).

Table 1 Observed Meridional d Spacings (Å) from Diffractometer Scans of Copoly(HBA/HNA) Fibers of Various Compositions and Calculated Meridional d Spacings for the Same Copolyesters.

		Observed d spacings (Å)			
25/75	*30/70*	*40/60*	*58/42*	*73/27*	*75/25*
7.97 ± 0.06	7.92	7.62	7.16	6.74	6.68
4.08 ± 0.04	4.05	4.11	—	—	—
2.83 ± 0.02	2.84	2.88	2.97	3.05	3.05
2.08 ± 0.01	2.08	2.08	2.07	2.07	2.07
		Calculated d spacings (Å)			
25/75	*30/70*	*40/60*	*58/42*	*73/27*	*75/25*
7.97	7.89	7.62	7.15	6.72	6.67
4.19	4.20	4.20	—	—	—
2.83	2.85	2.86	2.95	3.02	3.03
2.08	2.08	2.07	2.07	2.07	2.07

The X-ray scattering along the chain axis direction (corresponding to the meridian of the fiber diagrams) for copoly(HBA/HNA),[32–37] and for other wholly aromatic copolyesters,[38–40] can be modeled as the Fourier transform of the nearest neighbor probability function, $Q(z)$. A model of a short segment of a copoly(HBA/HNA) chain with a random sequence is shown in Figure 5.[41] The chain has an extended conformation, such that the axial advances per monomer are approximately constant for each monomer type, and are approximately equal to the distances between the terminal ester oxygens. The scattering intensity in the fiber axis direction, $I(Z)$, depends on the projection of the structure onto the chain axis (z), which is largely independent of conformation. For a nematic structure, *i.e.* random axial stagger of parallel chains, $I(Z)$ can be predicted from a consideration of a single chain averaged over all possible sequences. For a solid state structure with three-dimensional chain register, the scattering predicted for the single chain will be modified by the effects of intermolecular interferences. However, it has been shown that while these effects change the relative intensities of the maxima in $I(Z)$, their positions are largely unaffected.[41–43] Thus the sequence distribution and chain conformation in the three-dimensionally ordered structure can be investigated by comparison of the observed data with those predicted for a single extended infinite chain, or a finite chain averaged over all possible sequences.

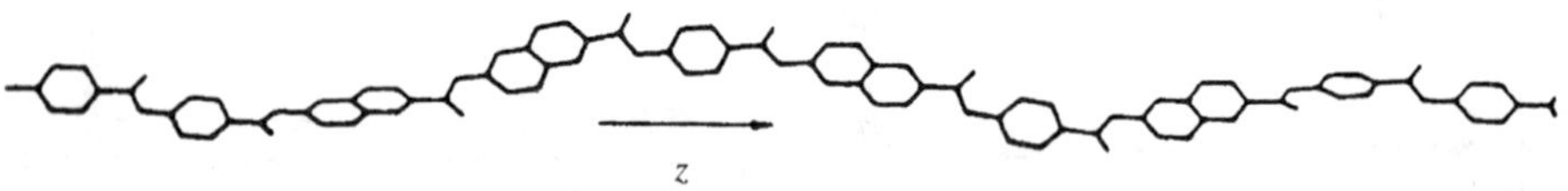

Figure 5 A model of a short chain segment of copoly(HBA/HNA) with a typical random sequence (after Biswas and Blackwell[41]).

$Q(z)$ defines the monomer nearest neighbor probability along the chain axis. Its Fourier transform is defined in terms of $H_1(Z)$, the Fourier transform of the first nearest neighbor component, which describes the probability that monomer A is followed by monomer B.[37] $I(Z)$ is given by:

$$I(Z) = \mathrm{Re}\left[\frac{1 + H_1(Z)}{1 - H_1(Z)}\right] \tag{3}$$

where Re designates the real component. $H_1(Z)$ is defined as follows:

$$H_1(Z) = \sum_{A}\sum_{B} F_{AB}(Z) p_A M_{AB} \exp(2\pi i Z z_B) \tag{4}$$

where the subscripts designate monomers A and B and the summation is over all AB monomer pairs.[37] $F_{AB}(Z)$ is the Fourier transform of the convolution of monomer A with monomer B:

$$F_{AB}(Z) = \sum_{A}\sum_{B} f_{j,A} f_{k,B} \exp(2\pi i Z(z_{j,A} - z_{k,B})) \tag{5}$$

where f is the scattering factor for the atom with axial coordinate z; the subscripts designate the jth atom of monomer A and the kth atom of monomer B, respectively, and the summations are over all atoms in monomers A and B. p_A is the mole fraction of monomer A; z_B is the axial advance of monomer B. M_{AB} defines the probability that monomer A is followed by monomer B. For a random sequence of copoly(HBA/HNA), $M_{AB} = p_B$, the molar ratio of the second monomer forming the pair, but deviates from this value if the sequence statistics are nonrandom. Examination of the molecular model in Figure 5 shows that the chains are only approximately linear: they have a sinuous character due largely to the 'crankshaft' HNA units. Consequently, the lengths of the monomers, z_B, are not constant, but are described by a relatively narrow distribution function, which can be derived experimentally by surveying molecular models.

Figure 4(b) shows the predicted $I(Z)$ for the same six monomer ratios for which we have presented the experimental data. The calculated intensity data have been corrected for Lorentz and polarization effects for ease of comparison with those observed. It can be seen that there is excellent agreement between the observed and calculated data, as is confirmed by comparison of the d spacings of the maxima in Table 1. The invariant peak at $d \approx 2.1$ Å arises because the lengths of the HBA and HNA monomers used in the model are 6.35 and 8.37 Å, respectively: 2.1 Å corresponds to the third order of one and the fourth order of the other, and thus a peak occurs at this position regardless of copolymer composition and is relatively sharp. The only notable discrepancy is the intensity of the first meridional maximum, which tends to be underestimated by this model. This is particularly true for the copolymers rich in HNA, *i.e.* the 25/75 and 30/70 copolymers, which exhibit very strong first meridional peaks that rival the intensity of the peak at $d \approx 2.1$ Å. This discrepancy is almost certainly due to interference effects due to chain packing, which is not considered in the above model. Modification of the M_{AB} terms in equation (4) leads to shifts and splitting of the calculated peaks,[37] and the good agreement is lost for all but minimal nonrandomness. Similar conclusions were drawn in studies of other wholly aromatic copolyesters, including copoly(HBA/DHN/TP)[38,39] and copoly(HBA/BP/PA).[40] Thus the X-ray data gave the first definition of the random microstructure for the melt synthesized and processed wholly aromatic LCPEs, and this conclusion was later confirmed by ^{13}C NMR.[44,45]

The width of the peak at $d \approx 2.1$ Å depends on the distribution of monomer lengths, and thus can be used to study the chain sinuosity by determining a persistence length for the extended chain in the solid state. Alternatively, this parameter can be derived as an axial crystallite width using the Scherrer equation if one assumes perfectly straight chains of finite length. The estimated axial persistence lengths for copoly(HBA/HNA) range from 9 to 13 residues,[33] which correlate well with the morphological evidence for 100 Å thick lamellae by electron microscopy.[46]

10.5.3 Three-dimensional Order

Three-dimensional models have been developed to attempt to explain the observed sharp equatorial and off-equatorial reflections.[41–43] The existence of a strong off-equatorial Bragg reflection at $d \approx 3.3$ Å on the first layer line (at an axial d spacing of ~14 Å, depending on the monomer ratio) requires a physical 'dimer repeat', which is due to an alternation of the mutual inclination of successive monomers along the chain by +60° and −60°. The observation of this and other off-meridional Bragg reflections also points to the existence of three-dimensional order, *i.e.* registration of adjacent sequences. We have shown that this can be explained by limited axial registration of nonidentical chains, which would arise as a result of stacking of aromatic rings or dipole–dipole interactions between adjacent carbonyl groups. The latter interactions are maximized if short sequences of 9–13 monomers are in register at their centers. X-ray simulations based on averaged chains showed that such a model is in fact more ordered than is observed, and the perfect registration at the centers of the lamellae needs to be replaced by a distribution about a central 'register plane', such that the actual registration is very minimal indeed. Windle and co-workers have proposed an alternative model in which identical nonrepeating sequences segregate to form nonperiodic lattices.[47,48] This also predicts the observed off-meridional diffraction maxima, but the model involving limited register plane is more feasible since it requires much less rearrangement of chains from the nematic melt. It is also more compatible with the observed lateral crystallite size and degree of crystallinity. The equatorial reflections observed for as-drawn HBA/HNA fibers ($d = 4.6$, 2.6, and 2.3 Å) are indexed as the 110, 020, and 220 reflections for a pseudohexagonal unit cell with dimensions $a = 9.2$ Å, $b = 5.2$ Å. However, annealing at 275 °C for 30 days results in the coexistence of a more dense orthorhombic form with dimensions $a = 7.6$ Å and $b = 5.7$ Å.[23]

X-ray analysis of the HBA/HNA copolymers under stress has provided information on the monomer conformations. Troughton *et al.*[49] have compared the tensile modulus of melt spun fibers with predicted values assuming either *cis* or *trans* conformations of HNA. The highest modulus occurs when the chains have restricted sinuosity, which occurs when a nonrandom distribution of conformations is present. Although macroscopically the 30/70 and 75/25 copolymers have the same modulus, X-ray analysis showed that the modulus of the 30/70 copolymer was smaller than that of the 75/25 copolymer. When all-*trans*-HNA was used, the predicted modulus was too high for the 30/70 copolymer and too low for 75/25 HBA/HNA. When *cis*-HNA was used exclusively, increasing the HNA content resulted in a decrease in modulus. Overall, the calculated moduli were lower than those observed, indicating that lateral packing plays a significant role in increasing the observed mechanical properties by creating additional shear forces that must be overcome during deformation. This group concluded that although the *trans*-HNA may be present in the HBA/HNA copolymers, *cis*-HNA is the predominant conformation.

10.5.4 Molecular Modeling of Arrays of Nonidentical Chains

Molecular mechanics modeling has been used to study the feasibility of packing nonidentical chains in a manner compatible with the observed X-ray data.[50] The torsion angles defining the chain conformation are shown in (**1**).

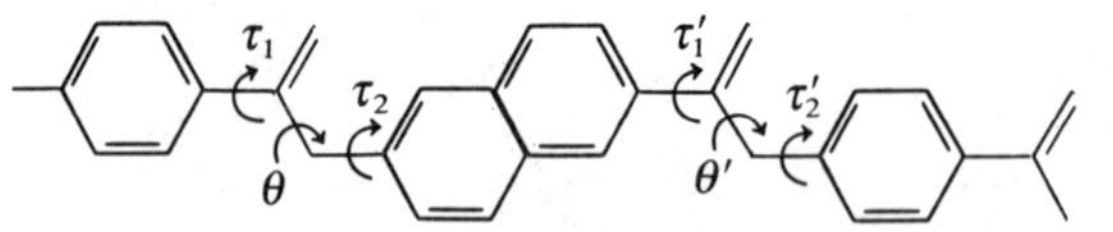

(**1**)

The energy minima are close to $\tau_1, \tau_1' = 0°, 180°$, $\theta, \theta' = 180°$, and $\tau_2, \tau_2' = \pm 60°, \pm 120°$, as seen in model compounds. The minima are relatively broad, and thus τ_1 and τ_2 can be changed significantly with only a relatively minor effect on the chain energy. The HNA residue was restricted to the shorter *cis* conformation ($\tau_1' = 180°$, as shown in (**1**)), as suggested by the X-ray data and also by the mechanical properties.[49] Thus considerable degeneracy is possible with regard to the mutual orientation of the carbonyls on adjacent residues of an individual chain. When the mutual inclinations of successive aromatic planes alternate between $+60°$ and $-60°$, two conformational extremes can be identified: conformation I with symmetric dispositions of the carbonyls to one side of the chain ('butterfly' conformation), and conformation II with random disposition of the carbonyls to both sides of the chain.

Chains of 12 residues were constructed from the dimer models, with random sequences selected using a random number generator, subject only to the overall monomer disposition in the arrays. In a particular array, all the chains had conformations of the same type (I or II). The chains were packed on an orthorhombic network with $a = 9.2$ Å, $b = 5.3$ Å, with their central ester oxygens in register along the fiber axis direction. Adjacent monomers in the register plane were packed in an approximate herringbone pattern. For conformation I, this condition could be achieved in two ways: either the 'butterflies' are oriented in the same direction along the *a* axis or are alternating. These two types of packing are designated A and B.

The arrays subjected to energy minimization consisted of 19 nonidentical chains of 12 monomers. Twenty different arrays were considered so as to explore the effects of chain sequences, starting conformations and packing. Energy minimization was done using the Tripos SYBYL package. The energy was calculated using the Tripos force field, as the sum of the bond stretching, bond angle bending, out-of-plane bending, van der Waals nonbonded interactions, and electrostatic energy contributions. Energy minimization for an array using a three-dimensional periodic boundary condition was not possible because the chains had different lengths, and the boundary problem was approached by a three-step procedure. First, the structure of the inner seven chains was optimized while the outer chains remained fixed. Thereafter, the central seven chains were held constant while the energy of the 12 outer chains was minimized. X-ray scattering by the arrays on the equator and first layer line was simulated by calculating the cylindrically averaged intensity transform.

For all 20 starting models the calculated potential energy was very high (in excess of $+4 \times 10^4$ kJ mol^{-1}), due to overlap of monomers on different chains. Minimization led to models which had potential energies in the range −5400 to −6000 kJ mol^{-1}, with most being closer to the lower figure. Similar calculations for homopoly(HBA), using the same lattice parameters, starting conformations, and packing models, led to minimized structures with energies in the range −5000 to −5400 kJ mol^{-1}. These figures are actually higher than those obtained for the copolymer models, but this probably reflects simply the different number of atoms (3021 for the homopolymer, 3363 for the 75/25 copolymer). For an equivalent poly(HNA) structure (4389 atoms), the energy is a little lower: −7400 kJ mol^{-1}. The significant point here is that the energies for both homopolymer models are not greatly different from those for the copolymer, and they demonstrate that the random chains of the copolymer can be packed on the observed lattice with relatively little difficulty. It means that there are no packing problems that would serve as the driving force for segregation of identical sequences.

The differences between the starting and final models occur almost entirely in the aromatic–ester torsion angles, τ_1 and τ_2. Even so, the averages for these angles are close to the energy minima for the isolated dimers. The conformational changes that occur are seen most easily in *c*-axis projections of a starting and minimized array, as shown in Figure 6. The straight lines are the projections of the aromatic units, which start out inclined by alternately +60° and −60°, but adjust to a distribution of orientations, such that the chain has an almost cylindrical projection. However, this effect is only appreciated in projections involving a number of residues (in this case 12 per chain). The standard deviations indicate that the changes from one residue to the next are relatively small. It is unlikely that any of the final models represent global energy minima, but rather than there are a very large number of almost equivalent minima. The energies of most of the minimized models are within 1.3 kJ mol^{-1} of the monomer and are probably representative of the many arrangements that occur in the quenched melt. Upon annealing, the conformation may change as a result of 180° flips of the aromatic and ester groups.

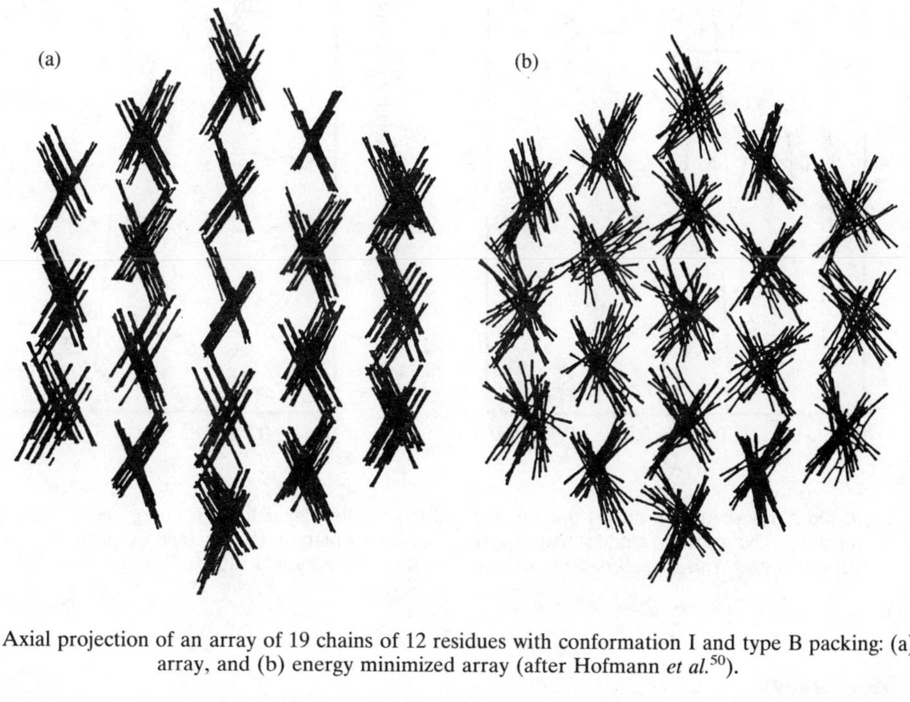

Figure 6 Axial projection of an array of 19 chains of 12 residues with conformation I and type B packing: (a) starting array, and (b) energy minimized array (after Hofmann *et al.*[50]).

Figure 7 shows the predicted X-ray scattering on the equator and first layer line for three arrays of 19 chains of 12 monomers, after energy minimization. The three starting models had chains with conformation II, and with conformation I in types A and B packing, respectively. The primary equatorial maxima indexed as $hk = 11/20$ and 02 for the two chains orthorhombic unit cell, which are seen in the experimental data, are predicted for all three models (Figure 7(a)). For the model with chains in conformation I and type B packing we also predict a significant maximum for $hk = 21$, which is not observed experimentally but arises in the calculations due to the differences between the projections of the two chains in the unit cell onto the *ab* plane. The other peaks are subsidiary maxima due to the small lateral size of the arrays. On the first layer line (Figure 7(b)), all three models generate the observed $hk = 21$ maximum at $d \approx 3.1$ Å that is characteristic of three-dimensional order. This is the only observed first layer line maximum predicted by the model containing chains in conformation II. For the models containing chains in conformation I, we also predict maxima indexed as $hk = 10$ for type A packing and $hk = 11/20$ for type B packing, which are unobserved experimentally. Thus we can conclude that the Bragg reflections seen on the equator and first layer line are predicted even by these small arrays of 19 chains of 12 monomers in different random sequences. Best agreement is obtained for the model in which the chains have type II conformation, *i.e.* random orientation of the ester carbonyls within an approximate herringbone packing of the monomers. It is reasonable that such randomness will be present in a structure quenched from the nematic melt, but will become more ordered on annealing when the second, more dense, orthorhombic phase is formed.

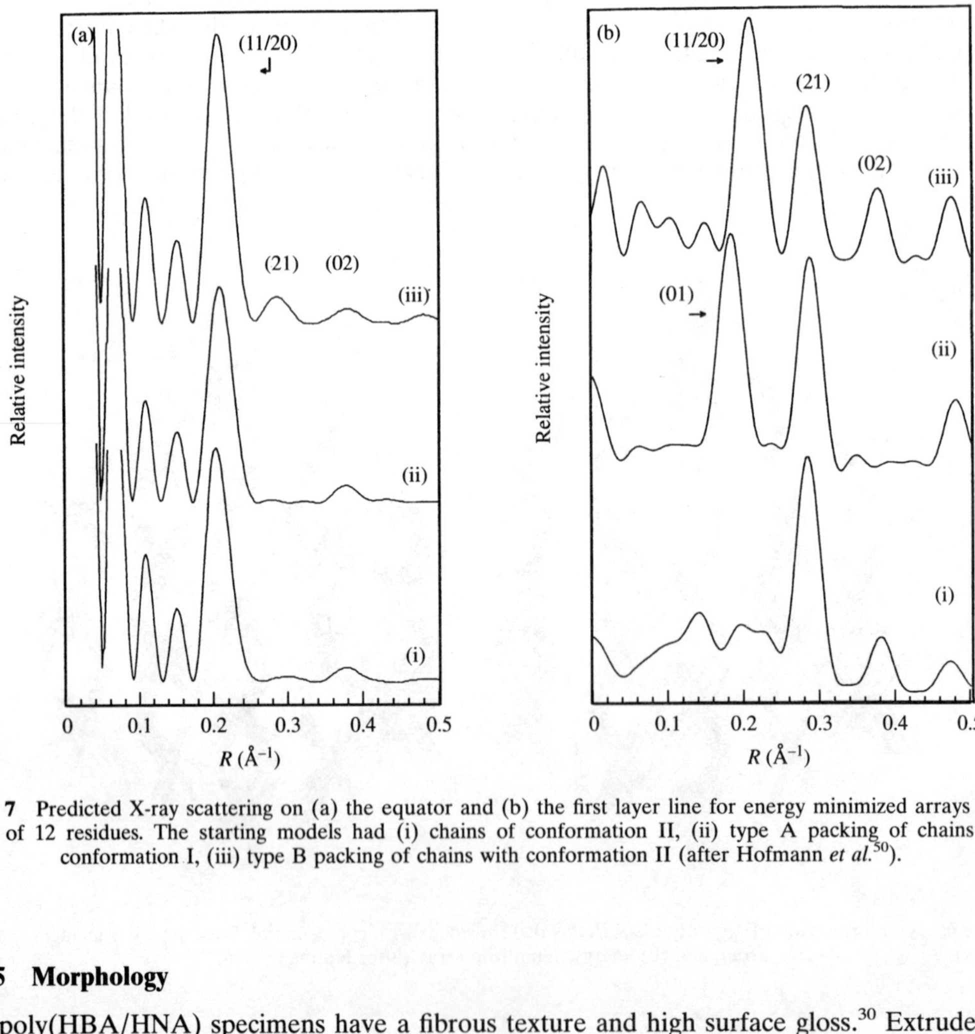

Figure 7 Predicted X-ray scattering on (a) the equator and (b) the first layer line for energy minimized arrays of 19 chains of 12 residues. The starting models had (i) chains of conformation II, (ii) type A packing of chains with conformation I, (iii) type B packing of chains with conformation II (after Hofmann *et al.*[50]).

10.5.5 Morphology

Copoly(HBA/HNA) specimens have a fibrous texture and high surface gloss.[30] Extruded or injection modeled specimens have a pronounced skin–core morphology. The birefringent skin contains highly aligned molecules, whereas there is less orientation in the core.[51] Transmission electron microscopy of the morphology of injection molded 73/27 copoly(HBA/HNA) showed that the defect density in the core was six orders of magnitude greater than in the skin.[52] Studies of both replicas of moldings and thin films revealed the existence of regular banding along the chain axis direction, consisting of 'crystalline' lamellae (~10 nm thick, ~100 nm wide), separated by less well ordered interlamellar regions, with an overall axial repeat distance of ~34 nm.[46] These data are consistent with the reported crystallinites of ~30% for unannealed specimens. Given the extended conformation of these polymers, a typical molecule is ~200 nm in length and therefore passes through ~6 ordered lamellae. This observation makes unlikely the lattice matching model proposed by Windle and co-workers[47,48] to account for the ordered regions. While it is possible that axial shifts could occur to match the sequences at one point on adjacent chains, it is very unlikely that similar matching would also occur at five other points separated by regular steps of 34 nm. However, the morphology is consistent with the limited axial registration proposed by this laboratory,[42,43] with the defects such as chain ends and cross-overs being concentrated in the interlamellar regions.

10.5.6 Solid State Relaxations

Several relaxations have been detected for copoly(HBA/HNA) in the solid state, using dynamical mechanical and dielectric relaxation methods.[53–55] The first, at −60 °C, is assigned as T_γ, and is attributed to the rotation of the phenylene rings; the activation energy is relatively small (<200 kJ mol^{-1}). The temperature at which this relaxation occurs is independent of composition, but is affected by the frequency of testing. Near to room temperature (~25–50 °C) a second relaxation occurs, T_β, which is thought to correspond to onset of rotation of the naphthalene rings. The temperature at which this relaxation occurs also depends on the test frequency and has a small activation energy, but unlike T_γ is composition dependent and increases with the HNA content. At ~100 °C, a third relaxation is detected, which is designated T_α, and is thought to correspond to increased rotational motions and segmental mobility, and hence is similar to T_g in glassy polymers. The activation energy is high (>400 kJ mol^{-1}) and the transition temperature is relatively independent of copolymer composition. Unlike the lower relaxations, T_α is independent of the test frequency. These relaxations can be seen in the plots of tensile tan δ against temperature shown for the 73/27 and 30/70 copolymers in Figure 8.[53]

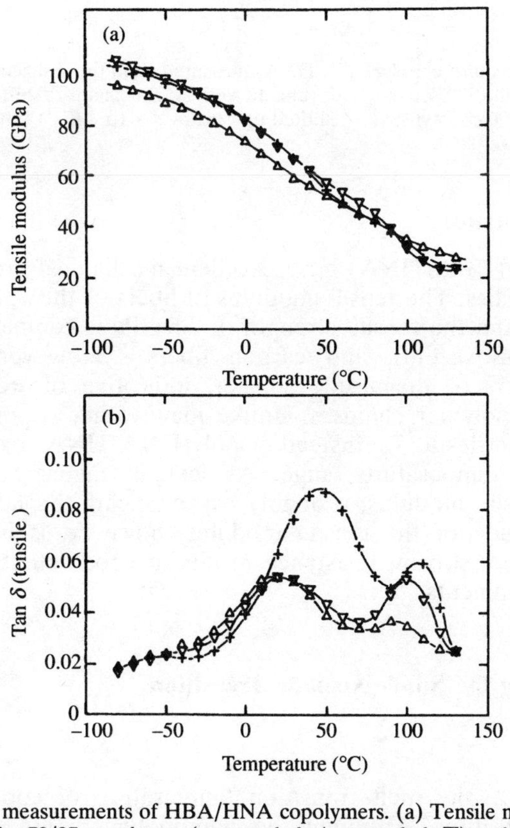

Figure 8 Dynamic mechanical measurements of HBA/HNA copolymers. (a) Tensile modulus and (b) tensile tan δ as a function of temperature for the 73/27 copolymer (unannealed, △; annealed, ▽) and the annealed 30/70 copolymer (+) (after Troughton *et al.*[53]).

10.5.7 Thermal Expansion

The thermal expansion of HBA/HNA fibers and crystallites has been studied using X-ray diffraction.[56–58] The coefficient of thermal expansion along the chain axis direction, α, was determined by Troughton *et al.*[56] from the small changes in d spacing for the diffraction maximum in the region of $d = 2.1$ Å. Their results are shown in Figure 9. α is negative, but is approximately constant at temperatures up to ~100 °C, at which point it decreases rapidly with increasing temperature. It appears that the chains are highly extended at low temperatures, and hence tend to contract in the axial direction as the structure becomes more disordered. Analysis of the lattice parameters a and b as a function of temperature is shown in Figure 10.[57] The crystallites expand laterally in the a dimension while the expansion in the b dimension is relatively slight. The change in thermal expansion coefficient above ~100 °C for the 75/25 copolymer on repeated thermal cycling (Figure 10(c)) is attributed to molecular rearrangement from the orthorhombic form to the less dense hexagonal packing. This is similar to the effect seen for the homopolymer of HBA, shown in Figure 10(d). Together, these studies indicate that the polymer chains become more sinuous with increasing temperature, causing the crystalline regions to expand laterally while contracting slightly in the chain axis direction.

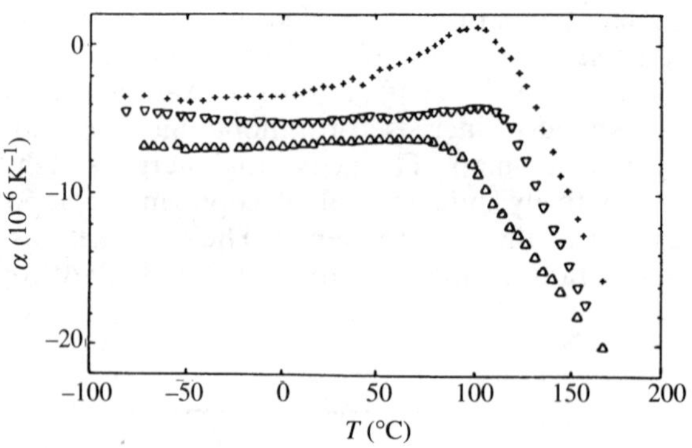

Figure 9 Axial thermal expansion of copoly(HBA/HNA) measured from the d spacing of the meridional reflection at ~2.1 Å. Results are shown for the 73/27 copolymer (unannealed, △; annealed, ▽, shifted upward by $2 \times 10^{-6}\,K^{-1}$) and the annealed 30/70 copolymer (+, shifted upward by $4 \times 10^{-6}\,K^{-1}$) (after Green *et al.*[56]).

10.5.8 Mechanical Properties

Preparations of copoly(HBA/HNA) have excellent mechanical properties, both in the form of fibers and molded plastics. The tensile modulus of fibers of the 73/27 mole ratio is ~65 GPa at room temperature,[53] and the tensile strength of 2.6 GPa is comparable to those for Kevlar, poly(p-phenylenebenzobisoxazole), and carbon fibers.[59] However, the shear modulus is approximately two orders of magnitude smaller, indicative of weak intermolecular forces between the random copolymer chains.[53] Unlike many other copolymers, there is no sharp decline in materials properties at T_α. Instead, copoly(HBA/HNA) exhibits a gradual decline in properties over a wide temperature range. As seen in Figure 8,[53] annealing leads to an improvement in the tensile modulus, probably by enhancing the lateral interactions between the chains. The magnitude of the tensile modulus, however, is independent of copolymer composition. The excellent solvent resistance results in good retention of modulus in a wide range of chemical environments.[30]

10.5.9 Factors Affecting the Solid–Nematic Transition

10.5.9.1 Composition

As shown in Figure 2, the melt transition temperature of copoly(HBA/HNA) depends primarily on the composition. Although the homopolymers are infusible, the copolymers consisting of 20–80% HBA melt in the range of ~250–350 °C. The minimum in T_m occurs at a

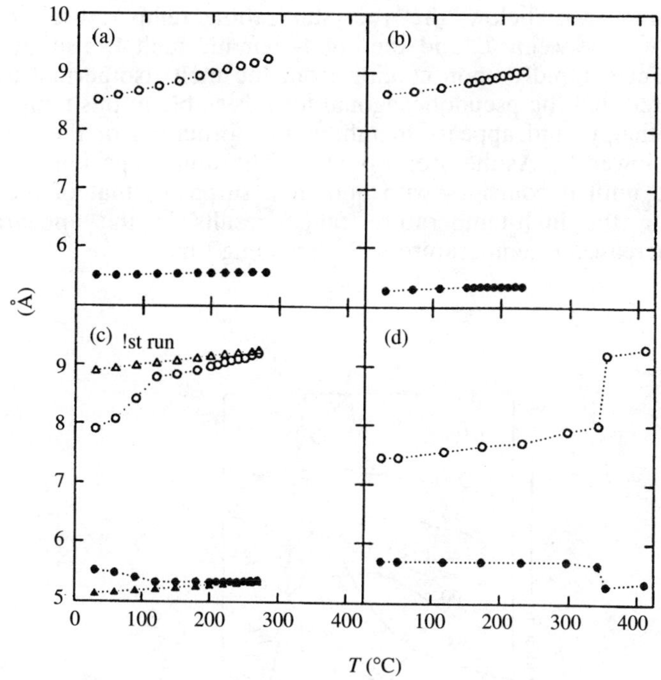

Figure 10 Lateral thermal expansion of copoly(HBA/HNA). The unit cell parameters *a* and *b* are indicated by filled and open symbols, respectively, for (a) 30/70, (b) 58/42, and (c) 75/25 copoly(HBA/HNA), and (d) poly(HBA) (after Flores *et al.*[57]).

composition of ~60/40 HBA/HNA.[30] A similar trend is observed for other wholly aromatic copolyesters. Copolymerization lowers T_m by introducing 'defects' in the homopolymer sequences, thereby reducing the packing efficiency of the chains.

10.5.9.2 Blending

Blends of copoly(HBA/HNA) have been examined by DeMeuse and Jaffe.[60] They found that T_m for a blend of two preparations with different monomer ratios was below that for the pure copolymer with the same total monomer ratio. In a 50:50 blend of the 30/70 and 75/25 copolymers, the observed $T_m = 236$ °C was lower than that of any HBA/HNA copolymer of a single composition. In this laboratory we have shown that this effect is understood in terms of the formation of a compatible blend of the two copolyester preparations.[61] We found that a melt blend of the 30/70 and 75/25 copolymers ($T_m = 303$ °C and 288 °C, respectively) at an overall monomer ratio of 60/40 had $T_m = 227$ °C, but this figure rose to 248 °C for the specimen compression molded for 60 minutes at 315 °C, the same as that for the 60/40 random copolymer. X-ray diffraction data showed that this is explained by transesterification, which will be discussed further below.

10.5.9.3 Annealing

The effects of annealing the HBA/HNA copolymers have been examined in this laboratory[23] and by other groups.[62] Annealing the 75/25 copolymer at 280 °C results in the appearance of a more dense orthorhombic form ($a = 7.6$ Å, $b = 5.7$ Å), which coexists with the pseudohexagonal form. Whereas the as-spun samples exhibited a single melt transition attributed to the less dense pseudohexagonal form, the annealed samples showed two endotherms which correlate with the two crystalline forms. Annealing experiments were performed on the 75/25 and 30/70 copolymers by Cheng in two distinct temperature ranges (Figure 11):[62] the low-temperature range was below the recrystallization temperature, T_c, and the high-temperature range was between T_c and the solid–nematic melt transition, T_m. The pseudohexagonal phase formed rapidly upon cooling from the melt. Isothermal experiments at low temperatures indicated that the pseudohexagonal form is stable in this temperature range, with constant T_m and enthalpy, and appears to inhibit the formation of the orthorhombic phase, which initially has a lower T_m. As the proportion of orthorhombic packing increases, T_m for this phase shifts upward until it coincides with and then surpasses that of the pseudohexagonal phase. Annealing in the high-temperature range results in the appearance of a single endotherm which increases in temperature with annealing time.

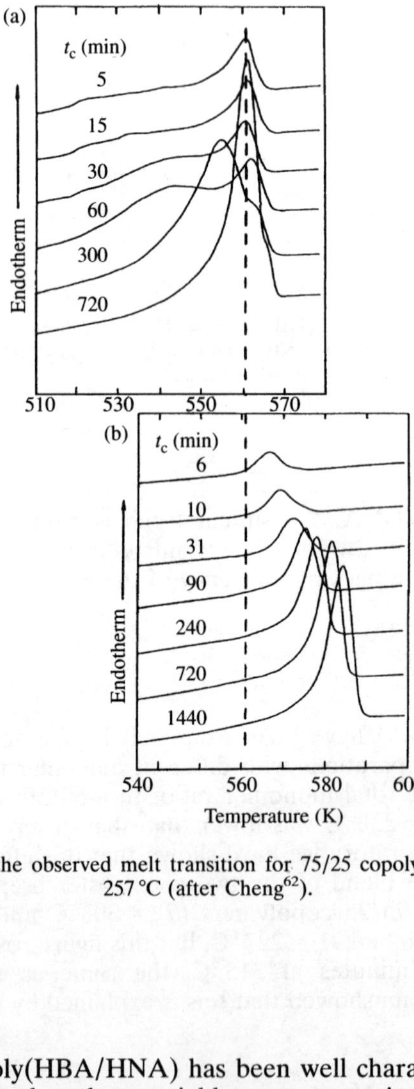

Figure 11 Effect of annealing on the observed melt transition for 75/25 copoly(HBA/HNA) at (a) 232 °C and (b) 257 °C (after Cheng[62]).

10.5.10 Melt Rheology

The melt rheology of copoly(HBA/HNA) has been well characterized.[63–65] Several features are common to LCPs in general, such as a yield stress, a transient negative first normal stress, and the absence of a zero shear viscosity. In contrast to observations for isotropic polymers, the slopes of the frequency-dependent dynamic moduli, $G'(\omega)$ and $G''(\omega)$, do not approach limiting values of ω^2 and ω, respectively, as the frequency approaches zero. Instead, $G'(\omega)$ levels off at low frequency, indicative of a more solid-like response, which may arise due to disclinations in the material.[65] In addition, a yield stress has been reported for copoly(HBA/HNA) preparations of different composition,[63] and is believed to arise due to the presence of unmelted crystalline domains. Annealing below T_m before measurement increases the yield stress, but preshearing the samples or holding them at high temperatures prior to testing eliminates this effect.[64,65] Under conditions of steady shear the viscosity exhibits a power law behavior over a wide range of shear rates, and no zero-shear viscosity is detected.[65] The viscosity of the nematic melt tends to be lower than that of an isotropic polymer due to molecular alignment in the direction of flow.

10.5.11 Transesterification in LCPEs

Transesterification is known to occur readily in isotropic polymer blends, and has been studied as a potential route to creating new polycondensates and as a method of increasing the miscibility of incompatible blends.[66–68] Three possible mechanisms of transesterification have been described by Kotliar:[66] alcoholysis, acidolysis, and ester–ester interchange. Although the rate constants of each of these reactions vary widely, depending on molecular structure and reaction conditions, their relative rates for low molecular weight esters generally follow the order:

$$k_{\text{alcoholysis}} \gg k_{\text{acidolysis}} \approx k_{\text{interchange}} \tag{6}$$

In polyesters, the number of ester linkages is so numerous that interchange reactions become more significant. However, due to the difficulties of distinguishing the mechanism of reaction in polymer blends, all three of the above mechanisms are collectively referred to as transesterification.

Transesterification of polyesters can result in the randomization of chains with ordered monomer sequences. Jin and co-workers[69] synthesized liquid crystalline copolyesters consisting of 50/25/25 copoly(HBA/2,7-DHN/TPA) with both random and repeating monomer sequences. Both were held in the nematic melt at 305 °C, and changes in sequence distributions were determined by analyses of triad content by ^{13}C NMR. Transesterification became apparent in the sequentially ordered copolymer after 5 min, but the reaction rate appeared to slow with time, and complete randomization was not achieved after 60 min. In contrast, no significant changes in triad content were detected in the random copolymer. This suggested the possibility of different monomer reactivities within the polyester chains.[69] However, both polymers increased in molecular weight over time, and the consequential increase in viscosity could also have tended to slow the reaction.

Lenz and co-workers reported the reverse effect, *i.e.* sequence reordering, for several copolyesters both below and above T_m,[70–72] also based on ^{13}C NMR data. Below T_m, random copolyesters transesterified to form more blocky copolymers that exhibited partial crystallinity,[70,71] in what the authors term a crystallization induced reaction (CIR). In the nematic phase, some random copolyesters were also reported to undergo CIR, but randomization was favored in the isotropic melt.[72] In both phases the reaction became more favorable in the presence of a transesterification catalyst and was hindered by increasing molecular weight.

The extent of transesterification in blends of non-LC polyesters has been investigated by a number of analytical methods. Devaux and co-workers[73] detected transesterification between bisphenol-A polycarbonate (PC) and poly(butylene terephthalate) (PBT) based on the solubility of the 50:50 blend in methylene chloride, a good solvent for PC but a nonsolvent for PBT. The unreacted blend showed ~50% solubility; blends undergoing esterification showed an initial decrease in solubility followed by a rapid increase to 100% solubility after ~30 min at 260 °C. IR spectroscopy and NMR analyses of the diads and triads indicated that transesterification leads to complete randomization after 200 min at 253 °C. The reaction rate was shown to be influenced by the presence of residual titanium catalysts in the blend, and was slowed by complexing the catalysts with phosphorus stabilizers. Small-angle neutron scattering (SANS) has also been used to determine transesterification kinetics in blends of deuterated and nondeuterated polyesters. For poly(ethylene terephthalate), an Arrhenius plot yielded an activation energy of 152 kJ mol^{-1} for the transesterification reaction[74] and a monomeric rate constant of $k = 1.87 \times 10^{-3}$ s at 280 °C.

Several of these methods also have been applied to study transesterification in wholly aromatic polyesters. Using SANS, an activation energy for transesterification was obtained for a more rigid terpolyester consisting of hydroquinone, isophthalic acid, and hydroquinone carbonate,[75] which was similar to that of PET.[74] The differences in rate constants for different molecular weight polymers suggested an activated end-group mechanism.[75] Economy and co-workers[76] report the rapid transesterification of mixed powders of poly(HBA) and poly(HNA) under pressure at 450 °C, based on NMR and thermal analysis. Application of

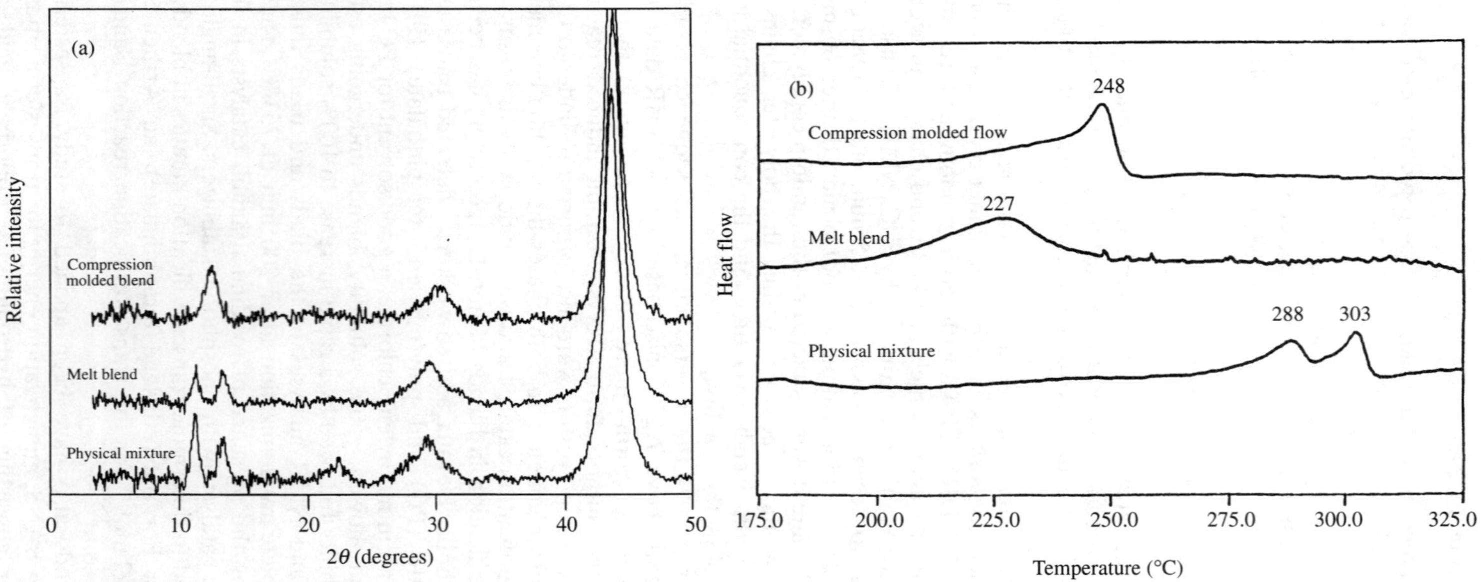

Figure 12 (a) $\theta/2\theta$ diffractometer scans along the fiber axis direction and (b) DSC scans for 30/70 and 75/25 copoly(HBA/HNA) fibers (overall 60:40 monomer mole ratio) prepared as a physical mixture, melt blend, and compression molded blend (after McCullagh *et al.*[61]).

heat and pressure to a 50:50 mole ratio mixture of the homopolymers resulted in a copolymer with $T_m \approx 254$ °C after <90 s at 450 °C, pointing to the formation of the random copolymer via transesterification. This was confirmed by ^{13}C NMR, which indicated that the diad content was approximately that of the random 50/50 copolymer. These authors also reported a reduction in the rate of transesterification between poly(HBA) and poly(HNA) for higher molecular weight preparations,[76] which they attributed to either the decreased number of end groups or a higher melt viscosity. However, SANS studies of a mesomorphic polyester[77] indicated that the rate of transesterification is independent of molecular weight in the isotropic melt. Economy has also suggested that sequence changes may occur when copoly(HBA/HNA) is annealed just below the solid to nematic transition.[45] In apparent conflict with the above data, DeMeuse and Jaffe interpret their X-ray data for copoly(HBA/HNA) blends to suggest that transesterification does not occur in the nematic state.[60]

In this laboratory we have studied the transesterification in blends of copoly(HBA/HNA) by X-ray methods and thermal analysis.[61] Specifically, we prepared a melt blend of the 75/25 and 30/70 copolymers, with an overall mole ratio of 60/40, and monitored the reaction that occurred in the melt at 315 °C. The positions of the intensity maxima along the chain axis direction are dependent on the monomer concentration: the first such maximum occurs at $d = 6.67$, 7.15, and 7.90 Å for the 75/25, 58/42, and 30/70 copolymers, respectively. Thermal analysis of these copolymers show single solid–nematic transitions at $T_m = 288$, 248, and 303 °C, respectively. All of these data are consistent with previously reported literature values.[62]

A diffractometer scan of a physical mixture of 75/25 and 30/70 copolymer fibers is shown in Figure 12(a). These data are essentially identical to those obtained by combining the scans of the component copolymers from Figure 4(a) according to the weight ratio in the mixture. A DSC scan of the physical mixture is shown in Figure 12(b). Thermal transitions occur at 303 and 288 °C, corresponding to the component copolymers. Figure 12(a) also shows the diffractometer scan for the initial melt-blended specimen. The *d* spacings of the observed maxima are within experimental error the same as those for the physical mixture. The decrease in certain peak heights, particularly at 7.91 Å, may be due to poorer three-dimensional packing

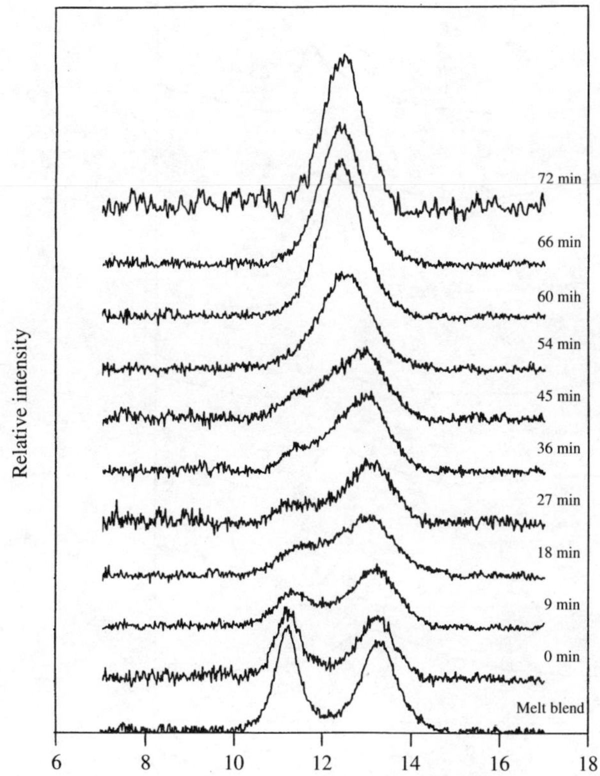

Figure 13 $\theta/2\theta$ diffractometer scans along the fiber axis direction for copoly(HBA/HNA) transesterification intermediates compression molded at 315 °C for the times indicated (after McCullagh *et al.*[61]).

of the 30/70 copolymer and an overall reduction of crystallinity in the melt blend. Figure 12(b) shows the DSC scan for the same specimen with a single T_m at 227 °C, well below that of the 58/42 copolymer (248 °C) and those of the component 30/70 and 75/25 copolymers. It appears that melt blending has resulted in the formation of a compatible blend, *i.e.* an intimate mixture at the molecular level, but there has not been significant transesterification during this process. Upon compression molding for 60 min at 315 °C, the first meridional maxima arising from the component copolymers are replaced by a single peak with a d spacing of 7.14 Å, as shown in Figure 12(a). This indicates a randomization of monomer sequences resulting from transesterification between the 30/70 and 75/25 copolymers. Thermal analysis indicates a shift in T_m to 248 °C (Figure 12(b)), consistent with that of a random copolymer with a 60/40 HBA/HNA composition.

Figure 13 shows $\theta/2\theta$ X-ray diffractometer scans along the fiber axis direction for melt-blended specimens subjected to compression molding at 315 °C for times ranging from 0 to 72 min. The maxima characteristic of the starting copolymers in the blend are seen to shift slowly toward each other as the molding time increases. The two-component peaks were resolved using a Gaussian–Lorentzian curve fitting program. It is seen that there is a steady but nonlinear change in d spacings until the two peaks converge after 60 min. The d spacings of the observed maxima of the final transesterification product are within experimental error those of the random 60/40 copolymer. DSC scans of the same specimens are shown in Figure 14, where it is observed that T_m shifts slowly from 227 to 248 °C. The single T_m at 248 °C matches the value for the random 58/42 copolymer and indicates a random copolymer of similar composition. There are no further changes in the meridional diffraction pattern or T_m after compression molding for times up to 72 min. We can conclude that transesterification has resulted in complete sequence randomization after 60 min at 315 °C in this specimen.

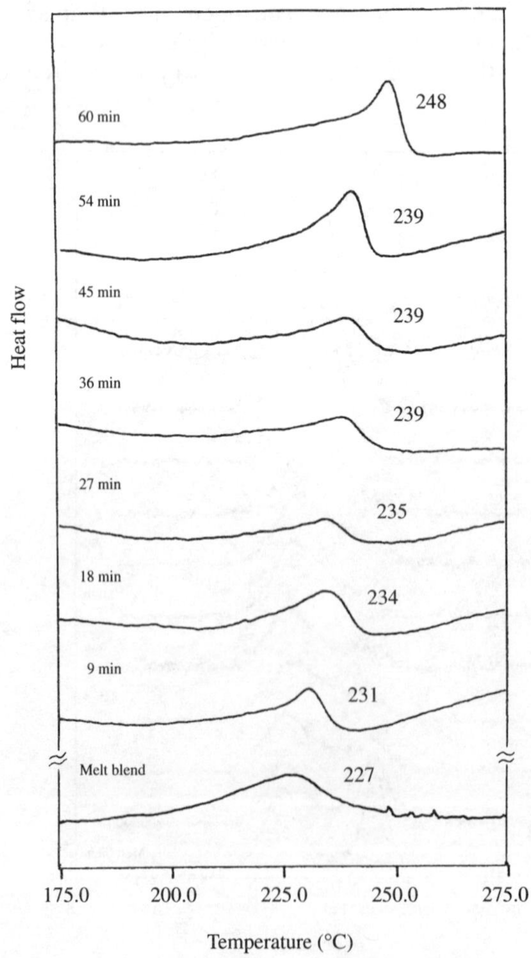

Figure 14 DSC scans of copoly(HBA/HNA) transesterification intermediates compression molded at 315 °C for the times indicated (after McCullagh *et al.*[61]).

From the above results it is clear that melt blending the 30/70 and 75/25 copolymers results first in the formation of a compatible blend, based on the single melt transition observed by DSC measurements. The X-ray data for the initial melt blend indicates that the specimen has the same copolymerization statistics as the physical blend, but the single T_m at lower temperature indicates the formation of a solution of one copolymer in the other. Further evidence comes from the time-dependent changes in the X-ray data for the compression molded melt blend. Transesterification leading to changes in sequence distribution of an incompatible blend would occur preferentially at the edges of the domains, resulting in copolymers of intermediate compositions in those regions. The diffraction maxima of these copolymers would occur between those of the initial copolymers, and the resulting pattern would show a filling in of the region between the two initial peaks, as the latter gradually decreased in intensity. Overall, one would expect to obtain one broad peak in the region $d = 8.5–6.3$ Å, which would gradually sharpen and become a more narrow peak at $d = 7.13$ Å when the copolymer fully randomized throughout the blend. For example, examination of the X-ray data of DeMeuse and Jaffe[60] shows a single broad maximum in the $d = 8.5–6.3$ Å region. These data probably reflect a partially transesterified specimen that had been insufficiently mixed to produce a compatible blend.

ACKNOWLEDGMENT

The work on LCPEs in this laboratory was supported by NSF under grant # DMR91-22227.

10.6 REFERENCES

1. F. Reinitzer, *Monatsh. Chem.*, 1988, **9,** 421.
2. D. Vorlander, *Z. Phys. Chem.*, 1923, **105,** 211.
3. C. Robinson, *Trans. Faraday Soc.*, 1956, **52,** 571.
4. F. C. Bawden and N. W. Pirie, *Proc. R. Soc. London., Ser. B*, 1957, **123,** 174.
5. S. L. Kwolek (DuPont), *US Pat.* 3 671 542 (1972).
6. H. F. Kuhfuss and W. J. Jackson (Eastman Kodak), *US Pat.* 3 804 805 (1974).
7. J. Economy and B. E. Nowak (Carborundum), *US Pat.* 3 759 870 (1973).
8. G. W. Calundann (Celanese), *US Pat.* 4 219 461 (1980).
9. J. Blackwell and A. Biswas, in 'Developments in Oriented Polymers — 2', ed. I. M. Ward, Elsevier Applied Science, New York, 1987.
10. H. Chedron, *Makromol. Chem., Macromol. Symp.*, 1990, **33,** 85.
11. G. W. Calundann, in 'High Performance Polymers: Their Origin and Development', eds R. B. Seymour and G. S. Kirshenbaum, Elsevier, New York, 1986.
12. H. Finkelmann, M. Happ, M. Portugall and H. Ringsdorf, *Makromol. Chem.*, 1978, **179,** 2541.
13. A.-I. Schneider, Ph.D. Thesis, Case Western Reserve University, 1991.
14. R. W. Lenz, *Faraday Discuss. Chem. Soc.*, 1985, **79,** 21.
15. K. Berger and M. Ballauff, *Mol. Cryst. Liq. Cryst.*, 1988, **157,** 109.
16. R. Stern, M. Ballauff and G. Wegner, *Makromol. Chem., Macromol. Symp.*, 1989, **23,** 373.
17. R. Stern, M. Ballauff, G. Lieser and G. Wegner, *Polymer,* 1991, **32,** 2096.
18. J. Economy, W. Volksen, C. Viney, R. Geiss, R. Siemens and T. Karis, *Macromolecules,* 1988, **21,** 2777.
19. G. Lieser, *J. Polym. Sci., Polym. Phys. Ed.*, 1983, **21,** 1611.
20. J. Blackwell, G. Lieser and G. A. Gutierrez, *Macromolecules,* 1983, **16,** 1418.
21. D. Y. Yoon, N. Masciocchi, L. E. Depero, C. Viney and W. Parrish, *Macromolecules,* 1990, **23,** 1793.
22. P. Iannelli, D. Y. Yoon and W. Parrish, *Macromolecules,* 1994, **27,** 3295.
23. Z. Sun, H.-M. Cheng and J. Blackwell, *Macromolecules,* 1991, **24,** 4162.
24. W. J. Jackson and H. F. Kuhfuss, *J. Polym. Sci.*, 1976, **14,** 2043.
25. A. E. Zachariades, J. Economy and J. A. Logan, *J. Appl. Polym. Sci.*, 1982, **27,** 2009.
26. L. Quach, E. Hornbogen, W. Volksen and J. Economy, *J. Polym. Sci., Polym. Chem. Ed.*, 1989, **27,** 775.
27. A. Muhlebach, J. Lyerla and J. Economy, *Macromolecules,* 1989, **22,** 3741.
28. H. Kromer, R. Kuhn, H. Pielartzik, W. Siebke, V. Eckhardt and M. Schmidt, *Macromolecules,* 1991, **24,** 1950.
29. G. W. Calundann (Celanese), *US Pat.* 4 161 470 (1979).
30. G. W. Calundann and M. J. Jaffe, 'Robert A. Welch Conferences on Chem. Res. Proc. Synth. Polymers', 1982, p. 247.
31. G. C. Rutledge, *Macromolecules,* 1992, **25,** 3984.
32. J. Blackwell, G. A. Gutierrez and R. A. Chivers, *Macromolecules,* 1984, **17,** 1219.
33. R. A. Chivers, J. Blackwell and G. A. Gutierrez, *Polymer,* 1984, **25,** 435.
34. R. A. Chivers, J. Blackwell, G. A. Gutierrez, J. B. Stamatoff and H. Yoon, in 'Polymeric Liquid Crystals', ed. A. Blumstein, Plenum, New York, 1985, pp. 153–166.
35. J. Blackwell, A. Biswas, G. A. Gutierrez and R. A. Chivers, *Faraday Discuss. Chem. Soc.*, 1985, **79,** 73.
36. J. Blackwell, R. A. Chivers, G. A. Gutierrez and A. Biswas, *J. Macromol. Sci., Phys.*, 1985, **B24,** 39.
37. A. Biswas and J. Blackwell, *Macromolecules,* 1987, **20,** 2997.
38. G. A. Gutierrez and J. Blackwell, *Macromolecules,* 1984, **17,** 2744.
39. G. A. Gutierrez, J. Blackwell and R. A. Chivers, *Polymer,* 1985, **26,** 348.

40. J. Blackwell, H.-M. Cheng and A. Biswas, *Macromolecules,* 1988, **21,** 39.
41. A. Biswas and J. Blackwell, *Macromolecules,* 1988, **21,** 3146.
42. A. Biswas and J. Blackwell, *Macromolecules,* 1988, **21,** 3152.
43. A. Biswas and J. Blackwell, *Macromolecules,* 1988, **21,** 3158.
44. A. Muhlebach, R. D. Johnson, J. Lyerla and J. Economy, *Macromolecules,* 1988, **21,** 3115.
45. J. Economy, R. D. Johnson, J. Lyerla and A. Muhlebach, in 'Liquid Crystalline Polymers,' ed. R. A. Weiss and C. K. Ober, American Chemical Society, Washington, 1990.
46. S. D. Hudson and A. J. Lovinger, *Polymer,* 1993, **34,** 1123.
47. S. Hanna and A. H. Windle, *Polymer,* 1988, **29,** 207.
48. S. Hanna, T. J. Lemmon, R. J. Spontak and A. H. Windle, *Polymer,* 1992, **33,** 3.
49. M. J. Troughton, A. P. Unwin, G. R. Davies and I. M. Ward, *Polymer,* 1988, **29,** 1389.
50. D. Hofmann, A.-I. Schneider and J. Blackwell, *Polymer,* 1994, **35,** 5603.
51. L. C. Sawyer and M. Jaffe, *J. Mater. Sci.,* 1986, **21,** 1897.
52. S. D. Hudson, A. J. Lovinger, S. K. Venkataraman, C. Liu and L. T. Manzione, *Polym. Eng. Sci.,* 1994, **34,** 1327.
53. M. J. Troughton, G. R. Davies and I. M. Ward, *Polymer,* 1989, **30,** 58.
54. M. H. Alhaj-Mohammed, G. R. Davies, S. Abdul Jawad and I. M. Ward, *J. Polym. Sci., Polym. Phys. Ed.,* 1988, **26,** 1751.
55. D. S. Kalika and D. Y. Yoon, *Macromolecules,* 1991, **24,** 3404.
56. D. I. Green, G. A. J. Orchard, G. R. Davies and I. M. Ward, *J. Polym. Sci., Polym. Phys. Ed.,* 1990, **28,** 2225.
57. A. Flores, F. Ania, F. J. Balta Calleja and I. M. Ward, *Polymer,* 1993, **34,** 2915.
58. S. Z. D. Cheng, J. J. Janimak, A. Zhang and Z. Zhou, *Macromolecules,* 1989, **22,** 4240.
59. W. W. Adams and R. K. Eby, *MRS Bull.,* 1987, **11,** 22.
60. M. T. DeMeuse and M. Jaffe, *Mol. Cryst. Liq. Cryst.,* 1988, **157,** 535.
61. C. M. McCullagh, J. Blackwell and A. M. Jamieson, *Macromolecules,* 1994, **27,** 2996.
62. S. Z. D. Cheng, *Macromolecules,* 1988, **21,** 2475.
63. D.-K. Yang and W. R. Krigbaum, *J. Polym. Sci., Polym. Phys. Ed.,* 1989, **27,** 819.
64. F. Cocchini, M. R. Nobile and D. Acierno, *J. Rheol.,* 1991, **35,** 1171.
65. S. M. Gusky and H. H. Winter, *J. Rheol.,* 1991, **35,** 1991.
66. A. M. Kotliar, *J. Polym. Sci.,* 1981, **16,** 367.
67. R. S. Porter and L.-H. Wang, *Polymer,* 1992, **33,** 2019.
68. R. S. Porter, J. M. Jonza, M. Kimura, C. R. Desper and E. R. George, *Polym. Eng. Sci.,* 1989, **29,** 55.
69. J.-I. Jin, J.-H. Chang, K. Hatada, K. Ute and M. Hotta, *Polymer,* 1992, **33,** 1374.
70. R. W. Lenz and S. J. Go, *J. Polym. Sci. A-1,* 1973, **11,** 2927; *J. Polym. Sci., Polym. Chem. Ed.,* 1974, **12,** 1.
71. R. W. Lenz and A. N. Sculer, *J. Polym. Sci., Polym. Symp.,* 1978, **63,** 343.
72. R. W. Lenz, J. Jin and K. A. Feichtinger, *Polymer,* 1983, **24,** 327.
73. J. Devaux, P. Godard and J. P. Mercier, *Polym. Eng. Sci.,* 1982, **22,** 229.
74. W. A. MacDonald, A. D. W. McLenaghan, G. McLean, R. W. Richards and S. M. King, *Macromolecules,* 1991, **24,** 6164.
75. J. Kugler, J. W. Gilmer, D. Wiswe, H.-G. Zachmann, K. Hahn and E. W. Fischer, *Macromolecules,* 1987, **20,** 116.
76. A. Muhlebach, J. Economy, R. D. Johnson, T. Karis and J. Lyerla, *Macromolecules,* 1990, **23,** 1803.
77. M. H. Li, A. Brulet, P. Keller, C. Strazielle and J. P. Cotton, *Macromolecules,* 1993, **26,** 119.

11

Recent Developments in the Molecular Theory of Elasticity

KENNETH J. SMITH, JR.

SUNY College of Environmental Science and Forestry, Syracuse, NY, USA

11.1 INTRODUCTION

Some materials are capable of sustaining enormous reversible tensile deformations, perhaps as much as 1000 per cent or even greater, while exhibiting tensile moduli of only small fractions of those characteristic of typical solids such as metals, ceramics, and glasses.[1-3] This phenomenon is generally referred to as rubber elasticity, and such materials are said to be rubber elastic. Rubber materials are polymers, without exception, but certain conditions must prevail: the macromolecules must be flexible at ambient temperature, and they must be tied to one another (crosslinked) at various points in order to prevent wholesale molecular slippage or flow under tension, otherwise reversibility is precluded.

Usually crosslinking is a purely chemical matter: functional groups within the molecules form intermolecular covalent bonds directly (such as ester or amide formation), or indirectly (via radical reactions), or through the use of an additive crosslinker (perhaps sulfur, as in vulcanization). Physical bonds, though lacking the stability of chemical covalent ones, are also possible. Crystalline and/or glassy regions can be effective but somewhat temperature sensitive. Crystallization bonding is possible with partially crystalline polymers below their melting points. Effective glass bonding is likely to require block copolymers, one part of which is rubbery and another glassy at ambient temperature. However, heterogeneous crosslinks, like crystallites and glassy zones, are temperature sensitive: above the melting point, or glass transition temperature, they disappear, along with elastic reversibility. Nevertheless, this property leads to easy fabrication as thermoset elastomers.

Elastic properties are slightly sensitive to the particular homogeneous chemical crosslinks. For example, peroxide generated crosslinks in natural rubber seem to affect elasticity a little differently than sulfur vulcanization of the same amount.[4] Clearly, large heterogeneous crosslinks must alter elastic responses considerably if, for no other reason, the 'crosslinks' by solidification convert appreciable fractions of rubbery material into solid, inelastic regions. However, our major concern herein is with small chemical crosslinks (homogeneous) and their import for network formation and consequent elasticity.

11.2 MOLECULAR FLEXIBILITY AND ELASTICITY

Long-range (rubber) elasticity is a phenomenon of intramolecular origin. 'Liquid' polymer molecules, above the melting point and glass transition temperature, upon crosslinking lose the capacity to flow: large molecular translations are prohibited by permanent molecular connectivity patterns established by crosslinks. As the crosslinks are few in number, internal liquid-like mobility of submolecular domains remains relatively unchecked. The 'solid' elastomer, incapable of large-scale molecular flow, therefore exhibits external dimensional stability simultaneously with localized internal liquidity or flexibility. Since liquidity implies an internal energy that is independent of shape (but not volume), van der Waals forces should play but a secondary role in determining equilibrium deformational forces and no role at all in an 'ideal' elastomer free of interunit interactional forces when the volume remains constant.

Elastic forces are usually ascribed to thermodynamic factors. Hence nomenclature such as 'entropic force' is often seen, implying that a deforming force decreases system entropy or, alternatively, decreases the number of configurations of the system. Geometrical configurations of a polymer molecule are referred to as conformations. It then follows that the conformations of a network chain, *i.e.* that portion of a polymer network subjoining two crosslinks, are determined by rotations about the covalent bonds making up a polymer chain. Those rotations of greatest import for elasticity pertain to bonds along the chain backbone. By far, most backbone bonds in a polymer adjoin neighboring ones at an obtuse angle, which by rotations (about single bonds) generate highly coiled chains whose geometry fluctuates as bond rotational angles continually change.

Altering rotational positions invariably depends upon the potential energies of bond rotations, which, of course, are sensitive to temperature in such a way that the potential energy decreases in importance as the temperature increases, as required by elementary statistical thermodynamics. Energy barriers are more easily overcome at high temperatures.

Now, exactly how does all of this translate into an elastic force? An exceptionally simple, vivid, and qualitative explanation can be sketched as follows. Consider a single polymer molecule as shown in Figure 1. An instantaneous view of a fluctuating molecule is traced by the continuous line meandering round about through space. Interstitial regions are not voids but filled with fluid. Consequently, there are thermodynamic interactions between polymer and fluid wherever they are in contact. At some points, however, the molecule folds back on itself through long-range loops that generate polymer–polymer contacts, or nonbonded interactions at various places. These interactions may differ thermodynamically from polymer–fluid interactions. However, if these long-range polymer–polymer contact interactions are equivalent to the corresponding polymer–fluid interactions, the fluid is said to be a theta solvent for the polymer.[5] In this instance, the thermodynamics of both kinds of interactions are identical, and a polymer unit is unable to recognize a neighboring contact as a polymer unit far removed along the molecule but folded back to cause an interaction or as a solvent contact. If the fluid (theta solvent) is replaced by pure polymer molecules of the same kind, nothing would change: the molecule in question cannot tell the difference. Therefore, one can say that a pure polymer in bulk is its own theta solvent,[3] and long-range intramolecular polymer–polymer interactions are indistinguishable from intermolecular polymer–polymer interactions. A conclusion is demanded that, in pure liquid polymer, long-range intramolecular interactions have no influence on molecular geometry: a particular polymer unit cannot know whether nonbonded contacts are with units of the same molecule or with other molecules. Then, the molecule in question has nothing to gain or to lose thermodynamically by altering its geometry as a result of its nonbonded contacts. This argument is often stated as an assumption of Flory,[25] who first enunciated it. Flory's reasoning is compelling far beyond a mere assumption. Indeed, it is now experimentally validated.[54–59]

An important distinction between pure liquid polymer and a polymer molecule dissolved in an extraneous theta solvent manifests itself upon crosslinking. A theta solvent (usually a low molecular weight fluid) does not impede or otherwise hamper conformational fluctuations of a polymer in a major way. However, with pure polymer, all units in the system are incorporated into long, linear molecules that cannot pass or 'cut' through one another. So long as the polymer is uncrosslinked, such linear connectivity may slow down the rate of conformational fluctuations but not prohibit any of them. Not so if crosslinking is imposed. Then, three-dimensional connectivity patterns are established throughout the network that must be permanently maintained. Since the chains can no longer work around their neighbors, and

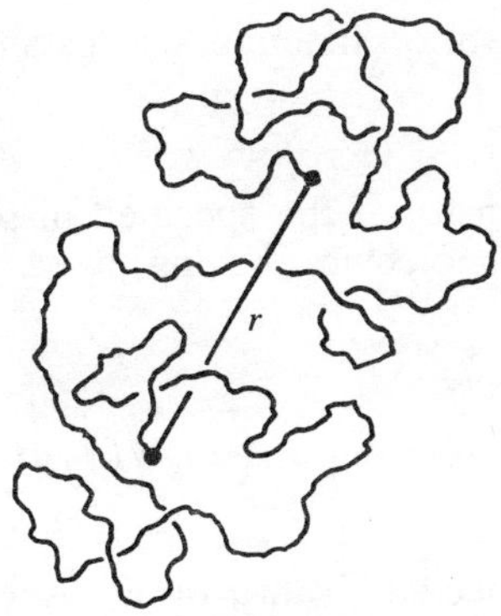

Figure 1 A random polymer molecule of end-to-end displacement r.

'cutting' through is physically not possible for real polymer molecules, a large number of chain conformations must be precluded by crosslinking.

Imagine now that we hold the two ends of the polymer as shown in Figure 1 at a fixed distance r. The polymer will then fluctuate through its store of conformations compatible with the fixed end-to-end displacement. If the displacement is increased, what will happen to the available conformations? They will decrease. This is clear once it is realized that if r is increased to the maximum value (a fully extended molecule), there can be only one conformation available. Alternatively, if the two ends are brought together, available conformations reach a maximum. Therefore, increasing r is accompanied by decreasing conformations, and therefore decreasing entropy.

Fluctuating conformations are generated by changes of bond rotational angles, which are random events. A polymer of end-to-end distance r has more conformations for smaller r than for larger. A random selection process for conformational change then requires that the end points move closer together because the greater probability lies on the side of the greater number of choices. A polymer whose ends are held at a fixed distance r must therefore exert a contractile force on its ends of a magnitude dependent upon the disparity between conformations of shorter and larger r. It is clear that this contractile force must increase as r increases. At the maximum end-to-end distance (a fully extended molecule), conformations of even greater r cannot exist, and all possible changes must be towards a smaller r. Hence, the contractile force at this point must be a maximum.

We can also establish qualitatively the influence of temperature on the contractile force. An increase in temperature increases the rate of bond rotation, or increases rotational momentum. The kinetic and rotational energies of the internal parts of the molecule are increased accordingly. This translates into a greater rate and momentum of conformational change. Increased momentum of interior chain parts increases the impact of the end points, which increases the net contractile force. The situation is completely analogous to a gas in a closed container. Gas pressure is determined by molecular collisions with the walls. Faster molecular velocities produced by increased temperature produce stronger collisions, *i.e.* greater pressure.

11.3 THE SINGLE MOLECULE (OR CHAIN), QUANTITATIVELY

The foregoing qualitative reasoning takes quantitative form rather easily. Volkenstein and Ptitsyn[6] define a polymer configurational, or conformational, integral (partition function) Z as

$$Z = \int \ldots \int \exp(-E(\boldsymbol{b})/kT)\pi_i \, \mathrm{d}\tau_i \qquad (i = 1, 2, \ldots, N) \tag{1}$$

where $E(\boldsymbol{b})$ is the conformational energy of the molecule including all rotational bonds in the chain backbone. The volume element $\mathrm{d}\tau_i$ is normalized. Integration is over all bonds, T is temperature, and k is the Boltzmann constant.

If the integrations of equation (1) are restricted by constraining the chain to a fixed end-to-end displacement $\boldsymbol{r}$, the partition function at constant $\boldsymbol{r}$, Z_r, emerges:

$$Z_r = ZW_0(\boldsymbol{r}) \tag{2}$$

where $W_0(\boldsymbol{r})$ is the probability density of the specified displacement $\boldsymbol{r}$.[6,7] This function is well known, even for real polymer molecules having fixed bond angles and restricted bond rotations.[8] It can be expressed as[9]

$$W_0(\boldsymbol{r}) = \int \delta(\boldsymbol{r}' - \boldsymbol{r})W(\boldsymbol{r}')\,\mathrm{d}\tau_{r'} \tag{3}$$

The symbolism is that $\boldsymbol{r}$ is a specified value of $\boldsymbol{r}'$. Actual integration is triple, over the coordinates of the volume element $\mathrm{d}\tau_{r'}$. The delta function is

$$\delta(\boldsymbol{r}' - \boldsymbol{r}) = \frac{1}{8\pi^3}\int \exp(i\boldsymbol{p}\cdot(\boldsymbol{r}' - \boldsymbol{r}))\,\mathrm{d}\tau_p$$

Again, the integration is triple over the coordinates of the volume element $\mathrm{d}\tau_p$. Equation (3) reads: the probability density $W_0(\boldsymbol{r})$ is given by the inverse Fourier transform of the Fourier transform of the identical probability density $W(\boldsymbol{r}')$. The prime merely denotes an integrated variable, as expressed by the integration of equation (3). Both angular integrations of equation (3) can be performed, giving

$$W_0(\boldsymbol{r}) = \frac{1}{2\pi^2}\int_0^\infty \left\langle \frac{\sin pr}{r} \right\rangle_0 \frac{\sin(pr)}{r}\,\mathrm{d}p \tag{4}$$

The symbol $\langle\ \rangle_0$ denotes an unconstrained average at theta conditions, that is, the 'unperturbed' situation, and p is the magnitude of the Fourier vector $\boldsymbol{p}$, which is of no importance since it is integrated away. However, that particular integration causes trouble; it cannot be done in closed form. Consequently, equation (4) is the last we will see of the complete, exact probability density function of a polymer molecule. From this point on, approximations must be introduced.

A common approximation is to expand the sine term enclosed by angle brackets and integrate term by term. The result is a series[10,11]

$$W_0(\boldsymbol{r}) = \left(\frac{3}{2\pi\langle r^2\rangle_0}\right)^{3/2} \exp\left(-\frac{3r^2}{2\langle r^2\rangle_0}\right)\left\{1 + \delta\left[5 - \frac{10r^2}{\langle r^2\rangle_0} + \frac{3r^4}{\langle r^2\rangle_0^2}\right] + \ldots\right\} \tag{5}$$

$$\delta = \frac{3}{40\langle r^2\rangle_0^2}[3\langle r^4\rangle_0 - 5\langle r^2\rangle_0^2] \tag{6}$$

The first term is the Gaussian approximation ($\delta = 0$), valid for flexible chains of small displacement r. The second term, involving δ, is the leading correction term for finite chain extensibility, that is, values of r at which the Gaussian approximation begins to break down. Highly stretched chains require additional terms (not shown) in the series. Nagai[8] employed a slightly different, though perhaps more logical, series expansion, and certainly more complicated. In any event, the rate of convergence of the series is determined by the ratio of r to the maximum length of the chain (contour length) and by the links in the chain, N: large N means rapid convergence. For example, consider a flexibly jointed chain, one whose bond angles are completely unrestricted. The second and fourth moments for a chain of N links, each of length b, are

$$\langle r^2\rangle_0 = Nb^2$$

$$\langle r^4\rangle_0 = \frac{5N^2b^4}{3}\left(1 - \frac{2}{5N}\right)$$

and

$$\delta = -\frac{3}{20N}$$

Accordingly, the distribution becomes[10]

$$W_0(\mathbf{r}) = \left(\frac{3}{2\pi Nb^2}\right)^{3/2} \exp\left(\frac{-3r^2}{2Nb^2}\right)\left\{1 - \frac{3}{20N}\left[5 - \frac{10r^2}{Nb^2} + \frac{3r^4}{N^2b^4}\right] + \ldots\right\} \quad (7)$$

The Hemlholtz energy, A, is related to the partition function Z_r:

$$A = -kT \ln Z_r \quad (8)$$

and the force along the displacement r is

$$f = \left(\frac{\partial A}{\partial r}\right)_{T,V}$$

Then from equation (5) we have, for a single chain,

$$f = \frac{3kT}{\langle r^2\rangle_0}\left[1 + \frac{20}{3}\delta - \frac{4\delta}{\langle r^2\rangle_0} r^2 + \ldots\right] r \quad (9)$$

In the Gaussian approximation ($\delta \to 0$) the response of force to deformation is Hookean:

$$f = \frac{3kT}{\langle r^2\rangle_0} r \quad (10)$$

Equations (9) and (10) validate the earlier qualitative argument that the chain contractile force increases as the end-to-end displacment, r, increases; and this force, at a given r, increases as the temperature increases.

The portion of this force originating from the bond rotational potential energy, labeled f_e, rather than from configurational (conformational) entropy is easily determined in the Gaussian approximation, equation (10):[7,12]

$$\frac{f_e}{f} = -T\left[\frac{\partial \ln(f/T)}{\partial T}\right]_{V,r} = T\frac{\mathrm{d} \ln\langle r^2\rangle_0}{\mathrm{d}T} \quad (11)$$

The quantity $\langle r^2\rangle_0$ is the mean-square unperturbed end-to-end displacement of the molecule; it is independent of material volume, V, and displacement, r, and its derivative is denoted as a total derivative to reflect this fact.

According to equation (11), the energy component of the force, f_e, is directly proportional to the force f. The constant of proportionality provides valuable insight into the unperturbed chain rotational potentials and their temperature dependencies, when coupled with reliable experimental data of f_e/f. A number of elaborate computations have been made that compare favorably with experimental results.[11,13,14]

The deformational energy, ΔE, of stretching the chain from r_* to r is obtained from equation (11), since at constant volume and temperature

$$\Delta E = \int_{r_*}^{r} f_e \, \mathrm{d}r = T\left(\frac{\mathrm{d} \ln\langle r^2\rangle_0}{\mathrm{d}T}\right)\Delta A$$

where ΔA, from equations (2) and (5) with $\delta = 0$, is

$$\Delta A = \frac{3kT}{2\langle r^2\rangle_0}(r^2 - r_*^2)$$

and the deformational entropy becomes

$$\Delta S = \left[\frac{\mathrm{d} \ln\langle r^2\rangle_0}{\mathrm{d}T} - \frac{1}{T}\right]\Delta A$$

The entropic contribution to the force, f_s, is

$$f_s = -T\left(\frac{\partial S}{\partial r}\right)_{T,V} = \left[1 - T\frac{\mathrm{d}\ln\langle r^2\rangle_0}{\mathrm{d}T}\right]f \tag{12}$$

which is usually somewhat larger than f_e. An 'ideal' elastomer is one for which there is no energy contribution, *i.e.* $f_e = 0$ and $f_s = f$, which means that all rotational conformations are of equal potential energy in an 'ideal' rubber. Most elastomers, as with gases, are not ideal.

11.4 THE RUBBER NETWORK: THE CLASSICAL VIEW

The foregoing commentary on the statistical behavior of single polymer chains was developed over several decades, beginning with the very early work of Kuhn[15,16] and James and Guth,[17,18] continuing through the refined molecular geometry calculations of Lifson[19] and Flory,[11] through the real-chain distributions of Nagai,[8] and advances still today. However, molecules rarely exist in isolation; rather, they are in the company of their own kind, and any attempt to investigate the behavior of one requires an accounting of the many. This is accomplished with the use of crosslinked networks.

Molecules in bulk or in a theta solvent have the probability distribution given by equation (5). Large subsections of the molecules will be distributed similarly. All of the molecules and parts thereof must be distributed according to the most probable distribution. Since crosslinking is a random chemical process joining the molecules together in a three-dimensional network structure, the chains running between crosslinks must also be distributed according to equation (5). Of course, the numbers of links in the chains must be distributed randomly, but an assumption is invariably made that the molecular weight of network chains is monodisperse, which greatly simplifies ensuing calculations but otherwise is of little consequence, except perhaps in a few special situations.

From the beginning, two distinct network theories of elasticity have existed. One by James and Guth[17,18,20,21] (JG) and the other by Kuhn,[15,16] Wall,[22–24] and Flory[2,25,26] (KWF). The results are similar in form but numerically different. For example, the functional relationship between stress and strain given by both theories is identical, but the modulus value in JG theory is one-half that of KWG (for tetrafunctional junctions).

James–Guth theory follows from a network partition function Z_N defined as

$$Z_N = \int \dots \int \pi_i Z_i \, \mathrm{d}\tau_j$$

Here Z_i is the partition function for the ith chain (of displacement r_i) and $\mathrm{d}\tau_j$ is the volume element of the jth crosslink (or junction). The integration over all junctions, connected by the network chains, is restricted to the connectivity pattern established by the crosslinking process. Now, the partition function does not contain nonbonded repulsive forces, which prevent interpenetration of matter into the space of others, and the cutting of chains through each other. As a result, a complete integration of Z_N, as it stands, leads to collapse of the network into a single point. To prevent this and to provide for dimensional stability, James and Guth assign some of the crosslinks to fixed positions on the surface of the elastomer, then integrate the interior ones subject to established connectivities. Integration of junction positions accounts for the Brownian motion, or fluctuations, of the junctions. At first glance this is reasonable; if the units of the single chains can fluctuate, why should not the junctions behave similarly? By allowing the junctions to do so, JG find for the deformational free energy ΔA

$$\Delta A = \frac{GkT}{2}\left[\frac{\langle r_\mathrm{m}\rangle^2}{\langle r^2\rangle_0}(\alpha_x^2 + \alpha_y^2 + \alpha_z^2) - 3\right] \tag{13}$$

where G is the number of chains, α_x is the extension ratio along the x axis, and $\langle r_\mathrm{m}\rangle$ is the mean chain displacement, which is proportional to the cube root of the unstretched volume at temperature T. The system described by equation (13) is referred to as a 'phantom' network because the chains pass (cut) freely through each other, but other than connectivity requirements do not interact in any fashion. However, are we really sure crosslinks actually fluctuate from their mean positions by an appreciable extent? In reality, real chains cannot transect. Since three or more chains converge to form a junction, material about the junction must entwine these chains and severely limit fluctuations to small regions, perhaps prohibiting them altogether. This is the view of Kuhn,[16] Wall,[22,23] and Wall and Flory,[24] who proposed that junctions are immobilized by entanglements of their constituent chains emanating from the crosslink with surrounding polymer material. Junctions, thus anchored to fixed positions in the network, are assumed to be displaced affinely upon deformation.

Such a network is simple and easy to handle. Its partition function is

$$Z_N = G! \frac{\pi Z_j^{G_j}}{\pi G_j!}$$

which leads to a deformational free energy of

$$\Delta A = \frac{GkT}{2}\left[\frac{\langle r^2\rangle_i}{\langle r^2\rangle_0}(\alpha_x^2+\alpha_y^2+\alpha_z^2)-3\right] - GkT\ln(\alpha_x\alpha_y\alpha_z) - \tfrac{3}{2}\,GkT\ln(\langle r^2\rangle_i/\langle r^2\rangle_0) \qquad (14)$$

This result and equation (15) are generally referred to as 'affine' models because junctions are displaced affinely with deformation. The quantity $\langle r^2\rangle_i$ is the mean-square displacement distance of the chains in the unstretched network; it may differ from $\langle r^2\rangle_0$, the mean-square displacement of the unperturbed chains.

Flory quickly backed away from this result, claiming it was valid only for a system of 'restrained' but unconnected chains and not for a genuine network.[27] However, he converted it to a network by adding to equation (14) a free energy of crosslinking, which he calculated. Then he found for the network

$$\Delta A = \frac{GkT}{2}\left[\frac{\langle r^2\rangle_i}{\langle r^2\rangle_0}(\alpha_x^2+\alpha_y^2+\alpha_z^2)-3\right] - \tfrac{1}{2}\,GkT\ln(\alpha_x\alpha_y\alpha_z) - \tfrac{3}{4}\,GkT(\langle r^2\rangle_i/\langle r^2\rangle_0) \qquad (15)$$

This result gained great importance, perhaps by default, as an alternative to the James–Guth equation (13).

Both equations (13) and (15) are in fact wrong. Of course, they are wrong because of their respective unrealistic assumptions, but they are also wrong on a more general account. Edwards[28] has put forth a principle, surely a correct one, that a network theory of elasticity must obey:

$$\lim_{\alpha_x\to 1}\left(\frac{\partial \Delta A}{\partial \alpha_x}\right)_{\alpha_y,\alpha_z} = 0$$

This is sometimes called the condition of stability. It is obvious that equation (13) fails this test; and equation (15) yields

$$\frac{\langle r^2\rangle_i}{\langle r^2\rangle_0} - \tfrac{1}{2} = 0$$

This cannot be right; the ratio $\langle r^2\rangle_i/\langle r^2\rangle_0$ is generally quite close to one and not to one-half. Much has been made of these clear failures[29] and perhaps rightly so, implying that the classical molecular approach to elasticity ought to be abandoned in favor of phenomenology, perhaps a hasty conclusion. Consider for a moment the 'non-network' equation (14) that has been left behind. The Edwards condition yields

$$\frac{\langle r^2\rangle_i}{\langle r^2\rangle_0} - 1 = 0 \qquad (16)$$

a perfect match! Perhaps then we should look more closely at Flory's divergence. Actually, someone else already has, and with great care.[30]

Hermans[30] calculated anew the free energy of crosslinking and concluded that equation (14) should be unaffected; that is, equation (14) as is ought to apply to networks, contrary to Flory's conclusion. This author is in agreement with Hermans. Flory's calculation[27] for crosslinking treats each chain end as independent from its other end, but this is a little hard to fathom. Hermans avoids this trick, leading him to a different result. Edwards' condition appears, therefore, to endorse Hermans in lieu of Flory, but the endorsement is not entirely unequivocal. The ratio appearing in equation (16) is exactly unity only at the temperature at which crosslinking takes place. At any other temperature, that ratio must differ slightly from unity because the two averages comprising the ratio depend differently upon temperature. The unperturbed average, $\langle r^2\rangle_0$, varies as rotational state populations vary, while $\langle r^2\rangle_i$ changes in accordance with thermal expansion (volume).

Uniaxial deformation along the x axis by an amount $\alpha_x = \alpha$ gives for the retractive force

$$f = \frac{GkT}{L_i}\frac{\langle r_{\mathrm{m}}\rangle^2}{\langle r^2\rangle_0}\left(\alpha - \frac{1}{\alpha^2}\right) \tag{17}$$

from equation (13) and

$$f = \frac{GkT}{L_i}\frac{\langle r^2\rangle_i}{\langle r^2\rangle_0}\left(\alpha - \frac{1}{\alpha^2}\right) \tag{18}$$

from equation (14) or (15). L_i is the unstretched sample length. The difference lies in the front factors, which are related for tetrafunctional junctions by the equation[31]

$$\frac{\langle r_{\mathrm{m}}\rangle^2}{\langle r^2\rangle_0} = \frac{\langle r^2\rangle_i}{\langle r^2\rangle_0} - \frac{1}{2} \tag{19}$$

Since $\langle r^2\rangle_i/\langle r^2\rangle_0 \approx 1$, $\langle r_{\mathrm{m}}\rangle^2/\langle r^2\rangle_0 \approx \frac{1}{2}$, and the modulus of equation (17) is about one-half that of equation (18). The phantom network of JG, because of junction fluctuations, is a softer network than that of KWF, but the functional relations between force and elongation are identical. So too are the thermoelastic ratios f_{e}/f for both equations:

$$\frac{f_{\mathrm{e}}}{f} = T\frac{\mathrm{d}\ln\langle r^2\rangle_0}{\mathrm{d}T} \tag{20}$$

Equation (20) requires an explanation. In the JG theory, $\langle r_{\mathrm{m}}\rangle$ accords with the network volume, but for KWF it is $\langle r^2\rangle_i$. Thus both quantities have been regarded as constant so long as the volume is unchanged. Differentiation with respect to temperature at constant volume and length of either equation then results in equation (20), which is identical to the behavior of a single, independent chain, shown by equation (11). It is clear that the energy contribution, f_{e}, to the network force must be independent of network factors and be determined solely by intramolecular rotational potentials in the Gaussian limit.

Beyond the Gaussian limit, approximate theories have been put forth, principally by Kuhn[16] and Wang and Guth,[32] and later by Treloar[33,34] and Smith.[9] The latter also shows that f_{e}/f is dependent upon deformation once the Gaussian region is exceeded (large elongations). We will not discuss this here since our interest is only in the Gaussian theory; we merely note that non-Gaussian effects due to finite extensibility have been conclusively demonstrated.[35–38]

None of the Gaussian theories are very accurate. For a given crosslinking, KWF predicts a force twice that of JG, shown in Figure 2. However, a real network follows neither curve. Instead, it seems to fall somewhere in the middle, shown by the dashed line. The experimental curve has been described (but not too successfully) in terms of a strictly phenomenological two-constant equation put forth by Mooney[39] and Rivlin.[40] An adequate molecular explanation was not available until recently, and it is generally thought to be associated with intermolecular obstructions caused by entanglements. Another evident theoretical failure is with composite networks. These are crosslinked networks stretched and crosslinked again before removing the force. According to Gaussian theory[41,42] the relaxed two-stage networks upon subsequent stretching are isotropic. In fact, they are not. Figure 3 illustrates the results for a composite stretched parallel (‖) to the original direction of orientation and perpendicular (⊥) to it.[43] These curves can be described by an empirical result based on the Mooney–Rivlin equation,[43] which again suggests that the failure of classical Gaussian theory stems from its omission of intermolecular obstructions that depend upon deformation.

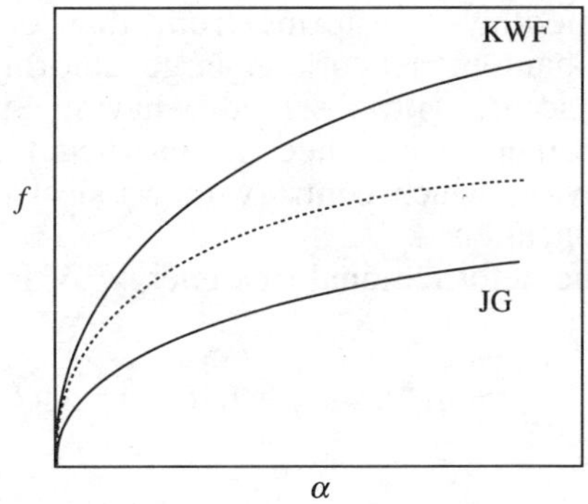

Figure 2 Force vs. deformation for KWF theory and JG theory. Dashed line shows behavior of a real network in the low deformation region.

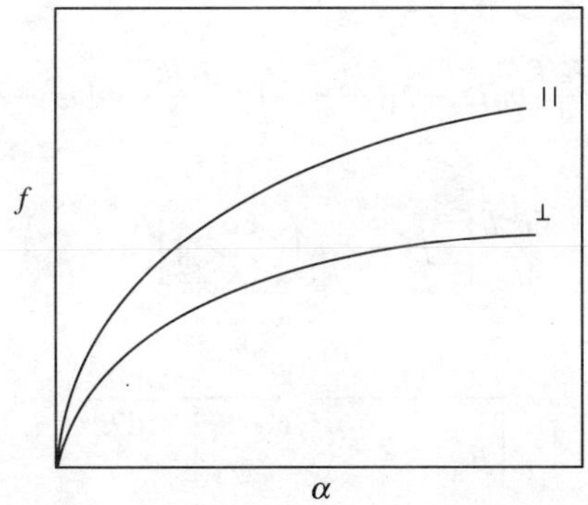

Figure 3 Force vs. deformation for a composite network in the direction parallel (‖) to original orientation and perpendicular (⊥).

11.5 CONTEMPORARY NETWORK THEORIES

Sometime in the mid-1970s it became apparent that elastomeric behavior varied from the affine model, equation (15), at low deformations to the phantom model, equation (13), at large deformations (in the absence of non-Gaussian finite extensibility).[44–46] Junction fluctuations, suppressed at low strains by isotropic entanglements, gain greater intensities from the anisotropy of larger strains. Since deformation of dry networks is almost a constant volume process, the constraint domains, isotropic in the unstretched system, deform anisotropically with strain, thereby permitting increased junction fluctuations in the stretch direction but not perpendicular to it. An alignment of chains along the stretch axis makes it easier for junctions to slide (migrate) along parallel channels, pulling their attached chains with them. In the transverse directions, fluctuations become even more difficult than in the isotropic system because the junction chains are more firmly encumbered by entanglements in transverse motion. Such ideas were incorporated into a new molecular theory of elasticity, of which there are two versions.

The first version was put forth by Ronca and Allegra[31] in 1975. Their work is original and elegant, but somewhat phenomenological. Starting with the James–Guth network partition function, a delta function is inserted that restricts junction fluctuations when deformations are small and permits them to increase as the extent of deformation increases, so that the network gains the characteristics of an unencumbered, phantom network as deformation increases without limit. Because such an explicit function is unknown, a general function is inserted so as to obtain a form of the network deformational free energy, which by minimization can be forced to yield up an acceptable explicit function. The results are remarkable. The deformational free energy smoothly transforms from that of the affine network at low deformations to that of the phantom network at large deformations, and the tensile force behaves similarly, in excellent accord with observed behavior. Such systems can no longer be described as 'Gaussian', but rather as modified or obstructed Gaussians. This is evidenced further by thermoelastic behavior, which contrary to classical elasticity theory is no longer found to be independent of elongation.

For a general deformation the deformational free energy, A, is

$$A = \frac{GkT}{8}\left[\eta f^* - 2 + \frac{6}{I_1}\ln I_1\right]I_1 - \frac{GkT}{2}\ln I_3 \tag{21}$$

where f^* is the junction functionality, $\eta = \langle r^2\rangle_i/\langle r^2\rangle_0$, $I_1 = \alpha_x^2 + \alpha_y^2 + \alpha_z^2$, and $I_3 = \alpha_x\alpha_y\alpha_z$. An unimportant additive constant to equation (21) is omitted. Consider a uniaxial deformation of amount $\alpha_x = \alpha$ at constant volume. Then

$$I_1 = \alpha^2 + 2/\alpha$$

$$I_3 = 1$$

and

$$A = \frac{GkT}{8}[\eta f^* - 2]\left(\alpha^2 + \frac{2}{\alpha}\right) + \frac{3GkT}{4}\ln\left(\alpha^2 + \frac{2}{\alpha}\right)$$

which yields

$$f = \frac{GkT}{4L_i}\left[\eta f^* - 2 + \left(\frac{6\alpha}{\alpha^3 + 2}\right)\right]\left(\alpha - \frac{1}{\alpha^2}\right) \tag{22}$$

and

$$\frac{f_e}{f} = \frac{(\eta f^* - 2)T}{\left[\eta f^* - 2 + \left(\frac{6\alpha}{\alpha^3 + 2}\right)\right]}\frac{d\ln\langle r^2\rangle_0}{dT} \tag{23}$$

These results are quite interesting. At low deformations, $\alpha \approx 1$, equation (22) reduces to the affine model; at large deformations, $\alpha \gg 1$, the phantom network model of James–Guth is recovered. This means that crosslink fluctuations increase as the anisotropy increases. However, the ratio f_e/f increases to a maximum value as $\alpha \to \infty$:

$$\frac{f_e}{f} \to T\frac{d\ln\langle r^2\rangle_0}{dT}$$

which is the same as equations (11) and (20). From equation (23) this is clearly the James–Guth value, not that of the affine model, which for $\alpha \approx 1$ reduces to

$$\frac{f_e}{f} \to \left(1 - \frac{2}{\eta f^*}\right)\frac{d\ln\langle r^2\rangle_0}{dT}$$

a result clearly at variance with that given by Flory, equation (20), for the affine model. The discrepancy is due to the differing interpretations of $\langle r_m\rangle$ and $\langle r^2\rangle_i$ and their temperature dependencies. So, we now have a major disagreement concerning the affine model. If Ronca–Allegra are correct, and this author thinks they are, or nearly so, all previous analyses, except perhaps those of highly swollen networks, could be wrong. The force has long been known to diverge considerably from both KWF and JG theories; now Ronca and Allegra tell us thermoelastic results are similarly wrong. The full consequences of this are not yet known, but a major reanalysis of thermoelastic data is necessary since much of this data has been used to gain information about rotational potential barriers, which are important for calculating characteristic ratios and polymer geometries in general.

Why are highly swollen networks exempt? Because reducing chain density reduces interchain obstructions and pushes the system towards the phantom network model. It is easy to extend the foregoing equations to swollen networks. If the network is first swollen by an amount v^{-1}, then stretched unidirectionally at constant volume by an amount α (referred to the swollen dimension), the deformational free energy becomes:

$$I_1 = v^{-2/3}\left(\alpha^2 + \frac{2}{\alpha}\right) \quad \text{and} \quad I_3 = v^{-1}$$
$$A_s = \frac{GkT}{8v^{2/3}}[\eta f^* - 2]\left(\alpha^2 + \frac{2}{\alpha}\right) + \frac{3GkT}{4}\ln\left(\alpha^2 + \frac{2}{\alpha}\right) \tag{24}$$

Unrequired constants are omitted. Therefore, we have

$$f = \frac{GkT}{4L_i v^{1/3}}\left[\eta f^* - 2 + \left(\frac{6\alpha v^{2/3}}{\alpha^3 + 2}\right)\right]\left(\alpha - \frac{1}{\alpha^2}\right) \tag{25}$$

$$\frac{f_e}{f} = T\left[1 + \frac{1}{(\eta f^* - 2)}\left(\frac{6\alpha v^{2/3}}{\alpha^3 + 2}\right)\right]^{-1} \frac{\mathrm{d}\ln\langle r^2\rangle_0}{\mathrm{d}T} \tag{26}$$

Here v is the volume fraction of polymer in the swollen network. For $v = 0.2$, $v^{2/3} = 0.342$, or $v = 0.5$, $v^{2/3} = 0.615$; the first reduces the correction terms to approximately a third, and the second to 61.5% of the unswollen value, not enough to eliminate it altogether but a significant reduction nevertheless. Isotropic swelling clearly decreases entanglement constraints and increases fluctuations, moving the system closer to the phantom network model.

The Ronca–Allegra theory is now 20 years old. Its results are relatively simple in form, devoid of ambiguous adjustable parameters, and in very reasonable agreement with experimental data. Even so, few have risen to carry its banner yet, but this will likely change.

The major reason for this dearth of interest is the availability of an alternate theory stemming from Flory.[44,45] Much of what Flory offers is the antithesis of Ronca–Allegra. Flory's equations are cumbersome, not simple and elegant. They also contain one or two adjustable parameters, thereby introducing ambiguity. However, ponderousness is only a nuisance if all else is right, and Flory's theory, especially as modified by others,[46] seems to be right, or nearly so.

Ronca–Allegra began with the James–Guth network theory and modified it by restricting junction fluctuations. Flory began with the affine model and modified it by removing restrictions to junction fluctuations. He did this by postulating that the entanglement domains in which the junctions are trapped deform affinely with deformation, so that fluctuations along the stretch axis are greatly increased as deformation increases. He further assumed that entanglements along the chains, away from the junctions, manifest themselves by further restricting junction flucturations; that is, the domains of constraint about a crosslink reflect the extent of entangled substituent chains with the surrounding medium as well as the lesser constraints along the contours of the chains. Flory was well aware that this assumption may falter. Actually it did, as shown by Erman and Monnerie[47] when they extended Flory's calculation to include a separate accounting of constraints all along the chains' contour lengths. Their work is an important contribution to elasticity theory. The major equations are:

$$A = \tfrac{1}{2}\zeta kT \sum_t \left\{\alpha_t^2 - 1 + \frac{G}{\zeta}[B_t + D_t - \ln(1 + B_t) - \ln(1 + D_t)]\right\} \tag{27}$$

where $\zeta = (1 - 2/f^*)G$ is the cycle rank, $t = x,y,z$, and

$$h(\alpha_t) = K_G[1 + (\alpha_t^2 - 1)\Phi]$$

$$B_t = h(\alpha_t)K_G(1 - \Phi)(\alpha_t^2 - 1)/[\alpha_t^2 + h(\alpha_t)]^2$$

$$D_t = \alpha_t^2 B_t / h(\alpha_t)$$

The force in simple elongation along the x direction becomes

$$f = \frac{\zeta kT}{L_i}\left(\frac{v_c}{v}\right)^{1/3}\left\{\alpha - \frac{1}{\alpha^2} + \frac{G}{\zeta}\left[\alpha K(\alpha_x^2) - \frac{1}{\alpha^2}K(\alpha_y^2)\right]\right\} \tag{28}$$

$$\alpha_x = (v_c/v)^{1/3}\alpha$$

$$\alpha_y = \alpha_z = (v_c/v)^{1/3}\alpha^{-1/2}$$

$$K(\alpha_t^2) = \frac{B_t\dot{B}_t}{1 + B_t} + \frac{D_t\dot{D}_t}{1 + D_t}$$

$$\dot{B}_t = B_t\left\{(\alpha_t^2 - 1)^{-1} - 2[\alpha_t^2 + h(\alpha_t)]^{-1} + \frac{K_G}{h(\alpha_t)}\frac{[\alpha_t^2 - h(\alpha_t)]\Phi}{[\alpha_t^2 + h(\alpha_t)]}\right\}$$

$$\dot{D}_t = B_t\left[h(\alpha_t)^{-1} - \frac{\alpha_t^2 K_G \Phi}{h(\alpha_t)^2}\right] + \frac{\alpha_t^2 B_t}{h(\alpha_t)}$$

The parameter Φ depends upon the way constraints affect chain fluctuations upon deformation. There are two cases. One is the constrained chain (CC) model, in which interior subchains (as opposted to crosslinks) fluctuate independently of deformation. The other is the modified constrained chains (MCC), in which interior chain points have strain dependent fluctuations:

$$\Phi = (1 - 2/f^*)^2 \text{ for the MCC model}$$

$$\Phi = \tfrac{1}{3}(1 - 2/f^*)^2(1 + 2/n) \text{ for the CC model}$$

where n is the number of subchains (subject to entanglement constraints) making up a network chain. The term $2/n$ can be neglected if n is large.

To avoid confusion, it is worthwhile to point out that Flory's original theory, and the altered version by Flory and Erman, involve junction fluctuations only. It is called the constrained junction model. The subsequent modification by Erman–Monnerie is based on subchain fluctuations (making up the entire network chains). These are constrained chain models, above. The parameter K_G is adjustable and must be determined from experimental data. It is a measure of the strength of constraints and ranges from zero (no constraints, plantom network) to infinity (infinite constraints, affine network). It is a function of the reference state only, independent of network volume and perhaps of temperature also. The symbol v is the volume fraction of polymer in the network and v_c is the volume fraction of polymer at the time of crosslinking.

These equations have been tested for several networks. In the case of *cis*-polyisoprene,[48] reasonably good results were found for the constrained junction, CC, and MCC models, with CC giving the best fit to experimental data. The constrained junction model works fairly well, but it requires another parameter in addition to K to obtain satisfactory results. Similar findings pertain to swollen networks.[49]

Thermoelastic behavior of these models, however, is ambiguous because it is not certain how K_G depends upon temperature. It is suggested that K_G may be given as[47,50]

$$K_G = I\langle r^2\rangle_0^{3/2}(\mu/V_0)$$

where μ is the number of crosslinks, V_0 is the volume at crosslinking, and I is a constant. Since $V_0 \propto \langle r^2\rangle_0^{3/2}$, K_G will be independent of T if the constant I is also. With such an assumption the ratio f_e/f was calculated[51] and found to vary with deformation in a fashion similar to the Ronca–Allegra theory. Since the Ronca–Allegra theory is quite definite about this point, it would seem that K_G is in fact independent of T, but this is not certain yet.

Erman and Flory[52] also extended the constrained junction theory to include strain birefringence, with good results.[53]

11.6 CONCLUSION

The Flory–Erman–Monnerie concepts have led to a comprehensive elasticity theory that seems to be in good accord with various experimental results. The theory is molecular, as opposed to phenomenological. Yet it does contain one or two adjustable parameters, depending on which version of the theory is considered. Contrast this with the Ronca–Allegra theory, which contains no adjustable parameters and is partially phenomenological but a molecular theory overall. Of course, adjustable parameters almost always improve agreement between theory and experiment: it is possible to fit any curve with a few parameters. A 'good fit' therefore should not discount one theory over the other at this stage of development. It is to be remembered that the Ronca–Allegra solution may not be unique, since other versions may surface, and the Ronca–Allegra treatment appears to be limited to junction fluctuations without accounting for subchain fluctuations. There is thus extensive room for further development.

With these latest developments in the molecular theory of rubber elasticity we can now be confident of a semiquantitative understanding of the phenomenon. Earlier, simple model networks — affine, phantom — were built upon the statistical characteristics of long, flexible polymer molecules. Crosslinking divides the molecules into submolecules, or chains, joining the crosslinks to form a network. Such chains follow the same statistical laws as the parent molecules. Each network chain was assumed to behave independently of all others, guided only by external forces or bulk dimensional changes. However, the phantom model did contain an element of interdependency by its insistence upon maintaining connectivity patterns established by crosslinking. Neither network was very satisfactory. The affine model failed to allow for junction fluctuations, and the phantom model allowed too many. The latest theories correct these deficiencies by incorporating interchain dependencies of the nature of entanglements or obstructions. Further refinements will no doubt be forthcoming. A final point: these new theories, good as they are, do not meet the Edwards condition of stability. This problem will have to be addressed sooner or later.

11.7 REFERENCES

1. L. R. G. Treloar, 'The Physics of Rubber Elasticity', Oxford University Press, Oxford, 1958.
2. P. J. Flory, 'Principles of Polymer Chemistry', Cornell University, Ithaca, NY, 1953.
3. Various texts on polymer science.
4. K. J. Smith, A. Greene and A. Ciferri, *Kolloid Z.*, 1964, **194,** 49.
5. P. J. Flory and W. R. Krigbaum, *J. Chem. Phys.*, 1950, **18,** 1086.
6. M. V. Volkenstein and O. B. Ptitsyn, *Zh. Tekh. Fiz.*, 1955, **25,** 649.
7. M. V. Volkenstein, 'Configurational Statistics of Polymer Chains', Wiley, New York, 1963.
8. K. Nagai, *J. Chem. Phys.*, 1963, **38,** 924.
9. K. J. Smith, Jr., *J. Polym. Sci. A-2,* 1971, **9,** 2119.
10. K. J. Smith, Jr., 'Polymer Science', ed. A. D. Jenkins, North Holland, Amsterdam, 1972, chap 5.
11. P. J. Flory, 'Statistical Mechanics of Chain Molecules', Hanser, New York, 1989.
12. P. J. Flory, C. A. J. Hoeve and A. Ciferri, *J. Polym. Sci.*, 1959, **34,** 337.
13. J.-P. Queslel and J. E. Mark, in 'Comprehensive Polymer Science', ed. G. Allen and J. C. Bevington, Pergamon, Oxford, 1989, vol. 2, chap. 9.
14. P. J. Flory, *Pure Appl. Chem.*, 1973, **33,** 1.
15. W. Kuhn, *Kolloid Z.*, 1936, **75,** 258; 1939, **87,** 3.
16. W. Kuhn and H. Grün, *J. Polym. Sci.*, 1946, **1,** 183.
17. H. M. James and E. Guth, *J. Chem. Phys.*, 1943, **11,** 455.
18. H. M. James and E. Guth, *J. Polym. Sci.*, 1949, **4,** 153.
19. S. Lifson, *J. Chem. Phys.*, 1958, **29,** 80.
20. H. M. James and E. Guth, *J. Chem. Phys.*, 1947, **15,** 669.
21. H. M. James and E. Guth, *J. Chem. Phys.*, 1953, **21,** 1039.
22. F. T. Wall, *J. Chem. Phys.*, 1942, **10,** 132, 485.
23. F. T. Wall, *J. Chem. Phys.*, 1943, **11,** 527.
24. F. T. Wall and P. J. Flory, *J. Chem. Phys.*, 1951, **19,** 1435.
25. P. J. Flory and J. Rehner, *J. Chem. Phys.*, 1943, **11,** 512.
26. P. J. Flory and J. Rehner, *J. Chem. Phys.*, 1943, **11,** 521.

27. P. J. Flory, *J. Chem. Phys.*, 1950, **18,** 108.
28. S. F. Edwards, *Proc. Phys. Soc.*, 1967, **92,** 9.
29. J. F. Douglas, in 'Elastomeric Polymer Networks', ed. J. E. Mark and B. Erman, Prentice Hall, Englewood Cliffs, NJ, 1992.
30. J. J. Hermans, *J. Polym. Sci.*, 1962, **59,** 191.
31. G. Ronca and G. Allegra, *J. Chem. Phys.*, 1975, **63,** 4990.
32. M. C. Wang and E. Guth, *J. Chem. Phys.*, 1952, **20,** 1144.
33. L. R. G. Treloar, *Trans. Faraday Soc.*, 1946, **42,** 77, 83.
34. L. R. G. Treloar, *Trans. Faraday Soc.*, 1954, **50,** 881.
35. L. Mullins, *J. Appl. Polymer Sci.*, 1959, **2,** 257.
36. K. J. Smith, Jr., A. Greene and A. Ciferri, *Kolloid Z.*, 1964, **194,** 49.
37. K. J. Smith, Jr. and D. Puett, *J. Appl. Phys.*, 1966, **37,** 346.
38. The last line of the abstract of Ref. 37 should read: '. . . non-Gaussian behavior does exist in rubber networks'. The phrase as printed is a misprint inconsistent with the findings reported in the body of the paper.
39. M. Mooney, *J. Appl. Phys.*, 1940, **11,** 582.
40. R. S. Rivlin, in 'Rheology: Theory and Applications', ed. F. R. Eirich, Academic Press, New York, 1956, **1,** 351.
41. J. P. Berry, J. Scanlan and W. F. Watson, *Trans. Faraday Soc.*, 1956, **52,** 1137.
42. R. D. Andrews, A. V. Tobolsky and E. E. Hanson, *J. Appl. Phys.*, 1946, **17,** 352.
43. A. Greene, K. J. Smith, Jr. and A. Ciferri, *Trans. Faraday Soc.*, 1965, **61,** 2772.
44. P. J. Flory, *Proc. R. Soc. London., Ser. A,* 1976, **351,** 351.
45. P. J. Flory, *J. Chem. Phys.*, 1977, **66,** 5720.
46. P. J. Flory and B. Erman, *Macromolecules,* 1982, **15,** 800.
47. B. Erman and L. Monnerie, *Macromolecules,* 1989, **22,** 3342.
48. F. Fontaine, C. Morland, C. Noel, L. Monnerie and B. Erman, *Macromolecules,* 1989, **22,** 3348.
49. F. Fontaine, C. Noel, L. Monnerie and B. Erman, *Macromolecules,* 1982, **22,** 3352.
50. B. Erman and P. J. Flory, *Macromolecules,* 1982, **15,** 806.
51. K. J. Smith, Jr., in 'Elastomeric Polymer Networks', ed. J. E. Mark and B. Erman, Prentice Hall, Englewood Cliffs, NJ, 1992.
52. B. Erman and P. J. Flory, *Macromolecules,* 1983, **16,** 1601.
53. B. Erman and P. J. Flory, *Macromolecules,* 1983, **16,** 1607.
54. R. G. Kirste, W. A. Kruse and J. Schelten, *J. Macromol. Chem.*, 1973, **162,** 299.
55. R. G. Kirste, W. A. Kruse and K. Ibel, *Polymer,* 1975, **16,** 120.
56. D. G. Ballard, J. Schelten and G. D. Wignall, *Env. Polym. J.*, 1973, **9,** 965.
57. J. P. Cotton, D. Decker, H. Benoit, B. Farnoux, J. Higgins, G. Jannink, R. Ober, C. Picot and J. des Cloizeaux, *Macromolecules,* 1974, **7,** 863.
58. M. Daoud, J. Cotton, B. Farnoux, G. Jannink, G. Sarma, H. Benoit, R. Duplessix, C. Picot and P. G. DeGennes, *Macromolecules,* 1975, **8,** 804.
59. M. Beltzung, C. Picot, P. Rempp and H. Hertz, *Macromolecules,* 1982, **15,** 1594.

12
Phenomenological Theory of Rubber Elasticity

O. H. YEOH

GenCorp Corporate Technology Center, Akron, OH, USA

12.1 INTRODUCTION

There are two different approaches to the study of rubber elasticity. First, the statistical or kinetic theory attempts to derive stress–strain properties of rubber from some idealized model of its structure. This theory is one of the cornerstones of our understanding of the macromolecular nature of rubber. The molecular approach is the subject of other reviews[1,2] in *Comprehensive Polymer Science.* Second, the phenomenological theory treats the problem from the viewpoint of continuum mechanics. The phenomenological approach, which is reviewed here, merely attempts to construct a mathematical framework to describe rubbery behavior so that stress analysis and strain analysis problems may be solved without reference to microscopic structure or molecular concepts.

It can be said that the pragmatic nature of the phenomenological theory makes it of greater interest to polymer engineers than to polymer scientists. Indeed, recent interest in the phenomenological theory stems largely from the desire to use finite element analysis in the engineering design of rubber components. Nevertheless, polymer scientists may also find this approach worthy of study. If nothing else, it provides a rational basis for characterizing the stress–strain behavior of rubber. It is clearly desirable that the stress–strain properties of the material be described in a form that is generally applicable (*i.e.* applicable to several different modes of deformation as opposed to just one), before attempts are made to interpret the behavior in terms of the molecular structure; otherwise, false deductions may result from incomplete or inadequate information.[3]

The phenomenological theory of rubber elasticity may also be viewed as an extension of the classical theory of elasticity. In the classical theory, all deformations are assumed to be small. While all solids exhibit behavior approximating that envisaged in the classical theory for some limited range, the small deformation assumption is very restrictive. Many materials, particularly vulcanized rubber, exhibit elastic behavior over large deformations. For these cases, the small deformation restriction must be relaxed. The study of large elastic deformations is sometimes referred to as finite elasticity theory.

12.2 THE MOONEY THEORY

According to Rivlin,[4] finite elasticity theory is almost as old as the classical elasticity theory itself. However, the first significant theory of large elastic deformations developed with particular reference to rubber appears to be that of Mooney.[5] Strictly, Mooney's theory was developed in two forms: a special and a general. In the special theory, Hooke's law is assumed to apply to shear deformation. In the general theory, this assumption of proportionality between stress and strain is relaxed and shear stress is merely assumed to be related to shear strain by a nonlinear stress–strain law. It is interesting to note that nearly all references to Mooney's theory cite just the special theory.[3]

12.2.1 Constant Shear Modulus

In developing his theory, Mooney[5] considered the deformation of an elemental volume of rubber such that the extension ratios (ratios of final to initial lengths) in the three principal coordinate directions are λ_1, λ_2, and λ_3 (see Figure 1). The material is assumed to be homogeneous, isotropic, elastic and incompressible. In the special version of his theory, Mooney further assumed that Hooke's law is obeyed in simple shear (*i.e.* shear stress is proportional to shear strain). Based upon these assumptions and mathematical arguments involving symmetry, Mooney showed that W, the elastic strain energy stored in unit volume of material, is given by

$$W = C_1(\lambda_1^2 + \lambda_2^2 + \lambda_3^2 - 3) + C_2(\lambda_1^{-2} + \lambda_2^{-2} + \lambda_3^{-2} - 3) \quad (1)$$

where C_1 and C_2 are material constants.

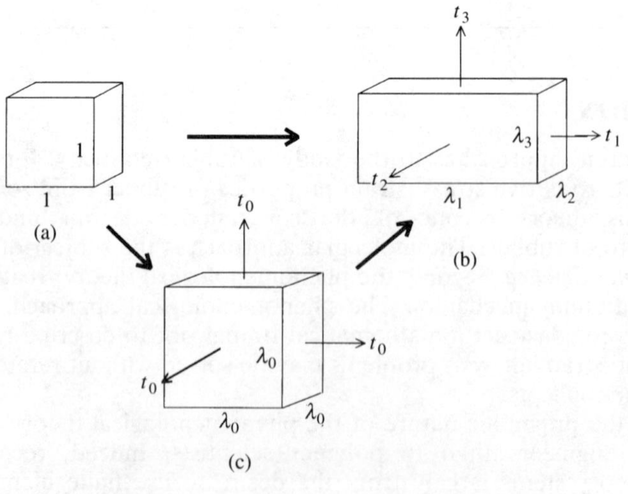

Figure 1 Pure homogeneous deformation of a unit cube: (a) undeformed state; (b) deformed state; (c) an intermediate state arising from uniform dilation such that $\lambda_0^3 = \lambda_1\lambda_2\lambda_3$.

Mooney also suggested that equation (1) may be conveniently rearranged by making the substitutions

$$C_1 = \tfrac{1}{4}(G + H) \quad \text{and} \quad C_2 = \tfrac{1}{4}(G - H) \quad (2)$$

yielding

$$W = \frac{G}{4}\sum_{i=1}^{3}\left(\lambda_i - \frac{1}{\lambda_i}\right)^2 + \frac{H}{4}\sum_{i=1}^{3}\left(\lambda_i^2 - \frac{1}{\lambda_i^2}\right) \tag{3}$$

In this form, it may be shown[5] that G is the shear modulus while H is a new constant not encountered in other theories of elasticity. To understand the significance of H, Mooney pointed out that for any deformation characterized by the extension ratios λ_1, λ_2, and λ_3, there exists an isochoric reciprocal deformation with the extension ratios

$$\lambda_1^* = 1/\lambda_1, \quad \lambda_2^* = 1/\lambda_2, \quad \lambda_3^* = 1/\lambda_3 \tag{4}$$

Now, if $H = 0$, then

$$W = \frac{G}{4}\sum_{i=1}^{3}\left(\lambda_i - \frac{1}{\lambda_i}\right)^2 = \frac{G}{4}\sum_{i=1}^{3}\left(\lambda_i^* - \frac{1}{\lambda_i^*}\right)^2 = W^* \tag{5}$$

where W^* is the strain energy for the reciprocal deformation. Thus, $H \neq 0$ is a measure of the departure from the symmetry represented by equation (5).

12.2.2 Variable Shear Modulus

In the general form of his theory, Mooney[5] assumed that the shear stress is an analytical function of the shear strain. Extending his arguments based upon symmetry, Mooney showed that the strain energy function may be written as

$$W = \sum_{n=1}^{\infty} [A_{2n}(\lambda_1^{2n} + \lambda_2^{2n} + \lambda_3^{2n} - 3) + B_{2n}(\lambda_1^{-2n} + \lambda_2^{-2n} + \lambda_3^{-2n} - 3)] \tag{6}$$

Alternatively, making substitutions analogous to equation (2), the strain energy function may be written as

$$W = \sum_{n=1}^{\infty}\frac{G_{2n}}{4n}\sum_{i=1}^{3}\left(\lambda_i^n - \frac{1}{\lambda_i^n}\right)^2 + \sum_{n=1}^{\infty}\frac{H_{2n}}{4n}\sum_{i=1}^{3}\left(\lambda_i^{2n} - \frac{1}{\lambda_i^{2n}}\right) \tag{7}$$

12.3 RIVLIN'S FORMULATION

The foundations of the phenomenological theory of rubber elasticity were largely laid by Rivlin.[6–9] The thrust of his contributions may be summarized[4,10] as follows.

12.3.1 Kinematics and the Strain Energy Function

The deformation of a body may be specified by the vector positions $\boldsymbol{X}$ and $\boldsymbol{x}$, referred to a fixed rectangular coordinate system, of a generic particle of the body in its undeformed and deformed states at time t, respectively. Thus

$$\boldsymbol{x} = \boldsymbol{x}(\boldsymbol{X}, t) \tag{8}$$

The deformation gradient, $\boldsymbol{F}$, is defined by

$$\boldsymbol{F} = \frac{\partial \boldsymbol{x}}{\partial \boldsymbol{X}} \tag{9}$$

If the material is perfectly elastic and the deformation takes place under isothermal conditions, then the strain energy density, W, is a function of $\boldsymbol{F}$ only. Since W is unaltered by a rigid body rotation, W depends on $\boldsymbol{F}$ through the symmetric tensor, $\boldsymbol{C}$, defined by

$$\boldsymbol{C} = \boldsymbol{F}^T \cdot \boldsymbol{F} \tag{10}$$

$\boldsymbol{C}$ is the right Cauchy–Green tensor (also called the Green deformation tensor).

Assuming further that the material is isotropic, the dependence of W on $\boldsymbol{C}$ must be through the three basic strain invariants which are scalar functions of $\boldsymbol{C}$ defined by

$$I_1 = \mathrm{tr}\,\boldsymbol{C} \tag{11}$$

$$I_2 = \tfrac{1}{2}[(\mathrm{tr}\,\boldsymbol{C})^2 - \mathrm{tr}\,\boldsymbol{C}^2] \tag{12}$$

$$I_3 = \det \boldsymbol{C} \tag{13}$$

When the deformation is expressed in terms of the principal extension ratios, λ_1, λ_2, and λ_3, then the strain invariants are simply given by

$$I_1 = \lambda_1^2 + \lambda_2^2 + \lambda_3^2 \tag{14}$$

$$I_2 = \lambda_1^2\lambda_2^2 + \lambda_2^2\lambda_3^2 + \lambda_3^2\lambda_1^2 \tag{15}$$

$$I_3 = \lambda_1^2\lambda_2^2\lambda_3^2 \tag{16}$$

Thus

$$W = W(I_1, I_2, I_3) \tag{17}$$

which may be expressed as the infinite series

$$W = \sum_{i+j+k=1}^{\infty} C_{ijk}(I_1 - 3)^i(I_2 - 3)^j(I_3 - 1)^k \tag{18}$$

where C_{ijk} are material constants and the three arguments, $(I_1 - 3)$, $(I_2 - 3)$, and $(I_3 - 1)$, have been chosen so that they assume the value of zero in the undeformed state. We note that I_3 is related to volumetric changes. If we restrict the discussion to incompressible materials, then equation (18) reduces to

$$W = \sum_{i+j=1}^{\infty} C_{ij}(I_1 - 3)^i(I_2 - 3)^j \tag{19}$$

where C_{ij} are material constants.

12.3.2 Some Particular Forms of the Revlin Strain Energy Function

The general Rivlin strain energy function is an infinite power series which is usually truncated by taking only the leading terms. Considering incompressible materials only and taking just the first term yields the neo-Hookean model

$$W = C_{10}(I_1 - 3) \tag{20}$$

The importance of the neo-Hookean model lies in the fact that the most elementary form of the statistical theory of rubber elasticity[3] gives the strain energy function

$$W = \tfrac{1}{2}NkT(I_1 - 3) \tag{21}$$

where N is the number of network chains per unit volume, k is Boltzmann's constant, and T is the absolute temperature. Equations (20) and (21) are of the same form and these provide a link between the experimentally determined coefficient, C_{10}, and molecular parameters. Unfortunately, we find that the neo-Hookean model is only valid for a relatively small range of strains.[3] The applicable strain range is particularly limited in tension, where the model is valid for strains up to about 30% only.

Taking the first two terms of equation (19) yields

$$W = C_{10}(I_1 - 3) + C_{01}(I_2 - 3) \tag{22}$$

Equation (22) is identical to equation (1) and it is usually referred to as the Mooney–Rivlin model. It represents the first-order approximation of equation (19). The Mooney–Rivlin model owes much of its popularity to the fact that it is usually quite successful in describing tensile data up to moderately large strains of about 100–200%. Tensile data are often the only type of experimental data available because of the relative inconvenience of performing experiments in other deformation modes. It is common to interpret C_{10} in terms of molecular parameters of the statistical theory and regard C_{01} as reflecting the deviations from ideal behavior. However, this interpretation is flawed. Rivlin and Saunders[9] have pointed out that the agreement between experimental tensile data and equation (22) is somewhat fortuitous. The Mooney–Rivlin model obtained by fitting tensile data is quite inadequate in other modes of deformation, especially compression.[9]

The Mooney–Rivlin model also fails to account for the upturn in the tensile stress–strain curve which is commonly observed at large strains. Tschoegl[11] suggested that this arises from not taking enough terms in truncating equation (19). He showed that the addition of a third term to the Mooney–Rivlin model enabled the fitting of tensile stress–strain data up to breaking strains. The third term is chosen empirically and, depending on the rubber, may be any of the second or third order terms.

Higher order approximations of the general Rivlin strain energy function have been used.[12–14] One variation suggested by Sato[15] chooses the terms based upon the order of the deformation (*i.e.* powers of λ_i) rather the order of the invariants. Thus, one popular form is the so-called third-order deformation approximation,[12–15] which includes all terms involving λ_i^2, λ_i^4, and λ_i^6, *viz.*

$$W = C_{10}(I_1 - 3) + C_{01}(I_2 - 3) + C_{11}(I_1 - 3)(I_2 - 3) + C_{20}(I_1 - 3)^2 + C_{30}(I_1 - 3)^3 \tag{23}$$

12.3.3 Stress

If the dependence of W on the Green deformation tensor, $\boldsymbol{C}$, is known, the Cauchy (true) stress tensor, $\boldsymbol{t}$, can be readily obtained by means of the relation[10]

$$\boldsymbol{t} = 2I_3^{-1/2}\left[\left(\frac{\partial W}{\partial I_1} + I_1\frac{\partial W}{\partial I_2}\right)\boldsymbol{B} - \frac{\partial W}{\partial I_2}\boldsymbol{B}^2 + I_3\frac{\partial W}{\partial I_3}\boldsymbol{\delta}\right] \tag{24}$$

where $\boldsymbol{B}$ is the left Cauchy–Green tensor (also called the Finger tensor) and $\boldsymbol{\delta}$ is the unit or identity tensor. $\boldsymbol{B}$ is defined by

$$\boldsymbol{B} = \boldsymbol{F} \cdot \boldsymbol{F}^T \tag{25}$$

If the material is incompressible, the stress is indeterminate to the extent of an arbitrary hydrostatic pressure, p, and we have

$$\boldsymbol{t} = 2\left[\left(\frac{\partial W}{\partial I_1} + I_1\frac{\partial W}{\partial I_2}\right)\boldsymbol{B} - \frac{\partial W}{\partial I_2}\boldsymbol{B}^2\right] - p\boldsymbol{\delta} \tag{26}$$

The constitutive equations (24) and (26) may be written in several alternative forms.[11,16–20] For the purpose of characterization of rubber vulcanizates it is sufficient to restrict the discussion to the principal extension ratios, λ_1, λ_2, and λ_3, and the principal Cauchy stresses, t_1, t_2, and t_3. In this case, the constitutive equations (24) and (26) reduce to

$$t_1 = 2I_3^{-1/2}\left[\lambda_1^2\frac{\partial W}{\partial I_1} + \lambda_1^2(\lambda_2^2 + \lambda_3^2)\frac{\partial W}{\partial I_2} + I_3\frac{\partial W}{\partial I_3}\right] \tag{27}$$

and

$$t_1 = 2\left[\lambda_1^2\frac{\partial W}{\partial I_1} + \lambda_1^2(\lambda_2^2 + \lambda_3^2)\frac{\partial W}{\partial I_2}\right] - p \tag{28}$$

and corresponding equations for t_2 and t_3.

12.3.4 Comparison with Experiment

We note that the stress–strain behavior is determined by the partial derivatives $\partial W/\partial I_1$, $\partial W/\partial I_2$, and $\partial W/\partial I_3$. Thus, characterization of rubber vulcanizates consists of elucidating the form of the strain energy function via these partial derivatives. For the present, we shall restrict our discussion to the incompressible case since this is a good approximation for most rubbers. Compressible materials will be discussed later in Section 12.5.

A convenient way to avoid consideration of the arbitrary hydrostatic pressure in incompressible materials is to take differences in principal stresses. With minor rearrangement, this gives equations of the form

$$\frac{t_1 - t_2}{\lambda_1^2 - \lambda_2^2} = 2\left[\frac{\partial W}{\partial I_1} + \lambda_3^2\frac{\partial W}{\partial I_2}\right] \tag{29}$$

For the specific case of simple tension or compression, where $\lambda_1 = \lambda$, $\lambda_2 = \lambda_3 = \lambda^{-1/2}$, and $t_2 = t_3 = 0$, equation (29) gives

$$\frac{\sigma_1}{\lambda - \lambda^{-2}} = 2\left[\frac{\partial W}{\partial I_1} + \frac{1}{\lambda}\frac{\partial W}{\partial I_2}\right] \tag{30}$$

where σ_1 is the engineering stress (which is referred to the undeformed state; note that $\sigma_1 = t_1/\lambda$).

For pure shear, $\lambda_1 = \lambda$, $\lambda_2 = 1$, $\lambda_3 = \lambda^{-1}$, and $t_3 = 0$, equation (29) gives

$$\frac{\sigma_1}{\lambda - \lambda^{-3}} = 2\left[\frac{\partial W}{\partial I_1} + \frac{\partial W}{\partial I_2}\right] \tag{31}$$

and

$$\frac{\sigma_2}{1 - \lambda^{-2}} = 2\left[\frac{\partial W}{\partial I_1} + \lambda^2 \frac{\partial W}{\partial I_2}\right] \tag{32}$$

Treloar[3] showed that simple shear is equivalent to pure shear plus a rotation. A simple shear deformation with an amount of shear (engineering shear strain) of γ is equivalent to a pure shear deformation with an extension ratio of λ where the relation between γ and λ is

$$\gamma = \lambda - \lambda^{-1} \tag{33}$$

and the simple shear stress, τ, is related to the amount of shear by

$$\frac{\tau}{\gamma} = 2\left[\frac{\partial W}{\partial I_1} + \frac{\partial W}{\partial I_2}\right] \tag{34}$$

For equal biaxial extension, $\lambda_1 = \lambda_2 = \lambda$, $\lambda_3 = \lambda^{-2}$, and $t_3 = 0$, equation (29) gives

$$\frac{\sigma_1}{\lambda - \lambda^{-5}} = \frac{\sigma_2}{\lambda - \lambda^{-5}} = 2\left[\frac{\partial W}{\partial I_1} + \lambda^2 \frac{\partial W}{\partial I_2}\right] \tag{35}$$

It should be noted that, in any single mode of deformation, the invariants I_1 and I_2 are intimately related and it is not possible to vary one without changing the other. Figure 2 shows the relation between the invariants for the deformation modes described above. In general, it is not possible to determine the partial derivatives, $\partial W/\partial I_1$ and $\partial W/\partial I_2$, separately from measurements in any one single deformation mode;[21] experiments in more than one deformation mode are necessary. The preferred method is general biaxial extension of a sheet of material where λ_1 and λ_2 are varied independently. Experiments of this type are described by Rivlin and Saunders,[9] James *et al.*,[12,13] Jones and Treloar,[22] Kawabata and Kawai,[21] and Chow and Cundiff.[14]

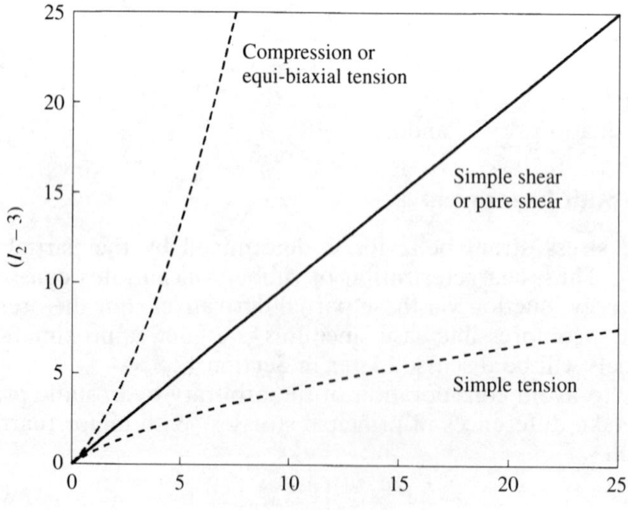

Figure 2 Relation between $(I_1 - 3)$ and $(I_2 - 3)$ for simple modes of deformation.

The required experiments are difficult and errors are correspondingly large, particularly at small deformations or when λ_1 and λ_2 are nearly equal.[9] Thus, it is not surprising to find that the literature contains conflicting results.[10] Table 1 summarizes some conclusions about the nature of $\partial W/\partial I_1$ and $\partial W/\partial I_2$ for unfilled vulcanizates of natural rubber or polyisoprene. The lack of consensus may arise from imperfect elasticity of the materials (even unfilled natural rubber vulcanizates, which are regarded as close to ideally elastic, display differences in the loading and unloading stress–strain curves, stress softening and permanent set) or anisotropy after stretching. McKenna and Zapas[25,26] suggested that there may also be effects due to different techniques of measurement or sample preparation. In any case, it appears that the behavior is complex, particularly at small strains, and that the Mooney–Rivlin model, which assumes constant $\partial W/\partial I_1$ and $\partial W/\partial I_2$, is incorrect.

Table 1 Dependence of $\partial W/\partial I_1$ and $\partial W/\partial I_2$ on Deformation for Unfilled Rubber Vulcanizates.

Authors	$\partial W/\partial I_1$	$\partial W/\partial I_2$
Rivlin and Saunders[9]	At large strains, $\partial W/\partial I_1$ is independent of I_1 and I_2	At large strains, $\partial W/\partial I_2$ is independent of I_1 but decreases with increasing I_2
Becker[23]	At small strains, $\partial W/\partial I_1$ decreases with increasing I_1 but increases with increasing I_2	At small strains, $\partial W/\partial I_2$ may be negative. It increases with increasing I_1 but decreases with increasing I_2
	At large strains, $\partial W/\partial I_1$ is independent of I_1 and I_2	At large strains, $\partial W/\partial I_2$ is independent of I_1 but decreases with increasing I_2
Jones and Treloar[22]	$\partial W/\partial I_1$ decreases monotonically with increasing I_1 and I_2	$\partial W/\partial I_2$ decreases monotonically with increasing I_1 and I_2
Kawabata *et al.*[24]	At small strains, $\partial W/\partial I_1$ increases with increasing I_1 but decreases with increasing I_2	At small strains, $\partial W/\partial I_2$ may be negative. It decreases with increasing I_1 but increases with increasing I_2
	At large strains, $\partial W/\partial I_1$ is independent of I_1 but decreases with increasing I_2	At large strains, $\partial W/\partial I_2$ is independent of I_1 but decreases with increasing I_2

The engineer who is concerned with the practical problem of characterizing a rubber material (usually filled with carbon black, which aggravates the problems of imperfect elasticity and anisotropy), so that he can solve design problems using finite element analysis, is apt to rely on regression analysis of some stress–strain data to obtain an approximate strain energy function for the material. Care should be taken in adopting this approach. It is a common misconception to assume that greater precision is always obtained by using strain energy functions with multiple high-order terms. While the additional coefficients do allow a more precise fit to the experimental data, they often result in unstable strain energy functions which predict physically unrealistic behavior under deformation conditions outside the range of the experimental data.[12–14] For this reason, it is usually recommended that the regression analysis be based upon stress–strain data obtained from two or more deformation modes so that the data cover a wider range of deformation. Nevertheless, higher-order strain energy functions show such poor ability to predict behavior outside the range of experimental data that Chow and Cundiff[14] recommended the use of the Mooney–Rivlin material model for finite element analysis. Indeed, the Mooney–Rivlin model appears to be the most widely used strain energy function in stress analysis.[27] Clearly, its simplicity is so valued that its inaccuracies are tolerated.

Some variations of the basic regression analysis are noteworthy. Haines and Wilson[28] suggested plotting W as a surface above a plane having $(I_1 - 3)$ and $(I_2 - 3)$ as rectangular Cartesian coordinates. The terms of the strain energy function are selected on the basis of an analysis of the shape of this surface. Also it is this surface, not the stress equations, which is the basis for the regression analysis. In this manner, Haines and Wilson obtained six-coefficient models to fit the stress–strain data of Treloar[29] and Jones and Treloar.[22]

As discussed above, strain energy functions containing high-order terms are often unstable, leading to predictions of physically unrealistic behavior under deformation conditions outside the range of the original data. Johnson *et al.*[30] considered the conditions for stable strain energy functions. By imposing the constraint that $C_{ij} \geq 0$ during the regression analysis, they obtained a stable three-coefficient model with the coefficients C_{10}, C_{01}, and C_{30}, that fits the data of Treloar.[29]

Yeoh[31] took a different approach. Following Gregory,[32] who had observed that $\partial W/\partial I_2$ is usually much smaller than $\partial W/\partial I_1$, Yeoh[31] assumed that $\partial W/\partial I_2 = 0$. Strictly, this is wrong, since the weight of evidence in the literature tells us that $\partial W/\partial I_2$ is finite even though there is lack of agreement about its sign or the nature of its dependence on deformation. However, $\partial W/\partial I_2$ appears to be sufficiently small in most cases that neglecting it is not expected to give rise to very serious errors. Equating $\partial W/\partial I_2$ to zero simplifies the problem considerably. It can be readily shown that the third-order deformation approximation (equation 23) of Rivlin's strain energy function reduces to the simple cubic equation

$$W = C_{10}(I_1 - 3) + C_{20}(I_1 - 3)^2 + C_{30}(I_1 - 3)^3 \tag{36}$$

Also, the reduced stress terms on the left hand side of equations (29)–(32) and equations (34) and (35) are now all equal to $2\partial W/\partial I_1$, which according to equation (34) is also the secant shear modulus. It is seen that a material obeying the cubic strain energy function has a strain-dependent shear modulus

$$\frac{\tau}{\gamma} = 2\frac{\partial W}{\partial I_1} = 2C_{10} + 4C_{20}(I_1 - 3) + 6C_{30}(I_1 - 3)^2 \tag{37}$$

Having declared I_2 irrelevant, it is no longer necessary to evaluate $\partial W/\partial I_1$ at fixed values of I_2. All available data may be pooled. We illustrate this using the general biaxial extension data of James and Green.[13] The reduced stress terms, $(t_1 - t_3)/(\lambda_1^2 - \lambda_3^2)$ and $(t_2 - t_3)/(\lambda_2^2 - \lambda_3^2)$, on the left hand side of equation (29), are plotted against $(I_1 - 3)$ in Figure 3 (note that t_3 is zero in this experiment). There is considerable scatter in the data but it is seen that a single curve may be drawn. The curve is a quadratic, as suggested by equation (37).

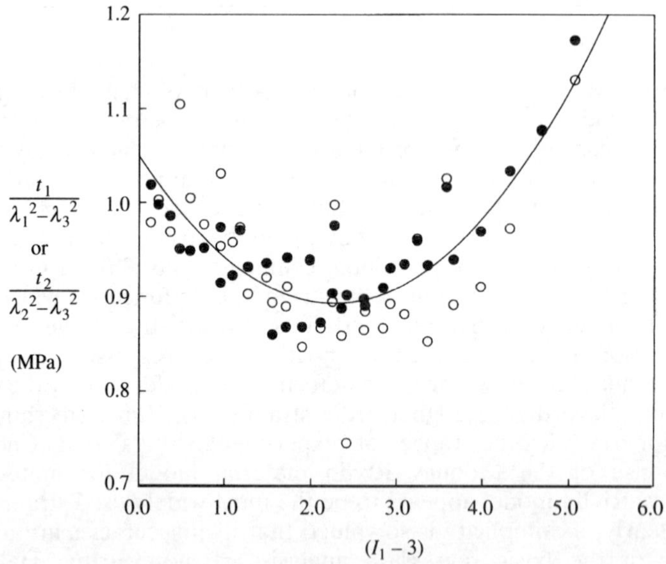

Figure 3 Reduced stress plotted against $(I_1 - 3)$. Biaxial data of James and Green[13] for an NR vulcanizate filled with 40 phr carbon black: ○, t_1; ●, t_2.

Noting that the amount of shear is related to I_1 by

$$\gamma^2 = (I_1 - 3) \tag{38}$$

it is seen that the cubic strain energy function represents a material whose shear modulus varies as the shear strain in a relatively simple manner. At small strains, the shear modulus decreases with increasing strain. At higher strains, the shear modulus increases with strain. This variation of shear modulus with strain is seen with most rubbers; so, usually, the C_{20}

coefficient in the cubic strain energy function is negative while C_{10} and C_{30} are positive. It is noteworthy that the constraint $C_{ij} \geq 0$ imposed by Johnson *et al.*[30] leads to a shear modulus which increases monotonically with strain. This fails to capture the decrease in shear modulus with increasing strain at small strains. The decrease can be so sharp that it is sometimes necessary to add an exponential term to the cubic strain energy function, yielding the modified form[33]

$$W = \frac{A}{B}[1 - \exp(-B(I_1 - 3))] + C_{10}(I_1 - 3) + C_{20}(I_1 - 3)^2 + C_{30}(I_1 - 3)^3 \tag{39}$$

Similar strain energy functions have been proposed by Othman and Gregory[34] and Davies *et al.*[35] to account for the sharp decrease in shear modulus at small strains.

The use of equations (36) or (39) has been shown[31,33] to permit the prediction of stress–strain behavior in different deformation modes from data obtained in one simple deformation mode such as simple tension. This circumvents the need for a rather difficult experiment in general biaxial extension. These equations are recommended as material models for the finite element analysis of rubber.

12.4 FORMULATION IN TERMS OF EXTENSION RATIOS

It may be argued that Rivlin's formulation in terms of the strain invariants is unnecessarily complicated. Formulation directly in terms of the principal extension ratios has some advantages in mathematical simplicity as well as being more intuitive. This was the original approach of Mooney.[3] Significant contributions to this approach were made by Valanis and Landel[36] and Ogden.[37]

12.4.1 The Valanis–Landel Hypothesis

Valanis and Landel[36] proposed as a hypothesis that the strain energy density, W, may be expressed as the sum of independent functions of the principal extension ratios, λ_1, λ_2, and λ_3, *viz.*

$$W = g(\lambda_1) + g(\lambda_2) + g(\lambda_3) \tag{40}$$

in which the function $g(\lambda_i)$ is by symmetry the same for each of the extension ratios. The symmetry arises from the assumption that the material is isotropic. In the Rivlin formulation, symmetry is embedded in the definitions of the three basic strain invariants. Equation (40) should not be taken to imply that the material is compressible. Indeed, the Valanis–Landel hypothesis was specifically proposed for incompressible materials so the relation

$$\lambda_1 \lambda_2 \lambda_3 = 1 \tag{41}$$

is implicitly assumed.

According to Rivlin and Sawyers,[38] the Valanis–Landel hypothesis is a special case of the Rivlin formulation. It is equivalent to an earlier form due to Carmichael and Holdaway,[39] who proposed that W be expressed as functions of (λ_i^2).

Based upon the observation that the Cauchy stress in simple tension is often found to be linearly related to the extension ratio up to strains of the order of 150%, Peng and Landel[40] obtained an explicit expression for $g(\lambda_i)$, *viz.*

$$g(\lambda_i) = E[\lambda_i - 1 - \ln \lambda_i - \tfrac{1}{6}(\ln \lambda_i)^2 + \tfrac{1}{18}(\ln \lambda_i)^3 - \tfrac{1}{216}(\ln \lambda_i)^4] \tag{42}$$

where E is the initial tensile modulus. This form appears attractive for use in finite element analysis since only one material constant needs to be determined. However, it does not appear to be popular. TEXPAC is the only commercial finite element program which uses this strain energy function.[27]

12.4.2 Ogden's Strain Energy Function

Ogden[37] proposed the strain energy function

$$W = \sum_{p=1}^{n} \frac{\mu_p}{\alpha_p}(\lambda_1^{\alpha_p} + \lambda_2^{\alpha_p} + \lambda_3^{\alpha_p} - 3) \tag{43}$$

In the Ogden formulation, the indices α_p need not be integers. Special cases of the Ogden formulation are:

$n = 1$, $\alpha_1 = 1$; this model was advocated by Varga[41] and by Bartenev and Nikiforov.[42]
$n = 1$, $\alpha_1 = 2$; this gives the neo-Hookean material model.
$n = 2$, $\alpha_1 = 2$ and $\alpha_2 = -2$; this gives the Mooney–Rivlin material model.

The Ogden strain energy function is a special case of the Valanis–Landel hypothesis. It has gained significant following among the finite element analysis community after its adoption in the commercial finite element codes, ABAQUS and MARC.

It is useful to understand the significance of the coefficients μ and α. For simplicity and compact printing, we shall consider an Ogden strain energy function with just one term and accordingly will drop the subscript. The arguments can be easily extended to Ogden models with n terms since each term has the same form.

We consider here shear deformation which occupies a central position in the Valanis–Landel hypothesis. For pure shear, $\lambda_1 = \lambda$, $\lambda_2 = 1$, and $\lambda_3 = \lambda^{-1}$, so we have

$$W = \frac{\mu}{\alpha}(\lambda^{\alpha} + \lambda^{-\alpha} - 2) \tag{44}$$

Simple shear is equivalent to pure shear plus a rotation.[3] A simple shear deformation with an amount of shear (engineering shear strain) of γ is equivalent to a pure shear deformation with an extension ratio of λ, where the relations between γ and λ are

$$\gamma = \lambda - \lambda^{-1}$$
$$\lambda = \frac{\gamma}{2} + \sqrt{1 + \frac{\gamma^2}{4}} \tag{45}$$

The simple shear stress, τ, is given by

$$\tau = \frac{\partial W}{\partial \gamma} = \frac{\partial W}{\partial \lambda} \cdot \frac{\partial \lambda}{\partial \gamma} \tag{46}$$

which after some manipulation yields

$$\tau = \mu\gamma\left(\frac{\lambda^{\alpha} - \lambda^{-\alpha}}{\lambda^2 - \lambda^{-2}}\right) \tag{47}$$

For small strains, equation (47) reduces to

$$\tau = \frac{\mu\alpha}{2}\gamma \tag{48}$$

Clearly, the small strain shear modulus, G_0, is given by

$$G_0 = \frac{\mu\alpha}{2} \tag{49}$$

and we may rewrite equation (47) in the form

$$\frac{\tau}{\gamma} = \frac{2G_0}{\alpha}\left(\frac{\lambda^{\alpha} - \lambda^{-\alpha}}{\lambda^2 - \lambda^{-2}}\right) \tag{50}$$

Equation (50) shows that the secant shear modulus, τ/γ, is a function of α and of the shear strain (which is reflected in λ). The form of the dependence is clearly shown in Figure 4, where the shear stress is plotted as a function of shear strain for materials with a small strain shear modulus, G_0, equal to 1 MPa but with different values for the index α. It is seen that when α takes the value of 2 (*i.e.* the material is neo-Hookean), the shear stress–strain curve is linear (*i.e.* the shear modulus is constant). When α is less than 2, the material softens with increasing strain. On the other hand, when α is greater than 2, the material stiffens with increasing strain. The form of equation (50) is such that the same behavior is obtained when α takes a negative value; it is the absolute value of α which determines whether the shear modulus decreases or increases with increasing strain.

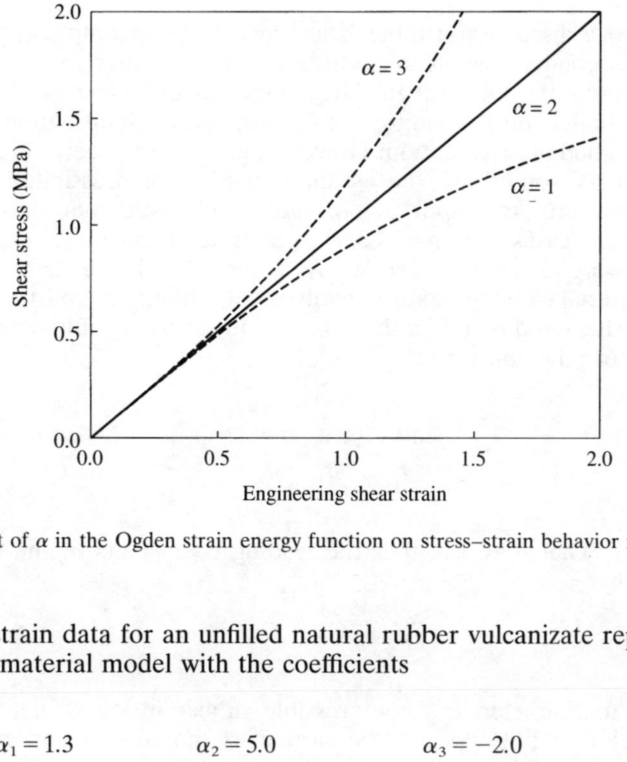

Figure 4 Effect of α in the Ogden strain energy function on stress–strain behavior in simple shear.

Using the stress–strain data for an unfilled natural rubber vulcanizate reported by Treloar,[29] Ogden[37] obtained a material model with the coefficients

$$\alpha_1 = 1.3 \qquad \alpha_2 = 5.0 \qquad \alpha_3 = -2.0$$

$$\mu_1 = 0.618\ \text{MPa} \qquad \mu_2 = 0.00118\ \text{MPa} \qquad \mu_3 = -0.00981\ \text{MPa}$$

Ogden[37] found that two terms were sufficient to describe the stress–strain properties in tension and shear. The third term is necessary to account for the behavior in compression. Jones and Treloar[22] performed biaxial strain experiments on another vulcanizate and obtained a similar material model with the coefficients

$$\alpha_1 = 1.3 \qquad \alpha_2 = 4.0 \qquad \alpha_3 = -2.0$$

$$\mu_1 = 0.69\ \text{MPa} \qquad \mu_2 = 0.01\ \text{MPa} \qquad \mu_3 = -0.0122\ \text{MPa}$$

The precise values of α_n are not critical. More sophisticated regression analysis techniques may be used if desired,[43] but little new insight is to be gained.

We can make some general statements about the nature of the coefficients. Considerations of stability and physically realistic response lead to the constraint[37] that

$$\mu_p \alpha_p > 0 \tag{51}$$

Also the value of α_1 is usually less than 2. This reflects the initial decrease in shear modulus with increasing strain. The value of α_2 is greater than 2 as it is necessary to account for the subsequent increase in shear modulus at large strains. Further grounds for expecting α_1 and α_2 to have values of this general magnitude are given by Ogden[37,44] on the basis of observed behavior of inflated rubber balloons. This combination of first and second terms gives the same general dependence of shear modulus on strain that Yeoh[31,33] describes as characteristic of rubbery behavior. However, it is not obvious how a simple Ogden strain energy function may be constructed to describe the behavior of rubbers (especially those containing reinforcing fillers) where the initial decrease in shear modulus is very sharp.

12.5 DEFORMATION INVOLVING CHANGES IN VOLUME

In much of the above discussion, rubber is assumed to be incompressible. Strictly, the bulk modulus of rubber is comparable in magnitude to that of other materials, so it is not less compressible. It is merely the fact that the shear modulus of rubber is so low compared to its bulk modulus that, under most loading conditions, shear deformation dominates and so incompressibility is a good approximation. However, in cases of triaxial stresses and constraints requiring rigid boundary conditions, the assumption of incompressibility will lead to errors. Also, cellular rubbers are an important category of elastic materials which are clearly compressible. For these cases it is necessary to consider the general strain energy function (equation (18)) involving all three invariants, I_1, I_2, and I_3. This leads to the stress equations (24) and (27) and material characterization involves determining the partial derivatives $\partial W/\partial I_1$, $\partial W/\partial I_2$, and $\partial W/\partial I_3$. It should be noted that in order to satisfy the initial unstrained conditions, it is necessary to impose the constraint[45]

$$\frac{\partial W}{\partial I_1} + 2\frac{\partial W}{\partial I_2} + \frac{\partial W}{\partial I_3} = 0 \tag{52}$$

when $\lambda_1 = \lambda_2 = \lambda_3 = 1$. That is to say that the leading coefficients of the Rivlin strain energy function are related by

$$C_{100} + 2C_{010} + C_{001} = 0 \tag{53}$$

While it is possible to characterize a compressible rubber in terms of $\partial W/\partial I_1$, $\partial W/\partial I_2$, and $\partial W/\partial I_3$, this approach does not appear to be particularly popular. An alternative approach due to Blatz and Ko[46] appears to be most commonly used.

12.5.1 The Blatz–Ko Model

Blatz and Ko[46] suggested the use of the invariants J_1, J_2, and J_3, defined by

$$J_1 = I_1 = \lambda_1^2 + \lambda_2^2 + \lambda_3^2 \tag{54}$$

$$J_2 = \frac{I_2}{I_3} = \lambda_1^{-2} + \lambda_2^{-2} + \lambda_3^{-2} \tag{55}$$

$$J_3 = I_3^{1/2} = \lambda_1\lambda_2\lambda_3 \tag{56}$$

implying the strain energy function

$$W = \sum_{i+j+k=1}^{\infty} D_{ijk}(J_1 - 3)^i(J_2 - 3)^j(J_3 - 1)^k \tag{57}$$

This leads to stress equations of the form

$$t_i = \frac{2}{J_3}\left(\lambda_i^2 \frac{\partial W}{\partial J_1} - \frac{1}{\lambda_i^2}\frac{\partial W}{\partial J_2}\right) + \frac{\partial W}{\partial J_3} \tag{58}$$

In order to satisfy the initial unstrained condition, we have to impose the requirement that

$$2\frac{\partial W}{\partial J_1} - 2\frac{\partial W}{\partial J_2} + \frac{\partial W}{\partial J_3} = 0 \tag{59}$$

when $\lambda_1 = \lambda_2 = \lambda_3 = 1$.

In general, the partial derivatives, $\partial W/\partial J_1$, $\partial W/\partial J_2$, and $\partial W/\partial J_3$, are functions of deformation. However, Blatz and Ko[46] proposed to take $\partial W/\partial J_1$ and $\partial W/\partial J_2$ as constants D_{100} and D_{010}, respectively, as a first approximation. Instead of considering $\partial W/\partial J_3$, it is convenient to relate J_3 to λ_i via the Poisson's ratio, ν, defined in terms of Hencky strain.[45] In this way, Blatz and Ko[46] obtained the strain energy function

$$W = D_{100}\left[(J_1 - 3) + \frac{1-2\nu}{\nu}(J_3^{-2\nu/(1-2\nu)} - 1)\right] + D_{010}\left[(J_2 - 3) + \frac{1-2\nu}{\nu}(J_3^{2\nu/(1-2\nu)} - 1)\right] \tag{60}$$

For a particular cellular polyurethane rubber, Blatz and Ko[46] found ν to be equal to 0.25 and D_{100} to be close to zero.

12.5.2 Penn's Formulation

Penn[47] proposed to separate the shear and dilation components of the deformation by the use of new invariants, L_1, L_2, and L_3, defined by

$$L_1 = \frac{I_1}{I_3^{1/3}} = \frac{\lambda_1^2}{\lambda_0^2} + \frac{\lambda_2^2}{\lambda_0^2} + \frac{\lambda_3^2}{\lambda_0^2} \tag{61}$$

$$L_2 = \frac{I_2}{I_3^{2/3}} = \frac{\lambda_1^2\lambda_2^2}{\lambda_0^4} + \frac{\lambda_2^2\lambda_3^2}{\lambda_0^4} + \frac{\lambda_3^2\lambda_1^2}{\lambda_0^4} \tag{62}$$

$$L_3 = I_3 = \lambda_1^2\lambda_2^2\lambda_3^2 = \lambda_0^6 \tag{63}$$

A convenient way to view this set of invariants is to regard the deformation as a two-step process as shown in Figure 1. The unit cube of rubber (Figure 1(a)) is first subjected to a uniform dilation to a cube of dimension λ_0 (Figure 1(c)). In the second step, the dilated cube is subjected to a shear deformation (without further change in volume), resulting in the final deformed state (Figure 1(b)). The deformation may be described in terms of I_1, I_2, and I_3 which are referenced to the original, undeformed state. Alternatively, the deformation may be described in terms of L_1, L_2, and L_3, which are referenced to the intermediate, dilated state. The advantage in using Penn's invariants lies in the fact that L_1 and L_2 are the shear or deviatoric strain invariants since they have the constant value of 3 for a pure dilation, while L_3 is the volumetric strain invariant. This approach was used by Yeoh[33] to study the stress–strain properties of rubber after swelling with a solvent.

Penn[47] postulated that the strain energy function may be separated as a sum of two functions

$$W = F(L_1, L_2) + G(L_3) \tag{64}$$

where F would involve only the shear part and G only the dilation part of the deformation. He tested this possibility but found that it cannot be true; for a solid natural rubber vulcanizate, $\partial G/\partial L_3$ was found to depend on L_1 and L_2 also. Subsequently, Fong and Penn[48] and Peng and Landel[49] derived strain energy functions for slightly compressible rubbers. There appears to have been no attempts to apply equation (64) to compressible elastic materials such as cellular rubbers.

12.5.3 Storåkers' Model

Storåkers[50] proposed a compressible version of Ogden's strain energy function

$$W = \sum_{p=1}^{n} \frac{\mu_p}{\alpha_p}\left[\lambda_1^{\alpha_p} + \lambda_2^{\alpha_p} + \lambda_3^{\alpha_p} - 3 + \frac{1}{\beta_p}(J_3^{-\alpha_p\beta_p} - 1)\right] \tag{65}$$

which has been adopted in the commercial finite element code ABAQUS. In this form, the small strain shear modulus is given by

$$G_0 = \sum_{p=1}^{n} \frac{\mu_p \alpha_p}{2} \tag{66}$$

and the corresponding bulk modulus is

$$K = \sum_{p=1}^{n} 2G_0(\beta_p + \tfrac{1}{3}) \tag{67}$$

If all β_i are equal to a constant value β, then it can be shown that β is related to an effective Poisson's ratio, ν, by

$$\nu = \frac{\beta}{(1 + 2\beta)} \tag{68}$$

12.6 CONCLUSION

To a first approximation, the stress–strain behavior of rubber may be described in terms of a strain-dependent shear modulus. This concept was advanced by Mooney in his general theory in 1940. So, in a sense, we have come full circle. A strain-dependent shear modulus contrasts with the elementary kinetic theory of rubber elasticity, which predicts a constant shear modulus. One view is to regard the kinetic theory as the ideal behavior and interpret the strain dependence of shear modulus as deviation from the ideal.

It appears that, for most rubbers, the shear modulus decreases with increasing strain at small strains. This dependence is quite strong. For filled rubber vulcanizates, this may be attributed to the breakdown of filler–filler or filler–rubber interaction.[31] However, the decrease in shear modulus with increasing strain is also evident in unfilled rubbers, though the dependence is usually not as strong as in filled systems.[33] Thomas[51] was the first to note that departures from the kinetic theory may be reconciled by means of a single additional term in the stored energy relation for a molecular chain. This additional term represents an attractive force which decreases in effect as the chain is extended. It appears that some similar mechanism has to be invoked to explain the observed strain dependence of shear modulus.

At large strains, the shear modulus increases with strain. This is explicable in terms of finite extensibility effects. When the assumption of Gaussian chain statistics in the kinetic theory of rubber elasticity is replaced by the assumption of Langevin chain statistics, the increase in shear modulus with strain is correctly predicted. It is interesting to note that a recent model by Arruda and Boyce[52] involving Langevin chain statistics leads to a strain energy function which involves I_1 only.

We started by pointing out that the objective of the phenomenological theory of rubber elasticity is to construct a mathematical framework to describe rubbery behavior so that stress analysis and strain analysis problems may be solved. This has been achieved. Most of the classical problems that are tractable analytically have been solved[4] and more complex problems are solvable using finite element analysis. Advances in both hardware and software have brought finite element analysis within the reach of most workers. Its widespread use is being hindered by the difficulty in material characterization in terms of a suitable energy function. The difficulty of this task is often underestimated and, in spite of recent progress, it remains a fundamental challenge.[53] Particular care should be taken in selecting a material model. Misleading results may arise when an inappropriate model is chosen and the regression analysis is based upon limited experimental data. Practical rubber materials deviate from the ideal behavior envisaged in the phenomenological theory and engineering judgement is necessary for useful results.

12.7 REFERENCES

1. J. E. Mark and J.-P. Queslel, in 'Comprehensive Polymer Science', ed. G. Allen and J. C. Bevington, Pergamon, Oxford, 1989, vol. 2, p. 271.
2. K. J. Smith, Jr., in 'Comprehensive Polymer Science', ed. G. Allen, Elsevier, Oxford, 1996, suppl. 2, chap 11.
3. L. R. G. Treloar, 'The Physics of Rubber Elasticity', 3rd edn., Clarendon Press, Oxford, 1975.
4. R. S. Rivlin, in 'Rheology: Theory and Applications', ed. F. R. Eirich, Academic Press, New York, 1956, vol. 1, p. 351.
5. M. Mooney, *J. Appl. Phys.*, 1940, **11,** 582.
6. R. S. Rivlin, *Philos. Trans. R. Soc. London, Ser. A,* 1948, **241,** 379.
7. R. S. Rivlin, *Proc. R. Soc. London,* 1949, **195,** 463.
8. R. S. Rivlin, *Philos. Trans. R. Soc. London, Ser. A,* 1949, **242,** 173.
9. R. S. Rivlin and D. W. Saunders, *Philos. Trans. R. Soc. London, Ser. A,* 1951, **243,** 251.
10. R. S. Rivlin, *Rubber Chem. Technol.*, 1992, **65** (3), G51.
11. N. W. Tschoegl, *J. Polym. Sci. A-1,* 1971, **9,** 1959.
12. A. G. James, A. Green and G. M. Simpson, *J. Appl. Polym. Sci.,* 1975, **19,** 2033.
13. A. G. James and A. Green, *J. Appl. Polym. Sci.,* 1975, **19,** 2319.
14. C. L. Chow and C. H. Cundiff, *Tire Sci. Technol.,* 1987, **15,** 73.
15. Y. Sato, *Rep. Prog. Polym. Phys. Jpn.,* 1973, **16,** 139.
16. A. E. Green and W. Zerna, 'Theoretical Elasticity', Oxford University Press, Oxford, 1954.
17. S. T. J. Peng and R. F. Landel, *J. Appl. Phys.,* 1975, **46,** 2599.
18. M. E. Gurtin, 'Topics in Finite Elasticity', Soc. Ind. Appl. Math., Philadelphia, 1981.
19. K. N. Morman, *Trans. ASME,* 1986, **53,** 726.
20. J. S. Chen, K. Satyamurthy and L. R. Hirschfelt, *Comput. Struct.,* 1994, **50,** 715.
21. S. Kawabata and H. Kawai, *Adv. Polym. Sci.,* 1977, **24,** 89.
22. D. F. Jones and L. R. G. Treloar, *J. Phys. D: Appl. Phys.,* 1975, **8,** 1285.
23. G. W. Becker, *J. Polym. Sci. C,* 1967, **16,** 2893.
24. S. Kawabata, M. Matsuda, K. Tei and H. Kawai, *Macromolecules,* 1981, **14,** 154.
25. G. B. McKenna and L. J. Zapas, *Polymer,* 1983, **24,** 1495.
26. G. B. McKenna and L. J. Zapas, *Rubber Chem. Technol.,* 1986, **59,** 130.
27. R. H. Finney and A. Kumar, *Rubber Chem. Technol.,* 1988, **61,** 879.
28. D. W. Haines and W. D. Wilson, *J. Mech. Phys. Solids,* 1979, **27,** 345.
29. L. R. G. Treloar, *Trans. Faraday Soc.,* 1944, **40,** 59.
30. A. R. Johnson, C. J. Quigley and J. L. Mead, *Rubber Chem. Technol.,* 1994, **67,** 904.
31. O. H. Yeoh, *Rubber Chem. Technol.,* 1990, **63,** 792.
32. M. J. Gregory, *Plast. Rubber: Mater. Appl.,* 1979, **4,** 184.
33. O. H. Yeoh, *Rubber Chem. Technol.,* 1993, **66,** 754.
34. A. Othman and M. J. Gregory, *J. Nat. Rubber Res.,* 1990, **5,** 144.
35. C. K. L. Davies, D. K. De and A. G. Thomas, *Rubber Chem. Technol.,* 1994, **67,** 716.
36. K. C. Valanis and R. F. Landel, *J. Appl. Phys.,* 1967, **38,** 2997.
37. R. W. Ogden, *Proc. R. Soc. London, Ser. A,* 1972, **326,** 565.
38. R. S. Rivlin and K. N. Sawyers, *Trans. Soc. Rheol.,* 1976, **20,** 545.
39. A. J. Carmichael and H. W. Holdaway, *J. Appl. Phys.,* 1961, **32,** 159.
40. S. T. J. Peng and R. F. Landel, *J. Appl. Phys.,* 1972, **43,** 3064.
41. O. H. Varga, 'Stress–Strain Behavior of Elastic Materials', Interscience, New York, 1966.
42. G. M. Bartenev and V. P. Nikiforov, *J. Polym. Sci., Part A-1,* 1975, **13,** 289.
43. E. H. Twizell and R. W. Ogden, *J. Austral. Math. Soc., Ser. B,* 1983, **24,** 424.
44. R. W. Ogden, in 'Mechanics of Solids', ed. H. G. Hopkins and M. J. Sewell, Pergamon, Oxford, 1982.
45. P. J. Blatz, *Rubber Chem. Technol.,* 1963, **36,** 1459.
46. P. J. Blatz and W. L. Ko, *Trans. Soc. Rheol.,* 1962, **6,** 223.
47. R. W. Penn, *Trans. Soc. Rheol.,* 1970, **14,** 509.
48. J. T. Fong and R. W. Penn, *Trans. Soc. Rheol.,* 1975, **19,** 99.
49. S. T. J. Peng and R. F. Landel, *J. Appl. Phys.,* 1975, **46,** 2599.
50. B. Storåkers, *J. Mech. Phys. Solids,* 1986, **34,** 125.
51. A. G. Thomas, *Trans. Faraday Soc.,* 1955, **51,** 569.
52. E. M. Arruda and M. C. Boyce, *J. Mech. Phys. Solids,* 1993, **41,** 389.
53. P. S. Oubridge, *Plastics, Rubber Composites: Process. Appl.,* 1993, **20,** 63.

13

Statistical Thermodynamics of Copolymers and their Blends

SEMION I. KUCHANOV

Keldysh Institute of Applied Mathematics, Moscow, Russia

and

SERGEY V. PANYUKOV

P. N. Lebedev Physics Institute, Moscow, Russia

13.1 INTRODUCTION

Unlike homopolymers, heteropolymers consist of macromolecules comprising more than one type of elementary unit. Any sample of a synthetic heteropolymer represents a mixture of an enormous (practically infinite) number of different individual chemical compounds. Even at the same degree of polymerization l the molecules vary in chemical composition and structure, *i.e.* in the fractions of units of different types and in the pattern of their arrangement. The number of conceivable isomers grows exponentially with l and for actual polymers essentially exceeds the Avogadro number. This is why it is impossible in principle to characterize a sample of such a polymer by setting the concentrations of individual chemical compounds because the vast majority of them will be represented by no more than a single macromolecule. There is a fundamental difference between this situation and that of homopolymers, where the number of molecules with any l in a sample is large enough to provide an adequate macroscopic description in terms of concentrations of homologues (l-mers). Consequently, polydisperse homopolymers and their blends can be described using traditional thermodynamics, although with a nominally infinite number of components which are homologues. Such a consideration, in view of the argument outlined above, fails in the case of heteropolymers so that when developing a rigorous thermodynamic theory of their solutions, alloys, and blends it is necessary to use nontraditional approaches.

One more inherent characteristic of heteropolymers, whose macromolecules comprise rather long blocks of units of the same type, is their ability to undergo microphase separation when, with the alteration of external conditions (*e.g.* temperature), thermodynamically stable microdomain structures appear in the system. The morphology of these phase-separated structures (*i.e.* their form, size, and spatial distribution of microdomains) depends not only on thermodynamic variables (temperature, pressure) but also on the molecular parameters of a sample of the particular block copolymer. The establishment of such dependencies is one of the most important problems in statistical thermodynamics of heteropolymers and requires for its correct solution the use of sophisticated methods of modern theoretical physics.

The present chapter is aimed at familiarizing the reader with progress in the theoretical interpretation of equilibrium states of heteropolymer systems, leaving out the problems of the kinetics of the phase separation process. These problems, together with the methods of blend preparation and the factors controlling their compatibility, have been covered in preceding volumes of 'Comprehensive Polymer Science' (Vol. 2, Chapter 5 and Vol. 7, Chapter 4), whereas only slight mention was made there of the quantitative theory of blends involving heteropolymers. The material of the present chapter extends rather than duplicates that published on polymer blends in earlier volumes. Special emphasis in this review will be placed on the analysis of theoretical concepts underlying the statistical thermodynamics of heteropolymers. A comparison of theoretical conclusions with the results of both computer simulations and experimental investigations can be found in a remarkable review by Binder.[1]

13.2 PECULIARITIES OF THE THERMODYNAMICS OF HETEROPOLYMERS

One of the fundamental problems of the thermodynamics of any system is associated with the construction of its phase diagram. Due to the polydispersity of real heteropolymers for their chemical length l and composition, their phase diagrams are expected to exhibit a number of characteristics distinguishing them from phase diagrams of solutions of low-molecular compounds. These characteristics are most readily elucidated by comparing solutions of monodisperse and polydisperse homopolymers where the molecules differ only in the number l of the units.

On the well-known phase diagram (Figure 1(a)), the areas of one- and two-phase solutions are separated by the binodal. This consists of two branches, one of which is situated to the left and the other to the right of the critical point. A segment of horizontal line, joining the points of its intersection with the binodal, is called the tie-line. Its terminal points lying on different binodal branches are conjugate in the sense that at a given temperature they correspond to the compositions of coexisting phases just at the moment when phase separation commences.

A crucially different situation takes place when dealing with a solution of a polydisperse homopolymer. Such a system belongs to the category of quasibinary ones since along with the solvent it contains only one quasicomponent, *e.g.* the monomer unit. In a binary heteropolymer or a mixture of two different homopolymers, the number of such quasicomponents will equal two, and in the general case it coincides with the number of types of monomer unit in the sample of interest. Therefore, for example, blends of two styrene copolymers with methyl methacrylate and with acrylonitrile have to be assigned to quasiternary systems. When constructing phase diagrams, the concentrations or fractions of monomer units might act as thermodynamic variables. However, it should be borne in mind that, unlike low-molecular solvents, they are not true thermodynamic components because the notion of chemical potential, for instance, proves to be physically meaningless for them.

Looking at Figure 1, one will become aware of qualitative differences in phase diagrams arising when polydispersity is taken into account. This factor is responsible for the split of a binodal into two curves. One of them, where the second phase is known to nucleate, is usually referred to as a cloud-point curve. Although it separates one- and two-phase regions like a binodal, this curve is distinguished by a number of specific features. The main one is the fact that, in the solvent with a specified fraction of polymer Φ, the value Φ' of this fraction in the first droplets of the precipitated phase corresponds in Figure 1(b) to the point of the intersection of the horizontal straight line with the dotted curve rather than with the solid one. The former, called the shadow curve, intersects the cloud-point curve at a critical point which divides each of these curves into two branches, corresponding to two different phases. These latter become identical, as in the case of a binary solution, only at the critical point which, however, is not now located at the maximum of the curve where the second phase nucleates. At this point, termed the precipitation threshold, allowance for the polydispersity factor results in an alteration of the type of phase transition which was found to be first order instead of second order. The spinodal where the homophase solution becomes thermodynamically unstable in a quasibinary system (Figure 1(b)) has a common tangent at the critical point with the cloud-point curve in the same was that it has such a common tangent with the binodal in a binary system (Figure 1(a)).

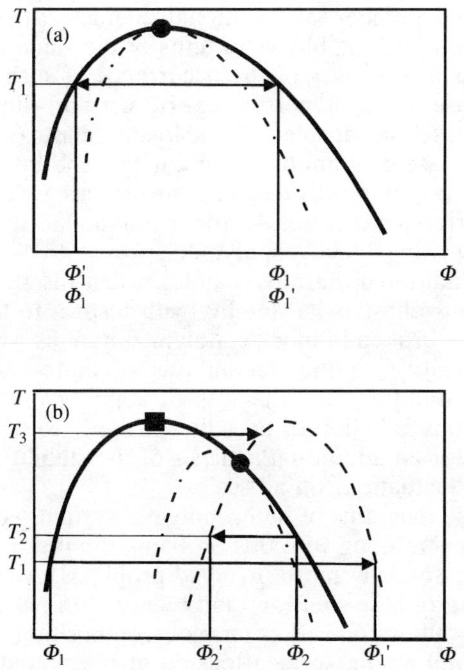

Figure 1 The shape of a two-dimensional phase diagram for binary (a) and quasibinary (b) systems describing, respectively, the solutions of a monodisperse and a polydisperse homopolymer in a low-molecular weight solvent (*T*, temperature; Φ, volume fraction of polymer in a solvent; full line, binodal (a) and cloud-point curve (b); dotted-dashed line, spinodal; dashed line, shadow curve; solid circle, critical point; solid square, precipitation threshold.

The region of the metastable states, located between the cloud-point curve and the spinodal, is of special significance for polymer systems as compared to low-molecular ones. This is because the phase separation in the polymer systems proceeds much more slowly than in low-molecular systems. Consequently, the lifetime of the metastable states may often far exceed the duration of a thermodynamic experiment and even the polymer operation time. In such cases, typical of concentrated solutions, melts, and blends of polymers, a metastable system is expected to exhibit an equilibrium behavior.

Apart from the above characteristics of phase diagrams, intrinsic to any polydisperse system, there are other characteristics inherent to more complicated diagrams comprising, for instance, regions where more than two phases coexist. The number of these regions, according to the Gibbs phase rule, generally speaking can be arbitrary, even for the above discussed quasibinary system, inasmuch as the number of its true thermodynamic components is formally infinite. Establishing the fractions of these components by means of the molecular weight distribution (MWD) function of a homopolymer sample provides an exhaustive description of its polydispersity. The dimensionality of a phase diagram will depend on how many independent stoichiometric parameters the MWD has.

If there is only one stoichiometric parameter, as in the exponential Flory distribution, the phase diagram will be a three-dimensional one. In this case the roles of spinodal, cloud-point curve, and shadow-curve will be performed by corresponding surfaces, the first two of which are tangential along the line of critical points. At a given value of the stoichiometric parameter, the two-dimensional sections of these surfaces, akin to those sketched in Figure 1(b), are of utmost importance for the investigation of demixing of solutions of a polymer with given MWD under variations of thermodynamic variables (temperature, volume, pressure). Fixing one of them makes possible the construction of other two-dimensional sections, which permits one to predict the conditions of phase separation in the course of polymer synthesis accompanied by the alteration of the MWD parameter. Even though this problem is of crucial practical significance for the manufacturing of polymers, it has not been properly elucidated in the literature so far.

With an increasing number of stoichiometric parameters, which define the polydispersity of the system, an increase of the dimensionality of its phase diagram leads to the emergence of multiple critical points. The question of the number of such points and their stability has been thoroughly examined only for solutions and blends of homopolymers.[2–9]

Phase diagrams feature a number of additional characteristics when a polymer contains macromolecules including rather long blocks of units of the same type. Then, provided specific conditions are met, there is a feasibility of the occurrence of such branches of the cloud-point curve (in the general case of a hypersurface of certain dimension), where inside the homogeneous principal phase the droplets of incipient phase (mesophase) possessing space periodicity will appear. A phase diagram of a solution or blend involving block copolymers with given molecular architecture may be divided into several areas, corresponding to mesophases differing in the type of periodic structure. As for the spinodal, the procedure of finding it for solutions and blends comprising block copolymers proves to be far more tedious than the procedure much used in traditional thermodynamics, when the sufficient condition of absolute stability of the homogeneous phase is its stability with respect to the fluctuations of an infinite space scale. This rule can be broken in block copolymer systems where the homogeneous phase appears to be unstable relative to the fluctuations of finite scale only. Thus, the regular approach to determining conditions for the loss of stability of the homogeneous state with respect to infinite fluctuations is believed, broadly speaking, to be insufficient to construct the whole spinodal. This calls for an additional analysis of the stability of this state in reference to the space inhomogeneous fluctuations on all scales.

The construction of phase diagrams of heteropolymer systems constitutes a problem of great practical importance. However, along with this problem, the thermodynamics of such systems is beset by a series of other essentially more involved problems. Among these are the derivation of equations describing heterophase solutions and blends with participation of copolymers. The solution of such equations allows one to determine compositions and volume fractions of all phases in equilibrium as well as the surface tension at the boundaries between them for any values of the thermodynamic and stoichiometric parameters of a system. Its complete thermodynamic characterization requires also the expression to be derived for the experimentally obtainable elastic scattered intensity of light, neutrons, or X-rays as a function of the magnitude of the scattering angle.

Here we attempt only to sketch the range of potential problems relevant to the statistical thermodynamics of heteropolymer systems. The material presented below will provide an understanding of the progress made towards their solution.

13.3 QUANTITATIVE CHARACTERIZATION OF HETEROPOLYMERS

In line with the arguments put forward in the preceding section, the traditional specification of an actual heteropolymer sample by a set of the concentrations of all its constituents makes no sense. Alternatively, it has been suggested[10] that such a sample be characterized by virtue of a probability measure $\mathcal{P}\{\sigma\}$ on a set $\{\sigma\}$ of all kinds of its macromolecules, *i.e.* stochastic sequences $\sigma = (\sigma_1, \sigma_2, \ldots, \sigma_l)$ of monomer units M_α $(\alpha = 1, \ldots, m)$. The symbol σ_τ (where $\tau = 1, 2, \ldots, l$) denotes here the type of unit situated at the τth position from the beginning of the polymer chain, so that, for instance, the terpolymer molecule $M_2M_1M_2M_3 \ldots M_1M_3$ corresponds to the sequence of integers $(2, 1, 2, 3, \ldots, 1, 3)$. Each of them may be thought of as the realization of a random process of conventional movement along the macromolecule during which the transitions between the states $0, 1, 2, \ldots, \alpha, \ldots, m$ occur at equal intervals of 'time' τ. Upon reaching the zeroth state, which corresponds to attaining the edge of the macromolecule, one will remain in such a state forever. That is why, as distinct from all the other ones, the latter is referred to as the absorbing one.

The above stochastic process with discrete time and discrete number of states bears in mathematics the name of a chain. Of particular importance are the Markov chains. Here the probability of the transition $\nu_{\alpha\beta}$ at any step is specified only by its initial (α) and final (β) states. The matrix of transition probabilities $\mathbf{Q}$ with elements $\nu_{\alpha\beta}$ and the vector of initial states $\boldsymbol{v}$ with components v_α equal to their probabilities provides an exhaustive description of a Markov chain. The probability of its arbitrary realization σ is defined by a simple equation (1), enabling one to find all the statistical characteristics of this chain.

$$\mathcal{P}(\sigma) = v_{\sigma_1} \nu_{\sigma_1\sigma_2} \nu_{\sigma_2\sigma_3} \cdots \nu_{\sigma_l 0} \tag{1}$$

The Markov chains owe their utmost importance in polymer science to the fact that they characterize the molecular structure of numerous commercially used heteropolymers. Among these are the products of many processes of radical copolymerization and polycondensation of an arbitrary number of monomers. The Markovian character of the sequence distribution in their macromolecules has been established both experimentally (largely by NMR spectroscopy) and theoretically, by means of the approaches of statistical chemistry of polymers.[10–12] These approaches allow one to express the elements of the matrix $\mathbf{Q}$ and the components of the vector $\boldsymbol{v}$ through kinetic and stoichiometric parameters of the reaction system; in other words, to relate any statistic characteristics of actual copolymers to the conditions of their synthesis.

A notorious particular case of the Markov chain is that when all rows of the matrix $\mathbf{Q}$ are the same. Here the probability of any transition $\nu_{\alpha\beta} = P_\beta$ does not depend on the initial state and thus the Markov chain is reduced to a sequence of independent trials with probabilities of a result $P_1, \ldots, P_m$. Such a stochastic process describes random copolymers, which are the simplest ones from the viewpoint of the characterization of their chemical structure among all statistical heteropolymers.

In the case of copolymers with a sequence distribution, which cannot be described by any Markov chain, it is frequently impossible to construct a probability measure $\mathcal{P}(\sigma)$(1), *i.e.* to find a molecular structure distribution (MSD) offering an exhaustive statistical description of the chemical structure. Here it is possible to determine only some of its statistical characteristics which carry limited (in comparison with the MSD) information on the chemical structure of a polymer sample. Naturally, within the scope of the present review only those which are relevant to the thermodynamics of heteropolymers are considered.

An individual heteropolymer molecule is most readily characterized either by the vector $\boldsymbol{l}$ with components $l_1, l_2, \ldots, l_m$ equal to the number of its monomer units M_1, $M_2, \ldots, M_\alpha, \ldots, M_m$ or by chemical size $l = l_1 + l_2 + \ldots + l_m$ and composition vector $\boldsymbol{\zeta}$ with components $\zeta_\alpha = l_\alpha / l$ $(\alpha = 1, \ldots, m)$. It is apparent that the distribution of macromolecules $f(l, \boldsymbol{\zeta})$ for size and composition (SCD) is equivalent to $f(\boldsymbol{l})$, but sometimes it appears to be more apt to describe those samples of high-molecular polymers where the function f arguments may be thought of as continuous variables rather than discrete ones. Several methods to introduce the SCD function are known in macromolecular chemistry. With one of them this function is equivalent to the probability of finding in a sample the molecule of a given size and composition, whereas with another it corresponds to the probability that the randomly chosen unit constitutes such a molecule. In these cases we are dealing with the number SCD (f_N) and with the weight SCD (f_W), respectively. These distributions

$$f_N(\boldsymbol{l}) = n(\boldsymbol{l})/\Pi, \qquad f_W(\boldsymbol{l}) = n(\boldsymbol{l})l/M,$$

where

$$\Pi = \sum_{\boldsymbol{l}} n(\boldsymbol{l}), \qquad M = \sum_{\boldsymbol{l}} n(\boldsymbol{l})l \tag{2}$$

are expressible in a simple way through the numbers $n(\mathbf{l})$ of molecules with given $\mathbf{l}$. The third type of the SCD

$$f(\mathbf{l})=n(\mathbf{l})/M, \qquad \sum_{\mathbf{l}} f(\mathbf{l})=\frac{\Pi}{M}\equiv Y, \qquad \sum_{\mathbf{l}} lf(\mathbf{l})=1 \tag{3}$$

differing from the distributions in equation (2) only by the normalization, is the most suitable when writing down thermodynamic formulae. Of great importance are the statistical moments of this SCD

$$X_\alpha\equiv\sum_{\mathbf{l}} l_\alpha f(\mathbf{l}), \qquad X_{\alpha\beta}\equiv\sum_{\mathbf{l}} l_\alpha l_\beta f(\mathbf{l}) \tag{4}$$

the first of which specifies the average composition of the sample whereas the second defines its polydispersity. With a knowledge of the generating function of the SCD

$$G(\mathbf{s})\equiv\sum_{\mathbf{l}} f(\mathbf{l})\prod_{\alpha=1}^{m} s_\alpha^{l_\alpha} \tag{5}$$

its statistical moments of any order can be found. Thus, quantities Y, X_α, $X_{\alpha\beta}$, introduced in formulae (3) and (4), are expressible through the generating function $G(\mathbf{s})$ and its derivatives

$$G_\alpha(\mathbf{s})\equiv\frac{\partial G(\mathbf{s})}{\partial \ln s_\alpha} \qquad G_{\alpha\beta}(\mathbf{s})\equiv\frac{\partial^2 G(\mathbf{s})}{\partial \ln s_\alpha\,\partial \ln s_\beta} \tag{6}$$

taken at point $\mathbf{s}=\mathbf{1}$ using the expressions

$$Y=G(\mathbf{1}), \qquad X_\alpha=G_\alpha(\mathbf{1}), \qquad X_{\alpha\beta}=G_{\alpha\beta}(\mathbf{1}) \tag{7}$$

The extension of equations (6) to statistical moments of arbitrary k will read

$$X_{\alpha\beta\ldots\omega}\equiv\sum_{\mathbf{l}} l_\alpha l_\beta\ldots l_\omega f(\mathbf{l})=G_{\alpha\beta\ldots\omega}(\mathbf{1}) \tag{8}$$

where $G_{\alpha\beta\ldots\omega}(\mathbf{s})$ stands for the kth derivative of the function $G(\mathbf{s})$ with respect to variables $\ln s_\alpha$, $\ln s_\beta, \ldots, \ln s_\omega$.

Currently available chromatographic methods enable one to separate some heteropolymers in accordance with both their size and composition in order to find experimentally their SCD. Alternatively, this distribution (or its generating function) is obtainable theoretically from the solution of corresponding material balance equations describing the chemical transformation during copolymer synthesis. Such solutions are presently known for several commonly accepted kinetic models of radical polymerization, polycondensation, and other ways of manufacturing polymer materials. The adoption of these solutions in thermodynamic calculations makes it possible to extend the application of the calculations from hypothetical model heteropolymers to real ones. Obviously, the SCD does not supply a complete description of a sample of a statistical heteropolymer, because this function carries no information on the pattern of alternation of units in polymer chains. In order to characterize quantitatively molecular structure in the statistical chemistry of polymers, two different approaches have been applied.

In the traditional approach, the microstructure of heteropolymers is specified in terms of a set of probabilities $P(U_k)$ of all conceivable sequences U_k, containing $k\geq 2$ successive units. For instance, the structure of binary copolymer chains may be roughly defined by the fractions (probabilities) of dyads M_1M_1, M_1M_2, M_2M_2. An increasingly more complete description of its microstructure is attainable by setting the probabilities of triads

$$U_3=M_1M_1M_1 \quad M_1M_1M_2 \quad M_1M_2M_1 \quad M_1M_2M_2 \quad M_2M_1M_2 \quad M_2M_2M_2 \tag{9}$$

tetrads, pentads, and so on.

As distinct from the probabilities of undirected sequences $P(U_k)$ obtainable experimentally, the probabilities of directed sequences $P\{U_k\}$ are normally derived using theoretical considerations. The probabilities mentioned coincide $P(U_k)=P\{U_k\}$ for symmetric sequences U_k (for example $M_1M_2M_1$) and differ by a factor of two $P(U_k)=2P\{U_k\}$ for asymmetric ones (such as $M_1M_1M_2$).

Today's experimental equipment allows one to determine by NMR spectroscopy, with an error of less than 10%, the fractions of sequences comprising up to 5–6 units. The theoretical problem of calculating such probabilities $P(U_k)$, *i.e.* the sequence distribution in copolymer chains, turns out to be trivial if these latter are Markovian.

However, modern methods of statistical chemistry permit this problem to be solved even for a number of non-Markovian heteropolymers[10,13] obtained, for example, in the course of the cooperative reaction of a chemical modification of a homopolymer such as the hydrolysis of poly(methyl methacrylate).

The second way to characterize quantitatively the chemical structure of heteropolymers is based on the introduction of chemical correlation functions (correlators). The simplest of them, $Y_{\alpha\beta}(k)$, is defined as the joint probability that two randomly chosen monomer units, separated along the macromolecule by an arbitrary sequence U_k, will be M_α and M_β. Using this definition, the expression for the two-point correlator can be written as

$$Y_{\alpha\beta}(k) \equiv Y^{(2)}_{\alpha\beta}(k) = \sum_{U_k} P\{M_\alpha U_k M_\beta\} = P\{M_\alpha X^k M_\beta\} \tag{10}$$

where X stands for a unit of an arbitrary type. The extension of formula (10) to the p-point correlation function is presented by the expression

$$Y^{(p)}_{\alpha\beta\ldots\omega}(k_1, \ldots, k_{p-1}) = P\{M_\alpha X^{k_1} M_\beta X^{k_2} \ldots X^{k_{p-1}} M_\omega\} \tag{11}$$

for the probability of finding $p \geq 2$ monomer units M_α, $M_\beta, \ldots, M_\omega$ separated along the polymer chain by arbitrary sequences consisting of $k_1, k_2, \ldots, k_{p-1}$ units, respectively.

The statistical description of the heteropolymers in terms of chemical correlators appears to be comprehensive, permitting one to calculate statistical parameters of polymer samples which characterize not only the microstructure but the polydispersity as well. In fact, it is readily seen that the correlators (11) at values $k_1 = k_2 = \ldots = k_{p-1} = 0$ coincide with the probabilities $P\{U_p\}$ occurring in the traditional method of describing sequence distribution in copolymers. On the other hand, statistical moments of the SCD may also be expressed through these correlators. For instance, the quantities $X_{\alpha\beta}$ in equation (4) were found to be related in a simple way to the two-point correlators in equation (10).[10,13]

As for the block copolymers, the samples obtained can have either determined or stochastic architecture, depending on the method of their synthesis. In the first case, unlike in the second one, the number of blocks n as well as their arrangement are the same for all macromolecules of a given sample, although they may differ in the length of the blocks involved. Among such heteropolymers, the most extensively studied both theoretically and experimentally are binary ones comprising two or three blocks

$$\underbrace{M_1 \ldots M_1}_{l_1}\underbrace{M_2 \ldots M_2}_{l_2} \equiv M_1^{l_1} M_2^{l_2} \tag{12a}$$

$$\underbrace{M_2 \ldots M_2}_{l_1}\underbrace{M_1 \ldots M_1}_{l_2}\underbrace{M_2 \ldots M_2}_{l_3} \equiv M_2^{l_1} M_1^{l_2} M_2^{l_3} \tag{12b}$$

Obviously, the statistical characterization of block copolymers of fixed architecture is substantially simpler because no sequence distribution problem arises. The only thing necessary to obtain a complete description of such a heteropolymer is to set the function of the distribution of blocks for lengths. This distribution is, naturally, prescribed by the synthesis conditions and may be either very narrow, if the blocks are being formed in the course of living polymerization via an anionic mechanism, or wider for radical or polycondensation mechanisms.

Among block copolymers of stochastic architecture, the simplest are segmented copolymers consisting of macromolecules with alternating blocks of units M_1 and M_2. A sample of such a copolymer in accordance with the conditions of its synthesis represents a mixture of macromolecules, the architecture of each of which can be unambiguously characterized by two integers $n = 1, 2, \ldots$ and $\alpha = 0, 1, 2$. The first of these indicates the overall number of blocks of a particular type contained in a randomly chosen molecule, while the second one specifies how many of them are located at its boundaries. The distribution f_{na} of molecules of any sample of segmented copolymer for stochastic variables n and a gives a complete statistical characterization of its architecture.

The question arises about the prospects of the solution of the above problem as applied to other types of block copolymers whose molecules may vary not only in the number of different blocks but also in the pattern of their arrangement along the polymer chain. It may be envisaged as a stochastic sequence of blocks comprising units of different kinds regardless of their lengths. Under such an architectural consideration of block copolymers with stochastic structure, the mathematical problem of its characterization is formally reducible to that of the microstructure of statistical copolymers discussed previously.

The present section provides a general idea of the information required to characterize completely or partially the chemical structure of a heteropolymer sample. This information proves to be indispensable for the calculation of thermodynamic parameters of a solution or a melt of a given sample. Here the set of necessary statistical characteristics of its molecular structure distribution depends on the particular thermodynamic problem to be solved. These characteristics can be determined either from experimental data or from theory. For this purpose the methods of statistical chemistry prove to be very efficient since they enable one to establish, within the framework of generally accepted kinetic models, the correlation between the statistical characteristics of samples of actual polymers and stoichiometric and kinetic parameters of the process of their synthesis. The use of information of this sort on the chemical structure of heteropolymers for thermodynamic calculations is of great value for the theoretical prediction of equilibrium properties of technologically important polymer systems.

To conclude this section, it is pertinent to point out the problem of the nomenclature of multicomponent polymers of complicated architecture. Even though their molecules are linear, a blend of several heteropolymer samples obtained under different conditions is being considered, and the elaboration of a universal system of notation suitable for practical use involves serious difficulties. The main difficulty is in the fact that each sample constituting a blend is by itself a mixture of an enormous set of individual compounds. Here a lot of credit must go to Sperling,[14,15] who called attention to this problem and devised for its solution the New General Nomenclature Scheme for polymer blends, graft copolymers, and interpenetrating polymer networks. Although this scheme, which adopts Group Theory concepts, furnishes no complete description of polymers, it nevertheless not only allows one to name accurately their known compositions but also to discover new ones. Schemes of such a kind are of particular significance in the development of computer software.

13.4 THERMODYNAMIC APPROACHES TO STATISTICAL COPOLYMERS

The very fact that macromolecules involve a great many units is responsible for a number of special features peculiar to the thermodynamic behavior of polymer systems when compared to that of low-molecular ones. Because such peculiarities have been observed in solutions and blends both of heteropolymers and homopolymers, a brief outline of homopolymers will be given before going on a detailed discussion of special features inherent to the more complicated heteropolymers. A retrospective journey into the history of the progression of theoretical ideas in the field of thermodynamics of statistical heteropolymers offers a convincing argument for first discussing homopolymers. The evolution of these ideas bears a close resemblance to that of the theory of miscibility of solutions and blends of homopolymers. Here we will indicate only the milestones and will refer the interested reader for details and references to a wealth of superb reviews and monographs.[16–44]

The necessary requirement for miscibility is that the Gibbs free energy of mixing

$$\Delta G = \Delta H - T\Delta S \tag{13}$$

be negative. To derive expressions for the entropy (ΔS) and the enthalpy (ΔH) of mixing of macromolecular systems, different models have been employed, the simplest of which was proposed independently by Flory and Huggins. Within the framework of this FH model, the ΔS is obtainable from a combinatorial consideration of the arrangements of macromolecules on a regular lattice, while for the heat of mixing it has been suggested that the van Laar relationship be used. As a result, for the entropy and the enthalpy of mixing (reduced to the net number of monomer units M) of a blend of two homopolymers $M_1^{l_1}$ and $M_2^{l_2}$ with volume fractions Φ_1 and Φ_2, familiar expressions are arrived at

$$\frac{\Delta S}{M} = \frac{\Delta S_c}{M} = -\frac{\Phi_1}{l_1}\ln\Phi_1 - \frac{\Phi_2}{l_2}\ln\Phi_2; \qquad \frac{\Delta H}{TM} = \chi_{12}\Phi_1\Phi_2 \tag{14}$$

Henceforward, for the sake of simplicity, the value of the gas constant is set equal to unity, so that the entropy is dimensionless, whereas the temperature T has the dimensionality of the energy. The sign of the dimensionless binary interaction parameter $\chi_{12} = (\gamma_{11} + \gamma_{22} - 2\gamma_{12})/T$, where γ_{ij} ($ij = 1, 2$) denotes the energy of the formation of a contact between units M_i and M_j, is of utmost importance. At positive χ_{12} the mixing of polymers is accompanied by the evolution of heat, while at negative χ_{12} the process is attended with heat absorption. For nonpolar polymers the Scatchard assumption, $\gamma_{ij} = \sqrt{\gamma_{ii}\gamma_{jj}}$, is frequently used, which makes the parameter $\chi_{12} = (\delta_1 - \delta_2)^2$ be positive, and an increase of the difference between the values of the solubility parameters δ_1 and δ_2 of polymers gives rise to an exothermic heat of mixing. Even a minor difference between δ_1 and δ_2 is enough for real homopolymers to be incompatible since their entropy of mixing (equation (14)) is small in comparison with the enthalpy in view of the large values of l_1 and l_2. The reciprocal value of the lower of them provides an estimate of the order of magnitude of the parameter χ_{12}, above which mixing is made impossible. Under this condition, entropic and enthalpic terms in equation (13) for the free energy become comparable. A similar estimate is obtainable from the expression $2\chi_{12} = (l_1^{-1/2} + l_2^{-1/2})^2$ for the interaction parameter value at the critical point.

The narrowness of the theoretically predicted interval of values of χ_{12} where the mixing of high-molecular homopolymers is feasible, gives the explanation for the commonly known incompatibility of the vast majority of them. This interval was found to expand markedly when going from blends to solutions of polymers, where the entropy of mixing due to the contribution from the solvent becomes comparable with the enthalpy at $\chi_{12} \approx 1$, as with solutions of low-molecular compounds. This conclusion can be inferred immediately from equation (14), where l_1 would be put equal to unity.

The FH theory, being the origin from which the thermodynamics of polymer systems emanated, in its first version rests on the following assumptions.

(i) Fluctuation effects are excluded from consideration. That is why, like any mean-field theory, this one does not work when dealing with dilute solutions or with phenomena occurring in the close vicinity of the critical point.

(ii) The system is supposed to be incompressible. This restriction precludes one from explaining the experimentally observed volume alteration under polymer mixing and leaves out effects induced by the pressure change.

(iii) The application of the rigid-lattice model. Within its framework, no account is taken of distinctions in size and shape of monomer units M_1 and M_2.

(iv) Random mixing approximation. This assumption, whereby the local composition of quasicomponents coincides with their average composition in a mixture, would be expected to be violated for some systems where specific interactions (either hydrogen bond or π-complex formation) are of special importance.

(v) Polydispersity is ignored.

The advancement of theoretical concepts in the field of thermodynamics of homopolymer solutions and blends was due to models developed by relaxing some of these five assumptions.

One possible trend is the extension of the applicability area of this FH theory by virtue of its generalization to polydisperse polymers whose free energy of mixing may be presented as

$$\frac{v\Delta G}{TV} = \sum_{\alpha}\sum_{l_\alpha} \frac{\Phi_\alpha(l_\alpha)}{l_\alpha} \ln \Phi_\alpha(l_\alpha) + \sum_{\alpha}\sum_{\beta<\alpha} \chi_{\alpha\beta}\Phi_\alpha\Phi_\beta$$

$$\Phi_\alpha = \sum_{l_\alpha} \Phi_\alpha(l_\alpha), \qquad \sum_{\alpha} \Phi_\alpha = 1, \qquad \Phi_\alpha(l_\alpha) = \Phi_\alpha f_w^\alpha(l_\alpha) \tag{15}$$

Here $\Phi_\alpha(l_\alpha)$ and Φ_α are volume fractions of the αth component (*i.e.* of the homopolymer molecule $M_\alpha^{l_\alpha}$ containing l_α units) and of the αth quasicomponent (*i.e.* of the unit M_α). In equation (15), the solvent is nominally regarded as a monodisperse polymer with molecular weight distribution $f_w^\alpha(l_\alpha) = \delta_{l_\alpha 1}$, where $\delta_{l_\alpha 1}$ represents the Kronecker delta-symbol ($\delta_{ij} = 0$ at $i \neq j$; $\delta_{ij} = 1$ at $i = j$). Proceeding from equation (15), which extends equation (14) to an arbitrary number of polydisperse polymers and solvents, general formulae for the cloud-point curve, spinodal, and critical points have been derived. Analogous results were later rederived in terms of continuous thermodynamics[43,45] where the molecular weight is thought of as a continuous variable, so that the operation of summing in expressions for thermodynamic characteristics is replaced by integration over this variable.

The FH model, being the simplest model in the thermodynamics of polymers, is not, naturally, a universal one. This model is, in principle, incapable of offering an explanation of a number of experimentally observed phenomena, such as, for instance, the volume alteration under mixing of the components or the negative sign of the entropy of mixing. Also, this model does not describe such important peculiarities of the topology of a polymer system phase diagram as the bimodal character of the cloud-point curve or the coexistence on it of the upper and lower critical solution temperatures.

These findings point to the necessity of relaxing the FH model assumptions, the most vulnerable of which is concerned with the neglect of the volume alteration under the mixing of polymers one with another or with the solvent. This effect is easy to understand in terms of the free volume, which is the difference between the true experimentally obtainable volume of a system and the intrinsic volume of its molecules. The latter can be calculated as a sum of the volumes of the atoms and their groups contained in these molecules by using experimental data obtained from X-ray analysis. Unlike the intrinsic volume, the free volume, being controlled by the intensity of interatomic interactions, does not exhibit the additivity property under the mixing of the liquids. The main distinction between a variety of models incorporating the compressibility of polymer mixtures (including solutions and blends) consists in the method of estimating the free volume. In terms of the lattice model, this is commonly done in one of the three following ways (see Figure 2).

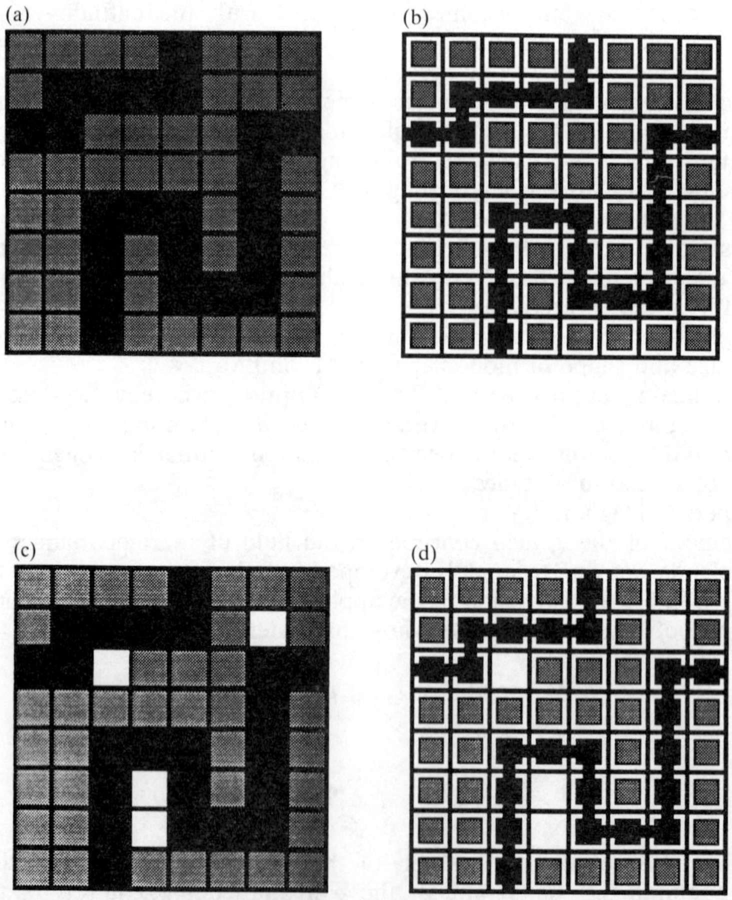

Figure 2 Four principal types of lattice model employed in the statistical thermodynamics of polymers: (a) Flory–Huggins model; (b) cell model; (c) hole model; (d) cell–hole model.

(i) Cell models. Every site of a lattice (termed a cell) is occupied by either a solvent molecule or a monomer unit, as in the case of the classical FH model. However, in contrast to the FH model, here such a quasicomponent does not occupy the whole volume of a cell but only its core. The remaining parts of such cells (which for these models are taken to be the same) constitute the free volume of a polymer system.

(ii) Hole models. As with the FH model, here the quasicomponent located in a cell occupies it completely. The distinction, however, lies in the fact that hole models admit the existence of vacant cells (holes), the total space inside which is identified with the free volume.

(iii) Cell–hole models. These combine the two above models and simultaneously take into consideration the contributions to the free volume from each of these models.

To the contraction of a mixture, induced by a pressure rise or a temperature alteration, in models of the first type there is an identical reduction of the volume of each cell with no change in their overall number. In contrast, for the second type of model, the contraction brings about a decrease in the concentration of holes with retention of the volume of each hole. In the third type of model, changes of both the cell size and the number of holes in liquids lead to the variation of the total volume.

In the course of the refinement of the statistical thermodynamics of polymer systems, repeated attempts have been made to relax the other restrictions of the FH approach while remaining within the framework of the lattice model. A proposal has been made to drop the rigid-lattice condition that each site has the same number of nearest neighbors and to assign interacting surface areas to the sites. The various species on the lattice, including the holes, are permitted to differ in surface area, in accordance with their molecular shape. Allowing for this factor yields a dependence of the heat of mixing ΔH on volume fractions of polymers, which differs from the quadratic dependence inherent to the FH model. This is formally reflected in the fact that the parameter χ_{12} in equation (14) becomes governed by Φ_1 or Φ_2. The same result has been arrived at using other models in which, in order to allow for deviations from additivity of pair interaction and the random mixing approximation, additional terms have been introduced, including association parameters, nonrandomness parameters, and others. Several approaches are also known in which attempts have been made to allow for the orientational ordering when calculating the combinatorial entropy of mixing of homopolymers. The relationships derived lead to expressions different from ΔS_c (equation 14), which include empirical chain flexibility parameters.

In the thermodynamics of polymer systems it is common practice to express the formulae for the free energy or the equation of state in terms of reduced quantities which are defined as dimensionless variables resulting from the division of each of the thermodynamic quantities (pressure, volume, temperature, entropy, and so on) by a corresponding parameter of the reduction. This is specified solely by the chemical nature of a substance and is believed to be identical for homologues of a particular polymer. When written down in terms of reduced parameters, the equation of state should have the same appearance for any arbitrary system belonging to the class in hand, for instance the blend of two homopolymers. The major problem encountered in the derivation of an equation of state written down in terms of reduced quantities is the absence of a rigorous algorithm capable of expressing the parameters of reduction for a mixture through those of its components. In order to cope with this difficulty within the framework of the equation-of-state theory (EST), resort is generally made to some semiempirical additivity rules, frequently differing for various models.

Currently available modifications of the EST, enabling one to elucidate many peculiarities of the thermodynamics of polymer systems, exhibit, however, a number of substantial shortcomings. Among these is the cumbersome appearance of the final formulae, which often incorporate too many empirical fitting parameters; this hampers general analysis and the revelation of qualitative features of phase diagrams.

13.4.1 Simplest Criterion of Polymer Blends Miscibility

Almost 40 years ago in a paper by Stockmayer *et al.*[46] it was pointed out that, when considering the solution of a binary monodisperse statistical copolymer in terms of the FH model, equation (14) for the enthalpy of mixing remains valid upon replacement of the binary interaction parameter with its modified value

$$\hat{\chi}_{12} = \chi_{13}\Phi_1 + \chi_{23}\Phi_2 - \chi_{12}\Phi_1\Phi_2 \tag{16}$$

Here, Φ_1 and Φ_2 are the volume fractions of monomer units M_1 and M_2 in the copolymer, and χ_{13}, χ_{23}, and χ_{12} are parameters of the interaction of these units with the solvent and one with another. Equation (16) remains true if we are dealing with a blend of a copolymer with a homopolymer, while its extension to a blend of two monodisperse copolymers with an arbitrary number of types of monomer units has been provided by Krause *et al.*[47] For a blend involving an arbitrary number n of such copolymers, the expression for the enthalpy of mixing can be presented as

$$\frac{\Delta H}{TM} = \sum_{i=1}^{n} \sum_{j=1}^{n} \hat{\chi}_{ij} \Phi_i \Phi_j \tag{17}$$

The value of the parameter $\hat{\chi}_{ij}$, which characterizes the volume interactions between molecules of the ith and jth copolymers whose volume fractions are Φ_i and Φ_j, is defined by the equation

$$\hat{\chi}_{ij} = \tfrac{1}{2} \sum_{\alpha=1}^{m} \sum_{\beta=1}^{m} \chi_{\alpha\beta} (\Phi^i_\alpha - \Phi^j_\alpha)(\Phi^j_\beta - \Phi^i_\beta) \tag{18}$$

where Φ^i_α represents a volume fraction of the units M_α in the ith copolymer. Use of equations (17) and (18) makes it possible to describe any heteropolymer blend in terms of the FH model.

Expressions for the parameters $\hat{\chi}_{ij}$ can be found in a number of papers[48–50] which have reported on the compatibility conditions for binary blends of a homopolymer with a two-component copolymer[48–58] or a mixture of two fractions of: (a) one such copolymer;[48,50] (b) two copolymers containing one identical type of monomer unit;[50] or (c) two copolymers with different pairs of monomer units.[49] In subsequent work,[51–56] analogous consideration is given to ternary blends of: (a) binary copolymers,[51,52] (b) two homopolymers and a copolymer composed of their monomer units,[53] as well as a binary blend of terpolymers with different composition obtained from the same triad of monomers.[55]

The case of an arbitrary blend of copolymers was examined by Paul and co-authors,[54–56] who made use of an algorithm which permitted them to write down a simple expression for the heat of mixing ΔH_m. The essence of this algorithm, applied for the calculation of ΔH_m within the mean-field theory formalism, is the need to subtract a certain term from the traditional one (the second item in the right-hand part of equation (15)). This term equals the heat of formation of copolymer molecules from monomer units[54] and takes into account intramolecular contacts between different types of unit. The same expression for ΔH_m can be written in quadratic form (equation 17) with respect to the volume fractions of macromolecules of different types. Coefficients (equation 18) of this form, having the sense of binary interaction parameters for macromolecule interactions, are readily expressed through analogous parameters of monomer units.[56]

Barlow and Paul[54] found compatibility conditions for heteropolymer blends by neglecting the entropic contribution to the free energy. The question of the compatibility of a particular blend is then answered by calculating its parameters χ_{ij} by equation (18) and subsequently determining the sign of ΔH by equation (17) for given values of $\Phi_1, \ldots, \Phi_i, \ldots, \Phi_n$. It is of primary importance to know at which ratio of components characterized by a set $\{\Phi_i\}$ the system will be homophasic. Mathematically, this problem reduces to the analysis of the quadratic equation (17) to find regions of parameters $\{\Phi_i\}$ where equation (17) for ΔH is negative.

Another problem in the thermodynamic examination of blends involving copolymers is in determining, at a given temperature, those regions of their composition values where the system remains homophasic. The size and shape of such miscibility regions of a particular polymer sample are termed[57] its miscibility map (MM). An exhaustive classification has been made[57] of all topologically conceivable kinds of such MMs depending on the Flory–Huggins parameters for a mixture of two monodisperse binary statistical copolymers having the same type of monomer unit. Later, the general principles of such a classification were formulated[58] for arbitrary blends with any numbers n and m of high-molecular components and types of their monomer units. Complete sets of topologically different MMs for the blend of a terpolymer with a homopolymer, as well as for two binary copolymers, have been presented. Besides, simple rules have been established[58] enabling one to determine the kind of system to which a particular system with known values of $\chi_{\alpha\beta}$ parameters belongs.

13.4.2 Account of Polydispersity and Microstructure of Heteropolymers

Scott[59] seems to have been the first person to examine the problem of thermodynamic compatibility of copolymers. He noticed that, for such a compatibility, a sufficiently narrow composition distribution of macromolecules in the sample is essential. Later, Koningsveld *et al.*,[60] proceeding from Scott's expression for the free energy of mixing for random copolymers varying in chemical composition and in chain length, derived an equation for the spinodal. Almost 30 years later, Leibler[61] managed within the framework of the Flory–Huggins model to obtain an expression for the spinodal and critical point for a blend of three fractions of the copolymer with different composition.

Koningsveld and Kleintjens[51] put forth a rather general, in terms of the lattice model, expression for the free energy of mixing of a low-molecular weight solvent with an arbitrary number of types of binary copolymer macromolecules differing in the degree of polymerization and the composition. This intriguing approach enabled them to take account of important factors which exert influence on the conditions of phase separation, such as copolymer composition inhomogeneity and the difference in dimensions of the monomer units and the molecules of the solvent.

A comparison of the experimental data on the compatibility of binary copolymers with conclusions from the Flory–Huggins theory demonstrates that, despite its simplicity, this theory can be used to estimate the conditions of demixing of solutions and blends containing heteropolymers.[27,52,60,62]

Another way of considering the 'liquid–liquid' equilibrium in solutions and blends of polydisperse random copolymers was proposed by German scientists[63–67] using a concept named 'continuous thermodynamics'. They analyzed the effect of the polydispersity of macromolecules on the cloud-point curve, the shadow curve, the spinodal, and the critical points, and also discussed the possibility of the existence of three-phase equilibrium regions. In spite of the impressive volume of thermodynamic information available in papers,[63–67] its practical value remains obvious because of the arbitrary character of the expression for the free energy of mixing for polydisperse copolymers, which the authors applied in both the initial[63,64] and the later[65–67] versions of their theory.

Another interesting formalism[68–73] allows one, even in the framework of the mean-field approximation, to explain the influence of the character of the sequence distribution in copolymers on the compatibility of their blends. This approach implies the need, when calculating the number of binary contacts between monomer units, to distinguish their pair interaction parameters depending on the types of units adjacent to a given one. In order to make final thermodynamic formulae suitable for practical application, the authors assumed that the majority of these parameters are equal to each other and confined themselves to the consideration of copolymers with a Markovian sequence distribution, which is characterized by a sole parameter θ. For many actual systems, this approximation seems to be good enough since it permits variation of the value of θ to embrace alternating, random, and block copolymers. In terms of such an approximation, the enthalpy of mixing is determined by effective values of molecular parameters of pair interactions; these comprise two groups of items. The first of them takes account of the composition distribution of copolymer macromolecules in the blend and looks just like that of the previous approach.[48–58] The addition of the second group of items, which take into consideration the character of the sequence distribution in monodisperse copolymers, enables the interpretation of known experimental data.

13.4.3 Further Development of the Theory

Despite the fact that the FH model finds rather wide utility in thermodynamic calculations of phase diagrams for many polymer systems, it proves to be incapable of explaining a considerable body of experimental data. This circumstance provided impetus to the development of more general theoretical approaches where some of the restrictions underlying the FH model are relaxed. For instance, to take the deviations from random mixing of monomer units into account, it has been proposed[74] to resort to a model which is a modification of that introduced earlier by Guggenheim[75] to calculate the enthalpy of mixing of low-molecular weight liquids. Using this modified Guggenheim quasichemical model, simple expressions for the enthalpy ΔH of multicomponent polymer systems have been obtained. A further development of the thermodynamic theory of solutions, melts, and blends containing copolymers allows for free-volume effects which are impossible to consider in terms of the FH theory. For this purpose,[76–80] use is normally made of different extensions of the equation-of-state approaches used for calculations of the thermodynamic characteristics of homopolymers.[20–22,30,31] An interested reader is referred to numerous reviews[81–90] which provide a detailed discussion of particular approaches. The discussion here will just highlight the main general features and will dwell briefly on the most significant approaches. As in the theory of low-molecular weight liquids, free volume effects in polymer systems may be considered in terms of either lattice models or off-lattice (continuous space) models. In the first case, almost all of the models, which differ slightly in detail, proceed from one of the versions of the compressible lattice, (b), (c) or (d), depicted in Figure 2. Along with a number of indisputable advantages, these models have one very serious shortcoming which is that, in terms of the lattice model, the correct account of the distinction in size and shape of monomer units involved in a copolymer macromolecule poses a rather serious problem. The above difficulties can be successfully overcome using the lattice cluster theory (LCT) introduced by Freed and Dudowicz.[91–94] Their generalization of the standard lattice model allows a single monomer unit to occupy several neighboring lattice sites, contrary to traditional approaches in which such an occupation is forbidden. Within the framework of such an extension of the Hole Model (Figure 2(c)), it becomes possible to develop a theory free simultaneously from the second and the third of the five assumptions underlying the FH theory. This result is also attainable by means of approaches based on current concepts of the theory of low-molecular weight liquids. A key role is played here by an off-lattice, microscopic statistical mechanical approach developed by Schweizer, Curro, and co-workers[95] for calculating thermodynamic characteristics of polymer liquids, including those comprising copolymers. The general methodology is known as the polymer reference interaction site model (PRISM) theory and is based on the RISM approach for low-molecular weight liquids.[96] Using such an approach, equations have been derived for which, upon a suitable closure approximation, either analytic[97,98] or numerical[99] solutions have been found for model binary polymer blends. Analytic PRISM theory has been recently extended by Schweizer[100] to the description of the thermodynamics of random copolymer alloys.

Because of the complexity of polymer liquids, their rigorous statistical mechanical treatment always presents serious difficulties so that progress in elaborating the thermodynamic theory of such liquids is related to the correct choice of the approximation employed. When examining theoretically solutions of low-molecular weight compounds, it is customary to consider as a reference system some simple model, permitting one to take into account as exactly as possible the excluded volume of a molecule and then to add a mean-field perturbation term allowing for the attractive interactions. Such an approximation rests on a difference in spatial scales of the short-range repulsion and long-range attraction between molecules. Often, as a reference system, an ensemble of hard spheres,[101] whose thermodynamic and correlation characteristics are well known, is chosen. When describing quantitatively equilibrium properties of polymer melts and blends, Prausnitz and co-workers[102] use as a reference system an athermal ensemble of hard-sphere chains, each of which can be pictured as a series of freely jointed tangent hard spheres. This model has been much studied[103] and, despite its simplicity, takes into account some significant features of real polymer liquids, including the excluded volume of monomer units and their connectivity. A perturbed hard sphere–chain (PHSC) theory appears rather attractive when performing thermodynamic calculations of real polymer systems inasmuch as it allows for distinctions in the size of different type monomer units[104] in a natural way and its input parameters (the effective hard-sphere diameters and attractive energy of pair interactions) are theoretically based functions of temperature. In a paper by Hino *et al.*,[105] an extension of the PHSC equation of state was presented for two types of copolymer mixtures; theoretical cloud-point curves and miscibility maps were also computed for binary random copolymer mixtures ($\mathrm{Mr}_1^{l_1}\mathrm{M}_2^{l_2} + \mathrm{M}_1^{l'_1}\mathrm{M}_2^{l'_2}$) and ($\mathrm{M}_1^{l_1}\mathrm{M}_2^{l_2} + \mathrm{M}_1^{l'_1}\mathrm{M}_3^{l'_3}$). The results of calculations for the first type are in good agreement with experimental data obtained in an investigation of the compatibility of blends involving two fractions of binary copolymer: poly(styrol-*co*-butadiene), poly(styrene-*co*-butyl methacrylate), or poly(butyl methacrylate-*co*-methyl methacrylate).[105] The efficiency of the use of the PHSC theory in the thermodynamics of blends with participation of copolymers has been supported in a recent paper by Prausnitz and co-workers.[106]

Analysis of the voluminous literature, including recent reviews,[81–90] supports the conclusion that 'there are no universal models for phase equilibrium calculations for copolymer systems available today'.[89] In Sections 13.5 and 13.6, material will be presented in terms of a unified approach which makes it possible to compensate to some extent for this deficiency. Although this theory has a rather universal character, it by no means refutes and even makes use of the main ideas of more generally accepted approaches. The indisputable advantage of such a theory is its ability to describe heteropolymers with any polydispersity and sequence distribution. This makes it possible within the framework of a review to discuss theoretical aspects of the thermodynamics of both statistical copolymers and block copolymers. On the other hand, when selecting molecular models, preference is given to those which could lead to final formulae simple enough to be applied for the treatment of experimental data. Such an approach allows one to follow all approximations used for obtaining the final results as well as to specify the area of their applicability for the description of particular systems.

13.5 LARGE-SCALE MEAN-FIELD THEORY

13.5.1 Homophasic Systems

In accordance with general ideas of statistical physics,[107] the interactions between units in terms of the mean-field approximation are taken into account via introduction of a certain external field $\boldsymbol{\eta}(\boldsymbol{r})$. Each component $\eta_\alpha(\boldsymbol{r})$ acts upon units M_α in the three-dimensional space point $\boldsymbol{r}$ forming a density distribution $\rho_\alpha(\boldsymbol{r})$ of such units. A standard procedure of statistical mechanics enables the dependence of $\boldsymbol{\rho}(\boldsymbol{r})$ on $\boldsymbol{\eta}(\boldsymbol{r})$ to be established. Further, by means of the self-consistency condition, it becomes possible to find another relation between these vectors which leads to a closed set of equations permitting the calculation of distributions $\rho_\alpha(\boldsymbol{r})$ $(\alpha = 1, \ldots, m)$.

In the present section, the situation will be discussed where scales of spatial change of densities (and, consequently, $\boldsymbol{\eta}(\boldsymbol{r})$) are large enough when compared with the geometrical dimensions of macromolecules. In this case, these latter can be considered as zero-dimensional points, presuming that the value $\eta_\alpha(\boldsymbol{r})$ of the field acting upon all units M_α of a macromolecule is the same. With such a consideration, the external field contribution to the energy of a macromolecule located in the point $\boldsymbol{r}$

$$\eta(\boldsymbol{l}; \boldsymbol{r}) = \boldsymbol{\eta}(\boldsymbol{r})\boldsymbol{l} = \sum_{\alpha=1}^{m} \eta_\alpha(\boldsymbol{r}) l_\alpha \tag{19}$$

is characterized only by the numbers l_α of units M_α of the macromolecule and does not depend on the pattern of their arrangement. Here the detailed chemical structure σ of polymer chains proves to be inessential, so that it is enough to distinguish them by vector $\boldsymbol{l}$ only. Consequently, within the framework of the large-scale mean-field theory (LSMFT), one may introduce the notion of macromolecule concentration $c(\boldsymbol{l}; \boldsymbol{r})$ and employ the methods of traditional thermodynamics by envisaging macromolecules as individual components.

Advances in the statistical thermodynamics of polymers are to a great extent due to the fact that the consideration of a polymer system with volume interactions can be reduced to that of two simpler auxiliary systems.[108] In the first of these, all inter-unit chemical bonds in the macromolecules of the system of interest are retained, as distinct from the physical interactions between them which are completely ignored. Conversely, in the second auxiliary system, such interactions are properly accounted for whereas all chemical bonds between units are taken to be broken, so that these latter are viewed as ordinary components of a low-molecular weight liquid. It turns out that any thermodynamic potential in terms of the mean-field approximation can be presented as a sum of the contributions of these systems of 'chemical bonds' and 'separate units'. Thus, for instance, the free energy of a spatially homogeneous homophasic system can be written down within the framework of the LSMFT as a sum of two items

$$F = F_{CB} + F^*, \quad \text{where} \quad F^* \equiv F_{SU} - F_{IG} = f^* V \tag{20}$$

the first of which describes an ideal gas of macromolecules placed in a volume V. Its free energy

$$F_{CB} = F_{in} - TS_C, \quad \text{where} \quad S_C = -V \sum_{l} C(\boldsymbol{l}) \ln \frac{C(\boldsymbol{l})}{e} \tag{21}$$

along with the contribution F_{in} from the internal degrees of freedom of macromolecules responsible for their conformation isomerism, takes into account also the combinatorial entropy S_C. The right-hand part of equation (20) equals the free energy F_{SU} of the system of separate units minus an analogous quantity F_{IG} for the ideal gas at the same value of densities ρ_α of units of all types. The concrete form of the function $f^*(\boldsymbol{\rho})$ depends on the model chosen for the liquid of separate units and is not of crucial importance for the theory of polymers. Here the free energy density f^* as well as the chemical potentials μ^*_α and the pressure P^*

$$\mu^*_\alpha = \frac{\partial f^*}{\partial \rho_\alpha}, \qquad P^* = \sum_\alpha \mu^*_\alpha \rho_\alpha - f^* \tag{22}$$

are functions of temperature T and the densities ρ_α of units

$$\rho_\alpha = \sum_{\boldsymbol{l}} C(\boldsymbol{l}) l_\alpha, \quad \text{where} \quad C(\boldsymbol{l}) = n(\boldsymbol{l})/V \tag{23}$$

which are known from the theory of low-molecular liquids.

When calculating thermodynamic parameters of a particular system, one can choose as functions f^*, μ^*_α, and P^* ones which have been derived within the framework of the simplest model of a lattice liquid

$$\frac{f^* v}{T} = (1-\Phi)\ln(1-\Phi) + \Phi - \frac{1}{T}\sum_\alpha \sum_\beta \gamma_{\alpha\beta}\Phi_\alpha\Phi_\beta \tag{24}$$

$$\frac{\mu^*_\alpha}{T} = -\ln(1-\Phi) - \frac{2}{T}\sum_\beta \gamma_{\alpha\beta}\Phi_\beta \tag{25}$$

$$\frac{P^* v}{T} = -\ln(1-\Phi) - \Phi - \frac{1}{T}\sum_\alpha \sum_\beta \gamma_{\alpha\beta}\Phi_\alpha\Phi_\beta \tag{26}$$

Here, as previously, $\gamma_{\alpha\beta}$ stands for the energy of the formation of a contact between units M_α and M_β whose volume fractions are Φ_α and Φ_β, whereas the quantity Φ

$$\sum_\alpha \Phi_\alpha \equiv \Phi = \frac{V_{\min}}{V} = \frac{vM}{V} = v_\rho = \frac{\rho}{\rho_{\max}}, \qquad \Phi_\alpha = \frac{\rho_\alpha}{\rho_{\max}} = v\rho_\alpha \tag{27}$$

is the fraction of V falling on the minimum volume $V_{\min} = vM$, which the M molecules of a liquid could occupy. The free volume of a liquid is considered to represent the difference between its actual volume V and the minimum one $V_{\min}$. The Φ in equation (27) is none other than the density of particles reduced to its maximum value $\rho_{\max}$, corresponding to the closest packing of molecules. Here we restrict ourselves to the simplest version of the hole model with the parameter of reducing $\rho_{\max} = 1/v$ where v is a volume of one cell, which has the physical meaning of the average close-packed volume of a particle in the fluid. Under this approximation, the empirical parameter v, having the same meaning as the excluded volume parameter when use is made of the nonlattice continuous model, is controlled only by the geometrical size of molecules.

For a system comprising r phases to be in thermodynamic equilibrium, the chemical potentials $\mu(\boldsymbol{l})$ of each component must be equal, as must the pressure P in all phases. Corresponding to equation (20) for the free energy are the following equations for these thermodynamic variables:

$$\mu(\boldsymbol{l}) \equiv \frac{\partial F}{\partial n(\boldsymbol{l})} = T\ln C(\boldsymbol{l}) + \sum_\alpha \mu^*_\alpha(\boldsymbol{\rho}) l_\alpha + \mu_{\text{in}}(\boldsymbol{l}) \tag{28}$$

$$P \equiv \sum_{\boldsymbol{l}} \mu(\boldsymbol{l}) C(\boldsymbol{l}) - f(\boldsymbol{\rho}) = T\sum_{\boldsymbol{l}} C(\boldsymbol{l}) + P^*(\boldsymbol{\rho}) \tag{29}$$

These, in parallel with equation (23), provide a description of the equilibrium states of multiphase polymer solutions and blends.

Applying equation (26) for a lattice fluid model of a low-molecular weight liquid, we arrive at the equation of state for a polymer liquid

$$\frac{vP}{T} = -\ln(1-\Phi) - (1-Y)\Phi - \frac{\gamma^{ef}}{T}\Phi^2, \qquad \text{where } \gamma^{ef} = \sum_{\alpha=1}^{m}\sum_{\beta=1}^{m} \gamma_{\alpha\beta}X_\alpha X_\beta \tag{30}$$

where Y and X_α have been determined earlier in equation (7). Equation (30) appears to be quite similar to that derived by Sanchez and Lacombe within the framework of the lattice fluid theory,[30,31] provided that parameters $v^*_{\alpha\beta}$ and $\varepsilon^*_{\alpha\beta}$ of their equation are substituted for v and $\gamma_{\alpha\beta}$, respectively. Unlike the simplified version of the lattice fluid model (equations (24)–(26)), the Sanchez–Lacombe treatment enables a correct explanation of the distinction in values of the excluded volume of different type units. Under such a treatment the parameter v is usually written down in quadratic form with phenomenological coefficients. No objective difficulties emerge if, instead of equations (24)–(26) for the description of the separate units system, recourse is made to those employed in the Sanchez–Lacombe theory and thus a more accurate account is taken of the excluded volume effect. However, for the sake of simplicity, we will confine our consideration to the simplest version of a lattice model which contains as few adjusting parameters as possible.

Within the framework of the LSMFT, general expressions have been derived for the enthalpy ΔH and the entropy ΔS of mixing

$$\frac{\Delta H}{M} = P\Delta V - \gamma^{ef}\Phi + \left\langle \sum_{\alpha=1}^{m}\sum_{\beta=1}^{m} \gamma_{\alpha\beta}\zeta_\alpha\zeta_\beta\Phi(\boldsymbol{l}) \right\rangle \tag{31}$$

$$\frac{\Delta V}{M} = v\langle 1 - 1/\Phi(\boldsymbol{l})\rangle \tag{32}$$

$$\begin{aligned}\frac{\Delta S}{M} = &-\sum_{\boldsymbol{l}} f(\boldsymbol{l})\ln[f_w(\boldsymbol{l})\Phi/\Phi(\boldsymbol{l})] \\ &+ \langle[1/\Phi(\boldsymbol{l}) - 1]\ln[1-\Phi(\boldsymbol{l})] - (1/\Phi - 1)\ln(1-\Phi)\rangle\end{aligned} \tag{33}$$

where angular brackets denote the procedure of averaging the quantity standing between them over weight SCD $f_w(\boldsymbol{l})$ (equation 2). The reduced density Φ of the blend with given values of statistical parameters Y and $\boldsymbol{X}$ can be obtained from the equation of state (30). For a pure component, characterized by vector $\boldsymbol{l} = l\boldsymbol{\zeta}$, the value $\Phi(\boldsymbol{l})$ can be calculated from the same equation (30) where Y and $\boldsymbol{X}$ should be replaced, respectively, by the reciprocal degree of polymerization $1/l$ and by the composition vector $\boldsymbol{\zeta}$ of this particular molecule.

13.5.2 Heterophasic Systems

13.5.2.1 Free energy

To provide an exhaustive description of a system involving r spatially homogeneous phases, it is necessary to set the values of their volumes V^i $(i = 1, \ldots, r)$ and concentrations $C^i(\boldsymbol{l})$ $(i = 1, \ldots, r)$ of all components with specified $\boldsymbol{l}$ in each ith phase. Given the values of these internal variables, the free energy is easily found by formula

$$F = \sum_{i=1}^{r} V^i\left\{f^*(\boldsymbol{\rho}^i) - T\sum_{\boldsymbol{l}} C^i(\boldsymbol{l})\ln[C^i(\boldsymbol{l})/e]\right\} + F_{in} \tag{34}$$

where components ρ^i_α of the vector $\boldsymbol{\rho}^i$ are related to $C^i(\boldsymbol{l})$ in any ith phase by the stoichiometric condition

$$\rho^i_\alpha = \sum_{\boldsymbol{l}} C^i(\boldsymbol{l}) l_\alpha \qquad (i = 1, \ldots, r; \alpha = 1, \ldots, m) \tag{35}$$

When in equilibrium, the chemical potential $\mu(\boldsymbol{l})$ (equation 28) of each macromolecule with given $\boldsymbol{l}$ has the same value in each phase. This circumstance along with the stoichiometric condition

$$\sum_{i=1}^{r} C^i(\boldsymbol{l}) V^i = n(\boldsymbol{l}) \tag{36}$$

permits one to exclude $\mu(\boldsymbol{l})$ and to express concentrations of molecules in each phase through $n(\boldsymbol{l})$

$$C^i(\boldsymbol{l}) = n(\boldsymbol{l}) e^i \Big/ \sum_{j=1}^{r} V^j e^j \tag{37}$$

where the following designations are used

$$e^i \equiv \exp\left\{ -\frac{1}{T} \sum_{\alpha=1}^{m} \mu^*_\alpha(\boldsymbol{\rho}^i) l_\alpha \right\} \tag{38}$$

Substitution of concentrations $C^i(\boldsymbol{l})$ from equation (37) in equations (29) and (31), with allowance being made for the equality of pressure in all phases, yields a closed set of $(m+1)r$ equations with respect to the same number of internal variables V^i and ρ^i_α. Between them there are $(m+1)$ stoichiometric relationships

$$\sum_{i=1}^{r} \rho^i_\alpha V^i = M X_\alpha \qquad (\alpha = 1, \ldots, m) \qquad \sum_{i=1}^{r} V_i = V \tag{39}$$

resulting from equations (35), (36), and (22), so that there will be only $(m+1)(r-1)$ independent internal variables. Volume fractions $y^i = V^i/V$ and concentrations of units $\boldsymbol{\rho}^i$ in all phases except the rth one are conveniently chosen as such variables whereas y^r and $\boldsymbol{\rho}^r$ can be determined from equations (39).

The equation (34) for free energy, in view of equations (22) and (37), will read

$$F = -T \sum_{\boldsymbol{l}} n(\boldsymbol{l}) \ln \sum_{i=1}^{r} y^i e^i - V \sum_{i=1}^{r} y^i P^*(\boldsymbol{\rho}^i) + T \sum_{\boldsymbol{l}} n(\boldsymbol{l}) \ln [n(\boldsymbol{l})/eV] + F_{\text{in}} \tag{40}$$

According to equation (40), the free energy of a multiphase system depends on $n(\boldsymbol{l})$, V, and T not only in an explicit but also in an implicit manner through the dependence of the concentrations $\boldsymbol{\rho}^i$ of units in different phases and their volume fractions y^i on the aforementioned external variables. However, when calculating the first derivatives of the free energy proceeding from equation (40), there is no need to take into account this implicit dependence since the function F reaches an extremum with respect to the internal variables y^i and ρ^i_α, *i.e.* there are the equalities

$$\frac{\partial F}{\partial y^i} = 0, \qquad \frac{\partial F}{\partial \rho^i_\alpha} = 0 \qquad (i = 1, \ldots, r-1; \alpha = 1, \ldots, m) \tag{41}$$

at any fixed values of the external variables. The above circumstance makes it possible to write out immediately the general expressions

$$\mu(\boldsymbol{l}) = \partial F / \partial n(\boldsymbol{l}) = -T \ln \left\{ V \sum_{i=1}^{r} y^i e^i / n(\boldsymbol{l}) \right\} + \mu_{\text{in}} \tag{42}$$

$$P = -\frac{\partial F}{\partial V} = \frac{T\Pi}{V} + \sum_{i=1}^{r} y^i P^*(\boldsymbol{\rho}^i) \tag{43}$$

for the calculation of the values of chemical potential and pressure in multiphase polymer systems, which can also be derived immediately from equations (28) and (29), provided that recourse is made to equation (37).

The thermodynamic description of an arbitrary multiphase polymer blend in terms of the LSMFT is reducible, as it appears from the above discussion, to the problem of the calculation of the internal parameters y^i and $\boldsymbol{\rho}^i$ from the solution of a set of several equations. Their number is defined along with r only by the number of types m of units, regardless of how many chemically individual constituents comprise the polymer solution. This is why such a mixture is called a 'quasi-m-component mixture'. It should be borne in mind that monomer units, unlike solvent molecules, are merely quasi-components and hence the thermodynamic significance in their terms has a number of fundamental peculiarities. One of these is the impossibility of introducing the concept of the chemical potential of a monomer unit, which by definition represents the work of transferring this unit from the given phase to infinity.[107] This procedure is not feasible in principle because any monomer unit can be transferred only together with the polymer molecule it enters. Another substantial peculiarity intrinsic to the thermodynamic consideration of quasi-components is the fact that the free energy (equation 40) of a multiphase system is not divided into terms with each dependent only on the density $\boldsymbol{\rho}^i$ of monomer units in the ith phase and its volume V^i. Clearly, the traditional thermodynamic approach is free from the above anomalies when the constituents are polymer molecules.

13.5.2.2 Phase diagram

When constructing a phase diagram of a solution of a heteropolymer with fixed SCD, the fractions of molecules of low-molecular weight solvents and the total fraction of monomer units in the polymer along with the temperature and volume (or pressure) are independent external variables. For the blends of several polymers (with given SCD for each of them), the system is defined by fractions of their units.

The nature of the variation of the equilibrium state of a polymer system with an arbitrary SCD, resulting from the change of any of its n parameters (thermodynamic or structural), can be predicted within the framework of the LSMFT. In the course of this change the point characterizing the equilibrium state of a system will drift within the n-dimensional hyperspace of the aforementioned external parameters. For such a drift, this point can intersect some specific $(n-1)$-dimensional hypersurfaces.

The most important one is a hypersurface, upon intersecting which the second liquid phase appears for the first time inside the homophasic system. Equations defining such a hypersurface of cloud-points have been derived[108]

$$\Phi'_\alpha = \Phi G_\alpha(\boldsymbol{s}), \qquad s_\alpha = \exp\{[\mu'_\alpha(\boldsymbol{\Phi}) - \mu'_\alpha(\boldsymbol{\Phi}')]/T\} \tag{44}$$

$$T\Phi G(\boldsymbol{s}) + vP^*(\boldsymbol{\Phi}') = T\Phi G(\mathbf{1}) + vP^*(\boldsymbol{\Phi}) = vP \tag{45}$$

where the components $G_\alpha(\boldsymbol{s})$ of the vector $\boldsymbol{G}(\boldsymbol{s})$ represent derivatives (equation 6) of the generating function $G(\boldsymbol{s})$ (equation 5) of the SCD of all the constituents including solvents.

If to turn to equations (44) for a vector $\boldsymbol{\Phi}' = v\boldsymbol{\rho}'$ with components equal to dimensionless concentrations of different type units, it will readily be seen that these equations always have the trivial solution

$$\Phi'_\alpha = U''_\alpha(\boldsymbol{\Phi}) \equiv \Phi_\alpha \equiv \Phi X_\alpha, \qquad (\alpha = 1, \ldots, m) \tag{46}$$

where the dependence of the total density of units Φ of an initial single-phase system on pressure and temperature at a given SCD can be found from the equation of state (30). Within a certain region of values of the parameters of the function SCD, pressure P and temperature T, equation (44) not only has the trivial solution but can also possess a nontrivial one $\Phi'_\alpha = U_\alpha(\boldsymbol{\Phi})$ which determines the dependence of the densities $\boldsymbol{\Phi}'$ of units in the first drops of the precipitated phase on their values $\mathbf{\Phi}$ in the bulk. Analogous dependencies of $\mathbf{\Phi}'$ on $\mathbf{\Phi}$ in the case of traditional thermodynamics are normally obtained[107] from the equality of the chemical potentials $\mu_\alpha(\mathbf{\Phi}') = \mu_\alpha(\mathbf{\Phi})$ of the constituents in both phases. These particular equations can be derived from general ones of equation (44) if the molecules of solvent are considered as units. Substituting every Φ'_α for the function $U_\alpha(\mathbf{\Phi})$ in equation (45) which denotes the equality of pressure of coexisting phases, we arrive at an equation for the hypersurface of cloud points. To each point $\mathbf{\Phi}$ located on it, there corresponds a concrete value $\mathbf{\Phi}'$ of the densities of units of all types in the precipitating phase, so that the manifold of all such points $\mathbf{\Phi}'$ will be the shadow hypersurface. Its particular case, namely the shadow curve, is depicted in Figure 1(b).

Another important specific hypersurface is the spinodal where the loss of the marginal stability of spatially homogeneous single-phase state appears. In terms of the application of traditional thermodynamics to this condition, there corresponds the reducing to zero of the determinant of the matrix $\mathbf{K}$ with elements $K_{\alpha\beta} = \partial^2 f/\partial\rho_\alpha\, \partial\rho_\beta = \partial\mu_\alpha/\partial\rho_\beta$, equal to the second derivatives of the density of free energy $f(\boldsymbol{\rho})$ with respect to the concentrations of particles of type α and β. This symmetrical matrix is a positively determined one (*i.e.* all its eigenvalues are positive) within that region of external parameters where small deviations $\Delta\rho_\alpha = \rho_\alpha - \bar{\rho}_\alpha \ll \bar{\rho}_\alpha$ of particle concentrations ρ_α from their average values $\bar{\rho}_\alpha$ do not disturb the system from thermodynamic equilibrium. At this region boundary, *i.e.* on the spinodal, the minimum eigenvalue in the spectrum of the matrix in hand becomes zero and then negative upon intersecting the spinodal hypersurface when some external parameter is being changed.

In the thermodynamics of polymer mixtures the matrix $\mathbf{K}$ will have as its elements

$$K(\mathbf{l}', \mathbf{l}'') \equiv \partial^2 f/\partial C(\mathbf{l}')\partial C(\mathbf{l}'') = \delta(\mathbf{l}', \mathbf{l}'')/C(\mathbf{l}') - \sum_{\alpha=1}^{m}\sum_{\beta=1}^{m} C_{\alpha\beta} l'_\alpha l''_\beta \tag{47}$$

where $\delta(\mathbf{l}', \mathbf{l}'')$ is the Kronecker delta symbol, whereas elements $C_{\alpha\beta}$ of the matrix $\mathbf{C}$ of direct correlation functions[109] are defined by the expression

$$C_{\alpha\beta} = -\frac{1}{T}\frac{\partial\mu^*_\alpha}{\partial\rho_\beta} = -\frac{1}{T}\frac{\partial^2 f^*}{\partial\rho_\alpha\, \partial\rho_\beta} \tag{48}$$

It was shown[108] that to find the spinodal there is no necessity to determine the minimum eigenvalue of the matrix $\mathbf{K}$ (equation 47); instead it will suffice to solve an analogous problem for the matrix $\mathbf{M}$

$$\mathbf{M} = \kappa^{-1} - \mathbf{C} \tag{49}$$

where the elements of matrix κ are proportional to the second-order statistical moments (equation 4) of SCD (equation 3)

$$\kappa_{\alpha\beta} \equiv \sum_{\mathbf{l}} C(\mathbf{l}) l_\alpha l_\beta = \frac{M}{V} X_{\alpha\beta} = \frac{\Phi}{v} X_{\alpha\beta} \tag{50}$$

The mathematical condition of the spinodal consists in the equality to zero of the determinant $|\mathbf{M}|$ of the matrix $\mathbf{M}$ (equation 49); generally speaking, this may be written down in three equivalent forms

$$\text{(a)}\ \ |\kappa^{-1} - \mathbf{C}| = 0, \qquad \text{(b)}\ \ |\mathbf{E} - \kappa\mathbf{C}| = 0, \qquad \text{(c)}\ \ |\mathbf{C}^{-1} - \kappa| = 0 \tag{51}$$

where $\mathbf{E}$ is the identity matrix with elements $\delta_{\alpha\beta}$.

Equation (49) makes sense only if the matrix κ is not a singular one, *i.e.* the determinant $|\kappa|$ is distinct from zero. This condition is certainly known to be fulfilled for all actual polymer samples and can be expected to be broken only when considering model polymer systems with degenerate SCD. In order to reveal this property of the distribution, one should write down the corresponding matrix with elements $l_{\alpha i}$, equal to the numbers of αth units in a molecule of

ith type. If the rank of the matrix turns out to be less than m, the SCD is degenerate whereas in the opposite case the determinant $|\kappa|$ never becomes zero. To find the spinodal of a model system with degenerate SCD where $|\kappa| \equiv 0$, recourse should be made to equation (51b).

Proceeding from equation (25), one can reduce the spinodal condition to a form suitable for practical calculation by dividing the matrix $\mathbf{M}$ (equation 49) into two items

$$\mathbf{M} = \mathbf{M}^R + \mathbf{M}^S, \qquad \text{where } M^R_{\alpha\beta} = (\kappa^{-1})_{\alpha\beta} - 2v\gamma_{\alpha\beta}/T, \qquad M^S_{\alpha\beta} = v/(1-\Phi) \tag{52}$$

which exhibit at $\Phi \to 1$ regular and singular behavior, respectively. The determinant of the matrix $\mathbf{M}$ is expressible through the sum D^R of all the elements of the matrix adjoint for $\mathbf{M}^R$ and its determinant, which enables one to write out equation (51a) as follows

$$|\mathbf{M}| = |\mathbf{M}^R| + \frac{v}{1-\Phi} D^R = 0 \tag{53}$$

The $(n-2)$-dimensional manifold (hyperline) of critical points on the cloud-point hypersurface can be found provided that the densities Φ'_α of units of all types within the incipient phase coincide with their values Φ_α within the principal phase. This means that equation (44) in all points of such a hyperline has a multiple root $\mathbf{\Phi}' = \mathbf{\Phi}$. Mathematically speaking, this implies that the determinant of the matrix $\mathbf{E} - \kappa\mathbf{C}$ equals zero, which agrees with the spinodal condition (51b). Thus, the manifold of common points on the cloud-point hypersurface and the spinodal hypersurface will contain the hyperline of critical points of a system.

The equation defining such a hyperline may be written down in a variety of ways, one of which looks as follows

$$L_3 = \sum_\alpha \sum_\beta \sum_\gamma \Gamma^{(3)}_{\alpha\beta\gamma} \psi_\alpha \psi_\beta \psi_\gamma = 0 \tag{54}$$

Here, along with the components of the eigenvector $\boldsymbol{\psi}$ of matrix $\mathbf{M}$ (equation 49), corresponding to its zeroth eigenvalue, the components $\Gamma^{(3)}_{\alpha\beta\gamma}$ of the third-order symmetric tensor $\mathbf{\Gamma}^{(3)}$ are also involved. These latter in accordance with the formula

$$\Gamma^{(3)}_{\alpha\beta\gamma} = -\sum_i \sum_j \sum_k \kappa^{(3)}_{ijk} (\kappa^{-1})_{i\alpha} (\kappa^{-1})_{j\beta} (\kappa^{-1})_{k\gamma} + \frac{f^{*(3)}_{\alpha\beta\gamma}}{T} \tag{55}$$

are expressed through the components

$$\kappa^{(3)}_{\alpha\beta\gamma} \equiv \sum_{\mathbf{l}} C(\mathbf{l}) l_\alpha l_\beta l_\gamma = \frac{M}{V} X_{\alpha\beta\gamma} = \frac{\Phi}{v} X_{\alpha\beta\gamma} \tag{56}$$

$$f^{*(3)}_{\alpha\beta\gamma} \equiv \partial^3 f^*(\boldsymbol{\rho}) / \partial\rho_\alpha \, \partial\rho_\beta \, \partial\rho_\gamma \tag{57}$$

of two other third-order symmetric tensors $\kappa^{(3)}$ and $\mathbf{f}^{*(3)}$. The first of these is characterized by statistical moments (equation 8) of the SCD whereas the second one is defined by the thermodynamic model chosen in the system of separate units. Condition (54) together with the spinodal condition (51) yields the equation of the hyperline of critical points of an arbitrary polymer mixture. Sometimes, when calculating this hyperline, it is more convenient to resort to an expression other than equation (54) for L_3 where the role of $\boldsymbol{\psi}$ is played by the eigenvector $\boldsymbol{\varphi}$ of the matrix $\kappa\mathbf{MC}^{-1} = \mathbf{C}^{-1} - \kappa$, while instead of $\mathbf{\Gamma}^{(3)}$ there appears the tensor $\mathbf{G}^{(3)}$ with components

$$G^{(3)}_{\alpha\beta\gamma} = \kappa^{(3)}_{\alpha\beta\gamma} - \sum_i \sum_j \sum_k \frac{f^{*(3)}_{ijk}}{T} (\mathbf{C}^{-1})_{i\alpha} (\mathbf{C}^{-1})_{j\beta} (\mathbf{C}^{-1})_{k\gamma} \tag{58}$$

With the knowledge of one of the vectors $\boldsymbol{\psi}$ or $\boldsymbol{\varphi}$, another can easily be found because $\boldsymbol{\varphi}$ is obtainable from $\boldsymbol{\psi}$ by multiplying the latter on the left by any of matrices κ^{-1} or $\mathbf{C}$.

When constructing phase diagrams of solutions of multicomponent heteropolymers, it is more convenient to reformulate the problem in terms of macromolecules only, having formally reduced the number of quasicomponents at the cost of an exclusion of the molecules of solvent. Under such an approach, the function $f(\boldsymbol{l})$ (equation 3) denotes the SCD of polymers which should be exclusively employed in all the foregoing formulae of the present section. However, all the characteristics of the system of separate units appearing in them now have to be replaced by their modified values. Thus, for instance, to calculate the spinodal using equations (51), one should substitute in them the matrix of direct correlation functions (equation 48) with the matrix

$$\hat{\mathbf{C}} = \mathbf{C}^{PP} + \mathbf{C}^{PS}\mathbf{K}^{-1}\mathbf{C}^{SP} \tag{59}$$

with the number of rows equal to that of the number of types of monomer unit. Into formula (59) there enter submatrices of $\mathbf{C}$ (equation 48)

$$\mathbf{C} = \begin{pmatrix} \mathbf{C}^{PP} & \mathbf{C}^{PS} \\ \mathbf{C}^{SP} & \mathbf{C}^{SS} \end{pmatrix} \qquad K_{ij} \equiv \frac{1}{T}\frac{\partial^2 f_{SU}(\boldsymbol{\rho})}{\partial\rho_i\,\partial\rho_j} = -C_{ij}^{SS} + \frac{\delta_{ij}}{\rho_i} \tag{60}$$

which take into account volume interactions of different types: (polymer + polymer), (polymer + solvent), and (solvent + solvent). In conformity with equation (59), in the reduced system the interactions of such a kind between monomer units proceed either directly (the first item) or through solvent molecules (the second item).

When calculating the critical point in a reduced system, the general formulae (54), (55), and (58) can be used by substituting $f^{*(3)}_{\alpha\beta\gamma}$ with its modified value

$$\hat{f}^{*(3)}_{\alpha\beta\gamma} = f^{*(3)}_{\alpha\beta\gamma} + \sum_i \sum_j \sum_k K^{(3)}_{ijk}\tau_{i\alpha}\pi_{j\beta}\pi_{k\gamma} \tag{61}$$

where the following designations are used

$$\pi_{i\alpha} \equiv \sum_j (\mathbf{K}^{-1})_{ij} C^{SP}_{j\alpha}; \qquad K^{(3)}_{ijk} = \frac{\partial^3 f_{SU}(\boldsymbol{\rho})}{T\,\partial\rho_i\,\partial\rho_j\,\partial\rho_k} \tag{62}$$

while summation is over all types of solvents.

13.5.3 Incompressible Systems

In the theory of polymer blends and solutions in widespread use is an approximation: for thermodynamic calculations, the compressibility of a system is neglected. This approximation in the framework of the lattice model is characterized by the following conditions

$$v \sum_{\alpha=1}^{m} \rho^i_\alpha = 1 \qquad (i = 1, \ldots, r) \tag{63}$$

replacing the conditions of the equality of pressures $P = P^i$ in all coexisting phases, which become physically meaningless if the conditions of equation (63) are implemented. It should be pointed out that the formal substitution of the relationship (63) into the expressions (25) and (26) leads to nonphysical divergencies not only for P^* but also for μ^*_α. However, their singular parts μ^S_α turn out to be inessential since they are canceled in equations (37). Consequently, the functions $\mu^*_\alpha = \mu^R_\alpha + \mu^S_\alpha$ which they contain should be substituted in terms of the incompressible system approximation for their regular parts

$$\frac{\mu_\alpha^{*R}}{T} \equiv R_\alpha(\boldsymbol{X}, Y) = \sum_\beta \chi_{\alpha\beta} X_\beta - \tfrac{1}{2} \sum_\alpha \sum_\beta \chi_{\alpha\beta} X_\alpha X_\beta - Y \tag{64}$$

resulting from eliminating the logarithmic terms in equations (25) and (26) with due regard for expression (30) for pressure. Substantially, within the framework of the approximation (63), the energies $\gamma_{\alpha\beta}$ of pairwise contacts of units are present in all formulae only as combinations $\chi_{\alpha\beta} = 1/T[\gamma_{\alpha\alpha} + \gamma_{\beta\beta} - 2\gamma_{\alpha\beta}]$, which are usually referred to as the binary interaction parameters.

Thus in the approximation under consideration, the state of the multicomponent multiphase system is defined by the set of internal variables y^i, X^i_α, Y^i ($i = 1, \ldots, r$; $\alpha = 1, \ldots, m$) which are connected with each other by $(m + r + 1)$ independent stoichiometry conditions

$$\sum_{i=1}^{r} y^i X^i_\alpha = X^0_\alpha \qquad \sum_{i=1}^{r} y^i = 1 \tag{65}$$

$$\sum_{\alpha=1}^{m} X^i_\alpha = 1 \qquad \sum_{i=1}^{r} y^i Y^i = Y^0 \tag{66}$$

Therefore, among $r(m + 2)$ internal system variables, the number of independent ones is $(r - 1)(m + 1)$. These can be found from the solution of the set of equations

$$X^i_\alpha = \sum_{\boldsymbol{l}} l_\alpha f(\boldsymbol{l}) \frac{e^i}{\sigma}; \qquad Y^i = \sum_{\boldsymbol{l}} f(\boldsymbol{l}) \frac{e^i}{\sigma} \tag{67}$$

where the following designations are used

$$e^i \equiv \exp\left[-\sum_\alpha R_\alpha(\boldsymbol{X}^i, Y^i) l_\alpha\right], \qquad \sigma \equiv \sum_{i=1}^{r} y^i e^i \tag{68}$$

while the components R_α of the vector $\boldsymbol{R}$ are defined by the expression (64). The very fact that among $r(m + 1)$ equations (67) the number of independent ones equals $(r - 1)(m + 1)$ testifies to the correct formulation of the mathematical problem of calculating the internal variables. To solve it, one is supposed to know, along with v and the interaction parameters $\chi_{\alpha\beta}$, only the explicit form of the SCD $f(\boldsymbol{l})$ (equation 3).

Generating the function $G(\boldsymbol{s})$ (equation 5) of this distribution prescribes the location in the phase diagram of the cloud-point hypersurface which can be found from the nontrivial solution of equations

$$X_\alpha = s_\alpha\, \partial G(\boldsymbol{s})/\partial s_\alpha, \qquad Y = G(\boldsymbol{s}) \tag{69}$$

$$s_\alpha = \exp[R_\alpha(\boldsymbol{X}^0, Y^0) - R_\alpha(\boldsymbol{X}, Y)] \tag{70}$$

Such a solution, as distinct from the trivial one

$$X_\alpha = X^0_\alpha = G_\alpha(\mathbf{1}), \qquad Y = Y^0 = G(\mathbf{1}) \tag{71}$$

can exist only on the hypersurface of the parameters of a system where a new phase appears for the first time.

Equations (67)–(70) can be derived by various means, the most elegant of which is realized in terms of the general approach set forth in Section 13.6.1. Allowance for the incompressibility condition (63) reduces here to the addition to the integrand of equation (96) of the functional delta function of argument $(\Sigma\rho_\alpha - \rho)$. Within the mean-field approximation, the function (97) should be minimized with regard to the aforementioned constraint. Instead of this, the unconstrained extremum of the modified functional $\mathscr{F}[\boldsymbol{\rho}]$ is conveniently sought by adding to the right-hand part of the expression (97) the item $T\int(\Sigma\rho_\alpha - \rho)w\mathbf{dr}$, where the Lagrange factor $Tw(\mathbf{r})$ can be found from the incompressibility condition (63). As a result, it turns out that in the limit of the compressibility absence, in equation (19) for the effective field, $\eta_\alpha(\mathbf{r})$ ought to be replaced by

$$\eta_\alpha(\mathbf{r}) = v \sum_\beta \chi_{\alpha\beta}\rho_\beta(\mathbf{r}) + w(\mathbf{r}) \tag{72}$$

In the case of spatially homogeneous phases, the components η_α of this field will equal $\mu_\alpha^{*\beta}$ and equation (72) transforms into equation (64).

Turning to equation (53), it is possible to write out the spinodal condition for incompressible systems

$$D \equiv \sum_\alpha \sum_\beta A_{\alpha\beta} = 0 \tag{73}$$

where $A_{\alpha\beta}$ are elements of matrix $\mathbf{A}$ adjoint for matrix $\mathbf{M}$ with elements

$$M_{\alpha\beta} = (\kappa^{-1})_{\alpha\beta} + v\chi_{\alpha\beta} \tag{74}$$

No difficulties arise in calculating the eigenvector $\boldsymbol{\psi}$ components entering into equation (54) which define the critical point, since, for incompressible systems, simple equations hold[108]

$$\psi_\alpha = \sum_\beta A_{\alpha\beta} \qquad \sum_\alpha \psi_\alpha = 0 \tag{75}$$

As for equation (54) for the determination of critical points, when dealing with an incompressible system, in equations (55) and (58) the second items may be put equal to zero, taking into consideration the second equation (75).

For any incompressible system, since all $\Phi(\boldsymbol{l})$ as well as the Φ equal unity, equations (31)–(33) become essentially simpler. Under such an approximation, $\Delta V = 0$, and the expression for the entropy of mixing ΔS contains only a combinatorial contribution, whereas the enthalpy of mixing ΔH, in a trivial manner

$$\frac{\Delta H}{MT} = -\sum_\alpha \sum_\beta \chi_{\alpha\beta}\lambda_{\alpha\beta} \tag{76}$$

is expressed through elements $\lambda_{\alpha\beta}$ of the correlation matrix λ of the composition vector $\boldsymbol{\zeta}$. By definition we have

$$\lambda_{\alpha\beta} \equiv \langle \zeta_\alpha \zeta_\beta \rangle - \langle \zeta_\alpha \rangle \langle \zeta_\beta \rangle \tag{77}$$

where angular brackets denote the averaging of the random value over the weight SDC (equation 2). If the system does not contain molecules of heteropolymers, the first item of equation (77) is equal to zero at $\alpha \neq \beta$ and the equation for the free energy of mixing of an incompressible blend of homopolymers reduces to the form presented in equation (15).

Equation (76) for the enthalpy of mixing of an incompressible heteropolymer system is similar to earlier presented expressions for ΔH (equation 17), and is distinguished by a more convenient notation for the calculation of actual polydisperse systems. The sum of the elements of each of m rows and columns of a symmetric matrix λ equals zero, so that among m^2 of its elements only $m(m-1)/2$ will be independent; this coincides with the number of parameters $\chi_{\alpha\beta}$.

For a melt of a polydisperse binary copolymer, equation (76) reduces to the form $\Delta H/MT = \chi_{12}\sigma^2$, which indicates that the enthalpy of mixing is proportional to the variation σ^2 of the composition distribution and becomes zero for a monodisperse sample. When dealing with a terpolymer, it appears more convenient to choose, as independent elements of the matrix λ, diagonal ones, $\lambda_{\alpha\alpha} = \sigma_\alpha^2$, by virtue of which the expression (76) becomes

$$\frac{\Delta H}{MT} = \sum_{\alpha=1}^{3} \chi_\alpha \sigma_\alpha^2, \quad \text{where} \quad 2\chi_\alpha \equiv \chi_{\alpha\beta} + \chi_{\alpha\gamma} - \chi_{\beta\gamma} \tag{78}$$

Here (α, β, γ) is a triple of indices among which no pair is identical.

13.5.4 Multiple Critical Points

At certain values of the external parameters of a system, critical points on its phase diagram become degenerate upon their convergence. Depending upon the degree of degeneration, they distinguish tricritical, tetracritical, and so on points which prescribe the topology of phase diagrams.[110,111] The problem of finding multiple critical points in macromolecular systems features a number of qualitative peculiarities associated with the polydispersity of polymer samples. For hompolymer systems, this problem was solved some time ago,[4] but for solutions and blends involving heteropolymers, until recent work[108] only some particular problems had been treated in the literature. The key ideas of this study connected with finding the critical points follow from the general approach (see Section 13.6.1) and comprise the following.

Since all critical points are located on the cloud-point hypersurface, it is sufficient to restrict the consideration to the linear contribution in a small volume $\Delta V'$ of an incipient phase to the alteration of the free energy ΔF. The proportionality coefficient

$$\frac{\Delta F(\tau)}{\Delta V' T} = \sum_{p=2}^{\infty} \frac{L_p}{p!} \tau^p \tag{79}$$

becomes zero at the hypersurface mentioned. The meaning of this formula is conveniently clarified by considering ΔF as a function of m variables $\Delta\rho_\alpha$ which denote deviations of the α-th-type unit concentration inside the nucleus of the incipient phase from their values within the principal phase. In the region of stability of the spatially homogeneous state, this function has a minimum at the point $\Delta\boldsymbol{\rho} = \mathbf{0}$. This stable node of hypersurface $\Delta F(\Delta\boldsymbol{\rho})$ is known to transform on the spinodal into a saddle to which the vanishing of the minimum eigenvalue Λ of matrix (49) mathematically corresponds. The normalized eigenvector $\boldsymbol{\psi}$ corresponding to Λ indicates the direction along which the loss of stability of the spatially homogeneous state of the system occurs. Because the remaining eigenvalues of this matrix are positive, the shape of the hypersurface $\Delta F(\Delta\boldsymbol{\rho})$ in the vicinity of the minimum will resemble a ravine. For the movement of the point along the hollow of this ravine (*i.e.* along a specific line, the location on which is characterized by parameter τ), the height of the moving point will change in accordance with equation (79). This specific line will pass through all the extrema of the function $\Delta F(\Delta\boldsymbol{\rho})$ lying in the vicinity of its principal minimum at point $\Delta\boldsymbol{\rho} = \mathbf{0}$. To the minima, appearing among these extrema, there will correspond other (meta)stable states of a system. As external parameters of the system change, the above minima shift and can be expected to merge with the main one at critical points. At an ordinary critical point, two minima merge; at a tricritical point, three merge, and so on. The mathematical condition for bifurcation inducing the merging of n minima of function $\Delta F(\Delta\boldsymbol{\rho})$ will be the vanishing of all coefficients of the expansion in series (79) up to the $(2n-1)$th one

$$L_2 = L_3 = \ldots = L_{2n-1} = 0, \qquad L_{2n} \neq 0 \tag{80}$$

Equations (80) represent general conditions for finding multiple critical points.

The coefficient $L_2 = \Lambda$ becomes zero on the spinodal, while the problem of the determination of all the remaining coefficients L_p of the expansion (79) in a power series was solved in general form for an arbitrary heteropolymer system.[108] It appears more convenient to present the final results in a diagram form rather than in an analytical one, when an appropriate analytical expression corresponds to each diagram element and a general algorithm of the construction of necessary phase diagrams is formulated. This can be accomplished in two equivalent ways, the first of which is based on the density functional (DF) formalism (see Section 13.6.1.2) while the second incorporates the random field (RF) formalism (see Section 13.6.1.1). Both these approaches make use of one and the same diagram technique for the calculation of the sought for values of L_p, to which there correspond in Figure 3 gray regular p-gons. To the same diagram elements depicted in Figure 3, different analytical expressions correspond in each approach.

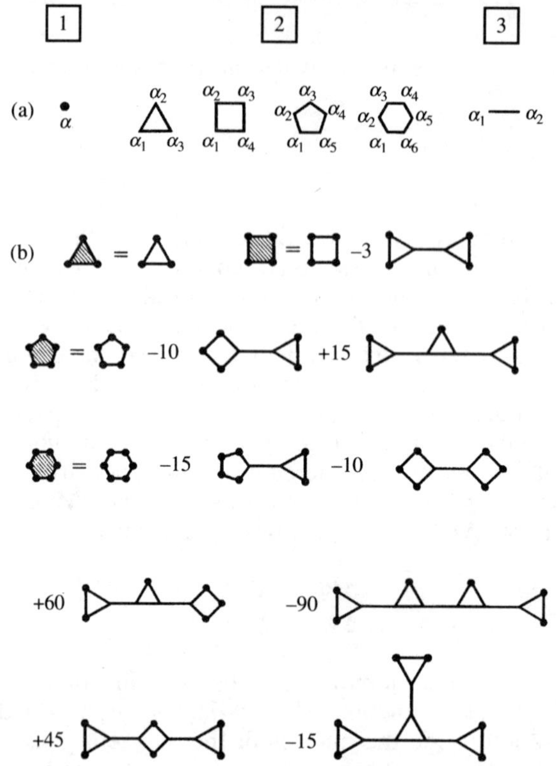

Figure 3 Diagram expressions relating the coefficients of the Landau expansion of free energy in formalisms of the density function (DF) and of the random field (RF).

In terms of the DF formalism, the elements of the diagram technique are depicted in Figure 3(a), where index α_i corresponds to the 'color' of the relevant ith apex. To elements 1, 2, and 3 there correspond, respectively, analytical expressions

$$\boxed{1}\quad \psi_\alpha \qquad \boxed{2}\quad \Gamma^{(p)}_{\alpha_1\ldots\alpha_p} = -\gamma^{(p)}_{\alpha_1\ldots\alpha_p} + f^{*(p)}_{\alpha_1\ldots\alpha_p}/T \qquad \boxed{3}\quad M^+_{\alpha_1\alpha_2} \tag{81}$$

where ψ_α are components of the eigenvector $\boldsymbol{\psi}$ of matrix (49) corresponding to its eigenvalue $\Lambda = 0$, while the meaning of the remaining designations will be elucidated below.

In order to write out an analytical expression for coefficient L_p, one should draw a set of all cactuses with number p of external apices comprising elements in Figure 3(a) in such a way as to provide for each apex to be joined to either element 1 or to one of the element 3 edges. Essentially, the junction is admissible of only apices of the same color, over which the summation is made. This may be conventionally equated to the color-erasing procedure resulting in the graphs with uncolored apices (Figure 3(b)). The integer appearing in front of each cactus is equal to the order of a group of its automorphisms and can be calculated by means of graph theory methods,[112] while the sign equals $(-1)^n$ where n represents the number of elements 3 in a cactus. The value of L_p of interest is obtainable by summing the contributions of all cactuses with an allowance for the weight coefficients appearing in front of them in the corresponding diagram.

It is easy to verify that the above-formulated diagram rules enabling one to find expressions for L_p are immediately conducive at $p = 3$ to expression (54) presented earlier, while for L_4, for instance, recourse to this diagram technique yields the following relationship

$$L_4 = \sum_{\{\alpha_i\}} \left[\Gamma^{(4)}_{\alpha_1\alpha_2\alpha_3\alpha_4} - 3 \sum_{\gamma\delta} \Gamma^{(3)}_{\alpha_1\alpha_2\gamma} M^+_{\gamma\delta} \Gamma^{(3)}_{\delta\alpha_3\alpha_4} \right] \prod_{i=1}^{4} \psi_{\alpha_i} \tag{82}$$

where summation is made over all sets of indices.

Having formulated the principle of the construction of the diagrams which permits the calculation of the unknown coefficients L_p of the expansion into power series (79), let us elucidate now the meaning of quantities (81) undefined so far. Among them are the elements $M^+_{\alpha\beta}$ of matrix $\mathbf{M}^+$ which is pseudoinverse to the matrix $\mathbf{M}$.

To obtain matrix $\mathbf{M}^+$ elements

$$M^+_{\alpha\beta} = \sum_{r=2}^{m} \Lambda_r^{-1} \psi^r_\alpha \psi^r_\beta \tag{83}$$

it is sufficient to find the complete set of eigenvalues Λ_r, and the corresponding eigenvectors $\boldsymbol{\psi}^r$ of the matrix $\mathbf{M}$ (equation 49), with the exception of the principal eigenvalue $\Lambda_1 \equiv \Lambda$ and its eigenvector $\boldsymbol{\psi}^1 \equiv \boldsymbol{\psi}$ which enter into the expression for the spinodal and the critical point.

To have the components of the symmetric tensor of pth order $\gamma^{(p)}$ expressed through the statistical moments (equation 8) of the SCD (equation 3), recourse should be made to the diagram technique similar to the one discussed above, whose elements are

$$\begin{array}{ccc} \boxed{1} & \boxed{2} & \boxed{3} \\ (\kappa^{-1})_{\alpha\beta} & \kappa^{(p)}_{\alpha_1 \ldots \alpha_p} & (\kappa^{-1})_{\alpha_1\alpha_2} \end{array} \tag{84}$$

Matrix κ is defined by equation (50). The symmetric tensors of pth order $\kappa^{(p)}$ as well as $f^{*(p)}/T$ entering into the expression for $\Gamma^{(p)}$ (equation 81), which characterize the system of chemical bonds and that of separate units, respectively, are as follows

$$\kappa^{(n)}_{\alpha_1 \ldots \alpha_n} \equiv \sum_{l} l_1 \ldots l_n C(\boldsymbol{l}), \qquad \frac{f^{*(n)}_{\alpha_1 \ldots \alpha_n}}{T} \equiv \frac{1}{T} \frac{\partial^n f^*(\boldsymbol{\rho})}{\partial \rho_1 \ldots \partial \rho_n} \tag{85}$$

As distinct from equation (81), to the element 1 there corresponds not a point but a line whose ends are marked by colors α and β. The first of these colors should be erased in accordance with the above described color-erasing procedure, while the second one should be retained. As a result, diagrams with colored apices are obtained similar to those depicted in Figure 4(a). The tensor $\gamma^{(p)}$, to which here a gray p-gon corresponds, can be found by summing the contributions of all cactuses with p external colored apices in accordance with the rules formulated for the diagram technique of equation (81). Thus, for instance, the diagram expression for $\gamma^{(5)}$ is presented in Figure 4(b), where Σ represents summation over all the different ways of coloring the cactus.

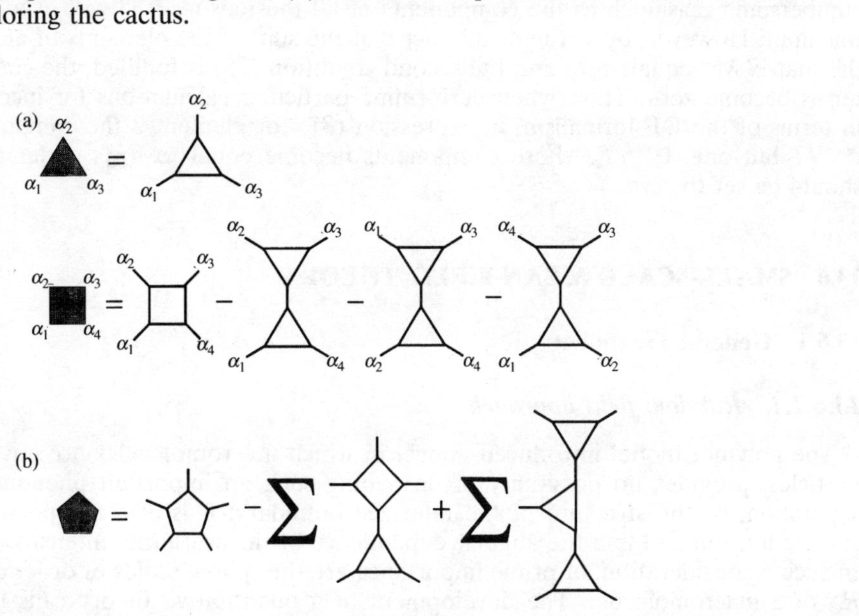

Figure 4 Elements of the diagram technique (a) and diagram expressions for the coefficient of expansion (equation 79).

When calculating the coefficients L_p in terms of the RF formalism, the diagram technique (81) remains the same as that discussed above with the only difference being that other quantities will now correspond to the diagram elements

$$\boxed{1} \qquad \boxed{2} \qquad \boxed{3}$$
$$\varphi_\alpha \qquad G^{(p)}_{\alpha_1\ldots\alpha_p}=(-1)^{p-1}\kappa^{(p)}_{\alpha_1\ldots\alpha_p}-B^{(p)}_{\alpha_1\ldots\alpha_p} \qquad G^{+}_{\alpha_1\alpha_2} \tag{86}$$

Here φ_α are the components of the eigenvector of the matrix $\mathbf{G}=\mathbf{C}^{-1}-\kappa$ corresponding to the eigenvalue $\Lambda=0$, whereas $G^{+}_{\alpha_1\alpha_2}$ represents the elements of the matrix $\mathbf{G}^{+}$ pseudoinverse to the matrix $\mathbf{G}$. The components of the tensor $\mathbf{B}^{(p)}$ with p colored external apices, describing volume interactions in the system of separate units, can be written down immediately using the diagram technique (84) under the following correspondence of diagram and analytical elements

$$\boxed{1} \qquad \boxed{2} \qquad \boxed{3}$$
$$(\mathbf{C}^{-1})_{\alpha\beta} \qquad f^{*(p)}_{\alpha_1\ldots\alpha_p}/T \qquad (\mathbf{C}^{-1})_{\alpha_1\alpha_2} \tag{87}$$

Which of the above discussed approaches is more appropriate to calculate the coefficients L_p is completely dictated by a researcher's requirements. When constructing phase diagrams of a concrete polymer sample investigated under different conditions (for instance, in various multicomponent solvents), preference should be given to the formalism (81). The reason is that the most tedious procedure of finding the explicit form of tensors $\gamma^{(p)}$ should be performed in this case only once. On the other hand, the advantages of the formalism (86) is especially pronounced in the case when thermodynamic properties of reaction systems are examined in the processes of polymer synthesis. In such systems, physical parameters (temperature, pressure) generally remain unaltered, whereas the SCD of the polymer formed changes continuously in the course of the process. The random-field formalism proves to be exceptionally convenient when considering the thermodynamics of incompressible systems since all the components of tensors $\mathbf{B}^{(p)}$ with $p\geq 3$ become zero in terms of such an approximation (63). The only exception is the tensor $\mathbf{B}^{(2)}\equiv\mathbf{B}$, whose components are determined by the simple expression

$$B_{\alpha\beta}=\bar{\rho}\left[(\chi^{-1})_{\alpha\beta}-\sum_{\gamma}(\chi^{-1})_{\alpha\gamma}\sum_{\delta}(\chi^{-1})_{\beta\delta}\Big/\sum_{\gamma}\sum_{\delta}(\chi^{-1})_{\gamma\delta}\right] \tag{88}$$

The DF formalism, if applied to limit transition to an incompressible system, is more cumbersome inasmuch as the components of all the tensors $\mathbf{f}^{*(p)}$ with $p\geq 2$ tend to infinity in this limit. However, by virtue of the fact that the sum of the elements of any row or column of the matrix $\mathbf{M}^{+}$ equals zero and the second condition (75) is fulfilled, the coefficients of singular terms become zero. Thus, when performing particular calculations for incompressible systems in terms of the DF formalism, in expression (81) for element 2 the components of all tensors $\mathbf{f}^{*(p)}/T$ but one, $\mathbf{f}^{*(2)}/T$, whose components become equal to $v\chi_{\alpha\beta}$ under this approximation, should be set to zero.

13.6 SMALL-SCALE MEAN-FIELD THEORY

13.6.1 General Treatment

13.6.1.1 Random field approach

The polymer model introduced above, in which macromolecules are envisaged as pointwise particles, provides no possibility of describing such an important phenomena as microphase separation, or the structure of the interface boundary; it is also not possible to estimate the surface tension and find the angular dependence of the scattering intensity in polymer systems. In such a consideration, of prime importance are the spatial scales of order of characteristic size R_G of a macromolecule. The development of a quantitative theory calls for new approaches capable of taking into account not only the chemical size and composition $\boldsymbol{l}$ of a macromolecule, but also its detailed microstructure σ.

The most general known approach of such a kind to the description of equilibrium properties of amorphous multicomponent polymer systems[113,114] makes it possible within the context of a unified treatment to take into consideration all the above mentioned effects for arbitrary polymer samples involving linear macromolecules of any microstructure. Such a high degree of universality within the framework of a unified approach becomes possible owing to the incorporation of the functional integration method extensively applied in contemporary theoretical physics. Nowadays this method is successfully employed also to tackle various problems of the theory of polymers.[115,116] Despite the fact that such current theoretical approaches might appear at first glance to be too sophisticated (to be more precise, too unfamiliar) for those who are not experts in this field, the resultant formulae normally prove to be rather simple and do contain only measurable parameters. The last circumstance opens fresh opportunities for experimental verification at the quantitative level of the main theoretical results. These latter will be discussed in subsequent sections, while here key ideas of the general approach[113,114] will be briefly formulated.

Dealing with this kind of problem, one is supposed to characterize the sample of interest by the numbers $n(\sigma)$ of macromolecules with a given microstructure σ. The synthesis conditions do not fix unambiguously the set $\{n(\sigma)\}$ of these numbers, but specify only the probability distribution of various sets $\{n(\sigma)\}$. So, to give a thermodynamic account of a system, it is necessary to perform the averaging of two types: over the Gibbs measure at fixed set $\{n(\sigma)\}$ and then over different sets $\{n(\sigma)\}$ with the measure specified by the synthesis conditions. As pointed out in the theory of disordered systems,[117] the free energy F of the system rather than its partition function is a self-averaging quantity. This means that the experimentally observed free energy $F(\{n(\sigma)\})$ of a particular (typical) sample coincides in the thermodynamic limit with the quantity F obtained by averaging $F(\{n(\sigma)\})$ over all sets $\{n(\sigma)\}$. To describe the phenomena occurring at space scales $R \leq R_G$, it is essential to choose a concrete conformational polymer model. For the sake of simplicity, we will discuss a free-joined model characterizing flexible chain polymers, which is defined by the conditional probability $\lambda(\mathbf{r})$ for the vector connecting by a chemical bond a pair of neighboring units to be equal to $\mathbf{r}$.

In the absence of volume interactions between units, the partition function of a macromolecule of a given microstructure σ with units located in the external field $\boldsymbol{H}^e(\mathbf{r})$ will read

$$Q(\sigma, [\boldsymbol{H}^e]) = \prod_n \int d\mathbf{r}_n \exp[-H^e_{\sigma_n}(\mathbf{r}_n)/T] \prod_k \lambda(\mathbf{r}_k - \mathbf{r}_{k+1}) \tag{89}$$

where the first product is over the monomer units and the second one is over the chemical bonds between neighboring units of a macromolecule. Inasmuch as the free energy F is estimated with an accuracy of the additive arbitrary constant, the quantity F_{CB} for such a system of chemical bonds is conveniently reckoned from its value at $\boldsymbol{H}^e = \mathbf{0}$

$$F_{\mathrm{CB}}(\{n(\sigma)\}, [\boldsymbol{H}^e]) = -T \sum_\sigma n(\sigma) \ln \{Q(\sigma, [\boldsymbol{H}^e])/V\} \tag{90}$$

Owing to self-averaging of the free energy, expression (90) coincides with the quantity $F_{\mathrm{CB}}[\boldsymbol{H}^e]$ arrived at by the replacement in equation (90) of $n(\sigma)$ by their averaged (in accordance with the synthesis conditions) values $\bar{n}(\sigma) = \mathcal{P}(\sigma)\Pi$ (see Section 13.3).

Each macromolecule makes an additive contribution to the expression (90) because we are obviously dealing with their ideal gas. No such additivity will be the case, of course, if the volume interactions are accounted for. It is common knowledge that the problem of finding the partition function of a system of interacting particles can be formally reduced to an analogous one for a system of noninteracting particles but residing in a random field $\boldsymbol{\eta}(\mathbf{r})$[117]

$$\exp[-F(\{n(\sigma)\}, [\boldsymbol{H}^e])/T] = \int D\boldsymbol{\eta} \exp\{-[F_{\mathrm{CB}}(\{n(\sigma)\}, [\boldsymbol{H}^e + \boldsymbol{\eta}]) + \Omega^*[\boldsymbol{\eta}]]/T\} \tag{91}$$

The function $-\Omega^*[\boldsymbol{\eta}]$ denotes the field $\boldsymbol{\eta}(\mathbf{r})$ energy. When density fluctuations in the system of separate units are disregarded, this function is defined by the relationships

$$\Omega^*[\boldsymbol{\eta}] = -\int d\mathbf{r} P^*(\boldsymbol{\rho}(\mathbf{r})), \qquad \eta_\alpha(\mathbf{r}) = \mu^*_\alpha(\boldsymbol{\rho}(\mathbf{r})) \tag{92}$$

Owing to self-averaging of the functional F_{CB}, it becomes possible to substitute in formula (91) the quantity $F_{CB}(\{n(\sigma)\}, [\boldsymbol{H}^e + \boldsymbol{\eta}])$ with its value averaged over its microstructure

$$F_{CB}[\boldsymbol{H}^e + \boldsymbol{\eta}] = -\Pi T \sum_{\sigma} \mathcal{P}(\sigma) \ln \{Q(\sigma, [\boldsymbol{H}^e + \boldsymbol{\eta}])/V\} \tag{93}$$

As a result, the right-hand part of equation (91) is no longer governed by the specific set $\{n(\sigma)\}$. This testifies to the fact that the allowance made for the volume interactions does not violate the self-averaging property of the free energy.

Given that the density fluctuations are neglected when calculating the integral (91), recourse can be made to the steepest descent method which yields

$$F[\boldsymbol{H}^e] = F_{CB}[\boldsymbol{H}^e + \boldsymbol{\eta}] + \Omega^*[\boldsymbol{\eta}] \tag{94}$$

where the field $\boldsymbol{\eta}(\mathbf{r})$ is obtainable from the extremum condition of the right-hand part of equation (94)

$$\rho_\alpha(\mathbf{r}) = \delta F_{CB}[\boldsymbol{H}^e + \boldsymbol{\eta}]/\delta\eta_\alpha(\mathbf{r}) \tag{95}$$

The above method appears to be highly convenient when treating effects associated with density distributions inhomogeneous in space. However, to deal with these effects, another approach is normally invoked in the literature,[118] consisting of the introduction of the density functional.

13.6.1.2 Density functional method

This method rests on the property that the partition function of a system situated within the external field is a generating function of the probability distribution of various space configurations of densities $\{\boldsymbol{\rho}(\mathbf{r})\}$

$$\exp(-F[\boldsymbol{H}^e]/T) = \int \mathrm{D}\boldsymbol{\rho} \exp\left[-(\mathcal{F}[\boldsymbol{\rho}] + \int \mathrm{d}\mathbf{r}\boldsymbol{H}^e(\mathbf{r})\boldsymbol{\rho}(\mathbf{r}))/T\right] \tag{96}$$

The function $\mathcal{F}[\boldsymbol{\rho}]$ denotes the conditional free energy of the specified configuration $\boldsymbol{\rho}(\mathbf{r})$

$$\mathcal{F}[\boldsymbol{\rho}] = \int \mathrm{d}\mathbf{r} f^*(\boldsymbol{\rho}(\mathbf{r})) - T\mathcal{S}_L[\boldsymbol{\rho}] \tag{97}$$

The function $f^*(\boldsymbol{\rho})$ has been determined earlier (equation 24), while the conditional free energy of a system of chemical bonds, $-T\mathcal{S}_L[\boldsymbol{\rho}]$, is expressed through the so called 'Lifshitz' entropy'[119] $\mathcal{S}_L(\boldsymbol{\rho})$

$$\exp \mathcal{S}_L[\boldsymbol{\rho}] = \int D\boldsymbol{H} \exp\left[\int \mathrm{d}\mathbf{r}\boldsymbol{H}(\mathbf{r})\boldsymbol{\rho}(\mathbf{r}) - F_{CB}[\boldsymbol{H}]/T\right] \tag{98}$$

Equation (96) for the free energy may be reduced to equation (91) if a series of consecutive procedures is carried out. The following procedure is necessary: substitute expression (97) for (98) and then (96) for (97); change the order of the integration over $\boldsymbol{\rho}(\mathbf{r})$ and $\boldsymbol{H}(\mathbf{r})$; neglect density fluctuations of the system of separate units to calculate, via the steepest descent method, the internal integral over $\boldsymbol{\rho}(\mathbf{r})$; and, finally, replace the variable of integration $\boldsymbol{H}(\mathbf{r}) = \boldsymbol{H}^e(\mathbf{r}) + \boldsymbol{\eta}(\mathbf{r})$.

The calculation of the integral (96) by means of the steepest descent method can be conducted in two stages. In the first stage, the Lifshitz entropy (98) should be found. The steepest descent method can be used to calculate the integral (97) under high densities when the number of units inside the coil proves to be sufficiently large. In this case the field $\boldsymbol{H}$ can be found from the extremum conditions

$$\rho_\alpha(\mathbf{r}) = \delta F_{CB}[\boldsymbol{H}]/\delta H_\alpha(\mathbf{r}) \tag{99}$$

$$T\mathcal{S}_L[\boldsymbol{\rho}] = \int \mathrm{d}\mathbf{r}\boldsymbol{H}(\mathbf{r})\boldsymbol{\rho}(\mathbf{r}) - F_{CB}[\boldsymbol{H}] \tag{100}$$

Substituting the obtained expression of the Lifshitz entropy into equation (96) and calculating the integral over $\boldsymbol{\rho}$ by virtue of the steepest descent method, we will obtain the final formula for the free energy

$$F[\boldsymbol{H}^e] = \mathscr{F}[\boldsymbol{\rho}] - \int \mathrm{d}\mathbf{r}\boldsymbol{H}^e(\mathbf{r})\boldsymbol{\rho}(\mathbf{r}) \tag{101}$$

This expression holds in the mean-field approximation where the dependence of density $\boldsymbol{\rho}$ on the field $\boldsymbol{H}^e$ is found from the equation resulting from relationship (98) upon the replacement in it of $\boldsymbol{H}(\mathbf{r})$ by the sum of the external field $\boldsymbol{H}^e(\mathbf{r})$ and the self-consistent field $\boldsymbol{\mu}^*(\boldsymbol{\rho})$.

It is readily seen that the final formulae for the free energy derived via the two different methods, RF and DF, coincide in the mean-field approximation if use is made of equations (92) and (22). It should be emphasized that this approximation substantially differs from that discussed in Section 13.5. The distinction consists in the markedly smaller scale of those spatial inhomogeneities which the small-scale mean-field theory (SSMFT) discussed here enables one to describe as distinct from the LSMFT.

13.6.2 Angular Dependence of Scattering Intensity

Static scattering of light or neutrons from polymer systems is a powerful tool for the investigation of their equilibrium properties.[120] The treatment of the data obtained by means of this method for heteropolymer blends and solutions provides important information on their chemical structure and also enables one to determine experimentally the location of the spinodal on a phase diagram. Because the scattering is known to be controlled by unit density fluctuations, their description becomes an important theoretical problem. In terms of the fluctuation theory, allowance is made for random deviations $\Delta\rho_\alpha(\mathbf{r}) = \rho_\alpha(\mathbf{r}) - \langle\rho_\alpha(\mathbf{r})\rangle$ ($\alpha = 1, \ldots, m$) of densities $\rho_\alpha(\mathbf{r})$ of units of all types from their average values $\langle\rho_\alpha(\mathbf{r})\rangle$ at the point $\mathbf{r}$ of three-dimensional space. The probabilities of such fluctuations $\Delta\boldsymbol{\rho}(\mathbf{r})$ are defined by the Gibbs measure, $\exp(-\mathscr{F}[\boldsymbol{\rho}]/T)$, where the function $\mathscr{F}$ has been introduced above (equation 97). By averaging over this probability distribution the product of density fluctuation at two arbitrary points $\mathbf{r}'$ and $\mathbf{r}''$, we can determine the two-point correlation function

$$S_{\alpha\beta}(\mathbf{r}', \mathbf{r}'') \equiv \langle\Delta\rho_\alpha(\mathbf{r}')\Delta\rho_\beta(\mathbf{r}'')\rangle = -T\frac{\delta^2 F[\boldsymbol{H}^e]}{\delta H^e_\alpha(\mathbf{r}')\delta H^e_\beta(\mathbf{r}'')} \tag{102}$$

where the second equality follows from equation (96). In the absence of an external field, when all $H^e_\alpha = 0$, the system is spatially homogeneous so that the elements in equation (102) of the correlation matrix $\mathbf{S}$ are dependent only upon the difference $\mathbf{r} = \mathbf{r}' - \mathbf{r}''$ of the three-dimensional vectors $\mathbf{r}'$ and $\mathbf{r}''$. Systems of such a kind, with $\boldsymbol{H}^e = \mathbf{0}$, will be discussed later.

When theoretical expressions for the Fourier component $\tilde{S}_{\alpha\beta}(q)$ of elements $S_{\alpha\beta}(\mathbf{r})$ of the correlation matrix are possible to obtain, the scattering intensity I at any angle θ of the incident radiation with the wavelength λ can be calculated by the known formula[121]

$$I(q) = \sum_\alpha \sum_\beta a_\alpha a_\beta \tilde{S}_{\alpha\beta}(q); \qquad q = \frac{4\pi}{\lambda}\sin\frac{\theta}{2} \tag{103}$$

where a_α is the scattering length of an α-type unit. In view of the spatial isotropy of the system, the matrix $\mathbf{S}$ and, consequently, the scattering intensity of equation (103) will be dependent only on the modulus q of the wave vector $\mathbf{q}$.

The calculation of the second derivative (102) of the function (101) leads in terms of the SSMFT to the simple matrix expression

$$\tilde{\mathbf{S}}^{-1}(q) = \tilde{\kappa}^{-1}(q) - \mathbf{C} \tag{104}$$

connecting the Fourier transform of the correlation matrix $\tilde{\mathbf{S}}$ and that of structural matrix $\tilde{\kappa}$. The latter represents, by definition, the second derivative of the functional $F_{\mathrm{CB}}[\boldsymbol{H}^e]$ with σ^2 of the composition distribution and becomes zero for a monodisperse sample. When dealing with a terpolymer it appears more convenient to choose, as independent elements of the matrix respect to the components of the field $\boldsymbol{H}^e$. From the explicit expression (90) for this function, the microscopic representation of matrix $\tilde{\kappa}$ elements follows

$$\tilde{\kappa}_{\alpha\beta}(q) = \frac{\Pi}{V}\sum_{\sigma}\mathscr{P}(\sigma)\left\langle\sum_{k}\sum_{l} e^{i\mathbf{q}(\mathbf{r}_k-\mathbf{r}_l)}\right\rangle_{\sigma} \tag{105}$$

Here angular brackets denote the procedure of averaging over the Gibbs measure of the conformational set of an individual macromolecule with the microstructure σ, while the inner summation is over all its units k and l whose types are $\sigma_k = \alpha$ and $\sigma_l = \beta$, respectively. Along with $\tilde{\kappa}$, the right-hand part of equation (104) comprises also the matrix of direct correlation functions **C** with elements (48) which characterize volume interactions within the system of separate units.

The elements of the structural matrix (105) are expressible through two-point chemical correlators (equation 10). To this end, it proves to be convenient at $\alpha = \beta$ to separate out from the inner double sum of equation (105) the items corresponding to the same units $k = l$ into an isolated term

$$\tilde{\kappa}_{\alpha\beta}(q) = \bar{\rho}_\alpha\delta_{\alpha\beta} + \tilde{\theta}_{\alpha\beta}(q) \tag{106}$$

where $\bar{\rho}_\alpha = \bar{\rho}X_\alpha$ is the density of α-type units averaged over space. The second term in formula (106) can be expressed in a trivial way[114]

$$\tilde{\theta}_{\alpha\beta}(q) \equiv \tilde{\theta}^{(2)}_{\alpha\beta}(q) = \bar{\rho}[W^{(2)}_{\alpha\beta}(x) + W^{(2)}_{\beta\alpha}(x)], \quad \text{where} \quad x = \tilde{\lambda}(q) \tag{107}$$

through the generating formation $W^{(2)}_{\alpha\beta}(x)$ of two-point chemical correlators (equation 10)

$$W^{(2)}_{\alpha\beta}(x) = \sum_{k=0}^{\infty} Y^{(2)}_{\alpha\beta}(k)x^{k+1} \tag{108}$$

in which the argument x is replaced by the Fourier transform $\tilde{\lambda}(q)$ of the probability $\lambda(r)$ to find a couple of neighboring monomer units of a macromolecule at a distance r from one another. As for the function $\tilde{\lambda}(q)$, it is well known that, regardless of the choice of the conformational microscopic model, this function is positive and alters as q increases between values $\tilde{\lambda}(0) = 1$ and $\tilde{\lambda}(\infty) = 0$, decaying at the characteristic scale $\sim 1/a$. Inasmuch as any function $\tilde{\lambda}(q)$ satisfying these conditions leads to the same final results, it is convenient to choose $\tilde{\lambda}(q) = \exp(-q^2a^2/6)$.

Apart from the above straightforward derivation of the fundamental expression (104), at least three more ways of obtaining it are available in the literature. The first of them goes back to papers by de Gennes,[122] Leibler,[123] and Eruchimovich,[124] who proceeded from the random phase approximation (RPA) method. Benoit and co-workers,[125] as well as Curro and Schweizer,[126] have extended the Ornstein–Zernike equation, known in the theory of low-molecular solvents, to macromolecular systems, whereas the third approach makes use of the Edwards Hamiltonian.[115,127,128] A brief overview of these three ways of deriving the fundamental relationship (104) is given in a review.[129] The fact that several formally different approaches are conducive to the same result comes as no surprise because all these approaches take into account fluctuations in the Gaussian approximation.

For the model of an incompressible lattice liquid (see Figure 2), a general expression for the Fourier transform of correlation matrix elements (102) can be presented as follows[130,131]

$$\tilde{S}_{\alpha\beta} = (\tilde{\mathbf{M}}^{-1})_{\alpha\beta} - \sum_{\gamma}(\tilde{\mathbf{M}}^{-1})_{\alpha\gamma}\sum_{\delta}(\tilde{\mathbf{M}}^{-1})_{\beta\delta}\Big/\sum_{\gamma}\sum_{\delta}(\tilde{\mathbf{M}}^{-1})_{\gamma\delta} \tag{109}$$

Here the dependence of the elements of matrix $\tilde{\mathbf{M}}$ on q is defined by equation (74) where κ is substituted by $\tilde{\kappa}$ (equation 105). Summation of the $\tilde{S}_{\alpha\beta}$ over the index α or β yields zero and thus the symmetric matrix $\tilde{\mathbf{S}}$ possesses only $m(m-1)/2$ independent elements. In a particular case, when the number m of monomer unit types equals two, a general equation (103) for the scattering intensity, taking due regard of equation (21), will look familiar[130,131]

$$I(q) = \frac{(a_1 - a_2)^2}{D(q)}, \qquad D(q) = \frac{\tilde{\kappa}_{11} + \tilde{\kappa}_{22} + 2\tilde{\kappa}_{12}}{-\tilde{\kappa}_{12}^2 + \tilde{\kappa}_{11}\tilde{\kappa}_{22}} - 2v\chi_{12} \tag{110}$$

Hence, no problem will be found in constructing for any polymer system the theoretical dependence of the scattering intensity on the angle θ, provided that it becomes possible to obtain for this system an explicit expression for the Fourier transform of the structural matrix $\tilde{\kappa}$ (equation 105). If so, the shape of the curve $I(q)$ for heteropolymer samples with different molecular structures may vary qualitatively. As an illustration, Figure 5 depicts two such dependences for the melt of a binary copolymer with a different character of sequence distribution in macromolecules. In the case of the statistical copolymer the curve $I(q)$ decays monotonically from a maximal value, $I(0)$, up to zero whereas for a block copolymer the appearance of the maximum on this curve at a certain $q = q^* \neq 0$ can be expected. At a specific value of a system parameter corresponding to the spinodal, the value $I(q)$ at one of these maxima will become infinity. According to which of them it occurs with, the spatially homogeneous state will lose its stability at the spinodal with respect to density fluctuations either with the zeroth wave vector or with a finite one $q = q^*$.

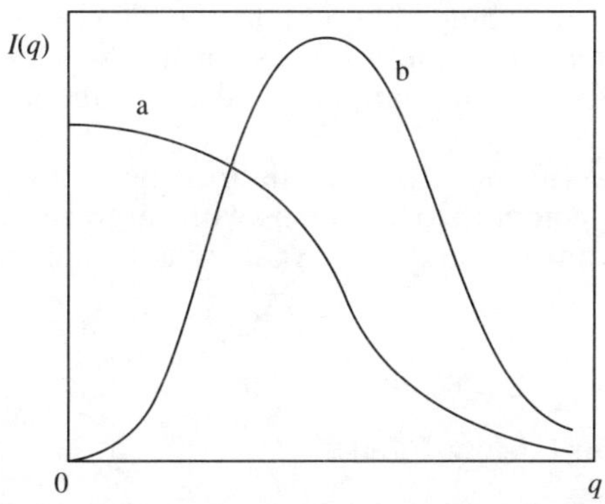

Figure 5 Characteristic forms of dependences of scattering intensity upon the modulus of the scattering wave vector q.

A simple model system where such transitions are possible has been theoretically studied in detail by Leibler.[131] He examined an incompressible melt of a diblock copolymer (equation 12a) with fixed numbers $l_1 = lX_1$, $l_2 = lX_2$ of monomer units M_1 and M_2 in each block and derived for this system at $l \rightarrow \infty$ simple expressions for the elements of the structural matrix

$$\begin{gathered} \tilde{\kappa}_{11} = lg(X_1, y), \qquad \tilde{\kappa}_{22} = lg(X_2, y) \\ \tilde{\kappa}_{12} = l[g(1, y) - g(X, y) - g(X_2, y)]/2 \end{gathered} \tag{111}$$

which are expressible through the Debye function

$$g(X, y) = \frac{2}{y^2}(Xy - 1 - e^{-Xy}), \qquad y = q^2 R_G^2, \qquad R_G^2 = \frac{a^2 l}{6} \tag{112}$$

where R_G represents a gyration radius of the Gaussian coil consisting of l monomer units. With a decrease in the angle of scattering θ, *i.e.* at $q \rightarrow 0$, the scattering power vanishes due to the system incompressibility[132] in accordance with the law $I(q) \simeq q^2$, while in the region of large q, function $I(q)$ tends to zero proportionally to $1/q^2$. At $q = q^* \simeq 1/R_G$ the scattering intensity has a maximum whose height increases as the parameter χ grows and, eventually, at a certain value $\chi = \chi_s$ it becomes infinity along with the correlation radius ξ of fluctuations in conformity with the asymptotical formulae

$$I(q^*) \simeq (\chi_s - \chi)^{-1} \qquad \xi \simeq al^{-1/2}(\chi_s - \chi)^{-1/2} \tag{113}$$

The influence of the compressibility effect in a system of interest on the angular dependence of the scattering intensity has been analyzed by Freed and co-workers,[133,134] who showed that this remains finite at $q = 0$ but retains the same form as in the case of incompressible systems. Within the area of large q, this curve exhibits asymptotical behavior $I(q) \sim 1/q^2$ only at rather high values of q. The compressibility effect acts upon the system behavior in the strongest way in the vicinity of phase transitions.

Of particular interest are studies[135,136] where the effect of heteropolymer polydispersity is accounted for. For this purpose it has been suggested[135] to use equation (111), characterizing a monodisperse diblock copolymer, by substituting in them the Debye function (112) by its number-averaged value. Later it was shown[136] that to this end the weight-averaged value of this function should be used. This assertion, proved by the authors[136] for a concrete model, follows in the general case directly from the definition (105) of the structural matrix.

The question has been theoretically examined[137,138] as to which alterations in scattering intensity cause the addition to the incompressible melt of a monodisperse diblock copolymer (equation 12a) of a homopolymer with units M_1. It was found[137] that with a low concentration of the latter these alterations are insignificant. However, as the homopolymer concentration increases, such a system can undergo a transition into a micellar phase.[138] A numerical calculation of the correlation matrix (104), which defines the scattering intensity (103), for a monodisperse melt of multiblock binary copolymers has been reported in a paper[139] in which the critical value of the quantity $\chi_s l$ was predicted to increase with the growth of the number of blocks in the macromolecule.

De Gennes[140] has analyzed the effect of polydispersity on the shape of curve $\boldsymbol{I}(q)$ for a mixture of macromolecules containing different length blocks. He has shown that, under a random distribution of the length of these blocks, such a structural disorder can lead to the situation where at certain values of the parameters the maximum disappears on the curve $I(q)$ at a finite value $q^* > 0$.

Explicit expressions for the function $I(q)$ for the treatment of experimental data on the light scattering in some particular compressible systems with three and four types of unit have been found by virtue of the diagram technique.[125] Analogous results are also obtainable using the RPA method.[141]

13.6.3 The Spinodal and the Lifshitz Point

The loss of stability of the spatially homogeneous state happens on a spinodal where the correlation functions (102) and the correlation length ξ diverge. The condition for the spinodal transition is, similarly to expression (104), the presence of the sole non-negative root of the equation

$$D(q) \equiv |\tilde{\kappa}^{-1}(q) - \mathbf{C}| = 0 \tag{114}$$

Generally speaking, this root can have a multiplicity of either 1 or 2. In the first case, the stability loss occurs at the zeroth wave vector ($q = 0$), while in the second case it takes place at the vector distinct from zero ($q = q^* \neq 0$). To two such cases there correspond two different regions on the spinodal hypersurface which are separated by a hyperline of the Lifshitz points, known from the thermodynamics of low-molecular compounds.[142] The root $q = 0$ of multiplicity 2 of equation (114) corresponds to these specific points. By the very definition of such a root, along with the function $D(q)$ its derivative must vanish as well. This condition together with equation (114) defines the region of the spinodal hypersurface where the spatially homogeneous state, remaining stable with respect to the fluctuations with the zeroth wave vector, loses its stability on the non-zeroth one. In order to find such spinodal regions in the case of incompressible systems, one should analyze instead of equation (114) the equation

$$\sum_{\alpha} \sum_{\beta} \tilde{A}_{\alpha\beta} = 0 \tag{115}$$

which ensues from equation (73), provided that in the latter $A_{\alpha\beta}$ are replaced by elements of matrix $\tilde{\mathbf{A}}$ adjoint for matrix $\tilde{\mathbf{M}}$.

The above condition for finding the spinodal can be reformulated in terms of the spectral problem of the matrix $\tilde{\mathbf{S}}^{-1}$. Because the determinant of any matrix is equal to the product of its eigenvalues, to the condition (104) there corresponds a vanishing of one of them, *i.e.* the minimum one, $\Lambda(q)$. This is obtainable from the condition of the existence of a nontrivial solution of the following set of homogeneous linear equations

$$\sum_{\beta} (\tilde{\kappa}^{-1}_{\alpha\beta} - C_{\alpha\beta})\tilde{\psi}_{\beta} = \Lambda\tilde{\psi}_{\alpha}, \qquad \sum_{\alpha} \tilde{\psi}^{2}_{\alpha} = 1 \tag{116}$$

Here the eigenvalue Λ and the components $\tilde{\psi}_{\alpha}$ of the corresponding normalized eigenvector are controlled by q because of the dependence on q of the matrix $\tilde{\kappa}$. It is readily noticed that Λ and $\tilde{\psi}_{\alpha}$ at $q=0$ reduce to the quantities λ and ψ_{α}, introduced in Section 13.5.2.2. As for the components of the eigenvector, they are expressed directly through cofactors $\tilde{A}_{\alpha\beta}(q)$ in the same way as in equation (75) when the compressibility is neglected.

It is noteworthy that the spinodal condition as well as the eigenvalues problem can be formulated in terms of the matrix $\mathbf{C}^{-1} - \tilde{\kappa}$, as has been done in Section 13.5. To do so, the matrix κ in all equations of that section should be substituted by $\tilde{\kappa}(q)$.

Thus, the problem of the theoretical determination of the spinodal with allowance for equations (114) and (115) reduces to finding the Fourier transform $\tilde{\kappa}$ of the structural matrix which is governed only by the microstructure and polydispersity of a polymer sample. A number of papers are available in the literature where this problem has been solved for some simple models, primarily in relation to the derivation of equations for the scattering intensity.[125–141]

The first who explicitly identified the Lifshitz point in polymer systems were, apparently, Broseta and Fredrickson.[143] They analyzed the dependence on q of the structure factor D (equation 110) of homogeneous melt $(A^{al} + B^{al} + A^{l}B^{l})$ comprising two different homopolymers and a diblock copolymer. As a result of such an analysis, the conditions were found for the existence of the Lifshitz point as well as its location on the phase diagram of the quasibinary system under consideration, characterized by four parameters: the volume fractions of the homopolymers; the ratio α of their degree of polymerization to that of the diblock copolymer; and the incompatibility degree $h = \chi\alpha l$ where χ is the Flory–Huggins parameter.

Comparatively recent theoretical works[113,114,144] have predicted the possibility of the existence of the Lifshitz point on the phase diagram of heteropolymer systems with the Markovian statistics of sequence distribution. In the study[144] the authors proceeded from the assumption of incompressibility of the system and monodispersity of a multiblock copolymer. They made recourse to the Edwards 'collective variables' method,[115] having taken volume interactions into account by means of supplementing the free energy by a quadratic with respect to the density of units item with a nonlocal parameter of their interaction. Within the framework of such a formal approach, where the parameter χ is dependent on the coordinates, they constructed a spinodal and showed the existence of the Lifshitz point on it. The very fact of its occurrence in terms of the model in hand is due to the nonlocal character of the interactions inasmuch as when this nonlocality scale approaches zero the Lifshitz point is known to vanish. However, it can appear in the traditional theory of heteropolymers whose units interact in a local manner provided that the system is compressible. For the first time this result has been arrived at in studies[113,114] where simple analytical expressions have been derived for the spinodal and the Lifshitz point for any Markovian binary copolymer with an arbitrary character of physical interactions. These expressions involve as parameters the elements of two matrices, of which the first describes the polymer chain microstructure and the second one is a matrix of direct correlation functions (48).

13.6.4 Critical Points

In melts and blends with the participation of block copolymers, there is a possibility of the existence of thermodynamically stable periodic spatial structures with a different symmetry of the crystal lattice. The free energy minimum corresponding to these structures can be located in the vicinity of the principal minimum which corresponds to the spatially homogeneous state and can even merge with it at the critical point under certain values of external parameters. These values will lie on the hyperline obtainable from two conditions (equation 80), which now should be supplemented with one more equation, $\partial\Lambda/\partial q = 0$, for determining the modulus of the wave vector $q = q^*$ of a superstructure. Expressions for coefficients L_2, L_3 in the case under consideration are as follows

$$L_2 = \Lambda(q^*), \qquad L_3 = \sum_{\{\mathbf{q}_i\}} \sum_{\{\alpha_i\}} \tilde{\Gamma}^{(3)}_{\alpha_1\alpha_2\alpha_3}(\mathbf{q}_1\mathbf{q}_2\mathbf{q}_3) \prod_{i=1}^{3} \tilde{\psi}_{\alpha_i}(\mathbf{q}_i) \tag{117}$$

where the external summation is over the reciprocal lattice primitive translations (RLPT) of the superstructure. The modulus of all these vectors is the same and equals q^*. This condition substantially restricts the number of potential types of d-dimensional superstructures which are normally classified by the number n of their RLPT vectors. The set of such structures includes at $d = 1$ a lamellar structure ($n = 1$), at $d = 2$ quadratic ($n = 2$) and triangular ones ($n = 3$), and at $d = 3$ the body-centred cubic structure ($n = 6$). Equation (117) differs from the analogous one of equation (54) in that the tensor $\tilde{\mathbf{\Gamma}}^{(3)}$ depends on the wave vectors, unlike $\mathbf{\Gamma}^{(3)}$. When allowance is made for this dependence, which extends also to the eigenvector with components $\tilde{\psi}_\alpha$, all the expressions for the coefficients L_p, describing the states with spatially homogeneous phases, are retained with minor modification. This modification suggests that the procedure of summing in all equations of Section 13.5.4 should be made not only over indices α_i of units but also over their wave vectors $\mathbf{q}_i$ belonging to the discrete set of n RLPTs of one of the above mentioned possible superstructures. With due regard for this modification, it is an easy matter to calculate the necessary coefficients L_p for a system to be in any spatially inhomogeneous state provided that the components of tensor $\tilde{\kappa}^{(p)}$ are known. They have a meaning of the Fourier transform of p-point correlation functions in a system of chemical bonds

$$\tilde{\kappa}^{(p)}_{\alpha_1\ldots\alpha_p}(\mathbf{q}_1 \ldots \mathbf{q}_p) = \frac{\Pi}{V} \sum_{\sigma} \mathscr{P}(\sigma) \left\langle \sum_{\sigma} \exp\left(i \sum_{j=1}^{p} \mathbf{q}_j \mathbf{r}_j \right) \right\rangle \tag{118}$$

Here, as in all the other equations for components of tensors $\tilde{\boldsymbol{\gamma}}^{(p)}(\{\mathbf{q}_i\})$ and $\tilde{\mathbf{\Gamma}}^{(p)}(\{\mathbf{q}_i\})$ of arbitrary order p, the sum of all the wave vectors $\mathbf{q}_1 + \ldots + \mathbf{q}_p$ equals zero. Note that if we put all $\mathbf{q}_j = \mathbf{0}$, the expression (118) for the quantity $\tilde{\kappa}^{(p)}$ can be put into that appearing in equation (84) for the quantity $\kappa^{(p)}$ and then the expansion coefficients L_p will reduce to those derived in Section 13.5.4.

As in the case $p = 2$ (equation 105), it turns out that the components of tensor $\tilde{\kappa}^{(p)}$ can be expressed in a simple manner[114] through generating functions of the chemical correlators of equation (11)

$$W^{(p)}_{\alpha_1\ldots\alpha_p}(x_1, \ldots, x_{p-1}) = \sum_{\{k_i\}} \mathrm{Y}^{(p)}_{\alpha_1\ldots\alpha_p}(k_1, \ldots, k_{p-1}) \prod_{i=1}^{p-1} x_i^{k_i+1} \tag{119}$$

where summing begins with $k_i = 0$. To do this we can, by analogy with equation (106), separate out into individual items the contributions of summing over repeating units.

A simple general algorithm can be put forward enabling one to write down immediately an analytical expression for any component (118) of the tensor $\tilde{\kappa}^{(p)}$ as a sum of several items. To any such item there corresponds a particular partition of p points, each of which is defined by an index α_j and a wave vector (momentum) $\mathbf{q}_j$, into r groups ($1 \le r \le p$), comprising only the points with the same indices. This index will characterize the given group, while the momentum attributed to it will be equal to the sum of the momenta of all points entering into this group. If each such group is now conventionally thought of as a point, there corresponds to this particular partition in terms of the algorithm suggested a particular r-point 'correlator' whose tensor indices and momenta characterize now not individual points but their groups. Having made a summation over contributions $\tilde{\boldsymbol{\theta}}^{(r)}$ from all such partitions we will arrive at the expression for the correlation tensor $\tilde{\boldsymbol{\kappa}}^{(p)}$. The use of this algorithm in the particular case $r = 3$ leads to the expression

$$\tilde{\kappa}^{(3)}_{\alpha_1\alpha_2\alpha_3}(\mathbf{q}_1\mathbf{q}_2\mathbf{q}_3) = \bar{\rho}_{\alpha_1}\delta_{\alpha_1\alpha_2}\delta_{\alpha_2\alpha_3} + \tilde{\theta}^{(2)}_{\alpha_1\alpha_2}(\mathbf{q}_1\mathbf{q}_2)\delta_{\alpha_2\alpha_3} + \tilde{\theta}^{(2)}_{\alpha_2\alpha_3}(\mathbf{q}_2\mathbf{q}_3)\delta_{\alpha_3\alpha_1} + \tilde{\theta}^{(2)}_{\alpha_3\alpha_1}(\mathbf{q}_3\mathbf{q}_1)\delta_{\alpha_1\alpha_2} + \tilde{\theta}^{(3)}_{\alpha_1\alpha_2\alpha_3}(\mathbf{q}_1\mathbf{q}_2\mathbf{q}_3) \tag{120}$$

where the Kroneker symbols take into consideration the similarity of all points of any group with respect to their type.

The expression for an arbitrary component $\tilde{\theta}^{(p)}_{\alpha_1\ldots\alpha_p}(\mathbf{q}_1\mathbf{q}_2 \ldots \mathbf{q}_p)$ of the symmetric tensor $\tilde{\boldsymbol{\theta}}^{(p)}$

represents up to the factor $\bar{\rho}$ the sum of $p!$ items, the first of which is $W^{(p)}_{\alpha_1\ldots\alpha_p}[\tilde{\lambda}(\mathbf{q}_1), \tilde{\lambda}(\mathbf{q}_1+\mathbf{q}_2),\ldots,\tilde{\lambda}(\mathbf{q}_1+\ldots+\mathbf{q}_{p-1})]$, while the remaining items are obtainable from it as a result of various transpositions of indices. Such a symmetrization procedure permits any component of a symmetric tensor $\tilde{\boldsymbol{\theta}}^{(p)}$ to be expressed through the components of a nonsymmetric one $\mathbf{W}^{(p)}$. When $p=2$ this algorithm leads to the expression (107), while for $p=3$ it gives

$$\tilde{\theta}^{(3)}_{\alpha_1\alpha_2\alpha_3}(\mathbf{q}_1\mathbf{q}_2\mathbf{q}_3)=\bar{\rho}[W^{(3)}_{\alpha_1\alpha_2\alpha_3}(b_1,b_3)+W^{(3)}_{\alpha_1\alpha_3\alpha_2}(b_1,b_2)+W^{(3)}_{\alpha_2\alpha_1\alpha_3}(b_2,b_3)$$
$$+W^{(3)}_{\alpha_2\alpha_3\alpha_1}(b_2,b_1)+W^{(3)}_{\alpha_3\alpha_1\alpha_2}(b_3,b_2)+W^{(3)}_{\alpha_3\alpha_2\alpha_1}(b_3,b_1)] \tag{121}$$

where use is made of the designation $b_i\equiv\tilde{\lambda}(\mathbf{q}_i)$. The above expression for L_3 (equation 54) of systems with spatially homogeneous phases follows from equations (118), (120), and (121) if all vectors $\mathbf{q}_i$ in them are put equal to zero.

When calculating the critical point, where the confluence occurs of the spatially inhomogeneous state with the homogeneous one, it is possible to apply the above reported relationships to incompressible systems, in which all components of the tensor $\mathbf{f}^{*(3)}$ would be equated to zero while the components $f^*_{\alpha\beta}/T$ are put equal to $v\chi_{\alpha\beta}$. The same rule works for the calculation of all the remaining coefficients L_p with values $p>3$.

13.6.5 Appearance of the Spatially Inhomogeneous Phase

In the theoretical construction of the cloud-point curves, it used to be assumed that the concentration of units inside the first drops of an incipient phase appearing in the system is the same at all points of these drops. In this case, the problem of finding the cloud-point curve hypersurfaces within the space of the external parameters of a system can be formulated in a trivial way in terms of the generating function $G(\boldsymbol{s})$ (equation 5) of the distribution of polymer molecules for size and composition (equation 3). Given this function, it becomes possible to use equations (144) and (145) to find the density of units of all types in the incipient phase at the moment of its appearance, as well as the values of the parameters at which this phase transition occurs.

The information on the statistical characteristics of a polymer sample contained in the generating function $G(\boldsymbol{s})$ proves to be insufficient to find those regions of the cloud-point curve hypersurface where the formation of the spatially inhomogeneous phase begins. It was demonstrated[114] that, to solve this problem, knowledge is required of generating the function

$$G[\boldsymbol{s}]=\frac{\Pi}{V}\sum_{\sigma}\mathcal{P}(\sigma)Q(\sigma;[-T\ln\boldsymbol{s}(\mathbf{r})]) \tag{122}$$

It has a meaning (up to the factor Π/V) of the partition function Q of an individual molecule with the chemical structure σ located in the external inhomogeneous field $\boldsymbol{H}^e(\mathbf{r})=-T\ln\boldsymbol{s}(\mathbf{r})$, which is averaged over the molecular structure distribution $\mathcal{P}(\sigma)$. The integration in expression (89) for Q, which should be inserted into the right-hand part of equation (122), is performed over the incipient phase volume V'.

The density distribution $\boldsymbol{\rho}(\mathbf{r})$ within the nucleus of the incipient phase can be determined by solving a set of equations

$$\rho_\alpha(\mathbf{r})=s_\alpha(\mathbf{r})\delta G[\boldsymbol{s}]/\delta s_\alpha(\mathbf{r}),$$
$$s_\alpha(\mathbf{r})=\exp\{[\mu^*_\alpha(\boldsymbol{\rho})-\mu^*_\alpha(\boldsymbol{\rho})]/T\} \tag{123}$$

where instead of the ordinary derivative entering into equation (144), here a variational one appears.

To determine cloud points, the expression should be written down for the alteration of the free energy ΔF of the system due to the formation of an incipient phase nucleus against the background of its spatially homogeneous state (principal phase), and then ΔF should be equated to zero. As a result, we arrive at the condition

$$\Delta F\equiv TG[\boldsymbol{s}]-\int_{V'}\mathrm{d}\mathbf{r}[P^*(\boldsymbol{\rho}(\mathbf{r}))-P]=0 \tag{124}$$

which in the case when the nucleus size is large compared with the superstructure period reduces to the condition of the equality of pressure in the incipient and principal phases.

Equations (123) and (124) extend (144) and (145) and become these latter in a particular case when the distribution $\boldsymbol{\rho}(\mathbf{r})$ of densities of quasicomponents does not depend upon $\mathbf{r}$.

13.7 THE INTERFACE PROBLEM

13.7.1 Central Ideas of the Self-consistent Field Approximation

In the thermodynamics of polymers, of primary importance are the problems of the description of the region between two incompatible homogeneous phases and that of establishing quantitative correlations between the surface tension and the statistical characteristics of a polymer sample. The interface tension σ is defined as the excess free energy of a system per unit surface A, which is a function of the distributions of quasicomponent density.

To solve the interface problem, it is normal to resort to various modifications of the self-consistent field method, the key ideas of which are easily understood if illustrated by the simplest example[145–147] of an incompressible melt of two immiscible monodisperse homopolymers with the number of monomer units l_1 and l_2. Initially macromolecule $M_\alpha^{l_\alpha}$ was considered in the absence of volume interactions, but situated in the external field $H_\alpha(\mathbf{r})$. The position of a unit in the polymer chain is conveniently characterized by a continuous variable $0 < s < l_\alpha$. The partition function of such a chain whose initial and final unit are at points $\mathbf{r}_0$ and $\mathbf{r}$ can be presented as an integral over trajectories $r(s)$[115]

$$Q_\alpha(l_\alpha; \mathbf{r}_0\mathbf{r}; [H_\alpha]) = \int_{\mathbf{r}_0}^{\mathbf{r}} D\mathbf{r}(s) \exp\left\{-\int_0^{l_\alpha} \left[\frac{\dot{\mathbf{r}}^2}{3a^2} + \frac{H_\alpha(\mathbf{r}(s))}{T}\right] \mathrm{d}s\right\} \tag{125}$$

This function is obtainable from the solution of the diffusion equation

$$\frac{\partial Q_\alpha}{\partial l_\alpha} = \frac{a^2}{6}\Delta Q_\alpha - \frac{H_\alpha(\mathbf{r})}{T} Q_\alpha, \qquad Q_\alpha(0; \mathbf{r}_0, \mathbf{r}; [H_\alpha]) = \delta(\mathbf{r}_0 - \mathbf{r}) \tag{126}$$

The density of monomer units at point $\mathbf{r}$ is expressed in a simple manner

$$\rho_\alpha(\mathbf{r}) = \frac{1}{Q_\alpha[H_\alpha]} \int_0^{l_\alpha} \mathrm{d}s q_\alpha(s; \mathbf{r}; [H_\alpha]) q_\alpha(l_\alpha - s; \mathbf{r}; [H_\alpha]) \tag{127}$$

through partition functions q_α and Q_α of a molecule with the number of fixed ends equal to 1 or 0. These quantities can be found by means of single and double integration of the partition function (125) over the coordinates of the terminal units. The procedure of closing a set of equations consists in choosing the expression for self-consistent fields $H_\alpha = T\eta_\alpha$ where η_α has been defined above (equation 72). In view of the fact that the field H_α depends only upon the sole coordinate x, this closed set of equations admits an exact analytical solution for the profile of dimensionless concentrations $\Phi_\alpha = v\rho_\alpha$ $(\alpha = 1,2)$

$$\Phi_1(x) = \tfrac{1}{2}\left[1 - \tanh\left(\frac{x}{d}\right)\right], \qquad \Phi_2(x) = \tfrac{1}{2}\left[1 + \tanh\left(\frac{x}{d}\right)\right] \tag{128}$$

where d stands for the interface width. Its dependence on the Flory parameter as well as the analogous dependence on it of surface tension

$$d = (2/6^{1/2})a\chi^{-1/2} \qquad \sigma = 6^{-1/2}\bar{\rho}aT\chi^{1/2} \tag{129}$$

can be arrived at proceeding from simple estimates.[148] To find these, let us consider a typical fluctuation resulting in the penetration of a chain segment involving z units of M_1 from the first-monomer-rich phase into the second-momomer-rich phase. The energy loss $T\chi z$ for the formation of a typical fluctuation is of order T, which yields the assessment $z \simeq 1/\chi$. The width of the interface is of the order of the Gaussian size of such a fragment $d \approx \alpha z^{1/2} \simeq \alpha\chi^{-1/2}$. When the interface energy is being evaluated, it is taken to be a product of two factors, the first of which is the number of units of such penetrating segments $\bar{\rho}dA$ and the second one is the energy of the volume interactions of each of them χT. As a result, the following estimate $\sigma \simeq \bar{\rho}aT\chi^{1/2}$ is arrived at. Analogous results can be also obtained within the framework of the Helfand lattice model.[149] The solution of these problems turns out to be possible if the finiteness of the lengths of macromolecules is neglected. In order to elucidate this approximation let us present the solution of equation (126) for the function $Q(l; \mathbf{r}_0\mathbf{r})$ as an expansion into a series

$$Q(l; \mathbf{r}_0\mathbf{r}) = \frac{1}{l}\sum_{k=0}^{\infty} \exp(-\varepsilon_k l)\psi_k(\mathbf{r}_0)\psi_k(\mathbf{r}) \tag{130}$$

in terms of the eigenfunctions of the equation

$$\left[-\frac{a^2}{6}\Delta^2 + H(\mathbf{r})\right]\psi_k(\mathbf{r}) = \varepsilon_k\psi_k(\mathbf{r}) \tag{131}$$

normalized to the number l of units of a molecule. In the limit $l \to \infty$, the main contribution into expansion (130) is made by the term with $k = 0$, which is the only one to be retained in the approximation of interest. As a result, the expressions for the partition function of a macromolecule and its monomer unit density will read

$$Q \simeq \exp(-\varepsilon_0 l), \qquad \rho(\mathbf{r}) = \psi_0^2(\mathbf{r}) \tag{132}$$

The condition of the applicability of this approximation is the presence of a finite gap between the two eigenvalues ε_0 and ε_1 of equation (131).

13.7.2 Heteropolymers with Strong Unit Interactions

A quantitative description of the interface based on the ideas formulated in the preceding subsection has been performed in a series of works by Noolandi *et al.*[150–152] They considered multicomponent solutions of homopolymers and block copolymers in a low-molecular solvent on the assumption of system incompressibility. This restriction is taken into account by the introduction of an external field w (equation 72) whose value is found from the condition that the overall unit density is constant at any point of the space. Joint numerical solution of the diffusion equations for every type of macromolecule and equations for self-consistent fields (equation 72) acting upon the units enabled the authors[150–152] to obtain many interesting results.

The effect has been examined[150] of the addition of a diblock copolymer (equation 12a) on the decrease of the interfacial tension of an immiscible blend of homopolymers $M_1^{l_1}$ and $M_2^{l_2}$ conditioned by the diminution of the entropy of mixing due to the displacement of molecules of these latter by block copolymer macromolecules from the region of the interface. Localization in it of copolymer junctions results in a decrease in the repulsion energy of different types of unit. This is responsible for the effect mentioned, which is known to be most strongly pronounced in blends of highly incompatible polymers. Another important factor, conducive to the reinforcement of the exclusion of homopolymers from the interface region as the molecular weight of the block copolymer is increased, is the increase in the enthalpy of mixing which is due to the penetration of αth type blocks into the region occupied by units of the αth homopolymer.

A curious consequence of the theoretical consideration is the fact that the interfacial tension is predicted to vanish for some concentration of block copolymer. To explain this effect it has been suggested[150] that only some of the block copolymer macromolecules find it energetically favorable to settle at the interface, while the remainder are randomly dispersed in the bulk phases.

Of particular interest is the question of the influence of solvent addition upon the properties of a system.[150] This question has been answered for the case when the solvent is athermic.[151] In this case the interface width in the limit of infinitely long chains of homopolymers grows exponentially, while the surface tension decreases drastically as the molecular weight of a copolymer increases because of localization of its molecules at the interface region. Special attention should be paid to work[152] where the general consideration of the interface problem was carried out for a blend containing two homopolymers, an arbitrary block copolymer, and solvent.

13.7.3 Heteropolymers with Weak Unit Interactions

The previous subsection was devoted to the discussion of the interface problem for blends of strongly incompatible polymers with a sharply pronounced boundary between the phases. Under changes of the external parameters of a system as it approaches the critical point, this boundary can spread because of the decay of volume interactions between units, resulting also in the diminution of the differences in composition of coexisting phases. A theoretical treatment of the interface of such relatively compatible polymers can be performed within the framework of the Landau theory (see Subsection 13.8.2.1).

Leibler[153] considered a melt of two monodisperse homopolymers M_1^l and M_2^l with the same degree of polymerization l and a symmetric block copolymer $M_1^{l/2}M_2^{l/2}$ where volume fractions of these three components are Φ_1^0, Φ_2^0, and Φ^0, respectively. In the vicinity of this system's critical point, which is the case at $\Phi_1^0 = \Phi_2^0 = 1/\chi l$, the Landau free energy was presented as an expansion in powers of two independent small parameters of order $(\Phi_1 - \Phi_2)$ and $(\Phi_3 - \Phi_3^0)$. Then ordinary theoretical analysis was performed in order to derive expressions for the surface tension σ. This simple expression looks like a difference of two terms $\sigma' - \sigma''$, corresponding to two different physical mechanisms which result in the decrease of interface tension when copolymer is added to a system. The first term, σ', reflects a corresponding diminution of the fraction of units M_1 and M_2 under the increase of Φ_3 in M_1-monomer-rich and M_2-monomer-rich phases, whereas the second term, σ'', represents the preferential location of copolymers at the interface. The contribution σ'' increases with increase in the incompatibility of polymers, remaining nevertheless small when compared to σ' at the vicinity of the critical point.

13.8 THEORY OF MICROPHASE SEPARATION

13.8.1 Introduction

Melts and solutions of block copolymers, as well as blends containing their macromolecules, can be often found in thermodynamically stable mesomorphic states. These are characterized by an inhomogeneous density distribution of monomer units in space. The basic mesomorphic states of polymer systems incapable of forming neither regular nor liquid crystals are micellar solutions and spatially periodic structures. Such superstructures, provided they are not formed in close vicinity to critical points, consist of blocks localized within microdomains periodically distributed in space. Usually this mesophase is one of three possibilities: (i) alternating layers (lamellae); (ii) cylinders on a two-dimensional lattice; and (iii) spheres on a three-dimensional lattice. Which of these particular structures is the case depends upon the statistical characteristics of the polymer sample as well as upon the conditions of observation. Much literature is devoted to the discussion of microdomain formation,[154–165] both from the theoretical and experimental standpoint.

A considerable body of experimental material is available which appears to be sufficient for the understanding of the fundamental peculiarities of this phenomenon. However, the elaboration of a general microscopic theory still remains a challenge for statistical thermodynamics. Such a theory, if available, should be able on the one hand to provide a recipe for constructing phase diagrams and, on the other hand, be able to predict the morphology of the phase-separated superstructure, *i.e.* the size, shape, and spatial distribution of microdomains. In the absence of a unified theory, two limit regimes for the formation of spatially periodic structures are generally considered.

In the first regime of strong incompatibility between different-type units of molecules of a binary block copolymer, microdomains are composed mainly of units of the same type, being separated by interfaces very thin in comparison with their sizes. The area of the applicability of this limit, the so-called strong segregation regime (SSR), covers polymer systems situated far enough from the critical points of their phase diagrams. Conversely, if a phase transition occurs under conditions close to the critical ones, the superstructure formation takes place under the weak segregation regime (WSR). Here the differences in concentrations of monomer units inside different type domains are known to be small and no sharp interface exists between them. Under such a regime, the size of a macromolecule is only slightly perturbed by volume interactions, unlike for the SSR where block copolymer chains are strongly extended, having the gyration radius R_G noticeably exceeding that of an unperturbed Gaussian coil.

Theoretical approaches normally employed in the statistical thermodynamics of superstructures formed under the two limit regimes generally differ markedly. A major success has been achieved here in the case of the WSR, where starting with a basic study by Leibler,[130] extensive use has been made of the Landau theory. Its mathematical apparatus has been thoroughly developed in modern theoretical physics, and so, once a particular problem of the description of equilibrium properties of a heteropolymer system is formulated in terms of the Landau theory, a further solution of this problem is a routine procedure. As for the SSR, attempts to develop an approach as universal as the Landau theory in the case of the WSR have not met with success so far, but, nevertheless, substantial progress has been made in describing superstructures within the framework of the mean-field approximation. The key idea of the latter, going back to pioneer works by Edwards[166] and Helfand[167] which deal with the thermodynamics of binary block copolymers, resides in a derivation under certain assumptions of self-consistent equations, enabling one to find the spatial distribution of concentrations of different types of monomer unit. These nonlinear equations are reminiscent of those describing nonstationary diffusion within a two-component gas with molecules capable of entering into chemical reactions. The role of the time, which appears as an independent variable in the latter equations, is performed in the former ones by the distance along the polymer chain reckoned from its end. The mathematical apparatus used in the theory to describe superstructures forming in the SSR resembles that employed when solving the interface problem.

13.8.2 Weak Segregation Regime

13.8.2.1 Introduction to the Landau theory

When considering phase transitions it is pertinent to introduce the parameter of order ψ equal to zero in the principal phase and distinct from it in the incipient one. For the solutions of low-molecular compounds one can choose as ψ, for instance, the difference in densities of the molecules in both phases. If these solutions contain m types of particles, the system will be defined by a set of m such variables ψ_α which can be thought of as components of the vector $\boldsymbol{\psi}$ being the m-component parameter of order. The latter can be either constant or a function of coordinates, depending upon the incipient phase symmetry. If it is spatially homogeneous, we are dealing with the first case; if periodic structure forms, the second case applies. The role of the parameter of order can be performed by other variables as well, which are known to become zero within the principal phase and are distinct from it within the incipient one.

When the phase transition occurs in the vicinity of the critical point, the parameter of order has a small value and thus the system free energy can be expanded into the power series of this parameter. The latter will be either an ordinary series or a functional one depending on whether the incipient phase is spatially homogeneous or not.

Under thermodynamic equilibrium, the free energy, envisaged as a function of the parameter of order, attains its minimum. This condition permits one to find the value of the parameter $\boldsymbol{\psi}$, and, consequently, the dependence of the free energy on the coefficients of expansion. Finding concrete expressions for them is, strictly speaking, beyond the scope of the traditional Landau theory, which is a phenomenological one. In order to specify these coefficients, it is necessary to take into account special features of the particular system within the framework of the corresponding microscopic theory. Peculiarities of this procedure when applied to heteropolymer systems will be discussed in the next subsubsection while here we will just outline briefly (supposing for the sake of simplicity that the parameter of order is one component) the central ideas of the Landau approach, indispensable for the understanding of the results arrived at in terms of the weak segregation theory.

The functional expansion of the free energy ΔF into a power series of the parameter of order $\psi(\mathbf{r})$ is normally written down as its Fourier transform $\tilde{\psi}(\mathbf{q})$. Here the expansion coefficients which are the functions of wave vectors become constants, provided that the function $\Delta F[\tilde{\psi}]$ is considered, as is customary, on the space of functions of particular form

$$\tilde{\psi}(\mathbf{q}) = \frac{\psi_n}{\sqrt{n}} \sum_{k=1}^{n} [\delta(\mathbf{q} - \mathbf{Q}_k) + \delta(\mathbf{q} + \mathbf{Q}_k)] \tag{133}$$

where the quantity in square brackets is the sum of two Dirac delta functions. Upon substitution of this expression into the function $\Delta F[\tilde{\psi}]$, the latter reduces to the ordinary Taylor series in powers of parameter ψ_n with coefficients controlled by n vectors $\{\mathbf{Q}_k\}$ which are the reciprocal lattice primitive translations (RLPT) of the superstructure formed. In terms of this approximation, all these vectors are generally taken to have the same modulus, equal to the value of the wave vector q^* where the loss of equilibrium of the spatially homogeneous state occurs at the critical point.

Among the coefficients $\{a_n^{(k)}\}$ of the expansion of the free energy $\Delta F[\psi_n]$, only that which appears in front of ψ_n^2 turns out to be independent of n. This coefficient along with $a_n^{(3)}$ is known to vanish at the critical point (not on the spinodal!). Provided that the coefficient $a_n^{(4)}$ is positive, when expanding the function $\Delta F(\psi_n)$ it is normal to restrict the expansion to the first three terms

$$\frac{\Delta F_n}{TM} = a^{(2)}\psi_n^2 + a_n^{(3)}\psi_n^3 + a_n^{(4)}\psi_n^4 \tag{134}$$

and omit the other terms. Having minimizes this expression with respect to ψ_n, it is possible to find the value whose substitution into equation (134) yields that of the free energy $\Delta F_n^{\min}$ reckoned from its value in the spatially homogeneous phase. The condition for the transition of a system from this phase to a superstructure is $\Delta F_n^{\min} = 0$. An analogous condition, but for the transition between two mesophases of different symmetry, is obtainable as a result of equating the values $\Delta F_n^{\min}$ of these phases. In order to find the condition for the loss of stability of any mesophase in terms of the approximation in question, it is necessary to substitute the value ψ_n obtained by minimization of ΔF_n into the equation $\partial^2\Delta F_n/\partial\psi_n^2 = 0$.

When the coefficient $a_n^{(4)}$ is either comparable in value to the dropped terms of the expansion (134) or negative, the next items should be retained in this expansion.

Elements of the Landau theory introduced here can be extended to the case when the parameter of order has more than one component. This appears to be rather promising when dealing with multicomponent systems.

13.8.2.2 Peculiarities of the Landau theory as applied to polymer systems

The number of components of the parameter of order $\boldsymbol{\psi}$ coincides with that of quasicomponents of a polymer system or is one less if the system is presumed to be incompressible. The circumstance that monomer units are linked to form polymer chains and, consequently, are incapable of moving independently in space, is responsible for the appearance of this fundamental characteristic. The latter consists in the impossibility (with the exception of degenerate model systems (see Section 13.5.2)) of presenting the free energy of a multiphase polymer system, considered as a function of parameters of order in different phases, as a sum of additive contributions from separate phases. This results in the appearance in the last item in the right-hand part of expression (134), as well as in all subsequent terms of the free energy expansion, of the items which are dependent upon the parameters of order in different phases. Such a nonlocality effect, inconceivable in principle in solutions of low-molecular compounds, was pointed out for the first time by Panyukov and Kuchanov.[168]

In this section, a general algorithm[114] will be formulated allowing one to find the coefficient of the expansion $\Delta F[\boldsymbol{\psi}]$ in a power series of the component ψ_α of the parameter of order $\boldsymbol{\psi}$. As ψ_α deviations can be chosen from average values of either the unit densities

$$\psi_\alpha(r) \equiv \Delta\rho_\alpha^{(r)} = \rho_\alpha(\mathbf{r}) - \bar{\rho}_\alpha \qquad \bar{\rho}_\alpha = \int \frac{d\mathbf{r}}{V} \rho_\alpha(\mathbf{r}) \tag{135}$$

or the components of the field $\boldsymbol{H}$ conjugate to these densities

$$\psi_\alpha(\mathbf{r}) \equiv \Delta H_\alpha^{(r)} = H_\alpha(\mathbf{r}) - \bar{H}_\alpha \qquad \bar{H}_\alpha = \int \frac{d\mathbf{r}}{V} H_\alpha(\mathbf{r}) \tag{136}$$

When the Landau expansion is performed with respect to the spatially homogeneous state, its coefficients are conveniently found in momentum rather than coordinate representation and their Fourier transforms $\tilde{\psi}_\alpha(\mathbf{q})$ appear instead of $\psi_\alpha(\mathbf{r})$.

In order to reveal specific features inherent to solutions and blends with the participation of macromolecules, it will suffice to consider only the system of chemical bonds, *i.e.* to find coefficients of the expansion of functions $F_{CB}[\boldsymbol{H}]$ and $S[\boldsymbol{\rho}]$ occurring in the random field formalism (Section 13.6.1.1) and the density functional formalism (Section 13.6.1.2), respectively

$$F_{CB}[\boldsymbol{H}] = F_{CB}[\bar{\boldsymbol{H}}] + V \sum_{p=2}^{\infty} \frac{(-1/T)^{p-1}}{p!} \prod_{i=1}^{p} \sum_{\{\mathbf{q}_i\}} \sum_{\{\alpha_i\}} \tilde{g}^{(p)}_{\alpha_1\ldots\alpha_p}(\mathbf{q}_1 \ldots \mathbf{q}_p) \Delta\tilde{H}_{\alpha_1}(\mathbf{q}_1) \ldots \Delta\tilde{H}_{\alpha_p}(\mathbf{q}_p) \tag{137}$$

$$S_L[\boldsymbol{\rho}] = S_L[\bar{\boldsymbol{\rho}}] - V \sum_{p=2}^{\infty} \frac{1}{p!} \prod_{i=1}^{p} \sum_{\{\mathbf{q}_i\}} \sum_{\{\alpha_i\}} \tilde{\gamma}^{(p)}_{\alpha_1\ldots\alpha_p}(\mathbf{q}_1 \ldots \mathbf{q}_p) \Delta\tilde{\rho}_{\alpha_1}(\mathbf{q}_1) \ldots \Delta\tilde{\rho}_{\alpha_p}(\mathbf{q}_p) \tag{138}$$

The summation over the momenta $\{\mathbf{q}_i\}$ in these expressions is over all vectors of the RLPT $\mathbf{Q}_1, \ldots, \mathbf{Q}_n$ of the superstructure forming in the incipient phase.

Let us discuss first the expansion (137) whose coefficients denote the Fourier transforms of functions of the response of a polymer system to an inhomogeneous external field $\boldsymbol{H}(\mathbf{r})$. The first two of them coincide with components of tensors $\tilde{\boldsymbol{\kappa}}^{(2)}$ and $\tilde{\boldsymbol{\kappa}}^{(3)}$, while for the components of tensor $\tilde{\mathbf{g}}^{(4)}$ the following expressions apply

$$\begin{aligned}\tilde{g}^{(4)}_{\alpha_1\alpha_2\alpha_3\alpha_4}(\mathbf{q}_1\mathbf{q}_2\mathbf{q}_3\mathbf{q}_4) &= \tilde{\kappa}^{(4)}_{\alpha_1\alpha_2\alpha_3\alpha_4}(\mathbf{q}_1\mathbf{q}_2\mathbf{q}_3\mathbf{q}_4)\delta(\mathbf{q}_1+\mathbf{q}_2+\mathbf{q}_3+\mathbf{q}_4) - [\tilde{\kappa}^{(4)}_{\alpha_1\alpha_2|\alpha_3\alpha_4}(\mathbf{q}_1\mathbf{q}_2\,|\,\mathbf{q}_3\mathbf{q}_4)\delta(\mathbf{q}_1+\mathbf{q}_2)\delta(\mathbf{q}_3+\mathbf{q}_4)\\ &+ \tilde{\kappa}^{(4)}_{\alpha_1\alpha_3|\alpha_2\alpha_4}(\mathbf{q}_1\mathbf{q}_3\,|\,\mathbf{q}_2\mathbf{q}_4)\delta(\mathbf{q}_1+\mathbf{q}_3)\delta(\mathbf{q}_2+\mathbf{q}_4) + \tilde{\kappa}^{(4)}_{\alpha_1\alpha_4|\alpha_2\alpha_3}(\mathbf{q}_1\mathbf{q}_4\,|\,\mathbf{q}_2\mathbf{q}_3)\delta(\mathbf{q}_1+\mathbf{q}_4)\delta(\mathbf{q}_2+\mathbf{q}_3)]\end{aligned} \tag{139}$$

where the components of tensor $\tilde{\kappa}^{(4)}$ have been determined above (equation 118). In parallel with these 'local' functions of the response, the tensor $\tilde{\mathbf{g}}^{(4)}$ also comprises nonlocal terms described by the expressions in square brackets in equation (139) which broke the locality of the response. The density distribution at a given point, as it follows from equation (139), was found to depend upon the values of the external field not only in the local vicinity of this point but also in the overall volume of a system.

A general algorithm enabling one to write an expression for the components of the symmetric tensor $\tilde{\mathbf{g}}^{(p)}(\{\mathbf{q}_i\})$ of any order involves three successive stages, each consisting in expressing the tensor through the other one. This procedure can be schematically presented as follows

$$\tilde{\mathbf{g}} \xrightarrow{(1)} \tilde{\boldsymbol{\kappa}} \xrightarrow{(2)} \tilde{\boldsymbol{\theta}} \xrightarrow{(3)} \mathbf{W} \tag{140}$$

The sum of wave vectors entering as arguments into the tensor $\tilde{\mathbf{g}}^{(p)}$, according to its definition, is zero at any p.

Stage 1. The set of p points, each being characterized by the index α and wave vector (momentum) $\mathbf{q}$, is partitioned into $\bar{\omega}$ groups in such a way that any νth group comprises the number of points $p_\nu \geq 2$ whose overall momentum equals zero. Every group makes an additive contribution to the expression for $\tilde{\mathbf{g}}^{(p)}$, which up to the factor $(-1)^{\bar{\omega}-1}$ will read

$$\tilde{\kappa}^{(p)}_{\alpha_1\ldots|\ldots\ldots|\ldots\alpha_p}\left(\underbrace{\mathbf{q}_1\ldots|}_{p_1\text{ terms}}\ \ldots\ldots\ \underbrace{|\ldots\mathbf{q}_p}_{p_{\bar{\omega}}\text{ terms}}\right) = \frac{\Pi}{\mathbf{V}} \sum_{\sigma} \mathcal{P}(\sigma) \prod_{\nu=1}^{\bar{\omega}} \left\langle \sum_{\sigma} \exp\left(i \sum_{j=1}^{p_\nu} \mathbf{q}_j \mathbf{r}_j\right) \right\rangle \tag{141}$$

where the designations of equation (118) are used. Here the product is over all groups of the partitioning in hand, so that summation in the exponent is made only over the points belonging to one and the same group. Vertical bars appearing in indices and arguments in equation (141) separate the points which pertain to different groups. Inasmuch as all points of any group are taken to be indistinguishable, the values of the tensor components remain the same under any transposition $(\alpha_i, \mathbf{q}_i)$ and $(\alpha_j, \mathbf{q}_j)$ inside this group.

Stage 2. Let us partition the points of every group mentioned, in accordance with the algorithm introduced in Section 13.6.4 to relate tensors $\tilde{\boldsymbol{\kappa}}$ and $\tilde{\boldsymbol{\theta}}$. As a result of such a procedure, any νth group will contain a number p'_ν of points less than p_ν. To each such partition there will correspond the contribution $\tilde{\boldsymbol{\theta}}^{(r)}(r = p'_1 + \ldots + p'_{\bar{\omega}}$, representing the overall number of points in a reduced system) to the expression for $\tilde{\boldsymbol{\kappa}}^{(p)}$. Summing the contributions from all possible partitions of such a kind yields the unknown value of component (141) of this tensor $\tilde{\boldsymbol{\kappa}}^{(p)}$. Retaining for $\tilde{\boldsymbol{\theta}}^{(r)}$ the designations used in equation (141) and proceeding from the above formulated algorithm, we will obtain in a particular case $p = 4$ the simple expression

$$\tilde{\kappa}^{(4)}_{\alpha_1\alpha_2\,|\,\alpha_3\alpha_4}(\mathbf{q}_1\mathbf{q}_2\,|\,\mathbf{q}_3\mathbf{q}_4) = \tilde{\theta}^{(4)}_{\alpha_1\alpha_2\,|\,\alpha_3\alpha_4}(\mathbf{q}_1\mathbf{q}_2\,|\,\mathbf{q}_3\mathbf{q}_4) \tag{142}$$

Stage 3. To have $\tilde{\boldsymbol{\theta}}^{(r)}$ expressed through $\mathbf{W}^{(r)}$, we can resort to the relationships derivable by virtue of the algorithm of Section 13.5.4 upon minor modification of the latter. In line with this algorithm, the tensor function $\mathbf{W}^{(r)}$ involves as arguments the functions $\tilde{\lambda}$ of the sum of momenta of different points which belong, generally speaking, to different groups. The set of all items in this sum is conveniently divided into subsets $\mathbf{q}_1, \ldots, \mathbf{q}_\nu, \ldots$, each containing the momenta of points of only one group. The above modification consists in replacing each function $\tilde{\lambda}$ by the product of factors $\tilde{\lambda}(\mathbf{q}_\nu)$ taken over all groups of type ν.

Having resorted to the above procedure (equation 140), it becomes possible to express the coefficients of the expansion (137) through the generating functions of chemical correlators (equation 119). The final expression for the coefficients of the expansion of free energy (equation 94), appearing in the Landau theory with the parameter of order (equation 136), for an arbitrary polymer mixture will read

$$\tilde{G}^{(p)}_{\alpha_1\ldots\alpha_p}(\mathbf{q}_1\ldots\mathbf{q}_p) = \tilde{g}^{(p)}_{\alpha_1\ldots\alpha_p}(\mathbf{q}_1\ldots\mathbf{q}_p) - \tilde{B}^{(p)}_{\alpha_1\ldots\alpha_p}\delta(\mathbf{q}_1 + \ldots + \mathbf{q}_p) \tag{143}$$

where the tensor $\tilde{\mathbf{B}}^{(p)}$ components have been determined earlier (equation 86). The fact that nonlocal terms in the coefficients (143) only appear beginning with the fourth-order terms with respect to the parameter of order comes as no surprise. This is concerned with the necessity when finding the free energy of a polydisperse polymer sample to average over the distribution $\mathcal{P}(\sigma)$ not the partition function of separate macromolecules but its logarithm. If we compare the expansion in the power series of the field $\Delta \boldsymbol{H}$ of this functional with that of the generating functional of separate molecules (equation 122), which can be obtained by averaging their partition function, it will be evident that the distinctions in coefficients commence just from the fourth-order terms. These distinctions are nothing but nonlocal items (equation 141), which are absent in equations generally applied to the construction of phase diagrams provided that the transitions between mesophases are not considered.

In the vast majority of studies on the weak segregation regime, use has been made of the Landau theory with the parameter of order (equation 135) corresponding to the density function formalism (Section 13.6.1.2). Coefficients of the expansion of this function into power series of $\Delta\boldsymbol{\rho}$ in a simple fashion (see equation 85)

$$\tilde{\Gamma}^{(p)}_{\alpha_1\ldots\alpha_p} = -\tilde{\gamma}^{(p)}_{\alpha_1\ldots\alpha_p}(\mathbf{q}_1\ldots\mathbf{q}_p) + (f^{*(p)}_{\alpha_1\ldots\alpha_p}/T)\delta(\mathbf{q}_1 + \ldots + \mathbf{q}_p) \tag{144}$$

are connected with the coefficients of the analogous expansion (138) of the Lifshitz entropy defined by equations (99) and (100). By substituting into the first of them the expansion (137) and solving the equation obtained by means of iterations, one can determine the dependence of the field $\boldsymbol{H}(\mathbf{r})$ on the density $\boldsymbol{\rho}(\mathbf{r})$ as an expansion into the power series of the parameter of order (135) Fourier transform. Insertion of this solution into equation (100) leads to the expansion (138), where the tensor $\tilde{\boldsymbol{\gamma}}^{(p)}$ components are determined using the diagram technique discussed in Section 13.6.4, provided that the tensors $\tilde{\boldsymbol{\kappa}}^{(p)}$ are substituted for $\tilde{\mathbf{g}}^{(p)}$. Apart from this distinction, there is another one residing in the fact that summing over momenta of all lines is made now under the condition $\mathbf{q} = \mathbf{0}$. Having unified the expansion

(138) with the contribution of volume interactions, one will arrive, in accordance with equation (97), at the expansion of functional $\mathscr{F}[\boldsymbol{\rho}]/T$ with coefficients (144).

A fundamental peculiarity of the Landau expansion of free energy in the case of polydisperse polymers consists of the presence of nonlocal terms in coefficients $\tilde{\boldsymbol{\Gamma}}^{(p)}$, caused by the nonlocality of the Lifshitz entropy. The role of these terms is conveniently illustrated for the derivation of tensor $\tilde{\boldsymbol{\gamma}}^{(4)}$ via the diagram technique generalizing that introduced in Sections 13.5.4 and 13.6.4. To have the nonlocal contribution to $\tilde{\boldsymbol{\gamma}}^{(4)}$ separated, it is pertinent to introduce the new diagram elements shown in Figure 6(a)

$$\begin{array}{ccc} \boxed{\text{I}} & \boxed{\text{II}} & \boxed{\text{III}} \\ \tilde{\kappa}^{(4)}_{\alpha_1\alpha_2\alpha_3\alpha_4}(\mathbf{q}_1\mathbf{q}_2\mathbf{q}_3\mathbf{q}_4) & \tilde{\kappa}^{(4)}_{\alpha_1\alpha_2\,|\,\alpha_3\alpha_4}(\mathbf{q}_1\mathbf{q}_2\,|\,\mathbf{q}_3\mathbf{q}_4) & \tilde{\kappa}^{-1}_{\alpha_1\alpha_2}(\mathbf{0})\delta(\mathbf{q}) \end{array} \tag{145}$$

In terms of this diagram technique, the expression (139) for $\tilde{\mathbf{g}}^{(4)}$ has the form presented in Figure 6(b). As for the expression for $\tilde{\boldsymbol{\gamma}}^{(4)}$, its diagram representation in Figure 6(c) differs from the traditional one (Figure 3) by the terms in square brackets.[168] They are mutually cancelled in the case of low-molecular weight solutions, whereas for polymer systems these terms annihilate each other only under the model assumption of a total absence of polydispersity. It was shown[114] that in this case such an annihilation takes place for all coefficients $\tilde{\boldsymbol{\gamma}}^{(p)}$ of the Landau expansion.

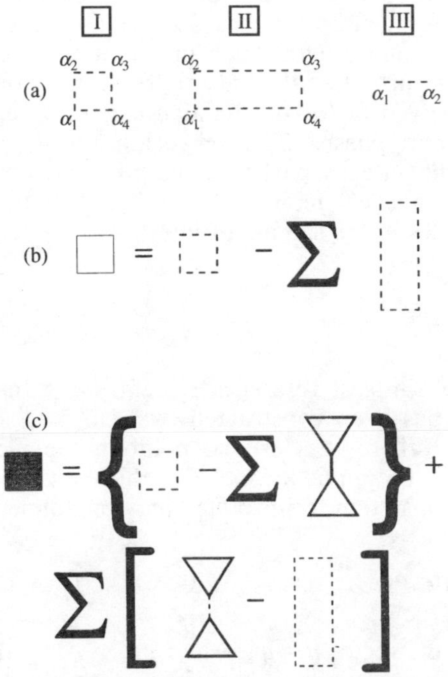

Figure 6 Elements of the diagram technique (equation 145), (a) for the description of a polydisperse system in terms of the formalism of the random field (b), and the density function (c).

The above-introduced Landau expansion has been formulated in terms of the small scale mean-field theory. When the characteristic scale of the cahnge of monomer unit densities is large compared to the size of an individual molecule, this expansion is noticeably simplified, reducing to a form appropriate for the description in terms of the large scale mean-field theory of multiphase polymer systems situated in the vicinity of a critical point. The coefficients of such an expansion can be found by two independent methods.

The first of them involves going from the free energy F (equation 34) to a new thermodynamic potential

$$\hat{F} = F + \sum_{\alpha=1}^{m}\sum_{i=1}^{r} h^i_\alpha\left(\rho^i_\alpha - \sum_{l} C^i(\mathbf{l})l_\alpha\right) \tag{146}$$

the minimization of which with respect to $C^i(\mathbf{l})$ leads to equations whose solution looks like equations (37) and (38) where $\mu_\alpha^*(\boldsymbol{\rho}^i)$ is replaced by $\mu_\alpha^*(\boldsymbol{\rho}^i) + h_\alpha^i$. In order to determine the dependence of the fields h_α^i on the densities ρ_α^i, one should substitute the expressions obtained for $C^i(\boldsymbol{l})$ into the stoichiometric condition of equation (35). By virtue of such an exclusion of concentrations $C^i(\boldsymbol{l})$ and fields h_α^i, it becomes possible to find an expression for thermodynamic potential $\hat{F}$ as a function of the densities $\{\boldsymbol{\rho}^i\}$ of monomer units in different phases ($i = 1, \dots, r$) and their volumes $\{V^i\}$. This potential is equal to the function of free energy F taken on a stepwise distribution of monomer unit densities describing the state of a system involving several spatially homogeneous phases. For low-molecular weight systems the function $\hat{F}[\{\boldsymbol{\rho}^i\}, \{V^i\})$ equals the sum of contributions $\hat{F}(\boldsymbol{\rho}^i, V^i)$ of all phases provided that surface effects are ignored. This property of additivity, as is easily shown using the above described procedure for deriving the expression for $\hat{F}$, is violated for real polymer systems. Despite the fact that such a procedure is possible to perform in principle, it proves to be inefficient for thermodynamic calculations of real polymer systems unless it is in the proximity of the critical point where monomer unit densities in each phase differ slightly from their average values. In this case, expansion of F into the power series of these deviations $\Delta\rho_\alpha$ will result in the sought for Landau expansion whose coefficients are, generally speaking, dependent in a nonlinear way on the volume fractions y^i of different phases.

Another, more straightforward way of deriving these Landau expansion coefficients suggests consideration of the function (138) on the space of stepwise distributions $\Delta\rho_\alpha(\mathbf{r})$ with the macroscopic linear sizes L^i of the phases. Because characteristic values of the modulus of the wave vectors $q^i \simeq 1/L^i$ on such a distribution are small compared to the inverse size of a macromolecule, the factors in front of δ-functions in expressions for tensor $\boldsymbol{\gamma}^{(p)}$ components may be taken outside a sign of the sums (integrals) provided that the values of all momenta are put equal to zero. Hence any sum (integral) over momenta is divided into the product of sums (integrals) of the individual groups, described above, each νth group containing p_ν terms. As a result, expressions will be arrived at for the Landau expansion coefficients as polynomials in volume fractions y^i of different phases. To each coefficient L_p corresponds the polynomial whose degree is equal to the integer part of a number $p/2$, while the coefficients of this polynomial depend on the statistical moments of the distribution for size and composition (equation 2) of macromolecules in the sample of interest.

13.8.2.3 Main results

The phase diagram of the simplest reference system, *i.e.* a melt of monodisperse diblock copolymers (equation 12a), has been constructed by Leibler.[130] In such a system, when the fraction of units M_1 is $f = \frac{1}{2}$ there is a critical point in the vicinity of which the Landau expansion of free energy can be applied. Since for the melt the incompressibility condition holds, it is sufficient for such a system to introduce only one dimensionless parameter of order $\psi(\mathbf{r})$

$$\rho_1(\mathbf{r}) = \bar{\rho}[f + \psi(\mathbf{r})], \qquad \rho_2(\mathbf{r}) = \bar{\rho}[1 - f - \psi(\mathbf{r})] \tag{147}$$

Near the spinodal at finite wave vector $q = q^*$, when $\chi - \chi_s \ll \chi_s$, the major contribution to the free energy functional is from a Fourier transform with this wave vector, and thus the parameter of order $\psi(\mathbf{r})$ can be chosen as

$$\psi(\mathbf{r}) = \frac{2}{\sqrt{n}} \psi_n \sum_{k=1}^{n} \cos \mathbf{Q}_k \mathbf{r} \tag{148}$$

The summation in expression (148) is over the vectors $\mathbf{Q}_k$ of the RLPT of the superstructure.

The substitution of the parameter of order (148) into the compression (147) for the densities of monomer units permits one to write down the free energy density, reckoned from the spatially homogeneous state as follows

$$F_n l/TV = 2\tau\psi_n^2 - \alpha_n\psi_n^3 + \beta_n\psi_n^4, \qquad \tau \equiv l(\chi_s - \chi) \tag{149}$$

The expansion coefficients α_n and β_n are controlled by both the symmetry of the superstructure and the parameter f; the coefficient α_n vanishes for symmetric polymers with $f = \frac{1}{2}$, while for a lamellar structure this is identically equal to zero. The equilibrium value of the amplitude ψ_n is found by minimization of the free energy (149)

$$\frac{F_1 l}{TV} = -\tau^2/\beta_1, \qquad \frac{F_n l}{TV} = \frac{27}{(64)^2}\frac{\alpha_n^4}{\beta_n^3}(1+\gamma_n^3)(1-3\gamma_n), \tag{150}$$

where at $n > 1$

$$\gamma_n = [1 - 64\beta_n\tau/9\alpha_n^2 l]^{1/2} \tag{151}$$

Conditions for a phase transition are obtainable from a comparison of the free energies of different microstructures and that of the homogeneous state. As a result, a phase diagram has been constructed (Figure 7) involving the regions of different microdomain structures for a block copolymer melt, the transitions between which at $f \neq \frac{1}{2}$ are the first-order phase transitions.

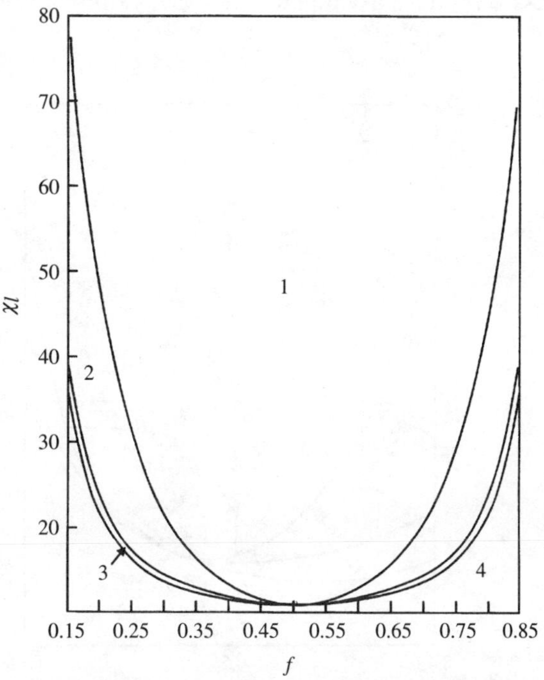

Figure 7 Phase diagram for a melt of a monodisperse diblock copolymer $M_1^{lf}M_2^{l(1-f)}$ (equation 12a).[130] Figures denote the type (symmetry) of phase: (1) lamellar ($d = 1$, $n = 2$); (2) planar triangular lattice ($d = 2$, $n = 3$); (3) body-centred cubic lattice ($d = 3$, $n = 6$); (4) spatially homogeneous.

In a paper by Fredrickson and Leibler[169] the properties were studied of the above reference system upon addition of a solvent good for both quasicomponents M_1 and M_2. To provide the description of such a system, along with the parameter of order (148), a second one, $\psi_s(\mathbf{r})$, should be introduced, proportional to the solvent density. In the vicinity of the spinodal it takes the form

$$\psi_s(\mathbf{r}) = \frac{2\psi_s}{\sqrt{n}}\sum_{k=1}^{n}\sum_{l=1}^{n}\cos(\mathbf{Q}_k + \mathbf{Q}_l)\mathbf{r} \tag{152}$$

For a lamellar structure this parameter of order is defined by the wave vector $2\mathbf{q}^*$, which corresponds to the displacement of the solvent from the regions enriched by monomer units to the interphase region. The amplitude ψ_s of the solvent inhomogeneities is a factor l^{-1} less than ψ_n. Owing to the smallness of this parameter, it appears possible to neglect the contribution of the solvent to the free energy of a system.

The main qualitative effect taking place in a concentrated solution ($1 - \Phi \ll 1$) of asymmetric polymers ($f \neq \frac{1}{2}$) is a splitting of the lines of the phase transitions shown in Figure 7 into the coexistence regions of corresponding phases. The widths of these regions, however, are too small to be observed experimentally. For symmetric polymers with $f = \frac{1}{2}$, the two-phase region has been obtained[169] at low temperatures.

A mixture of block copolymers, homopolymers, and solvent has been studied.[170,171] Numerous phase diagrams constructed in these papers show the presence of eutectic points and mesophase formation induced by the homopolymer. Unfortunately, the only possibility taken into consideration is the formation of the lamellar phase, for which the cubic term in the Landau expansion is absent in principle. That is because the phase transitions have been found to be second-order, which contradicts the above-discussed results.

A theoretical analysis[172] of copolymer melts of more complicated architecture, such as an asymmetric triblock copolymer (equation 12b), brought the authors[132] to the conclusion that the critical point is the case only at a particular value $f = f_c$ of a fraction of units M_1 being at the critical point. By virtue of the Landau expansion in the vicinity of the critical point, the phase diagram of the system has been constructed (see Figure 8). From the comparison of Figure 8 with Figure 7 as a reference system, it is evident that, as the parameter τ increases, the phase diagram becomes progressively more asymmetric and the widths of regions 2 and 3 increase.

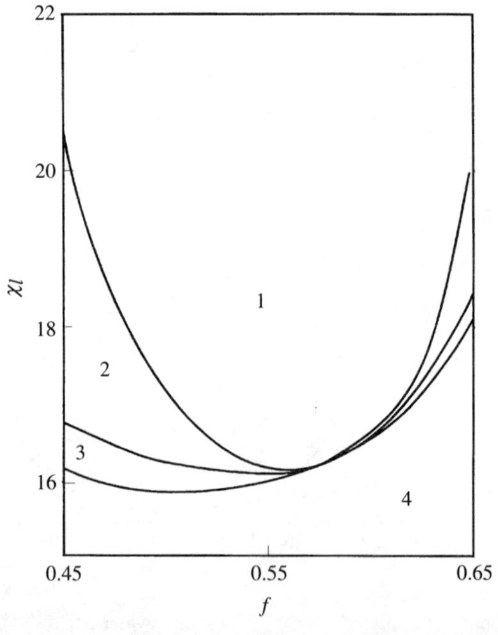

Figure 8 Phase diagram for a melt of a monodisperse triblock copolymer $M_1^{lf\tau}M_2^{l(1-f)}M_1^{lf(1-\tau)}$ (equation 12b) for the case $\tau = 0.25$.[130] Here τ denotes the symmetry parameter while the remaining designations are the same as in Figure 7.

Among the assumptions employed in the above-discussed studies on the theory of the weak segregation regime, the most vulnerable one is the approximation neglecting high-order harmonics of the parameter of order $\psi(\mathbf{r})$ (equation 148). Their influence on the value of the superstructure wave vector $\mathbf{q}^*$ has been elucidated.[173] Amplitudes of high-order harmonics of the Fourier series of $\psi(\mathbf{r})$ as well as the equilibrium value $x^* = q^{*2}R^2$ can be found from the minimum of the free energy function. For a lamellar structure, numerical calculations predict a linear dependence $\Delta x^* \approx 0.37(\chi_s - \chi)$, which is in a good agreement with experimental results obtained by Owens *et al.*[174] for a styrene–isoprene copolymer with $f = 0.4$. Analogous results have been arrived at for phases of other morphology.

Taking account of only the first harmonics in equation (148) provides no possibility for differentiating superstructures with the same first Bragg reflections pertaining to different spatial symmetry groups. Among them are the three-dimensional structures bcc (body-centered cubic) and bdd (bicontinuous double diamond) as well as a two-dimensional triangular structure and a three-dimensional hexagonal structure. Such alternative structures for the reference system have been investigated[175] with due regard for the contribution from higher-order harmonics. The phase diagram obtained here differs essentially from that presented in Figure 7 only within the region of values $0.47 < f < 0.52$, when the three-dimensional hexagonal structure has a free energy lower than that of lamellar, square, triangular, bcc, and simple cubic structures.

Dobrynin and Yerukhimovich[176] carried out a theoretical analysis of microphase separation in melts of binary copolymers of complicated architecture. By the numerical calculation, the phase diagrams were constructed in the space of parameters describing the chemical structure and temperature for asymmetric triblock and polyblock copolymers.

The allowance made for polydispersity can materially change phase diagrams of polymer systems. When polydispersity is weak enough, each of the above-considered lines of phase transitions on diagrams χ versus f expands to turn into an area of finite width where both phases coexist.[168] If polydispersity is strongly pronounced, new qualitative effects appear which are defined by nonlocal terms of the expansion of the Landau free energy function. The disregard of the latter when constructing the phase diagram led the authors of the paper[177] to incorrect results.

The simplest polydisperse system whose behavior was examined[178] is a melt of a statistically random heteropolymer with fraction $f = \frac{1}{2}$ of units M_1. On the assumption that the Flory parameter χ is dependent upon the wave vector **q**, the microdomain structure in such a system was shown to appear by a third-order continuous phase transition. The wave vector modulus $q^* \simeq (\chi_s - \chi)^{1/2}$ of such a superstructure in the vicinity of the spinodal at the zeroth wave vector $(\chi = \chi_s)$ is strongly controlled by the Flory parameter χ. The appearance of this superstructure is conditioned by the presence of randomly positioned M_1-rich and M_2-rich fragments which facilitate segregation.

This work is of prime importance for the thermodynamic theory of heteropolymers since here for the first time an expression has been derived for the Landau free energy, comprising nonlocal terms owing their existence to the polydispersity of a sample. Allowance for these terms provides a possibility of describing a superstructure whose period is strongly governed by temperature. This effect is due to the dependence $\chi(q)$ introduced by hand and is known to be absent in commonly accepted thermodynamic models of polymer melts where χ is a constant. This paper would be even more valid but for the restrictions, imposed by the incompressibility approximation, which do not permit an adequate description of experimentally observed macrophase separation in these systems. In order to describe the superstructures considered in the paper,[178] there is no need to use an arbitrary assumption about the dependence of the Flory–Huggins parameter χ on the wave vector q. It has been illustrated by authors[113,114] who, proceeding from the general approach (see Section 13.6.1) characterizing the thermodynamics of polymer systems with an arbitrary chemical structure, examined binary Markovian copolymers in detail. The period of such a structure in the proximity of the point $\chi = \chi_s$ of its appearance is large when compared to the characteristic spatial size $aN_b^{1/2}$ of one block whose average number of monomer units is N_b. For the sake of simplicity, the case is considered of a large number of blocks which corresponds to the infinite length limit of a macromolecule. Substituting expressions (119) for the generating functions of chemical correlators and relationships (147) and (148) for the parameter of order $\psi(\mathbf{r})$ into the general Landau expansion for free energy density we obtain

$$F_n N_b / TV = 2N_b(\chi_s - \chi + \delta a^2 Q^2)\psi_n^2 - \alpha_n \psi_n^3 + \beta_n \psi_n^4 + \gamma \psi_n^4 / a^2 N_b Q^2 \tag{153}$$

where coefficients γ and δ of the order of unity depend on the parameter f rather than on the structure symmetry.

The most essential qualitative difference of equation (153) from equation (149), which refers to monodisperse systems, is the presence of the additional last item in its right-hand part. The value of this item, representing the contribution from nonlocal effects, was found to increase approximately as Q^{-2} with the growth of the superstructure wave vector Q. This is due to the decrease in the probability of finding the density fluctuations with spatial scale $\mathbf{R} \approx \mathbf{Q}^{-1}$, which is large enough in comparison with the spatial size of block $aN_b^{1/2}$. Such fluctuations result from a local redistribution of macromolecules with the excess $l \simeq R^2/a^2$ of units of given type $\alpha = 1, 2$ at the scale $R \simeq Q^{-1}$ from the interphase areas into internal ones of domains of corresponding type α. Losses in the entropy of mixing induced by such a redistribution of chains are proportional to the number l of the excess units and are defined by a nonlocal term of the free energy (equation 153).

Obviously, the above-discussed physical mechanism, related to the polydispersity of the system of interest, is absent in the case of a homodisperse polymer.

The equilibrium value of a superstructure wave vector is obtainable by minimization of the free energy (equation 153) with respect to the value of Q which leads to $a^2N_bQ^2 = \sqrt{\gamma/2\delta}\,|\psi_n|$. Substituting this expression into equation (153) yields a free energy function which depends exclusively upon the superstructure amplitude ψ_n

$$F_nN_b/TV = 2N_b(\chi_s - \chi)\psi_n^2 + 2\sqrt{2\gamma\delta}\,|\psi_n|^3 - \alpha_n\psi_n^3 + \beta_n\psi_n^4 \qquad (154)$$

Here the coefficient in front of the cubic term is positive until the value $|\alpha_n|$ is not too large. Since the value $\alpha_n \simeq (f - \frac{1}{2})$ then, at $|f - \frac{1}{2}| < \Delta_c \simeq 1$, a nontrivial solution $\psi_n \neq 0$ minimizing the free energy (154) exists only under the condition $\chi > \chi_s$ of the loss of positive definiteness by the quadratic form of this function and remains continuous at the third-order phase transition point.

The equilibrium value of the superstructure amplitude can be found, in turn, by minimizing the free energy (154) with respect to ψ_n. Neglecting the last item in expression (154), which is small in comparison with the remaining terms in the vicinity of the transition point, it is possible to find the dependence of the superstructure parameters on the Flory parameter χ

$$|\psi_n| \sim N_b(\chi - \chi_s), \qquad Q \sim a^{-1}(\chi - \chi_s)^{1/2} \qquad (155)$$

More detailed analysis shows that an absolute minimum (154) is attainable on the solution with bcc symmetry ($n = 6$). The expression (148) with allowance for the dependence (155) for the parameter of order $\psi(\mathbf{r})$ becomes asymptotically exact within the limit $f \rightarrow \frac{1}{2}$. In the case $f \neq \frac{1}{2}$ this also involves high-order harmonics whose amplitude decreases as the parameter $|f - \frac{1}{2}|$ does.

When $\chi - \chi_s \geq \chi_s \simeq 1/N_b$ in the expansions (153) and (154), account should be taken for the fourth- and higher-order terms with respect to ψ_n. The amplitude of such a superstructure is of the order of overall density $\psi_n \simeq 1$, while its period is comparable with a block spatial size $aN_b^{1/2}$. To describe such a superstructure, the higher-order terms in the Landau expansion of the free energy should be taken into account.

13.8.3 Strong Segregation Regime

13.8.3.1 The Helfand–Edwards method

In the vast majority of theoretical papers devoted to periodic superstructures in systems far from the critical point, calculations are performed in terms of the self-consistent field aproximation. The mathematical apparatus employed in these papers is similar to that which was used for the solution of the interface problem (see Section 13.7). However, a distinction exists, since the solution of the diffusion equation (126) is searched in a periodic self-consistent field $H_\alpha(\mathbf{r})$. This solution can exhibit various symmetries corresponding to the type of the superstructure crystal lattice. Finding its period D constitues one of the crucial problems. The value D of a given superstructure is normally determined by minimization of the free energy. Comparing the equilibrium values of the free energy of different phases, it is possible to determine the values of parameters under which corresponding phase transitions occur. For the solution of this problem, it is normal to calculate the periodic profiles of quasicomponent concentrations necessary for realizing the self-consistency procedure.

The approach outlined in Section 13.7.1 is easily extended to the case of block copolymers. Thus, for instance, for macromolecules in equation (12a) we have

$$Q_{12}[\boldsymbol{H}] = \frac{1}{V}\int d\mathbf{r}\,q(l_1; \mathbf{r}; [H_1])q(l_2; \mathbf{r}; [H_2]) \qquad (156)$$

$$q(l_\alpha; \mathbf{r}; [H_\alpha]) = \int d\mathbf{r}_0 Q(l_\alpha; \mathbf{r}_0, \mathbf{r}; [H_\alpha]) \qquad (157)$$

where the integration in equation (156) is over the coordinate $\mathbf{r}$ of the junction point while in equation (157) it is over the coordinates of the end points of such a macromolecule. Its monomer unit densities are obtainable by differentiation of the logarithm of the partition function (156) with respect to external fields $H_\alpha(r)$

$$\rho_1(\mathbf{r}) = Q_{12}^{-1}[\boldsymbol{H}] \int_0^{l_1} \mathrm{d}s q(s; \mathbf{r}; [H_1]) \times \int \mathrm{d}\mathbf{r}' Q(l_1 - s; \mathbf{r}, \mathbf{r}'; [H_1]) q(l_2; \mathbf{r}'; [H_2]) \tag{158}$$

The density $\rho_2(\mathbf{r})$ of units M_2 is of similar form. Equations (156) and (158) completely define the free energy of the system in hand within the self-consistent field approximation. Thus the problem reduces to solving the diffusion equation (126) for each block influenced by given external fields $h_\alpha(\mathbf{r})$ and to finding self-consistent values of these latter.

When elaborating a theory of microdomain formation under the SSR ($D \gg d$), it is customary to assume that density profiles within the boundary region between two domains are similar to those found in the solution of the interface problem. In a series of papers by Helfand and Wasserman,[179–184] lamellar structure formation in melts of diblock and triblock copolymers (12a) and (12b) has been studied. Period D of such a structure is related to the widths D_1 and D_2 of domains of units M_1 and M_2 of the same volume in a simple way

$$D = D_1 + D_2 \tag{159}$$

where, neglecting the interface width, one can calculate values D_1 and D_2 by equations

$$\frac{D_1\bar{\rho}_1}{l_1} = \frac{D_2\bar{\rho}_2}{l_2} = \frac{D}{l_1/\bar{\rho}_1 + l_2/\bar{\rho}_2} \tag{160}$$

The theory of the strong segregation regime (SSR) developed in the studies referred to above is based on an approximation at which the copolymer chain partition function is factorable

$$Q_{12} \approx Q_1 Q_2 d/D \tag{161}$$

becoming a product of partition functions Q_α of each block and the interface contribution d/D. The system free energy, with an allowance for expressions (160) and (161), can be presented as follows

$$\frac{F}{\bar{\rho}VT} = \frac{2\sigma}{TD}\left(\frac{l_1}{\bar{\rho}_1} + \frac{l_2}{\bar{\rho}_2}\right) + \ln\left(\frac{D}{d}\right) - \ln Q_1 - \ln Q_2 - \chi\frac{(l_1/\bar{\rho}_1)(l_2/\bar{\rho}_2)}{(l_1/\bar{\rho}_1) + (l_2/\bar{\rho}_2)} \tag{162}$$

The first term represents the contribution of the interface, and the first logarithmic term describes the entropy loss due to the junction localization on it. The sum of the last three items in the right-hand part of equation (162) equals the free energy of mixing two homopolymers $M_1^{l_1}$ and $M_2^{l_2}$. Here the logarithmic terms provide an entropy contribution, while the last term takes into account the contribution of volume interactions.

The dependence of the values Q_α on the domain sizes D_α has been arrived at by means of a numerical solution of the diffusion equations (126). For the diblock copolymer (12a), this dependence can be successfully interpolated by the equation

$$-\ln Q_\alpha \approx \mathrm{const}(D_\alpha/al_\alpha^{1/2})^{2.5} \tag{163}$$

Substitution of equation (163) into equation (162) with due regard for relationships (160) enables one to find the explicit dependence of the free energy $F(D)$ on the superstructure period D which can be obtained by minimizing the function $F(D)$. In particular, at $\chi l = 37$, the expression for the period $D \approx 3.1al^{1/2}$ was found.

This theory has been extended to the description of domains of either spherical[185] or cylindrical[186] form. Further development of the theory for the SSR[187,188] was associated with the improvement of the numerical calculation method of solution of the diffusion equations (126). As a result, the authors of the paper[187] managed, with no recourse to the approximation of factorization (161), to describe not only the SSR for the solution of a diblock copolymer in an athermic solvent, but also to follow the transition from the WSR to the SSR. For the superstructure period in these two limit cases there was obtained the following expression

$$D \approx a\chi^p l^q \Phi_c^r \tag{164}$$

where Φ_c is the copolymer volume fraction. For the WSR, exponents in equation (164) coincide with those found in terms of the Landau theory

$$p \approx \tfrac{1}{3}, \qquad q \approx 0.8, \qquad r \approx 0.4 \tag{165}$$

By contrast, for the SSR they prove to be

$$p \approx 0.2, \qquad q \approx \tfrac{2}{3}, \qquad r \approx 0.22 \tag{166}$$

13.8.3.2 Modification of the Landau theory

At first glance it appears impossible to describe the SSR in terms of the Landau theory in view of the fact that the difference in densities $\Delta\rho_\alpha(\mathbf{r})$ is not small for this regime compared to $\bar{\rho}_\alpha$. Besides, the profiles $\rho_\alpha(\mathbf{r})$ are here very anharmonic, so that all coefficients of the expansion of this periodic function into Fourier series are of the same order.

The above difficulties can be surmounted provided that correct account is taken of normally dropped terms in the expansion of the Landau free energy of the parameter of order $\psi(\mathbf{r})$.[189–191] The central idea of these papers consists of separating the short-range F^S and the long-range F^L contributions in the free energy function

$$F[\psi] = F^S[\psi] + F^L[\psi] \tag{167}$$

The substantiation of such a separation of contributions lies with the existence of a hierarchy of spatial scales of the domains forming under the SSR. The first such a scale, d, defining the interface width, proves to be small in comparison with the Gaussian block size. Function F^S describes all effects responsible for system properties on such small scales. This function has a sharp minimum on configurations $\psi(\mathbf{r})$, defining the density profile between domains, and its value on such configurations

$$F^S[\psi] = \sigma A \tag{168}$$

is expressed through the interfacial tension σ and the total area of the interface A. The value of σ, entering into theoretical expressions as a phenomenological parameter, has not been calculated in these works.

The second scale, D, defining the characteristic size of domains, appears to be large compared to the Gaussian block size because of the strong extension of chains inside the domains. The contribution to the free energy due to the alteration $\psi(\mathbf{r})$ on this scale is described by function F^L. In view of the existence of hierarchy scales, $D \gg d$, the coarse-grained stepwise function $\bar{\psi}(\mathbf{r}) = a^{-3}\psi(\mathbf{r}/D)$ may be substituted into the function F^L. As a consequence of the Edwards screening of unit interactions in melts,[115] sufficiently long chains are asymptotically Gaussian ones. This result known for homopolymer chains has been verified by the investigation of the screening effect first in nonextended heteropolymer chains[128] and then with due regard for their extension.[192] Hence, to find the density functional F^L, it will suffice to make recourse to the Gaussian model of noninteracting chains. A scaling analysis[192] resting on the above presented considerations yields the following expression for the free energy function

$$\frac{F[\psi]}{V} = \frac{\sigma}{D}\left[\bar{A} + \sum_{n=2}^{\infty} \sum_{\mathbf{k}_1 \ldots \mathbf{k}_n} \frac{g_n(\bar{\omega}_n)}{k_1^2 + \ldots + k_n^2} \bar{\psi}_{\mathbf{k}_1} \ldots \bar{\psi}_{\mathbf{k}_n}\right] \tag{169}$$

where $\bar{A} = Da/V$ represents a dimensionless surface of the domain walls, $\bar{\psi}_{\mathbf{k}}$ stand for dimensionless Fourier transforms of the function $\bar{\psi}(x)$, and summing is over dimensionless wave vectors $\mathbf{k}_1, \ldots, \mathbf{k}_n$ of the reciprocal lattice of a superstructure. The domain characteristic size was found to be describable by the following scaling expression

$$D = (a^5\sigma)^{1/3} l^{2/3} \tag{170}$$

whereas the functions $g_n(\bar{\omega}_n)$ in equation (169) are controlled solely by the relative orientation $\bar{\omega}_n$ of the vectors $\mathbf{k}_1, \ldots, \mathbf{k}_n$. In practice these have been calculated only up to terms with $n = 4$. As a result, for fractions f' and f'' of units M_1 under transitions from bcc into a triangular structure and from this into a lamellar one, there have been calculated the following values: $f' = 0.194$ and $f'' = 0.35$. The exponent $\frac{2}{3}$ in equation (170) is in excellent agreement with the results of the numerical calculations in equation (166). It should be pointed out, however, that the evaluation of higher-order terms of series (169) changes this result materially. That is why another method of determination of function F^{L} scaling asymptotics has been devised[191] which is not concerned with the Landau expansion (169). Computer calculations of the parameter of the lamellar structure, carried out within the framework of this method, reproduce the results obtained by Semenov[193] as well as Helfand and Wasserman[181,182] in the SSR. This method verifies also the scaling dependence $D \simeq l^{2/3}$ (equation 170), which has been suggested to derive phenomenologically using the expression

$$F = \frac{\sigma}{n}\frac{2L}{D}L^2 + \frac{3}{2la^2}\left(\frac{D}{2}\right)^2 \tag{171}$$

for the free energy of a typical molecule with a degree of polymerization l. The first term in equation (171) characterizes the excess of its energy due to interface formation, while the second item stands for entropic losses by the chain under its extension up to a distance $D/2$ between its ends. Here n is the number of copolymer chains in the cube with linear dimension L and the factor $2L/D$ represents the number of interfaces. The minimum of function F (equation 171) is attained when the domain size is

$$D = 2(\sigma a^2/3\bar{\rho})^{1/3} l^{2/3} \tag{172}$$

where $\bar{\rho} = nl/L^3$ is the average density of monomer units. Equation (172) coincides with the result (equation 170) of more sophisticated calculations in the case of a polymer melt.

The dependence of the period D of a superstructure on the fraction f of units M_1 in a diblock copolymer molecule (see equation (12a)) is discussed in terms of the Landau theory,[194] where this dependence was shown to be of importance only under the WSR. Under the SSR, both scales, d and D, reach the asymptotic values which are not controlled by the parameter f. This inference correlates well with the result obtained by Wang and Safran.[195]

ACKNOWLEDGEMENTS

One of the authors (S. I. K.) expresses his gratitude to Professors K. Binder, G. H. Fredrickson, T. Hashimoto, L. Leibler, and J. M. Prausnitz, who provided him with the reprints of their papers. S. I. K. is very indebted to Professor S. Edwards for fruitful discussions on the application of field theory approaches to the thermodynamics of heteropolymers.

13.9 REFERENCES

1. K. Binder, *Adv. Polym. Sci.*, 1994, **112,** 181.
2. K. Šolc, *J. Polym. Sci.*, 1982, **20,** 1947.
3. K. Šolc, *Macromolecules,* 1983, **16,** 236.
4. K. Šolc, L. A. Kleintjens and R. Koningsveld, *Macromolecules,* 1984, **17,** 573.
5. K. Šolc and K. Battjes, *Macromolecules,* 1985, **18,** 220.
6. K. Šolc, *Macromolecules,* 1986, **19,** 1166.
7. K. Šolc, *Macromolecules,* 1987, **20,** 2506.
8. K. Šolc and Y. C. Yang, *Macromolecules,* 1988, **21,** 829.
9. K. Šolc, Y.-H. Huang and Y. C. Yang, *Macromolecules,* 1989, **22,** 4334.
10. S. I. Kuchanov, 'Methods of Kinetic Calculations in Polymer Chemistry', (in Russian), Khimia, Moscow, 1978.
11. 'Markov Chains and Monte Carlo Calculations in Polymer Science', ed. G. Lawry, Dekker, New York, 1970.
12. J. L. Koenig, 'Chemical Microstructure of Polymer Chains', Wiley, New York, 1980.
13. S. I. Kuchanov, in 'Mathematical Methods in Contemporary Chemistry', ed. S. I. Kuchanov, Gordon and Breach, New York, 1996, chap. 5, pp. 267–368.
14. L. H. Sperling, in 'Recent Advances in Polymer Blends, Grafts and Blocks', ed. L. H. Sperling, Plenum, New York, 1974, p. 93.
15. L. H. Sperling and E. M. Corwin, *Adv. Chem. Ser.*, 1979, **176,** 609.
16. P. J. Flory, 'Principles of Polymer Chemistry', Cornell University Press, New York, 1953.
17. H. Tompa, 'Polymer Solutions', Butterworth, London, 1956.
18. M. L. Higgins, 'Physical Chemistry of High Polymers', Wiley, New York, 1958.
19. I. Prigogine, 'Molecular Theory of Solutions', North-Holland, Amsterdam, 1957.
20. K. Kamide, 'Thermodynamics of Polymer Solutions', Elsevier, Amsterdam, 1990.

21. O. Olabisi, L. M. Roveson and M. T. Shaw, 'Polymer–Polymer Miscibility', Academic, New York, 1979.
22. L. A. Utracki, 'Polymer Alloys and Blends. Thermodynamics and Rheology', Hanser, Munich, 1989.
23. R. Koningsveld, in 'Polymer Science', ed. A. D. Jenkins, North-Holland, Amsterdam, 1972, vol. 5, p. 1047.
24. R. Koningsveld, *Br. Polym. J.*, 1975, **7,** 435.
25. R. Koningsveld and L. A. Kleintjens, *J. Polym. Sci., Ser. C,* 1977, **N61,** 221.
26. R. Koningsveld, M. H. Onclin and L. A. Kleintjens, in 'Polymer Compatibility and Incompatibility', ed. K. Šolc, Harwood, New York, 1982, p. 25.
27. R. Koningsveld and L. A. Kleintjens, in 'Polymer Blends and Mixtures', ed. D. J. Walsh, J. S. Huggins and A. Maconnachie, Nijhoff, Boston, 1985, p. 89.
28. E. J. Beckman, R. S. Porter, R. Koningsveld and L. A. Kleintjens, in 'Integration of Fundamental Polymer Science and Technology', ed. P. J. Lemstra and L. A. Kleintjens, Elsevier, Amsterdam, 1988, p. 197.
29. R. Koningsveld, *Pure Appl. Chem.*, 1989, **61,** 1051.
30. I. C. Sanchez, in 'Polymer Blends', ed. D. R. Paul and S. Newman, Academic, New York, 1978, chap. 3.
31. I. C. Sanchez, in 'Polymer Compatibility and Incompatibility', ed. K. Šolc, Harwood, New York, 1982, p. 59.
32. W. J. MacKnight and F. E. Karasz, in 'Comprehensive Polymer Science', ed. G. Allen, Pergamon, Oxford, 1989, vol. 7, p. 111.
33. D. R. Paul and J. W. Barlow, in 'Polymer Compatibility and Incompatibility', ed. K. Šolc, Harwood, New York, 1982, p. 1.
34. D. R. Paul, *Adv. Chem. Ser.*, 1986, **211,** 3.
35. D. R. Paul, J. W. Barlow and H. Keskkulla, in 'Encylopedia of Polymer Science and Engineering', Wiley, New York, 1988, vol. 12. p. 399.
36. S. Krause, *J. Macromol. Sci., Ser. C,* 1972, **7,** 251.
37. S. Krause, in 'Polymer Blends', ed. D. R. Paul and S. Newman, Academic, New York, 1978, chap. 2.
38. J. G. Curro, *J. Macromol. Sci., Rev. Macromol. Chem., Ser. C,* 1974, **11,** 321.
39. T. Kwei and T. Wang, in 'Polymer Blends', eds. D. R. Paul and S. Newman, Academic, New York, 1978, chap. 2.
40. J. W. Kennedy, in 'Macromolecular Chemistry', Royal Society of Chemistry, London, 1980, vol. 1, p. 296; 1984, vol. 3, p. 248.
41. R. Koningsveld, L. A. Kleintjens and M. H. Onclin, *J. Macromol. Sci., Ser. B,* 1980, **18,** 363.
42. D. J. Walsh, in 'Comprehensive Polymer Science', ed. G. Allen, Pergamon, Oxford, 1989, vol. 2, p. 135.
43. M. T. Rätzch and H. Kehlen, *Prog. Polym. Sci.,* 1989, **14,** 1.
44. S. Rostami, in 'Multicomponent Polymer Systems', ed. I. S. Miles and S. Rostami, Longman, London, 1992, p. 63.
45. R. L. Catterman, R. Bender and J. M. Prausnitz, *Ind. Eng. Chem., Process. Res. Dev.,* 1985, **24,** 194.
46. W. H. Stockmayer, L. D. Moore, M. Fixman and B. N. Epstein, *J. Polym. Sci.,* 1955, **26,** 517.
47. S. Krause, A. L. Smith and M. G. Duden, *J. Chem. Phys.,* 1965, **43,** 2144.
48. R. P. Kambour, J. T. Bendler and R. C. Bopp, *Macromolecules,* 1983, **16,** 753.
49. G. ten Brinke, F. E. Karasz and W. J. MacKnight, *Macromolecules,* 1983, **16,** 1827.
50. D. R. Paul and J. W. Barlow, *Polymer,* 1984, **25,** 487.
51. R. Koningsveld and L. A. Kleintjens, *Macromolecules,* 1985, **18,** 243.
52. J. M. G. Cowie, G. Li, R. Ferguson and I. J. McEwen, *J. Polym. Sci., Ser. B,* 1992, **30,** 1351.
53. D. Rigby, J. L. Lin and R.-J. Roe, *Macromolecules,* 1985, **18,** 2269.
54. J. W. Barlow and D. R. Paul, *Polym. Eng. Sci.,* 1987, **27,** 1482.
55. M. Nishimoto, H. Keskkula and D. R. Paul, *Polymer,* 1989, **30,** 1279.
56. G. R. Brannock and D. R. Paul, *Macromolecules,* 1990, **23,** 5240.
57. T. Schioni, F. E. Karasz and W. J. MacKnight, *Macromolecules,* 1986, **19,** 2274.
58. S. I. Kuchanov, *Polym. Networks Blends,* 1995, **5,** 191.
59. R. L. Scott, *J. Polym. Sci.,* 1952, **9,** 423.
60. R. Koningsveld, L. A. Kleintjens and G. Markert, *Macromolecules,* 1977, **10,** 1105.
61. L. Leibler, *Makromol. Chem., Rapid Commun.,* 1981, **2,** 393.
62. R. Koningsveld and W. J. MacKnight, *Makromol. Chem.,* 1989, **190,** 419.
63. M. T. Ratzsch, H. Kehlen and D. Browarzik, *J. Macromol. Sci., Ser. A,* 1986, **23,** 1349.
64. M. T. Ratzsch, D. Browarzik and H. Kehlen, *J. Macromol. Sci., Ser. A,* 1989, **26,** 903.
65. M. T. Ratzsch, C. Wohlfarth, D. Browarzik and H. Kehlen, *J. Macromol. Sci., Ser. A,* 1989, **26,** 1497.
66. M. T. Ratzsch, D. Browarzik and H. Kehlen, *J. Macromol. Sci., Ser. A,* 1990, **27,** 809.
67. M. T. Ratzsch, S. Enders, L. Tschersich and H. Kehlen, *J. Macromol. Sci., Ser. A,* 1991, **28,** 31.
68. A. V. Balazs, I. C. Sanchez, I. R. Epstein, F. E. Karasz and W. J. MacKnight, *Macromolecules,* 1985, **18,** 2188.
69. A. V. Balazs, F. E. Karasz, W. J. MacKnight, H. Ueda and I. C. Sanchez, *Macromolecules,* 1985, **18,** 2784.
70. J. van Hunsel, A. C. Balazs, R. Koningsveld and W. J. MacKnight, *Macromolecules,* 1988, **21,** 1528.
71. A. C. Balazs and M. T. Meuse, *Macromolecules,* 1989, **22,** 4260.
72. H.-J. Cantow and O. Schulz, *Polym. Bull.,* 1986, **15,** 449.
73. H.-J. Cantow and O. Schulz, *Polym. Bull.,* 1986, **15,** 539.
74. J. W. Barlow, C. H. Lai and D. R. Paul, in 'Multiphase Macromolecular Systems', ed. B. M. Culberton, Plenum, New York, 1989, p. 505.
75. E. A. Guggenheim, 'Mixtures', Clarendon Press, Oxford, 1952.
76. C. G. Payayiotou, *Makromol. Chem.,* 1987, **188,** 2733.
77. P. Dafniotis, E. Karayianni and C. Panayiotou, *Makromol. Chem.,* 1989, **190,** 1103.
78. T. Shiomi, H. Ishimatsu, T. Eguchi and K. Imai, *Macromolecules,* 1990, **23,** 4970.
79. T. Shiomi, T. Eguchi, H. Ishimatsu and K. Imai, *Macromolecules,* 1990, **23,** 4978.
80. T. Shiomi and K. Imai, *Polymer,* 1991, **32,** 73.
81. D. R. Paul, *Adv. Chem. Ser.,* 1986, **211,** 354.
82. F. E. Karasz and W. J. MacKnight, *Adv. Chem. Ser.,* 1986, **211,** 111.
83. R. J. Roe and D. Rigby, *Adv. Polym. Sci.,* 1987, **82,** 103.
84. W. J. MacKnight and F. E. Karasz, in 'Comprehensive Polymer Science', ed. G. Allen, Pergamon, Oxford, 1989, vol. 7, p. 111.

85. D. J. Walsh, in 'Comprehensive Polymer Science', ed. G. Allen. Pergamon, Oxford, 1989, vol. 2, p. 135.
86. M. T. Ratzsch and H. Kehlen, in 'Frontiers of Macromolecular Science', ed. T. Saegusa, T. Higashimura and A. Abe, Blackwell, Oxford, 1989, p. 76.
87. P. Munk, P. Hattam and Q. Du, *J. Appl. Polym. Sci., Appl. Polym. Symp.*, 1989, **43,** 373.
88. R. Koningsveld, W. H. Stockmayer and E. Nies, *Makromol. Chem., Macromol. Symp.*, 1990, **39,** 1.
89. C. Wohlfarth, *Makromol. Chem., Theory Simul.*, 1993, **2,** 605.
90. H. W. Kammer, J. Kressler and C. Kummerloewe, *Adv. Polym. Sci.*, 1993, **106,** 31.
91. J. Dudowicz and K. F. Freed, *Macromolecules,* 1991, **24,** 5076, 5112.
92. K. F. Freed and J. Dudowicz, *Theor. Chim. Acta,* 1992, **82,** 357.
93. J. Dudowicz and K. F. Freed, *J. Chem. Phys.*, 1992, **96,** 1644, 9147.
94. K. F. Freed and J. Dudowicz, *J. Chem. Phys.*, 1992, **97,** 2105.
95. K. S. Schweizer, *Macromolecules,* 1993, **26,** 6033, and references therein.
96. D. Chandler and H. C. Andersen, *J. Chem. Phys.*, 1972, **57,** 1930.
97. K. S. Schweizer and J. G. Curro, *Chem. Phys.*, 1990, **149,** 105.
98. K. S. Schweizer and J. G. Curro, *J. Chem. Phys.*, 1991, **94,** 3986.
99. K. S. Schweizer and J. G. Curro, *Phys. Rev. Lett.*, 1988, **60,** 809.
100. K. S. Schweizer, *Macromolecules,* 1993, **26,** 6050.
101. J. A. Barker and D. Henderson, *J. Chem. Phys.*, 1967, **47,** 4714.
102. Y. Song, S. M. Lambert and J. M. Prausnitz, *Chem. Eng. Sci.*, 1994, **49,** 2765, and references therein.
103. Y. C. Chiew, *Mol. Phys.*, 1990, **70,** 129.
104. Y. Song, S. M. Lambert and J. M. Prausnitz, *Macromolecules,* 1994, **27,** 441.
105. T. Hino, Y. Song and J. M. Prausnitz, *Macromolecules,* 1994, **27,** 5681.
106. T. Hino, Y. Song and J. M. Prausnitz, *Macromolecules,* 1995, **28,** 5717, 5725.
107. L. D. Landau and E. M. Lifshitz, 'Statistical Physics', Addison-Wesley, Reading, MA, 1969.
108. S. I. Kuchanov and S. V. Panyukov, *Macromolecules,* 1995, submitted for publication.
109. C. A. Croxton, 'Liquid State Physics – A Statistical Mechanical Introduction', Cambridge University Press, Cambridge, 1974.
110. A. Aharony, in 'Critical Phenomena', Lecture Notes in Physics, Springer, Heidelberg, 1983, vol. 186, p. 209.
111. J. Sivardiere, in 'Static Critical Phenomena in Inhomogeneous Systems', Lecture Notes in Physics, Springer, Heidelberg, 1984, vol. 206, p. 247.
112. F. Harary, 'Graph Theory', Addison-Wesley, Reading, MA, 1969.
113. S. V. Panyukov and S. I. Kuchanov, *Sov. JETP,* 1991, **72,** 368.
114. S. V. Panyukov and S. I. Kuchanov, *J. Phys. II,* 1992, **2,** 1973.
115. M. Doi and S. F. Edwards, 'The Theory of Polymer Dynamics', Clarendon Press, Oxford, 1986.
116. K. F. Freed, 'Renormalization Group Theory of Macromolecules', Wiley, New York, 1987.
117. J. M. Ziman, 'Models of Disorder', Cambridge University Press, Cambridge, 1979.
118. 'Density Functional Theory', ed. J. Keller and J. L. Gazquez, Lecture Notes in Physics, Springer, Heidelberg, 1983, vol. 187.
119. I. M. Lifshitz, A. Yu. Grosberg and A. R. Khokhlov, *Rev. Mod. Phys.*, 1978, **50,** 683.
120. I. A. Katime and J. R. Quintana, in 'Comprehensive Polymer Science', ed. G. Allen, Pergamon, Oxford, 1989, vol. 1, p. 103.
121. H. Benoit and D. Froelich, in 'Light Scattering from Polymer Solutions', ed. M. B. Huglin, Pergamon, London, 1972, p. 467.
122. P. G. de Gennes, 'Scaling Conceptions in Polymer Physics', Cornell University Press, Ithaca, NY, 1979.
123. L. Leibler, *Macromolecules,* 1982, **15,** 1283.
124. I. Ya. Erukhimovich, *Vysokomol. Soedin., Ser. A,* 1979, **21,** 427.
125. H. Benoit, M. Benmouna and W. Wu, *Macromolecules,* 1990, **23,** 1511.
126. J. G. Curro and K. S. Schweizer, *Macromolecules,* 1987, **20,** 1928.
127. M. Olvera de la Cruz and I. Sanchez, *Macromolecules,* 1986, **19,** 2051.
128. T. A. Vilgis and R. Borsali, *Macromolecules,* 1990, **23,** 3172.
129. T. A. Vilgis, M. Benmouna and H. Benoit, *Macromolecules,* 1991, **24,** 4481.
130. L. Leibler, *Macromolecules,* 1980, **13,** 1602.
131. I. Ya. Erukhimovich, *Vysokomol. Soedin., Ser. A,* 1982, **24,** 1942.
132. P. G. de Gennes, *J. Phys.*, 1970, **31,** 235.
133. W. E. McMullen and K. F. Freed, *Macromolecules,* 1990, **23,** 255.
134. H. Tang and K. F. Freed, *Macromolecules,* 1991, **24,** 958.
135. L. Leibler and H. Benoit, *Polymer,* 1981, **22,** 195.
136. K. M. Hong and J. Noolandi, *Polym. Commun.*, 1984, **25,** 265.
137. M. Olvera de la Cruz and I. C. Sanchez, *Macromolecules,* 1987, **20,** 440.
138. L. Leibler, H. Orland and J. C. Wheeler, *J. Chem. Phys.*, 1983, **79,** 3550.
139. T. A. Kavassalis and M. D. Whitmore, *Macromolecules,* 1991, **24,** 5340.
140. P. G. de Gennes, *Faraday Discuss. Chem. Soc.*, 1979, **68,** 96.
141. Y. Ijichi and T. Hashimoto, *Polym. Commun.*, 1988, **29,** 135.
142. R. M. Hornreich, M. Luban and S. Strikman, *Phys. Rev. Lett.* 1975, **35,** 1678.
143. D. Braseta and G. H. Fredrickson, *J. Chem. Phys.*, 1990, **93,** 292.
144. G. H. Fredrickson and S. T. Milner, *Phys. Rev. Lett.*, 1991, **67,** 835.
145. A. Dondos, *Makromol. Chem.*, 1971, **147,** 123.
146. A. Dondos, *J. Polym. Sci., Polym. Lett. Ed.*, 1971, **9,** 871.
147. E. Helfand and Y. Tagami, *J. Polym. Sci., Ser. B,* 1971, **9,** 741.
148. E. Helfand, in 'Polymer Compatibility and Incompatibility', ed. K. Šolc, Harwood, New York, 1983, p. 143.
149. E. Helfand, *J. Chem. Phys.*, 1975, **63,** 2192.
150. J. Noolandi and K. M. Hong, *Macromolecules,* 1982, **15,** 482.
151. J. Noolandi and K. M. Hong, *Macromolecules,* 1984, **17,** 1531.

152. T. A. Vilgis and J. Noolandi, *Macromolecules,* 1990, **23,** 2941.
153. L. Leibler, *Macromolecules,* 1982, **15,** 1283.
154. T. Inoue, T. Soen, T. Hashimoto and H. Kawai, in 'Block Copolymers', ed. S. L. Aggarwal, Plenum, New York, 1970, p. 53.
155. S. Krause, in 'Colloid and Morphological Behaviour of Block and Graft Copolymers', ed. G. E. Molau, Plenum, New York, 1971, p. 223.
156. S. Krause, in 'Block and Graft Copolymers', ed. J. J. Burke and V. Weiss, Syracuse University Press, Syracuse, NY, 1973, p. 143.
157. A. E. Skoulous, in 'Developments in Block Copolymers', ed. I. Goodman, Academic, New York, 1982, vol. 1, p. 81.
158. E. Gelfand, Z. R. Wasserman, in 'Developments in Block Copolymers', ed. I. Goodman, Academic, New York, 1982, vol. 1, p. 99.
159. T. Hashimoto, M. Shibayama, M. Fujimura and H. Kawai. in 'Block Copolymers', ed. D. J. Meier, Harwood, New York, 1983, p. 63.
160. B. R. M. Gallot, *Adv. Polym. Sci.,* 1978, **29,** 85.
161. R. A. Brown, A. J. Masters, C. Price and X. F. Yuan, in 'Comprehensive Polymer Science', ed. G. Allen, Pergamon, Oxford, 1989, vol. 2, p. 155.
162. T. Hashimoto, H. Tanaka and N. Iizuka, in 'Space–Time Organization in Macromolecular Fluids', ed. F. Tanaka, M. Doi and T. Ohta, Springer, Heidelberg, 1989, p. 2.
163. G. H. Fredrickson, E. Helfand, F. S. Bates and L. Leibler, in 'Space–Time Organization in Macromolecular Fluids', ed. F. Tanaka, M. Doi and T. Ohta, Springer, Heidelberg, 1989, p. 13.
164. M. D. Whitmore, in 'Space–Time Organization in Macromolecular Fluids', ed. F. Tanaka, M. Doi and T. Ohta, Springer, Heidelberg, 1989, p. 30.
165. Yu. D. Shivanov and Yu. K. Godovski, *Usp. Khim.,* 1988, **57,** 1713.
166. S. F. Edwards, *J. Phys. A,* 1974, **7,** 332.
167. E. Helfand, *Acc. Chem. Res.,* 1975, **8,** 295.
168. S. V. Panyukov and S. I. Kuchanov, *Sov. JETP Lett.,* 1992, **54,** 501.
169. G. H. Fredrickson and L. Leibler, *Macromolecules,* 1989, **22,** 1238.
170. J. Noolandi and K. M. Hong, *Polym. Bull.,* 1988, **7,** 561.
171. K. M. Hong and J. Noolandi, *Macromolecules,* 1983, **16,** 1083.
172. A. M. Mayes and M. Olvera de la Cruz, *J. Chem. Phys.* 1989, **91,** 7228.
173. A. M. Hayes and M. Olvera de la Cruz, *Macromolecules,* 1991, **24,** 3975.
174. J. N. Owens, I. S. Gancarz, J. T. Koberstein and T. P. Russell, *Macromolecules,* 1989, **22,** 3380.
175. M. Olvera de la Cruz, A. M. Mayes and B. W. Swift, *Macromolecules,* 1992, **25,** 944.
176. A. V. Dobrynin and I. Ya. Erukhimovich, *Vysokomol. Soedin., Ser. A,* 1991, **33,** 1100.
177. C. Burger, W. Ruland and A. N. Semenov, *Macromolecules,* 1990, **23,** 3339.
178. E. I. Shakhnovich and A. M. Gutin, *J. Phys.,* 1989, **50,** 1843.
179. E. Helfand, in 'Recent Advances in Polymer Blends, Grafts and Blocks', ed. L. H. Sperling, Plenum, New York, 1974, p. 141.
180. E. Helfand, *Acc. Chem. Res.,* 1975, **8,** 295.
181. E. Helfand, *Macromolecules,* 1975, **8,** 552.
182. E. Helfand and Z. R. Wasserman, *Macromolecules,* 1976, **9,** 879.
183. E. Helfand, *J. Chem. Phys.,* 1975, **62,** 999.
184. E. Helfand and Z. R. Wasserman, *Polym. Eng. Sci.,* 1977, **17,** 582.
185. E. Helfand and Z. R. Wasserman, *Macromolecules,* 1978, **11,** 960.
186. E. Helfand and Z. R. Wasserman, *Macromolecules,* 1980, **13,** 994.
187. M. D. Whitmore and J. Noolandi, *J. Chem. Phys.,* 1990, **93,** 2946.
188. M. D. Whitmore and J. D. Vavasour, *Macromolecules,* 1992, **25,** 2041.
189. T. Ohta and K. Kawasaki, *Macromolecules,* 1986, **19,** 2621.
190. K. Kawasaki, T. Ohta and M. Kohrogui, *Macromolecules,* 1988, **21,** 2972.
191. K. Kawasaki and T. Kawakatzu, *Macromolecules,* 1990, **23,** 4006.
192. K. Kawasaki, T. Ohta and M. Kohrogui, *Macromolecules,* 1986, **19,** 2621.
193. A. N. Semenov, *Sov. Phys. JETP,* 1985, **61,** 733.
194. T. Ohta and K. Kawsaki, *Macromolecules,* 1990, **23,** 2413.
195. Z. G. Wang and S. A. Safran, *J. Phys.,* 1990, **51,** 185.

14
Self-assembly and Morphology of Block Copolymer Systems

HIROKAZU HASEGAWA and TAKEJI HASHIMOTO
Kyoto University, Japan

14.1 BLOCK COPOLYMER SYSTEMS

The block copolymer systems discussed in this chapter are the following three systems: (i) neat block copolymers, (ii) blends of block copolymers and homopolymers, and (iii) blends of block copolymers and block copolymers. We shall focus primarily on recent advances in this field since about 1990. Before going into detailed discussion, each system will be briefly introduced.

14.1.1 Block Copolymers

The morphology of microphase separated structure (microdomain structure) in the segregation limit has been investigated for diblock copolymers of polystyrene (PS) and polyisoprene (PI) or polybutadiene (PB) (designated as SI or SB), for those of poly(ethylene-*alt*-propylene)

(PEP) and poly(ethylethylene) (PEE) (designated as PEP-PEE), for triblock copolymers such as PS-*block*-PI-*block*-PS (SIS) and PS-*block*-PB-*block*-PS (SBS), for linear ABC triblock copolymers such as PS-*block*-PI-*block*-poly(methyl methacrylate) (PMMA),[1] PS-*block*-PB-*block*-poly(4-vinylpyridine),[2] PS-*block*-PI-*block*-poly[(4-vinylbenzyl)dimethylamine] (SIA)[3] or SAI,[4] PI-*block*-PS-*block*-poly(2-vinylpyridine) (ISP)[5] or PIS,[6] PS-*block*-PB-*block*-PMMA,[7] and PS-*block*-poly(ethylene-*ran*-butene)-*block*-PMMA,[7] and for the ABC star block copolymer of PS, poly(dimethylsiloxane), and poly(*t*-butyl methacrylate).[8] We shall especially highlight new morphologies other than the classical morphologies, such as spheres in cubic lattices, hexagonally packed cylinders, and alternating lamellae.

The microphase transition, the so-called microphase separation transition (MST)[9] or order–disorder transition (ODT),[10] between the phase-separated state and the single-phase state has also been extensively investigated.[11] Recently, attention has been focused on the order–order transition (OOT),[12–15] *i.e.* the phase transition or morphology transition between different morphologies (*e.g.* the transition between the spheres and the cylinders,[12] or between the cylinders and the lamellae).[13,14]

14.1.2 Blends of Block Copolymers and Homopolymers

We will discuss binary blends of an A–B diblock copolymer and a homopolymer A (designated as HA) or B (HB) and ternary blends of A–B and HA and HB. A key parameter controlling the microdomain morphology of such systems is $r_A = N_{HA}/N_A$ and/or $r_B = N_{HB}/N_B$, where N_A and N_B are the degrees of polymerization (DP) of A and B block chains, respectively, and N_{HA} and N_{HB} are the DPs of HA and HB, respectively. In this chapter we shall consider the case of $N_A \approx N_B$ such that neat A–B diblock copolymers form the lamellar microdomain. Depending on the value r_A or r_B, we have a 'wet brush' criterion in the case of $r_A < 1$, $r_B < 1$, and a 'dry brush' criterion in the case of $r_A \approx 1$, $r_B \approx 1$. In the wet brush criterion, the lamellar morphology of an A–B block copolymer changes according to the volume fraction of the homopolymer(s) ϕ_H or the volume fraction of one of the components ϕ_A or ϕ_B (Section 14.5.2). On the other hand, in the dry brush criterion, upon increasing ϕ_H the morphology changes into vesicles composed either of bimolecular layers of A–B in the case of A–B/HA or A–B/HB blends, or monomolecular layers of A–B in the case of A–B/HA/HB ternary blends, the layers being dispersed in the homopolymer matrix as schematically shown in Figure 1 (Section 14.5.3). We define the wet-brush and dry-brush criteria in Section 14.5.1.

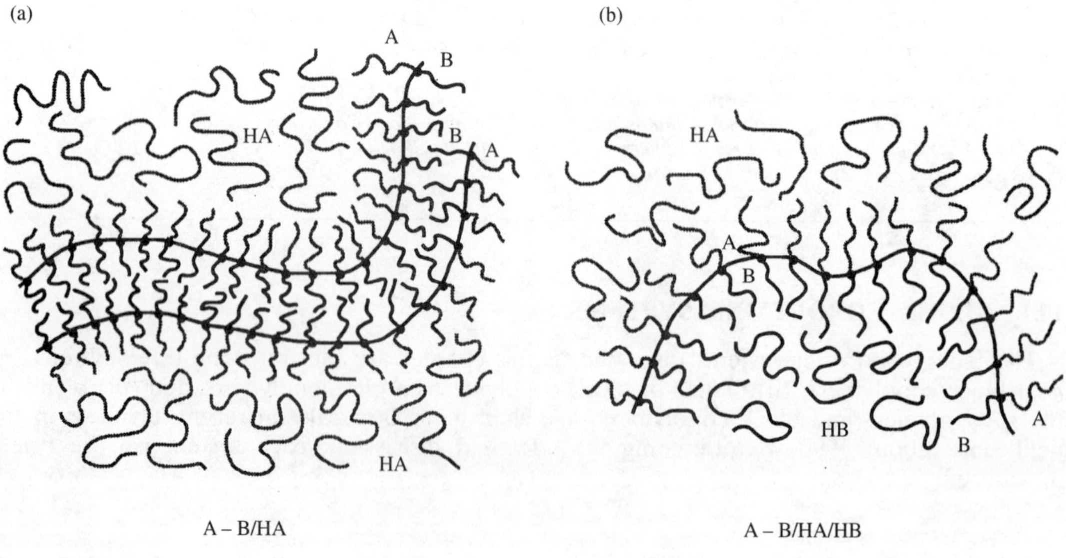

Figure 1 Schematic illustrations of two block copolymer systems, A–B/HA binary mixture (a) and A–B/HA/HB ternary mixture (b), in which A–B is a lamella-forming symmetric diblock copolymer and HA and HB are the corresponding homopolymers with a degree of polymerization comparable to those of the block chains of A–B ('dry brush' regime). In both cases the homopolymer content is much larger than the block copolymer content. (a) Bimolecular layers of A–B are formed and dispersed in the matrix of HA. (b) HA and HB are macrophase separated and a monomolecular layer of A–B is formed at the macrodomain interface.

We shall briefly discuss also the blend system of A–B/HC where HC denotes a homopolymer C such as poly(oxy-2,6-dimethyl-1,4-phenylene) (PPO),[16,17] poly(vinyl methyl ether) (PVME),[18] or poly(α-methylstyrene) (PαMeS),[19,20] and A–B is a SI, SB, or SBS block copolymer. In this case the segmental interaction parameters between C and A and between C and B, as well as the molecular weight (hereafter designated as MW) ratios between the block chains and the homopolymers, play an important role in controlling the morphology. Another important factor in the morphology of such systems is the interplay of two kinds of phase transitions: the microphase transition inherent in the block copolymer and the phase transition inherent in the polymer blends. We define the latter as the 'macrophase transition' in contrast to the microphase transition.

14.1.3 Blends of Two Block Copolymers A_α–B_α and A_β–B_β

Here we shall restrict our systems to a binary copolymer blend composed of segments A and B. The two block copolymers A_α–B_α and A_β–B_β have different MWs and different compositions $f_{A_i} = N_{A_i}/(N_{A_i} + N_{B_i})$ where N_{K_i} is the DP of the K block chain (K = A or B) in the *i* block copolymer ($i = \alpha$ or β). The system under consideration involves the microphase transition and the macrophase transition, as will be discussed in Section 14.6. If the macrophase separation occurs, two kinds of macrodomains coexist: one having the microdomain morphology inherent in the system rich in A_α–B_α, and the other having that rich in A_β–B_β. Intriguing experimental evidence was found that the macrophase separation occurs even in the system with $f_{A_i} \approx f_{B_i} \approx 0.5$ if the asymmetry in N_α and N_β is large, *e.g.* $N_\alpha/N_\beta \gtrsim 10$ (or $N_\alpha/N_\beta) \lesssim 0.1$) where $N_i = N_{A_i} + N_{B_i}$ ($i = \alpha$ or β) (Sections 14.6.2–14.6.4).

In the case in which only microphase separation occurs, two block copolymers A_α–B_α and A_β–B_β are mixed to form a single microdomain morphology. In this case a key physical factor is whether or not the chemical junction points of the two block copolymers are confined in the common interfaces between the coexisting microdomains composed of segments A and B. If they are confined in the interfaces, while keeping A_α and A_β block chains in A microdomains and B_α and B_β in B microdomains, the system should have much less translational entropy than A–B/HA systems, which makes a contribution of the conformational entropy to the free energy of the system much more important than in the A–B/HA systems. A unique 'co-surfactant effect' has been reported for the A_α–B_α/A_β–B_β systems with asymmetric composition f_{A_α} and f_{A_β}:[21,22] a small fraction of the block copolymer solubilized into the other block copolymer can effectively alter the curvature of the interface and hence the morphology of the block copolymer of the major component (Section 14.6.1). On the other hand, if the junctions of one copolymer, *e.g.* A_β–B_β, are not localized in the interfaces of A_α–B_α but rather distributed in the space of the microdomains A and/or B, the blend system A_α–B_α/A_β–B_β becomes analogous to A_α–B_α/HC where HC corresponds to A_β–B_β block copolymers which are solubilized into the A and/or B microdomain space.

14.2 THE MICROPHASE TRANSITION OF PURE BLOCK COPOLYMERS

Meier[23] and Helfand[24] are the pioneers who tried to develop theories of the microdomain structures. Leibler[9] established a firm theoretical basis for the microphase separation transition (MST) of block copolymers on the basis of Landau-type mean-field theory. His work stimulated experimental investigations of MST or ODT.[25–31] The experimental results obtained in turn provoked further theoretical explorations. The fact that it belongs to a new universality class other than the kinetic Ising universality class[32] has also attracted theoretical interest.

The transition had been thought to be a gradual transition and the experimental data obtained by small-angle X-ray scattering (SAXS) or small-angle neutron scattering (SANS) methods had been analyzed[25–31] on the basis of Leibler's Landau-type mean-field theory. Leibler's theory[9] predicts the following: (i) the SAXS or SANS profile of block copoolymer A–B shows a scattering maximum even in the disordered (single-phase) state and the maximum intensity I_m and the scattering vector q_m at the maximum further satisfy the following points; (ii) the inverse of the maximum intensity I_m^{-1} linearly decreases with the inverse of the absolute temperature T^{-1}; (iii) the characteristic wavelength D of the composition (or concentration) fluctuations as estimated from q_m

$$D = 2\pi / q_m \tag{1}$$

has the same temperature dependence as that of the unperturbed chain dimension $R_{g0}(T)$. Most of the experimental results were shown to obey approximately this prediction,[25–31] and the deviations of I_m^{-1} and D from the predictions were taken to be the onset of ODT or MST,[25–27,30,31] as schematically shown in Figure 2(a). Here the ODT temperature thus determined will be defined as the 'mean-field ODT temperature', $T_{ODT,MF}$. The system is thought to be in the ordered state (microphase-separated state) at $T < T_{ODT,MF}$ and in the disordered state at $T > T_{ODT,MF}$. In Figure 2(a), $D_r(T)$ shows the temperature dependence of D as predicted by the mean-field theory,

$$D_r(T) \equiv D(T_r) R_{g0}(T) / R_{g0}(T_r) \tag{2}$$

where T_r is a given reference temperature in the disordered state.

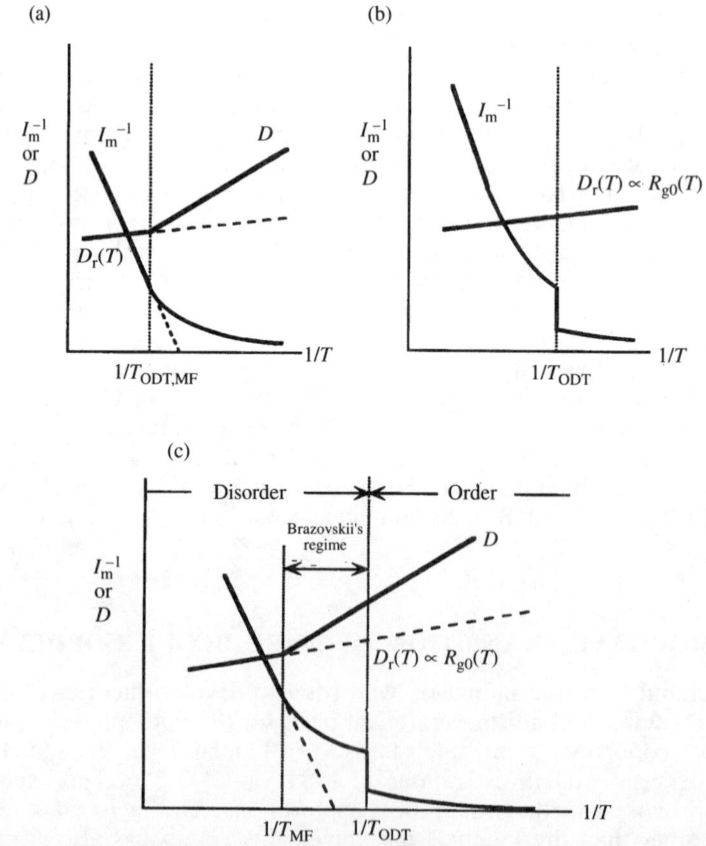

Figure 2 Temperature dependence of characteristic scattering parameters, *i.e.* the inverse maximum intensity I_m^{-1} and the characteristic length D determined from the scattering vector q_m at the first-order scattering maximum for A–B or A–B–A block copolymer having UCST (upper critical solution temperature)-type phase behavior: (a) a conventional view based upon a mean-field picture, (b) a new view, and (c) a unified view (after Sakamoto and Hashimoto[36]).

However, as a more elaborate theory which takes into account the effect of thermal noise (the Brazovskii effect)[33] on the ODT has been proposed by Fredrickson and Helfand[34] and a weakly first-order nature of the transition has been demonstrated, the temperature dependence of I_m and q_m (or D) has been reexamined more carefully. The experimental results obtained by Bates *et al.*[35] for PEP–PEE showed the following deviations from the mean-field prediction:[9] (i) the plot I_m^{-1} vs. T^{-1} in the disordered state is not linear but has a curvature and (ii) it shows a small discontinuous change at the ODT temperature as schematically shown in Figure 2(b). We shall define this temperature as T_{ODT}. Since the discontinuity of I_m^{-1} at T_{ODT} is small, it might be overlooked with the scattering data alone. However, the discontinuity was supported by T_{ODT} values separately determined from the low-frequency rheological behavior as a function of temperature and confirmed by the Fredrickson–Helfand theory. Here we should note a surprising report that D has the same temperature dependence as $R_{g0}(T)$ below and above T_{ODT} in the temperature range covered. This point has not been confirmed so far for SI and SB.[36]

For SI and SB this discontinuous change in I_m^{-1} at T_{ODT} was overlooked for a long time and confirmed only recently.[37–40] In particular, a high-temperature resolution SAXS experiment[39] unequivocally identified T_{ODT} with an accuracy of less than 1 °C, though this temperature T_{ODT} generally depends on the heating or cooling rate as is typical of a first-order phase transition.

Further elaborate experimental work[36] reveal a unified view as shown in Figure 2(c) on the temperature dependence of I_m^{-1} and D across the ODT temperature: we can 'operationally' define a temperature T_{MF} such that at $T > T_{MF}$ in the disordered state, the system approximately obeys the mean-field behavior as found in the earlier work (Figure 2(a)); namely $T_{ODT,MF}$ in Figure 2(a) corresponds to T_{MF} defined in this way but not to T_{ODT} itself; at $T_{ODT} < T < T_{MF}$, the system is still in the disordered state but the thermal noise effects strongly affect the concentration fluctuations in the disordered state so that the deviations (in I_m^{-1} and D) from the mean-field behavior become significant (Brazovskii's regime); at the ODT temperature, discontinuous changes in I_m^{-1} occur as shown in Figure 2(b) and the system is in the ordered state at $T < T_{ODT}$. T_{ODT} in Figure 2(c) corresponds to the T_{ODT} in Figure 2(b). It should be noted that, in the Brazovskii regime, I_m^{-1} (solid line) is much larger than that predicted by the mean-field theory (broken line). This indicates that the scattered intensity I_m and hence the concentration fluctuations in real systems are much suppressed compared with those predicted by the mean-field theory. It should be noted that D has a temperature dependence much greater than that of $D_r(T)$ or $R_{g0}(T)$, in contrast to the case of Figure 2(b), although D shows little change at T_{ODT} as in the case of Figure 2(b). It is expected that the discontinuity and the temperature gap $\Delta T \equiv T_{MF} - T_{ODT}$ will decrease with increasing N, the degree of polymerization of the block copolymer.[34,36] It is needless to say that, from the theoretical point of view, there is no temperature T_{MF} because the mean-field behavior is achieved only asymptotically. However, T_{MF} was operationally defined in such a way that above T_{MF} we can analyze practically our experimental data with good accuracy based on the mean-field theory.

The ordering mechanism below ODT or MST, *i.e.* whether or not it involves spinodal decomposition (SD) or nucleation growth (NG) and how it is different from the conventional SD or NG found in mixtures of HA and HB, is by and large left unexplored and merits future work. If the system has both OOT and ODT, we should be able to distinguish the two transitions. The clear distinction of them is not easy, and this should also merit future work. All these problems offer important research opportunities for a better understanding of the morphology of pure block copolymers. The dynamics of OOT and ODT should be a key physical factor in the morphology control of block copolymers.

14.3 RPA ANALYSIS OF THERMODYNAMIC STABILITY LIMITS FOR THE MICROPHASE TRANSITION AND THE MACROPHASE TRANSITION

Figure 3 shows schematically the microphase transition (or order–disorder transition) of pure A–B diblock copolymers (designated as A–B) (part a) and the macrophase transition between the two homopolymers in the binary mixtures of homopolymers HA and HB (designated as HA/HB) (part b), as well as the equilibrium structures both in the single-phase state (upper half) and the two-phase state (lower half).[41] In the two-phase state (or ordered state in the sense that it has a lower entropy), HA/HB separates into two macroscopic phases where the domain sizes parallel and normal to the interface ($D_{\parallel}$ and $D_{\perp}$, respectively) are much larger than the radius of gyration R_g of the individual polymer coils. On the other hand, A–B cannot separate into the macroscopic phases owing to the existence of a covalent bond between the A and B polymer chains; instead, it rather separates into submicroscopic domains with $D_{\perp}$ nearly equal to R_g.[9,23,24,42–49]

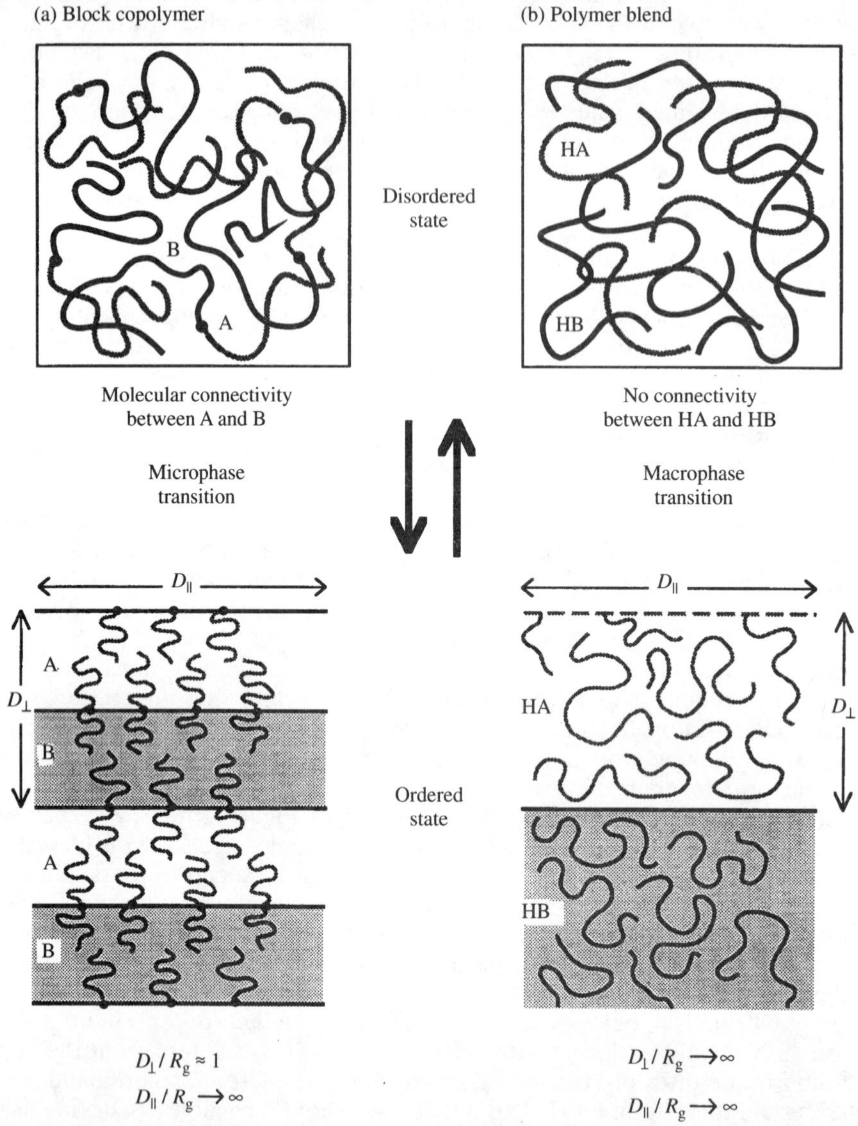

Figure 3 Comparison of concentration fluctuations of (a) A–B block copolymers and (b) HA/HB polymer blends in the single-phase (disordered) states (upper half) and the two-phase (ordered) states at thermal equilibrium (lower half). The phase transition between the two states, *i.e.* order–disorder transition ('microphase transition') for the block copolymers or liquid–liquid phase transition ('macrophase transition') for the polymer blends, occurs at $\chi_{AB}N = (\chi_{AB}N)_c$, where χ_{AB} is the Flory–Huggins interaction parameter between A and B and N is the degree of polymerization.

At the segregation limit the domains have to be arranged periodically in space with a long-range order characteristic of 'crystals' with a super-large lattice spacing D. The sketch in the bottom left of Figure 3 shows only one type of the domain morphology (*i.e.* lamella). The variety in the domain morphology depends on the volume fraction of one of the components A or B in A–B.[11,42,45,50,51] In contrast to the ordinary liquid–liquid phase transition of polymer mixtures, we denote the order–disorder transition of block copolymers as a 'microphase transition', and in contrast to this microphase transition we denote the ordinary phase transition as a 'macrophase transition'. Polymer mixtures containing a block copolymer at least as a component undergo in principle both the macrophase and microphase transitions. How can we determine thermodynamic stability limits of the mixtures for the macrophase and microphase transitions?

The spectra of thermal concentration fluctuations $\langle\eta_q^2\rangle_T$ of two-component polymer systems in the bulk disordered state have been extensively studied both theoretically[9,52–58] and experimentally[11,25,27–31,59–61] (where η_q is the amplitude of the Fourier mode of the fluctuations with the wave number q and $\langle\ \rangle_T$ stands for the thermal average). The spectrum can be determined by measuring the elastically scattered intensity, $I(q)$, of light, neutrons, and X-rays as a function of the magnitude of the scattering vector $q = (4\pi/\lambda)\sin(\theta/2)$, where λ and θ are the wavelength of incident beam and the scattering angle in the medium, respectively. The observable $I(q)$ or $\langle\eta_q^2\rangle_T$ is generally given by[9]

$$\langle\eta_q^2\rangle_T \sim I(q) \sim [S(q)/W(q) - 2\chi]^{-1} \tag{3}$$

for any two-component system on the basis of the mean-field random phase approximation (RPA),[9,52] where χ is the Flory–Huggins interaction parameter between A and B.[62] The functions $S(q)$ and $W(q)$ depend only on the conformational characteristics of polymers A and B and are composed of the terms $S_{ij}(q)$ (i,j = A or B),[9,52] which are the Fourier transforms of the density–density correlation for i and j monomers for Gaussian chains. It is important to note that $[S(q)/W(q) - 2\chi]$ is a thermodynamic (driving) force required for generating the q-Fourier mode of the composition fluctuations.

HA/HB has a maximum value of $I(q)$ or $\langle\eta_q^2\rangle_T$ at $q = 0$ and the stability limit as given by[9,52]

$$I^{-1}(q=0) \sim S(q=0)/W(q=0) - 2\chi_{\mathrm{s,macro}} = 0 \tag{4}$$

On the other hand, A–B has a maximum value of $I(q)$ or $\langle\eta_q^2\rangle_T$ at $q = q_{\mathrm{m}} \sim 1/R_{\mathrm{g}}$, R_{g} being the gyration radius of the block copolymer chain as a whole, and the stability limit as given by[9]

$$I^{-1}(q=q_{\mathrm{m}}) \sim S(q=q_{\mathrm{m}})/W(q=q_{\mathrm{m}}) - 2\chi_{\mathrm{s,micro}} = 0 \tag{5}$$

where $\chi_{\mathrm{s,macro}}$ and $\chi_{\mathrm{s,micro}}$ are the χ parameters at the spinodal points for the HA/HB and the A–B systems, respectively. In order to distinguish the stability limits at $q = 0$ and at $q = q_{\mathrm{m}}$, we define them as the stability limits for the macrophase transition and the microphase transition.[41,58] Polymer mixtures containing at least one block copolymer as a component can generally have both $\chi_{\mathrm{s,micro}}$ and $\chi_{\mathrm{s,macro}}$.

The functions $S_{ij}(q)$ (i,j = A or B) can be calculated for the binary A–B/HA and ternary A–B/HA/HB blends (see for example Tanaka and Hashimoto[63]) and hence we can predict $I(q)$, $\langle\eta_q^2\rangle_T$, $S(q)/W(q)$, and the stability limits $\chi_{\mathrm{s,macro}}$ and $\chi_{\mathrm{s,micro}}$. Figure 4 shows typical examples of $N_{\mathrm{AB}}S(q)/W(q)$ as a function of qR_{gAB} for the A–B/HA/HB systems with various block copolymer volume fractions ϕ_{AB}, where R_{gAB} and N_{AB} are R_{g} and the DP for the A–B diblock copolymer, respectively. The ternary systems to be considered here have the following parameters: $f_{\mathrm{A}} = 0.5$, $N_{\mathrm{HA}}/N_{\mathrm{AB}} = N_{\mathrm{HB}}/N_{\mathrm{AB}} = 0.5$, and $\phi_{\mathrm{HA}} = \phi_{\mathrm{HB}} = (1 - \phi_{\mathrm{block}})/2$ where ϕ_{HA} and ϕ_{HB} are volume fractions of HA and HB, respectively. Curve 1 is for the pure A–B diblock polymer ($\phi_{\mathrm{block}} = 1$), having $q_{\mathrm{m}} \approx 2/R_{\mathrm{gAB}}$ and $W(q_{\mathrm{m}})/S(q_{\mathrm{m}}) \approx 21 = 2(\chi N_{\mathrm{AB}})_{\mathrm{s,micro}}$, identical to Leibler's result.[9] Curve 6 is for the pure HA/HB blends ($\phi_{\mathrm{block}} = 0$), having $q_{\mathrm{m}} = 0$ and $W(q_{\mathrm{m}} = 0)/S(q_{\mathrm{m}} = 0) \approx 8 = 2(\chi N_{\mathrm{AB}})_{\mathrm{s,macro}}$, *i.e.* $(\chi N_{\mathrm{AB}}/2)_{\mathrm{s,macro}} = 2$ as predicted by classical thermodynamics.

It is important to note that the value q_{m} for the microphase transition shifts to a smaller value with decreasing ϕ_{block} and that there is a critical volume fraction ϕ_{c} of the block copolymer below which the nonzero value q_{m} for the microphase transition does not exist. This ϕ_{c} corresponds to a tricritical point called the Lifshitz point.[64] Similar features are obtained for the A–B/HA systems.

Let us examine below curve 3 with $\phi_{\mathrm{block}} = 0.6$. The value of χ changes with temperature. With increasing χ from zero, the value 2χ first reaches a minimum value of $N_{\mathrm{AB}}S(q)/W(q)$ at $q = q_{\mathrm{m}} \approx 1.4/R_{\mathrm{gAB}}$, and the minimum value is equal to $2(\chi N_{\mathrm{AB}})_{\mathrm{s,micro}} \approx 16.1$. This is the stability limit for the particular ternary mixture and the mode which becomes unstable has the wave number $q = q_{\mathrm{m}} \neq 0$. Hence this is the stability limit for the microphase transition. If the value (χN_{AB}) further increases slightly, the particular modes with q values between $q_{\mathrm{c,1}}$ and $q_{\mathrm{c,m}}$ can cause the thermodynamic driving force $[2\chi - S(q)/W(q)]$ to grow, as shown by the shaded

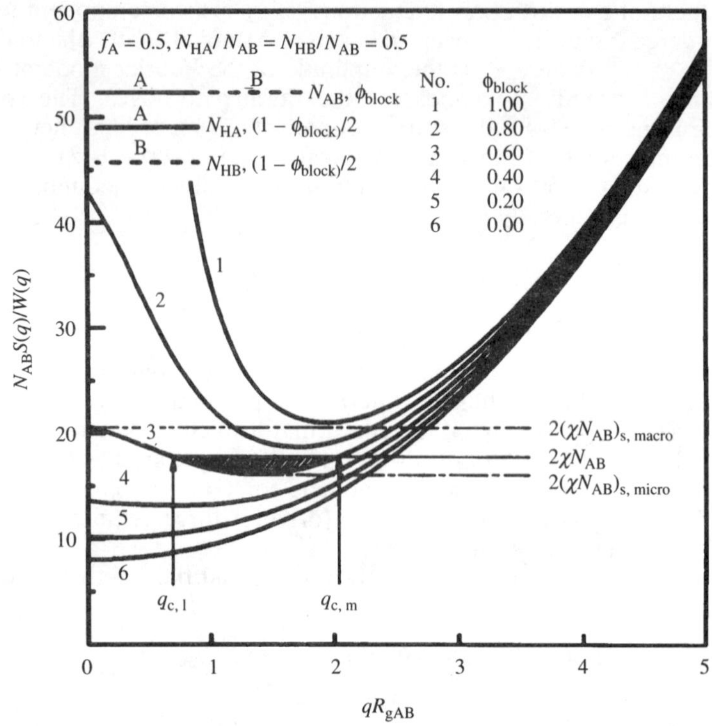

Figure 4 Function $N_{AB}S(q)/W(q)$ versus qR_{gAB} for the ternary mixtures of A–B block polymers with $f_A = 0.5$, HA and HB homopolymers, $N_{HA}/N_{AB} = N_{HB}/N_{AB} = 0.5$, and $\phi_{HA} = \phi_{HB} = (1-\phi_{block})/2$. Curve 1 is for pure A–B ($\phi_{block} = 1$) and curve 6 is for the homopolymer blend HA/HB ($\phi_{block} = 0$) (after Tanaka and Hashimoto[63]).

region drawn in Figure 4. Hence only the microphase transition can occur and the microdomain structure will be developed eventually at equilibrium. It is important to note that there is no possibility of phase separation into two macroscopic phases, and that the homopolymers are to be dissolved into the microdomains (the regime where only the microphase transition occurs (see Sections 14.5.1 and 14.5.2)).

With a further increase in χ, $2(\chi N_{AB})$ eventually reaches the value $S(q=0)/W(q=0) = 2(\chi N_{AB})_{s,macro}$ where the instability for the macrophase transition starts to occur. Let us consider the case where $\chi N_{AB} > (\chi N_{AB})_{s,macro}$. All the q-modes with $q < q_{c,m}$ have the thermodynamic potential to grow, where $q_{c,m}$ is the mode with the maximum wave number which can grow. In other words, there is no lower limit $q_{c,1}$. Consequently, in this regime both the microphase transition and the macrophase transition can take place in general. However, the detailed ordering or coarsening processes, and a change of the relative contribution of the two kinds of phase transition with time, are unknown and merit future experimental and theoretical work.

If ϕ_{block} is lower than ϕ_c, the minimum of $S(q)/W(q)$ occurs only at $q = 0$. For example, the curve for $\phi_{block} = 0.20$ corresponds to this case. If χN_{AB} is only slightly larger than $(\chi N_{AB})_{s,macro}$, the modes which can grow have very small q ($\ll 1/R_g$), *i.e.* $q_{c,m} \ll 1/R_g$, and hence only the macrophase transition can occur, at least in the beginning. This corresponds to the regime for the macrophase transition. It should be noted, however, that an increase in the local concentration fluctuation due to the macrophase transition may induce the microphase transition in the domains rich in block copolymer ('microphase transition induced by the macrophase transition'), as will be discussed later (Section 14.5.5). At large quench depth, *i.e.* the case where $\chi N_{AB} \gg (\chi N_{AB})_{s,macro}$, the mode q_m developed in the early stage spinodal decomposition may have a fairly large q value close to $1/R_{gAB}$.[65] However, the coarsening processes occurring in the late stage shift q_m to a smaller value with time[65–67] and eventually to zero at thermal equilibrium.

The stability limits have been analyzed as a function of the block copolymer composition ϕ_{block}, and typical results are given in Figure 5 for the same ternary system as discussed in

Figure 4. The solid and broken lines respectively show the stability limits for the macrophase transition $(\chi N_{AB})_{s,macro}$ and the microphase transition $(\chi N_{AB})_{s,micro}$. ϕ_c is the critical volume fraction of the block copolymer, below which only the macrophase transition has the stability limit.

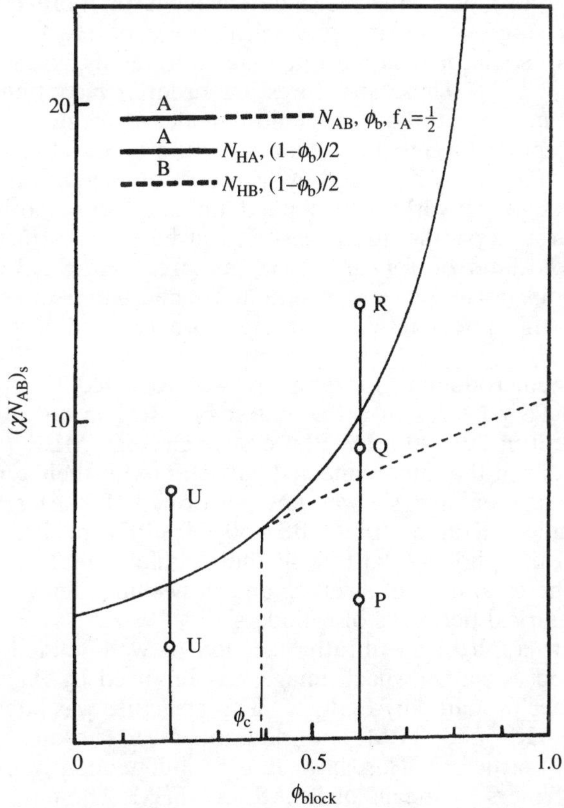

Figure 5 Stability limits for the macrophase transition (solid line) and microphase transition (broken line) for the same ternary mixtures of A–B/A/B given in Figure 4. At a point (*e.g.* the points designated as V, Q, and R) above the solid or broken lines $[\chi N_{AB} > (\chi N_{AB})_s]$, the mixtures are thermodynamically unstable. Upon increasing χN_{AB} from the disordered state (*e.g.* the points designated as U and P), the system encounters the macrophase separation first if $\phi_{block} < \phi_c$ (*e.g.* U → V), while the microphase separation occurs first if $\phi_{block} > \phi_c$ (*e.g.* P → Q). ϕ_c corresponds to a tricritical point called the Lifshitz point (after Tanaka and Hashimoto.[63]).

From Figure 5 we obtain a number of interesting results as summarized below. (i) $(\chi N_{AB})_{s,macro}$ increases with increasing ϕ_{block}, suggesting that the block copolymer suppresses the macrophase transition. This miscibility enhancement of the block copolymer is known as the compatibilizing effect of the block copolymer. (ii) Upon decreasing ϕ_{block} from unity, $(\chi N_{AB})_{s,micro}$ first decreases, suggesting that addition of the homopolymers enhances the microphase transition. (iii) In the case when $\phi_{block} < \phi_c$, at point U far below the $(\chi N_{AB})_{s,macro}$, the systems are in the disordered state and A–B, HA, and HB are mixed at the molecular level. (iv) On the other hand, at point V inside the spinodal limit the mixture is unstable and undergoes spinodal decomposition, resulting eventually in the macrophase separation. The effects of the block copolymer on the early stage spinodal decomposition and its subsequent coarsening processes of the ternary systems have been described elsewhere.[68,69] (v) In the case when $\phi_{block} > \phi_c$, at point P far below the spinodal point for the microphase transition, the system is in the disordered state. (vi) At point Q in the gap between the two stability limits the mixture can undergo only the microphase transition and regular, periodic microdomain structures would be formed at equilibrium.[70,71] HA and HB are expected to be solubilized primarily in the corresponding microdomains. (vii) At point R inside the spinodal of the macrophase transition, both the macrophase and microphase transitions occur in general. It would be obvious that the ordering processes would develop various interesting intermediate structures, as partly suggested elsewhere.[17,18,41]

Finally, it should be noted that the phase diagrams for the spinodal lines for the macrophase and microphase transitions are very sensitive to the DP of A–B, HA, and/or HB as well as f_A, the composition of A in A–B diblock copolymers.[41,63,70]

14.4 NEW ORDERED BICONTINUOUS MORPHOLOGY AND GRAIN BOUNDARY STRUCTURES

The morphologies of block copolymer systems have been discussed according to the classical classification proposed by Molau.[72] However, the existence of more complex structures was pointed out at an early stage of the morphological study of block copolymer systems.[73,74] Recently, the structures of some of such morphologies were analyzed in detail and continue to attract research interest.[50,51,77–79] These structures are ordered bicontinuous structures[50,51,77,79] and a mesh structure (or monocontinuous catenoid-lamellar structure).[78]

Matsuo and Sagaye[73] reported a 'piled network' structure in a cyclohexane-cast SBS triblock copolymer film (80 mol% PS). They observed by transmission electron microscopy (TEM) that the PB phase formed rough networks which piled up one after another in the PS matrix, tending to orient their surfaces parallel to the cast film surface. This structure is considered as a kind of mesh characterized later on for an SB/HS mixture.[78] Here HS designated homopolystyrene. Although they did not observe a bicontinuous microdomain structure, they had the insight to consider that this piled network structure was a modification of a bicontinuous structure.

The first bicontinuous microdomain morphology was reported by Aggarwal *et al.*[74] They observed interconnected PS particles in a PB matrix by TEM in an ultrathin section of SBS thermoplastic rubber (Kraton 101, 90 vol% PB) cast from THF/MEK (90/10) solution. They suggested short-range order in the interconnected particles (with their centre-of-masses packed in a square lattice) but further analysis was not performed. Folkes *et al.*[75] investigated the swelling effect of extruded Kraton 102 (SBS, 80 vol% PB) in hexane vapor by TEM. Well-oriented cylindrical morphology existing in the original sample was transformed into a complex network after the treatment of swelling and deswelling. They suggested the presence of Luzzati's highly symmetrical networks of cylinders.[80]

The first electron micrograph of an ultrathin section of well-defined ordered bicontinuous morphology that exhibited a 'wagon-wheel' image was reported by Aggarwal[76] for an SI star (or radial) block copolymer prepared by Fetters.[81] This structure was later analyzed by Thomas *et al.*[50] by means of SAXS and TEM and determined to be the ordered bicontinuous double-diamond (OBDD) structure. Hasegawa *et al.*[51] independently observed similar structures in SI diblock copolymers by means of SAXS and TEM. Thomas *et al.*[77] also reported a bicontinuous catenoid-lamellar structure. Recently, a new bicontinuous morphology, 'gyroid', has been reported for the high-temperature form of a lamellar-forming SI diblock copolymer having 37 wt% PS.[79]

The mesh and bicontinuous structures mentioned above are either equilibrium or nonequilibrium morphology. However, even for the nonequilibrium morphologies, such structures may be useful if we can develop the techniques to control and stabilize them. Therefore, it is worthwhile to investigate the possible structures and their self-assembling mechanism. Let us consider next why such a complex morphology is possible.

14.4.1 Microdomain Interfaces and Curved Surfaces

The microdomain morphology of block copolymer systems is determined by the shape of the interface between two phases, *i.e.* the local curvature of the interface (local aspect) and the topological continuity of the interface in space (global aspect). Therefore, the curvature of the interface is a key parameter to classify the microdomain morphologies. All kinds of curved surfaces should be considered for the microdomain morphologies.

Curved surfaces can be classified into three groups, namely, elliptic, parabolic, and hyperbolic surfaces, depending on the sign of their Gaussian curvature (K), which is a product of their two principal curvatures (k_1, k_2), *i.e.* $K = k_1 k_2$. An elliptic surface has a positive Gaussian curvature ($K > 0$), *i.e.* the surface is convex everywhere and forms a closed surface. Ellipsoids and spheres belong to this class. A parabolic surface has a zero Gaussian curvature ($K = 0$), *i.e.* at least one of the principal curvatures is zero. Cylinders belong to this class and so does a flat plane. A hyperbolic interface has a negative Gaussian curvature ($K < 0$), *i.e.* the surface is convex in one direction but concave in the orthogonal direction. Therefore, hyperbolic surfaces are the richest in morphology among these three groups. It is obvious that the classification of microdomain structures into spheres, cylinders, and lamellae (planes) ignored the existence of hyperbolic surfaces. It should be emphasized that hyperbolic surfaces are the ones that form new, complex, bicontinuous microdomain morphologies.

As for the microdomain interfaces, we can exclude self-intersecting hyperbolic surfaces. Since the microdomain structures are highly periodic, we will deal with only non-self-intersecting periodic hyperbolic surfaces below.

Simple examples of periodic hyperbolic surfaces are periodic minimal surfaces. Minimal surfaces are defined as the surfaces with zero mean curvature ($H = 0$). Mean curvature (H) is the average of the two principal curvatures, *i.e.* $H = (k_1 + k_2)/2$. Scherk's first surface is a doubly periodic minimal surface related to the grain boundary structure, which will be discussed in Section 14.4.5. Those related to the bicontinuous morphology are the triply periodic minimal surfaces. Mathematical formulations of more than 30 periodic minimal surfaces have been reported,[82,83] *i.e.* their Weierstrass functions are known. The repeat units of some well-known triply periodic minimal surfaces are shown in Figure 6.

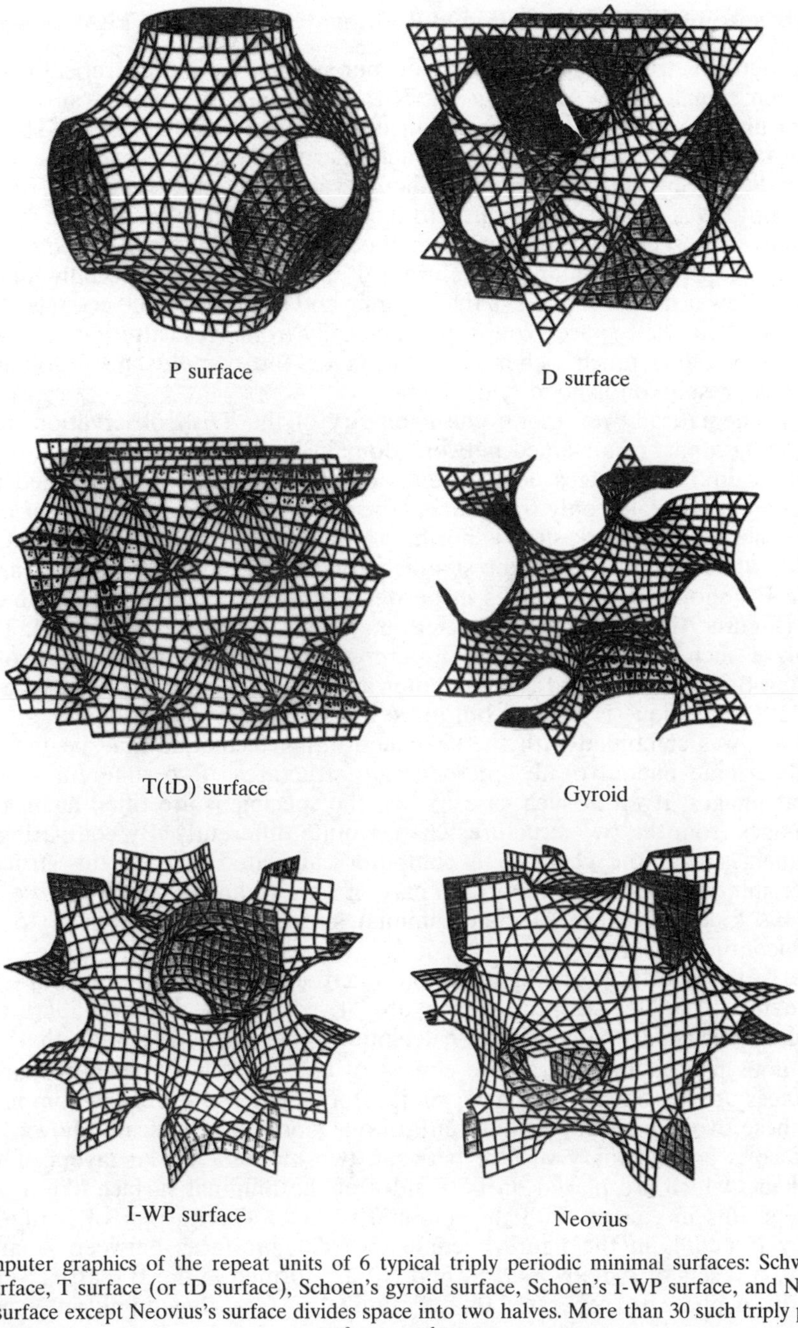

Figure 6 Computer graphics of the repeat units of 6 typical triply periodic minimal surfaces: Schwarz's P surface, Schwart'z D surface, T surface (or tD surface), Schoen's gyroid surface, Schoen's I-WP surface, and Neovius's surface. Each minimal surface except Neovius's surface divides space into two halves. More than 30 such triply periodic minimal surfaces are known.

However, those triply periodic minimal surfaces themselves cannot be the microdomain interfaces because they usually bisect the space, *i.e.* two phases on both sides of a triply periodic minimal surface have the equal volume fraction of 0.5. We know that a lamellar morphology results in such a case. A bicontinuous microdomain morphology must be characterized by a triply periodic hyperbolic interface of the microdomain which is equidistant from a triply periodic minimal surface ('parallel surface') or which has similar geometry to the parallel surface but with a constant mean curvature (cmc surface) (H = constant). Deformation of a periodic hyperbolic surface results in a periodic hyperbolic surface again. Since each of such periodic hyperbolic surfaces has a certain relationship with the interface of a microdomain morphology, a large number of microdomain morphologies is possible.

14.4.2 Importance of TEM Observation and Computer Analysis of TEM Images

Since microdomain structures of block copolymer systems have their repeat distances on the order of 10 nm, small-angle scattering of X-rays or neutrons and transmission electron microscopy of ultrathin sections are the major techniques to characterize the structures. These two techniques, scattering in reciprocal space and microscopy in real space, are complementary to each other. Scattering gives information about average size and symmetry of the structure, whilst microscopy gives information limited to the local area but related to the structural entity. Extinction rules are used in the scattering method to determine the symmetry (or the space group). However, pseudocrystalline structures of microdomains, especially of the complex ones, give much fewer diffraction peaks than atomic and small molecule crystals. Therefore it is not easy to determine their space groups unequivocally from the scattering data alone. On the other hand, TEM offers much richer information on the complex microdomain structures, provided that the resolution is good enough.

Figure 7(a) schematically represents the geometry of the TEM observation of an ultrathin section containing a part of a stained network domain in an unstained matrix. If it is assumed that the contrast in TEM images only depends on the thickness of the stained parts (l_{P_1}, l_{P_2}, *etc.*), the area which contains only the matrix appears bright (area composed of points like P_3) and the area where the network stem is normal to the section appears dark (area composed of points like P_1), while the area where the stem is parallel or tilted to the section (area composed of points like P_2) shows various shades in between (gray), depending on the thickness of the stained part (Figure 7(b)). Thus the projected image will be quite complicated. There are two ways to analyze such complex images: (i) stereoscopic observation of the two micrographs taken by tilting the specimen, and (ii) computer analysis of the TEM image with an adequate model. The latter technique is indirect but more objective than the former.

Computer analysis combined with the specimen-tilting technique may be the most reliable method to determine unequivocally microdomain structures. Two different structures might exhibit similar images. Even in such cases, when the specimens are tilted against the electron beam, the images from the two structures change quite differently. By comparing the changes in the real images with the changes in computer calculated images, the structures can be uniquely determined. The most important part of this technique is to choose an adequate model. Various kinds of triply periodic minimal surfaces may be utilized to consider the topology of bicontinuous structures.

As a model of the interfaces of a bicontinuous microdomain structure, we will consider a pair of surfaces named 'parallel surfaces' which are at a constant distance apart from a triply periodic minimal surface for simplicity in calculation. There is a suggestion that the interfaces of a bicontinuous microdomain structure consist of cmc surfaces,[77] but both parallel surfaces and cmc surfaces are considered to have a similar topology. Therefore the computer-simulated patterns for these two may not be so different. In our model schematically shown by Figure 8, a minimal surface is an imaginary surface between two monomolecular layers of A–B diblock copolymer chains which are placed on both sides of the minimal surface like a mirror image, with B block chains in contact with the minimal surface. Then a pair of parallel surfaces as designated by I_1 and I_2 in the figure become the real interfaces between A and B phases. Consequently, two sets of A networks, N_1 and N_2, are formed in the B matrix. Since these two networks of the A phase are separated by the minimal surface, they do not intersect each other throughout the space and are mirror images of one another.

Such a model agrees with the double-network structure of the reported bicontinuous

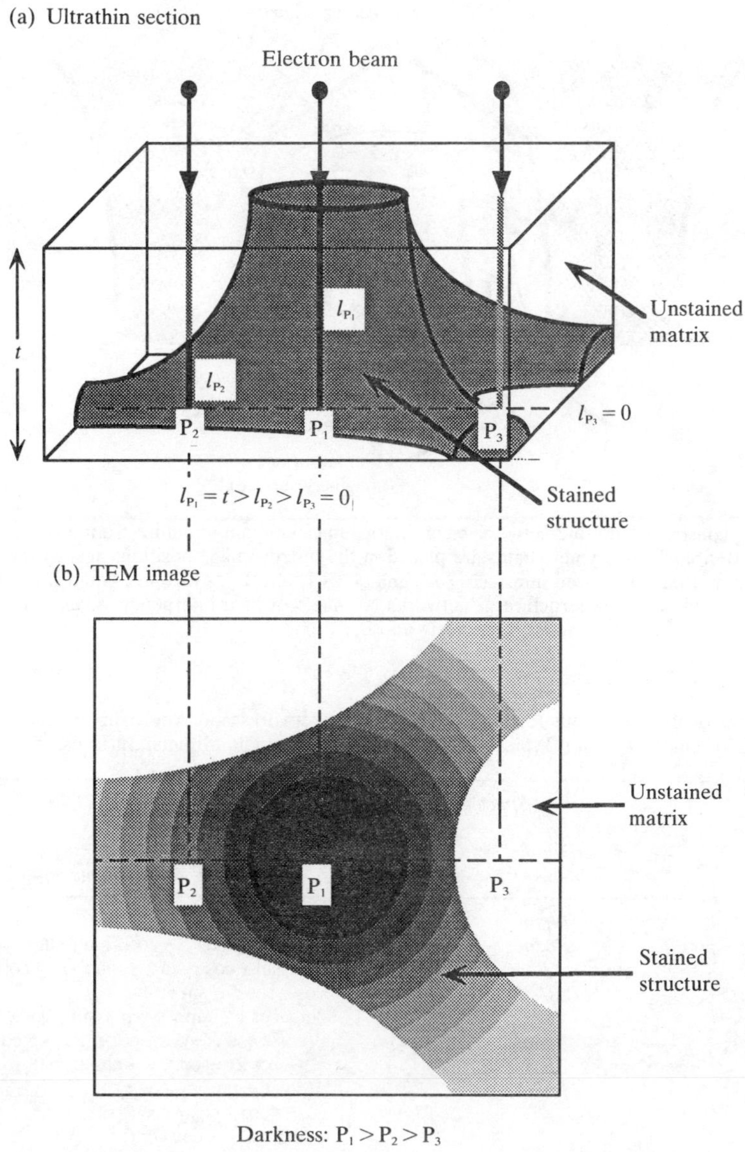

Figure 7 Schematic illustrations of (a) an ultrathin section with thickness *t* containing a potion of a network-forming microdomain stained uniformly and (b) the corresponding transmission electron microscopic image. The thickness of the stained phase along the direction of the electron beam varies at the positions designated by P_1, P_2, and P_3 in (a). The darkness in the image is proportional to the path length (l_{P_i}, $i = 1$, 2, or 3) of the electron beam through the stained phase. For $l_{P_1} = t > l_{P_2} > l_{P_3} = 0$ as shown in (a), the darkness in (b) decreases in order of $P_1 \rightarrow P_2 \rightarrow P_3$.

structures.[50,51,79] If necessary, a corresponding single-network structure can be readily constructed just by neglecting one of the networks of the double-network model. Calculation of the coordinates for a periodic minimal surface directly from the Weierstrass function is tedious in practice. Fortunately, most of the simple periodic minimal surfaces are reasonably approximated by simple trigonometric equations. Examples are given in Table 1 for the triply periodic minimal surfaces shown in Figure 6 and also for Scherk's first surface.[120] Their space groups for double- and single-network structures are listed as well. It should be noted that space groups are usually different for double- and single-network systems even though they are based on the same minimal surface because the latter has lower symmetry than the former.

The computer-calculated images of microdomain structures vary with the model chosen, observation direction (or sectioning angle designated by polar angle θ and azimuthal angle ϕ as shown in Figure 9), sectioning thickness and sectioning position with respect to the structural

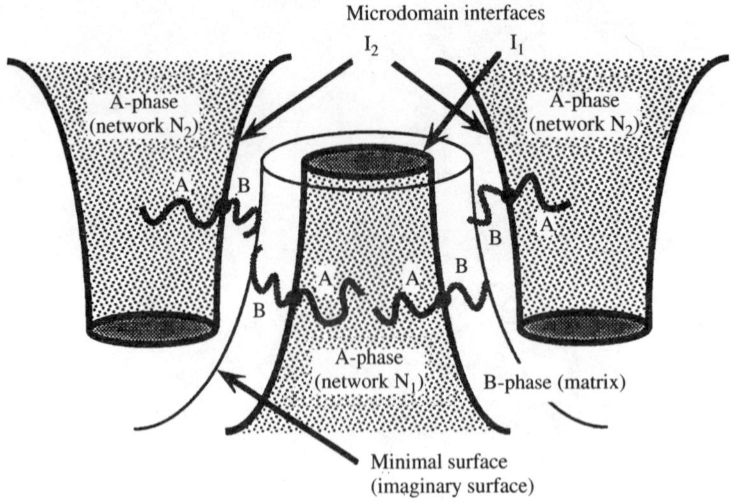

Figure 8 A model to construct a double-network bicontinuous microdomain structure from a triply periodic minimal surface. A pair of A–B diblock copolymer chains are placed in the mirror image positions against the minimal surface. The junction points form a pair of curved interfaces (designated as I_1 and I_2), which are equidistant from the minimal surface, resulting in a double-network structure of networks N_1 and N_2 which interpenetrate and are mirror images of one another.

Table 1 Space Groups of the Unit Cells for Double and Single Networks and Approximate Equations in Cartesian Coordinate *Oxyz* for Typical Triply and Doubly Periodic Minimal Surfaces.[120]

	Space group		
Minimal surface	*Double network*	*Single network*	*Approximate equation*
Schwarz's P surface	$Im\bar{3}m$	$Pm\bar{3}m$	$\cos x + \cos y + \cos z = 0$
Schwarz's D surface	$Pn\bar{3}m$	$Fd\bar{3}m$	$\cos x \cos y \cos z + \cos x \sin y \sin z$ $+ \sin x \cos y \sin z + \sin x \sin y \cos z = 0$
T surface (tD surface)	$P4_2/mm$	$I4_1/amd$	$\cos x \cos z - \sin y \sin z = 0$
Schoen's gyroid	$Ia\bar{3}d$	$I4_132$	$\sin x \cos y + \sin y \cos z + \sin z \cos x = 0$
Schoen's I-WP surface	$Im\bar{3}m$	$Im\bar{3}m$	$2(\cos x \cos y + \cos y \cos z + \cos z \cos x)$ $- \cos 2x - \cos 2y - \cos 2z = 0$
Neovius's surface	$Im\bar{3}m$	$Pm\bar{3}m$	$3(\cos x + \cos y + \cos z)$ $+ 4\cos x \cos y \cos z = 0$
Scherk's first surface			$\exp z \cos x - \cos y = 0$

unit, and tilt angle if the section is tilted with respect to the projection plane. In particular, the change of the image with sectioning angle is extraordinary. It is well known that a section of cylinders perpendicular to their axes gives a similar image as spheres and a parallel section gives a similar image as lamellae. However, the computer calculated images of such a simple structure of hexagonally packed cylinders turn out to be quite complex, depending on the sectioning angle as shown in Figure 10. For example, from the image of $\theta = 45°$ and $\phi = 15°$, it is difficult to imagine the cylinders. We learn the following from this: (i) a complex image does not necessarily arise from a complex structure; (ii) a single structure can generate a variety of images; (iii) analysis of a system with more than two structures coexisting is extraordinarily difficult.

14.4.3 Bicontinuous Morphology

Hasegawa *et al.*[51] investigated the microdomain morphology of toluene-cast films of a series of SI diblock copolymers with different compositions and MW by SAXS and TEM. Within their observation (number average MW $M_n = 2 \times 10^4$–1×10^6), there was no MW dependence

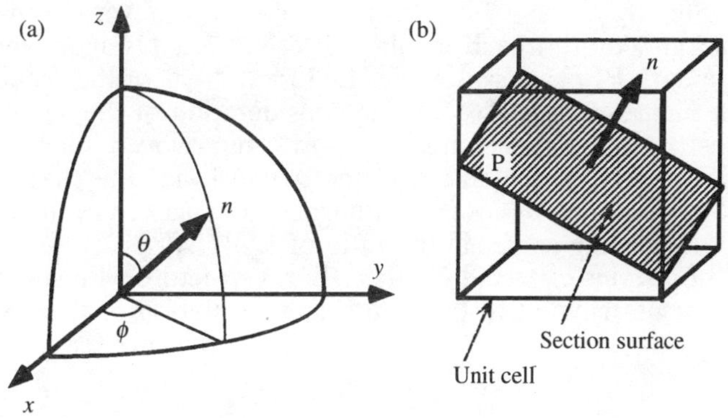

Figure 9 Cartesian coordinate *Oxyz* of the unit cell of the microdomain model (a) and the definition of the sectioning angle for the computer analysis of transmission electron microscopic images of a microdomain structure in terms of polar angle θ and azimuthal angle ϕ of the normal vector ***n*** of sectioning plane P (b).

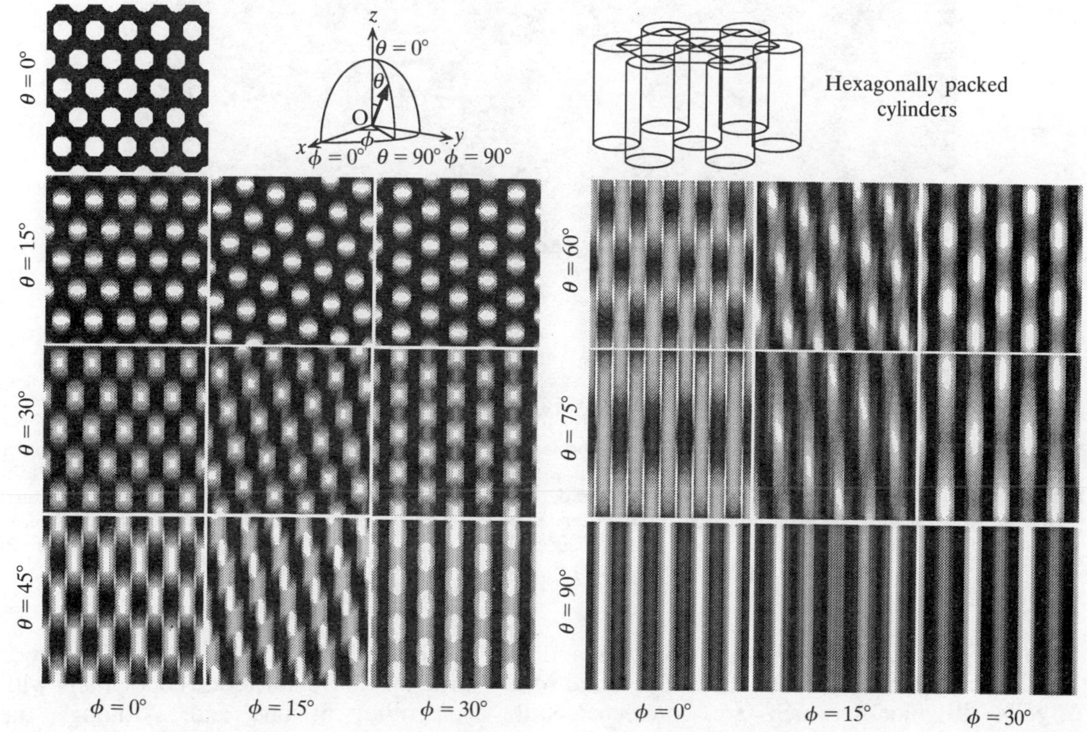

Figure 10 Dependencies of computer calculated images of hexagonally packed cylinders on sectioning angles θ and ϕ as defined in Figure 9 and the insertions. (210) and (100) planes of the hexagonal unit cell are parallel to x and y axes, respectively. The volume fraction of the cylindrical microdomains of the bright phase is 0.3 and the section thickness is 1.5 times the domain spacing (or unit cell length).

of the morphology, as expected for the morphology in the strong segregation limit. They found an ordered bicontinuous morphology, namely the double-diamond network structure composed of a tetrapod as a structural unit ('tetrapod-network' structure or 'OBDD-tetrapod'), appearing in the very narrow composition range of $0.62 \leq \phi_{PS} \leq 0.66$, ϕ_{PS} being the volume fraction of the PS component in SI. This composition range corresponds to that between lamellae and cylinders. Figure 11 shows a TEM picture of an ultrathin section (*ca.* 50 nm thick) of an SI diblock copolymer with $M_n = 1.86 \times 10^5$ and $f_{PS} = 0.66$ stained with osmium tetroxide. The picture shows a remarkably uniform and regular pattern called a 'wagon-wheel' pattern and the network nature of the dark PI phase is obvious. Different areas of the section showed different patterns, as shown in Figures 12(a)–12(d). From consideration of the TEM images alone, they concluded that the structure was composed of two interwoven diamond networks of PI microdomains in a PS matrix. The element of the network has the shape of a tetrapod. The SAXS profile of this structure resembles that of hexagonally packed cylinders, *i.e.* the scattering maxima appear at the scattering angles in the ratio of 1, $3^{1/2}$, $4^{1/2}$, $7^{1/2}$, *etc.* The scattering peaks at the $(3/2)^{1/2}$ and $2^{1/2}$ positions expected for the OBDD structure with space group *Pn3m* were not observed, which they attributed to the insufficient resolution of their SAXS apparatus for the structure.

Figure 11 A transmission electron micrograph obtained from an ultrathin section of toluene-cast film of an SI diblock copolymer (HY-10: $M_n = 1.64 \times 10^5$, $f_{PS} = 0.66$) stained with osmium tetraoxide, exhibiting a 'wagon-wheel' pattern with a hexagonal symmetry typically observed in the ordered bicontinuous double-network structures. The dark rings with a hexagonal shape correspond to the horizontal stems of one PI network as schematically represented by region P_2 in Figure 7; the darker spots in the centers of the rings probably correspond to the vertical stems of the other PI network as represented by region P_1 in Figure 7 (after Hasegawa *et al.*[51]).

The OBDD structure with PS networks and a PI matrix was not found in pure SI diblock copolymer systems, but Thomas *et al.*[50] found such a structure in SI star block copolymers with 67 vol% PI blocks which are connected with each other at one end. Although the three-dimensional bicontinuity was not shown directly by TEM due to their low arm MW ($M_n = 3.3 \times 10^4$), they could observe the scattering peak at the $(3/2)^{1/2}$ position. These two studies of the OBDD structures are complementary each other: the structural entity is better observed by TEM for the large microdomains of Hasegawa *et al.*,[51] while the symmetry is better observed by SAXS for the small microdomains of Thomas *et al.*[50] OBDD structures are observed also in SI/HS and SB/HS mixtures,[84,85] SI/PVME mixtures,[86] and ISP triblock copolymers (ordered tricontinuous double-diamond structure or OTDD structure).[87]

The OBDD structure can be modeled by Schwarz's D surface, one of the triply periodic minimal surfaces shown in Figure 6. The TEM pictures given in Figures 12(a)–12(d) are successfully reproduced by the computer calculation with the double-network model of the D surface, as shown in Figures 12(e)–12(h), respectively.

Recently, Hajduk *et al.*[79] found a new double-network bicontinuous morphology which can be modeled with Schoen's gyroid surface in an SI diblock copolymer ($M_n = 2.74 \times 10^4$ and $f_{PS} = 0.33$) after annealing at temperatures above 120 °C. They observed the first- and second-order scattering peaks at spacing ratios of $3^{1/2}:4^{1/2}$, which is inherent in the space group

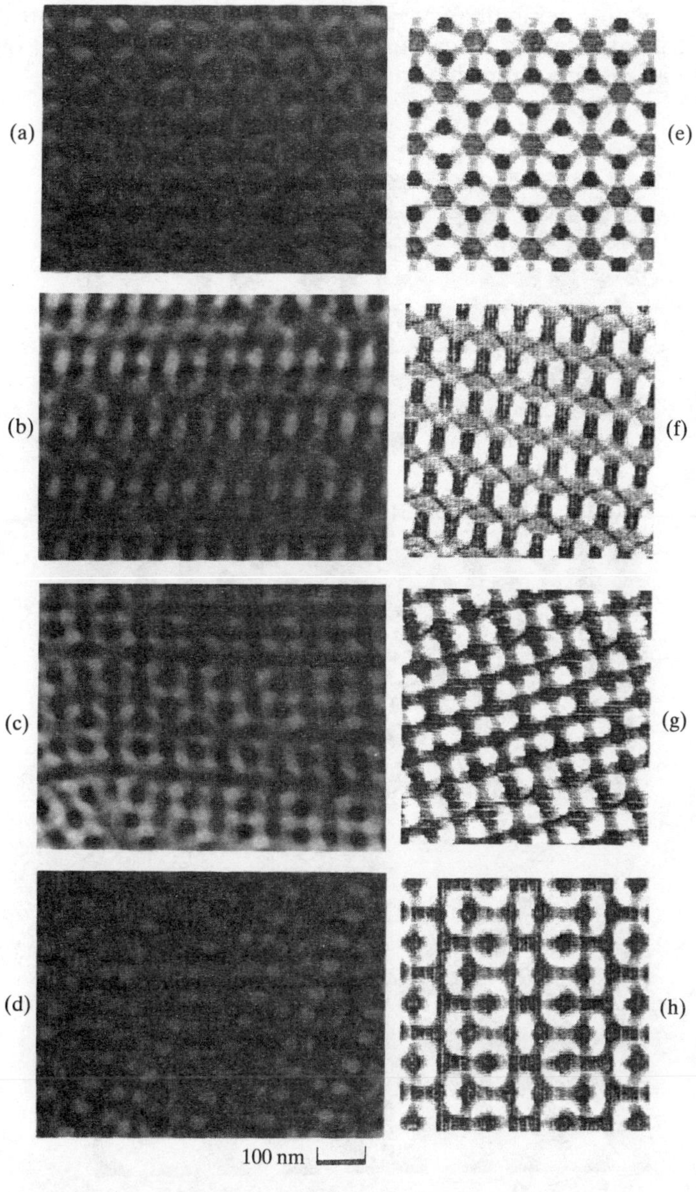

Figure 12 Transmission electron micrographs obtained from the same specimen as shown in Figure 11 exhibit a variety of patterns (a to d). These patterns can be reproduced fairly well by computer calculation with the double-network model based on a D surface as shown in parts (e) to (h).

*Ia*3*d* of the double-network structure of a gyroid surface. At temperatures below 115 °C this sample exhibited lamellar morphology. It is interesting that another bicontinuous morphology was found only at high temperatures in the composition range where the OBDD structure was expected. There is some argument about the stability of the OBDD structures.[88,89] Hajduk *et al.* reexamined by SAXS the same sample as the one Thomas *et al.*[50] used and obtained a completely different scattering profile exhibiting the gyroid symmetry.[90]

The structural element of the gyroid network is a tripod, in contrast to a tetrapod for the D surface (OBDD) network. Both structures have a cubic unit cell with threefold and fourfold symmetry. Thus the computer calculated images of the (111) sections of the double-network models with a D surface and gyroid both exhibit wagon-wheel patterns, as shown in Figures 13(a) and 13(b). Figure 13(c) also exhibits a wagon-wheel pattern but was obtained from the double-network model with a tD surface. However, if the section thickness is changed to 0.1 of the unit cell, the difference becomes remarkably clear, as shown in Figures 13(d)–13(f). So far, almost all of the structures exhibiting wagon-wheel patterns have been attributed to OBDD structures. Those structures should be carefully reexamined considering not only the D surface and gyroid but also all the other possible triply periodic minimal surfaces.

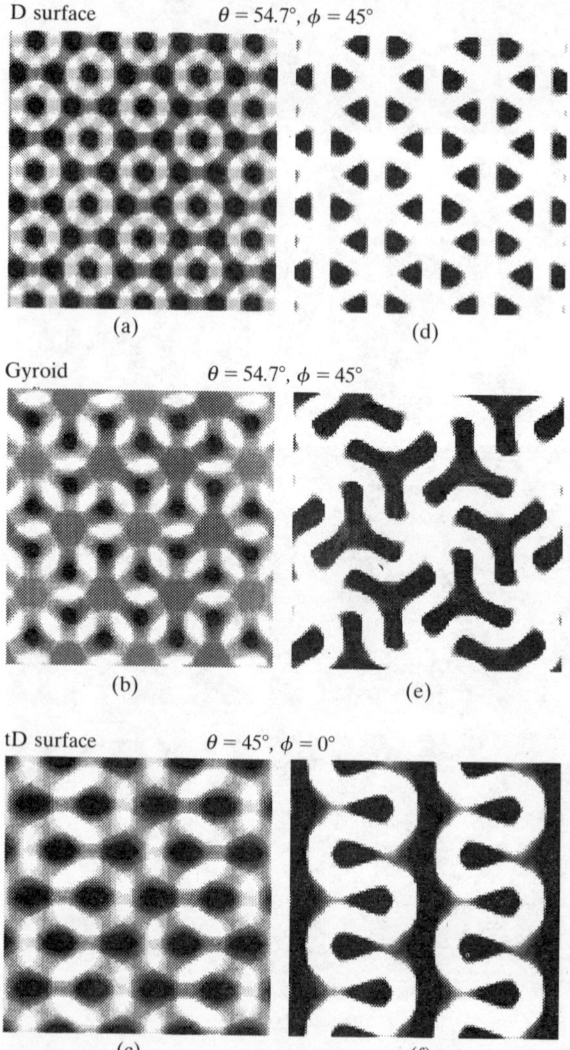

Figure 13 Comparison of computer-calculated TEM images for the double network structures based on a D surface (a and d), gyroid surface (b and c), and tD surface (c and f), all having the volume fraction of dark phase equal to 0.35. The left half (a to c) shows the sections with the thickness of one unit cell, exhibiting 'wagon-wheel' patterns, while the right half (d to f) with the thickness of 0.1 unit cell shows quite different patterns depending on the minimal surfaces. Sectioning angles θ and ϕ are given for each surface in the figure. The sectioning plane with $\theta = 54.7°$ and $\phi = 45°$ for the D surface and gyroid corresponds to the (111) plane of the cubic unit cell.

In addition to the above two morphologies, bicontinuous catenoid-lamellar structures in which both lamellar phases are perforated in a regular manner have been reported for SB[77] and PEP–PEE.[91] However, their structures are not yet fully characterized. Such structures can even be characterized with triply periodic minimal surfaces, *e.g.* an H surface,[82] which has a hexagonal arrangement of channels staggered between adjacent lamellae.

We believe many different ordered bicontinuous morphologies are possible in many kinds of block copolymer systems. It is important at this stage to find and characterize new ordered bicontinuous morphologies irrespective of whether they are equilibrium morphologies or not. Care should be taken not to persist in a single morphology and thus overlook interesting new morphologies. A TEM micrograph exhibiting a wagon-wheel pattern does not guarantee an OBDD or gyroid structure, and a bicontinuous structure does not necessarily exhibit a wagon-wheel pattern.

14.4.4 Mesh Structure

A 'mesh' structure was observed in a 70/30 wt%/wt% mixture of a SB radial block copolymer (K-Resin®, $M_n = 8.0 \times 10^4$ and 24 wt% PB) and HS ($M_n = 2.8 \times 10^3$) and characterized by Hashimoto *et al.*[78] The mesh structure in this system consists of alternating parallel sheets of the phase composed of PS block chains and HS in the matrix of PB phases, and the sheets of the major component (PS and HS) are interconnected by catenoidal channels of the same component traversing through the sheets of the minor component (PB). The perforated sheets of the PB phase appear as arrays of dark spots in the TEM image of a section perpendicular to the sheets (Figure 14), or are seen to be spot-like bright polystyrene microdomains dispersed in the dark PB matrix in the TEM image parallel to the sheets (Figure 15).

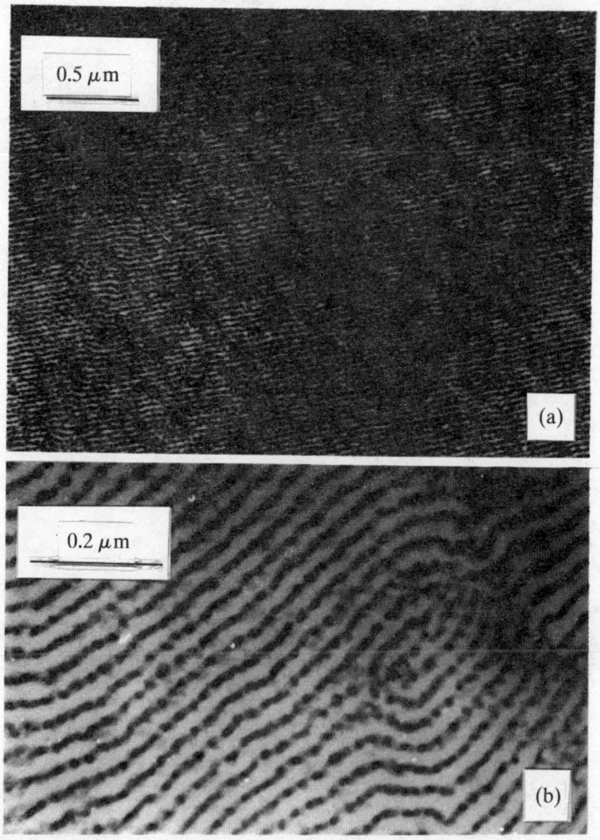

Figure 14 Transmission electron micrographs obtained from an ultrathin section of toluene-cast film of the 70/30 wt%/wt% mixture of the SB radial block copolymer (K-Resin: $M_n = 8.0 \times 10^4$, 24 wt% PB) and HS ($M_n = 2.8 \times 10^4$) cut perpendicular to the film surface and stained with osmium tetraoxide. The low magnification picture (a) exhibits lamellar-like stripes of dark and bright regions, but the high magnification picture (b) reveals that the dark PB 'lamellae' are composed of dark spots (after Hashimoto *et al.*[78]).

The mesh structure may be modeled by the modified P surface (see Figure 6 and Table 1). By disrupting the vertical connection of the channels of the P surface (a double network model in this case) at the positions between the sheets of the lateral channels, a mesh model shown in Figure 16 can be obtained. Such an operation can be done by just changing two parameters of the equation describing the Cartesian coordinates for the P surface.[78] An increase in the HS content to 40 wt% results in a change of the morphology to the bicontinuous network structure of PB in the PS and HS matrix ('strut' or 'sponge' structure) without long-range order. Since the neat block copolymer shows the lamellar phase, the mesh phase is considered to appear between the lamellar and the bicontinuous phases. The strut structure observed is very close to the distorted P surface, which may be designated also as interconnected cylinders of PB in the PS and HS matrix.

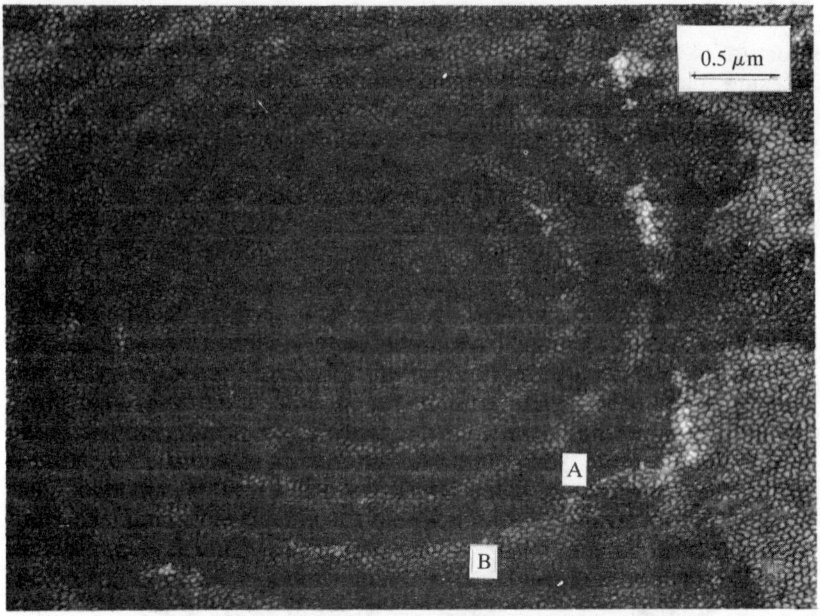

Figure 15 A transmission electron micrograph obtained from the same film as Figure 14 but with the ultrathin section cut parallel to the film surface. It exhibits a mesh-like pattern with the onion-like modulation in contrast to Figure 14, suggesting that the lamellar sheets of the dark PB phase are perforated like mesh. The onion-like modulation is due to the overlaps of two mesh sheets. Note that the mesh sheets tend to align with their normals parallel to the film normals. Thus the morphology appears different, depending on the sectioning direction (after Hashimoto *et al.*[78]).

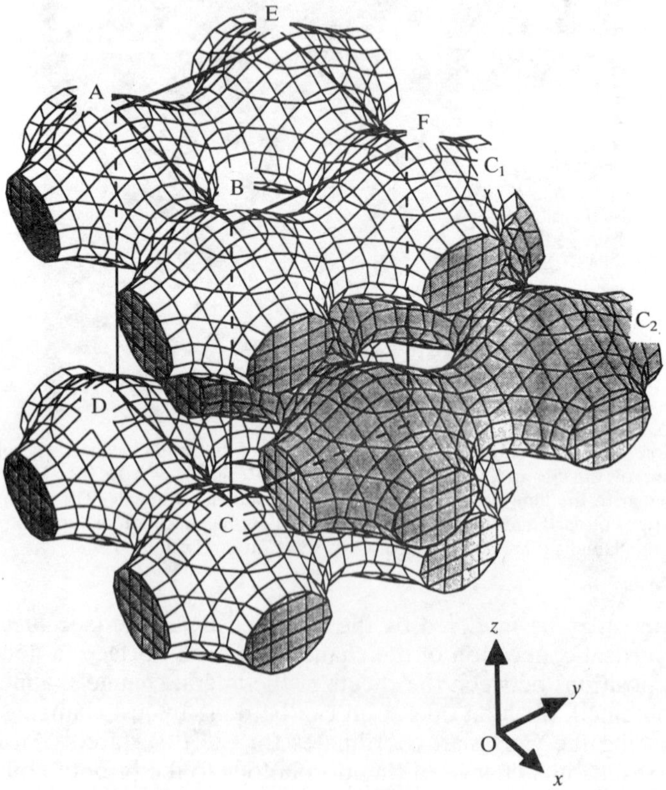

Figure 16 Computer graphics of a homogeneous global embedding of mesh surfaces with a constant mean curvature forming a three-dimensional lattice of body-centered tetragonal symmetry (the unit cell drawn with lines ABFE and ABCD). Thus the neighboring mesh sheets designated as C_1 and C_2 are staggered by half a unit cell. The graphics were generated by using the following equation with $a = 0.7$ and $b = 1.2$: $\cos x + \cos y + (1/a^2)\cos(z/a) = \pm b$ (after Hashimoto *et al.*[78]).

14.4.5 Grain Boundary Structures

The size of the grains, the coherent region of the microdomains, depends on the conditions of sample preparation and the segregation power of the constituent polymers in the block copolymers. Three-dimensional continuity of each microphase at the grain boundary is especially important for cylindrical and lamellar morphologies because it is expected to affect significantly the rheological[92] and diffusion properties.[93–95] The grain boundary is merely a surface with thickness of the order of the repeat distance of the microdomains. However, the orientation of the grain boundary in the ultrathin section varies indefinitely, giving rise to the variety of patterns in the TEM images.

Those images for the lamellar microdomains in PS-*block*-PEP[96] (SEP, $M_n = 17.7 \times 10^4$, $f_{PS} = 0.457$) and SB[97] diblock copolymers were analyzed by a computer generated image with the model of Scherk's first surface (Figure 17), one of the doubly periodic minimal surfaces. At the interface where two sets of orthogonal lamellae meet, a minimal surface is formed, as shown in Figure 17. Such an interface at the grain boundary is considered to minimize the local interfacial free energy. Figure 18 shows the computer-calculated TEM images of Scherk's first surface for various sectioning angles, some of which represent well the TEM images of the grain boundaries of lamellar microdomains in SEP stained with ruthenium tetraoxide (RuO_4)

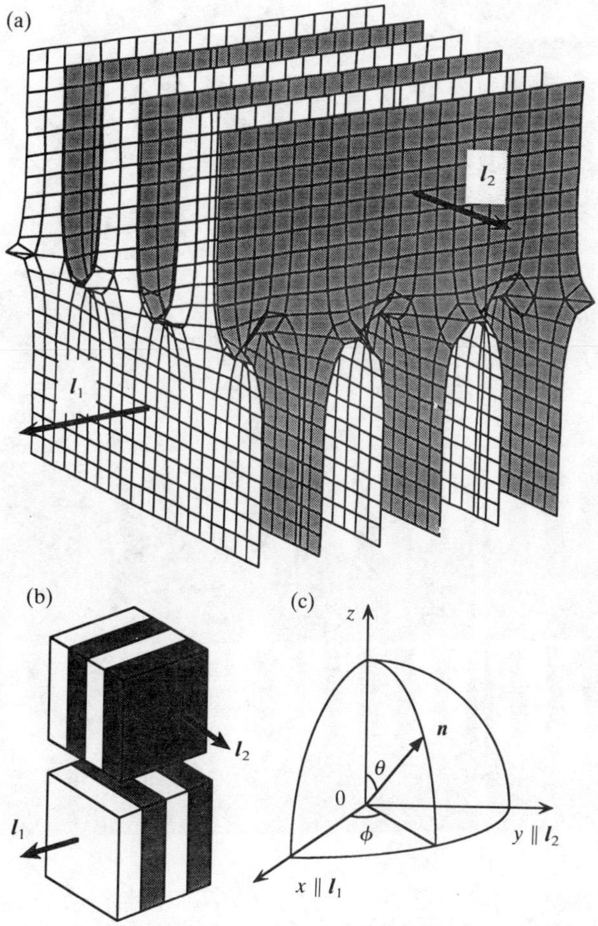

Figure 17 Computer graphic representation of Scherk's first surface (a), a typical doubly periodic minimal surface which serves as the model for the computer analysis of electron microscopic images of the boundary of two orthogonal lamellar grains as shown in (b). The Cartesian coordinates *Oxyz* and polar angle θ and azimuthal angle ϕ of the normal vector ***n*** of the sectioning plane are defined in (c) (after Nishikawa *et al.*[96]).

(Figure 19). Nonuniform staining with RuO_4 gives excess contrast to the lamellar interface and allows detailed analysis of the grain boundary structure. It was found that: (i) the deformed Scherk's first surface, twisted and/or elongated, is necessary for the explanation of some of the grain boundary images; (ii) although the lamellar microdomains locally have only two-dimensional continuity, the global structure has three-dimensional continuity through Scherk's first surface at the grain boundaries; (iii) the interfaces of the lamellar microdomains are flat in nature, and the lamellae can change their directions only through the normal or deformed Scherk's first surface at the grain boundaries.

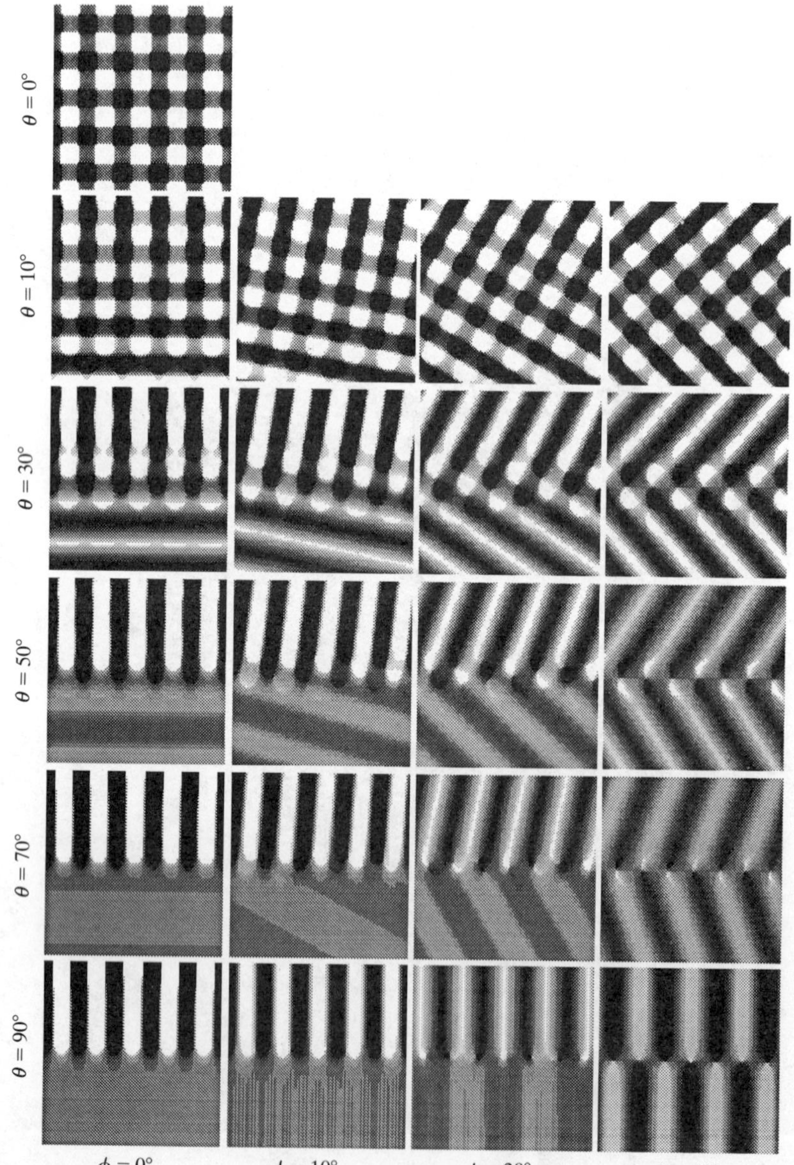

Figure 18 Computer generated images of ultrathin sections for grain boundaries of lamellar microdomains with Scherk's first surface model for various sectioning angles, θ and ϕ. Uniform staining within each domain is assumed and the section thickness was fixed to one repeat distance of the lamellae. $\theta = \phi = 0°$ corresponds to the section parallel to the grain boundary and $\theta = 90°$, $\phi = 0°$ corresponds to the section parallel to the lower-half lamellae (after Nishikawa *et al.*[96]).

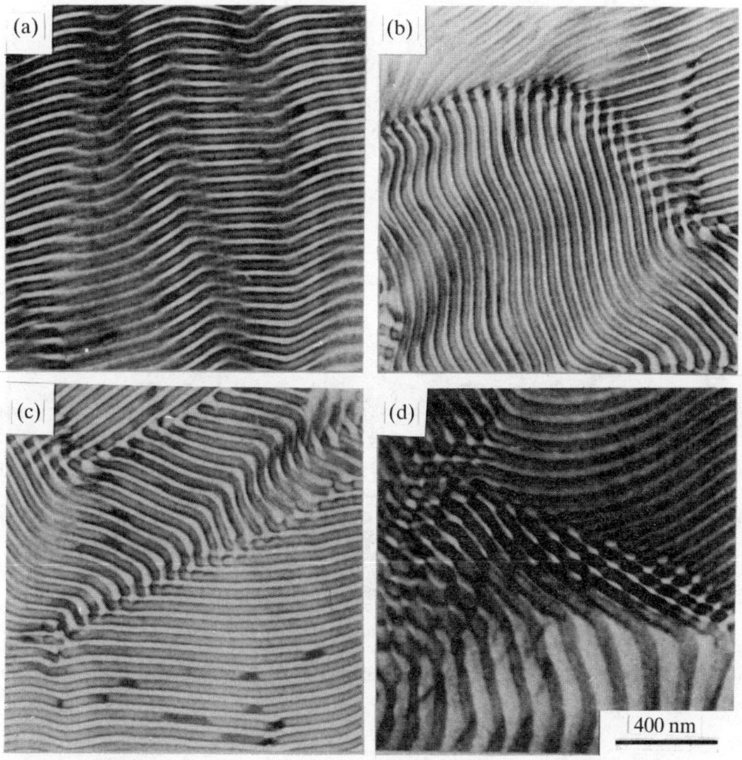

Figure 19 Transmission electron micrographs of the grain boundary areas of lamellar microdomains obtained from an ultrathin section of SEP diblock copolymer cast film stained with RuO_4 showing various images. The RuO_4 staining of the PS phases enhances the contrast at the microdomain interfaces and highlights the grain boundary structure. Except for the enhanced contrast at the interfaces, the grain boundary patterns in the micrographs are well accounted for by the computer generated images in Figure 18 which assume uniform staining (after Nishikawa *et al.*[96]).

14.5 BLOCK COPOLYMER/HOMOPOLYMER MIXTURES

Pattern formation in mixtures of an A–B block copolymer and an A homopolymer (HA) is a fascinating research subject, as introduced in Sections 14.1.2 and 14.3. We introduced the parameter r_A, which may control possible structures in a segregation limit.

Three extremes of the possible mixing states after phase separation for A–B/HA mixtures are illustrated in parts (b) to (d) in Figure 20, where part (a) shows the structure for pure A–B.[99,100] When $r_A \ll 1$ (part b), HA tends to be selectively solubilized into the A microdomain more or less uniformly, swelling the A block chains and hence expanding the average distance between the neighboring junctions of A–B from a_{J_0} to a_J (wet brush regime). This causes asymmetry in the average volumes occupied per single A and B block chains and hence changes the microdomain morphology. Where $r_A \approx 1$ (part c), HA tends to be still solubilized selectively into the A microdomains. However, HA does not significantly swell the A block chains and tends to localize in the middle of the A microdomains. Hence the average distance is not affected by HA ($a_{J_0} \approx a_J$) (dry brush regime). In this case, HA does not significantly affect the B microdomains in terms of their shape and size and in terms of the molecular conformations of the B block chains. When $r_A \gg 1$ (part d), A–B and HA undergo macrophase separation, forming microphase-separated A–B domains dispersed in the matrix of HA. We shall describe below experimental evidence and physical factors to support this behavior.

14.5.1 Wet Brush vs. Dry Brush

Polymer 'brushes' are long-chain molecules attached by one end to an interface at relatively high coverage and stretched away from the interface to avoid overlapping. Strongly

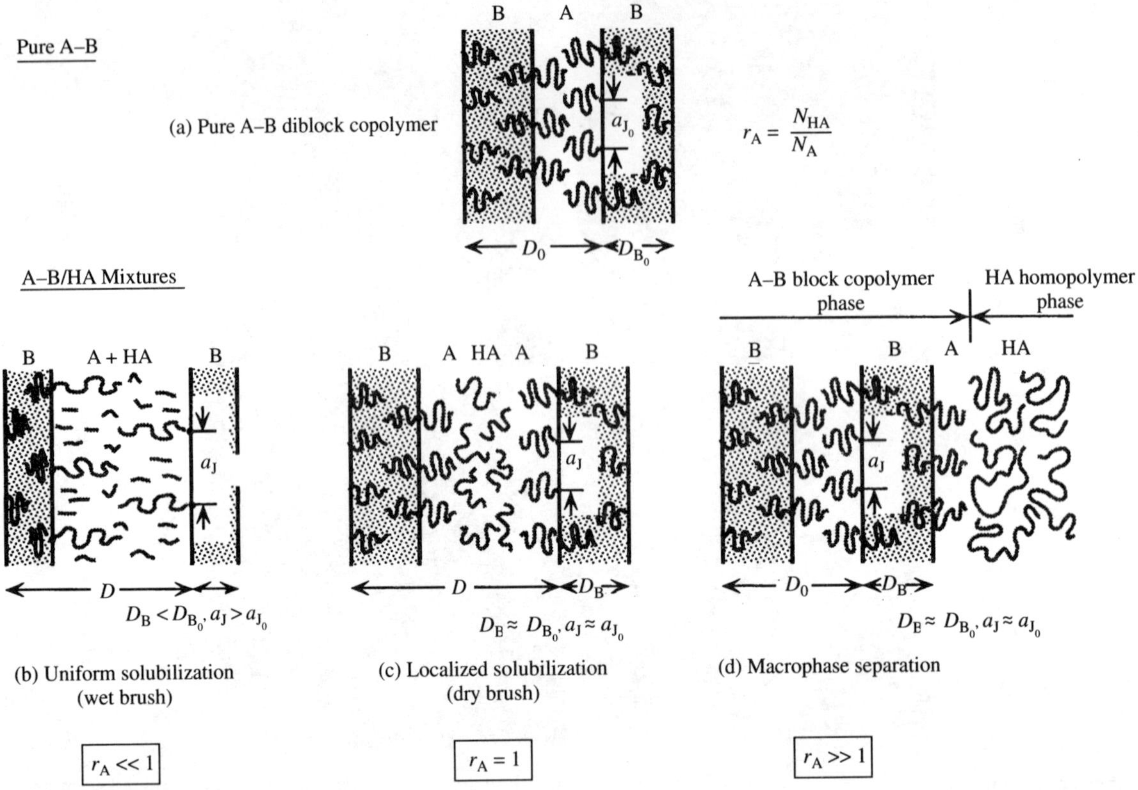

Figure 20 Possible states of mixing for a binary mixture of A–B/HA with $r_A = N_{HA}/N_A$. (a) Neat block copolymer A–B with domain spacing D_0, thickness of B domain D_{B_0} and average distance between neighboring junctions a_{J_0}. (b) Wet brush ($r_A \ll 1$). HA is uniformly solubilized in the A microdomain; swelling of A block chains by HA expands D_0 and a_{J_0} to D and a_J, respectively, and reduces D_{B_0} to D_B. Owing to the asymmetry in the effective volume of A and B block chains, the interface curvature and, hence, the morphology may change. (c) Dry brush ($r_A \approx 1$): HA is completely solubilized in the A microdomain but localized in the middle of the domain, expanding D_0 to D but D_B and a_J are unaffected by HA. (d) Macrophase separation between A–B and HA ($r_A \gg 1$).

segregated diblock copolymers are considered to be in such a physical situation. Because of this physical situation, polymer brushes behave differently from free polymer chains. The swelling behavior of brushes of polymer A with A homopolymer is quite different from that between two homopolymers A with different MW in free space.[98,99]

Hashimoto *et al.*[99] investigated the effect of the MW of the homopolymer on the microdomain structures by SAXS and TEM for given volume fractions of block copolymer (ϕ_{block}) in the mixtures of a given lamella-forming SI diblock copolymer (HY-8: $M_n = 3.16 \times 10^4$, $f_{PS} = 0.44$) and HS with various MWs ($M_n = 2.3 \times 10^3$–1.67×10^4). The systems are within the limit of the complete solubilization regime ($r_S = 0.15$–1.1, corresponding to r_A in the case of SI/HS) as shown in Figures 20(b) and 20(c), where only the microphase separation occurs. Figure 21(a) shows the ratio of the interdomain distances D/D_0 measured by SAXS as a function of ϕ_{block} for the mixtures composed of HY-8 and HS with various MWs (given by the two digits in the codes with the unit of 10^3 in the figure) and hence with various r_S, where D_0 is the interlamellar distance of pure SI ($D_0 = 26.7$ nm). The symbols S, C, and L in the figure designate, respectively, the spherical, cylindrical, and lamellar morphologies observed for the mixtures.

For a given r_S, D increases with increasing $\phi_H = 1 - \phi_{block}$. The volume fraction of PS microdomains as determined from SAXS increases with increasing ϕ_H at a rate as expected from the increases in the HS content, clearly indicating the solubilization effect (swelling of the PS microdomain with HA).[99] For a given ϕ_{block}, D increases with increasing r_S. It is a surprise to note that for $\phi_{block} = 0.5$, the morphology changes from the cylinders to the lamellae with increasing r_S, namely, the lamellar morphology for the neat SI is recovered with increasing r_S even for the SI/HS mixture with $\phi_H = 0.5$.

Figure 21(b) shows the corresponding change of a_J/a_{J_0} with ϕ_{block}. For a given r_S, a_J/a_{J_0} tends to expand with increasing ϕ_H and hence decreasing ϕ_{block}: the larger the value of r_S, the smaller

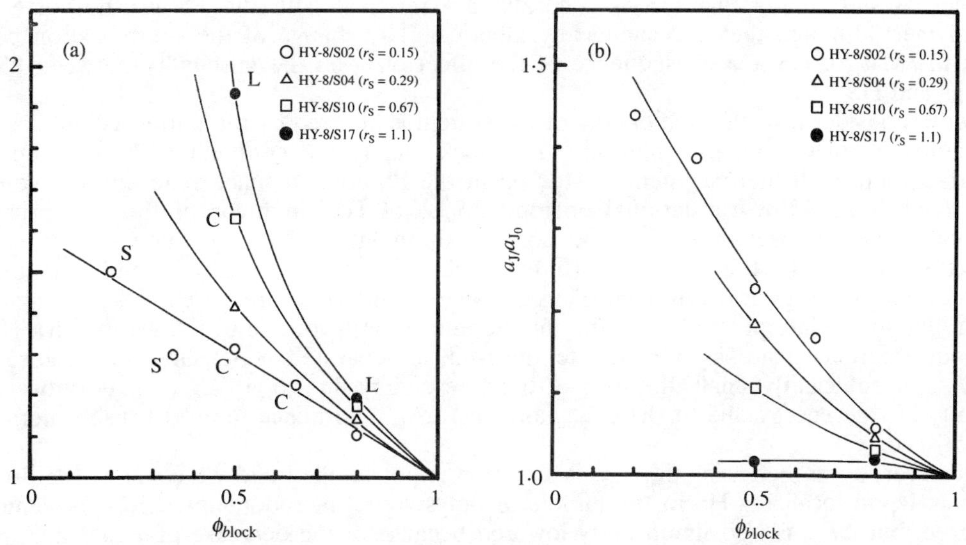

Figure 21 (a) Average interdomain distance relative to that for pure SI ($D_0 = 26.7$ nm), D/D_0, measured by SAXS, as a function of volume fraction of block polymer ϕ_{block} for the SI/HS mixture (SI: HY-8, $M_n = 3.16 \times 10^4$, $f_{PS} = 0.44$) with various HS having different MW (given by 2 digits after S in the figure) and hence having different r_S. Letters L, C, and S beside the data points designate lamellar, cylindrical, and spherical microdomain morphologies, respectively. (b) The average nearest-neighbor distance of chemical junction points along the interfaces of the microdomains relative to that for pure block polymer SI ($a_{J_0} = 2.00$ nm); a_J/a_{J_0} were plotted against the volume fraction of the block polymer ϕ_{block} for SI/HS blends (after Hashimoto *et al.*[99]).

the value of a_J/a_{J_0}. This implies that HS tends to cause the PS domains to swell laterally, *i.e.* in a direction parallel to the interface. For a given ϕ_H, a_J/a_{J_0} decreases and hence the degree of the lateral swelling decreases with increasing r_S. The lower the degree of the lateral swelling, the larger the expansion of the interdomain distance D/D_0. Figure 22 demonstrates these changes of a_J/a_{J_0} and D/D_0 with r_S or number-average MW of HS ($M_{n,HS}$) for $\phi_H = 0.5$.

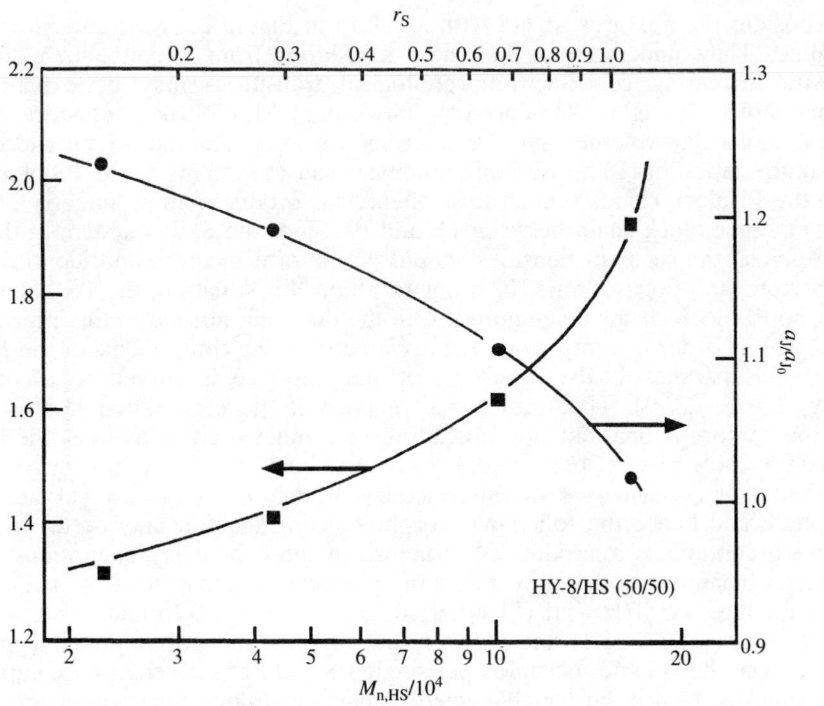

Figure 22 Replots of the D/D_0 and a_J/a_{J_0} data for $\phi_{block} = 0.5$ in Figure 21 as a function of MW of the homopolymer, $M_{n,HS}$, or r_S (after Hashimoto *et al.*[99]).

These results imply that HS is generally not miscible with the PS brushes of SI block copolymer, although they are chemically identical. The change of the solubilization behavior with the MW of HS is a consequence of the effect of r_S on the miscibility between PS block chains and HS.

Let us consider how the free energy of the system is affected by the uniform solubilization of HS into PS microdomains where the PS block chains are confined with one end at the interfaces and with their segments restricted in the PS domain space. The uniform solubilization (i) increases the translational entropy ΔS_{trans} of HS (in favor of the wet brush), (ii) decreases the conformational entropy ΔS_{conf} in both HS and PS by chain stretching in the direction normal to the interfaces (unfavorable to the wet brush),[103] and (iii) increases the interfacial energy ΔE_{int} by increasing the average interface area occupied per single block chain due to the lateral swelling of the microdomains (unfavorable to the wet brush). Here we set our reference state as a complete macrophase-separated state shown in Figure 20(d). If $M_{n,HS}$ is sufficiently small, the gain of the free energy due to (i) ΔS_{trans} may outweight the penalty of free energy due to (ii) ΔS_{conf} and (iii) ΔE_{int} and hence the wet brush criterion may be satisfied.

However, upon increasing $M_{n,HS}$ this situation is not fulfilled and we come to the dry brush regime. Upon localizing HS in the middle of polystyrene microdomains, ΔS_{trans} is significantly lowered, but ΔE_{int} is also significantly lowered because of the decrease of a_J/a_{J_0}. ΔS_{conf} is less intuitively obvious, because it may increase in a sense that the stretching of PS block chains and HS normal to the interface (caused by the uniform swelling) is relaxed upon the deswelling, but it may decrease in a sense that HS squeezed between the PS brushes may be compressed. Nevertheless, there may be a net increase in ΔS_{conf}. As $M_{n,HS}$ increases, the contribution of ΔS_{trans} to the free energy becomes increasingly trivial and hence the gains due to ΔE_{int} and ΔS_{conf} are expected to outweigh the penalty due to ΔS_{trans}. As a consequence, HS with larger r_S tends to be segregated more toward the middle of the PS microdomains. In the case when the density of brushes at the interface is fixed, the conformational entropy loss upon mixing with HS plays a dominant role in the miscibility of HS and PS brushes and hence on the solubilization of HS into PS microdomains. Thus, the miscibility of polymer brushes is distinctly different from that of the corresponding free chain.

14.5.2 Wet Brush

Let us consider the mixtures SI/HS with $r_S = 0.15$ in Figure 21, which are in the typical wet brush regime. They undergo morphological transitions from lamellae to cylinders and to spheres with increasing ϕ_H. Such morphological transitions may be explained with the illustrations shown in Figure 23. Part (a) shows pure SI diblock copolymer chains with a symmetrical molecular volume and the interface between PS and PI microdomains is flat. Uniform solubilization of HS into PS microdomains causes swelling of the PS block chains with HS, while the PI block chains remain unswollen. This produces an asymmetry of the volumes occupied per single block chain between PS and PI. Since the SI/HS system in the liquid state is incompressible, the segment densities should be uniform in space and identical to those for pure polystyrene and polyisoprene. In order to attain this situation, the PS block chains must stretch or the PI block chains be compressed in the direction normal to the interface, as shown in Figure 23(b), in order to compensate the asymmetry of the components of the R_g's of PS and PI block chains parallel to the interface, or the interface is curved to accommodate the asymmetry (Figure 23(c)). The former may happen if the segregation is strong enough to outweigh the conformational entropy loss. However, under usual situations, the latter (part c) is the more favorable picture, because it involves less conformational entropy loss.

Consequently, as ϕ_H increases, the interface has an increasing concave curvature toward the minority phase and hence the following morphological transition may occur: the minority PI phase forms a continuous microdomain structure of mesh or triply continuous network or a discrete microdomain structure of cylinders or spheres in the matrix of the majority phase. In the ternary mixtures of SI/HS/HI (PI homopolymer) with $r_S = 0.15$ and $r_I = 0.14$ (corresponding to r_A in the case of SI/HI) and with net polystyrene and polyisoprene segment-fractions kept symmetrical, the volumes occupied per single PS and PI block chains are kept symmetrical with increasing ϕ_H. Hence the lamellar microdomains with long-range order are conserved up to $\phi_H = 0.90$.[101] It is a surprise to note that a very minor fraction of the block copolymer SI added to the HS/HI blends can effectively avoid macrophase separation and keep the

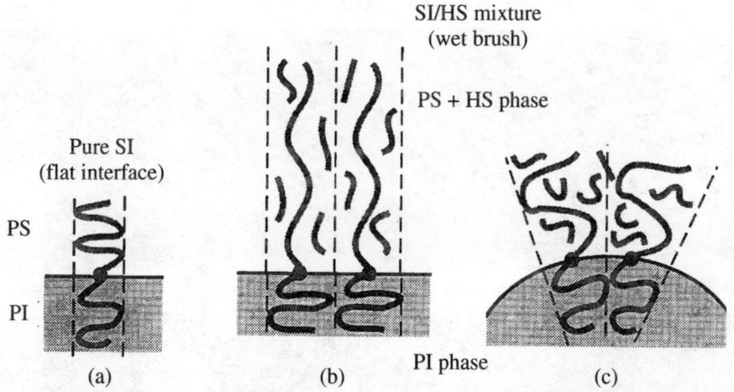

Figure 23 Schematic illustrations of chain packing. (a) Pure SI diblock copolymer chain with a symmetrical volume fraction, forming a flat interface. In the SI/HS mixture in the wet-brush regime, (b) swelling of PS block chains by uniformly solubilized HS causes stretching of PS block chains and/or compressing of PI block chains, resulting in a decrease in the conformational entropy, or alternatively, (c) a curved interface is introduced in order to attain the uniform distribution of the segments without significant stretching of the PS block chains or shrinking of the PI block chains.

microdomains with long-range order. It should be mentioned that the dependence of a_J/a_{J_0} on ϕ_H in the ternary system is exactly the same as that of the binary system.

Winey *et al.*[102] investigated over 130 diblock copolymer/homopolymer mixtures of SI/HS and SI/HB and obtained isothermal morphology diagrams. They observed the OBDD morphology in the mixtures with *ca.* 65 vol% overall PS content and $0.15 < r_S < 1.0$. They observed morphological transitions with increasing ϕ_H in order of lamellae → OBDD → cylinders → spheres → micelles. Hashimoto *et al.*[78] reported morphological transitions in order of lamellae → mesh → strut (interconnected cylinders) → spheres in SB radial block copolymer (K-Resin)/HS mixtures.

Hasegawa *et al.*[103] investigated the lateral and vertical components of the radius of gyration for a single block copolymer chain and those of a single homopolymer chain in the lamellar microdomain space formed by a mixture of SI and HS in the wet brush regime by means of contrast-matched small-angle neutron scattering (SANS). The result suggests that the homopolymer chains in the microdomain space as well as the block copolymer chains are more compressed in the direction parallel to the interface and more stretched in the direction perpendicular to the interface than the corresponding unperturbed polymer chains with the same MW. The block copolymer chains were found to undergo an isochoric affine deformation on addition of the homopolymers or with the change of thickness of the lamellar microdomains.

14.5.3 Dry Brush

Koizumi *et al.*[100] studied the morphology of a typical dry brush system of SI/HS mixtures with $r_S = 1.2$ by SAXS and TEM as a function of ϕ_H. Figure 24 shows the TEM pictures of the mixtures with increasing ϕ_H. The neat SI ($M_n = 1.0 \times 10^5$ and 50 wt% PS) has a lamellar morphology (part a). Upon increasing ϕ_H, the domain spacing D and its distribution increase, while the thickness of PI lamellae remains surprisingly constant and uniform. The latter seems to be typical of the dry brush systems. PI lamellae tend to bend and form lamellar, cylindrical, and spherical vesicles (part b–f), models of which are depicted in Figure 25.

It should be noted that such vesicular structures can be observed in the wet brush regimes in some special cases when an asymmetric block copolymer is used. Kinning *et al.*[104] investigated the morphology of mixtures of symmetric and asymmetric SB block copolymers and HS with large ϕ_H in the regimes of both wet and dry brush by TEM. In the dry brush regime with a symmetric block copolymer, they also observed spherical vesicles in the mixtures. Even in the wet brush regime ($r_S = 0.196$ and 0.382), they found lamellar and spherical vesicles in the mixtures of HS with the asymmetric SB diblock copolymers having $f_{PS} = 0.127$ and 0.269, for which spherical and cylindrical PS microdomains are expected in the pure copolymers, respectively.

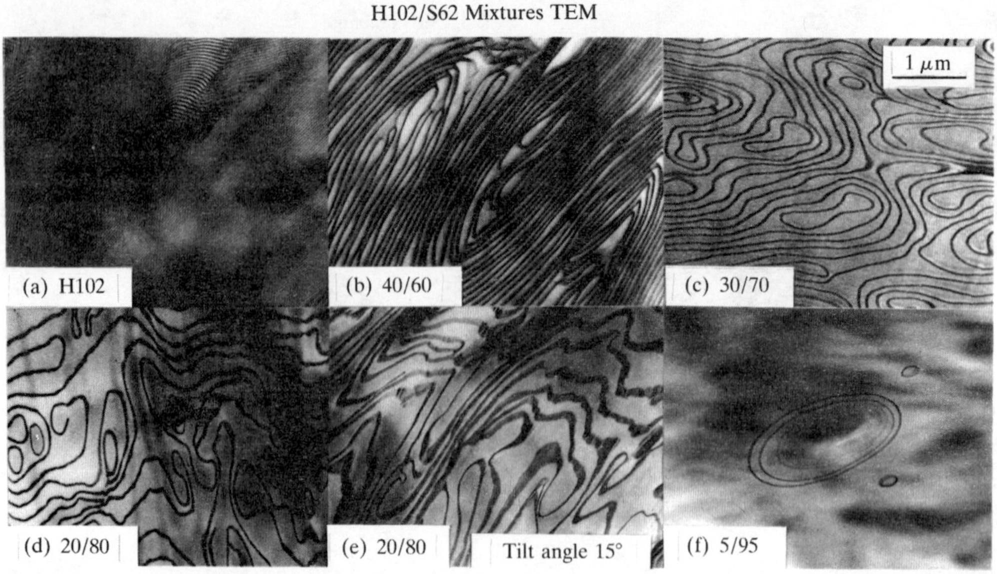

Figure 24 Transmission electron micrographs of SI/HS (H102/H62) mixtures (H102: $M_n = 1.0 \times 10^5$, $f_{PS} = 0.46$) having $r_S = 1.2$ (dry brush regime) slowly cast from toluene solutions: (a) neat SI diblock copolymer exhibiting lamellar microdomains, (b) SI/HS 40/60 wt%/wt% mixture (lamellar vesicles), (c) 30/70 mixture (lamellar and cylindrical vesicles), (d) 20/80 mixture (lamellar and cylindrical vesicles), (e) 20/80 mixture same as (d) but the specimen was tilted by 15°, showing the dark 'ribbons' of the bimolecular layers, and (f) 5/90 mixture (spherical vesicles). The PI lamellae with their normal vectors (set perpendicular to their interfaces) inclined with respect to the surface of the section increase their width and decrease their contrast in the micrograph as seen in (e) (after Koizumi *et al.*[100]).

Distribution of the homopolymer chains in microdomains can be measured by SANS[105,106] or by neutron reflectivity.[107] SANS with the deuterium labeling technique combined with SAXS[106] elucidated that the penetration depth of deuterated homopolystyrene chains (DS) ($M_n = 6.3 \times 10^4$) into the PS dry brush ($r_S = 1.2$) of SI diblock copolymer forming lamellar

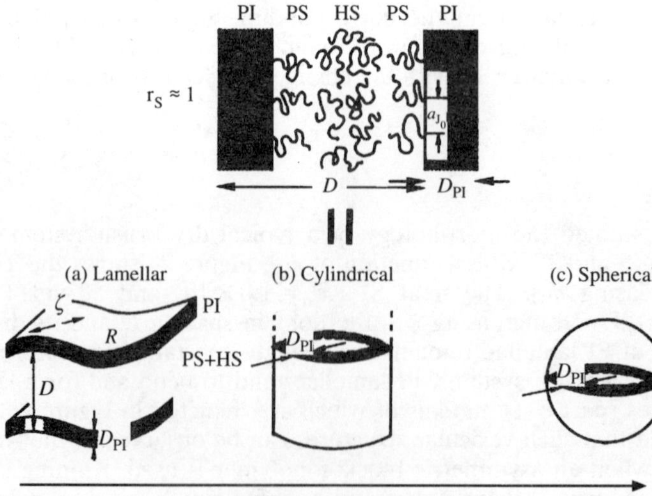

Figure 25 Model of vesicles for SI/HS mixtures in the dry brush regime ($r_S \equiv N_{HA}/N_A \approx 1$). With increasing volume fraction of HS ϕ_H in SI/HS, HS screens the interaction between the PS block chains emanating from the PI lamellae as shown in the upper half of the figure, resulting in lamellar vesicles (a). Further increase in ϕ_H causes a decrease in the persistence length ζ of the flat interface and an increase in the interface curvature R^{-1}, resulting in (b) cylindrical and (c) spherical vesicles. However, the thickness of the PI bimolecular layer D_{PI} remains unaffected (after Koizumi *et al.*[100]).

microdomains ($M_n = 1.0 \times 10^5$, 50 wt% PS) is 11 ± 1 nm and comparable to the unperturbed radius of gyration of PS or DS (7.0 nm). This result is also consistent with a theoretical prediction[108,109] (12.8 nm) and with the result of the neutron reflectivity measurement.[107]

14.5.4 Macrophase Separation Followed by Microphase Separation

Here we deal with the block copolymer/homopolymer mixtures which undergo macrophase transitions ($r_A \gg 1$). Since the structures formed in such mixtures are nonequilibrium and depend on the preparation processes, a variety of patterns is observed due to the interplay of macrophase and microphase transitions. Let us first consider the case in which a macrophase transition takes place prior to a microphase transition.

It may be worthwhile to investigate the thermodynamic stability limits for the real SI/HS mixtures whose morphologies are shown in Figures 24, 27, and 28. Upon increasing the interaction parameter χ beyond that at the spinodal point for the macrophase transition or the microphase transition, only specific modes of the thermal concentration fluctuations can obtain a driving force to grow to the ordered state. Figure 26 shows the spinodal lines for the macrophase transition (solid lines) and the microphase transition (broken lines) for the SI/HS mixtures having $r_S = 1.2$, 4.0, and 11.4.[100] The mixture with the larger r_S has less chance to encounter first the spinodal point for the microphase transition upon increasing χ from the disordered state, because the Lifshitz point ϕ_c increases with r_S. At $\phi_{block} > \phi_c$, the microphase transition should be encountered first during the ordering process so that HS may tend to be solubilized into the PS phase of a microdomain. A nice agreement between the stability analysis and experiments has been reported.[100]

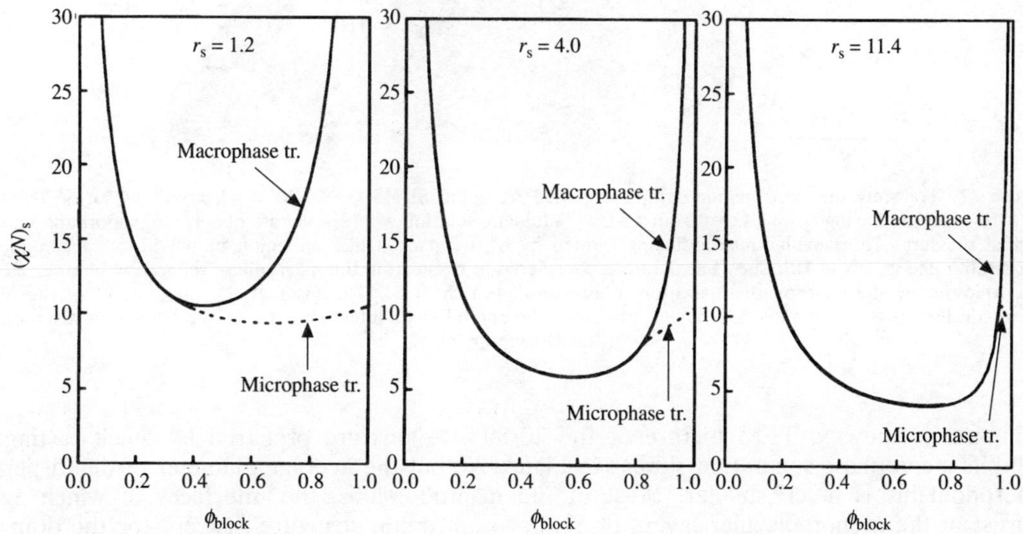

Figure 26 Stability limits $(\chi N)_S$ for macrophase transition (solid lines) and microphase transition (broken lines) calculated by RPA as a function of volume fraction of block copolymer ϕ_{block} for the mixtures of SI diblock copolymer with $M_n = 1.0 \times 10^5$ and $\phi_{PS} = 0.46$ (corresponding to H102) and HS having different r_S. (a) $r_S = 1.2$ (dry brush regime): the microphase separation occurs first for the compositions with $\phi_{block} \gtrsim 0.3$; (b) $r_S = 4.0$; and (c) $r_S = 11.4$; the macrophase separation occurs first for most of the compositions (after Koizumi *et al.*[100]).

The SI/HS mixtures having $r_S = 11.4$ were investigated by Koizumi *et al.*1[110] by TEM. These mixtures with compositions of 40/60 and 20/80 wt%/wt% first encounter the spinodal point for the macrophase transition when they are cast from solutions, as expected from the RPA results in Figure 26. Figure 27 shows a typical TEM picture of the 40/60 wt%wt% mixture prepared by slow casting. The SI-rich macrodomains composed of lamellar microdomains are not spherical but ellipsoidal and are dispersed in the matrix of HS. SAXS analysis suggests that the lamellar microdomains are surrounded by a dark layer of the PI phase, implying that a monomolecular layer of SI exists along the macrodomain interface between the HS- and SI-rich macrophases, as schematically shown in the inset. The following self-assembling mechanism was proposed. During the solvent evaporation process, the macrophase separation first occurs, generating droplets of SI-rich solution in the matrix of the HS-rich solution. The microphase separation is then followed, in the droplets of the SI-rich solution, from the periphery of the droplets toward their centers. The lamellae favor the flat interfaces, which drives deformation of the droplet into oblate ellipsoids as the net segregation power between PS and PI increases with the solvent evaporation.

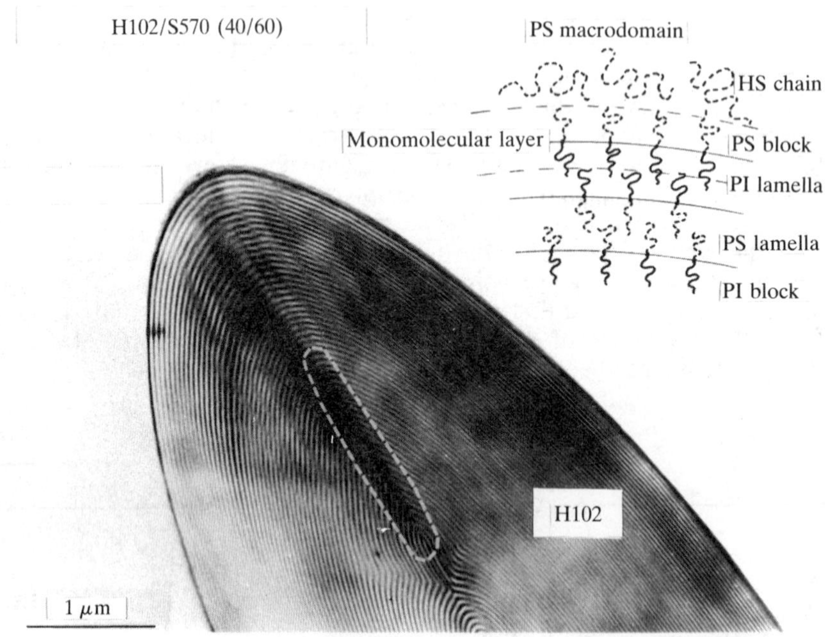

Figure 27 Transmission electron micrograph obtained from the SI/HS (H102/S570 40/60 wt%/wt%; S570: $M_n = 5.7 \times 10^5$) mixture having $r_S = 11.4$ cast from a 5.0 wt% toluene solution with a slow rate of solvent evaporation under a normal pressure. The lamellar microdomains formed by SI are packed like an onion in the SI-rich macrodomain dispersed in the matrix of HS. Such a structure is considered to result from the macrophase separation between SI and HS, followed by the microphase separation in the droplets rich in SI. The inset at the upper right of the figure schematically shows a monomolecular layer of SI along the microdomain interface between the HS and SI macrophases (after Koizumi *et al.*[110]).

Figure 28 shows a TEM picture of the 20/80 wt% mixture prepared by quick casting. It exhibits macrophase separation similar to Figure 27, but the average diameter of the dispersed macrodomains is much smaller. Most of the macrodomains, the interfaces of which again consist of the monomolecular layers of SI, have an ‘onion structure’, except for the domains marked by X–Z in the figure. The shape of the macrodomains rich in SI is not so elongated as that shown in Figure 27. At low magnification the size and distribution of these macrodomains are quite regular, suggesting macrophase separation via spinodal decomposition (SD). In fact, a spinodal ring was observed in the light scattering pattern of this specimen.

Koizumi *et al.*[110] also examined the SI/HS mixture having $r_S = 8.5$. The pure SI have the morphology of PI cylinders in a PS matrix. In this case, the following two structures were observed, depending on the solvent evaporation rate: a rapid evaporation develops the spherical microdomains of PI block chains, the centers of mass of which are dispersed in the continuous matrix of the HS with fractal-like spatial distribution; in the case of slow evaporation the self-assembled structure consists of ‘lens-like’ (oblate-ellipsoidal) macrodomains, in which cylindrical microdomains are packed hexagonally with their axes parallel to the axis of revolution of the oblate ellipsoid, dispersed in the continuous matrix of the HS. Thus the resulting morphology depends very much on the preparation process, *i.e.* the solvent evaporation rate in this case. In any case, during the solvent evaporation process the

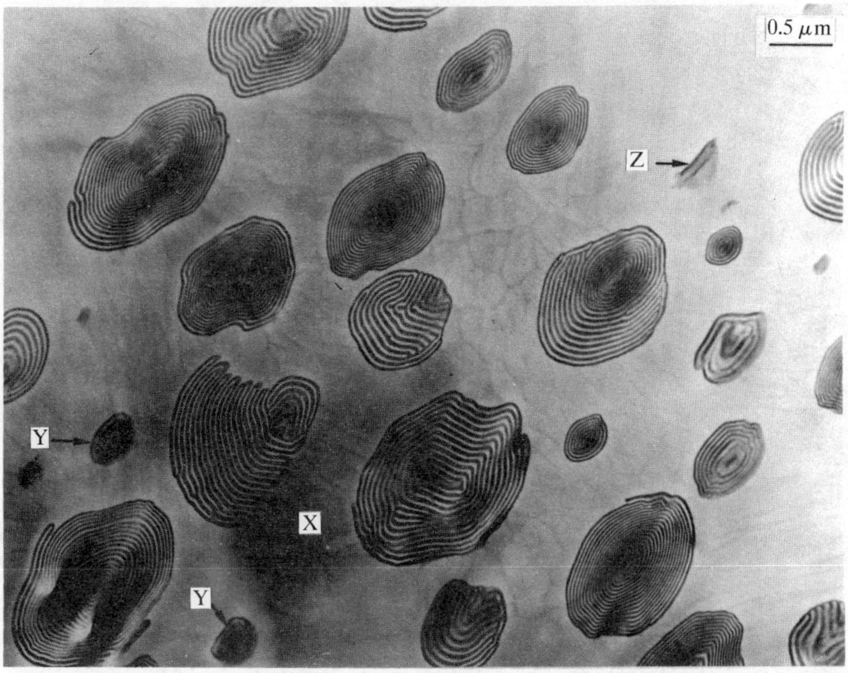

Figure 28 Transmission electron micrograph obtained from the binary mixture of SI/HS (H102/S570 20/80 wt%/wt%) cast from a 5.0 wt% solution in toluene by rapid evaporation under a low pressure. The SI macrodomains are much smaller than those in the slowly cast film (Figure 27) due to the microphase separation, and subsequent vitrification occurred before significant coarsening of the disordered SI macrodomains. Besides the 'onion structures', the vesicular structures marked X, Y, or Z are seen (after Koizumi *et al.*[110]).

macrophase separation occurs first, generating regions rich in SI and HS, and subsequently microphase separation, generating the microdomains in the macrophase rich in SI. In the lens-like macrodomains, the microphase separation was proposed to occur from the center toward the periphery of the SI-rich droplets which are first formed by the macrophase separation. The lens shape was explained in terms of interfacial free energy between HS and PS brushes emanating from the cylindrical microdomains.

The concentration fluctuations developed by the macrophase separation via spinodal decomposition between block copolymer and homopolymer can be frozen-in by vitrification and visualized through the spatial variation of the microdomain structures developed via the microphase separation subsequently occurring after the macrophase separation. The self-assembly discussed above indicates examples of such processes. Another example is shown below for SI/PPO mixtures.[17]

Figure 29 shows the TEM micrograph for the SI/PPO 80/20 wt%/wt% mixture with pure SI, having lamellar morphology. The micrograph shows the pattern in which the microdomain morphology changes spatially with the long period Λ_{macro} of the order of 5 μm: from the lamellar morphology (as in the regions marked A), through the cylindrical morphology (as in the regions marked D). The microdomains have the spacing of $\Lambda_{micro} = 100$ nm. This pattern can be explained as follows. The SD process may develop periodic concentration fluctuations of SI (or polymer α) and PPO (or polymer β) with a spacing of Λ_{macro} of the order of 1 μm as sketched in Figure 30(a), depending on the time spent for SD at a given phase separation condition.

If this macrophase transition is followed by the microphase transition, and if PPO and PS are mixed at the molecular level to form the microdomains of PS and PPO mixed with each other and segregated from the microdomains of PI, the self-assembling process may generate a modulated pattern or a pattern with a superlattice, as shown in Figure 30(b). Here the microdomain morphology with a short periodicity of Λ_{micro} of the order of 10 nm spatially varies with the long spacing of Λ_{macro} because SI and PPO form a wet brush when they are

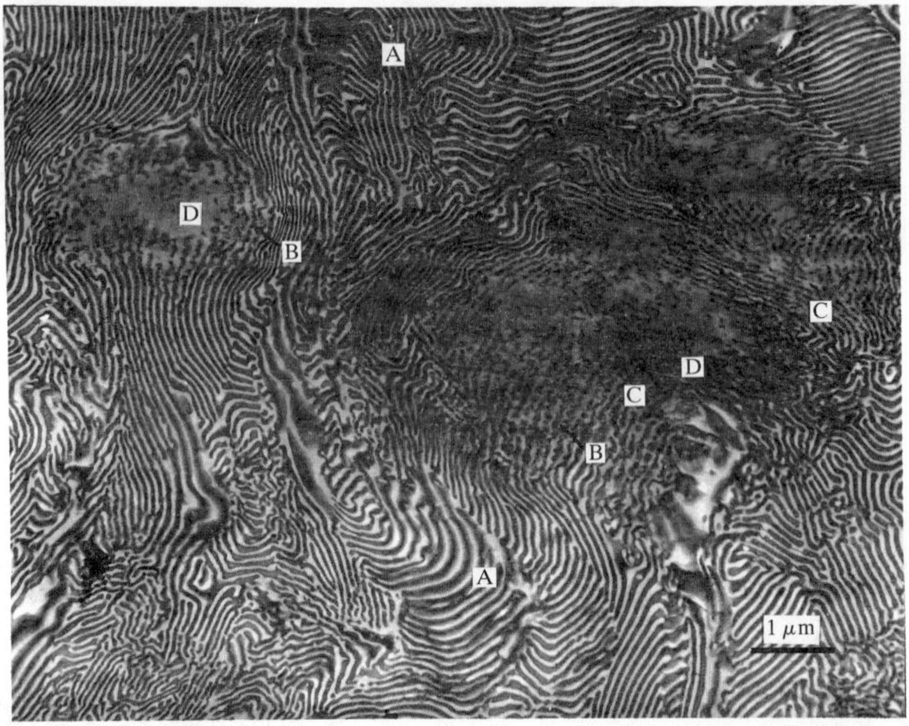

Figure 29 Electron micrograph of OsO_4-stained ultrathin section of SI/PPO 80/20 (wt%/wt%) mixture (SI: HY-12, $M_n = 5.24 \times 10^5$, $f_{PS} = 0.49$; PPO: $M_n = 1.17 \times 10^4$) slowly cast from toluene solution. Spatial variation in the microdomain morphology, such as lamellar morphology inherent in the neat SI (as in the regions marked A), mesh or bicontinuous (marked B), cylindrical (marked C), and spherical morphology (marked D), is considered to result from the concentration fluctuations produced by the SD between SI and PPO, followed by the microphase transition. PPO solubilized in the PS microdomain upon the microphase separation of SI swells the PS block chains and hence changes the microdomain morphology, according to its local concentration (after Hashimoto *et al.*[17]).

mixed and the microdomain morphology depends on the local concentration of SI or PPO, as discussed in Section 14.5.2. Such a process may generate a pattern comprised of spatially varying microdomain morphology. In the case of the lamellar-forming SI, the microdomain morphology changes from lamellae through cylinders to spheres with increasing concentration of PPO, as indicated in Figure 30(b). Mesh and/or bicontinuous morphology might exist between the regions with lamellar and cylindrical morphology. It should also be noted that the rapid evaporation of the solvent for the 80/20 mixture developed uniformly a strut morphology over the whole sample space. This implies that the mixture was quickly brought inside the spinodal line for the microphase separation through the gap between the two spinodal lines for the macrophase and microphase separations and hence the microphase separation had been completed before the macrophase separation proceeded. PPO was uniformly solubilized in the PS microdomain.

14.5.5 Microphase Separation Followed by Macrophase Separation

In certain A–B block copolymer/C homopolymer mixtures, temperature-induced macrophase separation can be observed after formation of a uniform microdomain structure. Such an example has been reported for the binary mixtures of SB radial block copolymer (K-Resin, $M_n = 1.45 \times 10^5$ and 75 wt% of PS) and PVME homopolymer ($M_n = 2.6 \times 10^4$).[18] The microphase separation was induced in a film preparation process using a solution cast method at room temperature. The as-cast film of the 50/50 wt%/wt% mixture exhibits a uniform strut morphology with a PB network in the matrix of PS and PVME mixed with each other, as expected from the TEM picture shown in Figure 31(a).

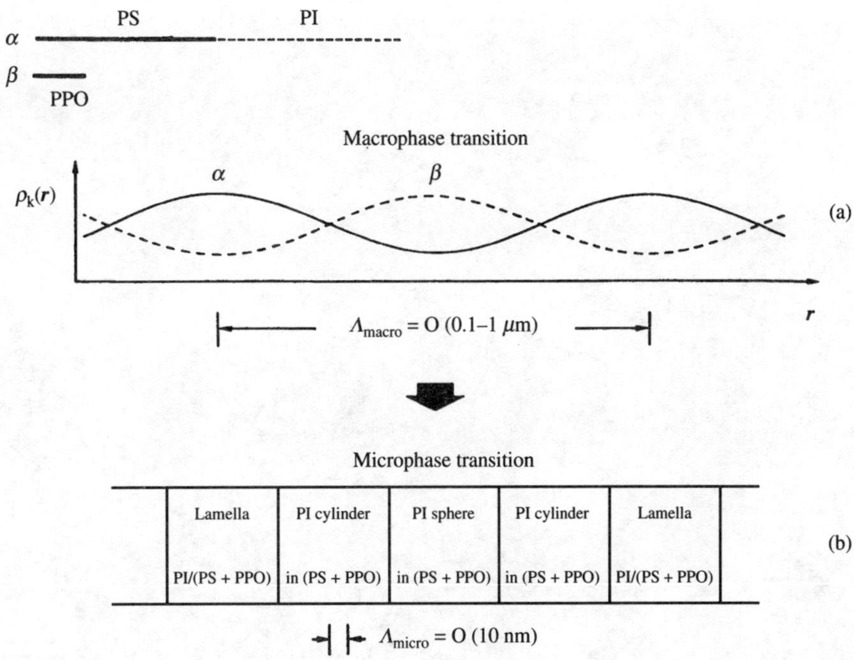

Figure 30 Interpretation of the micrograph in Figure 29. (a) Profiles of the periodic concentration fluctuations of polymer α or SI (solid curve) and polymer β or PPO (broken curve) with a spacing of Λ_{macro} (of the order of 1 μm) developed by the SD process in the macrophase transition. (b) Modulated pattern or a pattern with a superlattice of the spatially varying microdomains with a short periodicity of Λ_{micro} (of the order of 10 nm) superimposed over the periodic concentration fluctuation corresponding to (a). The morphology changes in space in order of lamellae → (mesh or bicontinuous) → cylinders → spheres depending on the local volume fraction of PPO solubilized in the PS microdomains (after Hashimoto *et al.*[17]).

As anticipated from the LCST-type phase behavior of PS/PVME binary homopolymer blends, heating this specimen at 102.5 °C induced macrophase separation between PS block chains and PVME homopolymer and a spinodal peak was developed in the light scattering profile. The TEM observation of the subsequently quenched specimen revealed a superlattice structure with two characteristic spacings as shown in Figure 31(b): Λ_{macro} between the PVME-rich regions (bright regions) and the SB rich-regions (the regions showing the periodic structure) developed via SD, and Λ_{micro} developed via the microphase separation in the SB-rich regions, both being generally time-dependent.

The morphology transition of the microdomains from strut (Figure 31(a)) to lamellae (Figure 31(b)) was also observed as the macrophase separation via SD progressed in the regions rich in the block copolymer for a further annealed sample. The dynamics of macrophase separation in the SB/PVME mixture was found to be much slower than that in PS/PVME binary homopolymer blends.

14.6 BLOCK COPOLYMER/BLOCK COPOLYMER MIXTURES

In Section 14.5 we showed that a variety of morphologies can be produced by blending homopolymers with block copolymers. The blends of two or more block copolymers have been considered as another variation of block copolymer/homopolymer blends.

14.6.1 Co-surfactant Effect

We will consider the mechanism of a morphology change in the block copolymer mixtures due to the co-surfactant effect below. Let us consider a mixture of a long diblock copolymer (A_α–B_α) with total DP N_α and composition f_{A_α} and a short diblock copolymer (A_β–B_β) with

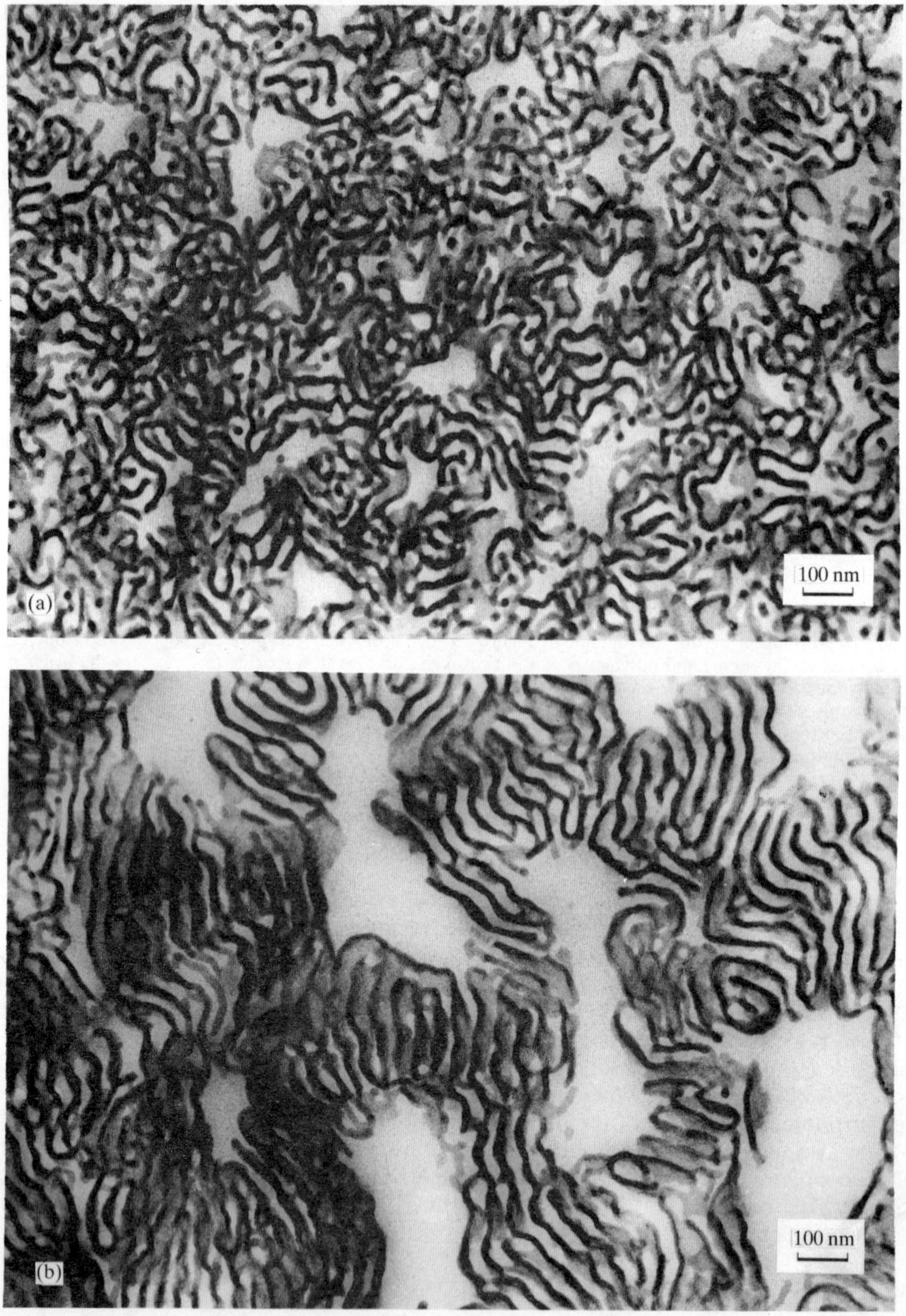

Figure 31 Transmission electron micrographs obtained from: (a) a 50/50 (wt%/wt%) binary mixture of SB/PVME (SB: K-Resin, $M_n = 1.45 \times 10^4$, 24 wt% PB; PVME: $M_n = 2.6 \times 10^4$) annealed at 111 °C (supercooling of $\Delta T = 18$ °C into the thermodynamically unstable state from the spinodal point) for 20 min exhibiting strut morphology, as observed also in the as-cast film. The small bright area is the region rich in PVME developed by SD. (b) The same mixture spinodally decomposed at $T = 111$ °C for 180 min. The macrophase separation between SB and PVME via SD further progressed and developed a superlattice structure with two characteristic spacings of Λ_{macro} of SI-rich and PVME-rich macrodomains and Λ_{micro} of PS and PB microdomains, and the microdomain morphology changed from strut to lamellae inherent in the neat SB (after Hashimoto *et al.*[18]).

N_β ($<N_\alpha$) and f_{A_β}. Here we employ the parameter r, the ratio of N_α and N_β ($r \equiv N_\alpha/N_\beta$) for the miscibility criterion, instead of the parameter r_A for block copolymer/homopolymer mixtures. We further consider the case where neat A_α–B_α prefers a morphology with curved interfaces ($f_{A_\alpha} > 0.5$), as schematically shown in Figure 32(a), and neat A_β–B_β with flat interfaces ($f_{A_\beta} \approx 0.5$), and B_α and B_β have nearly equal DP. We will consider the problem of chain packing under the following conditions: A_α–B_α and A_β–B_β are molecularly mixed to form a uniform microdomain structure with a single morphology; A_α and A_β block chains are segregated in A microdomains and B_α and B_β in B microdomains; and the junctions of both A_α–B_α and A_β–B_β are localized in the common interfaces.

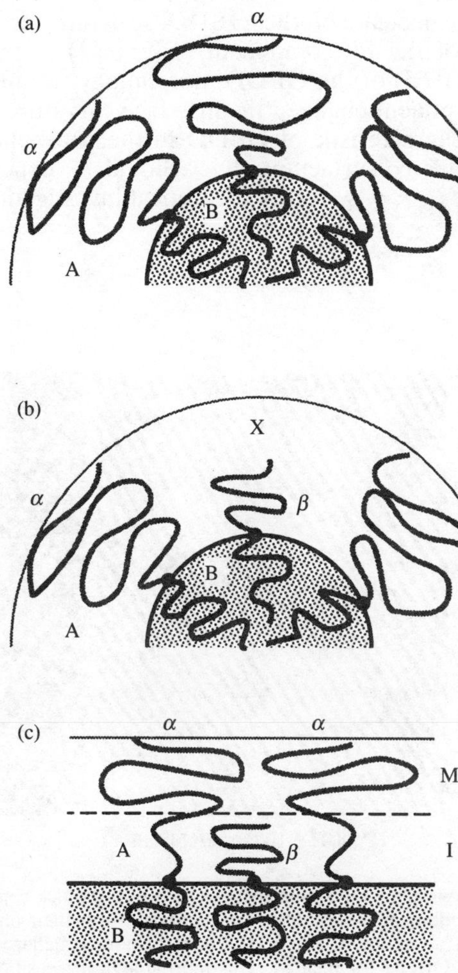

Figure 32 Schematic illustrations of the co-surfactant effect: (a) a long asymmetric diblock copolymer (A_α–B_α with $N_{A_\alpha} > N_{A_\beta}$) forming a microdomain structure with curved interfaces; (b) A_α–B_α blended with short symmetric A_β–B_β having $N_{A_\beta} \approx N_{B_\beta}$ and $N_{B_\alpha} \approx N_{B_\beta}$ without changing the original interfacial curvature of A_α–B_α, resulting in a segmental density dip (as in the region marked X) in the A microdomain; and (c) A_α–B_α/A_β–B_β blend with uniform segmental distribution in the A microdomain with an interface curvature smaller than that of (a). The long A_α chains and the short A_β chains tend to segregate, respectively, into regions M and I in the A microdomain space.

Under these conditions, replacing one A_α–B_α chain in the A_α–B_α microdomain shown in Figure 32(a) by one A_β–B_β chain without changing the interfacial curvature and without a significant change of conformation of α and β results in the situation as illustrated in Figure 32(b). Then the segmental distribution in the A microdomain becomes extremely inhomogeneous, giving rise to a density dip in region X in Figure 32(b). Such a situation is not favored. There may occur changes in the interfacial curvature and the conformations of the α and β chains to form the A microdomains with a uniform segmental distribution, as shown in

Figure 32(c), and hence the morphology may change. This implies a possibility that a small amount of block copolymer A_β–B_β changes the morphology of the majority block copolymer A_α–B_α and that the morphology of an A_α–B_α/A_β–B_β mixture does not depend necessarily on the net volume fraction of the A component of the mixture at all. This is a surprising effect, because we were used to the evidence that the morphology depended on the volume fraction of the A component in neat A–B block copolymers and in A–B/HA or A–B/HA/HB mixtures in the wet brush regime. This phenomenon is known as 'co-surfactant effect' in a surfactant system.

An example exhibiting such a co-surfactant effect has been reported by Hasihmoto *et al.* (Figure 33).[21] They mixed two SI diblock copolymers, a short lamellar-forming SI (Figure 33(a), HY-8: see Section 14.5.1) and a long OBDD-forming SI (Figure 33(c), HY-10: $M_n = 1.64 \times 10^5$, $f_{PS} = 0.66$) by 20/80 wt%/wt%. They found that the mixture exhibits a uniform lamellar microdomain structure instead of the OBDD structure, as shown in Figure 33(b), although the volume fraction of the PS component ($f_{PS} = 0.62$) in the mixture is still in the composition range ($0.62 \le f_{PS} \le 0.66$) of the OBDD morphology[51] in neat SI. The addition of a small amount of the minor component changed the interface curvature of the major component from the hyperbolic interface characteristic of OBDD to the parabolic interface characteristic of the lamellar morphology. This co-surfactant effect should be experimentally and theoretically explored as a function of f_{A_α}, r, χ_{AB}, *etc.* A theoretical investigation of the effect has been reported by Shi and Noolandi.[111]

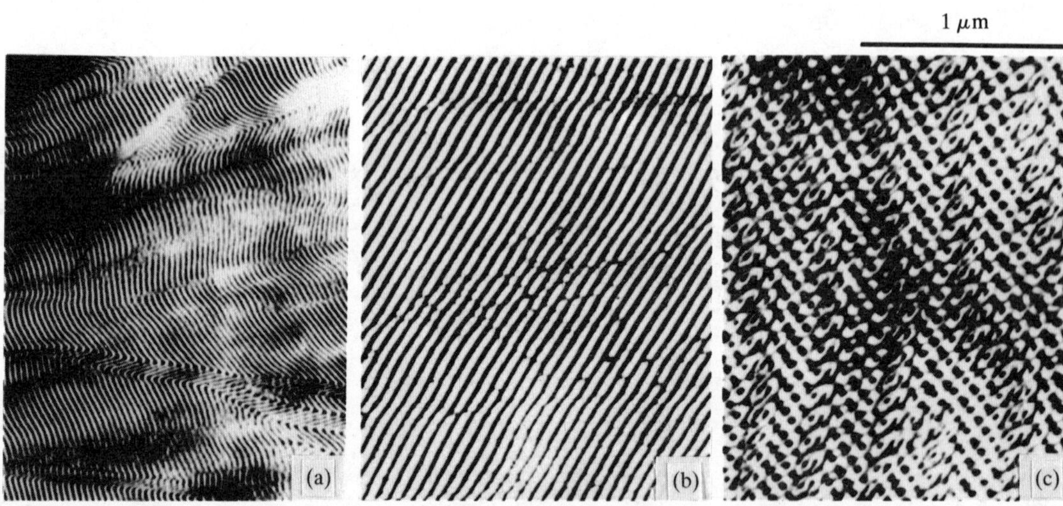

Figure 33 Transmission electron micrographs of ultrathin sections of two SI diblock copolymers, HY-8 (a) and HY-10 (c), exhibit lamellar and OBDD microdomain morphology in accordance with their composition $f_{PS} = 0.44$ and 0.66, respectively. However, a 20/80 wt%/wt% mixture of HY-8/HY-10 (b) exhibits lamellar morphology even though it has $f_{PS} = 0.62$, which is still in the limit of OBDD for neat SI. This morphological transition from OBDD to lamellae is considered as the co-surfactant effect, *i.e.* blending a small amount of the short symmetric diblock copolymer (HY-8) changed the interface curvature of the long asymmetric block copolymer (HY-10) and hence its microdomain morphology (after Hashimoto *et al.*[21]).

14.6.2 Mixtures of Lamella-forming Block Copolymers

The pioneering study on the structures of the mixtures of SI diblock and triblock copolymers by SAXS was reported by Hadziioannou and Skoulios.[112] They investigated the dependencies of domain spacing D and interfacial area per molecule S ($=a_J^2$) on blend composition within the limit that each block copolymer mixes with each other and forms a single microdomain structure. Their data were analyzed by Hashimoto[113] and it was shown that D is approximated by the 2/3 power law for the two block copolymers with a relatively small difference in $M_{n,\alpha}$ and $M_{n,\beta}$:

$$D \approx M_n^{2/3}, \quad M_n = x_\alpha M_{n,\alpha} + x_\beta M_{n,\beta} \tag{6}$$

where M_n is the number-average MW of the mixture, $M_{n,K}$ ($K = \alpha$ or β) is the number-average MW of the Kth component and x_K is the mole fraction of the Kth component ($x_\alpha + x_\beta = 1$).

The morphology of binary mixtures of nearly symmetrical SI diblock copolymers ($f_{PS} = 0.35$–0.69), the neat morphology of which are mostly lamellar, has been investigated by means of TEM and SAXS by Hashimoto *et al.*[21] The range of N_α (or N_β) employed in their study (1.03×10^2–1.24×10^4) was much wider than that of Hadziioannou and Skoulios[112] (2.07×10^2–1.33×10^3). Consequently, some of the mixtures exhibited macrophase separation. Figures 34 and 35 summarize the results of the structure analysis. If $r < 5$, α and β were found to be totally miscible on the molecular level at all compositions, forming a single microdomain morphology. Figure 34 shows a double-logarithmic plot of D vs. M_n for the mixtures exhibiting a single morphology. The result is consistent with the 2/3 power law (equation 6), except for the highest MW limit. The suppression of D to less than the predicted value may be due to the nonequilibrium effect.[114]

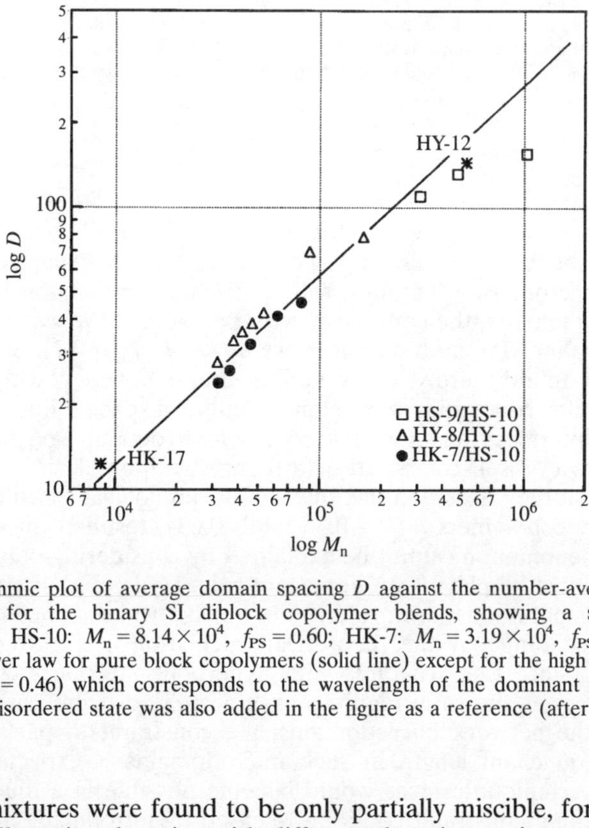

Figure 34 Double logarithmic plot of average domain spacing D against the number-average molecular weight M_n defined by equation (6) for the binary SI diblock copolymer blends, showing a single morphology (HS-9: $M_n = 1.03 \times 10^6$, $f_{PS} = 0.48$; HS-10: $M_n = 8.14 \times 10^4$, $f_{PS} = 0.60$; HK-7: $M_n = 3.19 \times 10^4$, $f_{PS} = 0.32$). The results are consistent with the 2/3 power law for pure block copolymers (solid line) except for the high MW limit. The D value for HK-17 ($M_n = 8.5 \times 10^3$, $f_{PS} = 0.46$) which corresponds to the wavelength of the dominant mode of the concentration fluctuations in the disordered state was also added in the figure as a reference (after Hashimoto *et al.*[21]).

When $r > 10$, the mixtures were found to be only partially miscible, forming macroscopically phase-separated lamellar microdomains with different domain spacings, as shown in Figures 35 and 37. Such macrophase separation is not expected to occur from homogeneous mixtures of α and β in the disordered state. The self-assembly of A_α–B_α/A_β–B_β involves macrophase and microphase separation as in the case of A–B/HA (Section 14.3). In the context of the mean-field theory, the spinodal point for the macrophase transition between single-phase disordered copolymers and two-phase but still disordered copolymers can be predicted by

$$\chi_{\text{eff,s}} = 1/N_\alpha \Phi_\alpha + 1/N_\beta (1 - \Phi_\alpha) \tag{7}$$

$$\chi_{\text{eff}} = (f_{A_\alpha} - f_{A_\beta})^2 \chi_{AB} \tag{8}$$

where χ_{eff} is the effective Flory segmental interaction parameter between two polymers α and β per segment, $\chi_{\text{eff,s}}$ is the χ_{eff} value at the spinodal point and corresponds to $\chi_{\text{s,macro}}$, and Φ_α is the volume fraction of copolymer α. Since the mismatch in the composition ($f_{A_\alpha} - f_{A_\beta}$) is small

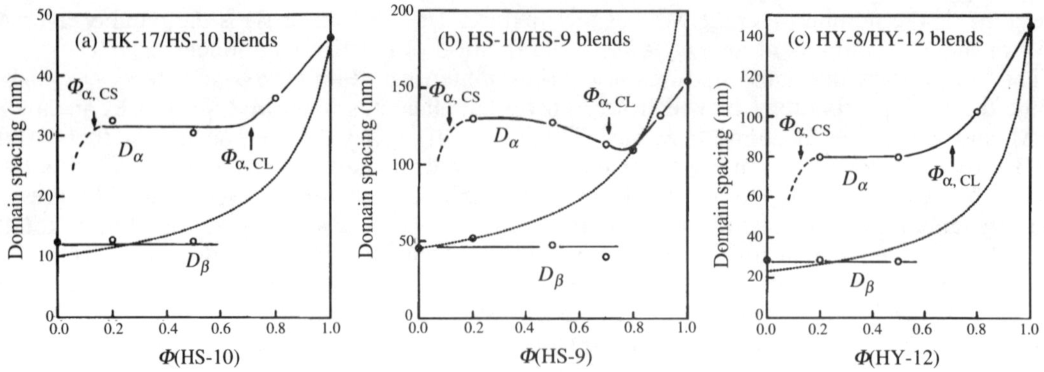

Figure 35 Plots of average domain spacing against the weight fraction of the block copolymer (upper lines) for the binary SI diblock copolymer blends α/β ($r_S = N_\alpha/N_\beta > 10$): (a) HK-17/HS-10, (b) HS-10/HS-9, and (c) HY-8/HY-12. Neat HK-17 is in the disordered state at $T > 298$ K and the other SIs have lamellar morphology. The mixtures show partial miscibility, forming the two coexisting macrophases of the lamellae with large and small spacings, D_α and D_β, or the lamellae with spacing D_α and the disordered phase with characteristic spacing D_β for HK-17/HS-10, in the composition range between the critical concentration $\Phi_{\alpha,CS}$ and $\Phi_{\alpha,CL}$. The lower curves are drawn according to the 2/3 power law. ($D = 0.025M_n^{2/3}$, $0.021M_n^{2/3}$, and $0.023M_n^{2/3}$ nm for (a) to (c), respectively) (after Hashimoto *et al.*[21]).

enough to fulfill the condition $\chi_{eff} \ll \chi_{eff,s}$ for all the mixtures shown in Figures 34 and 35, macrophase separation between the disordered copolymers is not expected to occur. A possible interpretation of the macrophase separation will be discussed in Section 14.6.3.

In the partially miscible pairs, the copolymer with the larger MW was found to solubilize the copolymer with the smaller MW up to about 30 wt% (*i.e.* $\Phi_{\alpha,CL} \approx 0.7$), but the copolymer with the smaller MW was found hardly to solubilize the copolymer with a larger MW (*i.e.* $\Phi_{\alpha,CS} < 0.2$). These results, concerning a problem of uniform space filling of the confined chains with one end at the interfaces and K-block chains in K-domain space (K = A or B), were successfully explained with a self-consistent field theory by Spontak.[115]

It should be noted that they observed the unusual morphological transition, *i.e.* mixing of two lamella-forming block copolymers ($f_{PS} = 0.35$ and 0.63) resulted in a new bicontinuous morphology. Such a phenomenon cannot be explained by considering only the volume fraction. A co-surfactant effect must be playing an important role. Hyde *et al.*[116] showed that hyperbolic interfaces (mesh and bicontinuous) are preferred in much a wider compositional range than the experimentally observed range by the theoretical calculation of the free energy for a single block chain. However, since the structures with hyperbolic interfaces characterized by triply periodic minimal surfaces as shown in Figure 6 have a broad distribution in the distance across the cross sections of the network microdomains, the constraint to pack uniformly the block chains having a uniform chain length in such microdomains is expected to be large. They suggested that the hyperbolic interfaces would become possible in a much wider composition range if the global packing constraint could be released by introducing a broad distribution in the block chain length; this may easily be attained by blending block copolymers. Schwier *et al.*[117] observed a bicontinuous morphology in the mixtures of three SB diblock copolymers and reported excellent mechanical strength, in contrast to the same system with cylindrical morphology.

14.6.3 Self-assembly in Mixtures of Immiscible Lamella-forming Block Copolymers

For the case of two lamella-forming block copolymers having partial miscibility, Hashimoto *et al.*[118] proposed an ordering mechanism, 'macrophase separations induced by microphase separation', which is a consequence of the fluctuation-induced segregation effect schematically shown in Figure 36. This fluctuation-induced segregation effect accounts for the following basic experimental observations: (i) macrophase separation giving rise to the β-rich domains as a minority phase for the 50/50 wt%/wt% mixture of HY-8/HY-12 (see Figure 37); (ii) a miscibility window in composition ($\Phi_\alpha < \Phi_{\alpha,CS}$, $\Phi_\alpha > \Phi_{\alpha,CL}$, in Figure 35); and (iii) the unique shape in the interface. Figure 38 shows the typical interfaces between two coexisting macrodomains. Beautiful commensurations of the two lamellae having different sizes at the macroscopic interfaces are observed (Figure 38). The difference in the domain sizes is compensated in various ways while the continuity of the respective domains at the macrointerface is maintained.

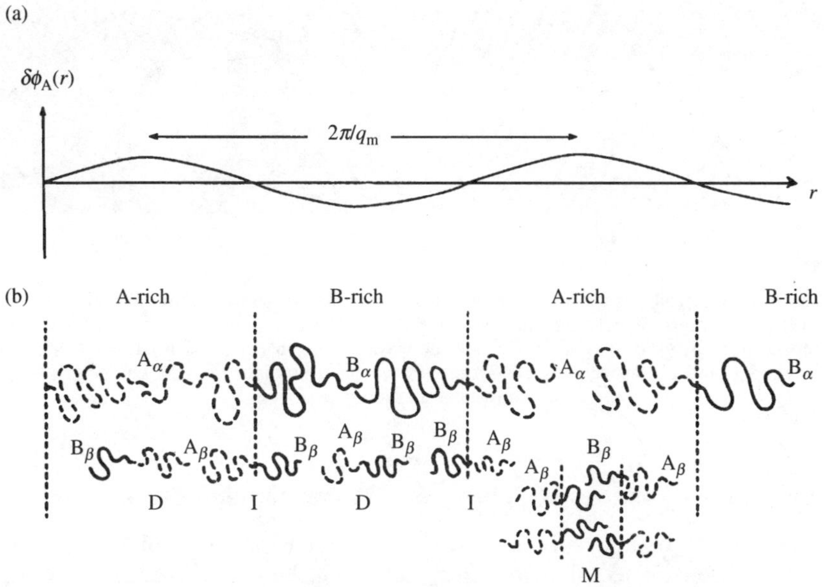

Figure 36 Schematic illustration showing the fluctuation-induced segregation effect: (a) when the thermodynamic driving force for the concentration fluctuations $\delta\phi_A(r)$ of A_α–B_α/A_β–B_β mixtures with $r = N_\alpha/N_\beta > 10$ become large, the plane wave concentration fluctuations are generated, and the wave number q_m of the dominant mode of concentration fluctuations is governed by the large copolymer A_α–B_α; (b) the small copolymers A_β–B_β marked D are put in the field rich in A_α or B_α and may feel as if they were put in the field of the homopolymers A_α or B_α; the repulsive interaction between A_α and A_β–B_β or between B_α and A_β–B_β drives (i) the segregation of A_β–B_β out of the A_α–B_α microdomains or (ii) the localization of A_β–B_β at the interface between A-rich and B-rich regions with A_β and B_β block chains confined in the corresponding domains as marked I, or (iii) the micellization into the lamellar domains as marked M (after Hashimoto *et al.*[118]).

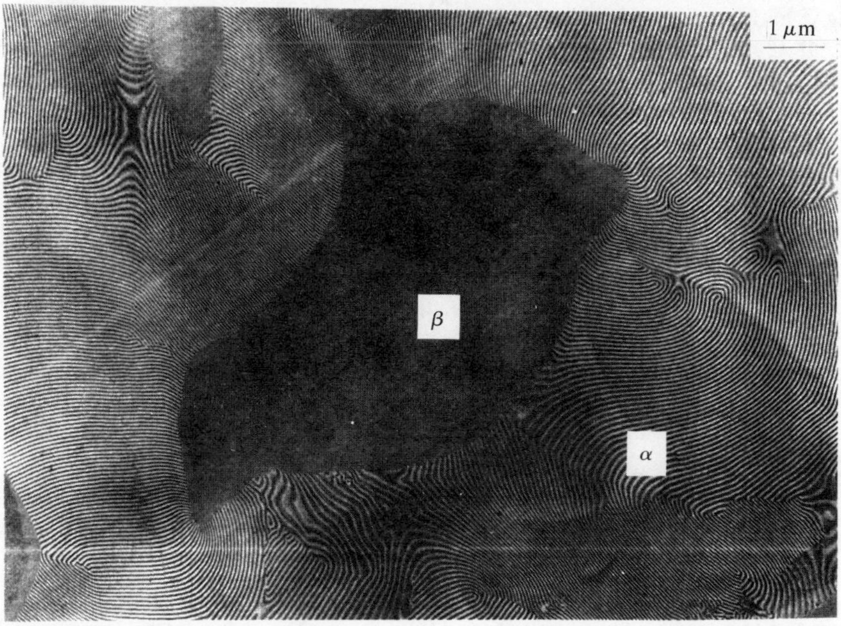

Figure 37 Transmission electron micrograph obtained for 50/50 wt%/wt% mixture of large (α: HY-12) and small (β: HY-8) lamella-forming SI block copolymers. The matrix is composed of the large lamellar (α-rich) microdomains, and the dispersed phase is composed of the small lamellar (β-rich) microdomains. The shape of the β-rich macrodomains are quite irregular (after Hashimoto *et al.*[118]).

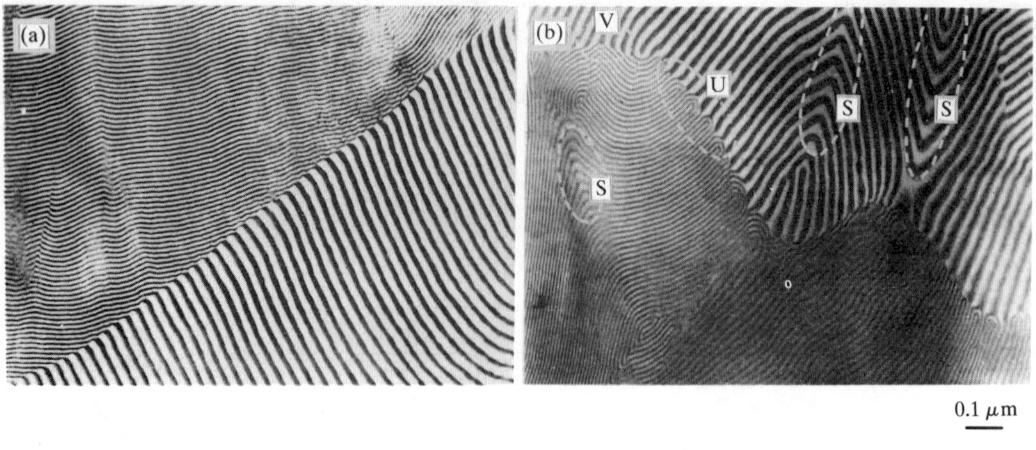

Figure 38 Transmission electron micrographs obtained from the same specimen as that in Figure 37. The micrographs focus on the macroscopic interface between the two coexisting macrodomains composed of thick and thin lamellae. Beautiful commensurations are shown in (a) (one-side inclination) and in (b) (V-shaped and U-shaped commensuration in regions marked V and U, respectively). S designates the grain boundary having characteristics of Scherk's first surface (after Hashimoto *et al.*[118]).

14.6.4 Self-assembly in Mixtures of Sphere- or Cylinder-forming Block Copolymers

Here we deal with binary mixtures of non-lamella-forming diblock copolymers with asymmetric compositions. The following three cases will be considered: (case i) $f_{A_\alpha} \approx f_{A_\beta}$ (and $\chi_{eff} \ll \chi_{eff,s}$) and r is so small that macrophase separation is not expected to occur so that α and β are mixed to form single microdomains; (case ii) $f_{A_\alpha} \approx f_{A_\beta}$ but r is so large that the fluctuation-induced segregation and, therefore, macrophase separation induced by microphase separation takes place; (case iii) the difference between f_{A_α} and f_{A_β} is so large that macrophase separation between disordered α and β occurs first ($\chi_{eff} > \chi_{eff,s}$) and this macrophase separation is followed by the microphase separation. Examples of these three cases for the binary mixtures of SI diblock copolymers were reported by Koizumi *et al.*[119]

Mixtures of sphere (composed of PI block chains)-forming SI diblock copolymer α, and cylinder (PI)-forming SI β, having $f_{PS_\alpha} \approx f_{PS_\beta} \approx 0.8$ and $r \approx 2.2$, or the mixtures of cylinder (PS)-forming α and cylinder (PI)-forming β, having $f_{PS_\alpha} \approx 1 - f_{PS_\beta} \approx 0.3$ and $r \approx 1.2$, were found to be mixed at the molecular level for all blending compositions and to form single domain morphologies where f_{PS_α} and f_{PS_β} designate volume fractions of the PS component in α and β, respectively (case i). Both mixture systems exhibited microdomain morphology transitions with blend composition, *i.e.* the morphology changed from PI spheres to PI cylinders for the former mixtures and from PS cylinders through alternating lamellae to PI cylinders for the latter, with increasing β.

In the case of the mixtures of two sphere (PI)-forming SI, having $f_{PS_\alpha} \approx f_{PS_\beta} \approx 0.8$ and $r \approx 7.4$, α and β were found to be partially miscible (case ii). They phase-separated into large PI spheres rich in α and small PI spheres composed of almost pure β for Φ_α, satisfying $\Phi_\alpha \leq \Phi_{\alpha,C} \approx 0.8$ (Figure 39). Since the composition difference between α and β is very small and $\chi_{eff} \ll \chi_{eff,s}$, the macrophase separation was proposed to result from the 'fluctuation-induced segregation effect'. However, this segregation does not extend to the macroscopic level in such a way that they form macrodomains packed with large spheres and macrodomains with small spheres. This is a big contrast to the case of the lamella-forming SI mixture shown in Figures 37 and 38.

In the case when $f_{PS_\alpha} \approx 1 - f_{PS_\beta} = 0.8$ and $r \approx 1.2$, the mixtures are found to form a modulated (or superlattice) structure (case iii). The mixture containing 30 wt% of β, for example, shows the macrophase-separated structure with β-rich macrodomains dispersed in an α-rich matrix. The α-rich macrodomains are composed of PI spherical microdomains in a PS matrix characteristic to pure α and the β-rich macrodomains are composed of PS spherical microdomains in a PI matrix characteristic to pure β.

Thus, there are numerous ways to control the self-assembly in block copolymer systems and a variety of patterns, either equilibrium or nonequilibrium, can be created.

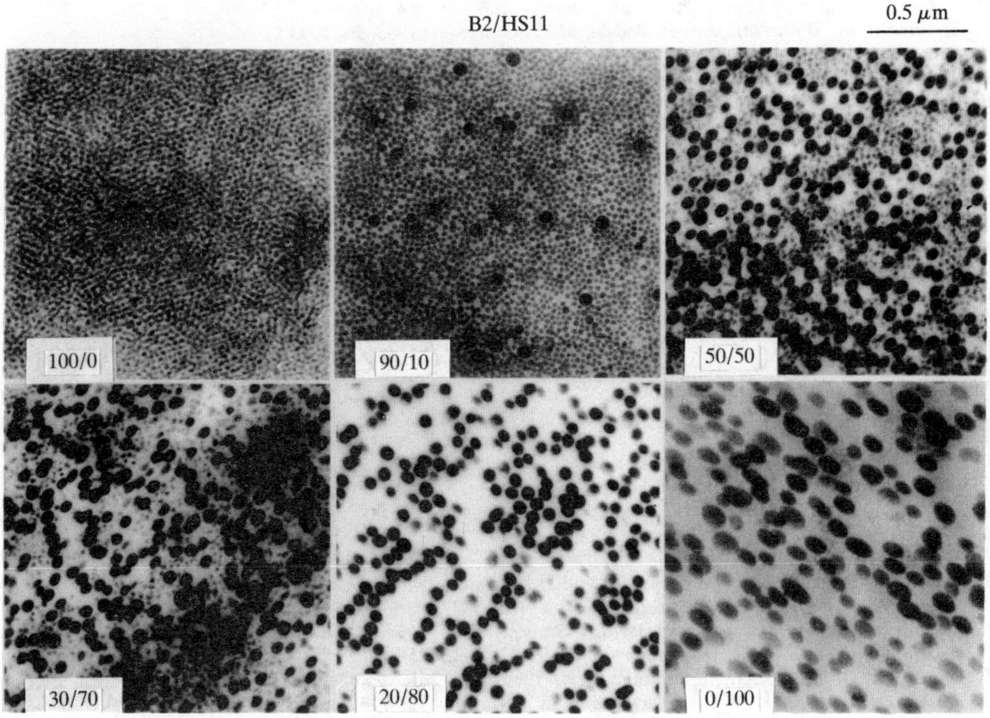

Figure 39 Transmission electron micrographs of ultrathin sections of the binary SI diblock copolymer blends β/α (β: B2, $M_n = 1.76 \times 10^5$, $f_{PS} = 0.83$; α: HS11, $M_n = 1.22 \times 10^6$, $f_{PS} = 0.78$) having $f_\alpha \approx f_\beta \approx 0.8$ and $N_\alpha/N_\beta \geq 7$. Both neat β and α show the PI spherical microdomains in the matrix of PS block chains. The mixture 20/80 shows only one kind of spherical microdomain whose size is between those of neat β and α, while the mixtures from 30/70 to 90/10 show the two kinds of spherical microdomains whose sizes are close to those of neat β and α (after Koizumi *et al.*[119]).

14.7 REFERENCES

1. G. Riess, M. Schlienger and S. Marti, *J. Macromol. Sci. Phys.*, 1980, **B17,** 355.
2. I. Kudose and T. Kotaka, *Macromolecules,* 1984, **17,** 2325.
3. M. Shibayama, H. Hasegawa, T. Hashimoto and H. Kawai, *Macromolecules,* 1982, **15,** 274.
4. Y. Matsushita, K. Yamada, T. Hattori, T. Fujimoto, Y. Sawada and M. Nagasawa, *Macromolecules,* 1983, **16,** 10.
5. Y. Mogi, H. Kotsuji, Y. Kaneko, K. Mori, Y. Matsushita and I. Noda, *Macromolecules,* 1992, **25,** 5408.
6. S. P. Gido, D. W. Schwark and E. L. Thomas, *Macromolecules,* 1993, **26,** 2636.
7. C. Auschra and R. Stadler, *Macromolecules,* 1993, **26,** 2171.
8. T. Fujimoto, H. Zhang, T. Kazama, Y. Isono, H. Hasegawa and T. Hashimoto, *Polymer,* 1992, **33,** 2208.
9. L. Leibler, *Macromolecules,* 1980, **13,** 1602.
10. T. Hashimoto, M. Shibayama, H. Kawai, H. Watanabe and T. Kotaka, *Macromolecules,* 1983, **16,** 361.
11. See, for example, review articles such as: T. Hashimoto, in 'Thermoplastic Elastomers', ed. N. R. Legge, G. Holden, H. E. Schroeder, Hanser, Munich, 1987 (2nd edn. in press, 1996); F. S. Bates and G. H. Fredrickson, *Annu. Rev. Phys. Chem.*, 1990, **41,** 525.
12. S. Sakurai, H. Kawada, T. Hashimoto and L. J. Fetters, *Macromolecules,* 1993, **26,** 5996.
13. S. Sakurai, T. Momii, K. Taie, M. Shibayama, S. Nomura and T. Hashimoto, *Macromolecules,* 1993, **26,** 485.
14. D. A. Hajduk, S. M. Gruner, P. Rangarajan, R. A. Register, L. J. Fetters, C. Honeker, R. J. Albalak and E. L. Thomas, *Macromolecules,* 1994, **27,** 490.
15. I. W. Hamley, K. A. Koppi, J. H. Rosedale, F. S. Bates, K. Almdal and K. Mortensen, *Macromolecules,* 1993, **26,** 5959.
16. P. S. Tucker, J. W. Barlow and D. R. Paul, *Macromolecules,* 1988, **21,** 1678, 2794, 2801.
17. T. Hashimoto, K. Kimishima and H. Hasegawa, *Macromolecules,* 1991, **24,** 5704.
18. T. Hashimoto, T. Izumitani and K. Ono, *Makromol. Chem., Macromol. Symp.*, 1995, **98,** 925.
19. C. D. Han, D. M. Baek, J. Kim, K. Kimishima and T. Hashimoto, *Macromolecules,* 1995, **25,** 3052.
20. K. Kimishima, T. Hashimoto and C. D. Han, *Macromolecules,* 1995, **28,** 3842.
21. T. Hashimoto, K. Yamasaki, S. Koizumi and H. Hasegawa, *Macromolecules,* 1993, **26,** 2895.
22. A.-C. Shi, J. Noolandi and H. Hoffmann, *Macromolecules,* 1994, **27,** 6661.
23. D. J. Meier, *J. Polym. Sci. C,* 1969, **26,** 81.
24. E. Helfand and Z. R. Wasserman, *Macromolecules,* 1976, **9,** 879; 1978, **11,** 960; 1980, **13,** 994; E. Helfand and Z. R. Wasserman, 'Developments in Block Copolymers', ed. I. Goodman, Applied Science, New York, 1982, chap. 4.

25. R. J. Roe, M. Fishkis and J. C. Chang, *Macromolecules,* 1981, **14,** 1091
26. T. Hashimoto, M. Shibayama and H. Kawai, *Macromolecules,* 1983, **16,** 1093.
27. W. C. Zin and R. J. Roe, *Macromolecules,* 1984, **17,** 183.
28. K. Mori, H. Hasegawa and T. Hashimoto, *Polym. J.,* 1985, **17,** 799.
29. F. S. Bates and M. A. Hartney, *Macromolecules,* 1985, **18,** 2478; F. S. Bates, *Macromolecules,* 1987, **20,** 2221.
30. T. Hashimoto, Y. Ijichi and L. J. Fetters, *J. Chem. Phys.* 1988, **89,** 2463; Y. Ijichi, T. Hashimoto and L. J. Fetters, *Macromolecules,* 1989, **22,** 2817.
31. J. N. Owens, I. S. Gancarz, J. T. Koberstein and T. P. Russell, *Macromolecules,* 1989, **22,** 3380.
32. See, for example: H. E. Stanley, 'Phase Transitions and Critical Phenomena', Oxford University Press, Oxford, 1971; S. K. Ma, 'Modern Theory of Critical Phenomena', Benjamin/Cummings, London, 1976.
33. A. Brazovskii, *Sov. Phys. JETP,* 1979, **41,** 85.
34. G. H. Fredrickson and E. Helfand, *J. Chem. Phys.,* 1987, **87,** 697.
35. F. S. Bates, J. H. Rosedale and G. H. Fredrickson, *J. Chem. Phys.,* 1990, **92,** 6255.
36. N. Sakamoto and T. Hashimoto, *Macromolecules,* 1995, **28,** 6825.
37. B. Stühn, R. Mutter and T. Albrecht, *Europhys. Lett.,* 1992, **18,** 427.
38. T. Wolff, C. Burger and W. Ruland, *Macromolecules,* 1988, **26,** 1707.
39. T. Hashimoto, T. Ogawa and C. D. Han, *J. Phys. Soc. Jpn.,* 1994, **63,** 2206.
40. G. Floudas, T. Pakula, E. W. Fischer, N. Hadjichristidis and S. Pispas, *Acta Polym.,* 1994, **45,** 176.
41. T. Hashimoto, H. Tanaka and H. Hasegawa, in 'Molecular Conformation and Dynamics of Macromolecules in Condensed Systems', ed. M. Nagasawa, Elsevier, Amsterdam, 1988, pp. 257–283.
42. T. Inoue, T. Soen, T. Hashimoto and H. Kawai, *J. Polym. Sci. A2,* 1969, **7,** 1283.
43. D. J. Meier, in 'Thermoplastic Elastomers', ed. N. R. Legge, G. Holden and H. E. Schroeder, Hanser, New York, 1987, pp. 269–301.
44. T. Hashimoto, M. Shibayama and H. Kawai, *Macromolecules,* 1980, **13,** 1237.
45. T. Hashimoto, M. Shibayama, M. Fujimura and H. Kawai, in 'Block Copolymers Science and Technology', ed. D. J. Meier, MMI Press/Harwood Academic, New York, 1983, pp. 63–108.
46. J. Noolandi and K. M. Hong, *Ferroelectrics,* 1980, **30,** 117.
47. G. Hadziioannou and A. Skoulios, *Macromolecules,* 1982, **15,** 258.
48. T. Ohta and K. Kawsaki, *Macromolecules,* 1986, **19,** 261; K. Kawasaki, T. Ohta and M. Kohrogui, *Macromolecules,* 1988, **21,** 2972; K. Kawasaki and T. Kawakatsu, *Macromolecules,* 1990, **23,** 4006.
49. A. N. Semenov, *Sov. Phys. JETP,* 1985, **61,** 733.
50. E. L. Thomas, D. B. Alward, D. J. Kinning, D. C. Martin, D. L. Handlin, Jr. and L. J. Fetters, *Macromolecules,* 1986, **19,** 2197.
51. H. Hasegawa, H. Tanaka, K. Yamasaki and T. Hashimoto, *Macromolecules,* 1987, **20,** 1651.
52. P. G. deGennes, 'Scaling Concepts in Polymer Physics', Cornell University Press, Ithaca, NY, 1979; P. G. deGennes, *J. Phys.* (*Paris*), 1970, **31,** 235.
53. A. D. LeGrand and D. G. LeGrand, *Macromolecules,* 1979, **12,** 450.
54. L. Leibler and H. Benoit, *Polymer,* 1958, **22,** 195.
55. M. Benmouna and H. Benoit, *J. Polym. Sci., Polym. Phys. Ed.,* 1983, **21,** 1227.
56. H. Benoit, W. Wu, M. Benmouna, B. Mozer, B. Bauer and A. Lapp, *Macromolecules,* 1985, **18,** 986.
57. M. Olvera de la Cruz and I. Sanchez, *Macromolecules,* 1986, **19,** 2051.
58. K. Mori, H. Tanaka and T. Hashimoto, *Macromolecules,* 1987, **20,** 381.
59. S. Nojima and R.-J. Roe, *Macromolecules,* 1987, **20,** 1866.
60. C. Herkt-Metzky and J. Schelten, *Phys. Rev. Lett.,* 1983, **51,** 896.
61. M. Shibayama, S. H. Yang, R. S. Stein and C. C. Han, *Macromolecules,* 1989, **18,** 2179.
62. P. J. Flory, 'Principles of Polymer Chemistry', Cornell University Press, Ithaca, NY, 1967.
63. H. Tanaka and T. Hashimoto, *Polym. Commun.,* 1988, **29,** 212.
64. G. H. Aharony and A. Bruce, *Phys. Rev. Lett.,* 1979, **42,** 462.
65. T. Hashimoto, M. Itakura and N. Shimidzu, *J. Chem. Phys.,* 1986, **85,** 6773.
66. T. Hashimoto, *Phase Transitions,* 1988, **12,** 47.
67. T. Hashimoto, in 'Materials Science and Technology', ed. R. W. Cahn, P. Haasen and E. J. Kramer, VCH, Weinheim, 1993, vol. 12.
68. R.-J. Roe and C.-M. Kuo, *Macromolecules,* 1990, **23,** 4635; D.-W. Park and R.-J. Roe, *Macromolecules,* 1991, **24,** 5324.
69. T. Hashimoto and T. Izumitani, *Macromolecules,* 1993, **26,** 3631.
70. T. Hashimoto, H. Tanaka and H. Hasegawa, *Macromolecules,* 1990, **23,** 4379.
71. H. Tanaka, H. Hasegawa and T. Hashimoto, *Macromolecules,* 1991, **24,** 240.
72. G. E. Molau, in 'Block Copolymers', ed. S. L. Aggarwal, Plenum, New York, 1970.
73. M. Matsuo and S. Sagaye, in 'Colloidal and Morphological Behavior of Block and Graft Copolymers', ed. G. E. Molau, Plenum, New York, 1971, p. 1.
74. S. L. Aggarwal, R. A. Livigni, L. F. Marker and T. J. Dudek, in 'Block and Graft Copolymers', ed. J. J. Burke and V. Weiss, Syracuse University Press, New York, 1973, p. 157; J. F. Beecher, L. Marker, R. D. Bradford and S. L. Aggarwal, *J. Polym. Sci. C,* 1969, **26,** 117.
75. M. J. Folkes, A. Keller and J. A. Odell, *J. Polym. Sci., Polym. Phys. Ed.,* 1976, **14,** 847.
76. S. L. Aggarwal, *Polymer,* 1976, **17,** 938.
77. E. L. Thomas, D. M. Anderson, C. S. Henkee and D. Hoffman, *Nature,* 1988, **334,** 598.
78. T. Hashimoto, S. Koizumi, H. Hasegawa, T. Izumitani and S. T. Hyde, *Macromolecules,* 1992, **25,** 1433.
79. D. A. Hajduk, P. E. Harper, S. M. Gruner, C. C. Honeker, G. Kim, E. L. Thomas and L. J. Fetters, *Macromolecules,* 1994, **27,** 4063.
80. V. Luzzati, A. Tardieu and T. Gulik-Krzywicki, *Nature,* 1968, **217,** 1028; V. Luzzati, T. Gulik-Krzywicki and A. Tardieu, *Nature,* 1968, **218,** 1031.
81. L. J. Fetters, 'MMI Press Symposium Series', ed. D. J. Meier, MMI Press, Midland, MI, 1983, Vol. 3, p. 17.
82. S. Andersson, S. T. Hyde, K. Larsson and S. Lidin, *Chem. Rev.,* 1988, **88,** 221.

83. E. Koch and W. Fischer, *Acta Crystallogr.*, 1990, **A46,** 33.
84. K. I. Winey, E. L. Thomas and L. J. Fetters, *Macromolecules,* 1992, **25,** 422.
85. R. S. Spontak, S. D. Smith and A. Ashraf, *Macromolecules,* 1993, **26,** 956.
86. R. Xie, B. Yang and B. Jiang, *Macromolecules,* 1993, **26,** 7097.
87. Y. Mogi, K. Mori, Y. Matsushita and I. Noda, *Macromolecules,* 1992, **25,** 5412.
88. S. Förster, A. Khandpur, M. F. Schulz and F. S. Bates, *Bull. Am. Phys. Soc.*, 1993, **38,** 605.
89. M. Matsen and M. Schick, *Macromolecules,* 1994, **27,** 4014.
90. D. A. Hajduk, P. E. Harper, S. M. Gruner, C. C. Honeker, E. L. Thomas and L. J. Fetters, *Macromolecules,* 1995, **28,** 2570.
91. I. W. Hamley, K. A. Koppi, J. H. Rosedale, F. S. Bates, K. Almdal and K. Mortensen, *Macromolecules,* 1993, **26,** 2859.
92. D. Scott, H. H. Winter, W. Gronsky, S. Okamoto and T. Hashimoto, *Macromolecules,* 1993, **26,** 7236.
93. D. Ehlich, M. Takenaka, S. Okamoto and T. Hashimoto, *Macromolecules,* 1993, **26,** 189.
94. D. Ehlich, M. Takenaka and T. Hashimoto, *Macromolecules,* 1993, **26,** 492.
95. M. C. Dalvi and T. P. Lodge, *Macromolecules,* 1993, **26,** 859.
96. Y. Nishikawa, H. Kawada, H. Hasegawa and T. Hashimoto, *Acta Polym.*, 1993, **44,** 192.
97. S. P. Gido, J. Gunther, E. L. Thomas and D. Hoffman, *Macromolecules,* 1993, **26,** 4506; S. P. Gido and E. L. Thomas, *Macromolecules,* 1994, **27,** 849, 6137.
98. P. G. deGennes, *Macromolecules,* 1980, **13,** 1069.
99. T. Hashimoto, H. Tanaka and H. Hasegawa, *Macromolecules,* 1990, **23,** 4378.
100. S. Koizumi, H. Hasegawa and T. Hashimoto, *Makromol. Chem., Macromol. Symp.*, 1992, **62,** 75.
101. H. Tanaka, H. Hasegawa and T. Hashimoto, *Macromolecules,* 1991, **24,** 240.
102. K. I. Winey, E. L. Thomas and L. J. Fetters, *Macromolecules,* 1991, **24,** 6182; 1992, **25,** 2645.
103. H. Hasegawa, H. Tanaka, T. Hashimoto and C. C. Han, *J. Appl. Crystallogr.*, 1991, **24,** 672.
104. D. J. Kinning, K. J. Winey and E. L. Thomas, *Macromolecules,* 1988, **21,** 3502.
105. C. V. Berney, P. Cheng and R. E. Cohen, *Macromolecules,* 1988, **21,** 2235.
106. S. Koizumi, H. Hasegawa and T. Hashimoto, *Macromolecules,* 1994, **27,** 7893.
107. A. M. Mayes, T. P. Russell, S. K. Satija and C. F. Majkrzak, *Macromolecules,* 1992, **25,** 6523.
108. K. R. Shull and K. I. Winey, *Macromolecules,* 1992, **25,** 2637.
109. K. R. Shull, A. M. Mayes and T. P. Russell, *Macromolecules,* 1993, **26,** 3929.
110. S. Koizumi, H. Hasegawa and T. Hashimoto, *Macromolecules,* 1994, **27,** 6532.
111. A.-C. Shi and J. Noolandi, *Macromolecules,* 1994, **27,** 2936.
112. G. Hadziioannou and A. Skoulios, *Macromolecules,* 1982, **15,** 267.
113. T. Hashimoto, *Macromolecules,* 1982, **15,** 1548.
114. K. Mori, H. Hasegawa and T. Hashimoto, *Polymer,* 1990, **31,** 2368.
115. R. J. Spontak, *Macromolecules,* 1994, **27,** 6363.
116. S. T. Hyde, A. Fogden and B. W. Ninham, *Macromolecules,* 1993, **26,** 6782.
117. C. E. Schwier, A. S. Argon and R. E. Cohen, *Polymer,* 1985, **26,** 1985.
118. T. Hashimoto, S. Koizumi and H. Hasegawa, *Macromolecules,* 1994, **27,** 1562.
119. S. Koizumi, H. Hasegawa and T. Hashimoto, *Macromolecules,* 1994, **27,** 4371.
120. S. T. Hyde, personal communication; S. Andersson, S. T. Hyde, B. W. Ninham, Z. Blum, S. Lidin, K. Larsson and T. Landh, 'The Language of Shape', Elsevier, Amsterdam, 1996.

15
Single-chain Condensed State of Flexible-chain Polymers

RENYUAN QIAN
Academia Sinica, Beijing, China

15.1 INTRODUCTION

The preparation of single-chain particles of a flexible-chain polymer dates back to some 50 years ago.[1] It was aimed at a direct observation of molar mass and molar mass distribution of polymers by electron microscopy (EM). The method employed a very dilute solution of concentrations $<10^{-3}\%$ by dropping a solution[1-3] or by atomizer spraying[4] a solution onto a substrate for EM. The amount of single-chain particles so prepared is usually so small that it is barely possible to carry out measurements of their physical properties. On the other hand, no difference has been clearly conceived between the condensed state of a single-chain particle and the condensed state of a conventional multi-chain bulk solid. It is perhaps proper to point out here that a polymer molecule is a unique case in that a single molecule could exist as a condensed state.

A polymer molecule with a molar mass of say 10^6 contains 10^4 monomer units in the chain for a monomer molar mass of 100. This single-chain particle resembles a condensed state of 10^4 monomer molecules. This is therefore equivalent to a condensed state of an assembly of a comparatively small number of small molar mass molecules. Assuming that the density of these single-chain particles does not deviate too far from the density of bulk solid samples, say around $1\ \mathrm{g\,cm^{-3}}$, we can calculate the diameter of the single-chain particle of $M = 10^6$ to be 14.7 nm. Consequently, single-chain particles are polymer nanoparticles. It is thus clear that for the study of single-chain particles it is essential to have a very high molar mass, *i.e.* $M \geq 1 \times 10^6$.

A flexible-chain polymer in its condensed state may be in a glassy state, a rubbery state, a flowing fluid state, or a crystalline state. Within the time scale of an ordinary laboratory experiment, a single-chain fluid state of a sample of single-chain particles will not last as the chains in the particles will gradually interpenetrate each other by segment diffusion to become a conventional multi-chain state consisting of interpenetrated Gaussian coils. A single-chain rubbery state could perhaps last long enough to perform some experiments. However, so far no paper has appeared in the literature in this regard. Thus the present article will be limited to the discussion of single-chain glasses and single-chain single crystals grown from the single-chain rubbery state.

It is to be noted that in the single-chain condensed state there will be no interchain entanglement; only intrachain entanglement is possible. The formation of a condensed state is necessarily a manifestation of an intersegmental cohesion which is an attractive intersegmental interaction. As the particle dimension is very small compared to the contour length of the chain, the molecular chain in a single-chain particle is necessarily folded in the sense that the chain has frequent chances to coil back. The chain morphology in the single-chain glass will not be a Gaussian coil. These are the main differences between the condensed state of a single-chain particle and that of a conventional multi-chain solid.

15.2 METHODS OF PREPARATION OF SINGLE-CHAIN PARTICLES

Experimental methods of preparation of single-chain particles can be categorized into two main routes. One is based on the use of a very dilute solution where the chains are isolated from each other, either by dropping the solution onto some substrate, or by atomizer spraying, or by spreading on a liquid surface, or by freeze drying. Because a very dilute solution has to be used, the amount of single-chain particles which can be produced is usually very small. The other route is by microemulsion polymerization of a monomer under the conditions that ideally only one monomer molecule per emulsion particle is initiated to undergo polymerization. This offers a method to prepare a sizable amount of single-chain particles.

15.2.1 Atomizer Spraying

The first attempt to obtain single-chain particles used a very dilute solution of polystyrene (PS) in benzene to which some MeOH was added to give a precipitating suspension. The suspension was dropped onto a SiO_2 substrate and single-chain particles were found after evaporation of the solvent.[1] Injecting droplets of a very dilute solution, 10^{-4}%, of poly(methyl methacrylate) of $M_w > 1 \times 10^7$ in acetone or of natural rubber in methylcyclohexane or *n*-propyl methyl ketone through a fine syringe needle onto a Formvar surface gave single-chain particles.[2,3] Fast evaporation of the solvent in the droplets, of the order of 10 s under vacuum,[3] was found to be essential to give individual particles instead of an aggregation of connected particles. Later, atomizer spraying of a very dilute solution in a poor solvent, for example PS in cyclohexane,[4] was introduced; it should give better results and it is more convenient to perform. The effects of various experimental factors on the performance of the solution spraying method to yield single-chains of an amorphous polymer in its tightly coiled conformation have been studied rather comprehensively by Richardson.[5,6]

When a good solvent is used, the sprayed droplets tend to be flattened into a thin film on evaporation of the solvent. The atomizer-sprayed droplets from a typical nebulizer show a distribution of diameters from a fraction of a micrometer to a few micrometers, while the majority of the droplets have a diameter around 2 μm.[4] A 10^{-4}% cyclohexane solution of PS with molar mass 2.5×10^6 when sprayed will have *ca.* 6 chains per droplet.[4] For the concentration values we shall hereafter simply express them in %, making no distinction of concentrations being in w/v or w/w as the solution density will not be too far from unity and only the order of magnitude of the values of the concentrations is of significance. During evaporation of the solvent the droplets will go from dilute to concentrated solutions and interpenetration of expanded coils would take place in a good solvent. Consequently a good solvent is neither useful for single-chain EM, nor for obtaining single-chain particles. A poor solvent consisting of a mixture of a good solvent and a nonsolvent is a convenient choice.[5] A nonsolvent less volatile than the good solvent is preferred so that during evaporation the polymer coil will be in the environment of the increasingly poorer solvent and the polymer chain will be in its tightly coiled conformation. It was found that a rapid evaporation of the droplets is an important requirement for a satisfactory production of single-chain particles[5] without interparticle aggregation. In the case of polyacrylamide, the change in the composition of a solvent/nonsolvent mixture, H_2O/n-PrOH, from 50/50 to 20/80 for spraying resulted in a change of morphology of the particles from branched filaments to compact spherical particles which result from agglomeration of the filaments.[7] For EM measurement of polymer molar mass, the production of well dispersed single-chain particles is essential. However, for harvesting a sample of single-chain particles some interparticle aggregation would be tolerable.

For the choice of solution concentration one has to compromise between good particle dispersion and the amount of harvested single-chain particles. For EM a concentration of 10^{-4} to 10^{-5}% is the usual practice. Richardson even expressed the view that when two polymer chains in tightly coiled conformation, *i.e.* in a poor solvent, come into contact with each other they immediately coalesce.[6] However, this has to be further confirmed, as has whether 'coalesce' means interpenetration of coils or just interparticle aggregation. For our present purposes of harvesting a sample, the single-chain particles come into contact and form an aggregate anyway in the harvested sample; what is essential is to have one chain per particle. In previous studies on different regimes in polymer solution concentrations the present author established a concept of a critical concentration c_s,[8,9] below which the polymer coils in solution are isolated and above which the polymer coils begin to shrink in size due to the repulsive interactions between segments of neighboring coils. The existence of c_s was also predicted from theoretical arguments.[10] Experimental estimation of c_s was carried out originally from excimer fluorescence of PS in dichloroethane. It has been further confirmed by precise size exclusion chromatography[11] of PS in toluene and from dynamic light scattering measurement of the diffusion coefficient of PS in toluene.[12] The value of c_s lies in the concentration range of 10^{-1} to 10^{-2}% and decreases weakly with increasing molar mass of PS. This signifies that when the solution concentration of PS is below 10^{-2}%, individual coils in the solution are isolated. When this solution is in a poor solvent, individual chains will be tightly coiled. On evaporation of the droplets of this solution there will be good chances to yield single-chain particles as the interpenetration of tightly coiled chains will not proceed readily. This was what was actually found to be the case from the results of a large number of experiments.[13] Thus for atomizer spraying a solution concentration of 10^{-3}% is a good choice.

The substrate onto which a very dilute polymer solution is sprayed should show sufficient wetting of the polymer,[6] otherwise chain aggregation tends to occur. Ideally the polymer and the substrate should match in interfacial properties so that the transition from polymer/solvent interface to polymer/solid substrate interface does not involve too many differences in interfacial interactions.[2] Evaporated carbon film and Formvar film are satisfactory for most polymers.

The form of the single-chain particles can deviate somewhat from being strictly spherical but may be in the form of prolate spheroids, especially when the molar mass is high,[4] or in the form of oblate spheroids or even thin film patches when the polymer particles are in a rubbery state, especially when there is some good solvent remaining in the particle. For rubbery polymers, Richardson[6] found that the method of spraying from a very dilute solution in a poor solvent failed to give spherical particles. However, this is perhaps not entirely true. One still has to find proper experimental conditions. For example, natural rubber in *n*-propyl methyl ketone at concentrations up to 8×10^{-4}% was dropped from a very fine syringe onto Formvar and gave single-chain spherical particles.[2] *cis*-1,4-Polybutadiene of 10^{-2}% in toluene–ethanol sprayed onto Formvar film gave single-chain spherical particles which lasted for a dozen minutes before they flattened down.[14] Here the nonsolvent ethanol is even slightly more volatile than the solvent toluene. Schulz *et al.*[2] mentioned a chain scission during atomizer spraying of a PS solution in toluene at 0.1% concentration. When poor solvent and a more dilute solution is used, as has been done in many laboratories, no chain degradation due to atomizer spraying has been noticed.

15.2.2 Surface Spread Film Patches

In the preparation of single-chain particles for EM, some authors have used the surface film technique by dropping a very dilute solution onto a liquid surface like H_2O.[15] More detailed studies were made by Kumaki,[16] who attempted to demonstrate the difference in the condensed state of single-chain particles as contrasted to the interpenetrated multi-chain state of conventional polymer glasses. Solutions of PS samples of $M_w = (3.8\text{–}8.4) \times 10^6$ in benzene at concentrations from 2×10^{-2} to 2×10^{-4}% were repeatedly dropped onto a water surface at

20 °C. The surface spread film was slightly compressed with a reduction of area of 10 times, which is still 5 times larger than the A_0 value of 2.6×10^3 cm^2 mg^{-1} extrapolated from the steepest slope of the spreading pressure vs. area plot to zero pressure. The value of A_0 was found to be independent of the molar mass of the sample. At the concentration of 10^{-2}%, noticeable agglomeration of particles was observed, but at concentrations below 10^{-3}%, there was no agglomeration. Transmission electron micrographs (TEM) showed that, for all concentrations studied, the morphology of the film patches obtained was in the form of strings of particles, probably formed due to an attractive interaction between the particles after they were produced on the water surface. These particles were shown to be single-chain particles. It was further shown that the observed particles were flattened spheroids with heights much less than their diameters. This flattening of spherical particles on a solid substrate or on a liquid surface occurs during evaporation of a polymer in a good solvent as it goes through the rubbery state before the solvent is completely driven off. Here, gravity may produce the flattening because of the low mechanical rigidity in the rubbery state, but adsorptive interactions between the rubbery polymer and the substrate surface may also play an important role.[17]

15.2.3 Freeze Drying

It seems to be a simple idea to fix the isolated chains of a very dilute solution in position by freezing the solvent on lowering the temperature and then subliming the solvent by vacuum pumping. In such a way the single-chain particles remaining after the solution is freeze dried should be able to be harvested. However, the processes involved are somewhat complicated due to possible phase separation of the solution, effects of stresses created during crystallization of the solvent, more pronounced effects of intersegmental cohesion at lower temperatures during freezing, and so on. All these factors will affect the chain segmental packing and the physical properties of the single-chain particles obtained by freeze drying. In many cases it has been found that complete removal of the solvent in the freeze drying process is rather difficult and the trace amount of remaining solvent in the freeze-dried single-chain particles may affect the DSC curves around and below T_g.[18]

The experimental preparation starts from a very dilute solution, 10^{-2} to 10^{-4}%, in a solvent which freezes at a temperature not too much lower than the ambient. For PS, for example, a solvent like cyclohexane[19,20] (m.p. 6.5 °C) has been used, and for poly(methyl methacrylate) (PMMA), *p*-xylene[17] (m.p. 14 °C) has been used. Rusher[21] combined dilute solution spraying with freeze drying of the solvent in the droplets to obtain single-chain particles and stressed the importance of complete removal of solvent, otherwise the particles will agglomerate on being warmed to room temperature. The solution should not undergo phase separation before solidification of the solvent, otherwise a concentrated phase would appear. Cooling baths of dry ice/acetone and liquid nitrogen are usually used. Fast cooling to complete solidification within 10 s leads to less chance of interparticle aggregation.[7] Wignall *et al.*[22] argued that during crystallization of the solvent the solution concentration in the vicinity of crystallized solvent will be much increased and this leads to agglomeration of chains. In order to suppress this solution concentration effect, it is necessary either to vitrify the solvent to a glass or to crystallize the solvent so fast that the polymer chains cannot stay clear of the crystallization front. Linear aggregation of particles was actually observed and this has been attributed to the crystal surface growth of the solvent where the crystal edges have a surface energy high enough to render the polymer coils concentrated there.[17]

For TEM observations of freeze dried particles it is particularly important to use carefully purified solvent free from any dust or foreign particles.[17] As very dilute solutions are to be used for freeze drying, a large volume of solvent is to be sublimed, leaving all the dust particles in the freeze dried sample.

There have been studies on polymer chain degradation during freeze–thaw cycling of very dilute solutions.[23–25] In all these investigations, molar masses were judged from the solution viscosity or size exclusion chromatography (SEC). It was recently found[26] that it was not chain degradation occurring during freeze thaw cycling but a change of conformation of the polymer in solution to more compact coils during cooling, which will even be retained in the solution when it is warmed to room temperature. This more tightly coiled conformation of the chains in the particle apparently takes a long time to relax at room temperature, even in the course of running a SEC. This rather long relaxation time must have its origin in the formation of some local cohesion between neighboring back-coiled chain segments during cooling the solution to low temperatures for freezing. Thus the reduced hydrodynamic volume of the more compact coil results in an apparent decrease of molar mass in solution viscosity or SEC measurements. For PS only in tetrahydrofuran solution, local cohesion between neighboring segments apparently does not form during freezing at liquid nitrogen temperature, so that no changes in solution viscosity or SEC curves were observed after repeated freeze–thaw cycles.[26] As to the change of conformation of the polymer in solution during cooling in the freeze drying process, there has been much discussion about the possibility of coil collapse[19,20] as the solvent power becomes lower in cooling to the freezing temperature of the solvent. However, these authors used solutions of rather high concentrations, 0.05% to over 1%, and so very probably the freeze dried samples they obtained are not single-chain particles. Small angle neutron scattering (SANS) experiments[22] on freeze dried PS from benzene solutions at 2.5–3.3% concentration showed no coil collapse and had normal values for the coil radius of gyration. A freeze dried sample of a 1:1 mixture of protonated and deuterated PS from a 0.8% solution in benzene showed interpenetrated coils, *i.e.* the freeze dried particles were not single-chain, by solid-state NMR cross polarization experiments.[27] For freeze dried single-chain particles, a more compact coil conformation seems to be a reality, as mentioned earlier in regard to the SEC experiments. One way to avoid the change of conformation on deep freezing is to use a solvent which has a melting point above room temperature. For PS and PMMA, naphthalene (m.p. 80.2 °C) has been used as the solvent for freeze drying[28] above room temperature.

15.2.4 Microemulsion Polymerization

Microemulsion polymerization is an emulsion polymerization in which the dispersion of a monomer in water is so fine that the droplets have a diameter less than 100 nm. Because of the small droplet size, the microemulsion looks optically transparent. The dispersion of monomer in water is aided by the use of an ionic surfactant, such as sodium dodecyl sulfate or cetyltrimethylammonium bromide, together with a co-surfactant which is usually a short-chain alcohol sparingly soluble in water. Oil in water microemulsions can only be formed over a small range of emulsifier concentrations at high emulsifier to monomer ratios.[29] The microemulsion is formed spontaneously without a requirement for vigorous stirring, high-pressure homogenization, or ultrasonication, and the system is believed to be thermodynamically stable.[29,30] Polymerization of a monomer/water microemulsion was first achieved in the early 1980s. For vinyl monomers like styrene and methyl methacrylate, oil-soluble initiators like azobisisobutyronitrile (AIBN)[31,32] or water-soluble initiators like potassium persulfate[31] have been used; γ-ray initiation[31,33] and photoinitiation with dibenzyl ketone as a photosensitizer[34] have also been used. The size of the microemulsion droplets can easily be measured by photon correlation spectroscopy (PCS) and the weight ratio, S, of surfactant to monomer has an important effect on the size of microemulsion droplets. The diameter of the droplets changes from 100 nm at $S = 0.2$ to 20 nm at $S \geq 3$.[35] The average molar mass and molar mass distribution of the polymerized latex particles are usually evaluated by SEC after dissolving the particles in a suitable solvent. The molar mass can be very high, of the order of 10^6, and the molar mass distribution was found to be very wide, with $M_w/M_n = 3$–5 for microemulsion polymerized PS particles.[30] Features of a microemulsion polymerization are no constant-rate interval and no gel effect in the polymerization kinetics, high molar mass, and low dependency of molar mass on the water-soluble initiator concentration. The polymerization locus lies primarily in the microemulsion droplets by capturing radicals generated in the aqueous phase. Diffusion of monomer from uninitiated microemulsion droplets to monomer-swollen growing polymer particles will proceed until all droplets disappear. The polymerized particle diameter was found to be larger than the microemulsion droplets before polymerization.[30] In some cases, bimodal distribution of latex particle size was found. It was suggested that larger particles are produced by the polymerization of microemulsion droplets and smaller particles by polymerization of monomer solubilized in micelles.[29] Inverse microemulsion polymerization of the acrylamide/toluene system was reported to yield a very high molar mass of 10^7 with a narrow particle size distribution.[36]

After the single-chain particles are prepared by one of the methods discussed in the previous sections, one has to judge whether the particles obtained are truly single-chain particles. Judgments frequently employed by workers in this field are as follows: (i) comparison of the average particle mass calculated from the average size of the particles determined by TEM or

PCS and an assumed density of the particles equal to the density of the bulk solid, with the average molar mass of the chains determined by dilute solution methods; (ii) comparison of the particle size distribution determined by TEM or PCS with the molar-mass distribution of the chains by SEC; (iii) comparison of a mixture of single-chain particles of two different molar masses and molar-mass distributions determined by SEC with the bimodal distribution of particle sizes determined by TEM or PCS. A good agreement indicates single-chain particles.

15.3 SINGLE-CHAIN AND PAUCI-CHAIN GLASSES

Because of the absence of interchain topological entanglement, the small size of the particles, and the large surface area, the single-chain polymer particles should have a macromolecular condensed state somewhat different from that of a conventional interpenetrated multi-chain polymer solid in either its glassy or rubbery states. At temperatures higher than T_g, enhanced segmental diffusion should cause the single-chain particles in a sample to become gradually interpenetrated with each other so that the sample eventually becomes an interpenetrated multi-chain polymer solid. However, owing to the experimental difficulties of obtaining a sizable amount of single-chain particles for the measurement of its physical properties, only very scanty experimental data are available up to the present time. In this regard, the preparation of single-chain particles by microemulsion polymerization is particularly advantageous compared with the other methods of preparation. The process of interchain penetration of single-chain particles to become an interpenetrated multi-chain polymer solid is a very interesting process to be explored. Solid-state NMR measurements of crossed nuclear polarization[27] on a mixture of single-chain protonated PS and single-chain deuterated PS at temperatures above T_g would yield valuable information.

15.3.1 Single-chain Polystyrene

Before the introduction of microemulsion polymerization to obtain single-chain PS (SCPS) samples, only a few experimental data about the density, thermal behavior in differential scanning calorimetry (DSC), and so forth were available in literature. For a PS sample of molar mass 5×10^4, a $5 \times 10^{-2}\%$ solution in cyclohexane was cooled to 7.5 °C and then rapidly frozen at liquid nitrogen temperature, the solvent was sublimed at −20 °C, and SCPS particles were collected and stored at −18 °C. An ultracentrifuge sedimentation run of the particles in a density gradient solution of EtOH–LiI–H_2O gave a value of the density of SCPS particles 0.03 g cm^{-3} lower than a conventional multi-chain PS sample[19] which has a density of 1.05 g cm^{-3}. Accumulated layers of SCPS film patches of $M_w = 3.84 \times 10^6$ was obtained by spreading a $2 \times 10^{-4}\%$ solution in benzene on a water surface and collected by horizontal lifting using a hydrophobic substrate like a Si single crystal plate, PMMA, or a quartz plate coated with a monolayer of Fe^{III} stearate.[37] Ellipsometric measurement on these multilayers of SCPS particles gave a refractive index of 1.45, much lower than the value of 1.60 for conventional PS, indicating a ~30% void in the multilayer.[37] Apparently this void volume fraction corresponds to that of close packed spheres. Consequently the lower refractive index obtained offers little information about the true refractive index of SCPS particles.

The SCPS particles have a dimension much smaller than the Gauss coil dimension in a θ-solvent or in a bulk solid. The macromolecular chain in a SCPS particle has been thought to be in a very compressed conformation. In a SCPS particle there is only one chain compared to a conventional PS solid, which in the space occupied by a Gauss coil would have dozens of interpenetrated chains. Consequently the space density of segments in a SCPS particle should actually be lower than that in a conventional PS solid. This explains the lower density value. The infrared spectrum of SCPS particles of $M_w = 4 \times 10^6$ obtained from microemulsion polymerization shows that the PS chains have a conformation corresponding to that of conventional PS at higher temperature.[33] It is to be noted that interchain penetration has important consequences on the density and chain conformation in a polymer solid. As to the issue of chain conformation, in single-chain (SC) particles obtained by the removal of solvent in a very dilute solution either by evaporation of sprayed droplets or surface spread film patches or by freeze drying, the compact conformation of the resulting SC particles is a manifestation of intersegmental cohesion of the single chain during the removal of solvent, whereas the compact conformation of a SC particle obtained from microemulsion polymerization is the result of confined chain growth in a monomer droplet during polymerization. The latter is really an experimental case of a self-avoiding walk confined in a sphere which has been analyzed by Monte Carlo simulation.[38]

A more drastic difference between SCPS and conventional multi-chain PS was observed in DSC measurements.[33] The SCPS was polymerized in a microemulsion by γ-rays at 60 °C, harvested by dilution of the latex with methanol, filtered, washed, and dried below 50 °C. The weight average diameter of the particles was 26 nm and $M_w = 4 \times 10^6$. DSC runs were carried out at a heating rate of 10 °C min^{-1} under a N_2 gas purge in the following sequences. The sample was first scanned from 30 to 130 °C (Figure 1, curve i), held at that temperature for 3 min, and then subjected to rapid air quenching to room temperature. The second heating run was from 30 to 150 °C (curve ii), the sample was held at that temperature for 3 min and then cooled at 1.25 °C min^{-1} to room temperature. The third run was a repetition of the second run to 150 °C but this was followed by a slower cooling rate of 0.31 °C min^{-1}. Then the fourth run was started (curve iv). The results are summarized in Figure 1. What is striking is the appearance of an exothermic peak in the glass transition region of curve i and the disappearance of this exothermic peak in curve ii. Sub-T_g annealing at 90 °C for 45 min of a SCPS sample did not remove the exothermic peak but it shifted to a temperature 2.8 °C higher (curve v). Annealing at 100 °C for 45 min removed completely the exothermic peak (curve vi). After the first scan to 130 °C, subsequent scans are normal: a simple endothermic step and the appearance of an endothermic peak (curve iv) due to slow cooling after the third scan. DSC runs on a solution cast film of SCPS in $CHCl_3$ (10%) showed no exothermic peak; as the SCPS particles go into solution the SCPS chain loses the original conformation prevailing in the particle and interchain penetration will occur during film formation.

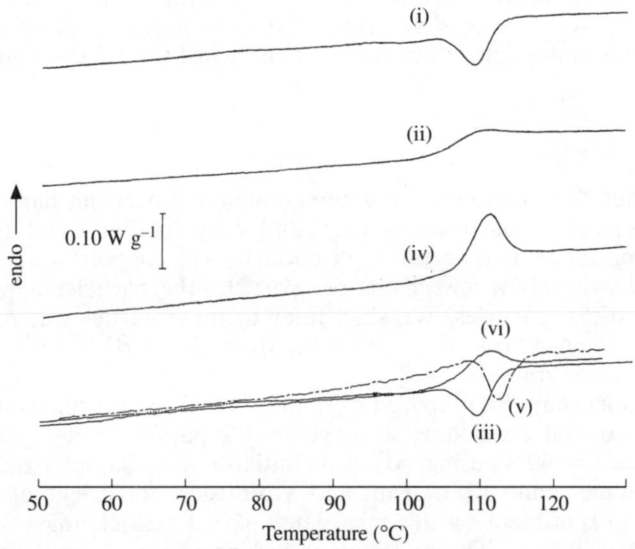

Figure 1 Differential scanning calorimetric measurements on a SCPS sample. Curves i to vi are explained in the text.

The exothermic peak in the DSC curve of as-polymerized SCPS has been interpreted as a manifestation of the formation of intrachain cohesional entanglements of the chain.[33] By cohesional entanglement, we refer to the local nematic interaction between neighboring segments in the polymer condensed state which may be interchain and/or intrachain. The cohesional entanglements are not topological in character, are more densely populated along a chain than topological entanglements, and play an important role in the physical properties of the polymer around and below its T_g. In the rubbery state ($T > T_g$), the cohesional entanglement spacing along the chain is necessarily longer than *ca.* 50–100 C—C bonds in the backbone chain needed for the rubbery elastic deformation. Rapid quenching of the sample from this rubbery state to temperatures below T_g does not allow sufficient time for the formation of new cohesional entanglements, so the quenched sample on being heated displays a simple step endothermic DSC curve. Slow cooling from the rubbery state or sub-T_g annealing of the quenched sample should both result in the formation of new cohesional entanglements along the chain. Consequently, on being heated, an endothermic peak will appear near T_g. This effect of sub-T_g annealing on the DSC curve of a flexible-chain polymer was first observed some 30 years ago.[39] It is to be expected that in SCPS particles the PS chain would have fewer cohesional entanglements as the chain conformation is kinetically controlled during polymerization. The local environment of the segments is more dilated and even macroscopically the particles have lower density. The exothermic process is the minimization of energy through intersegmental cohesion when the segmental mobility becomes sufficient at temperatures close to T_g. It is significant to note that the total enthalpy change involved in the exothermic peak is around 1 J g^{-1}, which is comparable to that involved in the endothermic peak of slowly cooled or sub-T_g annealed samples. It is reasonable to assume the same nature of these processes; one is the formation of cohesional entanglements, while the other is the destruction or melting of the cohesional disentanglements. A similar but much broader exothermic peak in a DSC curve was reported by Gaur and Wunderlich[40] in the case of PS latex particles of 85 nm diameter and was interpreted as some sort of stress release. The fact that a conventional multichain PS sample quenched from the rubbery state, where the cohesional entanglement spacing should be large, does not show an exothermic peak on the heating-up DSC run has its origin in the very slow process of cohesional entanglement formation for multichain samples as observed in the effect of sub-T_g annealing. In the case of a poly(vinyl chloride) (PVC) multichain sample after being annealed at 200 °C for 5 min and then quenched into ice water, the subsequent DSC heating-up run showed an exothermic peak around 65 °C (T_g 70 °C),[41] which disappeared on the second heating-up run after the first scan stopped and the temperature was maintained at 140 °C for 1 h. The exothermic peak may be of the same origin. For SCPS particles, the exothermic peak also disappeared on the second DSC run after the sample was heated in the first run to 130–150 °C. It still remains to be seen if some interchain penetration takes place at temperatures above T_g. Another possible origin of the exothermic peak of SCPS particles is the relaxation of nonequilibrium high conformational energy of the PS chain in the particle.

15.3.2 Pauci-chain Polystyrene

In experiments when the solution is not dilute enough, the particles harvested from spraying, surface spread film patches, or freeze drying, and particles harvested from microemulsion polymerization having molar masses not high enough, will contain more than one chain per particle. When there are only a few chains per particle, the particles apparently still possess some characteristics of SC particles. We shall refer to these particles as pauci-chain particles. At the moment, the demarcation and transition from SC particle properties to multichain properties have not been explored.

One example of pauci-chain PS prepared from microemulsion polymerization which has been carefully characterized and studied is a six-chain PS particle,[42] designated as 6CPS. The microlatex polymerized at 60 °C using AIBN as initiator was characterized by PCS to have a z-average hydrodynamic radius of 16.0 nm and a diffusion coefficient of 1.53×10^{-7} cm^2 s^{-1}. Static light scattering experiment on this microlatex gave a particle mass $M_w = 4.2 \times 10^6$ and a radius of gyration $R_z = 12.5$ nm. The refractive index increment determined for the microlatex was $dn/dc = 0.236$ mL g^{-1} at $\lambda_0 = 488$ nm, compared to that determined for a conventional PS latex of 22 nm radius of $dn/dc = 0.256$. The 8% reduction in the value of dn/dc for the 6CPS latex means that the 6CPS particles have a smaller refractive index than the multichain PS particles in the conventional latex. The latex particles were harvested by adding methanol, filtered, and repeatedly washed to yield the 6CPS particles. The harvested 6CPS in toluene solution showed a molar mass $M_w = 6.6 \times 10^5$ and $R_z = 49.5$ nm by static light scattering. This result indicates that the 6CPS particles contain 6.4 PS chains.

Density measurement on the 6CPS particles was carried out by Hg porosimetry at room temperature up to 3.6 kbar applied pressure. The experimental results are shown in Figure 2.[42] The pore volume penetrated by Hg under applied hydrostatic pressure signifies the interparticle space and space between the particle aggregates in the harvested 6PCS powder. From Figure 2 it is easy to assign the pore volume penetrated by Hg under pressures up to 10^2 bar as the space between aggregates and the pore volume penetrated by Hg under higher pressures as the intraparticle space. Above 1.8 kbar the pore volume did not change. On releasing the pressure, some expansion of the true sample volume could be noted in the figure, which was the result of reversible volume elasticity (compressibility) of the 6CPS particles. After correction for this compressibility at 1.8 kbar, the true volume of 0.3506 g of the 6CPS sample was 369 mm^3. This gives a density value for 6CPS particles as 0.95 $g\,cm^{-3}$ compared to a density of conventional multichain PS solid of 1.05 $g\,cm^{-3}$. This leads to a definite conclusion that SCPS and up to 6CPS particles from microemulsion polymerization have a lower density than the multichain PS solid. It is to be mentioned that the 6CPS did also show an exothermic peak in the first DSC run similar to SCPS discussed previously.

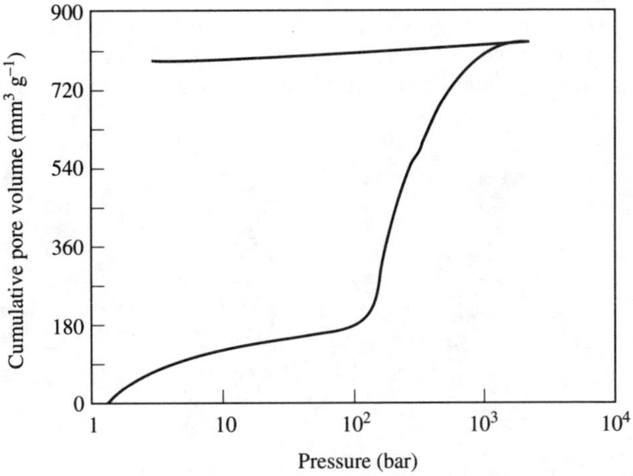

Figure 2 Hg porosimetric measurement on a 6CPS sample at room temperature.

CP/MAS solid-state ^{13}C NMR studies on the 6CPS sample showed that the spin–lattice relaxation times T_1 for C_{arom}, $\delta = 127.0$ ppm, and for C_{aliph}, $\delta = 40.0$ ppm, both conform to a two-exponential term decay to have the values 1.4 s, 37.6 s and 3.3 s, 48.8 s, respectively.[43] The longer lifetimes correspond to phenyl ring stacking for C_{arom} and a physical crosslinked system[44,45] through cohensional entanglements for C_{aliph}. The relative abundances of the longer lifetime components amount to 95 and 92%, respectively. As 95% of C_{arom} nuclei are stacked already in the 6CPS particles, the exothermic peak observed in the first run of DSC could not be assigned to the formation of phenyl ring stacking but to the formation of more cohesional entanglements of the chain. Only a few cohesional entanglements along the chain would be needed to observe 92% abundance of the long lifetime component of T_1 for C_{aliph}.

Positron annihilation lifetime spectroscopy (PALS) resolved in three decay times showed that $\tau_3 = 2.37$ ns for 6CPS,[42] compared to that of conventional PS found in the literature averaging at 2.08 ± 0.03 ns.[46–49] τ_3 is attributed to ortho-positronium pick-off annihilation in free volume. It signifies the mean free volume cavity radius if free volume is modeled as spherical cavities. The free volume cavity radius R could be estimated from the τ_3 value[50] to give $R = 3.2$ Å for 6CPS, compared to $R = 2.9$ Å for conventional PS. This is consistent with the fact that in SCPS the chain takes a conformation of higher temperature.

15.4 SINGLE-CHAIN SINGLE CRYSTALS

The possibility of forming a single crystal from a single-chain particle in its rubbery state is even more fascinating than the observation of a single-chain glass. Crystalline rod-like single-chain particles were first obtained from a 0.1% solution of cellulose tricarbanilate fraction of $M_w = 3.9 \times 10^6$ in $CHCl_3$/EtOH sprayed onto a carbon film.[51] They were probably single crystals, although no experimental evidence was presented. Truly single-chain single crystals were realized experimently only very recently. Up to the present moment, five examples are known; these are poly(ethylene oxide),[52] isotactic polystyrene,[53] *cis*-1,4-polybutadiene,[14] *trans*-1,4-polyisoprene,[54] and poly(vinylidene fluoride)[55] (PVF_2). Freeze-dried particles of PVF_2 obtained by pouring a 4×10^{-2}% solution in dioxane (m.p. 11.8 °C) into liquid nitrogen and subliming the solvent below its m.p. were found to be highly crystalline,[56] but it was not clear whether they were single-chain single crystals.

15.4.1 Poly(ethylene oxide)

Poly(ethylene oxide) (PEO) was the first example of a successful attempt to show the existence of single-chain single crystals. A sample of PEO of $M_n = 2.2 \times 10^6$ of broad molar mass distribution was dissolved in benzene at a concentration of 2×10^{-4}% and introduced drop by drop onto a hot water surface at 80 °C. As PEO is soluble in cold water but not in hot water, the solution spread on the hot water surface formed single-chain surface film patches. After being slightly compressed the patches were picked up by a Cu grid covered with collodion and reinforced with evaporated carbon for TEM. These single-chain PEO particles on the EM grid were heated to 80 °C for melting and then crystallized at 52 °C for 10 h. TEM images of these crystallized PEO single-chain particles having round to polygonal shapes were observed with a number average diameter of 7.6 nm, and selected area electron diffraction (ED) showed clearly a single-crystal diffraction pattern,[52] as shown in Figure 3.

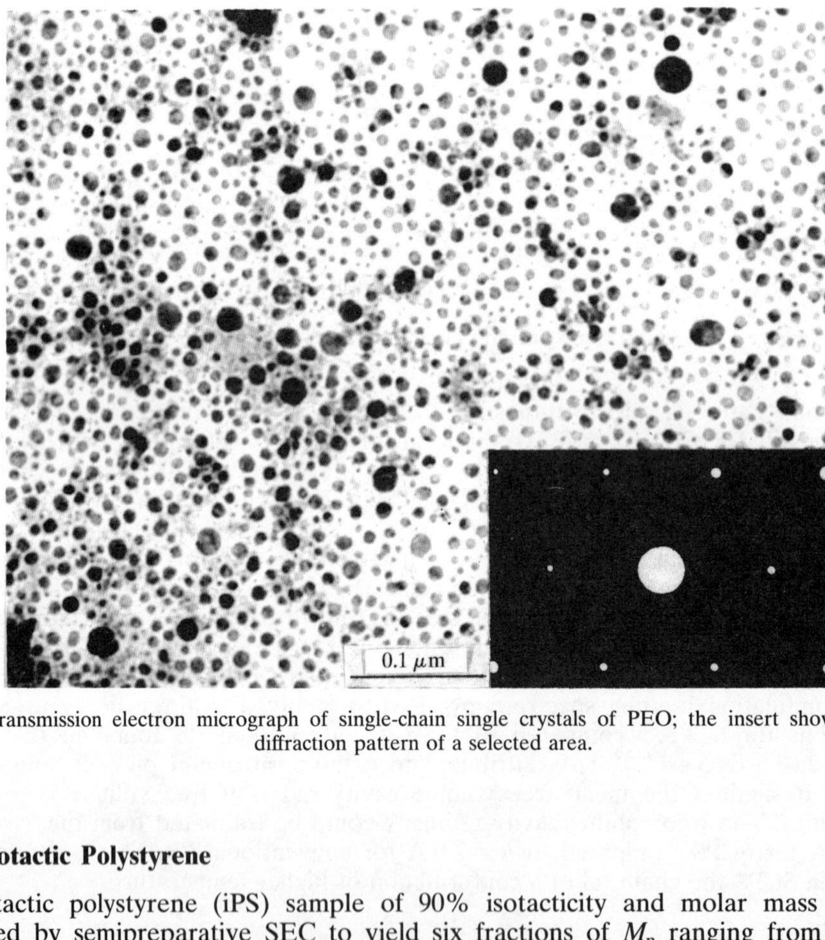

Figure 3 Transmission electron micrograph of single-chain single crystals of PEO; the insert shows the electron diffraction pattern of a selected area.

15.4.2 Isotactic Polystyrene

An isotactic polystyrene (iPS) sample of 90% isotacticity and molar mass 4×10^5 was fractionated by semipreparative SEC to yield six fractions of M_n ranging from 1.5×10^4 to 4.4×10^6. From a fraction of $M_w = 1.7 \times 10^6$, $M_n = 1.3 \times 10^6$ a 1×10^{-3}% solution in benzene was prepared and spread dropwise on a water surface at room temperature. After evaporation of the solvent, single-chain iPS surface film patches were picked up on a Cu grid for crystallization and TEM. Crystallization was carried out isothermally at 175 °C for 3 h. TEM images of single-chain iPS (SCiPS) before and after crystallization are shown in Figure 4.[53] After crystallization, the formation of polygonal shaped particles can be clearly seen. The thickness of the particles was found to be 10–20 nm while the average diameter was around 15 nm, compared to the average diameter of 28 nm before crystallization. For a longer crystallization time of 8 h, even more regularly shaped particles resulted. Some of them showed a tent-like morphology, revealing distinct sector boundaries as shown in Figure 5 and resembling single crystals of polyethylene grown from dilute solutions. Selected area ED patterns shown in Figure 6[53] confirm that these particles are single crystals. The macromolecular chain segments in the single crystals were standing on the substrate, if not precisely perpendicular to it. Scanning tunneling microscopy (STM) has also been applied to the study of single-chain iPS particles.[57]

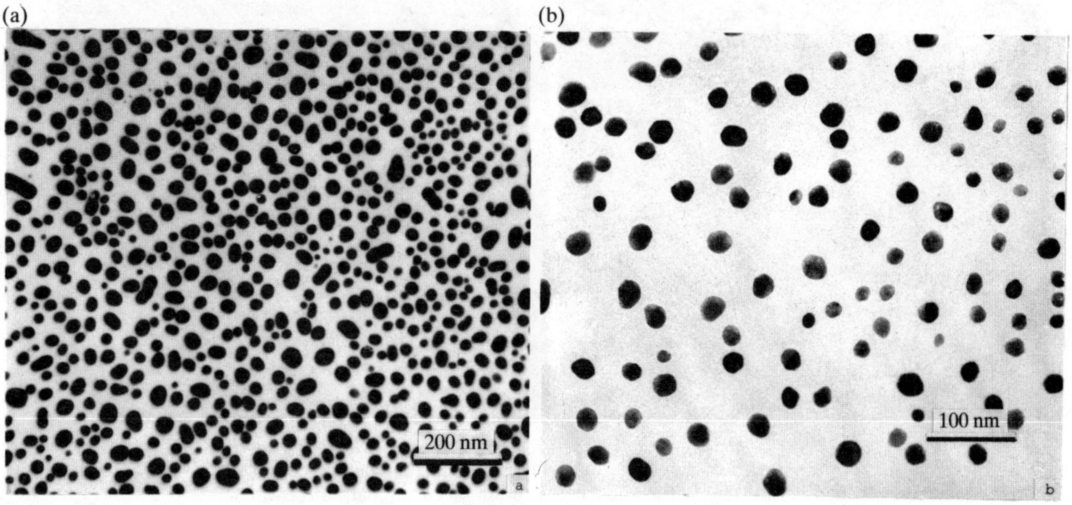

Figure 4 Transmission electron micrographs of (a) amorphous and (b) crystallized single-chain particles of iPS; crystallization at 175 °C, 3 h.

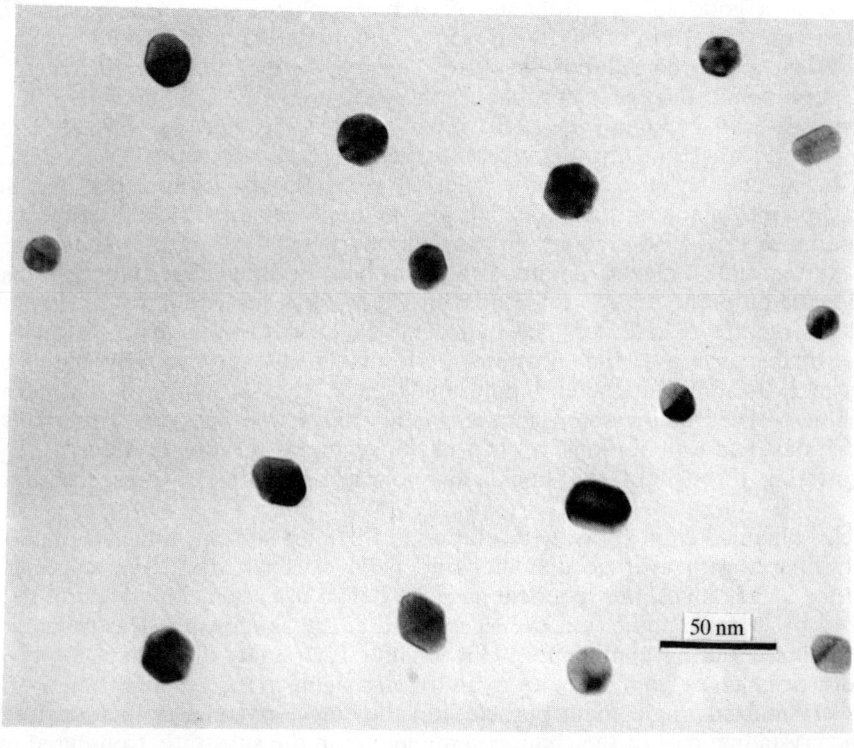

Figure 5 Transmission electron micrograph of single-chain single crystals of iPS crystallized at 175 °C, 8 h.

15.4.3 *cis*-1,4-Polybutadiene

A sample of *cis*-1,4-polybutadiene (cPBD), polymerized with a rare-earth catalyst system, which has high *cis* content of >98%, a very high molar mass, and no long-chain branching, was fractionated in toluene/ethanol. Solutions of a fraction of $M_n = 2.3 \times 10^6$ at a concentration of

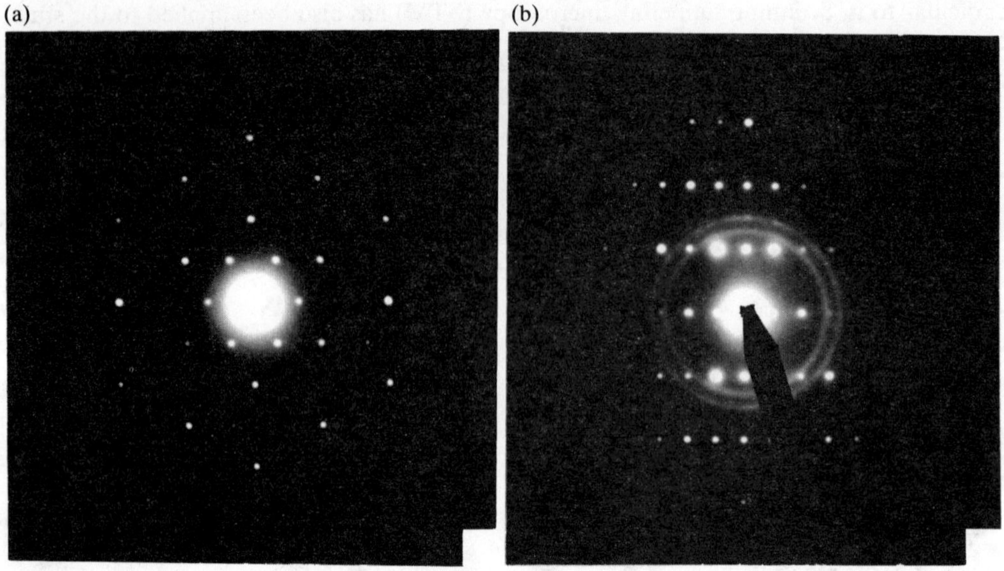

Figure 6 Electron diffraction patterns of a selected area of single-chain single crystals of iPS with the crystal *c*-axis parallel (a) and perpendicular (b) to the electron beam direction.

(1–2) $\times 10^{-2}$% in a good solvent (toluene) or a poor solvent (toluene/ethanol) (75/25 v/v) were atomizer sprayed onto evaporated carbon-reinforced Formvar film on a Cu grid for TEM.[14] When a toluene solution was used, the sprayed particles were irregular shaped patches of very small thickness, probably only fractions of a nanometer thick, which could not be measured by shadowing. cPBD has a $T_g = 160$ K and $T_m = 273$ K. The rubbery particles on the grid were put into the sample stage of the EM cooled to 143 K. The particles crystallized during cooling on the sample stage. It is interesting to note that the thin cPBD patches sprayed from a toluene solution became nonuniform in thickness, some part of the patch became much thicker, and many holes inside the patch appeared, as shown in Figure 7. The single-crystal ED patterns obtained with a field limiting aperture are also shown in Figure 7 with different crystal orientations. The reflections could be indexed from the known crystal structure of cPBD.[58] The molecular chain in the thin patch is most probably lying parallel to the surface of the substrate, while that in the single crystal stands up on the substrate surface to some height and then folds back, thus resulting in the formation of numerous holes. The quite sharp single-crystal diffraction pattern observed suggests that the positions and orientations of the chain segments in the crystalline region gave coherent scattering throughout the single-chain particle despite the presence of numerous holes.

The particles obtained from spraying a solution in the poor solvent, toluene/ethanol, were in the form of spheres with average diameter and height of *ca.* 20 nm. On standing at room temperature for longer times, the spherical particles began to flatten. The shape of the particles could be fixed by OsO_4 staining and the size measured by shadowing. The rubbery spherical particles crystallized during cooling to 143 K on the EM stage. The shape of the particles underwent drastic changes from a sphere to an irreglar shape after crystallization. Formation of holes in the crystallized single-chain particle and the single-crystal ED pattern are shown in Figure 8. The crystalline part of the chain is standing up on the substrate. Compared to the case of PEO and iPS, the crystallization rate of cPBD is much faster so that the morphology observed reflects processes occurring during the initial stage of crystallization. The outstanding ability for self-organization of a single macromolecular chain during crystallization is really striking. The crystallization process converts a molecular condensed state of somewhat close to a two-dimensional random walk in the thin patches or a confined random walk in a sphere on the substrate to a state of a single crystal, where the ordered chain segments are standing up on the substrate. After crystallization, numerous holes may be formed and very probably some amorphous regions may coexist.

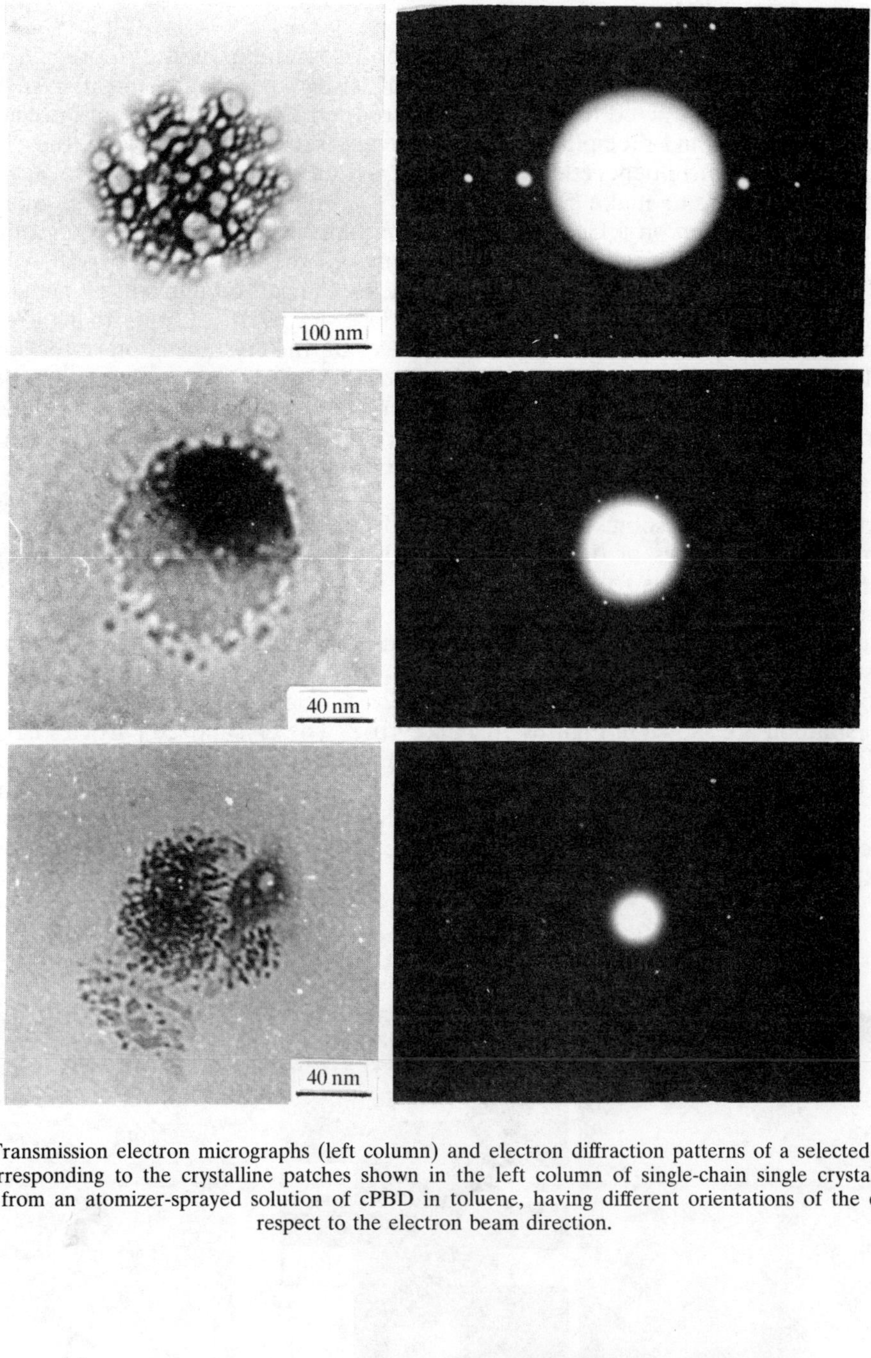

Figure 7 Transmission electron micrographs (left column) and electron diffraction patterns of a selected area (right column) corresponding to the crystalline patches shown in the left column of single-chain single crystals of cPBD crystallized from an atomizer-sprayed solution of cPBD in toluene, having different orientations of the crystal with respect to the electron beam direction.

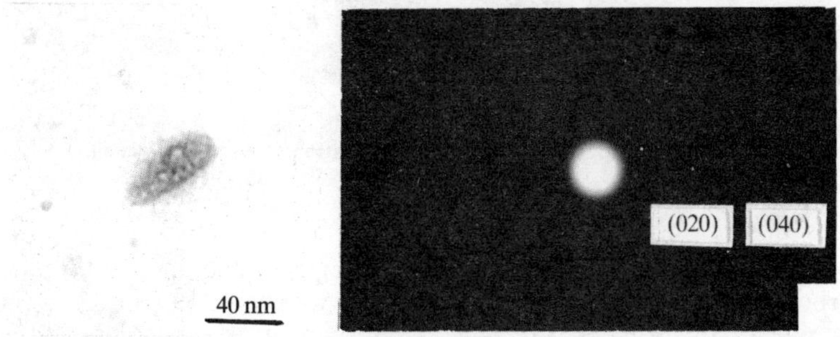

Figure 8 Transmission electron micrograph (a) and electron diffraction pattern (b) of a single-chain single crystal of cPBD crystallized from an atomizer-sprayed solution of cPBD in toluene/ethanol.

15.4.4 *trans*-1,4-Polyisoprene

Natural occurring gutta percha (GP) is *trans*-1,4-polyisoprene of high stereoregularity. Fallen leaves of *Eucommia ulmoides* were powdered and extracted with acetone to remove resins. Then gutta percha was extracted from these powdered leaves with toluene and precipitated by acetone. The extraction and precipitation were repeated several times to yield the GP sample. It was fractionated from toluene/ethanol and the fraction of $[\eta] = 8.17\,\mathrm{dL\,g^{-1}}$, $M_n = 3.7 \times 10^6$ was dissolved in $CHCl_3$ to make 5×10^{-3}–2×10^{-4}% solutions. The solution was atomizer sprayed onto a carbon film on a Cu grid. The carbon film on the grid was put on a filter paper which had been fully wetted with a proper amount of ethanol before the GP solution was sprayed. The ethanol on the filter paper could permeate the carbon film so that actually the GP solution was sprayed onto a non-solvent-coated surface. Experience shows that the amount of ethanol on the filter paper has important consequences during crystallization of the single-chain GP particles on the film. The rate of crystallization of GP is very fast, so that, when the solvent of the sprayed droplets evaporates, the GP chain crystallizes from its solvent-swollen rubbery state. Morphology observation and ED were carried out at 173 K.[55] When the filter paper was not wetted by ethanol, the EM image of the sprayed particles was irregularly shaped with an average linear size of *ca.* 120 nm and very low height. When the filter paper was wetted by a small amount of ethanol, some of the sprayed droplets showed the appearance of single crystals. In the surroundings of the thick part of the EM image of the single crystal, a faint image of the droplet remains to be seen, which is presumably the amorphous regions of the chain where the chain segments had not entered the single crystal. On increasing the amount of ethanol wetting the filter paper, some of the sprayed GP droplets could form polygonal-shaped single crystals.

The morphology and ED pattern of a crystallized single-chain particle sprayed from a 2×10^{-4}% solution are shown in Figures 9(a) and 9(b).[54] The crystallized particles have a shape close to circular cylinders. A typical particle has a diameter of 17 nm and a height of 11 nm, as shown in Figure 9(c). Some holes are also present in the crystalline region. The single-chain single crystals of GP are of the low-melting α-modification and the diffraction spots could be indexed according to the known crystalline structure of GP.[59] The GP macromolecular chain in the crystalline region is standing on the substrate.

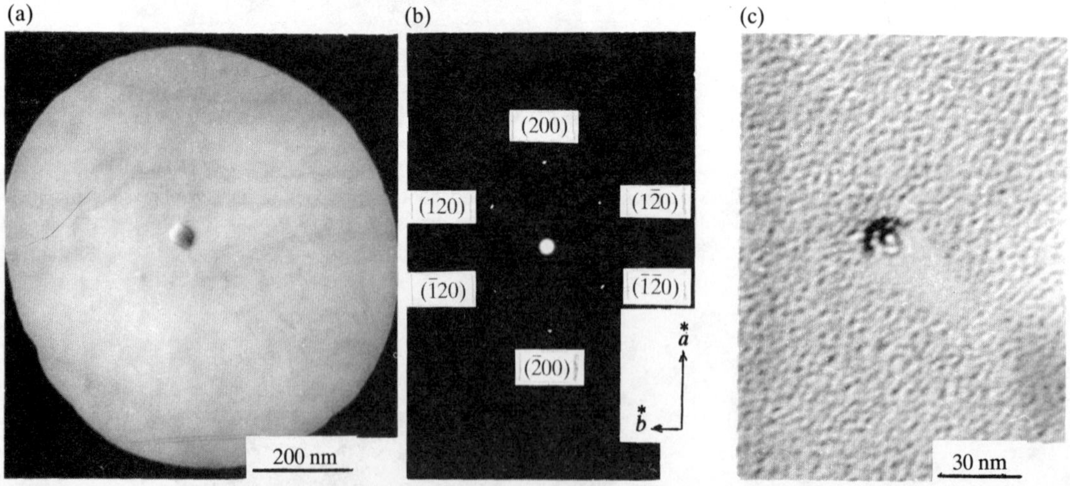

Figure 9 Transmission electron micrograph (a) and electron diffraction pattern of a selected area (b) of a single-chain single crystal of gutta percha. (c) Morphology under higher magnification of a single-chain single crystal of gutta percha with Pt shadowing at 15°.

With the proper amount of ethanol wetting the filter paper and the whole deck of sprayed droplets, the carbon film and filter paper was placed in a covered petri dish so as to slow down the evaporation rate of the solvent; GP single-chain single crystals of polygonal shape were observed. One example is shown in Figure 10(a) in the form of a hexagon. Sectoring of the hexagon is apparent from the figure. A selected area ED pattern is shown in Figure 10(b).

The diffraction spots form a regular hexagonal point lattice, with a Bragg spacing of 0.602 nm. If we draw three lines connecting every two opposite spots of the six inner hexagonal diffraction spots, the lines are perpendicular to the three pairs of the parallel side surfaces of the single crystal. The other next nearest diffraction spots from the electron beam centre show Bragg spacings of 0.354 and 0.301 nm. The diffraction pattern shows that the crystal is neither of the α-modification nor of the high-melting β-modification. The structure has yet to be interpreted.

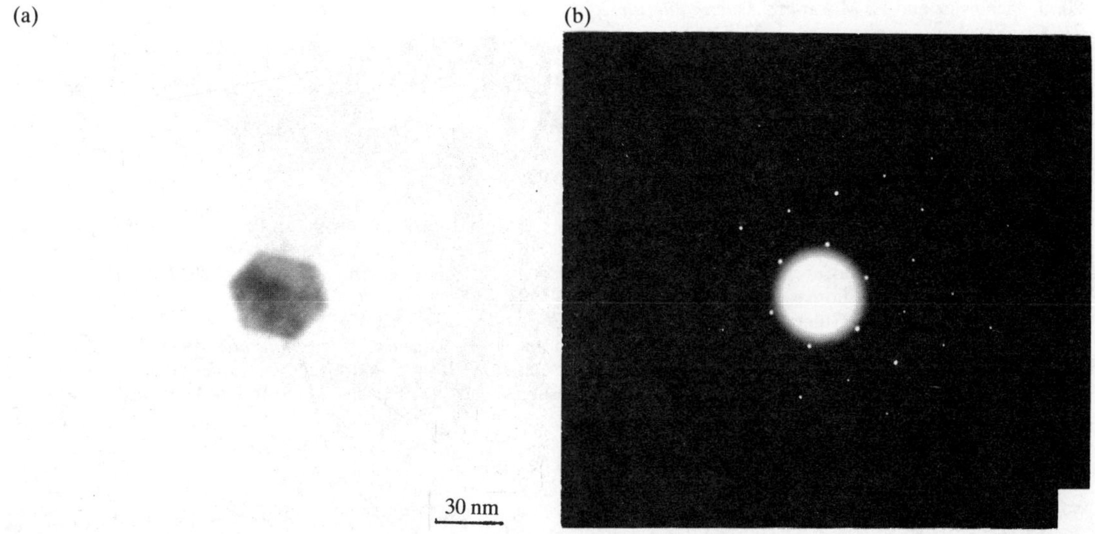

Figure 10 Transmission electron micrograph (a) and selected area electron diffraction pattern (b) of a single-chain single crystal of gutta percha crystallized at a slower evaporation of the solvent in the sprayed solution droplets.

It is clear that when the crystallization is fast the single-chain single crystal formed has no regular external surfaces, while a single-chain single crystal with polyhedral external surfaces can be formed when the crystallization is slow, so that there is enough time for the chain segments to rearrange into proper crystalline surfaces. This later stage of crystallization is probably not very much different from the later stage of crystallization of a multi-chain polymer solid. What is really interesting is the fact that the crystallization of a single-chain particle from its rubbery state reveals some insight on the initial stage of chain crystallization. All cases so far known show that the macromolecular chain in the single-chain single crystal is standing up from the substrate even when the rubbery state before crystallization may have the chain lying flat on the substrate surface. This implies that crystallization of a single chain is homogeneously nucleated and that the crystallization process has an outstanding self-organization ability.

Acknowledgement

Support by the National Basic Research Project – Macromolecular Condensed State, State Science and Technology Commission, People's Republic of China, for writing this chapter is gratefully acknowledged.

15.5 REFERENCES

1. R. F. Boyer and R. D. Heidenreich, *J. Appl. Phys.*, 1945, **16**, 621.
2. I. Barnikol, W. K. R. Barnikol, A. Beck, M. Campagnari-Terbojević, N. Jovanović and G. V. Schulz, *Makromol. Chem.*, 1970, **137**, 111.
3. G. Koszterszitz and G. V. Schulz, *Makromol. Chem.*, 1977, **178**, 2437.
4. B. M. Siegel, D. H. Johnson and H. Mark, *J. Polym. Sci.*, 1950, **5**, 111.
5. M. J. Richardson, *J. Polym. Sci.*, 1963, **C3**, 21.
6. M. J. Richardson, *Proc. R. Soc. London*, 1964, **A279**, 50.
7. D. V. Quayle, *Polymer*, 1967, **8**, 217.
8. R. Qian, in 'Macromolecules', ed. H. Benoit and P. Rempp, Pergamon, Oxford, 1982, p. 139.

9. R. Qian, T. Cao, S. Chen and F. Bai, *Zhongguo Kexue B,* 1983, 1080.
10. D. J. Lohse, *Macromolecules,* 1981, **14,** 1658.
11. R. S. Cheng and X. H. Yan, *J. Appl. Polym. Sci., Appl. Polym. Symp.,* 1991, **48,** 123.
12. C. Wu, *J. Polym. Sci. Part B: Polym. Phys.,* 1994, **32,** 1503.
13. R. Cheng, personal communication.
14. L. Liu, H. Li, Z. Jiang, E. Zhou and R. Qian, *Makromol. Chem., Rapid Commun.,* 1993, **14,** 747.
15. Th. G. F. Schoon and R. Kretschmer, *Koll. Z. Polym.,* 1964, **197,** 51.
16. J. Kumaki, *Macromolecules,* 1986, **19,** 2258.
17. G. Koszterszitz, W. K. R. Barnikol and G. V. Schulz, *Makromol. Chem.,* 1977, **178,** 1133.
18. L. Wu, personal communication.
19. G. Pouyet, A. Kohler and J. Dayantis, *Macromolecules,* 1981, **14,** 1126.
20. J. Jachowicz and H. Morawetz, *Macromolecules,* 1982, **15,** 828.
21. Ch. Ruscher, *J. Polym. Sci.,* 1967, **C16,** 2923.
22. H. R. Brawn and G. D. Wignall, *Macromolecules,* 1990, **23,** 683.
23. A. A. Berlin, E. A. Penskaya and G. I. Volkova, *J. Polym. Sci.,* 1962, **56,** 477.
24. F. Patat and W. Högner, *Makromol. Chem.,* 1964, **75,** 85.
25. V. Zysman, T. Q. Nguyen and H. H. Kausch, *J. Polym. Sci., Polym. Phys.,* 1994, **32,** 1257.
26. R. Cheng, personal communication, results to be published.
27. K. J. McGrath and C. M. Roland, *Macromolecules,* 1992, **25,** 1366.
28. A. R. Schultz and A. L. Young, *Macromolecules,* 1980, **13,** 663.
29. A. S. Dunn, in 'Comprehensive Polymer Science', ed. G. Allen and J. C. Bevington, Pergamon, Oxford, 1989, vol. 4, p. 222.
30. J. S. Guo, M. S. El-Aasser and J. W. Vanderhoff, *J. Polym. Sci,. Part A: Polym. Chem.,* 1989, **27,** 691.
31. S. S. Atik and J. K. Thomas, *J. Am. Chem. Soc.,* 1981, **103,** 4279.
32. P. L. Johnson and E. Gulari, *J. Polym. Sci., Polym. Chem. Ed.,* 1984, **22,** 3967.
33. R. Qian, L. Wu, D. Shen, D. H. Napper, R. A. Mann and D. F. Sangster, *Macromolecules,* 1993, **26,** 2950.
34. P. L. Kuo, N. J. Turro, C. M. Tseng, M. S. El-Aasser and J. W. Vanderhoff, *Macromolecules,* 1987, **20,** 1216.
35. M. Antonietti, W. Bremser, D. Müschenborn, C. Rosenauer, B. Schupp and M. Schmidt, *Macromolecules,* 1991, **24,** 6636.
36. F. Candau, Y. S. Leong and R. M. Fitch, *J. Polym. Sci., Polym. Chem. Ed.,* 1985, **23,** 193.
37. J. Kumaki, *J. Polym. Sci., Part B: Polym. Phys.,* 1990, **28,** 105.
38. D. Wu, D. Zhao and R. Qian, *Polymer,* 1986, **27,** 1087.
39. M. V. Volkenstein and Yu. A. Sharonov, *Vysokomol. Soedin.,* 1961, **3,** 1739.
40. U. Gaur and B. Wunderlich, *Macromolecules,* 1980, **13,** 1618.
41. K. H. Illers, *Makromol. Chem.,* 1969, **127,** 1.
42. C. Wu, K. K. Chan, K. F. Woo, R. Qian, X. Li, L. Chen, D. H. Napper, G. Tan and A. J. Hill, *Macromolecules,* 1995, **28,** 1592.
43. H. Feng, personal communication.
44. A. Charlesby, *Radiat. Phys. Chem.,* 1979, **14,** 919.
45. A. Charlesby and E. M. Jaroszkiewicz, *Eur. Polym. J.,* 1985, **21,** 55.
46. B. C. Groseclose and G. D. Loper, *Phys. Rev. A,* 1965, **137,** 939.
47. A. D. Kasbekar, P. L. Jones and A. Crowson, *J. Polym. Sci. Part A: Polym. Phys.,* 1989, **27,** 1373.
48. J. McGervey, N. Panigrahi, R. Simha and A. Jamieson, in 'Positron Annihilation', ed. P. C. Jain, R. M. Singru and K. P. Gopinathan, World Scientific, Singapore, 1985, p. 690.
49. P. J. O'Connor, C. W. Kocher, B. G. Landes, S. L. Anderson, P. T. DeLassus and E. A. Grulke, *Polym. Prepr. Am. Chem. Soc., Div. Polym. Chem.,* 1992, **33,** 308.
50. Q. Deng and Y. C. Jean, *Macromolecules,* 1993, **26,** 30.
51. H. Bittiger and E. Husemann, *Makromol. Chem.,* 1966, **96,** 92.
52. H. Bu, Y. Pang, D. Song, T. Yu, T. M. Voll, G. Czornyj and B. Wunderlich, *J. Polym. Sci. Part B: Polym. Phys.* 1991, **29,** 139.
53. H. Bu, E. Chen, S. Xu, K. Guo and B. Wunderlich, *J. Polym. Sci. Part B: Polym. Phys.,* 1994, **32,** 1351.
54. L. Liu, F. Su, H. Zhu, H. Li, E. Zhou, R. Yan and R. Qian, to be published.
55. E. Zhou, personal communication.
56. G. Xue, Y. Wang and Y. Lu, *Makromol. Chem., Rapid Commun.,* 1994, **15,** 719.
57. H. Bu, E. Chen, J. Yao, S. Xu and Y. Kuang, *Polym. Eng. Sci.,* 1992, **32,** 1209.
58. G. Natta and P. Corradini, *Angew. Chem.,* 1956, **68,** 615.
59. A. Keller and E. Martuscelli, *Makromol. Chem.,* 1972, **151,** 189.

16

Electrooptical Properties of Polymer–Liquid Crystal Mixtures

THEIN KYU

University of Akron, OH, USA

16.1 INTRODUCTION

In the search for novel electronic display technology, various liquid crystal optical materials and devices have been developed for use in flat panel displays.[1–3] The first generation liquid crystal device, which is much thinner and lighter and consumes less power than a conventional cathode-ray tube, was a twisted nematic (TN) cell. The TN device and its modified versions such as a supertwisted nematic liquid crystal (STN) have been utilized in hand-carried computer monitors. TN and STN devices have been customarily fabricated through tangential alignment of low molar mass liquid crystals (LC) on transparent conductive glass substrates by rubbing in one direction. The anchoring angle of the nematic directors may vary depending on the surface modification of the substrates. For instance, polyimide-coated substrates generally give a planar alignment whereas a silicate-treated surface yields a homeotropic alignment of the nematic directors. The two glass plates are then placed on each other such that the orientation of the LCs remains parallel, and subsequently the top plate is twisted to the orthogonal (90° for TN cells or 270° for STN cells) direction. The tangentially aligned liquid crystals in the TN cell change their orientation in a helical form from the bottom to the top plates, as depicted in Figure 1. In the absence of an external electric field, the device appears opaque as the polarized light cannot pass through the TN cell. When a modest electric voltage is applied across the TN cell, the LC directors in the bulk orient towards the direction of the applied electric field, *viz.*

normal to the glass substrates, thereby allowing incident light to transmit through the cell. Hence, the TN cells can be switched electrically between the transparent and the opaque states.

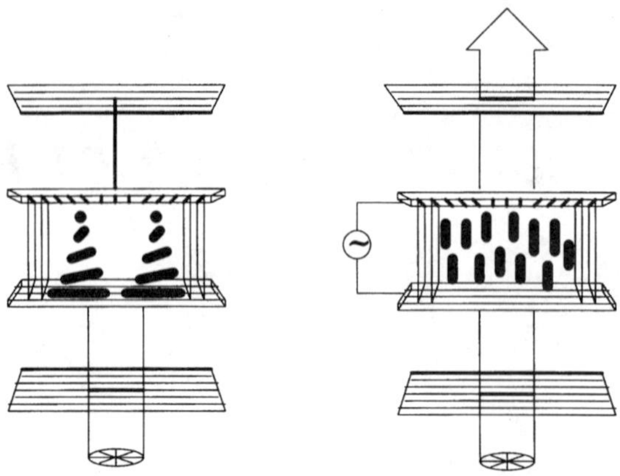

Figure 1 Schematic drawing of a twisted nematic liquid crystal cell in the absence of an external electric field and under the electric field.

The TN cell display devices have several advantages over the conventional television monitor, such as the low driving voltage required for operation, the reasonably short response time, and the light weight (it does not require a bulky cathode-ray tube for back-lighting of the screen). However, the slow recovery time of the nematic directors and the limited wide angle viewability are major obstacles to successful display applications. There have been several attempts to improve the electrical response (rise and decay) time by changing the director tilt angle and the twist angle of the cells through surface modification and stabilization. Another drawback with the TN cells is the difficulty of aligning the LCs uniformly over a large surface area of the substrates; therefore, the fabrication of TN cells for large area displays is presently not feasible.

In this connection, numerous efforts have been directed to the development of newer electrooptical devices.[4–19] The first light-shutter device patented by Hilsum[4] had micron-size soda spheres placed in a nematic liquid crystal layer sandwiched between two conductive glass substrates. The mismatch of the refractive indices between the soda glass spheres and the LCs gives rise to scattering of light such that the display screen appears opaque. When an external electric field is applied, the LC directors orient homeotropically towards the applied field. As the ordinary refractive index of the LC directors is close to the average refractive index of the glass spheres for normal incident light, the scattering of light can be reduced, thereby rendering the device transparent. Unfortunately, the contrast was not good enough to warrant further development for display applications. Craighead *et al.*[5] were the first to use polymers as a matrix material. A microporous polymer filter was filled with a nematic liquid crystal of positive dielectric anisotropy and sandwiched between two conductive substrates. Again, the contrast was not improved significantly enough for display applications.

Fergason[6] utilized an emulsification method by dispersing liquid crystal droplets in a water-based polymer medium, subsequently encapsulating them via polymerization, and creating micron-size liquid crystal capsules within the polymer matrix. It was perhaps the first electrically controllable device based on the principle of refractive index matching that could be switched successfully between the transparent and opaque states. Later, a polymer-dispersed liquid crystal (PDLC) was developed by Doane *et al.*[8–10] using a phase separation technique via polymerization and/or thermal quenching. The diverse choice of polymer types and phase separation procedures permits possible control of the domain morphology of PDLCs and ultimately their electrooptical properties.[8–15] There are some outstanding reviews on this subject that concentrate on thermal transitions, structure, and electrooptical properties of PDLCs.[13,16,17] The present review is aimed at elucidating the thermodynamics and kinetics of phase separation and related phenomena of pattern formation, since the understanding of the underlying mechanisms is a crucial part of the PDLC preparations for controlling morphology to improve the electrooptical properties of the PDLCs and related materials.

The PDLCs are basically inhomogeneous composite films comprising low molar mass liquid crystals (LC) which are dispersed in a continuum of a polymer matrix by forming micron-size domains. These domains can be of a droplet shape or an interconnected structure, depending on the kind of polymer matrix used and the way the PDLC films are prepared via phase separation. When the concentration of the phase-separated liquid crystal molecules exceeds its critical value, the LC molecules tend to self-align among themselves due to the large anisotropy of LC molecules by forming a nematic phase within the dispersed domains. The nematic liquid crystal domains strongly scatter light and thus the material appears opaque to the naked eye. For thermoset-based matrix binders, PDLCs are commonly sandwiched between two transparent conductive substrates such as indium–tin oxide (ITO) coated glasses. Alternatively, a PDLC may be prepared by dispersing LCs in some transparent matrix polymers and subsently sandwiching it between conductive flexible thermoplastic films such as ITO-coated poly(ethylene terephthalate) (PET) films.

With a modest external electric field, the sandwiched PDLC films can be switched from an opaque to a transparent state by aligning the liquid crystal molecules from their random configuration to the direction of the applied electric field (the film normal direction), or vice versa. It is common practice to match closely the refractive index of the matrix polymer to the ordinary component of the liquid crystal molecules. Therefore, the PDLC films are transparent in the on-state when the LC molecules orient in the direction of the electric field, as illustrated in Figure 2. Upon removal of the electric field, the LC molecules reorient randomly and revert to their original scattering state.

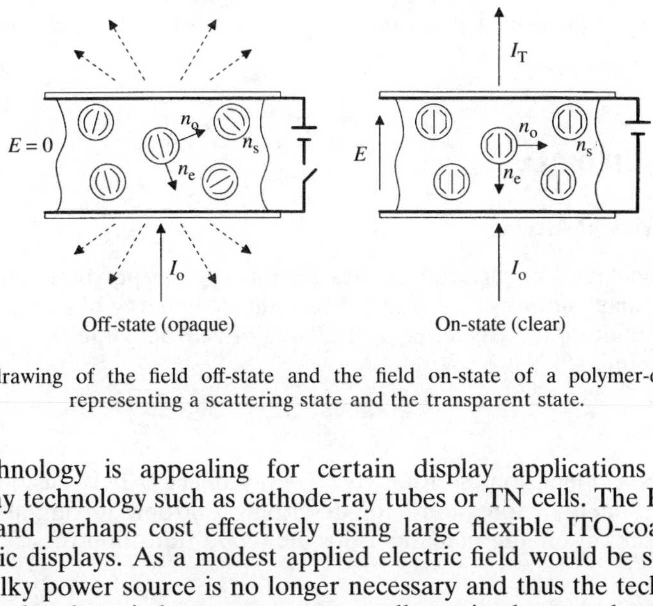

Figure 2 Schematic drawing of the field off-state and the field on-state of a polymer-dispersed liquid crystal representing a scattering state and the transparent state.

The PDLC technology is appealing for certain display applications compared to the conventional display technology such as cathode-ray tubes or TN cells. The PDLC films may be fabricated simply and perhaps cost effectively using large flexible ITO-coated PET films for large area electronic displays. As a modest applied electric field would be sufficient to operate the PDLCs, the bulky power source is no longer necessary and thus the technology is ideal for use in small-scale hand-carried computers as well as in large-scale flat-panel displays. Moreover, the power consumption can be reduced significantly. The lack of polarizers in PDLCs enhances the brightness in the forward view, and thus it can be used in the projection displays.

16.2 TYPES OF POLYMER-DISPERSED LIQUID CRYSTALS

Polymer-dispersed liquid crystals offer a relatively new category of optical materials which may be produced through various phase separation methods. To prepare PDLC films, low molar mass nematic liquid crystals are mixed with thermal or photo-curable prepolymers such as epoxy or methyl methacrylate or acrylate derivatives.[7–12] A mutual solvent such as toluene, tetrahydrofuran, or chloroform has been used to enhance mixing of liquid crystals with matrix polymers and crosslinking agents. The LC/prepolymer mixture generally remains in single phase at ambient temperature even after the solvent has been removed. The mixture is spread on a transparent ITO-coated glass slide, and then covered with another ITO-coated glass while the mixture is in the fluid state. Polymerization is then initiated by annealing at an elevated temperature, typically above the nematic–isotropic transition temperature of the LC utilized. When polymerization proceeds, the molecular weight of the prepolymer increases drastically. Concurrently, the entropy is reduced which, in turn, induces instability to the system. This instability has led to phase separation between the polymer liquid and the isotropic phase of LCs. Nematic ordering takes place subsequently within the dispersed domains, which may be in the form of spherical droplets or of percolated structures. This process is called polymerization-induced phase separation (PIPS). Alternatively, photopolymerization of prepolymers may be employed as a means of phase separating PDLC mixtures by subjecting the whole assembly to ultraviolet (UV) radiation. The photocrosslinking reaction of the matrix polymer induces phase separation while the fluid mixture is hardened. This photopolymerization-driven phase separation process is often categorized as a polymerization-induced phase separation.

Thermoplastics may be utilized as matrix polymers in producing a free-standing PDLC film. The preparation method requires use of a common solvent for dissolving polymers and mixing with LCs. Unlike the thermoset resins, the molecular weight of thermoplastics is rather high. Therefore, the LC molecules are segregated out from the matrix polymers during solvent evaporation, forming spherical droplets or irregular shape-domains. Such a process is called solvent-induced phase separation (SIPS). The two-phase structure thus formed is not thermally stable as it can be entrapped during the solvent evaporation. Thus, the PDLC films are generally subjected to heat treatment by annealing above the nematic–isotropic transition temperature of the LCs. The temperature may be lowered gradually at a natural cooling rate or instantaneously from a single phase to a two-phase temperature to induce phase separation; this process is generally termed thermally induced phase separation (TIPS). The length scale of the phase-separated domains varies from submicrons to a few tens of micrometers. The shape, size, uniformity, and dispersion depend on various thermodynamic parameters and kinetics of phase separation.

16.3 PHASE EQUILIBRIA

16.3.1 Free Energies of Mixing

A theoretical elucidation of phase equilibria for mixtures of polymers and liquid crystals has been attempted by many authors.[20–23] The Gibbs total free energy has been generally described in terms of a summation of the free energy of mixing of isotropic liquids, g^{i}, and the free energy of anisotropic ordering, g^{n}

$$g = g^{\mathrm{i}} + g^{\mathrm{n}} \tag{1}$$

where g represents a dimensionless total free energy density. It is customary to employ the Flory–Huggins free energy expression[24] for describing isotropic mixing and the Maier–Saupe free energy equation[25–27] for the nematic–isotropic phase transition of a nematic liquid crystal as

$$\begin{aligned} g &= \frac{\phi_1}{r_1}\ln\phi_1 + \frac{\phi_2}{r_2}\ln\phi_2 + \chi\phi_1\phi_2 + \frac{\phi_1}{r_1}\int f(\theta)\ln[4\pi f(\theta)]\,\mathrm{d}\Omega - \tfrac{1}{2}\nu\phi_1^2 s^2 \\ &= \frac{\phi_1}{r_1}\ln\phi_1 + \frac{\phi_2}{r_2}\ln\phi_2 + \chi\phi_1\phi_2 + \frac{\phi_1}{r_1}\ln\frac{1}{Z} + \tfrac{1}{2}\nu\phi_1^2 s^2 \end{aligned} \tag{2}$$

where Ω is the solid angle and ϕ_1 and ϕ_2 represent the volume fractions of components 1 (liquid crystal) and 2 (polymer), respectively; r_1 and r_2 are the numbers of lattice sites or segments occupied by one liquid crystal molecule and a single polymer chain, respectively; χ represents the intermolecular interaction, often known as the Flory–Huggins interaction parameter, which is generally assumed to be inversely proportional to the absolute temperature (T), *viz.*[24]

$$\chi = A + B/T \tag{3}$$

where A and B are constants associated primarily with the entropic and enthalpic contribution, respectively. The last two terms in equation (2) represent the contributions from the nematic ordering. Here, v is the Maier–Saupe quadrupole interaction parameter which has an inverse temperature dependence as[25]

$$v = 4.541\,\frac{T_{NI}}{T} \tag{4}$$

where T_{NI} is the nematic–isotropic (NI) transition temperature of the liquid crystals. s represents an orientational order parameter defined as

$$s = \tfrac{1}{2}(3\langle\cos^2\theta\rangle - 1) \tag{5}$$

The directional cosine can be determined by averaging the director orientations in terms of an orientation distribution function, $f(\theta)$, as

$$\langle\cos^2\theta\rangle = \int \cos^2\theta f(\theta)\,\mathrm{d}\cos\theta \tag{6}$$

$$f(\theta) = \frac{1}{4\pi Z}\exp\left[-\frac{u(\theta)}{kT}\right] \tag{7}$$

in which the partition function, Z, is given by

$$Z = \int \exp\left(-\frac{u(\theta)}{kT}\right)\mathrm{d}\cos\theta \tag{8}$$

where the pseudopotential of director orientations, $u(\theta)$, may be expressed in terms of a second-order Legendre polynomial by the equation[22,27]

$$\frac{u(\theta)}{kT} = -\tfrac{1}{2}m(3\cos^2\theta - 1) \tag{9}$$

where m is a mean field parameter characterizing the strength of a potential field. It is essential to determine the orientational order parameter for various blend compositions by means of the free energy minimization approach, and subsequently the free energy of the nematic ordering may be evaluated.

16.3.2 Chemical Potentials

At equilibrium, the chemical potentials of the individual component are, by definition, equivalent in every phase, *i.e.* the polymer-rich phase (α) and the liquid-crystal-rich phase (β) have the same chemical potential

$$\mu_1^{\alpha} = \mu_1^{\beta} \tag{10}$$

$$\mu_2^{\alpha} = \mu_2^{\beta} \tag{11}$$

As in the case of the total free energy, the total chemical potential of the PDLC has contributions from the nematic ordering and the isotropic mixing, *viz.* $\mu_j = \mu_j^{\mathrm{n}} + \mu_j^{\mathrm{i}}$. The chemical potentials of the nematic ordering, μ_j^{n}, can be expressed as

$$\mu_1^{\mathrm{n}} = -\ln Z + \tfrac{1}{2}vr_1\phi_1^2 s^2 \tag{12}$$

$$\mu_2^{\mathrm{n}} = \tfrac{1}{2}r_2 v\phi_1^2 s^2 \tag{13}$$

For the isotropic mixing, the chemical potentials, μ_j^{i}, may be deduced to be

$$\mu_1^{\mathrm{i}} = \ln \phi_1 + \left(1 - \frac{r_1}{r_2}\right)\phi_2 + \chi r_1 \phi_2^2 \tag{14}$$

$$\mu_2^{\mathrm{i}} = \ln \phi_2 + \left(1 - \frac{r_2}{r_1}\right)\phi_1 + \chi r_2 \phi_1^2 \tag{15}$$

By solving equations (10) and (11) simultaneously, the equilibrium temperature–composition phase diagram may be constructed.

16.3.3 Phase Diagram Calculations

Figure 3 depicts a theoretical phase diagram of a hypothetical PDLC system calculation based on the combined Flory–Huggins and Maier–Saupe free energies in which a liquid–liquid phase equilibrium was overlapped with a nematic–isotropic transition of the liquid crystal.[28] The phase diagram resembles essentially an upper critical solution temperature (UCST) in which the polymer liquid and the isotropic phase of the liquid crystal molecules are segregated. This liquid–liquid equilibrium region covers only a small temperature gap, below which nematic ordering takes place within the dispersed domains. At liquid-crystal-rich compositions, there is a narrow coexistence region where isotropic polymer liquids and nematic liquid crystals can coexist. Such a complex phase diagram of a PDLC system is often known as a 'tea-pot' phase diagram. At low temperatures or deep quenches, phase separation proceeds to the nematic phase limit where polymer chains can be excluded totally from the nematic domains. However, the isotropic (polymer-rich) phase may still contain some dissolved liquid crystal molecules. Recently, the existence of a spinodal line has been predicted within the narrow coexistence region of isotropic polymer fluids and nematic liquid crystals.

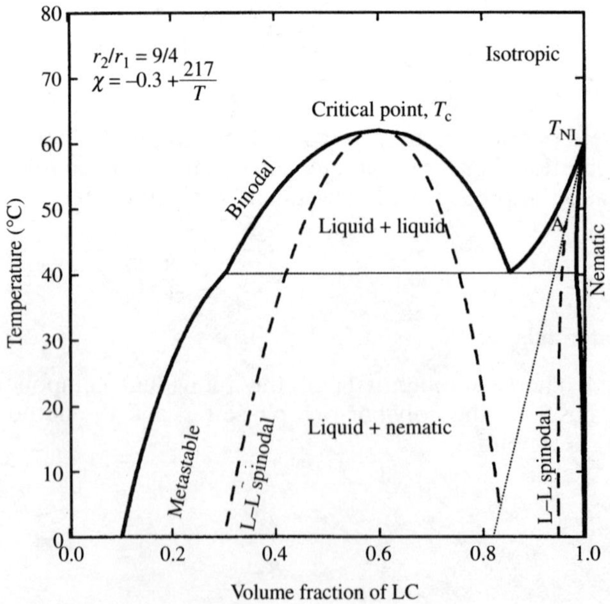

Figure 3 A calculated phase diagram for a hypothetical PDLC system with parameters employed in the calculations.

In Figure 4, the binodal curve calculated with a set of parameters r_2/r_1 and χ indicated in the figure is compared with the experimental cloud point phase diagram for blends of poly(methyl methacrylate) (PMMA)/monomeric liquid crystals.[29] The authors utilized the mixed eutectic main-chain liquid crystals (commercially known as E7) because a single component liquid crystal often shows a crystal–nematic transition in the proximity of the nematic–isotropic transition of the LC. Although r_1/r_2 and χ were used as adjustable parameters in their simulation, the authors[29] pointed out that χ at the spinodal point can be determined experimentally, for instance by small-angle neutron scattering. On the other hand, the value of r_1/r_2 will be dictated by the critical point which may be estimated, although by no means accurately, from the experimental cloud point phase diagram. Hence, their theoretical calculation could be tested rigorously with experiments. As can be seen in Figure 4, it certainly captures the experimental trends at least qualitatively.[29]

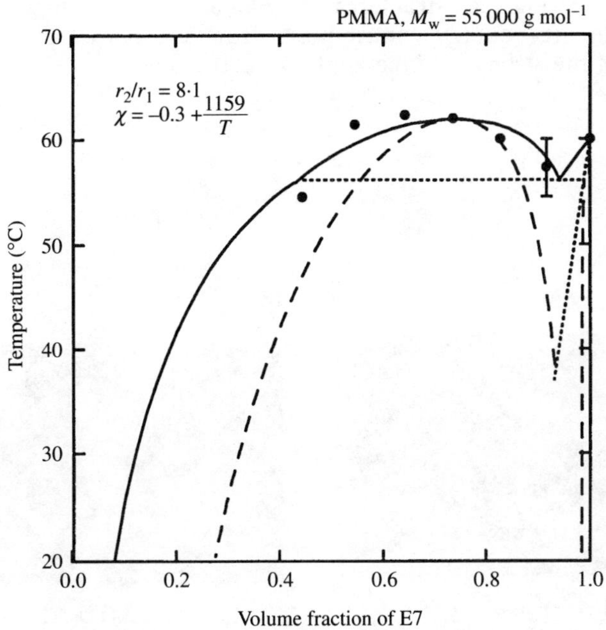

Figure 4 A comparison of a calculated phase diagram with the experimental cloud points and the NI transition of the PMMA/E7 system.

16.4 PATTERN FORMATION PHASE SEPARATION DYNAMICS

16.4.1 Thermally Induced Phase Separation

When a binary system is brought instantaneously from a single-phase to a two-phase system by thermal quenching, phase separation occurs through a series of heterogeneous states. The liquid–liquid phase separation can take place either through nucleation and growth (NG) or spinodal decomposition (SD). The SD is a diffusional process associated with phase separation driven by temporal concentration fluctuations. It is unstable at infinitesimal concentration fluctuations and thus occurs spontaneously without requiring any energy. On the other hand, the NG is stable at large concentration fluctuations but unstable against small concentration fluctuations. The initial stable fragment acts like a nucleus upon which coarsening occurs. The phase separation process is more complex in a PDLC system as the liquid–liquid phase separation has been generally accompanied by the nematic ordering.

16.4.1.1 Nucleation and growth

Nucleation and growth occur when a binary system is suddenly quenched from a stable single phase into a metastable state. In a majority of cases the NG process occurs through a heterogeneous process that results in a distribution of droplet sizes. Figure 5 exhibits the pattern formation of the off-critical 60/40 poly(*n*-butyl methacrylate) (PNBA)/E7 liquid crystals, following a thermal quench from a single phase regime (80 °C) to a metastability regime (ambient temperature).[29] A multiple droplet morphology, characteristic of a typical nucleation and growth, can be discerned. The average droplet size grows in diameter with elapsed time, but it ceases to grow further at a later stage. Concurrently, the nematic disclination develops within the preformed domains. These domains eventually transform into spherical droplets due to the difference in surface tension between the liquid crystals and the plasticized PNBA. When nematic LCs are confined to a small spherical cavity, a variety of unusual effects are introduced as a consequence of the interplay between the ordering interactions at the cavity wall and the elastic deformation of the LCs within the cavity. Notable features are the specific director configurations inside the cavity and the LC anchoring at the surface. The director ordering such structures as radial and axial configurations depends on the delicate balance of the free energy contributions from the elastic deformation of the LCs, surface anchoring, and the strength of the applied electric field.

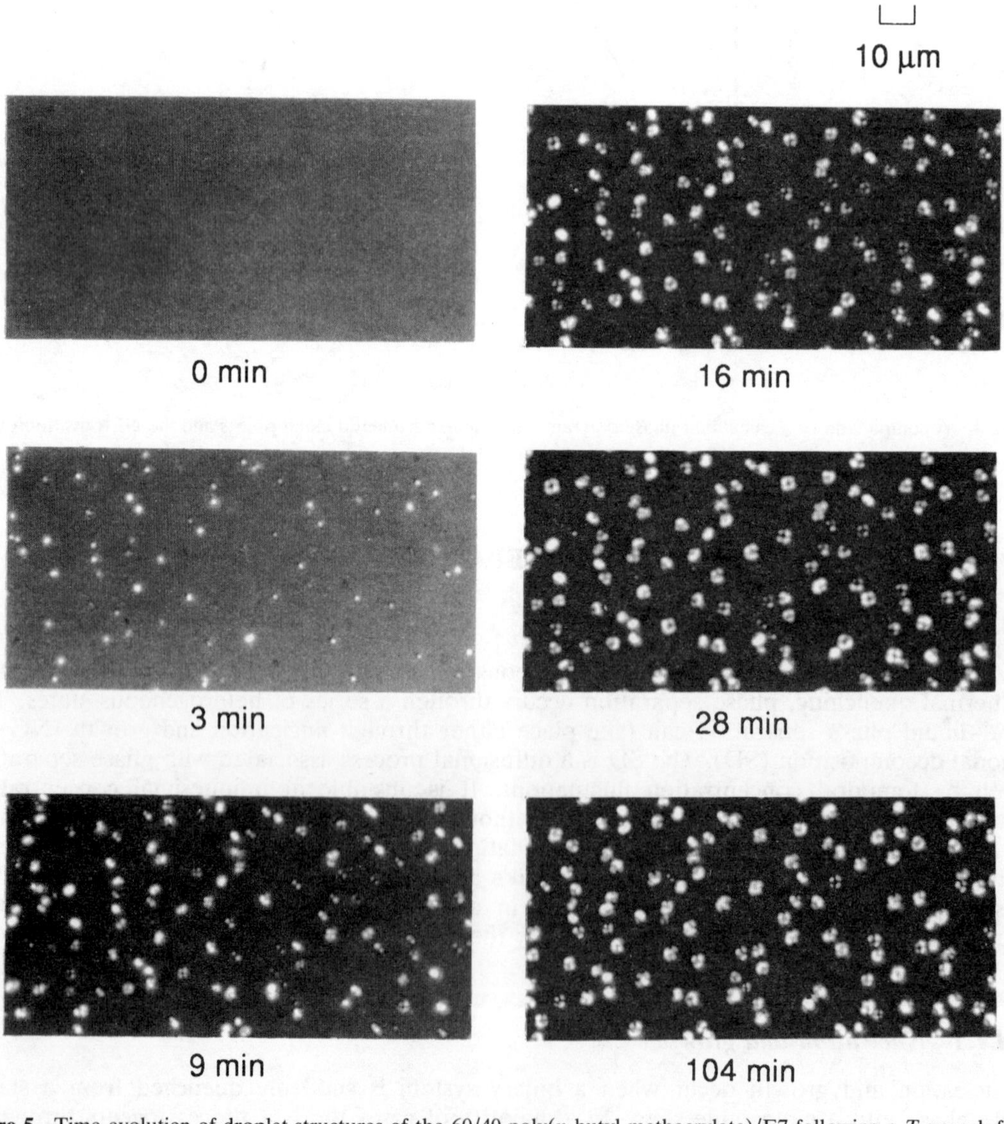

Figure 5 Time evolution of droplet structures of the 60/40 poly(*n*-butyl methacrylate)/E7 following a *T* quench from 80 °C to a metastable region (ambient temperature).

16.4.1.2 Nematic droplet configurations

The radial and axial director configurations are the most basic structures for PDLC systems.[30] In the radial configuration, the LC directors orient radially outward from a center such that their long axes are anchored perpendicular to the droplet wall (Figure 6). This configuration generally originates from a point singularity at the center of the droplet. In practice, most point defects are coupled; thus the radial configuration is rarely observed in PDLC materials. The other type is the axial configuration which is due to the equatorial line defect. The LC molecules are anchored normal to the surface, but the anchoring strength is not very large, and thus it can be influenced easily by the elastic torques of the LC. A modest electrical field is sufficient to transform the radial configuration to the axial. To describe this transition phenomenon, Erdmann *et al.*[30] calculated the stability limits of various PDLC droplet conformations by minimizing the free energy equation consisting of the Frank free energy[27] coupled with other contributions from the surface and the applied electric field, *viz.*

$$F = \tfrac{1}{2}\int [K_{11}(\operatorname{div}\boldsymbol{n}) + K_{22}(\boldsymbol{n}\operatorname{rot}\boldsymbol{n}) + K_{33}(\boldsymbol{n}\times\operatorname{rot}\boldsymbol{n})]\,\mathrm{d}V$$

$$-\tfrac{1}{2}\int \varepsilon_0\Delta\varepsilon(\boldsymbol{E}_l\cdot\boldsymbol{n})^2\,\mathrm{d}V + \int w_0(\xi-\xi_0)^2\,\mathrm{d}A \qquad (16)$$

where the first [16] term is made up of the usual contribution of splay, twist, and bend deformations with elastic constants K_{ii}, and $\boldsymbol{n}$ is the director orientation. The second term is the contribution arising from the applied field (E); E_l represents the local field and $\Delta\varepsilon$ is the dielectric anisotropy. The third term is associated with the surface anchoring where w_0 is the anchoring strength, and ξ and ξ_0 are the actual and preferred anchoring angles. Assuming $K = K_{11} = K_{22} = K_{33}$, equation (16) has been found to depend on the ratio of R/d_e where R is the radius of the cavity and $d_e = K/w_0$ is the surface extrapolation length. The authors[31] demonstrated that the structure becomes increasingly aligned with decreasing R/d_e. If the ratio

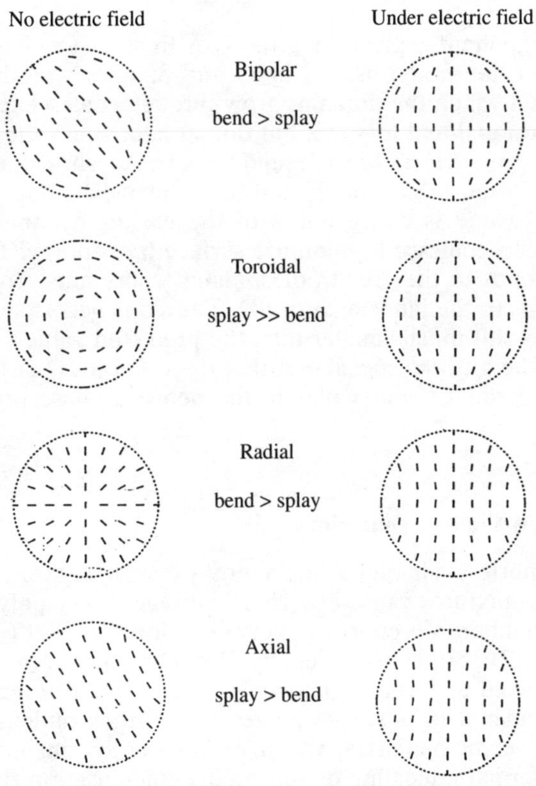

Figure 6 Various droplet configurations in the absence of an external electric field and under the electric field.

$R/d_e < 18$, the total free energy of the axial configuration is more stable compared to that of the radial configuration.[30] When a PDLC with a radial configuration is subjected to an electric field, a transition from the radial to the axial configuration has been predicted to take place. This configurational transition has also been experimentally observed for a polyurethane-based PDLC.

The dominant droplet structure found in PDLCs is the bipolar director configuration depicted in Figure 6. The bipolar structure is basically another type of axial configuration in which the LC directors are tangentially anchored at the droplet wall with a bipolar axis orienting randomly in the absence of an external field. When the bipolar structure is subjected to a modest electric field, the simulation predicts for the LCs of positive dielectric anisotropy that the bipolar axis would align readily in the direction parallel to the field.

16.4.1.3 Spinodal decomposition

Although the SD process has been extensively studied in polymer blends and metal alloys, there are only limited studies of SD for the PDLC systems. The SD process was observed in a PMMA/E7 PDLC system during thermal quenching from a single phase region (80 °C) to a two-phase regime (25 °C).[31] Figure 7 shows the time-evolution of the phase-separated domains as observed in a microscopic investigation. Tiny interconnected structures develop over the entire view of the microscope and grow as time elapses. These features resemble, although it is by no means a proof, the characteristics of phase separation by spinodal decomposition. Under the polarized configuration, no identifiable structure was discerned initially, but some birefringent entities developed with elapsed time that may be attributed to the nematic ordering within the preformed domains. Later, these birefringent entities form nematic disclinations with the strength of ±1. The disclinations further grow via annihilation while the phase-separated domains continue to grow.

In the light scattering studies, the time-evolution of scattering profiles showed that there was no discernible linear regime corresponding to the early stage of spinodal decomposition. The scattering results were subsequently analyzed in the context of the late stages of SD based on a power law scheme, *viz.*

$$q_m \propto t^{-\alpha} \tag{17}$$

where α is the kinetic exponent representing the growth rate. The exponent value varies from 0.21 to 1.0 depending on the model used. Binder and Stauffer[32] predicted $\alpha = 1/3$ based on a cluster dynamics model in which the domains grow through coalescence. Siggia[33] suggested that the initial growth is diffusional and $\alpha = 1/3$, but that at later times the domain growth proceeds via coalescence driven by surface tension and $\alpha = 1$. As shown in Figure 8, the growth exponent was estimated to be 1/3 at the 25 and 35 °C quenches, but it crosses over to a larger slope of −0.6. The 1/3 value is consistent with the cluster dynamics model of Binder and Stauffer[32] in which droplets coalesce to minimize surface free energy. The time scale covered by the initial slope corresponds to the growth of domains where anisotropy is absent. The second slope of −0.6 corresponds to the late stages of SD. The slope gets steeper slightly at a shallower quench to 45 °C, but it is still much smaller than the predicted value of −1 for the late stage of SD of isotropic fluids. The authors speculated that the differences in the effective diffusivity of coiled polymer and rigid rod LC molecules in the nematic phase probably effect the growth behavior.[31]

16.4.2 Reaction-induced Phase Separation

There are limited reports available in the literature with regards to the kinetics of phase decomposition of binary mixtures caused by chemical reaction or polymerization of monomers or prepolymers. The number of reports is very low for the PDLC systems.[34] Initially, the eutectic liquid crystal (E7), bisphenol-A epoxy (Epon 828), and a trifunctional thiol curing agent (Capcure) were mixed to form a homogeneous mixture. The mixture was annealed at an elevated temperature under a light scattering set-up. As the condensation reaction proceeds, the average molecular weight increases, which makes the system unstable. Phase separation takes place during isothermal annealing driven by the condensation reaction of the epoxy with Capcure. The kinetics of phase separation can be determined by time-resolved light scattering.

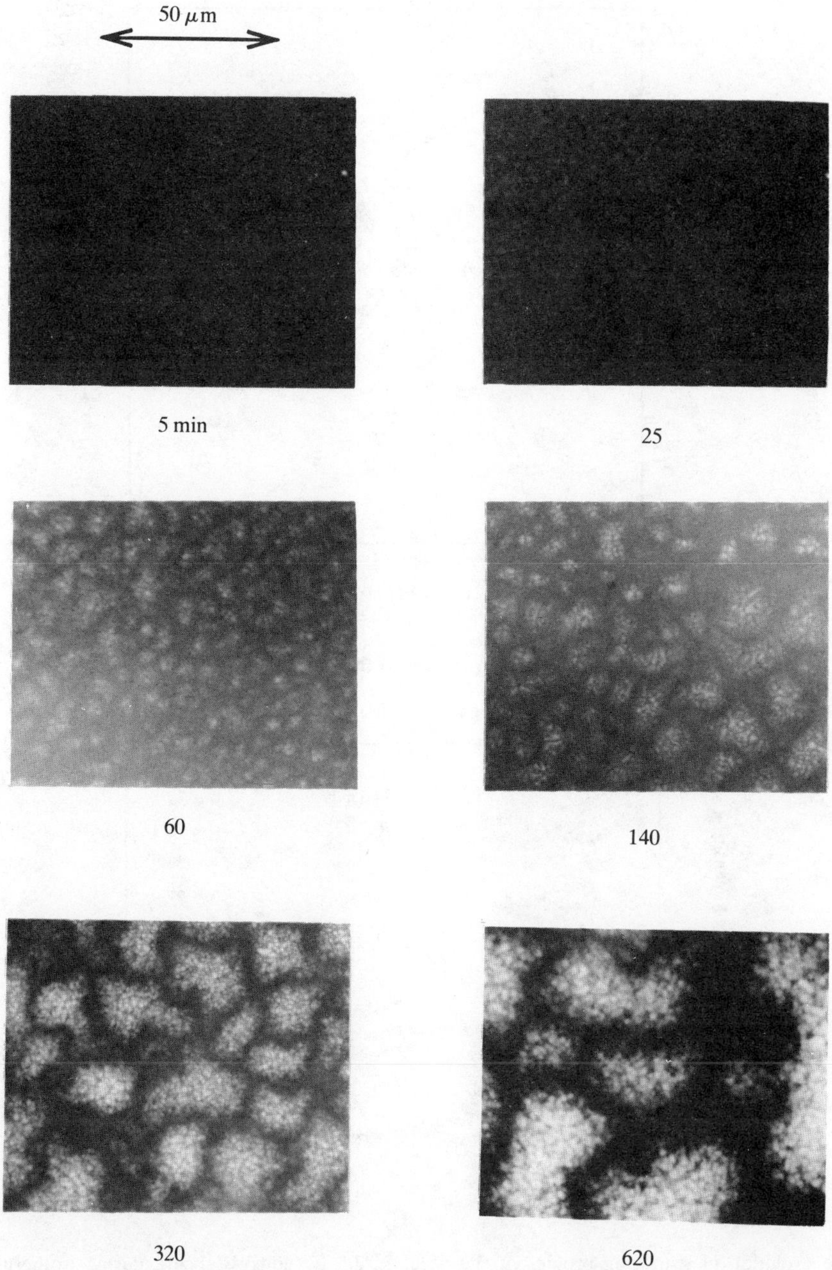

Figure 7 Time evolution of the spinodal structure of the 60/40 PMMA/E7 system following a T quench from a single phase (80 °C) to a unstable region (25 °C).

Figure 9 shows the time evolution of the structure factor of the 45/27.5/27.5 E7/epoxy/Capcure mixture at 62 °C. Although the polymerization may have started upon mixing, phase separation was not detected for some initial period, at least at the length scale of the wavelength of light. After this induction period, a scattering maximum appears and shifts to a lower scattering angle without revealing a linear regime corresponding to the early stage of spinodal decomposition where the scattering maxima remain stationary with time. The authors[34] analyzed the growth process in terms of a power law that has been customarily employed in the thermally induced phase separation.

Figure 10 shows the plot of $\ln q_m$ versus $\ln(t - t_0)$, where t_0 is the induction time. Although the data at short times can be approximated fairly well with a slope of $-1/3$, it should be cautioned that these results are susceptible to the accuracy of the t_0 measurement. At later

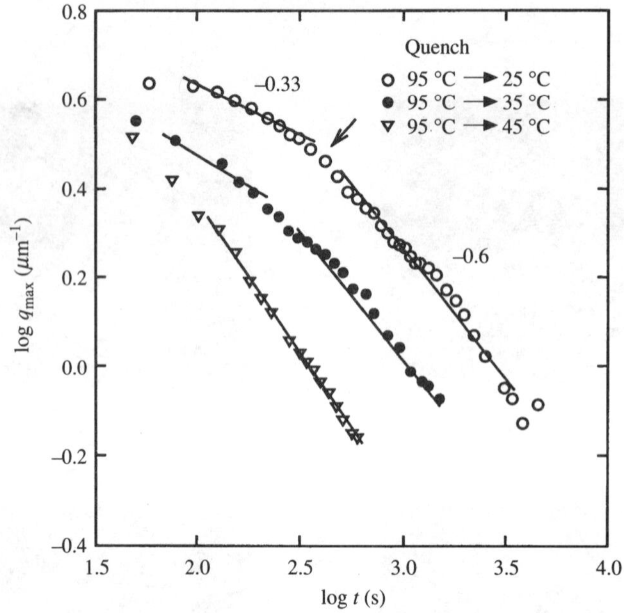

Figure 8 Double logarithmic plots of scattering wavenumber versus elapsed time for the 60/40 PMMA/E7 at several T quenches exhibiting the crossover of growth exponent from −1/3 to −0.6.

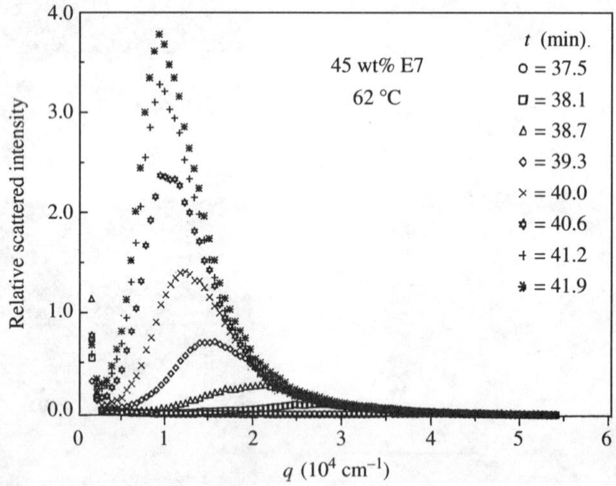

Figure 9 Time evolution of scattering profiles of the 45/27.5/27.5 E7/epoxy/Capcure during isothermal curing at 62 °C.

times, a cross-over of the exponent from −1/3 to −1 has been observed, which is in good agreement with Siggia's prediction[33] for the thermally quenched systems. The crosslinking reaction continues even though the domains can no longer grow further. The materials in both liquid-crystal-rich and -poor regions again become unstable and undergo phase separation for a second time within the preformed domains. This cascading phase separation phenomenon is strikingly similar to that reported for a two-step thermally quenched system.[35]

16.5 ELECTROOPTICAL PROPERTIES

The electrooptical properties of PDLC films depend strongly on their final morphology, which in turn can be influenced by various factors, such as the types of materials used and the way the films are prepared. The low driving voltage, fast switching time, a wide temperature range of operation, strong adhesion of the matrix binder to substrates, and stability against

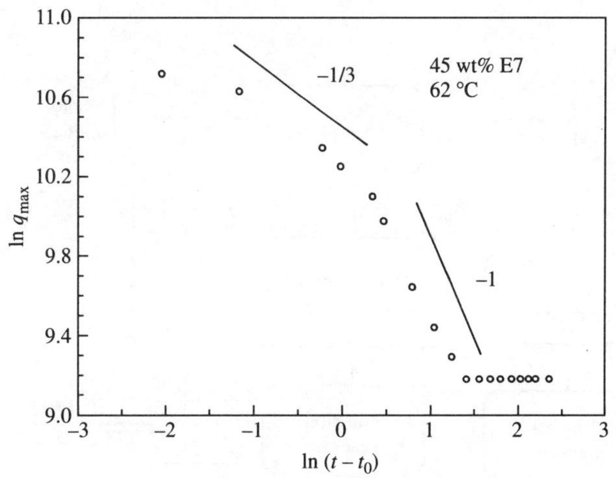

Figure 10 Double logarithmic plot of scattering wavenumber versus $(t - t_0)$ following reaction-induced phase separation of the 45/27.5/27.5 E7/epoxy/Capcure at 62 °C; t_0 is the induction time.

ultraviolet radiation are some of the desired properties. For a successful operation the PDLCs are expected to have maximum transmittance in the on-state and maximum opacity in the off-state. These criteria are, if not impossible, hard to meet.

The majority of optical transmittance measuring devices reported in the literature are home made. The details of the equipment set-ups may vary from one laboratory to another, but the basic principle is the same. Figure 11 illustrates a schematic diagram of an optical transmittance device used in the author's laboratory. It consists of a He–Ne laser light source, a photodiode (Hamamatsu), an oscilloscope (Tetronix), an alternating current (AC) switching device, an analog-to-digital (AD) converter (Quatec), and a microprocessor (IBM PC 2/30). The incident light (I_0) has been measured without placing any materials in the optical path. The optical transmittance of PDLCs in the on and off states has been determined in accordance with the equations

$$T_{on} = (I_{on} - I_{bkg})/(I_0 - I_{bkg}) \quad (18)$$

$$T_{off} = (I_{off} - I_{bkg})/(I_0 - I_{bkg}) \quad (19)$$

where I_{on} and I_{off} are the observed intensities at the zero scattering angle with and without applying an external electric field; I_{bkg} represents the intensity of the background due to the electronic noise.

Figure 12 shows the morphology change when the 40/60 PMMA-OH (poly(methyl methacrylate) with functional hydroxyl groups)/E7 PDLC is subjected to an alternating current (60 Hz and 90 V). In the off-state, the droplets are birefringent and scatter light. In the on-state, the domains are still discernible, but become very faint as the majority of liquid crystal directors orient to the applied field in the film normal direction. The LC directors at the internal droplet walls are unable to orient because of the surface anchoring. This director anchoring effect contributes to the lowering of optical transmittance in the on-state. Figure 13 depicts the change of optical transmittance driven by the AC voltage. The switching times, τ_{on} and τ_{off}, measured at 90% and 10% of the sample transmittance, are about 1.3 and 14.5 ms, respectively; these values are, if not better, comparable to the response times of a typical TN cell. Figure 14 exhibits the voltage dependence of the optical transmittance. A modest voltage of about 80 V (rms) is required to reach the limiting transmittance of about 80%. The transmittance falls off rapidly at the wide viewing angles.

To maximize the transmittance of a PDLC film in the on-state, it requires critical matching of refractive indices between liquid crystals and polymers, *i.e.* the polarizability of the matrix polymer (n_p) must be matched with the ordinary component (n_0) of the principal polarizability of the liquid crystal directors. It is usually obtained by carefully choosing polymers with matching refractive index or perhaps adjusting the refractive index of the matrix polymers

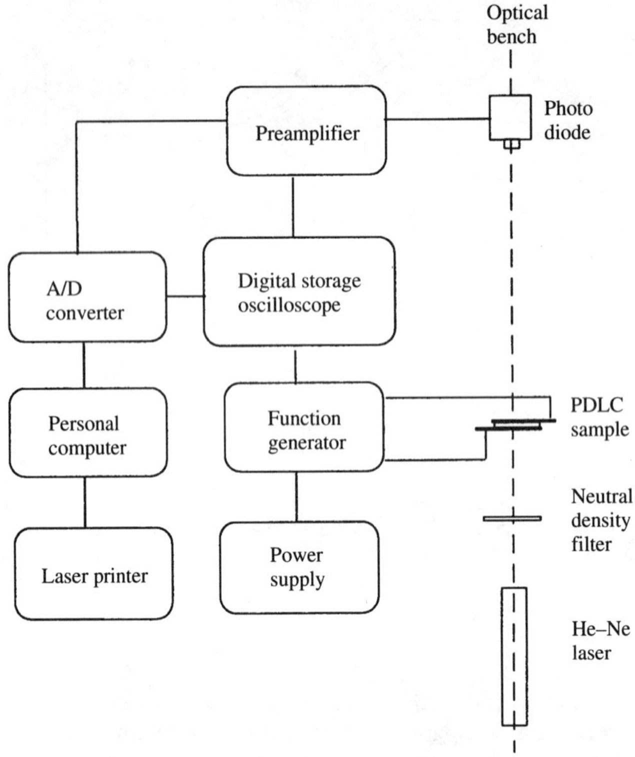

Figure 11 A schematic diagram of an electrooptical device.

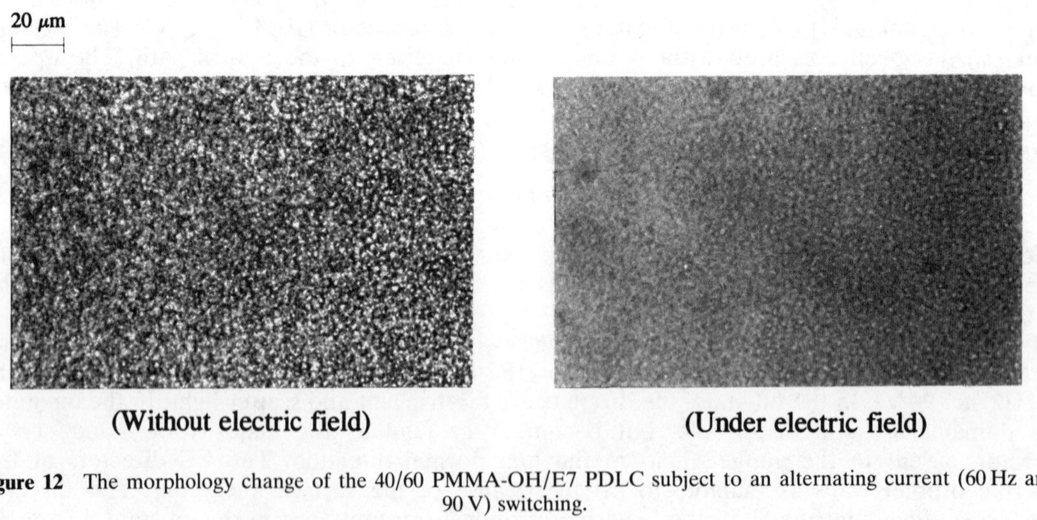

Figure 12 The morphology change of the 40/60 PMMA-OH/E7 PDLC subject to an alternating current (60 Hz and 90 V) switching.

through blending, although the miscibility problem could be severe for polymer blends. Nevertheless, a low molecular weight epoxy oligomer mixture with vastly different refractive indices has been shown to have complete miscibility, and thus may be used for refractive index matching.[13] To achieve the maximum opacity, the anisotropy of the liquid crystal component should be as large as possible so as to maximize the orientation fluctuation contribution to the scattering. The averaged length scales of the LC inclusions should be of the order of the wavelength of light to submicron sizes.

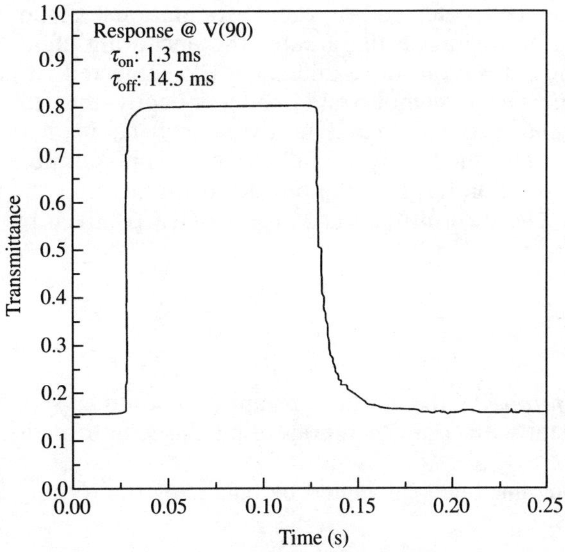

Figure 13 The change of optical transmittance as a function of rms voltage for characterizing the response times for the 40/60 PMMA-OH/E7.

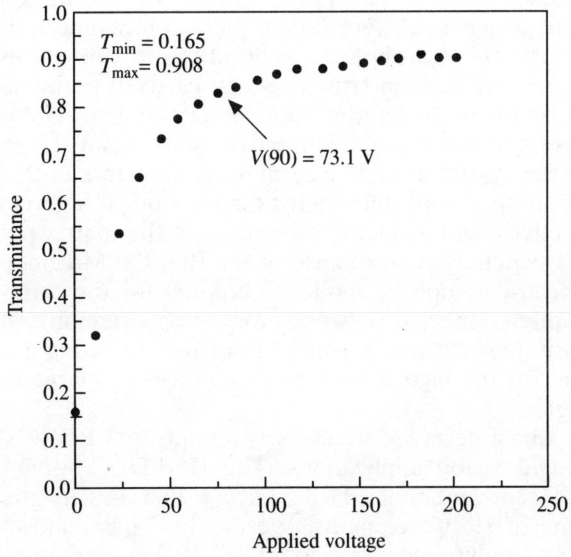

Figure 14 The dependence of optical transmittance on the driving voltage of the 40/60 PMMA-OH/E7.

The minimum voltage required for switching a PDLC shutter depends on the interaction between the LC directors and the field, *e.g.* the mobility of the LC as described by its effective viscosity and the orientational response of the LC director, which can be influenced by the configuration of the nematic directors and the confinement effect. The switching voltage and response time may vary depending on the shape, size, and distribution of the dispersed domains, as well as the topology of the LC directors within the droplets such as radial, axial, or bipolar configurations. Intuitively, the bipolar configuration would respond faster than the axial or radial configurations because only a rotation of the bipolar morphology is necessary. A larger domain size may be preferred for rapid orientational response of the LC directors because the smaller the domain size the greater the anchoring effect, which would make the director orientation and relaxation more difficult when subjected to the applied electric field. Drzaic[36] and Wu *et al.*[37] have demonstrated independently that the shape and size of the droplets indeed affect the response time. The dynamical theory of the director field inside a droplet has not been established fully for all director configurations. However, the bipolar model of director configuration has been employed to characterize the rise time and relaxation time of PDLC devices. The turn off-time (τ_{off}) for a PDLC is given by

$$\tau_{\text{off}} = \frac{\gamma_l a^2}{K(l^2 - 1)} \tag{20}$$

where γ_l *is the viscous torque,* K the elastic constant of a liquid crystal, and $l = a/b$ is the aspect ratio of the liquid crystal with a and b representing the length of the major and minor axes, respectively.

Similarly, the rise time has been calculated by balancing the torque densities, electric dipoles and viscosity as follows:

$$\tau_{\text{on}}^{-1} = \frac{1}{\gamma_l} \frac{9\varepsilon_0 \Delta\varepsilon V^2}{d^2(\rho_{\text{p}}/\rho_{\text{lc}} + 2)^2} + \frac{K(l^2 - 1)}{\gamma_l a^2} \tag{21}$$

It should be pointed out that the radial, axial, and bipolar configurations discussed thus far are idealized PDLC structures. In practice, the droplets in PDLC materials are strictly nonspherical. They may be flattened or distorted as the supporting matrix polymers are soft due to the plasticization by the LCs. In addition, phase separations commonly occur via spinodal decomposition, which could eventually yield interconnected domains or percolated channels such as network-like morphology as equilibrium can hardly be achieved in many PDLC systems. Such irregular domain structures can be fixed permanently by crosslinking the prepolymers with heat treatment or by photopolymerization with UV radiation. For irregularly shaped PDLC structures, the mismatch of refractive index is not the sole cause contributing to the scattering of light; the spatial distortion of nematic directors in the curved or distorted LC continuous channels is another important factor for the optical heterogeneity.

The electrical conductivity and dielectric properties of the matrix polymers can play a major role in determining the switching performance of the PDLCs. Miyamoto *et al.*,[38] who proposed a series-connected dielectric composite model to account for the complex dielectric properties of a PDLC having an interconnected network morphology, demonstrated that the electrooptical response, particularly the rise speed, can be improved by using a matrix polymer having a large dielectric constant for the high-frequency range or with a large electrical conductivity for the low-frequency range.

A recent innovation which deserves attention is the infrared PDLC shutter for video system, thermal sensor, and night vision applications. The IR PDLC systems operate on the same principle as that of the conventional PDLC, except that the length scale of the dispersed domains is in the range of IR wavelengths. West *et al.*[13] have shown that polymers such as PMMA, polystyrene, and poly(vinylpyrrolidone) (PVP) have acceptable transmittance in the IR range. The typical domain size (diameter) of the PVP/E7 mixture is in the range 10–40 μm, which can be controlled through the understanding of the kinetics of phase decomposition.

16.6 FUTURE TRENDS

Although the PDLC based displays are still in their infancy for commercialization, the related liquid crystal technology and sciences have made remarkable progress in recent years. As pointed out previously, the refractive index of the liquid crystal droplet in conventional PDLCs is angle dependent and can be matched with an isotropic polymer for incident light over a narrow range of view angles. This limited view angle presents a hazy appearance of the PDLC film in wide angle viewing and color tainting. Application of the conventional PDLC is limited thus far to projection displays and privacy windows.

An idea that is conceptualized for improving the wide angle viewing is the use of side-group liquid crystal polymers as a matrix with matching refractive indices. The dispersing LCs can be chosen as the mesogenic group for the LCP so that the extraordinary and ordinary polarizabilities between both LC mesogens are essentially the same. The nematic directors of the side-group LCP may be aligned homeotropically by an external electric or a magnetic field during the PDLC fabrication (Figure 15). The orientation of the side-group liquid crystals may be maintained via crosslinking or vitrification of the matrix. When an electric field is applied across the film, both directors of the side-chain liquid crystal group and the dispersing low molecular weight LC align parallel to the field direction. The refractive indices match closely for all incident angles; thus the film is haze-free in the on-state. Upon removal of the field, it reverts to the original scattering state.[17]

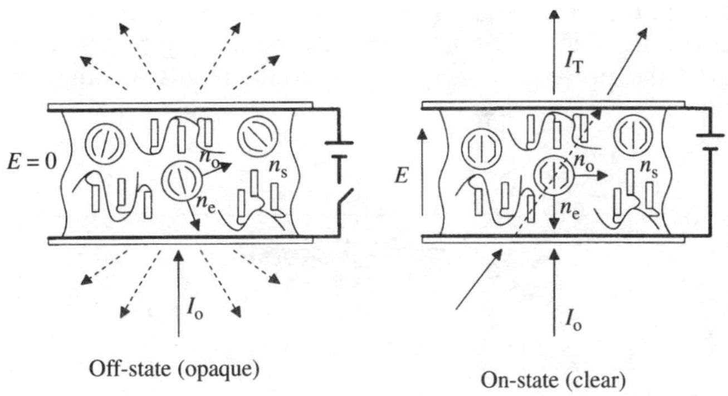

Figure 15 A schematic diagram of a side-group liquid crystalline polymer and low molar mass liquid crystal in the absence of an electric field and under an electric field.

Another development in liquid crystalline optical materials is the use of cholesteric liquid crystals (CLC) in reflective color displays and light shutters.[39–42] A CLC has a helical structure that reflects circularly polarized light at the wavelength of light corresponding to the cholesteric pitch in accordance with Bragg's formula $\lambda = nP \cos \beta$, where n is the refractive index, P the cholesteric pitch, λ the wavelength of the reflected light, and β the incident angle. Cholesteric liquid crystals can assume three different configurations depending on the surface treatment of the substrates and the electric field strength (Figure 16). In the preparation of the planar texture, the helical axis of the CLC is aligned perpendicular to the surface of the cell via the appropriate surface treatment. The planar texture represents the color reflecting state. When a modest voltage is applied, it transforms to a focal conic scattering state by forming cholesteric liquid crystal polydomains in which the helical axes orient randomly. If the applied voltage exceeds the critical voltage, the cholesteric texture is untwisted and the liquid crystal molecules align homeotropically. This homeotropic configuration corresponds to the transparent state.

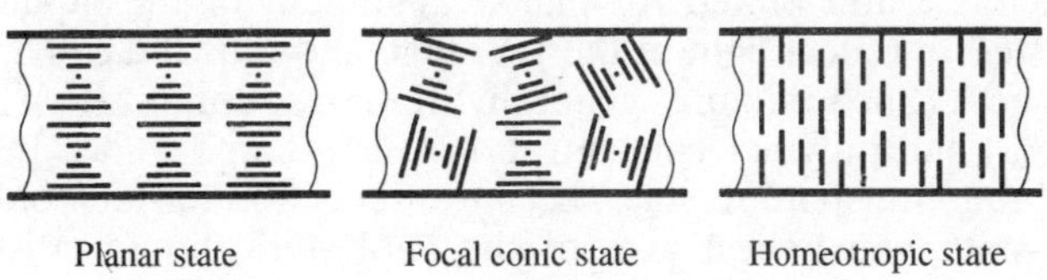

Figure 16 A polymer stabilized cholesteric texture (PSCT) with various director configurations showing the reflecting state (planar configuration), scattering state (focal conic), and transparent state (homeotropic alignment).

Recently, a polymer-stabilized cholesteric texture (PSCT) consisting of cholesteric liquid crystals and polymer has been developed for light shutter applications.[40,41] Surface treatment was not required for the normal mode PSCT because the focal conic cholesteric texture can be stabilized by the polymer network formation. The typical polymer concentration for the networks is about 3 wt%. The normal mode PSCTs are prepared by mixing a small amount of reactive monomer, chiral dopant, and photoinitiator in nonreactive low molar mass liquid crystals. The polymerization is then initiated *in situ* by UV radiation, leading to the formation of polymer gel networks in which LCs are contained. The addition of chiral dopants to nematic LCs causes the directors of the mesogenic molecules to adopt a helically twisted orientation. The chiral pitch of the resulting helicoidal structure is generally determined by the concentration (typically about 0.5 wt% for the normal mode) and the kind of chiral materials used. In the off-state, the CLC is stabilized in the focal conic scattering state. When a sufficient voltage is applied to the PSCT cell, the CLC is untwisted and the liquid crystal molecules align homeotropically. This homeotropic configuration represents the transparent state. The response of the PSCT device is, if not faster, comparable to that of PDLCs, but it requires a significantly lower operating voltage. As shown in Figure 17, the angular dependence of the optical transmittance of the PSCT is much wider relative to the conventional PDLC. Thus the normal mode PSCT appears haze-free at all viewing angles. Such PSCT may be used as light shutters, but its application to a large-area flat-panel display, however, remains a challenge since the support of the flexible ITO-coated film substrates by polymers is not strong. On the other hand, increasing the polymer concentration could result in phase separation and cause scattering of the light.

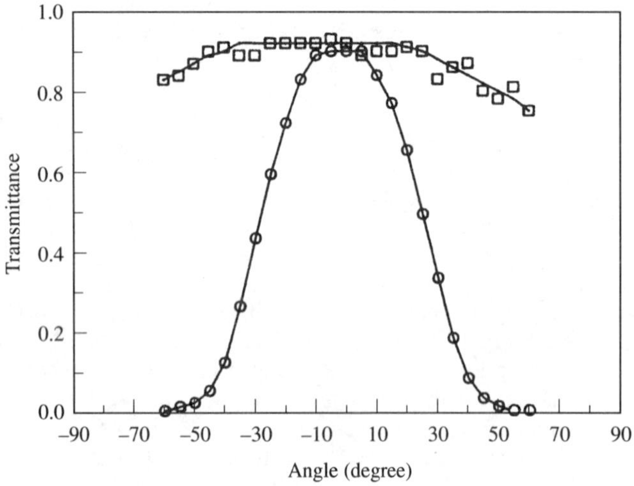

Figure 17 The angular dependence of the transmittance of the normal mode PSCT (square symbol) and of the PDLC (open circle) in their transparent state (after West *et al.*[41]).

Another interesting feature of the polymer-stabilized cholesteric textures (PSCT) is the reverse mode scattering mode cell, *i.e.* the cell is transparent in the off-state, but it becomes turbid in the field on-state.[40,41] The reverse mode PSCT is prepared using the same chiral nematics and other basic materials of the normal mode PSCT except that the glass substrates are treated with polyimide and buffed for planar alignment of the liquid crystals. The chiral pitch is cleverly adjusted to reflect light in the IR range and thus the cell appears transparent in the visible light range.[42] Such texture generally requires optimization of the chiral agent (typically about 7 wt%) and polymer concentrations (less than 5 wt%). In the off-state, such a PSCT texture is in the transparent state because reflection occurs only in the IR range. However, in the on-state, the helical axes of the cholesteric liquid crystals orient randomly similar to the focal conic state and therefore light is scattered. The reverse mode PSCT has a faster response (rise as well as decay times) than that of the normal mode PSCT and a smaller hysterisis. The slow response of the normal mode PSCT is not surprising in view of the fact that the homeotropically aligned polymer networks prevent the cholesteric liquid crystals from relaxing back to their original focal conic configuration. Hence, the reverse-mode PSCT has a greater potential for light valve applications and black-on-white viewing. Moreover, the cholesteric pitch of the PSCT may be controlled by adjusting the concentration of the chiral dopant mixtures such that the reflection occurs in the visible light range. A desired color spectrum corresponding to the controlled cholesteric pitch may be reflected from the PSCT cell. Hence, it may be feasible to devise a color display with the PSCT and this certainly warrants further consideration.

Another innovation which deserves attention is the surface-modified bistable cholesteric liquid crystal reflective display (SMBCD), in which cholesteric textures are aligned in a planar configuration via surface modification.[43] A near perfect planar alignment of cholesteric LC may be achieved by depositing cholesteric liquid crystal molecules on a rubbed polyimide-coated substrate or a poly(methyl methacrylate) coated surface. Photopolymerization is not required in SMBC display devices. In the off-state, the helical axis of the SMBCD is normal to the cell substrate and thus the incident light is reflected at a specular angle. Upon applying a modest external electric field, less than the threshold voltage, the planar alignment of the LC directors changes to the focal conic state. At this state, the PSCT device is opaque as the helical axis orientation orients randomly. The removal of the field reverts the system to its planar configuration through a conic relaxation. When the applied voltage exceeds the critical voltage, the focal conic alignment further transforms to the homeotropic state.

The successful application of the polymer stabilization effect in PSCT has invoked renewed interest in twisted nematics.[44,45] A polymer network (less than 3 wt%) can be aligned homeotropically in twisted nematic (TN) and supertwisted cells (STN) via *in situ* photopolymerization of reactive monomers under an applied electric field. The combined surface and polymer stabilization of the TN and STN cells has been shown to reduce the operational voltage and suppress the undesirable 'stripe' deformation of the STN cells.[45]

ACKNOWLEDGEMENT

The author is indebted to the NSF Science and Technology Center, Advanced Liquid Crystal Optical Materials (ALCOM), Grant No. DMR 89-20147.

16.7 REFERENCES

1. J. Cognard, *Mol. Cryst. Liq. Cryst.*, 1982, **78** (suppl.), 1.
2. E. Kaneko, 'Liquid Crystal TV Displays: Principles and Applications of Liquid Crystal Displays', KTK Scientific Publishers, Boston, MA, 1987.
3. B. Bahadur (ed.), 'Liquid Crystals: Their Application and Uses', World Scientific, Teaneck, NJ, 1990.
4. C. Hilsum, *Br. Pat.*, 1 442 360 (1976).
5. H. G. Craighead, J. Cheng and S. Hackwood, *Appl. Phys. Lett.*, 1982, **40**, 22.
6. J. L. Fergason, *Soc. Inf. Disp., Int. Symp. Digital Technol.*, 1985, **16**, 68; J. L. Fergason, A. Dalisa, S. Lu and P. S. Drzaic, *Soc. Inf. Disp. Dig.*, 1986, **86**, 126.
7. P. S. Drzaic, *J. Appl. Phys.*, 1986, **60**, 2142.
8. J. W. Doane, N. A. P. Vaz, B. G. Wu and S. Zumer, *Appl. Phys. Lett.*, 1986, **48**, 269.
9. J. W. Doane, G. Chidishimo and N. A. P. Vaz, *US Pat.* 4 688 900 (1987).
10. J. W. Doane, A. Golemme, J. L. West, J. B. Whitehead, Jr. and B. G. Wu, *Mol. Cryst. Liq. Cryst.*, 1988, **165**, 511.
11. T. Kajiyama, A. Miyamoto, H. Kikuchi and Y. Morimura, *Chem. Lett.* 1989, 813, 817; *J. Membr. Sci.*, 1982, **11**, 39.
12. A. Miyamoto, H. Kikuchi, Y. Morimura and T. Kajiyama, *New Polym. Mater.*, 1990, **2**, 27.
13. J. L. West in 'Polymer Liquid Crystals', ed R. A. Weiss and C. K. Ober, ACS Symposium Series 435, 1990, p. 475.
14. G. W. Smith and N. A. P. Vaz, *Liq. Cryst.*, 1988, **3**, 543.
15. N. A. P. Vaz, G. W. Smith and G. P. Montgomery, Jr., *Mol. Cryst. Liq. Cryst.*, 1987, **146**, 17.
16. G. W. Smith, *Int. J. Mod. Phys.*, 1993, **B7**, 4187.
17. J. W. Doane in 'Liquid Crystals: Their Applications and Uses', ed. B. Bahadur, World Scientific, Teaneck, NJ, 1990, p. 361.
18. D.-K. Yang, L. C. Chien and J. W. Doane, *Appl. Phys. Lett.*, 1992, **60**, 3102.
19. Y. Hirai, S. Niiyama, H. Kumai and T. Gunjima, *SPIE, Liq. Cryst. Disp. Appl.*, 1990, **2**, 1257.
20. R. L. Humphries and G. R. Luckhurst, *Proc. R. Soc. London*, 1976, **A352**, 41.
21. F. Brochard, J. Jouffroy and P. Levinson, *J. Phys. (Paris)*, 1984, **45**, 1125.
22. P. Palffy-Muhoray and J. J. de Bruyn, *Mol. Cryst. Liq. Cryst.*, 1985, **127**, 301.
23. V. K. Kelkar and C. Manohar, *Mol. Cryst. Liq. Cryst.*, 1986, **133**, 267.
24. P. J. Flory, 'Principles of Polymer Chemistry', Cornell University Press, Ithaca, NY, 1953.
25. W. Maier and A. Saupe, *Z. Naturforsch.*, 1959, **14a**, 882; 1960, **15a**, 287.
26. S. Chandrasekhar, 'Liquid Crystals', 2nd edn., Cambridge University Press, Cambridge, 1992.
27. P. G. de Gennes and J. Prost, 'The Physics of Liquid Crystals', 2nd edn, Clarendon Press, Oxford, 1993.
28. C. S. Shen and T. Kyu, *J. Chem. Phys.*, 1993, **102**, 556.
29. C. S. Shen, Ph.D. Dissertation, University of Akron, 1995.
30. J. H. Erdmann, S. Zumer and J. W. Doane, *Phys. Rev. Lett.*, 1990, **64**, 1907.
31. T. Kyu, I. Ilies and M. Mustafa, *J. Phys. IV*, 1993, **3**, 37.
32. K. Binder and D. Stauffer, *Phys. Rev. Lett.*, 1974, **33**, 1006.
33. E. Siggia, *Phys. Rev.*, 1979, **A20**, 595.
34. J. Y. Kim, C. H. Cho, P. Palffy-Muhoray, M. Mustafa and T. Kyu, *Phys. Rev. Lett.*, 1993, **71**, 2232.
35. C. Casagrande, P. Fabre, M. A. Guedau and M. Veyssie, *Europhys. Lett.*, 1987, **3**, 73.
36. P. S. Drzaic, *Liq. Cryst.*, 1988, **3**, 1543.

37. B. G. Wu, J. L. West and J. W. Doane, *J. Appl. Phys.*, 1987, **62,** 3925; *Mol. Cryst. Liq. Cryst.*, 1989, **5,** 1453.
38. A. Miyamoto, H. Kikuchi, S. Kobayashi, Y. Morimura and T. Kajiyama, *Macromolecules,* 1991, **24,** 3915.
39. P. O. Crooker and D. K. Yang, *Appl. Phys. Lett.*, 1990, **56,** 2529.
40. D. K. Yang and J. W. Doane, *Soc. Inf. Disp. Dig.*, 1990, 759; D. K. Yang, J. L. West, L. C. Chien and J. W. Doane, *J. Appl. Phys.*, 1994, **76,** 1331.
41. D. K. Yang, L. C. Chien and J. W. Doane, *Appl. Phys. Lett.*, 1992, **60,** 3102.
42. G. Solladie and R. G. Zimmerman, *Angew. Chem., Int. Ed. Engl.*, 1984, **23,** 348; *Mol. Cryst. Liq. Cryst.*, 1978, **41,** 109; *Mol. Cryst., Liq. Cryst.*, 1979, **56,** 1.
43. A. Mochizuki and S. Kobayashi, *Mol. Cryst. Liq. Cryst.*, 1983, **225,** 89.
44. E. Shimada and T. Uchida, *Jpn. J. Appl. Phys.*, 1992, **31,** 352.
45. P. J. Bos, J. A. Rahman and J. W. Doane, *Soc. Inf. Displ. Dig. Tech. Pap.*, 1993, **24,** 887.

17
Macromolecule–Metal Complexes

DIETER WÖHRLE and GÜNTER SCHNURPFEIL
Universität Bremen, Germany

17.1 FUNDAMENTALS OF MACROMOLECULE METAL COMPLEXES

In macromolecule metal complexes (MMCS), suitable compounds are combined to give materials with new and unusual properties. MMCs are organic or inorganic high molecular weight compounds incorporating metal/ion complexes/chelates or metal clusters. In order to appreciate the range of properties shown by MMCs, one has only to look at the following functions of natural MMC systems: metalloenzymes for catalysis; hemoglobin, myoglobin for gas transport; cofactors for electron interaction; the means of photosynthesis for energy conversion; metalloproteins and related systems for various functions. Selective properties are realized by special molecular arrangements. These arrangements may involve the primary structure (composition of a MMC), the secondary structure (steric orientation of a MMC unit), the tertiary structure (orientation of the whole MMC), or a quaternary structure (the interaction of different MMCs). The more detailed knowledge of biological MMCs has led in recent years to intensified research activity in the field of synthetic MMCs. MMCs can be classified into three main groups.

A type I MMC consists of a metal complex/chelate or a metal ion bound to the side chain of an organic polymer or on the surface of an inorganic high molecular weight compound. Preparation usually involves the following steps: functional reaction of a polymeric support; binding of the metal complex/chelate or metal ion; purification from unbound reagent (Scheme 1, equation (1)). Another method of synthesis is by polymerization of a vinyl-substituted metal complex (or of a ligand followed by metalation) (Scheme 1, equation (2)).

In a type II MMC, bifunctional or higher functional ligands and metal complexes/chelates are part of a polymer chain or network. Synthesis is carried out in the following steps: preparation of a low molecular weight bi/multifunctional ligand; interaction with a polyvalent metal ion; purification from starting materials (Scheme 1, equation (3)). Another method involves preparation of a low molecular weight bi/multifunctional metal complex/chelate; reaction with another bi/multifunctional reagent; purification from starting materials (Scheme 1, equation (4)). Yet another method is to first prepare a high molecular weight ligand and then interact it with a metal ion.

A type III MMC involves the physical interaction of a metal complex or metal cluster with an inorganic high molecular weight compound. Materials are obtained by different methods which can be exemplified as follows: preparation of a well defined polymer or a polymer with a developed support surface or cavity structure; interaction with a metal complex/chelate or formation of a metal cluster (in the presence of a stabilizing high molecular weight compound) (Scheme 1, (equation 5)).

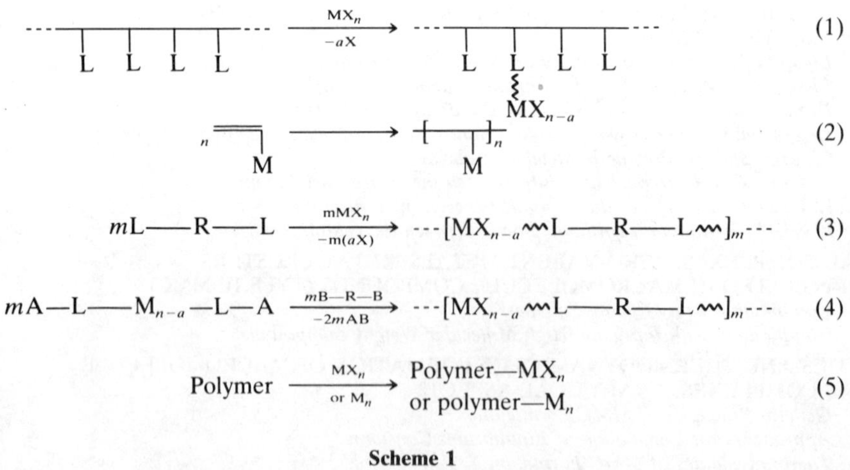

Scheme 1

The synthesis of a macromolecule metal complex needs careful design. Knowledge of the synthesis and, side reactions in the preparation, structure, and behavior of the comparable low molecular weight complex should provide experience for MMC preparation. Depending on the type of MMC, the polymer surroundings cause additional difficulties with structural uniformity and analytical characterization. For type I MMCs, chelation reactions are carried out either in homogeneous or heterogeneous phases which need more specific synthesis conditions. Starting from soluble precursors, type II MMCs are often insoluble and difficult to characterize. Metal complexes in a high molecular environment are often an equilibrium between monomer and aggregates and the stability of the metal cluster is influenced by the kind of polymer surroundings (type III). The nature of the solvent, the pH of the medium, the temperature, the ratio of reagents, and the reaction time have all to be considered in MMC formation. A few characteristic examples for the preparation of type I, II, and III MMCs are given in Sections 17.3–17.5 following a short overview of metal complexes in biological systems and of natural polymers in Section 17.2.

The nature and strength of the M—L bond in MMCs depend, in a similar way to low molecular weight metal complexes, on the ionization potential and electron affinity of the metal ion, and, in addition, on the Lewis base properties of the ligand groups. Kinetic and thermodynamic properties and the structure of the MMC must be analyzed. Compared with low molecular weight metal complexes, the ratio of metal to ligand (amount of loading of the macro ligand with metal ions) is more important for MMCs. Some kinetic and thermodynamic considerations are included in Section 17.6.

In macromolecule metal complexes the primary structure can be based on covalent, coordinate, or ionic bonds. These bonds are of high importance for chemical reactivity and physicochemical properties. The secondary, tertiary, and quaternary structures of MMCs are influenced also by the following kinds of secondary binding forces: anisotropic short-range charge-transfer interactions of medium strength; anisotropic short-range dipole–dipole interactions of weak strength; anisotropic short-range hydrogen bonds of medium strength; isotropic short-range hydrophilic interactions of weak strength; and isotropic long-range coulombic forces of medium strength. These interactions are dynamic in solution, but more fixed in the solid state, for example in a thin film. As a result of these various interactions, unusual characteristics are observed for MMCs. Some properties are reviewed in Section 17.7. Several publications summarize various topics in the area of macromolecule metal complexes: conference volumes on international symposia on macromolecule metal complexes,[1–4] monographs,[5–24] and selected reviews.[25–37]

No clear nomenclature for MMCs exists. Usually a rational nomenclature based on the metal salt/complex and macroligand is used for type I, *e.g.* a nickel(II) chloride complex with poly(4-vinylpyridine), a zinc(II) complex of 5,10,15,20-tetraphenylporphyrin covalently bonded to poly(methacrylic acid). IUPAC nomenclature of 'regular single-strand' and 'quasi single-strand' inorganic and coordination polymers of type II have been described.[38] For type III, names such as 'zinc(II) complex of phthalocyanine in poly(*N*-vinylpyrrolidone)' or 'platinum cluster in the cavities of faujasite X' are used.

The present situation and future aspects can be summarized as follows.

- Various metal complexes or, more recently, metal clusters of different types have been synthesized. Older work does not include detailed analytical characterizations.
- Besides organic polymers, more interest has recently been shown in inorganic high molecular weight carriers, including those with an organized two- or three-dimensional structure.
- Multiple interactions in macromolecule metal complexes exist which have been only basically analyzed.
- Also, only the basic kinetics and thermodynamics of macromolecule metal complex formation are known.
- Very interesting and promising properties have been described. Detailed analytical characterization of an MMC is needed to improve the properties of a newly synthesized material.

17.2 METAL BINDING BY NATURAL MACROMOLECULES (TYPE I MMC)

17.2.1 Metal Complexes in Living Systems

It is well known from bioinorganic chemistry that metal ions play an essential role in living systems.[15–17,39,40] Table 1 shows a selection of some important metals and their *in vivo* functions. Factors of importance for the structure, stability, and function are interactions of the metal ion or complex with directly surrounding neighbors and also quaternary structural arrangements. Beside the specific functions given in Table 1, certain toxic metals may replace nontoxic metal ions and the resulting complexes can be used as drugs (for example Pt compounds). Metal ions such as Ca, Mg, Na, K, and Mn take part in ionic or coordinative interactions whereas Pt, Hg, Cd and Pb usually from covalent bonds; Ni, Cu, and Zn show intermediate behavior.[41] In biological systems, metal ions can coordinate to a variety of bimolecules such as:[15–17,39,40] proteins, in particular to their side chains (N, O, S); nucleic acids, both at phosphates and at base N-donor atoms; carbohydrates, lipids (C—O, P—O); solids (bone, teeth, deposits like kidney stones, magnetite).

Table 1 Overview of Important Elements in Biology[a].

Element name	*Concentration in human species* (mg)	*Daily required* (mg)[b]	*Some comments about role in biology*
Calcium	10^6	800	Bones, teeth, muscle activation
Magnesium	4×10^4	350	Photosynthesis; nucleic acid processes
Iron	4500	10	Many enzymes; respiratory proteins
Zinc	2000	12	Hydrolytic enzymes; nucleic acid synthesis
Copper	100	2	Many enzymes; dioxygen transport
Manganese	20	3	Enzymatic activation, photosynthesis
Molybdenum	5	0.2	Many redox enzymes; nitrogenase (plants)
Cobalt	1	0.3	Vitamin B_{12}
Nickel	<0.1	<0.1	Role in hydrogenases
Vanadium	<0.1	<0.1	Role in bromoperoxidase

[a] After Reedijk.[39] [b] Generally accepted amounts for humans.

Besides the biological functions shown in Table 1, the role of metal ions can sometimes be limited to a structural function, for example holding together certain biomolecules. Examples are Ca^{2+} in thermolysin and Zn^{2+} in superoxide dismutase and alcohol dehydrogenase (for other examples, see Reedijk[15,39]). Living systems which use metal ions in several locations require an efficient means of transport of such metal ions: transferrin for iron transport in humans, albumin for copper transport, ferritin for iron storage. Efficient systems for removing excess toxic metal ions are known, *e.g.* metallothionines for Cd^{2+}, Zn^{2+}, and Cu^{2+}.[40] Examples of metal containing drugs are Au thiolates for arthritis, Bi salts for ulcers, Pt amine complexes as antitumor drugs,[14] Zn and Al complexes of phthalocyanines for photodynamic cancer therapy, and Tc coordination compounds for radiopharmaceutical imaging (see, for example, Kepler[14] and Reedijk[39]).

17.2.2 Metal Complexes with Natural Polymers

Every naturally occurring high molecular weight compound containing suitable donor groups of the N, P, O, or S type can coordinate with metal ions or metal complexes. Another possibility is to use substituents with ligand groups in order to give more efficient metal ion/complex binding.[5,7] Polysaccharides are most widely used for this purpose, especially cellulose (Scheme 2).[5,42] Chitin and its deacetylated derivative, chitosan, are also used for the preparation of macroligands and for metal ion binding.[5,42,43]

17.3 METAL COMPLEXES BOUND TO SYNTHETIC ORGANIC AND INORGANIC HIGH MOLECULAR COMPOUNDS (TYPE I MMC)

As shown in Scheme 1 (equations (1) and (2)), methods for the preparation of metal ions/complexes/chelates bound to a variety of synthetic polymers can be subdivided into two main routes: (a) reaction of a polymer bearing suitable ligand groups for binding metal ions/complexes/chelates; and (b) homo- or copolymerization of a vinyl monomer (or a monomer containing other polymerizable groups) bearing a metal complex/chelate or ligand as a metal complex/chelate precursor. Linear or crosslinked materials can be obtained using both routes. MMCs with a linear backbone are soluble and can be coated onto thin film devices. Crosslinked materials possess, depending on the amount of crosslinking and the method of copolymerization, pores of different type and size with a uniform crosslinked density.[44] One example is amorphous polystyrene crosslinked with divinylbenzene. Examples of nonporous materials are partially crystalline polymers like polyethylene and some inorganic carriers like silica gel. Ligand/metal ion/complex/chelate groups can be distributed in the

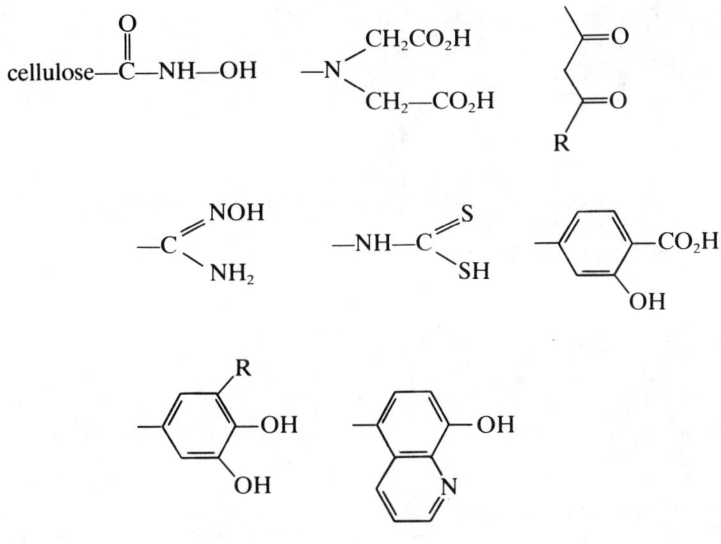

Scheme 2

whole polymer volume or localized only on the carrier surface, and connection to the carrier is possible via a direct bond or a spacer. Each possibility results in different reactivities (properties).[45,46]

17.3.1 Anchoring of Ligands or Metal Complexes at a Polymer Backbone

Various reactions are used for the functionalization of polymers by ligands or metal ions/complexes/chelates. The most employed method uses the immobilization of a ligand capable of metal ion complex binding.[4,5,7,37] Immobilized ligand groups may contain oxygen, nitrogen, sulfur, phosphorus, or arsenic donors. Besides open-chain amines, cyclic amines are also used. Other examples of chelating groups are pyridine-2-aldehyde, iminodiacetic acid, 8-hydroxyquinoline, hydroxylamine, bipyridyl, Schiff bases, Mannich bases, and porphyrin-type macrocycles. Most often chloromethylated polystyrene, either linear or with different degrees of crosslinking, is used as starting material. Scheme 3 shows some selected examples of ligands. Water-soluble polymers with chelate properties are formed by derivatization of linear polymers such as polyethyleneimine, polyvinylamine, poly(methacrylic acid), poly(acrylic acid), and poly(*N*-vinylpyrrolidone).[4,5,7,47–50] Other typical ligands are derived from phosphorus compounds like phosphines or phosphates on modified polystyrene for transition metal ion binding.[5,7,48] One example is binding of $Pd(PPh_3)Cl_2$ or $Rh(H)(PPh_3)(CO)$ on diphenylphosphinated polyethylene, $Bu(CH_2CH_2)_nPPh_2$, obtained by polymerization of ethylene with BuLi and quenching with Ph_2PCl.[49] An interesting approach using memory effects of three-dimensional polymers with catalytically active centers has been repoted.[50] Crosslinked polymers bearing phosphorylic, carboxylic, pyridine, amine, and imine functions have been used for the binding of Cu^{2+}, Ni^{2+}, and Co^{2+}. In the following, only some selected examples are shown. For the well-known metal ion binding at polycarboxylic acids, polyalcohols, polyamines, and polyvinylpyridines, see Ciardelli *et al.*[5] or Pomogailo and Uflyand.[7]

17.3.1.1 Diketones and keto carboxylic acids

Binding of metal acetylacetonates is achieved by the reaction of crosslinked chloromethylated polystyrene (DMF, 100 °C) with the formation of (**1**).[51] Rare earth (Eu^{III}) complexes of polystyrene-bound 1-carboxy-8-naphthoyl (**2**) are obtained by Friedel–Crafts acylation of the

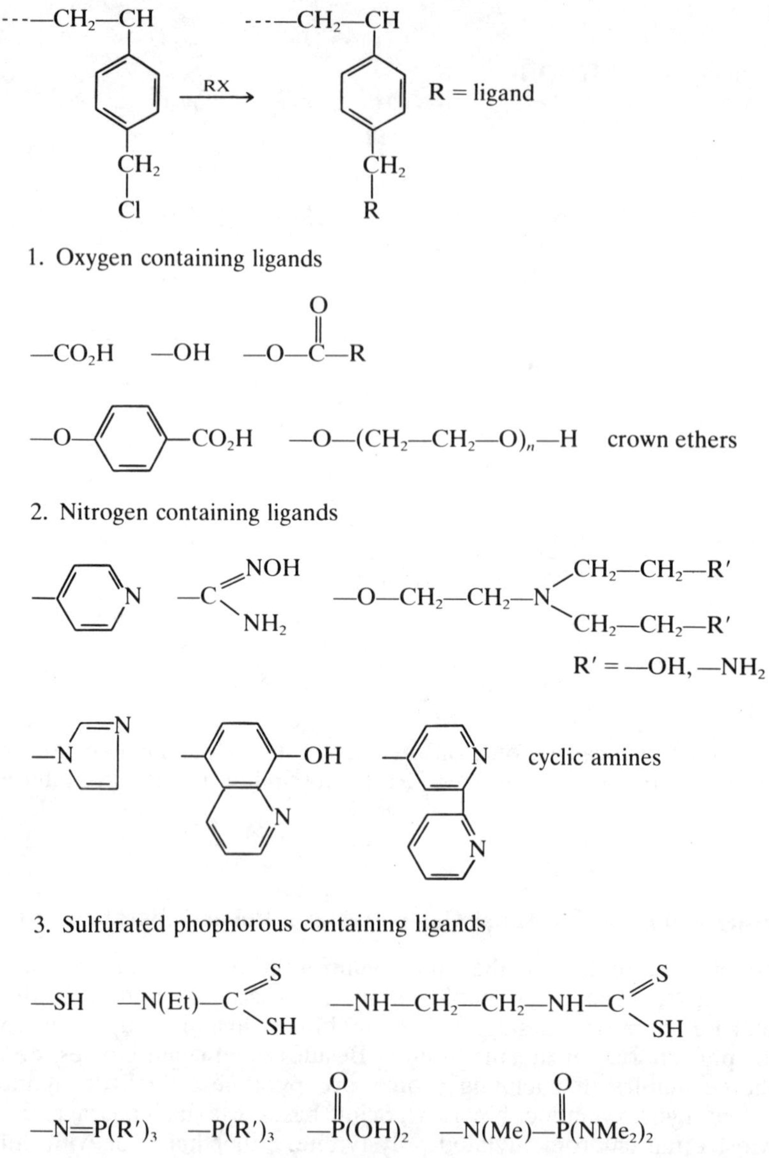

Scheme 3

corresponding naphthalenedicarboxylic anhydride with the polymer.[52] The luminescence properties of lanthanide ions with polycarboxylates have been investigated in detail.[53] The effects of the conformation of polymer chains on the electron transfer and luminescence behavior of Co^{II},Co^{III}-ethylenediamine complexes at polycarboxylates have been studied.[54]

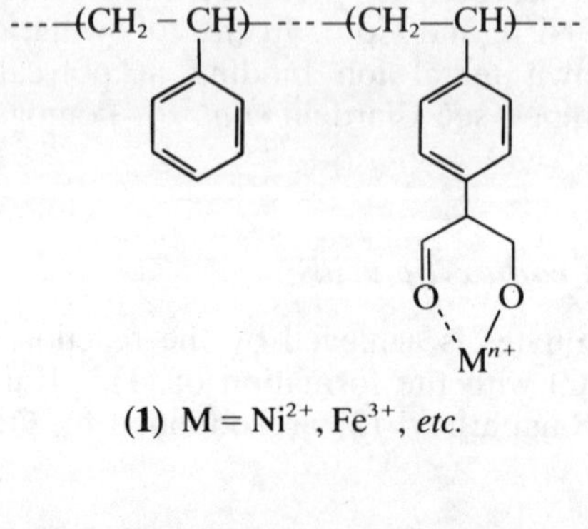

(1) $M = Ni^{2+}, Fe^{3+}$, *etc.*

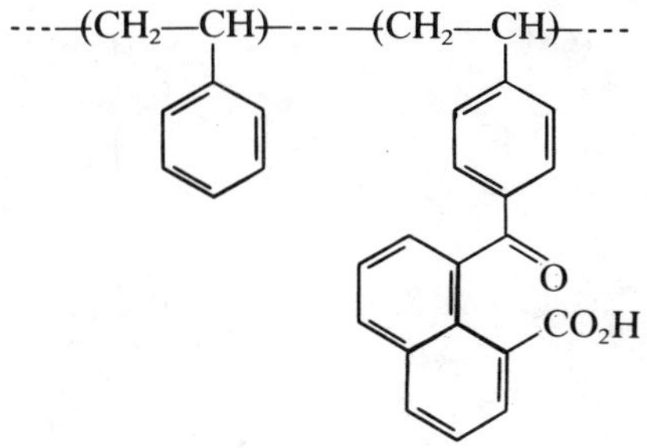

(2)

17.3.1.2 Ethers

Poly(oxyethylene)–metal salt complexes are of interest as solid polymer electrolytes. Complexes with Li^+, Na^+, K^+, Mg^{2+}, and Ba^{2+} have been investigated.[5,7,37,55] Preparation is carried out by direct interaction of the components in solution or, if crosslinked poly(oxyethylene) is used, by soaking in a solution of the metal salt. Poly(oxypropylene), poly(oxyethylene), and crosslinked phosphate esters and ethers have been used as the polymers.[56,57] Crown ether moieties on crosslinked polystyrene can be obtained by the reaction of the crosslinked chloromethylated polystyrene with hydroxy-substituted crown ethers (in THF in the presence of NaH).[58] The binding of alkali ions has been investigated. Crown ether moieties containing cinnamoyl groups (**3**), which can be photocrosslinked, are prepared by polymerization of the corresponding vinyl monomer containing cinnamoyl and crown ether groups.[59]

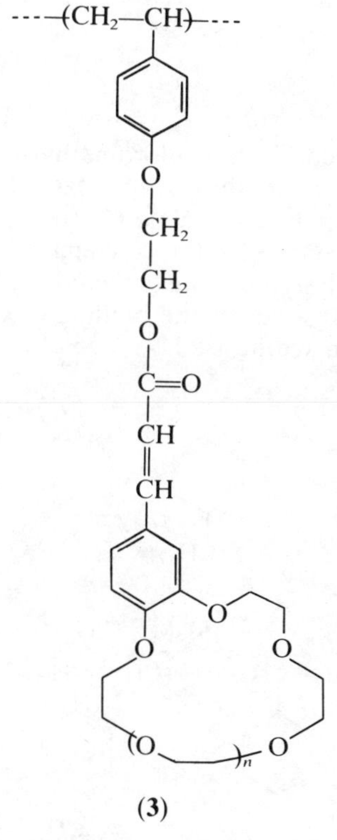

(3)

17.3.1.3 Amines

Open-chain and cyclic amines can coordinate with various metal ions. Poly(ethyleneimine), obtained from 2-methyloxazoline by ring-opening polymerization, has been investigated for Na^+ binding.[60] Various open-chain amines and amides and cyclic amines (**4**) and amides have been synthesized starting from crosslinked chloromethylated polystyrene.[61] The modified polymers contain up to 2.7 mmol g^{-1} of amine or amide groups. They have been investigated for the reversible binding of Co^{2+}, Ni^{2+} and Cu^{2+}. A crosslinked polystyrene with 2-amidooxime groups (**5**), prepared from crosslinked chloromethylated polystyrene by cyanoethylation and reaction with hydroxylamine, shows a good selectivity for the separation of UO_2^{2+} from seawater.[62] Amide-oxime polymers were also prepared from macroporous acrylonitrile–divinylbenzene copolymers by reaction with NH_2OH (around 2 mmol g^{-1} amide-oxine groups in the polymer);[63] their interaction with Cu^{2+} was studied.

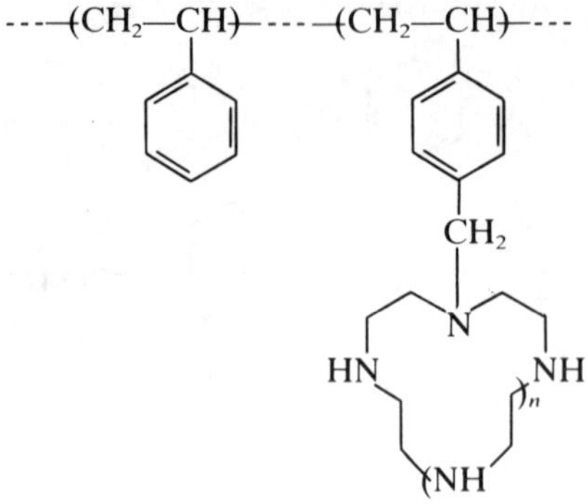

(4)

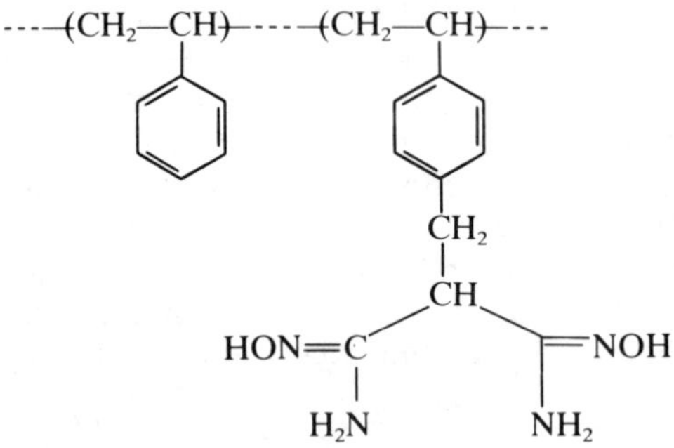

(5)

17.3.1.4 *Schiff bases*

Reaction of crosslinked polystyrene with 5-chloromethyl-2-hydroxybenzaldehyde followed by interaction with the Co^{II} chelate of the Schiff base from 2-hydroxybenzaldehyde and diaminomaleonitrile results in the polymer chelate (**6**) (content 0.2 mmol g^{-1} chelate centers).[64] This MMC was investigated as a catalyst for the conversion of quadricyclane to norbornadiene. Crosslinked chloromethylated polystyrene was reacted with N_2O_3 Schiff base ligands. The resulting macroligands were investigated for the binding of Co^{2+}, Mn^{2+}, and Fe^{2+} (**7**).[65] Cyclic Schiff base chelates have also been synthesized.[66]

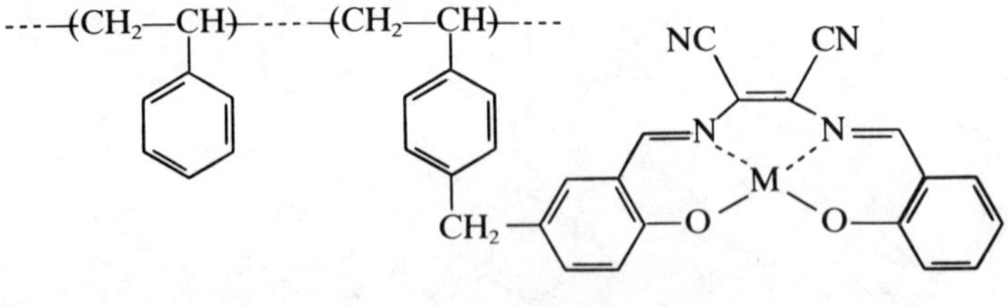

(**6**) $M = Co^{2+}$

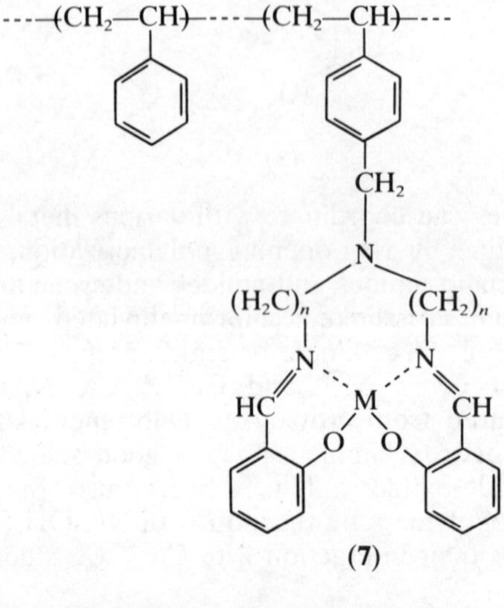

(**7**)

17.3.1.5 Pyridyl, bipyridyl, and other heterocycles

The excellent complexing ability of the pyridine group has led to many papers on the coordination of polymers bearing pyridyl or bipyridyl groups with metal ions like Ru^{2+}, Re^{2+}, and Co^{2+}.[5,7,37,67–69] Polymers and copolymers of vinylpyridine or *N*-vinylimidazole can easily interact by coordinative bonds in solution with a variety of transition metal salts, metal complexes, and macrocyclic metal chelates such as Schiff base chelates of the Co(salen) type, Co(dimethylglyoxime), or porphyrins like 5,10,15,20-tetraphenylporphyrin.[5,6,69–74] After film casting, the binding of oxygen and its separation in membranes was studied. For the coordinative interaction, by analogy to coordinative binding in low molecular weight complexes, the polymer must have groups with σ-donor or π-acceptor properties. In contrast to the monoaxial coordination of low molecular weight donors with Co complexes, the polymer donors can interact biaxially, resulting in crosslinking, a change of polymer conformation, and, therefore, different properties. Ru^{II}, Cu^{II}, and Cr^{III} complexes with 2,2′-bipyridyl and poly(4-vinylpyridine) (PVP) have been briefly reviewed.[5,7,37] *cis*-Ru^{II}(2,2′-bipyridyl)$_2^{2+}$ ($Ru(bpy)_2^{2+}$) reacts in methanol with PVP to give $[Ru(bpy)_2(PVP)_2]^{2+}$ and with PVP in the presence of pyridine (py) to give $[Ru(bpy)_2(PVP)(py)]^{2+}$.[75] A polymer complex containing $Ru(bpy)_3^{2+}$ pendant groups was prepared by the reaction of a lithium-substituted polystyrene with 2,2′-bipyridyl followed by interaction with *cis*-$Ru(bpy)_2^{2+}$.[76] Another example is the binding of 4,4′-dicarboxy-2,2′-bipyridyl to a copolymer of *p*-aminostyrene followed by reaction with *cis*-$Ru(bpy)_2^{2+}$ (**8**).[77] Other copolymers with pendant $Ru(bpy)_3^{2+}$ bound via a spacer or containing additional bound 4,4′-bipyridyl have been described. These materials are interesting as sensitizers for visible light energy conversion.

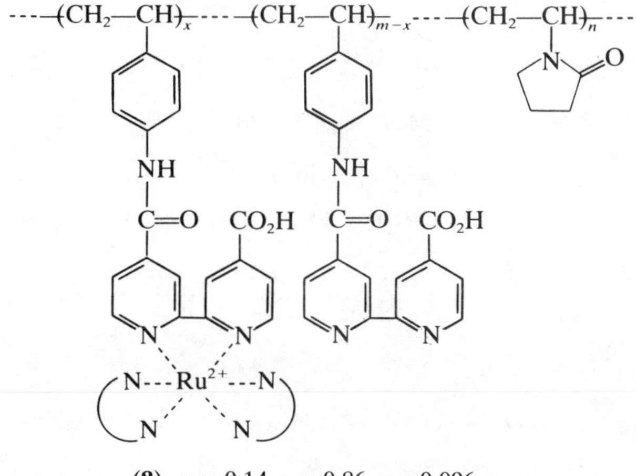

(**8**) $m = 0.14, n = 0.86, x = 0.096$

Polybenzimidazoles bearing cyanomethyl ligands were coordinated with $PdCl_2$, partly with $CuCl_2$ as cocomponent, and investigated for catalytic activity.[78] Several Pd^0 heteroarylene complexes were prepared by interaction of the polyheterocycles with $PdCl_2$ followed by reduction with $NaBH_4$ to Pd^0.[79] The catalytic activity was investigated.

17.3.1.6 Porphyrins, phthalocyanines, and related macrocycles

Tetraaminophenylporphyrin, tetraamino- and tetrahydroxyphenylphthalocyanine, and tetraaminonaphthalocyanine (for example (**9**)) were covalently bound to chloromethylated polystyrene, poly(methacrylic acid), or a copolymer of methacrylic acid and *N*-vinylpyrrolidone to obtain water-soluble positively, negatively, and uncharged polymers (**10a–c**) containing these macrocyclic moieties (loading of 0.003–0.06 mmol g^{-1} = 0.2–5.3 wt.%).[80] Crosslinking by the tetrasubstitutued porphyrins is avoided by using <2 mol% of them to 100 mol% of the reactive group of the polymer. These polymer porphyrins were investigated for electron- and photoinduced electron transfer.[80,81] Other examples of the covalent binding of different porphyrins and phthalocyanines have been reviewed.[25,31,34,82] The tetraacid chloride of Fe^{III}, Co^{III}, Cu^{II}, or Ni^{II} phthalocyanine has been bound to polystyrene or its copolymer with 2-vinylpyridine or to poly(γ-benzyl-L-glutamate).[83] Binding of porphyrins to methoxy-poly(oxyethylene) for use as photodrugs in the photodynamic therapy of cancer has been described.[84] The synthesis of covalent polymer-bound phthalocyanines by the reaction of polymer-bound dinitriles with low molecular weight dinitriles has been reviewed.[31]

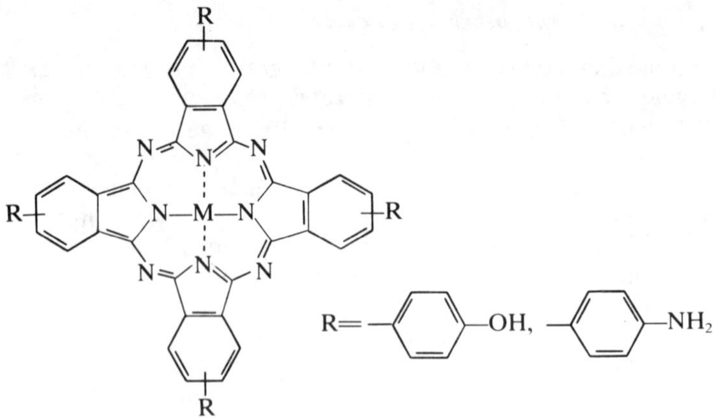

(**9**) $M = Zn^{2+}$

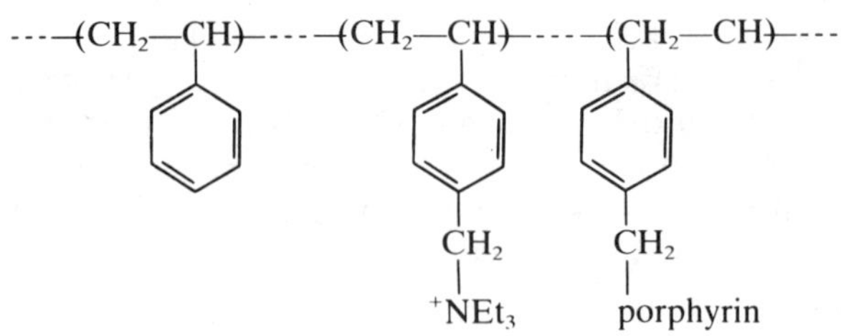

(**10a**)

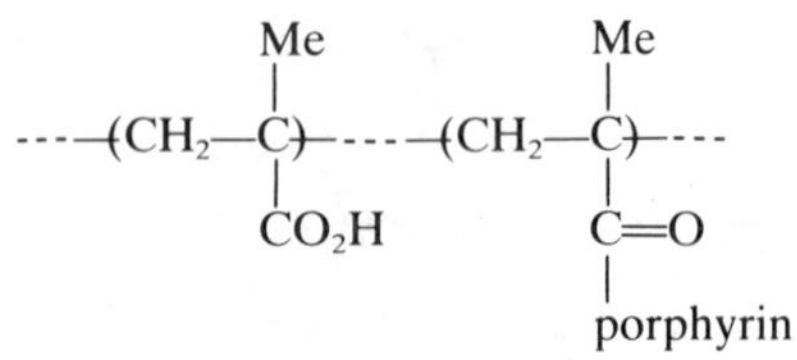

(**10b**)

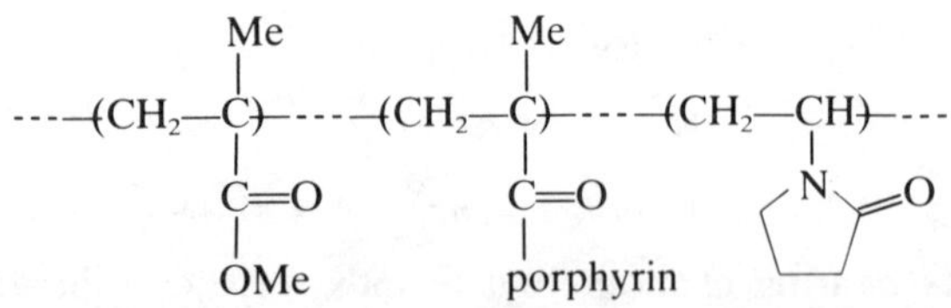

(**10c**)

17.3.2 Polymerization Reactions of Metal Complexes as Monomers

Monomers of metal complexes/chelates suitable for polycondensations or polyaddition reactions can be employed successfully for the preparation of MMCs. In most cases, the polymerizations are carried out in the presence of a comonomer to obtain polymers with sufficient solubility. For polymerizations either a vinyl-containing metal complex/chelate is directly converted into the MMC. In some cases, when chain transfer due to a transition metal ion in the core of a ligand occurs, polymerization of the ligand is preferable. The literature on this subject has been reviewed.[5,7,25,34] A few examples are given here.

17.3.2.1 Cyclic amines

Azacrown ether metal ion complexes were synthesized by copolymerization of vinyl azacrown ethers with styrene, acrylic acid, methacrylic acid, and *N*-vinylpyrrolidone.[85] The linear copolymers are soluble in some organic solvents or water. The coordination properties with transition metal ions were studied by UV/vis spectroscopy and ESR. Some of the ligands (*e.g.* (**11**)) exhibit high selectivity for Au ion binding.

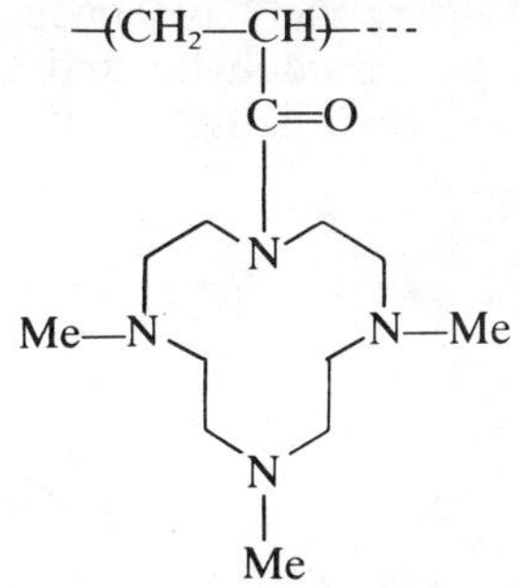

(**11**)

17.3.2.2 Bipyridyl

4-Methyl-4′-vinyl-2,2′-bipyridyl was copolymerized with styrene and then treated with *cis*-$Ru(bpy)_2Cl_2$ to form pendant $Ru(bpy)_3^{2+}$ (**12**).[5,7,33,67,68] In order to study ionic domains around the Ru complex, copolymers with acrylic acid were also prepared. Photophysical properties and photoinduced electron transfer were investigated, both in solution and as thin films. Photoluminescence properties were also studied for polysiloxane-pendant $Ru(bpy)_3^{2+}$ prepared from the corresponding substituted dihydroxysilanes.[86] Reductive electropolymerization of Fe^{II} and Ru^{II} complexes containing 2,2′-bipyridyl and other ligands in acetonitrile in the presence of Et_4NClO_4 (as electrolyte) on Pt results on films of <1 μm thickness.[87]

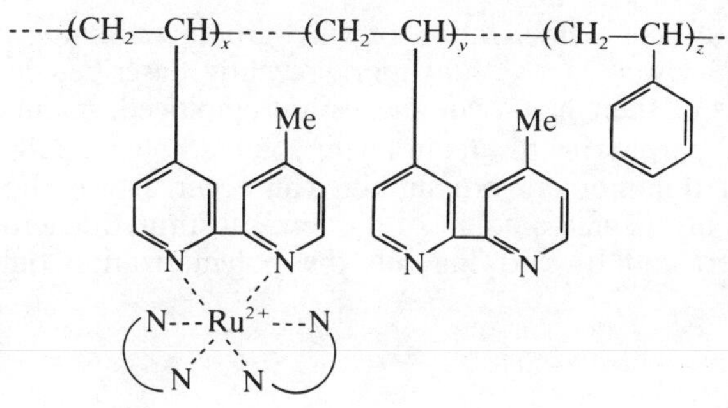

(**12**) $x = 0.062$, $y = 0.004$, $z = 0.934$

17.3.2.3 Metallocenes

Various strategies for the synthesis of metallocene monomers have been reported:[88] direct substitution of a cyclopentadienyl-metal compound with an alkene side chain; functionalization of a cyclopentadienyl with an unsaturated side chain followed by metallation; synthesis of an appropriately substituted cyclopentadienyl via ring closure followed by metalation. Examples of monomers, including (vinylcyclopentadienyl)tricarbonylmanganese, are shown in (**13a**) and (**13b**).[88,89] Copolymerizations with different comonomers were carried out. Binding of N_2 (at Mn complexes) was described.

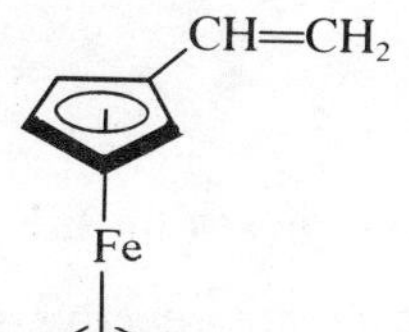

(**13a**)

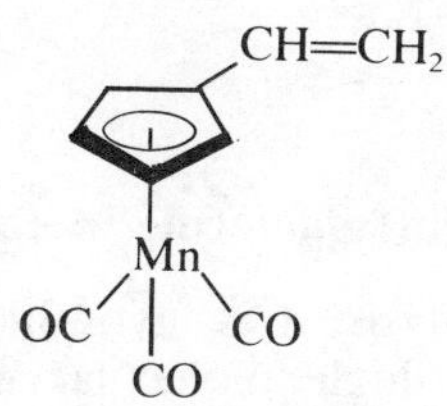

(**13b**)

17.3.2.4 Schiff base chelates

Polymeric cobalt(II) Schiff base chelates based on the radical copolymerization of 4,4′-divinylsalens with styrene or 4-vinylpyridine have been described.[70,90,91] One example of a polymer employing 2-hydroxy-5-vinylbenzaldehyde and 1,2-diaminoethane is shown in (**14**). The Co^{II} chelates were investigated for their activity in the binding of oxygen and for activation of the oxidation of 2,6-di-*t*-butylphenol.

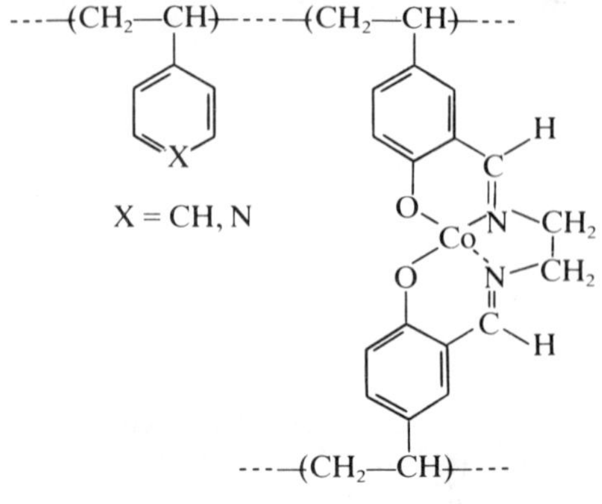

(**14**)

17.3.2.5 Porphyrins, phthalocyanines, and related macrocycles

The polymerization and copolymerization of vinyl-group-containing porphyrins and related compounds has been reviewed[25,31,34,92] and more recently described in detail.[93–98] Either the vinyl-substituted ligands or their metal complexes are employed, mainly in copolymerizations. Use of monosubstituted porphyrins results in linear soluble polymers like (**15**),[92,94,95,98] whereas with tetravinyl-substituted monomers, crosslinking can occur. It was shown that crosslinking is avoided by employing only a small amount of the tetrasubstituted macrocycle in comparison to the styrene comonomer, and by carrying out the polymerization only up to around 10% conversion.[97]

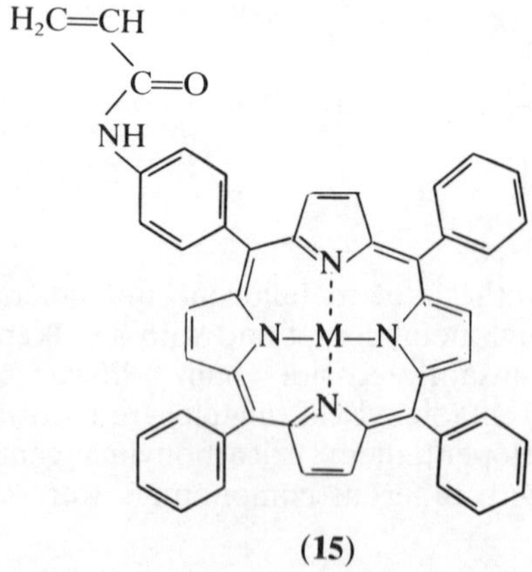

(**15**)

17.3.3 Immobilization of Macromolecule Metal Complexes on the Surface of Carriers

For purposes such as catalysis, it is favorable to create reactive sites on the surface of an organic polymer or on a high molecular inorganic carrier. This can be realized by polymerization of different monomers bearing ligand groups L for metal ion/complex or chelate binding (for example, by mechanical, chemical, or irradiation treatment of the carriers (equation 6)).[5,7]

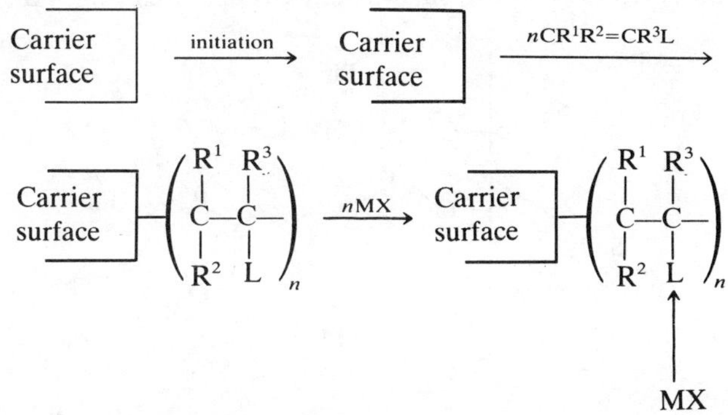

(6)

Gas-phase grafting is achieved by polymerization initiated by irradiation (γ-irradiation), accelerated electrons, or a low-pressure gas discharge.[5,7] Other possibilities use the immobilization of a ligand group capable of coordinative interaction and metal ion/complex/chelate binding or the direct covalent binding of a metal complex/chelate (equation 7).[5,7,99,100]

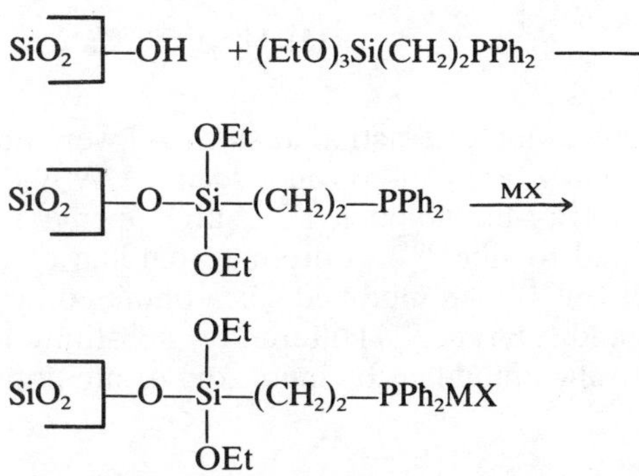

(7)

Such anchoring of MMCs allows the advantage of higher reaction rates for reactions at the metal complex centers and ease of separation from the reaction mixture for reuse. A few papers describe the grafting of polymers containing bound metal complexes onto the surface of organic polymers: polyethylene-*graft*-poly(methyl vinyl ketone)/ Schiff base with 2-aminophenol (**16**)[7,33,101] or salicylaldehyde hydrazide,[102] polyethylene-*graft*-poly(vinyl 1,3-diketone)[103] and polytetrafluoroethylene-*graft*-poly(acrylate) complexes with 2,2′-bipyridyl or 1,10-phenanthroline.[103]

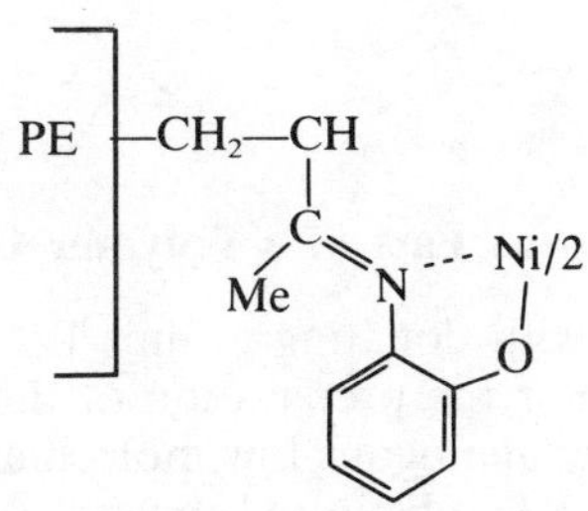

(**16**)

The immobilization of metal complexes or chelates on high molecular inorganic compounds has been investigated more frequently. The covalent binding has been described and reviewed.[5,104] Some examples are: the reactions of Cp_2ZrMe_2 (Cp = cyclopentadienyl) or diorgano-$ZrCl_2$ with silica gel and alumina (used, after treatment with $AlMe_3$, as catalysts for alkene polymerization); dichlorotitanium pyrocatecholate with silica gel; binding of a nickel P/O chelate at silica gel modified with tetrabenzyltitanium followed by binding of a nickel P/O chelate (**17**); and preparation of alumina-supported bis(arene)-Ti and tetra(neopentyl)-Zr.[105] The interest in this work is obtaining heterogeneous catalysts for alkene polymerization. Covalent binding of Co^{III}-tetraamino- and tetracarboxyphthalocyanine on silica gel of different surface areas (loading 25–1 μmol g^{-1}) was realized by the reaction of chloropropylsilyl or aminopropyl modified carriers with the substituted Co chelates (e.g. (**18**)).[106]

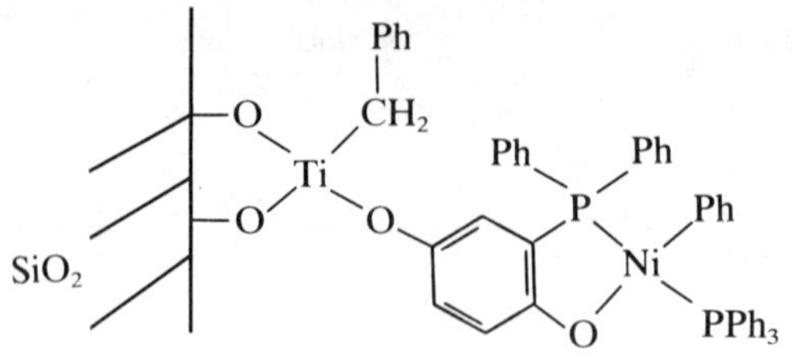

(**17**)

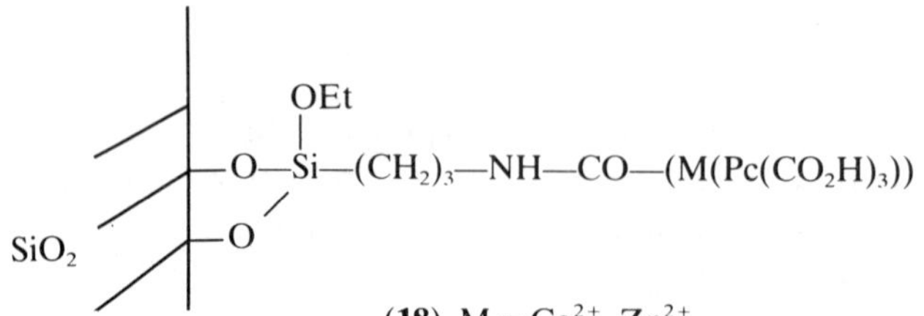

(**18**) M = Co^{2+}, Zn^{2+}

Ligands for transition metal ion interaction at silica gel were obtained by covalent connection of trialkoxysilanes containing a ligand group such as *N,N*-dimethyamino,[107] silica-grafted polymer-bound 3,3′,5,5′-tetra-*t*-butylbiphenyl-2,2′-diyl phenyl phosphite,[48] and trimethylene-phosphine covalently linked to silica.[108] Coordinate binding of Cu^{2+} by imidazole groups was realized by interaction of $CuCl_2$ with modified silica obtained by radical graft copolymerization with *N*-vinylimidazole and styrene.[70] Differently substituted Co^{II} phthalocyanines were coordinatively bonded to silica modified by imidazoyl or aminopropyl groups.[109]

17.4 METAL COMPLEXES AND METALS AS PART OF A POLYMER CHAIN OR NETWORK (TYPE II MMC)

The classical polymer backbone or network consists of C, N, O, S, P, and Si. However, most other elements in the Periodic Table, except the noble gases and the 12 elements of Groups 1 and 17 (including H), which are monovalent, can in principle act as chain/network forming or participating elements. In the following, a few characteristic examples are selected. More examples appear in reviews.[5,25,30,110]

17.4.1 Ligand of a Metal Complex as Part of a Polymer Chain or Network

In this case, a polymeric ligand with donating groups L binds a metal ion M similar to low molecular weight ligands (**19a, b**). For the preparation of these MMCs it is important to know the complexing characteristics of the analogous low molecular weight ligand. Two routes can be used: (i) reaction of a bifunctional substituted ligand (or bifunctional substituted metal complexes) with another bifunctional comonomer; and (ii) reaction of a bifunctional ligand or ligand precursor with a multivalent metal ion.

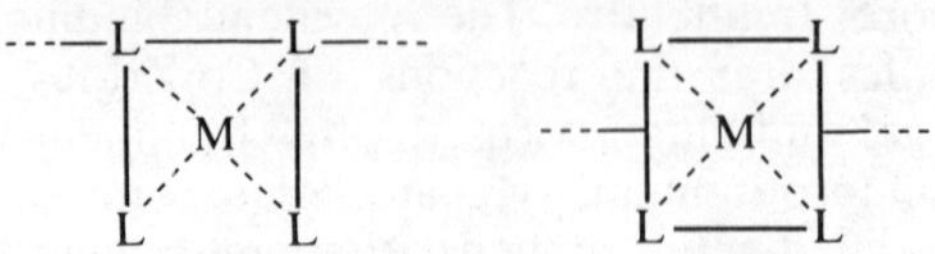

(**19a**) (**19b**)

Polymeric Schiff base chelates of the Co(salen) type are prepared, for example, by the reaction of a dihydroxy-substituted Schiff base ligand with a biscarboxylic acid derivative or by the reaction of bis(*o*-hydroxy aldehyde) with a diamine (for example (**20**)).[111–113] Thermotropic liquid crystalline polymers containing β-diketonato groups ($-RCOCH_2COR$) capable of metal ion binding (Cu^{2+}, Ni^{2+}) in the main chain of polyesters or in the side chain of polyacrylates have been described.[114] Binding of Pt^{4+} at poly(ethyleneimine) and other polymers containing 1,2-diimino moieties in the polymer chain were prepared for use in the chemotherapy of cancer.[115]

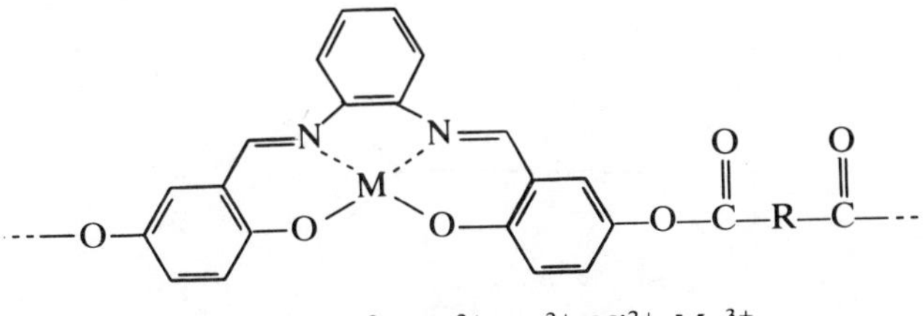

(**20**) $M = Co^{2+}, Cu^{2+}, Zn^{2+}, Ni^{2+}, Mn^{3+}$

Porphyrins, phthalocyanines, hemiporphyrazines, and tetraza[14]annulenes were investigated as cyclic ligands for polymer metal chelates.[25,31] Following the principles explained earlier, either tetrafunctional macrocycles can react with another bifunctional reagent or multifunctional ligand precursors are employed. One example for the first possibility is the reaction of 2,9,16,23-tetraaminophthalocyanine (in the presence of another diamine) with benzenetetracarboxylic dianhydride (or a similar tetracarboxylic acid derivative) at first to give polyamic acids in a solvent and then, by thermal treatment, to give films of polyimide derivatives.[116] The interfacial polycondensation technique to react tetraamino- or tetrahydroxyphenylporphyrins with tetrachlorocarbonylphenylporphyrins was used for thin film preparation.[117]

Bifunctional ligands or chelate precursors must be used to prepare polymeric macrocycles with, in the ideal case, a two-dimensional plane structure. Examples are polymeric phthalocyanines (**21**),[25,31,118–120] hemiporphyrazines,[25,31] and tetraaza[14]annulenes.[25,31,121] Different transition metal ions have been used. The formation of structurally uniform polymeric Cu^{II} phthalocyanines (see also Section 17.6.4[122]) with different molecular weights (up to ∞) was studied in detail, employing various tetracarbonitriles (equations 8 and 9).

$$2n\ C_{10}H_2N_4 + n\ Cu^0 \rightarrow [(C_{20}H_4N_8)^{2-}Cu^{2+}]_n \quad (8)$$

$$2n\ C_{10}H_2N_4 + n\ CuCl_2 \rightarrow [(C_{20}H_4N_8)^{2-}Cu^{2+}]_n + n\ Cl_2 \quad (9)$$

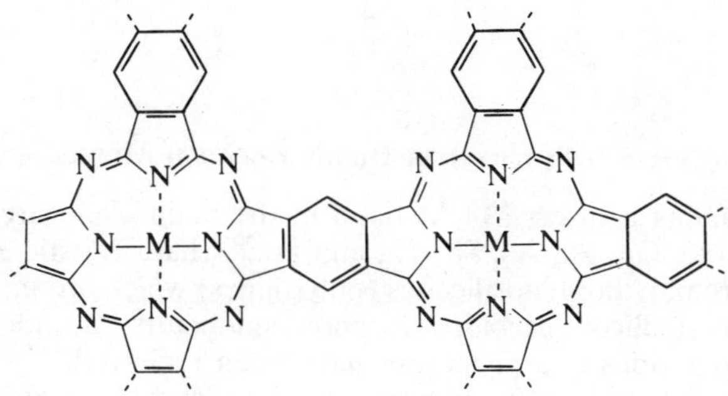

(**21**) $M = Cu^{2+}, Co^{2+}, Ni^{2+}$

Thin films of polymeric phthalocyanines were obtained by *in situ* synthesis from tetracarbonitriles on Co^0 or Ti^0 surfaces,[123] from porphyrins by the above-mentioned interfacial polycondensation,[117] by anodic electropolymerization of tetraaminophenyl-, tetrahydroxyphenyl-, or tetrapyrrolylphenylporphyrin or tetraaminophenylphthalocyanines,[124–126] or by plasma polymerization of evaporated low molecular weight phthalocyanines.[127] A few examples are known for the synthesis of one-dimensional ladder-type structured polymeric macrocycles. Bifunctional Diels–Alder reactions are used to prepare oligomeric ladder phthalocyanines and hemiporphyrazines.[128]

17.4.2 Ligand and Metal as Part of a Polymer Chain or Network

In this case, a bivalent ligand L and a metal ion M are responsible for the polymer backbone (**22**). Either inorganic or organic coordination polymers can be obtained.

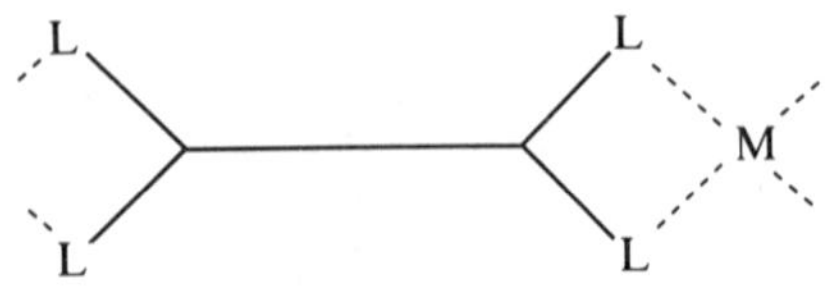

(22)

Examples of inorganic coordination polymers are AuI, $PdCl_2$, MoI_3 and AuCN (—C≡N→Au—C≡N→Au—) in the solid state.[8] Prussian blue ($Fe_4[Fe(CN)_6]_3 \cdot H_2O$) exhibits a three-dimensional network structure.[129] This material is interesting as a photosensing device, a rechargeable battery, and an electrochromic display material.[130]

Several examples of organic coordination polymers are known and have been reviewed.[5,30] Chelating groups are =O, =S, =NR, —NR_2, —O^-, —S^-, and NR^-. Most of these polymers are insoluble and it is difficult to characterize them analytically. Many of them exhibit high electronic conductivities as compressed powders. Examples which are reviewed in the above-mentioned literature are: (i) polymers from reaction of dicarboxylic acids with $SnCl_2$ or uranyl salts; (ii) polymers of Cu^{2+}, Ni^{2+}, Co^{2+}, Zn^{2+}, Cd^{2+}, Mo^{2+}, Mn^{2+}, VO^{2+}, and UO_2^{2+} with bis(*o*-hydroxy acids), *o,o*′-dihydroxybenzoquinones, -naphthoquinones, and -anthraquinones, 2,4-dihydroxybenzaldehyde, and formaldehyde; (iii) coordination polymers from dihydroxyquinolines, bis(anthranilic acid), and tetrahydroxyazomethines; (iv) MMCs with mixed N,S-ligands from rubeanic acid[131] (Cu complexes also as thin films on Cu electrodes by electrochemical polymerization[132]), 2,5-diamino-1,4-benzenethiol,[133] and butanethiooxamide;[134] and (v) coordination polymers of Fe^{2+}, Co^{2+}, Cu^{2+}, Pd^{2+}, and Pt^{2+} with tetrathiooxalate (**23**),[135] tetrathiosquarate, tetrathiofulvalenetetrathiolate, and benzene- and naphthalene-tetrathiolate.

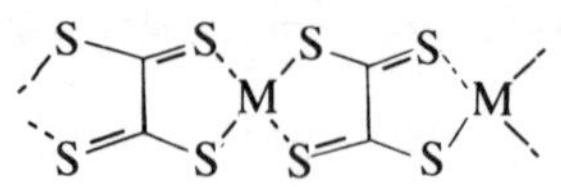

(23)

17.4.3 Homochain Polymers with Covalent Bonds Between Metals

Polymers (**24**) containing a direct M—M bond in the main chain are common with metals and semimetals like B, Si, Ge, Sn, As, Sb, Th, and Po.[30] Characteristic examples are shown in (**25**). Most examples are polydiorganosilicones (polysilanes) which are investigated for potential use as precursors for β-silicon carbide, as photoconductors, in nonlinear optics, and in microlithography. Preparations and properties have been reviewed.[30,136,137] A general method of preparation is a Wurtz-type reaction between dihalides and Na (equation 10).

$$n\ \text{hal—M(R}_x\text{)—hal} + 2n\ \text{A} \rightarrow \text{—[—M(R)}_x\text{—]}_n\text{—} + 2n\ \text{Ahal} \qquad (10)$$

R = alkyl, aryl ($x = 1, 2$); A = alkali metal

----M—M—M----

(24)

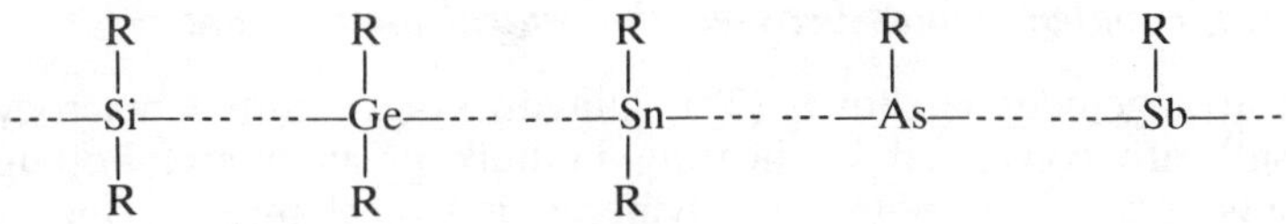

(25)

17.4.4 Heterochain Polymers with Covalent Bonds Between Metals and Another Element

The polymer backbone consists of a polar bond between a metal and another group, as shown in (**26**).

---M—X—M—X—M—X---

(26)

The most widely used method of preparation is by the reaction of bifunctional metal halides (mainly Cl) with a bifunctional Lewis base (diols, diamines, dihydrazines, dihydrazides, dithiohydrazines, dioximes, diamidoximes, dithiols, diacetylenes). Another method is by the polyaddition of a bifunctional metal hydride to bifunctional alkenes. These methods have been reviewed.[10,12,30,138] There is increasing interest in polymers with transition metals in the main chain, mainly σ-bonded to carbon atoms. CO, PR_3, or bulky organic substituents as ligands at the metal stabilize the M—C bond (e.g. (**27**)).[139] Well-used methods for the preparation of $[\text{—M—C}\equiv\text{C—Y—C}\equiv\text{C—}]_n$ are the Cu^I halide-catalyzed dehydrogenation of α,ω-bisethynyls, the oxidative coupling of bisethynyls, and alkynyl ligand exchange reactions.[5,30,140,141] Lyotropic nematic mesophases have been found.[142] Unit cell parameters and crystallite sizes have been determined.[142,143] Electronic spectra and luminescence show that electron conjugation is expanded over the whole polymer chain.[144] In addition, polymers with iron and two metals or semimetals (Pt, Si) in the main chain have been described.[145,146] Reactions of diacetylenes with $CpCo(PPh_3)_2$ leads to polyarylene cobalt–cyclopentadienylenes.[147]

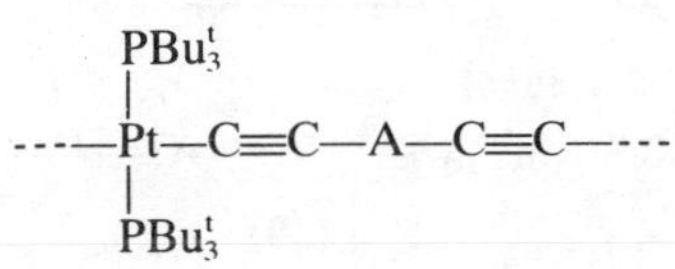

(27)

17.4.5 Cofacial Stacked Polymeric Metal Complexes

A metal ion M as part of a polymer chain is surrounded by a multivalent ligand. As a result, stacked arrangements (**28**) are realized with a face-to-face orientation of the ligands. Porphyrins, phthalocyanines, naphthalocyanines, etioporphyrin, tetraaza[14]annulene, hemiporphyrazines, and related macrocycles as ligands have been most intensively investigated. Some reviews summarize preparations and properties (conductivity, photoconductivity, NLO, electroluminiscent detectors) of these materials.[27,30,31,140,148–150] Several years ago, unsubstituted macrocycles were employed as ligands. These polymers are much less soluble. Recently, the use of substituted macrocycles has led to soluble polymers which are analytically easier to characterize. The self-organization in discotics of octasubstituted macrocycles giving liquid crystalline behavior has been described.[5,27,149]

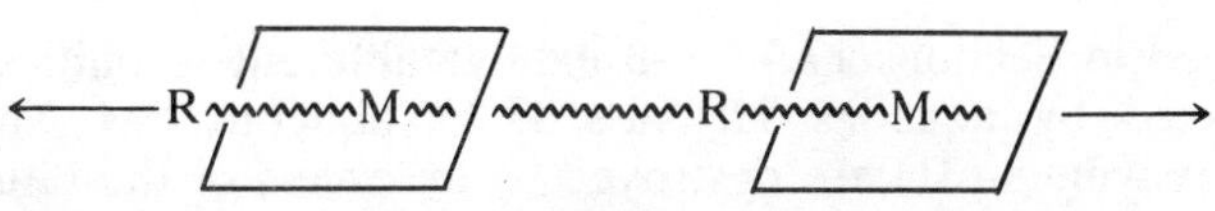

(28)

17.4.5.1 *Covalent/covalent bonds between the central metal atom*

The structural arrangement is shown (**29**). Dihydroxy-substituted macrocycles with Me = Si^{IV}, Ge^{IV}, or Sn^{IV} are converted by heating in bulk or in a high-boiling solvent to the μ-oxo-bridged polymers. The topotactic polymerization mechanism is mentioned in Section 17.6.2.[151] Depending on the reaction conditions, the dgeree of polymerization is <150. Use of octasubstituted phthalocyanines (oxyalkyl, methyleneoxyalkyl, crown ether, or ester groups) in $Si(OH)_2PcR_8$, partially with trifluoroacetyl as leaving group ($Si(OCOCF_3)_2PcR_8$), enhance the solubility of the polymers and favor polycondensation.[140,152–156] High molecular weight MMCs were obtained[153] and the rotational dynamic behavior of the polymers was investigated.[157] Langmuir–Blodgett films were prepared and liquid crystalline behavior of the polymers was observed.[153–156] Polymers with different bridges between the stacked macrocycles have been described.[27,31,140,149]

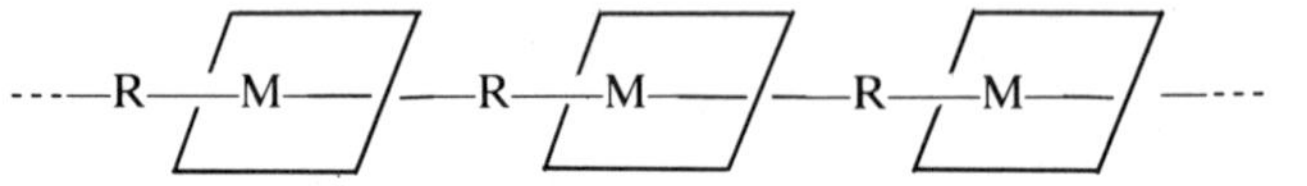

M = Si^{IV}, Ge^{IV}, Sn^{IV}
R = —O—, —S—, —O—R^1—O—, —C—C—, —C≡C—, —N=C=N—

(**29**)

17.4.5.2 *Covalent/coordinate bonds between the central metal ions*

As shown in (**30**), these stacked macrocycles contain Al^{III}, Ga^{III} or Cr^{III} with fluoro as the bridging atom or Co^{III}, Fe^{III}, Mn^{III}, Cr^{III}, or Rh^{III} with —C≡N, —S—C≡N, or —N_3 as the binding groups. Details for the preparations have been described.[27,31,140,149]

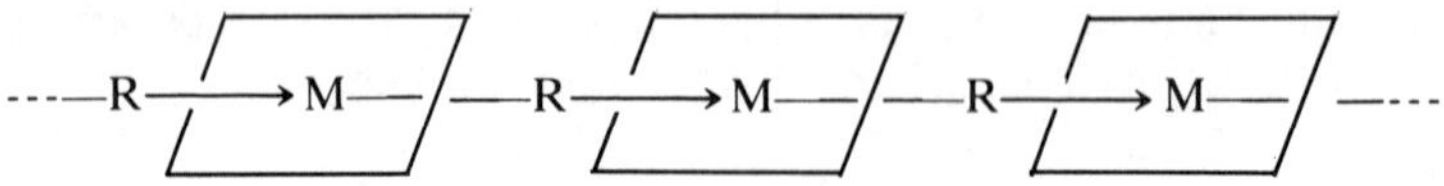

M = Al^{III}, Ga^{III}, Cr^{III} R = F^-
M = Fe^{III}, Co^{III}, Rh^{III}, Mn^{III}, Cr^{III} R = —C≡N^-, —S—C≡N^-, N_3^-

(**30**)

17.4.5.3 *Coordinate/coordinate bonds between the central metal ions*

In this case, macrocycles with transition metals ions in the +2 oxidation state and capable of hexacoordination, and neutral organic donors containing two groups or heteroatoms for coordination, are used in stoichiometric amounts as starting materials (arrangement (**31**)).[27,31,140,149] The degree of polymerization depends on the reaction conditions and is between 20 and 50.

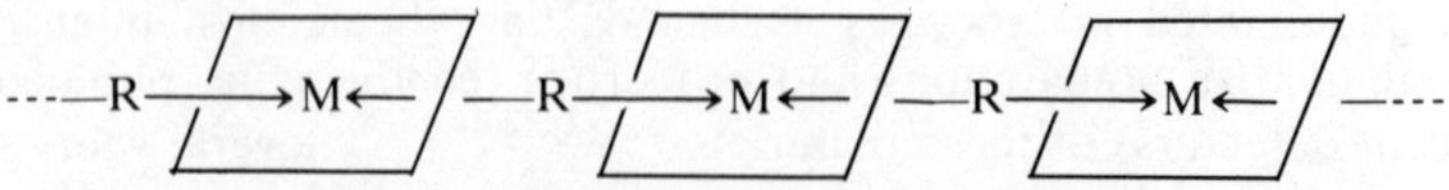

M = Fe^{II}, Co^{II}, Ru^{II}, Os^{II} R = pyrazine, tetrazine, triazine, *p*-diisocyanobenzene, *p*-phenylenediamine, fumaronitrile

(**31**)

The MMCs described in Sections 17.4.5.1–3 exhibit different stabilities. Whereas polymers (**29**) may be unaffected by aqueous HF at 100 °C, aqueous 2M NaOH at reflux, and concentrated H_2SO_4, polymers (**31**) are destroyed by an excess of the bifunctional ligand. The stability in general increases in the sequence: (**31**) < (**30**) < (**29**).

17.4.6 Metallocenes as Part of a Polymer Chain

The arrangement of bifunctional π-electron-rich charged aromatics connected by π-bonds in the main chain (**32**) has been known for a long time.[30,158–160] Polymers containing metallocenes are obtained by step-growth polycondensation from, for example, 1,1′-dilithioferrocenes with 1,1′-diiodoferrocenes ($M < 10\,000$).[30,158] Also, poly(1,1′-ruthenocenylenes) have been described.[158] Face-to-face polydecker sandwich complexes with a naphthalene spacer (**33**) are prepared from a monomer dianion by reaction with a transition metal ion.[159,161] Polydecker sandwich complexes can contain 2,3-dihydro-1,3-diborolo ligands and Rh^{2+}.[162] Several papers have reported on polyesters, polyamides, and polyurethanes containing metallocenylenes[136,158,163] and on polyferrocenylenes bridged by $(CH_2)_n$, SiR_2, or GeR_2.[158,160]

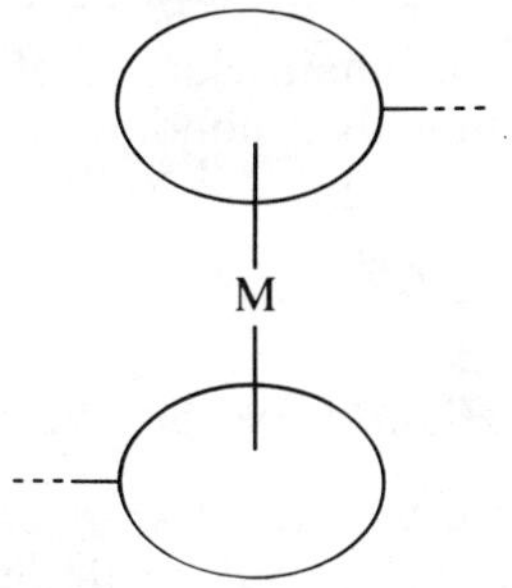

(32)

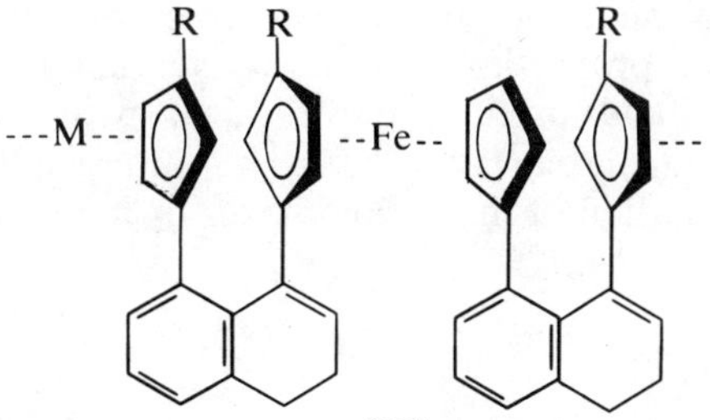

(33)

17.5 METAL COMPLEXES, ZEROVALENT METALS/METAL CLUSTERS INCORPORATED IN MACROMOLECULE COMPOUNDS (TYPE III MMC)

In this type, metal complexes or metal clusters are not covalently bonded to a polymer chain or are not part of a molecular chain/network. Interaction between the high molecular host and the guest may be caused by various secondary binding forces: physical adsorption by van der Waals interactions; coulombic or electrostatic interactions; charge-transfer or hydrophobic interactions; and coordinative or hydrogen binding.[6] These multiple and dynamic interactions influence the properties of the composite material. Some examples of type III MMCs are given below.

17.5.1 Combination with Organic Polymers

Every metal complex can be incorporated, either monomolecularly dissolved or as an aggregated particle, into an organic polymer. A solution of a polymer containing the metal complex is cast onto various conductive carriers (Pt, Au, ITO, graphite, *etc.*), or the metal complex is absorbed into a polymer film from solution and then this thin film is used as an electrode material. The methods are described by Kaneko and Wöhrle.[35] Examples are Ru, Re, and Co complexes such as $metal(bpy)_3^{2+}$, metal phthalocyanines, and Schiff base complexes.[5,6,67,68,72–74,164–166] The thickness of the films is of the order of 50 nm to a few μm. The apparent concentration is often as high as 0.1–5 M. Photophysical, photochemical, electrochemical, and photoelectrochemical properties have been investigated.

Considerable research effort has been directed to the study of metal clusters in homogeneous solutions in the presence of a protecting polymer or in a high molecular host matrix. The principal features of the synthesis of polymer-protected metal clusters of different size and the properties of such clusters in general have been reviewed recently.[167,168] The interest in these nanoscopic metal (and also semiconductor) materials, besides the specific activity and selectivity in catalysis, is in the quantum size effect (nonlinear optics, quantum size electronics).[168–172] The preparation is carried out by transition metal salt reduction with citrate or, more recently, by alcohol reduction and photoreduction.[170,173] Poly(*N*-vinylpyrrolidone) is one example of a protecting polymer. The particle size has a narrow distribution and is around 1–5 nm for Pd, Pt, and Rh clusters. Bimetallic clusters have been prepared.[6,173] General methods for the preparation of small metal particles (Pd, Ag, Au, Zn, and others) in various polymers include the vacuum codeposition of metal and vinyl monomers, followed by normal radical polymerization,[167,174] and the pyrolysis of a variety of MMCs at 300–1400 °C under an inert gas.[167,175] An interesting method is the pyrolysis of $(Ph_3P)_2Pt(C_2H_4)$ or $Co_2(CO)_8$ on poly(phenylenediacetylene) at 600 °C to obtain clusters of 1.6 nm average diameter.[176] Plasma polymers from plasma polymerization of monomers and metal sputtering have been mentioned.[177]

17.5.2 Combination with Inorganic High Molecular Weight Compounds

Metal complexes or metal clusters can interact with or are stabilized by contact with the surface of inorganic carriers of oxides such as silica gel, alumina, and titanium dioxide, and by encapsulation in molecular sieves like zeolites, aluminum phosphates, or layer silicates like hectorite clays. This interesting field exhibits, for the host systems, the advantage of high thermal and chemical stability and, for guests, specific carrier–metal complex/metal cluster interactions providing improved properties. A general overview has been given by Ciardelli *et al.*[5] Some additional reviews and original papers deal with the following topics: metal complexes or clusters on inorganic surfaces;[24,107,177–179] intercalation of metal complexes in silicate layers;[24,180] and encapsulation of metal complexes or metal clusters in the framework of molecular sieves.[172,180–184]

17.6 KINETICS AND THERMODYNAMICS OF FORMATION OF MACROMOLECULE METAL COMPLEXES; ANALYTICAL ASPECTS

17.6.1 General Principles of MMC Formation

As in low molecular weight metal complexes, the process of complex formation is accompanied by numerous complicated factors such as ion exchange equilibria, ligand conformational changes, influence in the change of the electrostatic potential, *etc.* The type and strength of the bonds formed between the metal and the ligand depend on the ionization potential of the metal ion, its electron affinity, and the donor properties of the ligand groups. For macromolecule metal complexes either in solution or in the solid state, various secondary binding forces are important and determine, besides the covalent and ionic bonds, the secondary, tertiary, and quaternary structures (see Section 17.1). In addition, specific polymer parameters like the degree of crosslinking, the distribution of ligands, and, in the case of insoluble polymers, the topography of a macroligand or protecting high molecular weight surrounding must be considered. Many unsolved problems exist in the field of the physical chemistry of complex formation, secondary binding forces, and composition and reactivity of MMCs because of their manifold structures. The present situation is best described in a recent monograph.[5]

17.6.2 Approaches for Calculation of Equilibrium Constants

Different models have been used to describe the interaction of metal ions with macroligands of type I and some type II (Scheme 1, equation 1) complexes.[5] In one model for linear macroligands, the polymer ligand L is the central particle and the metal ion/complex is added in a stepwise manner. In this case, the equilibrium constant will not depend on the molecular weight of the macroligand. A second model, which is based on the metal ion M as the central particle, is described by the Flory concept of infinitely large chains with the reactivity of the binding centers independent of their position in the polymer.[5] Another approach calculates the sequence equilibrium, which means that there are equilibrium constants for the metal ion binding at different positions on the macroligand.[185,186] The equilibrium is usually described by the equilibrium constant $\bar{K}$ of a macroligand containing metal ion-complexed repeating units (equations (11) and (12)),[5,185–189] where C_p and C_s are initial concentrations of polymer (expressed in repeating units) and metal salt, and α is the fraction of metal ion/complex not complexed by the polymer

$$\cdots\text{—L—L—L}_{n-1}\text{—L—} + M^+ \overset{\bar{K}}{\rightleftarrows} \text{—L—}\overset{\text{M}}{\text{L—L}_{n-1}}\text{—L—}\cdots \tag{11}$$

$$\bar{K} = \frac{[\text{M}(\text{—L}_n\text{—})]}{[\text{M}^+][\text{—L}_n\text{—}]} = \frac{1-\alpha}{\alpha(C_p/n - C_s(1-\alpha))} \tag{12}$$

Szymanski[185] argues that the right-hand side of equation (12) is not totally correct because the equilibrium concentration of the macroligand $[\text{—L}_n\text{—}] \neq (C_p/n - C_s(1-\alpha)$. The reason is that a sequence of $n+1$ vacant repeating units can consist of different but overlapping neighbor sequences of polymer units. Different lengths between noncomplexed sequences exist which influence each other and result in different equilibrium constants $k_1, k_2, k_3, \ldots, k_x$ (equations (13)–(15)).

$$\cdots\text{—L—L—L—L—L—L—L—}\cdots + \text{M}^+ \overset{k_1}{\rightleftharpoons} \cdots\text{—L—L—}\overset{\text{M}^+}{\text{L—L}}\text{—L—L—L—}\cdots \tag{13}$$

$$\cdots\text{—L—L—}\overset{\text{M}}{\text{L—L}}\text{—L—L—L—}\cdots + \text{M}^+ \overset{k_2}{\rightleftharpoons} \cdots\text{—L—L—}\overset{\text{M}^+}{\text{L—L}}\text{—}\overset{\text{M}^+}{\text{L—L}}\text{—L—}\cdots \tag{14}$$

$$\cdots\text{—L—L—}\overset{\text{M}}{\text{L—L}}\text{—}\overset{\text{M}}{\text{L—L}}\text{—L—}\cdots + \text{M}^+ \overset{k_3}{\rightleftharpoons} \cdots\text{—}\overset{\text{M}^+}{\text{L—L}}\text{—}\overset{\text{M}^+}{\text{L—L}}\text{—}\overset{\text{M}^+}{\text{L—L}}\text{—L—}\cdots \tag{15}$$

A theoretical model has been described which allows the determination of k_1 and k_2 on the basis of a numerical fit $\alpha = f(n, [\text{L}]_0, [\text{M}^+]_0, k_1, k_2)$ with $[\text{L}]_0$, $[\text{M}^+]_0$.[185] The validity of the model was tested for the interaction of Na^+ (as $\text{Na}^+\text{BPh}_4^-$) with poly(oxyethylene) in methanol. The best fit between measured and calculated values are found for $n = 1$ with $k_1 = 1.9\ \text{L mol}^{-1}$ and $k_1/k_2 = 3.5$. Cooperative effects which change the polymer chain conformation under complex formation must also be considered.[5] Bending of a polymer chain by coordination of different ligand groups of one polymer chain leads to an increase of macroligand reactivity (increase of formation constant in comparison to separated, *e.g.* low molecular weight, ligand centers). This has been discussed for metal binding at poly(oxyethylene) and poly(4-vinylpyridine).[190,191]

In the case of crosslinked macroligands, electrostatic factors significantly influence the composition, structure, and stability of a metal complex. Metal ion/complex binding can be described as mentioned before. In heterogeneous systems, when the ligand groups are mainly arranged on a surface with zero concentration in solution, diffusion and topological restrictions must be considered. At low binding-center concentrations, a Langmuir equation is valid for binding of a metal ion/complex (equation 16), where f_{max} is the maximum binding of the metal ion/complex by a macroligand.[5,191]

$$\frac{[\text{M}^+]}{[\text{M}]_{\text{bound}}} = \frac{1}{\bar{K}} + \frac{[\text{M}]}{f_{max}} \tag{16}$$

One example is the binding of Cu^{2+} by a crosslinked polymer containing bis(carboxymethyl)amino ligand groups with $\bar{K} = 3.5 \times 10^{-3}\ \text{L mol}^{-1}$ and $f_{\text{max}} = 0.075\ \text{mmol g}^-$.[192] For a nonporous solid matrix containing ligands grafted on a surface, the stability of the complex is independent of the degree of surface coverage, as shown for $CuCl_2$ or $PtCl_2$ on Aerosil from acetonitrile.[193]

Formation of type II MMCs obtained by the reaction of bi/multifunctional low molecular weight metal complexes with another bi/multifunctional ligand (Scheme 1, equation (4)) can be described by the usual rate constants, equilibrium, and kinetics as known for polycondensation or polyaddition reactions in macromolecular chemistry. Increasing insolubility results in chain termination and formation of oligomers.

Thermal polycondensation of dihydroxy(metallo)phthalocyanines in the solid state (equation (17); Section 17.4.5.1) is topotactic and under topochemical control, which means that well-defined intermolecular distances and interactions in the lattice control the reaction.[151] A kinetic study showed that the fraction of unreacted —OH end groups X over time does not obey first-order kinetics ($X = \exp(-k_1 t)$) but power-two-order kinetics ($X = \exp(-k_2 t^2)$).

$$n\,\text{HO—M(Pc)—OH} \rightarrow \text{HO—(M(Pc)—O—)}_n\text{H} + (n-1)\text{H}_2\text{O} \qquad \text{M} = \text{Si, Ge, Sn};\ n = 50\text{–}200 \tag{17}$$

17.6.3 Thermodynamics of MMC Formation

Beside the known thermodynamic changes during a reaction, the situation for the formation of MMCs is more complicated. Changes in, for example, the conformation of the macromolecules, chain flexibility, electrical charge, *etc.*, also influence the thermodynamic parameters such as ΔS.[5,194] The general expression for the reaction thermodynamics is shown in equation (18).

$$\Delta G = -RT \ln \bar{K} = \Delta H - T\Delta S \tag{18}$$

For the formation of a low molecular weight chelate, the chelate effect depends on the binding of the solvent and is of the order of −5 to −20 kJ mol^{-1} (mainly determined by entropic terms). The polymer chelate effect is more complicated and includes the above-mentioned parameters and also local, molecular, and supramolecular organizations of macromolecules.[7,195] With a low degree of chelation, ΔH values for macroligands and for low molecular weight ligands in the interaction with metal ions are comparable, but ΔS values are different (polymer chelate effect), as has been shown for the reaction of amines with Cu^{2+}.[196] For concentrated solutions as well as suspensions, interactions such as intermolecular or supramolecular organization must be considered and are determined by entropic terms. A more detailed discussion is given by Ciardelli *et al.*[5] Because of intermolecular interactions between the macroligand and the metal ion/complex, the temperature of ligand↔gel formation, T_{tr}, is influenced by the ultimate polymer concentration L^{ul}.[197] Above L^{ul}, the value of T_{tr} is independent of the concentration of the polymer and its molecular weight. In the case of Fe^{3+}–polyhydroxamic acid, infinite networks are formed when the probability of intermolecular metal binding is above 50%.[7,33]

Type II MMCs often form insoluble, more-or-less crystalline products. Therefore entropic terms arising from going from solution to a crystalline or amorphous precipitate must be considered. Entropic terms are also important for the stabilization of metal clusters or metal complexes/chelates in a high molecular weight surrounding (type III MMCs). During formation of MMCs, various thermodynamic side-effects driven by thermodynamically favored terms can occur. These include conformational changes, modification of functional groups, and macrochain breakage. Examples of conformational changes are: chain transformation in poly(oxyethylene)–transition metal complexes,[198,199] the double helix model of poly(oxyethylene)–alkali metal ion complexes,[200] and conformational modifications of poly(2-vinylpyridine) or poly(amido-amines) during complex formation.[201,202] It is important to mention here that chain destruction can occur in type I MMCs during their formation.[5,203,204]

17.6.4 Analytical Aspects

A detailed analysis is the fundamental prerequisite in the correlation of the structure and properties of new MMC materials. After preparation and isolation of an MMC, one has first to concentrate on the composition of the new material (primary structure). Well-known analytical methods can be used. For soluble MMCs, usual methods of molecular weight determination can be applied. Microcalorimetric studies allows measurement of the enthalpy of MMC formation. In some cases, complex formation constants can be determined by potentiometric or conductometric measurements.[5,7] More complicated is the investigation of the secondary, tertiary, and quaternary structure of MMCs either in solution or in the solid state. Each method (IR, UV/vis/NIR, Raman, acoustic, dielectric loss, X-ray, Mössbauer, ESCA, XAFS, magnetochemical, ESR, solution/solid NMR, *etc.*) provides some information on types I, II, and III MMCs.[5,7]

In nearly every case, some special analytical investigations have to be carried out. This is illustrated for polymer phthalocyanines of type II structure. Polymeric phthalocyanines (Section 17.4.1) are obtained by two-dimensional propagation from various tetracarbonitriles as bifunctional monomers. An ideal polymeric phthalocyanine has a regular planar structure which can be treated in a two-dimensional Cartesian coordinate system, allowing positive integers (propagation directions of the polymers are denoted by the letters x and y).[122] A model describing structural features such as the degree of polymerization, size, and shape of polymeric phthalocyanines has been discussed. Equation (19) correlates the number of macrocycles n (degree of polymerization) with the number of bridged monomers b and the number of end monomers e.

$$n = b/2 + e/4 \qquad (19)$$

Evaluation of some data (determination of the number of nitrile end groups and groups of bridged monomers by quantitative IR spectra) leads, depending on the kind of tetracarbonitrile and the reaction conditions, to values of $x = 4$ to infinity and $y = 1$ to infinity. In addition, it was shown that the structure of polymeric phthalocyanines exhibits fractal properties. They have a regular structure and four fractal dimensions for every size/shape/dilation combination.[122] This important mathematical model can serve as a polymer model for discussing basic fractals. Cofacial stacked polymeric phthalocyanines (Section 17.4.5) containing four substituents, and their possible isomers in such a stacking, have also been discussed mathematically.[183]

17.7 PROPERTIES

By combining high molecular compounds with a metal ion, metal complex/chelate, or metal cluster, new desired properties can be predicted for these compounds and composites. The principal properties are summarized below by selecting some aforementioned references. Also, other reviews[5–37] include information on these MMC properties: (a) binding of small molecules,[71–74] (b) catalysis,[39,45,46,48–50,64,70,71,78,79,83,98,104–109,167,169,170,173,179,181] (c) electron transfer, photoinduced electron transfer,[54,67–69,76,77,80,81,86,171,178] (d) luminescence,[53,54,86] (e) electrochemistry, photoelectrochemistry,[67,68,87,117,123–126,129,130,164,165,176] (f) conduction, photoconduction in solids,[27,127,131–135,144,148,149,155,166] (g) ionic conduction,[37,55–57] (h) liquid crystalline properties,[59,114,142,153–156] (i) medicine,[39,84,115] and (j) microelectronics, optoelectronics.[67,68,130,136,137,168–172,182,184]

17.8 REFERENCES

1. 'Macromolecule–Metal Complexes II', *J. Macromol. Sci., Chem.*, 1988, **A25,** Nos. 10 and 11.
2. 'Macromolecule–Metal Complexes III', *J. Macromol. Sci., Chem.*, 1990, **A26,** Nos. 2 and 3; 1990, **A27,** Nos. 9–11.
3. 'Macromolecule–Metal Complexes IV', *Makromol. Chem., Macromol. Symp.*, 1992, **59.**
4. 'Macromolecule–Metal Complexes V', *Macromol. Symp.*, 1994, **80.**
5. F. Ciardelli, E. Tsuchida and D. Wöhrle (eds.), 'Macromolecule Metal Complexes', Springer, Heidelberg, 1995.
6. E. Tsuchida (ed.), 'Macromolecular Complexes, Dynamic Interactions and Electronic Processes', VCH, New York, 1991.
7. A. D. Pomogailo and I. E. Uflyand, 'Macromolecular Metal Chelates', Nauka, Moscow, 1991.
8. F. G. A. Stone and W. A. G. Graham (eds.), 'Inorganic Polymers', Academic Press, New York, 1962.
9. N. H. Ray, 'Inorganic Polymers', Academic, London, 1978.
10. C. E. Carraher, Jr., J. E. Sheats and C. U. Pittmann, Jr. (eds.), 'Organometallic Polymers', Academic Press, New York, 1978.
11. C. E. Carraher, J. E. Sheats and C. U. Pittmann (eds.), 'Advances in Organometalic and Inorganic Polymer Science', Dekker, New York, 1982.
12. J. E. Sheats, C. E. Carraher and C. U. Pittmann (eds.), 'Metal-Containing Polymeric Systems', Plenum, New York, 1985.
13. Scientific journal: *J. Inorg. Organomet. Polym.*, Plenum, New York.

14. B. K. Kepler, 'Metal Complexes in Cancer Chemotherapy', VCH, New York, 1993.
15. J. Reedijk (ed.), 'Bioinorganic Catalysis', Dekker, New York, 1993.
16. H. Sigel (ed.), 'Metal Ions in Biological Systems', Dekker, New York (25 volumes).
17. T. G. Spiro (ed.), 'Metal Ions in Biology', Wiley, New York (7 volumes to 1985).
18. A. D. Pomogailo, 'Polymeric Immobilized Metallo-Complex Catalysts', Nauka, Moscow, 1988.
19. F. R. Hartley, 'Supported Metal Complexes, A New Generation of Catalysts', Reidel, Dordrecht, 1985.
20. V. I. Yemakov, B. N. Kuznetsov and V. A. Zakharov, 'Catalysis by Supported Complexes', Elsevier, Amsterdam, 1981.
21. V. I. Yermakov and V. Likholohov, 'Homogeneous and Heterogeneous Catalysis', VNU, Utrecht, 1986.
22. P. A. Jacobs, N. I. Jaeger, L. Kubelkova and B. Wichterlova (eds.), 'Zeolite Chemistry and Catalysts', Studies in Surface Science and Catalysis, Elsevier, Amsterdam, 1991, vol. 69.
23. H. Karge and J. Weitkamp (eds.), 'Zeolites as Catalysts and Detergent Builders', Studies in Surface Science and Catalysis, Elsevier, Amsterdam, 1989, vol. 46.
24. V. Ramamurthy (ed.), 'Photochemistry in Organized and Constrained Media', VCH, New York, 1991.
25. D. Wöhrle, *Adv. Polym. Sci.*, 1983, **50,** 45.
26. V. V. Korshak and N. M. Kozyreva, *Russ. Chem. Rev.*, 1985, **54,** 1091.
27. M. Hanack and M. Lang, *Adv. Mater.*, 1994, **6,** 819.
28. G. A. Ozin and C. Gil, *Chem. Rev.*, 1989, **89,** 1749.
29. D. C. Sherrington, *Pure Appl. Chem.*, 1988, **60,** 401.
30. D. Wöhrle, in 'Handbook of Polymer Synthesis', ed. H. Kricheldorf, Dekker, New York 1992, part B, p. 1133.
31. D. Wöhrle, in 'Phthalocyanines, Properties and Applications', ed. C. C. Leznoff and A. B. P. Lever, VCH, New York 1989, vol. 1, p. 55.
32. A. D. Pomogailo, *Russ. Chem. Rev.*, 1992, **61,** 133.
33. A. D. Pomogailo and J. E. Uflyand, *Adv. Polym. Sci.*, 1990, **97,** 61.
34. E. Tsuchida and H. Nishide, *Adv. Polym. Sci.*, 1977, **24,** 1.
35. M. Kaneko and D. Wöhrle, *Adv. Polym. Sci.*, 1988, **84,** 141.
36. M. Hanack, S. Deger and A. Lange, *Coord. Chem. Rev.*, 1988, **83,** 115.
37. M. Biswas and A. Mukherjee, *Adv. Polym. Sci.*, 1994, **115,** 89.
38. Nomenclature for regular single-strand and quasi-single-strand inorganic and coordination polymers: *Pure Appl. Chem.*, 1985, **57,** 151.
39. J. Reedijk, *Macromol. Symp.*, 1994, **80,** 95.
40. J. J. R. Frausto da Silva and R. J. P. Williams, 'The Biological Chemistry of the Elements', Oxford University Press, Oxford, 1991.
41. R. G. Pearson, *J. Am. Chem. Soc.*, 1963, **85,** 3553; R. G. Pearson, *Science*, 1966, **151,** 172.
42. R. A. A. Muzzarelli, 'Chitin', Pergamon, Oxford, 1977.
43. Y. An, D. Yuan, M.-Y. Huang and Y.-Y. Jiang, *Macromol. Symp.*, 1994, **80,** 257.
44. A. Guyot and M. Bartholin, *Prog. Polym. Sci.*, 1982, **8,** 277.
45. P. Hodge and D. C. Sherrington (eds.), 'Polymer Supported Reactions in Organic Synthesis', Wiley, New York, 1981.
46. N. Ise and I. Tabuski (eds.), 'Speciality Polymers', Iwanami Shoton, Tokyo, 1980.
47. K. Geckeler, G. Lange, H. Eberhart and E. Bayer, *Pure Appl. Chem.*, 1980, **52,** 1883.
48. P. W. N. M. van Leeuwen, T. Jongsma and G. Challa, *Macromol. Symp.*, 1994, **80,** 241.
49. E. A. Karakhanov, E. A. Runova, G. V. Berezkin and E. B. Neimerovets, *Macromol. Symp.*, 1994, **80,** 231.
50. A. A. Efendiev, *Macromol. Symp.*, 1994, **80,** 289.
51. C. P. Fei and T. H. Chan, *Synthesis*, 1982, 467.
52. Y. Ueba, K. J. Zhu, E. Banks and Y. Okamoto, *J. Polym. Sci., Polym. Chem. Ed.*, 1982, **20,** 1272; R. H. Joshi and M. N. Patel, *Macromol. Sci.*, 1983, **19A,** 919.
53. Y. Okamoto, *Makromol. Chem., Macromol. Symp.*, 1992, **59,** 83; Y. Okamoto and J. Kido, in 'Macromolecular Complexes', ed. E. Tsuchida, VCH, New York, 1991, p. 143.
54. Y. Kurimura, Y. Sairenchi and S. Nakayama, '*Makromol. Chem., Macromol. Symp.*, 1992, **59,** 199; Y. Kurimura, in 'Macromolecular Complexes', ed. E. Tsuchida, VCH, New York, 1991, p. 93.
55. S. Takeoka, K. Horiuchi, S. Yamagata and E. Tsuchida, *Macromolecules*, 1991, **24,** 2003.
56. C. K. Chiang, G. T. Davies, C. A. Harding and T. Takahashi, *Macromolecules*, 1985, **18,** 825.
57. J. M. R. Giles and M. P. Greenhall, *Polym. Commun.*, 1986, **27,** 360.
58. R. Sinta, B. Lamb and J. Smid, *Macromolecules*, 1983, **16,** 1382.
59. M. Shirai, A. Veda and M. Tanaka, *Makromol. Chem.*, 1985, **186,** 2519.
60. C. S. Harris, D. F. Shriver and M. A. Ratner, *Macromolecules*, 1986, **19,** 987.
61. V. Nicolaus and D. Wöhrle, *Angew. Makromol. Chem.*, 1992, **198,** 179.
62. H. Kise and H. Sato, *Makromol. Chem.*, 1985, **186,** 2449.
63. A. Boudakgi, J. Jezierska and B. N. Kolarz, *Makromol. Chem., Macromol. Symp.*, 1992, **59,** 343.
64. D. Wöhrle and P. Buttner, *Polym. Bull.*, 1985, **13,** 57.
65. H. Aeissen and D. Wöhrle, *Makromol. Chem.*, 1981, **182,** 2961.
66. R. S. Paredes, N. S. Valera and L. F. Lindoy, *Aust. J. Chem.*, 1986, **39,** 1071.
67. M. Kaneko and A. Yamada, *Adv. Polym. Sci.*, 1984, **55,** 1.
68. M. Kaneko and A. Yamada, in 'Metal Containing Polymeric Systems', ed. J. E. Sheats, C. E. Carraher, and C. U. Pittmann, Plenum, New York, 1985, p. 249.
69. T. Sasaki and F. Matsunaga, *Bull. Chem. Soc. Jpn.*, 1968, **41,** 2440.
70. G. Challa, W. Chen and J. Reedijk, *Makromol. Chem., Macromol. Symp.*, 1992, **59,** 59.
71. D. Wöhrle, H. Bohlen and J. K. Blum, *Makromol. Chem.*, 1986, **187,** 2081; D. Wöhrle, H. Bohlen, C. Aringer and D. Pohl, *Makromol. Chem.*, 1984, **185,** 669.
72. H. Nishide, M. Kuwahara, M. Ohyanagi, Y. Funada, H. Kawakami and E. Tsuchida, *Chem. Lett.*, 1986, 43.
73. E. Tsuchida, H. Nishide, M. Ohyanagi and O. Okada, *J. Phys. Chem.*, 1988, **92,** 6461; H. Nishide, M. Yuasa, E. Hasegawa and E. Tsuchida, *Macromolecules*, 1987, **20,** 1913.

74. K. Sugie, *Polym. Mater. Sci. Eng.*, 1988, **59,** 133.
75. T. Shimidzu, K. Izaki, Y. Akai and T. Iyoda, *Polym. J.*, 1981, **13,** 889.
76. M. Kaneko, S. Nemoto, A. Yamada and Y. Kurimura, *Inorg. Chim. Acta,* 1980, **44,** L289.
77. Y. Kurimura, N. Shinozaki, F. Ito, Y. Uratani, K. Shigehara, E. Tsuchida, M. Kaneko and A. Yamada, *Bull. Chem. Soc. Jpn.*, 1982, **55,** 380.
78. D. C. Sherrington and H.-G. Tang, *Macromol. Symp.*, 1994, **80,** 193.
79. A. L. Rusanov, M. E. Vol'pin, A. A. Beliy and L. G. Chigladze, *Macromol. Symp.*, 1994, **80,** 215.
80. D. Wöhrle, G. Krawczyk and M. Paliuras, *Makromol. Chem.*, 1988, **189,** 1001, 1013; D. Wöhrle and J. Gitzel, *Makromol. Chem., Rapid Commun.*, 1988, **9,** 229; D. Wöhrle, M. Paliuras and I. Okura, *Makromol. Chem.*, 1991, **192,** 819.
81. D. Wöhrle, J. Gitzel, G. Krawczyk, E. Tsuchida, H. Ohno, I. Okura and T. Nishisaka, *J. Macromol. Sci., Chem.*, 1988, **A25,** 1227.
82. K. Hanabusa and H. Shirai, in 'Phthalocyanines, Properties and Applications', eds. C. C. Leznoff and A. B. P. Lever, VCH, New York, 1989, vol. 2, p. 197.
83. H. Shirai, A. Maruyama, K. Kobayashi and N. Hojo, *Makromol. Chem.*, 1980, **181,** 575; S. Higaki, K. Hanabusa, H. Shirai and N. Hojo, *Makromol. Chem.*, 1983, **184,** 691; K. Hanabusa, C. Kobayashi, T. Koyama, E. Masuda, H. Shirai, Y. Kondo, K. Takemoto, E. Iizuka and N. Hojo, *Makromol. Chem.*, 1986, **187,** 753.
84. D. Wöhrle, A. Ardeshirpur, A. Heuermann, S. Müller, G. Graschew, H. Rinneberg, M. Kohl and J. Neukammer, *Makromol. Chem., Macromol. Symp.*, 1992, **59,** 17.
85. V. S. Psheshetskii, V. L. Lishinsky, A. I. Kokorin, L. A. Tsarkova, A. A. Rakhnianskaya and N. V. Pertsov, *Makromol. Chem., Macromol. Symp.*, 1992, **59,** 163.
86. K. Nagai, N. Nemoto, Y. Ueno, K. Ikeda and N. Takamiya, *Makromol. Chem., Macromol. Symp.*, 1992, **59,** 257.
87. C. R. Leidner and R. W. Murray, *J. Am. Chem. Soc.*, 1984, **106,** 1606.
88. E. J. Miller, M. J. Naughton, C. A. Weigelt, J. E. Bradt, J. A. Serth, C. L. Ofslager and W. J. O'Brien, *Makromol. Chem., Macromol. Symp.*, 1992, **59,** 135.
89. Y. Kurimura, Y. Uchino, F. Ohta, C. Saito, M. Koide and E. Tsuchida, *Polym. J.*, 1981, **13,** 247.
90. D. Wöhrle, H. Bohlen and G. Meyer, *Polym. Bull.*, 1984, **11,** 145; D. Wöhrle, H. Bohlen and G. Meyer, *Polym. Bull.*, 1984, **11,** 151.
91. Y. Fujii, K. Kikuchi, K. Matsutani, K. Ota, M. Adashi, M. Syoji, I. Haneishi and Y. Kuwana, *Chem. Lett.*, 1984, 1487.
92. E. Tsuchida, *J. Macromol Sci., Chem.*, 1979, **A13,** 545.
93. E. Tsuchida, H. Nishide, M. Yuasa, E. Hasegawa, K. Eshima and Y. Matsushita, *Macromolecules,* 1989, **22,** 2103; H. Nishide, M. Yuasa, E. Hasegawa and E. Tsuchida, *Macromolecules,* 1987, **20,** 1913.
94. M. Kamachi, H. Akimoto and S. Nozakura, *J. Polym. Sci., Polym. Lett. Ed.*, 1983, **21,** 693; A. Kajiwara and M. Kamachi, *Polym. J.*, 1989, **21,** 593; A. Kajiwara, K. Aramata, S. Nomura, Y. Morishima and M. Kamachi, *Chem. Lett.*, 1992, 95.
95. H. Hiroyoshi, T. Nakata and S. Komatsu, *Bull. Chem. Soc. Jpn.*, 1991, **64,** 2300.
96. J. F. van der Pol, E. Neeleman, R. J. M. Nolte, J. W. Zwikker and W. Drenth, *Makromol. Chem.*, 1989, **190,** 2727; J. F. van der Pol, E. Neeleman, J. C. van Miltenburg, J. W. Zwikker, R. J. M. Nolte and W. Drenth, *Macromolecules,* 1990, **23,** 155.
97. H. Eichhorn, M. Sturm and D. Wöhrle, *Makromol. Chem. Phys.*, 1995, **196,** 115.
98. M. Kimura, T. Dakeno, E. Adachi, T. Koyama, K. Hanabusa and H. Shirai, *Makromol. Chem. Phys.*, 1994, **195,** 2423.
99. M. R. Steele, P. M. Macdonald and G. A. Ozin, *J. Am. Chem. Soc.*, 1993, **115,** 7285.
100. M. C. Steigerwald and L. E. Brus, *Acc. Chem. Res.*, 1990, **23,** 183; Y. Wang, *Acc. Chem. Res.*, 1991, **24,** 133.
101. I. E. Uflyand, A. D. Pomogailo, N. D. Golubeva and V. N. Sheinker, in 'Proceedings of the XVII European Congress on Molecular Spectroscopy, Madrid', 1985, p. 152.
102. A. D. Pomogailo, I. E. Uflyand and N. D. Golubeva, *Kinet. Catal.*, 1985, **26,** 1104.
103. I. E. Uflyand, A. D. Pomogailo, M. O. Gorbumova, A. G. Starikov and V. N. Sheinker, *Kinet. Catal.*, 1987, **28,** 613; I. E. Uflyand, A. S. Kuzhanov, M. O. Gorbumova, V. N. Sheinker and A. D. Pomogailo, *React. Polym.*, 1989, **11,** 221.
104. F. Ciardelli, A. Altomare, G. Conti, G. Arribas, B. Mendez and A. Ismayel, *Macromol. Symp.*, 1994, **80,** 29.
105. S. D. Ittel, *J. Macromol. Sci., Chem.*, 1990, **A27,** 1133.
106. T. Buck, D. Wöhrle, G. Schulz-Ekloff and A. Andreev, *J. Mol. Catal.*, 1991, **70,** 256.
107. Y. Kurusu, *Makromol. Chem., Macromol. Symp.*, 1992, **59,** 313.
108. C. Pan and H. Zong, *Macromol. Symp.*, 1994, **80,** 265.
109. T. Buck, H. Bohlen, D. Wöhrle and G. Schulz-Ekloff, *J. Mol. Catal.*, 1993, **80,** 253.
110. A. K. Dey, *J. Indian Chem. Soc.*, 1986, **63,** 357.
111. M. Marcu, S. Lazarescu and G. E. Grigoriu, *Polym. Bull.*, 1986, **16,** 103.
112. W. Sawodny, R. Grünes and H. Reitzle, *Angew. Chem., Int. Ed. Engl.*, 1982, **21,** 775.
113. G. Manecke, W. E. Wille and G. Kossmehl, *Makromol. Chem.*, 1972, **160,** 111.
114. K. Hanabusa, T. Suzuki, T. Koyama and H. Shirai, *Makromol. Chem.*, 1991, **192,** 233; K. Hanabusa, T. Suzuki, T. Koyama and H. Shirai, *Makromol. Chem.*, 1992, **193,** 2149.
115. E. W. Neuse, *Macromol. Symp.*, 1994, **80,** 111.
116. B. N. Achar, G. M. Fohlen and J. A. Parker, *J. Polym. Sci., Polym. Chem. Ed.*, 1985, **23,** 801.
117. C. C. Wamser, R. R. Bard, V. Senthilathipan, V. C. Anderson, J. A. Yates, H. K. Lonsdale, G. W., Rayfield, D. T. Friesen, D. A. Lorenz, G. C. Stangle, P. van Eikeren, D. R. Baer, R. A. Ransdell, J. H. Golbeck, W. C. Babcock, J. J. Sandberg and S. E. Clark, *J. Am. Chem. Soc.*, 1989, **111,** 8485; C. C. Wamser *et al., Mol. Cryst. Liq. Cryst.*, 1991, **194,** 65.
118. D. Wöhrle, U. Marose and R. Knoop, *Makromol. Chem.*, 1985, **186,** 2209; D. Wöhrle and E. Preussner, *Makromol. Chem.*, 1985, **186,** 2189.
119. D. Wöhrle and B. Schulte, *Makromol. Chem.*, 1988, **189,** 1167; D. Wöhrle and B. Schulte, *Makromol. Chem.*, 1988, **189,** 1229.

120. D. Djurado, S. Tadlaoui, A. Hamwi and J. C. Cousseins, *Synth. Met.*, 1991, **41–43,** 295; M. S. Liao and K. T. Kuo, *Polym. J.*, 1993, **25,** 947.
121. D. Wöhrle and R. Müller, *Makromol. Chem.*, 1978, **179,** 2161.
122. G. Knothe and D. Wöhrle, *Makromol. Chem.*, 1989, **190,** 1573; G. Knothe, *Makromol. Chem., Theory Simul.*, 1992, **1,** 187.
123. D. Wöhrle, R. Bannehr, N. Jaeger and B. Schumann, *Angew. Makromol. Chem.*, 1983, **117,** 103; D. Wöhrle, B. Schumann, V. Schmidt, A. Yamada and K. Shigehara, *Ber. Bunsenges. Phys. Chem.*, 1987, **91,** 975.
124. K. A. Pressprich, S. G. Maybury, R. E. Thomas, R. W. Linton, E. A. Irene and R. W. Murray, *J. Phys. Chem.*, 1989, **93,** 5568.
125. H. Li and T. F. Guarr, *J. Chem. Soc., Chem. Commun.*, 1989, 832.
126. T. Malinski, A. Ciszewski, J. E. Bennett and J. R. Fish, *J. Electrochem. Soc.*, 1991, **138,** 2008.
127. Y. Osada, A. Mizumoto and H. Tsuruta, *J. Macromol. Sci., Chem.*, 1987, **A24,** 403.
128. C. Feucht, T. Linssen and M. Hanack, *Chem. Ber.*, 1994, **127,** 113 and unpublished results; M. Rack and M. Hanack, *Angew. Chem.*, in press.
129. M. Kaneko, X.-H. Hou and A. Yamada, *Bull. Chem. Soc. Jpn.*, 1987, **60,** 2523.
130. M. Kaneko, *J. Polym. Sci., Polym. Lett.*, 1986, **24,** 435; M. Kaneko, *J. Macromol. Sci., Chem.*, 1987, **A24,** 357.
131. R. N. Hurd, G. DeLa Mater, G. C. McElheny and J. V. Pfeifer, *J. Am. Chem. Soc.*, 1960, **82,** 4454; R. N. Hurd, G. DeLa Mater, G. C. McElheny, R. J. Turner and V. H. Wellingford, *J. Org. Chem.*, 1961, **26,** 3980.
132. G. Zotti, G. Mengoli and F. Decker, *Mol. Cryst. Liq. Cryst.*, 1985, **121,** 337.
133. O.-K. Kim, T. H. Yoon and D. J. McDermott, *J. Chem. Soc., Chem. Commun.*, 1989, 740.
134. M. Khandwe, A. Bajpai and U. D. N. Bajpai, *Macromolecules*, 1991, **24,** 5203.
135. J. R. Reynolds, J. C. W. Chien and C. P. Lillya, *Macromolecules*, 1987, **20,** 1184.
136. R. D. Miller, *Angew. Chem. Adv. Mater.*, 1989, **101,** 1773.
137. R. West, *J. Organomet. Chem.*, 1986, **300,** 327.
138. C. E. Carraher, Jr., *J. Chem. Educ.*, 1981, **58,** 921; C. E. Carraher, Jr., W. J. Scott and J. A. Schroeder, *J. Macromol. Sci., Chem.*, 1981, **A15,** 625.
139. P. J. Davidson, M. F. Lappert and R. Pearce, *Chem. Rev.*, 1976, **76,** 219; R. R. Schrock and G. W. Parshall, *Chem. Rev.*, 1976, **76,** 243.
140. 'Methoden der Organischen Chemie (Houben–Weyl)', Thieme, Stuttgart, 1987, vol. E20.
141. H. Lang, *Angew. Chem., Int. Ed. Engl.*, 1994, **33,** 547.
142. A. Abe, S. Tabata and N. Kimura, *Polym. J.*, 1991, **23,** 69.
143. A. E. Dray, R. Rachel, W. O. Saxton, J. Lewis, M. S. Khan, A. M. Donald and R. H. Friend, *Macromolecules*, 1992, **25,** 3473.
144. B. F. G. Johnson, A. K. Kakkar, M. S. Khan, J. Lewis, A. E. Dray, R. H. Friend and F. Wittmann, *J. Mater. Chem.*, 1991, **1,** 485.
145. B. F. G. Johnson, A. K. Kakkar, M. S. Khan and J. Lewis, *J. Organomet. Chem.*, 1991, **409,** C12.
146. S. Kotani, K. Shiina and K. Sonogashira, *Appl. Organomet. Chem.*, 1991, **5,** 417.
147. H. Nishihara, T. Shimura, A. Ohkubo, N. Matsuda and K. Aramaki, *Adv. Mater.*, 1993, **5,** 752; I. Tomita, A. Nishio, T. Igarashi and T. Endo, *Polym. Bull.*, 1993, **30,** 179.
148. C. W. Dirk, E. A. Mintz, K. F. Schoch and T. J. Marks, in 'Advances in Organometallic and Inorganic Polymer Science', ed. C. E. Carraher, J. E. Sheats and C. U. Pittmann, Dekker, New York, 1986, p. 275.
149. H. Schultz, H. Lehmann, M. Rein and M. Hanack, *Struct. Bonding*, 1991, **74,** 41.
150. T. J. Marks, *Angew. Chem., Int. Ed. Engl.*, 1990, **29,** 857.
151. K. G. Beltsios and S. H. Carr, *J. Polym. Sci., Part C, Polym. Lett.*, 1989, **27,** 355.
152. S. Schwiegk, H. Fischer, Y. Xu, F. Kremer and G. Wegner, *Makromol. Chem., Macromol. Symp.*, 1991, **46,** 211.
153. L. Dulog, A. Gittinger, S. Roth and T. Wagner, *Makromol. Chem.*, 1993, **194,** 493.
154. G. M. Crockett, A. J. Campbell and F. R. Ahmed, *Polymer*, 1990, **31,** 602.
155. A. P. M. Kentgens, B. A. Markies, J. F. van der Pol and R. J. M. Nolte, *J. Am. Chem. Soc.*, 1990, **112,** 8800.
156. A. Ferenz, R. Ries and G. Wegner, *Angew. Chem., Int. Ed. Engl.*, 1993, **32,** 1184.
157. S. Schwiegk, M. Werth, J. Leisen, G. Wegner and H. W. Spiess, *Acta Polym.*, 1993, **44,** 31.
158. E. W. Neuse and H. Rosenberg, 'Metallocene Polymers', Dekker, New York, 1970; E. W. Neuse, *J. Macromol. Sci., Chem.*, 1981, **A16,** 3.
159. M. Rosenblum, *Adv. Mater.*, 1994, **6,** 159.
160. I. Manners, *Adv. Mater.*, 1994, **6,** 68; M. Hnyene, A. Yassar, M. Escorne, A. Percheron-Guegan and F. Garnio, *Adv. Mater.*, 1994, **6,** 564.
161. H. M. Nugent, M. Rosenblum and P. Klemarczyk, *J. Am. Chem. Soc.*, 1993, **115,** 3848.
162. W. Sieher, *Russ. Chem. Rev.*, 1991, **60,** 784.
163. O. Nuyken, V. Burkhardt, T. Pöhlmann and M. Heberhold, *Makromol. Chem., Macromol. Symp.*, 1991, **44,** 195.
164. N. Oyama, M. Kitagawa, H. Inaba and K. Kawase, *Macromol. Symp.*, 1994, **80,** 337.
165. D. Schlettwein, N. I. Jaeger and D. Wöhrle, *Makromol. Chem., Macromol. Symp.*, 1992, **59,** 267; D. Schlettwein, M. Kaneko, A. Yamada, D. Wöhrle and N. I. Jaeger, *J. Phys. Chem.*, 1991, **95,** 1748.
166. K.-Y. Law, *Chem. Rev.*, 1993, **93,** 449.
167. A. D. Pomogailo, *Platinum Met. Rev.*, 1994, **38,** 60.
168. G. Schmid, *Chem. Rev.*, 1992, **92,** 1709.
169. G. A. Martin, *Catal. Rev.-Sci. Eng.*, 1988, **30,** 519; R. Burch, *Acc. Chem. Res.*, 1982, **15,** 24.
170. H. Hirai and N. Toshima, in 'Tailored Metal Catalysts', ed. Y. Iwasawa, Reidel, Dordrecht, 1986, p. 87.
171. M. Grätzel, 'Heterogeneous Photochemical Electron Transfer', CRC Press, Boca Raton, 1989.
172. G. A. Ozin and M. R. STeele, *Macromol. Symp.*, 1994, **80,** 45; G. A. Ozin, A. Kuperman and A. Stein, *Angew. Chem., Int. Ed. Engl.*, 1989, **28,** 359; G. A. Ozin, *Adv. Mater.*, 1994, **6,** 71.
173. N. Toshima, *J. Macromol. Sci., Chem.*, 1990, **A27,** 1225; M. Harada, K. Asakura and N. Toshima, *J. Phys. Chem.*, 1994, **98,** 2653; N. Toshima and Y. Wang, *Adv. Mater.*, 1994, **6,** 245.
174. G. Cardenas T. and C. Munoz D., *Makromol. Chem.*, 1993, **194,** 3377.
175. S. Miyanaga, H. Yasuda, A. Hiwara and A. Nakamura, *J. Macromol. Sci., Chem.*, 1990, **A27,** 1347.

176. T. X. Neenan, M. R. Callstrom and O. J. A. Schueller, *Macromol. Symp.*, 1994, **80,** 315.
177. R. Lamber, A. Baalman, N. I. Jaeger, G. Schulz-Ekloff and S. Wetjen, *Adv. Mater.*, 1994, **6,** 223.
178. M. Grätzel, *Comments Inorg. Chem.*, 1991, **12,** 93; B. O'Regan and M. Grätzel, *Nature,* 1991, **353,** 737.
179. D. Wöhrle, T. Buck, U. Hündorf, G. Schulz-Ekloff and A. Andreev, *Makromol. Chem.*, 1990, **190,** 961.
180. V. Ramamurthy, D. F. Eaton and J. V. Caspar, *Acc. Chem. Res.*, 1992, **25,** 299; S. L. Suib, *Chem. Res.*, 1993, **93,** 803.
181. D. E. De Vos, F. Thibault-Starzyk, P. P. Knops-Gerrits, R. F. Parton and P. A. Jacobs, *Macromol. Symp.*, 1994, **80,** 157; B. V. Romanovsky, *Macromol. Symp.*, 1994, **80,** 185.
182. D. Wöhrle and G. Schulz-Ekloff, *Adv. Mater.*, 1994, **6,** 875.
183. G. Knothe, *Makromol. Chem., Theory Simul.*, 1993, **2,** 917.
184. G. Schulz-Ekloff, in 'Comprehensive Supramolecular Chemistry', ed. J. M. Lehn, Pergamon, Oxford, 1996, vol. 7, in press.
185. R. Szymanski, *Makromol. Chem.*, 1991, **192,** 757; R. Szymanski, *Makromol. Chem.*, 1987, **188,** 2605.
186. K. Ono, H. Konami and K. Murakami, *J. Phys. Chem.*, 1979, **83,** 2665.
187. H. Awano, K. Ono and K. Murakami, *J. Macromol. Sci., Chem.*, 1989, **A26,** 567.
188. H. Awano, K. Ono and K. Murakami, *Bull. Chem. Soc. Jpn.*, 1982, **55,** 2530.
189. G. N. Arkhipovitch, S. A. Dubrovski, K. S. Kazanskii and A. N. Shupik, *Vysokomol. Soedin.*, 1981, **A23,** 1653.
190. A. S. Polinskii, V. S. Pshezhetskii and V. A. Kabanov, *Dokl. Akad. Nauk SSSR,* 1981, **256,** 129; A. S. Polinskii, V. S. Pshezhetskii and V. A. Kabanov, *Vysokomol. Soedin.*, 1982, **A25,** 72; A. S. Polinskii, V. S. Pshezhetskii and V. A. Kabanov, *Vysokomol. Soedin.*, 1985, **A27,** 2295.
191. L. Wong and J. Smid, *J. Am. Chem. Soc.*, 1977, **99,** 5637.
192. K. Arai and Y. Ogiwara, *J. Polym. Sci.*, 1986, **24,** 2027.
193. A. P. Filippov, *Eksp. Khim.*, 1983, **19,** 463.
194. A. D. Pomogailo, D. V. Sokol'skii and E. Baishiganov, *Dokl. Akad. Nauk SSSR,* 1974, **218,** 1111.
195. I. E. Uflyand, E. F. Vainshtein and A. D. Pomogailo, *Zh. Obshch. Khim.*, 1991, **61,** 1790.
196. R. Barbucci, M. J. M. Campbell and M. Casolaro, M. Nocentini, G. Reginato and P. Ferruti, *J. Chem. Soc., Dalton Trans.*, 1986, 2325.
197. W. W. Graessley, *Adv. Polym. Sci.*, 1974, **16,** 1.
198. H. Takadoro, Y. Chatani and T. Yoshihara, S. Tahara and S. Murahashi, *Makromol. Chem.*, 1964, **73,** 109.
199. R. E. Wetton, D. B. James and W. Whiting, *J. Polym. Sci., Polym. Lett. Ed.*, 1976, **14,** 577.
200. J. M. Parker, P. V. Wright and C. C. Lee, *Polymer,* 1981, **22,** 1305.
201. N. Higuchi, T. Hiraoki and K. Hikichi, *Polym. J.*, 1979, **11,** 139.
202. R. Barbucci and P. Ferruti, *Polymer,* 1979, **20,** 1061.
203. A. I. Kuzaev, A. D. Pomogailo and U. A. Mambetov, *Vysokomol. Soedin.*, 1981, **A23,** 213.
204. E. Roman, G. Valenzuela, L. Gargallo and D. Radic, *J. Polym. Sci., Polym. Chem. Ed.*, 1983, **21,** 2057.

18
Overview of Polymer Recycling Technologies

DURYODHAN MANGARAJ

Battelle Memorial Institute, Columbus, OH, USA

18.1 INTRODUCTION

With the unabated growth of plastics usage in our daily life, plastics recycling appears to be the most promising method to meet the future demands for polymeric materials. Starting in the 1970s, plastics recycling has grown slowly but steadily. The learning process has been tough, but the future is bright. It rests on safe grounds because it uses environmentally undesirable feedstock, conserves our limited fossil resources, uses less energy compared to virgin plastics, and, on the whole, solves our waste disposal problem.

It is a new and growing industry and does not have the well-defined structure of polymer synthesis and production. Unlike the latter, which derives its feedstock from centralized petrochemical refineries, plastic recycling obtains its feedstock from vastly diluted sources. Therefore this industry has some special features such as collection, cleaning, separation, restabilization, and reprocessing. In this article, we present the state of the art in plastics recycling science and technology. The topics addressed will include the source of raw materials; primary, secondary, and tertiary recycling; biodegradable and photodegradable polymers; the impact of emerging technologies on plastics recycling; and future trends and direction. The science and technology underlying identification, separation, compatibilization, and impact modification will be discussed briefly.

The information provided in this article has been taken from the open literature, especially the workshop proceedings of the Plastic Institute of America's short courses, the Proceedings of the Society of Plastic Engineers' (SPE) Annual Technical Conference, and Regional Technical Conferences. Two books, namely 'Plastic Recycling, Products, and Processes' edited by Erhig[1] and 'Emerging Technologies in Plastic Recycling', ACS Symposium Series 513,[2] have also been consulted. Whereas most publications treat plastics recycling as a waste disposal process, we consider recycling as an emerging technology which will exploit the nondegradability of plastics for multiple cycles of reprocessing and usage.

18.2 SOURCE AND FEEDSTOCK

An early estimate of plastics waste in the US was provided by Franklin Associates.[3] According to this estimate, the total solid waste, 320 billion pounds in 1988, was to reach 380 billion pounds by the year 2000. Plastics waste, which constituted only 7.2% of the total in 1988, would increase to approximately 11% during the same time span. Subsequently, Curlee[4] presented a detailed projection of plastics use in different product categories and predicted the volume of the post-consumer plastics waste in each category for the years 1995 and 2000. According to these estimates, the plastics waste in 1995 was to be 43.7 billion pounds, a 22% increase over 1990. Thermoplastics were projected to contribute the largest fraction, approximately 88.8% of the total, followed by thermoset (6.9%) and polyurethane (4.3%). Product packaging would contribute the largest fraction, approximately 45.1% of the total, followed by the consumer plastics at 10% and the construction industry at about 9%. The furniture, electrical, and electronics industries would contribute approximately 7% each, automobiles 4.9%, and other transport applications about 1.4%. Recently the consulting firms A. T. Kearney and Franklin Associates have estimated that solid waste generated in 1993 was approximately 413.8 billion pounds (Figure 1), of which plastics waste constituted 9.3% (*i.e.* 38 billion pounds).[5]

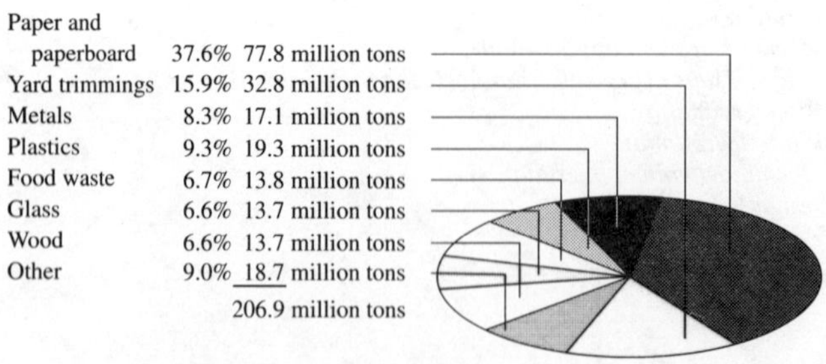

Figure 1 1993 total waste generation, before recycling.

The percentage of plastics waste recycled is the smallest amongst the major categories of post-consumer waste (Figure 2). It did, however, increase from approximately 1% in 1988 to about 4% in 1993. The amount of post-consumer plastic feedstock for individual plastics and the pattern of consumption is presented in Table 1 for the years 1993 and 1994.[6] This estimate falls short of 4% of the total plastics consumed (76.2 billion pounds) in 1994, because it excludes the PET recovered from X-ray films (55 billion pounds) and the loss of plastic waste at cleaning. It may be noted that the consumption of plastics and the generation of plastics waste is rising at a faster pace in building and construction than in other product categories. This is attributed to the higher rate of growth of plastics usage in this sector compared to the other sectors of plastics applications.[7]

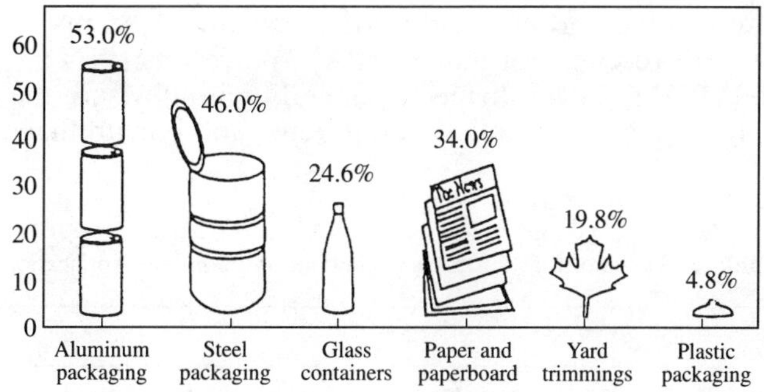

Figure 2 1993 recovery rates for selected components of municipal solid waste by percentage of total generation.

Table 1 Post-consumer Plastic Feedstock and Patterns of Consumption.

Post-consumer recyclate patterns of consumption[a]

End-use market	*Million lb* *1993*	*1994*	*End-use market*	*Million lb* *1993*	*1994*
Poly(ethylene terephthalate) bottle grade[b]			Low-density polyethylene		
Fibers	210	240	Film	95	102
Food bottles	25	20	Other	25	28
Nonfood bottles	35	40	Total LDPE	120	130
Strapping	25	35	Polystyrene		
Sheet	35	40	Foam packaging	14	16
Alloys, compounds	20	25	Loosefill	12	14
Export	25	50	Solid molding	8	8
All other	75	75	Other	6	7
Total PET	450	525	Total PS	40	45
High-density polyethylene[c]			Poly(vinyl chloride)		
Bottles[d]	120	150	Molded goods	7	8
Drainage pipe	65	95	Other	5	6
Film	50	45	Total PVC	12	14
Pallets	30	40	All other resins	60	75
Plastic lumber	35	35	Grand total	1162	1339
All other	180	185			
Total HDPE	480	550			

Post-consumer feedstock patterns of sourcing[e]

Resin/product	*Million lb* *1993*	*1994*	*Resin/product*	*Million lb* *1993*	*1994*
Poly(ethylene terephthalate)[f]			Low-density polyethylene	120	130
Soft drink bottles[g]	420	480	Polystyrene[j]	40	45
Custom bottles	30	45	Poly(vinyl chloride)[k]	12	14
Total PET	450	525	Polypropylene	25	30
High-density polyethylene			Other	35	45
Natural color bottles[h]	260	280	Grand total	1162	1339
Pigmented bottles	145	200			
Base cups	40	30			
All other[i]	35	40			
Total HDPE	480	550			

[a] Excludes typical 10–20% loss of collected volume that occurs during cleaning and reclaim steps. [b] Primarily PET bottles converted by chemical processes to monomer for reuse in drink bottles. [c] Includes bottles and nonpackaging film. [d] Includes base cups. [e] Excludes typical 10–20% losses of collected volume that occur during cleaning and reclaim steps. [f] Excludes 55 million lb of X-ray film, most of which is recycled in a chemical process after silver removal. [g] Base cups included in HDPE. [h] Typically, natural bottles are homopolymer for milk and water use; pigmented are copolymer for chemical and detergent use. [i] Includes packaging (primarily HDPE film) and nonpackaging. [j] Mostly PS foams. [k] Excludes 190–200 million lb reclaimed from vehicle battery cases.

The major fraction of plastic waste comes from packaging applications; it includes low-density and linear low-density polyethylene (LLDPE), high-density polyethylene (HDPE), polystyrene (PS), ethylene–vinyl acetate copolymer (EVA), poly(ethylene glycol terephthalate) (PET), poly(vinyl chloride) (PVC), and ethylene–vinyl alcohol copolymer (EVAL). The typical share of each of the plastics in packaging and automotive applications is presented in Table 2.[7] Plastic waste from automotive and allied industries constitute the major source of the waste from durable goods at present. Polyamide (PA), polyester (PET), polycarbonate (PC), poly(phenylene oxide) (PPO), polyurethane (PU), and their alloys are the major players in the automotive industry, followed by phenolics, polystyrene, and polyurethane.

Table 2 Types of Plastics in Packaging and Automobile Applications.

Automotive		*Packaging*	
Material	*% of total*	*Material*	*% of total*
Polyalkene	23.7	LD polyethylene	32.0
Polyurethane	19.4	HD polyethylene	31.0
Engineering plastics	10.9	Polystyrene	11.0
Polystyrene	10.4	Polypropylene	10.0
Poly(vinyl chloride)	10.0	Poly(ethylene terephthalate)	7.0
Unsaturated polyester	6.6	Poly(vinyl chloride)	5.0
Phenolic	1.5	Others	4.0
Others	17.5		

Although the lifecycle of the building materials is much longer (25 years) than that for packaging and automotive use, it is anticipated that a large volume of these materials including plastics will show up as waste in the coming years. It is interesting to note that a very large fraction of this waste will be natural polymers such as cellulose and concrete. The future challenge therefore lies in developing technologies which can combine them to manufacture useful products. Table 3 presents the consumption of some plastics in semidurable goods, particularly in appliances and electronics. ABS, poly(phenylene oxide) (PPO), polystyrene (PS), polycarbonate (PC), and PC alloys constitute the major components of the waste. Figure 3 presents Curlee's estimates of plastics waste for different polymers from this segment of plastics applications for the year 2000.

Table 3 Durable Plastics Consumed in Appliances, Electronics, and Equipments in 1994.

	ABS	*PS*	*SAN*	*Acetals*	*Nylon*	*PC*	*PP*	*Polyester*	*PPO*	*Total*
Appliances	269	162	14	19	22	63	242	14	36	841
Electronics	26	0	0	16	96	49	0	63	34	284
Business equipment and housing	81	0	0	0	0	38	0	0	27	146
Total	376	162	14	35	118	150	242	77	97	1271

In a recent update, Franklin Associates estimate that the total municipal solid waste will be approximately 391 million tons by the year 2000, compared to their earlier estimate of 380 million tons.[5]

18.3 PLASTICS RECYCLING TECHNOLOGY

Recycling of both post-consumer resin and industrial scrap has grown substantially during the past decade. Initiated under legislative pressure, recycling is now a profitable industry on its

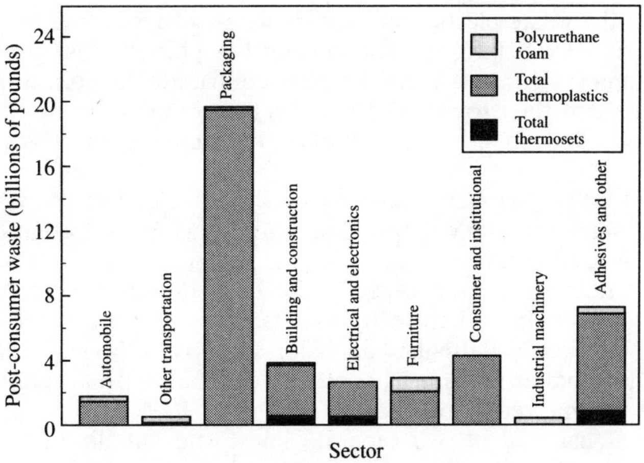

Figure 3 Post-consumer plastic waste, 2000.

own merit without the support of legislature or government handouts. Development of both the science and technology of recycling has largely supported this growth. Every area, including collection, transportation, sorting, cleaning, preprocessing and reprocessing, has seen innovation and invention. Significant among them are our knowledge of: the effect of contamination on processing and end application; the methodology for identifying and automatic separation of contaminants; restabilization of the post-consumer resin and industrial scrap to withstand multiple processing; compatibilization by both chemical and mechanical methods to prepare blends with properties close to those of virgin resin; chemical degradation into monomers for repolymerization, and methods of characterization of degraded products, degradation products, and development of standards for recycled products. Preprocessing of recycled products and development of sound methods for processing recycled materials have also contributed to the healthy growth of plastics recycling technology.

For convenience and clarification, the technology will be discussed under three headings, namely primary, secondary, and tertiary recycling. Each of the above developments will be discussed where they are most appropriate. In addition, the impact of emerging material and process technologies on plastics recycling will be discussed separately. At the end, the future of plastics will be discussed along with new directions for the technology.

18.3.1 Collection, Separation, and Handling

The sources of waste plastics is highly decentralized and diluted. For successful recycling, they need to be collected, transported to a central facility, cleaned, and sorted before they can be sold for reprocessing. The main points of collection are plastics manufacturing operations, municipal curbside, and junkyards. The manufacturing companies generate waste during resin production, fabrication, conversion, packaging, and assembling. The waste is either reprocessed at the site of operation or is sold in large quantity to waste processors. Most durable plastic products go to junkyards at the end of their service life. This includes automobiles, appliances, furniture, and fixtures, and waste from building and construction. The latter is expected to show a high rate of growth in the 1990s. The waste from these two sectors is automatically collected. However, systematic attempts are needed to collect post-consumer plastic waste. Thanks to legislation and incentive, a large infrastructure is being built to collect waste plastics alone and/or with other types of solid waste. More than nine states in the USA have enacted the bottle bill. This bill provides incentives, *e.g.* 5 to 10 cents per container for returning them to grocery stores or recycling centers. By the early 1990s, approximately 40 million people, 15% of the total population in the USA, had joined this program. The number of communities participating in curbside recycling increased from 1052 in 1988 to 2711 in 1990.[8]

Box-like household containers provided by the municipality or recycling companies have become handy to sort the waste plastics into different categories before their disposal. This type of source separation is very helpful in reprocessing the plastics waste, although it puts some burden on the consumer. The two factors which contribute to collection efficiency are the regularity of collection and the provision of a dedicated container to dispose of the waste. The participants are also asked to crush the bottles by stepping on them twice, to facilitate transportation.

Large vehicles, 15 to 40 cubic meter capacity, are used to collect the waste. They may have side-dumping bodies, semi-automatic top-loading bodies, or open-top rear dumps with movable divider bodies or rear-load compactor capability. Since plastics have a very large volume to weight ratio, either trucks with large capacity are used or smaller trucks are provided with compactors to reduce the volume before transportation. Compactors have been found better for densification of waste than flattener and baler systems, although both systems are being used at present. Compactors are simple to operate, accept all types of plastic products, and can reach a high level of volume reduction.

Usually the waste is handled in two separate lines, one handling heavy materials such as glass, steel, aluminum, *etc.*, and the other handling plastics, rubber, and newspaper. The newspaper is tipped off. The rest of the materials are manually sorted on an elevated platform and the containers are thrown into large storage bunkers beneath the platform. Commingled plastics are tipped into a continuous-feed floor conveyor, which is usually fitted with a rubber-like skirt on the receiving edge. This skirt minimizes breakage. The materials are then sent to a pre-sort station where the steel contaminants are separated by an overhead magnetic separator. The separator removes 90–95% of the steel. The rest of the material is then sent to a density separator, which splits the material into light and heavy fractions. This system is usually an air separating system, where the materials are blown to different distances from the center, depending on their density. Hydrocyclones are used in some cases to achieve better separation. Final separation is carried out prior to processing, using a variety of methods for distinguishing individual plastics or broad categories of plastics which are mutually compatible and processable. Table 4 summarizes the estimated collection levels of different product types, and the resins they are made of, from typical collection programs.

Table 4 Collection Levels of Different Product and Plastics Type in a Typical Collection Program.

		Estimated recovery level[a]		*Estimated density with step-on-it-program*	
Category sold to the public	*Resins collected*	lb $(HH\,yr)^{-1}$	(kg $(HH\,yr)^{-1}$)	lb yd^{-3}	(kg m^{-3})
Soft drink bottles	PET	5–11	(2.3–5.0)	30–45	(18–27)
Milk jugs	HDPE natural	5–10	(2.3–4.5)	25–30	(15–18)
Soft drink bottles & milk jugs	PET, HDPE natural	8–17	(3.6–7.7)	30–40	(18–24)
Beverage bottles	PET, HDPE natural, PVC clear	8–18	(3.6–8.2)	30–40	(18–24)
Detergent & bleach bottles	HDPE colored	1–2	(0.5–0.9)	35–50	(21–30)
All clean bottles	PET, HDPE natural, HDPE colored, PVC, PP, multilayer	15–24	(6.8–10.9)	40–50	(24–30)
All rigid containers	PET, HDPE, PS, PVC, PP, multilayer	20–30	(9.1–13.6)	40–50	(24–30)
All clean plastics	PET, HDPE, LDPE, PS, PVC, PP, multilayer	25–40	(11.3–18.1)	NA	NA

[a] HH yr = household year.

18.3.2 Primary Recycling of Plastics

In primary recycling, single polymers with the least amount of contaminants are reprocessed into products, the same as those produced from the virgin resin or other marketable products. For this, the material needs to be cleaned, separated from contaminants, restabilized, and processed by innovative methods to overcome the damage in reprocessing. Most plastics lose their impact resistance on reprocessing. The impact properties can be improved by using a suitable impact modifier. This section will therefore address the following topics: identification and separation of individual plastics; restabilization of regrinds and post-consumer resins; and reprocessing, including innovative processes and process changes. This will be followed by a detailed discussion of the recycling of individual plastics.

The plastics used for primary recycling are derived from manufacturing scraps, spills, purging materials, rejects, and off-spec products. Plastic containers such as beverage bottles, milk jugs, and detergent bottles constitute a large source of single plastics used in primary recycling. The third major source of raw materials for primary recycling are large parts of durable goods such as automotive bumpers, vertical and horizontal panels, doors, instrument panels, vinyl seat covers, computer housings, refrigerator panels and doors, *etc.* The building and construction industry will in future provide another large source of waste materials, such as those used in vinyl siding, roofing, windows, and insulated panels. Used flooring materials such as vinyl floors and carpets made from nylon, polyester, and polypropylene fibers will also provide a major source of polymers suitable for primary recycling. Another interesting source of individual polymers will be fishing gears made from polyamide and high-density polyethylene or polypropylene fibers.

18.3.2.1 Identification and separation

Primary recycling of plastics waste is accomplished in two steps, namely separation into individual plastics free from contamination, followed by reprocessing. Freeing from contaminants is very important because metallic contaminants degrade plastics during reprocessing, and polymeric contaminants produce phase-separated morphology with poor mechanical properties. Separation at the source is very helpful, but it has not been possible due to consumder apathy and an absence of a suitable code to differentiate one plastic from another. Hence, separation is carried out either at material recycling faciilties (MRF) or in plants before the material is fed into the processing equipment. Initially, the plastics were hand-sorted according to color, tint, and whitening characteristics when crushed, but this method was found to be very slow and inaccurate. Subsequently, gravitational separation methods were used because most plastics differ from metals and ceramics and from each other in respect to their specific gravity (Table 5).

Table 5 Specific Gravity of Waste Components.

Material type	*Specific gravity* (g cm^{-3})
Steel	7.86
Aluminum	2.7
Glass	2.4–2.5
PET	1.29–1.40
HDPE	0.94–0.96
LDPE	0.91–0.92
LLDPE	0.92–0.94
Polypropylene (PP)	0.90–0.91
Polystyrene (PS)	1.04
ABS	1.02–1.08
Polycarbonate (PC)	1.2
Polyamide (PA)	1.12–1.14

Substantial work has been done, particularly by the US Bureau of Mines, in developing gravitational methods for decontaminating and separating the plastics. This includes air classification, water elutriation, and an electrodynamic technique.[9] The latter method is used to separate nonmagnetic metals, which dissipate electrostatic charge rapidly compared to plastics, which hold charge for a while. The US Department of Agriculture has developed a method for separating plastics from paper by using a hot air stream in which the paper maintains its integrity, but the polymer film contracts and forms particles.[10] Selective wetting of specific plastics by special wetting agents has been used for separating the waste into individual plastic streams. Small air bubbles are introduced into the separation cell; these adhere to hydrophobic particles, causing them to float up to the surface, leaving the hydrophilic ones at the bottom. This technique can be adopted to separate plastics of similar density[11] but different polarity.

Dow Chemical has developed a gravitational technique based on chlorinated solvents and their mixtures.[12] Wehernburg at the Carnegie-Mellon Institute at Pittsburgh has developed a solvent separation process.[13] The Massachusetts Institute of Technology[9] has developed a vortex classifier in which the flakes move circularly in a radial, inward-flowing vertex and each plastic reaches a position of equilibrium due to the balance of gravitational and centrifugal forces. Newer methods are based on reflectance and absorption spectra of electromagnetic radiations such as X-ray, visible, and infrared light.

For gravitational separation, the waste stream, including dirty bottles, are granulated and reduced to flakes. Some of the labels and adhesives are removed in the size reduction process. These flakes are then moved to an air table where the flow of low-pressure air removes the loose dirt and label material from the flakes. The contaminated flakes are then agitated in a washing tank where detergent solutions are used at a selected temperature to remove the leftover dirt. Alkali such as sodium hydroxide (NaOH) is not used when cleaning PET because of the potential for hydrolysis. However, some manufacturers use NaOH along with surface-active agents. The crude polymer flakes are rinsed with fresh water and then moved to a flotation tank or hydrocyclone tank. Cyclones, both air and water, improve the separation forces by adding a centrifugal force to the gravitational force. A low-foam detergent is added to the solution to minimize capillary action that takes some lights to the bottom and *vice versa.* The concentration of solids also influences the efficiency of this process. The flakes, separated into heavy and light streams, are then rinsed, spin-dried, and hot-air dried.

The heavy fraction, which usually contains PET, PVC, and aluminum, is fed into a rotating drum, exposed to high-voltage electricity. In this electrostatic separation process, PET is separated from aluminum. The presence of PVC, even in very small quantity, is detrimental to PET reprocessing. It is identified and removed by an X-ray fluorescence spectra technique, initially developed by Tecoplas in Italy.[14]

For PET and HDPE bottle recycling, it is not necessary to granulate the bottles. They are presented to the inspection system with or without washing and are separated into different streams by color and type after their identification, manually or automatically. However, manual inspection is very slow and inaccurate. Hence, automatic separation systems are being adopted in most plants. Automatic separation systems use an array of IR, visible light, and X-ray fluorescence sensors to identify the different plastics and separate them into individual groups. Dinger has discussed how major improvements in sorting systems during the last five years have overcome the cost and quality hurdles and has described the potential new separation techniques likely to come in the near future.[15]

18.3.2.2 On-line sorting of plastic bottles

The automatic separation system, developed by Magnetic Separation System of Nashville, Tennessee, uses X-ray, IR, and visible spectra sensors for separating the post-consumer recyclate bottles or flakes into individual plastics and into different color groups. X-ray sensors, used for separating PVC, are very accurate and can operate at as high as 99% or better efficiency. IR and visible sensors are used to separate the colored bottles into individual polymers and color groups.

The separation system (Figure 4) essentially consists of a metering inclined conveyer, air knife, special disk screen, singulating infeed conveyor, and sensor module. A motor control system provides operator interface screens which control the sorting functions, including the number of bottles sorted into each fraction, ejection timing, and sort positions. Individual systems currently in use in Germany, Switzerland, and the United States are described in a paper by Kenny and Vaughan.[16] The systems are customized, based on the composition of the post-consumer recyclate and the end application of the separated streams. Some systems use X-ray and IR sensors in two locations to achieve better separation. In addition to sorting equipment, some systems also use equipment for breaking the bales and splitting the bottles into more than one stream for smooth operation. Grinders are used when the bottles have to be ground into flakes for further processing. Whereas PVC separation is accomplished at 99%. HDPE and PET separation is between 80 and 90%, depending on the level of contamination.

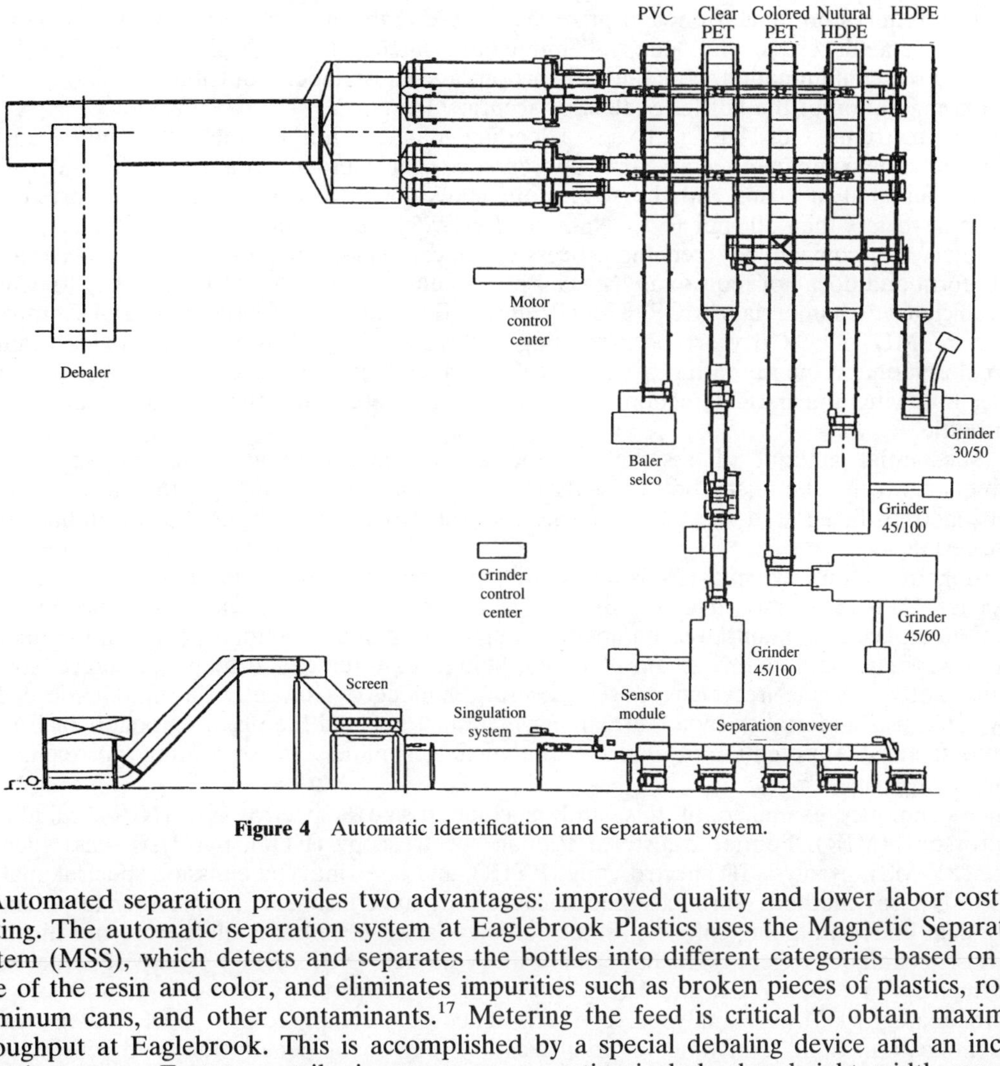

Figure 4 Automatic identification and separation system.

Automated separation provides two advantages: improved quality and lower labor cost for sorting. The automatic separation system at Eaglebrook Plastics uses the Magnetic Separation System (MSS), which detects and separates the bottles into different categories based on the type of the resin and color, and eliminates impurities such as broken pieces of plastics, rocks, aluminum cans, and other contaminants.[17] Metering the feed is critical to obtain maximum throughput at Eaglebrook. This is accomplished by a special debaling device and an incline metering system. Factors contributing to proper operation include clear height, width, spacing, belt speed, and incline angle. Proper presentation of the bottle to the sensor is critical. The bottles are split into four streams and two to three bottles are presented to the sensor per second, one at a time.

The primary identification sensor uses a multibeam, near-IR array to identify the bottles into three classes: Class 1, PVC, PET; Class 2, natural HDPE, PP; Class 3, mixed color HDPE and opaque containers. This sensor is also capable of separating colored PET from clear PET and PP from milk jug HDPE. The X-ray sensor identifies PVC, and a machine vision sensor system provides up to seven color classifications of the plastic bottles. After identification, the containers are ejected from the conveyors into appropriate collection stations using high-speed pulsed air nozzles. The motor control center (MCC) of the separation system controls motor protection, sequential slant up for the system, fault indication, and operation control. In addiiton, a touch screen input panel allows the operator to select any available sort to be directed to any ejection station. Visible light color sensors have been added which sort pigmented HDPE into different colors. The system also includes a decision cross-checking device between the primary sensor and the color sensor. This compares the decisions of the two sensors by comparing them with a logic file. The latter then provides correct identification in case there are discrepancies between the two decisions. The system has successfully operated for the last three to four years at a capacity of 5000 bottles h^{-1}.

The debaling system designed for Eaglebrook requires that the bales be presented to the debaling equipment in the same orientation as the original compression. This design feature requires less horsepower, reduces bottle clusters, and requires minimum energy. The debaling and declumping system incorporates a surge bin and metering conveyor to feed the screening system. The improved capacity and higher separation accuracy, due to increased metering efficiency, reduces bottle clusters and provides a more uniform feeding system. The separation efficiency depends on several factors. Timing and catcher bounceback accounts for 1–2% accuracy loss; contamination, container distortion, and loose labels contribute to about 3–4%, and nonsingulation of the bottles 5–10% of accuracy loss.

Asoma Instrument of Austin, TX, is a leading manufacturer of automated bottle sorting equipment. The company uses an X-ray fluorescence spectrophotometer sensor. The identification is completed in 10 ms and the separation takes about 20 s per bottle. The sorted PET streams have less than 50 ppm PVC. National Recovery Technology of Nashville, TN, uses a proprietary electromagnetic screening process which can handle the bottles either in crushed or whole form and does not require any special positioning or orientation of the bottle to achieve high efficiency. Chamberlain/MCR, Hunt Valley, MD, and Automation Industrial Control of Baltimore, MD, offer a 'paysort' bottle sorting system, which uses a sophisticated video camera and color monitor incorporating a strobe to detect and distinguish colors of post-consumer bottles following a near-IR detection system which also determines the primary resin found in each bottle.

A substantial amount of research is focused on microseparation techniques and on techniques which can reject bottles with trace amounts of harmful contaminant. Near-IR spectrometry is being used to separate bottles for household chemicals and ones with hazardous waste residues.

Sorting of automotive plastics is more difficult than sorting of plastics from packaging recyclates. Whereas only five to six polymers are used for packaging, post-consumer automotive plastics contain large numbers of engineering and commodity plastics, modified in various ways, including alloying and blending, filling, reinforcing, and foaming. Hence, sorting of automotive plastic recyclate poses several challenges. Recently, a systematic study, PRAVDA, was undertaken by a German car manufacturer and the plastic suppliers in Europe to investigate the potential of various analytical techniques in separating post-consumer automotive plastics.[18]

The techniques examined in this study include near-IR spectroscopy (NIR), middle-IR spectroscopy (MIR), Fourier transform Raman spectroscopy (FTR), pyrolysis mass spectrometry (PY-MS), pyrolysis IR spectroscopy (PYIR), and laser-induced emission spectral analysis (LIESA). X-ray methods were excluded because they have insufficient sensivitity to polymers, other than ones containing chlorine. Since commercial spectrophotometers were not available for most techniques except NIR, either laboratory models (MIR, FTR) or experimental stage instruments (PY-MS, PY-IR, and LIESA) were used in this study. A large number of parts (approximately 7000) were analyzed. The techniques were compared in respect to their success in identification, fault rate, time for identification, degree of penetration, and sensitivity to surface quality. The fault rate is the number of wrong identifications, given as percent. If the sum of the identification and fault rate is less than 100, the difference gives the rate of incomplete correct identification. The biggest stumbling block was the identification of black samples which could not be analyzed by NIR and FTR. MIR is the only technique which not only identified the black samples, but gave the highest identification rate. Some difficulties were experienced, however, in MIR analysis in the case of blends of two similar polymers such as PP/EPDM or nylon 6/nylon 66. The pyrolytic methods showed poorer identification rates and higher fault rates. The LIESA method is very fast and a remote technology, particularly for fast identification of heteroatoms. It is therefore suitable for identifying fillers, minerals, reinforcing fibers, pigments, flame retardants, and stabilizers specific to the individual plastic. The difficulty with MIR is that it is sensitive to surface micro-roughness and, hence, the samples need to be very smooth. Also, paint or surface coats on the part have to be removed for correct identification of the resin used for making the parts. Further, at this stage, no fiber optic or separated probe is available with MIR technology and, hence, the part has to be brought close to the spectrophotometer instead of the probe reaching the part. Another method of measuring efficiency is the level of contamination. Contamination of parts sorted by the MIR method was less than 1%, whereas contamination of parts sorted manually, using a Car Parts Dismantling Manual, is greater than 10–15%. When the level of contamination is high, further separation by swim–sink or hydrocyclone techniques are necessary.

The cost of a MIR spectrophotometer is approximately DM 100 000. The cost calculated for small dismantlers (dismantling less than 25 cars per day) is approximately DM 0.34 per kg and that for large dismantlers is somewhat less than DM 0.19. Manual sorting, on the other hand, would cost DM 0.71 and DM 0.23 per kg for small and large dismantlers, respectively. Spectrophotometric identification of plastics in automotive plastics waste therefore makes substantial economic sense.

18.3.2.3 Restabilization

Stabilizers are added to plastics to protect them from degradation during high-temperature, high-shear processing conditions, as well as to protect them against outdoor exposure in their service life. Hence, once the material goes through processing and a service life, it loses a substantial fraction of the stabilizers needed for it to survive degradation during subsequent processing and exposure to weather conditions. It is therefore important to add stabilizers and reactive compounds capable of compensating for the preceding damage. The need for restabilization of recyclates has been emphasized by a number of authors. Reactive stabilizers are needed to repair heavily damaged polymeric materials, particularly to improve the molecular weight.[19,20] Recently, Ciba-Geigy has developed a number of antioxidants (AO) and stabilizers for use with PP, HDPE, and PET.[21] The chemical structure and trade names of these AOs are given in Figure 5. Their efficiency was tested by multiple extrusion tests and oven aging followed by measurement of their melt flow index (MFI).

Polypropylene is most vulnerable among alkenic polymers because of its tertiary hydrogen atom. Phosphites, sterically hindered phenols, and their blends have been proven effective in prohibiting PP degradation and discoloration.[22] Polypropylene scrap, although slightly damaged during the first processing, becomes more damaged during the agglomeration process. A sample of PP scrap, when reextruded, showed an increase in the MFI, suggesting thermal degradation. However, when a blend of phenolic antioxidant, Irganox® 1010, and a phosphite, Irgafos® 168, was added to the scrap, the molecular weight degradation was substantially reduced. The melt flow index remained almost constant when the antioxidant Recyclostab® 411 was used in place of Irganox 1010 (Figure 6). The costabilizers present in Recyclostab compensate for preceding damage and form complexes with metallic impurities, rendering them harmless. The increase in stability is reflected in the compound's resistance to thermal aging. When the injection-molded sample was oven-aged at 135 °C, the unstabilized sample became completely brittle after 54 days, whereas the first and second stabilized samples survived for 80 and 92 days, respectively. If the scrap is to be used for outdoor applications, then a light stabilizer such as Tinuvin® 622 [poly(*N*-β-hydroxyethyl-2,2,6,6-tetramethyl-4-hydroxypiperidyl succinate)] has to be added to maintain mechanical properties at a high level. In the case of scrap polypropylene battery cases, although the stabilization was achieved by using Irganox 1010 and Irgafos 168, it was further improved by using 0.33% of Recyclostab 411 and 0.7% of a new recyclostab under development. The developmental product also provided the longest life to battery scrap against oven aging.

Polyethylene degradation depends on the catalyst package used in resin production. Whereas high-density polyethylene made by Ziegler–Natta catalyst undergoes molecular weight degradation, HDPE made using Phillips-type catalyst undergoes crosslinking. Low-density polyethylene undergoes both crosslinking and degradation. These harmful effects are overcome by incorporating blends of Irganox and Irgaphos stabilizers as well as Recyclostab. The latter is superior to the combination of Irganox and Irgaphos and is the stabilizer of choice when the history of the recyclate is unknown. Recyclostab 411 is the most effective stabilizer for low-density polyethylene (LDPE). It is usually accepted that films, when made with carbon black filled polyethylene, survive light degradation. In actual practice the effect of the black as an antioxidant is rather limited and one has to add HALS to survive the damage. This has been verified using black vineyard pipes made from a blend of LDPE/LLDPE.

Molecular weight is a critical criterion for blowing polystyrene foams. Hence, recycled polystyrene foam material has to be restabilized to avoid degradation below the critical molecular weight. Addition of a special stabilizer, Irganox® 245 (AO and costabilizer), is essential to prevent molecular weight degradation.

Poly(ethylene glycol terephthalate) (PET) is recycled in various ways: chemical, coextrusion, coinjection molding, and as fibers and films. The melt and color stability of PET recycled products is improved by using the phosphite stabilizer Ultranox® 626 as well as a small amount of the blend containing Irganox 1010 and Irgaphos 168. Addition of light stabilizer LS-1 to the

Irganox 1010
Pentaerythrityl tetrakis-3-(3,5-di-*t*-butyl-4-hydroxyphenyl)propionate

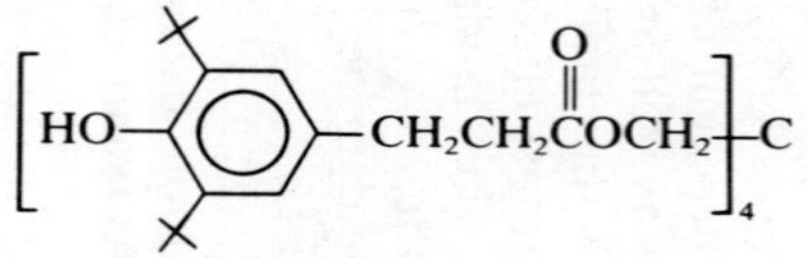

Recyclostab 411
(contains antioxidants and costabilizers)

Recyclostab developmental product
(contains antioxidants and costabilizers)

Recyclostab 421
(contains antioxidants and costabilizers)

Irganox 245
Triethylene glycol bis-3-(3-*t*-butyl-4-hydroxy-5-methylphenyl)propionate

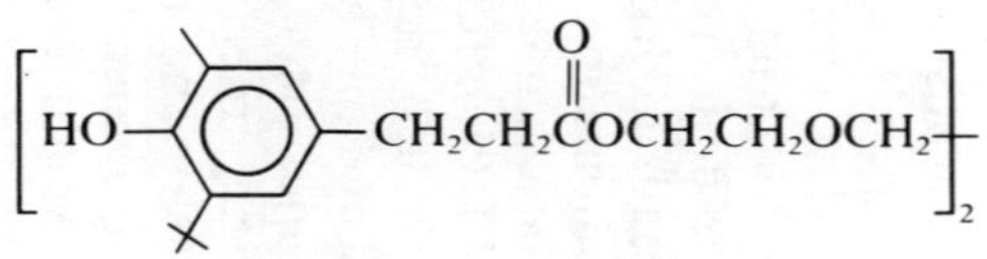

Irgafos 168
Tris(2,4-di-*t*-butylphenyl) phosphite

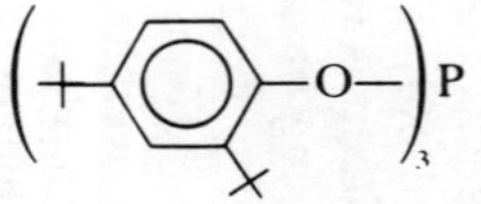

Ultranox 626
Bis(2,4-di-*t*-butylphenyl) pentaerythrityl diphosphite

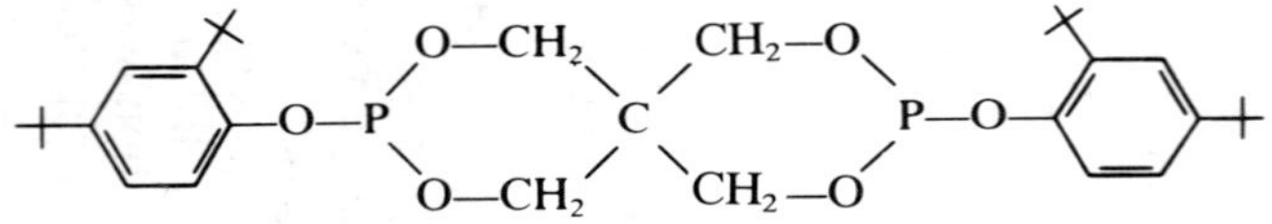

Sandostab PEPQ
Irgafos PEPQ
Tetrakis(2,4-di-*t*-butylphenyl)-4,4′-biphenylene diphosphonite

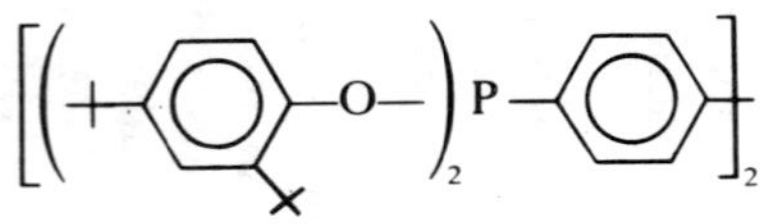

Irganox PS 802
Distearyl thiodipropionate

$$H_{37}C_{18}O{-}\underset{\underset{O}{\|}}{C}{-}CH_2CH_2{-}S{-}CH_2CH_2{-}\underset{\underset{O}{\|}}{C}{-}OC_{18}H_{37}$$

Recyclostab 811
(contains antioxidants and light stabilizers)

TINUVIN 622
Poly-(*N*-*β*-hydroxyethyl-2,2,6,6-tetramethyl-4-hydroxypiperidyl succinate)

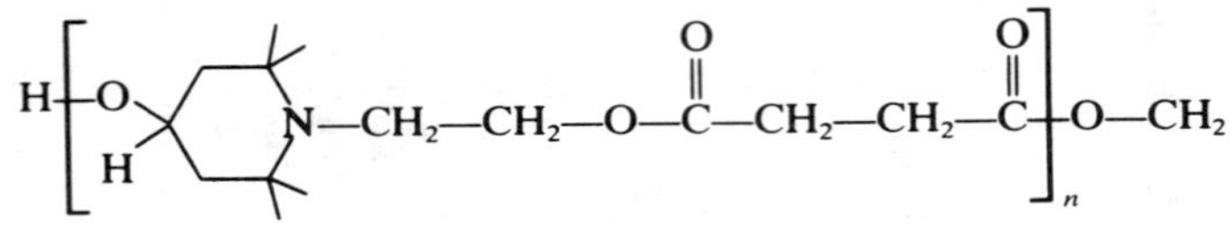

Figure 5 Chemical structures of antioxidants used for restabilization.

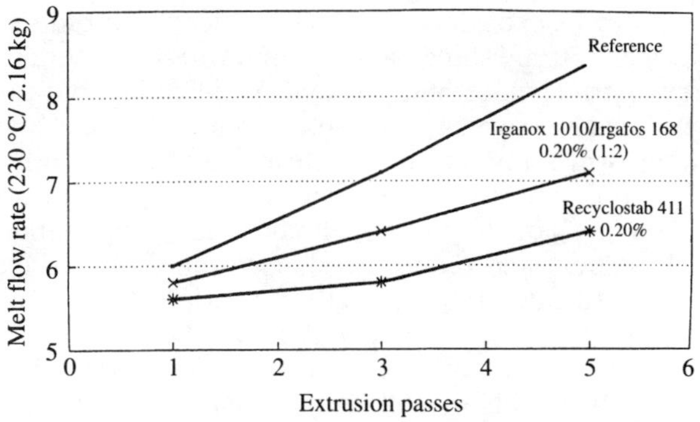

Figure 6 Change of melt flow index of polypropylene scrap (stabilized with different antioxidants) on repeated extrusion.

stabilizer packet (particularly Irganox 1010) is important for outdoor exposure of recycled PET products.

Specialty polypropylene copolymers are used in the automotive industry as bumpers, trims, *etc.* They are usually painted to maintain good aesthetics for the car. However, paint residue in PP or catalyst residue in EPDM causes severe molecular weight degradation during outdoor exposure. The loss of mechanical properties can be avoided by using a thiosynergist. A new stabilizer combination containing a reactive additive has been found very effective in stopping molecular degradation of PP.

Condensation polymers such as polyamides, polycarbonates, and polyesters undergo molecular weight loss due to hydrolysis during extrusion and in subsequent outdoor exposure. Hence, they are subjected to solid-phase condensation to restate their molecular weight. Ciba-Geigy has found a new reactive stabilizer to restate the molecular weight of polyamide by a reactive extrusion process.

Borg Warner Chemical recommends Ultranox® 626 phosphite antioxidant at 0.1% concentration to prevent discoloration and degradation of PET.[23] A 50:50 mixture of Irganox 1010-hindered phenol and Irgonox 168 hydrolytically stable phosphite is also recommended for stabilization of scrap PET. An FDA-approved hindered phenol stabilizer is available from Fairmount Chemical for stabilization of recycled HDPE and other polyalkenes. Penwalt corporation's Anoxyn® 422 acts as a synergist with most antioxidants in stabilizing plastic recyclates. ICI America, UniRoyal, and Dover Chemical Corporation also offer specialty antioxidants for recycled plastics. For multilayer barrier packaging it is important that the ethylene–vinyl alcohol barrier layer be preserved intact to maintain its efficiency and prevent discoloration. This can be accomplished by using Quantum Chemicals' Spetratect RM-11575 stabilizer, which stabilizes EVA against degradation.

As mentioned earlier, light stabilizers are essential, particularly as the recycled products are often used outdoors. DuPont offers Rylex NBC, a UV stabilizer which is very effective for HDPE recyclate products. Fairmont Chemical's Mixxim BB-100 is very effective in stabilizing PET up to 280 °C. However, this can only be used for nonfood applications.

18.3.2.4 Reprocessing

In primary recycling, processing is carried out using techniques commonly used for processing virgin resin such as extrusion, injection molding, blow molding, and thermoforming. The main difficulty in processing recyclates is their poor thermal stability and higher melt flow rate. This is usually taken care of by adding stabilizers and by restoring molecular weight, either through reactive additives or through solid stating. Since most recyclates are used in conjunction with large amounts of virgin resin, this does not pose serious difficulty. However, heat history adds up and recycling the same material over and over again can cause problems. It is therefore important that the processing cycles in reprocessing should be as short as possible, without affecting the product integrity and performance. In recent years, substantial development has taken place in extruder mixing. Twin screw extruders are available in both corotating, counter-rotating, intermeshing, and nonintermeshing configurations. It has been suggested that twin screw extruders are better suited for blending and processing of recyclates. One can select the type of machine which is suitable for the particular material or application.[24]

Intermeshing cororating twin screw extruders are preferred over other types of twin screw machines because they do not subject the recyclate to very high shear, as do counter-rotating twin screw extruders, and provide high throughput. This reduces both thermal and shear degradation because of low shear and reduced residence time. Vented extruders with feed and vacuum vents are desirable to add stabilizers at appropriate times and vent off the volatile byproducts. The screws in both cases are modular and the screw configuration should be adjusted to provide good dispersion. Extruders with tapering screws have been developed by Cincinnati Milacron to densify highly filled materials such as wood-fiber-filled plastic composites.

A substantial amount of work has been carried out in recent years on estimating the resistance of polymers toward multiple cycles of reprocessing, which simulates repeated recycling processes. Paul *et al.* have studied the blow molding behavior of recycled HDPE and PP with and without additional stabilizer.[25] Blow molding is a three-stage process: (1) parison formation, (2) inflation, and (3) cooling. The initial stage is critical to successful blow molding, since it controls the final part thickness distribution. This distribution is important, since excessive thickness leads to material weight and inadequate cooling, whereas an insufficient amount of material results in poor mechanical strength. As the parison comes through an annular die, it is affected by two phenomena, swell and sag. The die swell is caused by the relaxation of the melt and it is greater the broader the molecular weight distribution. The sag is caused by the gravitational forces that act on the suspended parison and it is greater the lower the melt strength (lower average molecular weight). The effect of recycling can be twofold, namely, chain scission resulting in lower molecular weight (low viscosity) and narrower molecular weight distribution or crosslinking leading to broader molecular weight distribution.

When the two polymers, HDPE and PP, were subjected to repetitive extrusion and regrinding processes (simulating recycling), it was found that whereas the viscosity of PP decreased with reprocessing, that of HDPE increased. This is in agreement with the fact that the tertiary hydrogen atom in PP is more prone to hydroperoxide formation and decomposition, whereas the few dangling chains in HDPE are more prone to mechanochemical degradation and crosslinking. It was found that increased reprocessing increases sag and reduces thickness for both HDPE and PP. The effect on sag is more predominant in the middle, thinning being more evident at the top. Use of additional stabilizer increases the die swell for polypropylene because it reverses the narrowing of molecular weight distribution by suppressing chain scission. In the case of HDPE, it suppresses crosslinking, maintaining narrow molecular weight distribution, leading to lower die swell.

Multilayer packaging made of polyethylene and polyamide layers is extensively used to package chemical, agricultural, and industrial products to provide adequate protection against moisture and solvents. The layers cannot be easily separated, making it impossible to separate the individual polymers. Hence, the resins in multilayer packaging are recycled as blends. Potente and Gao studied the recyclability of a blend of polyamide and polyethylene using an ethylene–acrylic acid ionomer as compatibilizer.[26] Dry blends of the three resins were extruded in a Leistiritz twin screw extruder with inter-screw distance of 30 mm and screw diameter of 34 mm. Screw configuration contained different quantities of right-handed screw elements and neutral kneading blocks. Prior to blending, the polyamide was vacuum dried at 80 °C for 6 hours. The mixture contained HDPE, PA6, and Ionomer at 80, 20, and 4 parts level. The screw speed was maintained at 100 rpm and the barrel temperature at 230 °C. Every compound was reextruded up to five times and both mechanical and flow properties were measured after each extrusion step. The second to fifth extrusion step was considered equivalent to recycling. It was found that the color of the extruded strands increased with every processing cycle, indicating shear degradation and oxidation. The elongation at yield and break decreased substantially after each extrusion step. After the fifth step, the tensile strength decreased by 14% and the elongation at break decreased by 71% compared to those for the original. This indicates that molecular degradation took place at each step, with the greatest decrease occurring at the fifth recycling step. There was no noticeable change in morhpology or thermal properties. It was concluded that although substantial loss in physical properties occurs due to molecular weight reduction with each cycle of reprocessing, the multilayer materials can be reprocessed a few times.

An earlier study by Dow Chemical demonstrated that in multilayer packaging materials containing PE, PP, PS, EVOH (ethylene–vinyl alcohol copolymer) and PVDC (poly(vinylidene chloride)), the tie layer acts as a compatibilizer in the recycling step.[27] The Center for Plastics Recycling at Rutgers University has developed a process for repelletizing PP/EVOH ketchup bottles and blow molding the pellets into monolayer and three-layer shampoo bottles.[28] The recycled material is often used as a middle layer in packaging, with virgin materials as the outer layers. Both extrusion and injection blow molding are used for making such multilayer packaging.

Coinjection molding is yet another method for using regrinds or reprocessed materials in recycling. In the coinjection process the resin component for the skin is injected into the mold followed by the injection of the core component.[29] The lamellar flow characteristics of polymeric material allows the process to take place in two individual steps in rapid sequence; this way the reground material is used as the core material with virgin material as the skin. The factors to consider in coinjection molding of recycled plastics are consistency of the flow property, such as melt index or viscosity, and compatibility of the skin and core materials required for interfacial adhesion. Consistency in viscosity can be adjusted by mixing with the desired amount of virgin resin of specific melt index. The interfacial adhesion can be enhanced by using an additive or compatiblizer to enhance the interaction between the core and skin material. Coinjection molding therefore offers the advantages of meeting the demand to increase the recycled content in a product, obtaining the combination of properties from two different materials, and building different properties into the skin and core. For example, in automotive horizontal panels one can use scratch-resistant material for the skin with a stiff material in the core. For exterior and interior trim, one can use a highly colored material or an easily paintable material for the skin only. In this case the recycled material does not need to be separated into individual plastics, but needs to be compounded to provide the necessary melt flow index and strong interfacial adhesion.

Some rules of thumb to be observed in coinjection molding include: skin and core material compatibility; extreme care for materials which are hard to process in injection molding; thin projections limited to skin only; core traveling the path of least resistance; foaming to be done only when the skin is 4 mm or more thick; and greater care in balancing flow fronts. Good knowledge of materials, processing, and equipment are the key to the success of the coinjection molding.

18.3.2.5 Testing

Testing and characterization of recyclates and recycled products are very important in recycling technology. Whereas characterization of recyclates is important for checking the quality of recyclate and its processing behavior, testing of recycled products is important for quality control and marketability. In fact, lack of standards on recycled products makes it difficult to sell either in the open market or to the government. Work is in progress under the stewardship of ASTM Committee D20 to develop standard test methods for recycled products, particularly plastic lumber. Recyclates are tested by the same standard methods as the virgin resin in respect to their thermal, rheological, and physical characteristics. However, these bulk tests do not indicate the type of contaminants or the type of volatile products they will generate during processing. Special analytical methods are being used for this purpose.

(i) Testing of recyclates

Ezrin, Wyatt *et al.* have discussed the use of a headspace GC/MS method as a tool to analyze volatiles and hazardous materials in recycled plastics, especially high-density polyethylene.[30] IR spectrophotometry in conjunction with GC/MS can quantify the levels of these contaminants. Cassel has discussed the use of thermal analysis techniques, including thermogravimetric analysis (TGA), differential scanning colorimetry (DSC), and thermomechanical analysis (TMA), to determine feedstock composition and to study the processing characteristics of PCR.[31] These techniques are relatively fast and automated. They indicate the temperature at which the polymer degrades and the glass transition and melting transition temperatures, and if the different materials in the recyclate are compatible. They also provide insight into the degree of crystallinity and rate of crystallization in the presence of contaminants which are important in PET recycling. Thermal analysis can also be used to select and optimize antioxidant and stabilizer concentration and study the role played by compatibilizer. Coupled with dynamic mechanical analysis (DMA), the thermal techniques can provide detailed knowledge of the viscoelastic and rheological properties of recyclates. Thermal analyses are therefore very handy and very useful in characterizing both the feedstock and the products in recycling.

(ii) Product testing

Computer tomography is a new technology to look into materials and products, particularly to view defects such as voids, cracks, concentration of impurities, and other internal features. The technique uses a 420-kV p X-ray (1.5 mm focal spot) tube to reconstruct images in diameters ranging from 3 to 50 cm. Reconstruction voxels (volume picture elements) are isotropic. Slice thicknesses range from 7.5 microns to 100 microns. Benner *et al.* have used this technique to look into the defects in recycled plastics.[32]

Plastic lumber samples made from HDPE, 20% glass fiber filled HDPE, recycled commingled plastics (New Jersey Curbside Tailings, with and without outdoor exposure), and LDPE were examined by optical microscopy.[33] It was found that most of the profiles have a foam core, possibly due to moisture present in the recycled plastics (Figure 7). The unfilled HDPE specimens failed due to separation of the core from the skin, whereas the glass reinforced specimens failed due to fracture at the bottom of the core. The foamed core appeared to be the dominant site of failure. This central region ruptures and tears away from the more dense region. Metallic impurities, which do not wet the polymer and create interfacial voids, occur in the skin and often initiate failure. Materials having smaller foam cores and greater homogeneity exhibit plastic deformation and less fracturing. It is recommended that elimination of contaminants and meticulous drying will help a great deal in improving the properties of recycled products, particularly plastic lumber. LDPE products are more homogeneous. In this case, the plastic wets the metallic contaminants, thus eliminating the interface. However, the compressive strength is lower because LDPE has approximately one-third the stiffness of HDPE.

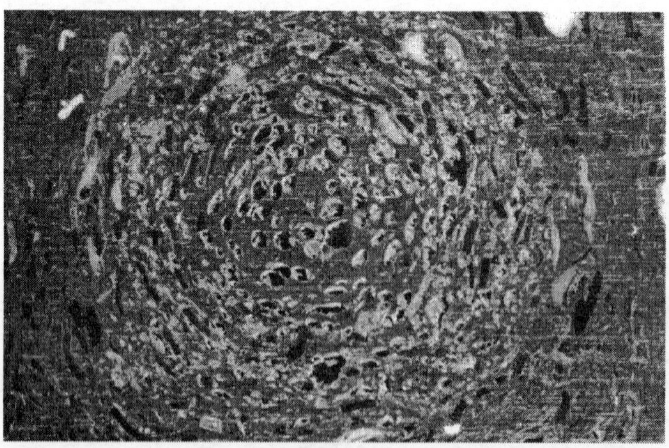

Figure 7 Typical cross-section of plastic lumbar showing cylindrical symmetry and various imperfections such as inclusions and voids.

As mentioned earlier, characterization and testing of plastic lumber is very important to improve the quality of the lumber and its potential for sale. Substantial work has been carried out in the last few years on testing plastic lumber using ASTM and other methods. The wear and abrasion behavior of two types of plastic lumber, one made from recycled polyalkene (Zahn Manufacturing) and the other from polyethylene and cellulose fibers (Riverite Manufacturing), were studied by Beatty *et al.*[34] They found the wear mechanism of both plastic lumbers are similar to that of virgin plastic samples. However, the wood fiber filled specimen exhibits better wear resistance than the lumber made from 100% recycled polyalkene.

Hill *et al.* have characterized plastic lumber posts by rectangular and circular cross-section using differential scanning calorimetry, NMR, and positron annihilation spectroscopy (PAS).[35] They came to the conclusion that there is material segregation in the lumber, low-density polyethylene (LDPE) moving to the skin and linear low-density PE staying in the core. This is irrespective of extrusion die geometry, circular or rectangular. The resins are possibly contaminated by poly(ethylene terephthalate) (PET). The polyethylene-to-sawdust ratio varies only within 10% throughout the cross-section. PALS measurement indicated a coreshell structure for the rectangular post, with fewer free volume sites in the skin (small I_3) than that in the core. The small free volume in the skin contributes to greater strength and the larger free volume in the core provides greater toughness, both properties being desirable for better performance of the lumber. Improved core–skin structure could be generated by designing specific geometry for the die.

Beatty has characterized thermal diffusivity of lumber samples made from recycled plastics, wood flour filled recycled plastics, and glass fiber filled recycled plastics.[36] The experimental setup included a heat blanket placed under a copper plate of 0.635 cm thickness. Small lumber specimens were placed over the copper plate. A liquid crystal temperature sensor was placed on the upper surface of the specimens. The temperature on the upper surface was measured for different thermal gradients. It was found that the thermal conductivity of the glass fiber filled and wood fiber filled lumber specimens are greater than that of the pure plastic lumber. Larger temperature gradients exhibited greater thermal diffusivity. However, this method could not be used to estimate equilibrium thermal conductivity due to the longer time required to inspect temperature change and edge effects of the specimen.

Yang *et al.* have measured flexural fatigue behavior of recycled plastics, using a Model VSS-40H Sheet Specimen Fatigue Testing machine supplied by Fatigue Dynamics.[37] The recycled commingled plastic materials provided both by Zahn Manufacturing and Cairo Manufacturing had a substantial amount of large voids. Hence, the samples were remolded by a pattern mold at 2.3 MPa and 180 °C to reduce the void size. Specimens cut from the recycled commingled plastic lumber were subjected to a fatigue test in flexural mode. The uncompatibilized commingled plastic lumber shows poor fatigue behavior due to high interfacial energy. The compatibilized commingled plastic exhibits fatigue behavior close to that of recycled high-density polyethylene. Both lumber specimens exhibited stress hardening and crack initiation before failure. At higher temperatures the plastic softened, requiring lower stress for failure, thereby lowering the fatigue strength of the specimen. At high frequency the modulus drops, probably due to frictional heating caused by the relative movement of the different phases. Similar behavior was exhibited by glass fiber reinforced polymeric composites. It is predicted that at a higher frequency the failure will be initiated by high-temperature ductility and not by fatigue crack initiation.

The same authors have also studied the fatigue behavior of virgin and recycled high-density polyethylene.[38] Both the specimens exhibited similar S–N (stress versus number of cycles) curves, but the recycled HDPE (RHDPE) lifetime was 60–70% shorter than that of virgin HDPE. This was attributed to lower molecular weight, broader molecular weight distribution, and smaller crystallinity of RHDPE compared to those for virgin HDPE. Unlike commingled plastics, both HDPE and RHDPE exhibit fatigue failure by cracking, not by a thermal mechanism.

Mason has used a STA/FTIR technique for studying the level of degradation in ABS pipe specimens made of virgin and virgin plus 10% and 20% regrind ABS.[39] In this technique the specimen is subjected to simultaneous thermogravimetric analysis (TGA) and differential scanning calorimetry (DSC). TGA indicates the types of endothermic decomposition and exothermic events that take place during degradation, which are simultaneously monitored by DSC due to the change in specific heat. FTIR, at the same time, measures the effluent gas and throws light onto the mechanism of degradation. The analysis of an ABS specimen showed that the material containing 10% regrind had 7.2% degradation of polybutadiene block whereas the specimen containing 20% regrind had approximately 11.8% degradation of the same block.

18.3.2.6 Recycling of individual plastics

(i) Poly(ethylene glycol terephthalate) (PET)

PET is a semicrystalline polyester, prepared from ethylene glycol (EG) and dimethyl terephthalate (DMT) by a polycondensation process, in the presence of an antimony catalyst.

Copolyesters which have smaller crystallinity but higher toughness are prepared by replacing DMT with isophthalic ester or EG with another glycol. The molecular weight of the polymer as measured by intrinsic viscosity (IV) varies between 0.58–0.67 $dL\,g^{-1}$. In 1970s a new technology, 'Solid Stating', helped in making a higher molecular weight PET having an IV close to 0.7 $dL\,g^{-1}$. The high-strength material could be blown into containers, which were later used to package beverages. Whereas the maximum crystallinity in PET is 55%, the bottle grade PET has a crystallinity close to 25%. The size of the crystallites are important for clarity. The crystallinity and the rate of crystallization are controlled by using a comonomer during synthesis and by adding nucleating agents during processing. The various applications of PET include bottles for soft drinks, films, straps, and fibers. Currently, the US recycles approximately a billion pounds of PET and HDPE bottles per year. Europe reclaims approximately 2.3 billion pounds, 7.5% of their consumption. Double digit growth rates are anticipated in both continents, especially with an impending initiative in Europe to reclaim 15% of waste PET per annum.[40]

PET bottles are made of a base cup of high-density polyethylene, a body of PET, and a cap of aluminum or polypropylene. Consequently, the reclamation technology involves separating these three different materials and recycling them separately. Most commercial PET bottle recovery processes are adapted from the earlier processes developed by DuPont and Goodyear.[41] Figure 8 provides the schematic of the Goodyear process for recycling PET bottles. This technique, however, is not very different from the recycling process recently developed at the Center for Plastics Recycling Research (CPRR) at Rutgers University in New Jersey. A detailed discussion of PET recycling and application at CPRR has been provided by Milgram.[42]

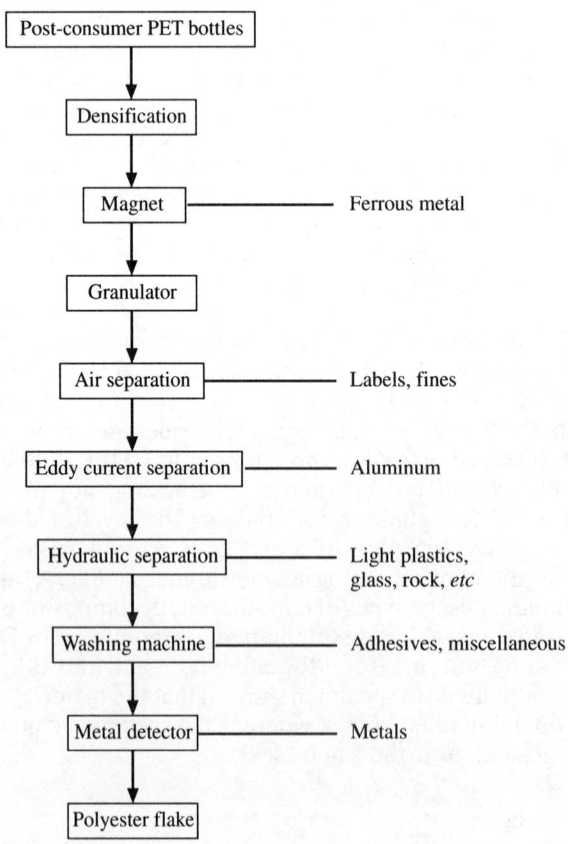

Figure 8 Process steps in recycling PET bottles.

Separation of individual constituents, particularly segregation by color, becomes easier if the whole bottle is available for recycling. An added incentive for separation by color is the fact that colorless RPET sells at a higher price than colored RPET. At St. Jude and Wellman, separation of base cups (colored HDPE) is accomplished by dissolving the glue line to separate the cups from the PET body; at Star Plastics, an automated mechanical system separates cap, base cup, and the main body of the bottles.

Off-the-shelf equipment is generally available for PET recycling. Shredders and granulators are supplied by Alpine American, Ball and Jewell, Conair, Cumberland Engineering, Foremost Machine Builders, Nelmor, Rapid Granulator, Allegheny Bottle Shredders, and Dresser Industries. Air separators are supplied by Conair, Cumberland, and Nelmor. Aidin Automation Corporation offers a commercial machine for separating base cups. Electrostatic separators are made by Carpco and Eriez Manufacturing. The latter uses a rare earth magnet that acts on traces of iron impurity present in aluminum and pulls it from the plastic stream. Cumberland and Lumus Development Corporation offers complete PET bottle recycling systems to new entrepreneurs. CPRR has also made its process available for licensing. Recently, a large number of material handling and size reduction equipments have been available to facilitate large-scale PET recycling.[43]

Currently, recycled PET is used for both food and noonfood containers, fibers, straps, and sheeting. Allied Signal has done considerable work during the last decade to develop a high-performance material, PETRA, from PET bottle recyclate.[44] PETRA has high heat deflection temperature, dimensional stability, strength, rigidity, chemical resistance, ease of processing, and excellent mechanical properties. PETRA is used in the automotive industry for grill opening reinforcement, panels, headlamp brackets, car door escutcheons, and window hardware due to its superior engineering properties. In addition to Allied Signal, General Electric and NCR Polymer in Chicago have developed engineering resin from recycled PET by alloying it with polycarbonate.[45] These alloys are used in the automotive industry for decorative body panels, trim, and spoilers. It was estimated that this market could reach as high as 30 million pounds in 1993.

Bakker has recently reviewed the various techniques used for PET recycling and the economic as well as the technical merits of each process.[46] The techniques can be broadly divided into two groups, namely mechanical and chemical. The mechanical recycled PET is available as flakes and pellets, the former being cheaper than the latter. Chemical recycling is carried out by glycolysis, methanolysis, and hydrolysis. Both the technologies have shown substantial progress, because of the bottle bills, economic advantage, and good markets for fibers and bottles.

Bottle to bottle recycling is very much regulated by the FDA. Bottles made from chemically recycled PET (CRPET) are allowed to come in contact with both solid and liquid food. Mechanically recycled PET (MRPET) can come into contact with food only if it is made in multilayers, the recycled PET being sandwiched by 1 mil (0.001 inch) layers of virgin PET on both sides. HDPE has not been allowed for use as barrier coatings for this application. Further, only some of the companies manufacturing multilayer bottles have been allowed to use them for liquid food packaging.

The difficulty with MRPET is inconsistency of intrinsic viscosity (IV), which makes it difficult to maintain process efficiency, container strength, and rigidity. The minimum intrinsic viscosity required for PET blow molding is close to 0.65. Recycled PETs often have IV much below the lower limit. The drop in intrinsic viscosity is caused by the presence of contaminants and alkali residue from cleaning steps, which enhance hydrolysis and molecular weight degradation. The hydrolysis reaction is autocatalytic, because terephthalic acid produced in the reaction catalyzes further degradation. MRPET with low IV and inconsistent IV cannot be used for making monolayer beverage bottles. This difficulty can be overcome by solid stating through transesterification, which adds to cost. Other difficulties of MRPET include a yellow cast in some batches due to label bleeding and the presence of contaminants in the flake. Although pelletization helps in contaminant removal and easy processing, it adds to the cost.

Coinjection technology has been used by Continental PET Technologies by a patented process to manufacture two-material three-layer (2M3L PET/MRPET/PET) bottles which have been allowed to package food materials. Both monolayer and three-layer bottles were tested for simulant migration by special extraction techniques. Whereas the monolayer bottles showed migration to the extent of 0.177 $mg\ mil^{-1}$, the 2M3L bottles showed no detectable migration. Currently the three-layer bottles are allowed to be used to package soft drinks and mineral water and as ambient-filled food products. Additional tests show that the simulant migration level is also very small for hot-filled food packaging, making these bottles suitable for additional food packaging applications. The difficulties encountered in the coinjection process are process sensitivity to IV variation, and contaminant; the need for drying; and controlling problems such as bridging in the feed hopper of the extruder and in conveying the flake to the injection molding machine. Melt filtration of MRPET can remove the contaminants, but experience has shown that some contaminants always escape filtration processes, especially PVC which appears in the feedstock as a black spot.

The loss of intrinsic viscosity on melt processing is a serious problem. The IV loss of the melt is higher the higher the temperature and the higher the concentration of moisture and contaminants. If large-scale recyclate content is enforced by law, then bottles will contain four or five times more RPET. This will substantially lower the melt viscosity, making it difficult to produce bottles with good performance characteristics. Whereas some plastics engineers consider this as a serious problem, others feel that the drop of IV after four cycles will be marginal and it will average out at accepted levels even after multiple processing.

Recyclate PET is available both in flakes and pellet form. Usually the flakes produced on grinding clean, dry, PET bottles are pelletized by an extruder. Pellets are more suitable to material handling and reprocessing than flakes.[47] Flakes require special material handling and reprocessing equipment for their processing into bottles. This is because the flakes have lower bulk density, 25–35 pounds per cubic foot compared to 52–60 pounds for pellets. The flakes also contain higher amounts of contaminants including glue, label scrap, metals, and vinyl. The contaminants reduce intrinsic viscosity and impart color to the flakes. It is therefore important to use a set of standard specifications for accepting the incoming raw materials, as broadly defined in Table 6. Standard test methods are evolving to monitor the contaminants. The largest single source of contamination is glue. It can be as high as 180 lbs in a 40 000 lb truck load of PET recyclate. The recyclate can be freed from glue during the normal washing and drying process. However, it recondenses on cooling as a waxy brown residue. It is recommended that the flakes be recrystallized prior to drying. Recrystallization prevents agglomeration as the material passes through the glass transition temperature. Recrystallization also reduces the residual moisture on the flake, thereby lowering the load on the drying and residual volatiles on the flake surface, including the glue, washing chemicals, and inks. Storage of flakes in conventional silos and storage bins cause bridging and rat-holing during extrusion, due to their different angle of repose. Hence special silos need to be used for storing flakes. Drying of the flakes should be carried out at relatively low temperatures, because at temperatures above T_g the flake curls and becomes sticky. This may cause erratic feeding and bridging in the hopper.

Table 6 Specifications for Post-consumer PET Recyclate.

Method	*Attribute*	*Specification*
	Intrinsic viscosity	0.71–0.80 dL g^{-1}
PTL1P604	CIE (L,A,B) color	3.0 max. 'b' value
PTLP601	Fines	0.4% max.
Pending	Flake pH	7.00 ± 1.5
	Flake residuals	Dye test, pending
PTLP603	Moisture content	0.8% max.
D-1505	Density	1.335 g cm^{-3}
D-1895	Bulk density	22–32 lb ft^{-3}
PTLP602	Floatable contamination	10 ppm max.
PTLP600	PVC contamination	none
	PETG contamination	≤10 ppm
	Metal contamination	≤10 ppm
	Green PET contamination	≤500 ppm

Since post-consumer PET comes from different sources, each having different catalyst systems, additives, and intrinsic viscosity, it is important that the material be homogenized and compatibilized before being pumped into the die. Hence, a longer extruder with 30:1 or higher length to diameter ratio should be employed to mix RPET. The mixing has to be carried out at low shear for a longer period at a compression ratio 4:1 or higher. Single-stage screw designs have been found to be particularly effective in promoting good mixing at low shear rate. Twin screw extruders, particularly corotating intermeshing extruders, can also be used for recycling RPET flakes.

Continuous melt filtration is essential in processing RPET. This eliminates contaminants from the melt stream. Special melt filtration systems for RPET are available. They can either be manual or automatic. Since RPET flakes contain some level of contamination that causes melt degradation, it is important that the final product design takes these factors into consideration in order to produce quality items.

Recycled PET pellets are, however, easier and simpler to use and can be mixed and extruded with equipment similar to that used for virgin resin. This is because RPET flakes and grinds become homogenized and decontaminated due to drying in the pelletization process. Crystallization and melt filtration takes place during pelletization. The intrinsic viscosity (IV) of the pellets are uniform, regardless of the initial range of IVs. In addition, many pelletizers perform custom compounding, tolling, and solid stating (further esterification to increase IV) to improve the properties and processability for RPET.

Recycled PET pellets are, therefore, best utilized for high-end applications including rigid packaging and injection molded products, whereas flakes may be used for low-end applications with sufficient care and PET processing expertise.

Golmanavich studied the degradation and recyclability of poly(butylene terephthalate) used for housing electronic components.[48] All samples contained 25% recyclate, but the recyclates came from different cycles of processing. The first series contained 25% of the recyclate from first grinding, the second series contained 18.5% first series recyclate and 6.5% second series recyclate. The tensile properties of five series of material were examined. No substantial difference was observed in pull strength. Some of the experiments showed product improvement when 25% of the regrind is added to the virgin resin.

(ii) Polyamide nylon 6, and nylon 66

Polyamide, like PET, is a crystalline polymer made by polycondensation of caprolactam or adipic acid with hexamethylenediamine. It is used for making fibers as well as engineering parts by conventional methods of extrusion and injection molding. The fibers are used for making carpets and yarns.

Plastic fishing gear is a major pollutant affecting marine life and disturbing the ecological equilibrium. An 1987 estimate shows that approximately 30 000 northern fur seals and 250 000 birds could die each year in nets left out after salmon fishing. Fishing gear lost in the sea was estimated at 135 000 metric tons a year, posing a serious threat to marine life and marine navigation. The problem is being dealt with by the US legislation in a number of ways, including methods to produce biodegradable yarns for fishing gear and recycling of the fishing gear material to value-added products.

Substantial work has been carried out at the Polymer Processing Institute at Hobeken, NJ, in developing methods to recycle nylon 6, nylon 66, and HDPE fibers that comprise the fishing nets.[49] This includes blending, reinforcement, and impact modification of the polymers to improve their properties, and use of chopped fibers from the nets to reinforce thermoplastic elastomers. The process steps in recycling include: collection, identification, and characterization; size reduction; cleaning; melt processing in the extruder; and blending of the three polymers if necessary.

Identification of plastics present in single and multifilament yarns was carried out using the polymer solution in 19–30% hydrochloric acid. IR spectroscopy was found most suitable for this study. Size reduction was carried out using a rotor grinder of approximately 300 hp. A 10 mm screen was placed at the discharge chute to separate larger particles. Sand was the most harmful contaminant for the fluff. Equipment containing a compressed air vibratory screen was used to remove the sand, followed by melt filtration in a short extruder. The chopped fibers were melted and pelletized by extruders provided with special feeder and densification equipment.

In addition to testing recycled HDPE, nylon 6, and nylon 66, the resins were impact modified using a Rohm Haas core-shell polymer, maleic anhydride-grafted EPDM rubber, and a maleic anhydride-modified SEBS copolymer. HDPE obtained from fishing gear was also grafted to maleic anhydride in the presence of dicumyl peroxide in a twin-screw extruder and was used as a compatibilizer for a nylon/HDPE blend.

The stiffness of HDPE was increased by 20% by using glass fiber reinforcement. Addition of MA-*g*-HDPE further increased stiffness and unnotched impact resistance. The best combination appears to be at 20% glass fibers and 20% MAH-*g*-HDPE. Use of a MA-*g*-EPDM reduced the brittleness and improved the gloss of nylon 66 molded products. This modifier also improved the impact resistance and flexural modulus of both nylon 6 and nylon 66 recyclates. Its effect on other properties was less significant. Addition of the impact modifier alleviated problems associated with sand contamination. A compound prepared by blending 33% nylon 6, 33% nylon 66 recyclates, and approximately 17% of HDPE recyclate and 17% MA-*g*-HDPE gave better mechanical properties and impact strength than any of the neat recyclates. Compounds of the recyclates modified by impact modifier and reinforced by glass fiber were found comparable to many virgin engineering plastics.

Chopped nylon 6 and nylon 66 fibers were used to reinforce an amorphous thermoplastic polyurethane (TPU). Fibers 8 cm long were compounded with the TPU in a corotating twin-screw extruder with screw configuration offering high total strain but low local stress. The mixing was carried out at 160–180 °C (below the melting point of the nylons) and the compound was pelletized. The pellets were injection molded after drying at 100 °C. Hardness, mechanical properties, abrasion resistance, and morphology were examined and were found to be improved on blending with the fibers.

Glass-reinforced nylon 66 is being currently used as under-the-hood parts in automobiles, including radiator endcaps, turbo coolers and air intake manifolds. The material can undergo extensive degradation on reprocessing due to depolymerization and oxidation of matrix resins, length reduction of glass fibers during attrition and compounding, and loss of stabilizers and thermal degradation during processing and reprocessing. Eriksson *et al.* have made a systematic investigation of the property change due to reprocessing of glass fiber reinforced nylon 66, in which they have measured fiber length, tensile and impact behavior, and thermal and thermooxidative stability.[50] They found that fiber attrition takes place during the first injection molding cycle and during the extrusion into pellets. Subsequent steps during recycling affect the fibers to a lesser extent, leading to a gradual decrease in fiber length. The distribution of fiber length becomes narrow, which helps in maintaining good mechancial strength. The tensile strength and impact resistance decreases rather slowly, and the interfacial adhesion between fiber and matrix remains unaffected. Thermal and oxidative stability decreases due to the consumption of stabilizers during reprocessing. The data show that the regrind from glass-reinforced nylon 66 parts can be recycled at least a few times when combined with virgin material.

(iii) Amorphous polycarbonate

Polycarbonate is made by polycondensation of bisphenol A with carbonyl chloride. It is largely amorphous with a high glass transition temperature. Its mechanical properties, particularly its impact resistance, are excellent. High clarity coupled with high impact resistance makes it an excellent choice for glazing, compact disks, and many automotive parts. A more highly amorphous polycarbonate copolymer (PC-HT35) made from 35 mol% bisphenol TMC (trimethylcyclohexanone bisphenol) and 65 mol% bisphenol A is currently used for reflectors for automobile headlamp assembly. The lens is usually made from bisphenol A polycarbonate (PC). The assembly, which contains an appreciable weight of plastic as lens and reflector in an automobile, is a good candidate for recycling. The hardware is easily removed. The presence of top coats, hard coats, and sealant as well as metals from the reflecting can however exert a substantial influence on the recyclability of this assembly.

Paul and Bier, in a recent work, studied the recyclability of a headlamp assembly based on a PC lens and a PC-HT35 reflector assembly.[51] They looked at seven different combinations of the polycarbonate resin and coating materials. The reground materials were molded and their tensile properties were measured. It was found that metallization by itself does not contribute to polymer (PC-HT35) degradation, but when combined with the base coat it can produce appreciable degradation. The silicone hard coat makes PC, the polymer of the lenses, highly vulnerable to degradation. The tensile properties of a blend of virgin PC-HT35 and virgin PC are much superior to the blend of the reflector and lens material. The polyurethane adhesive, however, exerts the worst influence. The properties of the blend containing the adhesive is very inferior and is unacceptable for any structural application. It is therefore recommended that the lens assembly should not use silicone or PU adhesive, but instead use another method of bonding such as heat, ultrasonic, or vibrational welding. Earlier work by GE had shown that polycarbonate does not lose its mechanical properties in multiple reprocessing and that clear water bottles can be used many times on reprocessing.

A substantial amount of polycarbonate is currently being used for the manufacturing of compact disks. In 19991 alone the world wide production of compact disks reached 400 million. With increasing use of CD-ROM and W-ROM technology and other optical media devices, a substantial amount of waste polycarbonate will become available as scrap and post-consumer resin.

Boudreau *et al.*[52] have reviewed the methods currently used for recovering PC from compact disks. Three techniques are used to remove contaminants, namely chemical separation, melt separation, and mechanical filtration. Chemical separation uses solvents to separate the metallic coating and the thermoset acrylic coating from polycarbonate. However, this process poses potential health problems due to harmful emission of organic vapor. Melt filtration is a bulk technique that does not involve hazardous chemicals. The product to be recycled is granulated into flake form, dried, and repelletized by extrusion. During processing, the screw forces the melt through a fine filter. However, whereas this technique is effective in separating aluminum, it is not capable of separating acrylic from PC. The additional heat history involved in this process also lowers the melt viscosity, which is usually low to start with. That makes recycled PC unsuitable for further processing. Mechanical separation does not involve chemical treatment or additional heating. In this technique, an abrasive medium scrapes the unwanted material from the polycarbonate substrate. The apparatus used for this technique is a rotating brush, situated above a conveyor belt that moves relative to the brush. The disks are fed one at a time onto the disk holders with the coated surface facing the brush. The disk holder advances while the rotating brush removes the coating from the PC substrate. A rotating knife removes the lacquer from the edge. Compressed air or water mist is used to remove the particles from the brush, thereby improving the efficiency of the stripping action. The disks are then granulated for further processing.

The authors compared the optical and mechanical properties of the regrind material from fully and partially stripped discs in comparison with virgin material. They found that whereas tensile modulus, strength, and elongation at break of the fully stripped material is comparable to those of the virgin material, those for partially stripped and unstripped material are very low. The material from the last two sources is very brittle and does not exhibit appreciable elongation of break. The transparency of the fully stripped material is only 12% lower, whereas the understripped and unstripped materials are opaque. The melt flow rate at 280 °C and the deflection temperature are not affected by stripping. However, the impact resistance of the stripped PC is substantially lower than that of the virgin material. The impact energy at break is an order of magnitude lower, probably due to contamination by the lacquer and aluminum. On the whole, the mechanically cleaned material has superior properties to recyclates cleaned either by the chemical or by the melt filtration process. Jones also studied the recyclability of polycarbonate from compact disk scrap.[53] They used regrind from injection molded CD grade polycarbonate (RPC) to blend with ABS, a bottle grade PC, and glass fibers. Whereas glass fibers improved stiffness and strength, the elongation at break was very small. The other two blends, 50:50 PC/ABS and RPC/PC (bottle grade), were found to have excellent srength, toughness, and impact resistance. Whereas RPC/PC blend has greater strength, RPC/ABS shows higher impact resistance.

(iv) Poly(phenylene oxide) (PPO)

PPO is prepared from 2,6-dimethylphenol through an oxidative coupling reaction. This polymer has a very high softening point of 210 °C and is very difficult to process. It is usually alloyed with polystyrene and modified polystyrene, to provide an alloy tradenamed Noryl®. The latter is available in many grades and is used in engineering applications such as instrument panels, seat backs, rear spoilers, and wheel covers in automobiles and as housing, keyboard, and printer bases in computers. GE has shown that Noryl can be recycled as roofing shingles after grinding and blending with virgin resin.[54]

(v) ABS

Acrylonitrile–butadiene–styrene (ABS) is used as a structural plastic in many applications in automotive components, computer housing, machine components, housing and construction, *etc.* Recycling of ABS is therefore an important concern in the industry. Michaeli and Dassow have studied the effect of contaminants and reprocessing conditions on the properties of recycled ABS.[55] They conducted their study using boiler control boxes which were collected and ground down by a conventional granulator. The regrind was divided into two parts. The first half was cleaned using industrial detergent and the other half was used without cleaning. Both batches were mixed with virgin ABS in different ratios, 0 to 100%, and their physical, rheological, morphological, and mechanical properties were measured. It was found that the blends of virgin and recycled ABS have deeper color than neat ABS. The difference in color increases with increasing percentages of regrind. The blends having cleaned regrind were more colored than the ones having uncleaned regrind. The batch containing 100% regrind was colored. There was no difference in density amongst the various blends. The melt flow index (MFI) of the blends and 100% regrind was higher than the MFI of the virgin material, showing that molecular degradation had occurred during production and aging. The MFI of cleaned and uncleaned batches did not differ appreciably, indicating that some of the lubricants present on the surface do not influence the MFI. Whereas the glass transition temperature (T_g) of the polybutadiene phase did not change by blending with regrind, that of the styrene–acrylonitrile phase decreased by 3 °C. The linear coefficient of thermal expansion was also higher for the blends (possibly due to molecular weight decrease), reaching a peak at 25% regrind. Recycling does not have any significant effect on morphology and does not influence phase separation. The short term mechanical properties, tensile strength, and modulus did not show significant differences on adding the regrind, although the yield strain decreased slightly. Under high speed tensile tear test, the strain at fracture and the volume-specific energy (a measure of fracture energy) of the cleaned material decreased significantly. The biggest drop came at 25% regrind content. In conclusion, the authors stated that ABS regrind can be added to the neat material without much loss of property except the impact resistance.

Potente and Gao carried out a similar study using lampshade parts made of ABS.[56] In addition to rheological properties, they also measured molecular weight and molecular weight distribution. They studied two types of regrind, one with artificial aging and one without. They observed an increase in melt flow index on regrinding and injection molding, possibly due to thermal and mechanical degradation. The shear dependence of the melt of the recycled material was slightly lower at low shear rate but almost equal to that of the virgin resin at higher shear rate. Gel permeation chromatography confirmed this by showing an increase in number average molecular weight and a decrease in weight average. The tensile strength of the virgin ABS and blends differed only by 3.7% and the modulus by 6%, but the elongation at break was reduced by 21% on blending with regrind. The izod impact strength also decreased by 18.5%. There was no difference between the aged and the unaged specimen. This study therefore confirms that ABS recyclate can be ground and added to virgin ABS for recycling.

(vi) TPE

In recent years, thermoplastic elastomers have found great application in the automotive industry. They provide easy processability and recyclability of the scrap compared to those of thermoset rubber. However, the recyclability of the used parts made from TPE has not been studied extensively. Advance Elastomer Systems, a manufacturer of thermoplastic vulcanizates (Santoprene and Geolast), have conducted a case study of recycling used automotive parts.[57] A rack and pinion steering gear bellow made of Santoprene was selected for this test, mainly because this part has been subjected to grease and car underbody environment for at least 10 years. The process steps involved in this recycling were cleaning, grinding, blending, predrying, and injection molding. Cleaning the TPE part was the most important step. After trying several methods, a hot water high-pressure and steam technique (110 °C at 100 bar pressure) was found to be the most effective method. Subsequently, the material was reprocessed alone and as a blend with Santoprene and a thermoplastic polyalkene, Vistaflex® 911B. The results are given in Table 7. It was found that the recyclate showed a minor loss of physical properties, but it can be effectively blended with 70 parts of virgin resin without any appreciable change in physical properties. The TPO blend with Vistaflex showed a greater improvement for the properties over those for neat resin.

Table 7 Physical Properties of TPO Blend with Polypropylene.

	Recyclate (% by weight)			
	0	30	70	100
	Virgin material[a] (% by weight)			
	100	70	30	0
Unaged (23 °C)				
Tensile (MPa)	9.1	9.3	11.1	14
Elongation (%)	650	630	570	560
100% Mod. (MPa)	7.6	7.6	7.7	7.5
Hardness (Sh. D)	37	38	38	37
Heat aged (125 °C/168 h)				
Tensile (% retention)	100	102	107	101
Elongation (% retention)	69	75	86	92
100% Mod. (% retention)	111	111	116	116
Hardness (% retention)	105	108	111	111
#3 Oil aged (125 °C/168 h)				
Tensile (% retention)	0	53	60	72
Elongation (% retention)	0	26	39	60
100% Mod. (% retention)	0	57	62	74
Hardness (% retention)	0	100	143	51
Weight gain (%)	340	160	90	50
Compression set % (100 °C/22 h)	85	80	70	65

[a] Vistaflex 911B.

(vii) Polyalkenes

Polyethylene, polypropylene, and their copolymers provide the bulk of polyalkenes used in the industry today. The density of polyethylene can be varied from 0.90 to 0.96 by suitable choice of polymerizing process. High-pressure polymerization in the presence of a radical catalyst gives low-density polyethylene, whereas stereospecific polymerization in the presence of chromia and Ziegler–Natta catalyst gives high-density polyethylene. Recent developments in metallocene catalysis have provided a means of making tailor-made ethylene–alkene copolymers with different densities and different mechanical properties.

Polyalkenes are commodity plastics, largely used for packaging. Whereas high-density polyethylene (HDPE) is used mostly for making rigid containers, low-density polyethylene (LDPE) and linear low-density polyethylene (LLDPE) are used for making films, straps, grocery sacks, packaging wraps, and in a variety of other applications. As a result, a large amount of PE and its copolymer goes as post-consumer resin into municipal solid waste. Both high- and low-density PE and PP waste are being recycled. Because of their different applications, recycling of the three polymers will be discussed separately.

(a) High-density polyethylene (HDPE). HDPE bottles are used in large quantity for retailing milk and nonfood items such as motor oil, detergent, bleach, agricultural chemicals, *etc.* Nearly four times as many HDPE containers are used as PET bottles. However, only 2.4% of the 3.3 billion pounds of HDPE is recycled compared to 20% of the 900 million pounds of PET used. HDPE containers are more suitable for recycling because they are plentiful, easily identified, and more forgiving in reprocessing than PET. Only 5% of the bottles are either coextruded or surface treated, a process that tolerates contaminants in the recyclate. The contaminants in HDPE containers are mainly mill oil and chemicals. Consequently, recycled containers are best salvaged by thorough cleaning followed by drying, granulating, and pelletizing. When thorough cleaning is not possible or the grades are either colored or filled with reinforcement (fiberglass or talc), the recycled materials are injection molded for use as flowerpots, pipes, and molded parts.

The technology used for PET recycling, including cleaning, grinding, and separation by air or hydrocyclone, can be used for reprocessing the HDPE waste. A second washing is often used in this process to remove contaminants before size reduction. The cleaned bottles are then converted into flakes of approximately 161–322 mm^2, using a shredder and granulator.

Turnkey recycling systems, such as the ones offered by John Brown of Providence, Rhode Island, and Sorema of Italy, are available for reprocessing both the HDPE and PET bottles from municipal solid waste. The main limiting factor for HDPE recycling is the fluctuating price of virgin resin. Another limiting factor is the retention of odor (sour milk odor) by polyethylene, in contrast to PET which does not retain any odor of the packaged material. Hence, the recycled HDPE is not used for food packaging. The major application of recycled HDPE is in making containers for industrial chemicals, motor oil, household chemicals such as detergent and floor polishes, and in drain pipes, leach tiles, and plastic lumber. Amberger Kaolinwerke has developed a process for producting high-quality PE pellets from household waste.[58] Recycled HDPE is also used to make plastic lumber from wood fiber composites beccause of its high melt strength, which prevents warpage during cooling. However, the largest application of recycled HDPE is blow molding of bottles, where the recyclate is used as a minor component (25–50%) of the HDPE compound.

HDPE bottle regrinds are generally limited to extrusion blow molding, because of their fractional melt index characteristics. However, the problems associated with recyclate blow molding include batch-to-batch variation in melt index, density, and resistance to environmental stress cracking. Alternatively, HDPE recyclate can be injection molded. However, this requires improving the melt flow properties of HDPE regrind, otherwise the molded parts exhibit warpage due to excessive orientation under high injection pressures. Lavieri in recent work has studied the possibility of increasing the melt flow index by adding low molecular weight HDPE grades (MFI 17, 38, and 65 g 10 min^{-1}) at the 10, 20, and 60% level.[59] The resin mixtures were dry blended and the blend was thoroughly melt mixed in a Leistnitz 34 mm counter rotating twin-screw extruder at 195 to 202 °C. Irregularities in extrudate appearance such as exfoliation were observed as the concentration and the melt index of the additive increased. The compounds were then injection molded using a 90 ton HPM machine. The effect of low molecular weight HDPE concentration on blend MFI is shown in Figure 9. The relation between blend MFI and additive concentration is linear, although the slope (rate of increase of MFI per unit percent of additive) increases with increased MFI of the additive. Warpage decreased substantially with increase in MFI. The tensile properties, however, showed a complex behavior. Whereas the tensile stress at yield decreased for all the three grades, the tensile strength at break decreased for 38 and 68 MFI but increased with 17 MFI grade. The average energy to break the samples increased with additive concentration for all three grades and decreased as the melt index increased. It is suggested that the compatibility of the blend decreases with an increase in the melt index of the low molecular weight HDPE additive.

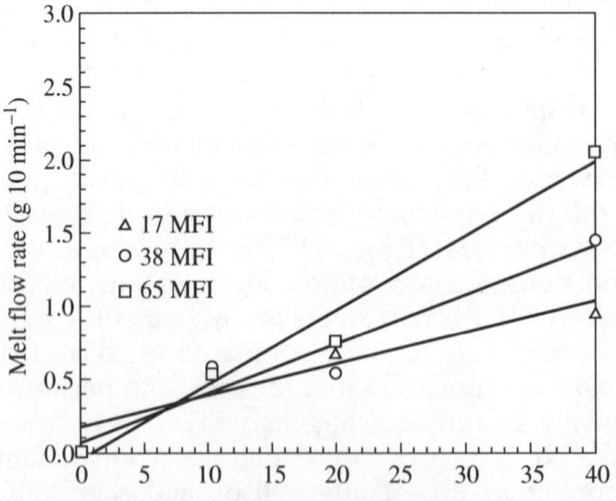

Figure 9 Effect of adding low molecular weight HPDE on the melt flow index of HDPE regrind.

Tripodis and Bradi examined the effect of adding different percentage post-consumer regrind on the properties of blow molded plastic bottles.[60] They used three extruders to prepare the parison in three layers and to include the recyclate and color concentrate in the middle layer. The parison was formed using a 12-station 24-cavity blow molding wheel. Bottles had a 0.91 L capacity with a height of 21.7 cm. The bottles were subjected to a top load test, section thickness measurement, and an environmental stress cracking resistance (ESCR) test. The weight of the bottles was varied between 47 and 51 g. The top load capacity is the highest for the 51 g/15% PCR system. However, the highest load was withstood by the 45 g/30% PCR bottle. The critical neck dimension was similar for all blends with a slightly better performance for 30% PCR. ESCR studies showed that bottles made with 30% regrind took a longer time to fail. This shows that a higher percentage of PCR can be used as the middle layer in multilayer packaging without affecting the mechanical properties.

Doyon *et al.* carried out extensive rheological studies of high- and low-density polyethylene samples after repeated extrusion and regrinding and found that the multiple extrusion process does not affect significantly either the resins or the blow molded films made out of them.[61] Rheological measurements indicated some degradation and crosslinking for the HDPE (degradation showing in the initial recycling cycles followed by crosslinking in the latter cycle), but no degradation for LDPE. The results of the rheological characteristics were complicated by alignment and relaxation of the side chains.

Shenoy and co-authors proposed a method for predicting the dependence of the melt viscosity of reprocessed PE.[62] They showed that the slope of MFI versus shear rate over MFI yielded a master curve for all grades of resin in a generic family including LDPE and PP, which can be used to estimate the melt viscosity of reprocessed PE. Gibbs examined the recyclability of HDPE milk jug recyclate from various samples of PCR and found a significant lot-to-lot variation of MFI.[63] The mechanical properties of PCR are comparable to those of virgin resins except that their elongation at break and ESCR are inferior.

Many states in the USA are enacting laws which mandate a 10% reduction in package weight or a 25% incorporation of post-consumer resin (PCR) in the packaging. Quite often a 10% reduction in package weight lowers the performance to below acceptance level, and so a 25% incorporation of PCR into new packaging is being accepted as the most viable method for meeting the requirements of the law.[64] However, the blends have poor environmental stress crack resistance (ESCR) and poor impact resistance. This is avoided by using virgin low density (0.92 in place of 0.94) copolymer as base resin. New copolymers are being produced which do not suffer the loss of ESCR on blending with 25% PCR.

Natural PCRs provide lower risk of product contamination because they are typically used to package milk and juices, in contrast to pigmented PCRs which are associated with packaging of laundry detergents, vegetable and mineral oil, *etc.* Hence a substantial amount of pigmented HDPE used in blow molding of bottles is available as recyclates. Recycling of pigmented PCR is accomplished by sorting them into three color groups, such as red–orange, yellow–blue green, and white. The color of the recycled product is obtained by color matching with an appropriate color concentrate. Multiple layer bottles are also blow molded from colored PCR, by incorporating the latter into the middle layer. This allows some moderation of hue and also prevents PCR from coming in direct contact with packaged materials. In the conventional blow molding process, trims and flashes from the finished product are often added back to the PCR. This has to be taken into consideration in designing the PCR product.

Among the two most recycled plastics, namely PET and HDPE, the latter retains larger amounts of volatile organic residues because of its lower solubility parameter, close to those of organic volatiles. Brahms and Ramaswami *et al.* have done a systematic study on the analysis of HDPE PCR for the volatiles.[65] Pigmented natural bottles were collected and color sorted. The pellets were extracted with hexane for 48 hours, and the extract was concentrated under a stream of nitrogen. One sample was also extracted with supercritical fluid. They found the presence of surfactants and long-chain alkylbenzenes in the extract. However, this may not be a problem if recycled HDPE is used for detergent bottles.

(b) High molecular weight polyethylene. Refuse carts, 30–100 gallon containers with two wheels, are used to collect municipal solid waste and these are semiautomatically or automatically dropped into garbage trucks. They are usually made out of high molecular weight polyethylene (HMWPE) and are designed to last 10 years or more. Lee *et al.* have investigated the recyclability of the plastics from refuse carts.[66] To analyze the problem successfully, they set up a list of performance criteria and identified standard and special tests to measure the criteria. The tests are automated lifting, sand drop, water drop, tensile properties, dent drop, and crash. Five series of materials were designed, one 100% virgin HMWPE and the rest containing a different percentage of recyclate. Containers were prepared using the experimental compounds and the latter were tested along with the control. No significant differences could be detected in squeeze, sand drop, tensile, dart drop, or crush tests. All specimens, from 25% recyclate to 100% recyclate, passed the specification. Subsequently, the specimens were

prepared from compounds which, in addition to virgin HMWPE, contained additional UV and heat stabilizers. It became apparent that specimens made from 50% and 75% recyclate had a sharp drop in toughness. When the specimens were exposed to UV light at 50 °C for 20 hours followed by a dark period of 4 hours, the compounds containing 50% and less recyclate maintained their performance adequately to last for about 12 years. The toughness of the blend was adequate.

(*c*) *Low-density and linear low-density polyethylene.* Substantial amounts of LDPE and LLDPE are used to make grocery bags, pallet wraps, packaging, and agricultural films. Resin manufacturers such as Mobil, Sunoco, Polypak, *etc.* have taken a leading role in recycling film material. Recycling of the used film stock is difficult because of high levels of contamination and the difficulty encountered in separating the bags. Early attempts to recycle the films using existing equipment were not successful because the layers of tightly baled stretch films adhere to one another, in effect laminating the contaminants within the bulk of the plastic. Further, the high specific surface area of the film stock allows the cleaning water and its suspended contaminants to adhere to the film. Additional drying, while adding cost, did not overcome the difficulty, nor could it break into the agglomerated layers of stretch film. Mobil Polymers of the USA, one of the largest manufacturers of stretch film, took on the challenge to overcome the difficulties. Mobil decided to recycle stretch film to stretch film, collected substantial information on the impact of contaminants on end applications, and carried out extensive research to find out novel methods to overcome agglomeration and adherence of contaminants to the film.[67] Additional care at the points of collection and statistical process control at different steps helped to overcome the contaminant problem. In the end, Mobil could produce quality recyclate materials for conversion into stretch films (Marketwrap®) and rigid containers. However, the recycled material could only be used in conjunction with virgin material. Blends of 10% and 25% stretch film post-consumer resin with high-strength virgin resin (NTX-095) provided functionality and mechanical properties of the level of the virgin resin. On the other hand, when post-consumer HDPE was added to copolymer NTX-095, both the dent impact and the mechanical strength of the latter deteriorated significantly.

Sikora and Bielinski examined the mechanical properties of reprocessed film grade PE containing about 5% of a mixture of agricultural fertilizer and humic soil.[68] The reprocessed film showed significant loss of modulus, tensile strength, and elongation at break. A similar study was carried out for blending LLDPE stretch film PCR with LDPE and ethylene–vinyl acetate copolymer (EVA).[69] With LDPE grade LLA-503 (suitable for films), the impact strength was unaffected by addition of 25% stretch PCR, while the tear strength in the machine direction decreased by about 20%. The overall tear balance (MD/TD ratio) improved significantly from 2.7 to 5.3 with LTA-051 (heavy gauge films) LDPE resin and the tear strength increased significantly in both directions but the impact resistance decreased by 12%. Creep resistance of both grades was not affected by the addition of stretch film PCR. With EVA resin, addition of PCR improved both tear strength and impact resistance, the maximum improvement of tear being at 20% PCR. The percentage of haze changed only marginally with an additional 30% PCR. In addition to strength and impact properties, the antiblocking properties of LDPE/PCR and LLDPE/PCR blends were measured. The stretch film PCR had 2% residual tackifier. Induced blocking increased in all cases on blending with PCR. However, addition of 0.6% antiblock concentrate reduced the blocking effect close to that of the virgin resin. It was therefore recommended that converters using 20% or more stretch film PCR should use additional amounts of antiblocking agents to overcome the effect of PCR addition.

LDPE and crosslinked LDPE are used for making insulating jackets for wires and cables. The used products and rejects in the cable factory are cut open for reclaiming copper, and the plastic is often hauled to landfills. Some of the firms recovering metal also recover LDPE. Bevis *et al.* have devised a rising current separation process for reclaiming PE and PVC separately from wire and cable jacket.[70] The electrostatic separator developed by the Bureau of Mines is effective in separating metallic contaminants from PE. Mangaraj *et al.* have done substantial work on reprocessing post-consumer crosslinked PE (XLPE) through *in situ* polymerization of the ground XLPE with vinyltoluene.[71]

The engineering properties and the performance of hot melt asphalt (HMA) for pavements can be greatly improved by adding about 5% recycled low-density polyethylene (LDPE). Novphalt®, the first LDPE modified HMA, was developed by the Felsinger group in Austria to provide a 10 year warranty on the performance of pavements in European countries. With the increased availability of recycled LDPE from used grocery bags, pallet wraps, agricultural films, and dry-cleaning bags, and the emphasis on the use of recycled materials in pavement construction as embodied in the Transportation Bill of 1991, a considerable amount of research is being carried out to improve HMA by adding recycled plastics and used tire products. At the 5% level, one mile of repaving of one lane highway will use approximately six tons of recycled LDPE. Recently, Tappeiner and Male have reviewed the LDPE-modified HMA technology and have emphasized the development of tests closely related to the use at the pavements to assess the benefits of LDPE modification.[72] These include creep test, fatigue test, wheel tracking test, and low-temperature cracking tests. The results indicate that the addition of LDPE improves the resistance to creeping, as is evident from the creep test, resistance to deformation (as measured by a GA DOT loaded wheel test), fatigue life as measured by methods used by Strategic Highway Research Program, and low-temperature cracking resistance as measured by a restrained thermal test. These improvements have been demonstrated practically by placing modified HMA on Route 75 near Sherman, Texas, where the section using modified asphalt has not shown any crack during the last 5 years of service. Modification of HMA increases the cost by 100% of the material, but only by 2–5% of the total cost of the pavement laying project. In fact, based on the performance improvements, extension of the service life, and reduced maintenance costs, the life cycle cost of modified HMA is about 20% lower than that of unmodified HMA. It has also been seen that the maintenance cost of airfield runways paved by modified HMA is approximately 25% less than those made by HMA. The authors have examined the effect of contamination of LDPE by HDPE, PP, PET, PVC, and PS on the performance of the modified HMA pavement. They have seen that up to 20% contamination by other plastics, the performance of modified HMA does not change.

(viii) Polystyrene

Polystyrene (PS) is made by polymerizing styrene monomer either in a batch suspension system in a stirred reactor or in a mass continuous process where the monomer is pumped through a series of heat exchangers until the conversion reaches a specified level. Radical initiators are used. Expanded polystryene (EPS) is manufactured in the batch suspension system by incorporating pentane or isopentane in the reactor. The EPS beads do not undergo extrusion pelletization like crystal polystyrene but they are expanded in heated molds to form to a particular shape. PS foams are also produced by extruding high molecular weight crystal PS with a nucleating agent and a foaming agent. High-impact polystryene is made by polymerizing styrene in the presence of polybutadiene rubber. Approximately 6.4 billion pounds of crystal and expandable PS is produced every year by eight major producers in the USA. These companies have formed a limited partnership company, the National Polystyrene Recycling Company (NPRC), to develop recycling technology for PS products.

Common applications of injection molded crystal PS include dust covers, cassette cases, windows for video cassettes, boxes for compact disks, office supplies, computer disks, tablewares, display cases, and medical accessories. Injection molded HIPS is used in electronic and appliance cabinets, business machine housings, video cassettes, furniture, and smoke detectors. Extruded PS products include luggage racks, door liners, *etc.* The biggest use of PS, however, is as foam for insulation panels, egg cartons, clam shells, and loose packing materials which are made out of EPS. Recycling of EPS therefore received the major attention, particularly because of its visibility in the solid waste stream.

A typical recycling operation on EPS involves removal of dirt by hand sorting, followed by granulation and washing in water with detergent on a vibrating screen. The wash water is removed through a strainer and the fluff is dried by a centrifugal drier or by conventional heat drying. Dewatering is also done in twin-screw extruders which provide clean and odor-free fluff. The latter is then densified using a densifier. Mackey *et al.*[73] have described a number of densifiers used for recycling EPS. The best ones are those which use either IR heating or mechanical compression to convert the fluff into granules without causing major degradation of the polymer. The recycled EPS has properties close to that of virgin EPS. The presence of a small amount of an internal plasticizer and rubber particles contributes to lower tensile strength but improves impact resistance.[81]

Expanded polystyrene is a superior packaging material because of its better cushioning effect, easy processability, and low cost. The amount of energy used in making EPS is almost half that used for alternative packaging materials such as corrugated cardboard. EPS is therefore used extensively in clam shells and packaging. However, disposal of EPS by landfill is difficult because of its lower density and slower degradability. Certain states in the USA have made it mandatory that EPS manufacturers use at least 25% post-consumer EPS in their new products. To meet this need, Arco Chemical has developed and produced Dylite R25950 containing 25% recycled EPS. Currently, this material is being used by a large number of manufacturers using a technology developed by Arco. Dylite R25950 has properties similar to natural EPS.

Mohareb *et al.* have carried out a systematic study of the density, compressive strength, and creep of EPS foams containing various percentages of post-consumer EPS.[74] They prepared 0.97 m × 0.81 m × 0.05 m molded slabs from which compression samples and creep samples as defined by ASTM standards were cut out and their properties measured using standard ASTM methods. It was found that the densities of the foam specimen containing some percentage of post-consumer EPS varied substantially. This was attributed to the difference in the method of incorporating the foaming agent pentane into the polystyrene. Whereas in natural EPS, pentane is added during polymerization of styrene, in the case of post-consumer EPS, pentane is injected into the beads. Hence, the properties were normalized by density before comparison. Blends containing EPS prepared by different processes behaved differently. Whereas the compressive strength and creep behavior of the 25% PCR blend is superior to that of natural, blends containing 50, 75, and 100% PCR have lower compressive strength and higher creep. The authors have developed a three-step procedure to predict the compressive strength of the foam produced from virgin EPS with different percentages of PCR and have demonstrated the accuracy of the procedure by further experimentation.

In a subsequent study, the same authors measured cushion curves both for virgin EPS and blends of virgin EPS with different percentages of recycled EPS.[75] The cushion curves are measured after subjecting the foam to a static compressive stress. Initially, the thickness decreases rapidly with stress, stays steady, and then increases very rapidly, giving box-shaped curves (Figure 10). It was found that under 8.27 kpm stress, the curves do not differ significantly, showing a similarity in performance. However, there is a significant decrease of cushioning performance under higher stress on blending recycled EPS with the virgin. This should be taken into account in designing packages where recycled EPS will be used partially or fully.

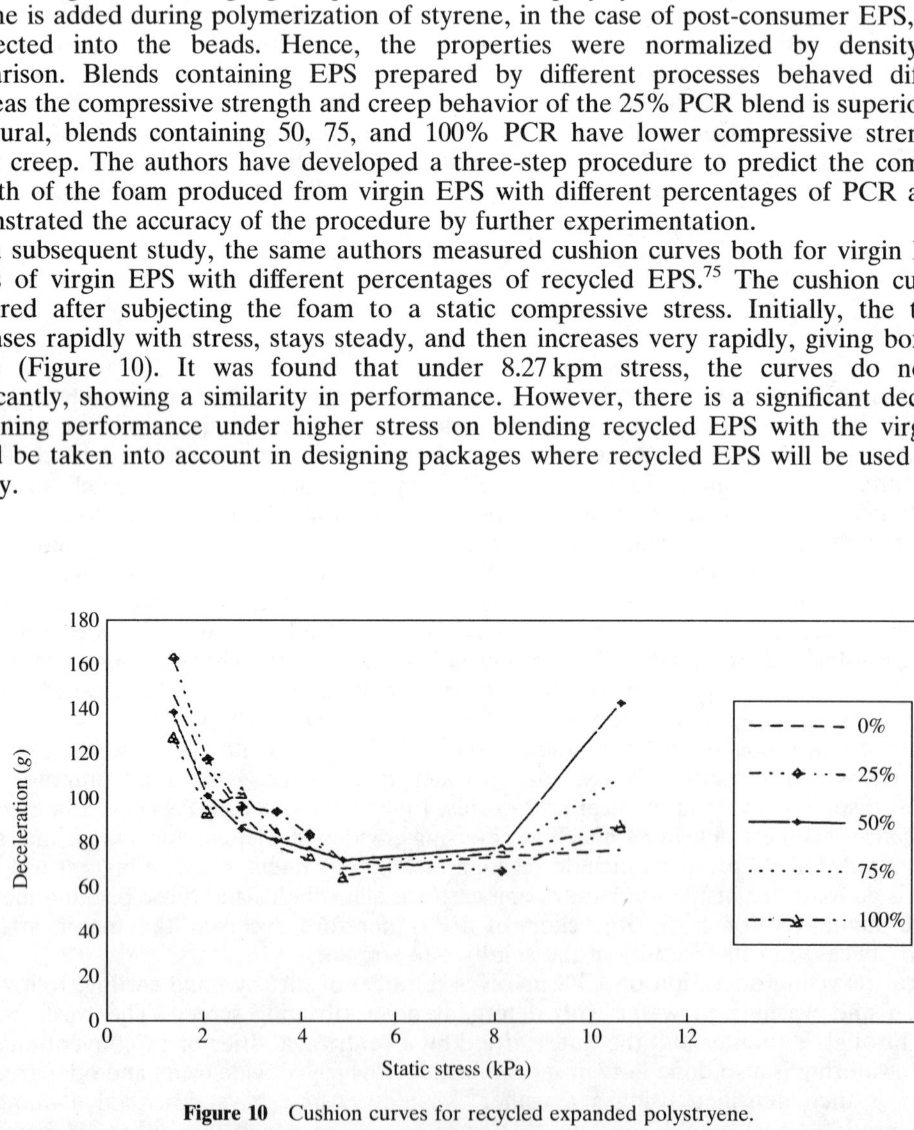

Figure 10 Cushion curves for recycled expanded polystyrene.

The recycled densified EPS is not allowed by the FDA for food packaging applications. However, it is used for making egg cartons, combs, key chain tags, coat hangers, picture frames, flowerpots, nursery trays, serving trays, *etc.* The foamed EPS is used as loose fill for packaging and foam core insulation boards. The revival of foam-core-panelized housing will provide a high-volume market for recycled EPS in the near future.

(ix) Polypropylene

Substantial amounts of polypropylene bale wrap and industrial fabrics are available as recyclate. Exxon Chemical has developed a facility to convert this waste into quality polypropylene pellets.[76] The major steps in this process include: loading the bales into a feed chute and removing wires and cardboard; cutting the bale into 4 to 6 inch slices by guillotine; grinding the slices into particles of half-inch diameter by a flexboard rotary grinder; hot water washing of the ground particles to remove nonpolymer impurities and hydrocycloning the washed particles to remove heavier contaminants; drying the product by a mechanical spin dryer followed by thermal drying; feeding the fluff to a 10-inch diameter vented extruder and extruding through continuous melt filtration equipment before feeding to the pelletizer; pelletization of clean polypropylene melt by a GALA underwater pelletizer; and blending recyclate with virgin resin (if necessary) to assure consistent quality. The recycled polypropylene is of very good quality. Currently it is being used to make geotextiles and landscaping fabrics. Closed loop recycling from bale wraps to bale wraps is under development. Attempts have also been made to recycle polypropylene by dissolving in a solvent and reprecipitating.

(x) Poly(vinyl chloride) (PVC)

PVC is largely prepared by suspension polymerization of vinyl chloride to obtain relatively porous particles of 130 μm diameter suitable for plasticization. PVC is used in nonplasticized (rigid) and plasticized grades and as plastisols (a fluid paste containing a high volume of plasticizer). Recycling of PVC uses the same process steps as recycling of polyalkenes, namely separation, size reduction, cleaning, drying, and pelletization.[77] PVC can be reprocessed by conventional methods, after adding a suitable amount of stabilizer and lubricant to the recyclate. Reprocessing of recycled PVC into packaging is difficult, because of loss of gloss, color, and clarity. Although PVC recyclate is quite suitable for making pipes, the use of recycled PVC in certain kinds of pipes has been restricted by code.

In the USA, PVC recycling is being developed under the auspices of the Vinyl Institute and major PVC producers. In Europe, GEOCOM, which combines the talent and support of PVC producers and users, is developing technology to recycle PVC. It may be mentioned that PVC bottles are used in large quantity in Europe to package distilled water. The bottles are collected at reverse vending machines, where they are flattened by heated roll, scraped off the roll with a knife, and bagged. The bags are sent to processing plants, where they are vibrated into cells of cleated conveyors and then passed in single file beneath an X-ray detector to remove the contaminants. Washing of PVC scrap is very important. Hot water, with or without caustic soda, and shear are used for cleaning PVC. Caustic soda removes labels and paper contaminants. This is followed by density separation to remove alkenic contaminants. PVC in this process sinks to the bottom. The pure PVC is then taken out, dried, extruded, and pelletized. During this process, PVC is melt-filtered using an automatic screen changer. A slide-through manual changer often introduces black specks in the pellets. Even then, some processors of sheet materials use manual filters to remove gels from the melt.

Removal of impurities is important in PVC reprocessing. A recent study has shown that the melt filtration efficiency depends on the temperature of the melt, the impurity level, and the overall quality of the recyclate. It has been shown that the impurity level should be less than 0.1%, preferably less than 0.05%. For acceptable performance, screens of 150 mesh size are recommended. Reformulation with virgin material is necessary to compensate for the loss of lubricant, stabilizer, and color. At low flow rates, the material stagnates at the low flow areas. Tooling for an extruder screen changer must therefore be designed to eliminate stagnation. Since PVC does not shrink as extensively as HDPE on cooling, it should be removed carefully from the breaker plates. The major imperfections in the bottle come from fibers, left-over PET, and adhesives. Unremoved oil and other impurities can change color, odor, and lubrication characteristics. Pigmented end-use products are more forgiving of color and gel content variation. Even with all these difficulties, clear PVC compounds have been formulated using 25% recyclate. The product EcoVinyl® Natural 25 is blow-molded at 1000 lbs per hour.[77]

Recycled PVC is not yet competitive with virgin, particularly due to the greater cost of transportation and the low cost of virgin material. In 1992, approximately 4.9 million tons of

PVC was sold, of which 60% went into housing and construction and a substantial amount into automotive and business equipment.[78] Only 2% went into packaging. It is estimated that 28–30% of PVC will enter into municipal solid waste in the near future.

Davidson Instrument Panel, a division of Textron, recycles its reject instrument panels which contain polyurethane foam (PU) layers and vinyl skin. Davidson has developed a process for careful removal of PU from vinyl and grinding the latter into flakes which are reprocessed for applications including boot heels, shoe soles, and calendered sheet. The PU foam is reclaimed and used as carpet underlay. The 'vinyl compatible' instrument panels which contain vinyl skin and a vacuum-formed vinyl substrate are easily recycled into interior trim, consoles, and door panels.

Wire and cable is another major application of vinyls. Usually the scrap cables are chopped, metal is removed, and the plastic insulation and jacket are disposed of in landfill. In the past, substantial work has been carried out in recycling wire and cable scrap, particularly in Europe. Ex Tech Plastics of Richmond, VA, has perfected a process by which they can make high-strength stretch film, 5–80 mil thick, by blow molding from 100% vinyl scrap. The Geon Company with Waxman and Sons have developed a process which recycles the vinyl waste into a variety of applications including floor mats, mud flaps, garden hoses, and electrical equipment. European Vinyl Corporation (EVC) operates a vinyl wire recycling plant where the vinyl from scrap wires is pulverized and pelletized to be converted into shoe soles, gaskets, furniture, food pads, *etc.*

Tecoplast in Italy is using vinyl bottle recyclates to manufacture sewer pipes, shipping tubes, and clips. Vermont Republic Industries has developed a technique for removing the end caps, clips, and labels from PVC containers and for recycling the vinyls to manufacture pipes. Cyrogrind of Australia, along with Geon, is developing a process to recycle PVC bottles to produce a compound with 25% recyclate for manufacturing storm water pipes. PVC pipes are also being collected in European countries, and being cleaned, sorted, and regranulated to manufacture new pipes. Vinyl windows are being recycled in Germany by Vekaplast (a major window producer) into nonvisible window parts. Turtle Group in Cleveland, OH, recycles vinyls into low end products such as medical trash containers, modular cushion floor tiles, urinal screens, and floor mats for restaurant kitchens, manufacturing shops, and horse barns.[79]

Separation of PVC from PET bottles and other contaminants is the major difficulty in PVC recycling. Although this can be done by visual observation, such as reading the resin code (1 for PVC, 3 for PET), observing the parison pinch-off scar, color (PVC slightly blue, PET colorless), and looking at the crease-whitening (only PVC whitens), the manual methods are very slow (60 per minute) and the accuracy is limited.[80] Even melt filtration does not separate PET from PVC properly. Technoplast Govini, an Italian company, first used X-ray transmission to identify chlorine.[15] Subsequently, ASCMA Instruments of Austin, TX, used X-ray fluorescence in a reflectance mode to identify singulated bottles. National Recovery Technologies (NRT) of Nashville developed the Pulsort® technology to eject each bottle automatically after its identification by X-rays. NRT's commercial lines operate in three capacities, up to 4500 pounds h^{-1} with a guaranteed purity up to 50 parts per million of PVC in PET. Magnetic Separation System (MSS) of Nashville, TN, has adopted X-ray technology along with a number of visible and IR sensors to separate PVC and to color sort the bottles. Automation Industrial Control (AIC) and MRC Chamberlain have developed a system based on near-IR spectroscopy to identify individual polymers and sort them using a vibratory conveyor. Both MSS and AIC systems require bottles to be singulated. NRT has recently expanded the technology to carry out full bottle sorting that does not require singulation. In the meantime, an assisted manual system has been developed to separate PVC from PET in which the identification is carried out by UV absorption. An optical technique is under development to separate PVC from HDPE, which works well with clear bottles. Other methods used to separate PVC from PET include hot belt separation, froth flotation, electrostatic separation, and separation procedures based on dielectric constant and density. Mangaraj and McGinnis have developed a process for separating PVC from PET based on selective swelling.[81]

18.3.2.7 Reinforcement of recycled plastics

The loss of physical properties of post-consumer recyclate can be compensated by adding reinforcing fillers and fibers.[82] Bockstedt of Polymer Composites has developed long glass fiber reinforced concentrates of HDPE, PP, and PET by using a pultrusion technique with subsequent chopping of the coated fibers. Inclusion of 5–15% of these concentrates increases the heat deflection temperature (HDT), reduces shrinkage, and enhances tensile strength and flexural modulus without affecting impact strength. This type of property enhancement has been utilized in producing special articles such as bakery trays with higher HDT, beer cases with higher flexural modulus, fiber crate racks with higher flex modulus and flatness, and utility cart wheels with precise control of roundness.

Although the physical properties of recycled thermoplastics can be greatly improved by addition of chopped fibers, fiber attrition in the melting zone of the extruder leads to loss of fiber length, leading to property loss. Thielen has recently developed a process which eliminates this difficulty.[83] He uses a Z-shaped-blade kneader with a bottom discharge screw which can be integrated to a blow molding machine. The extruder of the blow molding machine feeds the kneader with molten polymer, to which chopped glass fibers are added and mixed. The two components are mixed without shear and the molten mixture is discharged back into the blowhead. The first-in–first-out accumulator features a special die which allows good expulsion with satisfactory wall thickness. This process eliminates the potential for fiber damage.

Song and Suek *et al.* carried out a study of the recyclability of glass-reinforced flame retardant polystyrene by regrinding and remolding the injection molded parts five times.[84] To cover the most useful range of times recycled and the level of recyclate, they used a two-level factorial design of experiments. They measured strength, tensile modulus, flexural strength, flexural modulus, izod impact strength, and visual softening point for each material composition and drew response surfaces. They concluded that the key properties, except flame retardancy, are substantially reduced by the number of times recycled and the concentration of the recyclate. The number of times recycled exerts a greater influence as the concentration of the recyclate increases and the effect of the number of times recycled is greater than that of recyclate concentration for tensile and flexural strength. However, the opposite is true for other properties.

18.3.2.8 Wood fiber composites

Although all-plastic lumber provides excellent corrosion and weather resistance, it does not have the stiffness needed for structural applications. Further, the density of plastic lumber is higher, making it more expensive on a volume basis. Reinforcement of recycled polyalkenes can be done with sawdust and other types of cellulose fibers to provide improved stiffness, while lowering its density.

Schneider *et al.* of the Forest Products Laboratory prepared composites using four types of waste wood fiber, namely sawdust from oriented strand board, jute fibers bonded by polyester resin, Kenaf fibers, and fibers from waste newspaper.[85] Polypropylene with melt index 3 and 12 g $(10\,\mathrm{min})^{-1}$ was used as the binder and maleated polypropylene wax (Epolene A3002 and E43, Eastman Chemical) was used as the coupling agent. The fibers were hammer milled and sieved to produce 40–80 mesh particles. The waste newspaper fibers were granulated to approximately 6 mm long particles. The resin and the fibers were melt blended in a thermokinetic mixer and injection molded into test specimens. The strength and modulus were measured in tension and flexural mode along with notched and unnotched impact resistance. The results are summarized in Table 8.

It was concluded that small amounts of thermoset adhesive present in the waste wood flour do not have any influence on the properties, as long as particle sizes are small. The jute–polyester fiber gave inferior strength in spite of the higher aspect ratio of jute fiber, due to the presence of polyester particles which prevented wetting of the jute fibers by polypropylene. Kenaf core fibers gave similar properties to wood flour except unnotched impact energy, which was higher for Kenaf composite. Fibers from waste newspaper, on the other hand, provided superior strength and impact resistance, possibly due to their higher aspect ratio and better wetting by the binder resin.

However, Schneider *et al.*, in a study of composites made from old newspaper (ONP) waste and recycled high-density polyethylene, did not find any reinforcing effect from the paper fibers.[85] For this, the authors blended the resin and the ONP in an extruder, converted the resin into chips, and injection molded the chips to give the test specimen. The specimens were conditioned in both dry and wet atmospheres. The tensile strength and modulus were proportional to the amount of polyethylene, showing that neither the fiber nor the coupling agent contributed to the strength. Further, the wet conditioned specimens showed higher

Table 8 Mechanical Properties of Wood Flour/Waste Wood/PP Composites.[a]

System	*Izod impact energy* Notched (J m^{-1})	Unnotched (J m^{-1})	*Cantilever*[b] Modulus (GPa)	Strength (MPa)	*Flexural*[c] Modulus (GPa)	Strength (MPa)	*Tensile*[c] Modulus (GPa)	Strength (MPa)
Waste wood composite	25.9	99	2.61	75.0	3.82	61.0	4.29	38.2
Wood flour	25.7	88	2.48	70.7	3.98	62.1	4.62	37.4
% change[d] (wood flour to waste wood)	(1)	(13)	(5)	6	(−4)	(−2)	(−7)	(2)
Coefficient of variation (%)[e]	12	12	5		6	3	5	2

[a] 60/40 polypropylene/filler by weight with 5% E-43 by weight of filler. [b] Secant modulus and maximum strength. [c] Initial modulus and maximum strength. [d] Only the values without parentheses are significantly different by *t*-test at 95% confidence level. [e] Averaged over both systems.

strength than the dry conditioned specimens. This may indicate that the small amount of moisture on the fiber surface, in conjunction with the coupling agent, promotes the wetting of the fiber by polyethylene.

One of the main weaknesses of wood fiber–polyalkene composites is their lack of compatibility and interfacial adhesion. Cellulose is strongly hydrophilic whereas polyalkenes are hydrophobic. Hence, compatibilization plays a key role in producing high performance wood fiber composites. Oksman and Lindberg have studied the effect of compatibilizer on the mechanical properties of wood fiber composites.[86] The injection molded LLDPE compound contained 30% wood fiber and a series of compatibilizers including Surlyn 1702 (an ionomer), maleic anhydride-functionalized SEBS triblock copolymer, and polyethylene–maleic anhydride copolymer. The impact resistance and tensile strength showed improvement in the presence of SEBS copolymer. The ionomer showed improved tensile strength but had no effect on impact strength. Dynamic mechanical measurements showed that tan δ of polystryene in SEBS moved towards the cellulose T_g, indicating compatibilization. Scanning electron micrographs showed improved adhesion between LLDPE and the wood fiber surface. The maleic anhydride–PE copolymer did not provide any improvement in the interaction between the wood fiber and the LLDPE matrix.

Maldas and Kokta studied the effect of compatibilizers, including maleated PE, chlorinated PE and chlorosulfonated PE, on the mechanical properties and dimensional stability of Kenaf fiber-based wood fiber PE composites.[87] The composites contained 15–45% Kenaf fiber and 5–15% coupling agents. The fiber and PE were melt blended with recycled PE at 165 °C in a laboratory two-roll mill. The samples were remixed 5–10 times to improve dispersion. The compounds were then compression molded into dumbbell shaped specimens. Maleic anhydride–polyethylene copolymer showed the best performance as coupling agent among the compatibilizers studied. The tensile strength increased linearly with Kenaf fiber concentration, up to 45% in the dry state and up to 30% percent in the wet state. Whereas tensile strength and modulus decreased when exposed to boiling water, the elongation at break, toughness, and impact strength improved. The latter properties, however, decreased with an increase in fiber concentration. It is suggested that the acid group of maleated PE reacted with the OH groups of cellulose during the processing steps.

Anderson Window Corporation manufactures vinyl clad window components and in the process generates approximately 100 million pounds of sawdust each year. Whereas a large part of the sawdust is burnt to generate heat and power, the part that is contaminated with 5% PVC, generated during machining of the subsills, cannot be used for energy production. Anderson Company with the help of the Minnesota Office of Environmental Assistance has developed a process for using the dirty sawdust and recycled PVC to make reclaim composites.[88] The reclaimed material was used for manufacturing door subsills. Mobil Chemical is producing woodfiber based composite lumber using sawdust and recyclate from grocery sacks and plastic wraps.[89] The product 'Timberex' is used for industrial floors and marine pilings.

Advanced Environmental Recycling Technologies manufactures 'Moisture Shield' door and window sills from post-consumer PE and cedar wood chips.[90]

18.3.3 Secondary Recycling of Plastics

Secondary recycling involves processing the commingled plastic fraction of the waste which is not consumed by primary recycling. The two largest sources of commingled plastics are municipal solid waste (MSW) and automotive shredder residue (ASR). Once the plastics waste from building and construction is available in large quantities, this may provide another large source of commingled plastics waste. Municipal solid waste is largely dominated by packaging materials. After the separation of PET and HDPE bottles (approximately 75–80%), the rest of the waste is used in secondary recycling. This leftover portion is often called curbside tailing. Typical New Jersey curbside tailing (NJCT) consists of approximately 80% HDPE, 4% LDPE, 4% PP, 3% PS, 4% PET, and approximately 1% of other plastics.[91] However, the exact composition will vary from municipality to municipality and county to county. In addition, it contains several inorganic components including aluminum, calcium, sodium chloride, *etc.* Automotive shredder waste is not well defined. Based on the fraction of plastics used in a model year and a ten-year average life of a car, ASR in 1990 should contain approximately 19% PP, 12% PET, 10.6% PU foam, 7.6% PE, 6.64% ABS, 7.1% RIM polyurethane, and 5.9% PVC.[92] Thus, ASR contains a large number of plastics, each constituting a small fraction of the total instead of any one predominating over others. In addition, their processing temperature and properties vary over a wider range than those of curbside tailings. Hence, a certain amount of sorting and compatibilization is necessary for ASR waste.

18.3.3.1 Commingled plastics from municipal solid waste (MSW)

Commingled plastics from municipal solid waste are generally used for making products with large cross-sections so that the contaminants such as unmelted resin, glass, or aluminum do not seriously affect the mechanical or aesthetic properties of the product. Plastic lumber is a typical product from commingled plastics. A number of processes were developed in the 1980s to process commingled plastics; these may be categorized into the four basic types of Klobbie's intrusion design, the Reverzer process, continuous extrusion, and compression molding.[93]

Klobbie's equipment consists of a short extruder combined with several dies mounted on a rotating turret. The commingled plastics are mixed quickly with adiabatic heating and fed into the long die. The metallic and ceramic contaminants, as well as PET particles, are covered by molten PE. As one die is filled up, it moves out of position and allows the following die to be filled while the first die is being cooled in water. This overcomes the problem of cooling the profiles without allowing them to warp. The original patent included the use of foaming agents to foam the interior, which could be helpful in reducing the void size in the core. Advanced Recycling Technology of Belgium (ART) and Superwood International of Ireland produced equipment for Klobbie's process. ET is the trade name of the equipment supplied by ART. ET-1 has been used extensively in producing plastic lumber.

The Reverzer process allows the commingled plastics to soften in the hopper and then feeds this soft plastic into the extruder for uniform mixing. The output is flow molded, extruded, or compression molded. Both flow molding and extrusion are done under low pressure, but high pressure is used for compression molding. This versatile process accepts a large variety of mixtures and produces articles of various shapes.

Extrusion is often carried out in a short L/D extruder to minimize the thermal history. Cooling the profiles is the most difficult problem in this process. In the compression molding process, the plastics mixture is melted by frictional heat and the melt is scraped off by a knife into an extruder which produces roll-shaped profiles. The latter are then molded into different shapes by a compression molding machine. Instamelt Systems of Midland, TX, offers a rotary extruder which reduces residence time to approximately one fourth of the conventional extruder while providing homogeneous mix. This is claimed to be suitable for recycling plastics waste.[94]

Initially the plastics lumber produced by the ET-1 machine was very inhomogeneous with large voids in the core. This was attributed to the poor thermal conductivity of the plastics, which allows the skin to cool and solidify much earlier than the core. Subsequently, crystalline polyethylene in the core cools, shrinks, and creates large voids. This is reflected in nonuniform density and poor mechanical properties. The average compressive strength of approximately 22 MPa and the modulus of 790 MPa are much lower than the compressive strength and the modulus of wood. The density, which varies from 0.92 to 0.94 g cm^{-3} for plastics, is also greater than that of construction brand wood (~0.6 el% g cm^{-3}).[93]

Substantial work has been done by the Center for Polymer Recycling Research (CPRR) to improve the properties of plastic lumber.[95] Addition of recycled expanded polystryene (REPS) to commingled plastics at the 20–40% level improves both compressive strength and modulus. For example, a 35/65 REPS/NJCT mixture, when used for producing plastic lumber, improved the modulus from 1178 MPa to 2169 MPa. Examination by positron annihilation spectroscopy (PALS) showed uniform distribution of free volume sites (a measure of void size) across the cross-section. The voids were relatively smaller in size. Scanning electron microscopy showed co-continuous entangled morphology, indicating better adhesion between the phases. At higher polystyrene fractions, fibrillar structures appeared in the form of rods and plates. In other words, PS fibers reinforced PE in the extrusion process. However, neither the property improvement nor co-continuous morphology was observed when crystal PS was used in place of REPS. A similar study by the Rensselaer Polytechnic Institute (RPI) showed that addition of SEBS terpolymer to commingled plastics increased the tensile strain fivefold against a 14% decrease of tensile modulus.[96]

The properties of the commingled plastics can also be improved by good homogenization. Earlier work at the RPI showed that a mixture simulating commingled plastics showed substantial improvement in physical properties when mixed in a corotating intermeshing twin screw extruder.[96] CPRR used a 51 mm diameter counter-rotating twin-screw extruder from Welding Engineers, which is claimed to cause less degradation than a corotating nonintermeshing twin-screw extruder.[97] The Welding Engineers extruder had a proprietary Kleen Screen™ melt filter system and vacuum venting. The refined NJCT processed by this machine exhibited properties superior to virgin HDPE including impact resistance. Further, samples molded at higher temperatures showed higher impact resistance, possibly due to an annealing effect. Addition of 10% polystyrene to this compound showed significant improvement in flexural modulus, with slight decline in impact strength. The polystryene fraction existed as fibers oriented in the machine direction.

The RPI has also developed a selective dissolution system where the incompatible polymer systems like NJCT are dissolved in a common solvent to form a homogeneous mixture. This is followed by rapid devolatization under vacuum.[98] During this process, phase separation takes place by spontaneous decomposition, leading to microdispersion. This polymer material can then be processed quickly to obtain materials of high strength.

Mid Atlantic Systems in the USA provides a total turnkey operation for making plastic lumber starting from commingled plastics. In recent years, plastic lumber manufactured from commingled plastics has faced strong competition from lumber made from recycled HDPE bottle scrap and from woodfiber-filled polyethylene composites. Eaglebrook Products of Chicago manufactures plastic lumber from HDPE with consistent high quality and consistent cell structure. Foaming agent and heat stabilizer are added to HDPE recyclate to achieve resistance to weathering and uniform cell size.[99]

Beatty and co-workers have made an extensive evaluation of plastic lumber made from commingled plastics from a number of manufacturers. In general, they have observed large voids in the core.[100] Although the tensile strength of the core is slightly higher than that of the skin, the modulus is substantially lower. Both skin and core have poor impact resistance. The strength in the direction parallel to the machine is 25% greater than that in the transverse direction. Dynamic mechanical tests did not show any significant orientation effect in the mechanical properties of the skin. On accelerated aging at high temperatures (120–130 °C), the authors observed bubbling on the surface due to vaporization of residual water. Occasional cracks were caused by stress relaxation. Aging is strongly influenced by temperature. All samples gained weight when exposed to water, although the percentage increase varied widely.

One of the biggest difficulties in making plastic lumber is that the physical characteristics and strength vary over a wide range and there are currently no standard specifications to distinguish bad lumber from good. ASTM test procedures are inadequate, and do not address the composition and structure issue of plastic lumber profiles. A dog-bone specimen cut from one location on the lumber could vary widely in physical properties from a specimen taken from another location due to difference in voids and inclusions. Sachan has recently carried out measurements of strength and density on 13 samples of plastic lumber produced by 11 manufacturers.[101] The results illustrate the wide variation in properties (modulus, tensile strength, and density), as shown in Table 9. The modulus ranges from 262 to 1320 MPa, a 404% variation among 13 samples. The strength varied from 5.41 to 21.5 MPa. Further, high modulus does not coincide with high strength. The specific modulus (modulus/density) varies from 609 to 1680 MPa and specific strength from 15 to 24.6 MPa. The variations are attributed to different compositions of the plastic lumber tested and different levels of voids in the lumber due to a difference in processing variables and contaminants. Across a particular specimen the strength is high in the middle and low at both ends. The strength properties are, however, fairly constant for specimens produced by continuous extrusion. The modulus, however, varies randomly across the length. It is therefore important to grade the plastic lumber based on mechanical strength and composition in order to facilitate their marketing and application.

Table 9 Mechanical Properties of Plastic Lumber from Different Sources.

Sample	*Specific gravity*	*Modulus* (MPa)	*Specific modulus* (MPa)	*Yield stress* (MPa)	*Ultimate strength* (MPa)	*Specific strength* (MPa cm^3 g^{-1})
51A	0.2789	262	940	4.89	5.41	19.4
1B	0.7012	427	609	9.52	13.0	18.6
2D(br)	0.8630	588	682	11.5	16.0	18.5
2D(g)	0.8098	800	988	14.5	19.7	24.3
1E	0.8620	557	647	12.2	16.7	19.4
1F	0.7888	746	945	15.1	19.4	24.6
1J(b)	0.7534	643	854	13.1	16.3	21.6
1J(w)	0.9087	759	836	14.9	19.5	21.4
23L	0.7856	1320	1680	11.8	13.3	16.9
1M	0.5652	399	705	6.65	8.45	15.0
1S	0.9090	555	610	11.5	14.1	15.5
1T	0.8804	813	921	15.5	21.5	24.4
9U	0.7774	598	769	12.6	16.6	21.3

Practical tests performed on plastic lumber versus wood include nail pull-out across and parallel to the grain, screw pull-out, and percentage of strength loss at higher temperatures.[102] Nail pull-out tests are carried out on an 8 foot lumber section with nail penetration 1.5 inches using nails of 0.15 inch diameter. The results given in Table 10 show that the nail holding capacity of plastic lumber is greater than that of wood in both directions, perpendicular and parallel to the grain at room temperature. The nail holding capacity of wood in the direction parallel to the grain is 50% lower than the capacity in the perpendicular direction. In case of plastic lumber, the difference is marginal. However, the plastic lumber loses a large part of its strength at higher temperatures. The screw pull-out force of plastic lumber is smaller than that of wood.

Table 10 Properties of Plastic Lumber Compared to those of Wood.

	Nail pull-out force (MPa)		*Screw pull-out*	
	Across, rt[a]	*Parallel, rt*	*Across, 65 °C*	*Parallel, rt*
Wood	111–112	46	116	713
Plastic lumber	155–157	166	88	353
Difference	40–41%	261%	−43%	−50%

[a] rt = room temperature.

Substantial work has been carried out to study the harmful effects of curbside tailings (CT) on neat polymers recovered from a municipal waste stream (MSW). Earlier work at Quantum

Chemical showed that HDPE recovered from MSW had strength properties close to those of virgin HDPE. However, when contaminated with CT, the recyclates show lower elongation at break, lower impact resistance, unfavorable odor, and unacceptable levels of contamination. The contaminants can blow holes in the walls of the bottles made from recycled materials.[103]

VanNess *et al.* have studied the morphology of recycled PET and New Jersey post-consumer curbside tailings (NJCT).[105] The latter has greater than 95% by weight HDPE. PET and curbside tailings were mixed together in 25/75 and 5/95 ratios by a Welding Engineer nonintermeshing counter-rotating twin-screw extruder at 223–235 °C. The compounded pellets were injection molded at 177 and 271 °C into standard ASTM test bars. In another study, a complex recyclate containing 28% PET, 2% natural HDPE, 23% colored HDPE, 9% polystryene, 5.5% film, and 1% polypropylene were melt blended with and without compatibilizer and injection molded into test specimens under two sets of process conditions differing in temperature. The morphology of the blend specimens were studied by scanning electron microscopy. Both low- and high-temperature molded specimens were brittle and the morphology exhibited a complex layered structure: an outer layer of drawn rods and plates of PET in a HDPE matrix, an intermediate layer of a co-continuous phase, and an inner core of slightly oriented PET. When SEBS was used as compatibilizer, it was found to be uniformly distributed in the skin and core (Figure 11). The compatibilized blend contributed to the ductility. In subsequent work, the SEM specimens were subjected to dissolution by 1,1,2,2-tetrachloroethane to remove the PET phase and by toluene to remove the polystyrene phase. It was observed that the samples molded at lower temperature and slow injection speed displayed three-layered morphology with an intermediate layer exhibiting a more lamellar, co-continuous region. On the other hand, the specimen molded at higher mold temperature and fast injection speed did not exhibit the three-layered morphology. Instead it contained fibrous material (50 to 100 μm thick) throughout the skin and core with another region of alternating layers of PET and HDPE. This explained the delamination of the skin during flexural testing of the latter sample. The compatibilized blend containing PS and SEBS exhibited a co-continuous band bonded together by an intermediate layer of SEBS.

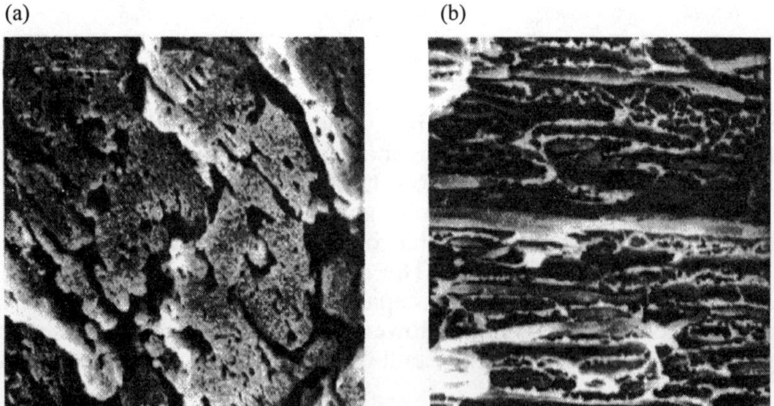

Figure 11 SEM pictures of HDPE/PET blend: (a) neat; (b) with SEBS compatibilizer.

18.3.3.2 Autoshredder residue (ASR)

A variety of engineering and commodity plastics are used in the automotive industry. The use of plastics in this industry is growing because of increased fuel efficiency, reduced manufacturing costs, and better aesthetics. It is estimated that approximately 320 lb of plastics is used in a 1995 car. More than 50% of this is thermoplastic. The rest is thermoset, including SMC, BMC, polyurethane, polyurea, and phenolics. Automotive recycling is carried out in two sequential steps: vehicle dismantling and shredding.[104] Whereas most dismantling is done in junkyards, shredding is carried out in a few facilities known as autoshredders. The latter use giant hammer mills to divide the autobody (without engine, fuel tank, battery, and catalytic converter) into fist-size pieces. Electromagnetic separation systems coupled with hydrocyclones are used to separate the pieces into three fractions: magnetic, noniron heavy, and light fluff. Plastics are largely present in the light fluff commonly known as autoshredder residue (ASR). An average car generates 500–800 lb of ASR, of which about 30% is plastic. ASR is cleaned and separated into different fractions by gravitational separation techniques. Estimated compositions of the plastic components of ASR are presented in Table 11.

Table 11 Approximate Composition of Autoplastics.

Material	*Density* ($g\,cm^{-3}$)	*Percentage present* 1980 *Car*	1990 *Car*
Polyurethane foam	0.5	13.2	10.6
Polypropylene	0.890	22.5	18.9
Polyethylene	0.920–0.960	5.5	7.6
ABS	1.06	8.8	6.6
Polyamide	1.14	2.5	4.0
Acrylic	1.18	2.6	2.0
Polyurethane RIM	1.20	8.8	7.1
PVC	1.34	18.1	5.9
Phenolic	1.38	2.0	1.9
Polyester SMC	1.8	13.2	12.1
Other	>1.0	2.7	22.8

Substantial work has been carried out by Plastics Institute of America to reprocess the plastics fraction of ASR into useful products.[106] The main difficulties in reprocessing ASR are the heterogeneity in plastics type and the presence of a large volume of filler, which results in poor melt flow characteristics. A variety of flow modifiers were examined.[107] A blend of 78:22 PP/LLDPE (Dowlex 2045), a flexible thermoplastic polyurethane (Tryin® 480-AR), phenoxy resin (Icar® PKHC 32201), and an ionomer (Surlyn® 1559) were found effective for increasing the MFI. The flexible PU and ionomer improved flexural strength and modulus. Addition of rigid polymers such as styrene–maleic anhydride copolymer (SMA) and rigid PU improved flexural modulus but lowered flexural strength. The phenoxy resin provided the best combination of properties, except impact resistance. Addition of chopped glass fiber to the mixture of ASR along with the flow modifiers mentioned above improved both strength and modulus. Several compatibilizers were examined including SBS and SEBS terpolymers, nitrile rubber, hexafluoroacetone-modified polystryene, and a chlorinated PE. The terpolymer appeared both to compatibilize and impact modify the ASR residue. Chlorinated PE gave the best combination of flexural strength, modulus, and impact properties (Tables 12 and 13).

In an attempt to mold ASR, Deanin and co-workers found that a blend of LLDPE and ASR, when injection molded in a machine with larger orifice, provided a homogeneous product with high surface gloss, high flexural strength[108] and modulus, and appreciable impact resistance. Both modulus and strength increased with an increase in applied pressure. Addition of a blowing agent, diazobisformamide, produced a foam of high strength and modulus. Deanin also used a number of thermoplastics, in combination with flow modifiers, to produce panels from 1990 fluff. ABS (Blendex 210 and 336) provided the best combination of strength, modulus, and impact resistance followed by polyethylene, both low and high density. PE wax and Acryloid 175 (Rohm and Haas) at the 10% level gave the highest increase in melt index but poor strength and modulus. High flow PP (Amoco) provided the best combination of MFI increase without jeopardizing mechanical strength.

A variety of thermoset materials, including phenolics, melamine, acrylics, and portland cement, were used as binders for ASR in attempts to make structural panels.[109] Maximum improvement in strength, modulus, and impact resistance was achieved when ASR was used in combination with a two-stage phenolic, melamine, and polyurethane. Injection molding of ASR gave products twice as much strength as compression molding. Thirty parts of binder for 100 parts of fluff appeared to provide the optimum improvement in properties. Tougher methods of mixing such as mastication and extrusion result in some loss of mechanical properties due to breakage of the fibrous structure of fluff.

Table 12 Effect of Hydrogen Bonding Additives in 1980 ASR Clean Light Fluff on the Property of Moldings.

Additives	*Percent*	*Melt rate* (g/10 min)	*Flexural modulus* (10^{-3} psi)	*Flexural strength* (10^{-3} psi)	*Izod impactstrength* (ft lb in^{-1}) *Unnotched*	*Notched*
Ionomer	0	0.21	1.02	0.68	0.82	0.59
	5	0.39	1.00	1.11	0.79	0.68
	10	0.47	0.81	1.14	1.09	1.04
	20	0.81	0.59	1.08	1.31	0.87
SMA	0	0.21	1.02	0.68	0.82	0.59
	5	0	1.45	1.21	0.68	0.66
	10	0	1.47	1.19	0.64	0.48
	20	0	1.28	1.31	0.53	0.48
Flexible PU	0	0.18	1.17	0.88	0.70	0.70
	5	0.43	1.26	1.15	0.89	0.66
	10	0.86	0.92	1.10	1.04	0.81
	20	1.41	0.47	0.99	1.43	0.89
Rigid PU	0	0.18	1.17	0.88	0.70	0.70
	5	0	1.13	0.71	0.65	0.53
	10	0.10	1.17	0.88	0.70	0.54
	20	0.28	1.10	0.99	0.83	0.57
Phenoxy	0	0.18	1.17	0.88	0.70	0.70
	5	0.25	2.43	1.47	0.91	0.41
	10	0.37	2.93	1.89	0.69	0.48
	20	0.79	2.31	2.00	1.14	0.48

Table 13 Effect of Elastomer Addition on Key Properties of ASR (1980).

Elastomer type	*Supplier*	*Trade name*	*Parts by wt.*	*Melt index* (g/10 min)	*Flexural* *Modulus*	*Strength*	*Izod impact* *Notched*	*Unnotched*
–	–	–	0	0.11	93 000	750	0.60	0.70
SBS block copolymer	Shell	Kraton	10	0.42	41 000	777	0.64	–
	Shell	Kraton	20	0.77	22 000	673	1.28	–
SEBS block copolymer	Shell	Kraton	10	0.51	55 000	1152	–	1.30
	Shell	Kraton	20	0.91	22 000	1075	–	2.20
Vinyl chloride/ vinyl acetate copolymer	–	–	10	–	38 000	640	0.9	16.5
	–	–	20	sticky	–	–	–	–
Nitrile rubber	Uniroyal	Pavacryl	10	–	25 600	589	0.8	1.56
	Uniroyal	–	20	–	25 300	443	0.88	1.7
Chlorinated PE	Dow	–	10	0.66	64 000	1266	1.0	–

A crosslinked acrylic system containing poly(methyl methacrylate), methyl methacrylate (90% level), and trihydroxymethylpentane trimethacrylate was used as a binder, with the hope that the liquid system will wet the fluff better than thermoplastic melt.[109] A radical catalyst (benzoyl peroxide) along with a promoter (dimethylaniline) was used to carry out the reaction. However, the composites did not have adequate strength or stiffness. The flexural strength dropped sharply with an increase in ASR content. A similar attempt to use a latex of butyl acrylate–methyl methacrylate copolymers along with portland cement did not give panels of good compressive strength. The latter decreased sharply with an increase in ASR content. Use of furan–formaldehyde resin as binder did not show any improvement in properties. It was concluded that none of the above resins wet the surface of ASR components, leading to brittle failure of the composites.[110]

Although this study provided some basis for recycling plastics waste from the automotive industry, substantial work has yet to be done for commercial exploitation of ASR. The automanufacturers in conjunction with the American Plastics Council (APC) are developing several methods, such as easy disassembly and dismantling, separating metal and ceramic parts from the plastic and elastomers, and recycling of SMC to reduce the volume of ASR.

18.3.4 Tertiary Recycling

In tertiary recycling, the polymer waste is degraded to produce useful chemicals, including monomers and oligomers. This is accomplished either by pyrolysis or hydrolysis. In pyrolysis, the physical and chemical decomposition of polymers is accomplished by heating in an oxygen-free or oxygen-poor atmosphere. In hydrolysis, the polymers are hydrolyzed to produce the low molecular weight monomers and oligomers. Less developed methods include programmed degradation and biological and photolytic degradation.

18.3.4.1 Pyrolysis

Pyrolysis is a viable method of recycling plastics waste, particularly when the waste components are commingled and contaminated. Early work on pyrolysis of municipal solid waste to produce fuel oil was carried out in the mid-1970s during the oil crisis. However, the cost of producing oil was economically unattractive. With the increase in oil price and tipping fees for landfill, pyrolysis may become an economical method to produce fuel oil from our garbage. Pyrolysis of organic materials may be defined as chemical decomposition at high temperature in the absence of oxygen. It is an endothermic process. External energy is therefore essential to initiate pyrolysis, but once it starts, the pyrolysis gases produce enough energy to sustain the high temperature. Pyrolysis is different from incineration in many respects. In incineration the organic material is combusted at a higher temperature, approximately 950–1000 °C compared to 700 °C or less for pyrolysis. Pyrolysis is self sustainable combustion whereas incineration requires additional energy for continuing combustion. There is very little emission in pyrolysis, whereas in incineration, extensive scrubbing of the gas is essential to retain harmful emissions from being discharged into the atmosphere. Further, whereas the residue after incineration cannot be reused as filler, the fly ash generated in pyrolysis can be safely used as filler.

Factors which affect the composition of pyrolysis products include residence time, reactor temperature, particle size of the feed material, and the gases and vapors present in the reactor. Whereas residence time and particle size increase solid content (*i.e.* char), higher temperature, steam and oxygen in the reactor increase the gaseous content and provide a higher yield of oil. Common pyrolysis reactors can be divided into three classes, namely shaft reactor, rotary kiln reactor, and fluidized bed reactor. Usually heat is provided either by partial combustion of water/fuel or through a heat exchange medium which is heated by a separate process. Presence of water in the reactor reduces char by reacting with hot carbon to produce methane. Predrying of the feedstock reduces energy input, heating time, and liquid cracking in the reactor. Huang and Dalton[110] have described the characteristics of various pyrolysis reactors including heating methods, product distribution, feed conditions, average reactor temperature, and state of development. With pyrolysis, both thermoplastic and thermoset waste can be thermally degraded in an oxygen-deficient atmosphere to produce monomers, fuel gas, fuel oil, and chemical feed stocks. This can be carried out in a fixed bed or fluidized bed. Pyrolysis has been used extensively in Japan and Europe and commercial plants are available from Michiame, Ebara, Ganet Research and Development Company, Barker-Colman, *etc.*

A few polymers such as poly(α-methylstyrene), polyperfluoroalkanes, and poly(methyl methacrylate) undergo thermal decomposition to produce monomers. The rest produce some monomers along with hydrocarbons and tar.[111] The products of pyrolysis can be directly fed into a heating furnace to provide heat in cement kilns and other manufacturing operations, or can be separated into monomers and different fuel oil fractions. In the 1980s, a number of petroleum manufacturers were experimenting on waste plastics as feedstock in their petroleum cracking plants.[112] They included the Japan Gasoline Company, Sanyo Electric Company, Nitsei Shipbuilding Company, Nikseie Petrochemical Industries, and Mitsubishi Heavy Industries in Japan and Chevron Chemical and Union Carbide in USA. Earlier, the Bureau of Mines had studied the pyrolysis of municipal solid waste containing plastics and had recovered 100–133 gallon of liquid per ton of refuse. Chromatographic analysis of the liquid showed a large percentage of aromatics including benzene, ethylbenzene, toluene, and xylene.[113] Roy *et al.* studied the pyrolysis of a model mixture of polyethylene and polystyrene, simulating the refuse fraction of food restaurants, and obtained decomposition products similar to those of pure PS and PE.[114]

In Germany, a reactor at Riehroel refinery (Bottop) is testing a thermal process in which unsorted municipal solid waste containing ~10% PVC and heavy oil is fed into a hydrogenator, yielding a light oil fraction at 97% efficiency.[115] The acid gases generated from chlorine-contamined oil are removed by an alkali injection system. Pilot plant studies are being carried out at BP Petrochemicals, Chevron, and Amoco refineries, where the objective is to replace a small part of the feedstock with municipal solid waste. The main difficulty is the presence of contaminants such as chlorine and nitrogen. However, when plastics waste is cracked in a coker, it produces 90% liquids compared to 80% in a normal refinery feedstock. Battelle Memorial Institute (Columbus, OH) has developed a fluidized bed pyrolysis unit that converts a mixture of LDPE, HDPE, PS, and PVC into a gas stream containing 90% ethylene, 27% methane, and 17% hydrogen. Unlike coker, Battelle's process accepts coarsely shredded solid waste and does not produce oxygenated compound such as dioxin. Wyane Technology Corporation (Victor, NY) has recently commercialized a process which takes scrap containing PE, PP, PS, wood pallets, cardboard boxes, *etc.*, and produces 100 barrels of oil per day. Pyrolyzing at 480 °C, the oil is separated into three cuts, the middle cut having a heat value and viscosity similar to #2 fuel oil. Another process that has gone commercial is that of the Fuji Recycle Industry (Aioi, Japan) in which polyalkene waste is converted to gasoline. In this process the plastic is first shredded and melted with a liquid feedstock which is thermally cracked at 400 °C, using an aluminum silicate catalyst. Fugi Recycle is also developing commercial pyrolysis units for converting polystyrene food trays and municipal mixed polyalkene waste into valuable fuel. Braslow *et al.* studied the pyrolysis of automotive shredder residue and obtained 53% volatiles including 34% liquid.[116] A noncatalytic process for pyrolyzing waste polypropylene to #2 and #6 fuel oils has been reported.[117]

Recently, the American Plastics Council has funded a large project with Conrad Industries of Centralia, WA, to develop and demonstrate the feasibiilty of pyrolysis as an effective recycling process.[118] This technology, once developed into commercial scale, may provide advantages that overcome problems associated with cleaning, shredding, and sorting and remove concerns over the quality of the end products. Furthermore, the technology may supplement the conventional recycling process, particularly as it can recycle the fraction that cannot otherwise be recycled. Both automotive and municipal solid waste disposal will benefit from Integrated Recycling's approach. Primary, secondary, and tertiary recycling can be used in tandem to achieve the highest recovery without sacrificing quality.[118]

The Conrad Recycling Process uses an auger kiln reactor that heats the plastics in the absence of oxygen to produce liquid petroleum, carbon, and gaseous products. Shredded plastics are fed into the system (Figure 12) through an air-lock device which prevents air coming into the system. The plastic then moves through an auger that reaches temperatures up to 540 °C. The carbon, 3–8% of the total waste, exits the system and is sold for the production of activated carbon and pigments and as filler for rubber products. The gases move into a condensation chamber where they provide a liquid product like crude. The latter, approximately 70–80% of this final output, is shipped to refineries for the production of monomers and other fine chemicals. Any gases that do not condense are used within the system to generate heat from the process.

A parametric study was carried out initially to identify process hurdles, key process parameters and to assess product values and markets. A base feed composition 60:20:20 of HDPE/PP/PS was selected, representing the post-consumer commingled plastics stream. Offspec resins were used to represent the damaged characteristics of post-consumer resin, while eliminating any feed composition variation. The overall yield and the composition of the product depend on the feed temperature. Lower temperatures produce a high grade of naphtha along with high yields of solid wax (incompletely depolymerized plastics). Reduction of the feed rate and slowing down the auger rotation increase the yield of liquid products and reduce

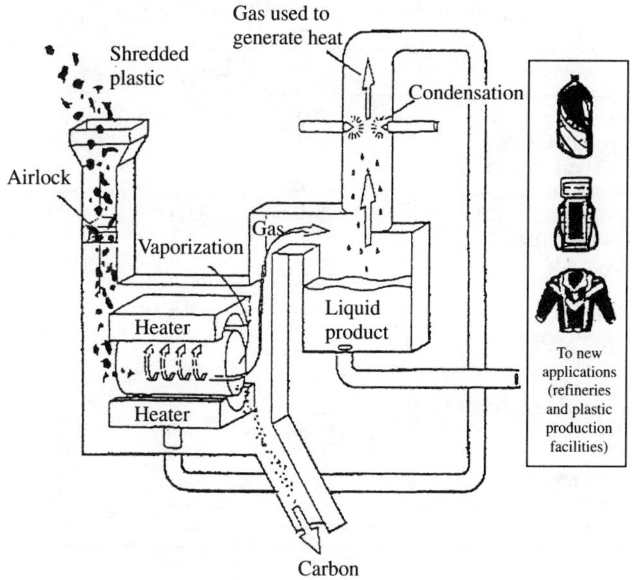

Figure 12 Conrad pyrolysis unit.

waste formation. Subsequently, by adjusting the temperature profile of the kiln, an increased feed rate was accommodated without losing the high yield of the liquid fraction. Use of larger units having larger kilns and recirculation of the first pyrolyzed product increase the residence time and eliminate the feed rate difficulty. The overall yield is given in Table 14, which shows that the maximum yield of the liquid fraction (79%) is obtained at around 540 °C kiln temperature. The composition of the liquid and gas is given in Table 15. The economics of this process works out at approximately $100 per ton tipping fee. This is comparable to the tipping fee for land filling in some of the northeastern states of the USA, but more costly compared to other states. However, it is possible to use smaller versions of the Conrad kiln at conventional recycling facilities to accept low-quality PCR.

Table 14 Product Yields from Conrad Study as a Function of Reactor Temperature.[a]

Oven temperature[b] (°C)	*Liquid* (%)	*Gas* (%)	*Solid* (%)
790	28	64	8
700	48	51	1
590	73	28	–
540	79	21	–

[a] Feed: 60% HDPE, 20% PP, 20% PS. [b] Retort temperature about 25–50 °C cooler than oven.

Thermoset materials such as sheet molding compounds (SMC), phenolics, urea, melamines, and polyurethane are not recyclable, because they do not fuse at high temperatures for reprocessing. Currently they are ground into fine powder and are used as filler in SMC and other thermoset compounds. With the increasing use of SMC in the automotive industry, recycling of SMC poses a serious problem. General Motors in the USA is therefore taking a lead in recycling of SMC through pyrolysis.[119] The pyrolysis unit used by GM consists of a feeding section, a pyrolysis section, and a condensation section. The feeding section has particle sizer, feed hopper, conveyer, air interlock, and other components. SMC waste is cut into small

Table 15 Product Analysis of Liquid Fraction.

	Base feed		
Oven Temp (°C)	700	650	590
Auger Temp (°C)	590	530	480
Partial oil analysis (%)			
Benzene	7.0	2.3	1.1
Toluene	20.0	11.0	9.1
Ethylbenzene	9.0	5.7	4.8
Styrene	20.0	17.0	14.7
C_{10}–C_{15}	24.5	26.0	19.9
C_{16}–C_{20}	8.6	9.9	14.4
C_{21}–C_{25}	3.0	3.5	6.4
C_{26}–C_{30}	1.7	1.7	4.5
C_{31}–C_{40}	1.1	1.5	4.2
Partial gas analysis (%)			
H_2	4.0	2.6	2.2
CH_4	18.0	13.6	10.5
Ethylene	19.4	18.1	13.8
Propylene	21.4	21.8	21.0
Isobutene	5.0	6.1	6.8

pieces, 4 inches or less. The smaller the size, the easier it is to feed and the shorter the residence time. The feed hopper has a vibrator and belt to convey the feed easily to the pyrounit inlet. The feed inlet and the air interlock are designed to prevent plug-up problems. The pyrolysis section consists of a retort, which is a hollow cylinder made from stainless steel to prevent sagging. The conveyer shaft in the retort has been specially designed to keep the feed material continuously flowing through the retort while being pyrolyzed, thereby preventing the build-up of char in the retort or on the conveyor interface. The pyroresidue is collected after it passes through the chamber. The pyrolyzed gases are continuously removed from the pyrolysis unit through a series of condensers where the pyro-oil or any other condensate is collected and removed. The control system maintains temperatures, pressure, vacuum, and feed rate at the desired levels. The pyrolyzed SMC residue has been successfully used to replace 50% calcium carbonate as filler for SMC formulations. The physical properties of SMC containing various levels of pyrolysis residue along with a virgin material are either superior or equivalent to that of virgin SMC. The bond adhesion characteristics such as lap shear, load, and percent of fiber torn at 27 and 82 °C are also superior. The gaseous byproduct from pyrolysis contains mainly aliphatic hydrocarbons from C_2 to C_9 along with CO, CO_2, O_2, H_2, and N_2 with a total Btu value of 1025. The process does not produce any harmful emissions, such as NO, SO_2, styrene, and nonmethane hydrocarbons. The pyrolysis oil contains 38.6% aromatic hydrocarbons, 24% α-alkenes, and 23% methylene, 11.3% alkenic, and 3.1% methyl hydrocarbons. The analysis of the pyroresidue showed that the heavy metal concentration is lower than the maximum allowable concentration in filler applications.

In addition to SMC, other polymer wastes including thermoplastics, paint sludge, wood chips, polyols, paint solvents, rubber tires, and automotive shredder residue have been successfully pyrolyzed.

18.3.4.2 Hydrolysis

Condensation polymers such as polyesters, polyamides, and polyurethanes can be depolymerized back to the original polymers. This process has been used to recycle condensation plastics such as PET, nylon, and polyurethane. The advantages of hydrolysis over pyrolysis are: control of product composition; less difficulty in separation of reaction products; and lower capital investment compared to pyrolyzer.

Food containers represent a large market for PET, including ~750 million lbs per year in soft drink bottles. The soft drink companies along with PET producers have obtained permission from the FDA for recycled PET to be used for soft drink packaging. This has encouraged a number of companies such as Hoeschst Celanese, Goodyear Tire and Rubber, DuPont, and Eastman Kodak in USA and Zimmer in Germany to develop PET depolymerization processes. Eastman Kodak's glycolysis process (Figure 13) gives a variety of products which can be used for making unsaturated polyester and polyurethane in addition to PET.[120] Furthermore, this process is much more forgiving (in respect of contaminant sensitivity) than either the Goodyear or DuPont processes. Another earlier process is based on hydrolysis followed by catalytic hydrogenation to obtain colorless diesters. Zimmer's glycolysis process produces low molecular weight oligomers which can be reprocessed into PET.[121] This process is claimed to be more economical than the complete depolymerization process. The presence of contaminants such as hot melt glue and PVC decreases the efficiency of the process. Hence they need to be removed before depolymerization.

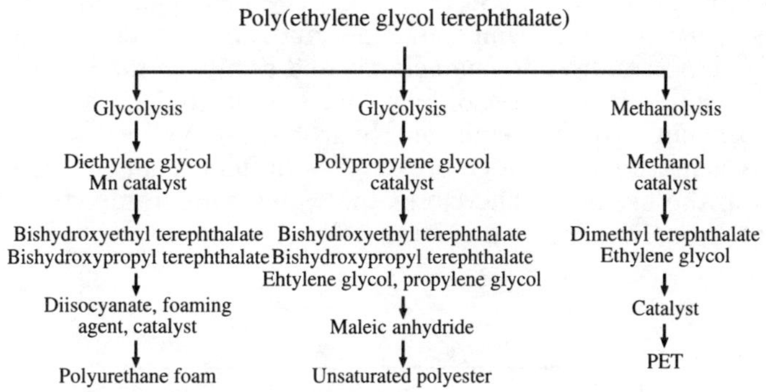

Figure 13 Eastman Kodak's glycolysis process.

Cleaning the PET bottles before methanolysis or glycolysis is very important, particularly if the recycled PET is to come into food contact. The FDA has defined tetracosane as a surrogate to spike the surface of the RPET chips and measure its residual concentration after cleaning. Komolprasert and Lawson have studied the influence of different cleaning process parameters on cleaning efficacy.[122] In addition to tetracosane, they have used a number of hydrocarbons as surrogate. Thirteen different washing conditions using different levels of NaOH along with different levels of Triton X and different rpm of the agitator were used for cleaning. They found that at various combinations of mixer speed, temperature, and NaOH concentration, only the temperature and the mixer speed showed any significant effect on the yield. The sodium hydroxide concentration has no effect on the cleaning efficiency in the presence of 1% of Triton X-100. The optimal washing condition was 600 rpm at 90 °C. A number of dimensionless parameters, including Reynolds number (*R*), Fraude number (*F*), Power number (*P*), and Weber number (*We*), were calculated and it was found that statistically *F, P,* and *We* and their interactions have a significant effect on cleaning efficiency. An increase in percent of PET weight per batch also appeared to decrease cleaning efficiency. Approximately 80–97% of surrogates were removed in the washing process.

Methanolysis of poly(ethylene glycol terephthalate) (PET), developed by DuPont and Eastman Kodak, is accomplished by reaction of methyl alcohol and ethylene glycol (EG) with PET.[123] The process is exactly the mirror image of the polymerization process, where bis(hydroxyethyl terephthalate) (BHET), formed by the condensation of dimethyl terephthalate (DMT) and ethylene glycol (EG), undergoes step growth condensation to form PET. The reverse reaction follows the following steps: $PET_{(n)} + (n-1)EG \rightarrow n\,BHET$; $BHET + 2\,MeOH \rightarrow DMT + 2\,EG$, where n indicates the number of repeating units in the PET chain. PET can also be hydrolyzed in the molten state to give terephthalic acid and ethylene glycol, but this process faces formidable technical and quality problems. Hence depolymerization is carried out primarily by converting chopped PET sections into a glycol-rich medium of oligomers or by adding methanol to molten PET under pressure. Esterification catalysts are used. Details of reaction steps and technology are considered proprietary. Ethylene glycol and dimethyl terephthalate are recovered along with diethylene glycol, dimethyl isophthalate and cyclohexyldimethanol. The contaminant level, particularly poly(vinyl chloride) (PVC) or poly(vinylidene chloride) (PVDC), has to be kept below 0.5% for product acceptance. The economics of methanolysis is not favorable, due to the high cost of the methanolysis plant.

Chemical recycling of PET has also been carried out by reactive extrusion. This is an efficient continuous process, in which better control can be exerted on reaction conditions. Further, reactant characteristics can be held constant and are not changed from time to time. Provision of vents and feeding ports provides easy removal of byproducts and addition of catalysts, *etc.*, in appropriate time. The twin-screw extruder used to carry out the reaction is simple and easy to handle. Earlier work has shown that polyurethane can be successfully hydrolyzed by reactive extrusion to provide recyclable polyethers and diamines. This has been accomplished by pumping water at high temperature into molten polymer. Kamal *et al.* used this technique to hydrolyze PET to produce a high level of hydrolysis.[124] The reaction was carried out in a 25 mm Berstorff ZE 25 fully instrumentated corotating twin-screw extruder with a L/D ratio 23:1. The extruder was starve-fed by the hopper and a positive displacement pump was used to supply water under pressure. Kodapak PET copolymer 12,440 was used for this experiment. A tracer of black PET granule was added to monitor residence time. The melting range of the resin was from 230 to 240 °C. The extent of reaction was monitored by estimating carboxyl concentration using an end-group analysis technique. Hence, observed melting point change has been used to characterize end groups in the hydrolysis product. It was observed that under starve-feeding conditions, the use of cold or hot saturated water with molten polymer does not provide significant hydrolysis. However, a significant amount of depolymerization is achieved using high-pressure saturated steam injected against the back pressure generated in the extruder. The closer the water temperature to that of the melt, the more effective is the hydrolysis reaction (Figure 14).

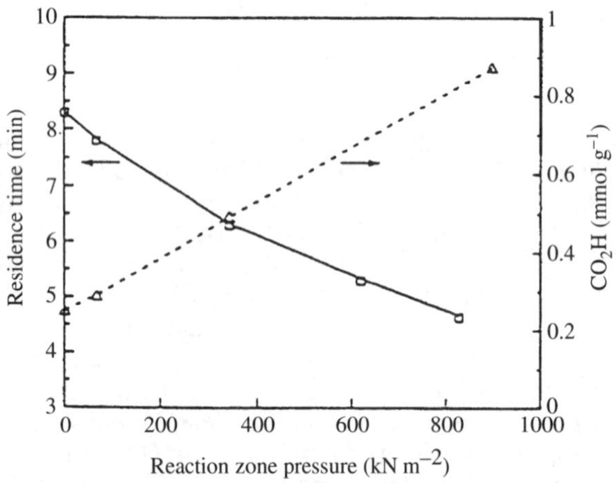

Figure 14 Residence time and carboxyl-group concentration against reaction zone pressure. Reaction temperature, 300 °C; screw speed, 10 rpm; PET feed rate, 0.61 kg h^{-1}; H_2O/PET ratio, 1.67:1.

Synergistics Industries in Canada has recently received a patent for a degradative transesterification process for PET which can be used to produce materials useful as plasticizers.[125] The PET is reacted with a high molecular weight alcohol in the presence of a catalyst such as tin(II) oxalate at a temperature of 160–260 °C. For example, with 2-ethylhexanol the final product is dioctyl isodecyl terephthalate.

(i) *Polyurethane ((PU)*

Hydrolysis of polyurethane to glycols, polyols, and diamines, and use of the polyols back into foam production, has been extensively studied.[126] When low-density PU foam was mixed with superheated steam, toluenediamine and propylene glycol were formed. The rate of formation of toluenediamine was pseudo-first order. The yields of diamine and glycol were 85% and 90% of the theoretical yields. Earlier work by GM on commercial development of PU hydrolysis was not cost effective.[127] Subsequently, Leverkusen developed a continuous process for hydrolyzing PU, using a twin-screw extruder (Figure 15).[128] The shredded foam is crushed and kneaded into a pulp and hydrolyzed at 185–210 °C in a nitrogen atmosphere in the presence of an organometallic catalyst. Polyols produced in this process were extracted and were used in producing foam. Alternatively, the reaction is carried out in the presence of a glycol solvent to produce low molecular weight polyols. Foams produced from the polyols have mechanical properties similar to those produced from virgin polyol.[128]

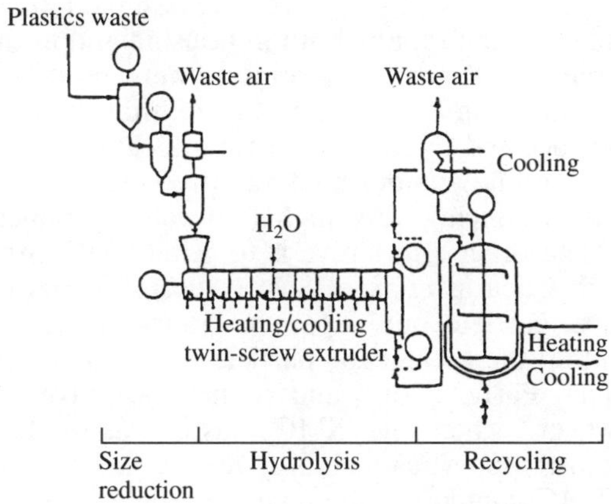

Figure 15 Schematic diagram of a polyurethane glycolysis unit.

The Federal Government of West Germany and the plastic producers worked on a closed-loop recycling process for polyurethane foam.[129] The hydrolysis reactor consisted of a specially equipped twin-screw extruder. Water was injected at 300 °C into a sealed reaction chamber against prevailing internal pressure. Recycled PU foam was fed continuously from the rear and the reaction was completed in 5 minutes. Borek and Peabody have developed a process in which the foam is dissolved in dicarboxylic acids and is subjected to depolymerization at 250 °C.[130] The products are used for coatings, adhesives, and reforming foams. The Ford Motor Company in Europe have successfully developed a process for hydrolyzing PU to glycols and use the latter to make PU foam.

(ii) Polyamide (PA)

Chemical recycling of nylon to caprolactam has also been developed as a closed-loop recycling process for waste nylon carpet. Originally developed by BASF, this process is being evaluated by an alliance of BASF and Shredtech.[131] Carpets are made of multicomponent construction, consisting of face fibers and primary and secondary backing. The face fibers are largely made of nylon. Both nylon 6 and 66 with smaller amounts of cotton, polypropylene, wool, and acrylic are used for making face fibers. The backing is usually made of polypropylene with calcium carbonate filler. Styrene–butadiene rubber latex is used as an adhesive. In addition, the carpets also contain smaller amounts of dye, pigments, antistatic fibers, and surfactants. For hydrolysis, the carpets are collected, identified as to their face fiber content, and subjected to a mechanical process for removing the fibers from the backing, adhesive, and filler. The resulting nylon fibers are then depolymerized by a proprietary process back to caprolactam. The latter is purified and used to produce nylon, which is subsequently used to weave carpets. Nylon 6 is separated from other impurities by a solvent dissolution process: 4.2 M hydrochloric acid dissolves nylon 6, leaving other components intact. Recently, Rockville Industries of Baltimore, MD, has developed a near-IR spectrophotometer which identifies all types of resin used for carpet facing.[131] Nylon fibers after separation are divided into different batches before they are subjected to depolymerization. The crude caprolactam obtained from depolymerization is collected, purified by distillation, and repolymerized into nylon 6 at BASF's Enka site. The nylon 6 obtained in the process is converted into yarn and tufted into test carpet. The carpet is then tested for fuzzing, tetrapod wear, Vetterman wear, xenon light fastness, and cracking. The carpet made from recycled nylon has properties very close to the carpets made from virgin nylon yarn except that the latter have slightly higher delamination strength. The properties of caprolactam from recycled carpet are comparable to pure caprolactam. The results show that the hydrolytic process is an effective method of recycling nylon resin and nylon fibers from the carpet.

18.3.5 Recycling of Thermoset Materials

The use of thermoset materials such as styrene-crosslinked unsaturated polyester, polyurethane, and phenolics is increasing rapidly both in housing and in the automotive industry. In 1995, approximately 423 million pounds of thermosets were used by the automotive industry, compared to 242 million pounds in 1990. Recycling of thermosets poses a serious problem because they cannot be reprocessed by melt processing techniques. The presence of crosslinks does not allow them to fuse and flow under heat and pressure.

SMC can be incinerated, producing very high fuel value products, and pyrolyzed to give monomers and chemicals. The other alternative is to grind SMC into fine powder and reuse it as filler in BMC compound. Cucuras *et al.* of SMC Automotive Alliance and Jute of Owen's Corning Fiber Glass have carried out separate laboratory evaluation of this recycling process.[132,133] It was found that the greater the particle size of the SMC powder, the greater the property loss of the product. Particles of $\frac{1}{8}$ and $\frac{1}{16}$ inch size gave approximately 30% loss in physical properties. However, when the SMC waste was milled into fine powder and incorporated into BMC, most properties including tensile flex modulus and impact resistance improved. When used in SMC compound the coarse particles gave lower tensile strength and modulus but higher elongation at break and impact resistance compared to fine particles (Table 16). Furthermore, when ground SMC powder was used as filler in a polypropylene compound, tensile and flex modulus improved but the elongation at break and impact resistance went down. Mangaraj *et al.* at Battelle have developed a proprietary process for surface activating the SMC powder and using it as filler in a SMC formulation.[134] Their results indicate that both tensile strength and modulus increase appreciably on addition of chemically activated SMC powder.

Table 16 Mechanical Properties of SMC Formations Using SMC Waste.

		Tensile		*Elongation*	*Flexural*		*Izod impact* (ft lb)	
Sample no.	*Composition weight (%)*	*Strength* (psi $\times 10^3$)	*Modulus* (psi $\times 10^6$)	*at break* (%)	*Strength* (psi $\times 10^3$)	*Modulus* (psi $\times 10^6$)	*Notched*	*Unnotched*
Control	Resin/$CaCo_3$/SMC 21.1/57.4/0	4.05	1.90	0.44	14.03	1.53	5.06	6.76
Compound coarse	21.5/52.0/5.8	2.33	1.43	0.68	9.95	1.42	5.06	5.20
Compound fine	21.5/52.0/5.8	2.51	1.84	0.22	10.36	1.48	3.91	5.13
Coarse	21.6/46.5/11.6	2.11	1.33	0.34	7.97	1.37	2.96	3.42
Fine	21.6/46.5/11.6	2.33	1.48	0.14	9.26	1.34	4.14	5.63

Polyurethane is another major thermoset which is difficult to recycle. However, it has several advantages over polyalkenes in automotive applications, because of its potential for reaction injection molding, parts integration, and better aesthetics. In recent years, several technologies have been developed to recycle thermoset polyurethanes. The technologies can generally be divided into two categories, namely regrinding and using the regrind as a filler in polyurethane reaction injection molding (RIM), and glycolysis of scrap PU and the reuse of glycols in polyurethane RIM. Research at Dow Chemical has shown that both filled and unfilled RIM scrap can be commingled with RIM compound as feedstock.[135] Different processes use different levels of regrind, depending on the requirements of the final product. For example, for RIM fascia, where surface finish is very important, only 5% regrind is incorporated. The two other critical issues are particle size and moisture content. The particle size should be under 250 μm and the moisture content should be maintained at less than 0.25%. Diethyltoluenediamine (DETDA), used in RIM formulations as a catalyst, is not fed in the same stream as PU regrind, because it has adverse effects on the chemical properties of the end product. If an inorganic filler is needed, it can be introduced either with the polyol stream and/or with the main stream containing diamine catalyst additives. The product manufactured in this process has properties similar to virgin PU. In this process the recyclate content is kept at a low level to avoid high viscosity and to achieve a Class A surface. Class A surface finish cannot usually be accomplished with regrind mixed thermoplastics. It is anticipated that, with the development of tools that can accommodate shrink rate, higher levels of regrind can be used in this process, thus alleviating the difficulties of in-factory scrap disposal. Another benefit of this process is the reduction of urethane chemicals in the feed stream.

Krauss Maffei in Europe has developed another process which is based on pre-glycolysis of the scrap before feeding into an RIM machine.[136] The process uses RIM scrap from old cars as well as rejects and sprues from RIM processing. The material is pulverized into 200 μm size particles which are reacted with an appropriate amount of glycol (mainly dipropylene glycol) to undergo metathesis and produce a mixture of glycols. Excess glycols are taken out and new polyols are mixed to maintain proper stoichiometry in the polyol stream for RIM processing. This polyol stream has approximately half regrind and half glycolized material. It is then used without an additive in the RIM process to manufacture thick RIM products without any blemish. This process can reuse 20% recyclate, saving 16% raw materials from RIM feedstock.

Bayer has developed a compression molding process that uses pulverized RIM scraps and polyurethane parts mixed with iron oxide pigment to provide a uniform black color.[137] The material is then dried to a minimum moisture content and compression molded into automotive parts. Bayer is also investigating other methods of recycling including glycolysis, use of powdered PU as a filler, and manufacturing chemicals by controlled pyrolysis of polyurethane foam. Ground polyurethane has also been recycled for other automotive applications. The Toyota Motor Company pulverized polyurethane bumper (particle size less than 1 μm) and added that to a polyurethane adhesive to produce a repair material for damaged bumpers.[138] Mitsubishi has patented a process for using pulverized PU bumper scrap to make backing for nylon carpet.[139] The same company also patented a process to manufacture sound insulator and floor covering from ground urethane waste.

18.3.6 Recycling of Tires

Approximately 275 million tires are thrown away each year in the USA.[140] The majority of these cast-off tires, numbering as many as three billion, are sitting around in stockpiles in different locations in the country. In addition to their ugly look, they pose serious fire hazards which cost millions of dollars to contain. Toxic fumes from these tires contain carcinogens, such as polynuclear hydrocarbons and dioxin. Hence, tire disposal is a serious problem throughout the USA. Almost every state has now imposed a tire tax, 1–5 dollars per tire, and the money collected from this tax is being used to develop technology and facilities to recycle scrap tires.

Figure 16 gives a schematic representation of how scrap tires are being disposed of. Of the 49% being reused, 33% of the tires are burnt for energy and 13% for retreading. Only 2% is being used for civil engineering applications and 2% as crumb rubber. Incineration of scrap tires is not only unwise but also dangerous since it gives SO_2, nitrogen oxide fumes, and toxic emissions of heavy metals and volatile organic compounds. Retread companies take tires which are less worn out and retread them. Approximately 30–33 million tires are salvaged by the retread industry. However, this market is shrinking because the steel belted tires manufactured today last too long and are difficult to retread. Some tires are split or pinched for making products such as gaskets, shoe soles, and dock bumpers. Devulcanization of tire rubber is carried out by heating rubber crumbs in the presence of lime and other chemicals. The devulcanized or reclaim rubber is added to virgin rubber in formulating rubber compounds. The use of reclaim rubber has also declined in recent years. A process for the devulcanization of rubber by ultrasonics is under investigation.[141] Devulcanization is not a viable process for tire recycling because of its high cost. Recycling tire rubber into value added products appears to be more lucrative. In this process, tires are made into crumb rubber by grinding using special granulators at room temperature or at cryogenic temperature in the presence of liquid nitrogen. Solid state shear grinding appears to be a viable process for making rubber powder. The latter process is very costly. The crumb rubbers are used along with virgin rubber to make a variety of products, including mud flaps for trucks, core for solid tires, wheel chucks, pickup truck bed liners, bridge bearing parts, *etc.*[142] Crumb rubbers are also mixed with plastics and the rubber plastic composites are used to make garbage cans, flowerpots, and curbside collection bins. Roadside sound barrier walls is another large application of crumb rubber. Formulations used in these applications contain approximately 40–50% of crumb rubber. Tires are also chipped and shredded and are used in making products such as sports mats and industrial filters.

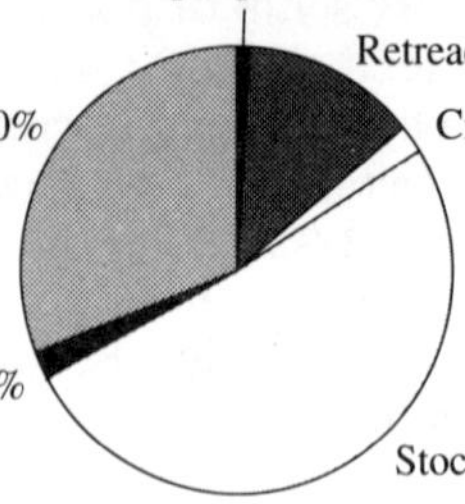

Figure 16 Used tire consumption.

The largest use of scrap tires is in rubberized asphalt. Typical formulations contain a large amount of aggregate with a small amount of asphalt. Crumb rubber is introduced as part of the binder or aggregate. The compound, often known as Crumb Rubber Modified Asphalt (CRM), has outperformed conventional asphalt. CRM-built roads last longer and require less maintenance. Because of its elastic nature, it can sustain the dynamic stress caused by traffic without showing cracks or dents. Since water does not accumulate in cracks, the roads are not affected by the freeze–thaw effect in winter. CRM's initial cost is approximately 20% higher than conventional asphalt. However, the savings on labor for installation and the reduced maintenance pay off the extra cost in a short time. The cost of CRM has been dropping since 1992, when the patent on CRM expired. Besides cost, the other drawback of CRM is that it cannot be recycled, particularly when the rubber content is high. Tires are also pyrolyzed, generating pyro-oil which can be cracked to give valuable chemicals.

18.4 IMPACT OF EMERGING TECHNOLOGIES ON PLASTICS RECYCLING

As mentioned earlier, plastics recycling technology is intimately related to a number of polymer technologies such as identification and sorting, restabilization, blending and reprocessing, impact modification, and compatibilization as well as the design of plastics parts for recycling. All these technologies have emerged in recent years and they are growing very fast. Their growth in coming years will substantially impact plastics recycling technology. In addition, biodegradable and photodegradable polymers are entering into the packaging industry. Since most of the plastics waste is generated from post-consumer packaging, emergence of these polymers will influence the recycling industry as well as the technology. Life-cycle analysis is also becoming an important feature of the environmentally conscious manufacturing technology. It is therefore important to discuss some of these technologies briefly in order to assess their impact on the future of plastics recycling technology. In Section 18.1, we have already discussed plastics identification, sorting, restabilization, and processing of recycled plastics. In this section we will briefly review reactive processing, compatibilization, biodegradable and photodegradable polymers, design for recyclability, and lifecycle analysis as related to plastics recycling.

18.4.1 Reactive Processing

Reactive processing is a generic technology where low molecular weight materials are converted into macromolecules and the latter are processed into a product in one step. A number of processes, including reaction injection molding (RIM), reinforced RIM, sheet and bulk molding, pultrusion, and reactive extrusion (RE), are included in this generic technology. RIM and RE are the most important, so far as recyclying is concerned.

In RIM, the monomers/oligomers and the additives are carefully metered and injected under high pressure into the mixing head and the mixture is immediately transferred into the mold. As the mold is closed, the reaction starts and gelation takes place. In recent years, polyurea

RIM has replaced polyurethane RIM in many automotive applications. In addition, RIM systems based on nylon and cyclopentadiene technology are available. RIM provides several advantages for recycling. The low viscosity of the monomer/oligomer allows good distributive mixing of the recyclate at low energy input. The monomers wet the surface of the recyclate and thereby provide good interphase adhesion. In structural RIM (SRIM), the aspect ratios of the fibers are not reduced as much as in the case of standard extrusion or injection molding. Hence, both of these technologies will be useful in recycling post-consumer polymer waste. A RIM process has been successfully used for recycling waste PU foam.

Reactive extrusion, on the other hand, involves continuous *in situ* polymerization, including graft and block copolymerization, in the extruder. Availability of twin-screw extruders with feed and vent ports, gear pump, modular screw configuration, automatic screen changer, and online monitoring of composition and flow characteristics of the extruders have helped this technology to break new frontiers in blending and alloying technology. Reactive extrusion is particularly suitable for forming block and graft polymers *in situ* while blending recyclable mixtures. Polymer recycling technology will therefore benefit from the growth of reactive processing. Already, reactive extrusion is being used for blending recyclates and for hydrolytic degradation of PET and PU.

18.4.2 Compatibilization

Plastics available from post-consumer sources are always contaminated or commingled with other plastics. Since polymers in general are immiscible with each other, phase separation takes place, leading to loss of physical properties. This loss can be restored to a large extent by using a compatibilizer, which acts as a glue at the phase interface. It reduces interfacial tension and promotes mutual interaction between the two phases. The net result is a fine dispersion of one phase in the other and good stress transfer between the phases, leading to higher mechanical performance. In short, compatibilizers do not make the two phases miscible, but they help them stay together in harmony and thereby produce a synergistic improvement in the properties of the blend. Mangaraj *et al.* have studied the changes in free volume size in compatible and incompatible polymer blends and have shown that the average size of the free volume decreases and the number of free volume sites increases for compatibilized blends.[143] Compatibilizers are basically high molecular weight surfactants. One part of the compatibilizer chain mixes, interacts, or reacts with one phase and the other part does the same for the second phase. Hence copolymers, block and graft polymers, and polymers with functional groups act as compatibilizers. Very often, surface active agents such as silanes and organic titanates act as compatibilizers. Theoretically, addition of a compatibilizer reduces the enthalpy of mixing, thereby reducing the free energy of mixing to zero or negative. The thermodynamic principles involved in the miscibility and compatibility of polymers have been discussed by Stein[144] and Coleman *et al.*[145]

Reduction of phase size by a mechanochemical method can also assist compatibility by increasing the surface to volume ratio, which leads to strong interfacial interaction. Hence, a substantial amount of work has been carried out to develop compatibility for blends of virgin materials as well as those for polymers present in solid waste. The state of the art in using compatibilizers for recycling has been reviewed by Mangaraj and co-workers in a recent Battelle report on the compatibilization of polymer blends.[146] Briefly speaking, compatibilizers are copolymers, graft or block copolymers, or polymers with reactive or interactive side groups which, when added in small quantity, improve the desired properties of a binary, ternary, or quaternary blend. The same result can be achieved by good dispersion when an increase in surface area of the dispersed phase improves interfacial adhesion.

Chotourou *et al.* observed that the properties of a post-consumer polyethylene–wood fiber blend is enhanced by the presence of phenol–formaldehyde resin on the fiber surface.[147] Beatty and co-workers used graft polymerization of methyl methacrylate for compatibilizing post-consumer plastics waste.[148] Copolymers such as ethylene–vinyl acetate, ethylene–acrylic acid, and ethylene–maleic anhydride compatibilize the plastic fraction of MSW and improve their mechanical properties. SEBS terpolymer (Kraton G1650) or maleic anhydride grafted SEBS (Kraton F1901X), when added to solid waste, also acts as a compatibilizer and enhances the properties of the blend.[149] Dagli *et al.* used acrylic acid–ethylene copolymer as a compatibilizer for blending nylon with polypropylene (the two basic resins used for yarn in fishing gear) and found that it enhances the properties of the blend.[150] SEBS terpolymer also acts as an excellent compatibilizer for polystyrene–polyalkene blends. Xanthos *et al.* have successfully used SEBS and MA grafted SEBS as a compatibilizer for New Jersey Curbside Tailings and found that the anhydride modified terpolymer (Kraton 1901X) is superior to SEBS terpolymer as a compatibilizer.[151] La Mantia and Curto used photooxidized greenhouse polyethylene bags to compatibilize PE with PS and found substantial improvement in their mechanical properties and morphology.[152] The Plastic Institute of America has used a large number of compatibilizers, flow enhancers, and impact modifiers to improve the properties of ASR, as discussed earlier. Table 17 presents a list of compatibilizers currently available in the market, some of which have already have been used as compatibilizers for scrap reclaim.[153]

Table 17 Compatibilizers Used for Polymer Blending Including Commercially Available Compatibilizers.

Compatibilizers	*Materials*	*End properties*
MA-modified TP elastomers	PP/nylon	Increased toughness
MA-modified alkenic copolymers	a) Alkenes/EVOH, PS/EVOH, PE/polyester b) PP/nylon	a) Scrap reclaim, combined barrier properties with inexpensive materials b) Increased toughness
Polyacrylic imide	a) Alkenes/nylon b) Alkenes/polycarbonate	b) Moisture and chemical resistance, paintability
AA-modified PP	Alkenes/PET	Scap reclaim
Peroxypolymers	EPR/engineering resins	Increased toughness
Silanes	a) PPE/nylon b) PET/EPR, PET/styrenics	a) Combined crystalline and amorphous properties b) Higher elongation, impact resistance
Reactive PS	Syrenics/nylon, styrenics/alkenes, styrenics/polycarbonate	Increased toughness, scrap reclaim
SMA	Polycarbonate/nylon	Increased toughness
Phenolics	Polycarbonate/SMA, polycarbonate/ABS	Improved weld strength, higher impact resistance
Polycaprolactone	a) PVC/PS b) Polycarbonate/SAN	a) Higher HDT, scrap reclaim b) Increased toughness
CPEs	PVC/PE	Higher HDT, scrap reclaim
PS graft copolymers	a) PS/PE b) PS/nylon	a) Scrap reclaim b) Increased toughness
Syrenic TP elastomers	PS/alkenes	Higher elongation, impact resistance

Immiscible blend morphology appears to depend on several material and process parameters, including component ratio, compatibilizer concentration, surface tension, component viscoelastic properties, degree of mixing, material addition protocol, and screw design. The degree of mixing depends on the screw design and residence time. The residence time depends on a balance of feed rate and shear rate as given by the rpm of the extruder. Adewole *et al.* carried out a systematic study of the effect of material and process variables on the micromorphology and mechanical properties of recycled HDPE and recycled PP blend in the presence of maleated SEBS terpolymer (Kraton F1901X).[149] They found that both dispersed and co-continuous phase structures could be achieved by mixing at 300–400 rpm. The material variables, on the other hand, did not exert a significant influence on either the micromorphology or the mechanical properties. Subsequently they used optimization techniques to obtain the optimum processability window for a given balance of properties. They used an orthogonal rotatable response surface design, in three variables, namely rpm, HDPE, and Kraton concentration. They evaluated flexural modulus, notched izod impact resistance, elongation at break, tensile strength, and heat distortion temperature. Analysis showed that elongation at break is significantly affected by rpm, but rpm does not have any effect on flexural modulus, impact, tensile strength, or HDT (heat deflection temperature). This result allows adjustment of the operating conditions to include a wide range of possible compositions which satisfy the predesired performance criteria. For a property profile of flexural modulus 1100 MPa, izod impact greater than 65 J m^{-1}, elongation at break beteeen 125–175%, tensile strength greater than 28 MPa, and HDT greater than 90 °C, they found that at 300 rpm screw speed, the desired balance of properties and cost appears as a narrow triangular region. At 400 rpm, there are two operating windows, one for HDPE below 36% and the other above 46% (Figure 17). The region for lower HDPE level requires higher levels of Kraton whereas the region for higher HDPE level requires a lower amount of Kraton. At 500 rpm, the operating window expands to include the entire range of HDPE, generally with lower levels of Kraton. Since the cost of Kraton is very high compared to that of post-consumer HDPE and PP and the cost of compounding is approximately the same as that of processing virgin resin, these maps can very well provide the optimum processing window for making a cost-effective compatibilized blend of recycled PP and HDPE.

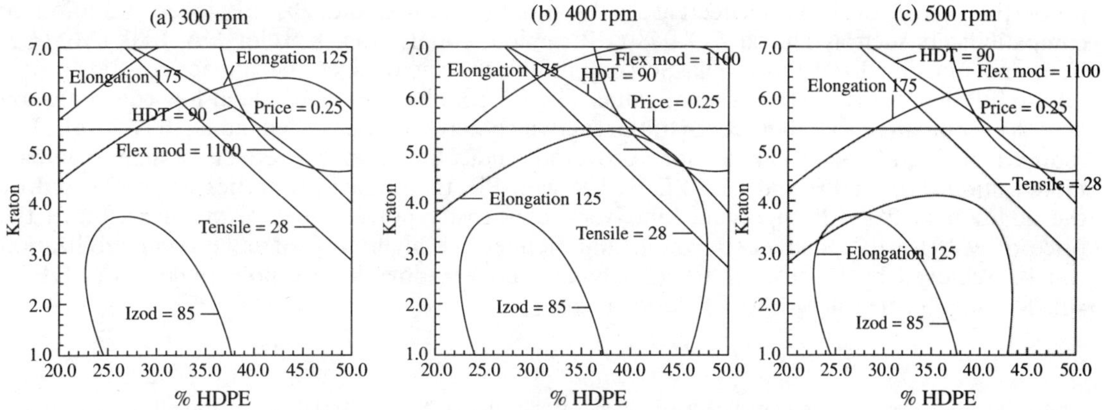

Figure 17 Operating windows of HPDE/NJCT blends.

Starke *et al.* have discussed the importance of good dispersion and chemical interaction for stabilizing mixtures of plastics waste and have experimented on various methods for achieving property improvements by good mixing and compatibilization.[137] According to the authors, good impact properties are obtained when the size of the particles (agglomerates) in each phase is smaller than a critical size. Large particles are unstable and they initiate fracture under dynamic stress. Very often, even small particles can cause fracture initiation due to excessive elongation under stress. Agglomeration of fine particles can produce larger particles, which cause fracture initiation and delamination. The critical particle size is approximately 1 μm for HDPE and 3 μm for PVC. Hence, the agglomerates in the feedstock have to be reduced by dispersion. The shear stress γ_T at the agglomerate surface is given by the relation $\gamma_T = (2.15$ to $2.5)\gamma_F$, where γ_F is the shear stress in a fluid medium. The latter is related to shear rate by the relation $\gamma_F = \eta_F \gamma$, where η_F is viscosity of the medium and γ is the shear rate in the fluid. For systems with large viscosity differences, dispersion in the relevant temperature range can only be achieved at high shear rates. Thermal sensitivity of plastics also requires that the mixing time should be short. The authors have devised a short disk-type dispersing unit and have successfully used this equipment to mix PVC and HDPE. The stress is minimum at the nip and increases exponentially towards the wall. Hence dispersing mainly occurs in the wall.

The authors have investigated the compatibilizing efficiency of many products, including solid solvents and radical precursors. They found that elongation at break and toughness are substantially improved by adding bitumen as a compatibilizer to plastics in municipal solid waste. This is attributed to the solubilization of PE and PVC by certain components of the bitumen. Addition of solid solvents therefore achieves compatibilization by mutual solubilization. Whereas random copolymers are not very effective as compatibilizers, block and graft polymers are capable of compatibilizing. The authors found SEBS block polymer to compatibilize blends of PE and PS very effectively. Addition of only 1% block copolymer reduced the particle size from 20 to 4 μm. Surface active agents can promote wetting and energy exchange at the phase boundaries. This requires modification of interfacial tension by introducing dipole/dipole interaction. The authors used an ethylene propylene glycol type surfactant and achieved property enhancement by adding only 1.5% of the surfactant to the blend.

Thus, there are many ways to enhance compatibility between the components of MSW recyclates, including intimate mixing, solubilization, and compatibilization with block copolymers, surfactants, and radical precursors. Polyethylene is the major component of the plastics portion of municipal solid waste and is always contaminated with other polymers in spite of sorting. The contaminants can substantially affect the mechanical properties of recycled PE, particularly the impact strength and elongation at break. Hope *et al.* carried out a systematic study of the effect of small amounts of PP, HIPS (high impact polystyrene), PVC, and PET on the loss of physical properties of HDPE, LDPE, and LLDPE and the possibility of recovering the performance by the addition of compatibilizer.[154] Regrind PET flake/pellet mixture was used in place of neat PET resin. The results are summarized in Table 18. It may be seen that the properties of HDPE were not affected by the presence of LDPE even at the 20% level. However, they are substantially affected by the presence of PP, HIPS, and PVC at the 10% level and PET at the 2% level. Whereas moderate addition of PP to LDPE does not affect the properties, those of LLDPE decrease substantially even at the 2% PP level. Addition of compatibilizers such as Dutral CO 058 (EP random copolymer), Exxelor VA 1801 (MAH-*g*-EP), and functionalized thermoplastic compatibilizers can recover the impact resistance of a HDPE/PP blend close to the original value. The EP copolymer is effective at low concentration (2–5%), compared to other compatibilizers which have to be added at a 10% level. The reduced elongation at break is not recovered, indicating that the effect is mainly impact modification. For LDPE and LLDPE blends with PP, the impact properties are recovered by the addition of 2% EP copolymer and very-low-density polyethylene (Norsoflex). Since the price of MAH-*g*-EP, SEBS copolymer, and Bennet are high, cost effective compatibilization can be achieved by the use of EP copolymer and very-low-density polyethylene (VLDPE), which is similar in structure to EP copolymer.

Table 18 Mechanical Properties of HDPE/PP Blends Containing 20% PP by Weight and Various Compatibilizers.

Compatibilizer	*Grade/description*	*Addition level* (pph)	*Breaking strain* (1%)	*Impact strength* (kJ m^{-2} at 23 °C)
None	–	–	15	4
VLDPE	Norsoflex MW 1960	2	15	4
MAH-*g*-EP copolymer	Exxelor VA 1801	2	15	3
		4	15	4
		6	30	5
		10	20	8
EP random copolymer	Dutral CO 058	2	15	6
		4	15	6
		6	10	8
		10	10	15
EP block copolymer	EP D60R	2	15	4
S-EB diblock copolymer	Kraton G1701	2	15	5
Radical initiator	Dicumyl peroxide	0–1	20	4
Functional thermoplastics	Bennett G25	5	15	5
		10	20	6
	Bennett GR110	5	15	5
		10	20	9
	Bennett GR125	5	40	4

18.4.2.1 Mechanical compatibilization

In addition to chemical compatibilization, mechanical compatibilization appears to work very successfully in making viable products from commingled plastics. This has been accomplished by advanced elastic deformation grinding,[153] which converts the plastic mass into a fine powder of controlled size and size distribution. The pulverization is carried out in a corotating twin-screw extruder designed and manufactured by Berstroff Corporation of Germany. Usually, the extrusion is carried out at 0.25–0.3 MPa pressure with very high shear, combined with extensive cooling to maintain temperature at between 15–60 °C.

Depending on the location, the configuration of the screw elements, and the quantity of shearing, the resin flakes are transposed into elongated fibrous shapes and then further transformed into powder of different coarseness. The powder typically has a size less than 500 μm. The throughput as well as the coarseness can be varied using the screw profiles and processing parameters such as rpm, pressure, temperature, and feed rate. Although the powder formation mechanism is not fully understood, it is apparent that solid state pulverization achieves better homogenization, resulting in a light gray color instead of dark shades of brown or gray seen in conventional recycling. The powder, when injection molded, provides high tensile strength and good impact resistance ($10.5–16\,J\,m^{-1}$) for all types of polymer combination. Electron spin resonance studies show an increase in radical concentration during pulverization, particularly for PP and LDPE. The spectra for LDPE corresponds to peroxy radicals (ROO·). In other words, solid state pulverization under high shear conditions breaks chemical bonds, forming reactive sites in different polymers. This probably results in the formation of a small quantity of block polymers incorporating repeating units from different individual polymers. These graft and block copolymers then serve as compatibilizers for the commingled plastics.

Pulverization of shredded tires (5–6 mm long) by a solid state shear extrusion (SSSE) process provides particles of 40 μm size with a sharp particle size distribution. When polyethylene is added to rubber and the mixture is subjected to SSSE, the surface area of the particles increases dramatically from about 2500 to 5500 μm^2 and reaches a peak at around 5% polyethylene.[155] Scanning electron micrographs of the fracture surface indicated good wetting of rubber particles by polyethylene. When polyurethane (PU) foam was subjected to SSSE, it was pulverized to very fine powder. The powder size varied from 17–32 μm for flexible foams to 96–170 μm for rigid foams. The shape of the particles, however, did not show much change. TGA and DSC data did not indicate any change in PU characteristics. When the pulverized powders were used as fillers in polystyrene and PVC, the physical properties of the resins showed higher values compared to the unfilled matrix.

Rihai *et al.* have also applied the solid state shear extrusion process to pulverized rubber from automotive tires.[156] On adding a minor component, polyethylene, they found that the roundness of the particles increases up to 5% and beyond 5% it decreases sharply. The roundness is maximum at the 5% level. Hardness of the rubber particles is also maximum at this composition. It was suggested that PE acts almost as a catalyst in the pulverization of rubber. Similar experiments with wood fiber and polyethylene showed that the particle size of the wood particles become almost 20 times smaller than regular sawdust. This led to a substantial enhancement of the Young's modulus of wood fiber–rubber–PE composite.

One may conclude from the above examples that the properties of the recycled products can be greatly improved by addition of a compatibilizer or compatibilizer combinaton as well as by proper mixing and pulverization through solid state shear extrusion.

18.4.3 Biodegradation, Photodegradation, and Mechanochemical Degradation

18.4.3.1 Biodegradation

Biodegradable polymers are being considered as environmentally benign substitutes for polymers currently used in packaging. Many microorganisms produce hydrolase, an enzyme capable of hydrolyzing aliphatic polyesters, polyethers, polyamides, and polyurethane. However, long chain length and the presence of side groups and aromatic moieties prevent this degradation. Hence, biodegradable polymers such as low molecular weight carbohydrates (*e.g.* starch) and aliphatic polyesters (*e.g.* polylactides, lactide–glycolide copolymers, and butyrate–valerate copolymers) have the potential to enter into the packaging market.[157] Substantial research has been carried out by DuPont, Cargill, Dai Nippon Ink & Chemicals, Mitsui Toatsu Chemicals, and Shimodzu Corporation on the commercial development of polylactides and lactide copolymers. The Battelle Memorial Institute has done considerable research and holds many patents on polylactide and copolymers.[158] However, there are formidable difficulties which have prevented the commercial development of biodegradable polymers, particularly in obtaining D- and L-lactides in the desirable ratio and in controlling the rate of crystallization of lactide polymers. Imperial Chemical Industries has developed poly(isobutyrate–valerate) copolymers using microbiological techniques. Although these are used in making containers, seed cases, and mulches, they have not found any application in packaging as yet because of the high cost of production.

A number of companies in North America and Japan are on the verge of introducing polylactides and copolymers on a commercial scale.[159] Cargill of Minneapolis has announced the introduction of Eco PLA, a polylactide copolymer for applications in packaging and semistructural products. The mechanical properties of Eco PLA are comparable to those of polystyrene (Table 19). Like PS, Eco PLA can be toughened to provide higher impact resistance. Its melt rheology is similar to that of polystyrene. It is crystal clear, glossy, stiff, and brittle and has higher moisture barrier properties as well as environmental stress crazing resistance. Its low heat-distortion temperature (around 55 °C at 66 psi) and slow crystallization rate can be improved by copolymerization and nucleation. A variety of biodegradable repeating units are available for copolymerization, including glycolate, tartrate, mandelic acid derivatives, and caprolactone. Eco PLA needs to be dried to 150 ppm moisture level before processing. Eco PLA can be extruded, injection molded, and thermoformed using currently available machines and is quite suitable for extrusion coating. In addition to Cargill, Ecochem at Newport, DE, Chronopol Company at Golden, CO, and Shimadzu in Japan are planning pilot plant production of polylactides in the near future.

Table 19 Mechanical Properties of Eco PLA Molded Sheet.

	Injection molded		*Sheet*			
Property	*PLA neat*	*Filled*	*Toughened*	*Neat*	*PLA-A*	*HIPS-A*
Ultimate tensile strength (psi)	7700	8400	3600	7700	6100	3100
Ultimate elongation at break (%)	4.1	2.1	4.0	5.0	7.2	25
Tensile strength at yield (psi)	8900	8700	4800	8700	7500	3400
Tensile modulus (psi $\times 10^3$)	500	1000	300	510	–	260
Notched izod impact (ft lb in^{-1})	0.3	0.3	3.9	0.2	0.3	0.68
Tear strength (lb)	—	–	–	0.11	0.11	0.3

Earlier, St. Lawrence of Canada and Ampacet of the US were promoting starch-based concentrates which, when blended with polyethylene and polypropylene, would make the latter biodegradable.[160] The hypothesis is that, while biodegrading starch, the bacteria will degrade the polymers or reduce their molecular weight to a point where they will become biodegradable. However, the products were not commercially successful, mainly due to their questionable performance. DuPont and Air Products manufacture poly(vinyl alcohol) (PVOH) under the trade names Airvol and Elvanol, respectively. These polymers are used in packaging and bagging, particularly because of their oxygen barrier capability. PVOH is both water soluble and biodegradable. Warner Lambert Company has announced the introduction of an all-starch polymer 'Novon', which can be biodegraded by both aerobic and anaerobic processes in the presence of moisture.[161] This will replace gelatin in making pharmaceutical capsules and will be used for single service packaging in food, medical, and other fields. Novon's mechanical properties are in between those of polyethylene and polystyrene with a melting point between 170 and 200 °C. It can be filled, colored, and compounded like common polymers and its solubility in water is adjustable. Warner Lambert has over 20 patents covering processes for destructuring starch without eliminating water, which contributes to improved mechanical properties. The destructuring process was carried out in a twin-screw extruder. Warner Lambert is building its pilot plant in Switzerland, at its Capragel Division, to commercialize Novon.

18.4.3.2 Photodegradation

Many polymers degrade in UV light, either through photochemical or photophysical processes.[162] For photochemical decomposition, the UV-sensitive moiety has to be a structural unit in the main chain of the polymer. Monomers like vinyl alkyl ketones induce photodegradation. Alternatively, a photosensitizer, such as 4-chlorobenzophenone, phenylacetophenone, or benzophenone, can be added to the polymer. This will absorb the UV and transfer the energy to the main chain and will lead to molecular degradation. The amount of ketone needed for successful photodegradation is very small. Polysar of Canada and Eco Corporation of the USA have started a joint venture to produce and market a light-degradable concentrate which, when added to a base polymer, will degrade it in a small time. The product will go by the trade name Ecolyte. This is essentially a copolymer of a vinyl ketone and styrene. The concentrate is usually added in a 10:1 or 20:1 ratio. The approximate cost is \$2.56 per pound. The photodegradable resin can be reprocessed many times, like ordinary resins. Table 20 provides failure times for polyethylene film in the presence of photosensitive additives. Combinations of metal-organic compounds and UV sensitizers lower the degradation time substantially.

Table 20 Photodegradation of Polyethylene Film in the Presence of Photosensitive Additives.

Metal organic compound (A)	*Aromatic carbony compound (B)*	*Failure times* (h)		
		A	*B*	*A + B*
Zirconium neodecanoate	4-Chlorobenzophenone	234	214	142
Cobalt pentanedione	Phenylacetophenone	311	236	42
Iron(II) stearate	Benzophenone	115	236	98

18.4.3.3 Mechanochemical degradation

When long-chain molecules are subjected to rapid deformation under adequate stress, they are very likely to degrade, forming chain radicals of lower molecular weight. These chain radicals can be suitably end capped to provide oligomers, which can be repolymerized under appropriate conditions. Porter and Casale[163] have discussed mechanochemical degradation in great detail. One of the subtle ways of introducing rapid deformation and mechanochemical degradation is by ultrasonics. When chain molecules are subjected to high-frequency vibration, they undergo rapid compression and decompression. The latter can generate negative pressure strong enough to break the cohesive forces of liquids and gases, causing bubble formation. The rapid formation, growth, and collapse of these bubbles set up shock waves and generate localized hot spots as well as a high level of stress. This phenomenon is known as cavitation. The high temperature and stress generated during cavitation causes chain molecules to break down into fragments. Earlier work has shown that the viscosity of a dilute solution of polystyrene decreases substantially as the solution is exposed to ultrasonic vibration.[163] Work is in progress at Akron University for developing a process for devulcanization of rubber by using ultrasonic vibration.

18.4.4 Product Design

At present, proper product design is used for providing high performance, good processability, and low cost. These are functions within the service life of a product. Recycling of the post-consumer product into valuable raw material after the first cycle of use is over was not a part of product design. With the increasing cost of waste disposal and soaring prices of virgin resin, it makes great sense to design a part or a product for recyclability.

To design a part which can be easily recyclable one has to look for easy dismantling, easy identification and sorting, easy reprocessing, compatibility of the resins used for a product, and inclusion of long-life additives to go through multiple processing without damage. The automotive industry has already taken steps to design for disassembling. This is particularly useful for parts recycling. Use of a bar code with information on the resin type, manufacturing company, and year of production in large parts will make it easier to identify the resin or resin blends used for a particular part. In addition, minimizing the number of resins and using compatible resins will help a great deal in reprocessing the post-consumer waste either for closed-loop recycling or for value added products. One has therefore to evaluate the recyclability of each part and use lifecycle analysis before a part is designed for production.

Recyclability of a particular plastic can be defined in two ways. Technically, it may mean how easily a material can be recycled while still maintaining most of its key properties. However, in practice it will depend on its demand and marketability. Souza-Couch *et al.*[164] have defined a series of factors for estimating recyclability. They are: existence of a published price; annual volume of the post-consumer material sold; specific legislation demanding recycling for the polymers; and difference in PCR price *vis-a-vis* virgin material price. Existence of a published price for post-consumer flakes (PCF) shows that there is a market demand for the material. The volume sold is another indication of the marketability of the PCF and often reflects the time it has taken for the demand to grow. Legislation forces the infrastructure needed to create the demand, and penalties motivate industrial and consumer compliance. However, for this to happen, the legislation has to be specific. The price difference between the virgin and recyclate has the major influence on marketability. The larger the difference, the greater is the possibility that the recyclate can be sold at a profit. The authors have proposed a quantitative estimate of recyclability by taking the above factors into consideration.

As mentioned earlier, material compatibility is an important issue in designing for recyclability. For example, in conventional packaging a printed paper label is adhered to the bottle with an adhesive. This not only adds foreign materials to the bottle resin but introduces the potential for color bleeding. Therefore, special machinery and chemical stripping agents are required to remove the label, adhesive, and ink from the resins to be recycled. On the other hand, if labels made from HDPE laminates are applied to the HDPE container in a mold without adhesives, the bottle can be reground and recycled without washing and label removal steps. In addition to getting rid of foreign materials, production line speed may be increased because the in-mold HDPE label takes less time to cool in the mold than the time taken for attaching the paper label. This concept was verified by reprocessing and measuring the properties of the HDPE material from the bottle and label resin, unlabeled and labeled bottles, and the bottle after use.[165] Tensile properties including percentage elongation at break, impact resistance, and viscosity were measured using standard methods. The label resin had lower tensile strength, lower percentage elongation, and lower impact resistance than bottle resin. Whereas the two bottle samples (with and without the label) showed almost no difference in tensile strength, the elongation at break differed by 17% and the impact resistance by 3.8%. The difference in viscosity was found to be insignificant. HDPE compounds containing 5%, 10%, and 15% label resin were used for blow molding of bottles. It was found that, whereas compounds with 5% addition of label resin allowed production of quality bottles, compounds with 10% and 15% label resin produced lower quality bottles. Since the label weight is much less than 5% it was concluded that bottles with similar-resin labels can be used in bottle production without impairing the mechanical properties.

Finally, with the increasing emphasis on recyclability of plastics, it is important that the products made from plastic be properly designed to facilitate post-consumer recycling. The history of PET recycling is a good example of the difficulties encountered in PET bottle recycling. Back in the 1970s when PET recycling started, high levels of adhesives used for gluing the HDPE base cup to the PET bottle and the presence of the aluminum caps were the main hurdles in PET recycling. As the adhesive level decreased and polypropylene (PP) replaced aluminum for caps, the higher level of PP caused the recycled material to be brittle. Also, the presence of even a small amount of PVC caused loss of viscosity and color degradation. Bleeding label problems, where the ink dissolves in water at temperatures of 70 °C and above in detergent solution, discolors the transparent flake and prevents it from being converted into bright-colored products. All this requires that design engineers be careful while designing a polymeric product which will be subsequently recycled. The Association of Post-Consumer Recyclers (APR) is carrying out substantial research to develop guidelines for the packaging industry, so that a poor choice of materials or design that would hinder recycling is avoided.[166] For example, for high-density milkjug bottles, the caps should be made of either HDPE or PP with a melt index similar to the melt index of the HDPE used for making the main body of the bottles.

18.4.5 Lifecycle Analysis

Every material impacts on the environment not only at the time of its disposal but at earlier stages of its life, including raw material processing, product manufacture, and application. It is therefore important that the service life of the material be lengthened as much as possible to reduce the total impact on the environment while getting the best use of the material. Multiple recycling is therefore important, not only in maximizing the economic return from the material but also for minimizing the total environmental impact. Since plastics withstand the vagaries of nature successfully and have a life expectancy greater than most metals, it is desirable that a lifecycle approach should be taken to lengthen the service life by repeated recycling. Currently there are differing methodologies for lifecycle analysis. However, topics such as recyclability, the socioeconomic drive for recycling, and hierarchy of application are important to plastics recycling.

Maximizing the lifecycle of polymers may be considered as the key to recycling. This can be achieved once the performance of the polymer is known after each lifecycle. Alternatively, if the factors contributing to the performance are known and are measured as a function of aging through repeated use, then one can devise a strategy to recycle polymers by taking some corrective measures or by incorporating them into virgin resin. Many engineering resins such as PC, ABS, HDPE, and nylon do not undergo substantial property loss over successive regrinding.

Shriver *et al.* have developed a methodology for material lifecycle analysis which uses the factors contributing to performance to predict the properties after repeated recycling. This is essential to devise guidelines for recycling.[167] They used polycarbonate (Lexan 141, GE) as a test material and measured molecular weight (MW) and MW distribution, melt flow index (MFI), viscosity as a function of shear rate, impact resistance, and color. Based on the analysis of the changes that occur on ten successive regrinds, they have developed a three-axis nomograph which uses a mechanical property, melt index, and percentage of virgin material in the blend for estimating recyclability. MFI is shown to be a good indicator of performance level. Therefore, the nomograph based on MFI can be used to estimate the minimum virgin material required to blend with a recyclate to restore the original performance with respect to a particular property.

Beatty emphasized a hierarchy of applications of plastic products as related to the residual properties at the end of each lifecycle and the economic viability of the reuse.[148] According to him, we are in the first lifecycle of many uses of plastics and when we complete two or three iterations of this sequence of the lifecycle there will be sufficient interaction between society and industry which will result in harmony between the reapplications of used plastics and social needs. The recycling methods such as chemical and mechanical, *etc.,* should be taken into account along with the cost of operation and environmental effects versus the benefits in terms of actual applications.

18.5 FUTURE TRENDS AND DIRECTION

The plastics recycling industry went through a rough time in the last decade during its learning process. About 1000 European, 400 North American, and a handful of Japanese recyclers had to fight to survive. The anticipated government and legislative support did not materialize. The procurement mechanism, raw material quality, and reprocessing technology were not sufficiently advanced to make quality recycled plastics for acceptance in the market place. Record low prices of virgin resin and worldwide restructuring of economics also contributed to the loss of many recyclers. Despite these tribulations, the industry sees several reasons for optimism.[168] The integration of plastics recycling into the mainstream of developed economics based on the profit motive shows maturity, which is a signal for healthy growth. Recyclates of engineering resins are being viewed as a major source of feedstock for select niches in the market. The success story of PET recycling has demonstrated that those companies able to innovate and invent will be winners in the plastics recycling market.

The high growth of PET recycling is due to profit. The cost of producing RPET is approximately 37 cents per lb as against the virgin resin price of 55–60 cents per lb. RPET, therefore, provides a cost avoidance of 18–28 cents per lb to fiber manufacturers such as Wellman, Martin Colorlife, PureTech, *etc.* RPET is penetrating into visible, high-fashion applications for knit, fleece, and woven textiles where it will fetch even higher profit. This success is attributed to quality improvement and faster recycling through the use of electronic polymer recognition and automatic sorting. The current recycling rate of PET, approximately 4.5% in the USA and 7.5% in Europe, is going to reach double digit figures. In Europe, the use of virgin PET is growing at a very high rate. The laws are changing in the UK and Italy, the major users of PET bottles, in favor of legislating a recyclability content. The possiblity of monolayer bottles containing a significant RPET being used for food packaging is being seriously considered both in the US and Europe. Similar potential for large scale recycling of clear and mixed color HDPE bottles and LDPE stretch wraps, grocery sacks, and agricultural films is very bright. An appreciable quantity of polyalkenes will also be used in making woodfiber-based products which will meet the market demand created by the vanishing wood supply for building and construction. More and more recyclates will go into multilayer packaging, which is growing due to improvement in both extrusion and injection molding technology. A further increase will take place by economies of scale as the recycling operations grow bigger and bigger. Market acceptance is growing and the public image that recycled products are second class materials is gradually diminishing as both producers and consumers see the high performance of recycled goods.

The durable sector of plastics use, such as automobiles, building and construction, business machine housing, and appliances will bring a substantial amount of used plastics into the market in the near future. These robust plastics do not lose much of their strength during their service life. They are easier to procure since they all come back to some sort of junkyard instead of entering into the trash stream. Provisions such as design for disassembly, bar coding the plastic type, a worldwide dismantling handbook, and worldwide use of standard ISO test methods will facilitate the recovery of plastics from durable goods. Identification and sorting systems for separating individual plastics from durables are under development. Both automotive and computer companies are showing deep interest in closed-loop recycling and in using post-consumer resins for new automobiles. Even design changes are being made to accommodate recycled plastics when necessary. All these are strong signs for a healthy demand for structural plastics recyclate in the future. Furthermore, increased use of plastics recycled from durable goods is being supported by a growing number of key groups and consortia, both in the US and in Europe. Wigtsky, in a recent article, has described some of these efforts and their payoffs.[169] These groups are supporting recycling in many ways, including problem solving, assistance in applied engineering, processing, market assessment, material characterization, application validation for recyclate use, product optimization, testing, and demonstration and selection of equipment and processes. Comprehensive programs are being funded by the American Plastics Council (APC), the Automotive Alliance for Recycling, and the US Council of Automotive Research. High priority is given to rapid dismantling technology, rapid material identification and sorting with portable optical devices based on IR spectroscopy, and materials liberation/size recognition technology.[170] The Argonne National Laboratory, with a large grant from APC, is working on a cost-effective separation system for plastics in automotive shredder residue. Similarly, the Vinyl Institute is sponsoring research and development of PVC recycling, particularly from the construction industry.

It is also being predicted that the cost of collection and reprocessing of recyclates will go down. The aura of uncertainty is vanishing. Processors now know how to stabilize, compatibilize, and mobilize plastics for a high throughput, while still producing high performance parts. Both destructive and nondestructive tests are being established to achieve quality control, thereby reducing rejects. On the whole, the science and technology of plastics recycling is maturing.

However, further technological innovations are needed for devising methods and developing materials for: easy collection and transportation of plastics waste; sorting and classifying into single plastics or individual groups of plastics based on application, compatibility, and processing; resisting thermal degradation and rejuvenating damaged macromolecular chains; restoring properties by compatibilization and impact modification; developing new methods of processing at minimum expense of energy and time; and developing cost-effective processes for depolymerization. The synthetic polymer industry has taken more than half a decade to invade the territories of metals and ceramics. We hope recycling of plastics waste will mature and recycled plastics will be used in place of virgin plastics in a much shorter time frame.

18.6 REFERENCES

1. 'Plastics Recycling, Products and Processes', ed. R. J. Ehrig, Hanser, New York, 1992.
2. 'Emerging Technologies in Plastics Recycling', ACS Symposium Series 513, ed. G. D. Andrews, and P. M. Subramanian, American Chemical Society, Washington, 1992.
3. A. M. Thayer, *Chem. Eng. News,* 1989, **67**(5), 7.
4. R. Curlee and S. Das, *Mater. Soc.,* 1991, **15**(1), 41.
5. W. E. Franklin, B. J. Sauer and M. A. Franklin, *Soc. Plast. Eng. Annu. Tech. Conf.* (*ANTEC*) *Proc.*, 1995, 3504.
6. Anon., *Mod. Plast.,* Jan. 1995, 67.
7. D. Mangaraj, *Indian J. Technol.,* 1993, **31,** 468.
8. P. Dinger, in 'Plastic Recycling, Products and Processes', ed. R. J. Ehrig, Hanser, New York, 1992, p. 30.
9. R. J. Sperber and S. L. Rosen, *Polym. Plast. Technol., Eng.,* 1974, **3**(2), 215.
10. J. F. Laundrie, Publication, #227004-9, US Department of Agriculture, 1972.
11. K. Saitoh, I. Nagano and S. Izumi, in 'Proceedings of the Fifth Mineral Waste Utilization Symposium, 1976, p. 322.
12. 'Plastics Recycling, Products and Processes', ed. R. J. Ehrig, Hanser, New York, 1992, p. 56.
13. R. H. Wehrenburg II, *Mater. Eng.,* Mar. 1979, 34.
14. E. Rasmussen, *US Pat.* 4 578 184 (1986); G. Carlo, *US Pat.* 4 884 386 (1989).

15. P. W. Dinger, 'Modern Plastics Encyclopedia', McGraw Hill, New York, 1995, p. 50.
16. G. R. Kenny and D. Vaughan, *Soc. Plast. Eng., Reg. Tech. Conf. (RETEC) Proc.*, Nov. 1994, 59.
17. G. R. Kenny, *Soc. Plast. Eng. RETEC Proc.*, June 1993, 34.
18. K. Vornberger and B. Willenber, *Soc. Plast. Eng., RETEC Proc.*, Nov. 1994, 130.
19. R. Pfaender, H. Herbst, K. Hoffman, and F. Steik, presented at the 16th Annual Conference on Advances in the Stabilization and Degradation of Polymers, Luzern, 1994, p. 217.
20. J. Ogando, *Plast. Technol.*, July 1991, 51; R. Monk, *ibid.*, Jan. 1991, 26.
21. R. Pfaender, *Soc. Plast. Eng., RETEC Proc.*, Nov. 1994, 19.
22. A. R. Patel and S. D. Paster, *US Pat.* 5 149 774 (1992).
23. G. R. Smoluk, *Mod. Plast.*, Feb. 1988, 87.
24. W. Thiele and M. Knights, *Plast. Technol.*, Apr. 1995, 75.
25. J. H. Paul, C. L. Prestangen and A. L. Simons, *Soc. Plast. Eng., ANTEC Proc.*, 1994, 3002.
26. H. Potente and X. Gao, *Soc. Plast. Eng., ANTEC Proc.*, 1994, 3050.
27. Anon., *Plast. Technol.*, 1990, **36**(1), 78.
28. Anon., *Plast. Technol.*, 1990, **36**(1), 11.
29. R. Derry, *Plast. Rubber Proc. Appl.*, 1990, **13**(4), 221.
30. M. Ezrin, E. D. Wyatt, G. Lavigne and A. Garton, *Soc. Plast. Eng., ANTEC Proc.*, 1994, 2922.
31. B. Cassel, *Mod. Plast.*, May 1993, 65.
32. R. Bennett, D. Faron and B. J. Vaughan, *Soc. Plast. Eng., ANTEC Proc.*, 1993, 882.
33. L. Cao, R. M. Ramer and C. L. Beatty, 'Emerging Technologies in Plastics Recycling', ACS Symposium Series 513, American Chemical Society, Washington, 1991, p. 88.
34. M. A. El Rahaman, W. Lee, R. Cannaday and C. L. Beatty, *Soc. Plast. Eng., ANTEC Proc.*, 1994, 1846.
35. A. J. Hill, M. D. Zipper, T. J. Bastow and C. L. Beatty, *Soc. Plast. Eng., ANTEC Proc.*, 1994, 1870.
36. T. M. Joseph and C. L. Beatty, *Soc. Plast. Eng., ANTEC Proc.*, 1994, 1867.
37. S. G. Yang, D. Bennet and C. L. Beatty, *Soc. Plast. Eng., ANTEC Proc.*, 1994, 1852.
38. S. G. Yang, D. Bennet and C. L. Beatty, *Soc. Plast. Eng., ANTEC Proc.*, 1994, 1855.
39. J. W. Mason, *Soc. Plast. Eng., ANTEC Proc.*, 1994, 2935.
40. R. D. Leaversuch, *Mod. Plast.*, July 1994, 48D.
41. L. L. Gaines and A. M. Wolsky, *Conserv. Recycling*, 1983, **6**(1/2), 11.
42. J. Milgrom, in 'Plastics Recycling, Products and Processing', ed. R. J. Ehrig, Hanser, New York, 1992, p. 45.
43. R. D. Leaversuch, *Mod. Plast.*, May 1993.
44. J. Pyper, *Soc. Plast. Eng., RETEC Proc.*, June 1993, 49.
45. J. Milgrom, *Beverage World*, Jun. 1985, 86.
46. M. Bakker, *Soc. Plast. Eng., ANTEC Proc.*, 1994, 2944.
47. D. W. Hayward, *Soc. Plast. Eng., RETEC Proc.*, Nov. 1994, 111.
48. J. L. Golmanavich, *Soc. Plast. Eng., ANTEC Proc.*, 1994, 2893.
49. S. S. Dagli, S. K. Dey, R. Tupil and M. Xanthos, *Soc. Plast. Eng., ANTEC Proc.*, 1994, 3017.
50. P. A. Eriksson, K. Eriksson and J. A. E. Manson, *Soc. Plast. Eng., ANTEC Proc.*, 1994, 2918.
51. W. G. Paul and P. N. Bier, *Soc. Plast. Eng. Eng., ANTEC Proc.*, 1994, 3013.
52. K. Boudreau, R. Malloy and A. Tayebi, *Soc. Plast. Eng., ANTEC Proc.*, 1993, 869.
53. J. Jones, *Soc. Plast. Eng., ANTEC Proc.*, 1993, 2865.
54. R. H. Burnett and G. A. Baum, 'Plastics Recycling, Products and Processes', ed. R. J. Ehrig, Hanser, New York, 1992,
p. 165.
55. W. Michaeli and J. Dassow, *Soc. Plast. Eng., ANTEC Proc.*, 1994, 2994.
56. H. Potente and X. Gao, *Soc. Plast. Eng., ANTEC Proc.*, 1994, 2989.
57. C. P. Raeder and R. C. Weglin, *Soc. Plast. Eng., ANTEC Proc.*, 1994, 3026; B. Bennyhoff, *Soc. Plast. Eng., RETEC Proc.*, June 1993, 213.
58. R. Derry, *Plast. Rubber Proc. Appl.*, 1990, **13**(4), 221.
59. N. Lavieri, *Soc. Plast. Eng., ANTEC Proc.*, 1994, 2856.
60. C. Tripodis and C. W. Brady, *Soc. Plast. Eng., ANTEC Proc.*, 1994, 2871.
61. J. F. Doyon, P. G. Lafleur and P. J. Carreau, *Soc. Plast. Eng., ANTEC Proc.*, 1994, 2876.
62. A. V. Shenoy, D. R. Saini and V. M. Nadkarni, *Polymer*, 1983, **24,** 722.
63. M. Gibbs, *Plast. Eng.*, 1990, **46**(7), 57.
64. J. E. Hiltner, *Soc. Plast. Eng., RETEC Proc.*, Nov. 1994, 81.
65. J. Brahms, S. Ramaswami, *et al., Soc. Plast. Eng., RETEC Proc.*, Nov. 1994, 100.
66. N. Lee, A. B. Strong and R. L. Walling, *Soc. Plast. Eng., ANTEC Proc.*, 1993, 866.
67. M. Shaver, *Soc. Plast. Eng., RETEC Proc.*, June 1993, 183.
68. R. Sikora and M. Bielinski, *Kunststoffe*, 1988, **78,** 335.
69. J. H. Paul, C. L. Prestangen and A. L. Simons, *Soc. Plast. Eng., ANTEC Proc.*, 1994, 3002.
70. M. Bevis, N. Irving and P. Allan, *Conserv. Recycling*, 1983 **6**(1/2), 3.
71. D. Mangaraj, Battelle Columbus Division, Patent Disclosures, 1981.
72. W. Tappeiner and A. Male, *Soc. Plast. Eng., ANTEC Proc.*, 1993, 860.
73. G. A. Mackey, R. C. Westphal and R. Coughanour, 'Plastics Recycling, Products and Processes', ed. E. J. Ehrig, Hanser, New York, 1992, p. 118.
74. N. Mohareb, T. Hornberger, T. Piree and T. Hight, *Soc. Plast. Eng., ANTEC Proc.*, 1994, 2979.
75. N. Mohareb, *et al. Soc. Plast. Eng., ANTEC Proc.*, 1994, 2979.
76. R. Prioleau, *Soc. Plast. Eng., RETEC Proc.*, June 1993, 163.
77. W. F. Caroll, Jr., R. G. Elcik and D. Goodman, 'Plastics Recycling, Products and Processing', ed. R. J. Ehrig, Hanser, New York, 1992, p. 133.
78. F. Drause, *Soc. Plast. Eng., RETEC Proc.*, June 1993, 171.
79. T. B. Norton, paper presented at the Environmentally Conscious Manufacturing Seminar, Battelle Memorial Institute, April 13, 1995.

80. W. F. Caroll, Jr., *Soc. Plast. Eng., ANTEC Proc.*, 1994, 2949.
81. D. Mangaraj and V. McGinnis, Battelle Columbus Division, Patent Disclosure, 1992.
82. R. J. Bockstedt, *Soc. Plast. Eng., ANTEC Proc.*, 1994, 2998.
83. M. Thielen, *Soc. Plast. Eng., ANTEC Proc.*, 1994, 2964.
84. J. Song, L. Suek and M. L. Moran, *Soc. Plast. Eng., ANTEC Proc.*, 1994, 2970.
85. J. P. Schneider, G. E. Myers, C. M. Clemons and B. W. English, *Soc. Plast. Eng., ANTEC Proc.*, 1994, 1840.
86. K. Oksman and H. Lindberg, paper presented at the Wood Fiber–Plastic Composites Conference, Madison, Wisconsin, May 1–2, 1995.
87. D. Maldas and B. V. J. Kokta, *Reinf. Plast. Compos.*, 14 May 1995, 58.
88. M. Deaner and K. Heikkilla, paper presented at the Wood Fiber–Plastic Composites Conference, Madison, Wisconsin, May 1–2, 1995.
89. Anon., *Mod. Plast.*, Jan. 1993, 19; *Plast. Technol.*, Mar. 1993, 94.
90. Anon., *Reuse, Recycle, Mon. Newsl.*, 1993, **23**(12), 90.
91. M. Applebaum, D. R. Morrow, J. J. Nosker *et al. Soc. Plast. Eng., RETEC Proc.*, Oct. 1990, 17.
92. D. Mangaraj, *Indian J. Technol.*, 1993, **31,** 468.
93. K. E. Van Ness and T. J. Nosker, 'Plastics Recycling, Products and Processes', ed. R. J. Ehrig, Hanser, New York, 1992, p. 193.
94. Brochure on Extraordinary Extruder, Instamelt Systems Inc., PO Box 1714, Midland, TX 79702.
95. T. Nosker, R. N. Renfree and D. R. Morrow, *Plast. Eng.*, Feb. 1990, 33.
96. S. S. Swint, K. D. Webber and C. I. Chung, *Soc. Plast. Eng., RETEC Proc.*, Oct. 1990; Ref. 1, p. 228.
97. M. D. Applebaum, K. E. Van Ness, T. J. Nosker *et al. Soc. Plast. Eng., ANTEC Proc.*, 1991.
98. J. C. Lynch and E. B. Nauman, 'Recyling of Commingled Plastics, Proceedings of 10th International Co-extrusion Conference', 1989, p. 99; Ref. 1, p. 228.
99. L. Kleres, *Soc. Plast. Eng., RETEC Proc.*, Nov. 1994, 121.
100. L. Cao, R. M. Ramer and C. L. Beatty, 'Emerging Technologies in Plastics Recycling', ed. G. D. Andrews and P. M. Subramanian, ACS Symposium Series 513, American Chemical Society, Washington, 1992, p. 88.
101. R. Sachan, *Soc. Plast. Eng., RETEC Proc.*, Nov. 1994, 158.
102. I. Liedermooy, C. L. Beatty and R. M. Ramer, 'Emerging Technologies in Plastics Recycling', ed. G. D. Andrews and P. M. Subramanian, ACS Symposium Series 513, American Chemical Society, Washington, 1992, p. 113.
103. M. L. Gibbs, *Soc. Plast. Eng., ANTEC Proc.*, 1990.
104. A. Golovoy, S. S. Labana and R. A. Pet, *Soc. Plast. Eng., RETEC Proc.*, Nov. 1994, 29.
105. K. E. VanNess, C. C. Edwards and T. J. Nosker, *Soc. Plast. Eng., ANTEC Proc.*, 1994, 3064.
106. Plastics Institute of America Inc., 'Secondary Reclamation of Plastics Waste Research Report', Phase I, 1987; Phase II, 1988.
107. G. J. DeAngelis, B. Porter and R. D. Deanin, *Soc. Plast. Eng., ANTEC Proc.*, 1985, 1316.
108. R. Deanin and M. J. Curry, 'Secondary Reclamation of Plastics Waste Research Report', Plastics Institute of America, Phase I, 1987, p. 257.
109. W. J. Crawford and J. A. Manson, 'Secondary Reclamation of Plastics Waste Research Report', Plastics Institute of America, Phase I, 1987, p. 185; *Polym. Prepr.*, 1983, **24**(2), 432.
110. C. J. Huang and C. Dalton, 'Energy Recovery from Solid Waste', NASA Report No. CR 2526, 1975.
111. E. H. Immergut and J. Brandrup, 'Polymer Handbook', 2nd ed., Wiley, New York, 1974.
112. R. D. Leaversuch, *Mod. Plast.*, July 1991, 40.
113. W. S. Sanner, C. Ortaglio *et al.*, US Bureau of Mines Report No. 7428, 1970.
114. M. Roy, A. L. Rollins and H. P. Schreiber, *Polym. Eng. Sci.*, 1978, **18,** 721.
115. F. K. Shelly and S. Moore, *Chem. Eng.*, July 1992, 30.
116. J. Braslaw, R. L. Gealer and R. C. Wengfield, *Polym. Prepr.*, 1983, **24**(2)**,** 434.
117. J. Bhatia and R. A. Rossi, *Chem. Eng.*, 1982, **89**(20), 58.
118. M. Meszsaros, *Soc. Plast. Eng., RETEC Proc.*, Nov. 1994, 246.
119. K. Agarwal, *Soc. Plast. Eng., RETEC Proc.*, Nov. 1994, 232.
120. S. S. Labana *et al. Plast. Rubber Proc. Appl.*, 1990, **13**(4)**,** 229; D. Cornell, *Soc. Plast. Eng., RETEC Prepr.*, June 1993, 107; H. Grusche *et al., US Pat.* 3 403 115 (1968).
121. Anon, *Plast. News*, Dec. 10, 1990; *Reuse–Recycle*, Jan. 1991, 10.
122. V. Komolprasert and A. Lawson, *Soc. Plast. Eng., ANTEC Proc.*, 1994, 2906.
123. K. Foughy and S. Moore, *Chem. Eng. News*, July 1992, 35.
124. M. R. Kamal, R. A. Liafook and T. Yalcunyuva, *Soc. Plast. Eng., ANTEC Proc.*, 1994, 2896.
125. W. Macomis, Battelle Memorial Institute, personal communication.
126. L. R. Mahoney, S. H. Weine and F. C. Ferris, *Environ. Sci. Technol.*, 1974, **8,** 135; J. Leidner, 'Plastics Waste, Recovery of Economic Value', Dekker, New York, 1981.
127. Anon., *Chem. Eng.*, Feb. 1976.
128. E. G. Leverkusen, *Kunststoffe,* 1978, **68**(5).
129. K. W. Steiner, *Kunststoffe,* 1984, **74**(4), 186.
130. T. R. Borek and D. W. Peabody, *US Pat.* 2 937 151 (1960); B. Davis, *Plast. News*, April 24, 1989.
131. R. Sferrazza, A. Sarian and G. Shorel *et al. Soc. Plast. Eng., RETEC Proc.*, Nov 1994, 268; Chem. Eng. News 1993, **71**(12), 28.
132. SMC Automotive Alliance, 1275 K St. NW, Suite 400, Washington, DC 20005, Catalog 184; C. N. Cucuras, A. M. Flax, W. D. Graham and G. N. Heart, Society of Automotive Engineers International Congress and Exposition, Detroit, MI, February 25, 1991, paper #910387.
133. R. B. Jute and W. D. Graham, Technical Paper, Owens Corning Fiberglass, 1991.
134. D. Mangaraj and M. Luttinger, Battelle Columbus Division, Patent Disclosure, 1989.
135. W. J. Farrisey, R. E. Morgan, R. L. Tabor and M. Zawisza, 'Emerging Technologies in Plastics Recycling', ACS Symposium Series 513, American Chemical Society, Washington, 1991, p. 273.
136. W. J. Farrisey, R. E. Morgan, *et al.* 'Emerging Technologies in Plastics Recycling', ACS Symposium Series 513, American Chemical Society, Washington, 1991, p. 272.

137. L. Starke, Z. Funke and N. Kolzch, *Kuntstoffe,* 1992, **82**(4), 349.
138. *Japan Kokai,* 58/98374 [83/98374] (11 June 1983).
139. *Japan Kokai,* 58/42650 A2 [83/42650] (12 March 1983).
140. R. Dimino, *Resour. Recycling,* Dec. 1994, 14.
141. A. Isayev, paper presented at the Annual Meeting of the Ultrasonic Industry Association, October 19, 1994; D. Mangaraj, paper presented at the International Conference on Plastics and Rubber Business Planning, Dusseldorf, Germany, Oct. 31, 1989.
142. A. G. Bukens, *Conserv. Recycling,* 1977, 368.
143. D. Mangaraj, B. Mayo *et al. Soc. Plast. Eng., ANTEC Proc.,* 1988; D. Mangaraj, B. Mayo and J. Pfau, Proceedings of the Ultralloy Conference, Houston, Texas, Nov. 29, 1990.
144. R. S. Stein, 'Emerging Technologies in Plastics Recycling', ACS Symposium Series 513, American Chemical Society, Washington, 1991, p. 39.
145. M. M. Coleman, L. F. Graf and P. C. Painter, 'Specific Interactions and the Miscibility of Polymer Blends', Techneconomic, 1991, p. 495.
146. R. Markham, D. Mangaraj and A. Parson, 'Final Report on 'Compatibilization of Polymer Blends', A Multi-Client Program', Battelle Columbus Division, 1994.
147. H. Chotourou, R. Ridel and A. Aitkadi, *Soc. Plast. Eng., ANTEC Proc.,* 1991, 600.
148. C. L. Beatty and G. A. Staab, paper presented at the 4th Chemical Congress of North America, New York, August 25–30, 1991.
149. A. A. Adewole, K. Dackson and M. D. Wolkowitz, *Soc. Plast. Eng., ANTEC Proc.,* 1994, 3044.
150. S. S. Dagli, M. Xanthos and H. A. Biesenberger, 'Emerging Technologies in Plastics Recycling', ACS Symposium Series 513, American Chemical Society, Washington, 1991, p. 241.
151. M. Xanthos, T. J. Nosker and K. E. Van Ness, 'Compalloy 92: Proceedings of Sixth International Congress on Compatilibzers', ed. J. A. Snediker, April 1992, p. 59.
152. F. P. La Mantia and D. Curto, *Polym. Degradation Stab.,* 1992, **36**(2), 131.
153. Anon, *Plast. Technol.,* Feb. 1989, 67.
154. P. S. Hope, J. G. Bonner and A. C. Miles, *Plast. Rubber Compos. Proc.,* 1994, **22**(3), 147.
155. K. Khalt, *Soc. Plast. Eng., RETEC Proc.,* Nov. 1994, 72.
156. A. Rihai, J. Arastoopour *et al., Soc. Plast. Eng., ANTEC Proc.,* 1993, 891.
157. J. Ogando, *Plast. Technol.,* Aug. 1992, 60.
158. E. S. Lipinsky and R. G. Sinclair, *Chem. Eng. Prog.,* Aug. 1986, 26.
159. M. H. Naitov, *Plast. Technol.,* Mar. 1995, 16.
160. A. M. Thayer, *Chem. Eng. News,* Jun. 25, 1990, 7.
161. B. Mill, *Plast. World,* Mar. 1990, 12.
162. W. Schnabell, 'Polymer Degradation, Principles and Practical Application', Hanser/Macmillan, New York, 1981; R. Leaversuch, *Mod. Plast.,* Aug. 1987, 52.
163. R. S. Porter and A. Casale, 'Polymer Stress Reactions', Academic Press, New York, 1978, 79.
164. L. Souza-Couch, L. Hornberger *et al., Soc. Plast. Eng., ANTEC Proc.,* 1994, 2868.
165. L. A. Dzeskiewicz and R. E. Farrel, *Soc. Plast. Eng., ANTEC Proc.,* 1994, 3055.
166. S. Babinchak, *Soc. Plast. Eng., RETEC Proc.,* Nov. 1994, 50.
167. M. E. Shriver, K. A. Beiter and K. Ishij, *Soc. Plast. Eng., ANTEC Proc.,* 1994, 2910.
168. R. Leaversuch, *Mod. Plast.,* July 1994, 48D; J. Powell, *Resour. Recycling,* May 1995, 23.
169. V. Wigtsky, *Plast. Eng.,* Dec. 1994, 10.
170. J. Raykirby and D. Pitts, *Soc. Plast. Eng., RETEC Proc.,* Nov. 1994, 41, S. Labana, *Soc. Plast. Eng., RETEC Proc.,* June 1993, 181.

19

Physics and Chemistry of Microwave Processing

FABRIZIO PARODI
Saiag SpA, Torino, Italy

19.1 INTRODUCTION

Radio waves and microwaves are nonionizing, electromagnetic radiation with frequencies conventionally ranging from 3 kHz to 300 MHz and from 300 MHz to 300 GHz, respectively (corresponding to wavelengths in vacuum of 10^5 to 1 m and 1 to 10^{-3} m, respectively).

Except for magnetic metals, their alloys, and conductive polymers, good electronic conductors interact with radio waves and microwaves by synchronous generation of superficial, high-frequency alternating currents and electromagnetic reirradiation with minimum energy losses.[1–4] In other words, they behave as wave reflectors, a characteristic which is the basis of radar detection of metal objects. Insulators or dielectrics, having no or little electrical conductivity, by contrast, tend to dissipate the electric field energy internally as heat *via* cyclically reversing polarization and dielectric relaxation phenomena.[5–8] Such energy dissipation varies from very weak to strong, resulting in materials which range from being almost transparent to being fairly opaque to waves. This absorption of radio waves and microwaves by dielectrics with material heating is commonly termed dielectric heating.

Other energy transfer processes from this spectrum of electromagnetic radiation to matter with internal heating[8] are electric conduction losses in electrolytic solutions,[9] ionic conductors in general, and poor electronic conductors, as well as magnetic relaxation, resonance, and eddy current dissipation phenomena in magnetic materials, *etc.* In all these cases, heating tends to proceed extremely rapidly, simultaneously, and uniformly throughout the entire volume of material, largely irrespective of its thermal conductivity. It is also very efficient in terms of energy conversion yield, occurs with little external energy losses, and can be activated or interrupted instantaneously by simply switching on or off the irradiation device. Moreover, combined materials having different electromagnetic characteristics may be heated at different rates with impressive selectivity. These unique features make radio wave and microwave heating powerful and versatile techniques for both laboratory and industrial processing of materials. This approach, indeed, offers striking advantages over any conventional heating method known, where heating itself proceeds from the surface to the core of objects. It is used particularly to process many dielectric materials having poor thermal conductivity, such as natural and synthetic polymers and resins, ceramics, glasses, foods, wood, minerals, textiles, and so on.

19.2 GENERAL ASPECTS

19.2.1 Dielectric Heating of Materials

The dielectric heating technique dates back to the last 20 years of 19th century, when it was used in the therapeutic method of diathermy,[10,11] consisting of the internal warming of the human body or its parts by exposure to electromagnetic fields (hyperthermy). For decades the 0.3–3 MHz range of medium frequency radio waves, or MF band (typically around 0.6 MHz), was used. After the advent of radio communications technology, medical diathermy was implemented with the adoption of higher frequencies (10–100 MHz, more energetic and penetrant in biological tissues) around the first quarter of this century: neodiathermy, ultra-diathermy, Marconi-therapy in Italy, *etc.*, and worldwide referred to as shortwave diathermy or shortwave therapy.[12] About 20 years later, microwave technology was introduced in this field, leading to the development of microwave diathermy.[13,14]

Heating of dielectrics with radio waves, to which the definition of dielectric heating is often confined, is usually called radiofrequency or high-frequency heating. It began to find industrial applications in the early 1940s, and underwent rapid growth during the next two decades. In this technology, frequencies are normally in the 3–30 and 30–300 MHz intervals of high and very-high frequencies, or HF and VHF bands, respectively, although the 10–100 MHz range is preferred in most cases.

In contrast to radiofrequency (RF) heating, industrial uses of microwave (MW) heating were introduced only in the 1960s, when reliable, high-power, and reasonably cheap MW generators became available. Development was comparatively slow, and is still in progress. Irradiation frequencies in the 300–3000 MHz and 3–30 GHz ranges are employed: ultra-high frequency (UHF) and super-high frequency (SHF) bands, respectively. Frequencies in the 300–3000 MHz range are almost universally adopted, and, for such reason, MW heating is often referred to as UHF heating.

Communications and radar purposes dominate both civil and military uses of RF and MW spectra.[15] Thus, in order to minimize radio interferences, only a few precise frequencies with narrow tolerance are allocated under national and international regulations for industrial, scientific, and medical power applications (the so-called ISM bands). Frequencies worldwide allowed by the International Telecommunications Union (ITU) are: 13.560 ± 0.007 MHz, 27.120 ± 0.163 MHz, 40.68 ± 0.02 MHz, 2450 ± 50 MHz (the frequency used in almost all domestic MW ovens), 5800 ± 75 MHz, and 24125 ± 125 MHz. Besides these, the intermediate MW frequencies of 896 ± 10 and 915 ± 13 MHz (extensively used since the beginning of the technology) are allocated to the UK and the Americas, respectively. A few other frequencies are presently under consideration, and can be employed with special permission of local administrations. All other frequencies can be used for laboratory as well as for industrial heating purposes provided that radio emissions are suppressed, by careful shielding of the entire apparatus, below very low limiting levels fixed or recommended by the International

Special Committee on Radio Interference (CISPR). Even with the free radiation ISM bands, RF and MW leakages are subject to severe restrictions and controls by inspective authorities because of the health hazards derived from prolonged or repeated exposure to high-frequency electromagnetic fields.

As expected, due to the identical energy absorption principles by which both are governed, RF and MW heating techniques are very similar from the standpoint of general features and applications, and are interchangeable with each other for many purposes. The RF and MW techniques, however, require use of substantially different equipment, *i.e.,* high-frequency generators and devices for materials irradiation (or applicators). In RF heating (more specifically, below 100 MHz), the so-called 'open wire circuit' arrays are used as energy applicators. In other words, the material being processed is placed or slitted either between or nearby a pair (or pairs) of electrodes fed with a high-frequency oscillatory voltage. The comparatively low frequencies employed and the uniformity of the electromagnetic field, which may be imagined to cross the whole material, mean that RF heating can be particularly homogeneous, even in very large and thick objects.

Applicators for MW technology above 500 MHz, by contrast, include resonant cavities, travelling wave devices, horns, their hybrids, and other complex devices (supplied by a MW generator through a waveguide or a coaxial cable), where the material may be visualized as being immersed in, or surrounded by, the UHF or SHF field. The field intensity distribution in the applicator is typically, and often remarkably, inhomogeneous, according to regular or irregular patterns. The intensity itself decays appreciably from the surface to the interior of objects, as sharply as the energy absorption of the material becomes stronger and the radiation frequency higher. The resulting tendency to uneven heating can be minimized by a variety of methods, including: (i) proper applicator design; (ii) rapid, repeated change of the field intensity patterns in the applicator (resonance mode switching or randomization); (iii) transit, rotation, or spatial revolutions of objects through the field; and (iv) joint use of MW irradiation and conventional heating sources.

Heating with frequencies in the spectral region around the conventional border between radio waves and microwaves (roughly, in the 100–500 MHz band) presents features intermediate between those of RF and MW technologies, and implies hybridization of the related circuit arrays and applicators. The lack of good design criteria for such equipment is the reason why this portion of the electromagnetic spectrum is not often used for industrial heating purposes.

19.2.2 Radiofrequency and Microwave Processing of Materials: Processing of Polymers, Resins, and Related Composites

The energy supplied to an absorbing material by electromagnetic radiation is proportional to the frequency. Hence, the more expensive MW technology results in superior heating rates (orders of magnitude higher at comparable field strength) compared to the RF technique. This advantage is particularly valuable and has been exploited in a large number of practical cases involving the processing of objects of not excessive thickness. Furthermore, MW technology allows virtually complete freedom regarding the shape of objects to be processed, whereas comparatively simple geometries are imposed by RF applicators.

In RF heating, the electromagnetic field strengths, and thus the heating rates, are enhanced by increasing the voltage to the electrodes, and this represents a major limitation of the RF technology when compared to the MW one. Accordingly, maximum applicable voltages are imposed by the inherent electrical properties (*i.e.* dielectric strength, electric bulk and surface conductivities) of the material being processed. Above such critical voltage values, arcing between electrodes may occur across the material or may track its surface; this results in damage to the object undergoing treatment and in significant energy waste, absent in MW systems. Because of the above considerations, MW technology is gaining more and more favor in most fields where dielectric heating is used, and appears to offer a still inexhausted source of industrial opportunities.

Several books and review articles give general descriptions of RF[16–20] and MW[20–28] technologies for materials processing, including the basic principles and features, applications and economics, equipment design and plant arrays, radio interference, as well as health and safety issues. Papers have been devoted to the analysis of plant and processing economics of both techniques[29,30] in comparison with conventional heating methods.

Technical advantages and versatility make RF and MW heating suitable for many established and potential uses; these have been overviewed for the MW technology.[31–33] Dielectric materials to which RF and/or MW heating have been successfully applied include: foods[34] (RF/MW baking, drying and freeze-drying, pasteurization/sterilization, tempering of frozen products); wood[35,36] (RF drying, bending, gluing); textiles[37] (RF/(MW) drying, dye fixation); leather[20] (RF/MW drying); paper and board[20] (RF moisture profiling, drying of coatings and adhesives); ceramics[38,39] (MW setting, drying, firing, sintering, joining); pharmaceuticals[20] (MW drying); hazardous wastes[40] (MW heating); polymers, resins, and related materials.

Processing of polymers and resins was among the earliest industrial uses of dielectric heating; it was first used in the 1950s and gained importance in the next decades. Applications include: RF drying/hardening of aqueous thermosetting resins, or drying of polymer emulsions, for wood gluing (in the manufacture of wood laminates and panels);[35,36] RF welding of plastics[41,42] RF preheating in compression molding of thermoplastics. Two reviews[43,44] have been devoted to general aspects of these and other applications of RF heating to polymeric materials (precure of resin binders in the fabrication of particle and fiber boards, molding and foaming of PVC plastisols, *etc.*).

A remarkable achievement of the late 1960s was the adoption of MW power technology for the continuous vulcanization of extruded rubbers and the discontinuous vulcanization of molded rubber articles.[45] The MW vulcanization of rubber compounds, which has been used industrially since the early 1970s, is the most important application of MW heating to polymeric materials in terms of the number of installed plants and the volume of products processed worldwide each year.

In the 1970s, the technology of MW heat-drawing, reviewed some years ago,[46] was developed for the manufacture of high-modulus polymeric materials (in particular, of polyoxymethylene rods). Many other applications of MW heating were introduced in the 1970s and 1980s, including the rapid bonding of polymeric films, sheets, and various components, fast curing of thermosetting resins, *etc.*

In the mid-1980s, MW technology revealed, almost simultaneously in polymer and in organic chemistry, an outstanding potential beyond the known advantages of heating rapidity, efficiency, and selectivity. MW energy showed the capacity to cause strong acceleration and/or selective promotion of many chemical processes; this was further highlighted by the adoption of pulsed irradiation methods.

In the last 10 years, MW technology applied to polymer processing, and particularly to polymerization processes, has become the subject of a great deal of academic and industrial research, with a huge body of scientific and patent literature being generated. Among the numerous papers, some recent reviews summarize the advances in MW curing of thermosetting resins and composites,[47–49] the basic principles of MW processing of thermoplastic polymers[50] and polymeric materials in general,[51,52] and the uses of MW irradiation in welding techniques for polymers and composites.[53]

19.3 FUNDAMENTALS OF ELECTROMAGNETIC HEATING OF POLYMERS, RESINS, AND RELATED COMPOSITES

As mentioned in the introduction, a variety of energy absorption (or energy loss) mechanisms can be involved in, or exploited for, the RF or MW heating of materials, by interaction with the electric and/or the magnetic components of the high-frequency electromagnetic field. Since neat, ordinary organic polymers and thermosetting resins are dielectrics with no or negligible magnetic response, coupling with the electric field component and energy absorption *via* dielectric relaxation phenomena can be considered the principal electromagnetic heating mechanism for this family of materials.

However, numerous practical reasons, including the necessity of providing polymers and resins with tailored mechanical, thermomechanical, electrical, and/or magnetic characteristics (or with enhanced RF or MW heating attitudes), have led to a progressively widespread adoption of polymer blends and polymer matrix composites. In such materials, a continuous polymeric phase embeds polymeric or inorganic constituents of amorphous, mono- or polycrystalline nature, with different morphology and dimensions, simply dispersed (powders, microbeads, flakes, milled or short fibers) or distributed in random or regular arrays (long or continuous fibers). Composite structures consisting of parts or layers of different polymeric materials, or comprising both polymers and inserts, wires, films, plates, *etc.*, of metals, ceramics, or inorganic glasses, are also of great technological importance. In polymer blends and, in particular, in polymer matrix composites and composite structures, multisymmetrical properties, morphology, and/or spatial arrangements of additives, inserts, or parts, often contribute appreciably to high degrees of microscopic and/or macroscopic electromagnetic anisotropy. Although focused on isotropic materials, the following discussion will also give some attention to such anisotropic systems.

With respect to both their electric and magnetic properties, inorganic constituents of composites and composite structures can be classified into the following main categories.

(a) Electrically conductive materials: diamagnets (carbon black and graphite, Cu, Zn and related alloys, Pb, Ag, Au, *etc.*); paramagnets (Al, Mg, stainless steel, Pt, *etc.*); ferromagnets (Fe, Co, Ni, their alloys, alloys of them with a variety of non-ferromagnetic elements, etc.); antiferromagnets (*e.g.* Cr).

(b) Dielectric and semiconducting materials: diamagnets (*e.g.* oxides and salts of diamagnetic elements and most paramagnetic ones, such as SiO_2, ZnO, $CaCO_3$, $MgCO_3$, Al_2O_3, silicate glasses, and ceramics in general); paramagnets (most salts of ferromagnetic metals); ferromagnets (*e.g.* CrO_2); antiferromagnets (*e.g.* hematite (α-Fe_2O_3)); ferrimagnets (maghemite (γ-Fe_2O_3), magnetite (Fe_3O_4), and ferrites in general); ferroelectrics (perovskites such as $BaTiO_3$, $PbTiO_3$ and Pb zirconate–titanate, *etc.*); antiferroelectrics (*e.g.* $PbZrO_3$).

In addition to the dielectric loss of the polymer matrix, the presence of inorganic or polymeric additives with different electrical and/or magnetic properties leads to a wide variety of energy loss processes, with heat generation, including: interfacial electric polarization losses and electric conduction losses in conductor-filled polymers; eddy current dissipations in metal parts of composite structures; magnetic relaxation and resonance losses in magnetic constituents of composites and composite structures. When used either as matrices or additives, moreover, conductive and ferroelectric polymers exhibit strong RF/MW absorptions through their unusual electric conduction and dielectric relaxation loss capacities, respectively.

The overall electromagnetic response of any polymer blend, composite, or composite structure is the resultant of the contributions of each constituent according to its own electromagnetic characteristics, as well as by possible cooperative phenomena between different constituents (such as that of interfacial polarization, mentioned above).

Although the dielectric relaxation loss properties of neat polymers and resins is the major factor, a comprehensive treatment of RF and MW heating of polymeric materials must consider the overall world of electric and magnetic responses of matter, whose marvelous, almost perfect symmetry of behavior is sketched in Scheme 1.

ELECTRIC BEHAVIOR	MAGNETIC BEHAVIOR
–	Diamagnetic
Paraelectric	Paramagnetic
Ferroelectric	Ferromagnetic
Antiferroelectric	Antiferromagnetic
(Ferrielectric)	Ferrimagnetic

Scheme 1

19.3.1 Interactions of Condensed Matter with Electric and Magnetic Fields: Basic Concepts

The macroscopic effects of electric polarization[6,7,54–56] and magnetization[57–59] exerted by steady electric and magnetic fields on isotropic dielectrics with no electrical conductivity are given, in the SI system, by the classical equations (1) and (2).

$$\boldsymbol{D} = \varepsilon_0 \boldsymbol{E} + \boldsymbol{P} \tag{1}$$

$$\boldsymbol{B} = \mu_0 \boldsymbol{H} + \boldsymbol{M} \tag{2}$$

In such equations: the $\boldsymbol{D}$ and $\boldsymbol{P}$ vectors, whose values are expressed in coulombs per square meter ($C\,m^{-2}$), are the electric displacement, and the electric volume polarization or electrization (*i.e.* the average electric dipole moment per unit volume), respectively, achieved in (by) a material subjected to an electric field of strength or intensity represented by the vector $\boldsymbol{E}$ (values in volts per meter ($\mathrm{V\,m^{-1}}$)); the $\boldsymbol{B}$ and $\boldsymbol{M}$ vectors, expressed in tesla (T), are the magnetic induction or magnetic flux density, and the magnetic volume polarization or magnetization (*i.e.* the average magnetic dipole moment per unit volume), respectively, produced in the material by a magnetic field of strength or intensity represented by the vector $\boldsymbol{H}$ (values in amperes per meter ($\mathrm{A\,m^{-1}}$)); ε_0 (the dielectric constant or permittivity of free space) $= 8.854 \times 10^{-12}$ farads per meter ($\mathrm{F\,m^{-1}}$); μ_0 (the magnetic permeability of free space) $= 4\pi \times 10^{-7}$ henries per meter ($\mathrm{H\,m^{-1}}$).

The electric displacement $\boldsymbol{D}$ and the electric volume polarization $\boldsymbol{P}$ are proportional to the electric field strength $\boldsymbol{E}$ as in equations (3a) and (3b), where ε_s and $\chi_{e,s}$ are the static relative dielectric constant and the static electric volume susceptibility of the material. Such dimensionless parameters are interrelated, through equation (1), as in equation (3c).

$$\boldsymbol{D} = \varepsilon_0 \varepsilon_s \boldsymbol{E} \tag{3a}$$

$$\boldsymbol{P} = \varepsilon_0 \chi_{e,s} \boldsymbol{E} \tag{3b}$$

$$\varepsilon_s = 1 + \chi_{e,s} \tag{3c}$$

Analogous considerations apply to magnetization phenomena as in equations (4), with μ_s and $\chi_{m,s}$ being the static relative magnetic permeability and the static magnetic volume susceptibility of the medium subjected to the magnetic field.

$$\boldsymbol{B} = \mu_0 \mu_s \boldsymbol{H} \tag{4a}$$

$$\boldsymbol{M} = \mu_0 \chi_{m,s} \boldsymbol{H} \tag{4b}$$

$$\mu_s = 1 + \chi_{m,s} \tag{4c}$$

Values of the magnetic volume susceptibility in the SI system as hereinafter considered can be obtained through equation (5), from those usually given in electromagnetic c.g.s. units.

$$\chi_m \ (\mathrm{SI}) = 4\pi\chi_m \ (\mathrm{c.g.s.\ emu}) \tag{5}$$

ε_s, μ_s, $\chi_{e,s}$ and $\chi_{m,s}$ depend in different ways on temperature, and for ferro- or antiferroelectrics and for ferro-, antiferro-, or ferrimagnets are also functions of the strength of the applied electric or magnetic field. In good electrical conductors such as metals or graphite, whose ε_s is extremely high (in principle, $\varepsilon_s \rightarrow \infty$) and having specific d.c. conductivity σ above $10^4\,\Omega^{-1}\,\mathrm{m}^{-1}$, steady electric fields cause continuous charge flow instead of polarization. Therefore, in isotropic conductors, for which equation (2) is adequate, equation (1) must be replaced with Ohm's law (6), where $\boldsymbol{J}$ is the electric current density vector (in $\mathrm{A\,m^{-2}}$), parallel to, and in the same direction as, $\boldsymbol{E}$.

$$\boldsymbol{J} = \sigma \boldsymbol{E} \tag{6}$$

Most organic polymers and resins are dielectrics. These are characterized by small, finite electrical conductivities (typically, $10^{-18} < \sigma < 10^{-5}\,\Omega^{-1}\,\mathrm{m}^{-1}$) and the polarization phenomenon described by equation (1) is accompanied by a weak electric current flow according to equation (6). In anisotropic materials, the scalar parameters ε_s, μ_s, $\chi_{e,s}$, and $\chi_{m,s}$ become 3rd rank tensors ($\tilde{\boldsymbol{\varepsilon}}_s$, $\tilde{\boldsymbol{\mu}}_s$, $\tilde{\boldsymbol{\chi}}_{e,s}$, and $\tilde{\boldsymbol{\chi}}_{m,s}$) with *ij*-components equal to the symmetrical *ji* ones. Similarly, σ becomes a 2nd rank tensor $\tilde{\boldsymbol{\sigma}}$. Tensorial expressions of these parameters require the replacement of equations (1)–(4) and (6) with the generalized forms (7)–(9).

$$\boldsymbol{D} = \varepsilon_0 \tilde{\boldsymbol{\varepsilon}}_s \boldsymbol{E} = \varepsilon_0 \boldsymbol{E} + \varepsilon_0 \tilde{\boldsymbol{\chi}}_{e,s} \boldsymbol{E} \tag{7}$$

$$\boldsymbol{B} = \mu_0 \tilde{\boldsymbol{\mu}}_s \boldsymbol{H} = \mu_0 \boldsymbol{H} + \mu_0 \tilde{\boldsymbol{\chi}}_{m,s} \boldsymbol{H} \tag{8}$$

$$\boldsymbol{J} = \tilde{\boldsymbol{\sigma}} \boldsymbol{E} \tag{9}$$

Electric and magnetic polarization phenomena imply energy transfer from an external electric or magnetic field to matter, and energy storage into matter itself. According to equations (1) and (2), or (7) and (8), $\boldsymbol{D}$ and $\boldsymbol{B}$ quantify the overall displacements of electric and magnetic charges caused by the electric and magnetic forces $\boldsymbol{E}$ and $\boldsymbol{H}$, respectively, each of them as the vector sum of two contributions: (a) that corresponding to the electric or magnetic potential in the free space occupied by the material subjected to the external field ($\varepsilon_0\boldsymbol{E}$ or $\mu_0\boldsymbol{H}$); and (b) that produced within the material by its electric or magnetic polarization according to equations (3b) and (4b) or their generalizations for anisotropic materials. Hence, the electric or magnetic work w_e or w_m done on (*i.e.* the energy supplied to) a unit volume of a material ($\mathrm{J\,m^{-3}}$) by a steady electric or magnetic field of strength $\boldsymbol{E}$ or $\boldsymbol{H}$ (applied under quasi-static conditions at constant volume) can be calculated by integration of equations (10) and (11).

$$\delta w_e = \boldsymbol{E}\,\mathrm{d}\boldsymbol{D} = \varepsilon_0 \boldsymbol{E}\,\mathrm{d}[\boldsymbol{E}\varepsilon_s(T, \boldsymbol{E})] \tag{10}$$

$$\delta w_m = \boldsymbol{H}\,\mathrm{d}\boldsymbol{B} = \mu_0 \boldsymbol{H}\,\mathrm{d}[\boldsymbol{H}\mu_s(T, \boldsymbol{H})] \tag{11}$$

In the cases of isochoric polarization phenomena in isotropic dielectrics whose ε_s and μ_s are independent of the field strength (and supposed to have a linear dependence on temperature), w_e and w_m are given by equations (12) and (13).

$$w_e = \varepsilon_0 \int_0^{\boldsymbol{E}} \int_T^{T+\Delta T} \boldsymbol{E}\,\mathrm{d}[\boldsymbol{E}\varepsilon_s(T)] = \tfrac{1}{2}\varepsilon_0 \mathrm{E}^2\left[\varepsilon_s(T) + \left(\frac{\partial \varepsilon_s}{\partial T}\right)_{\mathrm{V}} \Delta T\right] \tag{12}$$

$$w_m = \mu_0 \int_0^{\boldsymbol{H}} \int_T^{T+\Delta T} \boldsymbol{H}\,\mathrm{d}[\boldsymbol{H}\mu_s(T)] = \tfrac{1}{2}\mu_0 H^2\left[\mu_s(T) + \left(\frac{\partial \mu_s}{\partial T}\right)_{\mathrm{V}} \Delta T\right] \tag{13}$$

The variations of internal energy associated with electric and magnetic polarization (or depolarization) can be considered through equations (14) and (15), indicating the reversible character of quasi-static polarization (and depolarization) processes under adiabatic ($\delta Q = 0$) conditions.

$$\mathrm{d}U_e = \delta Q + \delta w_e \tag{14}$$

$$\mathrm{d}U_m = \delta Q + \delta w_m \tag{15}$$

The temperature coefficients of the dielectric constant (with the exception of most polymers) and permeability ($\partial\varepsilon_s/\partial T$) and ($\partial\mu_s/\partial T$) are usually negative. This accounts for the progressively lower (reversible) field energy transfer to matter at increasing temperatures, due to thermal agitation opposing the atomic, molecular, or supramolecular ordering associated with polarization. Similar thermodynamic considerations can be applied, and appropriate equations can be drawn, for polarization processes at constant pressure.

In contrast to what was said previously, the electric currents induced within isotropic, real dielectrics, semiconductors, and conductors always imply (irreversible) energy dissipations by the Joule effect, expressed, at constant temperature and per unit volume of material ($\mathrm{W\,m^{-3}}$), by equation (16).

$$W_\sigma = \sigma E^2 \tag{16}$$

The various types of electric and magnetic behavior given in Scheme 1 correspond to different values and different types of dependence of ε_s, $\chi_{e,s}$, and μ_s, $\chi_{m,s}$ on temperature, or also, in complex ways, on both the electric or magnetic field strength and the history of a material.

19.3.1.1 Diamagnets

As is well known, the diamagnetism[57–59] of a material, a property having no correspondence in electricity, is its tendency to oppose, *i.e.* to reduce the strength of, an external magnetic field acting on the material itself. As a general property of matter, it can be associated with, and often masked by, other magnetic behavior (*e.g.* para- or ferromagnetism). Diamagnetism always makes a small, negative contribution, χ_{dia}, to the overall $\chi_{m,s}$ of any substance. This contribution is almost independent of both temperature and field strength. Diamagnetic materials, *i.e.* those where diamagnetism dominates or is the sole magnetic response, are characterized by $\chi_{m,s}$ and μ_s values just slightly less than 0 and less than 1, respectively. Their weakly negative interactions with a magnetic field, according to equations (2), (10), and (13), are negligible with regard to energy exchanges with the field itself.

19.3.1.2 Paraelectrics and paramagnets

Paraelectric and paramagnetic substances are those undergoing weak to appreciable temporary electric or magnetic polarization when subjected to an electric or magnetic field, respectively. They are characterized by positive values of $\chi_{e,s}$ or $\chi_{m,s}$ (and thus by ε_s or $\mu_s > 1$) which are independent of the applied field strength, implying a linear polarization response. Paraelectricity and paramagnetism relate to different polarization phenomena at the molecular or atomic level.

(i) Paraelectrics

Most materials referred to as dielectrics[6,7] exhibit paraelectric behavior, so that the electric polarization of true paraelectrics is normally called dielectric polarization. This term is also used incorrectly to include any other kind of electric polarization in nonconductors or semiconductors, such as the electrization phenomena in ferroelectrics mentioned in Section 3.1.3.

Dielectric polarization of homogeneous materials proceeds by three mechanisms:[54–56]

(a) electronic polarization ($\boldsymbol{P}_e$), consisting of the induction of atomic dipoles through shifts of electrons with respect to the atomic nuclei;

(b) atomic (or ionic) polarization ($\boldsymbol{P}_a$), that is, the reciprocal displacement of different atomic nuclei of a molecule from their original, relative positions, with induction of molecular dipoles;

(c) orientational polarization ($\boldsymbol{P}_\mu$), consisting of the alignment of permanent dipoles of a molecule with an external electric field.

With respect to the orientation or disorientation of permanent dipoles, electronic and atomic polarization and depolarization are much quicker (at least a hundred times faster) and are practically temperature-independent phenomena. The total polarization is given by the vector sum of the above three contributions as in equation (17). The contributions of electronic and atomic polarization will be considered together as distortion polarization ($\boldsymbol{P}_d$).

$$\boldsymbol{P} = \boldsymbol{P}_e + \boldsymbol{P}_a + \boldsymbol{P}_\mu \tag{17}$$

The detailed molecular description of electric polarization phenomena[60–62] is very complex and beyond the scope of this chapter, so that only the basic aspects are summarized here in a simplified form. If an isotropic, homogeneous substance consisting of simple dipolar molecules with a permanent dipole moment $\hat{\boldsymbol{\mu}}$ (expressed in C m) is considered, the total polarization is given by equation (18), where N_A is the Avogadro number, $\hat{\boldsymbol{\mu}}_d$ the average dipole moment induced by the field by distortion polarization, and $\langle\hat{\boldsymbol{\mu}}\rangle$ the average dipole moment in the field direction resulting under the field itself by orientation of permanent molecular dipoles, ρ the density (kg m^{-3}), and M the molecular weight (kg mol^{-1}).

$$\boldsymbol{P} = \boldsymbol{P}_d + \boldsymbol{P}_\mu = \frac{\rho(T)}{M} N_A[\hat{\boldsymbol{\mu}}_d + \langle\hat{\boldsymbol{\mu}}\rangle] \tag{18}$$

$\hat{\boldsymbol{\mu}}_d$ and $\langle\hat{\boldsymbol{\mu}}\rangle$ can be expressed as in equations (19) and (20), where $\tilde{\boldsymbol{\alpha}}$ is the molecular distortion polarizability (3rd rank) tensor (F m^2), $\mathscr{L}(x)$ the Langevin function, and $\boldsymbol{E}_L$ the strength of the internal or local field really acting on molecules, which is different from the applied field due to the electrostatic effects exerted by the surrounding matter itself.

$$\hat{\boldsymbol{\mu}}_d = \tilde{\boldsymbol{\alpha}}\boldsymbol{E}_L \tag{19}$$

$$\langle\hat{\boldsymbol{\mu}}\rangle = \hat{\boldsymbol{\mu}}\,\mathscr{L}\left(\frac{\hat{\boldsymbol{\mu}}\boldsymbol{E}_L}{kT}\right) \approx \frac{\hat{\mu}^2\boldsymbol{E}_L}{3kT} \tag{20}$$

The last term of equation (20) is a reasonable approximation for $\langle\hat{\boldsymbol{\mu}}\rangle$, obtained by reducing the Langevin function to the first term of its power series expression. The molecular polarizability $\tilde{\boldsymbol{\alpha}}$ is a temperature-independent parameter, so that $\boldsymbol{P}_d$ is influenced by temperature only slightly through the substance density. Equation (20), by contrast, quantifies the disordering effects of thermal agitation in opposing dipole orientation, *i.e.* the strong and negative influence of temperature on orientational polarization.

A critical issue is the knowledge of $\boldsymbol{E}_L$, and several theoretical models have been developed for its calculation. The classical Lorentz approach predicts $\boldsymbol{E}_L$ by its relationship to the external field strength $\boldsymbol{E}$ as in equation (21).

$$\boldsymbol{E}_L = \frac{\varepsilon_s + 2}{3}\boldsymbol{E} \tag{21}$$

If $\tilde{\boldsymbol{\alpha}}$ reduces to a 2nd rank tensor (*i.e.* to its *ii*-components), the scalar $\alpha = \frac{1}{3}(\alpha_{11} + \alpha_{22} + \alpha_{33})$ may be introduced into equation (19), and the Debye formula (22) can be deduced from equations (18)–(21). It interrelates the ε_s of a substance with temperature, its molecular polarizability, and its permanent dipole moment.

$$\frac{\varepsilon_s - 1}{\varepsilon_s + 2}\frac{M}{\rho(T)} = \frac{N_A}{3\varepsilon_0}\left(\frac{\alpha_{11} + \alpha_{22} + \alpha_{33}}{3} + \frac{\hat{\mu}^2}{3kT}\right) \tag{22}$$

Equation (22) is a rough approximation, in that it neglects any intermolecular interactions. Another approach for the calculation of $\boldsymbol{E}_L$ is that of Onsager, which leads to equation (23), where ε_∞ is the so-called instantaneous, high-frequency, or optical (relative) dielectric constant of the substance. ε_∞ expresses the contribution of the sole very fast (instantaneous) distortion polarizability to ε_s, and is thereby related to α through the Debye–Mossotti equation (24). Conversely, the quantity $(\varepsilon_s - \varepsilon_\infty)$ gives the contribution of the dipole orientation polarizability. It is known that ε_∞ equals the square of the refractive index of the substance at far-infrared frequencies. A good approximation is often given by the square of the easily accessible refractive index measured in visible light (usually at the sodium D line) or obtained from the molar refraction of the substance (calculated, for organic compounds, from tabulated atomic, group, and structural contributions).

$$\varepsilon_s - \varepsilon_\infty = \frac{\varepsilon_s(\varepsilon_\infty + 2)^2}{2\varepsilon_s + \varepsilon_\infty}\frac{\rho(T)}{M}\frac{N_A\hat{\mu}^2}{9\varepsilon_0 kT} \tag{23}$$

$$\frac{\varepsilon_\infty - 1}{\varepsilon_\infty + 2}\frac{M}{\rho(T)} = \frac{N_A\alpha}{3\varepsilon_0} \tag{24}$$

At least at a certain extent, equation (23) considers intermolecular dipolar interactions and is thus applicable to a number of simple liquids and even solids, provided that strong intermolecular forces (strong dipolar interactions, hydrogen- and π-bonds) are not present. A far better description is given by the well-known Fröhlich–Kirkwood equation (25).

$$\varepsilon_s - \varepsilon_\infty = \frac{\varepsilon_s(\varepsilon_\infty + 2)^2}{2\varepsilon_s + \varepsilon_\infty}\frac{\rho(T)}{M}\frac{N_A g\hat{\mu}^2}{9\varepsilon_0 kT} \tag{25}$$

$$g = \frac{(\hat{\mu})^2}{\hat{\mu}^2} \tag{26}$$

Equation (25) is a simple modification of the Onsager formula (23) by introduction of the Kirkwood correlation or reduction factor g (equation (26)). It accounts for a different effective or apparent dipole moment $(\hat{\boldsymbol{\mu}})$ of molecules (or also dipole moment 'in the condensed phase') to be used in place of their intrinsic, permanent dipole moment $\hat{\boldsymbol{\mu}}$ in the absence of intermolecular interactions (as measured in the gas phase, and often called 'vacuum dipole moment') hereinabove considered. This implies that $(\hat{\boldsymbol{\mu}})$ values of molecules of a liquid or solid substance differ from the $\hat{\boldsymbol{\mu}}$ values in the gas phase as indicated by equation (27), where ε_s and

ε_∞ are the experimental values for the substance itself in the condensed state under consideration.

$$\langle\hat{\mu}\rangle = \hat{\mu}\frac{(\varepsilon_\infty + 2)(2\varepsilon_s + 1)}{3(2\varepsilon_s + \varepsilon_\infty)} \tag{27}$$

The Kirkwood factor g (even when significantly different from unity) expresses the effects of the (nearest) neighbors of a dipolar molecule on its own dipole moment in condensed matter. It depends markedly on temperature and on so many parameters (number, shape, geometrical disposition, permanent dipole moment, and tensor components of the polarizability of surrounding molecules, *etc.*) that it cannot realistically be calculated *a priori.* Hence, its meaning remains unfortunately almost completely conceptual to quantify the experimental deviations from the Onsager equation (23). Whenever available (for thermally stable compounds of reasonably low molecular weight), experimental dipole moment values in the gas phase can be used for rough calculations of ε_s and $(\varepsilon_s - \varepsilon_\infty)$ through equations (23) and (24). For substances of high molecular weight (such as many organic compounds, oligomers, and polymers), vacuum molecular dipole moments can only be calculated from the molecular configuration or configuration distribution as vector sums of the individual characteristic dipole moments of the constituent atomic groupings. Table 1 collects typical values (or narrow ranges of variation) of the dipole moment in the gas phase of some important dipolar groups, obtained from selected $\hat{\mu}$ data of small molecules.[63–65]

Table 1 Dipole Moments of Various Chemical Groups (from Dipole Moments of Aliphatic Organic Compounds in the Gas Phase).[a,b]

Group	*Dipole moment* $\hat{\mu} \times 10^{30}$ (C m)	*Group*	*Dipole moment* $\hat{\mu} \times 10^{30}$ (C m)
—HN—CO—NH—	15.3–17.0[c]	—Cl	6.7–7.2
$—SO_2—$	~15[c]	—F	6.3–6.8
—C≡N	13.1–13.7	—COOH	5.8
—N=C=O	~13	—COO—	5.7–5.9
—CO—NH—	12.3–12.7	—OH	5.5–5.7
$—NO_2$	11.7–12.5	—SH	~5.2
$—ONO_2$	~10.5	$—NH_2$	~4.2
—CO—	~9.8	—O—	3.8–4.3
—Br	6.8–7.3	—OCOO—	~3.7

[a] Due to conjugation, the same groups exhibit appreciably different and scattered values when linked to aromatic nuclei (higher values for groups exerting a −M effect, lower values in the opposite case). [b] Values deduced from Refs. 63–65. [c] Values in organic solutions ($\langle\hat{\mu}\rangle$).

At room temperature, values of ε_s for inorganic dielectrics are typically in the range 4–12, whereas the values for organic liquids range from around 2 for hydrocarbons to high values for highly dipolar and strongly hydrogen-bonded substances, *e.g.* 10–40 for alcohols and glycols, 78.5 for H_2O at 25 °C, 109 for formamide at 20 °C, *etc.* In practically all cases, the temperature coefficient of ε_s is negative, as expected from the Debye formula (22).

(ii) Polymers

Like the majority of organic compounds, common organic polymers and resins are good dielectrics,[66–69] with paraelectric, weakly semiconducting and diamagnetic properties. At room temperature, indeed: (i) σ values of pure products are in the range of 10^{-16} to $10^{-12}\,\Omega^{-1}\,m^{-1}$ (easily increased by several orders of magnitude by ionic impurities); (ii) ε_s values typically vary between 2.0 and 4.5, with higher values (up to around 8) being exhibited by strongly hydrogen-bonded products such as cellulose and its derivatives, poly(vinyl alcohol), urea–, melamine– and phenol–formaldehyde resins, *etc.*; (iii) $\chi_{m,s}$ values are in the -0.5 to -1.5×10^{-5} range found for most liquid and solid organic substances.[70] Exceptions to such general picture include: electrically conductive polymers, *i.e.* polymers (normally still paraelectric and diamagnetic) having electronic conductivity and characterized by values of σ from those normal for inorganic semiconductors (10^{-5} to $10^4\,\Omega^{-1}\,m^{-1}$) up to those for poor metallic conductors ($\sim 10^5\,\Omega^{-1}\,m^{-1}$);[71–73] ionic-conducting polymer systems (σ up to $1\,\Omega^{-1}\,m^{-1}$ and over);[74,75] ferroelectric polymers;[76] some (rather exotic) iron-containing and noniron-containing (ferro)magnetic polymers.[77]

Owing to the continuous motion and conformational changes of polymer chains, the vector sum of moments of dipolar groups brought by a polymer backbone (*i.e.* the overall dipole moment of a macromolecule) varies continually and rapidly. For this reason, the dipole moment of a polymer is given by the statistical average of all possible instantaneous dipole moment values, through which mean value, appropriately calculated, the Fröhlich–Kirkwood equation is again applicable.[66,78] Furthermore, it is worth noting that in polymers (in bulk and in solution) both inter- and intramolecular interactions of dipolar groups are involved, though intermolecular forces play (especially in solution) a comparatively minor role. In the simplest case of flexible-chain amorphous polymers, the Fröhlich–Kirkwood equation can be written as equation (28), where $(\hat{\boldsymbol{\mu}})$ is the dipole moment of the repeat unit in the condensed phase calculated from the dipole moments of chemical groups as previously mentioned and *via* equation (27), N is the number of repeat units per unit volume, and g is the Kirkwood factor expressing the intramolecular angular correlations among dipole vectors of repeat units as in equation (29).

$$\varepsilon_s - \varepsilon_\infty = \frac{\varepsilon_s(\varepsilon_\infty + 2)^2}{2\varepsilon_s + \varepsilon_\infty} N \frac{g(\hat{\mu})^2}{9\varepsilon_0 kT} \tag{28}$$

$$g = \frac{(\hat{\mu})^2 + \sum_k \sum_{k'} \langle(\hat{\boldsymbol{\mu}}_k)(\hat{\boldsymbol{\mu}}_{k'})\rangle}{(\hat{\mu})^2} \tag{29}$$

In equation (29), $\langle(\hat{\boldsymbol{\mu}}_k)(\hat{\boldsymbol{\mu}}_{k'})\rangle = (\hat{\mu})^2\langle\cos\vartheta_{kk'}\rangle$ is the mean square dipole moment resulting from the vector sum of dipoles of repeat units k and k', with $\vartheta_{kk'}$ the angle between the two dipole vectors, averaged over all molecules. The term $(\sum_k \sum_{k'}\langle(\hat{\boldsymbol{\mu}}_k)(\hat{\boldsymbol{\mu}}_{k'})\rangle)/(\hat{\mu})^2$ accounts for the reciprocal enhancement ($g > 1$) or cancellation ($g < 1$) of dipole moments of repeat units within the polymer chain, due to reciprocal conformational arrangements and specific configurational arrangements (tacticity) of the repeat units themselves imposing certain reciprocal orientations ($\vartheta_{kk'}$ values) of their dipole moments. A remarkable fact is that, in contrast to almost all other dielectrics, increasing temperatures generally lower the strength of the dominating intramolecular interactions, and thus cause ε_s of polar polymers to have a positive dependence on temperature through a markedly positive dependence of the g factor on the temperature itself.

(iii) Interfacial electric polarization

This is the important polarization mechanism generally referred to as space-charge polarization or as the Maxwell–Wagner or Maxwell–Wagner–Sillars (MWS) effect.[79–81] It occurs in heterogeneous dielectrics, consisting of a dielectric matrix and a dispersed phase with relative dielectric constants and d.c. conductivities $\varepsilon_{s,1}$, $\varepsilon_{s,2}$ and σ_1, σ_2, respectively, whenever $\varepsilon_{s,1}\sigma_2 \neq \varepsilon_{s,2}\sigma_1$. This phenomenon, consisting of an electric charge accumulation at the interface between phases, typically occurs in dielectrics (*e.g.* in all ordinary polymers) containing inclusions of conductors such as metals, graphite, or carbon black ($\varepsilon_{s,2} \gg \varepsilon_{s,1}$), water droplets ($\varepsilon_{s,2} > \varepsilon_{s,1}$), or gas bubbles or voids ($\sigma_2 \ll \sigma_1$). Water can also act in a different manner, by dissolving and dissociating ionic substances and thereby becoming a dispersed phase with appreciable (ionic) conductivity ($\sigma_2 \gg \sigma_1$). In dielectrics with conductive inclusions, the applied electric field causes electric charges to flow within the inclusions and opposite charges to accumulate on their opposite surfaces. The field thus converts the inclusions themselves into microscopic electric dipoles which contribute to the molecular polarization of the dielectric matrix.

(iv) Paramagnetism

Paramagnetism derives from atoms, molecules, or ions possessing a permanent magnetic moment associated with unpaired electron spins (atoms or ions of the majority of transition metals and radicals). Paramagnetic polarization consists of the alignment of such magnetic dipoles which have random orientations under normal conditions. Paramagnets[57,58] are those materials whose paramagnetism prevails over the diamagnetism general to all matter, *i.e.* the positive contribution of paramagnetism χ_{para} (paramagnetic susceptibility) to the overall $\chi_{m,s}$ ($=\chi_{dia}+\chi_{para}$) is such that $\chi_{para}>|\chi_{dia}|$. As a rule, the paramagnetic susceptibility smoothly decreases with temperature according to the well-known Curie–Weiss law. Values of $\chi_{m,s}$ at room temperature for most paramagnets are in the range 10^{-5} to 10^{-1} ($1<\mu_s<2$). The coupling with magnetic fields is thus very weak, as in the case of diamagnets.

19.3.1.3 Ferroelectrics and ferromagnets

Conceptually, ferroelectricity and ferromagnetism are two almost equivalent types of response to electric and magnetic fields, respectively. Ferroelectrics[82–86] and ferromagnets[57–59,87–89] are crystalline materials achieving strong, partially to almost entirely permanent electrization or magnetization under the influence of electric or magnetic fields, respectively, even of modest intensity. Such materials exhibit abnormally high values of $\chi_{e,s}$ and $\chi_{m,s}$ (and thus of ε_s and μ_s) which depend in a complicated manner on temperature and on electric or magnetic field strength, as well as on their previous electrization or magnetization history, according to their hysteretic behavior, expressed by well-known electric or magnetic polarization–depolarization–reverse polarization loops.

Polarization phenomena in ferroelectrics and ferromagnets are based on crystal domains of isooriented permanent atomic, ionic, or molecular dipoles (*i.e.* of spontaneously and permanently polarized material) called Weiss domains in ferromagnets, with linear dimensions of the order of microns, up to millimeters in ferroelectrics. Two polarization mechanisms may be involved: (a) enlargement of domains already aligned with the external field, by displacement of domain boundaries (or Bloch walls), in weak fields; (b) rotation of domain dipoles in the field direction, in strong fields. Due to such mechanisms, the polarization of ferroelectrics and ferromagnets shows marked saturation effects at certain field strength values, above which $\boldsymbol{P}$ or $\boldsymbol{M}$ cannot increase further.

Starting from the depolarized state, $\boldsymbol{D}$ or $\boldsymbol{B}$ increases with the applied field strength according to a sigmoidal curve, which is the locus of the $\boldsymbol{E}$–$\boldsymbol{D}$ or $\boldsymbol{H}$–$\boldsymbol{B}$ limits of hysteresis loops, up to saturation. Accordingly, ε_s or μ_s increases from an initial value for $\boldsymbol{E}$ or $\boldsymbol{H}\rightarrow 0$ ($\varepsilon_{s,i}$ and $\mu_{s,i}$, respectively), reaches a maximum, and then decreases again to level off at a constant value at saturation. In ferromagnets, whose μ_s normally decreases with temperature, the dependence of $\boldsymbol{B}$ and μ_s on the magnetic field strength at constant temperature is expressed, for weak fields, by the simple Rayleigh law (equations (30)).

$$\boldsymbol{B}=\mu_0(\mu_{s,i}\boldsymbol{H}+rH^2) \tag{30a}$$

$$\mu_s(\boldsymbol{H})=\mu_{s,i}+r\boldsymbol{H} \tag{30b}$$

Inorganic ferroelectrics display $\varepsilon_{s,i}$ values from 10^2 up to 10^4 and over, while $\mu_{s,i}$ of ferromagnets, commonly in the range 10^2 to 10^4, can reach values of the order of 10^5. Values of μ_s at saturation are of the order of 10^3 to 10^6. As the temperature is increased, at the critical Curie point T_C, variable from material to material, any ferroelectric or ferromagnet undergoes a sharp transition to paraelectric or paramagnetic behavior. Above T_C, ε_s or μ_s decays rapidly with temperature.

Whereas ferromagnetic behavior pertains almost exclusively to inorganic substances, ferroelectric properties are exhibited by such a rather common organic material as poly(vinylidene fluoride) (PVDF), a semicrystalline polymer whose ferroelectric characteristics have been studied since 1969.[76,90] At room temperature it shows an $\varepsilon_{s,i}$ value (in the range 10–20) 2.5–5-fold the values known for normal paraelectric polymers, and maximum ε_s values of 50–60; its T_C (above the melting point) is extrapolated to be ~200 °C. Copolymers of vinylidene fluoride with up to about 50% mol of tri- or tetrafluoroethylene are also ferroelectric, with T_C values ranging between room temperature and their melting points.

19.3.1.4 Antiferroelectrics and antiferromagnets

In these rather uncommon, crystalline materials, intrinsic ionic, molecular, or atomic dipoles are arranged in the crystal lattice in antiparallel directions. At absolute zero, dipoles are exactly compensated with one another, thus forming crystal domains (formed by interpenetrating sublattices with opposite dipole orientations) with overall zero electric or magnetic moment. Raising the temperature liberates more and more elementary dipoles from such self-balanced arrangements, causing domains to achieve a nonzero, and progressively stronger, dipole moment. Susceptibilities of antiferroelectrics[82–86] and antiferromagnets[57,59,87] are of the same order of magnitude of those of paraelectrics and paramagnets. In contrast to these, however, the susceptibilities of antiferroelectrics and antiferromagnets increase upon heating up to maximum values at a temperature, variable from material to material, known as the antiferroelectric or antiferromagnetic Curie point (or Néel point T_N). Above T_N, they attain paraelectric or paramagnetic properties, and the relevant values of $\chi_{e,s}$ or $\chi_{m,s}$ decrease again with temperature. Antiferromagnetism is exhibited by Cr up to 35 °C, by α-Mn, and by some rare-earth metals, as well as by several compounds of Fe, Co, Ni, Cr, Mn, and other transition elements, up to widely variable temperatures (*e.g.* up to 34 and ~680 °C) for Cr_2O_3 and α-Fe_2O_3, respectively). Their χ_{para} values at $T > T_N$ decrease with temperature according to a Curie–Weiss-type equation.

19.3.1.5 Ferrimagnets

In these crystalline materials, magnetic dipoles are arranged in antiparallel directions as in antiferromagnets, though always with only partial reciprocal compensation, and thus with a nonzero vector sum of moments. Both at the microscopic and the macroscopic level, ferrimagnetism[57–59] is a magnetic behavior very similar to ferromagnetism. Ferrimagnets, indeed, are characterized by a Weiss domain microstructure, magnetization mechanisms and hysteresis loops, high values of susceptibility and permeability ($\mu_{s,i}$ of the order of 10^2 to 10^4), and a Curie transition to paramagnetic materials, as in ferromagnets. Major differences are: (i) ferrimagnets are nonconductors or semiconductors (with respect to ferromagnetic materials, which are, with a few exceptions such as CrO_2, metals or metal alloys); (ii) above T_C, $\chi_{m,s}$ decays with increasing temperature more sharply than according to the Curie–Weiss law, but approaches asymptotically this behavior as the temperature becomes higher and higher.

Ferrimagnets[87,88,91,92] are usually classified as: (a) spinels, *i.e.* γ-Fe_2O_3 and compounds of general formula $MO \cdot Fe_2O_3$, with M a divalent cation such as Fe^{2+} (magnetite), Ni^{2+}, Co^{2+}, Zn^{2+}, Mg^{2+}, *etc.*; (b) magnetoplumbites $MO \cdot 6Fe_2O_3$, where M is a divalent ion such as Ba^{2+}, Sr^{2+}, Pb^{2+}; (c) magnetic garnets $3M_2O_3 \cdot 5Fe_2O_3$, with M a trivalent rare-earth ion. Very fine grinding of ferri- and ferromagnets fractionates the Weiss domains and cancels the ferro- or ferrimagnetic properties, thus yielding peculiar paramagnetic powders (useless, *e.g.* for magnetic information storage[93]) which have outstandingly high values of susceptibility (superparamagnets).[57,88,93]

19.3.2 Energy Loss Phenomena in Alternating Electromagnetic Fields at Radio and Microwave Frequencies

Electric and magnetic polarization of a material under a steady electric or magnetic field (and the associated energy supply to, and energy storage into, the material itself) can ideally be treated as reversible processes, to which the static (or equilibrium) susceptibilities, relative dielectric constant, and permeability of the considered material refer. Obviously, reversibility considerations apply also to depolarization processes, implying field energy return upon removal of the external field.

Whereas it is possible to speak in terms of reversible processes for paraelectrics and paramagnets, polarization and depolarization phenomena in ferroelectrics and ferro-/ferrimagnets are characterized by a high degree of irreversibility. They proceed in fact in low fields as successions of microscopic discrete $\boldsymbol{D}$ or $\boldsymbol{B}$ variations with $\boldsymbol{E}$ or $\boldsymbol{H}$ (the Berkhausen effect in ferromagnets), linked to discrete (nonequilibrium) electric or magnetic domain growth steps. The mechanism of domain dipole rotation, prevailing in high fields, progressively accentuates the reversibility of polarization and depolarization processes in these materials as the applied field strength increases towards their saturation values. Other inherently irreversible processes are the electric conduction phenomena, with their energy losses by the Joule effect, under both steady and variable fields.

Conversely, all, either transient or cyclic, electric and magnetic polarization and depolarization processes with finite rates of variation of the external field strength are partially irreversible. An important point is that all electric and magnetic polarization and depolarization phenomena, according to the variety of mechanisms summarized in Section 19.3.1, are hindered and slowed by internal frictions on the atomic, molecular, or supramolecular scale, conceptually equivalent to the internal viscous effects in mechanical deformations of materials (as, in particular, in polymeric viscoelastic solids and liquids).

In practice, materials always take time and consume energy (internally dissipated as heat) to overcome internal constraints opposing electric or magnetic polarization and depolarization, analogous to any mechanical strain and strain release. That is, they can comply with external force fields (electric, magnetic, or mechanical) only through those processes known as electric, magnetic, and mechanical relaxations. In other words, electric and magnetic polarization (like mechanical strains) correspond to a reversible (elastic or reactive) field energy storage into a material, whereas electric and magnetic (or mechanical) relaxation processes correspond to the irreversible (viscous or resistive) field energy losses occurring to reach any state itself of polarization (or mechanical strain).

19.3.2.1 Paraelectric (dielectric) and paramagnetic relaxation

The interactions of permanent electric and magnetic dipoles with each other and with the surrounding matter determine and govern electric and magnetic relaxation phenomena in paraelectrics (paraelectric relaxation, normal for most dielectrics and therefore commonly called dielectric relaxation) and in paramagnets (paramagnetic relaxation), respectively. Obviously, such relaxation processes pertain also to ferroelectric and ferro- and ferrimagnetic materials above their Curie temperature, as well as to antiferroelectrics and antiferromagnets beyond their Néel point. Phenomenological and molecular aspects of dielectric relaxation are surveyed in the vast literature on dielectrics,[5–7,54–56] including polymers,[66–68] mentioned above, and treated in specific monographs for dielectrics in general,[94] liquids,[62] solids,[95] and polymers.[78,96–99] The book by McCrum *et al.*,[96] in particular, deals with the parallelism of dielectric and mechanical relaxations in polymers. Dissertations on paramagnetic relaxation can be found in the general literature on magnetism and in encyclopedias.[57,59,100]

Relaxation phenomena retard the alternating (cyclically reversing) electric or magnetic polarization caused in materials by an alternating electric or magnetic (or an electromagnetic) field. At any instant, the value of $\boldsymbol{P}$ or $\boldsymbol{M}$, and hence of $\boldsymbol{D}$ or $\boldsymbol{B}$, differs from that expected from equations (1)–(4) for any instantaneous $\boldsymbol{E}$ or $\boldsymbol{H}$ value of the applied alternating field. The $\boldsymbol{D}$ or $\boldsymbol{B}$ oscillation lags behind that of $\boldsymbol{E}$ or $\boldsymbol{H}$, with a frequency-dependence governed by an appropriate relaxation function $\Phi(t)$. In the simplest, almost ideal cases this function is expressed, or can be approximated, by equation (31), with $\tau\,[=\tau(T)]$ the relaxation time proper to the relaxation process considered at a given temperature.

$$\Phi(t) = \exp\left(-\frac{t}{\tau}\right) \tag{31}$$

In sinusoidally alternating fields of frequency ν and angular frequency ω ($\omega = 2\pi\nu$), $\boldsymbol{E}$ and $\boldsymbol{H}$ vary with time according to equations (32) and (33), where $\boldsymbol{E}_0$ and $\boldsymbol{H}_0$ are the relevant oscillation amplitudes.

$$\boldsymbol{E}(t) = \boldsymbol{E}_0 \cos \omega t = \boldsymbol{E}_0 \exp(j\omega t) \tag{32}$$

$$\boldsymbol{H}(t) = \boldsymbol{H}_0 \cos \omega t = \boldsymbol{H}_0 \exp(j\omega t) \tag{33}$$

The corresponding electric displacement and magnetic flux density in the material subjected to the alternating field follow equations (34) and (35), with δ [$\delta = \delta(\omega, T)$] the phase lag in $\boldsymbol{D}$

and $\boldsymbol{B}$ with respect to $\boldsymbol{E}$ and $\boldsymbol{H}$ due to the relaxation phenomenon, and $\boldsymbol{D}_0$ and $\boldsymbol{B}_0$ the amplitudes of $\boldsymbol{D}$ and $\boldsymbol{B}$ oscillations.

$$\begin{aligned}\boldsymbol{D}(t) &= \boldsymbol{D}_0 \cos(\omega t - \delta_e) \\ &= \boldsymbol{D}_0 \cos \omega t \cos \delta_e + \boldsymbol{D}_0 \sin \omega t \sin \delta_e \\ &= \boldsymbol{D}_0 \exp[j(\omega t - \delta_e)]\end{aligned} \tag{34}$$

$$\begin{aligned}\boldsymbol{B}(t) &= \boldsymbol{B}_0 \cos(\omega t - \delta_m) \\ &= \boldsymbol{B}_0 \cos \omega t \cos \delta_m + \boldsymbol{B}_0 \sin \omega t \sin \delta_m \\ &= \boldsymbol{B}_0 \exp[j(\omega t - \delta_m)]\end{aligned} \tag{35}$$

(i) Dielectric relaxation

Using the expressions (36a–c), equation (34) can be written as (36d), with $\varepsilon'(\omega)$, $\varepsilon''(\omega)$, and $\varepsilon^*(\omega)$ being the (relative) dielectric constant, the dielectric loss factor, and the (relative) complex dielectric constant, respectively, of the material considered, at the frequency ω at constant temperature.

$$\boldsymbol{D}_0 \cos \delta_e(\omega) = \varepsilon_0 \varepsilon'(\omega) \boldsymbol{E}_0 \tag{36a}$$

$$\boldsymbol{D}_0 \sin \delta_e(\omega) = \varepsilon_0 \varepsilon''(\omega) \boldsymbol{E}_0 \tag{36b}$$

$$\frac{\boldsymbol{D}(t)}{\varepsilon_0 \boldsymbol{E}(t)} = \varepsilon'(\omega) - j\varepsilon''(\omega) = \varepsilon^*(\omega) \tag{36c}$$

$$\begin{aligned}\boldsymbol{D}(t) &= \varepsilon_0 \varepsilon'(\omega) \boldsymbol{E}_0 \cos \omega t + \varepsilon_0 \varepsilon''(\omega) \boldsymbol{E}_0 \sin \omega t \\ &= \varepsilon_0 \varepsilon^*(\omega) \boldsymbol{E}_0 \exp(j\omega t)\end{aligned} \tag{36d}$$

For each material, as indicated above for δ, ε' and ε'' are also functions of temperature, although this dependence is hereinafter generally considered implicit and is omitted for simplicity. Of the two additive terms in equation (36d), the former refers to the reversible (in-phase) component of polarization, and the latter to the dissipative (out-of-phase) one. The ratio $\varepsilon''(\omega)/\varepsilon'(\omega) = \tan \delta_e(\omega)$ is the well-known dielectric loss tangent or dissipation factor, *i.e.* the ratio between the energy dissipated (absorbed from the field) and that reversibly stored and released per cycle of alternating polarization at the frequency ω. The lowest frequencies of IR radiation (10^{12} Hz) represent the upper limit for the alternating orientational polarization phenomena and the associated relaxation processes. Above such frequencies, indeed, only the much faster distortion polarization–depolarization can follow the rapid, cyclic variations of the external field strength. In the frequency range from zero (steady fields) to such a limit, ε' and ε'' vary with increasing frequency in different ways from material to material, and also as functions of temperature. ε' always decreases monotonically with ω from the highest value ε_s $[=\varepsilon'(0)]$ to level off at the value ε_∞ (dielectric dispersion), whereas $\varepsilon'' \to 0$ as $\omega \to 0$, as for $\omega \to$ far-IR frequencies. $\varepsilon'(\omega)$ and $\varepsilon''(\omega)$ are related to ε_s, ε_∞, and $\Phi(t)$ through equation (37), whose integration for $\Phi(t)$ according to equation (31) leads to the Debye formulas (38).

$$\frac{\varepsilon^*(\omega) - \varepsilon_\infty}{\varepsilon_s - \varepsilon_\infty} = 1 - j\omega \int_0^\infty \Phi(t) \exp(-j\omega t)\, dt \tag{37}$$

$$\varepsilon^*(\omega) = \frac{\varepsilon_s - \varepsilon_\infty}{1 + j\omega\tau} + \varepsilon_\infty \tag{38a}$$

$$\varepsilon'(\omega) = \frac{\varepsilon_s - \varepsilon_\infty}{1 + \omega^2\tau^2} + \varepsilon_\infty \tag{38b}$$

$$\varepsilon''(\omega) = (\varepsilon_s - \varepsilon_\infty) \frac{\omega\tau}{1 + \omega^2\tau^2} \tag{38c}$$

Figure 1 exemplifies the dependence of ε' and ε'' on $\ln \omega$ according to such a model, predicting a symmetrical dielectric loss peak with a maximum ε'' (the highest energy dissipation capacity of a material) given by $(\varepsilon_s - \varepsilon_\infty)/2$, at the frequency $\omega_{max} = 1/\tau$.

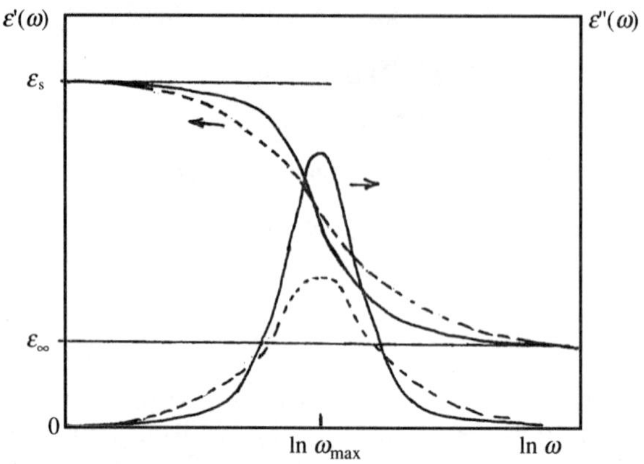

Figure 1 (Relative) dielectric constant and dielectric loss factor of a dielectric substance vs. frequency: for an ideal Debye dielectric (——); for a real dielectric (- - - -).

As shown by equation (38c), the energy dissipation capacity of a material at any frequency is proportional to the quantity $(\varepsilon_s - \varepsilon_\infty)$, related to the integral of the overall relaxation curve ε'' vs. $\ln \omega$ as in equation (39), and known as the dielectric relaxation strength of the material.

$$\varepsilon_s - \varepsilon_\infty = \frac{2}{\pi}\int_{-\infty}^{+\infty} \varepsilon''(\omega)\, \mathrm{d} \ln \omega \tag{39}$$

In accordance with equations (23), (25), and (28), $(\varepsilon_s - \varepsilon_\infty)$ is linked to the molecular structure (in particular, it is proportional to the square of the molecular dipole moment, or to the square of the dipole moment and the number of repeat units per unit volume in polymers), and to inter- and intramolecular interactions through the appropriate g factor, as well as (in a complex way) to temperature (with a direct proportionality to T^{-1} and through the dependence of both density and the g factor on the temperature itself). Yet, according to equation (38c), temperature influences $\varepsilon''(\omega)$ through the temperature dependence of the relaxation time as in equation (40a), where h is the Planck constant, and ΔG^*, ΔS^*, and ΔH^* are the molar free enthalpy, molar entropy, and molar enthalpy of activation, respectively, for the specific relaxation phenomenon.

$$\tau = \frac{h}{kT}\exp\left(\frac{\Delta G^*}{RT}\right) = \frac{h}{kT}\exp\left(-\frac{\Delta S^*}{R}\right)\exp\left(\frac{\Delta H^*}{RT}\right) \tag{40a}$$

In other words, heating makes relaxation phenomena easier and faster, with a progressive decrease of τ with temperature dominated by a proportionality to the function $\exp(T^{-1})$. Over not too broad temperature intervals, the preexponential and the entropic factors can be combined into a practically temperature-independent term τ_0. Equation (40a) can thus assume the approximate Arrhenius form (40b), where τ_0 has the meaning of a relaxation time for $T \rightarrow \infty$.

$$\tau \approx \tau_0 \exp\left(\frac{\Delta H^*}{RT}\right) \tag{40b}$$

Equations (40) account for the regular, marked shifts of relaxation peaks (*i.e.* of energy absorption bands) of dipolar compounds to higher frequencies with increasing temperatures. In particular, equation (40b) describes, through a linear decrease of $\ln \tau$, the common linear increase of $\ln \omega_{max}$ (or $\ln \nu_{max}$) with decreasing T^{-1}.

At room temperature, low-viscosity dipolar liquids typically show dielectric loss maxima at MW frequencies, with relaxation times of the order of 10^{-11} to 10^{-10} s (for example,

$\nu_{max} \simeq 4$ GHz and $\tau \simeq 3.5 \times 10^{-11}$ s for 1-bromooctane at 25 °C, $\nu_{max} \simeq 15$ GHz and $\tau \simeq 1 \times 10^{-11}$ s for H_2O at 25 °C, *etc.*). At lower temperatures, τ can be considerably prolonged, according to equations (40), with relaxation peaks being shifted down to radio and even audio frequencies (*e.g.* $\nu_{max} \simeq 3$ kHz and $\tau \simeq 6 \times 10^{-5}$ s for glycerol at −50 °C, $\nu_{max} \simeq 0.3$ kHz and $\tau \simeq 5 \times 10^{-4}$ s for *n*-propanol at −140 °C, *etc.*). The Debye equations (38c) and (38d) can be combined to obtain equation (41), correlating, through a semicircular curve of radius $(\varepsilon_s - \varepsilon_\infty)/2$ in an Argand diagram ε'' vs. ε' as in Figure 2, the (relative) dielectric constant and the dielectric loss factor of a substance at any frequency at constant temperature. Values of $\varepsilon'(\omega)$ and $\varepsilon''(\omega)$ of most low-viscosity dipolar liquids fit this equation well, demonstrating a good agreement with the Debye equations (38).

$$\left[\varepsilon'(\omega) - \frac{\varepsilon_s + \varepsilon_\infty}{2}\right]^2 + [\varepsilon''(\omega)]^2 = \left(\frac{\varepsilon_s - \varepsilon_\infty}{2}\right)^2 \tag{41}$$

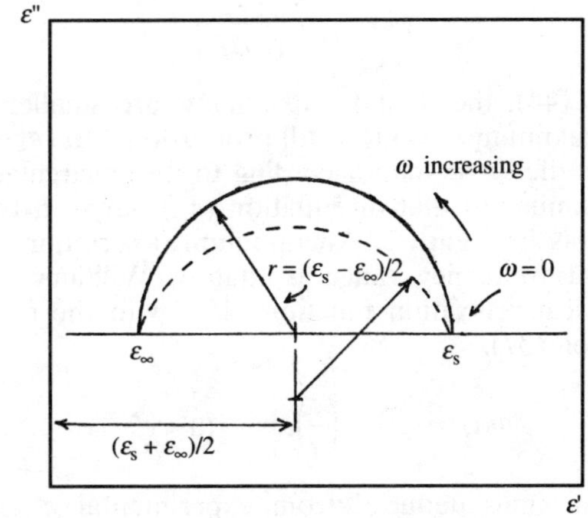

Figure 2 Argand plot of the dielectric loss factor vs. the (relative) dielectric constant for a dielectric substance: according to the Debye model (——); for a real dielectric (– – –).

The Debye single-relaxation time model is usually inadequate to describe dielectric dispersion curves and loss peaks (as exemplified in Figure 1, both much broader and flatter then expected from equations (38)) of viscous liquids and amorphous solids in general, polymers, polymer solutions, and most oligomeric substances. In these cases, where inter- and/or intramolecular interactions play an important role, models considering multiplicities of associated relaxation microphenomena are much more preferable. Therefore, complex relaxation functions based on discrete or continuous, multiple-relaxation time distributions according to equations (42a) and (42b), respectively, are suitable, with f_i the (relative) probability of relaxation events characterized by a relaxation time value τ_i in histographic τ distributions, and $\varphi(\tau)$ the (normalized) function defining each continuous τ distribution.

$$\Phi(t) = \sum_i f_i \exp\left(-\frac{t}{\tau_i}\right), \quad \text{with} \quad \sum_i f_i = 1 \tag{42a}$$

$$\Phi(t) = \int_0^\infty \varphi(\tau)\exp\left(-\frac{t}{\tau}\right)\mathrm{d}\tau, \quad \text{with} \quad \int_0^\infty \varphi(\tau)\,\mathrm{d}\tau = 1 \tag{42b}$$

Introduction of equations (42) into (37) leads to multiple-relaxation time expressions for $\varepsilon^*(\omega)$ (equations (43)).

$$\varepsilon^*(\omega) = \left[(\varepsilon_s - \varepsilon_\infty)\sum_i \frac{f_i}{1 + j\omega\tau_i}\right] + \varepsilon_\infty \tag{43a}$$

$$\varepsilon^*(\omega) = \left[(\varepsilon_s - \varepsilon_\infty)\int_0^\infty \frac{\varphi(\tau)}{1 + j\omega\tau}\mathrm{d}\tau\right] + \varepsilon_\infty \tag{43b}$$

Unfortunately, both discrete and continuous τ distributions cannot be known *a priori.* To overcome this lack of knowledge, semiempirical equations can be derived from the Debye equations (38), still based on the single-relaxation time approach, but allowing a better description of dielectric properties of viscous liquids, polymers, and oligomeric products over wide frequency ranges.[78] Equations worth mentioning are the Cole–Cole (44a), Davidson–Cole (44b), and Havriliak–Negami (44c) equations, which are generalizations of the Debye equation (38a), where w and x are empirically adjustable parameters, positive and ≤ 1, and $\bar{\tau}$, simply calculated from the experimental loss peak as $\bar{\tau} = 1/\omega_{\max}$, is the most probable relaxation time over the entire relaxation process considered.

$$\varepsilon^*(\omega) = \frac{\varepsilon_s - \varepsilon_\infty}{1 + (j\omega\bar{\tau})^w} + \varepsilon_\infty \tag{44a}$$

$$\varepsilon^*(\omega) = \frac{\varepsilon_s - \varepsilon_\infty}{(1 + j\omega\bar{\tau})^x} + \varepsilon_\infty \tag{44b}$$

$$\varepsilon^*(\omega) = \frac{\varepsilon_s - \varepsilon_\infty}{[1 + (j\omega\bar{\tau})^w]^x} + \varepsilon_\infty \tag{44c}$$

According to equations (44), the w and x exponents are smaller when the loss peaks are broader and flatter, and maximum ε'' values, still proportional to $(\varepsilon_s - \varepsilon_\infty)$, are lower. Equation (44a) fits the experimental dielectric data according to a semicircular ε'' vs. ε' curve (the well known Cole–Cole plot), similar to that of equation (41) but with the center located beneath that of the Debye curve, as in Figure 2. An optimum description of dielectric properties of polymers in a very broad frequency range is that of Williams and Watts, based on the introduction of the empirical relaxation function (45) (with the exponent y normally in the range 0.4–0.6) into equation (37).

$$\Phi(t) = \exp\left[-\left(\frac{t}{\bar{\tau}}\right)^y\right] \qquad (0 < y \leq 1) \tag{45}$$

As expected, relaxation times deduced from experimental $\varepsilon''(\omega)$ data for viscous, low-molecular weight liquids, and for oligomers, polymers, and polymer solutions are appreciably longer at comparable temperatures than those for low-viscosity liquids, because of the stronger frictional interactions of the molecules. Dielectric loss maxima in the RF domain, with τ values of the order of 10^{-6} s, are typical, for example, of glassy, polar high polymers and are at variance with what is advisable for their efficient MW heating. In general, however, heating at temperatures above the glass transition temperature T_g, or plasticization with solvents or compatible low-T_g polymers or oligomers, causes τ to decrease to such values that relaxation peaks can move into the MW region, greatly enhancing the MW heatability of the material. In reality, the dielectric behavior of polymers and polymer solutions is much more complex than hereinabove considered. A plurality of polarization and relaxation modes, to which markedly different τ values or τ ranges are normal, is typically displayed over the entire zero-to-far-IR frequency range, causing a plurality of relaxation peaks normally characterized by a different temperature dependence, as described in Section 19.3.2.2.

(ii) Paramagnetic relaxation

Energy loss phenomena associated with paramagnetic relaxation are treated similarly to those of dielectric relaxation. Equations (46) equivalent to (36) can be written, with $\mu'(\omega)$, $\mu''(\omega)$, and $\mu^*(\omega)$ the (relative) permeability, the magnetic loss factor, and the (relative) complex permeability, respectively, of a material at the frequency ω, and $\tan\delta_m(\omega) = \mu''(\omega)/\mu'(\omega)$.

$$\boldsymbol{B}_0 \cos\delta_m(\omega) = \mu_0\mu'(\omega)\boldsymbol{H}_0 \tag{46a}$$

$$\boldsymbol{B}_0 \sin\delta_m(\omega) = \mu_0\mu''(\omega)\boldsymbol{H}_0 \tag{46b}$$

$$\frac{\boldsymbol{B}(t)}{\mu_0\boldsymbol{H}(t)} = \mu'(\omega) - j\mu''(\omega) = \mu^*(\omega) \tag{46c}$$

$$\begin{aligned}\boldsymbol{B}(t) &= \mu_0\mu'(\omega)\boldsymbol{H}_0\cos\omega t + \mu_0\mu''(\omega)\boldsymbol{H}_0\sin\omega t \\ &= \mu_0\mu^*(\omega)\boldsymbol{H}_0\exp(j\omega t)\end{aligned} \tag{46d}$$

The complex magnetic susceptibility $\chi^*_m(\omega)$ as in equation (47), with its in-phase and out-of-phase components $\chi'_m(\omega)=\mu'(\omega)-1$ and $\chi''_m(\omega)=\mu''(\omega)$, respectively, is commonly used to define the response of paramagnets to alternating fields.

$$\chi^*_m(\omega)=\chi'_m(\omega)-j\chi''_m(\omega) \tag{47}$$

Expressions for $\mu^*(\omega)$, $\mu'(\omega)$, and $\mu''(\omega)$ equivalent to the Debye equations (38) are applicable to monocrystalline substances, and are often reasonable for polycrystalline materials such as metals. Paramagnetic relaxation peaks can range from radio to MW frequencies, depending on the material considered and the temperature, increasing temperatures shifting them to higher frequencies. Unfortunately, data of χ''_m for paramagnetic substances at various frequencies and temperatures are not easy to find in the literature. A general consideration, however, is that, due to the very low values of χ'_m for ordinary paramagnets (see Section 19.3.1.2(iv)), even appreciable $\tan\delta_m$ values imply relatively low χ''_m values which are, in most cases, negligible from the standpoint of magnetic field energy losses. The superparamagnetic powders cited in Section 19.3.5.1 may be exceptions.

19.3.2.2 Dielectric properties of polymeric substances at radio and microwave frequencies

The RF heating technology and the techniques of conventional dielectric spectroscopy using impedance bridges and analyzers[101] up to mid-radio frequencies (10^6 to 10^7 Hz) have a relatively long history and well established uses. For such reasons, the dielectric properties of matter as a function of both temperature and frequency in the 10 to 10^5 Hz range of audio frequencies have been studied for a long time, and extensive literature is available and reviewed (or simply summarized) in several publications for a great variety of dielectrics,[5–7,16–18,20,56,61,94,95] including polymers.[66–69,78,96–99] By contrast, dielectric properties of materials at high radio frequencies and at MW frequencies (*i.e.* in the range 10^6 to 10^{10} Hz and over) must be measured by delicate and comparatively less accessible techniques. With the exception of high-frequency impedance analyzers covering the 1 MHz to 1 GHz interval, appropriate techniques generally need the use of noncommercial instrumentations or instrumentation assemblies. Techniques include:

(a) Resonant cavity techniques, for measurements at fixed resonant frequencies, including the re-entrant cavity method (usually for the range 10^8 to 10^9 Hz) and various resonant MW cavity perturbation methods (CPM), typically applied up to 10 GHz.

(b) MW reflection, transmission, and reflection/transmission methods, for measurements over broad MW bands and often simply referred to as broadband techniques, *i.e.* (i) coaxial line methods, for frequencies in the 100 MHz to 12 GHz interval; (ii) waveguide methods (typically applied in the range 3–60 GHz), including time domain spectroscopy (TDS) and, in particular, time domain reflectometry (TDR), conveniently used for frequencies from 10^7 to 10^{10} Hz; (iii) free space methods, suitable over frequency ranges extending to the 30–300 GHz band of extremely high frequencies (EHF).

Basic aspects of such methods, most of which are also suitable for measurements of complex magnetic permeability, are treated in the vast literature on the theory and practice of MW physics[102–104] and MW measurements.[105–109] Surveys of methods can be found in publications on MW processing,[25] dielectrics in general,[55,56,60,61,63] and dielectric properties of polymers.[66,78,98] The above MW measuring techniques are not described in this chapter, but literature references regarding their applications to polymeric materials are worth mentioning here.

Uses of CPM[110,111] for measurements of the dielectric constant of polymeric products at fixed MW frequencies were reported many years ago for films and fibers,[112] and later for rubber compounds.[113–115] Applications of CPM for the monitoring of epoxy/amine curing processes have also been described.[116,117] A fast and precise modified cavity perturbation method was developed some years ago,[118] and used to study the kinetics of radical polymerizations.[119–121] The TDR technique[122,123] has been applied to monitor polymerization processes such as amine-curing of epoxy resins[124–126] and *n*-butyl acrylate polymerization.[126,127] Waveguide methods[128] have been proposed for measuring complex dielectric constant and conductivity data of thin specimens,[129] including polymer films.[130] A simple method to obtain dielectric constant data of polymers at GHz frequencies is based on the measurement of the wavelength on the surface of cylindrical polymer rods connected to a MW circuit.[131] Among the numerous applications of coaxial line methods are those which use open-ended coaxial probes for studying thermosetting resins.[132,133] A newer broadband measuring technique recently developed is coherent MW transient spectroscopy (COMITS).[134] It belongs to the category of free space methods, and is suitable for a range of high MW frequencies as wide as from 14 to 140 GHz. Its application for studying the dielectric properties of a variety of polymers has been reported.[135]

In spite of the advances of the last two decades in measuring techniques and instrumentation, the body of data on dielectric properties of materials in general and polymers in particular at frequencies above 100 MHz is still far from extensive. Values of the complex dielectric constant of materials at MW frequencies as a function of either frequency or temperature are given in an impressive variety of publications, including several books on dielectrics[6,7,56] and industrial MW heating technology.[20,21,23,25] A collection of data for a broad range of substances (including many types of natural and synthetic rubbers) at frequencies from 10^4 to 10^{10} Hz was compiled more than 20 years ago.[136] A systematic study of the complex dielectric constant of a series of polar high polymers as a function of temperature at the fixed MW frequency of 8.6 GHz dates back to 30 years ago.[137] Investigations on dielectric properties of polar and nonpolar polymers at MW frequencies have multiplied since then and were reviewed[138] in 1985, while continuing with undiminished interest up to the present time.

(i) *Polarity and dielectric relaxation strength of polymers*

From the standpoint of their arrangement with respect to the polymer chain, two major types of dipolar groups can be identified in polar polymers.[66]

(a) dipolar groups forming part of a side grouping possessing rotational mobility with respect to the polymer backbone. Familiar examples are the ester groupings of poly(meth)acrylates and poly(vinyl ester)s, the ether groupings of poly(vinyl ether)s, and the —CH_2Cl unit in polyepichlorohydrin.

(b) Dipolar groups rigidly joined to, or forming part of, the polymer backbone. Examples of the former are innumerable: —Cl in poly(vinyl chloride) (PVC), poly(vinylidene chloride) (PVDC), polychloroprene [1,4-poly(2-chloro-1,3-butadiene)] and poly(4-chlorostyrene), —F in PVDF, —CN in polyacrylonitrile (PAN), *etc.*; examples of the latter are —O— and —S— in polyethers and polythioethers, respectively, —COO— and —OCOO— in polyesters and polycarbonates, —CONH—, —HN—COO— and —HN—CO—NH— in polyamides, polyurethanes, and polyureas, —SO_2— and —CO— in aliphatic polysulfones and polyketones (alkene–sulfur dioxide and alkene–carbon monoxide copolymers), as well as in poly(arylene ether sulfones) and poly(arylene ether ketone)s, and so on. The dipole moment may be either perpendicular to the backbone (as in PVC, PVDC, PVDF, PAN, polyoxymethylene (POM), poly(ethylene oxide), *etc.*), or directed transversely to the backbone but forming an angle different from 90° with it (in most polymers).

The different characteristics of dipolar groups in polymers imply different types of dependence of ε_s and $(\varepsilon_s - \varepsilon_\infty)$ on temperature, as well as the plurality of distinct dipolar relaxation phenomena in alternating fields mentioned in Section 19.3.2.1(i).[66,68,99] In flexible-chain amorphous polymers with dipoles originating solely from side groupings (dipolar groups of type (a)) with high rotational mobility (*e.g.* poly(allyl ester)s and polyacrylates), the dipoles themselves are free to orientate with electric fields irrespective of the backbone mobility, similarly to low-molecular weight substances. Analogously to them, and based on the concepts expressed in Section 19.3.1.2(i), ε_∞ is nearly constant with temperature (see, for example, equation (24)), whereas their ε_s and $(\varepsilon_s - \varepsilon_\infty)$ values decrease monotonically with temperature (for regular dielectrics according to equations (22) and (23), or to the simple equation (48) from Debye and Fröhlich[56]) both below and above T_g as sketched in Figure 3(a), due to the effect of increasing thermal agitation.

$$\left(\frac{\partial \varepsilon_s}{\partial T}\right)_V = -\frac{1}{T}(\varepsilon_s - \varepsilon_\infty) \tag{48}$$

In amorphous polymers with dipolar groups of type (b), the backbone mobility governs the possibilities of dipoles aligning with an external electric field. This implies that, in the glassy state, orientational polarization is strongly hindered. Their ε_s, $\varepsilon'(\omega)$, and $(\varepsilon_s - \varepsilon_\infty)$ values are thus lower than expected on the basis of the strength and concentration of dipolar groups, as discussed in Section 19.3.1.2(ii) (values of the Kirkwood g factor much less than one in equation (28)). Through an increasing mobility of dipolar groups and a decreasing strength of intra- and intermolecular interactions, however, the orientational polarizability and therefore ε_s, $\varepsilon'(\omega)$, and $(\varepsilon_s - \varepsilon_\infty)$ values of such polymers tend to increase strongly with temperature (rapidly increasing g values) as T_g is approached. At T_g the polymer chains achieve a high flexibility, with high degrees of rotational and translational freedom being achieved by the individual chain segments (activation of their micro-Brownian motion regime), and with correspondingly high levels of orientational mobility being imparted to dipoles rigidly attached to them. Above T_g, increasing temperatures cause micro-Brownian motions of chain segments to increase in rapidity and amplitude. Consequently, progressively faster and larger fluctuations in the directions of dipoles oppose more and more strongly, and then overcome, the orientational effects of electric fields, thus causing ε_s, $\varepsilon'(\omega)$, and $(\varepsilon_s - \varepsilon_\infty)$ to decrease with temperature as in regular dielectrics mentioned above (Figure 3(b)).

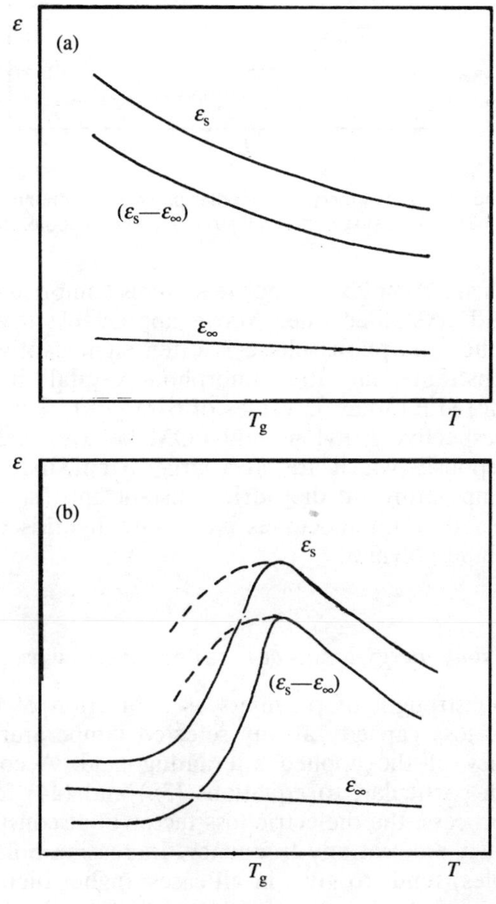

Figure 3 Variations of ε_s, ε_∞, and $(\varepsilon_s - \varepsilon_\infty)$ of amorphous polar polymers with temperature: (a) polymers with dipoles originating from side groups with good rotational mobility; (b) polymers with dipoles rigidly attached to the backbone.

Most polymers and thermosets (crosslinked thermosetting resins) conform more or less precisely to this latter picture. Behavior intermediate between those typical for polymers having dipolar groups of types (a) and (b) are often displayed (see, for example, the dotted curves of Figure 3(b)) by polymer blends, polymers with dipolar groups of both types, and polymers bearing dipolar side-groups with modest (inherently small or sterically hindered) rotational mobility (*e.g.* poly(vinyl ester)s, poly(methacrylate)s and poly(ethacrylate)s).

As exemplified in Figure 4, crystalline polymers may show dielectric behavior analogous to those discussed above for amorphous materials. Most frequently, the crystal structure partially locks the dipoles and hinders their free orientation in applied electric fields, thus diminishing their ε_s, $\varepsilon'(\omega)$, and $(\varepsilon_s - \varepsilon_\infty)$ values. Heating progressively unlocks the dipoles, making positive the temperature coefficients of ε_s, $\varepsilon'(\omega)$, and $(\varepsilon_s - \varepsilon_\infty)$ up to the melting point T_m, as depicted in Figure 4 (curve c). Above T_m, the disordering effect of thermal agitation prevails over the orientational freedom achieved by the dipoles, causing again the well-known monotonical decrease of ε_s, $\varepsilon'(\omega)$, and $(\varepsilon_s - \varepsilon_\infty)$ with temperature.

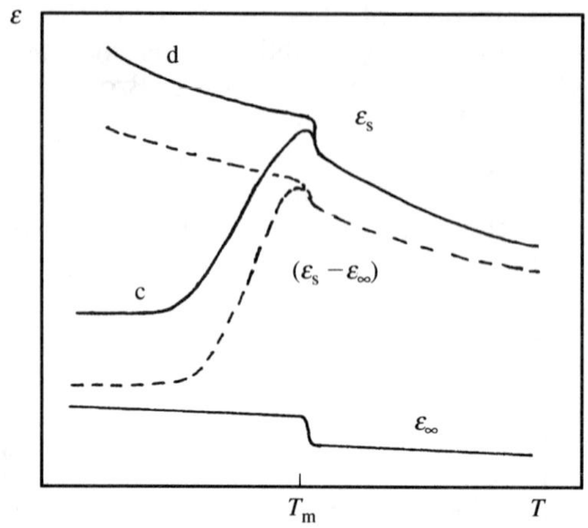

Figure 4 Variations of ε_s, ε_∞, and $(\varepsilon_s - \varepsilon_\infty)$ of a crystalline polar polymer with temperature. Curve c: crystal structure exerting partial locking of dipoles. Curve d: crystal structure where dipoles have rotational mobility.

Semicrystalline polymers are complex composite systems comprising an amorphous (glassy or rubbery) phase and a solid crystalline one. Above appreciable dgerees of crystallinity (say, >30%), the behavior of the amorphous phase is often significantly perturbed by crystallites acting as relaxation constraints at the amorphous–crystal boundaries. For example, poly(ethylene terephthalate) (PET) has T_g values of 67 and 81 °C in wholly amorphous and in semicrystalline samples, respectively; nylons and POM behave analogously. The two phases give additive dielectric responses which are, at a large extent, independent of each other. In terms of the effects of temperature on dielectric constant and $(\varepsilon_s - \varepsilon_\infty)$, the contributions of phases smoothly overlap and result in curves which are hybrids of those arising from the amorphous and the crystalline polymer.

(ii) Dielectric relaxations and energy losses at radio and microwave frequencies

The dielectric relaxation strength of polymers as a function of temperature defines their overall potential dielectric loss capacity at any selected temperature over a certain interval, regardless of the frequency of the applied alternating field. According to the concepts of Section 19.3.2.1(i), and, in particular, to equations (38) and (44) for dielectrics displaying a single dipolar relaxation process, the dielectric loss factor ε'' at constant temperature is always expected to increase with $(\varepsilon_s - \varepsilon_\infty)$ at any frequency. Increasing polarity (in terms of strength and concentration of dipoles) tends to give, in all cases, higher dielectric relaxation strengths. On the other hand, $(\varepsilon_s - \varepsilon_\infty)$ expresses the number and strength of dipoles really capable of undergoing polarization and dipolar relaxation (*i.e.* dielectric loss) processes. The above considerations of the dependence of $(\varepsilon_s - \varepsilon_\infty)$ on temperature show how, as anticipated in Section 19.3.2.1(i), the dielectric loss attitudes ε'' at any frequency will be modest to quite small at temperatures below T_g for most amorphous and semicrystalline polymers, even for those having a very high polarity. Polymers whose backbone bears either no or reciprocally compensated dipoles ($(\langle\hat{\mu}\rangle = 0$, or $\langle\hat{\mu}\rangle \neq 0$ and $g = 0$, respectively, in equation (28)) are known to be nonpolar polymers. According to equations (28), (38), and (44) they exhibit low, nearly constant, ε_s, $\varepsilon'(\omega)$, and $(\varepsilon_s - \varepsilon_\infty)$, and negligible $\varepsilon''(\omega)$, over the entire range of temperatures at which they are stable, at any frequency. This is the case for linear (high-density) polyethylene (HDPE) and poly(tetrafluoroethylene) (PTFE), representative polymers of these two types.[67,68,138] Indeed they show (*e.g.* at 23 °C) modest, almost invariable, $\varepsilon'(\omega)$ and extremely low $\varepsilon''(\omega)$ values from audio to MW frequencies (10 to 10^9 Hz): for HDPE, $\varepsilon' = 2.26$, $\varepsilon'' \leq 4.5\text{–}11 \times 10^{-4}$; for PTFE, $\varepsilon' = 2.1$, $\varepsilon'' = 2\text{–}6 \times 10^{-4}$. Very low dielectric dissipations are also reported over wide temperature ranges: $\tan \delta_e < 0.006$ from -80 to 110 °C for HDPE; $\tan \delta_e < 5 \times 10^{-4}$ from -160 to 160 °C for PTFE.

The category of nonpolar polymers conventionally includes also a number of polymers (among which are many important plastics and rubbers) containing very weak dipolar groupings (not listed in Table 1) such as aliphatic >CH—CH_3 and Ar—CH_2— (characterized by $\hat{\mu}$ values around $0.40–0.45 \times 10^{-30}$ and 1.2×10^{-30} C m, respectively), *cis* and *gauche* alkenic —CH=CH— structures, >CF—CF_3 groups, and so on. Polymers of these types include: homo- and copolyalkenes; polydienes; alkene–diene copolymers and alkene^{I}-alkene^{II}–diene terpolymers (*e.g.*, ethylene–propylene–diene rubbers (EPDM); polystyrene (PS) and diene–styrene copolymers; copolyperfluoroalkenes).

Table 2 gives ε' and ε'' values at room temperature for representative nonpolar polymers at frequencies from 10^4 to 10^9–10^{10} Hz. As expected, they display, like all other polymers of this type over the same wide frequency interval, no or negligible decreases of ε' and always very low values of ε'' (of the order of 10^{-3}). Similar properties are also shown by PS: ε' decreasing from 2.55 to 2.49, and maximum ε'' of around 4×10^{-3}, from 1 kHz to 1 GHz.

Table 2 Dielectric Properties of Some Nonpolar Polymers at Radio and Microwave Frequencies at Room Temperature (20–25 °C).

	Frequency (Hz)							
Polymer	10^4	10^5	10^6	10^7	10^8	10^9	3×10^9	10^{10}
Polypropylene[a,f]								
ε'	2.4	–	2.4	–	–	2.4	–	–
$\varepsilon'' \times 10^3$	<1.2	–	<1.2	–	–	1.2–4.5	–	–
Butyl rubber[b,g]								
ε'	2.37	2.36	2.35	2.35	2.35	–	2.35	–
$\varepsilon'' \times 10^3$	6.4	3.4	2.4	2.4	2.4	–	2.4	–
Butadiene–styrene rubber[c,g]								
ε'	2.5	2.5	2.5	2.45	2.45	–	2.45	–
$\varepsilon'' \times 10^3$	2.5	4.5	9.5	16.9	17.4	–	10.8	–
Hevea rubber[d,g]								
ε'	2.4	2.4	2.4	2.4	2.4	–	2.15	–
$\varepsilon'' \times 10^3$	3.4	3.4	4.3	7.7	12.0	–	6.4	–
FEP[e,f]								
ε'	–	2.1	–	2.1	–	2.1	–	2.1
$\varepsilon'' \times 10^3$	–	0.5	–	~1	–	2.3	–	1.5

[a] Average values for semicrystalline, isotactic polymer (crystallinity ~60%). [b] Isobutene–isoprene copolymer (99:1–98:2). [c] Butadiene–styrene copolymer (75:25 w/w). [d] Purified, natural rubber (pale crepe): >98% *cis*-1,4-polyisoprene. [e] Tetrafluoroethylene–hexafluoropropene copolymer (90:10–85:15 w/w). [f] Data from: J. Brandrup and E. H. Immergut (eds.), 'Polymer Handbook', 3rd edn., Wiley, New York, 1989. [g] Data from Ref. 136.

In the theoretically nonpolar HDPE and PTFE, some polarity can easily be caused by dipolar impurities and/or dipolar end-groups, or may be associated with the normal chain folds in lamellar crystallites.[138] In HDPE, as in all the feebly polar polymers mentioned above with the exception of perfluorinated ones, slight oxidation such as that accompanying the usual processing may create some additional polarity, slightly increasing the dielectric loss capacities with respect to the levels already presented. In all cases, neat nonpolar polymers show very weak coupling tendencies with electric fields, and must be described as extremely poor RF and MW absorbers.

In contrast, ε' and ε'' values at room temperature for some important polar polymers over the similar frequency range of 10^3–10^{10} Hz are collected in Table 3. This shows the significantly stronger dielectric loss capacities, with ε'' values at least one order of magnitude higher than those for nonpolar polymers. Such data exemplify, through ε'' values typically 100-fold higher over the entire RF and UHF spectra examined, the superior dielectric relaxation strength of rubbers (*e.g.* polychloroprene) and semicrystalline polymers with the amorphous phase in the rubbery state (PVDF), previously discussed.

Conversely, the same data show the relatively low loss capacities of such considerably polar materials as poly(methyl methacrylate) (PMMA), PET, and nylon-6 at temperatures (23–30 °C) appreciably below T_g, as mentioned above. From the literature on dielectric spectroscopy of polymers at audio and subsonic (<10 Hz) frequencies,[66,78,96–98] the typical dielectric response

Table 3 Dielectric Properties of Some Polar Polymers at Radio and Microwave Frequencies at Room Temperature (23–26 °C).

Polymer	T_g (°C) T_m (°C)		*Frequency* (Hz) 10^3	10^4	10^5	10^6	10^7	10^8	10^9	10^{10}
Polychloroprene[a]	−48	ε'	–	6.2	6.1	5.7	4.7	3.4	2.84[b]	–
	115	ε''	–	0.205	0.183	0.541	0.940	0.544	0.136[b]	–
PMMA[c]	105	ε'	3.2	3.0	2.9	2.8	2.75	2.7	–	2.57[d,h]
	–	ε''	0.140	0.100	0.082	0.070	0.048	0.032	–	0.018[d,h]
PET[e,h]	81	ε'	3.25	–	–	3.0	–	–	2.8	–
	260	ε''	0.016	–	–	0.048	–	–	0.008	–
Nylon-6[f,h]	63	ε'	3.5	3.4	3.4	3.3	3.2	3.2	3.1	–
	229	ε''	0.035	0.048	0.065	0.079	0.096	0.096	0.062	–
Cellulose[g,h]		ε'	7.2	7.0	–	–	5.8	5.6	–	–
		ε''	0.145	0.21	–	–	0.46	0.39	–	–
PVDF[i]	≈ −50	ε'	7.5–13.2[h]	–	–	6.0–7.6[h]	–	3.3[j]	2.9[j]	2.6[j]
	171	ε''	0.10–0.25[h]	–	–	0.9–1.7[h]	–	0.54[j]	0.3[j]	0.09[j]

[a] 1,4-poly(2-chloro-1,3-butadiene), essentially amorphous; data from Ref. 136. [b] Values at 3 GHz. [c] Poly(methyl methacrylae), atactic (amorphous); data from: M. G. Broadhurst and A. J. Bur, *J. Res. Natl. Bur. Stand.*, 1965, **69C**, 165. [d] Values at 30 GHz. [e] Poly(ethylene terephthalate), semicrystalline: Mylar A (Du Pont) biaxially oriented film. [f] Semicrystalline, injection-molded specimens; values at 30 °C. [g] Pure, dry; values at 20 °C. [h] Data from: J. Brandrup and E. H. Immergut (eds.), 'Polymer Handbook', 3rd edn., Wiley, New York, 1989. [i] Poly(vinylidene fluoride), semicrystalline. Due to its ferroelectric characteristics, ε' and ε'' may vary within intervals depending on its previous electrization and thermal history. [j] Data from: T. Wentink, *J. Appl. Phys.*, 1961, **32**, 1063.

of a generic, amorphous polar polymer is known at constant temperature as a function of frequency (Figure 5(a)), and at constant frequency as a function of temperature (Figure 5(b)).

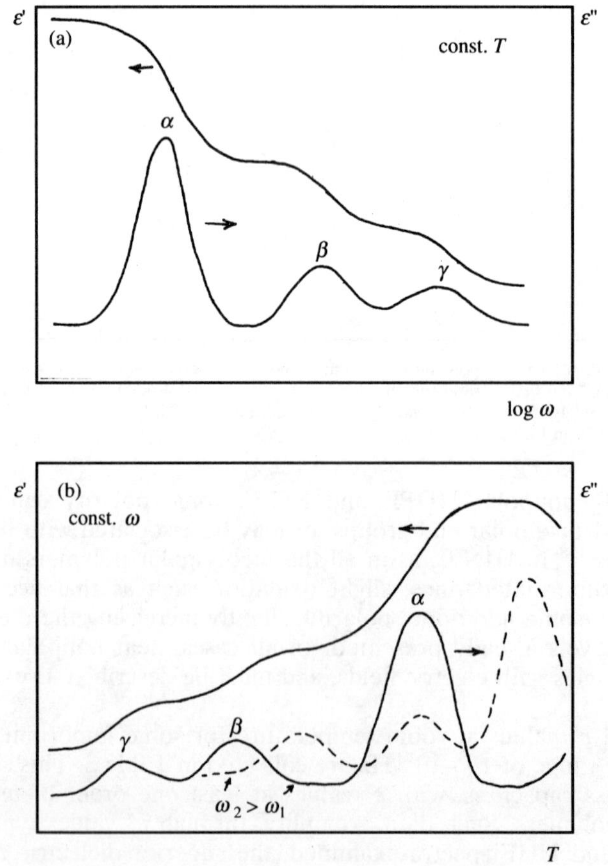

Figure 5 Dielectric relaxations and dielectric properties of an amorphous polar polymer: (a) as a function of frequency at constant temperature; (b) as a function of temperature at constant frequency.

These diagrams depict, through a succession of descending steps in ε' and a corresponding series of ε'' peaks, the typical multiplicity of dielectric relaxations associated with the different possible modes of orientational polarization and related relaxation phenomena, occurring in distinct ranges of frequency and temperature. It is known, as well, that different relaxation processes may occur at the same frequencies and temperatures, with partial overlapping of related loss peaks. In some cases, two distinct relaxation phenomena can merge, at suitably high frequencies and temperatures, into a single (particularly strong) relaxation process.

In the vast number of publications on dielectric properties of polymers, the various relaxations are labeled in different ways, but the labeling used in Figure 5 is perhaps the one most commonly used.[96] The α relaxation at the lowest frequencies, and at the highest temperatures before the normal continuous ε' values decrease with temperature, is associated with the glass transition of the polymer, *i.e.* to the movements of the main chain *via* cooperative motions of a multiplicity of chain segments, as previously discussed. It has the maximum intensity with respect to the other relaxations by virtue of the great energy dissipations caused by simultaneous inter- and intramolecular frictions of a large number of chain segments. The subsidiary, weaker relaxations β, γ, *etc.*, appearing at progressively higher frequencies and lower temperatures, are associated with motions of single structural elements and do not involve the segmental mobility of the whole polymer backbone. Examples include: rotations of dipolar side-groups, conformational changes of side-groups, local motions of individual segments of the main chain, *etc.*

Rare and/or weak dipoles also impart to nonpolar polymers a dielectric behavior qualitatively equivalent to that just described. For the majority of polymers in the glassy state, the weak secondary relaxations fall in the audio and radio frequency ranges, and only the high-frequency tail of one of them may extend into the lowest portion of the MW spectrum. Moreover, the much stronger α relaxations occur only at subsonic frequencies.[138] This accounts for the modest dielectric dissipations of glassy polar polymers at radio and MW frequencies discussed above, and shown for some specific cases in Table 3. At any fixed temperature, a partial dielectric relaxation strength $(\Delta\varepsilon')$ can be assigned to each relaxation. The sum of the $(\Delta\varepsilon')$ values of the various relaxations as in equation (49) over the entire frequency range from 0 to far-IR radiations gives the overall $(\varepsilon_s - \varepsilon_\infty)$ of a polymeric substance at the temperature considered, according to equation (39).

$$\varepsilon_s - \varepsilon_\infty = (\Delta\varepsilon')_\alpha + (\Delta\varepsilon')_\beta + (\Delta\varepsilon')_\gamma + \ldots \tag{49}$$

Table 4 reports $(\Delta\varepsilon')$ values for the α relaxation at different temperatures (of the α-peak maxima at 100 kHz) for a variety of amorphous polar polymers. The data are in qualitative agreement with the polarity of the different polymers, based on their structures and the $\hat{\mu}$ values of Table 1.

Table 4 Dielectric Relaxation Strength and Apparent Activation Energy of the α Relaxation, and Activation Enthalpy of the β Relaxation, for Several Amorphous Polar Polymers.

Polymer	T_g (°C)	$(\Delta\varepsilon')_\alpha$	*α Relaxation*[a] at T (°C)	E_a (kJ mol^{-1})[b]	*β Relaxation*[c] ΔH^* (kJ mol^{-1})
Poly(vinyl chloride)	85	5.60	120	380	63
Poly(vinyl acetate)	30	5.00	90	210	42
Poly(ether sulfone)[d]	190	4.80	224	570	–
Polychloroprene	−48	3.45	0	190	–
Poly(methyl methacrylate)[e]	105	2.25	145	100	79
Poly(ethyl methacrylate)[e]	66	1.60	129	170	58.5
Poly(butyl methacrylate)[e]	20	1.50	102	130	–
Polycarbonate[f]	150	1.15	182	550	205
Poly(phenylene ether)[g]	210	0.50	251	430	–

[a] Data taken from Ref. 50b. [b] Apparent activation energies calculated from temperatures of maximum ε'' at 1, 10, and 100 kHz. [c] Data taken from Refs. 66 and 68. [d] Poly[propane-2,2-bis(4-phenyl) ether sulfonyl-bis(4-phenyl) ether]. [e] At the temperature T, α and β relaxations merged. [f] Poly[propane-2,2-bis(4-phenyl) carbonate]. [g] Poly(2,6-dimethyl-1,4-phenylene ether).

By adoption of the appropriate $(\Delta\varepsilon')$ value in place of $(\varepsilon_s - \varepsilon_\infty)$, the various relaxation processes can be described by the Debye formulas (38) or, with a far better fit of experimental data as mentioned in Section 19.3.2.1(i), by semiempirical equations such as (44a)–(44c) or

(45). Plotting of ε' and ε'' data in an Argand diagram typically shows the various relaxations as in Figure 6, generally as semicircular ε'' vs. ε' Cole–Cole curves. Partial superposition of relaxation peaks implies a partial intersection of the relevant semicircular ε'' vs. ε' curves.

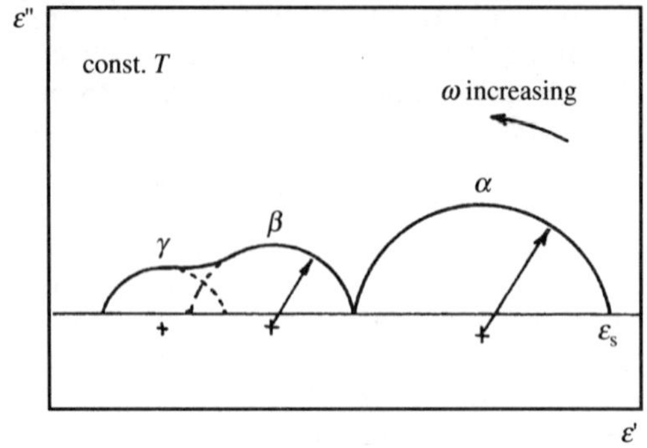

Figure 6 Argand plot of the dielectric loss factor vs. the (relative) dielectric constant for an amorphous polar polymer.

As mentioned previously, dielectric loss peaks in viscous liquids and in polymers are broader in frequency than expected from the Debye model, which predicts a half-height width of 1.14 decades. Typically, α relaxations display half-widths of around two frequency decades, whilst those of subsidiary relaxations are commonly of several decades. Besides that, and most importantly, α relaxation processes are characterized by a temperature dependence markedly different from that of the subsidiary ones. In all cases, the relaxation times for α, β, $\gamma, \ldots$ processes are expected to decrease with temperature, as extensively discussed in Section 19.3.2.1(i), with a monotonical shift of all relaxation peaks to higher-frequency regions (*i.e.* towards MW frequencies) at progressively higher temperatures.

Owing to the nature of relaxation processes of individual dipolar elements of the polymer structure, equivalent to those in low molecular weight substances, τ values of β, $\gamma, \ldots$ relaxations (and thus the values of their maximum loss frequency ω_{max}) follow the temperature dependence described by equation (40a). Generally, experimental data are in satisfactory agreement with the simple Arrhenius-type equation (40b), thus allowing an easy calculation of the activation energy values for the different relaxations and reliable extrapolations of their frequency migration with temperature. By contrast, because of their cooperative character, the dependence of τ on temperature above T_g for α relaxations does not follow a simple Arrhenius equation, but it is well described by equation (50), having the form of the Williams–Landel–Ferry (WLF) equation for mechanical relaxation phenomena.[139] In equation (50), τ_T and τ_{T_g} are the relaxation times at the temperatures T and T_g, respectively.

$$\log_{10}\frac{\tau_T}{\tau_{T_g}} = -\frac{c_1(T - T_g)}{c_2 + T - T_g} \tag{50}$$

This means that the frequency values ω_{max} (or ν_{max}) of maximum ε'' of loss peaks for α and subsidiary β, $\gamma, \ldots$ relaxations increase with temperature in different ways, according to the WLF and Arrhenius equations, respectively, as exemplified in the log ν_{max} vs. T^{-1} diagram of Figure 7.

From the WLF equation, in fact, log ν_{max} values of α relaxations increase with decreasing T^{-1} by following curved lines (in place of the straight lines for β, $\gamma, \ldots$ relaxations) which might be said to arise from Arrhenius-governed processes with activation energy progressively decreasing with temperature. Therefore, only apparent activation energies (progressively varying with temperature) can be defined for α relaxations. In agreement with their cooperative nature, α relaxations exhibit (apparent) activation energies greater than those of subsidiary relaxations, as shown in Table 4 for several polymers. This implies a steeper temperature dependence of ν_{max} of α relaxations with respect to the β, $\gamma, \ldots$ ones, thus causing α peaks to migrate towards MW frequencies on increasing temperature more rapidly than peaks for subsidiary relaxations.

Semicrystalline polymers evidence a dielectric behavior similar to that of amorphous ones. Figures 8(a) and 8(b) show the dielectric spectra of a representative semicrystalline polar

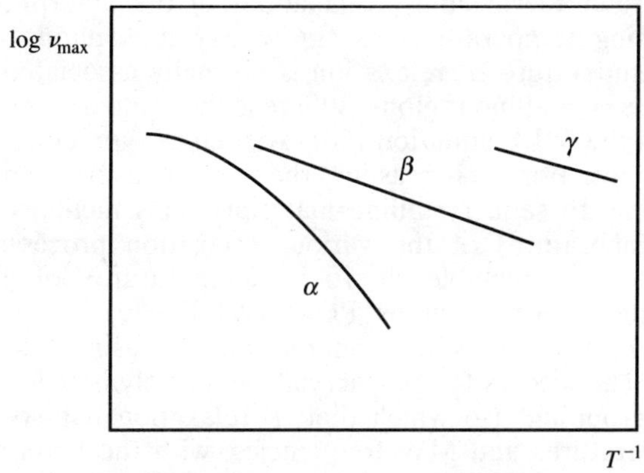

Figure 7 Dependence on T^{-1} of the logarithm of the maximum dielectric loss frequency for the α, β, and γ relaxations of an amorphous polar polymer.

polymer at constant temperature as a function of frequency, and at constant frequency as a function of temperature, respectively, with the typical relaxations labeled in the most commonly accepted way.

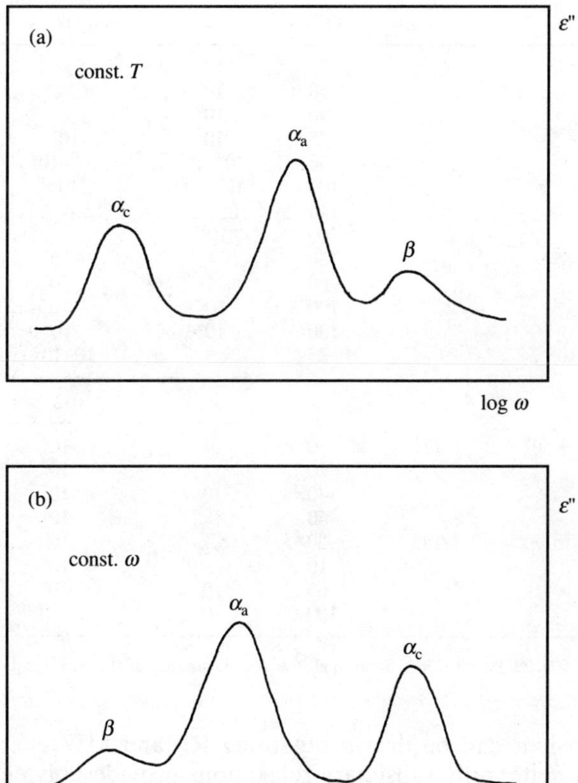

Figure 8 Dielectric relaxations and dielectric loss factor of a semicrystalline polar polymer: (a) as a function of frequency at constant temperature; (b) as a function of temperature at constant frequency.

The principal α_a and α_c peaks pertain to relaxations of the amorphous and crystalline phases, with relative intensities proportional to the relative weight amounts. Broader and weaker β, γ, ... peaks refer to subsidiary relaxations of the amorphous phase at increasing frequencies and decreasing temperatures, as in wholly amorphous polymers. The lowest-frequency and highest-temperature α_c relaxation is normally associated with (noncooperative) chain reorientations in the crystalline regions. Whereas the temperature dependence of ν_{max} for the α_a relaxation follows the WLF equation (50), Arrhenius-type equations are appropriate to the α_c, β, γ, ... relaxations. For such reasons, the same considerations expressed above for amorphous polymers apply to semicrystalline materials. This includes the frequency location (and migration with temperature) of the various relaxation processes, particularly that at subsonic frequencies (or undetectable at room temperature) of the α_c peak of most semicrystalline plastics, such as PET, nylons, POM, PVDF, *etc.*

In several polymers, α and β relaxations merge, on increasing temperature, into a single, joint relaxation process. This occurs for polyacrylates and polymethacrylates, whose rotational motions of side (ester) groupings (to which their β relaxation is associated) can combine, at appropriately high temperatures and MW frequencies, with the main-chain motions of the α relaxation. This gives place to an intense $(\alpha + \beta)$ relaxation, as reported in Table 4 for various polymethacrylates. Table 5 reports the approximate frequencies of maximum dielectric loss ν_{max} (Hz) for different relaxations at increasing temperatures for several amorphous and semicrystalline polymers: PVDC, poly(methyl acrylate) (PMA), PVC, PMMA, PVDF, and POM.

Table 5 Approximate Frequencies of Maximum Loss (ν_{max}) for Different Dielectric Relaxations of Polar Polymers at Various Temperatures.[a]

Polymer	T_g (°C)	T_m (°C)	*T* (°C)	ν_{max} (Hz) *for Relaxation* α	β	γ
PVDC	−18	–	25	10^3	–	–
			50	10^5	–	–
			90	10^6	–	–
PMA	~5	–	25	10^2	10^8	–
			50	10^4–10^5	10^8–10^9	–
			100	10^8–10^9	10^9	–
			~140	10^9–10^{10}	(α & β relax. merge)	
			200	10^{10}	$(\alpha + \beta)$	–
PVC	85	–	30	–	~10^4	–
			60	<1	10^5	–
			100	10^3	~10^6	–
			130	10^6	–	–
PMMA	105	–	25	≪1	10–10^2	10^6
				α_c	α_a	β
PVDF	~ −50	171	0	–	10^5	~10^6 [b]
			20	<1	10^6	~10^7 [b]
			40	~10	10^7	~10^8 [b]
			60	~10^2	~10^8	–
POM	−90 to −25	183	−30	–	10^6	–
			10	–	10^8	–
			60	10^3	10^9	–
			100	10^6	10^9–10^{10}	–

[a] Data taken from Refs. 50b, 78, 96, 98, 138. [b] Small high-frequency shoulder of the α_a relaxation.

Some general conclusions can be drawn regarding RF and MW energy absorption of neat polymers. In the glassy state, only subsidiary relaxations provide polymers with RF absorption capability; exclusively from such subsidiary relaxations, a contribution to the MW absorption capacity can be expected by virtue of their considerable width in frequency. Heating towards T_g shifts minor relaxations to progressively higher frequencies in the MW region, with a progressive increase of dielectric absorption thus occurring at GHz frequencies. Only appreciably beyond T_g (say, 50–100 °C above), the strong α relaxation (or α_a in semicrystalline products) moves into the MW region to make a large contribution to the MW energy absorption capacity of the material. Obviously, even higher temperatures are required for the intense α_c relaxation to play an analogous role in semicrystalline polymers. Only low-T_g polar rubbers, such as polychloroprene and butadiene–acrylonitrile copolymers, show maximum

α-loss at the highest radio frequencies, or at the low edge of the MW region, at room temperature; paradoxically, heating may reduce their UHF heatability by shifting the α relaxation to very high (SHF or EHF) MW frequencies.

The effects of heating on the variations of ε'' at radio and MW frequencies (in the direction of appreciable increases of ε'' with temperature from below to above T_g) are shown in Table 6 for nylon-6,6. For various polar polymers, Table 7 shows the evolution of ε' and ε'' with temperature at the single SHF frequency of 8.6 GHz.

Table 6 Evolution of Dielectric Properties of Nylon-6,6 at Radio and Microwave Frequencies with Temperature.[a]

					Frequency (Hz)			
T (°C)		10^3	10^4	10^5	10^6	10^7	10^8	10^9
−30	ε'	3.1	3.1	3.0	3.0	3.0	3.0	3.0
−30	ε''	0.033	0.033	0.039	0.050	0.048	0.030	0.015
30	ε'	3.5	3.4	3.4	3.2	3.1	3.1	3.0
30	ε''	0.044	0.061	0.073	0.080	0.079	0.068	0.041
90	ε'	8.9	7.6	6.2	5.0	4.0	3.4	3.2
90	ε''	1.29	1.0	0.90	0.80	0.52	0.275	0.14

[a] $T_g = 34$ °C; $T_m = 254$ °C. Semicrystalline, injection-molded specimens. Data taken from: J. Brandrup and E. H. Immergut (eds.), 'Polymer Handbook', 3rd edn., Wiley, New York, 1989.

Table 7 Evolution of Dielectric Properties at 8.6 GHz of Some Polar Polymers with Temperature.[a]

					Temperature (°C)		
Polymer	T_g (°C) T_m (°C)		*20*	*60*	*100*	*140*	*180*
Poly(vinyl acetate)	30	ε'	3.1	3.15	3.25	3.4	–
	–	ε''	0.04	0.055	0.09	0.21	–
PVC	85	ε'	3.05	~3.1	3.15	3.3	–
	–	ε''	0.02	0.03	0.06	0.13	–
Polycarbonate[b]	150	ε'	2.76	2.79	2.81	2.82	2.87
	–	ε''	0.019	0.024	0.029	0.036	0.062
POM	~ −60	ε'	2.7	2.8	2.9	3.15	3.6
	175[c]	ε''	0.05	0.11	0.19	0.26	0.31

[a] Data taken from Ref. 137. [b] Poly[propane-2,2-bis(4-phenyl) carbonate]. [c] Experimental T_m of the polymer sample studied.

As mentioned at the end of Section 19.3.2.1(i), an alternative way to decrease the temperatures at which the α relaxation process shifts to high radio or low MW frequencies is by the addition of organic plasticizers.[138] In PVC, for example, either polar or nonpolar plasticizers can easily lower T_g to room temperature and cause α and β relaxations to merge, with marked increases of ε'' at 10^7–10^9 Hz at room temperature. Common dipolar plasticizers such as dialkyl phthalates and phosphoric acid esters can additionally contribute their own RF and MW loss capacities.

19.3.2.3 Interfacial polarization losses

Interfacial (or MWS) polarization mentioned in Section 19.3.1.2(iii) involves relaxation phenomena in alternating electric fields analogous to, and often indistinguishable from, those of authentic dielectric relaxations.[7,55,56,66,79–81] Interfacial relaxation in a composite dielectric leads to a decrease of ε' with frequency and a relaxation peak superimposed on the normal $\varepsilon'(\omega)$ and $\varepsilon''(\omega)$ curves of the neat dielectric matrix discussed above, *i.e.* it gives additive contributions $\varepsilon'_{MWS}(\omega)$ and $\varepsilon''_{MWS}(\omega)$ to the overall $\varepsilon'(\omega)$ and $\varepsilon''(\omega)$ of the whole material under consideration. As in dielectric relaxation, any interfacial relaxation phenomenon is characterized by a proper relaxation time τ_{MWS} (or τ_{MWS} distribution).

An evaluation of such a dissipation process has been given for the simple model system consisting of a dielectric matrix (such as a nonpolar polymer) of static (relative) dielectric constant $\varepsilon_{s,1}$, and with negligible conductivity and dielectric relaxation strength $(\varepsilon_{s,1} - \varepsilon_{\infty,1})$, containing a small volume fraction ϕ of dispersed spherical conductive particles with dielectric constant $\varepsilon_{s,2} \approx \varepsilon_{\infty,2}$ and d.c. conductivity σ_2.[80] $\varepsilon'_{MWS}(\omega)$ and $\varepsilon''_{MWS}(\omega)$ for the composite dielectric can be calculated by Wagner's equations (51), where $\varepsilon_{s\,1,2}$ and $\varepsilon_{\infty\,1,2}$ are the static and the instantaneous (relative) dielectric constants, respectively, of the composite material (matrix 1 + dispersed phase 2) itself.

$$\varepsilon'_{MWS}(\omega) = \frac{\varepsilon_{s1,2} - \varepsilon_{\infty 1,2}}{1 + \omega^2 \tau^2_{MWS}} + \varepsilon_{\infty 1,2} \tag{51a}$$

$$\varepsilon''_{MWS}(\omega) = (\varepsilon_{s1,2} - \varepsilon_{\infty 1,2}) \frac{\omega \tau_{MWS}}{1 + \omega^2 \tau^2_{MWS}} \tag{51b}$$

where:

$$\varepsilon_{s1,2} = \varepsilon_{s,1}(1 + 3\phi)$$

$$\varepsilon_{\infty 1,2} = \varepsilon_{s,1}\left[1 + \frac{3\phi(\varepsilon_{s,2} - \varepsilon_{s,1})}{2\varepsilon_{s,1} + \varepsilon_{s,2}}\right]$$

$$\varepsilon_{s1,2} - \varepsilon_{\infty 1,2} = \frac{9\phi \varepsilon^2_{s,1}}{2\varepsilon_{s,1} + \varepsilon_{s,2}}$$

These equations correspond to the Debye formulas (38b) and (38c), where ε_∞ and the dielectric relaxation strength $(\varepsilon_s - \varepsilon_\infty)$ of the homogeneous dielectric are replaced by the $\varepsilon_{\infty\,1,2}$ and the relaxation strength $(\varepsilon_{s\,1,2} - \varepsilon_{\infty\,1,2})$, respectively, of the composite dielectric. In this representative system, the relaxation time for the interfacial relaxation process, independent of the volume fraction of the conducting phase, is expressed by equation (52).

$$\tau_{MWS} = \frac{\varepsilon_0(2\varepsilon_{s,1} + \varepsilon_{s,2})}{\sigma_2} \tag{52}$$

Based on current ε_s values of polymer and resin matrices, equation (52) predicts that either weakly conducting inclusions (such as semiconductors and ionic conductors, with moderate σ and low ε_s) or dispersed metallic conductors (with very high σ and ε_s) imply τ_{MWS} values well above 10^{-10} s, with MWS relaxation maxima at audio or radio frequencies. The phenomenon of MWS relaxation is quite complex, as it is influenced by morphological and geometrical factors (shape, surface characteristics, and orientation of dispersed particles, *etc.*). For instance, elongation of conductive particles in the electric field direction broadens the loss peaks and shifts them to frequencies lower than those expected with spherical inclusions. As shown by equation (52), moreover, the temperature dependence of τ_{MWS} is a function of the temperature coefficients of dielectric constants and conductivities of the coexisting phases, implying that a case-by-case evaluation of the frequency shifts of relaxation peaks with temperature is necessary.

19.3.2.4 Electric conduction losses

Electric conduction and related energy losses, as discussed from a general standpoint under steady fields in Section 19.3.1, are in polymeric materials a wide and complex subject. Different conduction mechanisms, of the two fundamental electronic[140] and ionic[141] types, may be concomitant in a single material, whose specific conductivity thus results as the sum of the contributions given by each conduction phenomenon and by the relevant conductive species. This is often the case in polymeric materials, where both the electronic and ionic conductivity of neat polymers and resins,[66–69] as well as the electronic and/or ionic conductivity of different phases in polymer blends and polymer matrix composites (*e.g.* in conductor-filled polymers[66,72,142]), contribute to the overall conductivity and the associated electromagnetic energy dissipations. Electrical conductivity may or may not depend appreciably on the applied field frequency, whereas it always is more or less strongly influenced by temperature. Although the conductivity of many materials is negligibly to only slightly variable with frequency, and d.c. values σ can often be considered even up to MW frequencies, its expression as a frequency function $\sigma(\omega)$ is of more general validity. In a good conductor or a semiconductor, a sinusoidally alternating electric field induces an oscillating current whose instantaneous density is expressed by equation (53).

$$[\boldsymbol{J}(t)]_\sigma = \sigma(\omega)\boldsymbol{E}(t) = \sigma(\omega)\boldsymbol{E}_0 \exp(j\omega t) \tag{53}$$

In real dielectrics (displaying weak electronic semiconductivity and/or ionic conductivity) the total current density $\boldsymbol{J}(t)$ induced by the field results from two contributions: the conduction current density of equation (53) and a dielectric displacement current density expressed by equation (54), as shown in equation (55).

$$[\boldsymbol{J}(t)]_{\boldsymbol{D}} = j\omega \boldsymbol{D}(t) = j\omega\varepsilon_0\varepsilon^*(\omega)\boldsymbol{E}(t) \tag{54}$$

$$\begin{aligned} \boldsymbol{J}(t) &= [\boldsymbol{J}(t)]_\sigma + [\boldsymbol{J}(t)]_{\boldsymbol{D}} \\ &= \sigma(\omega)\boldsymbol{E}(t) + j\omega\varepsilon_0\varepsilon^*(\omega)\boldsymbol{E}(t) \\ &= \sigma(\omega)\boldsymbol{E}(t) + j\omega\varepsilon_0[\varepsilon'(\omega) - j\varepsilon''(\omega)]\boldsymbol{E}(t) \end{aligned} \tag{55}$$

Rearrangement of equation (55) into equation (56), by grouping the two energy dissipation contributions of conduction and dielectric relaxation of permanent dipoles, leads to equation (57), which gives the contribution $\varepsilon''_\sigma(\omega)$ of conduction losses to the overall loss factor of a material.

$$\boldsymbol{J}(t) = j\omega\varepsilon_0\left\{\varepsilon'(\omega) - j\left[\varepsilon''(\omega) + \frac{\sigma(\omega)}{\omega\varepsilon_0}\right]\right\}\boldsymbol{E}_0 \exp(j\omega t) \tag{56}$$

$$\varepsilon''_\sigma(\omega) = \frac{\sigma(\omega)}{\omega\varepsilon_0} \tag{57}$$

Owing to the often weak (or neglected) frequency dependence of σ, ε''_σ is commonly referred to as the d.c. conductivity loss factor. The total conductivity $\sigma(\omega)$ results as the sum $\sigma_e(\omega) + \sigma_i(\omega)$ of the two contributions of electronic and ionic conductivities, respectively.

(i) Electronic conductivity

Electronic conductivities comparable to those of inorganic metallic conductors (metals and nonmetals) such as graphite and carbon black) and inorganic semiconductors[143] are shown only by a few particular types of organic compounds (molecular semiconductors)[144] and by those polymeric products classified as conductive polymers,[71–73] including undoped and doped conjugated-chain polymers (polyacetylenes, polypyrrole, polythiophenes, partially pyrolyzed PAN, poly(sulfur nitride), *etc.*) and polymeric radical-cation salts.[145] Apart from such types of polymeric products, both nonpolar and polar ordinary polymers can be considered to be very weak semiconductors, like the majority of organic molecular substances.[146,147] A peculiarity of polymers, however, is that intramolecular conduction phenomena may easily prevail over intermolecular ones. In semiconducting materials in general, two major mechanisms of electronic conduction are recognized.[66,68,143] These are intrinsic conduction and extrinsic conduction.

Intrinsic conduction occurs by drift of free charge carriers. Charge carriers (free electrons and electron vacancies or holes) are generated by thermal excitation of electrons to current-carrying states and conductivity rapidly rises with temperature. The simple dependence of σ_e on temperature given by equation (58) is followed, with ΔE the activation energy for the charge carrier generation and $\xi(T)$ a slowly varying function of temperature.

$$\sigma_e(T) = \xi(T)\exp\left(-\frac{\Delta E}{2kT}\right) \tag{58}$$

Semiconductors where intrinsic conduction dominates (intrinsic semiconductors) include: pure to lightly doped, crystalline and polycrystalline inorganic semiconductors, molecular semiconductors, polymeric semiconductors such as poly(sulfur nitride), undoped or lightly doped conjugated-chain polymers such as polyacetylene and polypyrrole, *etc.*

Extrinsic conduction is operated by sparse, localized charge carriers, either hopping over, or tunneling through, energy barriers between adjacent, randomly distributed sites. Charge carriers for hopping and tunneling conduction come from impurities, *i.e.* electrons and holes provided by electron donors and electron acceptors, or ions. Both conduction phenomena occur between sites of charge separation at the atomic, molecular or supramolecular level, up to microclusters or microdomains, *e.g.* doping sites in the crystal lattice of crystalline semiconductors, doping microregions in amorphous semiconductors, doping centers along the main chain of conductive polymers, *etc.* Hopping conduction is normal for inorganic, amorphous semiconductors. Tunneling is typical of highly doped, crystalline and polycrystalline (inorganic) semiconductors. An important case is that of electron tunneling between conductive particles in conductor-filled polymers.[148,149] Hopping and tunneling conductions may often be concomitant in the same material, and their relative importance depends primarily on the energy-barrier characteristics and temperature. Semiconductors where hopping and/or tunneling mechanisms largely prevail are known as extrinsic semiconductors. Extrinsic conduction is also a thermally activated process; this implies an Arrhenius-type increase of conductivity with temperature according to equation (59), where $\Delta E_x(T)$ is the appropriate activation energy (increasing with temperature and related to the mobility gap the charge carriers must overcome) and $\sigma_{e,0}$ an infinite-temperature conductivity.

$$\sigma_e(T) = \sigma_{e,0} \exp\left[-\frac{\Delta E_x(T)}{2kT} \right] \tag{59}$$

As mentioned earlier, ordinary organic polymers display quite feeble conductivities, and may be considered as weak semiconductors from two standpoints. Highly pure polymers may be thought to be intrinsic semiconductors with values of the activation energy ΔE in equation (58) so high as to give extremely low values of σ_e. Polymers containing small but appreciable amounts of impurities of a different kind, by contrast, may be ascribed to the class of (weak) extrinsic semiconductors. Traces of substances acting as electron donors or acceptors, or (as very common) ionic compounds, may in fact cause a hopping conductivity of great importance. In both cases, rapid increases of conductivity with temperature can be expected from equations (58) and (59). Increases of d.c. conductivity with temperature of an exponential type are shown in Figure 9 for various polymeric materials.[67] Only the conductivity of PTFE exhibits a substantial insensitivity to temperature up to its very high melting point (327 °C). Crosslinked and semicrystalline polymers tend to have lower conductivities than linear and amorphous ones, because of the lower mobilities of charge carriers predictable in network structures and crystalline regions.

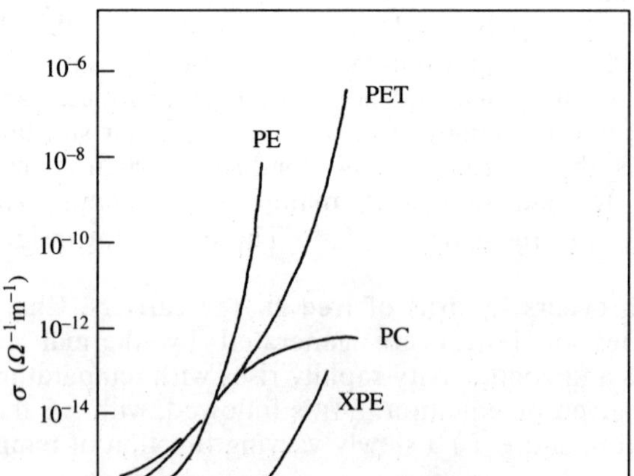

Figure 9 Variation of the d.c. conductivity with temperature for different polymeric materials: polyethylene (PE); poly(ethylene terephthalate) (PET); poly[propane-2,2-bis(4-phenyl) carbonate] (PC); crosslinked polyethylene (XPE); poly(tetrafluoroethylene) (PTFE).

The frequency dependence of conductivity in inorganic and organic (polymeric) semiconductors is a rather intriguing matter. Normally, conductivities vary only slightly from zero up to

MHz frequencies, showing general increases well below one order of magnitude from the d.c. values to the $\sigma_e(\omega)$ ones at the highest radio frequencies. For this reason, d.c. conductivity values can be used for computation of conduction losses in RF heating without incurring any large errors. At MW frequencies, on the contrary, considerable deviations may easily arise, with conductivities often being very different (in many instances several orders of magnitude higher) from d.c. values. Unfortunately, only a few $\sigma_e(\omega)$ data are available in the literature, so that values at the frequencies, at the temperatures, and for the materials of interest must be measured case by case. Nevertheless, some qualitative criteria of general applicability can be given. The overall electronic conductivity $\sigma_e(\omega)$ can be split, as in equation (60), into two additive terms:[150,151] (i) the d.c. conductivity σ_e [$=\sigma_e(0)$], expressing the contribution of charge carriers flowing through continuous conduction (or percolation) paths across the material; (ii) an a.c. conductivity $\sigma_\sim(\omega)$ related to the conduction phenomena by transitions of charge carriers between different states.

$$\sigma_e(\omega) = \sigma_e + \sigma_\sim(\omega) \tag{60}$$

The conductivity $\sigma_\sim(\omega)$ at constant temperature is governed by the proportionality expressed in equation (61), where the exponent s varies from material to material, and can also change with frequency itself in the same material.[152]

$$\sigma_\sim(\omega) \propto \omega^s \qquad (0 \leq s \leq 2) \tag{61}$$

Intrinsic conduction phenomena seem to be weakly dependent on frequency ($0 \leq s < 0.5$), as reported for intrinsic molecular semiconductors,[144] intrinsic polymeric semiconductors such as poly(sulfur nitride), and iodine-doped polyacetylene,[138] whose $\sigma_e(\omega)$ values at frequencies of tens of GHz are not so different from d.c. σ_e (less than tenfold higher). In amorphous semiconductors, characterized by hopping, but also tunneling, the exponent s typically ranges from 0.5 to 1.[151,153] At constant temperature, a charge carrier hopping phenomenon is characterized by a hopping frequency (or frequency distribution), to which a hopping relaxation time τ_x (or distribution of hopping relaxation times $\varphi'(\tau_x)$) can be associated. This leads to the Debye-type equation (62a), or to the corresponding equation (62b) for systems with a relaxation time distribution, for the frequency dependence in $\sigma_\sim(\omega)$.

$$\sigma_\sim(\omega) \propto \frac{\omega^2 \tau_x}{1 + \omega^2 \tau_x^2} \tag{62a}$$

$$\sigma_\sim(\omega) \propto \int_0^\infty \frac{\omega}{1 + \omega \tau_x} \varphi'(\tau_x)\, d\tau_x \tag{62b}$$

Equation (62a) simply predicts a quadratic proportionality of $\sigma_\sim$ to ω, and s values varying from 1 to 2 are observed at GHz frequencies in polymeric semiconductors.[152] Equations (62) help in discussing the similarities between hopping phenomena of electric charge carriers and relaxation processes of permanent dipoles. The former phenomena, in fact, can be regarded as involving alternating orientations of electron-rich/electron-poor pairs under the action of an alternating field, responsible for energy losses equivalent to those of dielectric relaxation in dipolar substances (with analogous loss peaks being exhibited).

Hopping phenomena are also given by impurity–vacancy pairs in crystalline ionic solids,[141] ionic impurities in polymers,[78] and by ionic centers sparse but immobilized in a polymeric structure. The above concepts are applicable in all these cases. In some respects, the tunneling conduction is similar to intrinsic conduction, with a weak dependence of $\sigma_\sim$ on ω ($s < 0.5$). The modest variations of conductivity from zero to GHz frequencies in heavily-doped crystalline semiconductors and in certain conductor-filled polymers,[152] where tunneling conduction dominates, might lead to this conclusion. It could also explain the s values, which are much less than 2 in amorphous inorganic semiconductors despite the predictions of equation (62a), as being due to a significant participation of tunneling conduction in such materials. Nearly constant conductivities of heavily-doped conductive polymers up to 10^{10} Hz frequencies could also be interpreted as being linked to intramolecular tunneling mechanisms of conduction.

(ii) Ionic conductivity

Conductive polymers containing ionic centers, such as electrochemically oxidized and protonated polypyrrole and polyaniline, display both electronic and ionic conductivity. Strong ionic conductivity is possessed by the so-called polymeric electrolytes, *i.e.* microporous polymers or polyelectrolytes swollen with electrolyte solutions and solid solutions of inorganic salts in ion-solvating polymers (namely, in poly(alkylene oxide)s).[74,75] In contrast to these types of polymer systems, perfectly pure, ordinary polymers and resins have *per se* no ionic conductivity. A weak to appreciable ionic conductivity, however, is commonly exhibited by them because of the frequent presence of ionic substances of varied nature and source, *e.g.* catalysts employed in resin and polymer syntheses, hardening catalysts in thermosetting resins, residual byproducts of many polycondensation processes, impurities in starting monomers, and so on.

For such reasons, ionic conduction losses may play a nonnegligible role in RF and MW heating of most resins and polymeric materials, although their chemical structure intrinsically does not involve ionic centers. Electronic and ionic conductivities are measured jointly; they act altogether as RF and MW loss sources, and their separate contributions to the total conductivity are not immediately distinguishable. In current resin systems and polymers, the contribution of ionic conductivity is in most cases notably stronger than the electronic one. With respect to the values and the variation of conductivity with temperature of polymeric products shown in Figure 9, for example, nylons commonly exhibit strong increases of conductivity on heating, largely attributable to ionic conduction phenomena.[67] Ion-hopping phenomena as causes of subsidiary dielectric relaxation peaks in polymers (appearing as high-temperature and low-frequency relaxations, typical, *e.g.* in PVDF) have been mentioned in the preceding section. Ionic conductivity can require a frequency-dependent form of the type of equation (60). For the most important free-ion conduction, however, the relatively long relaxation time of the Debye–Hückel cloud surrounding the ions makes the second, frequency-dependent term practically negligible ($\sigma_i(\omega) \approx \sigma_i$) at MW frequencies.[141]

The dependence of σ_i on temperature, by contrast, is always significant and is worth discussing in some detail. Each ionic compound A_aB_b ($A_aB_b \rightleftharpoons aA^{b+} + bB^{a-}$) dissolved in an organic medium (resin or polymer) gives to the overall conductivity an additive contribution expressed by equation (63), with ι the degree of dissociation of the ionic compound itself, $[A_aB_b]_0$ its original concentration (molecules m^{-3}), m_{A^+} and m_{B^-} the electric mobilities (*i.e.* the average drift velocities in an electric field of unit strength, in m^2 s^{-1} V^{-1}) of the cation and anion species produced by dissociation, and e the electron charge (1.602×10^{-19} C).

$$\sigma_i = \iota[A_aB_b]_0 abe(m_{A^+} + m_{B^-}) \qquad (63)$$

In the simplest case of an ionic compound $AB \rightleftharpoons A^+ + B^-$ in an organic medium of reasonably low ε_s, such as most organic polymers and resins, σ_i can be calculated, with some approximation, by equation (64), where K_0 is the preexponential factor of equation (65), expressing the Arrhenius temperature dependence of the equilibrium constant of the AB dissociation, and ΔE_{AB} the energy for separation of the ions in a medium of unit ε_s.

$$\sigma_i = K_0^{0.5}[AB]_0^{0.5} e(m_{A^+} + m_{B^-}) \exp\left(-\frac{\Delta E_{AB}}{2\varepsilon_s kT}\right) \qquad (64)$$

$$K = K_0 \exp\left(-\frac{\Delta E_{AB}}{\varepsilon_s kT}\right) \qquad (65)$$

Equation (64) indicates how σ_i increases with the electrolyte concentration, with temperature and with the ε_s of the medium. In general, the dependence of σ_i on temperature is more complicated than the Arrhenius-type one suggested by equation (64). A good empirical description of the temperature dependence of σ_i in resins and amorphous polymers above and moderately below their T_g is given by a Vogel–Tammann–Fulcher (VTF) equation (66) for diffusion-controlled processes, with B proportional to the activation energy and $T_0 = T_g - 50$.

$$\sigma_i(T) = \sigma_{i,0} \exp\left(-\frac{B}{T - T_0}\right) \qquad (66)$$

Since ion mobilities are strictly related to the viscosity of the medium, the ionic conductivity at constant temperature decreases monotonically with increasing molecular weight in liquid to rubbery oligomeric systems and linear or branched polymers, as evidenced in the course of polymerization processes. When crosslinking reactions are involved, *i.e.* in curing processes of liquid thermosetting resins, the conductivity progressively decreases with the degree of conversion from an initial value and levels off at gelation.[154]

(iii) Conductor-filled polymers and resins

Polymers and resins filled with particles of fibers of electronic conductors[66,72,142] can reach high values of conductivity, and thus find many applications such as lightweight conductors in electronics and in electrical and electromechanical devices (electromagnetic radiation shields, *etc.*). Powders or flakes of Al, Cu, brass, Ag, Ni, Fe, and graphite, carbon black, milled, chopped and long carbon fibers, specialty metal-coated fibers, conductive polymers, *etc.*, can be used as the conducting phase. Besides these tailored conductive composites, carbon black-filled rubbers and plastics and carbon fiber-reinforced thermosets and plastics are important materials, in that their broad spectrum of properties generally includes at least a certain degree of electrical conductivity.

A percolation model applies satisfactorily to electronic conduction in these materials, *i.e.* through conductive paths formed by elongated aggregates of particles or fibers in electrical contact with each other, carrying electrons within and through the material. The conductivity σ_e of such composites strongly depends on: (i) the overall volume fraction ϕ of the dispersed conductor; (ii) the conductivity, morphology, and topology of particles or fibers, progressively higher values of conductivity being imparted by fillers of higher inherent conductivity $\sigma_{e,c}$, higher aspect ratio (from spheres to flakes, to needles, to fibers), and by the increasing formation of aggregates (eminent in certain cases, *i.e.* with carbon black particles, known to agglomerate spontaneously into fiber-like arrangements or structures[155]); (iii) the quality of interparticle contacts (surface oxidation of metal fillers being increasingly detrimental to σ_e, from the negligible oxidation of noble metals to that of electropositive metals such as Zn and Al).

Typically, σ_e of composites increases with ϕ as in Figure 10. At low ϕ values the weak conductivity of the polymer or resin matrix remains substantially unaltered until, for a certain critical ϕ value, the creation of conductive paths takes place (percolation threshold). After that, σ_e increases sharply with ϕ as the number of interparticle or interfiber contacts rapidly becomes higher and higher. The increase of conductivity with ϕ then subsides, approaching values of metallic conductivity 2–4 orders of magnitude below that of the conductive filler.

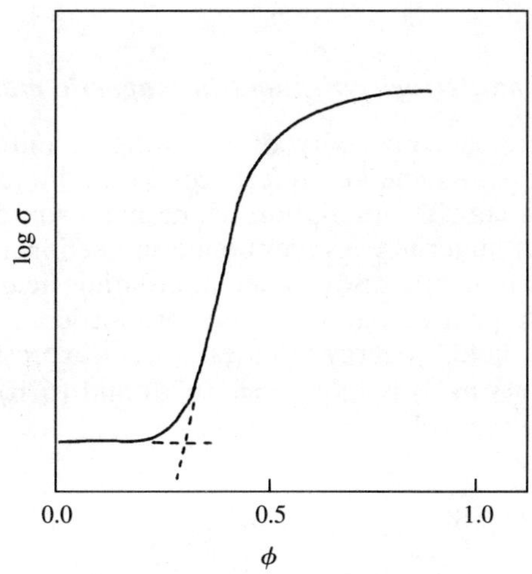

Figure 10 Dependence of the d.c. conductivity of a conductor-filled polymer on the volume fraction of the dispersed conductive phase.

Generally, two conduction mechanisms are involved: (a) non-Ohmic conduction by electron tunneling across interparticle or interfiber gaps of up to some tens of Å;[148,149] (b) Ohmic conduction *via* physical interparticle contacts. The former mechanism determines the conduction regime in the low-conductivity range of filler concentrations, whilst the latter dominates in high-conductivity composites. The morphology and topology of the dispersed conductor govern the critical ϕ value for the percolation threshold, which is low when the aspect ratio of particles is high, and the formation and elongation of their aggregates are favored. Extremely low values of the critical ϕ (even $<0.1\%$) are imparted by dispersed charge-transfer salts[145] with dendritic morphology.[72] Preferential orientations of elongated particles, particle aggregates, or fibers, with even strong degrees of electrical anisotropy, are often encountered. In these cases, the composite conductivity varies from a maximum in the direction of preferential orientation to much lower values in orthogonal directions. In fiber-filled and fiber-reinforced polymers, these differences may be particularly pronounced.

The dependence of conductivity on temperature in conductor-filled polymers varies greatly from case to case (either positive or negative, or even neutral), due to the many parameters by which it is controlled, *e.g.* the temperature coefficient of $\sigma_{e,c}$ (positive for graphite and carbon black, negative for most metals and alloys), the coefficients of thermal expansion of conductor and polymer matrix (expanding or shrinking the insulating interparticle or interfiber gaps), *etc.* The frequency dependence of the conductivity of conductor-filled systems is expected to be even more complicated than that of neat polymers, because of the great variety of parameters involved. At MW frequencies, in fact, such materials may show appreciable deviations of conductivity from their d.c. values, and some recent studies have addressed this subject. At filler concentrations below the percolation threshold, conductive inclusions play a minimal role, the characteristics of the very weak semiconductor of the neat polymer matrix largely prevail, and the concepts expressed in Section 19.3.2.4(i) are applicable. In the percolation threshold region, where composites manifest strong semiconductivity by electron tunneling, $\sigma_{\sim}(\omega)$ of equation (60) has been calculated for two-component semiconductors,[156] and verified to be proportional to $\omega^s (s = 0.4\text{–}0.5)$ over the frequency range from 10^6 to 10^{10} Hz in composites of doped substituted polyacetylene filled with graphite.[152]

The conductivity at frequencies from 8 to 12 GHz of thermosets (epoxy and unsaturated polyester resins) filled with metal powders or flakes has recently been studied.[157,158] Percolation phenomena are not observable when the conductor is perfectly dispersed and conductive structures are thus not formed.[157] In the frequency range examined, the conductivity above the percolation threshold does not show a marked dependence on frequency.[158] The conductivity as a function of frequency from 0.13 to 18 GHz has been investigated for composites comprising a dielectric polymer matrix and a conductive polymer: polypyrrole in PVC,[159,160] and polypyrrole in butyl rubber, epoxy resin, and PTFE.[161] These studies have indicated that the simple relation $\sigma(\omega) \propto \omega^t$ is applicable above the percolation threshold, and t gradually decreases from ~ 1 to level off at 0 for ϕ increasing from its critical value. This shows a vanishing influence of frequency on conductivity as materials become better and better Ohmic conductors, *i.e.* as the d.c. conductivity term in equation (60) increases and becomes predominant.

19.3.2.5 Energy losses in ferro-/antiferroelectrics and in magnetic materials

Because of the extent and extreme complexity of the subject, only an overview of the fundamental aspects is given here, with some key references to the literature being cited for a deeper insight. This section gives a concise description of the main sources of electromagnetic energy loss in the various classes of materials. As mentioned in Section 19.3.1, almost all these materials undergo cyclical polarization processes under alternating fields following hysteresis loops of different shape and area, implying hysteresis losses proportional to the loop area itself. Such (electric or magnetic) field energy losses are expressed in a general form per unit volume of material ($\mathrm{W\,m^{-3}}$) by equations (67a) and (67b) for ferroelectrics[82–86] and ferro-/ferrimagnets,[57,58,87,89] respectively.

$$W_h = \frac{\omega}{2\pi} \oint \mathbf{D}(\mathbf{E})\, d\mathbf{E} \tag{67a}$$

$$W_h = \frac{\omega}{2\pi} \oint \mathbf{B}(\mathbf{H})\, d\mathbf{H} \tag{67b}$$

These materials follow hysteresis cycles, and exhibit related energy losses, with alternating fields of any oscillation amplitude $\boldsymbol{E}_0$ or $\boldsymbol{H}_0$. Only certain antiferroelectrics and antiferromagnets, by contrast, follow hysteresis loops only above suitable values of $\boldsymbol{E}_0$ or $\boldsymbol{H}_0$, below which they behave similarly to paraelectrics and paramagnets, respectively. Except for antiferromagnetic metals and the majority of ferromagnets, all the subject materials are dielectrics or semiconductors. In ferromagnetic metals and alloys, alternating magnetic fields are well known to induce intense electric eddy currents (or Foucault currents),[59,89] accompanied by correspondingly strong energy dissipations by the Joule effect (eddy current losses). The eddy current generation in electronic conductors and semiconductors by oscillating magnetic fields is the basis of induction heating metallurgy,[162] and represents the principal source of energy dissipations in ferromagnetic materials in their common electrical and electromechanical applications. The calculation of eddy current losses is made particularly difficult by their strong dependence on both geometrical factors and material anisotropy. For the present purposes, however, it is sufficient to express by equation (68) the main parameters and their qualitative influence on the power dissipated per unit volume (W_F) *via* this loss mechanism, at constant temperature. In this equation, ζ is a shape factor inherent to the object considered, d is the smallest thickness of material perpendicular to the magnetic flux, and $\mu'(\omega, \boldsymbol{H}_0)$ and $\sigma(\omega)$ are the permeability and the electric conductivity of the material itself at the frequency ω and for the field oscillation amplitude $\boldsymbol{H}_0$.

$$W_F \propto \zeta d^2 \omega^2 \mu'(\omega, \boldsymbol{H}_0)\sigma(\omega) \tag{68}$$

According to equation (68), strong eddy current losses are expected at radio and, particularly, at MW frequencies, especially in ferromagnets by virtue of their very high μ' and σ. Energy losses orders of magnitude lower are displayed in diamagnetic and paramagnetic conductors, proportionally to their far lower permeabilities. Appreciable eddy current losses at MW frequencies may be given by conductive antiferromagnets due to their significant μ' (especially, around the Néel point), and by ferrimagnets,[163–165] having low conductivities accompanied by very high values of permeability. The eddy current circulation is restricted to a surface layer of objects (skin effect) which becomes thinner as the frequency and the material permeability increase,[162] *i.e.* the surface layer has a thickness of millimeters to fractions of a millimeter at radio frequencies, and of microns at MW frequencies, in ferromagnetic metals.

The third and often predominant category of energy dissipations in magnetic materials is that of residual losses, comprising a variety of magnetic relaxation and resonance phenomena. In ferromagnets, various relaxation processes are recognized to accompany magnetization and demagnetization,[57,87,89,166] together with ferromagnetic resonance phenomena.[87–89,166–168] Resonance lines at radio and MW frequencies are broadened by magnetic relaxation,[163,166] giving place to magnetic spectra (μ' and μ'' vs. log ω) analogous to dielectric ones. Similar relaxation processes and resonance phenomena are exhibited in antiferromagnets[57,87,169] and ferrimagnets.[57,87,163–165,169] Figure 11 depicts a typical magnetic spectrum at constant field strength for a ferrite material, where the maximum μ'' corresponds to a broadened resonance loss.

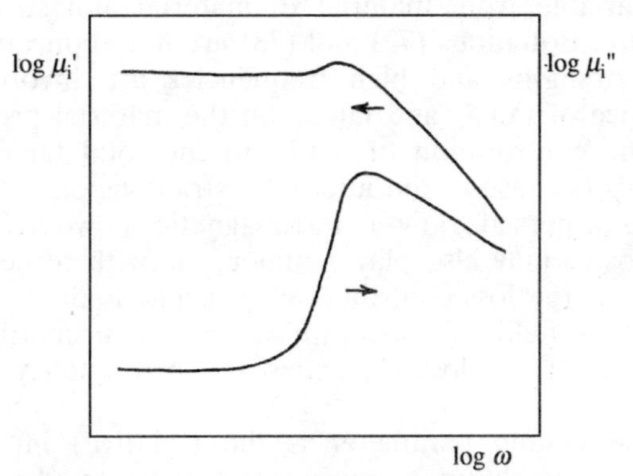

Figure 11 Variation of the (relative) complex initial permeability of a typical ferrite material with frequency.

The three types of loss phenomena described above give additive contributions to a total magnetic loss tangent $\tan\delta_m$ and to an overall (or effective) magnetic loss factor μ''_{eff} of a material as in equations (69) and (70).

$$\tan\delta_m(\omega, \boldsymbol{H}_0) = \frac{\mu''_{eff}(\omega, \boldsymbol{H}_0)}{\mu'(\omega, \boldsymbol{H}_0)} = \tan\delta_h + \tan\delta_F + \tan\delta_r \tag{69}$$

$$\mu''_{eff}(\omega, \boldsymbol{H}_0) = \mu''_h(\omega, \boldsymbol{H}_0) + \mu''_F(\omega, \boldsymbol{H}_0) + \mu''(\omega, \boldsymbol{H}_0) \tag{70}$$

The calculation of the hysteresis loss tangent $\tan\delta_h$ and of the eddy current loss tangent $\tan\delta_F$ is extremely difficult because of the complex behavior of magnetic materials. An exact calculation of $\tan\delta_h$ is possible only in simplified cases, *i.e.* with low field strengths, when minor hysteresis loops are followed, and the Rayleigh formulas (71) (generalizations of equations (30)) can be used to express the permeability of a material as a function of $\boldsymbol{H}_0$, with $\mu'_i(\omega)$ the initial permeability at the frequency ω (permeability at the frequency ω for $\boldsymbol{H}_0 \rightarrow 0$).

$$\boldsymbol{B}_0 = \mu_0[\mu'_i(\omega)\boldsymbol{H}_0 + rH_0^2] \tag{71a}$$

$$\mu'(\omega, \boldsymbol{H}_0) = \mu'_i(\omega) + r\boldsymbol{H}_0 \tag{71b}$$

Magnetic relaxation and resonance processes determine at the frequency ω an angular phase lag $\delta_r(\omega)$ between $\boldsymbol{B}$ and $\boldsymbol{H}$ (see Section 19.3.2.1(ii)), creating an elliptical $\boldsymbol{B}$–$\boldsymbol{H}$ relation superimposed on the parabolic hysteresis loop. If residual (relaxation + resonance) losses dominate, elliptical hysteresis loops are followed, whereas parabolic loops are shown in the opposite case. The parabolic loop component of power loss linked to pure hysteresis can thus be calculated from equation (67b) according to equation (72). The hysteresis loss tangent value is correspondingly given by equation (73), showing also a reasonable approximation for this parameter under fields of low strength ($\mu'(\omega, \boldsymbol{H}_0) \approx \mu'_i(\omega)$).

$$W_h = \frac{\omega}{2\pi}\frac{4\mu_0 r \mathrm{H}_0^3}{3} \tag{72}$$

$$\tan\delta_h(\omega, \boldsymbol{H}_0) = \frac{4r\boldsymbol{H}_0}{3\pi\mu'(\omega, \boldsymbol{H}_0)} \approx \frac{4r\boldsymbol{H}_0}{3\pi\mu'_i(\omega)} \tag{73}$$

The eddy current loss tangent can be calculated at low frequencies according to equation (74).

$$\tan\delta_F(\omega) = \pi\zeta d^2\omega\mu_0\mu'(\omega, \boldsymbol{H}_0)\sigma(\omega) \tag{74}$$

The value of $\tan\delta_r$ derives from contributions of all the relaxation and resonance loss processes; it is highly variable from material to material and strongly dependent on both frequency and temperature. Equations (72) and (73) are not applicable in most practical cases, when remarkable field strengths and high frequencies are involved, but they are useful to describe the dependence of $\tan\delta_h$ and $\tan\delta_F$ on the material properties, object geometry, and field parameters. The contribution of $\tan\delta_F$ to the total $\tan\delta_m$ is small in finely subdivided magnetic materials because of the internal restrictions on the eddy current circulation, and it is negligible in dispersed, low-σ ferrimagnetic powders. Under not particularly strong fields, hysteresis losses may also play a minor role with respect to residual losses. This means that the magnetic energy loss contribution of magnetic fillers dispersed in nonmagnetic dielectrics under RF or MW fields of moderate intensity is primarily linked to their intrinsic magnetic relaxation and resonance loss capacities at the frequency considered, quantified by $\mu''(\omega, \boldsymbol{H}_0)$.

A useful and easily accessible parameter is the (relative) initial complex permeability $\mu_i^*(\omega) = \mu'_i(\omega) - j\mu''_i(\omega)$, attainable by measurements on dispersed powders at vanishingly low magnetic field strengths. Values of μ_i' and μ_i'' at radio and MW frequencies at room temperature for various ferrimagnetic materials (γ-Fe_2O_3, nickel ferrite, a manganese ferrite, and the yttrium–iron garnet) are given in Table 8. It reports also their typical σ values at room temperature, as well as the relevant Curie temperature, above which their magnetic field energy dissipation capacities disappear with the transition to a simple paramagnetic behavior.

Table 8 (Relative) Complex Dielectric Constant and (Relative) Initial Complex Permeability of Some Inorganic Paraelectric, Ferroelectric, and Ferrimagnetic Materials at Radio and Microwave Frequencies at Room Temperature.

Material	T_C (°C)	σ (Ω^{-1} m^{-1})		*Frequency* (Hz) 10^6	10^7	10^8	10^9	3×10^9
Muscovite (ruby mica)[a]	–	10^{-13}	ε'	–	5.4	–	–	5.4
			ε''	–	0.0016	–	–	0.0016
$CaCO_3$ (marble)[a]	–		ε'	–	9.0	–	–	9.0
			ε''	–	0.33	–	–	0.22
$BaTiO_3$[b]	130		ε_i'	1400	1300	1150	890	550[c]
			ε_i''	28	43.5	73	133	220[c]
γ-Fe_2O_3[d]	747		μ_i'	–	–	–	11	2.8
			μ_i''	–	–	–	~13	4
$NiO\cdot Fe_2O_3$[e]	590	10^{-2}	μ_i'	–	–	10	2.7	~1.5
			μ_i''	–	–	2.1	3.1	1.75
$Mn_{1.42}Fe_{1.58}O_4$[e]	222	$\sim10^{-2}$	μ_i'	–	–	33	~0.15	–
			μ_i''	–	–	~18.5	2.4	–
$3Y_2O_3\cdot 5Fe_2O_3$[e]	277	10^{-8}	μ_i'	115	115	35	1.5	–
			μ_i''	1.5	60	27	6.5	–

[a] Data taken from Ref. 25. [b] Polycrystalline ceramics. Data taken from: K.-H. Hellwagge (ed.), 'Landolt-Börnstein – Numerical Data and Functional Relationships in Science and Technology', Springer, Berlin, 1981, vol. III/16a. [c] Values at 10 GHz. [d] Data taken from Ref. 163. [e] Data taken from Ref. 164.

Besides magnetic losses, nonconducting magnetic materials under electromagnetic fields can suffer sometimes high electric field energy losses by dielectric relaxation. Indeed, ferrite materials show $\varepsilon'(\omega)$ and $\varepsilon''(\omega)$ values comparable to, and in many cases much higher than, those of many inorganic dielectrics, as exemplified in Table 9 for the two ferrites $NiO\cdot Fe_2O_3$ and $MgO\cdot Fe_2O_3$.

Table 9 Dielectric and Magnetic Properties of the Ferrites $NiO\cdot Fe_2O_3$ and $MgO\cdot Fe_2O_3$ (Quenched from 1200 °C) at 4.5 GHz and Room Temperature.[164]

	$NiO\cdot Fe_2O_3$	$MgO\cdot Fe_2O_3$
ε'	13.40	9.66
ε''	3.52	0.17
μ_i'	0.26	0.80
μ_i''	3.46	0.97

Hysteresis, dielectric relaxation, and electric conduction losses are the causes of electric field energy dissipations in ferroelectrics. Equations similar to (69) and (70), where the eddy current loss term is obviously absent, express the total electric loss tangent $\tan\delta_e(\omega, \boldsymbol{E}_0)$ and the overall electric loss factor $\varepsilon''_{eff}(\omega, \boldsymbol{E}_0)$ of a material (equation (75)).

$$\varepsilon''_{eff}(\omega, \boldsymbol{E}_0) = \varepsilon''_h(\omega, \boldsymbol{E}_0) + \varepsilon''(\omega, \boldsymbol{E}_0) + \frac{\sigma(\omega)}{\varepsilon_0\omega} \tag{75}$$

Considerations similar to those for ferro- and ferrimagnets are applicable to ferroelectric hysteresis losses, and to the meaning of the (relative) complex dielectric constant $\varepsilon^*(\omega, \boldsymbol{E}_0)$ and the (relative) initial complex dielectric constant $\varepsilon_i^*(\omega) = \varepsilon_i'(\omega) - j\varepsilon_i''(\omega)$ for $\boldsymbol{E}_0 \to 0$. Typical dielectric properties of the ferroelectric $BaTiO_3$[170] at room temperature at frequencies from 10^6 to 10^{10} Hz are reported in Table 8, in comparison with those of two representative inorganic dielectrics (muscovite and $CaCO_3$). The values of ε'' at GHz frequencies exemplify the extremely strong dielectric loss capacities of the ferroelectric material.

19.4 MICROWAVE HEATING AND PROCESSING OF POLYMERIC MATERIALS

Subjects covered in this section are the theoretical and experimental aspects of MW heating of neat polymers, thermosetting resins, and related composites as functions of their own electrical and magnetic properties at such frequencies, as well as a survey of applications of MW heating itself for processing of polymeric materials. In the latter respect, however, design and performance of MW heating equipment (including MW generators) are not covered; this section focuses on the intrinsic physical, physicochemical, and chemical response of polymeric materials and polymerizing systems to MW irradiation. Descriptions of equipment design and performance can be found in various books on industrial MW heating[20–23,25] and RF/MW engineering in general.[171] Surveys are given in several reviews on the applications of microwaves to chemical technologies.[24,26,43]

MW generators (or MW tubes) are of various types and power output capacities;[24,102,104] types include disk-seal triodes, klystrons, magnetrons, travelling-wave tubes (TWT), backward wave oscillators (BWO), and other less commonly used devices based on similar operation principles. Magnetrons are the most widely employed generators for laboratory and industrial MW processing of materials (as well as being used in domestic MW ovens). Magnetrons of nominal power from 0.5 to 30 kW and over are commercially available. High-power klystrons (up to 50 kW) have also been used conveniently in industrial MW heating plants. TWT generators, traditionally confined to scientific and military uses, have found recent interesting applications for materials processing.

MW applicators are of many different types, and an updated comprehensive description of their theory and technology is given by Metaxas and Meredith.[25] Despite their variety, most MW applicators belong to three main categories, whose principles are explained in textbooks on fundamentals of MW physics and MW devices.[1–4,102–104] These are travelling wave applicators, single-mode resonant cavity applicators, and multimode resonant cavity applicators.

Travelling wave applicators consist of hollow metal tubes along which waves, supplied by the generator in a certain position, propagate and are partially absorbed by the material being treated inside, and their residue is finally dissipated in a highly MW-absorbing termination. The material undergoing MW irradiation may flow continuously in the form of a strip, bar, pipe, *etc.*, axially to the tube (axial travelling-wave applicators), or transversely to it, describing a serpentine, through slots (serpentine, meander, or folded-guide applicators).

Single-mode resonant cavity applicators comprise metallic containers of exactly determined geometry and dimensions (usually cylindrical or parallelepipedal) where a MW signal, injected by a coaxial cable or a waveguide, gives place to a standing wave regime *via* a multiple-reflection mechanism by the cavity walls. Various resonance modes can be selected, with correspondingly different (inhomogeneous but regular) spatial distributions of the electric and magnetic field intensities being attainable. These applicators, enabling extremely high (local) field intensities, are not practical for treatment of large specimens, but they are particularly useful and versatile for laboratory investigations and whenever selective processing or strong irradiation of thin samples (fibers, films, *etc.*) is required.

Multimode resonant cavity applicators consist of metal boxes (currently rectangular) where the MW radiation injected through a waveguide gives rise to a great number of simultaneous resonance modes, with a complex and normally irregular (and also scarcely controllable) distribution of the MW field intensity. Most of the MW furnaces presently used (including all domestic MW ovens) are of this type, because of the simplicity, versatility, and virtual absence of limits to dimensions and MW power. Closed multimode ovens are usable for discontinuous materials treatments. Open ovens are easily employed as tunnel furnaces for continuous-flow processes. As anticipated in Section 19.2.1, many methods can be used to reduce, and often to almost eliminate, the MW heating inhomogeneities linked to the uneven distribution of the field intensity (*e.g.* the rotating plate solution commonly adopted in domestic ovens).

In recent years, many efforts have been made to enhance the knowledge and technical potential of single-mode[172–176] and controlled multimode[172,175] resonant cavity applicators. Numerous other types of MW applicators not ascribable to one of the above principal categories, or hybrids of them, have been devised to tailor the MW irradiation to specific applications (*e.g.* to fibers, films, profiles of complex cross-section, *etc.*). Due to their variety and complexity, only a few of them can be mentioned here: ridged waveguide applicators, travelling resonator applicators, dielectric loaded applicators, helix applicators, single-mode resonant cavities with elliptical cross-section, radiator (*i.e.* nonresonant) applicators (such as horn applicators and waveguide radiators), and so on.

19.4.1 Volumetric Radiofrequency and Microwave Heating of Neat and Filled Polymers and Resins

The major aspects of the effects of RF and MW irradiation to consider in this context as functions of frequency and strength of the applied field, and of the electric and magnetic properties of the irradiated material, are: (i) the electromagnetic power dissipated (or lost) per unit volume in, or absorbed by, the material; (ii) the consequent heating rate of the material itself; (iii) the penetration depth of the electromagnetic radiation into the material considered; and (iv) the MW heating profiles and temperature-distribution patterns in objects of defined shape and dimensions.[20–26]

The time-average power flux associated with an electromagnetic radiation propagating in space is expressed by the Poynting vector $\mathbb{P}$ ($W\,m^{-2}$) as in equation (76), *i.e.* by the vector product of the mean electric and magnetic field strengths $\bar{\boldsymbol{E}}$ and $\bar{\boldsymbol{H}}$, or the half vector product of the oscillation amplitudes of the electric and magnetic field strengths.

$$\mathbb{P} = \bar{\boldsymbol{E}} \wedge \bar{\boldsymbol{H}} = \tfrac{1}{2}\boldsymbol{E}_0 \wedge \boldsymbol{H}_0 \tag{76}$$

From $\mathbb{P}$ and Maxwell's equations it is possible to calculate, as in equation (77), the time-average power absorbed per unit volume ($W\,m^{-3}$) by a material from an incident electromagnetic radiation of frequency ω and intensity constant in time and uniform over the material volume considered, at constant temperature and in the absence of physical and chemical transformations (*i.e.* with constant electric and magnetic properties of the material itself).

$$\begin{aligned} W &= \omega\varepsilon_0\varepsilon''_{\text{eff}}(\omega, \boldsymbol{E}_0)\bar{\mathrm{E}}^2 + \omega\mu_0\mu''_{\text{eff}}(\omega, \boldsymbol{H}_0)\bar{\mathrm{H}}^2 \\ &= \tfrac{1}{2}\omega\varepsilon_0\varepsilon''_{\text{eff}}(\omega, \boldsymbol{E}_0)\mathrm{E}_0^2 + \tfrac{1}{2}\mu_0\mu''_{\text{eff}}(\omega, \boldsymbol{H}_0)\mathrm{H}_0^2 \end{aligned} \tag{77}$$

In equation (77), $\varepsilon''_{\text{eff}}(\omega, \boldsymbol{E}_0)$ is the effective (or total) electric loss factor of the material, *i.e.* the sum of the ε'' contributions of all possible electric field energy losses at the frequency ω and for the electric field strength amplitude $\boldsymbol{E}_0$, as in equation (78): hysteresis, dielectric relaxation and electric conduction losses (discussed in Sections 19.3.2.2(ii), 19.3.2.4, and 19.3.2.5), as well as interfacial polarization losses in multiphase materials (Section 19.3.2.3). Obviously, in polymer blends, composites, and composite structures, the various material components and interfacial polarization phenomena make their own contributions to the overall $\varepsilon''_{\text{eff}}$.

$$\varepsilon''_{\text{eff}}(\omega, \boldsymbol{E}_0) = \varepsilon''_{\text{h}}(\omega, \boldsymbol{E}_0) + \varepsilon''(\omega, \boldsymbol{E}_0) + \frac{\sigma(\omega)}{\varepsilon_0\omega} + \varepsilon''_{\text{MWS}}(\omega) \tag{78}$$

Analogously, $\mu''_{\text{eff}}(\omega, \boldsymbol{H}_0)$ is the effective (total) magnetic loss factor of the material considered, discussed in Section 19.3.2.5 (equation (70)). With reasonably low field strengths, electric and magnetic hysteresis losses are minimized, and the related loss factor contributions may be neglected. Consistent with what was mentioned in Section 19.3.2.5, in ordinary polymers and resins filled with dispersed ferromagnetic metal particles below the percolation threshold or with ferrimagnetic powders, the eddy current loss term $\mu''_{\text{F}}(\omega)$ is also very small and often negligible. In such common situations, the expressions of $\varepsilon''_{\text{eff}}$ and μ''_{eff} reduce to equations (79a) and (79b), respectively.

$$\varepsilon''_{\text{eff}}(\omega, \boldsymbol{E}_0) = \varepsilon''(\omega, \boldsymbol{E}_0) + \frac{\sigma(\omega)}{\varepsilon_0\omega} + \varepsilon''_{\text{MWS}}(\omega) \tag{79a}$$

$$\mu''_{\text{eff}}(\omega, \boldsymbol{H}_0) = \mu''(\omega, \boldsymbol{H}_0) \tag{79b}$$

In all ordinary paraelectric and diamagnetic polymers, polymer blends, resins, and their composites with nonmagnetic fillers or fibers, $\mu' \approx 1$ and $\mu''_{\mathrm{eff}} \approx 0$, so that only electric losses are involved and equation (77) reduces to the first additive terms, with $\varepsilon''_{\mathrm{eff}}$ being expressed by the further simplified equation (80).

$$\varepsilon''_{\mathrm{eff}}(\omega) = \varepsilon''(\omega) + \frac{\sigma(\omega)}{\varepsilon_0 \omega} + \varepsilon''_{\mathrm{MWS}}(\omega) \tag{80}$$

From equation (77) it is simple to deduce the expression (81) for the instantaneous heating rate at constant volume and in adiabatic conditions of a material (of specific heat at constant volume $C_v(T)$) subjected to an electromagnetic radiation of determined frequency and intensity.

$$\left(\frac{\mathrm{d}T}{\mathrm{d}t}\right)_{V,S} = \frac{\omega}{2\pi\rho(T)C_v(T)}[\varepsilon_0\varepsilon''_{\mathrm{eff}}(\omega, \boldsymbol{E}_0, T)\bar{\mathrm{E}}^2 + \mu_0\mu''_{\mathrm{eff}}(\omega, \boldsymbol{H}_0, T)\bar{\mathrm{H}}^2] \tag{81}$$

Since, for condensed matter, $C_p(T) \approx C_v(T)$, the more easily accessible specific heats at constant pressure can be introduced in equation (81) in place of those at constant volume. Yet, $(\mathrm{d}T/\mathrm{d}t)_{V,S}$ may be considered approximately equal to $(\mathrm{d}T/\mathrm{d}t)_{\mathrm{p},S}$ in most practical cases. Equation (81), reduced to its first electric loss term for materials containing no magnetic components, shows a direct variation of the heating rate to the applied radiation frequency, square field strength, and loss factor. This is the basis for predictions of the RF and MW heatability of any material. As a practical criterion, one may define different degrees of heatability of materials suffering only electrical losses according to different ranges of $\varepsilon''_{\mathrm{eff}}$ values at the frequency and temperature considered: feeble or negligible heating, $\varepsilon''_{\mathrm{eff}} < 0.01$; modest heating, $0.01 < \varepsilon''_{\mathrm{eff}} < 0.08$; fair heating, $0.08 < \varepsilon''_{\mathrm{eff}} < 0.2$; good heating, $0.2 < \varepsilon''_{\mathrm{eff}} < 1$; very strong heating, $1 < \varepsilon''_{\mathrm{eff}}$. Obviously, magnetic losses can be considered analogously, in addition to the electric ones, through the μ''_{eff} value.

Equation (81) shows that the adiabatic heating rate is a complex function of temperature *via* the temperature dependence of density, specific heat, and effective electric and magnetic loss factors of the material. Temperature–time curves for electromagnetic adiabatic heating can be calculated by integration of the same equation. The temperature dependence at radio and MW frequencies of the loss factor contributions linked to the various loss phenomena in polymeric materials have been discussed in Section 19.3.2. Accordingly, general and rapid increases of both dielectric relaxation and electric conduction losses with temperature are normal for polymeric substances, with progressive and remarkable increases of RF and MW heatability with temperature being displayed by most polar polymeric materials. Owing to the very low conductivities in neat ordinary polar polymers, appreciable contributions to $\varepsilon''_{\mathrm{eff}}$ can be expected only from dielectric losses, especially at high (MW) frequencies because of the dominant proportionality to ω^{-1} in the expression for the electric conduction loss factor (equation (57)). According to the representative data of Tables 3, 6, and 7, and Figure 9, this is true over broad temperature intervals, despite the rapid increases of conductivity with temperature itself. Only at high temperatures, at which ionic conductivity may become important (*e.g.* in nylons and PET above 200 °C), conduction losses can contribute significantly to $\varepsilon''_{\mathrm{eff}}$. Different and even reversed ratios between the contributions of dielectric relaxation and conduction losses are expected in conductive and conductor-filled polymers.

Among the initial studies on MW heating of neat polymers, it is worth mentioning the one, already cited,[113] on a series of commercial elastomers ranging from nonpolar (natural rubber, EPDM, butadiene–styrene rubbers) to polar ones (polychloroprene, butadiene–acrylonitrile rubber, *etc.*), at the ISM frequency of 2.45 GHz. According to their ε'' values at best slightly greater than 0.01 at this frequency over the entire temperature interval 25–100 °C, nonpolar rubbers displayed under a MW field of constant strength only low heating rates of the order of 10^{-1} °C s^{-1}. By comparison, polar rubbers, with ε'' variable from ~0.08 to ~0.6 over the same temperature range, showed heating rates of the order of 5–10 °C s^{-1} under the same experimental conditions. More recently, systematic investigations have been conducted on the MW heating at 2.45 GHz of a range of polar polymers under continuous irradiation in single-mode cavities and in waveguide applicators: synthetic polar elastomers (polychloroprene and styrene–acrylonitrile rubbers),[50b] glassy amorphous polymers (various polyacrylates and polymethacrylates, poly(vinyl acetate) (PVAc), *etc.*),[50b,177–179] and some semicrystalline plastics (nylons, PET, poly(4,4′-oxydiphenylene ether 1,4-phenylene ketone (PEEK)).[180–182]

From equation (81), the variation of the electromagnetic heating rate with temperature under constant frequency and field strength is proportional to the temperature variation of the effective electric loss factor $(d\varepsilon''_{eff}/dT)_\omega$. In other words, the variation of $(dT/dt)_{\omega,\bar{E}}$ with temperature for any polymeric substance is expected to follow the derivative of its dielectric relaxation spectrum at fixed frequency (Figures 5(b) and (8b)). According to the general description in Section 19.3.2.2 of dielectric relaxation processes in polymers at radio and MW frequencies, the strong α or α_a relaxations can contribute to give high ε'' and $(dT/dt)_\omega$ values at MW frequencies only beyond T_g, at temperatures which increase as the frequency increases.

Figure 12 shows typical continuous MW heating curves for amorphous and semicrystalline polymers at constant frequency and field intensity under nearly adiabatic conditions, together with the related ε'' vs. temperature curve. The temperature increases slowly with time, and $(dT/dt)_{\omega,\bar{E}}$ is modest and almost constant with temperature until a critical temperature T_{crit} ($>T_g$), variable from polymer to polymer and increasing with frequency, is reached. At T_{crit}, the α or α_a relaxation begins to be involved. As the temperature increases further, ε'' rapidly increases, implying a progressive, nearly proportional, increase of the heating rate. This means that, after having reached T_{crit}, heating becomes faster and faster, *i.e.* a temperature runaway occurs, with dramatic risks of overheating, and polymer degradation, up to combustion. The process may be controlled by modulation of the MW field intensity.

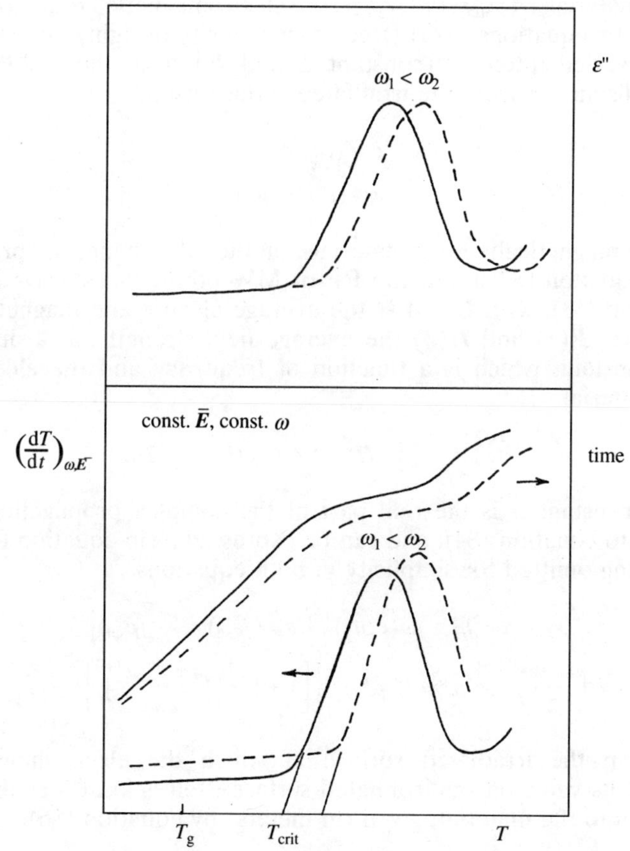

Figure 12 Dielectric loss factor and heating rate vs. temperature, and variation of temperature with time, for an amorphous polar polymer under continuous microwave irradiation in quasi-adiabatic conditions.

Table 10 reports T_{crit} values for some amorphous and semicrystalline polymers at the irradiation frequency of 2.45 GHz. T_{crit} is low when T_g is low, so that polar rubbers exhibit T_{crit} values around or below room temperature. This agrees with what was said in Section 19.3.2.2 about their having good MW heatability already at room temperature. Weakly polar polymers manifest a similar, though less marked, behavior. The difference ($T_{crit} - T_g$) is small when the apparent activation energy of the α or α_a relaxation is low (see Table 4).[178]

Table 10 Critical Runaway Effect Temperature (T_{crit}) of Various Polymers in Microwave Heating at 2.45 GHz.[a]

Polymer	T_g (°C)	T_m (°C)	T_{crit} (°C)
PVAc	30	–	65
Nylon-12	37	178	85
Nylon-6	63	229	105
PVC	85	–	113
PMMA	105	–	185

[a] Data taken from Refs. 178, 179, and 182.

The above approach interprets the kinetics of adiabatic RF and MW heating with the intensity of the electromagnetic field supposed uniform throughout the material. This assumption, acceptable for thin specimens and inherently low-absorbing or low-density materials, may be largely incorrect for thick objects under MW fields, and is totally inapplicable to conductors and other highly lossy materials (*e.g.* ferroelectrics and ferromagnets) both at radio and MW frequencies.[16–18,20–25,162] This can be evaluated by an analysis of the propagation and attenuation of electromagnetic plane waves through condensed matter.[3–5] In free space and in nonabsorbing materials ($\varepsilon''_{eff} \approx 0$, $\mu''_{eff} \approx 0$), electromagnetic radiations propagate with velocities expressed by equations (82a) (free space velocity of light) and (82b), respectively, in both cases with unvaried intensity (constant $\bar{\boldsymbol{E}}$ and $\bar{\boldsymbol{H}}$) from the irradiated surface to the interior, up to any distance z from the irradiated surface itself.

$$c = (\varepsilon_0 \mu_0)^{-0.5} \tag{82a}$$

$$v = (\varepsilon_0 \varepsilon' \mu_0 \mu')^{-0.5} \tag{82b}$$

In electrically and magnetically lossy materials, on the other hand, the propagation velocity is still that given by equation (82b), but the RF or MW power flux decays exponentially with z according to equation (83), with $\bar{\boldsymbol{E}}$ and $\bar{\boldsymbol{H}}$ the average electric and magnetic field strengths on the irradiated surface, $\bar{\boldsymbol{E}}(z)$ and $\bar{\boldsymbol{H}}(z)$ the average field strengths at a distance z, and α the field attenuation constant, which is a function of frequency and the electrical and magnetic properties of the material.

$$\mathbb{P}(z) = \bar{\boldsymbol{E}}(z) \wedge \bar{\boldsymbol{H}}^*(z) = \bar{\boldsymbol{E}} \wedge \bar{\boldsymbol{H}} \exp(-2\alpha z) \tag{83}$$

The attenuation constant α is the real part of the complex propagation constant γ of the radiation according to equation (84), and can be expressed as in equation (85), the dependence of ε and μ on ω being omitted for simplicity in both equations.

$$\gamma = \alpha + j\beta = j\omega(\varepsilon_0\mu_0)^{0.5}[(\varepsilon' - j\varepsilon''_{eff})(\mu' - j\mu''_{eff})]^{0.5} \tag{84}$$

$$\alpha = \omega\left(\frac{\varepsilon_0\mu_0}{2}\right)^{0.5}(\varepsilon'\mu' - \varepsilon''_{eff}\mu''_{eff})^{0.5}\left\{\left[1 + \left(\frac{\varepsilon'\mu''_{eff} + \varepsilon''_{eff}\mu'}{\varepsilon'\mu' - \varepsilon''_{eff}\mu''_{eff}}\right)^2\right]^{0.5} - 1\right\}^{0.5} \tag{85}$$

The distance from the irradiated surface at which the electromagnetic power flux is attenuated to e^{-1} of its value on the irradiated surface itself is known as the penetration depth D_p of the radiation into the material, given (in meters) by equation (86).

$$D_p = \frac{1}{2\alpha} \tag{86}$$

For nonmagnetic materials ($\mu' \approx 1$, $\mu''_{eff} \approx 0$) such as neat ordinary polymers, equation (85) reduces to equation (87), implying a correspondingly simplified expression of D_p.

$$\alpha = \omega\left(\frac{\varepsilon_0\varepsilon'\mu_0\mu'}{2}\right)^{0.5}\left\{\left[1 + \left(\frac{\varepsilon''_{eff}}{\varepsilon'}\right)^2\right]^{0.5} - 1\right\}^{0.5} \tag{87}$$

Equations (85) and (86) show how the intensity attenuation of an electromagnetic radiation travelling through a material increases, *i.e.* its penetration capacity into the material itself decreases, as the frequency of the radiation and ε', μ', $\varepsilon''_{\text{eff}}$, and μ''_{eff} of the material at the frequency considered become higher. The penetration depth gives an indication of when the attenuation of the electromagnetic field intensity must be taken into consideration: this is when the thickness of the object is of the same order of magnitude as D_p, or higher. Table 11 reports D_p values calculated by equations (87) and (86) at frequencies from 10^4 to 10^{10} Hz for a hypothetical (representative) polymer having $\varepsilon' = 3$ and $\varepsilon''_{\text{eff}} = 0.01$, 0.1, and 1 (with both ε' and $\varepsilon''_{\text{eff}}$ assumed to be independent of frequency).

Table 11 Penetration Depth of Radio Waves and Microwaves into a Non-magnetic Material of (Relative) Dielectric Constant $\varepsilon' = 3.0$ and Different (Effective) Electric Loss Factor (Penetration Depths in Meters).[a]

$\varepsilon''_{\text{eff}}$	*Frequency* (Hz) *10^4*	*10^6*	*10^8*	*10^{10}*
0.01	2.7×10^5	2.7×10^3	27	2.7×10^{-1}
0.1	8.2×10^4	8.2×10^2	8.2	8.2×10^{-2}
1	8.4×10^3	84	8.4×10^{-1}	8.4×10^{-3}

[a] Values calculated by equations (86) and (87).

Minimum D_p values of the order of one meter at 100 MHz for such highly lossy materials as the one with $\varepsilon''_{\text{eff}} = 1$ show that no great attention has generally to be paid to the field uniformity across objects in practical RF processing of polymers. D_p values of the order of from centimeters to millimeters at GHz frequencies for significantly to highly lossy materials, by contrast, indicate that field intensity decays must be considered in MW processing of polymer samples of any appreciable thickness. Even smaller D_p values (below one millimeter) can be calculated at MW frequencies for conductive polymers, ionic-conducting polymer systems, and conductor-filled polymers with σ exceeding $1\,\Omega^{-1}\,\text{m}^{-1}$, as well as in polymers filled with inorganic ferroelectrics, leading to $\varepsilon''_{\text{eff}}$ values $\gg 1$ (see Table 8). Similarly low D_p values can also be expected according to equation (85) in polymers containing magnetic fillers, by virtue of their strong μ''_{eff} contributions (as in Table 8).

In summary, increasing electric and/or magnetic loss capacities, enabling materials to undergo proportionally faster and faster MW heating, are responsible for a correspondingly stronger and stronger attenuation of the MW field intensity across their thickness, with a tendency to progressively steeper decreases of heating rate from the irradiated surface to the core of objects. The above features make the temperature versus thickness profiles in thick objects undergoing MW irradiation appreciably uneven, and the prediction of the related time evolution a very difficult task. Indeed, each MW heating step determines, *via* a certain temperature profile created across the sample thickness, a corresponding profile of values of density, specific heat, and electrical and magnetic properties of the material. This distribution of density values and thermal, electrical, and magnetic properties determines in turn a certain profile of MW field intensity, a certain heating rate distribution, and then a new temperature profile across the material thickness, and so on. The MW heating thus proceeds as a continuous succession of (continuously changing) temperature profiles (transient temperature profiles) across the object thickness.

Further computational problems arise when nonadiabatic conditions are considered. Obviously, MW-heated bodies immersed in cool air lose more heat as their surfaces become hotter, both by irradiation (governed by the Stephan–Boltzmann equation) and convection, according to the temperature of the gas stream and an appropriate heat transfer coefficient.[25] Only numerical methods of calculation are viable to model such complex situations. Many studies based on finite-difference, finite-element, and boundary element methods of calculation have been devoted to analyze electromagnetic power and temperature distributions in different objects under MW heating.[183] Computer simulations of transient temperature profiles in plane slabs of homogeneous materials with temperature-dependent dielectric properties and constant specific heat and density were published some years ago.[184,185] A recent study[183] predicts by a finite-element calculation the transient temperature profiles in homogeneous and isotropic slabs (of materials with both temperature-dependent electrical and thermal properties) immersed in air at room temperature, and irradiated on their opposite faces with a homogeneous MW field. When applied to nylon-6,6 slabs, the model predicts the inner temperature runaways proper to the MW heating of polymeric substances under conditions approaching adiabatic ones, as discussed above.

A finite-element modeling of MW heating profiles in cylindrical and parallelepipedal polymer samples, with and without air convection, was published a few years ago.[186] Inhomogeneous space distributions of the MW field (as normally encountered in almost all MW applicators, including the multimode resonant cavity ovens) create transverse transient temperature-distribution patterns in addition to the transient temperature profiles across the sample thickness. A three-dimensional, finite-difference time-domain method has been developed for the numerical simulation of MW heating in such very common cases.[187]

19.4.2 Applications of Microwave Heating

Practical and potential applications of MW heating technology to polymeric materials, as described in the patent and scientific literature, are numerous and spread over a wide field. The survey presented here is based on a subdivision into a few main areas of interest; this is more an attempt to manage a huge collection of literature documents than a scientific criterion.

19.4.2.1 Applications of purely physical microwave processing

Rapid, intense, and selective heating of thin polymer layers is one of the most valuable features of MW irradiation. The selective MW melting of interlayers of thermoplastic polar polymers as hot-melt adhesives has been disclosed in many patents[188–195] for quick bonding (in times of seconds) of thermally degradable or shrinkable fabrics, paper, polymer films, and so on. Fast and selective heating has been claimed for the rapid embedding of thermally shrinkable fibers into polar polymers,[196] and the differential stretching of multilayer films of polar and nonpolar polymers.[197] A great variety of polar polymers have been considered, including PVC, PVDC, PVAc, polyacrylates, and related copolymers, cellulose esters, polyurethanes, *etc.*

Other applications of fast MW heating of highly polar polymers described in the patent literature are for heating of preforms for blow molding of acrylonitrile and methacrylonitrile copolymers,[198] and for quick sealing (in tens of milliseconds) of flexible containers made of ethylene–carbon monoxide copolymers and ethylene–carbon monoxide–acrylic acid terpolymers, under electromagnetic irradiation with frequencies from 1 MHz to 10 GHz.[199,200] Thin layers of conductive polymers, strongly and rapidly heatable by MW irradiation, have been proposed for different purposes: (a) promotion of almost instantaneous MW melting of polymer seals of vial lids;[201] (b) tapes or powders of polypyrrole and polyaniline *p*-toluenesulfonates (with σ of 2.2×10^3 and $9\times10^2\ \Omega^{-1}\,m^{-1}$, respectively) as MW-absorbing interlayers for fast MW welding (in times of a few seconds) of nonpolar or weakly polar thermoplastic polymers.[202]

The MW heatability of polar polymers to be used as fast 'MW adhesives' can be enhanced by the addition of highly dipolar organic liquids, as in the case of PVC, PVDC, and related copolymers by introduction of small amounts of sulfolane.[203] The accelerated formation of polymer films from solutions of polar or nonpolar polymers in organic solvents of different polarity under MW heating has also been studied.[204] Organic and inorganic high-ε'' additives have been used to enable the RF and MW processing of nonpolar or weakly polar polymers of extremely high melt-viscosity (*e.g.* ultra-high molecular weight polyethylene) and high-melting polymers such as poly(phenylene sulfide),[205] *e.g.* *N*-ethyltoluenesulfonamides,[206] ZnO, bentonite clay, alkali and alkaline-earth metal aluminosilicates.[207] Other MW-absorbing additives disclosed for similar purposes are compounds such as Ti or Zr hydroxyethylphosphonates.[208] More recently, it has been claimed that inorganic fillers of high ε' and low $\varepsilon''_{\mathrm{eff}}$ (>10 and <0.05, respectively, such as Al_2O_3, TiO_2, $CaTiO_3$, *etc.*) enhance *via* synergistic effects the MW heatability of alkene–carbon monoxide copolymers and related terpolymers with other vinyl monomers.[209] The complex dielectric constant at radio and MW frequencies of polymer–mineral filler systems can be predicted from the ε' and ε'' of the constituents according to the logarithmic Lichtenecker formula (88), with ϕ the volume fraction of the filler and the indexes 1 and 2 referring to the polymer matrix and the filler, respectively.[210]

$$\log_{10}\varepsilon^* = (1-\phi)\log_{10}\varepsilon_1^* + \phi\log_{10}\varepsilon_2^* \qquad (88)$$

In accordance with what was mentioned in the previous subsection and, particularly, in Section 19.3.2.5, the strong dielectric loss capacities of inorganic ferroelectrics and the strong magnetic (and often also dielectric) loss attitudes of ferrimagnets can be exploited by using them as fillers to make MW-heatable nonpolar polymers, or to enhance the MW heatability of polar ones. This approach has been claimed to produce polymeric adhesives or coatings ultra-rapidly meltable by MW irradiation, comprising polar polymers (*e.g.* ethylene–vinyl acetate copolymers) containing high percentages of $BaTiO_3$, γ-Fe_2O_3, $BaO{\cdot}18Fe_2O_3$, *etc.*[211,212] Heat-resistant blends of aromatic poly(ether sulfones)s, poly(ether ketone)s, or polyimides and inorganic ferroelectric fillers ($BaTiO_3$, $BaZrO_3$, *etc.*) or ferrites (*e.g.* Ba ferrites) have been devised as heavy-duty MW-absorbing materials.[213] Studies on modeling of the electromagnetic properties of MW-absorbing composites of polymers and Fe_3O_4[214] or magnetic fillers in general[215] have also been published.

An important method to achieve or enhance the MW heatability of polymers is the introduction of conductive fillers. For example, the addition of carbon black powder was claimed years ago to enable the MW sintering of such a MW-transparent polymer as PTFE.[216] The introduction of carbon black, graphite powder, or carbon fibers has also been disclosed for fast MW-activatable adhesives[217] similar to those mentioned above, and MW heating of thermoplastics.[218] As stated earlier, conductive fillers can give to the overall $\varepsilon''_{\rm eff}$ of a material the two contributions of electric conduction and interfacial polarization (MWS) losses. Both contributions are worth considering in this respect, although the former is expected to play a dominant role in UHF and SHF heating because of the normal location of MWS relaxations in the RF region (Section 19.3.2.3). The use of carbon blacks as conductive fillers in polymers and the relevant effects on their MW heatability have been studied since the beginning of MW processing of polymeric materials, within the best established area of rubber vulcanization.[113–115] Several types and many grades of carbon black are known, differing in particle size and morphology, surface area, conductivity, *etc.*:[155] furnace blacks (the most widely used ones in the rubber industry), 'medium thermal' blacks, and extra-conductive blacks.

The increases of $\varepsilon''_{\rm eff}$ and MW heating rate (dT/dt) caused by the dispersion of carbon black into a polymer can be considerable and sometimes even excessive for practical purposes.[113] At the same concentration, different types and grades of black can produce greatly different enhancements of $\varepsilon''_{\rm eff}$ and MW heatability. Under the same experimental conditions of continuous MW irradiation at 2.45 GHz, for example, 33.3% by weight of various blacks increased the MW heating rate of natural rubber from ~0.2 °C s^{-1} to values ranging from 0.8 to 8 °C s^{-1}.[113] A systematic study of the dielectric properties and MW heatability of a great variety of EPDM–carbon black compounds at 2.5, 3.2, and 4.0 GHz has clarified the general influence of important parameters on the MW response of carbon black-filled polymers.[219] The $\varepsilon''_{\rm eff}$ and MW heating rate of a composite increase with the black fraction ϕ according to the general trend shown in Figure 10 for σ of conductor-filled polymers. The critical ϕ value for the percolation threshold decreases with increasing conductivity and surface area of the carbon black. At constant ϕ, log $\varepsilon''_{\rm eff}$ and log (dT/dt) increase linearly with the surface area of the black and temperature, respectively.

Further investigations on the RF (10 Hz to 800 kHz) and MW (2.45 GHz) responses of conductor-filled polymers have been performed on the MW-promoted curing of epoxy resins, filled with conductive powders of different nature: copper,[220,221] iron,[222] carbon black,[223,224] and aluminum.[225,226] From determinations of the power absorbed by the different materials (with various filler amounts), several important conclusions have been drawn, particularly regarding the influence of the composite micro-morphology. MWS relaxation losses have been detected within the RF region (*e.g.* with a peak centered at ~1 kHz in copper-filled materials), so that, at 2.45 GHz, the overall electric losses result in practice from the sole contributions of relaxation losses of the resin matrix and the electronic conduction losses induced by the filler. Below the electrical percolation threshold, conduction losses comprise (relatively modest) eddy current losses at the surfaces of isolated particles and particle aggregates (or clusters) of the conductor, and interparticle and inter-/intracluster conduction phenomena by tunneling and/or hopping mechanisms.

The number and dimensions of particle clusters tend to increase with the filler concentration. At a suitably high filler concentration, the formation of a continuous network of interparticle contacts takes place, corresponding to the percolation threshold of the material and its transition from a dielectric to a semiconductor (*e.g.* slightly below 20% by weight, or 3.1% by volume, of copper). No percolation, however, could be observed in Al-filled products, because of the insulating layer of alumina particles covering this electropositive metal. As a general behavior, the absorbed power increases with the filler amount up to a maximum for a critical concentration, variable from filler to filler, above which it decreases. This critical filler concentration, corresponding to considerable clustering, differs from that for the percolation threshold. This phenomenon has been interpreted by considering multiple reflections of waves by the surfaces of conductive particles through the dielectric resin matrix. Multiple reflections lengthen the path of waves within the material, implying enhanced power absorption by the resin *via* dielectric relaxations. In clusters, inner particles cannot reflect waves to the dielectric matrix, because they are shielded by the outer ones, and only cluster shells can contribute to the multiple-reflection phenomenon. The power absorption is thus expected to increase with the number of reflecting particles (*i.e.* with the filler concentration), until the subtraction of particles resulting from the possibility of reflecting waves by clustering becomes dominant.

Further investigations on the MW properties of particulate metal-filled polymers have focused on the influence of filler concentration and conductivity, particle size, and morphology: cured polyester and epoxy resins filled with a variety of metal powders (Al, Cu, doped Si, Ag, Ni) at 8.5–11.5 GHz,[227] silicone rubber-Al fibers composites at 2–25 GHz,[228] and iron powder-filled polypropylene at 2400 MHz.[229] Electric percolation does not take place when particles are thoroughly dispersed, *i.e.* when they are well separated from each other and clustering is modest or negligible. In other cases, the onset of percolation occurs, accompanied by a rapid ε''_{eff} increase, at filler concentrations which are lower when the particles are smaller. Dielectric constants follow the Wu–McCullough equation (89),[227] with ε_1^* the (relative) complex dielectric constant of the polymer matrix, and $\langle h \rangle$ an average shape factor of conductive particles, *e.g.* $h = \frac{2}{3}$ for spheres, and $h \rightarrow 1$ for inclusions with aspect ratio $\rightarrow \infty$ (*i.e.* fibers or chain structures of particles).

$$\varepsilon^* = \frac{(1 - \langle h \rangle)(1 - \phi) + \langle h \rangle \phi}{(1 - \phi)^2 (1 - \langle h \rangle)} \varepsilon_1^* \tag{89}$$

In Fe-filled polypropylene, both electrical and magnetic losses, increasing with the Fe content, contribute to an increasingly stronger MW heatability, reaching high levels above the percolation threshold (at around 35% by volume of Fe), mainly by virtue of the strong contribution of conduction losses.

In recent years, ordinary polymers filled with the conductive polymers mentioned in Section 19.3.2.4(ii) have achieved importance as excellent MW absorbers.[230] Contrary to what was said in Section 19.3.2.3, these composites seem to exhibit broad MWS relaxation losses at MW frequencies.[160] This is indeed predictable from equation (52) based on the peculiar characteristics of conductive polymers (having rather high σ associated with rather low ε') with respect to metals. For example, equation (52) applied to a polypyrrole-filled PVC (with σ and ε' of polypyrrole being $1 \times 10^2\ \Omega^{-1}\ m^{-1}$ and 300, respectively) predicts a MWS relaxation maximum at 5.8 GHz, in good agreement with a relaxation experimentally observed around 5 GHz.[160]

An interesting application of MW heating to polymers is MW heat-drawing to obtain high-modulus rods, tubes, and fibers; the fundamental and practical aspects have been reviewed.[46] This technique consists of the tensile drawing of semicrystalline polymers under MW irradiation which heats selectively the amorphous regions of the material with minimal heating and thermal disordering of crystalline regions (either already existing or as they are produced in the course of the drawing process itself). This facilitates the attainment of high degrees of orientation and crystallinity, and particularly high values of tensile modulus of the drawn materials.

The most successful application of the method has been for the fabrication of ultra-high modulus rods and tubes of POM with diameters of 1–2 mm.[231–233] The MW heat-drawing was accomplished in a cylindrical travelling waveguide applicator, operating at 2.45 GHz. It yielded POM rods with values of tensile modulus at room temperature varying from ~20 to 63 GPa, and T_m from 183 to 188 °C, for draw ratios from 10 to 33, respectively. The new technique led to gains in tensile modulus of 40–50% over the values attainable by conventional drawing procedures.

A detailed analysis of the influence of processing variables (draw ratio, temperature, electrical field strength, *etc.*), molecular weight, dimensions of the precursor rod, *etc.*, on the mechanical, thermal, and morphological properties of the resulting materials has been conducted.[232–236] It was estimated that a maximum tensile modulus for POM of ~72 GPa[234] could be achieved by MW heat-drawing. The reasons for the efficiency of this technique are understandable on the basis of the considerations on dielectric properties of semicrystalline polymers given in Section 19.3.2.2(ii). The frequency values for the relaxation peaks at increasing temperatures reported for this polymer in Table 5 show how the irradiation at GHz frequencies tends to heat selectively the amorphous regions (α_a relaxation) in POM at temperatures above 60 °C with respect to crystalline ones (α_c relaxation). Only in the proximity of T_m do the two relaxations merge, with the possibility of optimum selective heating and modulus increases under MW irradiation being ensured at temperatures between 100 and 160 °C.

The MW heat-drawing technique failed with PVDF, even though a similar behavior can be predicted for this polymer from the data of Table 5,[237] and gave minimal modulus gains with poly(vinyl alcohol),[238] PET,[239] and its blends with a liquid-crystalline copolyester [poly(ethylene terephthalate-*co*-*p*-hydroxybenzoate)].[240] Appreciable rigidity enhancements (~10%) were obtained with ethylene–vinyl alcohol copolymers.[238] Further applications of MW heating have been found for the faster and more efficient drying of polymer suspensions,[241] polymer solutions,[242] and solid polymers.[243,244]

19.4.2.2 Rubber vulcanization

As mentioned previously, rubber vulcanization is by far the most important industrial application of MW energy to polymeric materials. The feature of MW processing that has been of the greatest interest since the beginning of the technology is the ability to heat large and thick samples rapidly and uniformly, *e.g.* in tire manufacturing, either to perform molding or to promote vulcanization of rubber compounds.[111–114] A number of papers have been published on the uses of MW heating in the rubber industry: discontinuous preheating, prior to compression or injection-molding, in closed, multimode resonant cavity ovens,[245–248] and continuous vulcanization of extrudates of various cross-section, sheets, *etc.*, passing through waveguide-type applicators, MW resonant cavities, or the especially-developed devices mentioned in Section 19.4.[245–252] The main features of a variety of MW applicators for continuous rubber vulcanization are discussed by Schindler.[247] Generally, a hot air stream is coupled with MW irradiation, primarily to compensate for the heat losses from the surfaces of the MW-heated bodies and thus to achieve homogeneous internal temperature profiles. The two ISM frequencies around 900 Mz and that of 2.45 GHz are universally adopted.

Sulfur vulcanization is the most suitable one for MW processing. As was mentioned in Section 19.3.2.2 and in Section 19.4.1, only polar rubbers are directly MW processible. Carbon black-filled rubber compounds, which form the majority of rubber materials manufactured because of the reinforcing effects exerted by this class of fillers,[155] normally possess a sufficient MW heatability, which can be enhanced by use (or further addition) of highly conductive blacks.[111] Ordinary white mineral fillers such as clays, $CaCO_3$, *etc.* (employed in place of carbon black for lightly colored materials), cannot impart, because of their low ε'' and σ, any marked MW heatability to nonpolar elastomers. These rubbers, however, can achieve good MW heatability by being blended with polar ones (in particular, polychloroprene or butadiene–acrylonitrile copolymers), or other polar organic substances such as chlorinated paraffins, hydroxyalkylamines and related salts with fatty acids, glycols, glycerol, poly(oxyalkylene glycols,) and so on.[111,247] Silicas probably give the highest dielectric loss contributions among white mineral fillers. Rubbers filled with various metal powders have been considered as MW shields, and their dielectric properties have been studied over the broad 0.4–40 GHz range.[253] A curious phenomenon caused by MW heating is the blow-out of thick blocks of vulcanized rubbers at temperatures typically >180 °C.[254] MW blow-out is linked to a strong increase of internal pressure, principally due to the generation of volatile products.

19.4.2.3 Polymerization processes and 'athermal' microwave effects

In 1969, it was shown that it was possible to achieve fast radical polymerization at moderate temperatures (and in the absence of radical initiators) of vinyl monomers (*e.g.* styrene, alkyl (meth)acrylates, *etc.*) emulsified in water or as aqueous solutions ((meth)acrylic acid), under the influence of a pulsed RF field.[255] Similarly, the use of MW irradiation was claimed for the continuous and rapid peroxide-crosslinking of a variety of rubbers (including EPDM, butyl, and natural rubber) and plastics (polyethylene, PVC, *etc.*) containing small amounts of metal powders.[256] In the second half of the 1970s, the application of MW energy to some polymerization processes was disclosed not only as an extremely efficient heating method, but also as a way of inducing profitable reaction rate enhancements: (a) fast curing at moderate temperatures of epoxy binders for solid propellants;[257] (b) polycondensation of aliphatic dicarboxylic acids and diamines or ω-aminocarboxylic acids (*e.g.* the polycondensation of 12-aminoundecanoic acid was shown to require only 5 min of MW heating to afford a nylon-12 compared with times >100 min using conventional heating);[258] (c) synthesis of amino resins in times of less than a half of those normally needed.[259] In subsequent years, MW heating has been used to promote a great variety of both step and chain polymerization processes. A number of patent disclosures and scientific papers have addressed this subject, so we can now consider the existence and growth of a true body of literature on the MW activation of polymerization processes in general.

(i) Step polymerizations under microwave irradiation

Either continuous or pulsed MW irradiation (in practically all cases, at the standard frequency of 2.45 GHz) has been used to promote many types of di- and multifunctional polymerizations, generally with more or less pronounced reaction rate increases (and often with modified thermal and/or mechanical properties of the resulting polymers) with respect to the same processes under conventional heating.

(a) Polycondensation processes. Reaction accelerations have been described in the syntheses of polyester resins[260] and in the continuous post-polymerization of PET.[261] MW-accelerated syntheses of aromatic polyamides have also been studied recently.[262,263] MW irradiation has been found to be an excellent method to generate high-temperature-resistant polyimide films (for various applications in electronics) by cyclocondensation of polyamic acid precursors.[264] The kinetics of the thermal and MW-assisted imidization stage in solution for the process of Scheme 2 have been analyzed in detail.[265]

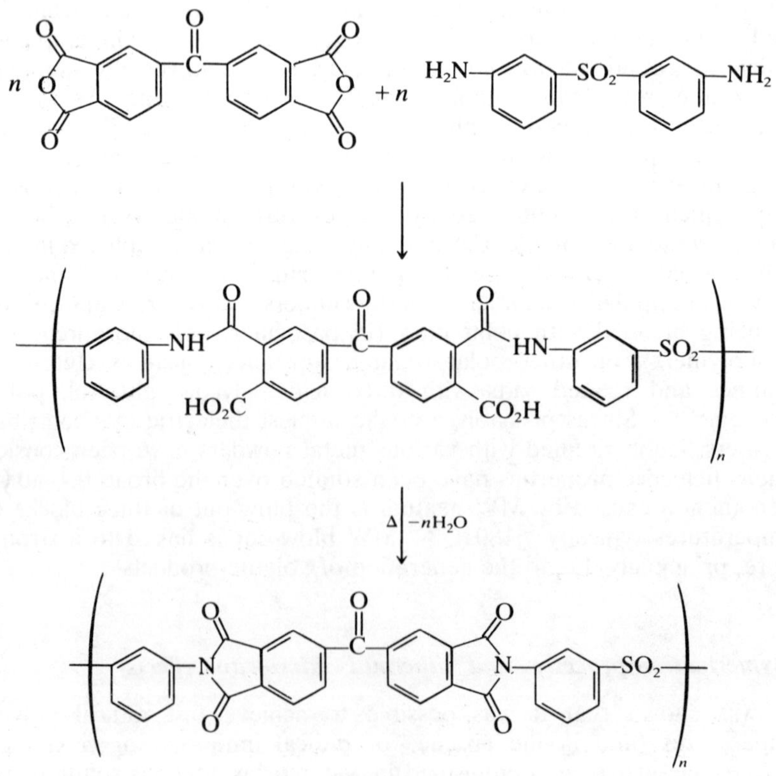

Scheme 2

A comparison of the first-order velocity constants at the same temperatures has shown the MW treatment to induce rate enhancements of 20–40 times, and also to lower the apparent activation energy of the reaction to approximately one half that of the conventional thermal process. With respect to solution processing, MW accelerations are found in solid thin films only in the rubber state, *i.e.* at temperatures maintained above the T_g of the polymer, gradually increasing along with the progress of the imidization process. In the glassy state, MW and thermal reactions proceed at comparable rates.[266,267] Other applications of MW heating to polycondensation processes are those for the manufacture of ladder polysiloxanes usable as photoresists and coatings,[268] and for the fast crosslinking of trialkoxysilylalkylated polyalkenes *via* formation of siloxane bridges.[269–271]

(b) Polyaddition processes. More than 15 years ago, MW irradiation was shown to be an excellent method for the rapid and selective heating of thin polyurethane coatings on weakly MW-heatable substrates. This not only effects prompt solvent evaporation, but also accomplishes the rapid polymerization of systems of blocked polyisocyanates and polyols, much faster (typically in 3–5 min) than with conventional heating.[272,273] The polyisocyanate–polyol polymerization has been studied under continuous and pulsed irradiation with different pulse regimes, at the constant frequency of 2.45 GHz and with constant average irradiation power.[274] MW-synthesized polyurethanes displayed superior hardness levels with respect to thermally cured materials. Further investigations on the MW-assisted synthesis of polyurethanes have since been conducted.[275]

The MW-promoted curing of epoxy resins has been by far the most extensively studied type of polymerization process. Various epoxy–amine systems have been the subjects of detailed investigations, in particular those comprising the liquid diglycidyl ether of bisphenol A (DGEBA) and different amine curing agents: 4,4′-diaminodiphenylmethane (DDM), 4,4′-diaminodiphenyl sulfone (DDS), *m*-phenylenediamine (*m*PDA), 4,4′-diamino-3,3′-dimethyldicyclohexylmethane (3DCHM), and dicyanodiamide (DCDA). MW curing under continuous irradiation has been monitored in comparison with thermal curing in conventional hot air ovens for reactive systems with a stoichiometric epoxy/amine ratio: DGEBA–DDM,[276–278] DGEBA–DDS,[279–281] and DGEBA–3DCHM.[282] In all cases, the temperature–time profiles followed by the materials during their continuous MW irradiation at constant power are of the type shown in Figure 13. The temperature–time curve can be divided into three regions: (I) simple MW heating of the resin mixture with no reaction; (II) rapid temperature increase caused by the onset of the exothermic curing reaction; (III) slow, continuous cooling, due to heat losses by convection and irradiation associated with a progressively decreasing reaction heat generation and a progressively lower MW heatability of the material undergoing curing. The decrease of MW heatability of the material is characterized by the typical time evolution of its complex dielectric constant with curing, linked to the disappearance of the dipolar reactive epoxy and primary/secondary amine groups, as well as to the variation of physical properties of the material itself, changing from a viscous liquid to a crosslinked rubber and finally a glass.[116,124–126,279]

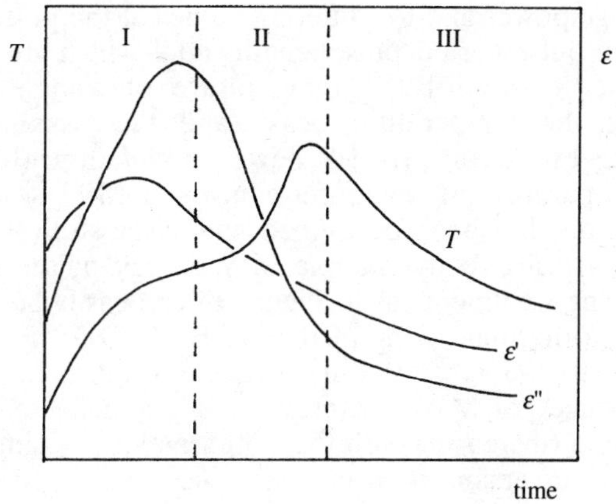

Figure 13 Typical variation of temperature and (relative) complex dielectric constant of a thermosetting resin with microwave cure time under continuous irradiation at constant power.

A higher irradiation power obviously enhances heating and therefore accelerates the entire curing cycle, as depicted in Figure 14. As seen in Section 19.4.2.1, the addition of conductive fillers (Cu, Al, and carbon black powders) increases the MW heatability of epoxy systems, and hence accelerates their curing at constant MW power.[220,222–224,226]

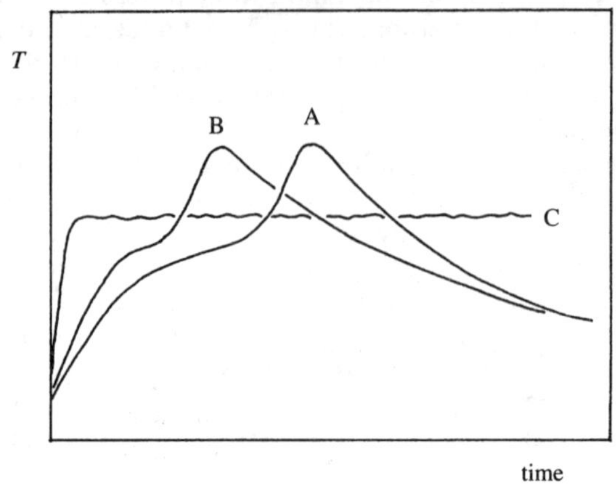

Figure 14 Temperature of a thermosetting resin vs. microwave cure time at fixed frequency. Curves A and B: under continuous irradiation at two power levels (B > A). Curve C: under computer-controlled pulsed irradiation.

All these studies have highlighted the capacity of MW irradiation for promoting curing processes of epoxy resins much faster than those feasible under conventional heating, especially whenever thick samples (opposing conventional external heating because of their low thermal conductivity) are to be treated. Materials cured by continuous MW irradiation have also shown higher elastic moduli when compared to thermally cured ones.[283] MW processing by the technique of pulsed irradiation with a fixed radiation frequency (known in shortwave therapy since the early 1950s[12]) has been applied to epoxy resin systems.[283] Pulse regimes are defined by the peak pulse power, the pulse frequency (or pulse period) and the pulse duration (or pulse length). The effects of pulsed MW (2.45 GHz) curing on the chemical structure of amine-crosslinked epoxy resins have been investigated for the DGEBA–DDS[284,285] and DGEBA–DCDA[285] systems. In both cases, MW treatments with low pulse periods (<10 ms) have been shown to catalyze the epoxy homopolymerization, *i.e.* to promote the formation of bis-aliphatic ethers, whereas only epoxy–amine reactions are promoted under irradiation with pulse periods of >50 ms (Scheme 3).

Studies of the pulsed MW curing of the DGEBA–DDM system[286,287] have displayed temperature–time profiles of the type described above for continuous irradiation (Figure 13), and, under the same average irradiation power, reaction rate enhancements with respect to continuous irradiation. These studies have also revealed the importance of the pulse regime adopted. At a fixed average power and for different values of the pulse period, pulse length can be varied to find optimum pulse period/pulse length ratios which give the highest curing rates. Two methods are viable to control the temperature of samples undergoing MW cure, particularly to eliminate the temperature peak (and the risks of material degradation) associated with the curing exotherm: (a) MW power modulation; (b) on–off switching of the irradiation power. An apparatus for computer-controlled MW power switching (computer-controlled pulsed irradiation) has been developed and successfully tested with the DGEBA–DDS system.[288,289] By delivering to the sample MW energy pulses with computer-controlled frequency, isothermal curing treatments as in Figure 14 can easily be performed.

The isothermal MW and thermal curing of the DGEBA–DDS system have been compared at temperatures from 125 to 195 °C.[290,291] Curing rates are much faster than those of thermal processing in the early stages of MW treatment, and gradually tend to become equal as a certain degree of conversion (increasing with the isothermal processing temperature adopted) is approached, after which an inversion of curing rates occurs. Further investigations on the MW and thermal curing kinetics under isothermal conditions of DGEBA–DDS and DGEBA-*m*PDA at temperatures from 125 to 145 and 80 to 120 °C, respectively,[292,293] have given substantially different results: much faster curing (with much shorter gelation and vitrification times) and higher final T_g of the resulting materials (particularly for DGEBA–DDS) under MW irradiation. Epoxy resin systems with MW processibility enhanced by addition of conductive fillers[294] and a variety of additives with high MW absorption capacity (including

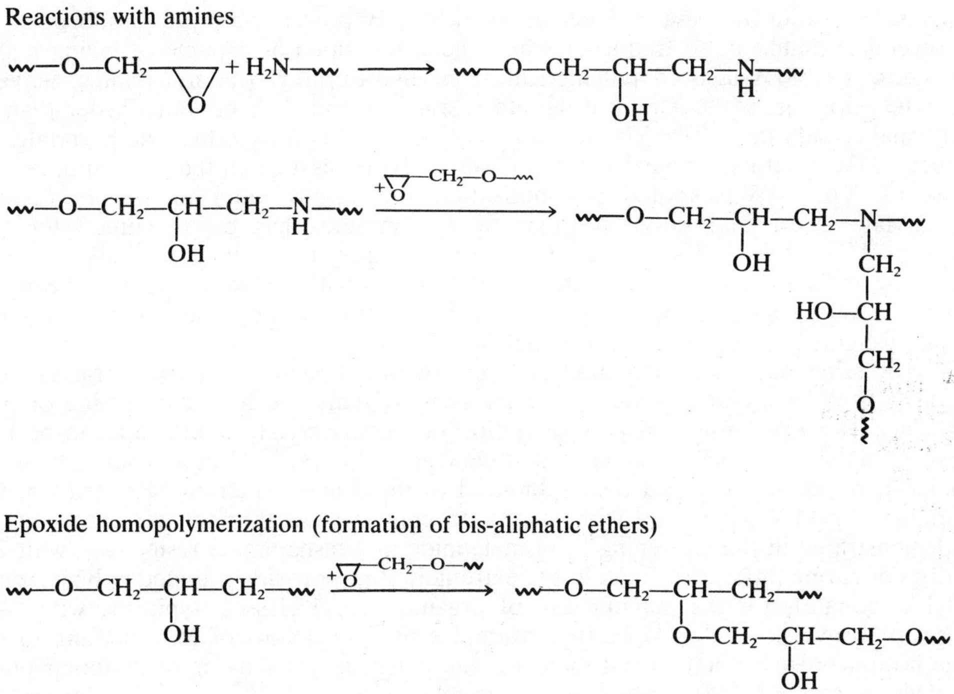

Scheme 3

polyelectrolytes)[295] have been the subjects of patent disclosures. Ferro- and ferrimagnetic fillers with suitably low Curie point have been claimed as MW curing enhancers, capable of limiting the thermal excursion of resins undergoing MW processing[296] to maximum temperatures near the T_C of the filler.

The thermal and the continuous MW curing of DGEBA–DDS resins toughened with amino-terminated bisphenol-A polysulfones has been studied.[297–301] Also in this case, MW heating has allowed curing times much shorter (minutes versus hours) than those required by conventional heating. Moreover, MW-processed materials have displayed lower volume fractions and smaller particle diameters of the separated (dispersed) polysulfone phase with respect to thermally processed samples, though with equivalent toughness levels. Similarly, the thermal and the MW-promoted cure of the cyanate ester resin 2,2-bis(4-cyanatophenyl)propane toughened with a hydroxy-terminated poly(arylene ether sulfone) has been studied from the standpoint of reaction kinetics (at least 10-fold faster under continuous MW irradiation) and of the morphology of the resulting two-phase materials.[302] MW irradiation has also been used to accelerate the curing of resin systems of aromatic polyisocyanates and epoxides, affording high-T_g isocyanurate–oxazolidone thermosets for heavy-duty electrical devices and structural composites.[303,304]

(ii) Chain polymerizations under microwave irradiation

The effects of MW energy on the promotion of a variety of radical polymerization processes are analogous to those exerted on step polymerizations, with considerable accelerations with respect to conventional thermal processes normally being exhibited. The 2.45 GHz frequency has always been used for both practical applications and scientific investigations in this field. The fast, peroxide-initiated curing of unsaturated polyester resins under continuous MW irradiation was studied at the end of the 1970s,[276] and was soon claimed for embedding or coating of electrical devices[305] (with the advantage of processing times of a few minutes). Temperature–time profiles are similar to those discussed above for curing of epoxy resin systems (Figure 13).[276] Higher irradiation power levels and/or higher initiator concentrations accelerate the process and anticipate the exothermic peak.

The MW curing of unsaturated polyester resins by pulsed irradiation has been investigated in comparison with that under continuous irradiation, with faster kinetics resulting from pulsed treatments.[306] Also in this case, at constant average MW power and pulse period/pulse length ratio, several optimum pulse frequencies have been identified as capable of giving maximum curing rates. The MW-assisted polymerization of hydroxyalkyl (meth)acrylates, acrylic and methacrylic acids, *etc.* without radical initiators, was also found to be much faster than under conventional conditions.[277,307] The polymerization of 2-hydroxyethyl methacrylate under continuous MW treatment revealed rates easily 3–10 times higher than the simple thermal processes.[308] The MW-promoted polymerization of various vinyl monomers (especially, (meth)acrylate esters and salts) in times of 1–2 minutes has been claimed for specific applications.[309,310] The MW-assisted and thermal copolymerizations in bulk at constant temperature of methyl and 2-hydroxyethyl methacrylates with azo initiators have been studied in detail.[311] At the same temperatures, far higher polymerization rates and almost double molecular weights have been achieved under MW heating.

MW processing has also been disclosed for the attainment of interpenetrating polymer networks[312] from epoxy/unsaturated polyester resin systems, *via* selective heating of the two components. The MW-promoted polymerization of (meth)acrylate monomers can be used to improve mechanical properties of hardened cement moldings,[313,314] impregnated with a fluid monomer–peroxide mixture and then submitted to rapid and penetrant MW heating. Curing accelerations by MW irradiation with respect to conventional thermal treatments have also been demonstrated in the hardening of bismaleimide and bisnadimide resins (*e.g.* with 20-fold reductions of curing times),[315,316] and in the extremely slow peroxide-initiated polymerization of bis(allyl carbonate)s for the manufacture of organic optical glasses (typically, with 2–3-fold increases of conversion rate).[317] Further examples of applications of MW heating to radical polymerization processes have been those for the peroxide-crosslinking of nonpolar polymers (polyethylene and EPDM) with specialty dipolar peroxides,[318,319] for the fabrication of crosslinked polyalkene foams,[320] and the quick crosslinking of vinyl-substituted silicone rubbers.[321]

(iii) 'Athermal' microwave effects

The reaction rate enhancements observed for a great variety of di- and multifunctional polymerization processes (both with chain- and step-growth mechanisms) invite conclusions to be drawn about the marked catalytic capacities of MW irradiation. Catalytic effects are self-evident in chain polymerizations of vinyl monomers, which are easily performed without radical initiators, and in the curing of a broad range of thermosetting systems under isothermal conditions, when MW experiments are compared to conventional thermal processes. Despite the clear scientific importance and the exciting technological perspectives of such results, however, this section has merely presented experimental facts without reporting related interpretations. In contrast to the well-established theoretical principles of MW heating given earlier, specific MW activation phenomena in polymer chemistry (from the standpoint either of reaction accelerations or reaction selectivity) do not currently have theoretical justifications, and simple speculative explanations are presently given. Moreover, this subject still has a controversial character, as some authors deny, at least under the experimental conditions used, the existence of any catalytic influence of MW treatments, and even maintain that polymerization kinetics are faster under conventional conditions.[290,291]

The definition of athermal MW effects was introduced in the 1970s in MW diathermy[14] (analogously to the corresponding one referring to RF effects used in shortwave therapy[12]) to include all the biochemical activation phenomena of electromagnetic irradiation which are unpredictable on the basis of the faster and more efficient heat transfer of this type of energy supply. In therapeutic diathermy, MW and RF athermal effects have been debated for a long time; their reality is also denied[11] in spite of much experimental support, and investigations are still in progress.[12,14]

Since the mid-1980s, MW heating has been used to perform a variety of organic syntheses: Diels–Alder and ene reactions, Claisen rearrangements, hydrolysis of esters and amides, esterification of carboxylic acids, oxidation of alkylbenzenes, Williamson synthesis of ethers, *etc.* In most cases, dramatic reductions of reaction times (reaction rate enhancements of up to 1000 times) and also higher yields have been obtained with respect to conventional procedures, although in some other cases no reaction accelerations have been observed. Several review articles, summarizing and discussing these works, have been published in recent years.[322–330] Although much evidence seems in favor of even strong catalytic influences of microwaves, some authors tend to disclaim athermal effects, and explain anomalous reaction kinetics by the superheating of reaction mixtures caused by MW irradiation.[325–327,330] Actually, most organic syntheses have been conducted in home MW ovens under poorly controlled conditions, so that only a little is known about the electromagnetic field intensity applied and the reaction temperature profiles during experiments.

Studies in polymer chemistry, by contrast, have been performed under far better controlled conditions and are more diagnostic of athermal MW effects. General conclusions which can be drawn are that less reactive systems display higher MW enhancements of reaction rate, and that MW accelerations become weaker as temperature increases.[51,52] MW effects in polymer chemistry have been tentatively explained in terms of an improved diffusion capacity of reactive species through the medium induced by MW irradiation, as well as by a selective superheating of reacting polar groups (thus achieving a superior reactivity) with respect to the bulk,[265–267] a molecular-scale model equivalent to the one of hot spots at the microscopic level introduced in MW processing of ceramics.[38,39]

19.5 REFERENCES

1. E. C. Jordan, 'Electromagnetic Waves and Radiating Systems', Prentice-Hall, Englewood Cliffs, NJ, 1950.
2. P. Fleury and J.-P. Mathieu, 'Courants Alternatifs – Ondes Hertziennes', Editions Eyrolles, Paris, 1958.
3. V. Rojansky, 'Electromagnetic Fields and Waves', Prentice-Hall, Englewood Cliffs, NJ, 1971.
4. D. K. Cheng, 'Field and Wave Electromagnetics', Addison-Wesley, Philadelphia, PA, 1983.
5. A. R. Von Hippel, 'Dielectrics and Waves', Wiley, New York, 1954.
6. J. C. Anderson, 'Dielectrics', Reinhold, New York, 1964.
7. A. R. Von Hippel, 'Dielectric Materials and Applications', 2nd edn., The Technology Press of MIT and Wiley, New York, 1966.
8. J. White, in 'Transactions of the International Microwave Power Institute', Clifton, VA, 1973, vol. 1, p. 40.
9. J. B. Hasted, 'Aqueous Dielectrics', Chapman & Hall, London, 1973.
10. W. R. Adey, *Physiol. Rev.,* 1981, **61,** 436.
11. J. Low and A. Reed, 'Electrotherapy Explained: Principles and Practice', Butterworths-Heinemann, London, 1990.
12. S. S. Kitchen and C. J. Partridge, *Physiotherapy,* 1992, **78,** 243.
13. B. O. Scott, *Physiotherapy,* 1965, **51,** 143; 1968, **54,** 150; B. H. Brown and S. G. Johnson, *Physiotherapy,* 1975, **61,** 171.
14. S. S. Kitchen and C. J. Partridge, *Physiotherapy,* 1991, **77,** 647.
15. W. J. Weisz, in 'Encyclopedia of Physical Science and Technology', ed. R. A. Meyers, Academic Press, Orlando, FL, 1987, vol. 11, p. 679.
16. G. M. Brown, C. N. Hoyles and R. A. Bierwirth, 'Theory and Applications of Radio-Frequency Heating', Van Nostrand, New York, 1947.
17. L. L. Langton, 'Radio-Frequency Heating Equipment', Pitman, New York, 1949.
18. F. Lauster, 'Elektrowärmetechnik', Teubner, Stuttgart, 1968.
19. D. Bialod, in 'Radio Frequency/Radiation and Plasma Processing', ed. P. N. Cheremisinoff, O. G. Farah and R. P. Ouellette, Technomic, Lancaster, PA, 1985, chap. 1.
20. P. Hulls and J.-M. De Hoe (eds.), 'Dielectric Heating for Industrial Processes', Union Internationale d'Electrotherapie, Paris, 1992.
21. H. Puschner, 'Heating with Microwaves', Springer, New York, 1966.
22. E. C. Okress (ed.), 'Microwave Power Engineering', Academic Press, New York, 1968, vols. 1 and 2.
23. D. A. Copson, 'Microwave Heating', 2nd edn., Avi, Westport, CT, 1975.
24. J. M. Osepchuk, in 'Kirk-Othmer Encyclopedia of Chemical Technology', 3rd edn., ed. H. F. Mark, D. F. Othmer, C. G. Overberger and G. T. Seaborg, Wiley, New York, 1981, vol. 15, p. 494.
25. A. C. Metaxas and R. J. Meredith, 'Industrial Microwave Heating', Peregrinus, London, 1983.
26. R. F. Schiffmann, in 'Encyclopedia of Chemical Processing and Design', ed. J. J. McKetta and W. A. Cunningham, Dekker, New York, 1989, vol. 30, p. 202.
27. W. H. Sutton, *MRS Bull.,* 1993, **18** (11)**,** 22.
28. J. G. P. Binner and T. E. Cross, in 'The Encyclopedia of Advanced Materials', ed. D. Bloor, R. J. Brook, M. C. Flemings and S. Mahajan, Pergamon, Oxford, 1994, vol. 3, p. 1620.
29. K. W. Peterson, J. R. Button and D. A. Annett, in Ref. 19, chap. 3.
30. M. R. Sanio and P. S. Schmidt, *Mater. Res. Soc. Symp. Proc.,* 1988, **124,** 337.
31. Ref. 22, vol. 2.
32. N. W. Lord, in Ref. 19, chap. 2.
33. R. D. Smith, *Mater. Res. Soc. Symp. Proc.,* 1991, **189,** 383.
34. R. V. Decareau, 'Microwaves in the Food Processing Industry', Academic Press, Orlando, Florida, 1985.
35. J. Pound, 'Practical RF Heating for the Wood Industry', Heywood, London, 1957.
36. J. Pound, 'Radio Frequency Heating in the Timber Industry', Spon, London, 1973.
37. S. M. Burkinshaw, *Colour. Annu.,* 1986, 83.
38. D. E. Clark, D. C. Folz, R. L. Shulz, Z. Fathi and A. D. Cozzi, *MRS Bull.,* 1993, **18** (11)**,** 41.
39. R. Silberglitt, I. Ahmad, W. M. Black and J. D. Katz, *MRS Bull.,* 1993, **18** (11)**,** 47.
40. S. J. Oda, *Mater. Res. Soc. Symp. Proc.,* 1994, **347,** 371.
41. H. P. Zade, 'Heatsealing and High-Frequency Welding of Plastics', Temple, London, 1959.

42. E. C. Stanley, 'High Frequency Plastic Sheet Welding', Radyne, Wokingham, Berks, UK, 1961.
43. M. Rothstein, in 'Encylopedia of Polymer Science and Technology', ed. H. F. Mark, N. G. Gaylord and N. M. Bikales, Wiley, New York, 1966, vol. 5, p. 1.
44. T. L. Wilson, in 'Encyclopedia of Polymer Science and Engineering', 2nd edn., ed. H. F. Mark, N. M. Bikales, C. G. Overberger, G. Menges and J. I. Kroschwitz, Wiley, New York, 1986, vol. 5, p. 1.
45. C. Akyel and E. Bilgen, *Energy,* 1989, **14,** 839; B. Krieger, *Polym. Mater. Sci. Eng.,* 1992, **66,** 339; B. Krieger, *Mater. Res. Soc. Symp. Proc.,* 1994, **347,** 57.
46. K. Nakagawa and M. Amano, in 'High Modulus Polymers', ed. A. E. Zachariades and R. S. Porter, Dekker, New York, 1988, chap. 15.
47. J. Mijović, *Polym. News,* 1989, **14,** 370; 1990, **15,** 50.
48. J. Mijović and J. Wijaya, *Polym. Compos.,* 1990, **11,** 184.
49. M. C. Hawley, J. Wei and V. Adegbite, *Mater. Res. Soc. Symp. Proc.,* 1994, **347,** 669; J. Wei and M. C. Hawley, *Polym. Mater. Sci. Eng.,* 1995, **72,** 10.
50. (a) T. C. Ward and M. Chen, *Polym. Mater. Sci. Eng.,* 1992, **66,** 335; (b) M. Chen, E. J. Siochi, T. C. Ward and J. E. McGrath, *Polym. Eng. Sci.,* 1993, **33,** 1092.
51. D. A. Lewis, *Mater. Res. Soc. Symp. Proc.,* 1992, **269,** 21; D. A. Lewis and J. M. Shaw, *MRS Bull.,* 1993, **18** (11), 37.
52. M. T. DeMeuse and C. L. Ryan, *Adv. Polym. Technol.,* 1993, **12,** 197; J. B. Wei, T. Shidaker and M. C. Hawley, *Trends Polym. Sci.,* 1996, **4,** 18.
53. A. Benatar, in Ref. 28, vol. 2, p. 1214.
54. J. C. F. Böttcher, 'Theory of Electric Polarization', Elsevier, Amsterdam, 1952.
55. W. F. Brown, Jr., in 'Handbuch der Physik', ed. S. Flügge, Springer, Berlin, 1956, vol. XVII, p. 1.
56. H. Fröhlich, 'Theory of Dielectrics', 2nd edn., Clarendon, Oxford, 1958.
57. A. H. Morrish, 'The Physical Principles of Magnetism', Wiley, New York, 1965.
58. S. V. Vonsovskii, 'Magnetism', Israel Program for Scientific Translations, Jerusalem, 1974, vol. 1.
59. J. K. Watson, 'Applications of Magnetism', Wiley, New York, 1980.
60. C. P. Smyth, 'Dielectric Behavior and Structure', McGraw-Hill, New York, 1955.
61. N. Hill, W. E. Vaughan, A. H. Price and M. Davies, 'Dielectric Properties and Molecular Behavior', Van Nostrand, New York, 1969.
62. P. Madden and D. Kivelson, *Adv. Chem. Phys.,* 1984, **56,** 467.
63. J. R. Partington, 'An Advanced Treatise on Physical Chemistry', Longmans, London, 1954, vol. 5, sect. XI.
64. J. W. Smith, 'Electric Dipole Moments', Butterworths, London, 1955.
65. A. L. McClellan, 'Tables of Experimental Dipole Moments', Rahara Enterprises, El Cerrito, CA, 1989.
66. A. R. Blythe, 'Electrical Properties of Polymers', Cambridge University Press, Cambridge, 1979.
67. K. N. Mathes, in 'Encyclopedia of Polymer Science and Engineering', 2nd edn., ed. H. F. Mark, N. M. Bikales, C. G. Overberger, G. Menges and J. I. Kroschwitz, Wiley, New York, 1986, vol. 5, p. 507.
68. C. C. Ku and R. Liepins, 'Electrical Properties of Polymers: Chemical Principles', Hanser, New York, 1987.
69. J. I. Kroschwitz, 'Electrical and Electronic Properties of Polymers: A State-of-the-Art Compendium', Wiley, New York, 1988.
70. D. M. Van Krevelen, 'Properties of Polymers', 3rd edn., Elsevier, Amsterdam, 1990, chap. 12.
71. J. E. Frommer and R. R. Chance, in Ref. 67, vol. 5, p. 462.
72. D. Bloor, in 'Comprehensive Polymer Science', ed. C. Booth and C. Price, Pergamon, Oxford, 1989, vol. 2, chap. 22.
73. R. H. Friend, in 'Encyclopedia of Advanced Materials', ed. D. Bloor, R. J. Brook, M. C. Flemings and S. Mahajan, Pergamon, Oxford, 1994, vol. 1, p. 467.
74. J. Owen, in Ref. 72, vol. 2, chap. 21.
75. J. M. G. Cowie, in Ref. 73, vol. 2, p. 1195.
76. H. S. Nalwa, *J. Macromol. Sci., Rev. Macromol. Chem. Phys.,* 1991, **C31,** 341; R. G. Kepler, in Ref. 73, vol. 2, p. 811.
77. H. Ishida, in Ref. 67, 1989, suppl. vol., p. 446.
78. G. Williams, in Ref. 72, vol. 2, chap. 18.
79. J. C. Maxwell, 'Electricity and Magnetism', Clarendon, Oxford, 1892, vol. 1, p. 452.
80. K. W. Wagner, *Arch. Elektrotechnol.,* 1914, **2,** 371; W. A. Yager, *Physics,* 1936, **7,** 434.
81. R. W. Sillars, *Proc. R. Soc. London,* 1939, **A169,** 66.
82. E. Fatuzzo and W. J. Merz, 'Ferroelectricity', North-Holland, Amsterdam, 1967.
83. H.-G. Unruh, in 'Ullmann's Encyclopedia of Industrial Chemistry', 5th edn., ed. L. Kaudy, J. F. Rounsaville and G. Schulz, Verlag Chemie, Weinheim, 1987, vol. A10, p. 309.
84. Y. Xu, 'Ferroelectric Materials and Their Applications', North-Holland, Amsterdam, 1991.
85. S.-J. Jang, in 'Kirk-Othmer Encyclopedia of Chemical Technology', 4th edn., ed. J. I. Kroschwitz and M. Howe-Grant, Wiley, New York, 1993, vol. 10, p. 413.
86. A. J. Bell, in Ref. 73, vol. 2, p. 816.
87. Ref. 58, vol. 2.
88. R. S. Tebble and D. J. Craik, 'Magnetic Materials', Wiley, London, 1969.
89. E. P. Wohlfarth (ed.), 'Ferromagnetic Materials', North-Holland, New York, 1989–92, vols. 1–3.
90. R. Hayakawa and Y. Wada, *Adv. Polym. Sci.,* 1973, **11,** 1.
91. A. Goldman, 'Modern Ferrite Technology', Van Nostrand-Reinhold, New York, 1990.
92. F. X. N. M. Kools and D. Stoppels, in Ref. 85, vol. 10, p. 381.
93. G. Bate, in Ref. 73, vol. 3, p. 2236; J. C. Lodder, in Ref. 85, 1995, vol. 14, p. 339.
94. V. Daniel, 'Dielectric Relaxation', Academic Press, New York, 1967.
95. A. K. Jonscher, 'Dielectric Relaxation in Solids', Chelsea Dielectric Press, London, 1983.
96. N. G. McCrum, B. E. Read and G. Williams, 'Anelastic and Dielectric Effects in Polymeric Solids', Wiley, New York, 1967.
97. F. E. Karasz (ed.), 'Dielectric Properties of Polymers', Plenum, New York, 1971.

98. P. Hedvig, 'Dielectric Spectroscopy of Polymers', Wiley, New York, 1977.
99. S. Matsuoka, in Ref. 67, vol. 5, p. 23.
100. J. G. Park, in 'Encyclopaedic Dictionary of Physics', ed. J. Thewlis, R. C. Glass, D. J. Hughes and A. R. Meetham, Pergamon, Oxford, 1962, vol. 5, p. 289.
101. J. R. McDonald (ed.), 'Impedance Spectroscopy', Wiley, New York, 1987.
102. G. J. Wheeler, 'Introduction to Microwaves', Prentice-Hall, Englewood Cliffs, NJ, 1963.
103. R. N. Ghose, 'Microwave Circuit Theory and Analysis', McGraw-Hill, New York, 1963.
104. F. E. Gardiol, 'Introduction to Microwaves', Artech House, Dedham, MA, 1984.
105. M. Wind and H. Rapaport (eds.), 'Handbook of Microwave Measurements', Polytechnic Institute of Brooklyn/Interscience, New York, 1955.
106. E. L. Ginzton, 'Microwave Measurements', McGraw-Hill, New York, 1957.
107. M. Sucher and J. Fox (eds.), 'Handbook of Microwave Measurements', Brooklyn Polytechnic Press, Brooklyn, NY, 1963, vol. 2.
108. T. S. Laverghetta, 'Microwave Measurements and Techniques', Artech House, Dedham, MA, 1976.
109. P. Guillon, *Mater. Res. Soc. Symp. Proc.*, 1994, **347,** 203.
110. E. F. Labuda and R. C. Le Craw, *Rev. Sci. Instrum.*, 1961, **32,** 391.
111. M. A. Rzepecka, *J. Microwave Power,* 1973, **8,** 3; D. Couderc, M. Giroux and R. G. Bosisio, *J. Microwave Power,* 1973, **8,** 69.
112. T. M. Shaw and J. J. Windle, *J. Appl. Phys.*, 1950, **21,** 956.
113. J. Ippen, *Rubber Chem. Technol.*, 1971, **44,** 294.
114. H. F. Schwarz, R. G. Bosisio, M. R. Wertheimer and D. Couderc, *J. Microwave Power,* 1973, **8,** 303.
115. B. Terselius and B. Ranby, *J. Microwave Power,* 1978, **13,** 327.
116. J. Jow, M. C. Hawley, M. Finzel and T. Kern, *Polym. Eng. Sci.*, 1988, **28,** 1450.
117. K. R. Baker, E. Marand and J. D. Graybeal, *Polym. Mater. Sci. Eng.*, 1992, **66,** 474.
118. M. Martinelli, P. A. Rolla and E. Tombari, *IEEE Trans. Microwave Theory Tech.*, 1985, **MTT-33,** 779.
119. M. Martinelli, P. A. Rolla and E. Tombari, *IEEE Trans. Instrum. Meas.*, 1985, **IM-34,** 417.
120. C. Carlini, F. Ciardelli, P. A. Rolla and E. Tombari, *J. Polym. Sci., Polym. Phys. Ed.,* 1987, **25,** 1253.
121. C. Carlini, P. A. Rolla and E. Tombari, *J. Appl. Polym. Sci.*, 1990, **41,** 805.
122. R. H. Cole, S. S. Mashimo and P. Winsor, IV, *J. Phys. Chem.*, 1980, **84,** 786.
123. R. H. Cole, J. C. Berberian, S. S. Mashimo, G. Chryssikos, A. Burns and E. Tombari, *J. Appl. Phys.*, 1989, **66,** 793.
124. S. Carrozzino, G. Levita, P. A. Rolla and E. Tombari, *Polym. Eng. Sci.*, 1990, **30,** 366.
125. A. Livi, G. Levita and P. A. Rolla, *J. Appl. Polym. Sci.*, 1993, **50,** 1583.
126. C. Carlini, G. Levita and P. A. Rolla, *Polym. Mater. Sci. Eng.*, 1992, **66,** 472.
127. C. Carlini, A. Livi, P. A. Rolla and D. Fioretto, *J. Non-Cryst. Solids,* 1994, **172–174,** 569.
128. H. Sobol and J. J. Hughes, *IEEE Trans. Microwave Theory Tech.*, 1967, **MTT-15,** 377; Susanta Sen, P. K. Saha and B. R. Nag, *Rev. Sci. Instrum.*, 1979, **50,** 1594.
129. D. C. Dube and R. Natarajan, *J. Appl. Phys.*, 1973, **44,** 4927; *J. Phys.*, 1974, **E7,** 256.
130. D. C. Dube, *Bull. Mater. Sci.*, 1984, **6,** 1075.
131. K. K. S. Jamwal and A. Dhar, *Rev. Sci. Instrum.*, 1981, **52,** 767.
132. F. Legros, A. Fourrier-Lamer, D. Le Pen and A. Gourdenne, *Eur. Polym. J.*, 1984, **20,** 1057; 1986, **22,** 331, 335.
133. K. R. Baker and J. D. Graybeal, *Polym. Mater. Sci. Eng.*, 1992, **66,** 470.
134. Y. Pastol, G. Arjavalingam, J.-M. Halbout and G. V. Kopcsay, *Appl. Phys. Lett.*, 1989, **54,** 307; G. Arjavalingam, Y. Pastol, J.-M. Halbout and G. V. Kopcsay, *IEEE Trans. Microwave Theory Tech.*, 1990, **MTT-38,** 615.
135. W. M. Robertson, Y. Pastol, G. Arjavalingam, J.-M. Halbout and G. V. Kopcsay, *Mater. Res. Soc. Symp. Proc.*, 1991, **214,** 87.
136. W. R. Tinga and S. O. Nelson, *J. Microwave Power,* 1973, **8,** 23.
137. V. Frosini, E. Butta and M. Calamia, *J. Appl. Polym. Sci.*, 1967, **11,** 527.
138. A. J. Bur, *Polymer,* 1985, **26,** 963.
139. J. D. Ferry, 'Viscoelastic Properties of Polymers', 2nd edn., Wiley, New York, 1970, chap. 11.
140. F. J. Blatt, 'Physics of Electronic Conduction in Solids', McGraw-Hill, New York, 1968.
141. A. B. Lidiard, in 'Handbuch der Physik', ed. S. Flügge, Springer, Berlin, 1957, vol. XX, p. 246.
142. R. H. Norman, 'Conductive Rubbers and Plastics', Applied Science, London, 1970; A. T. Ponomarenko, V. G. Shevchenko and N. S. Enikolopyan, *Adv. Polym. Sci.*, 1990, **96,** 125.
143. H. F. Wolf, 'Semiconductors', Wiley, New York, 1971.
144. J. Simon, M. Bouvet and P. Bassoul, in 'Encyclopedia of Advanced Materials', ed. D. Bloor, R. J. Brook, M. C. Flemings and S. Mahajan, Pergamon, Oxford, 1994, vol. 3, p. 1680.
145. P. Day, in Ref. 144, vol. 1, p. 417.
146. H. Kallmann and M. Silver (eds.), 'Electrical Conductivity in Organic Solids', Interscience, New York, 1961.
147. J. J. Brophy and J. W. Buttrey (eds.), 'Organic Semiconductors', McMillan, New York, 1962.
148. P. Sheng, *Phys. Rev. B,* 1980, **21,** 2180.
149. R. D. Sherman, L. M. Middleman and S. M. Jacobs, *Polym. Eng. Sci.*, 1983, **23,** 36.
150. R. M. Hill and A. K. Jonscher, *J. Non-Cryst. Solids,* 1979, **32,** 53.
151. P. Extance, S. Elliot and E. A. Davis, *Phys. Rev. B,* 1985, **32,** 8148.
152. A. Fourrier-Lamer, A. Fizazi, N. Belhadj-Tahar, A. Fontanille and A. Soum, *Synth. Met.*, 1988, **24,** 95.
153. S. Elliot, *Philos. Mag.*, 1977, **36,** 1291.
154. W. Y. A. Tajima, *Polym. Compos.*, 1982, **3,** 166.
155. G. Kraus, *Adv. Polym. Sci.*, 1971, **8,** 155; Z. Rigbi, *Adv. Polym. Sci.* 1980, **36,** 21.
156. A. I. Ivon and I. M. Chernenko, *Sov. Phys. Semicond.*, 1981, **15** (2), 152.
157. Y.-S. Ho, *Diss. Abstr. Int. B,* 1991, **52** (2), 1009.
158. S. M. Musameh, M. K. Abdelazeez, M. S. Ahmad, A. M. Zihlif, M. Malinconico, E. Martuscelli and G. Ragosta, *Mater. Sci. Eng.*, 1991, **B10,** 29.
159. L. Olmedo, C. Deleuze, P. Hourquebie and F. Jousse, *Rev. Compos. Nouv. Matér.*, 1991, **1,** 123.

160. L. Olmedo, C. Deleuze and M. Dufort, *49th Annu. Tech. Conf. – Soc. Plast. Eng.*, 1991, Conference Proceedings, p. 774; L. Olmedo, P. Hourquebie and F. Jousse, *Adv. Mater.*, 1993, **5,** 373.
161. P. Hourquebie, L. Olmedo and C. Deleuze, *Springer Ser. Solid-State Sci.*, 1992, **107,** 125.
162. W. J. Cable, 'Induction and Dielectric Heating', Reinhold, New York, 1954; M. G. Lozinskii, 'Industrial Applications of Induction Heating', Pergamon, Oxford, 1969.
163. B. Lax and K. J. Button, 'Microwave Ferrites and Ferrimagnetics', McGraw-Hill, New York, 1962.
164. W. H. von Aulock (ed.), 'Handbook of Microwave Ferrite Materials', Academic Press, New York, 1965.
165. E. C. Snelling, 'Soft Ferrites – Properties and Applications', Iliffe, London, 1969.
166. C. W. Haas and H. B. Callen, in 'Magnetism', ed. G. T. Rado and H. Suhl, Academic Press, New York, 1963, vol. 1, chap. 10.
167. R. W. Damon, in Ref. 166, vol. 1, chap. 11.
168. S. V. Vonsovskii (ed.), 'Ferromagnetic Resonance', Pergamon, New York, 1966.
169. S. Foner, in Ref. 166, vol. 1, chap. 9.
170. D. M. Smyth, in Ref. 144, vol. 1, p. 209.
171. T. A. Johnk, 'Engineering Electromagnetic Fields and Waves', Wiley, New York, 1975.
172. J. Asmussen, Jr., H. H. Lin, B. Manring and R. Fritz, *Rev. Sci. Instrum.*, 1987, **58,** 1477.
173. J. Jow, M. C. Hawley and M. C. Finzel, *Rev. Sci. Instrum.*, 1989, **60,** 96.
174. E. B. Manring and J. Asmussen, Jr., *Ceram. Trans.*, 1991, **21,** 159.
175. J. Asmussen, *Polym. Mater. Sci. Eng.*, 1992, **66,** 341.
176. E. B. Manring and J. Asmussen, Jr., *Polym. Mater. Sci. Eng.*, 1992, **66,** 476.
177. M. Chen, J. E. McGrath and T. C. Ward, *Polym. Mater. Sci. Eng.*, 1989, **60,** 443.
178. M. Chen, M. A. Zumbrum, J. C. Hedrick, J. E. McGrath and T. C. Ward, *Mater. Res. Soc. Symp. Proc.*, 1991, **189,** 431.
179. M. Chen, E. J. Siochi, T. C. Ward and J. E. McGrath, *Polym. Eng. Sci.*, 1993, **33,** 1110.
180. H.-F. Huang, *J. Microwave Power,* 1976, **11,** 305.
181. M. T. DeMeuse, *Polym. Mater. Sci. Eng.*, 1992, **66,** 428.
182. M. Chen, J. W. Hellgeth, E. J. Siochi, T. C. Ward and J. E. McGrath, *Polym. Eng. Sci.*, 1993, **33,** 1122.
183. K. G. Ayappa, H. T. Davis, E. A. Davis and J. Gordon, *AIChE J.*, 1991, **37,** 313.
184. C. De Wagter, *J. Microwave Power,* 1984, **19,** 97.
185. J. P. Joly and I. Turner, *J. Microwave Power EM Energy,* 1990, **25,** 3.
186. A. Joly, J. P. Joly, E. Karmazsin and J.-L. Veau, *J. Therm. Anal.*, 1993, **39,** 1285.
187. M. F. Iskander, *MRS Bull.*, 1993, **18 (**11**),** 30.
188. T. Murayama, K. Honda, T. Kasugai and N. Minagawa (Fuji Photo Film Co.), *Ger. Pat.* 2 108 555 (1971) (*Chem. Abstr.*, 1972, **76,** 26 607).
189. Asahi-Dow, Ltd., *Jpn. Pat.* 83 104 279 (1983) (*Chem. Abstr.*, 1984, **100,** 8405).
190. Seiren Co., Ltd., *Jpn. Pat.* 83 191 279 (1983) (*Chem. Abstr.*, 1984, **100,** 87 211).
191. H. Yogo and M. Muto (Asahi Chem. Ind. Co.), *Jpn. Pat.* 87 15 386 (1987) (*Chem. Abstr.*, 1987, **106,** 215 438).
192. M. Nomura and M. Furukawa (Seinen Co.), *Jpn. Pat.* 87 33 883 (1987) (*Chem. Abstr.*, 1987, **106,** 157 861).
193. S. Yoshida and M. Mabuchi (Nissan Motor Co.; Kawashime Textile Mills, Ltd.), *Jpn. Pat.* 87 57 988 (1987) (*Chem. Abstr.*, 1987, **107,** 24 524).
194. M. Mimura (Teijin Cordley, Ltd.), *Jpn. Pat.*, 88 85 170 (1988) (*Chem. Abstr.*, 1988, **109,** 39 338).
195. T. Ohayashi and H. Hiraoka (Hiraoka and Co.), *Jpn. Pat.* 88 92 779 (1988) (*Chem. Abstr.*, 1988, **109,** 56 511).
196. T. A. Brooke and P. M. Robinson (Dunlop Rubber Co.), *US Pat.* 3 639 190 (1972) (*Chem. Abstr.*, 1972, **76,** 155 021).
197. C. R. Ashcraft (Mobil Oil Corp.), *Eur. Pat. Appl.* 113 203 (1984) (*Chem. Abstr.*, 1985, **102,** 46 968).
198. Monsanto Co., *Fr. Pat.* 2 182 010 (1973) (*Chem. Abstr.*, 1974, **81,** 64 817).
199. G. M. Lancaster and J. A. Allen (Dow Chem. Co.), *World Pat.* 85/1294 (1985) (*Chem. Abstr.*, 1985, **103,** 88 613); *World Pat.* 86/7012 (1986) (*Chem. Abstr.*, 1987, **107,** 41 231).
200. E. R. Peterson, *Mater. Res. Soc. Symp. Proc.*, 1994, **347,** 697.
201. H.-R. Han, L. Christensen and P. H. LePere (3M Co.), *Eur. Pat. Appl.*, 461 868 (1991) (*Chem. Abstr.*, 1992, **117,** 9409).
202. P. Kathirgamanathan, *Polymer,* 1993, **34,** 3105.
203. W. R. Rinker and R. S. Varga (The B. F. Goodrich Co.), *US Pat.* 4 253 898 (1981) (*Chem. Abstr.*, 1981, **94,** 176 369).
204. H. Jullien and H. Valot, *Polymer,* 1983, **24,** 810.
205. A. K. Thorsrud, *Mater. Res. Soc. Symp. Proc.*, 1988, **124,** 195.
206. A. K. Thorsrud (Phillips Petroleum Co.), *US Pat.* 4 840 758 (1989).
207. A. K. Thorsrud (Phillips Petroleum Co.), *US Pat.* 4 968 726 (1990).
208. J. R. Ellis and M. B. Dines (Occidental Research Corp.), *US Pat.* 4 721 738 (1988) (*Chem. Abstr.*, 1988, **108,** 187 928).
209. E. R. Peterson and G. M. Lancaster (Dow Chem. Co.), *World Pat.* 92/13 023 (1992) (*Chem. Abstr.*, 1993, **118,** 60 771).
210. G. Vicq and A.-M. Bottreau, *J. Mol. Liq.*, 1988, **37,** 181.
211. K. Kamiyoshi, M. Hotsuta and T. Shiobara (Sekisui Chem. Co.), *Jpn. Pat.* 79 161 645 (1979) (*Chem. Abstr.*, 1980, **92,** 182 225); *Jpn. Pat.* 86 171 783 (1986) (*Chem. Abstr.*, 1986, **105,** 154 421).
212. P. R. Dufour and F. E. Nix (Stamicarbon B.V.), *Neth. Pat.* 90 714 (1991) (*Chem. Abstr.*, 1992, **116,** 22 412).
213. P. Ittemann, G. Heinz and W. Eberle (BASF A.G.), *Ger. Pat.* 4 003 908 (1991) (*Chem. Abstr.*, 1991, **115,** 209 479).
214. V. K. Varadan, V. V. Varadan, Y. Ma and W. F. Hall, *IEEE Trans. Microwave Theory Tech.*, 1986, **MTT-34,** 251.
215. X. Wang, Y. Sheng, Z. Qiu, J. Qian, H. Lu and G. Hu, *Nanjing Daxue Xuebao, Ziran Kexue,* 1992, **28,** 551 (*Chem. Abstr.*, 1994, **120,** 64 897).
216. T. M. Cadle, L. D. Manton and J. P. Marriott, *Br. Pat.* 1 340 503 (1973) (*Chem. Abstr.*, 1974, **80,** 146 659).
217. H. D. Hellmann and R. Krieger (Uzin-Werk Georg Utz GmbH und Co. K.-G.), *Ger. Pat.* 3 802 881 (1989) (*Chem. Abstr.*, 1991, **111,** 135 673).
218. A. Piller, *Ger. Pat.* 4 036 876 (1992) (*Chem. Abstr.*, 1992, **117,** 113 161).

219. N. Probst and J. Iker, *Kunststoffe,* 1984, **37** (5), 385.
220. R. G. Raj, *Polym. Mater. Sci. Eng.*, 1986, **55,** 49.
221. Y. Baziard, S. Breton, S. Toutain and A. Gourdenne, *Eur. Polym. J.*, 1988, **24,** 633.
222. R. G. Raj, *Polym. Mater. Sci. Eng.*, 1987, **57,** 537.
223. A. Bouazizi and A. Gourdenne, *Rev. Gen. Electr.*, **1988,** (5), 27.
224. A. Bouazizi and A. Gourdenne, *Eur. Polym. J.*, 1988, **24,** 889.
225. Y. Baziard, S. Breton, S. Toutain and A. Gourdenne, *Eur. Polym. J.*, 1988, **24,** 521.
226. Y. Baziard and A. Gourdenne, *Eur. Polym. J.*, 1988, **24,** 873, 881.
227. Y. S. Ho and J. J. Kramer, *Mater. Res. Soc. Symp. Proc.*, 1988, **124,** 161.
228. A. N. Lagarkov, S. M. Matitsin and A. K. Sarychev, *Polym. Mater. Sci. Eng.*, 1992, **66,** 426.
229. J. P. Ludman and C. H. Stern, *Mater. Res. Soc. Symp. Proc.*, 1994, **347,** 717.
230. A. Piller, *Ger. Pat.* 4 006 134 (1991) (*Chem. Abstr.*, 1992, **116,** 22 233).
231. K. Nakagawa, O. Maeda and S. Yamakawa, *J. Polym. Sci., Polym. Lett. Ed.*, 1983, **21,** 933.
232. K. Nakagawa, T. Konaka and S. Yamakawa, *Polymer,* 1985, **26,** 84.
233. T. Konaka, K. Nakagawa and S. Yamakawa, *Polymer,* 1985, **26,** 462.
234. Y. Takeuchi, F. Yamamoto, K. Nakagawa and S. Yamakawa, *J. Polym. Sci., Polym. Phys. Ed.*, 1985, **23,** 1193.
235. Y. Takeuchi, K. Nagakawa and F. Yamamoto, *Polymer,* 1985, **26,** 1929.
236. K. Nagakawa and T. Konaka, *Polymer,* 1986, **27,** 1030, 1037, 1553.
237. K. Nakagawa and M. Amano, *Polym. Commun.*, 1986, **27,** 310.
238. M. Amano and K. Nakagawa, *Polym. Commun.*, 1987, **28,** 119.
239. M. Amano and K. Nakagawa, *Polymer,* 1986, **27,** 1559.
240. M. Amano and K. Nakagawa, *Polymer,* 1987, **28,** 263.
241. A. J. Park (BP Co.), *Ger. Pat.* 2 612 572 (1976) (*Chem. Abstr.*, 1976, **85,** 193 548).
242. Mitsubishi Chem. Ind. Co., *Jpn. Pat.* 81 135 524 (1981) (*Chem. Abstr.*, 1982, **96,** 53 265).
243. N. P. Vasilakos and F. Magalhaes, *J. Microwave Power,* 1984, **19,** 135.
244. M. Oota, K. Ishikawa and R. Hayashi (Kawata Mfg. Co.), *Jpn. Pat.* 89 163 006 (1989) (*Chem. Abstr.*, 1989, **111,** 234 291); *Jpn. Pat.* 89 163 007 (1989) (*Chem. Abstr.*, 1989, **111,** 234 292); M. Oota and K. Ishikawa (Kawata Mfg. Co.), *Jpn. Pat.* 89 163 008 (1989) (*Chem. Abstr.*, 1989, **111,** 234 290).
245. Anon., *Rubber J.*, 1970, **152** (10), 43.
246. R. A. Shute, *J. Microwave Power,* 1971, **6,** 193.
247. U. Schindler, *Elastomerics,* 1978, **110** (5), 42.
248. I. J. Chabinsky, *Elastomerics,* 1983, **115** (1), 17.
249. D. Anders, *Rubber J.*, 1970, **152** (3), 19.
250. R. G. Bosisio, C. F. Weissfloch and M. R. Wertheimer, *Rubber World,* 1972, **167,** 66.
251. R. J. Meredith, *J. Elast. Plast.*, 1976, **8,** 191.
252. C. S. Wang, *Rubber Chem. Technol.*, 1984, **57,** 134.
253. B. G. M. Helme, *Mater. Res. Soc. Symp. Proc.*, 1991, **189,** 517.
254. A. N. Gent and M. Hindi, *J. Appl. Polym. Sci., Appl. Polym. Symp.*, 1989, **44,** 255.
255. J. W. Vanderhoff (The Dow Chem. Co.), *US Pat.* 3 432 413 (1969) (*Chem. Abstr.*, 1969, **70,** 97 422).
256. E. O. Forster (Esso Research and Engineering Co.), *Ger. Pat.* 1 809 807 (1969) (*Chem. Abstr.*, 1969, **71,** 102 939).
257. D. C. Sayles (Secretary of the Army of the USA), *US Pat.* 3 972 967 (1976) (*Chem. Abstr.*, 1976, **85,** 162 932).
258. J. G. Dolden and A. C. Sturt, *Br. Pat.* 1 534 151 (1978) (*Chem. Abstr.*, 1979, **91,** 57 806).
259. Y. Taniyama (Otomo Kasei K.K.), *Jpn. Pat.* 78 109 535 (1978) (*Chem. Abstr.*, 1979, **90,** 24 455).
260. E. J. Zalewski (Schenectady Chem., Inc.,) *Ger. Pat.* 3 036 314 (1981) (*Chem. Abstr.*, 1981, **95,** 8928).
261. H. Ishihara and K. Kurita (Toyobo Co.), *Jpn. Pat.* 85 144 331 (1985) (*Chem. Abstr.*, 1986, **104,** 34 532).
262. K. H. Park, S. Watanabe, M. Kakimoto and Y. Imai, *Polym. J.*, 1993, **25,** 209.
263. S. Watanabe, K. Hayama, K. H. Park, M. Kakimoto and Y. Imai, *Makromol. Chem., Rapid Commun.*, 1993, **14,** 481.
264. Hitachi, Ltd., *Jpn. Pat.* 84 184 226 (1984) (*Chem. Abstr.*, 1985, **102,** 79 976).
265. D. A. Lewis, T. C. Ward, J. D. Summers and J. E. McGrath, *Polym. Prepr., Am. Chem. Soc., Div. Polym. Chem.*, 1988, **29** (1), 174; D. A. Lewis, J. D. Summers, T. C. Ward and J. E. McGrath, *J. Polym. Sci., Polym. Chem. Ed.*, 1992, **30,** 1647.
266. D. A. Lewis, S. J. Lamaire and T. L. Nunes, *Mater. Res. Soc. Symp. Proc.*, 1994, **347,** 681.
267. D. A. Lewis, T. L. Nunes and E. Simonyi, *Polym. Mater. Sci. Eng.*, 1995, **72,** 40.
268. H. Morimoto and H. Yoshioka (Shin-Etsu Chem. Ind. Co.), *Jpn. Pat.* 90 107 638 (1990) (*Chem. Abstr.*, 1990, **113,** 79 264).
269. F. Gimpel, *US Pat.* 4 529 750 (1985) (*Chem. Abstr.*, 1985, **103,** 124 521); F. Gimpel and C. Brichta, *Eur. Pat. Appl.* 149 782 (1985) (*Chem. Abstr.*, 1985, **103,** 196 912).
270. F. Gimpel and C. Brichta, *US Pat.* 4 680 319 (1987) (*Chem. Abstr.*, 1988, **109,** 38 897).
271. G. Bressan, G. Barbero, C. Troglia and C. Brichta (EniChem Augusta Ind., Srl), *Eur. Pat. Appl.* 475 064 (1992) (*Chem. Abstr.*, 1992, **116,** 215 738).
272. A. J. Berteaud, H. Valot and H. Jullien (ANVAR), *Fr. Pat.*, 2 458 323 (1981).
273. A. J. Berteaud, H. Jullien and H. Valot, *Rev. Gen. Electr.*, **1981,** (11), 826.
274. H. Jullien and H. Valot, *Polymer,* 1985, **26,** 506.
275. B. Silinski, C. Kuzmycz and A. Gourdenne, *Eur. Polym. J.*, 1987, **23,** 273.
276. A. Gourdenne, A.-H. Maassarani, P. Monchaux, S. Assudre and L. Thourel, *Polym. Prepr., Am. Chem. Soc., Div. Polym. Chem.*, 1979, **20** (2), 471.
277. A. Gourdenne, *2nd Int. Conf. React. Process. Polym.*, 1982, Conference Proceedings, p. 23.
278. Q. Le Van and A. Gourdenne, *Eur. Polym. J.*, 1987, **23,** 777.
279. J. Jow, M. C. Hawley, M. Finzel, J. Asmussen, Jr., H.-H. Lin and B. Manring, *IEEE Trans. Microwave Theory Tech.*, 1987, **MTT-35,** 1435.
280. D. A. Lewis, J. C. Hedrick, J. E. McGrath and T. C. Ward, *Polym. Prepr., Am. Chem. Soc., Div. Polym. Chem.*, 1987, **28** (2), 330.
281. S. M. Singer, J. Jow, J. D. DeLong and M. C. Hawley, *SAMPE Q.*, 1989, **20,** 14.

282. C. Jordan, J. Galy, J. P. Pascault, C. Moré and M. Delmotte, *Polym. Mater. Sci. Eng.*, 1992, **66,** 380.
283. E. Karmazsin, P. Satre and J. F. Rochas, *Thermochim. Acta,* 1985, **93,** 305.
284. F. M. Thuillier, H. Jullien and M.-F. Grenier-Loustalot, *Polym. Commun.*, 1986, **27,** 206; *Makromol. Chem., Macromol. Symp.*, 1987, **9,** 57.
285. F. M. Thuillier and H. Jullien, *Makromol. Chem., Makromol. Symp.*, 1989, **25,** 63.
286. N. Beldjoudi, A. Bouazizi, D. Douibi and A. Gourdenne, *Eur. Polym. J.*, 1988, **24,** 49.
287. N. Beldjoudi and A. Gourdenne, *Eur. Polym. J.*, 1988, **24,** 53, 265.
288. J. Jow, *Diss. Abstr. Int. B,* 1989, **50** (3)**,** 1040.
289. J. Jow, J. D. DeLong and M. C. Hawley, *SAMPE Q.*, 1989, **20,** 46.
290. J. Mijović and J. Wijaya, *Macromolecules,* 1990, **23,** 3671.
291. E. Marand, K. R. Baker and J. D. Graybeal, *Macromolecules,* 1992, **25,** 2243.
292. J. Wei, M. C. Hawley and M. T. DeMeuse, *Polym. Mater. Sci. Eng.*, 1992, **66,** 478.
293. J. Wei, M. C. Hawley, J. D. DeLong and M. T. DeMeuse, *Polym. Eng. Sci.*, 1993, **33,** 1132.
294. A. Gourdenne and Y. Baziard (Electricité de France), *Fr. Pat.* 2 555 188 (1985) (*Chem. Abstr.*, 1986, **104,** 89 773).
295. S. Amano and Y. Sasaki (NOK Corp.), *Jpn. Pat.*, 89 95 174 (1989) (*Chem. Abstr.*, 1989, **111,** 135 723).
296. W. G. Clark and R. E. Shannon (Westinghouse Electric Corp.), *Eur. Pat. Appl.* 498 998 (1992) (*Chem. Abstr.*, 1992, **117,** 214 071).
297. J. C. Hedrick, D. A. Lewis, T. C. Ward and J. E. McGrath, *Polym. Prepr., Am. Chem. Soc., Div. Polym. Chem.*, 1988, **29** (1)**,** 363.
298. D. A. Lewis, J. C. Hedrick, G. D. Lyle, T. C. Ward and J. E. McGrath, *Mater. Res. Soc. Symp. Proc.*, 1988, **124,** 181.
299. S. C. Liptak, S. P. Wilkinson, J. C. Hedrick, W. D. Joseph, T. C. Ward and J. E. McGrath, *Polym. Prepr., Am. Chem. Soc., Div. Polym. Chem.*, 1990, **31** (2), 320.
300. S. C. Liptak, S. P. Wilkinson, J. C. Hedrick, T. C. Ward and J. E. McGrath, *ACS Symp. Ser.*, 1991, **475,** 364.
301. S. C. Liptak, A. Gungor, T. C. Ward and J. E. McGrath, *Polym. Mater. Sci. Eng.*, 1992, **66,** 387.
302. S. A. Srinivasan, S. S. Joardar, D. B. Priddy, Jr., T. C. Ward and J. E. McGrath, *Mater. Res., Soc. Symp. Proc.*, 1994, **347,** 705.
303. M. T. DeMeuse and F. Parodi (EniChem SpA), *Eur. Pat. Appl.* 598 433 (1994) (*Chem. Abstr.*, 1995, **122,** 11 496).
304. F. Parodi, R. Gerbelli and M. T. DeMeuse (Eniricerche SpA; EniChem SpA), *Eur. Pat. Appl.* 655 472 (1995) (*Chem. Abstr.*, 1995, **123,** 229 292).
305. Mitsubishi Electric Corp., *Jpn. Pat.* 84 152 919 (1984) (*Chem. Abstr.*, 1985, **102,** 46 819); *Jpn. Pat.* 84 221 360 (1984) (*Chem. Abstr.*, 1985, **102,** 205 521).
306. L. Chan and A. Gourdenne, *Polym. Mater. Sci. Eng.*, 1992, **66,** 382.
307. P. Gagne (Societé Civile Immobiliere FIBAR), *Fr. Pat.* 2 463 795 (1981) (*Chem. Abstr.*, 1982, **96,** 36 191).
308. M. Teffal and A. Gourdenne, *Eur. Polym. J.*, 1983, **19,** 543.
309. F. Boardman and J. M. Lesniak (Johnson & Johnson Patient Care, Inc.), *Eur. Pat. Appl.* 295 438 (1988) (*Chem. Abstr.*, 1989, **110,** 155 442).
310. A. Nishizawa and T. Kondo (Victor Co. of Japan, Ltd.), *Jpn. Pat.* 91 281 606 (1991) (*Chem. Abstr.*, 1992, **116,** 195 096).
311. J. Palacios, J. Sierra and M. Paz Rodriguez, *Polym. Prepr., Am. Chem. Soc., Div. Polym. Chem.*, 1991, **32** (3), 244; *New Polym. Mater.*, 1992, **3,** 273.
312. A. Gourdenne and P. Heintz (Electricité de France), *US Pat.* 4 468 485 (1984).
313. M. Onofusa, S. Kobayashi and E. Tazawa (Mitsui Toatsu Chem., Inc.; Taisei Constr. Co.), *Jpn. Pat.* 74 10 926 (1974) (*Chem. Abstr.*, 1974, **80,** 146 704).
314. J. Hetz, J. Chang and M. Brodwin, *Mater. Res. Soc. Symp. Proc.*, 1994, **347,** 711.
315. J. C. Hedrick, D. A. Lewis, G. D. Lyle, S. D. Wu, T. C. Ward and J. E. McGrath, *Polym. Mater. Sci. Eng.*, 1989, **60,** 438.
316. J. C. Hedrick, D. A. Lewis, T. C. Ward and J. E. McGrath, *Mater. Res. Soc. Symp. Proc.*, 1991, **189,** 421.
317. M. T. DeMeuse, C. L. Ryan, Jr. and F. Parodi (Ist. G. Donegani SpA; EniChem SpA), *US Pat.* 5 237 032 (1993) (*Chem. Abstr.*, 1994, **120,** 246 613).
318. H. Meyer, M. Dorn and H. Seidl (Peroxid-Chemie GmbH), *Ger. Pat.* 3 018 321 (1981) (*Chem. Abstr.*, 1982, **96,** 163 676).
319. W. C. Endstra, *Caoutch. Plast.*, **1983,** (630), 35.
320. K. Matsumoto, N. Kotani, N. Ida and Y. Miyano (Sanwa Kako Co.), *Eur. Pat. Appl.* 526 872 (1993) (*Chem. Abstr.*, 1993, **119,** 74 106).
321. T. Saruyama and H. Kobayashi (Dow Corning Toray Co.), *Jpn. Pat.* 90 242 856 (1990) (*Chem. Abstr.*, 1991, **114,** 63 626).
322. D. M. P. Mingos and D. R. Baghurst, *Chem. Soc. Rev.*, 1991, **20,** 1.
323. R. J. Giguere, *Mater. Res. Soc. Symp. Proc.*, 1992, **269,** 387.
324. A. Loupy, *Spectra Anal.*, 1993, **22** (175)**,** 33.
325. D. M. P. Mingos, *Chem. Ind.* (*London*), 1994, **15,** 596.
326. D. M. P. Mingos, *Res. Chem. Intermed.*, 1994, **20,** 85.
327. R. Gedye, K. Westaway and F. Smith, *Polym. Mater. Sci. Eng.*, 1995, **72,** 8.
328. G. Majetich and R. Hicks, *Radiat. Phys. Chem.*, 1995, **45,** 567.
329. J. Berlan, *Radiat. Phys. Chem.*, 1995, **45,** 581.
330. K. C. Westaway and R. N. Gedye, *J. Microwave Power EM Energy*, 1995, **30,** 219.

Epilogue

RETROSPECT

We bring this series to an end after nine volumes. The main text, confined to seven volumes, has been supplemented by two additional volumes. Here topics that were not included in the main text are dealt with and other topics already included for which recent developments called for further authoritative review are to be found.

With this Second Supplement the editors feel that the task of creating a 'Comprehensive Polymer Science' treatise is reasonably complete. Developments in polymer science in the next decade are likely to be incremental, stepping out from the science base that we have described and set in wider horizons of physics and chemistry.

The chemistry of polymer synthesis is linked equally to free radical, anionic and cationic and condensation mechanisms with coordination catalysts becoming increasingly important. The underlying theory of the physics of polymers, developed principally by Edwards and de Gennes, embracing the concept of connectivity has a comprehensive reach.

The corresponding applications of polymer technology are growing at a much faster pace than is the science base. None of these applications, important though many of them are, involve new principles of polymer science. Thus in its present form our compendium should provide the essential background from which the reader will be able to follow and understand these new advances.

There has been a global shift in the practice of polymer science in the period covered by the series. Early discoveries in the 1930s and 1940s were uniquely located in Europe (including Russia) and the USA. The creation of an international research community benefitted greatly from the emigration of European scientists and technologists to the USA. Since the turn of the 1970s there has been a significant growth in the contribution from Asian scientists, especially from Japan. To some extent this has marked a decline in the UK and US contributions since the middle of the 1980s, leading to a more balanced global effort. This is manifest too in the uniformly international spread of the corresponding industries.

PROSPECT

Looking into the future, we have to recognize the growing impact of the interface that polymer science shares with the burgeoning life sciences. We know that nature has a way of controlling tightly either the stereochemistry of its macromolecules or the polydispersity of an assembly of chains. It would seem that development in the coming decades of polymer chemistry will address the same issues. Only in very special instances have we been able to prepare polymers with Poisson distributions. Control of stereoregular sequence lengths within a polymer chain is a topic for the future. Recent developments in metallocene catalysts have reawakened the prospect of tailoring stereoblocks and this will stimulate immediately the discovery of new chemical systems aimed at solving these problems.

The physics of polymers will develop as new diagnostic tools evolve. There is already evidence that the characterization and understanding of thin polymer films and polymers at interfaces is of considerable importance in the development of new polymeric materials and new applications of polymers.

Finally, we expect that chemistry and physics will join to provide polymer materials of novel and controlled structure for increasingly important applications in electronic and electrooptical devices and in biotechnology. Monomers can be organized at interfaces into regular physical arrays waiting for suitable physical or catalytic means to turn them into polymer chains. Or, polymer chain structures may be designed to facilitate certain novel solid state morphologies. In our view, the emphasis in the future will be on what we may call 'Precision Polymerization and Self Assembly of Polymers'. Such developments will need participation of research scientists in disciplines other than polymer science. We hope that the present nine-volume work 'Comprehensive Polymer Science' will be valuable to these scientists for an overview of topics of polymer science basic to their research.

GEOFFREY ALLEN
London, UK
April 1996

Subject Index